D0161272

Environmental Microbiology

Third edition

p462

Environmental Microbiology
Third edition

Ian L. Pepper

Charles P. Gerba

Terry J. Gentry

AMSTERDAM • BOSTON • HEIDELBERG • LONDON • NEW YORK • OXFORD
PARIS • SAN DIEGO • SAN FRANCISCO • SINGAPORE • SYDNEY • TOKYO
Academic Press is an imprint of Elsevier

BPL
QR100
P463
2015

Academic Press is an imprint of Elsevier
525 B Street, Suite 1800, San Diego, CA 92101-4495, USA
32 Jamestown Road, London NW1 7BY, UK
225 Wyman Street, Waltham, MA 02451, USA

Copyright © 2015, 2009, 2000 Elsevier Inc. All rights reserved.

No part of this publication may be reproduced, stored in a retrieval system, or transmitted in any form
or by any means electronic, mechanical, photocopying, recording or otherwise without the prior written
permission of the publisher.

Permissions may be sought directly from Elsevier's Science & Technology Rights Department in Oxford, UK:
phone (+44) (0) 1865 843830; fax (+44) (0) 1865 853333; email: permissions@elsevier.com. Alternatively,
visit the Science and Technology Books website at www.elsevierdirect.com/rights for further information.

Notice
No responsibility is assumed by the publisher for any injury and/or damage to persons or property as a matter
of products liability, negligence or otherwise, or from any use or operation of any methods, products, instructions
or ideas contained in the material herein. Because of rapid advances in the medical sciences, in particular,
independent verification of diagnoses and drug dosages should be made.

British Library Cataloguing-in-Publication Data
A catalogue record for this book is available from the British Library

Library of Congress Cataloging-in-Publication Data
A catalog record for this book is available from the Library of Congress

ISBN: 978-0-12-394626-3

For information on all Academic Press publications
visit our website at elsevierdirect.com

Printed and bound in the United States of America

15 16 17 18 19 10   9 8 7 6 5 4 3 2 1

Working together
to grow libraries in
developing countries

www.elsevier.com • www.bookaid.org

This book is dedicated to my friend, who was always a "Blaze" of glory!

Ian Pepper

This book is dedicated to my wife and sons Peter and Phillip for all their support

Charles Gerber

This book is dedicated to my grandparents, Coy and Sybal, who gave me the opportunity and encouragement to pursue my dreams

Terry Gentry

Contents

Part II
Microbial Environments

Part III
Detection, Enumeration, and Identification

8 Environmental Sample Collection and Processing

Ian L. Pepper and Charles P. Gerba

9 Microscopic Techniques

Timberley M. Roane and Ian L. Pepper

10 Cultural Methods

Ian L. Pepper and Charles P. Gerba

11 Physiological Methods

Raina M. Maier and Terry J. Gentry

Part IV
Microbial Communication, Activities, and Interactions with Environment and Nutrient Cycling

14   Microbial Source Tracking

Channah Rock, Berenise Rivera and Charles P. Gerba

Part V
Remediation of Organic and Metal Pollutants

18 Microorganisms and Metal Pollutants

Timberley M. Roane, Ian L. Pepper and Terry J. Gentry

19 Microbial Diversity and Interactions in Natural Ecosystems

Terry J. Gentry, Ian L. Pepper and Leland S. Pierson III

20 Microbial Communication: Bacteria/Bacteria and Bacteria/Host

Leland S. Pierson III, Raina M. Maier and Ian L. Pepper

21 Bioinformation and 'Omic Approaches for Characterization of Environmental Microorganisms

Emily B. Hollister, John P. Brooks and Terry J. Gentry

Part VI
Water- and Foodborne Pathogens

22 Environmentally Transmitted Pathogens

Charles P. Gerba

23 Indicator Microorganisms

Charles P. Gerba

24 Risk Assessment

Charles P. Gerba

Part VII
Wastewater Treatment and Disinfection

Everywhere on Earth, microorganisms are abundant, even within our own human bodies. Astonishingly, an adult human contains ten times as many microbial cells as mammalian cells, carrying around approximately 1.25 kg of microbes. Bacteria are found all over our skin, in our mouths, up our noses, and within every part of our interior bodies, particularly within the gastrointestinal tract. Outside of ourselves, microorganisms including viruses, bacteria, fungi, and protozoa proliferate in every known environment, from the frozen arctic regions to the warm tropics. We define "Environmental Microbiology" as the study of microbes within all earthly habitats, and their beneficial and detrimental effects on human health and welfare. As such, environmental microbiology involves the effect or impact of microbes on human activities, either directly or indirectly. Examples of the role of microbes in our daily lives are shown in Table 1.1 (Chapter 1), which illustrates that life without environmental microbes is impossible.

Environmental microbiology is not only a dynamic field of science; it is transformative, constantly changing as new microbes are discovered and new aspects of microbial activities are understood. Ultimately microbes impact the quality of our environment, which subsequently impacts human health. In this textbook we invite you to join us on a journey into the extraordinary world of how microbes impact our environment and how we put them to use to enhance the world we live in.

This third edition has eight subject areas: (i) foundation chapters that provide adequate background for subsequent advanced chapters; (ii) chapters on microbial environments; (iii) chapters on the microbial methods and technologies utilized to study microbes and their activities; (iv) chapters on biochemical and industrial transformations, including microbial transport; (v) chapters on bacterial diversity and communication; (vi) chapters on pathogens found within water and food and their potential risks; (vii) chapters on waste- and potable-water treatment and distribution; and finally (viii) chapters on urban microbiology and global emerging issues.

This textbook is designed for a senior-level undergraduate class, or a graduate level class in environmental microbiology. Because environmental microbiology is constantly evolving, new beneficial and pathogenic microbes have emerged since the second edition was published in 2009 (Table 1.2, Chapter 1). Likewise, new technologies have been developed that aid our understanding of microbes. These recent discoveries are documented in Information Box 1.2 (Chapter 1).

Also, there are a number of changes in the third edition. The authors are Professors Pepper, Gerba, and Gentry, who collectively have expertise over a broad spectrum of environmental microbiology. Key contributions to the text were also made by nine colleagues who have collaborated with the authors at The University of Arizona and Texas A & M University. Overall, there are five new chapters in the third edition, and seven chapters that have undergone significant change or modification. All of the remaining chapters were updated with respect to references, and many new graphics have been added.

Ian Pepper
Charles Gerba
Terry Gentry

Instructors: For lecture slides and answers to the end of chapter tests, please visit the instructor website at http://textbooks. elsevier.com/web/product_details.aspx?isbn=9780123946263

Ian L. Pepper Ph.D., The Ohio State University, 1975. Currently Professor of Environmental Microbiology. Dr. Pepper's diverse research interests are reflected in the fact that he is Fellow of The American Association for the Advancement of Science, The American Academy of Microbiology, the Soil Science Society of America, and the American Society of Agronomy. He is also Director of the National Science Foundation Water and Environmental Technology Center at the University of Arizona. Dr. Pepper has been active in the area of soil molecular ecology as well as waste utilization including biosolids and effluent reuse. More recently he has been pursuing research on real-time monitoring of microbial contaminants in potable water, and "smart water distribution systems."

Terry J. Gentry Ph.D., The University of Arizona, 2003. Currently Associate Professor and Director of the Soil and Aquatic Microbiology Laboratory at Texas A&M University. Dr. Gentry is an environmental microbiologist specializing in the development and use of molecular technologies to enhance the detection and remediation of environmental contamination. This includes the use of genotypic methods to detect and identify microbial contaminants from animal, human, and natural sources and also the characterization of microbial populations and communities contributing to applied processes such as the bioremediation of organic and metal contaminants and ecosystem restoration and sustainability. Dr. Gentry teaches an undergraduate course in soil and water microbiology and a graduate course in environmental microbiology.

Charles P. Gerba Ph.D., University of Miami, 1973. Currently, Professor of Microbiology. Dr. Gerba is a Fellow of the American Academy of Microbiology. He is recipient of the A.P. Black Award from the American Water Works Association for outstanding contributions to Water Science, and the McKee Award from the Water Environment Federation for outstanding contributions to groundwater protection. He has an international reputation for his methodologies for pathogen detection in water and food, pathogen occurrence in households, and risk assessment.

Contributing Authors

 **John P. Brooks** USDA ARS Mississippi State, MS

 **Julia W. Neilson** The University of Arizona, Tucson, AZ

 **Marilyn J. Halonen** The University of Arizona, Tucson, AZ

 **Deborah T. Newby** Idaho National Laboratory, Idaho Falls, ID

 **Emily B. Hollister** Baylor College of Medicine, Houston, TX

 **Leland S. Pierson III** Texas A & M University, College Station, TX

 **Raina M. Maier** The University of Arizona, Tucson, AZ

 **Timberley M. Roane** University of Colorado, Denver, CO

 **Channah Rock**   The University of Arizona, Maricopa, AZ

 **Berenise Rivera**   The University of Arizona, Tucson, AZ

 **Virginia I. Rich**   The University of Arizona, Tucson, AZ

 **Hye-Weon Yu**   The University of Arizona, Tucson, AZ

Part I

Review of Basic Microbiological Concepts

Chapter 1

Introduction to Environmental Microbiology

Ian L. Pepper, Charles P. Gerba and Terry J. Gentry

1.1 ENVIRONMENTAL MICROBIOLOGY AS A DISCIPLINE

We define "environmental microbiology" as the study of microbes within all habitats, and their beneficial and detrimental impacts on human health and welfare. Environmental microbiology is related to, but also different from, "microbial ecology," which focuses on the interactions of microorganisms within an environment such as air, water or soil. The primary difference between the two disciplines is that environmental microbiology is an applied field in which we attempt to improve the environment and benefit society. Environmental microbiology is also related to many other disciplines (Figure 1.1).

Microorganisms occur everywhere on Earth. An adult human body contains 10 times as many microbial cells as mammalian cells, consisting of approximately 1.25 kg of microbial biomass (Wilson, 2005). Although the study of microbial inhabitants of humans resides within clinical microbiology, it was the discovery of environmental pathogenic microorganisms that invaded the human body that resulted in the beginning of environmental microbiology. These roots were enabled by the work of Louis Pasteur and Robert Koch, who developed the Germ Theory of Disease in the 1870s, following which, the presence of waterborne human pathogens then became the initial focus of environmental microbiology. In developed countries, applied environmental studies related to drinking water and wastewater treatment dramatically reduced bacterial waterborne disease. However, other microbial agents such as viruses and protozoa, which are more resistant to disinfection than enteric bacteria, still cause problems, resulting in water quality continuing to be a major focus in environmental microbiology. There is an estimated 20,000,000 cases of illness per year due to drinking contaminated water (Reynolds *et al.*, 2008). The largest waterborne outbreak of disease in the United States occurred in 1993, when over 400,000 people became ill and around 100 died in Milwaukee, Wisconsin, due to the protozoan parasite *Cryptosporidium* (Eisenberg *et al.*, 2005). In developing countries, poor sanitation resulting from a lack of water and wastewater treatment still results in millions of deaths annually.

Controlling the contamination of our food supply also continues to be a concern; and the Centers for Disease Control estimates that in the United States each year there are 48 million cases with 128,000 people hospitalized and 3000 deaths. The third most deadly outbreak of foodborne infection in the United States occurred in 2011, when 29 persons died from *Listeria* contamination of cantaloupe. Information Box 1.1 documents some of these foodborne outbreaks.

Until the middle of the 20th century, industrial chemicals in the United States were routinely disposed of by dumping them into sewers, soils, rivers or oceans, without regard to the pollution that this caused, or the subsequent adverse

I.L. Pepper, C.P. Gerba, T.J. Gentry: Environmental Microbiology, Third edition. DOI: http://dx.doi.org/10.1016/B978-0-12-394626-3.00001-6
© 2015 Elsevier Inc. All rights reserved.

FIGURE 1.1   Environmental microbiology interfaces with many other disciplines.

Information Box 1.1   Large-scale Food Recalls Due to Foodborne Outbreaks

Food Recall	Year	Organism	# People Involved
Turkey and chicken products	2002	Listeria	120
Green onions	2003	Hepatitis A	600
Spinach	2006	E. coli O157:H7	200
Canned meat products	2007	Clostridium botulinum	4
Canned beef	2007	E. coli O157:H7	14
Fresh salsa	2008	Salmonella enterica	1442
Peanut butter	2009	Salmonella	22,500
Eggs	2010	Salmonella	2000
Cantaloupe	2011	Listeria	29
Strawberries	2012	Norovirus	11,200

---

ecological and human health effects. This all changed in the 1960s when concern over a toxic dump was highlighted by Rachel Carson's landmark book *Silent Spring*. In essence, this resulted in the birth of the environmental movement in the United States, and a new field of study for environmental microbiology known as "bioremediation." Many chemicals discharged into the environment without regard to the consequences have been shown to result in adverse human health impacts. However, since hydrocarbons, chlorinated solvents and most pesticides are organic in nature, they can

potentially be degraded by heterotrophic microorganisms including bacteria and fungi. The field of bioremediation within environmental microbiology involves enhancing and optimizing microbial degradation of organic pollutants, resulting in environmental cleanup and reduced adverse human health effects. The efficacy of bioremediation was demonstrated in 1989, when the *Exxon Valdez* oil tanker spilled approximately 11 million gallons of crude oil into Prince William Sound. Optimization of bioremediation was a major factor in cleaning up and restoring Prince William Sound. Bioremediation has also been shown to be critically important in cleaning up the more recent 2010 Gulf of Mexico oil spill (see Chapter 31).

Also in the 20th century, soil microbiology, a component of environmental microbiology, became important as a means to enhance agricultural production. Studies on the rhizosphere (the soil surrounding plant roots), and specific studies on root-microbial interactions involving nitrogen fixing rhizobia, and mycorrhizal fungi that enhanced phosphorus uptake, were all utilized to improve plant growth. Other studies of plant growth-promoting bacteria that reduced the incidence of plant pathogens were also effective in aiding the "Green Revolution," which resulted in stunning increases in crop yields throughout the United States and in many parts of the world. Overall, these fundamental study areas have helped shape the current discipline of environmental microbiology, and all affect our everyday life.

1.2 MICROBIAL INFLUENCES ON OUR DAILY LIVES

Some of the influences that microorganisms have on our daily lives are shown in Table 1.1. These influences can be summarized in terms of:

- The overall health of the planet
- What infects us
- What heals us
- What we drink
- What we eat
- What we breathe

1.2.1 Overall Health of the Planet

Life on Earth depends on the biogeochemical cycles that are microbially driven. For example, carbon dioxide is removed from the atmosphere during photosynthesis by both plants and photosynthetic microbes. The result of this process is that carbon dioxide is converted into organic carbon building blocks as plant or microbial biomass, which ultimately results in the formation of organic matter. Fortunately, this organic matter is ultimately degraded by microorganisms via respiratory processes, which again release carbon dioxide into the atmosphere. Without microbial respiration, a vast array of organic matter would accumulate. Similar biogeochemical processes exist for all other elements, and are also driven by microorganisms. All life on Earth is dependent on these

biogeochemical cycles. In addition, these cycles can benefit human activity, as in the case of remediation of organic and metal pollutants, or be detrimental, as in the formation of nitrous oxide which can deplete Earth's ozone layer (Ravishankara *et al.*, 2009).

A major indirect effect of environmental microbes may be the influence of soil microbes on global warming. However, currently there is still debate about the net impact of microbes on this process (Rice, 2006). Soils can be a source of "greenhouse gases" such as carbon dioxide, methane and nitrous oxide due to microbial respiration, or they can be a sink for carbon due to enhanced photosynthetic activity and subsequent carbon sequestration. Although the debate has yet to be resolved, it is clear that even relatively small changes in soil carbon storage could significantly affect the global carbon balance and global warming. In turn, many scientists believe that continued global warming will ultimately have catastrophic impacts on human health via extreme weather events and natural disasters.

1.2.2 What Infects Us

Humans are subject to microbial attack from a plethora of pathogens that can be viral, bacterial or protozoan in nature (Table 1.2). Likewise, the route of exposure is variable and can be through ingestion or inhalation of contaminated food, water or air, or from contact with soils or fomites. The infections resulting from microbial pathogens can be mild to severe, or even fatal. In extreme

TABLE 1.1 Microbial Influences on Our Daily Lives

Activity	Environmental Matrix	Impact	Microorganisms
Municipal wastewater treatment	Wastewater	Waterborne disease reduction	*E. coli* *Salmonella*
Water treatment	Water	Waterborne disease reduction	Norovirus *Legionella*
Food consumption	Food	Foodborne disease	*Clostridium botulinum* *E. coli* O157:H7
Indoor activities	Fomites	Respiratory disease	Rhinovirus
Breathing	Air	Legionellosis	*Legionella pneumophila*
Enhanced microbial antibiotic resistance	Hospitals	Antibiotic resistant microbial infections	Methicillin resistant *Staphylococcus aureus*
Nutrient cycling	Soil	Maintenance of biogeochemical cycling	Soil heterotrophic bacteria
Rhizosphere/Plant interactions	Soil	Enhanced plant growth	Rhizobia Mycorrhizal fungi
Bioremediation	Soil	Degradation of toxic organics	*Pseudomonas* spp.

TABLE 1.2 Emerging Environmentally Transmitted Microbial Pathogens and Biological Agents

Agent	Type	Mode of Transmission	Why Important	Disease/Symptoms
Adenovirus	Virus	Water Air Fomites	Most resistant waterborne agent to UV light	Respiratory; gastroenteritis; eye, ear infections
Toxigenic *E. coli* (O157:H7)	Bacterium	Foodborne Waterborne	Virulence increasing	Enterohemorrhagic fever, kidney failure
Cryptosporidium	Protozoan	Waterborne Foodborne	Resistance to chlorination	Gastroenteritis
Norovirus	Virus	Waterborne Foodborne Fomites	Low infectious dose	Gastroenteritis
Prions	Protein	Cows/Humans	Very stable in the environment	Variant Creutzfeldt—Jakob disease
Naegleria fowleri	Protozoan	Water	Causes fatal brain disease via swimming and drinking water	Brain encephalitis

cases, pandemics can occur, as in the case of the 1918 – 1919 influenza pandemic, which spread worldwide and killed more people than the number that died in the First World War (Brundage and Shanks, 2008). More recently, concern has centered on the potential for a pandemic originating from avian influenza (H5N1) virus (Malik Peiris *et al.*, 2007). Overall, every person on Earth has experienced some form of infection, and every location on Earth can be a source of infections. For example, hospitals that are designed to house patients recovering from various maladies can be a source of methicillin resistant *Staphylococcus aureus* (MRSA).

1.2.3 What Heals Us

Although numerous microbes are pathogenic to humans, many others provide a treasure chest of natural products critical to maintaining or improving human health. The earliest classes of compounds to be discovered were the antibiotics. Antibiotics are compounds produced by environmental microorganisms that kill or inhibit other microorganisms. The first discovered antibiotic was penicillin isolated from the soil-borne fungus *Penicillium* by Sir Alexander Fleming in 1929. Later, Selman Waksman discovered streptomycin in 1943, a feat for which he received the Nobel Prize. This antibiotic was isolated from *Streptomyces griseus*, and, since then, soil actinomycetes have been shown to be a prime source of antibiotics.

In addition to bacteria, fungi are also a source of natural products that aid human health. In particular, endophytes, which are microbes that colonize plant roots

without pathogenic effects, are a rich source of novel antibiotics, antimycotics, immunosuppressants and anticancer agents (Strobel and Daisy, 2003). Microtubule-stabilizing agents (MSA) such as paclitaxel have been isolated from endophytic fungi associated with species of the yew tree (*Taxus* spp.). Because paclitaxel acts as a cell poison that arrests cell division, it has become a highly potent anticancer agent (Snyder, 2007). Endophytes have also been shown to have useful applications in agriculture and industry (Mei and Flinn, 2010). A new technology known as "genomic mining" has resulted in new discoveries of useful natural products. These molecular technologies are allowing for the identification of new drug products that result from gene clusters that are not normally expressed under laboratory conditions (Gross, 2009). These new approaches bode well for future sources of new natural products that will improve human health.

1.2.4 What We Drink

Environmental microbes also influence the quality of the water we drink, both directly and indirectly. Direct adverse effects can include the contamination of surface water or groundwaters with pathogenic microorganisms. Microbes can also exacerbate chemical contamination of water, as in the case of arsenic. Specifically, some soil microbes utilize arsenate as a terminal electron acceptor under anaerobic conditions, thus converting arsenate to arsenite which is a more toxic and mobile species that is more likely to contaminate groundwater (National Research Council, 2007). On the other hand, microbes can also indirectly protect water quality, such as through degradation of toxic organics in the Critical Zone which protects groundwater (Lin, 2010).

1.2.5 What We Eat

Soil is a fundamental requirement for food production since the vast majority of food grown for human or animal consumption is derived from soil. Soil in close proximity to the plant roots is known as the rhizosphere, which contains vast numbers of soil microorganisms essential for plant growth. Without rhizosphere organisms, plant growth is severely repressed, since beneficial microbes enhance nutrient uptake. In addition, specific soil bacteria known as rhizobia fix atmospheric nitrogen into ammonia for leguminous plants, and mycorrhizal fungi enhance plant uptake of phosphate. Adverse effects of microorganisms on what we eat also include contamination with pathogenic microbes and microbial toxins (Information Box 1.1).

1.2.6 What We Breathe

Microbes can be aerosolized through both natural and human activities. Humans influence the transport of aerosolized microbes through a variety of activities including, for example, land application of wastes. The introduction of cooling towers and hot showers also creates a route for human exposure to aerosolized *Legionella* bacteria which can result in life-threatening infections. With as much as 80% of our time now spent indoors, air quality in these environments can also result in "sick building syndrome" and asthma attacks. Microbial derived allergens are also readily transported into and through the air. Mycotoxins produced by soil fungal molds including *Aspergillus*, *Alternaria*, *Fusarium* and *Penicillium* can cause a variety of health problems. For example, aflatoxin produced by *Aspergillus flavus* is a potent carcinogen (Williams *et al*., 2004).

1.3 ENVIRONMENTAL MICROBIOLOGY IN 2014

Issues continue to emerge where solutions depend on an understanding of environmental microbiology. For example, an outbreak of avian bird flu raised concerns about a worldwide pandemic, with little hope of quickly developing a vaccine. Better information was needed on how this virus spread through the environment from one person to another, in order to develop successful interventions. It became evident that little was known about how important routes of transmission occurred (i.e., air vs. fomites vs. water), and how they influenced transmission of the

FIGURE 1.2 Oil surrounds the site of the *Deepwater Horizon* oil spill in the Gulf of Mexico near the coast of Louisiana, Monday, May 31, 2010. Photo Source: Jae C. Hong, Association Press (ID #100531020566).

influenza virus. It was critical that these exposure routes be better understood so that appropriate environmental controls could be developed. In 2010, the Gulf of Mexico oil release following an explosion on an oil rig devastated the economies of local communities, but intrinsic and enhanced bioremediation was significant in mitigating the hazard (Figure 1.2) In 2011, a new virulent strain of *E. coli* (O104:H4) was the source of a foodborne microbial outbreak in Germany that killed more than 50 people (Rasko *et al*., 2011).

On a positive note, new techniques and methodologies are aiding our efforts to contain adverse microbial contaminants (Information Box 1.2). New molecular techniques including qPCR (quantitative polymerase chain reaction) are allowing for "near real-time" detection of pathogens. Microbial source tracking now allows us to pin-point sources of microbial contamination. New sensors are allowing us to monitor microbial water quality in real time. Advances in quantitative microbial risk assessment are now allowing us to determine whether particular activities such as land application of biosolids and animal manures are "safe." New molecular ecological techniques, including next-generation genome sequencing technologies, are allowing us to create better estimates of microbial diversity in the environment, and exploit that diversity for new sources of natural products. Advances in DNA synthesis and transplantation technologies are enabling the construction of "synthetic microorganisms" and may revolutionize our approach to characterizing and mining the genomes of environmental microorganisms. Self-sanitizing surfaces will potentially provide for proactive disinfection of fomites that will reduce microbial infections.

Overall, the field of environmental microbiology is mature, yet evolving, and well situated to deal with the variety of microbial issues that face today's (and tomorrow's) society. Join us on the exciting journey as we examine the state of the science.

Information Box 1.2   State-of-the-Art Microbial Methodologies and Techniques

Technique	Purpose	Reference
High-throughput DNA sequencing	Rapid, large-scale sequencing of microbial genomes and communities	Novais and Thorstenson, 2011
"OMICS"	Molecular estimates of diversity and function	Jansson et al., 2012
qPCR	Near real-time detection of pathogens	van Frankenhuyzen et al., 2011
Real-time Sensors	Detection of contaminants in potable water	Miles et al., 2011
Aptamer sensors	Detection of specific microbes	Song et al., 2012
Microbial source tracking	Determine source of pathogens	Staley et al., 2012
Quantitative microbial risk assessment	Evaluation of potential microbial hazards	Haas et al., 2014
Self-sanitizing surfaces	Proactive disinfectants	Sattar, 2010
Synthetic microbial cells	Creation of microbial cells with entirely synthetic genomes	Gibson et al., 2010

QUESTIONS AND PROBLEMS

1. Describe five ways in which environmental microbiology directly affects you today.
2. Identify the most recent waterborne and foodborne outbreaks in the United States.
3. Based on the current U.S. population, how many food and drinking water infections can you expect to contract in a 70-year lifetime?
4. Identify and describe two major bioremediation projects within 1000 miles of your home town.
5. Identify two new natural products that are now being utilized to treat cancer.

REFERENCES AND RECOMMENDED READING

Brundage, J. F., and Shanks, G. D. (2008) Deaths from bacterial pneumonia during 1918–19 influenza pandemic. *Emerg. Infect. Dis.* **14** (8), August 2008.

Eisenberg, J. N. S., Lei, X. D., Hubbard, A. H., Brookhart, M. A., and Colford, J. M., Jr. (2005) The role of disease transmission and conferred immunity in outbreaks: analysis of the 1993 *Cryptosporidium* outbreak in Milwaukee, Wisconsin. *Am. J. Epidemiol.* **161**, 62–72.

Gibson, D. G., Glass, J. L., Lartigue, C., *et al.* (2010) Creation of a bacterial cell controlled by a chemically synthesized genome. *Science* **329**, 52–56.

Gross, H. (2009) Genomic mining: a concept for the discovery of new bioactive natural products. *Curr. Opin. Drug Discov. Dev.* **12**, 207–219.

Haas, C. N., Rose, J. B., and Gerba, C. P. (2014) "Quantitative Microbial Risk Assessment," Second edition. John Wiley & Sons, New York, NY.

Jansson, J. K., Neufeld, J. D., Moran, M. A., and Gilbert, J. A. (2012) Omics for understanding microbial functional dynamics. *Environ. Microbiol.* **14**, 1–3.

Lin, H. (2010) Earth's critical zone and hydropedology: concepts, characteristics and advances. *Hydrol. Earth Syst. Sci.* **14**, 25–45.

Malik Peiris, J. S., deJong, M. D., and Guan, Y. (2007) Avian influenza virus (H5N1): a threat to human health. *Clin. Microbiol. Rev.* **20**, 243–267.

Mei, C., and Flinn, B. S. (2010) The use of beneficial microbial endophytes for plant biomass and stress tolerance improvement. *Recent Pat. Biotechnol.* **4**, 81–95.

Miles, S. L., Sinclair, R. G., Riley, M. R., and Pepper, I. L. (2011) Evaluation of select sensors for real-time monitoring of *Escherichia coli* in water distribution systems. *Appl. Environ. Microbiol.* **77**, 2813–2816.

National Research Council (2007) "Earth Materials and Health," The National Academies Press, Washington, DC.

Novais, R. C., and Thorstenson, Y. R. (2011) The evolution of Pyrosequencing® for microbiology: from genes to genomes. *J. Microbiol. Methods* **86**, 1–7.

Rasko, D. A., Webster, D. R., Sahl, J. W., Bashir, A., Boisen, N., Scheutz, F., *et al.* (2011) Origins of the *E. coli* strain causing an outbreak of hemolytic-uremic syndrome in Germany. *N. Engl. J. Med.* **365**, 709–717.

Ravishankara, A. R., Daniel, J. S., and Portmann, R. W. (2009) Nitrous oxide (N_2O): the dominant ozone-depleting substance emitted in the 21st Century. *Science* **326**, 123–125.

Reynolds, K. A., Mena, K. D., and Gerba, C. P. (2008) Risk of waterborne illness via drinking water in the United States. *Rev. Environ. Contam. Toxicol.* **192**, 117–158.

Rice, C. W. (2006) Introduction to special section on greenhouse gases and carbon sequestration in agriculture and forests. *J. Environ. Qual.* **35**, 1338–1340.

Sattar, S. A. (2010) Promises and pitfalls of recent advances in chemical means of preventing the spread of nosocomial infections by environmental surfaces. *Am. J. Infect. Control* **38**, S34–S40.

Snyder, J. P. (2007) The microtubule-pore gatekeeper. *Nat. Chem. Biol.* **3**, 81–82.

Song, K.-M., Lee, S., and Ban, C. (2012) Aptamers and their biological applications. *Sensors* **12**, 612–631.

Staley, C., Reckhow, K. H., Lukasik, J., and Harwood, V. J. (2012) Assessment of sources of human pathogens and fecal contamination in a Florida freshwater lake. *Wat. Res.* **46**, 5799–5812.

Strobel, G., and Daisy, B. (2003) Bioprospecting for microbial endophytes and their natural products. *Microbiol. Molec. Biol. Rev.* **67**, 491–502.

van Frankenhuyzen, J. K., Trevors, J. T., Lee, H., Flemming, C. A., and Habash, M. B. (2011) A review: molecular pathogen detection in biosolids with a focus on quantitative PCR using propidium monoazide for viable cell enumeration. *J. Microb. Methods* **87**, 263–272.

Williams, J. H., Phillips, T. D., Jolly, P. E., Stiles, J. K., Jolly, C. M., and Aggarwal, D. (2004) Human aflatoxicosis in developing countries: a review of toxicology, exposure, potential health consequences and interventions. *Am. J. Clinical. Nutr.* **80**, 1106–1122.

Wilson, M. (2005) "Microbial Inhabitants of Humans," Cambridge University Press, New York, NY.

Chapter 2

Microorganisms Found in the Environment

Ian L. Pepper and Terry J. Gentry

Microorganisms other than viruses can be defined as free-living organisms that are so small that they cannot be seen with the naked eye. Generally, this size range is less than 100 μm, but defining microbes just in terms of size can be confusing since some microbes can be seen with the naked eye and are greater than 100 μm in size. Examples of larger microbes include some protozoa, and bacteria such as *Epulopiscium fishelsoni*. Mushrooms are certainly large enough to be seen with the naked eye, and yet are classified as fungi. Viruses also complicate the picture since, although they are certainly small (10−100 nm), they are not free-living and do not metabolize.

Despite these anomalies, microbes found in the environment are generally thought to consist of: Bacteria (including actinomycetes); Archaea; Fungi; Protozoa; Algae; and Viruses. Microbes indigenous to the environment, which includes soil, water and air, are characterized as being able to adapt to variable environmental conditions such as temperature, redox potential, pH, moisture regime and pressure. This differentiates them from microbes found within the human body which exist under much more constant conditions, and which normally do not survive when introduced into the environment. Microorganisms are also capable of existing under oligotrophic (low nutrient) conditions, essentially living under conditions of starvation. These characteristics allow microbes to be found in every habitat imaginable including deserts and jungles, and even under Arctic conditions.

In this chapter we introduce the different types of microbes found in the environment including their structural features and some of their major functions and impacts, not only on human health and welfare, but also on the environment. To put the importance of microbes into perspective, it is interesting to realize that they first appeared on Earth approximately 4 billion years ago, and have been critical to the formation of current global conditions, including the presence of free molecular oxygen which first appeared around 2.5 billion years ago. Additional information and background on these types of microbes can be found in textbooks in the recommended reading listed at the end of this chapter.

2.1 CLASSIFICATION OF ORGANISMS

Until the 1970s, classification of macro- and microorganisms was based primarily on physiological differences with anywhere from two to six major kingdoms proposed for categorizing life as we know it. However, in the 1970s,

I.L. Pepper, C.P. Gerba, T.J. Gentry: Environmental Microbiology, Third edition. DOI: http://dx.doi.org/10.1016/B978-0-12-394626-3.00002-8
© 2015 Elsevier Inc. All rights reserved.

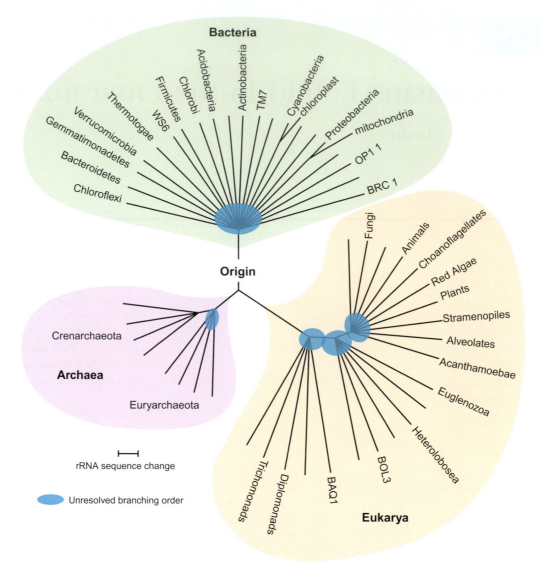

FIGURE 2.1   The three domain tree of life. Classification is based on the ribosomal RNA gene.

techniques became available to allow examination of nucleic acids, including ribosomal RNA (rRNA), which is a highly conserved structure used for synthesis of proteins in living things. Based on analysis of 16S rRNA, Carl Woese identified an entirely new group of organisms—the Archaea (Woese and Fox, 1977)—which eventually led to the modern classification of living organisms into a three domain system consisting of Archaea, Eukarya and Bacteria (Figure 2.1). Of these, the Bacteria and Archaea are termed prokaryotes, and the Eukarya are known as eukaryotes. Eukaryotic microbes other than algae and fungi are collectively called protists. Within the Eukarya are fungi, protozoa, algae, plants, animals and humans.

2.2 PROKARYOTES

Prokaryotes are the simplest of organisms and are characterized by the lack of a true nucleus and membrane-bound

cell organelles, such as mitochondria or chloroplasts. The prokaryotes consist of two separate large groups, the Bacteria and the Archaea. The structural features of prokaryotes are shown in Information Box 2.1.

2.2.1 Bacteria

The bacteria are the least complex of the living microorganisms but offer the greatest metabolic flexibility and have the greatest diversity. They dominate numerous environmental processes critical not only to humans, but also to the environment (such as nitrogen fixation); however, they also include some of the most notorious human, animal and plant pathogens. It is estimated that there are more than 50 bacterial phyla based on the analysis of the conserved 16S rRNA sequence (Schloss and Handelsman, 2004). Approximately half of these phyla have not yet been cultured. Thus, we know relatively little about the majority of

Information Box 2.1   Prokaryotic Cell Structure and Function

Location	Structure	Function
Cytoplasm	Chromosome	Information storage and replication (DNA).
	Plasmid	Extrachromosomal DNA that often confers a competitive advantage to the cell, e.g., antibiotic resistance.
	Ribosomes	Protein synthesis (rRNA and protein).
Cell envelope	Cell membrane	Selectively permeable layer found in all bacteria that allows import and export of nutrients, toxins and waste products. Composed of a phospholipid bilayer with proteins that serve as ions channels, proton pumps and receptors.
	Cell wall	Rigid, permeable structure that confers shape and protection to cells.
	Periplasmic space	Involved in nutrient acquisition, electron transport, and alteration of substances toxic to the cell. Especially important in Gram-negative cells.
	Outer membrane	A second semipermeable membrane found only in Gram-negative cells. The outer leaflet of the outer membrane contains lipopolysaccharide (LPS) molecules.
	Lipopolysaccharide (LPS)	Found anchored into the outer leaflet of the outer membrane in Gram-negative cells. This negatively charged molecule helps mediate interactions of the cell with the environment. The molecule is an endotoxin and antigenic.
	Teichoic acids	Found anchored into the peptidoglycan wall of Gram-positive cells. This negatively charged molecule helps mediate interactions of the cell with the environment, e.g., adhesion. Teichoic acids are antigenic.
	S-Layer	Monomolecular protein layer on the exterior of cells that can provide protection against phage, act as a barrier to entry of high molecular weight molecules, help stabilize the cell, and act as an adhesion site for exoproteins. The S-layer is associated with the LPS in gram-negative bacteria, with peptidoglycans in Gram-positive bacteria, with the lipid membrane in Gram-negative Archaea, and with Pseudomurein or Methanochondroitin in Gram-positive Archaea.
Cell exterior	Glycocalyx	A heterogeneous layer of polysaccharides, protein, and DNA that encapsulates the cell and provides protection against predation and desiccation. A diffuse irregular layer is known as a slime layer and a more defined distinct layer is known as a capsule.
Appendages	Flagella	Long appendages that impart motility to a cell.
	Fimbriae or pili	Hollow fine protein structures that aid in adhesion to other cells and surfaces.

environmental bacteria, and the discussion that follows pertains to cells that have been successfully cultured.

Bacteria grown in the laboratory average $0.5-1$ μm in diameter and $1-2$ μm in length and have the basic composition shown in Table 2.1. They are generally characterized by high rates of replication (*Escherichia coli* can replicate by binary fission in less than 10 minutes), high surface area-to-volume ratio and genetic malleability. They have a single large circular chromosome located in the cytoplasm and there is no compartmentalization of the cell (Figure 2.2). The relative simplicity of the bacterial cell allows it to rapidly respond and adapt to changing environmental conditions.

Actinomycetes are technically classified as bacteria, but are unique enough that they are discussed and frequently cited as an individual group. What distinguishes actinomycetes from other typical bacteria is their tendency to branch into filaments or hyphae that structurally resemble the hyphae of fungi, only smaller in nature. Overall, actinomycetes are Gram-positive organisms that are highly prevalent in soils. Actinomycetes are important antibiotic producers, and are also responsible for the production of geosmin, which can cause odor problems in potable water.

2.2.2 Bacterial Cell Envelope

Those bacteria that have been cultured can be structurally separated into two major groups based on their cell envelope architecture: Gram positive or Gram negative (Figure 2.3). This major architectural difference helps dictate strategies for survival in the environment. For example, the thick cell wall of Gram-positive bacteria, such as in *Bacillus* and *Clostridium*, helps them withstand the harsh physical conditions found in soil environments. On the other hand, the more complex architecture of the cell envelope in Gram-negative bacteria such as *Pseudomonas* and *Shewanella* seems to aid these microbes in interacting with mineral surfaces and solutes in the environment to obtain required nutrients for metabolism.

Starting from the interior side of the cell envelope, both types of bacteria have a cytoplasmic membrane that is impermeable to many of the nutrients the cell needs for growth and energy production (Figure 2.4). Consequently, embedded throughout the cytoplasmic membrane are membrane-spanning proteins specific for the transport of molecules into and out of the cell. These proteins turn the cytoplasmic membrane into a semi-permeable structure that separates the cytoplasm from the exterior of the cell.

TABLE 2.1 Overall Macromolecular Composition of an Average *E. coli* B/r Cell

Macromolecule	Percentage of Total Dry Weight	Weight per Cell (10^{15} × Weight, Grams)	Molecular Weight	Number of Molecules per Cell	Different Kinds of Molecules
Protein	55.0	155.0	4.0×10^4	2,360,000	1050
RNA	20.5	59.0			
23S rRNA		1.0	1.0×10^6	18,700	1
16S rRNA		16.0	5.0×10^5	18,700	1
5S rRNA		1.0	3.9×10^4	18,700	1
Transfer		8.6	2.5×10^4	205,000	60
Messenger		2.4	1.0×10^6	1380	400
DNA	3.1	9.0	2.5×10^9	2.13	1
Lipid	9.1	26.0	705	22,000,000	4
Lipopolysaccharide	3.4	10.0	4346	1,200,000	1
Peptidoglycan	2.5	7.0	$(904)_n$	1	1
Glycogen	2.5	7.0	1.0×10^6	4360	1
Total macromolecules	96.1	273.0			
Soluble pool	2.9	8.0			
Building blocks		7.0			
Metabolites, vitamins		1.0			
Inorganic ions	1.0	3.0			
Total dry weight	100.0	284.0			
Total dry weight/cell		2.8×10^{-13}g			
Water (at 70% of cell)		6.7×10^{-13}g			
Total weight of one cell		9.5×10^{-13}g			

Adapted with permission from Neidhardt et al. (1990).

Other important functions of the cytoplasmic membrane and its embedded proteins are electron transport and energy generation for the cell, as well as biosynthesis of structural molecules and secondary metabolites such as antibiotics that are exported from the cell.

Moving on to the exterior of the cell envelope, both types of bacteria have a cell wall made of peptidoglycan that is external to the cytoplasmic membrane. One important function of the cell wall is to maintain the shape and integrity of the cell giving rise to various bacterial morphologies ranging from the bacillus (rod) and coccus (round) to the spirillum (twisted), vibrio (comma-shaped) and even stalked bacteria. The cell wall is composed of repeating units of *N*-acetylmuramic acid (NAM) and *N*-acetylglucosamine (NAG) attached to each other through peptide crosslinking (Figure 2.3). This NAM-NAG network forms a rigid porous structure that freely allows molecules of <15,000 MW to gain access to or

diffuse away from the cytoplasmic membrane. In Gram-negative bacteria, the cell wall is a thin NAM-NAG layer sandwiched between the periplasmic space and the outer membrane (Figure 2.3). The periplasmic space is well defined and contains transport proteins, signaling proteins and degradative enzymes that support growth and metabolism. Continuing the journey toward the exterior of the Gram-negative cell envelope, there is a second membrane called the outer membrane that is attached to the cell wall by lipoproteins. The inner leaflet of the outer membrane is structurally similar to the cytoplasmic membrane, while the outer leaflet contains immunogenic lipopolysaccharides (LPS) that extend out from the cell into the environment. LPS confers a negative charge to the cell, and have both antigenic (causes an immune response) and endotoxic (potentially toxic to humans and animals) properties. The outer membrane has a variety of functions. It acts as a diffusion barrier against large molecules such as

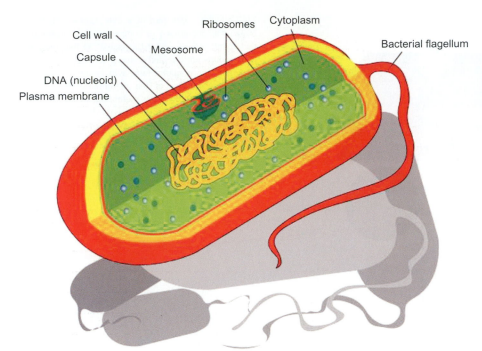

FIGURE 2.2 The components of a bacterial cell are illustrated. All bacteria have the cellular DNA dispersed in the cytoplasm, ribosomes, cell membranes, a cell wall and a capsule. Some bacteria do not have flagella, some have a single flagellum and some have multiple flagella. Not all bacteria have pili, but bacterial pathogens do, and the pili allow them to attach to host cells.

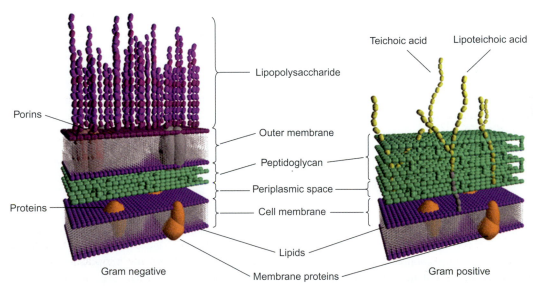

FIGURE 2.3 A comparison of the cell envelope of Gram-negative and Gram-positive bacteria. The Gram-negative cell envelope is characterized by two membranes, the inner membrane and the outer membrane, which are separated by a thin layer of peptidoglycan called the cell wall and the periplasmic space. On the exterior side of the cell wall, lipopolysaccharide molecules stretch out and mediate cell interactions with the environment. The Gram-positive cell wall is characterized by a single cell membrane that is interior to a thick layer of peptidoglycan (cell wall). In this case, teichoic acids stretch out from the cell and mediate interactions with the environment. Drawing courtesy Wayne L. Miller, McGill University.

antibiotics; it contains phage receptors and is involved in the process of conjugation (DNA exchange); it has specific nutrient uptake systems, e.g., for iron, vitamins and sugars; it contains passive diffusion pores that allow entry of low molecular weight substrates; and, finally, it provides protection for periplasmic proteins.

In Gram-positive bacteria, the cell wall is made up of many stacked layers of peptidoglycans to form a thick structure. In addition, there are covalently bound negatively charged teichoic acids, polymers of glycerol or ribitol joined by phosphate groups that extend out from the surface of the cell wall. They are antigenic and help

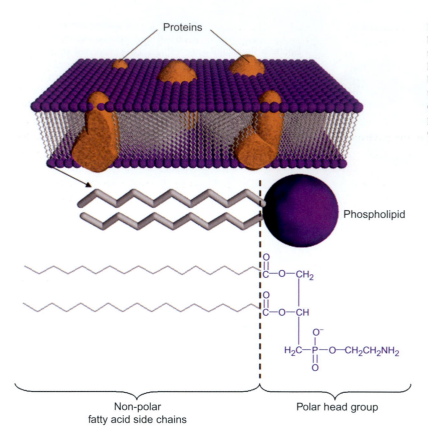

Proteins

Phospholipid

Non-polar
fatty acid side chains

Polar head group

FIGURE 2.4 Structure of a cytoplasmic cell membrane. The membrane is comprised of phospholipids that spontaneously orient themselves so that the polar head groups are directed to the exteriors of the membrane and the nonpolar fatty acid groups are directed toward the interior of the membrane. Proteins are embedded throughout the membrane to aid in transport of molecules into and out of the cell. Drawing courtesy Wayne L. Miller, McGill University.

mediate interactions of the cell, e.g., surface adhesion, with the environment and other microorganisms. To the interior of the cell wall, there is a periplasmic space (much less well defined than for Gram-negative bacteria), which has recently been identified in several Gram-positive microbes, and is thought to be involved in peptidoglycan synthesis (Matias and Beveridge, 2006).

2.2.3 Bacterial Cytoplasm

Cell replication and protein synthesis is centered in the cell cytoplasm, a complex gel-like matrix composed of water, enzymes, nutrients, wastes and gases, as well as ribosomes responsible for protein synthesis, a single circular chromosome and a varied number of small circular auxiliary plasmids ranging up to several thousand base pairs (kbp) (Figure 2.2). The chromosome is localized in the cytoplasm in a region called the nucleoid. Bacterial chromosomes average 4 million base pairs (Mbp) in size and encode for several thousand genes (see Information Box 2.2). It is an amazing feat of packaging that allows the chromosome to fit into the nucleoid region. When stretched out, the chromosome is approximately 1.3 mm in length in comparison to the cell which is 1−2 μm. Thus, the cell uses a number of tightly regulated proteins including small nucleoid-associated ("histone-like")

> **Information Box 2.2   Chromosome Size and Environmental Niche**
>
> The environment selects for the fittest microorganisms and thus each bacterium has only the necessary genes for survival (to have additional genes simply requires additional energy to maintain them). Here we compare two bacteria, *Pseudomonas aeruginosa* (a generalist) and *Nitrosomonas europaea* (a specialist). *P. aeruginosa* is a versatile Gram-negative chemoheterotroph found in soil, marshes, and marine habitats as well as on plants and animals, for which it is an opportunistic pathogen. The *P. aeruginosa* genome is 6.3 Mbp with 5570 predicted genes. Approximately 10% of these genes are transcriptional regulators or environmental sensors so that the bacterium can quickly respond to its environment. Compare this to *N. europaea*, which is a Gram-negative chemoautotroph that specializes in the oxidation of ammonia to nitrite. *N. europaea* is found in soil, freshwater, and sewage and in polluted environments with elevated levels of ammonia. The *N. europaea* chromosome is 2.8 Mbp with 2715 predicted genes. Thus, the specialist has a chromosome that is less than half the size of the generalist.

proteins and structural maintenance of chromosomes (SMC) complexes to coil, bend and ultimately condense the chromosome so that it packs into the cell, but yet is available for replication, translation (synthesis of messenger RNA) and recombination (gene rearrangement)

(Thanbichler and Shapiro, 2006). This feat is even more impressive when one considers that in an actively growing cell, there may be from two to four copies of the chromosome being actively replicated at the same time.

Plasmids are DNA sequences that are separate from the chromosome. Normally, plasmids encode genes that are not mandatory for cell growth and division but that often make the cell more competitive in a particular niche in the environment. Plasmids are often only retained if there is a selective pressure, such as the presence of an antibiotic to maintain a plasmid which confers antibiotic resistance. The relationship between plasmids and the chromosome is complex because some plasmids can integrate into the chromosome during

replication and function as part of the chromosome. During later replications, this process can be reversed, with the plasmid DNA being excised and allowed to function as a self-replicating entity within the cell. Some of the more important functions of plasmids are shown in Information Box 2.3. Plasmids are autonomous in that the plasmid copy number, or number of identical plasmids per cell, is normally independent of the number of chromosome copies. Plasmids are also expendable, meaning that the accessory genetic element is not essential to the growth of the organism in its normal environment. Plasmids range in size from 10 to 1000 kbp, and a bacterium can harbor a single or several different plasmids with variable copy numbers. Plasmid types are shown in Information Box 2.4.

Ribosomes are the other distinctive feature of the cytoplasm. Ribosomes transcribe messenger RNA into proteins that carry on the basic metabolism of the cell. Ribosomes are composed of a large (50S) and a small (30S) subunit that each contain both ribosomal RNA (rRNA) and proteins. The importance of the ribosomal RNA is illustrated by the highly conserved nature of regions of the gene that encodes for the rRNA. In fact, because it has a combination of both highly conserved and highly variable regions, the 16S rRNA gene that encodes for the 16S rRNA component of the small (30S) subunit of the ribosome is currently used for classification of the Bacteria and the Archaea.

2.2.4 Bacterial Glycocalyx

Finally, the exterior of the cell can have some important features. Some bacteria have an extracellular layer composed primarily of polysaccharide, but which can also contain proteins and even nucleic acids known as extracellular or eDNA. This layer is called a glycocalyx, also known as

---

Information Box 2.3   Bacterial Plasmid Functions

Plasmid Type	Function	Reference
Resistance plasmids	Antibiotic resistance	Brooks et al., 2007
	Mercury resistance	Smit et al., 1998
Degradative plasmids	2,4-D degradation	Newby et al., 2000
Plant-interactive plasmids	Tumor induction for crown gall disease nodule formation and nitrogen fixation in rhizobia	Cho and Winans, 2007, Rogel et al., 2001
Fertility plasmids	Conjugative plasmids that contain *tra* genes	Van Biesen and Frost (1992)
Col plasmids	Code for the production of colicins or proteins that kill other bacteria	Riley and Gordon, 1992
Virulence plasmids	Code for toxins in pathogenic bacteria	Sayeed and McClane, 2007
Cryptic plasmids	Unknown	Srivastava et al., 2006

Information Box 2.4   Types of Plasmids

Low-copy-number plasmids. Plasmids larger than 10 kbp that have one or two copies per cell.

High-copy-number plasmids. Smaller plasmids (<10 kbp) that have 10 to 100 copies per cell.

Relaxed plasmids. Plasmids whose replication does not depend on initiation of cell replication. Therefore these plasmids can be amplified (i.e., copy number increased) relative to cell number.

Stringent plasmids. These plasmids are dependent on cell replication, and plasmid replication is synchronized with replication of the bacterial chromosome. Thus, stringent plasmids have low copy numbers that cannot be amplified. Since cells growing rapidly may have three or four chromosomes, stringent plasmids can still be present with copy numbers greater than one per cell.

Conjugative plasmids. Self-transmissible plasmids that can be transferred from one bacterial cell to another during conjugation.

The cells can be of the same species or of different species. Conjugative plasmids are normally large and contain transfer genes known as **tra genes**.

Nonconjugative plasmids. These do not contain *tra* genes and are not self-transmissible. However, some plasmids can transfer to other cells by "mobilization" to other conjugative plasmids, although not all nonconjugative plasmids are mobilizable. In the process of mobilization, transfer of non-conjugative plasmids relies on the *tra* genes of the conjugative plasmids.

Incompatible plasmids. Plasmids vary in their ability to coexist within the same cell. Incompatible plasmids cannot exist together and give rise to incompatibility groups (Inc groups). Compatible plasmids belong to different Inc groups and vice versa.

a slime layer (more diffuse and irregular) or capsule (more defined and distinct) (Figure 2.5). The resulting sticky layer provides protection against desiccation, predation, phagocytosis, and chemical toxicity, such as from antimicrobials, and acts as a means of attachment to surfaces. Glycocalyx-producing bacteria, such as *Pseudomonas* spp., are often found associated with biofilms and microbial mats (Section 6.2.4). This material has been found to bind metals and is being used commercially in the binding and removal of heavy metals from industrial waste streams (Chapter 18).

2.2.5 Bacterial Appendages

Several accessory structures extend from the cell envelope out into the environment surrounding the cell. These appendages are not present in all bacterial types, but they are common, and they typically aid bacteria with either motility or attachment to surfaces. The flagellum (plural flagella) is a complex appendage used for motility (Figure 2.6). Motility is important in aiding a bacterial cell to move short distances (μm) toward nutrients

(positive chemotaxis) and away from potentially harmful chemicals (negative chemotaxis). Pili and fimbriae are any surface appendages that are not involved in motility. Fimbriae (singular fimbria) are numerous short surface appendages. The fimbriae aid in attachment of cells to surfaces, and so are important for initial colonization for biofilm formation and also for cell attachment to initiate an infection process. Pili (singular pilus) are normally less numerous than fimbriae but are longer. They are found only on Gram-negative bacteria, and are involved in a mating process between cells known as conjugation (Figure 2.7). In this process the exchange of DNA is facilitated by a pilus forming a connection between two cells. Conjugation in environmental bacteria is important because it enhances microbial diversity, often allowing specific populations to "fit" better in their environmental niche. It has recently been discovered that some bacteria also form extracellular filaments that are conductive (nanowires). During anaerobic respiration (Section 2.2.8),

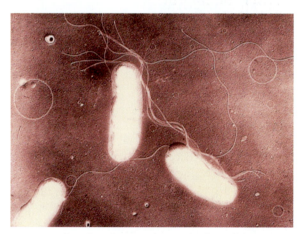

FIGURE 2.6 Scanning electron micrograph of a soil bacterium with multiple flagella. The circles are detached flagella that have spontaneously assumed the shape of a circle.

Slime Layer

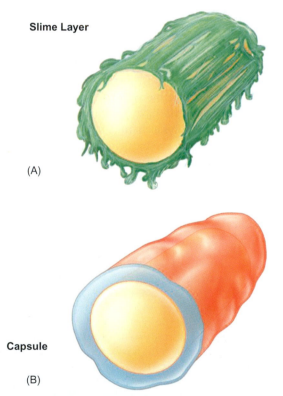

(A)

Capsule

(B)

FIGURE 2.5 The surface of bacteria is covered with a sticky coating called a glycocalyx. The glycocalyx is composed of polysaccharides (sugars) and proteins. The bacterial glycocalyx has two forms: (A) a loose slime layer and (B) a rigid capsule. Capsules are found on many pathogenic (disease-causing) bacteria including *Streptococcus pneumoniae*, which causes a respiratory infection of the lungs. The glycocalyx has several functions including: protection, attachment to surfaces and formation of biofilms.

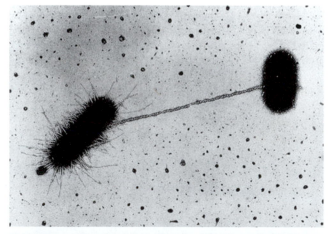

FIGURE 2.7 Bacterial cells exchanging DNA by conjugation using a pilus.

the nanowires transfer electrons directly from the bacterial cells to solid-phase terminal electron acceptors (e.g., iron oxides) in a process termed direct extracellular electron transfer (DEET) (Chapters 3 and 18).

2.2.6 Bacterial Endospores

Some Gram-positive bacteria, such as *Bacillus* and *Clostridium* spp., produce endospores, multi-layered structures capable of withstanding adverse conditions including radiation, UV light, heat, desiccation, low nutrients and chemicals, to ensure the survival of the cell. Endospores are environmentally significant because they can remain in a metabolically dormant state for long periods of time only to germinate and reactivate when conditions become favorable for growth.

2.2.7 Genetic Information Transfer

Perhaps the most unique ability of bacteria is their ability to quickly respond to changing environmental conditions. This can be attributed to their rapid growth and the flexibility of the bacterial chromosome. Bacteria readily incorporate new DNA into their genome through homologous recombination. Homologous recombination involves the alignment of two DNA strands of similar sequence, a crossover between the two DNA strands and a breaking and repair of the DNA at the crossover point to produce an exchange of material between the two strands. The acquisition of new DNA generally occurs via lateral or horizontal gene transfer by one of three mechanisms—conjugation, transduction or transformation—which allow for the exchange of chromosomal and plasmid DNA. The relative importance of these DNA transfer mechanisms is still not known but all have been shown to occur in the environment. Variations of these three methods can be used in the laboratory to genetically modify an organism.

Conjugation relies on direct cell-to-cell transfer of conjugative plasmid DNA through a protein pilus (Figure 2.7). The pilus is extended from a donor cell (containing a conjugative plasmid) to a recipient cell (lacking a conjugative plasmid). The conjugative plasmid is similar to other plasmids in that it can code for a variety of nonessential genes, such as antibiotic resistance or degradation genes. Unlike other plasmids, however, conjugative plasmids also code for the *tra* genes, genes coding for the production of the sex pilus. When a donor cell encounters a recipient cell, the pilus is formed and allows for the replication and transfer of a copy of the conjugative plasmid from donor to recipient. Upon receipt of the plasmid, the recipient now becomes a donor cell capable of spreading the plasmid and its corresponding genes to another recipient cell. Conjugation is thought to require a high cell concentration to increase the odds of encounter between

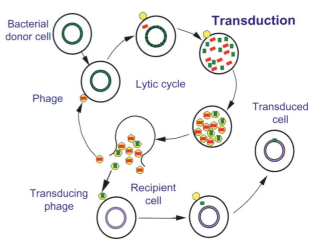

FIGURE 2.8   Transduction is the transfer of host bacterial DNA to a recipient bacterium via a bacteriophage.

compatible donor and recipient cells. Conjugation, since it is dependent on plasmids, is thought to play a significant role in the rapid transfer of plasmid encoded genes, e.g., antibiotic resistance, among bacterial populations.

Transduction occurs due to the accidental packaging of cellular genetic material during bacteriophage replication inside of its host cell (Figure 2.8). Transducing viruses sacrifice some of their own genome in place of the host's genetic material, resulting in a virus that can still infect a recipient cell but can no longer replicate. When a transducing virus infects a recipient cell, the host's genetic material is incorporated into the recipient's genome. Since the infecting transducing virus is replication defective, the recipient cell continues to grow and metabolize like normal with the acquisition of new genes. The genetic material picked up by transducing viruses reflects a variety of genes, some useful for the recipient organism and others not. Transduction can occur at lower cell concentrations since the process relies on viruses as carriers of genetic information. This process is extensively used in biotechnology for the introduction of genes into cells.

Transformation occurs when a bacterial cell obtains free DNA from its surrounding medium (Figure 2.9). When cells die, they readily lyse releasing cellular contents including chromosomal and plasmid genetic material. Much of this material is rapidly degraded by nucleases in the environment, but some can be adsorbed onto the surfaces of soil particles and organic matter, which can protect the DNA from degradation for long periods of time. Approximately one in every thousand cells is thought to be competent or capable of transporting DNA directly into the cell. Competency is the ability of a cell to transport DNA from its external environment inside the cell and is dependent on the stage of growth and the cell concentration. For example, an exponentially growing culture of 10^7-10^8 cells/mL of *Streptococcus pneumoniae* secretes a competency protein that initiates the production of several other proteins that convert the cell into one capable of taking up external DNA, e.g., through the production of DNA-specific transport proteins. The uptake of DNA is random; however, when it occurs, the DNA can become incorporated into the genome of the organism increasing genetic diversity.

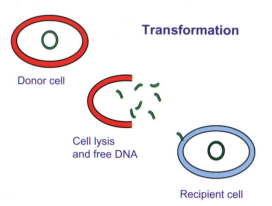

Transformation

Donor cell

Cell lysis
and free DNA

Recipient cell

FIGURE 2.9 Transformation is the uptake of free DNA released from a dead or dying cell by a recipient cell.

2.2.8 Bacterial Metabolism

Full consideration of bacterial metabolism is provided in general microbiology textbooks. Here we simply summarize the four major types of metabolism based on the source of energy and carbon used for growth (Table 2.2). Energy can be obtained from light through photosynthesis (phototroph), or from the oxidation of organic or inorganic chemicals (chemotroph). Carbon is obtained either from carbon dioxide (autotroph) or from organic compounds such as glucose (heterotroph or organotroph). Thus, chemoheterotrophs (chemoorganotrophs) use organic compounds both for energy and for carbon, chemoautotrophs (chemolithotrophs) obtain their energy from the oxidation of inorganic compounds and their carbon from carbon dioxide, photoautotrophs obtain energy from light and fix carbon from carbon dioxide and, finally, photoheterotrophs obtain energy from light and carbon from organic compounds.

There are two ways in which bacteria can harvest energy to use for building new cell material, respiration and

TABLE 2.2 Metabolic Classification of Bacteria

Metabolism	Electron Donor (Terminal Electron Acceptor)	Carbon Source	Metabolism Type	Products
			Chemoheterotroph	
Respiration (aerobic)	organic compounds (O_2)	organic compounds	*Pseudomonas, Bacillus*	CO_2, H_2O
(anaerobic)	(e.g., NO_3, Fe^{3+}, SO_4^{2-})[a]		*Micrococcus, Geobacter, Desulfovibrio*	CO_2, NO_2^-, N_2O, N_2, Fe^{2+}, S, S^{2-}
Fermentation (anaerobic only)	organic compounds (organic acids)	organic compounds	*Escherichia, Clostridium*	CO_2, organic acids, alcohols
			Chemoautotroph or Chemolithotroph	
Chemolithotrophy (aerobic) (anaerobic)[a]	H_2, S^{2-}, NH_4^+, Fe^{2+} (O_2) (NO_3)	CO_2	Hydrogen bacteria, Beggiatoa, Planctomycetes	H_2O, SO_4^{2-}, NO_2^-, Fe^{3+}
			Photoautotroph	
Photosynthesis (oxygenic)	Light + H_2O ($NADP^+$)	CO_2	*Cyanobacteria*	O_2
(anoxygenic)	Light + H_2S (bacteriochlorophyll)	CO_2	bacteria including: purple sulfur bacteria e.g., *Chromatium*; purple nonsulfur bacteria, e.g., *Rhodospirillum*; green nonsulfur bacteria, e.g., *Chloroflexus*; Heliobacteria, e.g., *Heliobacterium*	S^0
			Photoheterotroph	
Photoheterotrophy	light + H_2S (bacteriochlorophyll)	organic compounds	many purple nonsulfur bacteria purple sulfur bacteria to a limited extent	S^0

[a]*The majority of chemoautotrophs are aerobic. However, there have been several descriptions of anaerobic chemoautotrophs including those that participate in ammonia oxidation (Anammox) using NO_2^- as the terminal electron acceptor (from the order Planctomycetes, e.g., Brocadia anammoxidans) and those that participate in sulfur oxidation using NO_3^- as a terminal electron acceptor (Thiobacillus denitrificans).*

fermentation. In respiration, the cell uses a combination of substrate level phosphorylation (provides a small amount of ATP) and oxidative phosphorylation, which combines the tricarboxylic acid cycle (TCA cycle) and electron transport chain (provides a large amount of ATP) to generate ATP energy and reducing power. Key for respiration is the terminal electron acceptor (TEA) that is used to receive electrons from the electron transport chain. Under aerobic conditions, the TEA is oxygen which maximizes the amount of energy produced; a net total of 38 moles of ATP per mole glucose metabolized. Under anaerobic conditions, an alternate TEA such as NO_3^-, Fe^{3+}, SO_4^{2-} or CO_2 is used. However, the relative amount of energy that can be derived from the reduction of alternate terminal electron acceptors is less than for oxygen. Thus, respiration under anaerobic conditions is always less efficient than under aerobic conditions (less energy is produced). It should be noted that some alternate TEAs are very close to oxygen in the amount of energy produced, most notably nitrate. Thus, many bacteria are facultative anaerobes in that if oxygen is present they use it as a TEA, but if it is not present, they can use NO_3^- instead. A good example of a facultative genus is *Pseudomonas*. Although facultative anaerobes can use either oxygen or nitrate, the range of TEAs they can use is limited. For example, the sulfate-reducers are obligate anaerobes that specialize in using sulfate as a TEA.

Fermentation is an anaerobic process that uses only substrate-level phosphorylation with a net generation of 2 moles of ATP per mole glucose (thus there is no use of the electron transport chain or need for an external electron acceptor). Instead, electrons are shunted among organic compounds usually ending in the production of organic acids or alcohols resulting in very small amounts of energy. Thus, in fermentation, the end products include a combination of CO_2 and organic acids and alcohols. Fermentation is a process that has been taken advantage of in the manufacture of alcoholic beverages and a variety of other food products (vinegar, olives, yogurt, bread, cheese).

In considering metabolism in terms of this textbook, it is important to consider the environment and which type of respiration or photosynthesis will occur given the local availability of oxygen. In fact, it must be recognized that there is not simply a presence or absence of oxygen in the environment but rather a continuum of concentrations. As a result, within a single tiny niche, there may be both aerobic and anaerobic respiration occurring (as well as fermentation). Each of these types of metabolism has energy requirements and energy outputs that can be calculated. These calculations are shown in Chapter 3.

Overall, because of the ubiquitous nature of bacteria and their impact on human health and welfare, the habitats, functions and importance of bacteria are described in detail throughout this textbook.

2.2.9 The Archaea

The archaeans are microbes that look somewhat similar to bacteria in size and shape under the light microscope, but they are actually genetically and biochemically quite different. They appear to be a simpler form of life, and may in fact be the oldest form of life on Earth. Archaeans were originally thought only to inhabit extreme environments, leading to their being described as extremophiles, but more recently they have been shown to exist in a variety of normal or nonextreme environments. For example, Fan *et al.* (2006) identified similar archaeans from nonextreme environments in four Chinese and American pristine soils. The significance of the presence of archaeans in nonextreme environments has yet to be determined.

Interestingly, some aspects of archaean cell structure and metabolism are similar to those of bacterial cells. However, there are key differences in genetic transcription and translation that are actually more similar to those of the eukaryotes than to bacteria. Some of the key structural differences between archaeans and bacteria are identified in Table 2.3. Also of interest is the fact that the archaean lipids are based on the isoprenoid side chain, which is a 5-carbon unit that is also present in rubber. Archaeans have tough outer cell walls that contain different kinds of amino acids and sugars than those found in bacteria. The cell membranes are also different with glycerol-ether lipids

TABLE 2.3 Structural Comparison of Archaea and Bacteria

Structure	Archaea	Bacteria
Lipid	Glycerol-based phospholipid but stereochemistry of the glycerol is opposite that of bacteria and eukaryotes	Glycerol-based phospholipid
Membranes	Composed of glycerol-ether lipids	Composed of glycerol-ester lipids
Cell wall	Lacks peptidoglycan, contains surface layer proteins or S-layer	Peptidoglycan and S-layer
Flagella	Resembles Type IV pili	Resembles Type III pili
Chromosome	One circular chromosome	One circular chromosome

rather than the glycerol-ester lipids found in bacteria. The ether bonds are chemically more resistant than ester bonds and may help archaeans survive extreme environments.

2.2.9.1 Archaean Habitats

Many archaeans are extremophiles that can survive either hot or cold temperatures, or extreme salinity, alkalinity or acidity (Pikuta et al., 2007). Nonextreme archaeans have been found in a variety of environments including soil, seawater or even sewage. Originally it was thought that no pathogenic archaea existed (Cavicchioli et al., 2003). However, archaeans were later linked to clinical infections (Vianna et al., 2006). Archaeans are usually placed into three groups based on habitat. The two major divisions of archaea are the Crenarchaeota (mostly thermophiles) and the Euryarchaeota (mostly haloarchaeans and methanogens). The halophiles or haloarchaeans exist in saline environments. In contrast, methanogens live in anaerobic environments and produce methane. Methanogens can be found in low temperature environments, which is in contrast to thermophiles which are located in high temperature environments such as hot springs in Yellowstone Park (Information Box 2.5). Archaeans are also found in high numbers in cold marine environments (Giovannoni and Stingl, 2005).

2.2.9.2 Archaean Function

Many archaeans remain nonculturable, and this, coupled to the relatively short period of time since the discovery of many archaeans, means that information is limited on archaean physiology, function and the impact on global biochemical cycles. Despite this their widespread presence and major roles in extreme environments suggest that they are likely to be critically important to nonextreme ecosystems also. For example, it has recently been demonstrated that archaeans capable of nitrification are widely distributed (Santoro et al., 2010), and may actually control nitrification processes (instead of bacteria) in some ecosystems. Archaea have also been implicated as mediators of horizontal gene transfer between archaeans and bacteria (Nelson et al., 1999). Clearly, information on the archaea will increase dramatically in the near future, particularly information on archaeans found in nonextreme environments.

2.2.10 The Planctomycetes, Verrucomicrobia and Chlamydiae Superphylum

Planctomycetes, Verrucomicrobia and Chlamydiae (PVC) form a discrete phylum of the domain Bacteria, and possess very distinctive features that set them apart from other phyla. Structurally, the cell walls do not contain peptidoglycan, a universal component of most prokaryotic microbes. Also, they are capable of intracellular compartmentalization of cells, utilizing internal membranes and even membrane-bound nucleoids (Fuerst and Sagulenko 2011). Further, utilizing a process associated with internal vesicle formation, they have the ability to take up proteins from external solutions, similar to eukaryotic endocytosis (Lonhienne Thierry et al., 2010). Overall, the origin of these eukaryotic characteristics and traits are currently being debated, including the concept that the Planctomycetes may have evolved from a eukaryotic-like last universal common ancestor (LUCA) (Fuerst and Sagulenko, 2011). Thus Planctomycetes could be evolutionary intermediates between prokaryotes and eukaryotes.

Planctomycetes have been identified from all corners of the world, but are particularly prevalent in soils, fresh waters and marine environments (Buckley et al., 2006).

Information Box 2.5   Thermophilic Archaeans—How Do They Survive?

Hyperthermophilic archaeans are those that can grow in temperatures above 80°C. *Pyrococcus furiosus* can grow in temperatures from 73 to 103°C, with an optimum growth temperature of 100°C. Originally isolated from heated marine sediments on Vulcano Island, Italy, *P. furiosus* is an anaerobic bacterium capable of fermenting proteins and carbohydrates and has one of the fastest growth rates of 40 minutes observed among the hyperthermophiles. In order to survive in such extreme temperatures, structural modifications of the cell are required. Heat stabilization of proteins occurs through key amino acid substitutions that prevent protein denaturation under high temperatures. Membranes have increased concentrations of saturated fatty acids, which increase the hydrophobicity and stability of the membrane. Many hyperthermophiles have no fatty acids in their membranes, substituting them with long chain (e.g., C_{40}) hydrocarbons bonded together by ether linkages. These hydrocarbon membranes, conversely, form very temperature stable monolayers instead of the bilayers observed with fatty acid-based membranes.

Hyperthermophiles also have to prevent their DNA chromosome from melting under the high temperature. This is thought to be accomplished through the use of a reverse DNA gyrase enzyme that introduces positive supercoils into DNA. Nonhyperthermophilic bacteria use DNA gyrase to introduce the negative supercoils found in their DNA. Positive supercoiling prevents DNA denaturation. Additionally, some hyperthermophiles use heat-stable DNA binding proteins that can increase the temperature required to melt DNA by several-fold. Obviously, other structural modifications exist to stabilize lipids and small molecules in the cell. What is the upper temperature limit such modifications can afford? Because of the instability of ATP beyond 150°C, scientists think this is the upper limit for life. Only future research will tell if this is the case.

In addition, they have been identified as dominating the biofilm microbial communities associated with the surfaces of brown kelp seaweed in marine waters (Bengtsson and Øvreåes, 2010). Specifically, depending on the time of year, 24–53% of all bacteria within the kelp biofilm were identified as Planctomycetes. Two potential roles for Planctomycetes within biofilms include C and N nutrient cycling. Planctomycetes possess genes that encode enzymes for C_1 transfer, but the role of these genes is currently under debate (Chistoserdova *et al.*, 2004). Some Planctomycetes are capable of carrying out anammox reactions, which involves the anaerobic oxidation of ammonium to dinitrogen using nitrite as an electron acceptor as carbon dioxide is reduced. This chemoautotrophic metabolism occurs within a membrane-bound cell compartment called the anammoxosome, in some ways similar to eukaryotic mitochondria. Thus, biofilms with Planctomycetes could have potential for the removal of nitrogen from wastewaters (Kartal *et al.*, 2010). "Overall Planctomycetes challenge our concept of the bacterial cell and of a prokaryote as a structural cell type, as well as our ideas about origins of the eukaryotic nucleus" (Fuerst, 2010).

2.3 EUKARYOTES

Eukaryotes are more complex than prokaryotes and contain a true nucleus and membrane-bound cell organelles. Important groups of environmental eukaryotic microorganisms include fungi, protozoa and algae.

2.3.1 Fungi

While bacteria may represent the most abundant microorganisms in terms of numbers of individuals, the fungi, which are a physically larger group of eukaryotic microorganisms,

have the greatest biomass. In a landmark paper, 1.5 million fungal species were estimated to exist (Hawksworth, 2001) with only 7% of them identified so far (Crous *et al.*, 2006). Traditionally, the identification of fungi has been based on morphology, spore structure and membrane fatty acid composition. However, more recent estimates using high-throughput sequencing methods suggest that as many as 5.1 million fungal species exist (Blackwell, 2011).

Fungi are ubiquitous and primarily found in the soil environment where they can adapt to a variety of conditions and have a primary role as decomposers. As with bacteria, some fungi are pathogenic to both humans and plants (in fact, economically, fungi are the most important plant pathogens). Other fungi are important in industrial processes involving fermentation, and in biotechnology for the production of antimicrobial compounds (Table 2.4). Metabolically, fungi are chemoheterotrophs. Most fungi are obligately aerobic, but yeasts are facultative anaerobes and the zoosporic fungi found in ruminants are obligately anaerobic. These anaerobic fungi generally ferment sugars and in doing so produce a variety of useful by-products, such as ethanol, acetic acid and lactic acid, making them important commercially for production of many staple foods (e.g., yogurt, cheese, bread, pickles) and alcoholic products such as beer and wine.

In addition to their primary metabolism which supports biosynthesis and energy production, fungi are known for producing secondary metabolites (compounds produced during the stationary phase of growth). These secondary metabolites have revolutionized medicine, biotechnology and agriculture. For example, fungi are responsible for such antimicrobials as penicillin produced by *Penicillium notatum*, cephalosporin produced by *Cephalosporium acremonium* and griseofulvin produced by *Penicillium griseofulvum*. While the fungal production of antimicrobials under *in situ* conditions is not well

TABLE 2.4 Examples of Fungi and their Role in the Environment

Fungus	Common Environments	Role
Molds		
Rhizopus spp.	Spoiled food; soil; crops	Degradation; plant diseases, e.g., rice seedling blight
Penicillium spp.	Spoiled food; soil	Degradation; antibiotic (penicillin) production
Mushrooms		
Polyporus squamosus	Dead trees and plant material	Decomposition
Cryptococcus neoformans	Soil; can be airborne	Degradation; causes *Cryptococcus* (infection of the lungs and central nervous system in humans)
Yeasts		
Saccharomyces cerevisiae	Fruits; soils; aquatic environments	Fermentation; degradation
Candida albicans	Normal microbiota of animals	Causes candidiasis (yeast infections) of the skin or mucous membranes

understood, it is hypothesized that they help reduce competition from other microorganisms for nutrients.

2.3.1.1 Fungal Structure

Fungal membranes and cell walls are complex structures that act as selectively permeable barriers and protective outer barriers, respectively. The composition of these two structures varies somewhat among genera, in part due to the large variation in behaviors and life cycles, habitats and physiologies seen in the fungi. As eukaryotes, fungi have membrane-bound organelles in addition to a cytoplasmic membrane composed of a phospholipid bilayer with interspersed proteins for transport and degradation. Fungal membranes can be quite complex with structural and compositional differences observed in organelle membranes and in the life cycle stages. In addition to phospholipids, fungal membranes can include sterols, glycolipids and sphingolipids, which can be used for fungal identification as the ratio, type and amount of lipids can be species specific.

Fungal cell walls are multilayered structures composed of chitin, the glucose derivative *N*-acetylglucosamine (Figure 2.10). Fungal cell walls may also contain cellulose, galactosans, chitosans and mannans. Other cell wall components include proteins and lipids. Similarly to bacteria, the fungal cell wall lies outside of the cytoplasmic membrane, protecting the membrane from damage. The cell wall additionally provides the scaffolding for the fairly complex three-dimensional structures characteristic of some fungi, e.g., mushrooms.

2.3.1.2 Fungal Diversity

Fungi can be divided into three general groups based on morphological descriptions: molds, mushrooms and yeasts (Figure 2.11). Molds, such as *Aspergillus*, *Penicillium*,

Rhizopus and *Pilobolus*, are filamentous fungi which are found in many fungal phyla. Each filamentous fungal cell is called a hypha (pl. hyphae), which grows in mass to form tufts of hyphae or mycelia. Some hyphae extend out from the mycelium to form aerial hyphae responsible for the formation of asexual spores or conidia ranging from 1 to 50 μm in diameter. The fuzzy appearance of mold colonies is due to the aerial hyphae and the color of fungal colonies is the result of the coloration of the spores. Some molds produce sexual spores as the result of sexual reproduction. While not as resistant as bacterial spores, both asexual and sexual spores can be resistant to extreme temperatures, desiccation and chemicals, and are in large part responsible for the widespread occurrence of molds.

The mushrooms are part of the Basidiomycota, which are filamentous fungi that form the large fruiting bodies referred to as mushrooms. Aerial mycelia come together to form the macroscopic mushroom, whose main purpose is dispersal of the sexual basidiospores found underneath the cap. The rest of the mushroom fungus is below ground as a mycelium that extends outward for nutrient absorption. Both molds and mushrooms are important decomposers of natural products, such as wood, paper and cloth, as discussed below. However, both groups of fungi can additionally produce sticky extracellular substances that bind soil particles to each other to form stable soil aggregates that reduce soil erosion (Chapter 4). In some cases, fungi are thought to play a more important role in controlling erosion than plants.

The yeasts are unicellular fungi that are able to ferment under anaerobic conditions. Most important are *Saccharomyces* and *Candida*, which are members of the Ascomycota. While the yeasts do not produce spores, they are prolific in sugary environments, and are particularly associated with fruits, flowers and sap from trees. With a few exceptions where sexual reproduction occurs, yeasts

FIGURE 2.10　Structure of a fungal cell wall.

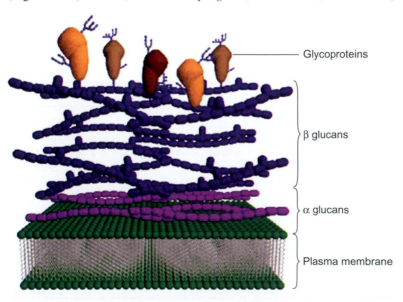

Glycoproteins

β glucans

α glucans

Plasma membrane

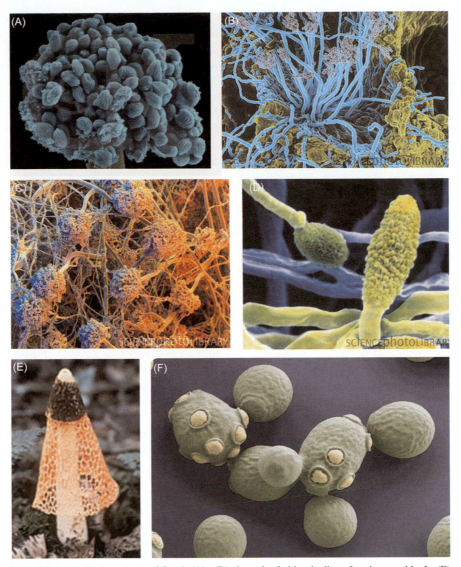

FIGURE 2.11   Various types of fungi. (A)–(D) show the fruiting bodies of various molds. In (E) is illustrated the Maiden Veil fungus or *Dictophora indusiata* within the *Basidiomycota* (mushroom). In (F) shown the budding yeast *Saccharomyces cerevisiae*.

reproduce by budding where the daughter cell forms as an outgrowth from the mother cell, eventually pinching off as a single cell. The number of bud scars left behind with each budding event can be used to estimate the number of replication cycles a particular yeast cell has undergone. While commonly found in the environment, yeasts, especially *Saccharomyces*, play a significant role in commercial applications, including the production of food, alcoholic and medicinal products. Some yeasts, e.g., *Candida*, can cause vaginal, oral and respiratory infections, and can experience a filamentous stage during pathogenesis.

Some fungi, usually members of the Ascomycota, establish symbiotic relationships with algae and cyanobacteria to form lichens. These are extremely important because lichens encourage mineral weathering through the secretion of organic acids that degrade rocks and other inorganic surfaces. The fungus:phototroph relationship actually occurs when fungal haustoria (hyphal projections) penetrate the algal cell wall. Both organisms have to be nutritionally deprived to establish their relationship. In exchange for the oxygen and organic carbon provided by the alga, the fungus provides water and minerals in addition to protection from harsh environmental conditions.

2.3.1.3 Fungal Environmental Aspects

Fungi are chemoheterotrophic microorganisms that rely on simple sugars for carbon and energy. However, simple sugars are limiting in many environments due to intense competition. Consequently, many fungi secrete extracellular enzymes (exoenzymes) to break down complex polymers to simple carbon compounds for cell utilization. Often referred to as saprophytic, fungi are extremely important in the degradation and recycling of dead plant, insect and animal biomass, especially the complex polymers associated with these organisms, e.g., cellulose and lignin found in plants, and chitin found in insects. The filamentous fungi and mushrooms are especially adapted to a saprophytic lifestyle due to the large surface area provided by their hyphae. Because of their unique ability to degrade complex polymers, fungi have been found to have the ability to degrade a variety of environmental contaminants making them important in waste degradation and recycling (Chapters 16 and 17). For example, the yeast-like fungus *Aureobasidium pullulans* has been found to degrade polyvinyl chloride (PVC) containing plastics (Webb *et al.*, 2000), and filamentous fungi such as *Penicillium, Stachybotrys, Allescheriella* and *Phlebia* can degrade the aromatic hydrocarbons associated with petroleum products and agricultural pesticides (Boonchan *et al.*, 2000; D'Annibale *et al.*, 2006). One special fungus is *Phanerochaete chrysosporium*, which is described further in Information Box 2.6.

A second important environmental group of fungi are the mycorrhizae. Mycorrhizae form symbiotic relationships with many plants. Through their increased surface area, mycorrhizae can increase the absorptive area of a plant's roots by hundreds of thousands of times, help to prevent desiccation of the roots and increase uptake of nutrients, especially phosphates. In return, the plant provides the fungus with sugars made during photosynthesis. Mycorrhizal fungi are present in 92% of all plant families studied and include ectomycorrhizal fungi (Wang and Qiu, 2006) and endomycorrhizal fungi (Rinaldi *et al.*, 2008).

Endophytes are defined as microbes that live within plants and include fungal species; they are frequently a source of natural products. For example, recently a novel endophytic fungus was isolated from wild pineapples in the Bolivian Amazon basin. This endophyte produces volatile organic compounds with antibiotic properties capable of killing plant pathogens such as *Pythium ultimum* and human pathogens such as *Mycobacterium tuberculosis* and *Staphylococcus aureus* (Mitchell *et al.*, 2010) (see also Section 19.4).

Phenotypically similar to both fungi and protozoa, slime molds produce spores but move with amoeba-like gliding motility. Phylogenetically, slime molds are more related to the amoeboid protozoa than the fungi. There are two types of slime molds. The cellular slime molds are composed of single amoeboid cells during their vegetative stage, while the vegetative acellular slime molds are comprised of plasmodia, amorphous masses of protoplasm. Both forms can be found in moist environments on decaying organic matter where they consume bacteria and other microorganisms via phagocytosis. Environmental factors, such as nutrients or stress, can trigger cell accumulation and differentiation into fruiting bodies for the production and dispersal of spores. The spores can later germinate into vegetative amoeboid cells. The consensus of individual vegetative cells coming together and forming fruiting bodies, implying cell-to-cell communication involving chemical signals, is of much interest to scientists (see also Chapter 20).

2.3.2 Protozoa

The 18S rRNA-based classification being used for eukaryotic microorganisms has revealed fundamental

---

Information Box 2.6　The White-Rot Fungi: The Ultimate Fungus?

Fungi, in general, are known for their degradative abilities, and much of the conversion of recognizable organic matter into unrecognizable organic matter is attributed to fungal activity. One group of fungi, however, far exceeds other groups of fungi in their ability to degrade recalcitrant and xenobiotic compounds. These fungi are known as the white-rot fungi. The majority of white-rot fungi are basidiomycetes and include members such as *Phanerochaete chrysosporium* and *Trametes versicolor*. The white-rot fungi are especially known for their ability to degrade lignin, a structurally complex component of wood, as a result of a secondary metabolic process that yields no energy for the fungus. Lignin seems to be accidentally degraded as white-rot fungi release extracellular enzymes to access the polysaccharides tied up in the lignin. Because these enzymes are oxidases, the process of lignin degradation is aerobic. White-rot fungi are the only identified organisms capable of significant lignin degradation. The three primary enzymes responsible for degradation include lignin peroxidase, manganese peroxidase, and laccase. The nonspecificity of these enzymes for their substrate target allows them to act on other substrates, including pollutants, to facilitate their degradation. Consequently, white-rot fungi are being actively pursued in bioremediation. To date, white-rot fungi have been noted to degrade munitions waste, for example, TNT (2,4,6-trinitrotoluene); pesticides, for example, DDT (1,1,1-trichloro-2,2-bis(4-chlorophenyl)ethane) and lindane; polychlorinated biphenyls used as plasticizers and in hydraulic fluids; polyaromatic hydrocarbons, benzene homologues found in fossil fuels; synthetic dyes such as azo dyes and triphenylmethane used in textiles and plastic; synthetic polymers such as plastics; and toxic wood preservatives such as creosote and pentachlorophenol (PCP) (Kullman and Matsumura, 1996; Pointing, 2001). Widespread in the environment and able to withstand harsh environmental conditions, white-rot fungi are being examined for *in situ* and *ex situ* remediation applications.

genetic differences among the protozoa. Consequently, it has emerged that the single-celled protozoa are a poly-phyletic (developed from more than one ancestral type) group of eukaryotic microorganisms that are grouped together based on similar morphological, physiological, reproductive and ecological characteristics. All protozoa rely on water, and as such they are most commonly observed in freshwater and marine habitats, although some are terrestrial in moist soils and others are exclusively found in the gastrointestinal tracts of animals. However, some protozoa are saprophytic, some are parasitic and some are photosynthetic. Protozoa are important environmentally because they serve as the foundation of the food chain in many aquatic ecosystems, making up a large part of the plankton fed on by aquatic animals (see also Chapter 6).

2.3.2.1 Structure and Function

Cell morphology among the protozoa is extremely diverse, perhaps due the absence of a cell wall and wide range in size (5 μm to 1 mm) (Figure 2.12). In protozoa, cell rigidity is provided, in part, by a gelatinous cytoplasmic material called the ectoplasm located just inside the cell membrane. The ectoplasm is where the cilia and flagella are anchored in motile protozoans. The cell membrane and the ectoplasm together are known as the pellicle. Finally, inside the pellicle is the fluid endoplasm which contains organelles. Similar to other microorganisms, the cell membrane is a phospholipid bilayer with interspersed proteins for nutrient and waste transport.

Within the protozoa cytoplasm, vacuoles serve a variety of functions. Phagocytic vacuoles participate in the digestion of food; contractile vacuoles maintain osmolarity for protozoa living in hypotonic environments; and secretory vacuoles contain enzymes for various cell functions. Also

apparent in the cytoplasm are multiple nuclei. In some protozoa, the nuclei are identical, while in others, such as the Ciliophora, there is a macronucleus and a micronucleus. The larger macronucleus is associated with cell growth and metabolism. The smaller micronucleus is diploid and is involved in genetic recombination during reproduction and regeneration of the macronucleus. Many genetically identical copies of each nucleus can exist in a cell.

Some protozoa form cysts or oocysts as part of a complex life cycle. Similar to spores in bacteria and fungi, cysts can increase the survival of the organism. Giardia, for example, produces cysts that persist under environmental conditions until transmission to an animal host occurs, at which point the cyst will undergo excystation to produce a vegetative cell resulting in giardiasis, a diarrheal disease. The ability to form resilient cysts substantially increases the likelihood of exposure, and thus the possibility of disease. In the last few decades, Giardia and another cyst-forming protozoan Cryptosporidium have been linked to outbreaks of waterborne illness.

Many protozoa are distinguished based on their structural morphology and mechanism of motility. Morphological characterizations are based on colony formation (single existence or in colonies), swimming style (sedentary or motile), external structures (naked, shelled or scaled) and pigmentation. Most protozoa are motile and are divided into taxonomic groups based on their mechanism of motility. For example, the Mastigophora use flagella; the Ciliophora use cilia (hair-like structures that extend outward from the cell membrane); the Sarcodina use ameboid motility; and the Apicomplexa are nonmotile.

2.3.2.2 Protozoan Environmental Aspects

Protozoa have a number of important ecological roles (Information Box 2.7). Many protozoa are chemoheterotrophic using either aerobic respiration or fermentation.

 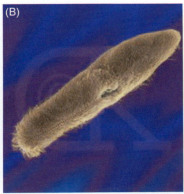 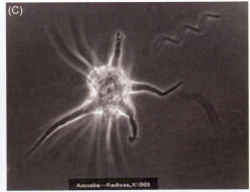

FIGURE 2.12 Morphological diversity seen among the protozoa. (A) Infection of an intestinal cell with *Cryptosporidium* (an Apicomplexan), specifically a sporozoite is attached to an intestinal epithelial cell with several merozoites emerging. For details on the life cycle see Chapter 22. (B) Paramecia are members of the Ciliophora. They are commonly found in freshwater environments where they feed on bacterial cells. (C) *Amoeba radiosa* is a member of the Sarcodina. It is found in many different environments and obtains its food by surrounding and engulfing it.

Information Box 2.7   Roles of Protozoa in Environmental Microbiology

- Serve as the base of the food chain in aquatic systems
- Population control through the predation of bacteria, algae, and even other protozoa
- Human and vertebrate parasites, food-borne and water-borne disease
- Degradation of complex organic materials, such as cellulose
- Symbioses with some animals, such as termites and ruminants

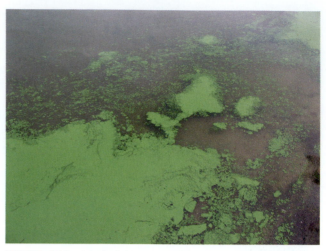

FIGURE 2.13   An algal bloom leaves a green scum-like substance on the waters of Lake Zurich, Switzerland. Courtesy Dr. Jennifer L. Graham, U.S. Geological Survey.

Interestingly, anaerobic and microaerophilic protozoa do not contain true mitochondria (as found in other eukaryotes). Instead, they rely on membrane-bound structures called hydrogenosomes for energy production. Hydrogenosomes lack many of the citric acid cycle enzymes normally associated with mitochondria and use protons as terminal electron acceptors forming molecular hydrogen instead of water.

Protozoa have an important role in the degradation and cycling of organic matter in the environment. These protozoa make and release a variety of extracellular enzymes for the degradation of polymers such as cellulose from plants and peptidoglycan from bacterial cell walls. Some bacteriovorous protozoa release such enzymes to aid in feeding on bacteria. Other protozoa engulf their nutrients via phagocytosis (engulfment) which are then degraded by digestive enzymes stored in phagocytic vacuoles. The ability to degrade large molecules contributes to the complex relationships these organisms have with animals. In fact, protozoa are responsible for up to one-third of fiber digestion in ruminants and contribute half the microbial mass in the rumen. In the anaerobic environment of the rumen, protozoa carry out fermentation producing organic acids and alcohols.

Water quality is an area where protozoa are having an increasingly important impact. Outbreaks of disease have been attributed to protozoa in drinking and recreational waters. The three most commonly reported protozoa affecting water quality are *Giardia* (*Mastigophora*), *Cryptosporidium* (*Apicomplexa*) and *Toxoplasma* (*Apicomplexa*). Originating from infected humans and animals, the cysts and oocysts of these organisms can withstand the water temperatures and salinities encountered to survive long periods of time in the environment (Fayer *et al*., 2004). They can also withstand chlorine disinfection (see also Chapters 22 and 25). Other important protozoa include the so-called brain-eating amoeba such as *Naegleria fowleri* and *Balamuthia mandrillaris*. These organisms can be found in water or soil, and result in brain encephalitis which is often fatal in humans (Niyyati *et al*., 2009) (see also Chapter 22).

2.3.3 Algae

Algae are a group of eukaryotic oxygenic photosynthetic microorganisms that contain chlorophyll *a* (as seen in plants). Algae range from single-celled organisms to complex multicellular organisms like seaweeds (Figure 2.13). Inhabiting a wide range of habitats from aquatic environments (freshwater, marine and brackish) to soils and rocks, only inadequate light or water seems to limit the presence of algae. Algae are most commonly found in saturated environments either suspended (planktonic), attached to surfaces or at the air—water interface (neustonic). Endolithic algae can be found in porous rock or as surface crusts on desert soils. Algae are often the predominant microorganisms in acidic (<pH 4) habitats, as seen with the red alga *Cyanidium* that can grow at <pH 2. Generally free-living, some algae have symbiotic relationships with fungi (lichens), mollusks, corals and plants, and some algae can be parasitic. Classification of algae is complex involving numerous cellular properties. For example, algae can be grouped based on cell wall chemistry, cell morphology, chlorophyll molecules and accessory pigments, flagella number and type of insertion in the cell wall, reproductive structures, life cycle and habitat. Based on cell properties, algae include the green algae (*Chlorophyta*), the euglenoids (*Euglenophyta*), the dinoflagellates (*Dinoflagellata*), the golden-brown algae, diatoms (*Chrysophyta*), the brown algae (*Phaeophyta*) and the red algae (*Rhodophyta*). However, similar to the protozoa, 18S rRNA criteria reveal a phylogenetically diverse group.

2.3.3.1 Cell Structure

Algae can be unicellular, colonial (occurring as cell aggregates) or filamentous, resulting in great diversity in overall cell morphology. Algal cell walls surround cytoplasmic membranes and are thin and rigid but vary in their composition. They generally contain cellulose with a variety of other polysaccharides including pectin, xylans and alginic acid. Some walls are calcareous containing calcium carbonate deposits. Chitin (a polymer of *N*-acetylglucosamine) may also be present in some algae. The euglenoids, however, differ from other algae by lacking cell walls. For diatoms, the cell wall is composed of silica giving rise to fossils. Other cell wall-associated structures include gelatinous capsules outside the cell wall for adhesion and protection, and flagella arranged in different patterns on the cell for motility.

All algae also contain membrane-bound chloroplasts containing chlorophyll *a* and other chlorophylls, such as chlorophyll *b*, *c* or *d*. Some contain differently colored pigments called xanthophylls, which can give rise to differently colored algae. Many algae also contain pyrenoids, which serve as sites for storage and synthesis of starch. Starch is one of the many types of carbohydrate storage that algae use to support respiration in the absence of photosynthesis. Other types of storage molecules include paramylon (β-1,2-glucan), lipids and lammarin (β-1,3-glucan).

2.3.3.2 *Algal Environmental Aspects*

Algae are primarily oxygenic photoautotrophs although a few are chemoheterotrophic using simple organic compounds, for example acetate, to help support cell metabolism. Oxygenic photosynthesis produces oxygen as a waste product obtaining energy from the breakdown of water. The production of oxygen is one of the desirable effects of algal growth in some aquatic systems. Oxygenic photosynthesis by algae is also responsible for primary production (production of organic matter) in many aquatic habitats. Thus, primary production by algae sustains the food web in many aquatic environments, and is equivalent to the role plants play in terrestrial systems.

Found in a variety of disparate habitats, the versatility of algal reproduction is partially responsible for their success. Algae can reproduce sexually and asexually with sexual reproduction involving the formation of eggs within structures called oogonia and sperm within antheridia. The egg and sperm fuse forming a diploid zygote resulting in a vegetative algal cell. Asexually, algae reproduce through binary fission or fragmentation, where fragments of filamentous algae break off and continue to grow. Binary fission is especially prevalent among the single-celled algae. Finally, some algae can produce spores (e.g., zoospores or aplanospores) that can germinate into fully functioning vegetative cells.

An interesting characteristic of some algae, especially the coastal dinoflagellates, is the production of secondary metabolites. Many of these metabolites are in the form of toxins released extracellularly. For example, the dinoflagellates *Gymnodinium* and *Gonyaulax* species can produce the neurotoxin saxitoxin that paralyzes muscles of the respiratory system in vertebrates. The toxin itself, potent at nanogram (ng) concentrations, does not harm shellfish; however, shellfish accumulate the toxin making them dangerous for consumption. Ciguatera is a disease resulting from the consumption of fish that have ingested or have accumulated the toxin of the dinoflagellate *Gambierdiscus toxicus*. The toxin itself survives cooking and can cause diarrhea and central nervous system disorders. Toxin accumulation in fish and shellfish generally occurs during blooms of dinoflagellates (algal blooms). Unfortunately, the occurrence of toxic algal blooms is on the rise nationally and internationally. This is due to a number of contributing factors, the most prevalent of which is the increasing nutrients, especially nitrogen and phosphorus, that are being released into coastal waters from sewage and agricultural runoff.

2.4 VIRUSES

Viruses are a unique group of biological entities that can infect eukaryotic or prokaryotic organisms. Although some viruses do contain a few enzymes, they are obligate parasites that have no metabolic capacity and rely on host metabolism to produce viral parts that self-assemble. Viruses consist of nucleic acid encapsulated within a protein coat known as the capsid of variable size (Figure 2.14) and morphology (Figure 2.15). Viral nucleic acids can consist of single- or double-stranded DNA or RNA. The replication of a virus can be described in five steps: (1) adsorption; (2) penetration; (3) replication; (4) maturation; and (5) release (Figure 2.16). All viruses share a common mechanism of replication at the molecular level, but different viruses replicate at varying rates. For example, prokaryotic viruses known as bacteriophage or phage infect bacteria and often replicate rapidly, in minutes, whereas a typical animal virus replicates in hours to days. Coliphages are viruses that infect coliform bacteria such as *Escherichia coli* (Figure 2.17). All viruses begin infection by adsorption to the host via specific receptors and injection of the nucleic acid or uptake of the total virus particle into the cell. The cycle then goes into what is known as the eclipse phase, a period of time during which no virus particles can be detected because of release and incorporation of the nucleic acid in the host cell machinery. Finally, new viral components are produced, assembled and released from the host by disruption of the cell or budding at the cell membrane surface. The latter release mechanism is less destructive to the host cell and may support a symbiotic condition between the virus and the host.

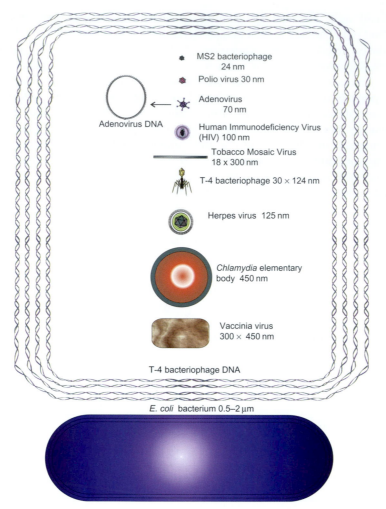

FIGURE 2.14   Comparative sizes of selected viruses in comparison to a bacterial cell and nucleic acids.

2.4.1 Infective Nature of Viruses

Outside their hosts, viruses are inert objects, incapable of movement. Thus, these tiny infectious agents require a vehicle, such as air or water, for transport. Once in contact with a potential host, viruses find their way into target cells using specific receptor sites on their capsid or envelope surfaces. This is why viruses of bacteria or plants do not normally infect humans and vice versa. Once viruses invade host cells and replicate, they can invade neighboring cells to continue the infection process. Infection of a cell by a virus is often but not always debilitating to the cell's regular functions. Thus, viral infection may be asymptomatic or may cause acute, chronic, latent or slow infections, or may cause cell death.

In bacteria, some viral infections appear to cause no immediate harm to the host cell. Therefore, the phage is carried by the host. These carrier hosts, however, may still be sensitive to other phage populations. This condition is known as lysogeny. In a lysogenic phage, also known as a temperate phage, the nucleic acid is integrated with the chromosome of the host, persisting indefinitely, and is transmitted to host descendants or daughter cells. This stable, noninfectious form of the virus is known as a prophage. The phage may remain latent for many generations and then suddenly be mobilized and initiate replication and eventually cause host lysis. Typically, only a portion of temperate phage becomes lysogenic, while other members of the population remain virulent, multiplying and lysing host cells. Similar to lysogenic bacteriophage, animal retroviruses integrate their nucleic acid into the cell chromosome, producing persistent infections. Such a cycle is typical of herpesvirus infections in humans. In fact, the herpesvirus may be passed from grandparent to grandchild, remaining dormant through two generations. Half of the human population is estimated to be infected by the age of 1 year, and up to 85% of the population is

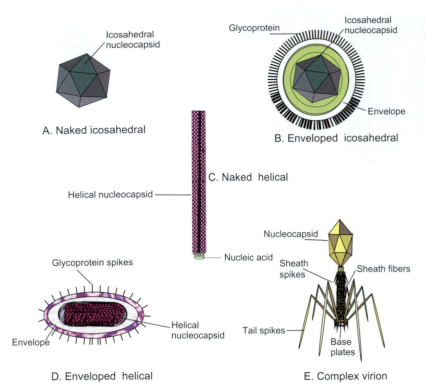

FIGURE 2.15 Simple forms of viruses and their components: (A) Naked icosahedral viruses resemble small crystals; (B) the enveloped icosahedral viruses are made up of icosahedral nucleocapsids surrounded by the envelope; (C) naked helical viruses resemble rods with a fine regular helical pattern in their surface; (D) enveloped helical viruses are helical nucleocapsids surrounded by the envelope; and (E) complex viruses are mixtures of helical, icosahedral and other structural shapes.

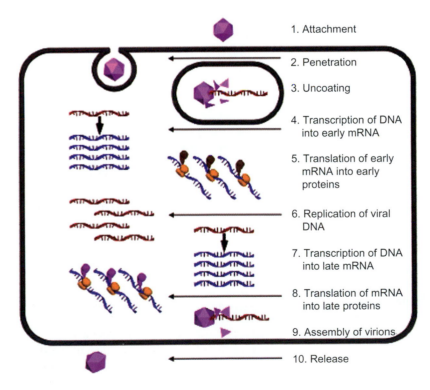

FIGURE 2.16 The basic steps of virus multiplication. Representation of an icosahedral DNA virus showing the main steps including adsorption, penetration, replication, maturation and release.

seropositive by puberty. Most latent animal viruses, such as mumps and measles viruses, lack the ability to lyse the host cell or prevent host cell division. Infection is therefore ensured by the production of infected daughter host cells. In latent infections, an equilibrium is reached between host and parasite until a nonspecific stimulus, such as compromised host immunity, evokes active infection.

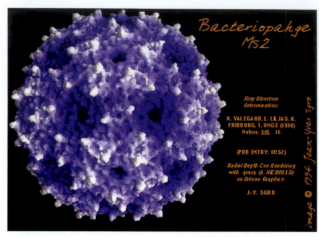

FIGURE 2.17 The bacteriophage MS2 infects the *E. coli* bacterium. Courtesy of Dr. Karin Valegard.

2.4.2 Prokaryotic Viruses

2.4.2.1 Structure and Life Cycle of Prokaryotic Viruses

The interactions of bacteriophages with a prokaryotic host are diverse and outlined in Information Box 2.8. Lytic phage are predators of prokaryotes. In contrast, lysogenic and chronic infections are actually a parasitic interaction that could be described as mutualism (Weinbauer, 2004). In the lytic cycle, the number of virions released per cell is known as the burst size. Figure 2.18 shows a bacterial cell visibly infected with phage. Overall, the size of phage usually ranges from 30 to 60 nm. The morphology of environmental phage varies considerably. Typically, phage consist of a head and a tail held together by a connector, but other forms can be cubic, spindle, filamentous or pleomorphic. Further on the metabolic state of phage and methods of viral infection and replication are given in Section 2.4.3 on eukaryotic viruses.

2.4.2.2 Prokaryotic Virus Environmental Aspects

Prokaryotic phage are abundant in a variety of environments (Table 2.5), and phage populations appear to be correlated with bacterial populations. Interestingly, estimates of phage populations vary depending on the methodology utilized in a manner similar to that for bacteria. When direct microscopic counts are made utilizing transmission electron microscopy, counts are two to three orders of magnitude higher than when traditional viable plaque counts are used (Ashelford *et al.*, 2003). Virus to bacteria ratios (VBR) are often used to illustrate the large number of phage in a sample. For example, large diverse populations of phage exist in marine environments (Sogin *et al.*, 2006) resulting in VBRs in aquatic systems that often average around 10 but sometimes exceed 100. Parada *et al.*

Information Box 2.8 Host/Phage Interactions

Lytic infection. Lytic or virulent phages redirect the host metabolism toward the production of new phages which are released as the host cell lyses.

Lysogenic infection. Phages nucleic acid material of the temperate or lysogenic phage remains dormant within the host as **prophage**. Prophage are replicated along with the host until the lytic cycle is induced.

Chronic infection. Infected host cells constantly release phage progeny by budding or extrusion without lysing the host cell.

Pseudolysogenic infection. Phages multiply in only a fraction of the infected host cells. Also known as the phagecarrier state.

Mode of Infection	Phage Type
Infection via pili or flagella	F-specific phage
Infection via recognition of outer host layer or polysaccharide capsule	Capsule phage
Infection via cell wall	Somatic phage

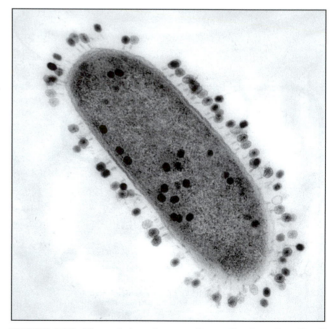

FIGURE 2.18 Transmission electron micrograph of an *E. coli* cell being infected by several T4 phages. Courtesy of J. Wertz.

(2007) reported marine virus to picoplankton ratios ranging from 9 at 100 m depth to 110 at a depth of 3500 to 5000 m (see Chapter 6). Picoplankton are aquatic microorganisms found in the size range 0.2 to 2 μm including small protozoa and bacteria. In comparison to aquatic systems, VBRs in terrestrial systems appear to be far more variable— ranging from <1 to several thousand (Ashelford *et al.*, 2003; Srinivasiah *et al.*, 2008).

TABLE 2.5 Incidence of Phages in Different Environments

Environment	Number of Phages (g^{-1} or mL^{-1})	Virus to Bacteria Ratio (VBR)
Marine		
Oligotrophic	$1.4 \times 10^6 - 1.6 \times 10^7$	3−110
Mesotrophic and eutrophic	$2.8 \times 10^6 - 4.0 \times 10^7$	
Freshwater		
Oligotrophic	$4.2 \times 10^6 - 4 \times 10^7$	3−9
Mesotrophic	$5.3 \times 10^6 - 1.4 \times 10^8$	
Eutrophic	$5.6 \times 10^6 - 1.2 \times 10^9$	0.1−72
Terrestrial		
Forest soil	$1.3 - 4.2 \times 10^9$	5−12
Agricultural soil	$1.5 \times 10^8 - 1.1 \times 10^9$	0.04−3346

Data compiled from Srinivasiah et al. (2008) and Williamson et al. (2005).

Information Box 2.9   Role of Phage in Environmental Microbiology

- Control of bacterial populations
- Control of specific bacterial pathogens
- Control of marine cyanobacteria
- Interactions with food web processes
- Interactions with biogeochemical cycles
- Enhanced prokaryotic diversity via horizontal gene transfer

Given these large numbers, phages play critical roles in the environment (Information Box 2.9). The direct role of phages is to control bacterial populations by induction of the lytic cycle, which causes bacterial cell lysis. Without phage as a controlling factor, bacterial populations could increase significantly. Controlled studies by Wommack and Colwell (2000) showed that additions of phage to bacterial populations resulted in a 20−40% decrease in bacterial numbers. Overall, phage and protozoa (nanoflagellates) are the two major predators of prokaryotes in marine waters. Control of bacterial populations occurs in both marine and soil environments and can be general or very specific. For example, bacteriophage SF-9 is known to specifically lyse *Shigella dysenteriae* Type 1 (Faruque *et al.*, 2003). Phage have also been explored for use in the biological control of fish disease. In this case, two different phage

were used to control the bacterium *Pseudomonas plecoglossicida*, the causative agent of bacterial hemorrhagic ascites disease in cultured ayu fish (*Pleoglossus altiuelis*) (Park *et al.*, 2000).

Viruses are also influential in controlling marine cyanobacteria. Such viruses are known as cyanophage, which infect numerically dominant primary producers such as the marine cyanobacteria *Prochlorococcus* and *Synechococcus* (Sullivan *et al.*, 2006). Due to the large numbers of bacteria that are lysed daily in marine waters, phages are important in both food web processes and biogeochemical cycles (Figure 2.19). Significant amounts (6−26%) of the carbon fixed by primary producers enter the dissolved organic matter (DOM) pool via virus-induced lysis at different trophic levels. In addition, it has been estimated that phage contribute from 1 to 12.3% of the total dissolved DNA in samples from freshwater, estuarine and offshore oligotrophic environments (Jiang and Paul, 1995).

Phages also influence bacterial diversity by mediating horizontal gene transfer through the process of transduction. Two types of transduction are known to be undertaken by phage. For generalized transduction, bacterial host genetic material is packed in error into the capsids of virulent phages, which is subsequently transferred into a new recipient host following infection. In specialized transduction, a host sequence is excised along with the prophage and subsequently transferred to a host. Information on transduction rates in natural ecosystems is limited but it is believed that generalized transduction may be more important, because phage-encapsulated DNA is protected from degradation (Weinbauer, 2004).

2.4.3 Eukaryotic Viruses

Eukaryotic viruses infect humans and other animals, plants and other eukaryotic microorganisms including algae and fungi. Eukaryotic viruses are also ubiquitous and are readily found in marine and soil environments. Most interest in eukaryotic viruses has centered on human and plant pathogens. Human viruses cause ailments to almost every part of the human body and include smallpox, mumps, measles, meningitis, hepatitis, encephalitis, colds, influenza and diarrhea. Information on fate and transport of some the human viruses is given in Chapter 22; however, a detailed discussion of infectious disease is beyond the scope of this book. Examples of important animal viruses include the Rhabdoviridae, which causes rabies in dogs, and the Apthovirus, which causes foot and mouth diseases in cows. Studies on plant pathogenic viruses have focused on those affecting major agricultural crops including tobacco, potatoes and tomatoes, but yet again a discussion of plant pathology is not warranted here.

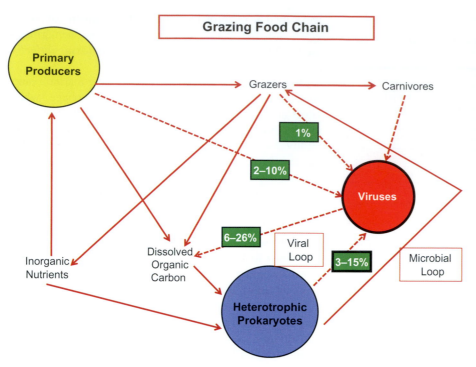

FIGURE 2.19 Pelagic food chain model and virus-mediated carbon flow. The dotted lines point to virus-mediated pathways. All values are in terms of the flux of carbon fixed by primary producers (100%). Only data for viruses are shown. The data indicate that between 6% and 26% of the carbon fixed by primary producers enters the DOC pool via virus-induced lysis at different trophic levels. Adapted from Weinbauer (2004).

Studies on eukaryotic viruses that affect algae have been mostly limited to the marine environment. Here, viruses have been identified as important agents in controlling phytoplankton populations (Brussard, 2004). Eukaryotic viruses not only aid in the control of algal blooms, but are also now being recognized as important agents that affect fluxes of energy, nutrients and dissolved organic matter in marine waters. Similar to the prokaryotic viruses, interest in eukaryotic viruses is now increasing rapidly as it becomes clear that algal viruses are important in the biogeochemical cycles. In recent years, 13 viral infections of marine microalgae have been reported (Brussard, 2004). As an example, diatoms are the major phytoplankton group that play a role in maintaining oxygen levels in the atmosphere and in the carbon cycle that sustains primary production in aquatic environments. The most abundant diatom is *Chaetocerus*, which was recently reported to have been infected by a previously unknown virus (Nagasaki *et al.*, 2005).

Eukaryotic viruses that infect fungi are known as mycoviruses. Although mycoviruses have been identified in all major fungal families, information on environmental aspects of these viruses is limited. An example of a mycovirus with economic repercussions is the causative agent of La France Disease, which affects the edible common mushroom *Agaricus bisporus* (Romaine and Schlagnhaufer, 1995). Viral infections of yeasts have also been reported (Schmitt and Neuhausen, 1994). Undoubtedly, numerous mycoviruses are present in soil environments that most likely influence control of fungal populations and

biogeochemical cycling, but currently data on such incidence are limited. One intriguing recent example from Yellowstone National Park involves a fungus, *Curvularia protuberata*, and a plant, *Dichanthelium lanuginosum*, that only grow at high temperature (65°C) when in a mutualistic association. Research has shown that this association involves a third partner, a mycovirus. In the absence of viral infection, the fungus does not confer thermal tolerance on the plant (Márquez *et al.*, 2007).

2.5 OTHER BIOLOGICAL ENTITIES

2.5.1 Viroids

Viroids are sub-viral particles that have virus-like properties but are not viruses. Viroids are infectious circular single-stranded RNAs from 250 to 400 nucleotides long that replicate in their plant host and are transmitted from host to host through mechanical means, e.g., through a wound or through transmission of contaminated pollen or in ovules. Viroids have been linked to over 16 plant diseases, including potato spindle-tuber disease and citrus exocortis disease, and appear to be highly homologous to each other indicating a common evolutionary origin. Viroids lack a protein capsid and the RNA of the viroid contains no protein-coding genes. As a result, the mechanism of disease for viroids remains unclear. In a given infected cell, however, hundreds to thousands of viroid copies may be present.

2.5.2 Prions

2.5.2.1 Structure of Prions

Prions are found in all humans and consist totally of protein (Figure 2.20). Abnormal prions are infectious protein that can destroy brain tissue, giving it a spongy appearance. Diseases caused by prions are termed trans-missible spongiform encephalopathy (TSE) diseases. TSE diseases are thought to be the active agent of prion disease TSEs including: the agent of "mad cow disease" (or bovine spongiform encephalopathy—BSE) in cattle; scrapie in sheep; Creutzfeldt–Jakob disease in humans; kuru in humans; and chronic wasting disease (CWD) in wild deer and elk.

TSE diseases are transmissible from host to host of a single species or from one species to another, e.g., cow to humans. Normal prions are found in the human body and are known as PrP^c where:

$$PrP = \text{prion protein}$$
$$c = \text{cellular}$$

PrP^c have three-dimensional configurations that are easily digested by proteases. The secondary structure of PrP^c is dominated by α helices (Figure 2.21). The abnormal prions are known as PrP^{sc} where:

$$SC = \text{scrapie}$$

The primary structure of PrP^{sc} is similar to that of PrP^c (amino acid sequences) but the secondary structure is dominated by β sheet conformations (Figure 2.21). PrP^{sc} are not easily degraded by proteases. Of critical importance is the fact that when PrP^{sc} comes into contact with PrP^c it converts the PrP^c into PrP^{sc}. Therefore, although abnormal prions do not replicate, there exists a mechanism to increase the numbers of the abnormal form. When numbers of PrP^{sc} in the brain exceed a critical threshold, a TSE illness results.

2.5.2.2 Prion Environmental Aspects

BSE gained notoriety in Britain where almost 200,000 cases in cattle were detected by 2005. BSE has also been confirmed in cattle in many European countries as well as Canada and the United States (Sreevatsan and Michel, 2002). However, of greater concern to the United States is chronic wasting disease (CWD) that infects deer and elk. To date, there is no evidence of transmission of CWD from elk to humans, but there is concern about the potential for such transfer. However, a recent publication has stated that the risk if any of transmission of CWD to humans is low (Belay et al., 2004).

Recently, studies have demonstrated that prions are sorbed by mica, montmorillonite and other natural soils (Rigou et al., 2006). There is evidence both for the eventual degradation of prions in the environment (Gale and Stanfield, 2001) and that prions can remain infectious even after sorption to soil minerals (Johnson et al., 2006), so the question of prion survival is still an open one. Other concerns were that prions were considered to be capable of surviving conventional wastewater treatment, especially treatment that utilized mesophilic (normal temperature) digestion instead of thermophilic (high temperature) digestion (Kirchmayr et al., 2006). However, early studies utilized technologies for detection that did not distinguish between infectious and noninfectious prions. More recent studies utilizing a standard scrapie cell assay that only detects infectious prions show that prions do not survive meosphilic or thermophilic wastewater treatment. Further, lime treatment of Class B biosolids to produce Class A biosolids is very effective in eliminating infectious prions (Miles et al., 2013).

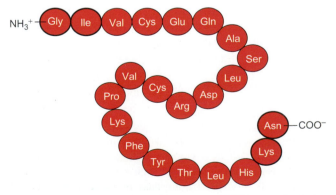

FIGURE 2.20   Primary amino acid sequence of PrP^c (a normal prion).

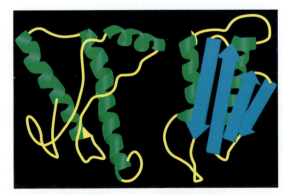

FIGURE 2.21   Secondary structures of PrP^c and PrP^{sc}.

QUESTIONS AND PROBLEMS

1. For the following groups of microorganisms—viruses, bacteria, actinomycetes, archaea, fungi, protozoa and algae—identify:
   a. Which can be pathogenic to humans?
   b. Which can be pathogenic to plants?
   c. Which can degrade organic compounds?
   d. Which can be autotrophic and heterotrophic? Give samples at the genus or species level.
2. What are some similarities and differences between:
   a. Fungi and protozoa?
   b. Archaeans and bacteria?
   c. Viruses and prions?

REFERENCES AND RECOMMENDED READING

Recommended General Microbiology Texts

Madigan, M. T., and Martinko, J. M. (2006) "Brock Biology of Microorganisms," Prentice Hall, New Jersey.

Prescott, L. M., Harley, J. P., and Klein, D. A. (2005) "Microbiology," McGraw Hill, New York.

Recommended Texts on Archaea

Garrett, R. A., and Klenk, H.-P. (2007) "Archaea: Evolution, Physiology, and Molecular Biology," Blackwell Publishing, MA.

Recommended Texts on Fungi

Gadd, G. M. (2006) "Fungi in Biogeochemical Cycles," Cambridge University Press, New York.

Jennings, D. H. (1995) "The Physiology of Fungal Nutrition," Cambridge University Press, Cambridge.

Kuhn, P. J., Trinci, A. P. J., Jung, M. J., Goosey, M. W., and Copping, L. G. (1990) "Biochemistry of Cell Walls and Membranes in Fungi," Springer-Verlag, Berlin.

Mountfort, D. O., and Orpin, C. G. (1994) "Anaerobic Fungi," Marcel Dekker, New York.

Recommended Texts on Viruses

Voyles, B. A. (2002) "The Biology of Viruses," McGraw Hill, New York.

Williamson, K. E. (2011) Soil phage ecology: abundance, distribution, and interactions with bacterial hosts. In "Biocommunication in Soil Microorganisms. Soil Biology, vol. 23, Part 1," Springer-Verlag, Berlin, pp. 113–136.

Chapter References

Ashelford, K. E., Day, M. J., and Fry, J. C. (2003) Elevated abundance of bacteriophage infecting bacteria in soil. *Appl. Environ. Microbiol.* **69**, 285–289.

Belay, E. D., Maddox, R. A., Williams, E. S., Miller, M. W., Gambetti, P., and Shonberger, L. B. (2004) Chronic waste disease and potential transmission to humans. *Emerg. Infect. Dis.* 10.

Bengtsson, M. M., and Øvreås, L. (2010) Planctomycetes dominate biofilms on surfaces of the kelp *Laminaria hyperborea. BMC Microbiol.* **10**, 261.

Blackwell, M. (2011) The Fungi: 1, 2, 3…5.1 million species? *Am. J. Bot.* **98**, 426–438.

Boonchan, S., Britz, M. L., and Stanley, G. A. (2000) Degradation and mineralization of high-molecular-weight polycyclic aromatic hydrocarbons by defined fungal-bacterial cocultures. *Appl. Environ. Microbiol.* **66**, 1007–1019.

Brooks, J. P., Rusin, P. A., Maxwell, S. L., Rensing, C., Gerba, C. P., and Pepper, I. L. (2007) Occurrence of antibiotic-resistant bacteria and endotoxin associated with the land application of biosolids. *Can. J. Microbiol.* **63**, 616–622.

Brussard, C. P. (2004) Viral control of phytoplankton populations—a review. *J. Eukaryot. Microbiol.* **51**, 125–138.

Buckley, D. H., Huangyutitham, V., Nelson, T. A., Rumberger, A., and Thies, J. E. (2006) Diversity of planctomycetes in soil in relation to soil history and environmental heterogeneity. *Appl. Environ. Microbiol.* **72**, 6429.

Cavicchioli, R., Curmi, P., Saunders, N., and Thomas, T. (2003) Pathogenic archaea: do they exist? *Bioessays* **25**, 1119–1128.

Chistoserdova, L., Jenkins, C., Kalyuzhnaya, M. G., Marx, C. J., Lapidus, A., Vorholt, J. A., *et al.* (2004) The enigmatic planctomycetes may hold a key to the origins of methanogenesis and methylotrophy. *Molec. Biol. Evol.* **21**, 1234–1241.

Cho, H. B., and Winans, S. C. (2007) TraA, TraC and TraD autorepress two divergent quorum-regulated promoters near the transfer origin of the Ti plasmid of *Agrobacterium tumefaciens. Molec. Microbiol.* **63**, 1769–1782.

Crous, P. W., Rong, I. H., Wood, A., Lee, S., Glen, H., Botha, W., *et al.* (2006) How many species of fungi are there at the tip of Africa? *Mycology* **55**, 13–33.

D'Annibale, A., Rosetto, F., Leonardi, V., Federici, F., and Petruccioli, M. (2006) Role of autochthonous filamentous fungi in bioremediation of a soil historically contaminated with aromatic hydrocarbons. *Appl. Environ. Microbiol.* **72**, 28–36.

Fan, H., Fairley, D. J., Rensing, C., Pepper, I. L., and Wang, G. (2006) Identification of similar non-thermophilic Crenarchaeota in four Chinese and American pristine soils. *Biodiv. Sci.* **14**, 181–187.

Faruque, S. M., Chowdhury, N., Khan, R., Hasan, M. R., Nahar, J., Islam, M. J., *et al.* (2003) *Shigella dysenteriae* Type 1-specific bacteriophage from environmental waters in Bangladesh. *Appl. Environ. Microbiol.* **69**, 7028–7031.

Fayer, R., Dubey, J. P., and Lindsay, D. S. (2004) Zoonotic protozoa: from land to sea. *Trends Parasitol.* **20**, 531–536.

Fuerst, J. A. (2010) Beyond prokaryotes and eukaryotes: planctomycetes and cell organization. *Nat. Educ.* **3**, 44.

Fuerst, J. A., and Sagulenko, E. (2011) Beyond the bacterium: planctomycetes challenge our concepts of microbial structure and function. *Nat. Rev. Microbiol.* **9**, 403–413.

Gale, P., and Stanfield, G. (2001) Towards a quantitative risk assessment for BSE in sewage sludge. *Appl. Microbiol.* **91**, 563–569.

Giovannoni, S. J., and Stingl, U. (2005) Molecular diversity and ecology of microbial plankton. *Nature* **427**, 343–348.

Hawksworth, D. L. (2001) The magnitude of fungal diversity: the 1.5 million species estimate revisited. *Mycol. Res.* **105**, 422–1432.

Jiang, S. C., and Paul, J. H. (1995) Viral contribution to dissolved DNA in the marine environment as determined by differential centrifugation and kingdom probing. *Appl. Environ. Microbiol.* **61**, 317–325.

Johnson, C. J., Phillips, K. E., Schramm, P. T., McKenzie, D., Aiken, J. M., and Pederson, J. A. (2006) Prions adhere to soil minerals and remain infectious. *PLOS Pathogens* **2**, www.plospathogens.org.

Kartal, B., Kuenen, J. G., and van Loosdrecht, M. C. M. (2010) Engineering. Sewage treatment with anammox. *Science (New York, N.Y.)* **328**, 702–703.

Kirchmayr, R., Reichi, H. E., Schildorfer, H., Braun, R., and Somerville, R. A. (2006) Prion protein: detection in "spiked" anaerobic sludge and degradation experiments under anaerobic conditions. *Wat. Sci. Technol.* **53**, 91–98.

Kullman, S. W., and Matsumura, F. (1996) Metabolic pathways utilized by *Phanerochaete chrysosporium* for degradation of the cyclodiene pesticide endosulfan. *Appl. Environ. Microbiol.* **62**, 593–600.

Lonhienne Thierry, G. A., Sagulenko, E., Webb, R. I., Kuo-Chang, L., Franke, J., Devos, D. P., *et al.* (2010) Endocytosis-like protein uptake in the bacterium *Gemmata obscuriglobus*. *Proc. Natl Acad. Sci. U.S.A.* **107**, 12883–12888.

Márquez, L. M., Redman, R. S., Rodriguez, R. J., and Roossinck, M. J. (2007) A virus in a fungus in a plant: three-way symbiosis required for thermal tolerance. *Science* **315**, 513–515.

Matias, V. R., and Beveridge, T. J. (2006) Native cell wall organization shown by cryo-electron microscopy confirms the existence of a periplasmic space in *Staphylococcus aureus*. *J. Bacteriol.* **188**, 1011–1021.

Miles, S. L., Sam, W., Field, J. A., Gerba, C. P., and Pepper, I. L. (2013) Survival of infectious prions during wastewater treatment. *J. Res. Sci. Technol.* In press.

Mitchell, A. M., Strobel, G. A., Moore, E., Robison, R., and Sears, J. (2010) Volatile antimicrobials from *Muscodor crispans*, a novel endophytic fungus. *Microbiol* **156**, 270–277.

Nagasaki, K., Tomaru, Y., Takao, J., Nishida, K., Shirai, Y., Suzuki, H., *et al.* (2005) Previously unknown virus infects marine diatom. *Appl. Environ. Microbiology* **71**, 3528–3535.

Neidhardt, F. C., Ingraham, J. L., and Schaechter, M. (1990) "Physiology of the Bacterial Cell: A Molecular Approach," Sinauer Associates, Sunderland, MA.

Nelson, K. E., Clayton, R. A., Gill, S. R., Gwinn, M. L., Dodson, R. J., Haft, D. H., *et al.* (1999) Evidence for lateral gene transfer between archaea and bacteria from genome sequence of *Thermotoga maritima*. *Nature* **399**, 323–329.

Newby, D. T., Josephson, K. L., and Pepper, I. L. (2000) Detection and characterization of plasmid pJP4 transfer to indigenous soil bacteria. *Appl. Environ. Microbiol.* **66**, 290–296.

Niyyati, M., Lorenzo-Morales, J., Rezaeian, M., Martin-Navarro, C. M., Haghi, A. M., MacIver, S. K., *et al.* (2009) Isolation of *Balamuthia mandrillaris* from urban dust, free of known infectious involvement. *Parasitol. Res.* **106**, 279–281.

Parada, V., Sintes, E., van Aken, H. M., Weinbauer, M. G., and Herndl, G. J. (2007) Viral abundance, decay, and diversity in the meso- and bathypelagic waters of the North Atlantic. *Appl. Envir. Microbiol.* **73**, 4429–4438.

Park, S. C., Shimamura, I., Fukunaga, M., Mori, K.-I., and Nakai, T. (2000) Isolation of bacteriophages specific to a fish pathogen, *Pseudomonas plecoglossicida*, as a candidate for disease control. *Appl. Environ. Microbiol.* **66**, 1416–1422.

Pikuta, E. V., Hoover, R. B., and Tang, J. (2007) Microbial extremophiles at the limits of life. *Crit. Rev. Microbiol.* **33**, 183–209.

Pointing, S. B. (2001) Feasibility of bioremediation by white-rot fungi. *Appl. Microbiol. Biotechnol.* **57**, 20–33.

Rigou, P., Rezaei, H., Grosclaude, J., Staunton, S., and Quiquampoix, H. (2006) Fate of prions in soil: adsorption and extraction by electrolution of recombinant ovine prion protein from montmorillonite and natural soils. *Envir. Sci. Technol.* **40**, 1497–1503.

Riley, M. A., and Gordon, D. M. (1992) A survey of Col plasmids in natural isolates of *E. coli* and an investigation into the stability of Col-plasmid linages. *J. Gen. Microbiol.* **138**, 1345–1352.

Rinaldi, A. C., Comandini, O., and Kuyper, T. W. (2008) Ectomycorrhizal fungal diversity: separating the wheat from the chaff. *Fung. Diver.* **33**, 1–45.

Rogel, M. A., Hernandez-Lucas, I., Kuykendall, D., Balkwill, D. L., and Martinez-Romero, E. (2001) Nitrogen-fixing nodules with *Ensifer adhaerens* harboring *Rhizobium tropici* symbiotic plasmids. *Appl. Environ. Microbiol.* **67**, 3264–3268.

Romaine, C. P., and Schlagnhaufer, B. (1995) PCR analysis of the viral complex associated with La France disease of *Agaricus bisporus*. *Appl. Environ. Microbiol.* **61**, 2322–2325.

Santoro, A. E., Casciotti, K. L., and Francis, C. A. (2010) Activity abundance and diversity of nitrifying archaea and bacteria in the central California Current. *Environ. Microbiol.* **12**, 1989–2006.

Sayeed, S., and McClane, B. A. (2007) Virulence plasmid diversity in *Clostridium perfringens* Type D isolates. *Infect. Immun.* **75**, 2391–2398.

Schloss, P. D., and Handelsman, J. (2004) Status of the microbial census. *Microbiol. Mol. Biol. Rev.* **68**, 686–691.

Schmitt, M. J., and Neuhausen, F. (1994) Killer toxin-secreting double-stranded RNA mycoviruses in the yeasts *Hanseniaspora uvarum* and *Zygosaccharomyces* bailii. *J. Virol.* **68**, 1765–1772.

Smit, E., Wolters, A., and van Elsas, J. D. (1998) Self-transmissible mercury resistance plasmids with gene-mobilizing capacity in soil bacterial populations: influence of wheat roots and mercury addition. *Appl. Environ. Microbiol.* **64**, 1210–1219.

Sogin, M. L., Morrison, H. G., Huber, J. A., Welch, D. M., Huse, S. M., Arrieta, J. M., *et al.* (2006) Microbial diversity in the deep sea and underexplored "rare biosphere. *Proc. Natl Acad. Sci. U.S.A.* **103**, 12115–12120.

Sreevatsan, S., Michel, F.C. Jr., (2002). Prion diseases (*Spongiform encephalopathies*): an overview. In: Michel Jr., and R. F. Rynk (Eds.), "International Symposium, Composting and Compost Utilization" 2002, Columbus, OH.

Srinivasiah, S., Bhavsar, J., Thapar, K., Liles, M., Schoenfield, T., and Wommack, K. E. (2008) Phages across the biosphere: contrasts of viruses in soil and aquatic environments. *Res. Microbiol.* **159**, 349–357.

Srivastava, P., Nath, N., and Deb, J. K. (2006) Characterization of broad host range cryptic plasmid pCR1 from *Corynebacterium renale*. *Plasmid* **56**, 24–34.

Sullivan, M. B., Lindell, D., Lee, J. A., Thompson, L. R., Bielawoki, J. P., and Chisholm, S. W. (2006) Prevalence and evolution of core photosystem II genes in marine cyanobacterial viruses and their hosts. *PloS Biol.* **4**(8), e234.

Thanbichler, M., and Shapiro, L. (2006) Chromosome organization and segregation in bacteria. *J. Struc. Biol.* **156**, 292–303.

Van Biesen, T., and Frost, L. S. (1992) Different levels of fertility inhibition among F-like plasmids are related to the cellular concentration of fin mRNA. *Mol. Microbiol.* **6**, 771–780.

Vianna, M. E., Conrads, G., Gomes, B. P. F. A., and Horz, H. P. (2006) Identification and quantification of Archaea involved in primary endodontic infections. *J. Clin. Microbiol.* **44**, 1274−1282.

Wang, B., and Qiu, Y. L. (2006) Phylogenic distribution and evolution of mycorrhizas in land plants. *Mycorrhiza* **16**, 299−363.

Webb, J. S., Nixon, M., Eastwood, I. M., Greenhalgh, M., Robson, G. D., and Handley, P. S. (2000) Fungal colonization and biodeterioration of plasticized polyvinyl chloride. *Appl. Environ. Microbiol.* **66**, 3194−3200.

Weinbauer, M. G. (2004) Ecology of prokaryotic viruses. *FEMS Microbiol. Rev.* **28**, 127−181.

Williamson, K. E., Radosevich, M., and Wommack, K. E. (2005) Abundance and diversity of viruses in six Delaware soils. *Appl. Environ. Microbiol.* **71**, 3119−3125.

Woese, C., and Fox, G. (1977) Phylogenetic structure of the prokaryotic domain: the primary kingdoms. *Proc. Natl. Acad. Sci. U.S.A* **74**, 5088−5090.

Wommack, K. E., and Colwell, R. R. (2000) Virioplankton: viruses in aquatic ecosystems. *Microbiol. Mol. Biol. Rev.* **64**, 69−114.

Bacterial Growth

Raina M. Maier and Ian L. Pepper

Microorganisms conduct a series of highly organized chemical reactions that collectively are known as metabolism. There are several thousand potential reactions in a microbial cell, many of which are utilized to make new cells. These reactions are known as growth metabolism. Other reactions are termed nongrowth reactions, and are needed for cellular activity such as maintenance of intracellular metabolite pools, repair of cellular structures, motility and response to environmental stress (Schaechter et al., 2006). In the laboratory, we can manipulate conditions so that cells are undergoing growth metabolism most of the time. In the environment it is a different story—most microorganisms are in a nongrowth state, simply surviving and awaiting new nutrient sources.

Overall, metabolism is a complex process involving numerous anabolic (synthesis of cell constituents and metabolites) and catabolic (breakdown of cell constituents and metabolites) reactions. Ultimately, these biosynthetic reactions result in cell division as shown in Figure 3.1. In a homogeneous rich culture medium, under ideal conditions, a cell can divide in as little as 10 minutes. In contrast, it has been suggested that cell division may occur as slowly as once every 100 years in some subsurface terrestrial environments. Such slow growth is the result of a combination of factors including the fact that most subsurface environments are both nutrient poor and heterogeneous. As a result, cells are likely to be isolated, and cannot share nutrients or protection mechanisms, and therefore have slower growth rates.

Most information available concerning the growth of microorganisms is the result of controlled laboratory studies using pure cultures of microorganisms. There are two approaches to the study of growth under such controlled conditions: batch culture and continuous culture. In a batch culture, the growth of a single organism or a group of organisms, called a consortium, is evaluated using a defined medium to which a fixed amount of substrate (food) is added at the outset. In continuous culture, there is a steady influx of growth medium and substrate such that the amount of available substrate always remains the same. Growth under both batch and continuous culture conditions has been well characterized physiologically and also described mathematically. This information has been used to optimize the commercial production of a variety of microbial products including antibiotics, vitamins, amino acids, enzymes, yeast, vinegar and alcoholic beverages. These materials are often produced in large batches (up to 500,000 liters), also called large-scale fermentations.

Unfortunately, it is difficult to extend our knowledge of growth under controlled laboratory conditions to an understanding of growth in natural soil or water environments, where enhanced levels of complexity are encountered (Figure 3.2). This complexity arises from a number

I.L. Pepper, C.P. Gerba, T.J. Gentry: Environmental Microbiology, Third edition. DOI: http://dx.doi.org/10.1016/B978-0-12-394626-3.00003-X
© 2015 Elsevier Inc. All rights reserved.

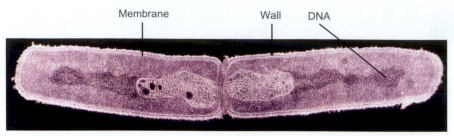

FIGURE 3.1 Gram positive *Bacillus subtilis* undergoing cell division. Reprinted with permission from Madigan and Martinko *et al.* (2006).

FIGURE 3.2 Compare the complexity of microbial growth in a flask and growth in a soil environment. Although we understand growth in a flask quite well, we still cannot always predict growth in the environment.

of factors, including an array of different types of solid surfaces, microenvironments that have altered physical and chemical properties, a limited nutrient status and consortia of different microorganisms all competing for the same limited nutrient supply (see Chapter 4). Thus, the current challenge facing environmental microbiologists is to understand microbial growth in natural environments. Such an understanding would facilitate our ability to predict rates of nutrient cycling (Chapter 16), microbial response to anthropogenic perturbation of the environment, microbial interaction with organic and metal contaminants (Chapters 17 and 18) and survival and growth of pathogens in the environment (Chapter 22). In this chapter, we begin with a review of growth under pure culture conditions, and then discuss how this compares to growth in the environment.

3.1 GROWTH IN PURE CULTURE IN A FLASK

Typically, to understand and define the growth of a particular microbial isolate, cells are placed in a liquid medium in which the nutrients and environmental conditions are controlled. If the medium supplies all nutrients required for growth and environmental parameters are optimal, the increase in numbers or bacterial mass can be measured as a function of time to obtain a growth curve. Several distinct growth phases can be observed within a growth curve (Figure 3.3). These include: the lag phase; the exponential or log phase; the stationary phase; and the death phase. Each of these phases represents a distinct period of growth that is associated with typical physiological changes in the cell culture. As will be seen in the following sections, the rates of growth associated with each phase are quite different.

3.1.1 The Lag Phase

The first phase observed under batch conditions is the lag phase in which the growth rate is essentially zero. When an inoculum is placed into fresh medium, growth begins after a period of time called the lag phase. By definition, the lag phase transitions to the exponential phase after the initial population have doubled (Yates and Smotzer, 2007). The lag phase is thought to be due to the physiological adaptation of the cell to the culture conditions. This may involve a time requirement for induction of specific messenger RNA (mRNA), and subsequent protein synthesis to meet new culture requirements. The lag phase may also be due to low initial densities of organisms that result in dilution of exoenzymes (enzymes released from the cell), and of nutrients that leak from growing cells. Normally, such materials are shared by cells in close proximity. But when cell density is low, these materials are diluted and are not taken up as easily. As a result, initiation of cell growth and division, and the transition to exponential phase growth may be delayed.

The lag phase usually lasts from minutes to several hours. The length of the lag phase can be controlled to some extent because it is dependent on the type of medium as well as on the initial inoculum size. For example, if an inoculum is taken from an exponential phase culture in trypticase soy broth (TSB), and placed into fresh TSB medium at a concentration of 10^6 cells/ml under the same growth conditions (temperature, shaking speed), there will be no noticeable lag phase. However, if the inoculum is taken from a stationary phase culture,

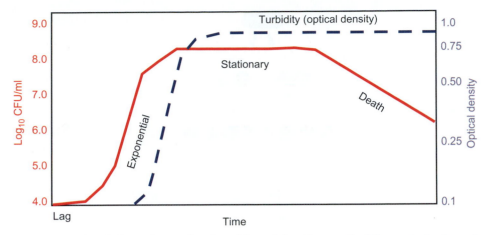

FIGURE 3.3 A typical growth curve for a bacterial population. Compare the difference in the shape of the curves in the death phase (colony-forming units versus optical density).

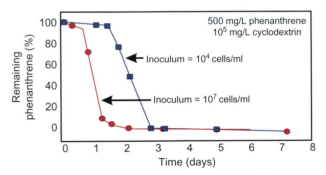

FIGURE 3.4 Effect of inoculum size on the lag phase during degradation of a polyaromatic hydrocarbon, phenanthrene. Because phenanthrene is only slightly soluble in water and is therefore not readily available for cell uptake and degradation, a solubilizing agent called cyclodextrin was added to the system. The microbes in this study were not able to utilize cyclodextrin as a source of carbon or energy. Courtesy E.M. Marlowe.

there will be a lag phase as the stationary phase cells adjust to the new conditions, and shift physiologically from stationary phase cells to exponential phase cells. Similarly, if the inoculum is placed into a medium other than TSB, for example, a mineral salts medium with glucose as the sole carbon source, a lag phase may be observed while the cells adjust physiologically to synthesize the appropriate enzymes for glucose catabolism.

Finally, if the inoculum size is small, for example 10^4 cells/ml, and one is measuring activity, such as disappearance of substrate, a lag phase will be observed until the population reaches approximately 10^6 cells/ml. This is illustrated in Figure 3.4, which compares the degradation of phenanthrene in cultures inoculated with 10^7 and with 10^4 colony-forming units (CFU) per ml. Although the degradation rate achieved is similar in both cases (compare the slope of each curve), the lag phase was 1.5 days when a low inoculum size was used (10^4 CFU/ml), in contrast to only 0.5 day when the higher inoculum was used (10^7 CFU/ml).

3.1.2 The Exponential Phase

The second phase of growth observed in a batch system is the exponential phase. The exponential phase is characterized by a period of exponential growth—the most rapid growth possible under the conditions present in the batch system. During exponential growth, the rate of increase of cells in the culture is proportional to the number of cells present at any particular time. There are several ways to express this concept both theoretically and mathematically. One way is to imagine that during exponential growth the number of cells increases in the geometric progression 2^0, 2^1, 2^2, 2^4 until, after n divisions, the number of cells is 2^n (Figure 3.5). This can be expressed in a quantitative manner as:

$$X = 2^n X_0 \qquad \text{(Eq. 3.1)}$$

where:

X_0 = initial concentration of cells
X = concentration after time t
n = number of generations or cell division

From Eq. 3.1 it follows that:

$$\ln X = n \ln 2 + \ln X_0 \qquad \text{(Eq. 3.2)}$$

During the exponential phase of growth, if the number of cells initially, and at any particular time thereafter, is known, the number of generations can be calculated from Eq. 3.3.

$$n = \frac{\ln X - \ln X_0}{0.693} \qquad \text{(Eq. 3.3)}$$

Example Calculation 3.1 shows that if one starts with a low number of cells, exponential growth does not initially produce large numbers of cells. However, as cells accumulate after several generations, the number of new cells with each cell division increases dramatically.

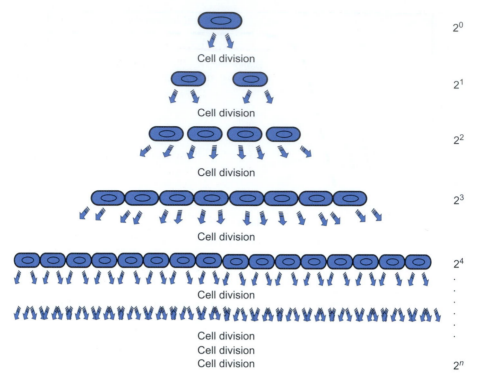

FIGURE 3.5 Exponential cell division. Each cell division results in a doubling of the cell number. At low cell numbers the increase is not very large; however, after a few generations, cell numbers increase explosively.

Example Calculation 3.2 demonstrates how the number of generations and the mean generation time can be calculated.

As long as a culture is in the exponential or logarithmic phase, the culture is said to be undergoing balanced growth.

3.1.2.1 Mean Generation Time Versus Specific Growth Rate

Two terms that are used to describe growth in the exponential phase are generation time and specific growth rate. The generation time refers to the time needed for cell doubling while the specific growth rate is the maximum growth rate that can be achieved given the environmental conditions present (unlimited substrate, temperature, etc.). When substrate becomes limiting or toxic by-products build up, cells will leave the exponential phase and, correspondingly, the specific growth rate will decrease.

In mathematical terms in the exponential phase:

$$\frac{dX}{dt} = \mu X \qquad \text{(Eq. 3.4)}$$

where:

dX/dt = change in cell number X during time t

μ = specific growth rate expressed as reciprocal time (hours^{-1})

Example Calculation 3.1   Calculation of the Number of Cells in a Pure Culture

Problem: If one starts with 10,000 (10^4) cells in a culture that has a generation time of 2 h, how many cells will be in the culture after 4, 24, and 48 h?

From Eq. 3.1, $X = 2^n X_0$, where X_0 is the initial number of cells, n is the number of generations, and X is the number of cells after n generations.

After 4 h, n = 4h/2 h per generation = 2 generations:

$$X = 2^2(10^4) = 4.0 \times 10^4 \text{ cells}$$

After 24 h, n = 12 generations:

$$X = 2^{12}(10^4) = 4.1 \times 10^7 \text{ cells}$$

After 48 h, n = 24 generations:

$$X = 2^{24}(10^4) = 1.7 \times 10^{11}$$

This represents an increase of less than one order of magnitude for the 4-h culture, four orders of magnitude for the 24-h culture, and seven orders of magnitude for the 48-h culture.

Example Calculation 3.2   Calculation of Mean Generation Time

Following a dilution and plating experiment, the following data were obtained:

At the beginning of exponential growth:

$$t_0 = 0$$
$$X_0 = 1000 \text{ cells/ml}$$

At time $t = 6$ hours:

$$X = 16,000 \text{ cells/ml}$$

Using Eq. 3.3:

$$n = \frac{\ln X - \ln X_0}{0.693}$$

$$n = \frac{\ln 16,000 - \ln 1000}{0.693}$$

$$\therefore n = \frac{9.7 - 6.9}{0.693}$$

And

$$n = \frac{1.204}{0.693} = 4 \text{ generations}$$

$\therefore$ Since there are 4 generations in 6 hours, the mean generation time $= 6/4 = 1.5$ hours.

By integration:

$$X/X_0 = e^{\mu t} \qquad \text{(Eq. 3.5)}$$

Taking the natural log (ln) of both sides:

$$\ln X/X_o = \ln X - \ln X_o = \mu t$$

$$\therefore \mu = \frac{\ln X - \ln X_0}{t} \qquad \text{(Eq. 3.6)}$$

Thus, the specific growth rate in exponential phase is the slope of the growth curve. See Example Calculation 3.3 for an illustration of how to determine specific growth rates.

To calculate the generation time (g), we consider the special case where X_0 is doubled. In the simplest case we can consider when one cell becomes two cells and $X = 2$ while $X_0 = 1$:

$$\therefore \mu = \frac{0.693 - 0}{g}$$

$$\therefore \mu = \frac{0.693}{g} \qquad \text{(Eq. 3.7)}$$

3.1.3 The Stationary Phase

The third phase of growth is the stationary phase. The stationary phase in a batch culture can be defined as a state of no net growth, which can be expressed by the following equation:

$$\frac{dX}{dt} = 0 \qquad \text{(Eq. 3.8)}$$

Although there is no net growth in stationary phase, cells still grow and divide. Growth is simply balanced by an equal number of cells dying. There are several reasons why a batch culture may reach stationary phase. One reason is that the carbon and energy source or an essential nutrient becomes limiting. When a carbon source is used up it does not necessarily mean that all growth stops. This is because dying cells can lyse and provide a recycled source of nutrients. Growth resulting from dead cells is called endogenous metabolism. Endogenous metabolism occurs throughout the growth cycle, but can be best observed during the stationary phase when growth is measured in terms of oxygen uptake or evolution of carbon dioxide. Thus, in many growth curves such as that shown in Figure 3.6, the stationary phase can actually show a small amount of growth. Again, this growth occurs after the substrate has been utilized, and reflects the use of dead cells as a source of carbon and energy. A second reason that the stationary phase may be observed is that waste products build up to a point where they begin to inhibit cell growth or are toxic to cells. This generally occurs only in cultures with high cell density. Regardless of the reason why cells enter the stationary phase, growth in the stationary phase is referred to as unbalanced growth because it is easier for the cells to synthesize some components than others. As some components become more and more limiting, cells will still keep growing and dividing as long as possible. As a result of this nutrient stress, stationary phase cells are generally smaller and rounder than cells in the exponential phase. Ultimately, since the reuse of some cell components is not 100% efficient, more cells die than new cells are produced, and the culture will enter the death phase.

3.1.4 The Death Phase

The final phase of the growth curve is the death phase, which is characterized by a net loss of culturable cells. Even in the death phase there may be individual cells that are metabolizing and dividing, but more viable cells are lost than are gained so there is a net loss of viable cells. The death phase is often exponential, although the rate of cell death is usually slower than the rate of

Example Calculation 3.3   Calculation of Specific Growth Rate

Problem: The following data were collected using a culture of *Pseudomonas* during growth in a minimal medium containing salicylate as a sole source of carbon and energy. Using these data, calculate the specific growth rate for the exponential phase.

Time(h)	Culturable Cell Count (CFU/ml)
0	1.2×10^4
4	1.5×10^4
6	1.0×10^5
8	6.2×10^6
10	8.8×10^8
12	3.7×10^9
16	3.9×10^9
20	6.1×10^9
24	3.4×10^9
28	9.2×10^8

The times to be used to determine the specific growth rate can be chosen by visual examination of a semilog plot of the data (see figure). Examination of the graph shows that the exponential phase is from approximately 6 to 10 hours. Using Eq. 3.6, which describes the exponential phase of the graph, one can determine the specific growth rate for this *Pseudomonas*. (Note that Eq. 3.4 describes a line, the slope of which is μ, the specific growth rate.) From the data given, the slope of the graph from time 6 to 10 hours is:

$$\mu = \frac{\ln 10^9 - \ln 10^5}{10 - 6} \, 2.3 \text{ hour}^{-1}$$

It should be noted that the specific growth rate and generation time calculated for growth of the *Pseudomonas* on salicylate are valid only under the experimental conditions used. For example, if the experiment were performed at a higher temperature, one would expect the specific growth rate to increase. At a lower temperature, the specific growth rate would be expected to decrease.

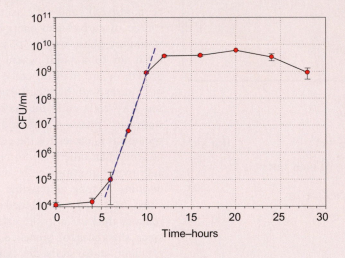

growth during the exponential phase. The death phase can be described by the following equation:

$$\frac{dX}{dt} = -k_d X \qquad \text{(Eq. 3.9)}$$

where k_d = the specific death rate.

It should be noted that the way in which cell growth is measured can influence the shape of the growth curve. For example, if growth is measured by optical density instead of by plate counts (compare the two curves in Figure 3.3), the onset of the death phase is not readily apparent. Similarly, if one examines the growth curve measured in terms of carbon dioxide evolution shown in Figure 3.6, again it is not possible to discern the death phase. Still, these are approaches commonly used to measure growth, because normally, the growth phases of most interest to environmental microbiologists are the lag phase, the exponential phase and the time to the onset of the stationary phase.

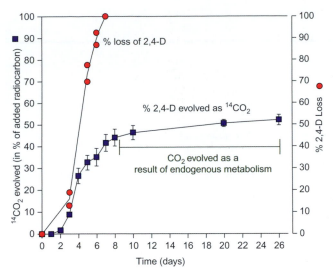

FIGURE 3.6 Mineralization of the broadleaf herbicide 2,4-dichlorophe-noxyacetic acid (2,4-D) in a soil slurry under batch conditions. Note that the 2,4-D is completely utilized after 6 days but the CO_2 evolved continues to rise slowly. This is a result of endogenous metabolism. *From* Estrella *et al.* (1993).

3.1.5 Effect of Substrate Concentration on Growth

So far we have discussed each of the growth phases and have shown that each phase can be described mathematically (see Eqs. 3.1, 3.8 and 3.9). One can also write equations to allow description of the entire growth curve. Such equations become increasingly complex. For example, one of the first and simplest descriptions is the Monod equation, which was developed by Jacques Monod in the 1940s:

$$\mu = \frac{\mu_{max}S}{K_s + S} \qquad \text{(Eq. 3.10)}$$

where:

μ = the specific growth rate (1/time)
μ_{max} = the maximum specific growth rate (1/time) for the culture
S = the substrate concentration (mass/volume)
K_s = the half-saturation constant (mass/volume) also known as the affinity constant

Equation 3.10 was developed from a series of experiments performed by Monod. The results of these experiments showed that at low substrate concentrations, growth rate becomes a function of the substrate concentration (note that Eqs. 3.1 to 3.9 are independent of substrate concentration). Thus, Monod designed Eq. 3.10 to describe the relationship between the specific growth rate and the substrate concentration. There are two constants in this equation, μ_{max}, the maximum specific growth rate, and K_s, the half-saturation constant, which is defined as the substrate concentration at which growth occurs at one-half the value of μ_{max}.

Both μ_{max} and K_s reflect intrinsic physiological properties of a particular type of microorganism. They also depend on the substrate being utilized and on the temperature of growth (see Information Box 3.1). Monod assumed in writing Eq. 3.10 that no nutrients other than the substrate are limiting and that no toxic by-products of metabolism build up.

As shown in Eq. 3.11, the Monod equation can be expressed in terms of cell number or cell mass (X) by equating it with Eq. 3.4:

$$\frac{dX}{dt} = \frac{\mu_{max}S\,X}{K_s + S} \qquad \text{(Eq. 3.11)}$$

The Monod equation has two limiting cases (see Figure 3.7). The first case is at high substrate concentration where $S \gg K_s$. In this case, as shown in Eq. 3.12, the specific growth rate μ is essentially equal to μ_{max}. This simplifies the equation and the resulting relationship is zero order or independent of substrate concentration:

$$\text{for } S \gg K_s: \quad \frac{dX}{dt} = \mu_{max}X \qquad \text{(Eq. 3.12)}$$

Under these conditions, growth will occur at the maximum growth rate. There are relatively few instances in which ideal growth as described by Eq. 3.12 can occur. One such instance is under the initial conditions found in pure culture in a batch flask when substrate and nutrient levels are high. Another is under continuous culture conditions, which are discussed further in Section 3.2. It must be emphasized that this type of growth is likely to be rare under natural conditions in a soil or water environment, where either substrate or other nutrients are commonly limiting.

The second limiting case occurs at low substrate concentrations where $S \ll K_s$ as shown in Eq. 3.13. In this case there is a first-order dependence on substrate concentration (Figure 3.7):

$$\text{for } S \ll K_s: \quad \frac{dX}{dt} = \frac{\mu_{max}S\,X}{K_s} \qquad \text{(Eq. 3.13)}$$

As shown in Eq. 3.13, when the substrate concentration is low, growth (dX/dt) is dependent on the substrate concentration. Since the substrate concentration is in the numerator, as the substrate concentration decreases, the rate of growth will also decrease. This type of growth is typically found in batch flask systems at the end of the growth curve when almost all substrate has been consumed. This is also the type of growth that would be more typically expected under the conditions found in a natural environment, where substrate and nutrients are limiting.

The Monod equation can also be expressed as a function of substrate utilization given that growth is related to substrate utilization by a constant called the cell yield (Eq. 3.14):

$$\frac{dS}{dt} = -\frac{1}{Y}\frac{dX}{dt} \qquad \text{(Eq. 3.14)}$$

Information Box 3.1   The Monod Growth Constants

Both μ_{max} and K_s are constants that reflect:

- The intrinsic properties of the degrading microorganism
- The limiting substrate
- The temperature of growth

The following table provides representative values of μ_{max} and K_s for growth of different microorganisms on a variety of substrates at different temperatures and for oligotrophs and copiotrophs in soil.

Organism	Growth Temperature (°C)	Limiting Nutrient	μ_{max} (L/h)	K_s (mg/L)
Escherichia coli	37	Glucose	0.8–1.4	2–4
Escherichia coli	37	Lactose	0.8	20
Saccharomyces cerevisiae	30	Glucose	0.5–0.6	25
Pseudomonas sp.	25	Succinate	0.38	80
Pseudomonas sp.	34	Succinate	0.47	13
Oligotrophs in soil			0.01	0.01
Copiotrophs in soil			0.045	3

Source: Adapted from Blanch and Clark (1996), Miller and Bartha (1989), Zelenev et al. (2005).

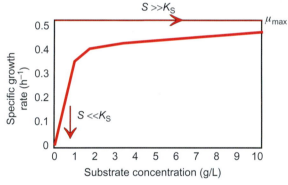

FIGURE 3.7 Dependence of the specific growth rate, μ, on the substrate concentration. The maximal growth rate, $\mu_m = 0.5\ h^{-1}$ and $K_s = 0.5$ g/L. Note that μ approaches μ_{max} when $S >> K_s$ and becomes independent of substrate concentration. When $S << K_s$, the specific growth rate is very sensitive to the substrate concentration, exhibiting a first-order dependence.

where Y = the cell yield (mass/mass). The cell yield coefficient is defined as the unit amount of cell mass produced per unit amount of substrate consumed. Thus, the more efficiently a substrate is degraded, the higher the value of the cell yield coefficient (see Section 3.3 for more details). The cell yield coefficient is dependent on both the structure of the substrate being utilized and the intrinsic physiological properties of the degrading microorganism. As shown below, Eqs. 3.11 and 3.14 can be combined to express microbial growth in terms of substrate disappearance:

$$\frac{dS}{dt} = -\frac{1}{Y}\frac{\mu_{max}S\,X}{K_s + S} \qquad \text{(Eq. 3.15)}$$

Figure 3.8 shows a set of growth curves constructed from a fixed set of constants. The growth data used to generate this figure were collected by determining protein as a measure of the increase in cell growth (see Chapter 11). The growth data were then used to estimate the growth constants μ_{max}, K_s and Y. Both Y and μ_{max} were estimated directly from the data. K_s was estimated using a mathematical model that performs a nonlinear regression analysis of the simultaneous solutions to the Monod equations for cell mass (Eq. 3.11) and substrate (Eq. 3.14). This set of constants was then used to model or simulate growth curves that express growth in terms of CO_2 evolution and substrate disappearance. Such models are useful because they can help to: (1) estimate growth constants such as K_s that are difficult to determine experimentally; and (2) quickly understand how changes in any of the experimental parameters affect growth without performing a long and tedious set of experiments.

3.2 CONTINUOUS CULTURE

Thus far, we have focused on theoretical and mathematical descriptions of batch culture growth, which is currently of great economic importance in terms of the production of a wide variety of microbial products. In contrast to batch culture, continuous culture is a system that is designed for long-term operation. Continuous culture can be operated over the long term because it is an open system (Figure 3.9), with a continuous feed of influent solution that contains nutrients and substrate. It also contains a continuous drain of effluent solution that has cells, metabolites, waste products and any unused nutrients and substrate. The vessel that is used as a growth container in continuous culture is called a bioreactor or a chemostat. In a chemostat, one can control the flow rate and maintain a constant substrate concentration, as well as provide continuous control of pH, temperature and oxygen levels. This allows control of the rate of growth, which can be used to optimize the production of specific microbial products. For example, primary metabolites or growth-associated products, such as ethanol, are produced at high flow or dilution rates which stimulate cell growth. In contrast, a secondary metabolite or nongrowth-associated product such as an antibiotic is produced at low flow or dilution rates which maintain high cell numbers. Chemostat cultures are also being used to aid in the study of the functional genomics of growth, nutrient limitation and stress responses at the whole-organism level. The advantage of the chemostat in such studies is the constant removal of secondary growth effects that may mask or alter subtle physiological changes under batch culture conditions (Hoskisson and Hobbs, 2005).

Dilution rate and influent substrate concentration are the two parameters controlled in a chemostat to study microbial growth or to optimize metabolite production. The dynamics of these two parameters are shown in Figure 3.10. By controlling the dilution rate, one can control the growth rate (μ) in the chemostat, represented in this graph as doubling time (recall that during the exponential phase the growth rate is proportional to the number of cells present). By controlling the influent substrate concentration, one can control the number of cells produced or the cell yield in the chemostat since the number of cells produced will be directly proportional to the amount of substrate provided. Because the growth rate and the cell number can be controlled independently, chemostats have been an important tool for the study of the physiology of microbial growth. They are also useful in the long-term development of cultures and consortia that are acclimated to organic contaminants that are toxic and difficult to degrade. Chemostats can also produce microbial products more efficiently than batch fermentations. This is because a chemostat can essentially hold a culture in the exponential phase of growth for extended periods of time. Despite these advantages, chemostats are not yet widely used to

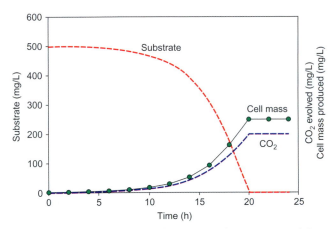

FIGURE 3.8 This figure shows the same growth curve expressed three different ways; in terms of substrate loss, in terms of CO_2 evolution and in terms of increasing cell mass. The parameters used to generate these three curves were: $\mu_{max} = 0.29\ h^{-1}$, $K_s = 10\ mg/L$, $Y = 0.5$, initial substrate concentration = 500 mg/L and initial cell mass = 1 mg/L. In this experiment, cell mass was measured and so the data points are shown. The data for CO_2 evaluation and substrate loss were simulated using a model and so the data are shown using dashed lines.

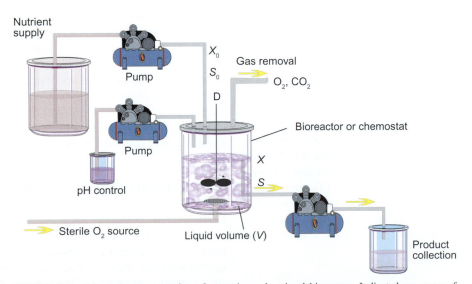

FIGURE 3.9 Schematic representation of a continuously stirred bioreactor. Indicated are some of the variables used in modeling bioreactor systems. X_0 is the dry cell weight, S_0 is the substrate concentration and D is the flow rate of nutrients into the vessel.

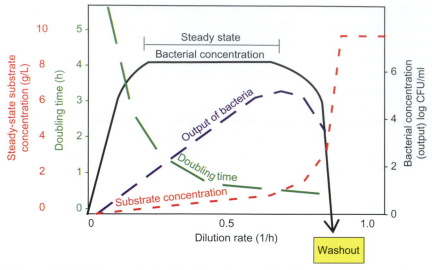

FIGURE 3.10 Steady-state relationships in the chemostat. The dilution rate is determined from the flow rate and the volume of the culture vessel. Thus, with a vessel of 1000 ml and a flow rate through the vessel of 500 ml/h, the dilution rate would be 0.5 L/h^{-1}. Note that at high dilution rates, growth cannot balance dilution and the population washes out. Thus, the substrate concentration rises to that in the medium reservoir (because there are no bacteria to use the inflowing substrate). However, throughout most of the range of dilution rates shown, the population density remains constant and the substrate concentration remains at a very low value (that is, steady state). Note that although the population density remains constant, the growth rate (doubling time) varies over a wide range. Thus, the experimenter can obtain populations with widely varying growth rates without affecting population density. Adapted with permission from Madigan and Martinko (2006).

produce commercial products because it is often difficult to maintain sterile conditions over a long period of time.

In a chemostat, the growth medium undergoes constant dilution with respect to cells due to the influx of nutrient solution (Figure 3.9). The combination of growth and dilution within the chemostat will ultimately determine growth. Thus, in a chemostat, the change in biomass with time is:

$$\frac{dX}{dt} = \mu X - DX \qquad \text{(Eq. 3.16)}$$

where:

 X = the cell mass (mass/volume)
 μ = the specific growth rate (1/time)
 D = the dilution rate (1/time)

Examination of Eq. 3.16 shows that a steady state (no increase or decrease in biomass) will be reached when $\mu = D$. If $\mu > D$, the utilization of substrate will exceed the supply of substrate, causing the growth rate to slow until it is equal to the dilution rate. If $\mu < D$, the amount of substrate added will exceed the amount utilized. Therefore, the growth rate will increase until it is equal to the dilution rate. In either case, given time, a steady state will be established where:

$$\mu = D \qquad \text{(Eq. 3.17)}$$

Such a steady state can be achieved and maintained as long as the dilution rate does not exceed a critical rate,

D_c. The critical dilution rate can be determined by combining Eqs. 3.10 and 3.17:

$$D_c = \mu_{max}\left(\frac{S}{K_s + S}\right) \qquad \text{(Eq. 3.18)}$$

Looking at Eq. 3.18, it can be seen that the operation efficiency of a chemostat can be optimized under conditions in which $S >> K_s$, and therefore $D_c \approx \mu_{max}$. But it must be remembered that when a chemostat is operating at D_c, if the dilution rate is increased further, the growth rate will not be able to increase (since it is already at μ_{max}) to offset the increase in dilution rate. The result will be washing out of cells and a decline in the operating efficiency of the chemostat. Thus, D_c is an important parameter because if the chemostat is run at dilution rates less than D_c, operation efficiency is not optimized; whereas if dilution rates exceed D_c, washout of cells will occur as shown in Figure 3.10.

3.3 GROWTH IN THE ENVIRONMENT

How is growth in the natural environment related to growth in a flask or in continuous culture? There have been several attempts to classify bacteria in soil systems on the basis of their growth characteristics and affinity for carbon substrates. The first was by Sergei Winogradsky (1856−1953), the "father of soil microbiology," who introduced the ecological classification system of autochthonous versus zymogenous organisms. The former metabolize slowly in soil, utilizing slowly released nutrients from soil organic matter as a substrate. The latter are adapted to intervals of dormancy when substrate availability is low, or to rapid growth following the addition of fresh substrate or amendment to the soil. In addition to these two categories, there are the allochthonous organisms, which are organisms that are freshly introduced into soil and usually survive for only short periods of time (Figure 3.11).

Current terminology distinguishes soil microbes as either oligotrophs, those that prefer low substrate concentrations, or copiotrophs, those that prefer high substrate concentrations. This is similar to the concept of r and K selection. Organisms that respond to added nutrients with rapid growth rates are designated as r-strategists, while K-strategists have low but consistent growth rates and numbers in low nutrient environments. In reality, a soil community normally has a continuum of microorganisms with various levels of nutrient requirements ranging from obligate r-strategists, or copiotrophs, to obligate K-strategists, or oligotrophs. Typical maximum growth rates (μ_{max}) and affinity constants (K_s) for these two groups of microbes are given in Information Box 3.1.

When considering oligotrophic microbes in the environment, it is unlikely that they exhibit the stages of growth observed in batch flask and continuous culture.

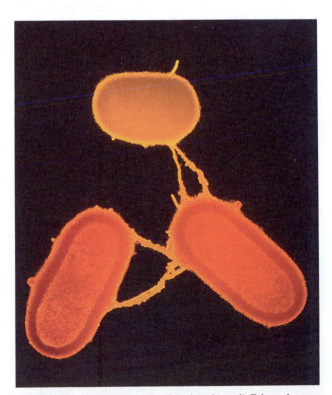

FIGURE 3.11 Conjugating cells of *Escherichia coli*. False-color transmission electron micrograph (TEM) of a male *E. coli* bacterium (bottom right) conjugating with two females. This male has attached two F-pili to each of the females.

These microbes metabolize slowly, and as a result have long generation times. Frequently, they use energy obtained from metabolism simply for cell maintenance. On the other hand, copiotrophic organisms may exhibit high rates of metabolism and perhaps exponential growth for short periods, or may be found in a dormant state. Note also that exponential growth does not normally persist for long periods of time in the environment. Rather, bacteria frequently alternate between periods of growth and nongrowth, i.e., constantly entering and leaving the stationary phase (Schaechter *et al.*, 2006). Dormant cells are often rounded and small (approximately 0.3 μm) in comparison with healthy laboratory specimens, which range from 1 to 2 μm in size. Dormant cells may become viable but nonculturable (VBNC) with time because of extended starvation conditions or because cells become reversibly damaged (Roszak and Collwell, 1987). VBNC microbes are thus difficult to culture because of cell stress and damage. Specific microbes may also be difficult to culture for several reasons, which are discussed in Section 10.3.1. New approaches for enhanced cultivation of soil bacteria are defined in Section 10.3.1.1.

Both of these cases contribute to the fact that direct counts from environmental samples, which include all viable cells, are often one to two orders of magnitude higher than culturable counts, which include only cells capable of growth on the culture medium used. When a soil culture is plated on a solid medium, a subset of the community which is dominated by copiotrophs quickly takes advantage and begins to metabolize actively. In a sense, this is similar to the reaction by microbes in a batch flask when nutrients are added. Studies utilizing both culture-dependent and culture-independent methodologies on the same samples have shown that culturing bacteria from soil actually selects for less abundant species, or members of the "soil rare biosphere" (Shade *et al.*, 2012) (see also Case Study 4.2 and Section 10.3.1.2). Thus, these microbes can exhibit the growth stages described in Section 3.2 for batch and continuous culture, but the pattern of the stages is quite different as described in the following sections.

3.3.1 The Lag Phase

The lag phase observed in a natural environment can be much longer than the lag phase normally observed in a batch culture. In some cases, this longer lag phase may be caused by very small numbers of initial populations that are capable of metabolizing the added substrate. Note also that chemicals which humans may consider to be organic contaminants can be a useful source of substrate for growth for those microbes with the necessary set of enzymes to degrade the contaminant. In this case, neither a significant disappearance of the contaminant nor a significant increase in cell numbers will be observed for several generations. Note that in a pure culture, the transition between the lag and exponential phase is defined to have occurred after the initial population has doubled. However, this is a troublesome definition to impose in the environment, where it is difficult to accurately measure the doubling of a small subset of the microbial community that is responding to the addition of nutrients including contaminants. Alternatively, populations capable of metabolizing the substrate may be dormant or injured, and require time to recover physiologically and resume metabolic activities. Further complicating growth in the environment is the fact that generation times are usually much longer than those measured under ideal laboratory conditions. This is due to a combination of limited nutrient availability and environmental conditions such as temperature or moisture that are not optimal for specific microbes. Thus, it is not unusual to observe lag periods of months or even years after an initial application of a newly synthesized anthropogenic pesticide, for significant degradation to be observed. However, once an environment has been exposed to a new pesticide and developed a community for its degradation, the degradation of succeeding pesticide applications will occur with shorter and shorter lag periods. This phenomenon is called acclimation or adaptation, and has been observed with successive applications of many pesticides including the broad-leaf herbicide 2,4-dichlorophenoxyacetic acid (2,4-D) (Newby *et al.*, 2000).

A second explanation for long lag periods in the environment is that a community with microbes genetically capable of utilizing a specific carbon source may not initially be present within the existing population. This situation may require a mutation or a gene transfer event to introduce appropriate degradative genes into indigenous microbes. For example, one of the first documented cases of gene transfer in soil was the transfer of the plasmid pJP4 from an introduced organism into an indigenous soil population. The plasmid transfer resulted in rapid and complete degradation of the herbicide 2,4-D within a microcosm (Case Study 3.1). In this study, gene transfer to indigenous soil recipients was followed by growth and survival of the transconjugants at levels significant enough to affect degradation. There are still few such studies, and the likelihood and frequency of gene transfer in the environment are topics that are currently under debate.

3.3.2 The Exponential Phase

In the environment, the second phase of growth, exponential growth, occurs for only very brief periods of time following addition of a substrate. Such substrate might be crop residues, vegetative litter, root residues or contaminants added to or spilled into the environment. Following substrate addition, it is the zymogenous cells, many of which are initially dormant, that can most quickly respond to added nutrients. Upon substrate addition, these dormant cells become physiologically active and briefly enter the exponential phase until the substrate is utilized, or until some limiting factor causes a decline in substrate degradation. Thus, in many environmental samples, bacteria alternate between short periods of exponential or balanced growth and subsequent nongrowth or unbalanced growth (Schaechter *et al.*, 2006). Thus, such bacteria may constantly leave and re-enter the stationary phase. As shown in Table 3.1, culturable cell counts rapidly increase one to two orders of magnitude in response to the addition of 1% glucose. In this experiment, four different soils were left untreated or were amended with 1% glucose and incubated at room temperature for 1 week.

Because nutrient levels and other factors, e.g., temperature or moisture, are seldom ideal, it is rare for cells in the environment to achieve a growth rate equal to μ_{max}. Thus, rates of growth in the environment are slower than growth rates measured under laboratory conditions. This is illustrated in Table 3.2, which compares the metabolism or degradation rates for wheat and rye straw in a laboratory environment with degradation rates in natural environments. These include: a Nigerian tropical soil that undergoes some dry periods; an English soil that is exposed to a moderate, wet climate; and a soil from Saskatoon, Canada, that is subjected to cold winters and dry summers. As shown in Table 3.2, the relative rate of straw degradation under laboratory conditions is twice as fast as in the Nigerian soil, eight times faster than in the English soil and 18 times faster than in the Canadian soil. This example illustrates the importance of understanding that there can be large differences between degradation rates in the laboratory and in natural environments. This understanding is crucial when attempting to predict degradation rates for contaminants in an environment.

3.3.3 The Stationary and Death Phases

Stationary phase in the laboratory is a period of time where there is active cell growth that is matched by cell death. In batch culture, cell numbers increase rapidly to levels as high as 10^{10} to 10^{11} CFU/ml. At this point, either the substrate is completely utilized or waste metabolites inhibit further growth. Recall that most cells are only likely to achieve an exponential phase for brief periods of time because of nutrient limitations and environmental stress. Rather, they are likely to be either dormant or in a maintenance state. Cells that do undergo growth in response to a nutrient amendment will quickly utilize the added food source. However, even with an added food source, cultural counts rarely exceed 10^8 to 10^9 CFU/g soil except perhaps on some root surfaces. At this point, cells will either die or, in order to prolong survival, they

Case Study 3.1   Gene Transfer Experiment

Biodegradation of contaminants in soil requires the presence of appropriate degradative genes within the soil population. If degradative genes are not present within the soil population, the duration of the lag phase for degradation of the contaminant may range from months to years. One strategy for stimulating biodegradation is to "introduce" degrading microbes into the soil. Unfortunately, unless selective pressure exists to allow the introduced organism to survive and grow, it will die within a few weeks as a result of abiotic stress and competition from indigenous microbes. DiGiovanni *et al.* (1996) demonstrated that an alternative to "introduced microbes" is "introduced genes." In this study the introduced microbe was *Ralstonia eutrophus* JMP134. JMP134 carries an 80-kb plasmid, pJP4, that encodes the initial enzymes necessary for the degradation of the herbicide 2,4-D. A series of soil microcosms were set up and contaminated with 2,4-D. In control microcosms, there was slow, incomplete degradation of the 2,4-D over a 9-week period (see figure). In a second set of microcosms, JMP134 was added to give a final inoculum of 10^5 CFU/g dry soil. In these microcosms, rapid degradation of 2,4-D occurred after a

1-week lag phase and the 2,4-D was completely degraded after 4 weeks. The scientists examined the microcosm 2,4-D-degrading population very carefully during this study. What they found was surprising. They could not recover viable JMP134 microbes after the first week. However, during weeks 2 and 3 they isolated two new organisms that could degrade 2,4-D. Upon closer examination, both organisms, *Pseudomonas glathei* and *Burkholderia caryophylli*, and found to be carrying the pJP4 plasmid! Finally, during week 5 a third 2,4-D degrader was isolated, *Burkholderia cepacia*. This isolate also carried the pJP4 plas-mid. Subsequent addition of 2,4-D to the microcosms resulted in rapid degradation of the herbicide, primarily by the third isolate, *B. cepacia*. Although it is clear from this research that the pJP4 plasmid was transferred from the introduced microbe to several indigenous populations, it is not clear how the transfer occurred. There are two possibilities: cell-to-cell contact and transfer of the plasmid via conjugation or death, and lysis of the JMP134 cells to release the pJP4 plasmid which the indigenous populations then took up, a process called transformation.

Soil + 2,4-D

Slow, incomplete degradation of 2,4-D over a one-week period.

Soil + 2,4-D + JMP134

Complete degradation of 2,4-D in 4 weeks. JMP134 was not recovered but plasmid pJP4 was found in 3 indigenous microbes.

Alcaligenes eutrophus JMP134

pJP4 plasmid

There are two possible mechanisms of gene transfer which may explain these results.

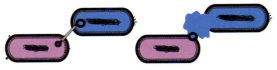

A. Plasmid transfer via conjugation

B. Cell lysis and uptake of plasmid via transformation

TABLE 3.1 Culturable Counts in Unamended and Glucose-Amended Soils[a]

Soil	Unamended (CFU/g soil)	1% Glucose (CFU/g soil)	Log Increase
Pima	5.6×10^5	4.6×10^7	1.9
Brazito	1.1×10^6	1.1×10^8	2.0
Clover Springs	1.4×10^7	1.9×10^8	1.1
Mt. Lemmon	1.4×10^6	8.3×10^7	1.7

Courtesy E.M. Jutras.
[a]*Each soil was incubated for approximately 1 week and then culturable counts were determined using R2A agar.*

can re-enter the dormant phase until a new source of nutrients becomes available. Thus, stationary phase periods of growth are likely to be very short, similar to that for exponential growth. In contrast, the death phase in environmental samples can certainly be observed, at least in terms of culturable counts. Once added nutrients are consumed, both living and dead cells become prey for protozoa that act as microbial predators. Bacteriophage can also infect and lyse significant portions of the living bacterial community. Dead cells are also quickly scavenged by other microbes in the vicinity, which reuse available carbon and nitrogen substrate. Thus, culturable cell numbers increase in response to nutrient addition (see Table 3.1), but will decrease again just as quickly to former background levels after all nutrients have been utilized.

3.4 MASS BALANCE OF GROWTH

During growth there is normally an increase in cell mass which is reflected in an increase in the number of cells. In this case, one can say that the cells are metabolizing substrate under growth conditions. However, in some cases, when the concentration of substrate or some other nutrient is limiting, utilization of the substrate occurs without production of new cells. In this case, the energy from substrate utilization is used to meet the maintenance requirements of the cell under nongrowth conditions. The level of energy required to maintain a cell is called the maintenance energy (Neidhardt et al., 1990).

Under either growth or nongrowth conditions, the amount of energy obtained by a microorganism through the oxidation of a substrate is reflected in the amount of cell mass produced, or the cell yield (Y). As discussed in Section 3.1.5, the cell yield coefficient is defined as the amount of cell mass produced per amount of substrate consumed. Although the cell yield is a constant, the value of the cell yield is dependent on the substrate being utilized. In general, the more reduced the substrate, the larger the amount of energy that can be obtained through its oxidation. For example, it is generally assumed that approximately half of the carbon in a molecule of sugar or organic acid will be used to build new cell mass, and half will be evolved as CO_2, corresponding to a cell yield of approximately 0.4. Note that the glucose ($C_6H_{12}O_6$) molecule is partially oxidized because the molecule contains six atoms of oxygen. One can compare this to a very low cell yield of 0.05 for pentachlorophenol, which is highly oxidized due to the presence of five chlorine atoms, and a very high cell yield of 1.49 for octadecane, which is completely reduced (Figure 3.12). As these examples show, some substrates support higher levels of growth and the production of more cell mass than others.

We can explore further why there are such differences in cell yield for these three substrates. As microbes have evolved, standard catabolic pathways have developed for common carbohydrate- and protein-containing substrates. For these types of substrates, approximately half of the carbon is used to build new cell mass. This translates into a cell yield of approximately 0.4 for a sugar such as glucose (see Example Calculation 3.4). However, since industrialization began in the late 1800s, many new molecules have been manufactured for which there are no standard catabolic pathways. Pentachlorophenol is an example of such a molecule. This material has been commercially produced since 1936, and is one of the major chemicals used to treat wood and utility poles. To utilize a molecule such as pentachlorophenol, which appeared in the environment relatively recently on an evolutionary scale, a microbe must alter the chemical structure to allow use of standard catabolic pathways. For pentachlorophenol, which has five carbon–chlorine bonds, this means

TABLE 3.2 Effect of Environment on Decomposition Rate of Plant Residues Added to Soil

Residue	Half-life (days)[a]	μ[b]	
		(1/days)	Relative Rate[c]
Wheat straw, laboratory	9	0.08	1
Rye straw, Nigeria	17	0.04	0.5
Rye straw, England	75	0.01	0.125
Wheat straw, Saskatoon	160	0.003	0.05

From Paul and Clark (1989).
[a]The half-life is the amount of time required for degradation of half of the straw initially added.
[b]μ is the specific growth rate constant.
[c]The relative rate of degradation of wheat straw under laboratory conditions is assumed to be 1. The degradation rates for straw in each of the soils were then compared with this value.

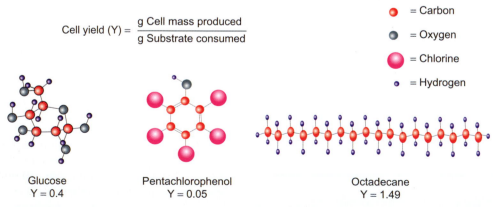

$$\text{Cell yield (Y)} = \frac{\text{g Cell mass produced}}{\text{g Substrate consumed}}$$

● = Carbon
● = Oxygen
● = Chlorine
● = Hydrogen

Glucose
Y = 0.4

Pentachlorophenol
Y = 0.05

Octadecane
Y = 1.49

FIGURE 3.12 Cell yield values for various substrates. Note that the cell yield depends on the structure of the substrate.

Example Calculation 3.4   Conversion of Substrate Carbon into Cell Mass and Carbon Dioxide during Growth

Problem: A bacterial culture is grown using glucose as the sole source of carbon and energy. The cell yield value is determined by dry weight analysis to be 0.4 (i.e., 0.4 g cell mass was produced per 1 g glucose utilized). What percentage of the substrate (glucose) carbon will be found as cell mass and as CO_2?

Assume that you start with 1 mole of carbohydrate ($C_6H_{12}O_6$, molecular weight = 180 g/mol):

$$\text{(substrate mass)(cell yield)} = \text{cell mass produced}$$
$$(180\ g)(0.4) = 72\ g$$

Cell mass can be estimated as $C_5H_7NO_2$ (molecular weight = 113 g/mol):

$$\text{mol cell mass} = \frac{72\ g\ \text{cell mass}}{113\ g/\text{mol cell mass}} = 0.64\ \text{mol cell mass}$$

In terms of carbon,
For cell mass: (0.64 mol cell mass)(5 mol C/mol cell mass)(12 g/mol C) = 38.4 g carbon
For substrate: (1 mol substrate)(6 mol C/mol substrate)(12 g/mol C) = 72 g carbon
The percentage of substrate carbon found in cell mass is

$$\frac{38.4\ g\ \text{carbon}}{72\ g\ \text{carbon}}(100) = 53\%$$

and by difference, 47% of the carbon is released as CO_2.

Question

Calculate the carbon found as cell mass and CO_2 for a microorganism that grows on octadecane ($C_{18}H_{36}$), Y = 1.49, and on pentachlorophenol (C_6HOCl_5), Y = 0.05.

Answer

For octadecane, 93% of the substrate carbon is found in cell mass and 7% is evolved as CO_2. For pentachlorophenol, 10% of the substrate carbon is found in cell mass and 90% is evolved as CO_2.

---

that a microbe must expend a great deal of energy to break the strong carbon–halogen bonds before the substrate can be metabolized to produce energy. Because so much energy is required to remove the chlorines from pentachlorophenol, relatively little energy is left to build new cell mass. This results in a very low cell yield value.

In contrast, why is the cell yield so high for a hydrocarbon such as octadecane? Octadecane is a hydrocarbon typical of those found in petroleum products (see Chapter 17). Because petroleum is an ancient mixture of molecules formed on early Earth, standard catabolic pathways exist for most petroleum components, including octadecane. The cell yield value for growth on octadecane is high because octadecane is a saturated molecule (the molecule contains no oxygen, only carbon–hydrogen bonds). Such a highly reduced hydrocarbon stores more energy than a molecule that is partially oxidized such as glucose (glucose contains six oxygen molecules). This energy is released during metabolism, allowing the microbe to obtain more energy from the degradation of octadecane than from the degradation of glucose. This in turn is reflected in a higher cell yield value.

3.4.1 Aerobic Conditions

Under aerobic conditions, microorganisms metabolize substrates by a process known as aerobic respiration. The complete oxidation of a substrate under aerobic conditions is represented by the mass balance equation:

$$(C_6H_{12}O_6) + 6(O_2) \rightarrow 6(CO_2) + 6(H_2O) \quad \text{(Eq. 3.19)}$$
$$\text{substrate} \quad \text{oxygen} \quad \text{carbon dioxide} \quad \text{water}$$

In Eq. 3.19, the substrate is a carbohydrate such as glucose, which can be represented by the formula $C_6H_{12}O_6$. Oxidation of glucose by microorganisms is more complex than shown in this equation because some of the substrate carbon is utilized to build new cell mass, and is therefore not completely oxidized. Thus, aerobic microbial oxidation of glucose can be more completely described by the following, slightly more complex, mass balance equation:

$$a(C_6H_{12}O_6) + b(NH_3) + c(O_2) \rightarrow$$
$$\text{substrate} \quad \text{nitrogen source} \quad \text{oxygen}$$
$$d(C_5H_7NO_2) + e(CO_2) + f(H_2O) \quad \text{(Eq. 3.20)}$$
$$\text{cell mass} \quad \text{carbon dioxide} \quad \text{water}$$

where a, b, c, d, e and f represent mole numbers.

It should be emphasized that the degradation process is the same whether the substrate is readily utilized (glucose) or only slowly utilized as in the case of a contaminant such as benzene. Equation 3.20 differs from Eq. 3.19 in two ways: it represents the production of new cell mass, estimated by the formula $C_5H_7NO_2$, and in order to balance the equation, it has a nitrogen source on the reactant side, shown here as ammonia (NH_3).

The mass balance equation has a number of practical applications. It can be used to estimate the amount of oxygen or nitrogen required for growth and utilization of a particular substrate. This is useful for wastewater treatment (Chapter 25), for production of high value microbial products (e.g., antibiotics or vitamins) and for remediation of contaminated sites (see Chapter 17 and Example Calculation 3.5).

3.4.2 Anaerobic Conditions

The amount of oxygen in the atmosphere (21%) ensures aerobic degradation for the overwhelming proportion of the organic matter produced annually. In the absence of oxygen, organic substrates can be mineralized to carbon dioxide by fermentation or by anaerobic respiration, although these are less efficient processes than aerobic respiration (see Information Box 3.2). In general, anaerobic metabolism is restricted to water-saturated niches such as sediments, isolated water bodies within lakes and oceans, and microenvironments in soils. Anaerobic degradation requires alternative electron acceptors, either an organic compound for fermentation, or one of a series of inorganic electron acceptors for anaerobic respiration (Table 3.3). During anaerobic respiration, the terminal electron acceptor used depends on availability, and follows a sequence that corresponds to the electron affinity of the electron acceptors. Examples of alternative electron acceptors in order of decreasing electron affinity are: nitrate (nitrate-reducing conditions); manganese (manganese-reducing conditions); iron (iron-reducing conditions); sulfate (sulfate-reducing conditions); and carbonate (methanogenic conditions). More recently, additional terminal electron acceptors have been identified, among them arsenate, arsenite, selenate and uranium IV (Stolz et al., 2006). These may be important in environments where they can be found in abundance.

Often, under anaerobic conditions, organic compounds are degraded by an interactive group or consortium of microorganisms. Individuals within the consortium each carry out different, specialized reactions that together lead to complete mineralization of the compound (Stams et al., 2006). The final step of anaerobic degradation is methanogenesis, which occurs when other inorganic electron acceptors such as nitrate and sulfate are exhausted. Methanogenesis results in the production of methane and is the most important type of metabolism in anoxic freshwater lake sediments. Methanogenesis is also important in anaerobic treatment of sewage sludge, in which the supply of nitrate or sulfate is very small compared with the input of organic substrate. In this case, even though the concentrations of nitrate and sulfate are low, they are of basic importance for the establishment and maintenance of a sufficiently low electron potential that allows

proliferation of the complex methanogenic microbial community.

Mass balance equations very similar to that for aerobic respiration can be written for anaerobic respiration. For example, the following equation can be used to describe the transformation of organic matter into methane (CH_4) and CO_2:

$$C_nH_aO_b + \left(n - \frac{a}{2} - \frac{b}{4}\right)H_2O$$
$$\rightarrow \left(\frac{n}{2} - \frac{a}{8} + \frac{b}{4}\right)CO_2 + \left(\frac{n}{2} - \frac{a}{8} + \frac{b}{4}\right)CH_4 \quad \text{(Eq. 3.21)}$$

where n, a and b represent mole numbers.

Note that after biodegradation occurs, the substrate carbon is found either in its most oxidized form, CO_2, or in its most reduced form, CH_4. This is called disproportionation of organic carbon. The ratio of methane to carbon dioxide found in the gas mixture which results from anaerobic degradation depends on the oxidation state of the substrate used. Carbohydrates are converted to approximately equal amounts of CH_4 and CO_2. Substrates that are more reduced such as methanol or lipids produce relatively higher amounts of methane, whereas substrates that are more oxidized such as formic acid or oxalic acid produce relatively less methane.

QUESTIONS AND PROBLEMS

1. Draw a growth curve of substrate disappearance as a function of time. Label and define each stage of growth.
2. Calculate the time it will take to increase the cell number from 10^4 CFU/ml to 10^8 CFU/ml assuming a generation time of 1.5 h.
3. From the following data calculate the mean generation time:
   a. At the beginning of exponential growth when time $t = 0$, initial cell concentration = 2500 per ml
   b. At time t = 8 hours cell concentration = 10,000 per ml
4. Differentiate between metabolism during growth and nongrowth conditions.
5. You are given a microorganism that has a maximum growth rate (μ_{max}) of 0.39 h^{-1}. Under ideal conditions (maximum growth rate is achieved), how long will it take to obtain 1×10^{10} CFU/ml if you begin with an inoculum of 2×10^7 CFU/ml?
6. Is there any way to increase the growth rate observed in Question 3?
7. Write the Monod equation and define each of the constants.

Information Box 3.2   Biological Generation of Energy

Growth and metabolism requires energy, which is usually stored and transferred in the form adenosine triphosphate (ATP). As with any chemical reaction, metabolic reactions are subject to the second law of thermodynamics which says:

In a chemical reaction, only part of the energy is used to do work. The rest of the energy is lost as entropy. For any chemical reaction the free energy ΔG is the amount of energy available for work.

For the reaction $A + B \rightleftharpoons C + D$, the thermodynamic equilibrium constant, K_{eq}, is defined as

$$K_{eq} = \frac{[C][D]}{[A][B]}$$

Case 1: If the formation of products (C and D) is favored, then K_{eq} will be positive. That is, if $[C][D] > [A][B]$ then $K_{eq} > 1$. In this case, the logarithm of K_{eq} is also positive; for example, if $K_{eq} = 2$, then $\log K_{eq} = 0.301$.

Case 2: If product formation is not favored, then K_{eq} will be less than 1. That is, if $[A][B] > [C][D]$ then $K_{eq} < 1$. In this case, $\log K_{eq}$ will be negative; for example, if $K_{eq} = 0.2$, $\log K_{eq} = -0.699$.

Thus, for a reaction to have a positive K_{eq} and proceed as written (from left to right), the energy of the products must be lower than the energy of the reactants, meaning that overall, energy is released during the reaction.

The relationship between K_{eq} and the Gibbs free energy change (ΔG^0) is given by

$$\Delta G^0 = -RT \log K_{eq}$$

where R is the universal gas constant and T the absolute temperature (K).

ΔG^0 is negative when $\log K_{eq}$ is positive and the reaction will proceed spontaneously.

ΔG^0 is positive when $\log K_{eq}$ is negative and the reaction will not proceed as written.

Thus we can use ΔG values for any biochemical reaction mediated by microbes to determine whether energy is liberated for work and how much energy is liberated. Soil organisms can generate energy via several mechanisms, which can be divided into three main categories: photosynthesis, respiration, and fermentation.

1. **Photosynthesis**

$$2H_2O + CO_2 \xrightarrow{\text{light}} \underset{\text{biomass}}{CH_2O} + O_2 + H_2O$$

$$\Delta G^0 \simeq +115 \text{ kcal/mol}$$

Note that ΔG^0 is positive and so this reaction is not favorable. It requires reaction energy supplied by sunlight. The fixed organic carbon is then metabolized to generate energy via respiration. Examples of photosynthetic soil organisms are *Rhodospirillum*, *Chromatium*, and *Chlorobium*.

2. **Respiration**

    a. *Aerobic heterotrophic respiration:* Many bacteria function as either obligate or facultative aerobic chemoheterotrophs including *Pseudomonas*, *Bacillus*, and *E. coli*.

$$\underset{\text{succinic acid}}{C_4H_6O_4} + 3.5O_2 \leftrightarrow 4CO_2 + 3H_2O$$

$$\Delta G^0 (25°)C - 1569.25 \text{ kJ/mol}$$

b. *Anaerobic heterotrophic respiration:* Anaerobic chemoheterotrophs utilize alternate terminal electron acceptors and organic compounds from the electron donor. The examples below use the same electron donor (succinic acid, $C_4H_6O_4$) with three different terminal electron acceptors (nitrate, iron, sulfate).

$$C_4H_6O_4 + 2.8NO_3^- + 2.8H^+ \leftrightarrow 1.4N_2 + 4CO_2 + 4.4H_2O$$
$$\Delta G^0 (25°C) = -1507.59 \text{ kJ/mol}$$

$$C_4H_6O_4 + 14Fe^{3+} + 4H_2O \leftrightarrow 4CO_2 + 14Fe^{2+} + 14H^+$$
$$\Delta G^0 (25°C) = -891.00 \text{ kJ/mol}$$

$$C_4H_6O_4 + 1.75SO_4^2 + 3.5H^+ \leftrightarrow 4CO_2 + 1.75H_2S + 12H_2O$$
$$\Delta G^0 (25°C) = -257.60 \text{ kJ/mol}$$

c. *Aerobic autotrophic respiration:* The reactions carried out by the obligately aerobic chemoautotrophs *Nitrosomonas* and *Nitrobacter* are known as nitrification:

$$NH_3 + 1.5O_2 \rightarrow H^+ + NO_2^- + H_2O \text{ (\textit{Nitrosomonas})}$$
$$\Delta G^0 (25°C) = -267.50 \text{ kJ/mol}$$

$$NO_2^- + 0.5O_2 \rightarrow NO_3^- \text{ (\textit{Nitrobacter})}$$
$$\Delta G^0 (25°C) = -86.96 \text{ kJ/mol}$$

The following two reactions are examples of chemoautotrophic sulfur oxidation:

$$H_2S + 0.5O_2 \rightarrow S^0 + H_2O \text{ (\textit{Beggiatoa})}$$
$$\Delta G^0 (25°C) = -217.53 \text{ kJ/mol}$$

$$S^0 + 1.5O_2 + H_2O \rightarrow SO_4^{2-} + 2H^+ \text{ (\textit{Thiobacillus thiooxidans})}$$
$$\Delta G^0 (25°C) = -532.09 \text{ kJ/mol}$$

The next reaction involves the chemoautotrophic degradation of carbon monoxide:

$$CO + 0.5O_2 \rightarrow CO_2 \quad \Delta G^0 (25°C) = -274.24 \text{ kJ/mol}$$

d. *Anaerobic autotrophic respiration:* *Thiobacillus denitrificans* can utilize nitrate as a terminal electron acceptor.

$$S^0 + 1.2NO_3^- + 0.4H_2O \rightarrow SO_4^{2-} + 0.6N_2$$
$$\Delta G^0 (25°C) = -505.59 \text{ kJ/mol}$$

3. **Fermentation**

$$\underset{\text{ethanol}}{C_2H_5OH} + CO_2 \rightarrow \underset{\text{acetic acid}}{1.5CH_3COOH} \quad \Delta G^0 (25°C) = -27.45 \text{ kJ/mol}$$

All of the preceding reactions illustrate how organisms mediate reactions involved in biogeochemical cycling of carbon, nitrogen and sulfur (Chapter 14). The $\Delta G^0 (25°C)$ values provided were compiled from Amend and Shock (2001).

Example Calculation 3.5   Leaking Underground Storage Tanks

The Environmental Protection Agency (EPA) estimates that more than 1 million underground storage tanks (USTs) have been in service in the United States alone. Over 436,000 of these have had confirmed releases into the environment. Although regulations now require USTs to be upgraded, leaking USTs continue to be reported at a rate of 20,000/year and a cleanup backlog of > 139,000 USTs still exists. In this exercise we will calculate the amount of oxygen and nitrogen necessary for the remediation of a leaking UST site that has released 10,000 gallons of gasoline. To simplify the problem, we will assume that octane (C_8H_{18}) is a good representative of all petroleum constituents found in gasoline. We will use the mass balance equation to calculate the biological oxygen demand (BOD) and the nitrogen demand:

$$a(C_8H_{18}) + b(NH_3) + c(O_2) \rightarrow d(C_5H_7NO_2) + e(CO_2) + f(H_2O)$$

octane   ammonia   oxygen   cell mass   carbon dioxide   water

In this equation, the coefficients a through f indicate the number of moles for each component. To solve the mass balance equation we must be able to relate the amount of cell mass produced to the amount of substrate (octane) consumed. This is done using the cell yield Y where

$$Y = \frac{\text{mass of cell mass produced}}{\text{mass of substrate consumed}}$$

Literature indicates that a reasonable cell yield value for octane is 1.2. Using the cell yield we can calculate the coefficient d. We will start with 1 mole of substrate ($a = 1$) and use the following equation:

$$d(\text{MW cell mass}) = a(\text{MW octane})(Y)$$
$$d(113 \text{ g/mol}) = 1(114 \text{ g/mol})(1.2)$$
$$\therefore d = 1.2$$

We can then solve for the other coefficients by balancing the equation. We start with nitrogen. We know that there is one N on the right side of the equation in the biomass term. Examining the left side of the equation, we see that there is similarly one N as ammonia. We can set up a simple relationship for nitrogen and use this to solve for coefficient b:

$$b(1 \text{ mol nitrogen}) = d(1 \text{ mol nitrogen})$$
$$b(1) = 1.2(1)$$
$$b = 1.2$$

Next we balance carbon and solve for coefficient e:

$$e(1 \text{ mol carbon}) = a(8 \text{ mol carbon}) - d(5 \text{ mol carbon})$$
$$e(1) = 1(8) - 1.2(5)$$
$$e = 2.0$$

Next we balance hydrogen and solve for coefficient f:

$$f(2 \text{ mol hydrogen}) = a(18 \text{ mol hydrogen}) + b(3 \text{ mol hydrogen}) - d(7 \text{ mol hydrogen})$$
$$f(2) = 1(18) + 1.2(3) - 1.2(7)$$
$$f = 6.6$$

Finally we balance oxygen and solve for coefficient c:

$$c(2 \text{ mol oxygen}) = d(2 \text{ mol oxygen}) + e(2 \text{ mol oxygen}) + f(1 \text{ mol oxygen})$$
$$c(2) = 1.2(2) + 2(2.0) + 6.6(1)$$
$$c = 6.5$$

Thus, the solved mass balance equation is $1(C_8H_{18}) + 1.2(NH_3) + 6.5(O_2) \rightarrow 1.2(C_5H_7NO_2) + 2.0(CO_2) + 6.6(H_2O)$. Now we use this mass balance equation to determine how much nitrogen and oxygen will be needed to remediate the site. First we convert gallons of gasoline into moles of octane using the assumption that octane is a good representative of gasoline. Recall that we started with 10,000 gallons of gasoline:

Convert to liters (L) : 10, 000 gallons (3.78 L/gallon) = 3.78×10^4 L gasoline
Convert to grams (g) : 3.78×10^4 L gasoline (690 g gasoline/L) = 2.6×10^7 g gasoline in the site
$$\text{Convert to moles}: = \frac{2.6 \times 10^7 \text{ g gasoline}}{114 \text{ g octane/mol}} = 2.3 \times 10^5 \text{ mol octane in the site}$$

Now we ask, how much nitrogen is needed to remediate this spill? From the mass balance equation we know that we need 1.2 mol NH_3/mol octane (see coefficient b).

$$(1.2 \text{ mol } NH_3/\text{mol octane})(2.3 \times 10^5 \text{ mol octane in the site})(17 \text{ g } NH_3/\text{mol}) = 4.7 \times 10^6 \text{ g } NH_3$$
$$\Rightarrow 4.7 \times 10^6 \text{ g}(1 \text{ kg}/1000 \text{ g})(2.2046 \text{ lb/kg}) = 10000 \text{ lb or 5 tons of } NH_3!$$

Finally we ask, how much oxygen is needed to remediate this spill? From the mass balance equation we know that we need 6.5 mol O_2/mol octane (see coefficient c).

$$(6.5 \text{ mol } O_2/\text{mol octane})(2.3 \times 10^5 \text{ mol octane in the site}) = 1.5 \times 10^6 \text{ mol } O_2$$

A gas takes up 22.4 liters/mol, but remember that air is only 21% oxygen.

$$(1.5 \times 10^6 \text{ mol } O_2)(22.4 \text{ L/mol air})(\text{L mol air}/0.21 \text{ mol } O_2) = 1.6 \times 10^8 \text{ L air}$$
$$1 \text{ cubic foot of air} = 28.33 \text{ L}$$
$$\Rightarrow 1.6 \times 10^8 \text{ L air (1 cubic foot}/28.33 \text{ L gas}) = 5.5 \times 10^6 \text{ cubic feet of air or enough air to fill a football field to a height of 100 ft!}$$

From Pepper *et al.*, 2006.

TABLE 3.3 Relationship Between Respiration, Redox Potential and Typical Electron Acceptors and Products[a]

Type of Respiration	Reduction reaction Electron Acceptor → Product	Reduction Potential (V)	Oxidation Reaction Electron Donor → Product	Oxidation Potential (V)	Difference (V)
Aerobic	$O_2 - H_2O$	+0.81	$CH_2O - CO_2$	−0.47	−1.28
Denitrification	$NO_3 - N_2$	+0.75	$CH_2O - CO_2$	−0.47	−1.22
Manganese reduction	$Mn^{4+} - Mn^{2+}$	+0.55	$CH_2O - CO_2$	−0.47	−1.02
Nitrate reduction	$NO_3 - NH_4^+$	+0.36	$CH_2O - CO_2$	−0.47	−0.83
Sulfate reduction	$SO_4^{2-} - HS^-, H_2S$	−0.22	$CH_2O - CO_2$	−0.47	−0.25
Methanogenesis	$CO_2 - CH_4$	−0.25	$CH_2O - CO_2$	−0.47	−0.22

Adapted from *Zehnder and Stumm (1988)*.
[a]*Biodegradation reactions can be considered a series of oxidation−reduction reactions. The amount of energy obtained by cells is dependent on the difference in energy between the oxidation and reduction reactions. As shown in this table, using the same electron donor in each case but varying the electron acceptor, oxygen as a terminal electron acceptor provides the most energy for cell growth and methanogenesis provides the least.*

---

8. There are two special cases when the Monod equation can be simplified. Describe these cases and the simplified Monod equation that results.

9. List terminal electron acceptors used in anaerobic respiration in the order of preference (from an energy standpoint).

10. Define disproportionation.

11. Define the term critical dilution rate, D_c, and explain what happens in continuous culture when D is greater than D_c.

12. Compare the characteristics of each of the growth phases (lag, log, stationary and death) for batch culture and environmental systems.

13. Compare and contrast the copiotrophic and oligotrophic style of life in the environment.

REFERENCES AND RECOMMENDED READING

Amend, J. P., and Shock, E. L. (2001) Energetics of overall metabolic reactions of thermophilic and hyperthermophilic Archaea and Bacteria. *FEMS Microbiol. Rev.* **25**, 175−243.

Blanch, H. W., and Clark, D. S. (1996) "Biochemical Engineering," Marcel Dekker, New York.

DiGiovanni, G. D., Neilson, J. W., Pepper, I. L., and Sinclair, N. A. (1996) Gene transfer of *Alcaligenes eutrophus* JMP134 plasmid pJP4 to indigenous soil recipients. *Appl. Environ. Microbiol.* **62**, 2521−2526.

Estrella, M. R., Brusseau, M. L., Maier, R. S., Pepper, I. L., Wierenga, P. J., and Miller, R. M. (1993) Biodegradation, sorption, and transport of 2,4-dichlorophenoxyacetic acid in saturated and unsaturated soils. *Appl. Environ. Microbiol.* **59**, 4266−4273.

Hoskisson, P. A., and Hobbs, G. (2005) Continuous culture—making a comeback? *Microbiology* **151**, 3153–3159.

Madigan, M. T., and Martinko, J. M. (2006) "Brock Biology of Microorganisms," Prentice Hall, New Jersey.

Miller, R. M., and Bartha, R. (1989) Evidence from liposome encapsulation for transport-limited microbial metabolism of solid alkanes. *Appl. Environ. Microbiol.* **55**, 269–274.

Neidhardt, F. C., Ingraham, J. L., and Schaechter, M. (1990) "Physiology of the Bacterial Cell: A Molecular Approach," Sinauer Associates, Sunderland, MA.

Newby, D. T., Gentry, T. J., and Pepper, I. L. (2000) Comparison of 2,4-dichlorophenoxyacetic acid degradation and plasmid transfer in soil resulting from bioaugmentation with two different pJP4 donors. *Appl. Environ. Microbiol.* **66**, 3397–3407.

Paul, E. A., and Clark, F. E. (1989) "Soil Microbiology and Biochemistry," Academic Press, San Diego, CA.

Roszak, D. B., and Collwell, R. R. (1987) Survival strategies of bacteria in the natural environment. *Microbiol. Rev.* **51**, 365–378.

Schaechter, M., Ingraham, J. L., and Neidhardt, F. C. (2006) "Microbe," ASM Press, Washington, DC.

Shade, A., Hogan, C. S., Klimowicz, A. K., Linske, M., McManus, P. S., and Handelsman, J. (2012) Culturing captures members of the soil rare biosphere. *Environ. Microbiol.* **14**, 2247–2252.

Stams, A. J. M., de Bok, F. A. M., Plugge, C. M., van Eekert, M. H. A., Dolfing, J., and Schraa, G. (2006) Exocellular electron transfer in anaerobic microbial communities. *Environ. Microbiol.* **8**, 371–382.

Stolz, J. E., Basu, P., Santini, J. M., and Oremland, R. S. (2006) Arsenic and selenium in microbial metabolism. *Ann. Rev. Microbiol.* **60**, 107–130.

Yates, G. T., and Smotzer, T. (2007) On the lag phase and initial decline of microbial growth curves. *J. Theoret. Biol.* **244**, 511–517.

Zehnder, A. J. B., and Stumm, W. (1988) Geochemistry and biogeochemistry of anaerobic habitats. In "Biology of Anaerobic Microorganisms" (A. J. B. Zehnder, ed.), Wiley, New York, pp. 1–38.

Zelenev, V. V., van Bruggen, A. H. C., and Semenov, A. M. (2005) Modeling wave-like dynamics of oligotrophic and copiotrophic bacteria along wheat roots in response to nutrient input from a growing root tip. *Ecol. Model.* **188**, 404–417.

Part II

Microbial Environments

Chapter 4

Earth Environments

Ian L. Pepper and Terry J. Gentry

4.1 EARTH'S LIVING SKIN

Soil is the thin veneer of material that covers much of Earth's surface. This fragile part of Earth's skin is frequently less than a meter thick, yet is absolutely vital for human life. It has a rich texture and fragrance and teems with plants, insects and microorganisms. Young and Crawford (2004) described it as "the most complicated biomaterial on the planet." The complexity of soil is driven by two components: the abiotic soil architecture and biotic diversity which is driven and supported by large amounts of energy from the sun through photosynthesis. Integrated together these components result in amazing physical, chemical and biological heterogeneity among soils globally.

The abiotic portion of all soils consists of inorganic particles of different size ranges, notably sands, silts and clays. Not only are the size ranges different, but the shapes and morphology of these particles also differ. This results in different specific surface areas of the particulates, with the smaller clays having larger surface areas per unit of mass than the silts and sands. Surface area in turn impacts the surface chemistry of the soil in question, as well as the rates of chemical reactions and

transformations. Under the influence of the soil biota, the different sized inorganic particles combine to form secondary aggregates. Pore spaces within the aggregate structure (intraaggregate pore space) and between the aggregates (interaggregate pore space) are crucial to the overall soil architecture (Figure 4.1). The soil architecture in turn is critical for the regulation of water movement and retention, gas exchange and microsite redox potentials within the soil. Totally enclosed pores within aggregates can have much lower redox potentials than open pores between aggregates. The resulting heterogeneity that develops means that both aerobic and anaerobic microorganisms can exist in very close proximity to one another.

Soils also contain biotic components (e.g., plant vegetation, decaying residues, stable soil humus and soil organisms), which add to the soil matrix complexity and architecture. Plant vegetative growth originates in soils, and following the death of plants, senesced organic vegetation is returned to the soil where it is degraded by heterotrophic soil microorganisms. Nutrients released during degradation are utilized by soil microbes and by new vegetation. Inorganic substrates such as ammonium, nitrate or sulfate are subject to autotrophic microbial

I.L. Pepper, C.P. Gerba, T.J. Gentry: Environmental Microbiology, Third edition. DOI: http://dx.doi.org/10.1016/B978-0-12-394626-3.00004-1
© 2015 Elsevier Inc. All rights reserved.

FIGURE 4.1 Soil architecture resulting from secondary aggregate formation with intraaggregate and interaggregate pore space. *Source:* Pepper (2014).

Information Box 4.1   The Five Soil Forming Factors

Parent material. The rock and mineral base from which soil is formed through weathering.

Climate. Precipitation and temperature are particularly important in weathering of parent material.

Organisms. Plants, animals, and microbes add organic matter and aid in decomposition and nutrient cycling that are part of the weathering process.

Topography. In particular the site slope angle and length.

Time. Essential for the soil weathering process; soils generally form more rapidly in warm environments than in cold ones.

transformations. Some organic residues are incorporated into the organic backbone of soils known as humus. Degradation of organic substrates also results in microbial gums and slimes, which together with fungal hyphae enhance the process of binding primary inorganic particles into secondary aggregates. Microbial populations proliferate in soil, with billions of bacteria and fungi coexisting in close proximity. Other biological entities include phage and protozoa which are important for the control of bacterial populations. The diversity of these microbes with respect to substrate utilization (organic versus inorganic) and redox requirements (aerobic versus anaerobic) results in diverse microbial communities capable of coexisting in microsite niches within the heterogeneous soil matrix. The microbial populations mediate innumerable biochemical transformations within soils. Despite their very large numbers, microbes occupy less than 1% of the total soil surface area, about the same land area on Earth occupied by humans (Young and Crawford, 2004).

The soil colloidal matrix, which consists of micronsized particles including inorganic, organic and biological entities, dominates soil architecture. Soil architecture, in turn, controls soil chemical and biochemical transformations and soil diversity. The diversity of soil is characterized by physical and temporal heterogeneities across all measured scales from nm to km, and is probably the driving force for the microbial diversity that we see in soil (Young and Crawford, 2004). Diversity estimates of the number of bacterial species in soil range from 2000 to 8.3 million per gram of soil depending on the methodologies utilized (Roesch *et al.*, 2007). Regardless of the true estimate, the microbial diversity within soil is clearly enormous and greatly impacts soil health and ultimately human health (Pepper, 2014).

4.2 PHYSICOCHEMICAL CHARACTERISTICS OF THE EARTH ENVIRONMENT

4.2.1 Earth Environments

4.2.1.1 Soil

Soil is the weathered end product of the action of climate and living organisms on soil parent material with a particular topography over time. We refer to these factors as the five soil-forming factors (Information Box 4.1). The soil weathering process can take decades to millions of years depending on the soil-forming factors involved. The physical and chemical characteristics of soils are discussed in detail in Section 4.2.2. The major difference between a surface soil and the subsurface is that in the subsurface, the parent material has generally not been weathered by climate. In addition, microbial numbers are much lower in subsurface environments than in soil because of reduced inputs of plant residues that function as substrate for heterotrophic microbes.

4.2.1.2 Vadose Zone

The vadose zone is defined as the subsurface unsaturated oligotrophic environment that lies between the surface soil and the saturated zone. The vadose zone contains mostly unweathered parent materials and has a very low organic carbon content (generally <0.1%). Thus, the availability of carbon and micronutrients is very limited compared with that in surface soils. The thickness of the vadose zone varies considerably. When the saturated zone is shallow or near the surface, the unsaturated zone is narrow or sometimes even nonexistent, as in a wetland area. In contrast, there are many arid or semiarid areas of the world where the unsaturated zone can be hundreds of meters thick. These unsaturated regions, especially deep

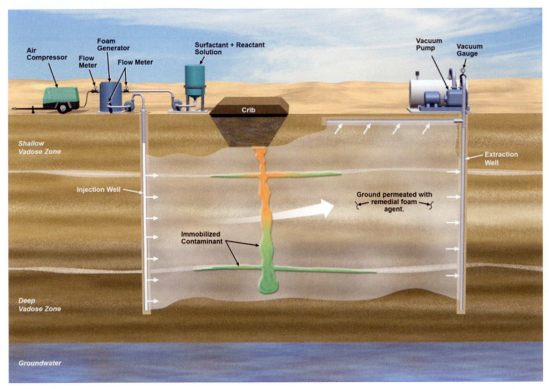

FIGURE 4.2 Delivery of remedial solutions through a heterogeneous deep vadose zone to remove contaminants. *Source:* Pacific Northwest National Laboratory.

unsaturated regions, may receive little or no moisture recharge from the surface, and normally have limited microbial activity because of low nutrient and/or moisture status. However, these regions are receiving more attention from a microbiological perspective, because pollutants that are present from surface contamination must pass through the vadose zone before they can reach groundwater (Figure 4.2).

4.2.1.3 Saturated Zone—Aquifers

The saturated zones that lie directly beneath the vadose zone are commonly called aquifers and are composed of porous parent materials that are saturated with water. Like the vadose zone, aquifers are generally oligotrophic environments. The boundary between the vadose zone and the saturated zone is not a uniformly distinct one, because the water table can rise or fall depending on rainfall events. The area that makes up this somewhat diffuse boundary is called the capillary fringe (Figure 4.3). Aquifers serve as a major source of potable water for much of the world. For example, in the United States, approximately 50% of the potable water supply currently comes from aquifers.

There are several types of aquifers, including shallow table aquifers and intermediate and deep aquifers that are

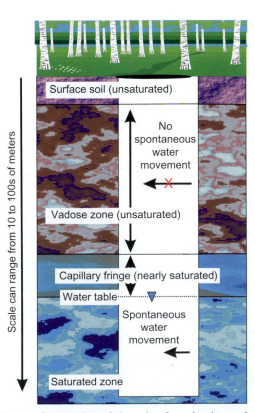

FIGURE 4.3 Cross-section of the subsurface showing surface soil, vadose zone, capillary fringe and saturated zone.

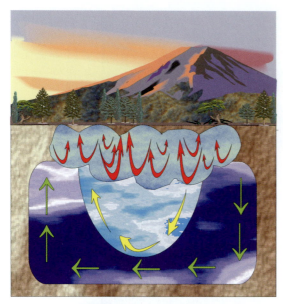

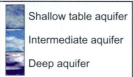

Shallow table aquifer

Intermediate aquifer

Deep aquifer

FIGURE 4.4 Shallow, intermediate and deep aquifer systems. Arrow thickness indicates the relative flow rates (the thicker the arrow, the faster the flow rate) in the different aquifer systems. Adapted from Chapelle (1993) and reproduced by permission of Wiley, New York.

separated from shallow aquifers by confining layers (Figure 4.4). Confining layers are composed of materials such as clay that have very low porosity. Such layers allow little water movement between shallow and deeper aquifers. Of these different types of aquifers, shallow aquifers are most closely connected to the Earth's surface and have the highest organic carbon content. They receive water from rainfall events and provide recharge to adjacent streams or rivers. In addition, shallow aquifer systems are very active with rapid groundwater flows (meters per day), and hence usually remain aerobic. Confined aquifers within 300 m of the surface soil are termed intermediate aquifers. These have much slower flow rates, on the order of meters per year. It is this aquifer system that supplies a major portion of drinking and irrigation water. Deep aquifers, those more than 300 m in depth, are characterized by extremely slow flow rates (meters per century). Because so little water flow occurs, these aquifers are usually anaerobic. Deep aquifers are not directly recharged or affected by surface rainfall events.

4.2.1.4 Saturated Zone—Wetlands

Wetlands are important ecosystems throughout the world in areas that have a temperate climate and include

Information Box 4.2 Lindow Man

"Lindow Man" was buried for 2000 years in an English peat bog and his body was discovered in August, 1984. This was an exciting find because his body was so well preserved that scientists could tell that his last meal consisted primarily of cereal grains. Scientists also discovered that "Lindow Man " was murdered (he was hit on the head, strangled, and his throat was cut for good measure). It is thought that a number of factors were important for the preservation of " Lindow Man's " body: the acidic pH, the absence of oxygen and the presence of antimicrobials in the peat, along with the presence of peat components such as sphagnan that reacted with collagen tissue in the body and basically tanned it (Painter, 1991). All of these factors aided in suppressing microbial activity that normally acts to degrade dead tissue.

swamps, marshes and bogs. Such areas are saturated for most or all of the year because the water table is at or above Earth's surface. These ecosystems are of increasing interest to environmental microbiologists for their potential to treat polluted waste streams such as sewage effluent (Chapter 25) and acid mine drainage.

One important example of wetlands are bogs, which are extensive worldwide, covering 5 to 8% of the terrestrial surface. Bogs are composed of deep layers of waterlogged peat and a surface layer of living vegetation. The peat layers are composed of the dead remains of plants that have accumulated over thousands of years. Thus, in these extensive areas, the production of plant material has consistently exceeded the rate of decomposition of plant material. There are several reasons for this. Because these areas are completely submerged, the limited dissolved oxygen in the water is quickly used up, resulting in extensive anaerobic regions. Under anaerobic conditions, the rate and extent of decomposition of organic material is much lower. A second factor is that many bogs become highly acidic (pH 3.2 to 4.2) as a result of the growth of sphagnum mosses that are an integral part of these areas. The combination of anaerobic and acidic conditions suppresses the growth of most microorganisms that are essential for plant or animal decomposition (Information Box 4.2).

Canada has the most extensive bog system in the world, with 129,500,000 hectares or 18.4% of its land area composed of bogs. Harvesting of peat is an important industry in Canada (peat moss for gardening) and in Ireland and Finland (for production of energy). However, the mining and use of peat sources globally returns fixed and sequestered carbon back to the atmosphere as carbon dioxide. In fact, it is estimated that peat bogs hold three times the amount of fixed carbon than rainforests do and so their use and destruction is likely one factor that is contributing to global warming.

4.2.2 The Solid Phase

All soils and subsurface environments are three-phase systems consisting of: (1) a solid or mineral inorganic phase that is often associated with organic matter; (2) a liquid or solution phase; and (3) a gas phase or atmosphere. Soil properties are dependent on the specific composition of each of these phases, which are discussed in the following sections.

4.2.2.1 Primary Particles and Texture

Typically, a soil contains 45 to 50% solids on a volume basis (Figure 4.5). Of this solid fraction, 95 to >99.9% is the mineral fraction. Silicon (47%) and oxygen (27%) are the two most abundant elements found within the mineral fraction of Earth's crust. These two elements, along with lesser amounts of other elements, combine in a number of ways to form a large variety of minerals. For example, quartz is SiO_2 and mica is $K_2Al_2O_5[Si_2O_5]_3Al_4(OH)_4$. These are primary minerals that are derived from the weathering of parent rock. Weathering results in mineral particles that are classified on the basis of three different sizes: sand, silt and clay (Information Box 4.3). The distribution (on a percent by weight basis) of sand, silt and clay within a porous medium defines its texture. Soils predominated by sand are considered coarse textured while those with higher proportions of silt and clay are known as fine textured.

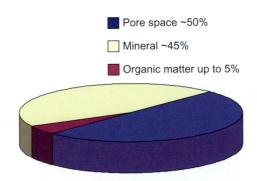

■ Pore space ~50%

□ Mineral ~45%

■ Organic matter up to 5%

FIGURE 4.5 Three basic components of a porous medium, such as a typical surface soil, on a volume basis.

Information Box 4.3 Primary Mineral Size Classifications

Sand	0.05 to 2 mm
Silt	0.002 to 0.05 mm
Clay	<0.002 mm (2 μm)

4.2.2.2 Soil Architecture

Soil particles do not normally remain as individual entities. Rather, they aggregate to form secondary structures or soil architecture (Information Box 4.4). These structures occur because microbial gums, polysaccharides and other microbial metabolites bind the primary particles together. In addition, particles can be held together physically by fungal hyphae and plant roots. These secondary aggregates, which are known as peds, can be of different sizes and shapes, depending on the particular soil (Figure 4.6). Soils with even modest amounts of clay usually have well-defined peds, and hence a well-defined soil structure. These aggregates of primary particles usually remain intact as long as the soil is not disturbed, for example, by plowing. In contrast, soils that are primarily sand with low amounts of clay have less well-defined soil structure.

In between the component mineral particles of a porous medium are voids known as pore space. These pores allow movement of air, water and microorganisms through the porous medium. Pores that exist between aggregates are called interaggregate pores, whereas those within the aggregates are termed intraaggregate pores (Figure 4.7). Pore space may be increased by plant roots, worms and small mammals, whose root channels, worm holes and burrows create macro openings. These larger openings can result in significant aeration of soils as well as preferential flow of water through these large pores where flow is the easiest.

Texture and structure are important factors that govern the movement of water, contaminants and microbial populations in porous media. Of the three size fractions that make up a porous medium, clay particles are particularly dominant in determining the physical and chemical characteristics. For example, clays, which are often composed of aluminum silicates, add both surface area and charge to a soil. As shown in Table 4.1, the surface area of a fine clay particle can be five orders of magnitude larger than the surface area of a 2-mm sand particle. To put this into a microbial perspective, the size of a clay particle is similar to that of a bacterial cell. Clays affect not only the surface area of a porous medium but also the average pore size (Figure 4.8). Although the average pore size is smaller in a clay soil, there are many more pores than in a sandy soil, and as a result the total amount of pore space is larger in a fine-textured (clay) soil than in a coarse-textured (sandy) soil. However, because small pores do not transmit water as fast as larger pores, soils with higher clay content will slow the movement of any material moving through it, including air, water and microorganisms (Figure 4.9). Often, fine-textured regions or layers of materials, e.g., clay lenses, can be found in sites composed primarily of coarser materials, creating very heterogeneous environments. In this case, water will prefer to travel through the coarse material and flow

Information Box 4.4   The Importance of Aggregation — Cryptobiotic Crusts a Special Case

Aggregation is an extremely important factor for soil sustainability because aggregated soils resist water and wind erosion. One can find a special case of aggregate formation in semiarid and arid areas of the world called cryptobiotic crusts. Cryptobiotic crusts are highly specialized communities of cyanobacteria, mosses, and lichens that form a surface crust of soil particles bound together by organic materials. These crusts are important for soil stability and protect against erosion processes. Unfortunately, these crusts are slow growing and fragile and easily destroyed by hikers and large animals. The pictures show (right) cryptobiotic crusts growing on the Colorado Plateau (United States) and (below) a close-up of a piece of crust showing the aggregate architecture.

From United States Geological Survey, 2006.

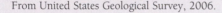

FIGURE 4.6 Soil structure results from secondary aggregates known as peds.

around the fine-textured lens. However, once water moves into a clay lens it will be retained more tenaciously than within sandy materials because of the smaller pore spaces.

For microorganisms, which are much larger than individual water molecules, a fine-textured horizon or lens will inhibit bacterial movement either into or out of the region. Such heterogeneity poses great difficulties when trying to remove contaminants because some finely textured regions are relatively inaccessible to water flow or to microorganisms—a process also known as micropore exclusion (Section 15.1). Thus, contaminants trapped within very small pores may remain there for long periods of time, acting as a long-term "sink" of contaminant that diffuses out of the pores very slowly with time.

4.2.2.3 Soil Profiles

The process of soil formation generates different horizontal layers, or soil horizons, that are characteristic of that particular soil. It is the number, nature and extent of these horizons that give a particular soil its unique character. A typical soil profile is illustrated in Figure 4.10. Generally, soils contain a dark organic-rich layer, designated as the O horizon, then a lighter colored layer, designated as the

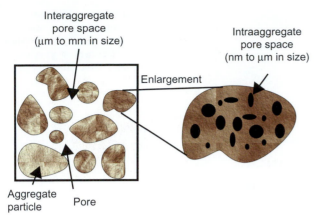

FIGURE 4.7 Pore space. In surface soils, mineral particles are tightly packed together and even cemented in some cases with microbial polymers forming soil aggregates. The pore spaces between individual aggregates are called interaggregate pores and vary in size from micrometers to millimeters. Aggregates also contain pores that are smaller in size, ranging from nanometers to micrometers. These are called intraaggregate pores.

A horizon, where some humified organic matter accumulates. The layer that underlies the A horizon is called the E horizon because it is characterized by eluviation, which is the process of removal or transport of nutrients and inorganics out of the A horizon. Beneath the E horizon is the B horizon, which is characterized by illuviation. Illuviation is the deposition of the substances from the E horizon into the B horizon. Beneath the B horizon is the C horizon, which contains the parent material from which the soil was derived. The C horizon is generally unweathered parent material and marks the transition between a soil and the vadose zone. Although certain diagnostic horizons are common to most soils, not all soils contain each of these horizons.

4.2.2.4 Cation-Exchange Capacity

The parameter known as cation-exchange capacity (CEC) arises because of the negative charge associated with clay

TABLE 4.1 Size Fractionation of Soil Constituents

Specific Surface Area Using a Cubic Model	Soil Mineral Constituents	Size	Organic and Biologic Constituents
0.0003 m²/g	Sand Primary minerals: quartz, silicates, carbonates	2 mm	Organic debris
0.12 m²/g	Silt Primary minerals: quartz, silicates, carbonates	50 μm	Organic debris, large microorganisms Fungi Actinomycetes Bacterial colonies

3 m²/g	Granulometric clay
	Microcrystals of primary minerals
	Phyllosilicate
	  Inherited: illite, mica
	  Transformed: vermiculite, high-charge smectite
	  Neoformed: kaolinite, smectite
	Oxides and hydroxides
		2 μm	Amorphous organic matter
			Humic substances
			Biopolymers
			Small microorganisms
			Bacteria
			Fungal spores
			Large viruses

30 m²/g	Fine clay Swelling clay minerals Interstratified clay minerals Low range order crystalline compounds	0.2 μm	Small viruses

Adapted from Robert and Chenu (1992).

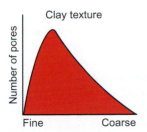

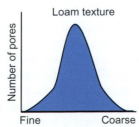

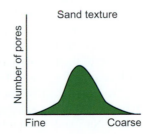

FIGURE 4.8 Typical pore size distributions for clay-, loam- and sand-textured horizons. Note that the clay-textured material has the smallest average pore size, but the greatest total volume of pore space.

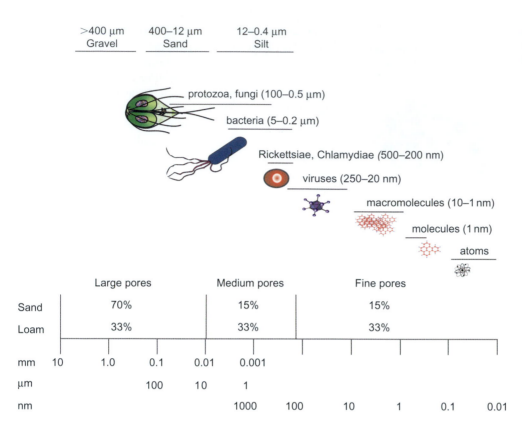

FIGURE 4.9 Comparison of sizes of bacteria, viruses and molecules with hydraulic equivalent diameters of pore canals. Adapted from Matthess *et al.* (1988).

particles and organic matter (Information Box 4.5). Clays are negatively charged for one of two reasons:

1. **Isomorphic substitution:** Clay particles exist as inorganic lattices composed of silicon and aluminum oxides. Substitution of a divalent magnesium cation (Mg^{2+}) for a trivalent aluminum cation (Al^{3+}) can result in the loss of one positive charge, which is equivalent to a gain of one negative charge. Other substitutions can also lead to increases in negative charge.
2. **Ionization:** Hydroxyl groups (OH) at the edge of the lattice can ionize, resulting in the formation of negative charge:

$$Al - OH = Al - O^- + H^+$$

These are also known as broken-edge bonds. Ionizations such as these usually increase as the pH increases, and are therefore known as pH-dependent charge.

The many functional groups of organic matter, such as carboxyl moieties, are also subject to ionization, and can contribute to the total pH-dependent charge. The clay and organic particles that participate in creating CEC are generally very small, <1 μm in diameter, and due to their small size are referred to as soil colloids. Because of their small size, these colloids offer extensive surface area for CEC to occur.

How does the process of cation exchange work? Common soil cations such as Ca^{2+}, Mg^{2+}, K^+, Na^+ and H^+, which exist in the soil solution, are in equilibrium with cations on exchange sites. If the concentration of a cation in the soil solution is changed, for example, increased, then that cation is likely to occupy more exchange sites, replacing existing cations within the site (Figure 4.11). Thus, a monovalent cation such as K^+ can replace another monovalent cation such as Na^+, or two K^+ can replace one Mg^{2+}. Note, however, that when working with charge equivalents, one milliequivalent of K^+ replaces one milliequivalent of Mg^{2+}. Cation exchange ultimately depends on the concentration of the cation in soil solution and the adsorption affinity of the cation for the exchange site. The adsorption affinity of a cation is a function of its charge density, which in turn depends on its total charge and the size of the hydrated cation. The adsorption affinities of several common cations are given in the following series in decreasing order:

$$Al^{3+} > Ca^{2+} = Mg^{2+} > K^+ = NH_4^+ > Na^+$$

Highly charged small cations such as Al^{3+} have high adsorption affinities. In contrast, monovalent ions have lower affinities, particularly if they are highly hydrated such as Na^+, which increases the effective size of the cation. The extensive surface area and charge of soil colloids (clays + organic material) are critical to microbial activity since they affect both binding or sorption of solutes and microbial attachment to the colloids.

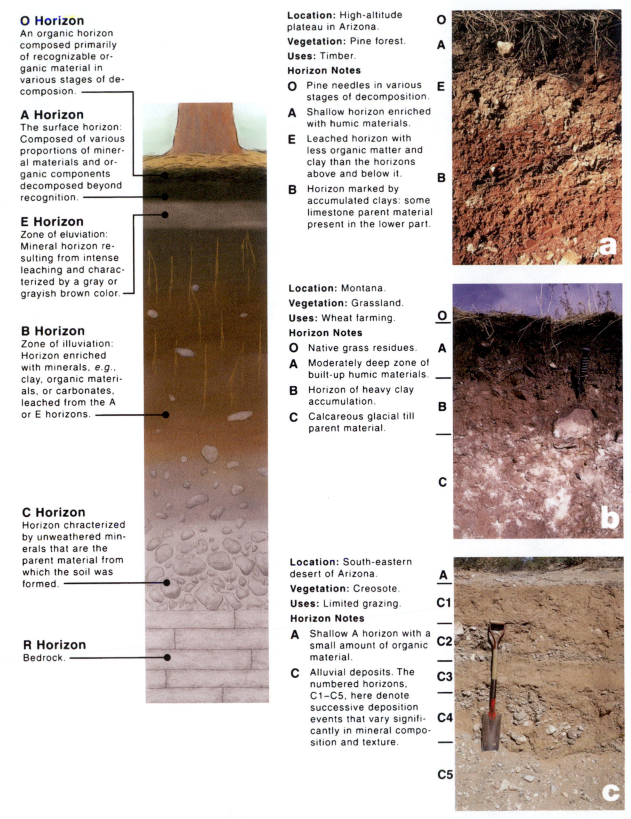

O Horizon
An organic horizon composed primarily of recognizable organic material in various stages of decomposition.

A Horizon
The surface horizon: Composed of various proportions of mineral materials and organic components decomposed beyond recognition.

E Horizon
Zone of eluviation: Mineral horizon resulting from intense leaching and characterized by a gray or grayish brown color.

B Horizon
Zone of illuviation: Horizon enriched with minerals, *e.g.*, clay, organic materials, or carbonates, leached from the A or E horizons.

C Horizon
Horizon chracterized by unweathered minerals that are the parent material from which the soil was formed.

R Horizon
Bedrock.

Location: High-altitude plateau in Arizona.
Vegetation: Pine forest.
Uses: Timber.
Horizon Notes
O Pine needles in various stages of decomposition.
A Shallow horizon enriched with humic materials.
E Leached horizon with less organic matter and clay than the horizons above and below it.
B Horizon marked by accumulated clays: some limestone parent material present in the lower part.

Location: Montana.
Vegetation: Grassland.
Uses: Wheat farming.
Horizon Notes
O Native grass residues.
A Moderately deep zone of built-up humic materials.
B Horizon of heavy clay accumulation.
C Calcareous glacial till parent material.

Location: South-eastern desert of Arizona.
Vegetation: Creosote.
Uses: Limited grazing.
Horizon Notes
A Shallow A horizon with a small amount of organic material.
C Alluvial deposits. The numbered horizons, C1–C5, here denote successive deposition events that vary significantly in mineral composition and texture.

FIGURE 4.10 Typical soil profiles illustrating different soil horizons. These horizons develop under the influence of the five soil-forming factors and result in unique soils. From Pepper *et al.* (2006).

Information Box 4.5   Cation Exchange Capacity in Soil

The total amount of negative charge in soil is usually measured in terms of equivalents of negative charge per 100 g of soil and is a measure of the potential CEC. A milliequivalent (meq) is one-thousandth of an equivalent weight. Equivalents of chemicals are related to hydrogen, which has a defined equivalent weight of 1. The equivalent weight of a chemical is the atomic weight divided by its valence. For example, the equivalent weight of calcium ion is 40/$2 = 20$ g. The five most common exchangeable cations in soil are Ca^{2+}, Mg^{2+}, K^+, Na^+, and Al^{3+}.

In soil, a CEC of 15—20 meq per 100 g is considered to be average, whereas a CEC above 30 is considered high. Soils with low CEC (sandy, low organic matter) often have limited nutrient content because they cannot hold cations tightly and therefore these nutrients are leached during precipitation events. Thus, soils with low CEC generally do not support plant growth as well as those with higher CEC.

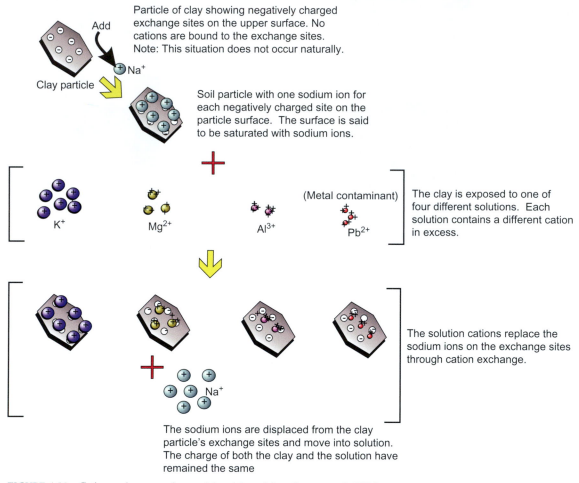

FIGURE 4.11   Cation exchange on clay particles. Adapted from Pepper *et al.* (2006).

Sorption is a major process influencing the movement and bioavailability of essential compounds and pollutants in soil. The broadest definition of sorption is the association of organic or inorganic molecules with the solid phase of the soil. For inorganic charged molecules, cation exchange is one of the primary mechanisms of sorption (Figure 4.11). Generally, positively charged ions, for example, calcium (Ca^{2+}) or lead (Pb^{2+}), participate in cation exchange. Since sorbed forms of these metals are

in equilibrium with the soil solution, they can serve as a long-term source of essential nutrients (Ca^{2+}) or pollutants (Pb^{2+}) that are slowly released back into the soil solution as the soil solution concentration of the cation decreases with time.

Attachment of microorganisms can also be mediated by the numerous functional groups on clays (Figure 4.12). Although the clay surface and microbial cell surface both have net negative charges, clay surfaces are neutralized

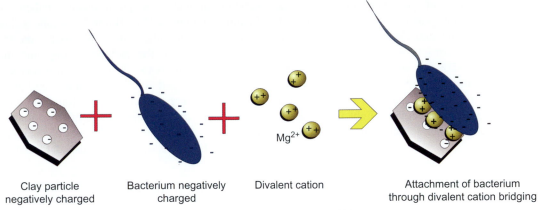

Clay particle   Bacterium negatively   Divalent cation   Attachment of bacterium
negatively charged   charged   through divalent cation bridging

FIGURE 4.12   Attachment of a bacterial cell to a clay particle via cation bridging.

by the accumulation of positively charged counterions such as K^+, Na^+, Ca^{2+}, Mg^{2+}, Fe^{3+} and Al^{3+}. Together, these negative and positive surface charges form what is called the electrical double layer. Similarly, microbes have an electrical double layer. The thickness of the clay double layer depends on the valence and concentration of the cations in solution. Higher valence and increased cation concentrations will shrink the electrical double layer. Because the double layers of the clay particles and microbial cells repel each other, the thinner these layers are, the less the repulsion between the clay and cell surfaces. As these repulsive forces are minimized, attractive forces such as electrostatic and van der Waals forces allow the attachment of microbial cells to the surface (Gammack *et al.*, 1992). As a result, most microbes in terrestrial environments exist attached to soil colloids, rather than existing freely in the soil solution (see Section 15.1.3).

4.2.2.5 Soil pH

Soil pH affects the solubility of chemicals in soils by influencing the degree of ionization of compounds and their subsequent overall charge (Information Box 4.6). The extent of ionization is a function of the pH of the environment and the dissociation constant (pK) of the compound. Thus, soil pH may be critical in affecting transport of potential pollutants through the soil and vadose zone.

In areas with high rainfall, basic cations tend to leach out of the soil profile; moreover, soils developed in these areas have higher concentrations of organic matter, which contain acidic components and residues. Thus, such soils tend to have decreased pH values (<5.5) and are acidic in nature. Soils in arid areas do not undergo such extensive leaching, and the concentrations of organic matter are lower. In addition, water tends to evaporate in such areas, allowing salts to accumulate. These soils are therefore alkaline, with higher pH values (>8.5). Neutral soils range from pH 6 to 8.

Information Box 4.6   pH

pH is defined as the negative logarithm of the hydrogen ion concentration:

$$pH = -\log[H^+]$$

Usually, water ionizes to H^+ and OH^-:

$$HOH \rightleftharpoons H^+ + OH^-$$

The dissociation constant (K_{eq}) is defined as

$$K_{eq} = \frac{[H^+][OH^-]}{[HOH]} = 10^{-14} \text{ mol/L}$$

Since the concentration of HOH is large relative to that of H^+ or OH^-, it is normally given the value of 1. Therefore, $[H][OH^-] = 10^{-14}$ mol/L. For a neutral solution, $[H^+] = [OH^-] = 1 \times 10^{-7}$ mol/L and

$$pH = -\log[H^+] = -(-7) = 7$$

A pH value of less than 7 indicates an acid environment while a pH value greater than 7 indicates an alkaline environment.

4.2.2.6 Organic Matter

Organic matter in soil is defined as a combination of: (1) live biomass, including animals, microbes and plant roots; (2) recognizable dead and decaying biological matter; and (3) humic substances, which are heterogeneous polymers formed during the process of decay and degradation of plant, animal and microbial biomass (Figure 4.13). Soil organic matter contents range from less than 1% in hot arid climates that have low plant residue inputs, to 5% in cooler more humid areas with large plant inputs. In contrast, subsurface environments usually contain only very small amounts of organic matter, $<0.1\%$. This is due to an absence of plant residues and other macroorganisms as well as a smaller numbers of microorganisms.

The humic fraction of organic matter is a stable nutrient base that serves as a slow release source of carbon and energy for the autochthonous (indigenous), slow-growing microorganisms in soil (see Section 3.3). The turnover rate is 2—5% per year. Humic substances have extremely complex structures that reflect the complexity and diversity of organic materials produced in a typical soil. They range in molecular weight from 700 to 300,000. An example of a humic acid polymer is shown in Figure 4.14.

Overall, humus has a three-dimensional, spongelike structure that contains both hydrophobic (water-hating) and hydrophilic (water-loving) regions. Thus, the humus molecule folds so that the hydrophilic portions face the charged exterior water or mineral phases and the hydrophobic portions are attracted toward the interior of the molecule. This hydrophobic interior provides a favorable environment for solutes that are less polar than water. This means that humus can sorb nonpolar solutes from the general soil

solution through a sorption process called hydrophobic binding (Figure 4.15). Humic substances also contain numerous hydrophilic functional groups, the most important of which are the carboxyl group and the phenolic hydroxyl group, both of which can become negatively charged in the soil solution. As noted earlier, these functional groups are similar to those found on clays, and can contribute to the pH-dependent CEC of soil and participate in sorption of solutes and attachment of microorganisms by cation exchange, as shown in Figures 4.11 and 4.12.

4.2.3 The Liquid Phase

4.2.3.1 Soil Solution Chemistry

The soil solution is a constantly changing matrix composed of both organic and inorganic solutes in aqueous solution. The composition of the liquid phase is extremely important for biological activity in a porous medium,

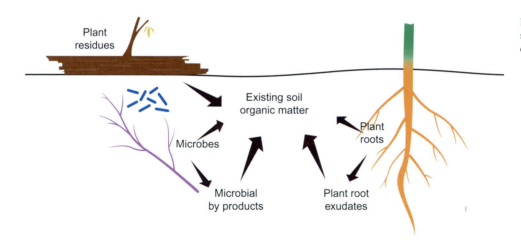

FIGURE 4.13 Schematic representation of the formation of soil organic matter.

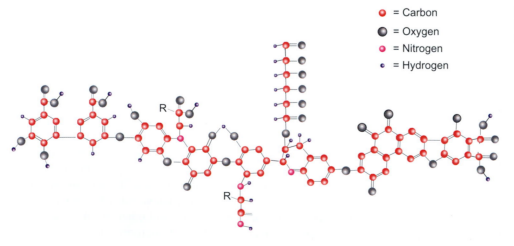

= Carbon
= Oxygen
= Nitrogen
= Hydrogen

FIGURE 4.14 Humus polymer. R can represent various functional groups.

because microorganisms are approximately 70% water and most require high levels of water activity (>0.95) for active metabolism. Indeed, all microorganisms, even attached ones, are surrounded by a water film from which they obtain nutrients and into which they excrete wastes. Thus, the amount and composition of the liquid phase ultimately controls both microbial and plant growth.

The soil solution composition reflects the chemistry of the soil as well as the dynamic influx and efflux of solutes in response to water movement. Water movement results from rainfall or irrigation and affects mineral weathering, organic matter formation and decomposition (Figure 4.16). This composition is also altered by anthropogenic activities, such as irrigation, fertilizer and pesticide addition, and chemical spills. The chemistry of the soil affects not only the composition of the soil solution, but also the form and bioavailability of nutrients which are those in the soil solution (Figure 4.16). For example,

as illustrated in Table 4.2, the form of the common cations found in a soil varies as a function of soil pH. As shown in this table, most cations are found in a more soluble form in acidic environments. In some cases, as for magnesium and calcium, this results in extensive leaching of the soluble form of the cation, leading to decreased concentrations of the nutrient. For many metallic cations, this can lead to increased metal toxicity (see Section 18.6). In other cases, as for iron and phosphate, a slightly acidic pH provides optimal availability of the element. Overall, the pH range that supports maximum microbial and plant activity is between 6.0 and 6.5.

4.2.3.2 Soil Water Potential

Water is the primary solvent in porous medium systems, and water movement is generally the major mechanism responsible for the transport of chemicals and

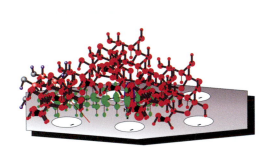

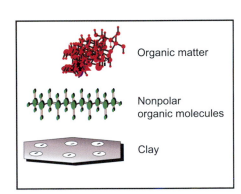

FIGURE 4.15 Hydrophobic sorption mechanisms. Nonpolar organic molecules tend to sorb to organic matter that is associated with solid mineral surfaces by diffusing into the sponge-like interior of organic matter molecules. Adapted from Schwarzenbach *et al.* (1993); reproduced by permission of John Wiley and Sons, Inc.

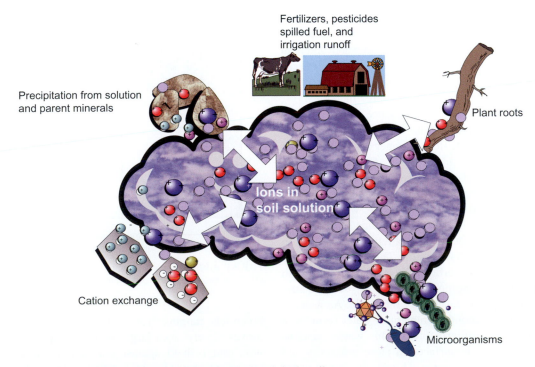

FIGURE 4.16   Paths of dissolution and uptake of minerals in the soil.

TABLE 4.2 The Form of Common Cations Found in Acid and Alkaline Soils

Cation	Acid Soils (Low pH)	Alkaline Soils (High pH)
Na^+	Na^+	Na^+, $NaHCO_3^0$, $NaSO_4^0$
Mg^{2+}	Mg^{2+}, $MgSO_4^0$, organic complexes	Mg^{2+}, $MgSO_4^0$, $MgCO_3^0$
Al^{3+}	organic complexes, AlF^{2+}, $AlOH^{2+}$	$Al(OH)_4^-$, organic complexes
Si^{4+}	$Si(OH)_4^0$	$Si(OH)_4^0$
K^+	K^+	K^+, KSO_4^-
Ca^{2+}	Ca^{2+}, $CaSO_4^0$, organic complexes	Ca^{2+}, $CaSO_4^0$, $CaHCO_3^+$
Mn^{2+}	Mn^{2+}, $MnSO_4^0$, organic complexes	Mn^{2+}, $MnSO_4^0$, $MnCO_3^0$, $MnHCO_3^+$, $MnB(OH)_4^+$
Fe^{2+}	Fe^{2+}, $FeSO_4^0$, $FeH_2PO_4^+$	$FeCO_3^+$, Fe^{2+}, $FeHCO_3^+$, $FeSO_4^0$
Fe^{3+}	$FeOH^{2+}$, $Fe(OH)_3^0$, organic complexes	$Fe(OH)_3^0$, organic complexes
Cu^{2+}	Organic complexes, Cu^{2+}	$CuCO_3^0$, organic complexes, $CuB(OH)_4^+$, $Cu[B(OH)_4]_4^0$
Zn^{2+}	Zn^{2+}, $ZnSO_4^0$, organic complexes	$ZnHCO_3^+$, $ZnCO_3^0$, organic complexes, Zn^{2+}, $ZnSO_4^0$, $ZnB(OH)_4^+$
Mo^{5+}	$H_2MoO_4^0$, $HMoO_4^-$	$HMoO_4^-$, MoO_4^{2-}

Adapted from Sposito (1989).

microorganisms. Water movement in a porous medium depends on the soil water potential, which is the work per unit quantity necessary to transfer an infinitesimal amount of water from a specified elevation and pressure to another point somewhere else in the porous medium. The soil water potential is a function of several forces acting on water, including matric and gravitational forces. Soil water potential is usually expressed in units of pressure (pascals, atmospheres or bars). Values of the matric contribution are negative because the reference is generally free water, which is defined to have a soil water potential of zero. Since the matric force decreases the free energy of water in the soil solution, the soil water potential becomes increasingly negative as this force increases.

In the saturated zone, the presence of a regional hydraulic gradient usually results in a general horizontal flow of water. In contrast, flow is generally downward in the unsaturated zone. In an unsaturated zone, where the soil pores are not completely filled with water, there are several incremental forces that affect water movement. These are related to the amount of water present (Figure 4.17). In very dry soils, there is an increment of adsorbed water that exists as an extremely thin film on the order of angstroms (Å) in width. This thin film is held very tightly to particle surfaces by surface forces with soil water potentials ranging from -31 to $-10,000$ atm. As a result, this water is essentially immobile. As water is added to this soil, a second increment of water forms as a result of matric or capillary forces. This water exists as bridges between particle surfaces in close proximity and can actually fill small soil pores. This results in soil water

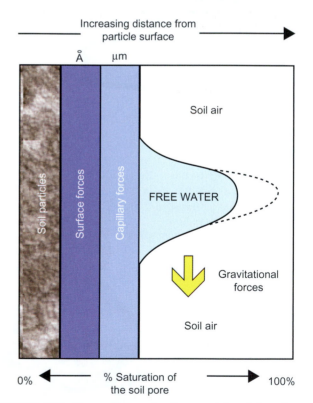

FIGURE 4.17 The continuum of soil water. Adapted from Dragun (1988). Reproduced with permission of ASP.

potentials ranging from -0.1 to -31 atm. This water moves slowly from larger to smaller pores in any direction, and is held against gravitational forces. The next increment of water added, free water, can be removed by gravitational forces. The soil water potential for free water ranges from 0 to -0.5 atm. Although we have classified these different types of water in categories, in porous media they actually occur as a continuum rather than with sharply defined boundaries.

The amount of water present in the pore space of a medium is another important parameter in understanding the level and type of microbial activity in an environment. The optimal environment for active aerobic microbial growth in a porous medium is one in which water is easily available but the medium is not completely saturated. Why is this? As stated earlier, microorganisms obtain nutrients from the water phase surrounding them, so water is absolutely necessary for active microbial growth. Generally, microbial activity in the soil is greatest at -0.1 atm. As the water potential becomes less negative, the soil becomes saturated. In a completely saturated environment, oxygen, which governs aerobic microbial activity, may become limiting because of its limited solubility (9.3 mg/L at 20°C and 1 atm pressure) in water. Because the diffusion of oxygen through water is slow, once the available dissolved oxygen is used up, it is not replenished rapidly. As the water potential becomes more negative than -0.1 atm, water becomes less available because it is held tightly by matric and capillary forces.

4.2.4 Soil Atmosphere

4.2.4.1 Constituents of Soil Atmosphere

Soil and the atmosphere are in direct contact; therefore, the soil atmosphere has the same basic composition as air: nitrogen, oxygen and carbon dioxide. As shown in Table 4.3, there is little difference between the atmospheres in a well-aerated surface soil and in the air. However, plant and microbial activity can greatly affect the relative proportions of oxygen and carbon dioxide in soils that are not well aerated, that are far removed from the soil surface or that have undergone a recent flooding

event due to heavy rains or irrigation. For example, in a fine clay or a saturated soil, oxygen can be completely removed by the aerobic activity of respiring organisms. During this respiration process, carbon dioxide (CO_2) is evolved, eventually resulting in elevated CO_2 levels. These changes affect the redox potential of the porous medium, which affects the availability of terminal electron acceptors for aerobic and anaerobic microbes.

4.2.4.2 Availability of Oxygen and Soil Respiration

The amount of oxygen in the atmosphere (21%) allows aerobic degradation of the overwhelming proportion of the organic matter produced annually. In the absence of oxygen, organic substrates can be mineralized to carbon dioxide by fermentation or by anaerobic respiration, although these are less efficient processes than aerobic respiration. Thus, the oxygen content of soil is vital for aerobic activity, which depends on oxygen as a terminal electron acceptor during degradation of organic compounds.

Soil moisture content controls the amount of available oxygen in a soil (Information Box 4.7). In soils saturated with water, all pores are full of water and the oxygen content is very low. In dry soils, all pores are essentially full of air, so the soil moisture content is very low. Since aerobic microorganisms require both oxygen and water, soils at field capacity (moist but drained soils), which have moderate soil moisture and optimize both air (oxygen) and moisture, will maximize aerobic microbial metabolism. It is important to note, however, that even in field capacity soils, low-oxygen concentrations may exist in certain isolated pore regions (Figure 4.18). These wet but unsaturated regions of soil can quickly go anaerobic due to microbial activity since the rate of oxygen diffusion through water is roughly 10,000 times slower than

TABLE 4.3 Soil Atmosphere

Location	Composition (% volume basis)		
	Nitrogen (N$_2$)	Oxygen (O$_2$)	Carbon Dioxide (CO$_2$)
Atmosphere	78.1	20.9	0.03
Well-aerated surface soil	78.1	18–20.5	0.3–3
Fine clay or saturated soil	>79	≈0–10	Up to 10

Information Box 4.7   Oxygen in the Gas and Water Phases of Soil

Oxygen can be found either in the soil atmosphere or dissolved in the soil solution, but the relative solubility of oxygen in water is low:

Compare 9.3 mg O$_2$ per liter water to approximately 1300 mg O$_2$ per liter air.

Microorganisms obtain oxygen from the water phase. Therefore as the dissolved oxygen is utilized by aerobic respiration, oxygen will be driven from the soil atmosphere into the water phase until it is used up, finally creating an anaerobic environment. In the environment, saturated soils will hold less oxygen (due to its limited solubility) than unsaturated ones will and, therefore, are more prone to developing anaerobic conditions.

FIGURE 4.18 Gradient of oxygen concentrations in a typical soil aggregate. Adapted from Sextone *et al.* (1985).

FIGURE 4.19 The harsh soil environment found at the White Sands National Monument still supports plant and microbial life. *Source:* Pepper (2014).

through air and much of the oxygen will be used by other microorganisms as it slowly diffuses into the interior portions of soil aggregates. Thus, anaerobic microsites will exist even in aerobic soils that support transformation processes carried out by facultative anaerobes and strict anaerobes. This is an excellent example of how soil can function as a discontinuous environment of great diversity.

4.2.4.3 Oxygen and Respiration in the Subsurface

Overall activity levels in subsurface environments are lower than in soils. This is primarily because there are low amounts of organic carbon to serve as substrate in subsurface environments. Thus, the vadose zone, because it has low organic carbon and is unsaturated, is generally aerobic. However, due to the heterogeneous nature of the subsurface, anaerobic zones can occur particularly in clay lenses. In contrast, oxygen availability in saturated zones of the subsurface varies considerably, depending on availability of organic carbon.

4.3 SOIL AS A MICROBIAL ENVIRONMENT

Given the physical and chemical characteristics just described, just what kind of environment is soil for its microbial inhabitants? The short answer is that the soil environment, like most, is a competitive one. Those microorganisms that are best adapted to the stresses of the soil environment are most successful. The stresses found are both biotic, including competition from other microbes, and abiotic, including the physical and chemical characteristics of the environment. However, despite such stresses, even the harshest soil environments will support plant and microbial life (Figure 4.19).

4.3.1 Biotic Stresses

Since indigenous soil microbes are in competition with one another, the presence of large numbers of organisms results in biotic stress factors. Competition can be for substrate, water or growth factors. In addition, microbes can secrete allelopathic substances (inhibitory or toxic), including antibiotics, that harm neighboring organisms. Finally, many organisms are predatory or parasitic on neighboring microbes. For example, protozoa graze on bacteria, and viruses infect both bacteria and fungi. Because of biotic stress, nonindigenous organisms that are introduced into a soil environment often survive for very short periods of time (days to several weeks) unless there is a specific selective niche. This effect has important consequences for survival of pathogens and for other organisms introduced to aid biodegradation or for biological control.

4.3.2 Abiotic Stresses

4.3.2.1 Light

Sunlight does not penetrate beyond the top few centimeters of the soil surface. Phototrophic microorganisms are therefore limited to the top few centimeters of soil. At the surface of the soil, however, such physical parameters as temperature and moisture fluctuate significantly throughout the day and also seasonally. Hence, most soils provide a somewhat harsh environment for photosynthesizing microorganisms. Some phototrophic organisms, including algae, have the ability to switch to a heterotrophic respiratory mode of nutrition in the absence of light. Such "switch-hitters" can be found at significant depths within soils. Lichens, a mutualistic association between

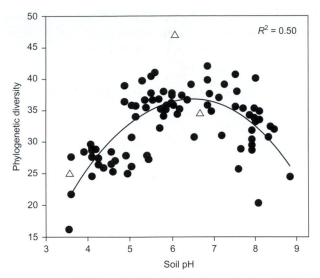

FIGURE 4.20   Soil bacterial diversity in soil samples from North and South America is strongly correlated with soil pH. From Lauber *et al.* (2009).

Case Study 4.1   Soil pH is a Major Determinant of Bacterial Diversity

Fierer and Jackson (2006) used terminal restriction fragment length polymorphism (T-RFLP) analysis to determine which environmental factors were related to bacterial diversity in 98 soils from across North and South America. Interestingly, bacterial diversity was not correlated with diversity of vegetation, site temperature or latitude. The best predictor of bacterial diversity was soil pH ($r^2 = 0.70$) with the highest diversity occurring in soils with a near-neutral pH. Similarly, Lauber *et al.* (2009) later used bar-coded pyrosequencing to investigate the bacterial communities in 88 soils from North and South America. They also found that overall diversity was also correlated with pH ($r^2 = 0.50$) and was highest in soils with a near-neutral pH. Furthermore, they found that bacterial community composition was strongly correlated with soil pH ($r = 0.79$), largely due to changes in populations of *Acidobacteria*, *Actinobacteria* and *Bacteroidetes*. Although these results suggest that soil pH is one of the main drivers of bacterial diversity at a continental scale, other factors such as vegetation type and soil C may be more important at regional or local scales.

algae or cyanobacteria and fungi, are common in extremely harsh environments. In this association the fungus provides protection from desiccation and the phototroph provides energy from photosynthesis.

4.3.2.2 Soil Moisture

The availability of water is critical for microbial activity. Typically, optimal microbial activity occurs at -0.1 atm, which is the transition between capillary water and free water (Figure 4.17). As a group, the fungi are most desiccation resistant, followed by actinomycetes and finally the Bacteria.

4.3.2.3 Soil Temperature

Soil temperatures vary widely, particularly near the soil surface. Most soil populations are resistant to wide fluctuations in soil temperature although soil communities can be psychrophilic (prefer $< 20°C$), mesophilic (prefer $20-45°C$), thermophilic ($45-90°C$) or hyperthermophilic ($> 90°C$) depending on the geographic location of the soil. Most soil organisms are mesophilic because of the buffering effect of soil on soil temperature, particularly at depths beneath the soil surface.

4.3.2.4 Soil pH

Undisturbed soils usually have soil pH values within the range of $6-8$, and most soil organisms have pH optima within this range. There are, of course, exceptions to this rule, as exemplified by *Thiobacillus thiooxidans*, an organism that oxidizes sulfur to sulfuric acid and has a pH optimum of $2-3$. Interestingly, soil pH appears to be

a major determinant of bacterial diversity (Figure 4.20; Case Study 4.1).

4.3.2.5 Soil Texture

All soils contain microbial communities regardless of the soil texture. However, soils with a mixture of sand, silt and clay particles offer a more favorable habitat for organisms because they hold more nutrients (Section 4.2.2.4) and provide for better water (Section 4.2.3) and air flow (Section 4.2.4) than do pure sands or clays. Microbial communities found in pure sands or clays are lower in numbers and activity.

4.3.2.6 Soil Nutrients

Carbon and nitrogen are generally the most important limiting nutrients that are found in soils although any limiting nutrient will reduce microbial activity. Since both carbon and nitrogen are usually present in low concentrations, growth and activity of soil organisms are slow. In fact, many organisms exist in soil under semi-starvation conditions and hence are dormant. The major exception to this is the plant rhizosphere, where root exudates maintain much higher microbial numbers and activity.

4.3.2.7 Redox Potential

Redox potential (E_h) is the measurement of the tendency of an environment to oxidize or reduce substrates. An aerobic soil, which is an oxidizing environment, has an E_h of $+800$ mV; an anaerobic soil, which is a reducing environment, has a negative E_h which can reach -300 mV. Oxygen is found in soils at a redox potential of about $+800$ mV. When soil is placed in a closed container, oxygen is used by aerobic organisms as a terminal electron acceptor until all of it is depleted. As this process occurs, the redox potential of the soil decreases, and other compounds can be used as terminal electron acceptors. Table 4.4 illustrates the redox potential at which various substrates are reduced, and the activity of different types of organisms in a soil.

4.4 MICROORGANISMS IN SURFACE SOILS

Surface soils are occupied by indigenous populations of bacteria (including actinomycetes), archaeans, fungi, algae and protozoa. In general, as the size of these organisms increases from bacteria to protozoa, the number present decreases. It is also known that there may be phages or viruses present that can infect each class of organism, but information on the extent of these infectious agents in surface soils is limited (see also Chapter 2). In addition to these indigenous populations, specific microbes can be introduced into soil by human or animal activity. Human examples include the deliberate direct introduction of bacteria as biological control agents or as biodegradative agents. Microbes are also introduced indirectly as a result of application of animal manures or sewage sludge to agricultural fields (see Chapter 26). Animals introduce microbes through bird droppings and animal excrement. Regardless of the source, introduced

TABLE 4.4 Redox Potential at which Soil Substrates are Reduced

Redox Potential (mV)	Reaction	Type of Organism and Metabolism
$+800$	$O_2 \rightarrow H_2O$	Aerobes, aerobic respiration
$+740$	$NO_3^- \rightarrow N_2, N_2O$	Facultative anaerobes, nitrate reduction
-220	$SO_4^{2-} \rightarrow S^{2-}$	Anaerobes, sulfate reduction
-300	$CO_2 \rightarrow CH_4$	Anaerobes, methanogenesis

TABLE 4.5 Characteristics of Bacteria, Actinomycetes, and Fungi

Characteristic	Bacteria	Actinomycetes	Fungi
Numbers	Most numerous	Intermediate	Least numerous
Biomass	Bacteria and actinomycetes have similar biomass		Largest biomass
Degree of branching	Slight	Filamentous, but some fragment to individual cells	Extensive filamentous forms
Aerial mycelium	Absent	Present	Present
Growth in liquid culture	Yes—turbidity	Yes—pellets	Yes—pellets
Growth rate	Exponential	Cubic	Cubic
Cell wall	Murein, teichoic acid, and lipopolysaccharide	Murein, teichoic acid, and lipopolysaccharide	Chitin or cellulose
Complex fruiting bodies	Absent	Simple	Complex
Competitiveness for simple organics	Most competitive	Least competitive	Intermediate
Fix N	Yes	Yes	No
Aerobic	Aerobic, anaerobic	Mostly aerobic	Aerobic except yeast
Moisture stress	Least tolerant	Intermediate	Most tolerant
Optimum pH	6–8	6–8	6–8
Competitive pH	6–8	>8	<5
Competitiveness in soil	All soils	Dominate dry, high-pH soils	Dominate low-pH soils

organisms rarely significantly affect the abundance and distribution of indigenous populations.

The following discussion is an overview of the dominant types of microbes found in surface soils, including their occurrence, distribution and function. Overall soil microorganisms are critical in a variety of areas including surface soil formation, nutrient cycling (Chapter 16), bioremediation (Chapter 17) and land application of municipal wastes (Chapter 26).

4.4.1 Bacteria

Bacteria are almost always the most abundant organisms found in surface soils in terms of numbers (Table 4.5). Culturable numbers vary depending on specific environmental conditions, particularly soil moisture and temperature. Culturable bacteria can be as numerous as 10^7 to 10^8 cells per gram of soil, whereas total populations (including viable but nonculturable organisms) can exceed 10^{10} cells per gram. In unsaturated soils, aerobic bacteria usually outnumber anaerobes by two or three orders of magnitude. Anaerobic populations increase with increasing soil depth but rarely predominate unless soils are saturated and/or clogged.

Indigenous soil bacteria can be classified on the basis of their growth characteristics and affinity for carbon substrates. As explained in Section 3.3, two broad categories of bacteria are found in the environment, those that are K-selected or autochthonous and those that are r-selected or zymogenous. The former metabolize slowly in soil, utilizing slowly released soil organic matter as a substrate. The latter are adapted to intervals of dormancy and rapid growth, depending on substrate availability, following the addition of fresh substrate or amendment to the soil.

Bacteria are also classified according to diversity or the different types present. An intriguing question that has not yet been completely answered is: "How many different bacteria are there in soils?" Traditionally, this has been determined using culture techniques, but most recently, estimates of diversity have been made based on DNA sequencing in combination with statistical approaches. In this case, diversity is indicated by the number of operational taxonomic units (OTU) where each OTU theoretically represents a different bacterial population in the community (Chapter 19). These approaches are providing estimates of diversity at $> 10,000$ species (OTUs) of bacteria per gram of soil (Roesch et al., 2007). It is important to remember that most of these populations are not easily cultured and we are just beginning to develop methods to study the viable but difficult-to-culture

TABLE 4.6 Dominant Culturable Soil Bacteria

Organism	Characteristics	Function
Arthrobacter	Heterotrophic, aerobic, Gram variable. Up to 40% of culturable soil bacteria.	Nutrient cycling and biodegradation.
Streptomyces	Gram-positive, heterotrophic, aerobic actinomycete. 5–20% of culturable bacteria.	Nutrient cycling and biodegradation. Antibiotic production, e.g., *Streptomyces scabies*.
Pseudomonas	Gram-negative heterotroph. Aerobic or facultatively anaerobic. Possess wide array of enzyme systems. 10–20% of culturable bacteria.	Nutrient cycling and biodegradation, including recalcitrant organics. Biocontrol agent.
Bacillus	Gram-positive aerobic heterotroph. Produce endospores. 2–10% of culturable soil bacteria.	Nutrient cycling and biodegradation. Biocontrol agent, e.g., *Bacillus thuringiensis*.

TABLE 4.7 Examples of Important Autotrophic Soil Bacteria

Organism	Characteristics	Function
Nitrosomonas	Gram negative, aerobe	Converts $NH_4^+ \rightarrow NO_2^-$ (first step of nitrification)
Nitrobacter	Gram negative, aerobe	Converts $NO_2^- \rightarrow NO_3^-$ (second step of nitrification)
Thiobacillus	Gram negative, aerobe	Oxidizes $S \rightarrow SO_4^{2-}$ (sulfur oxidation)
Thiobacillus denitrificans	Gram negative, facultative anaerobe	Oxidizes $S \rightarrow SO_4^{2-}$; functions as a denitrifier
Thiobacillus ferrooxidans	Gram negative, aerobe	Oxidizes $Fe^{2+} \rightarrow Fe^{3+}$

TABLE 4.8 Examples of Important Heterotrophic Soil Bacteria

Organism	Characteristics	Function
Actinomycetes, e.g., *Streptomyces*	Gram positive, aerobic, filamentous	Produce geosmins "earthy odor," and antibiotics
Bacillus	Gram positive, aerobic, spore former	Carbon cycling, production of insecticides and antibiotics
Clostridium	Gram positive, anaerobic, spore former	Carbon cycling (fermentation), toxin production
Methanotrophs, e.g., *Methylosinus*	Aerobic	Methane oxidizers that can cometabolize trichloroethene (TCE) using methane monooxygenase
Cuprivadus necator	Gram negative, aerobic	2,4-D degradation via plasmid pJP4
Rhizobium	Gram negative, aerobic	Fixes nitrogen symbiotically with legumes
Frankia	Gram positive, aerobic	Fixes nitrogen symbiotically with nonlegumes
Agrobacterium	Gram negative, aerobic	Important plant pathogen, causes crown gall disease

microbes in soil (Chapter 8). Tables 4.6–4.8 identify some of the culturable bacterial genera that are known to dominate typical surface soils and other bacterial genera that are critical to environmental microbiology. Of course, these lists are by no means all inclusive. A very important point that follows is that any methodology that relies on characterizing environmental organisms via a procedure involving culture may in fact obtain a very small subsection of the total population that may not be representative of the majority of the community (Figure 4.21; Case Study 4.2).

4.4.2 Actinomycetes

Actinomycetes are prokaryotic organisms that are classified as bacteria, but are unique enough to be discussed as an individual group. Actinomycete numbers are generally one to two orders of magnitude smaller than the total bacterial population (Table 4.5). They are an important component of the bacterial community, especially under conditions of high pH, high temperature or water stress. Morphologically, actinomycetes resemble fungi because of their elongated cells that branch into filaments or hyphae. These hyphae can be distinguished from fungal hyphae on the basis of size with actinomycete hyphae much smaller than fungal hyphae (Figure 4.22). Characteristics and unique functions of actinomycetes are shown in Information Box 4.8. One distinguishing feature of this group of bacteria is that they are able to utilize a great variety of substrates found in soil, especially some of the less degradable insect and plant polymers such as chitin, cellulose and hemicellulose. Although originally recognized as soil microorganisms, it is now being recognized that marine actinomycetes are also important. Specifically, marine actinomycetes have been shown to possess novel secondary metabolites that add a new

dimension to microbial natural products (Jensen *et al.*, 2005) that have been discovered within soil actinomycetes (Chapter 19).

4.4.3 Archaea

Once thought to occur primarily in extreme environments such as thermal springs or hypersaline soils, culture-independent techniques have revealed that archaeans are actually widespread in nature. Although archaeal populations can be very large ($>10^8$ per gram of soil), they are typically two or more orders of magnitude less numerous than bacteria. The Archaea contribute to multiple soil processes including the biogeochemical cycling of C, N and S. For example, they have major roles in nitrification (ammonia oxidation) and methanogenesis (Chapter 16). Numerous studies have documented large populations of ammonia-oxidizing archaeans (AOA) in a variety of ecosystems. In general, AOA appear to be more important to ammonia oxidation in environments that have lower levels of N, such as natural ecosystems and pastures. In contrast, although there are also often high levels of AOA in managed ecosystems such as agricultural fields, nitrification in these environments having higher levels of N actually appears to be dominated by ammonia-oxidizing bacteria (AOB) (Taylor *et al.*, 2010; Verhamme *et al.*, 2011). However, even in these managed systems, AOA may be of greater relative importance in the subsoil and subsurface environments due to lower nutrient levels and pHs. In contrast to nitrification, which is performed by both Archaea and Bacteria, methanogenesis is solely an archaeal process and is critically important to the global C cycle with some researchers estimating that $>40\%$ of global methane emissions originate from soils and associated wetlands.

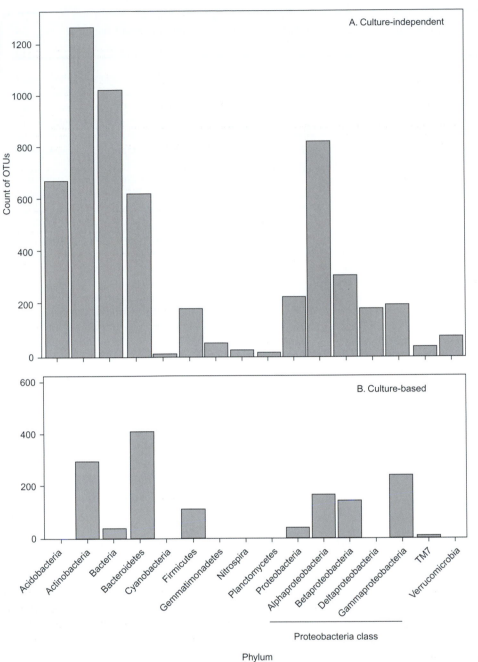

FIGURE 4.21 Impact of culture-based and -independent methods on characterization of a soil bacterial community. Note the greater number of phyla and operational taxonomic units (OTUs) detected using the culture-independent approach. From Shade *et al.* (2012).

4.4.4 Fungi

Fungi other than yeasts are aerobic and are abundant in most surface soils. Numbers of fungi usually range from 10^5 to 10^6 per gram of soil. Despite their lower numbers compared with bacteria, fungi usually contribute a higher proportion of the total soil microbial biomass (Tables 4.5 and 4.9). This is due to their comparatively large size; a fungal hypha can range from 2 to 10 μm in diameter. Because of their large size, fungi are more or less restricted to the interaggregate regions of the soil matrix. Yeasts can metabolize anaerobically (fermentation) and

are less numerous than aerobic mycelium-forming fungi. Generally, yeasts are found at populations of up to 10^3 per gram of soil. Because of their reliance on organic sources for substrate, fungal populations are greatest in the surface O and A horizons, and numbers decrease rapidly with increasing soil depth. As with bacteria, soil fungi are normally found associated with soil particles or within plant rhizospheres.

Fungi are important components of the soil with respect to nutrient cycling and especially decomposition of organic matter, both simple (sugars) and complex (polymers such as cellulose and lignin). The role of fungi

Case Study 4.2   Comparison of Culture-Based and Independent Approaches for Characterizing a Soil Bacterial Community

Shade *et al.* (2012) used both culture-based and -independent methods to characterize the bacterial community in an orchard soil from Wisconsin, U.S.A. For culture-independent analysis, DNA was extracted from soil and sequenced using a 16S rRNA gene pyrosequencing approach (Chapters 13 and 21). For culture-based analysis, soil bacteria were first grown on rhizosphere isolation medium and then characterized by 16S rRNA gene pyrosequencing. Over 37,000 sequences were obtained with each approach. The results for the two methods were strikingly different, with a much greater number of bacterial phyla and operational taxonomic units (OTUs) detected using the culture-independent approach (Figure 4.21). The culture-based approach indicated that the community was dominated by *Bacteroidetes*, *Proteobacteria* and *Actinobacteria*. In contrast, the culture-independent approach indicated that the community was dominated by *Actinobacteria*, *Proteobacteria* (with increased levels of *Alphaproteobacteria* and *Deltaproteobacteria* and decreased levels of *Gammaproteobacteria*), *Acidobacteria* and *Bacteroidetes*. Approximately 90% of the bacteria detected using the culture-independent approach were not detected using the culture-based approach. For example, no *Acidobacteria* were detected using the culture-based method, but they comprised >10% of the bacterial community based upon the culture-independent results and represented the third most abundant phylum. Numerous other studies using culture-independent methods have also found that *Acidobacteria* may comprise ≥ 30% of many environmental bacterial communities. Prior to use of culture-independent methods,

the abundance and distribution of *Acidobacteria* in the environment were unknown due to the inability or difficulty to grow these organisms on traditional media. Current evidence based upon a limited number of isolated representatives and genome sequence data indicates that *Acidobacteria* can metabolize a wide variety of substrates and they have a competitive advantage in low C and pH environments.

What is perhaps most surprising about the Shade *et al.* (2012) study is that approximately 60% of the bacteria detected using the culture-based approach were not detected using the culture-independent approach. This indicates that the culture-based method captured rare members of the soil community that were missed with the sequencing approach and highlights an important limitation of the scale of sequencing currently used in most studies for characterizing bacterial communities. Due to costs associated with sequencing, many studies only sequence ≤ 10,000 bacteria per sample, thus only detecting the most dominant organisms (unless a targeted sequencing approach is used). For illustration, even if the 40,000 most abundant bacteria in a soil sample were characterized out of a total community of 1 million bacteria (a very conservative estimate), this would represent only the top 4% of the community. In other words, 96% of the bacterial community would be missed! Fortunately, rapid advances in sequencing technologies are allowing more thorough sequencing of these communities and thus increasing our ability to detect and characterize these rare members of the soil biosphere using culture-independent methods.

in decomposition is increasingly important as the soil pH declines because fungi tend to be more acid tolerant than bacteria (Table 4.5) Some of the common genera of soil fungi involved in nutrient cycling are *Penicillium* and *Aspergillus*. These organisms are also important in the development of soil structure because they physically entrap soil particles with fungal hyphae (Figure 4.23). As well as being critical in the degradation of complex plant polymers such as cellulose and lignin, some fungi can also degrade a variety of pollutant molecules. The best-known example of such a fungus is the white rot fungus *Phanerochaete chrysosporium* (Information Box 2.6). Other fungi, such as *Fusarium* spp., *Pythium* spp. and *Rhizoctonia* spp., are important plant pathogens. Still others cause disease; for example, *Coccidioides immitis* causes a chronic human pulmonary disease known as "valley fever" in the southwestern deserts of the United States. Finally, note that mycorrhizal fungi are critical for establishing plant–fungal interactions that act as an extension of the root system of almost all higher plants. Without these mycorrhizal associations, plant growth as we know it would be impossible.

4.4.5 Algae

Algae are typically phototrophic and thus would be expected to survive and metabolize in the presence of a light-energy source and CO_2 for carbon. Therefore, one would expect to find algal cells predominantly in areas where sunlight can penetrate, the very surface of the soil. However, one can actually find algae to a depth of 1 m because some algae, including the green algae and diatoms, can grow heterotrophically as well as photoautotrophically. In general, though, algal populations are highest in the surface 10 cm of soil (Curl and Truelove, 1986). Typical algal populations close to the soil surface can range from 5000 to 10,000 per gram of soil, but where a visible algal bloom has developed there can be millions of algal cells per gram of soil.

Algae are often the first to colonize surfaces in a soil that are devoid of preformed organic matter. Colonization by this group of microbes is important in establishing soil formation processes, especially in barren volcanic areas, desert soils and rock faces. Algal metabolism is critical to soil formation in two ways: algae provide a carbon input

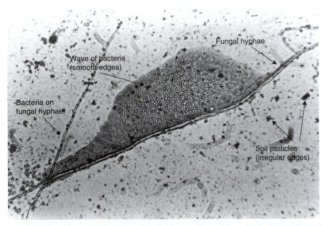

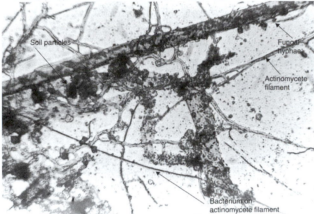

FIGURE 4.22 Comparison of soil bacteria, actinomycetes and fungi viewed under a light microscope. From Pepper *et al.* (2006).

TABLE 4.9 Approximate Range of Biomass of Each Major Component of the Biota in a Typical Temperate Grassland Soil

Component of Soil Biota	Biomass (tons/ha)
Plant roots	Up to 90 but generally about 20
Bacteria	1—2
Actinomycetes	0—2
Fungi	2—5
Protozoa	0—0.5
Nematodes	0—0.2
Earthworms	0—2.5
Other soil animals	0—0.5
Viruses	Negligible

From Killham (1994).

Information Box 4.8   Characteristics and Functions of Actinomycetes

Characteristics

Structure	Prokaryotic
Size	1—2 μm diameter
Morphology	Filamentous lengths of cocci
Gram stain	Gram positive
Respiration	Mostly aerobic, can be anaerobic
Habitat	Soil or marine
Abundance, marine isolates	5—40 CFU/ml
Abundance, soils	$10^6 - 10^8$/g

Functions

- Source of natural products and antibiotics, e.g., streptomycin
- Produce geosmin, the compound which gives soil and water a characteristic earthy odor
- Capable of degradation of complex organic molecules
- Capable of biological nitrogen fixation with species of the non-legume-associated *Frankia*

---

through photosynthesis and as they metabolize, they produce and release carbonic acid, which aids in weathering the surrounding mineral particles. Further, algae produce large amounts of extracellular polysaccharides, which also aid in soil formation by causing aggregation of soil particles (Killham, 1994).

Populations of soil algae generally exhibit seasonal variations with numbers being highest in the spring and fall. This is because desiccation caused by water stress tends to suppress growth in the summer and cold stress affects growth in the winter. Four major groups of algae are found in soil. The green algae or the Chlorophyta, for example, *Chlamydomonas*, are the most common algae found in acidic soils. Also widely distributed are diatoms such as *Navicula*, which are members of the Chrysophycophyta. Diatoms are found primarily in neutral and alkaline soils. Less numerous are the yellow—green algae such as *Botrydiopsis*, which are also members of the Chrysophycophyta, and the red algae (Rhodophycophyta, e.g., *Prophyridium*). In addition to these algal groups, there are the cyanobacteria (e.g., *Nostoc* and *Anabaena*), which are actually classified as bacteria but have many characteristics in common with algae. The cyanobacteria participate in the soil-forming process discussed in the previous paragraph, and some cyanobacteria also have the capacity to fix nitrogen, a nutrient that is usually limiting in a barren environment. In temperate soils the relative abundance of the major algal groups follows the order green algae > diatoms > cyanobacteria > yellow—green algae. In tropical soils the cyanobacteria predominate.

TABLE 4.10 Average Length and Volume of Soil Protozoa Compared with Bacteria

Group	Length (μm)	Volume (μm³)	Shape
Bacteria	<1−5	2.5	Spherical to rod shaped
Flagellates	2−50	50	Spherical, pear shaped, banana shaped
Amoebae			
Naked	2−600	400	Protoplasmic streaming, pseudopodia
Testate	45−200	1000	Build oval tests or shells made of soil
Giant	6000	4×10^9	Enormous naked amoebae
Ciliates	50−1500	3000	Oval, kidney shaped, elongated and flattened

From Ingham (1998).

4.4.6 Protozoa

Protozoa are unicellular, eukaryotic organisms that range up to 5.5 mm in length, although most are much smaller (Table 4.10). Most protozoa are heterotrophic and survive by consuming bacteria, yeast, fungi and algae. There is evidence that they may also be involved, to some extent, in the decomposition of soil organic matter. Because of their large size and requirement for large numbers of smaller microbes as a food source, protozoa are found mainly in the top 15 to 20 cm of the soil. Protozoa are usually concentrated near root surfaces that have high densities of bacteria or other prey. Soil protozoa are flatter and more flexible than aquatic protozoa, which makes it easier to move around in the thin films of water that surround soil particle surfaces as well as to move into small soil pores.

There are three major categories of protozoa: the flagellates, the amoebae and the ciliates (Chapter 2). The flagellates are the smallest of the protozoa and move by means of one to several flagella. Some flagellates (e.g., *Euglena*) contain chlorophyll, although most (e.g., *Oicomonas*) do not. The amoebae, also called rhizopods, move by protoplasmic flow, either with extensions called pseudopodia or by whole body flow. Amoebae are usually the most numerous type of protozoan found in a given soil environment. Ciliates are protozoa that move by beating short cilia that cover the surface of the cell. The protozoan population of a soil is often correlated with the bacterial population, which is the major food source present. For example, increases in protozoan populations often occur shortly after a proliferation of soil bacteria, as would result during the bacterial degradation of organic pollutants like 2,4-D. Numbers of protozoa reported range from 30,000 per gram of soil from a nonagricultural temperate soil to 350,000 per gram of soil from a maize field to 1.6×10^6 per gram of soil from a subtropical area.

4.5 DISTRIBUTION OF MICROORGANISMS IN SOIL

In surface soils, culturable microorganism concentrations can reach 10^8 per gram of dry soil, although direct counts are generally one to two orders of magnitude larger. These diverse microorganisms have been estimated to represent >10,000 species of bacteria alone (Roesch *et al.*, 2007). In addition, there are substantial populations of fungi, algae and protozoa. In general, microbial colonies are found in a nonuniform "patchlike" distribution on soil particle surfaces. This "patchlike" distribution of microorganisms in unsaturated soils results in increased microbial diversity as compared to saturated soils which are more highly connected and allow for more competitive interactions between microorganisms to occur (Treves *et al.*, 2003). Despite the large number of microorganisms found, they make up only a small fraction of the total organic carbon and a very small proportion of the soil volume (0.001%) in most soils.

In surface soils, microbial distribution is also dependent on soil texture and structure. As soils form, microbes attach to a site that is favorable for replication. As growth and colony formation take place, exopolysaccharides are formed, creating a "pseudoglue" that helps in orienting adjacent clay particles and cementing them together to form a microaggregate (Figure 4.23). Although the factors that govern whether a given site is favorable for colonization are not completely understood, several possible factors have been identified that may play a role including nutrient availability and surface properties. In addition, in surface soils, pore space seems to be an important factor. Pore spaces in microaggregates with neck diameters less than 6 μm have more activity than pore spaces with larger diameters, because the small pore necks protect resident bacteria from protozoal predation. Pore space also controls water content to some extent. Larger pores drain more quickly than smaller pores, and therefore the interior of a small pore is generally wetter and more conducive to microbial activity. It has further been suggested that Gram-negative bacteria prefer the interior of microaggregate pore space because of the increased moisture, whereas Gram-positive bacteria, which are better adapted to withstand dry conditions, tend to occupy the microaggregate exteriors.

Most microorganisms in Earth environments are attached. It has been estimated that approximately 80 to 90% of the cells are sorbed to solid surfaces and the remainder are free-living. As stated earlier, attached microbes are found in patches or colonies on particle

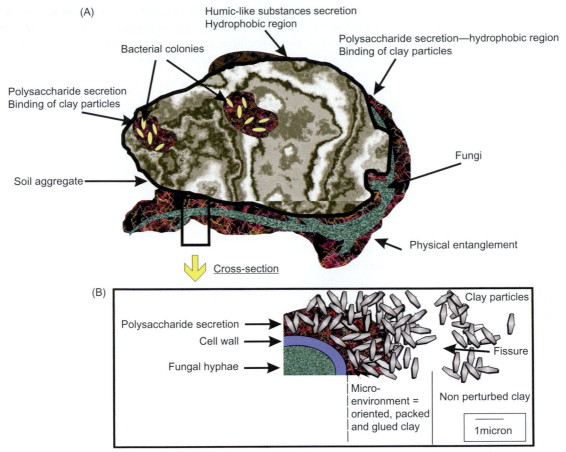

FIGURE 4.23   Microbially mediated aggregation: (A) schematic representation of the binding and stabilization of a soil aggregate by microorganisms; (B) detail of the microenvironment in the vicinity of a fungus. Adapted with permission from Robert and Chenu (1992).

surfaces. Attachment and growth into colonies confer several advantages for microorganisms, bacteria in particular (Gilbert *et al.*, 1993). Attachment can help protect bacteria from protozoal predation. Attachment and colony formation can also help provide localized concentrations of nutrients that are contained in and recycled among the attached cells within the colony rather than being diluted into the general environment. This is especially important in oligotrophic environments. Another advantage of colony formation is that a microbial colony can alter the immediate microsite environment surrounding the cell, such as pH, to optimize growth conditions. Finally, genetic exchange can occur much more frequently within a colony than between isolated cells in a soil environment.

Although free-living cells are less common, they are an important mechanism for the dispersion of microorganisms. As nutrient supplies at a particular surface site are consumed, microorganisms need a mechanism by which cells can disperse to new sites that may have additional food supplies. Fungi spread via spores released from fruiting bodies or via hyphal extension. Bacteria, which undergo only simple cell division, need a different

mechanism of dispersal, namely, release of free-living daughter cells. In fact, there is evidence that bacterial cells at the surface of a colony undergo changes in their surface properties that cause the release of a newly formed daughter cell after cell division. As these free-living daughter cells grow, their surfaces undergo a chemical change that makes attachment at a new site more favorable.

4.6 MICROORGANISMS IN SUBSURFACE ENVIRONMENTS

In subsurface environments, the same patchlike distribution of microbes exists that is found in surface soils. Culturable counts range from essentially zero to 10^7 per gram of dry soil depending on the depth and type of porous medium. Direct counts generally range from 10^5 to $> 10^7$ cells per gram of porous medium. Thus, the difference between culturable and direct counts is often much larger in the subsurface than in surface soils. This is most likely due to the presence of viable but nonculturable microbes (VBNC) (Chapter 3). These microbes exist as a result of the nutrient-poor status of subsurface environments, which is directly reflected in their low organic matter content. When subsurface cells are examined, they are rarely dividing and contain few ribosomes or inclusion bodies. This is not surprising considering the nutrient-limited conditions in which subsurface microbes live. Recall that environmental bacteria have diverse, specific, nutritional needs and thus may be difficult to culture on traditional media. Further, they often exist under adverse conditions and as a result may be sublethally injured. Such injured bacteria cannot be cultured by conventional methods. It has been estimated that 99% of all soil organisms may be VBNC. Likewise, most culture-based methods only enumerate heterotrophic microorganisms. As C concentrations decrease in the subsurface, the relative number of autotrophic microorganisms typically increases and these organisms may be missed when using typical culture-based methods.

The weathered component minerals (which serve as a source of micronutrients) and organic matter (which serves as a carbon and nitrogen source) are two of the primary differences between surface soils and subsurface materials as environments for microorganisms. These differences in nutrient content are reflected in a higher and more uniform distribution of microbial numbers and activity in surface soil environments. The other major factor impacting microbial density and activity in surface and subsurface environments is water content. Areas that have high recharge from rainfall and water flow tend to have both higher microbial numbers and activity.

4.6.1 Microorganisms in Shallow Subsurface Environments

Although the microorganisms of surface soils have been studied extensively, the study of subsurface microorganisms is relatively new, beginning in earnest in the 1980s. Complicating the study of subsurface life are the facts that sterile sampling is problematic and many subsurface microorganisms are difficult to culture. Some of the initial studies evaluating subsurface populations were invalidated by contamination with surface microbes. As a result, study of subsurface organisms has required the development of new tools and approaches for sterile sampling (Chapter 8) and for microbial enumeration and identification. For example, it has been demonstrated that rich media are not suitable for culturing subsurface organisms that are adapted to highly oligotrophic conditions and that viable counts from these environments on less-rich media often produce microbial counts one order of magnitude or more higher than those produced on richer media (Chapter 8).

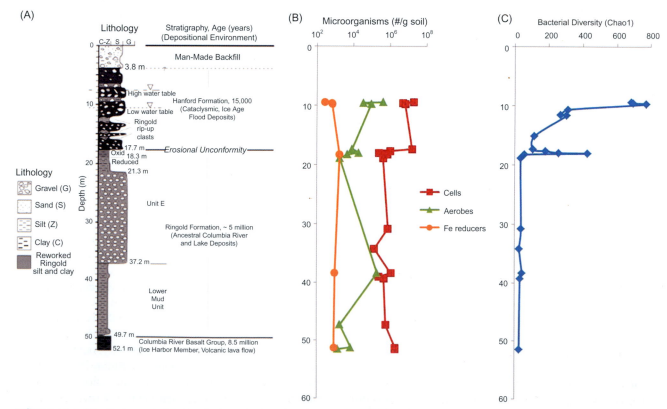

FIGURE 4.21 Distribution and diversity of microorganisms in a shallow subsurface environment. Adapted from Lin *et al.* (2012a,b).

Case Study 4.3   Microbial Counts in the Shallow Subsurface

One of the most thorough characterizations of microorganisms in the shallow subsurface environment was conducted by Lin *et al.* (2012a,b) on samples from the Hanford Site 300 Area near Richland, WA, USA. This area represents a geologically and hydrologically complex system, and is of interest due to the extensive amount of nuclear-related research (much in response to environmental contamination) that has been conducted in the area. The authors collected cores from a 52-m deep borehole at the site and used a large variety of microbial methods including DNA sequencing, quantitative PCR (qPCR), phospholipid fatty acid analysis (PLFA) and culture-based methods to characterize the microbial communities.

The water table at the site fluctuated from 7.5 to 10.5 m below the surface and became a reduced (anaerobic) environment around 18.3 m, shortly after transitioning from coarse- to fine-textured sediment (Figure 4.24). Microbial counts and diversity were greatest in the upper, aerobic region of the saturated zone and decreased dramatically in the anaerobic zone. In general, culturable counts were about two orders of magnitude lower than those obtained with direct methods (qPCR or PLFA). The region surrounding the aerobic/anaerobic interface appeared to be a zone of active biogeochemical redox cycling. Although the levels of metal-reducing bacteria were low overall, there was an increase in the proportion of the bacterial community comprised by organisms using alternate electron acceptors such as iron and sulfate in anaerobic samples with some reaching maximum levels around the aerobic/anaerobic interface. Populations of archaea also varied with depth, but likewise reached maximum levels ($\approx 8\%$ of community) at the aerobic/anaerobic interface. Additionally, the authors detected 13 novel orders of *Deltaproteobacteria* in the samples. This group contains many members known to be involved in transformations of metals, and these novel organisms may prove to be important to the remediation of metal and radionuclide contaminants at Hanford and other contaminated sites (Chapter 18).

Because subsurface microbiology is still a developing field, information is limited in comparison with that for surface microorganisms. Yet there is still enough information available to know that many subsurface environments, once thought to contain very few if any microorganisms, actually have significant and diverse populations of microorganisms. In particular, shallow subsurface zones, specifically those with a relatively rapid rate of water recharge, have high numbers of microorganisms. The majority of these organisms are bacteria, but protozoa and fungi are also present. In general, both microbial numbers and diversity decrease with depth in shallow subsurface systems, especially once the environment becomes anaerobic (Figure 4.24; Case Study 4.3). Total numbers of bacteria, as measured by direct counts, tend to remain fairly constant, ranging between 10^5 and 10^7 cells per gram throughout the profile. For comparison, numbers in surface soils range from 10^9 to 10^{10} cells per gram. This decrease in numbers is directly correlated with the low amounts of inorganic nutrients and organic matter in subsurface materials. Subsurface eukaryotic counts are also lower than surface counts by several orders of magnitude. Low eukaryotic counts are a result of low organic matter content but, perhaps more importantly, result from removal by physical straining by small soil pores as they move downward (Chapter 15). A final point to be made is that both prokaryotic and eukaryotic counts are highest in portions of the subsurface containing sandy sediments. This does not mean that clayey regions are not populated, but that the numbers tend to be lower. This may also be due to exclusion and physical straining of microorganisms by small pores in clay-rich media.

Numbers of culturable bacteria in subsurface environments generally show more variability than those from direct counts (molecular-based or microscopic methods). Thus, the difference between direct and cultural counts in the subsurface is often greater than the difference in surface soils (one to two orders of magnitude). Several factors may explain the larger difference between direct and cultural counts in the subsurface. First, because nutrients are much more limiting in the subsurface, a greater proportion of the population may be in a nonculturable state. Second, the physiological and nutritional requirements of subsurface organisms are not well understood. Therefore, even though we know that a dilute nutrient medium is better than a rich medium, the type of dilute nutrient medium used may still not be appropriate for many environmental microorganisms (Chapter 8).

4.6.2 Microorganisms in Deep Subsurface Environments

Until relatively recently, it was thought that the deep subsurface environments contain few if any microorganisms because of the extreme oligotrophic conditions found there. However, recent research has shown that microorganisms can be found to a depth of > 3 km below Earth's surface! Interest in this area began in the 1920s, when increased consumption of oil led to increased oil exploration and production. Upon examination of water extracted from deep within oil fields, Edward Bastin, a geologist at the University of Chicago, found that significant levels of hydrogen sulfide and bicarbonate were present. The presence of these materials could not be explained on a chemical basis alone, and Bastin suggested that sulfate-reducing bacteria were responsible for the hydrogen sulfide and bicarbonate found in the drilling water. Subsequently, Frank Greer, a microbiologist at the University of Chicago, was able to culture sulfate-reducing bacteria from water extracted from an oil deposit that was hundreds of meters below Earth's surface. Bastin and Greer suggested that these microorganisms were descendants of organisms buried more than 300 million years ago during formation of the oil reservoir. However, their suggestions were largely ignored because the sampling techniques and microbial analysis techniques available at the time could not ensure that the bacteria were not simply contaminants from the drilling process.

Other research hinted at the existence of subsurface microorganisms, most notably the work of Claude Zobell. But not until the 1980s, with the growing concern over groundwater quality, did several new efforts address the questions of whether subsurface microorganisms exist and what range and level of microbial activity occur in the subsurface. Agencies involved in these new studies included the U.S. Department of Energy, the U.S. Geological Survey, the U.S. Environmental Protection Agency, the German Federal Ministry of the Interior (Umbeltbundesamt), the Institute for Geological Sciences (Wallingford, England) and the Water Research Center (Medmanham, England). A number of new techniques were developed to facilitate the collection of sterile samples from deep cores in both the saturated and the unsaturated zone (Chapter 8), and a great deal of information has been generated concerning the presence and function of microorganisms in deep subsurface environments (Fredrickson and Onstott, 1996).

4.6.2.1 Microorganisms in the Deep Vadose Zone

Several studies have looked at deep cores in the unsaturated zone. In one of the first such studies, Frederick Colwell (1989) collected a 70-m core from the eastern Snake River Plain, which is a semiarid, high desert area in southeastern Idaho. Table 4.11 shows a comparison of bacterial numbers in the surface and subsurface samples from this site. Following the pattern described in Section 4.6.1 for shallow subsurface environments, the direct counts from deep

subsurface samples remained high, declining by only one order of magnitude in comparison with the surface samples. In contrast, culturable counts declined by four orders of magnitude to less than 100 colony-forming units per gram of sediment. The majority of the isolates from the subsurface in this study were Gram positive and strictly aerobic. In contrast, in surface soils Gram-negative bacteria are more numerous. The subsurface atmosphere was found to be similar to ambient surface air in most samples, suggesting that the subsurface was aerobic.

Subsequent studies have largely confirmed these findings and have added some new information. In general, microbial numbers and activities are higher in paleosols (buried sediments) that have had exposure to Earth's surface and plant production. These materials tend to have associated microorganisms and nutrient reserves, albeit at low concentrations, that can maintain very slow-growing populations for thousands of years. These later studies also suggested that there are some vadose zone materials, most notably massive basalt samples collected by the Idaho National Engineering Laboratory, that lack viable microorganisms and do not show any detectable metabolic activity. In summary, our present understanding of the deep vadose zone is limited. However, it appears that there are areas of the vadose zone that contain microbes that may be stimulated to interact with environmental contaminants, whereas other areas of the vadose zone simply act as a conduit for the downward transport of contaminants.

It must be emphasized that although microbes are present in deep vadose zones, rates of metabolic activity are much lower than rates in surface soils. This is illustrated in Figure 4.25, which depicts metabolic activity in a range of surface and subsurface environments. Metabolic activity is expressed as the rate of CO_2 production and was estimated by groundwater chemical analysis and geochemical modeling. As can be seen in this figure, the difference between the

TABLE 4.11 A Comparison of Microbial Counts in Surface and 70-m Unsaturated Subsurface Environments

Sample Site	Direct Counts (counts/g)	Viable Counts (CFU/g)[a]
Surface (10 cm)	2.6×10^6	3.5×10^5
Subsurface basalt-sediment interface (70.1 m)	4.8×10^5	50
Subsurface sediment layer (70.4 m)	1.4×10^5	21

[a]CFU, colony-forming units.

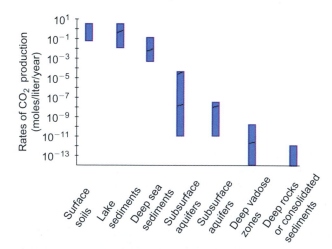

FIGURE 4.25 Ranges of rates of *in situ* CO_2 production for various surface and subsurface environments, as estimated by groundwater chemical analyses and geochemical modeling. Adapted from Kieft and Phelps (1997).

Case Study 4.4   Deep Probe—How Low Can Life Go?

In 1987, the US Department of Energy sponsored the drilling of several deep boreholes (0.5 km) in Cretaceous sediments (70 million to 135 million years old) in South Carolina, near the Savannah River nuclear materials processing facility. A team of scientists used a sophisticated sampling device to ensure that the core samples taken from the boreholes were not contaminated with microorganisms from other parts of the borehole or the surface. The samples were collected from depths of almost 0.5 km and then shipped to several laboratories, where microbial analyses were initiated immediately. Microbial analysis of core materials showed diverse and numerous populations of microbes, with total counts ranging from 10^6 to 10^7 cells per gram of sediment. Culturable counts were much lower, ranging from 10^3 to 10^6 colony-forming units per gram of sediment from samples taken from 350 to 413 m in a permeable, saturated Middendorf sediment, to nondetectable to 10^4 colony-forming units per gram of sediment from a low-permeability Cape Fear sediment (450 to 470 m) (Fredrickson *et al.*, 1991). The most abundant culturable forms in these samples were aerobic or facultatively anaerobic chemoheterotrophs. Results of these analyses have helped confirm the theory that subsurface bacteria are ubiquitous, although their abundance varies considerably, depending on the site characteristics.

More recently, between 2001 and 2006, a group of collaborating scientists took a series of water samples from depths of 0.72 to > 3 km below Earth's surface in the Witwatersrand Basin in central South Africa (Gihring *et al.*, 2006). Total microbial numbers in the samples were estimated to be as low as 10^3 cells/ml.

Diversity was also low as shown by analysis of the 16S rRNA gene, which generated only an average of 11 bacterial OTUs from all the samples. Compare this to surface soils that have up to 6300 OTUs! Interestingly, these researchers found growth substrates in the samples such as methane, ethane, propane, butane, and acetate and H_2 that were clearly not being used by the microbial community (Kieft *et al.*, 2005). They speculate that there must be factors other than these electron donors that ultimately limit growth, such as the limitation of an inorganic nutrient like iron or phosphate. Growth limitations are also evidenced by the small average cell diameter of 0.3 μm.

What types of microorganisms were present in these samples and how do they compare to what are found in surface soils? The bacteria identified in the Savannah River study were dominated by the Gram-negative divisions *gamma-Proteobacteria*, *beta-Proteobacteria*, and *alpha-Proteobacteria*, and the Gram-positive division *Actinobacteria* (Balkwill and Boone, 1997). The Witwatersrand samples were dominated by *beta-*, *gamma-*, and *alpha-Proteobacteria*, followed by the Gram-positive spore-forming division *Firmicutes*. Each of these sample sites also had bacteria that did not match closely to any bacteria identified to date. The pattern of bacteria found in these deep subsurface samples when compared to that found in surface soils is surprisingly similar. Most uncultured soil bacterial libraries are dominated by *Proteobacteria* (predominantly *alpha-Proteobacteria*), *Acidobacteria* (known to be difficult to culture), and *Actinobacteria* (Janssen, 2006).

---

rates of CO_2 production in a surface soil and in the deep vadose zone is at least nine orders of magnitude.

4.6.2.2 Microorganisms in the Deep Saturated Zone

Intermediate and deep aquifers are characterized by low rates of recharge and groundwater flow that create a habitat for microorganisms different from that in shallow aquifers. Samples taken from deep cores (Case Study 4.4) have generally shown that there are lower numbers and a more limited diversity of microorganisms in deep saturated zones than in surface soils. However, the types of organisms detected included a wide range of aerobic and facultatively anaerobic chemoheterotrophs; denitrifiers; methanogens; sulfate-reducers; sulfur-oxidizers; nitrifiers; and nitrogen-fixing bacteria. Low numbers of unicellular cyanobacteria, fungi, and protozoa have also been detected in some samples from 0.5 km depth (Balkwill and Boone, 1997). Culture-based analysis of some of these samples has shown that the bacteria are able to metabolize simple sugars, organic acids and even complex polymers such as the storage product β-hydroxybutyric acid. Thus, subsurface microbes exhibit diverse metabolic capabilities. What is interesting about these data is that they suggest that subsurface microbes reflect in large part what is found in surface soils.

Scientists have also discovered microbial life at even greater depths (0.72 to > 3 km below Earth's surface) in water samples from deep gold mines in South Africa (Kieft *et al.*, 2005; Gihring *et al.*, 2006). Interestingly, microbial numbers in the samples were estimated to be quite low ($\approx 10^3$ cells/mL) despite the presence of relatively high concentrations of growth substrates such as methane, ethane, acetate and H_2. The researchers speculated that other factors, such as the lack of inorganic nutrients or alternate electron acceptors, may ultimately be limiting microbial growth in these deep environments. However, our knowledge of microbial communities in these environments is likely biased by the methodologies used to characterize them—based largely on our greater relative understanding of microbial communities in surface soils. Additional work on these deep subsurface environments will undoubtedly yield exciting discoveries that will only deepen our astonishment at the complexity of microbial life on Earth.

REFERENCES AND RECOMMENDED READING

Recommended Readings

Amend, J. P., and Teske, A. (2005) Expanding frontiers in deep subsurface microbiology. *Palaeogeogr. Palaeoclimatol. Palaeoecol.* **219**, 131−155.

Atlas, R. M., and Bartha, R. (1998) "Microbial Ecology," fourth ed. Benjamin Cummings, Redwood City, CA.

Brady, N. C., and Weil, R. R. (2007) "The Nature and Properties of Soils," 14th ed. Prentice Hall, Upper Saddle River, NJ.

Chapelle, F. H. (2001) "Ground-Water Microbiology and Geochemistry," second ed. John Wiley & Sons, New York.

Chapter References

Balkwill, D. L., and Boone, D. R. (1997) Identity and diversity of microorganisms cultured from subsurface environments. In "The Microbiology of the Terrestrial Deep Subsurface" (P. S. Amy, and D. L. Haldeman, eds.), CRC Press, Boca Raton, FL, pp. 105–117.

Chapelle, F. H. (1993) "Ground-Water Microbiology and Geochemistry," Wiley, New York.

Colwell, F. S. (1989) Microbiological comparison of surface soil and unsaturated subsurface soil from a semiarid high desert. *Appl. Environ. Microbiol.* **55**, 2420–2423.

Curl, E. R., and Truelove, B. (1986) "The Rhizosphere," Springer-Verlag, New York, pp. 105.

Dragun, J. (1988) "The Soil Chemistry of Hazardous Materials," Hazardous Materials Control Research Institute, Silver Spring, MD.

Fierer, N., and Jackson, R. B. (2006) The diversity and biogeography of soil bacterial communities. *Proc. Natl. Acad. Sci. U.S.A.* **103**, 626–631.

Fredrickson, J. K., and Onstott, T. C. (1996) Microbes deep inside the earth. *Sci. Am.*, 68–73.

Fredrickson, J. K., Balkwill, D. L., Zachara, J. M., Li, S-M. W., Brockman, F. J., and Simmons, M. A. (1991) Physiological diversity and distributions of heterotrophic bacteria in deep Cretaceous sediments of the Atlantic Coastal Plain. *Appl. Environ. Microbiol.* **57**, 402–411.

Gammack, S. M., Paterson, E., Kemp, J. S., Cresser, M. S., and Killham, K. (1992) Factors affecting the movement of microorganisms in soils. In "Soil Biochemistry" (G. Stotzky, and J.-M. Bollag, eds.), vol. 7, Marcel Dekker, New York, pp. 263–305.

Gihring, T. M., Moser, D. P., Lin, L.-J., Davidson, M., Onstott, T. C., Morgan, L., *et al.* (2006) The distribution of microbial Taxa in the subsurface water of the Kalahari Shield, South Africa. *Geomicrobiol. J.* **23**, 415–430.

Gilbert, P., Evans, D. J., and Brown, M. R. W. (1993) Formation and dispersal of bacterial biofilms in vivo and in situ. *J. Appl. Bacteriol. Symp.*(Suppl. 74), , 67S–78S.

Ingham, E. R. (1998) Protozoa and nemadoes. In "Principles and Applications of Soil Microbiology" (D. M. Sylvia, J. J. Fuhrmann, P. G. Hartel, and D. A. Zuberer, eds.), Prentice-Hall, Upper Saddle, NJ, pp. 114–131.

Janssen, P. H. (2006) Identifying the dominant soil bacterial taxa in libraries of 16S rRNA and 16S rRNA genes. *Appl. Environ. Microbiol.* **72**, 1719–1728.

Jensen, P. R., Mincer, T. J., Williams, P. G., and Fenicel, W. (2005) Marine actinomycete diversity and natural product discovery. *Antonie van Leeuwenhoek* **87**, 43–48.

Kieft, T. L., and Phelps, T. J. (1997) Life in the slow lane: activities of microorganisms in the subsurface. In "The Microbiology of the Terrestrial Deep Subsurface" (P. S. Amy, and D. L. Haldeman, eds.), CRC Lewis, Boca Raton, FL, pp. 137–163.

Kieft, T. L., McCuddy, S. M., Onstott, T. C., Davidson, M., Lin, L. H., Mislowack, B., *et al.* (2005) Geochemically generated, energy-rich substrates and indigenous microorganisms in deep, ancient ground-water. *Geomicrobiol. J.* **22**, 325–335.

Killham, K. (1994) "Soil Ecology," Cambridge University Press, Cambridge.

Lauber, C. L., Hamady, M., Knight, R., and Fierer, N. (2009) Pyrosequencing-based assessment of soil pH as a predictor of soil bacterial community structure at the continental scale. *Appl. Environ. Microbiol.* **75**, 5111–5120.

Lin, X., Kennedy, D., Fredrickson, J., Bjornstad, B., and Konopka, A. (2012a) Vertical stratification of subsurface microbial community composition across geological formations at the Hanford site. *Environ. Microbiol.* **14**, 414–425.

Lin, X., Kennedy, D., Peacock, A., McKinley, J., Resch, C. T., Fredrickson, J., *et al.* (2012b) Distribution of microbial biomass and potential for anaerobic respiration in Hanford site 300 area subsurface sediment. *Appl. Environ. Microbiol.* **78**, 759–767.

Matthess, G., Pekdeger, A., and Schroeder, J. (1988) Persistence and transport of bacteria and viruses in groundwater—a conceptual evaluation. *J. Contam. Hydrol.* **2**, 171–188.

Painter, T. J. (1991) Lindow Man, Tollund Man, and other peat-bog bodies—the preservative and antimicrobial action of sphagnam, a reactive glycuronoglycan with tanning and sequestering properties. *Carbohyd. Polym.* **15**, 123–142.

Pepper, I. L. (2014) The soil health:human health nexus. *Crit. Rev. Environ. Sci. Technol.* In press.

Pepper, I. L., Gerba, C. P., and Brusseau, M. L. (2006) "Environmental and Pollution Science," second ed. Elsevier Science/Academic Press, San Diego, CA.

Robert, M., and Chenu, C. (1992) Interactions between soil minerals and microorganisms. In "Soil Biochemistry" (G. Stotsky, and J.-M. Bollag, eds.), vol. 7, Marcel Dekker, New York, pp. 307–418.

Roesch, L. F. W., Fulthorpe, R. R., Riva, A., Casella, G., Hadwin, A. K. M., Kent, A. D., *et al.* (2007) Pyrosequencing enumerates and contrasts soil microbial diversity. *ISME J.* **1**, 283–290.

Schwarzenbach, R. P., Gschwend, P. M., and Imboden, D. M. (1993) "Environmental Organic Chemistry," John Wiley & Sons, New York.

Sextone, A. J., Revsbech, N. P., Parkin, T. B., and Tiedje, J. M. (1985) Direct measurement of oxygen profiles and denitrification rates in soil aggregates. *Soil Sci. Soc. Am. J.* **49**, 645–651.

Shade, A., Hogan, C. S., Klimowicz, A. K., Linske, M., McManus, P. S., and Handelsman, J. (2012) Culturing captures members of the soil rare biosphere. *Environ. Microbiol.* **14**, 2247–2252.

Sposito, G. (1989) "The Chemistry of Soils," Oxford University Press, New York.

Taylor, A. E., Zeglin, L. H., Dooley, S., Myrold, D. D., and Bottomley, P. J. (2010) Evidence for different contributions of archaea and bacteria to the ammonia-oxidizing potential of diverse Oregon soils. *Appl. Environ. Microbiol.* **76**, 7691–7698.

Treves, D. S., Xia, B., Zhou, J., and Tiedje, J. M. (2003) A two-species test of the hypothesis that spatial isolation influences microbial diversity in soil. *Microb. Ecol.* **4**, 20–28.

Verhamme, D. T., Prosser, J. I., and Nicol, G. W. (2011) Ammonia concentration determines differential growth of ammonia-oxidising archaea and bacteria in soil microcosms. *ISME J.* **5**, 1067–1071.

Young, I. M., and Crawford, J. W. (2004) Interactions and self-organization in the soil-microbe complex. *Science* **304**, 1634–1637.

Chapter 5

Aeromicrobiology

Ian L. Pepper and Charles P. Gerba

5.1 INTRODUCTION

In the 1930s, F.C. Meier coined the term aerobiology to describe a project that involved the study of life in the air (Boehm and Leuschner, 1986). Since then, aerobiology has been defined by many as the study of the aerosolization, aerial transmission and deposition of biological materials. Others have defined it more specifically as the study of diseases that may be transmitted via the respiratory route (Dimmic and Akers, 1969). Despite the variations in definition, this evolving area is becoming increasingly important in many aspects of diverse fields including public health, environmental science, industrial and agricultural engineering, biological warfare and space exploration.

This chapter introduces the basics of aerobiology, including the nature of aerosols and the fundamentals of the aeromicrobiological (AMB) pathway. The remainder of the chapter focuses on a subset of the science that we shall term aeromicrobiology. Aeromicrobiology, as defined for the purpose of this text, involves various aspects of intramural (indoor) and extramural (outdoor) aerobiology, as they relate to the airborne transmission of environmentally relevant microorganisms, including viruses, bacteria, fungi, yeasts and protozoans.

5.2 AEROSOLS

Particles suspended in air are called aerosols. These pose a threat to human health mainly through respiratory intake and deposition in nasal and bronchial airways. In addition, soil or dust particles can act as a "raft" for biological entities known as bioaerosols (Brooks et al., 2004). Smaller aerosols travel further into the respiratory system and generally cause more health problems than larger particles. For this reason, the United States Environmental Protection Agency (USEPA) has divided airborne particulates into two size categories: PM_{10}, which refers to particles with diameters less than or equal to $10 \,\mu m$ (10,000 nm), and $PM_{2.5}$, which are particles less than or equal to $2.5 \,\mu m$

I.L. Pepper, C.P. Gerba, T.J. Gentry: Environmental Microbiology, Third edition. DOI: http://dx.doi.org/10.1016/B978-0-12-394626-3.00005-3
© 2015 Elsevier Inc. All rights reserved.

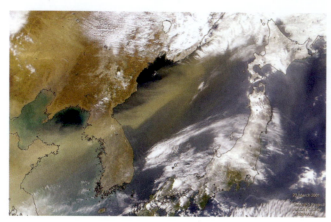

FIGURE 5.1 Mongolian dust over the Sea of Japan. Image provided by NASA.

(2500 nm) in diameter. For this classification, the diameter of aerosols is defined as the aerodynamic diameter:

$$d_{pa} = d_{ps}(\rho_p/\rho_w)^{\frac{1}{2}} \qquad \text{(Eq. 5.1)}$$

where:

d_{pa} = aerodynamic particle diameter (μm)
d_{ps} = Stokes' diameter (μm)
ρ_p = particle density (g/cm^{-3})
ρ_w = density of water (g/cm^{-3})

Atmospheric particulate concentration is expressed in micrograms of particles per cubic meter of air (μg/m^3). The USEPA established a National Ambient Air Quality Standard (NAAQS) for PM$_{10}$ of 150 μg/m^{-3} averaged over a 24-hour period, and 50 μg/m^{-3} averaged annually. More recently, separate standards for PM$_{2.5}$ of 65 μg/m^{-3} for 24 hours and 15 μg/m^{-3} annually have been introduced.

Symptoms of particulate matter inhalation include: decreased pulmonary function; chronic coughs; bronchitis; and asthmatic attacks. The specific causal mechanisms are poorly understood. One well-documented episode occurred in London in 1952, when levels of smoke and sulfur dioxide aerosols, largely associated with coal combustion, reached elevated levels due to local weather conditions. Over a 10-day period, approximately 4000 deaths were attributed to cardiovascular and lung disorders brought on or aggravated by these aerosols.

Airborne particles can travel great distances. Intense dust storms during 1998 and 2001 in the Gobi desert of western China and Mongolia (Figure 5.1) elevated aerosol levels to concentrations near the health standard in western North America several thousand miles away.

Smaller particles tend to travel greater distances than large particles. Stokes' law (Eq. 5.2) is used to describe the fall of particles through a dispersion medium, such as air or water:

$$V = [D^2 \times (\rho_p - \rho_1) \times g]/18\rho \qquad \text{(Eq. 5.2)}$$

Information Box 5.1   Influence of Particle Size on Velocity of Deposition of Particles in Air, Calculated Using Stokes' Law

Particle Diameter (mm)	Particle Type	Rate of Fall in Air (cm s^{-1})
1	Sand	7880
0.1	Silt	79
0.001	Clay	7.9×10^{-5}
0.002	Clostridial spore	0.016

From Pepper et al., 2006.

where:

V = velocity of fall (cm/s^{-1})
g = acceleration of gravity (980 cm/s^{-2})
D = diameter of particle (cm)
ρ_p = density of particle (density of quartz particles is 2.65 g/cm^{-3})
ρ_1 = density of dispersion medium (air has a density of about 0.001213 g/cm^{-3}; water has a density of about 1 g/cm^{-3})
ρ = viscosity of the dispersion medium (about 1.83×10^{-4} poise or g cm^{-1}s^{-1} for air; 1.002×10^{-2} poise for water)

Using Stokes' law, we can calculate the rate of fall of particles in air (Information Box 5.1). Small particles are thus a greater concern than larger particles for several reasons. Small particles stay suspended longer and so they travel further and stay suspended longer. This results in an increased risk of exposure. Small particles also tend to move further into the respiratory system, exacerbating their effects on health. Stokes' law explains why we can expect viruses to persist as a bioaerosol longer than bacteria, which are much larger.

5.3 NATURE OF BIOAEROSOLS

Biological contaminants include whole entities such as bacterial and viral human pathogens. They also include airborne toxins, which can be parts or components of whole cells. In either case, biological airborne contaminants are known as bioaerosols, which can be ingested or inhaled by humans.

Bioaerosols vary considerably in size, and composition depends on a variety of factors including the type of microorganism or toxin, the types of particles they are associated with such as mist or dust, and the gases in which the bioaerosol is suspended. Bioaerosols in general range from 0.02 to 100 μm in diameter and are classified on the basis of their size. The smaller particles (<0.1 μm in diameter) are considered to be in the nuclei mode, those ranging from 0.1 to 2 μm are in the accumulation mode and larger

Information Box 5.2   Possible Modes of Respiratory Transmission of Influenza A

Direct Contact
Transmission occurs when the transfer of microorganism results from direct physical contact between an infected or colonized individual and a susceptible host.

Indirect Contact
Transmission occurs by the passive transfer of microorganisms to a susceptible host via inanimate contaminated object or fomite.

Droplet
Transmission occurs via large droplets (≥5 μm diameter) generated from the respiratory tract of the infected individual during coughing or sneezing, talking or during procedures such as suctioning or bronchoscopy. These droplets are propelled a distance of less than 1 m through the air, and are deposited on the nasal or oral mucosa of the new host or in their immediate environment. These large droplets do not remain suspended in the air and true aerosolization does not occur.

Airborne
Transmission occurs via the dissemination of microorganisms by aerosolization. Organisms are contained in droplet nuclei (airborne particles less than 5 μm that result from the evaporation of large droplets), or in dust particles containing skin cells and other debris that remain suspended in the air for long periods of time. Microorganisms are widely dispersed by air currents and inhaled by susceptible hosts.
See also Section 5.6.2 and Case Study 5.2.

The composition of bioaerosols can be liquid or solid, or a mixture of the two, and should be thought of as microorganisms associated with airborne particles, or as airborne particles containing microorganisms. This is because it is rare to have microorganisms (or toxins) that are not associated with other airborne particles such as dust or water. This information is derived from particle size analysis experiments, which indicate that the average diameter of airborne bacterial particles is greater than 5 μm (Fengxiang *et al.*, 1992). By comparison, the average size of a soil-borne bacterium, 0.3 to 1 μm, is less than one-fifth this size. Similar particle size analysis experiments show the same to be true for aerosolized microorganisms other than bacteria, including viruses.

5.4 AEROMICROBIOLOGICAL PATHWAY

The aeromicrobiological pathway describes: (1) the launching of bioaerosols into the air; (2) the subsequent transport via diffusion and dispersion of these particles; and finally (3) their deposition. An example of this pathway is that of liquid aerosols containing the influenza virus launched into the air through a cough, sneeze or even through talking. These virus-associated aerosols are dispersed by a cough or sneeze, transported through the air, inhaled, and deposited in the lungs of a nearby person, where they can initiate a new infection (Figure 5.3). Traditionally, the deposition of viable microorganisms and the resultant infection are given the most attention, but all three processes (launching, transport and deposition) are of equal importance in understanding the aerobiological pathway.

5.4.1 Launching

The process whereby particles become suspended within Earth's atmosphere is termed launching. Because

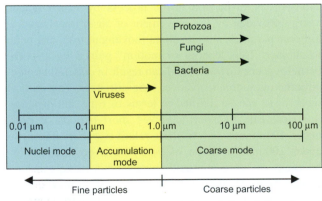

FIGURE 5.2 Diagrammatic representation of the relative sizes of bioaerosols. The depictions of the various kinds of organisms are indicative of their potential sizes when associated with airborne particles (rafts). The terminologies used to describe the various sizes of the bioaerosols are also indicated.

FIGURE 5.3 A cough or sneeze launches infectious microbes into the air. Anyone in the vicinity may inhale the microbes, resulting in a potential infection.

particles are considered to be in the coarse mode (Committee on Particulate Control Technology, 1980). As shown in Figure 5.2, particles in nuclei or accumulation mode are considered to be fine particles and those in coarse mode are considered coarse particles.

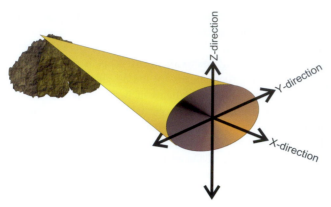

FIGURE 5.4 Schematic representation of the type of bioaerosol distribution expected from a point source, given three planes of diffusion: (1) the X-direction is the mean direction in which the wind is blowing; (2) the Y-direction is the lateral diffusion; and (3) the Z-direction is the vertical diffusion.

bioaerosols must be launched into the atmosphere to be transported, it is important to understand this process. The launching of bioaerosols is mainly from terrestrial and aquatic sources, with greater airborne concentrations or atmospheric loading being associated with terrestrial sources than with aquatic sources. A recent model estimated that the total global emission of bacteria containing particles to the atmosphere to be 7.6×10^{23} to 3.5×10^{24} (Burrow *et al.*, 2009). Some researchers speculate that there may even be atmospheric sources of bioaerosols in addition to terrestrial and aquatic ones. This phenomenon is related to the limited potential for microorganisms to reproduce while airborne. This, however, is an area of aeromicrobiology for which there is little available information.

Launching into the surface boundary layers can include, but is certainly not limited to, diverse mechanisms such as: air turbulence created by the movement of humans, animals and machines; the generation, storage, treatment and disposal of waste material; natural mechanical processes such as the action of water and wind on contaminated solid or liquid surfaces; and the release of fungal spores as a result of natural fungal life cycles.

Airborne particles can be launched from either point, linear or area sources. A point source is an isolated and well-defined site of launching such as a pile of biosolid material, before it is applied over a field. Point sources tend to display a general conical-type dispersion (Figure 5.4). Point sources can be further defined on the basis of the type of launching phenomenon: (1) instantaneous point sources, for example, a single event such as a sneeze; or (2) continuous point sources, from which launching occurs over extended periods of time, such as a biosolid pile.

In contrast to point sources, linear sources and area sources involve larger, less well-defined areas. When considered on the same size scale, linear and area sources display more particulate wave dispersion as opposed to the conical type of dispersion displayed by point sources.

FIGURE 5.5 A linear bioaerosol source using the example of the release of biological warfare agents. This is an illustration of an instantaneous linear bioaerosol release.

Linear and area sources can also be divided into instantaneous and continuous launching points of origin. For example, an instantaneous linear source might be a passing aircraft releasing a biological warfare agent (Figure 5.5). A continuous area source might be exemplified by release of bioaerosols from a large field that has received an application of biosolids or animal manures.

5.4.2 Transport

Transport or dispersion is the process by which kinetic energy provided by the movement of air is transferred to airborne particles, with resultant movement from one point to another. This "energy of motion" gained by airborne particles is considerable, and can result in dissemination of airborne microorganisms over long distances. Transport of bioaerosols can be defined in terms of time and distance. Submicroscale transport involves short periods of time, under 10 minutes, as well as relatively short distances, under 100 m. This type of transport is common within buildings or other confined spaces. Microscale transport ranges from 10 minutes to 1 hour, and from 100 m to 1 km, and is the most common type of transport phenomenon. Mesoscale transport refers to transport in terms of days and distances up to 100 km, and in macroscale transport, the time and distances are extended even further. Because most microorganisms have limited ability to survive when suspended in the atmosphere, the most common scales considered are the submicroscale and microscale. It should be noted, however, that some viruses, spores and spore-forming bacteria have been shown to enter into mesoscale and even macroscale transport.

As bioaerosols travel through time and space, different forces act upon them such as diffusion, inactivation and ultimately deposition. Diffusion is the scattering and/or dissipation of bioaerosols in response to a concentration gradient as well as gravity, and is generally aided by airflow and atmospheric turbulence. The amount of turbulence associated with airflow, and thus the relative amount of diffusion that may occur in association with particulates such as bioaerosols, can be estimated using the method of Osbert Reynolds. Reynolds found that factors associated with mean wind velocity, the kinetic viscosity of the air and the relative dimension of the interfering structures could provide an indication of the amount of turbulence associated with linear airflow. Without turbulence, airborne particles from a point source would travel in a concentrated stream directly downwind. The Reynolds equation is written as follows:

$$\text{Reynolds number} = \frac{\text{velocity} \times \text{dimension}}{\text{viscosity}} \quad \text{(Eq. 5.3)}$$

Consider, for instance, a situation in which there are relatively high winds (500 cm/sec) that are passing over a small bush (24 cm). Because the occurrence of frictional turbulence associated with an object depends on the wind velocity being high enough, and the object it is flowing over being large enough, we find that at normal air viscosity (0.14 cm^2/sec) the Reynolds number (Re) becomes:

$$\text{Re} = \frac{500 \text{ cm/sec} \times 24 \text{ cm}}{0.14 \text{ cm}^2/\text{sec}} = 85,700 \quad \text{(Eq. 5.4)}$$

The limiting value for the Reynolds equation is usually considered to be 2000, with values above this number indicating turbulent conditions. The higher this value, the higher the relative turbulence of the airflow, and the greater the microorganism-associated particle diffusion that occurs per unit time. In the preceding example, one would expect a great deal of turbulence around items such as a bush, which would increase the diffusion rates of passing bioaerosols.

When dealing with particulate transport over time and distance, Tayler (1915) indicated that diffusion during horizontal transport could be viewed as an increase in the standard spatial deviation of particles from the source over time. What does this mean? For an instantaneous point source under the influence of a mean wind direction, spread would be a standard spatial deviation from a linear axis (x) extending from the source (origin) in the mean direction of wind flow, with diffusion caused by turbulence occurring in the lateral (y) and vertical (z) axes (Figure 5.4). The standard deviation of particulate diffusion cannot be considered constant over a particular spatial orientation, but is instead dependent on the time taken to reach the particular distance. Mathematical models that attempt to estimate the transport of airborne particles use this basic premise as a foundation for predictions. To picture this concept, imagine standing at the door of a room,

where someone is holding a smoking candle. If there is no air current in the room the smoke will still eventually reach you at the door, but it will be very diffuse as it is also spreading in every other direction. However, if there is a fan behind the person holding the smoking candle and this fan is pointed at the door, then the smoke from the candle will be carried by this air current. It will travel the same distance as it did before, but it will travel faster, undergo less diffusion and as a result be more concentrated when it reaches you. This is the principle of time-dependent diffusion as indicated by Tayler's theory.

5.4.3 Deposition

The last step in the aeromicrobiology pathway is deposition. An airborne bioaerosol will eventually leave the turbulence of the suspending gas and will ultimately be deposited on a surface by one or a combination of interrelated mechanisms. These mechanisms are discussed in the following sections and include: gravitational settling; downward molecular diffusion; surface impaction; rain deposition; and electrostatic deposition. These processes are linked in many ways, and even though viewed separately, they all combine to create a constant, if not steady, deposition of particles.

5.4.3.1 Gravitational Settling

The main mechanism associated with deposition is the action of gravity on particles. The force of gravity acts upon all particles heavier than air, pulling them down and essentially providing spatial and temporal limitations to the spread of airborne particles. Steady-state gravitational deposition (Figure 5.6) in the absence of air movement can be described in very simplistic terms by Stokes' law, which takes into account gravitational pull, particle density, particle diameter and air viscosity (Section 5.2).

5.4.3.2 Downward Molecular Diffusion

Downward molecular diffusion, as indicated by the name, can be described as a randomly occurring process caused by natural air currents and eddies that promote and enhance the downward movement of airborne particulates (Figure 5.7). These random movements exist even in relatively still air and tend to be in the downward direction because of gravitational effects. As a result, measured rates of gravitational deposition tend to be greater than those predicted by the Stokes equation. The increase in the rate of deposition is due to the added effects of downward molecular diffusion. Molecular diffusion is also influenced by the force of the wind. Molecular diffusion-enhanced deposition rates tend to increase with increasing wind speed and turbulence.

FIGURE 5.6 Schematic representation of gravitational settling, which is a function of Earth's gravitational pull, particle density, particle diameter and the viscosity of air. This figure does not take into account random air movement. Stokes' equation was developed to give an estimate of the terminal velocity achieved by particles as a function of gravitational settling.

FIGURE 5.7 Schematic representation of downward molecular diffusion, a naturally occurring process caused by the air currents and eddies that promotes and enhance gravitational settling of airborne particles. Although molecular diffusion can occur in any direction, due to the effects of gravity the overall trend of the process results in net downward movement and deposition.

5.4.3.3 Surface Impaction

Surface impaction is the process by which particles make contact with surfaces, such as leaves, trees, walls and computers. With impaction there is an associated loss of kinetic energy. In nature, it is rare to find flat, smooth surfaces on which wind currents are unobstructed. Thus, surface impaction is a very critical factor influencing transport and deposition, especially for bioaerosols.

Impaction potential is the relative likelihood that an airborne object will collide with another object in its path. Impaction does not necessarily result in permanent deposition, however. Once a particle collides with an object, it has the potential to bounce. Bouncing off a surface causes the particle to reenter the air current at a lower rate, which can have one of two effects: (1) it can allow subsequent downward molecular diffusion and gravitational settling to occur, resulting in deposition on another nearby surface; or (2) it can allow the particle to escape the surface and once again reenter the air current. Studies have shown that impaction is influenced by the velocity and size of the particle, as well as the size and shape of the surface it is approaching.

5.4.3.4 Rain and Electrostatic Deposition

Rainfall and electrostatic charge also can affect deposition. Rainfall deposition occurs as a condensation reaction between two particles (raindrop and bioaerosol), which combine and create a bioaerosol with a greater mass, which settles faster. This can be described mathematically using the Stokes equation. In the example presented in Information Box 5.1, a clostridial spore alone has a calculated terminal velocity of 0.016 cm/sec. The same spore (bioaerosol), if it condensed with another particle such as a water droplet, has a greater mass and thus a greater terminal velocity. For instance, if the clostridial spore were to condense with a water droplet that doubled the bioaerosol density from 1.3 to 2.6 g/cm^3, the terminal velocity would be increased from 0.016 to 0.032 cm/sec. The overall efficiency of rain deposition also depends on the spread area of the particle plume. Larger, more diffuse plumes undergo stronger impaction than smaller, more concentrated plumes. Rain deposition is also affected by the intensity of the rainfall. The heavier the rainfall, the greater the overall rates and numbers of the condensation reactions, and the greater the subsequent increase in rain deposition.

Electrostatic deposition also condenses bioaerosols, but is based on electrovalent particle attraction. All particles tend to have some type of associated charge. Microorganisms typically have an overall negative charge associated with their surfaces at neutral pH. These negatively charged particles can associate with other positively charged airborne particles, resulting in electrostatic condensation. The major phenomenon occurring may be a coagulation effect between particles (much like the condensation of the clostridial spore with the water droplet), which would increase the bioaerosol mass and enhance deposition. It might also be assumed that as an electromagnetically charged bioaerosol comes into close proximity with an electromagnetically charged surface, electroattractive or electrorepulsive influences may be present.

5.5 MICROBIAL SURVIVAL IN THE AIR

The atmosphere is an inhospitable climate for microorganisms mainly because of desiccation stress. This results in a limited time frame in which microbes can remain biologically active. Many microorganisms, however, have specific mechanisms that allow them to be somewhat resistant to the various environmental factors that promote loss of biological activity. Spore-forming bacteria, molds, fungi and cyst-forming protozoa all have specific mechanisms that protect them from harsh gaseous environments, increasing their ability to survive aerosolization. For organisms that have no such specific mechanisms, the survival in aerosols can often be measured in seconds. In contrast, organisms with these mechanisms can survive indefinitely.

As a result, viability is highly dependent on the environment, the amount of time the organism spends in the environment and the type of microorganism. In addition, microbes may be viable but nonculturable (Chapter 3), but for simplicity in this chapter we will use the term viable rather than the term culturable. Many environmental factors have been shown to influence the ability of microorganisms to survive. The most important of these are relative humidity and temperature. Oxygen content, specific ions, UV radiation, various pollutants and AOFs (air-associated factors) are also factors in the loss of biological activity. Each of these factors is discussed in the following sections.

The loss of biological activity can be termed inactivation and can generally be described using the following equation:

$$X_t = X_0 e^{-kt} \qquad \text{(Eq. 5.5)}$$

where:

X_t represents the viable organisms at time t
X_0 is the starting concentration
k is the inactivation constant, which is dependent on the particular species of microorganisms as well as a variety of environmental conditions

5.5.1 Relative Humidity

The relative humidity or the relative water content of the air has been shown to be of major importance in the survival of airborne microorganisms. Wells and Riley (1937) were among the first to show this phenomenon, indicating that as the relative humidity approaches 100%, the death rate of Escherichia coli increases. In general, it has been reported that most Gram-negative bacteria associated with aerosols tend to survive for longer periods at low to mid levels of relative humidities, with enhanced decay at relative humidities above 80% (Brooks et al., 2004). The opposite tends to be true for Gram-positive bacteria, which tend to remain viable longer in association with high relative humidities (Theunissen et al., 1993). Thus, the ability of a microorganism to remain viable in a bioaerosol is related to the organism's surface biochemistry. One mechanism that explains loss of viability in association with very low relative humidity is a structural change in the lipid bilayers of the cell membrane. As water is lost from the cell, the cell membrane bilayer changes from the typical crystalline structure to a gel phase. This structural phase transition affects cell surface protein configurations and ultimately results in inactivation of the cell (Hurst et al., 1997). In general, Gram-negative bacteria react unfavorably to desiccation, whereas Gram-positive cells are more tolerant of desiccation stress (Mohr, 2001).

Early studies by Loosli et al. (1943) showed that the influenza virus was also adversely affected by an increase in relative humidity. More recent work suggests that viruses possessing enveloped nucleocapsids (such as the influenza virus) have longer airborne survival when the relative humidity is below 50%, whereas viruses with naked nucleocapsids (such as the enteric viruses) are more stable at a relative humidity above 50% (Mohr, 2001). It should be noted that viruses with enveloped nucleocapsids tend to have better survival in aerosols than those without. Some viruses are also stable in the AMB pathway over large ranges of relative humidity, which makes them very successful airborne pathogens.

5.5.2 Temperature

Temperature is a major factor in the inactivation of microorganisms. In general, high temperatures promote inactivation, mainly associated with desiccation and protein denaturation, and lower temperatures promote longer survival times (Mohr, 2001). When temperatures approach freezing, however, some organisms lose viability because of the formation of ice crystals on their surfaces. The effects of temperature are closely linked with many other environmental factors, including relative humidity.

5.5.3 Radiation

The main sources of radiation damage to microorganisms including bacteria, viruses, fungi and protozoa are the shorter UV wavelengths and ionizing radiation such as

X-rays. The main target of UV irradiation damage is the nucleotides that make up DNA. Ionizing radiation or X-rays cause several types of DNA damage, including single strand breaks, double strand breaks and alterations in the structure of nucleic acid bases. UV radiation causes damage mainly in the form of intrastrand dimerization, with the DNA helix becoming distorted as thymidines are pulled toward one another (Freifelder, 1987). This in turn causes inhibition of biological activity such as replication of the genome, transcription and translation.

Several mechanisms have been shown to protect organisms from radiation damage. These include association of microbes with larger airborne particles, possession of pigments or carotenoids, high relative humidity and cloud cover, all of which tend to absorb or shield bioaerosols from radiation. Many types of organisms also have mechanisms for repair of the DNA damage caused by UV radiation. An example of an organism that has a radiation resistance mechanism is *Dienococcus radiodurans*. *D. radiodurans* is a soil bacterium that is considered the most highly radiation-resistant organism that has yet been isolated. An important component of its radiation resistance is the ability to enzymatically repair damage to chromosomal DNA. The repair mechanism used by these bacteria is so highly efficient that much of the metabolic energy of the cell is dedicated exclusively to this function.

5.5.4 Oxygen, OAF and Ions

Oxygen, open air factors (OAFs) and ions are environmental components of the atmosphere that are difficult to study at best. In general, it has been shown that these three factors combine to inactivate many species of airborne microbes. Oxygen toxicity is not related to the dimolecular form of oxygen (O_2), but is instead important in the inactivation of microorganisms when O_2 is converted to more reactive forms (Cox and Heckley, 1973). These include superoxide radicals, hydrogen peroxide and hydroxide radicals. These radicals arise naturally in the environment from the action of lightning, UV radiation, pollution, etc. Such reactive forms of oxygen cause damage to DNA by producing mutations, which can accumulate over time. The repair mechanisms described in the previous section are responsible for control of the damaging effects of reactive forms of oxygen.

Similarly, the open air factor (OAF) is a term coined to describe an environmental effect that cannot be replicated in laboratory experimental settings. It is closely linked to oxygen toxicity, and has come to be defined as a mixture of factors produced when ozone and hydrocarbons (generally related to ethylene) react. For example, high levels of hydrocarbons and ozone can cause increased inactivation rates for many organisms, probably because of damaging effects on enzymes and nucleic acids (Donaldson and Ferris, 1975). Therefore, OAFs have been strongly linked to microbial survival in the air.

The formation of other ions, such as those containing chlorine, nitrogen or sulfur, occurs naturally as the result of many processes. These include the action of lightning, shearing of water and the action of various forms of radiation that displace electrons from gas molecules, creating a wide variety of anions and cations not related to the oxygen radicals. These ions have a wide range of biological activity. Positive ions cause only physical decay of microorganisms, e.g., inactivation of cell surface proteins, whereas negative ions exhibit both physical and biological effects such as internal damage to DNA.

5.6 EXTRAMURAL AEROMICROBIOLOGY

Extramural aeromicrobiology is the study of microorganisms associated with outdoor environments. In the extramural environment, the expanse of space and the presence of air turbulence are two controlling factors in the movement of bioaerosols. Environmental factors such as UV radiation, temperature and relative humidity modify the effects of bioaerosols by limiting the amount of time that aerosolized microorganisms will remain viable. This section provides an overview of extramural aeromicrobiology that includes: aerosolization of indigenous soil pathogens; influenza pandemics; the spread of agricultural pathogens; the spread of airborne pathogens associated with waste environments; and important airborne toxins.

5.6.1 Aerosolization of Indigenous Soil Pathogens

Geo-indigenous pathogens are those found in soils that are capable of metabolism, growth and reproduction (Pepper *et al.*, 2009). These are found in all soils and include both prokaryotic and eukaryotic organisms, many of which are spore formers. Such spores can potentially be aerosolized and cause human infections. *Bacillus anthracis* is a bacterial geo-indigenous pathogen that causes lethal disease in humans via pulmonary, gastrointestinal or cutaneous modes of infection (Gentry and Pepper, 2002). The organism is found worldwide and, because it is a spore former, can remain viable in soil for years.

Studies have shown the potential for anthrax to be disseminated by aerosols. Turnbull *et al.* (1998) found airborne concentrations of anthrax spores as high as 2.1×10^{-2} CFU L^{-1} of air, and airborne movement as far as 18 m from a contaminated carcass in Etosha National Park, Namibia. However, the majority of samples taken were negative, and the number of spores collected in positive samples was very low, making airborne contraction of disease at a distance from the carcass unlikely. A more serious outbreak in humans resulting from a *B. anthracis* aerosol is described in Case Study 5.1.

Important fungal geo-indigenous pathogens include *Coccidioides immitis* and *Histoplasma capsulatum*. *Coccidiodes immitis* is a soil-borne fungi that causes a respiratory illness known as Valley Fever. It preferentially

grows in the soils of semiarid regions of the Southwest United States, including California, Arizona, New Mexico and Texas (Baptista-Rosas *et al.*, 2007). Symptoms can be mild to fatal. *Histoplasma capsulatum*, another fungus causing respiratory infections, is found worldwide in soils, but, in the United States, it is endemic to southeastern and midwestern states (Deepe and Gibbons, 2008). Histoplasmosis can be asymptomatic or mild, but the infections can be very serious or even fatal for immuno-compromised individuals.

Case Study 5.1   Anthrax

In 1979, an anthrax outbreak occurred in Sverdlovsk, in the then U.S.S.R., due to the accidental release of a bioaerosol from a military microbiological facility (Meselson *et al.*, 1994). At least 66 people died as a result of the release. Human anthrax cases extended 4 km along an axis to the south of the military facility and livestock cases extended up to 50 km in the same direction. The geographic distribution of human and animal cases was consistent with meteorological patterns existing when the accidental release was believed to have occurred. There has been no indication that human anthrax cases have occurred in Sverdlovsk since 1979.

Case Study 5.2   The Spanish Influenza Pandemic of 1918

This pandemic affected approximately one-third of the world population at that time, with 3−6% dying (Barry, 2004). The pandemic lasted from January 1918 to December 1920, the responsible virus being H1N1. This was the first outbreak resulting from H1N1, with the second epidemic occurring in 2009. Although the pandemic did not originate in Spain, the term "Spanish flu" was coined due to the severity of the infections in Spain. It is believed that the pandemic began in Haskell County, Kansas, before spreading rapidly to Europe. Estimates of the total number of deaths range from 50 to 100 million worldwide with 500,000 to 675,000 deaths in the U.S.A. (Barry, 2004).

5.6.2 Influenza Pandemics

Influenza pandemic is the term given to an epidemic of an influenza virus that occurs on a worldwide scale with a resultant infection of a large proportion of the human population. Known colloquially as the "flu," influenza is an infectious disease of birds and mammals caused by an RNA virus of the family *Orthomyxoviridae*. Influenza can cause the common flu symptoms of muscle ache, headache, coughing, weakness and fatigue, or pneumonia which can be fatal.

Avian influenza refers to a large group of influenza viruses that primarily affect birds, but have the potential to adapt and infect humans. An influenza pandemic occurs when an avian influenza virus adapts into a strain that is contagious among humans and that has not previously circulated within humans. Such adaptations can be devastating, as illustrated in Table 5.1.

Influenza virus transmission among humans can occur via four mechanisms: by direct contact with infected individuals; by indirect contact with contaminated objects of fomites; by inhalation of droplets that contain the virus; or by inhalation of aerosolized virus. Interestingly, despite 70 years of research since the influenza A virus was discovered, there is still debate about the modes of influenza transmission, specifically whether influenza is mainly transmitted via true bioaerosols, or by droplets, or by direct or indirect contact (Brankston *et al.*, 2007).

TABLE 5.1 History of Major Influenza Pandemics

Name of Pandemic	Period	Deaths	Influenza Subtype
Asiatic (Russian) flu	1889−1890	1 million	Unknown
Spanish flu	1918−1920	Up to 50 million	H1N1
Asian flu	1957−1958	2 million	H2N2
Hong Kong flu	1968−1969	1 million	H3N2
Swine flu	2009−2010	≅18,000	H1N1

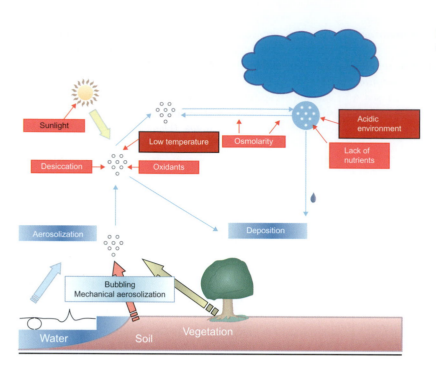

5.6.3 Microbiology in the Clouds

Recent studies have suggested that microbes can potentially affect meteorological processes. In particular, some microorganisms, called ice nucleators, efficiently catalyze ice formation and may play a role in the formation and precipitation within clouds (Chistner, 2012). Based on recent studies, 95% of ice nuclei are biological particles and at least 40% originate from bacteria. Microorganisms are present in both clouds and fog. The abundance of culturable bacteria and fungi in clouds varies with the season, with greater numbers occurring in the summer and fall. While only 1% of the bacteria and 50% of the fungi in clouds are culturable, studies suggest that the majority are metabolically active (Delort et al., 2010). Bacterial numbers range from 10^3 to 10^4/ml compared to fungal numbers of 10^2 to 10^4/ml. The cloud environment is a harsh environment with UV light irradiation, desiccation, low temperatures and other factors potentially adversely affecting microbes (Figure 5.8). Microorganisms may modify this environment by metabolizing organic compounds, and also by playing a role in cloud chemistry and physics, but much additional research is needed because the cloud environment is difficult to study.

5.6.4 Agriculture

Numerous plant pathogens are spread by the aeromicrobiological pathway (Information Box 5.3). Contamination of crops and animals via bioaerosols has a large worldwide economic impact. Rice and wheat are two of the major staple crops that are paramount to world food security. Major pathogens of such crops are the wheat rust fungi. These spore-forming fungi cause some of the most devastating diseases of wheat and other grains. In 1993, one type of wheat rust (leaf rust) was responsible for the loss of over 40 million bushels of wheat in Kansas and Nebraska alone. Even with selective breeding for resistance in wheat plants, leaf rust continues to have major economic impacts. The high concentration of wheat in areas ranging from northern Texas to Minnesota and up into the Dakotas makes this whole region highly susceptible to rust epidemics.

Spores of wheat rust are capable of spreading hundreds if not thousands of kilometers through the atmosphere (Ingold, 1971). The airborne spread of rust disease has been shown to follow a predictable trend, which starts during the fall with the planting of winter wheat in the southern plains. Any rust-infected plant produces thousands of spores, which are released into the air (Figure 5.9) by either natural atmospheric disturbance or mechanical disturbance during the harvesting process. Once airborne, these spores are capable of long-distance dispersal, which can cause downwind deposition onto other susceptible wheat plants. The generation time of new spores is measured in weeks, after which new spores are again released from vegetative fungi into the AMB pathway. For example, during the harvest of winter wheat in Texas, the prevailing wind currents are from south to north, which can allow rust epidemics to spread into the maturing crops farther north in Kansas and up into the young crops in the Dakotas (Figure 5.10). This epidemic spread of wheat rust and the resulting economic destruction produced are

Information Box 5.3   Examples of Airborne Plant Pathogens

Fungal Plant Disease	Pathogen
Dutch Elm disease	*Ceratocystis ulmi*
Potato late blight	*Phytophthora infestans*
Leaf rust	*Puccinia recondite*
Loose smut of wheat	*Ustilago tritici*
Downy mildew	*Pseudoperonospora humuli*
Maize rust	*Puccinia sorghi*
Powdery mildew of barley	*Erysiphe graminis*
Southern corn leaf blight	*Helminthosporium maydis*

The figure shows the airborne spread of late blight of potato that caused the 1845 epidemic known as the Irish potato famine. *Phytophthora infestans* spread from Belgium (mid-June) throughout Europe by mid-October. Famine related deaths are estimated from 750,000 to 1,000,000. Economic devastation from this famine caused the population of Ireland to decrease from approximately 8 million to 4 million from 1840 to 1911.

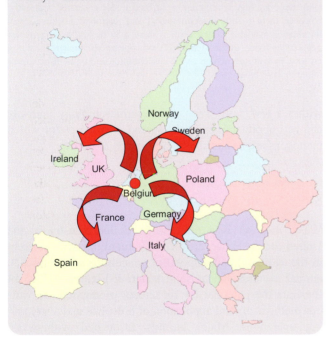
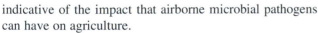

FIGURE 5.9   Field of wheat highly infected by phytopathogenic wheat rust. The field is being harvested by a hay machine, which is releasing a cloud of rust spores into the aeromicrobiological pathway. These spores can spread thousands of miles and infect other crops downwind, causing catastrophic losses to wheat crops.

FIGURE 5.10   The arrow indicates the northern path of wheat rust infections as spread by the aeromicrobiological pathway. The wheat rust infection begins in the winter harvest in Texas and spreads northward with the prevailing wind currents. The epidemic spread of these phytopathogens infects maturing crops in Kansas and then moves up into the young crops in the Dakotas.

indicative of the impact that airborne microbial pathogens can have on agriculture.

A factor that complicates the control of such diseases is that chemical treatment for the control of pathogens is viewed as undesirable. This is because many pesticides have long half-lives and their residence in an ecosystem can be extremely harmful. Therefore, instead of using wheat rust fungicides, attempts are being made to breed strains of wheat that are more resistant to the fungi. Another method used for controlling phytopathogenic (plant pathogenic) fungi is spore monitoring as a disease control strategy. In this approach, the life cycle of the fungi, especially the release of spores, is monitored, and

fungicide application is timed to coincide with spore release. This approach minimizes use of harmful chemicals. Thus, efficient aeromicrobiology pathway sampling, monitoring, detection and modeling have the ability to aid in the control of airborne pathogens.

The airborne spread of pathogenic microorganisms is also highly important in the animal husbandry industry (Information Box 5.4). The occurrence of foot-and-mouth disease is an example of the importance of bioaerosols in the spread of airborne disease (Case Study 5.3). It has long been thought that bioaerosol spread is linked primarily to respiratory pathogens, but there is growing evidence that gastrointestinal pathogens are also important in airborne transmission of disease among animals. One example of bioaerosol spread of a gastrointestinal pathogen is

Information Box 5.4 Examples of Airborne Animal Pathogens

Animal Disease	Pathogen
Bacterial diseases	
Tuberculosis	*Mycobacterium bovis*
Brucellosis	*Brucella* spp.
Fungal diseases	
Aspergillosis	*Aspergillus* spp.
Coccidioidomycosis	*Coccidioides immitis*
Viral diseases	
Influenza	Influenza virus
Rabies	Rhabdoviridae
Foot-and-mouth disease	Aphthovirus

Case Study 5.3 The United Kingdom Foot and Mouth Crisis 2001

Outbreaks of foot and mouth disease have occurred worldwide including multiple occurrences within the U.S.A. However, one of the more devastating outbreaks occurred in the U.K. in the spring and summer of 2001. Foot and mouth viruses are from the *Aphthovirus* genus of the family Picornaviridae, and are single-stranded RNA viruses. The 2001 U.K. outbreak was due to a type O pan-Asia strain that was thought to have arisen from infected meat illegally imported into the U.K. Transmission of the virus can occur via direct contact or via bioaerosols (Grubman and Baxt, 2004). Overall, 2000 cases were reported throughout Britain, resulting in the culling of 7 million sheep and cattle, costing the U.K. ≅ $16 million.

FIGURE 5.11 Application of secondary treated wastewater onto agricultural lands. This method is highly efficient at conserving water and has been shown to improve the fertility of soils. Due to the presence of pathogens in wastewater, and the nature of these land application systems, there are high concentrations of bioaerosols generated. Currently, however, there is little epidemiological and microbial risk assessment information available to determine if there may be health concerns for populations living in the vicinity of such operations, though there is a growing base of information on the concentration and types of pathogens found in these bioaerosols.

within a few days. Thus, the aeromicrobiology pathway can be important even in the spread of diseases for which pathogens are not normally considered airborne.

5.6.5 Waste Disposal

Waste disposal is a multibillion dollar industry in the United States. However, there are many hazards inherent in the treatment and disposal of wastewater (Figure 5.11), animal manures and biosolid material. Figures 5.12−5.14 illustrate the potential for bioaerosol production via various methods of land application of biosolids and also loading operations. Major hazards associated with waste effluents are pathogenic microorganisms including bacteria, viruses, protozoa and helminths. Wastewater treatment plants utilize activated sludge and trickling filter systems, and all of these treatment processes potentially create relatively large amounts of aerosols, which have been shown to include pathogenic microorganisms. Other aspects of the treatment process such as composting and land disposal are also associated with the generation of aerosols containing pathogenic microorganisms.

One of the primary methods for the disposal of biosolids and manure is agricultural land application. The major concern associated with the aerosolization process in relation to waste disposal operations is the exposure of waste disposal workers to pathogenic microorganisms (occupational risk). In addition, nearby population centers are also potential

transmission of *Salmonella typhimurium* among calves that are housed individually in small pens (Hinton *et al.*, 1983). The potential for bioaerosol spread of this pathogen was recognized because the initial symptoms resembled those of pneumonia and appeared randomly within these animals, two factors that are not characteristic of oral transmission. Oral transmission generally occurs sequentially from one pen to the next, whereas aerial transmission can carry organisms past nearby pens, infecting calves randomly. Furthermore, Wathes *et al.* (1988) showed that *S. typhimurium* could survive for long periods in an airborne state, and calves and mice exposed to aerosolized *S. typhimurium* developed symptoms, proving that gastrointestinal pathogens could be spread by aerosolization. Finally, Baskerville *et al.* (1992) showed that aerosolized *Salmonella enteritidis* could infect laying hens. These hens showed clinical symptoms and were shedding the test strain of salmonellae in their feces

FIGURE 5.12 Land application of liquid biosolids via a spray application, and collection of air samples via biosamplers.

FIGURE 5.13 Land application of liquid biosolids via a sprinkler system.

FIGURE 5.14 Land application of cake biosolids via a slinger in Solano County, California.

exposure risks (community risk). The potential for aerosolization of pathogens from land application of biosolids has become a nationally debated issue. A major national study on aerosolization from land application in the United States was conducted by Brooks *et al.* (2005a,b). This study showed that occupational risks of infection from bioaerosols was greater than for offsite communities, where risks were minimal (Brooks *et al.*, 2012) (Case Study 5.4). Baertsch *et al.* (2007) used DNA-based microbial source tracking to measure aerosolization during land application.

5.6.6 Important Airborne Toxins

Microbial toxins can also be airborne. For example, a toxin from *Clostridium botulinum* (botulinum A toxin) is a potential biological warfare agent (Amon *et al.*, 2001). Botulinum toxin is a neurotoxin that is normally associated with ingestion of contaminated food. However, the lethal dose is so small that aerosolization can also be a means of dissemination. The lethal dose for botulinum toxin by inhalation is 0.3 μg, with death occurring 12 hours after exposure. Death is due to asphyxiation caused by the paralysis of respiratory muscles. Another toxin produced by bacteria is staphylococcal enterotoxin. On occasion, this toxin can be fatal with the lethal dose estimated to be 25 μg by inhalation. The symptoms include cramping, vomiting and diarrhea, which occur within 1 hour of exposure by aerosolization.

An important airborne toxin is lipopolysaccharide (LPS) (Hurst *et al.*, 1997). Lipopolysaccharide is derived from the outer membrane of Gram-negative bacteria. It is also referred to as endotoxin and is a highly antigenic biological agent that, when associated with airborne particles such as dust, is often associated with acute respiratory symptoms such as chest tightness, coughing, shortness of breath, fever and wheezing. Due to the ubiquity of Gram-negative bacteria, especially in soil, LPS is considered by some to be the most important aerobiological allergen. LPS (Figure 5.15) has three major components: a lipid A moiety, which is a disaccharide of phosphorylated glucosamines with associated fatty acids; a core polysaccharide; and an O-side chain. The lipid A moiety and the core polysaccharide are similar among Gram-negative bacteria, but the O-side chain varies among species and even strains. It is the O-side chain that is responsible for the hyperallergenic reaction. There are many sources associated with the production of high levels of LPS, such as cotton mills, haystacks, sewage treatment plants, solid waste handling facilities, swine confinement buildings, poultry houses, and even homes and office buildings. LPS is liberated when Gram-negative bacteria in these environments are lysed but can also be released when they are actively growing.

In soils, bacterial concentrations routinely exceed 10^8 per gram and soil particles containing sorbed microbes can

---

Case Study 5.4   Occupational and Community Risks of Infection from Bioaerosols Generated During Land Application of Biosolids

In 2002, the National Research Council (NRC) issued a report titled, *Biosolids Applied to Land: Advancing Standards and Practices*. One of the recommendations of this report was the need to document and evaluate the risk of infections from bioaerosols generated during land application. In response to this the University of Arizona conducted a national study to evaluate occupational and community risks from such bioaerosols. Overall, more than 1000 aerosol samples were collected and analyzed for bacterial and viral pathogens (Tanner *et al.*, 2005; Brooks *et al.*, 2005b).

The study was undertaken in two parts. First, the emission rate of pathogens generated during loading of biosolids from trucks into spreaders, and also during land application of biosolids, was evaluated. This was assumed to be direct exposure to biosolid workers on-site, that is, occupational risk, since there was no pathogen transport required to for exposure. For community risk, fate and transport of pathogens was taken into account, since residents live off-site, allowing for natural attenuation of pathogens due to environmental factors such as desiccation and ultraviolet light.

Based on exposure data gathered on-site during land application, occupational risk of infection from Coxsackie virus A21 was determined using a one-hit exponential model: $P_i = 1 - \exp(-rN)$, where:

P_i = the probability of infection per work day,

r = parameter defining the probability of an organism initiating infection = 0.0253 for Coxsackie A21, and

N = number of pathogens inhaled per day

The annual risk of infection can be calculated from the daily risk using

$$P_{year} = 1 - (1 - P_i)^d$$

where d = number of days exposed per year.

Occupational Risk of Infection

Annual risk of infection for Coxsackie virus A21 during loading operations was 2.1×10^{-2}.

This risk suggests that approximately 1 worker per 50 is likely to be infected with Coxsackie virus A11 working on-site over the course of 1 year.

Community Risk of Infection

Risk was calculated for a distance of 30 m from a land application site assuming 6 days of land application annually, and 8 h exposure duration.

Annual risk of infection for Coxsackie virus A21 during loading was 3.8×10^{-5}.

Annual risk of infection for Coxsackie virus A21 during land application was 2.1×10^{-5}.

These data imply that community risks of infection are minimal. As a comparison, for drinking water a 1:10,000 risk of infection per hear is considered acceptable (Haas *et al.*, 2014).

---

be aerosolized, and hence act as a source of endotoxin. Farming operations such as driving a tractor across a field have been shown to result in endotoxin levels of 469 endotoxin units (EU) per cubic meter as measured by the *Limulus* amebocyte assay. These values are comparable to those found during land application of biosolids operations (Table 5.2) (Brooks *et al.*, 2006). Daily exposures of as little as 10 EU/m^{-3} from cotton dust can cause asthma and chronic bronchitis. However, dose response is dependent on the source of the material, the duration of exposure and the frequency of exposures (Brooks *et al.*, 2004). The data in Table 5.2 illustrate that endotoxin aerosolization can occur during both wastewater treatment and land application of biosolids. However, the data also show that endotoxin of soil origin resulting from dust generated during tractor operations results in similar amounts of aerosolized endotoxin (see also Section 26.3.2).

5.7 INTRAMURAL AEROMICROBIOLOGY

The home and workplace are environments in which airborne microorganisms create major public health concerns.

In comparison with the extramural environment, intramural environments have limited circulation of external air and much less UV radiation exposure. Indoor environments also have controlled temperature and relative humidity, which are generally in the ranges that allow extended microbial survival. Thus, these conditions are suitable for the accumulation and survival of microorganisms within many enclosed environments, including office buildings, hospitals, laboratories and even spacecraft. In this section, we will consider these three diverse areas as examples of current topics related to intramural aeromicrobiology. Again, it should be noted that this section does not cover all aspects of intramural aeromicrobiology, but instead attempts to show the wide diversity of the science.

5.7.1 Buildings

Many factors can influence bioaerosols and therefore how "healthy" or how "sick" a building is. These include: the presence and/or efficiency of air filtering devices, the design and operation of the air circulation systems, the health and hygiene of the occupants, the amount of clean outdoor air

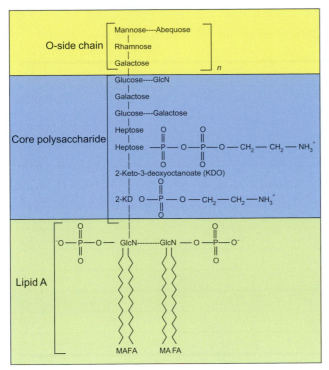

FIGURE 5.15 Schematic structural representation of the lipopolysaccharide from *Salmonella typhimurium*. The number of repeating units (*n*) in the side chain varies from 10 to 40. The sugars found in the side chain vary among bacterial species, whereas the composition of the core polysaccharide is usually the same. There is a molecule of beta-hydroxymyristic acid (MA), which is a14-carbon fatty acid attached to each N-acetylglucosamine (GlcN) residue. Other fatty acids (FA) are attached to these residues as well.

circulated through the building, the type of lighting used, the ambient temperature in the building, and the relative humidity (Information Box 5.5).

Some pathogens are uniquely adapted for survival and transmission in the intramural environment. One good example of such an organism is *Legionella pneumophilia*, the causative agent of both Legionnaires' disease and Pontiac fever. Legionnaires' disease or legionellosis is a pneumonia that causes disease in up to 5% of those exposed. Of those who contract the disease, up to 39% die from the infection. Pontiac fever is associated with flu-like symptoms and affects up to 100% of those exposed, although it is generally not associated with mortality. The causative agent of both diseases is a poorly staining, Gram-negative bacillus called *L. pneumophila*. This organism is named in association with the first highly characterized outbreak of the disease, which occurred in 1976 at an American Legion convention in Philadelphia.

Legionella spp. are ubiquitous in the environment. They are found in association with lakes, ponds, compost and streams, and have even been found in deep terrestrial subsurface environments. In addition to natural reservoirs, there are many human-made systems within which

legionellae can find a niche. These include cooling towers, evaporative condensers, plumbing systems, whirlpools, shower heads and hot-water faucets (Bollin *et al.*, 1985). In the case of the American Legion convention, the reservoir for the organism that caused the outbreak was a poorly maintained cooling tower, which provided optimal conditions for *Legionella* proliferation. Because of the poor design of the air circulation system at the convention, this proliferation led to the subsequent aerosolization and spread of the organisms throughout the building.

What conditions promote the proliferation of *Legionella* spp.? Stagnant water and temperatures in the range of 35–46°C are factors that can lead to the rapid multiplication of background levels of *Legionella* spp. Another interesting aspect of the ecology of *Legionella* is that they can grow intracellularly within cyanobacteria and protozoa. How can growth and spread of *Legionella* spp. be avoided? Several strategies can be used. In the maintenance of hot-water plumbing systems, operating temperatures should be greater than 50°C. All potential places where water can stagnate in water pipes should be avoided. For cooling towers, the recommendations involve the installation of ozonization units, dry convective heat exchange designs and the avoidance of any design that could potentially mix the wet system with the supply air. Biocidal agents such as chlorine or copper can also be effective when used regularly at low levels.

5.7.2 Hospitals and Laboratories

Hospitals and microbiology laboratories are the two indoor environments with perhaps the greatest potential for the aerosolization of pathogenic microorganisms. Hospitals, because they are centers for the treatment of patients with diseases, have a high percentage of individuals, including patients and staff, who are active carriers of infectious, airborne pathogens. Of particular concern are neonatal wards, surgical transplant wards and surgical theaters, all critical areas where the control of nosocomial infection is imperative. Illustrating this point is a study by Portner *et al.* (1965) that evaluated airborne microbial concentrations in surgical theaters, industrial clean rooms, typical industrial manufacturing areas and a horizontal laminar flow clean room designed for the space industry. The surgical theater had by far the highest counts of pathogenic airborne microbial contaminants, followed by the industrial manufacturing area, the industrial clean room and finally the laminar flow room, which had the lowest counts of airborne microbes.

Because microbiology laboratories often handle pathogens, procedures have been developed and refined to protect laboratory workers. However, even under the strictest of conditions, aerosolization events may occur. In 1988, for

TABLE 5.2 Aerosolized Endotoxin Concentrations Detected Downwind of Biosolids Operations, a Wastewater Treatment Plant Aeration Basin, and a Tractor Operation

Sample Type	# of Samples Collected	Distance From Site (m)	Aerosolized Endotoxin (EU m[a])			
			Avg	Median	Minimum	Maximum
Controls						
Background	12	NA	2.6	2.49	2.33	3.84
Biosolids operations						
Loading	39	2–50	343.7	91.5	5.6	1807.6
Slinging	24	10–200	33.5	6.3	4.9	14.29
Biosolids pile	6	2	103	85.4	48.9	207.1
Total operation	33	10–200	133.9	55.6	5.6	623.6
Wastewater treatment plant						
Aeration basin	6	2	627.3	639	294.4	891.1
Nonbiosolids field						
Tractor	6	2	469.8	490.9	284.4	659.1

[a]EU m = Endotoxin units per m³.

Information Box 5.5   Molds in Buildings

In moist environments within buildings mold and bacteria can proliferate rapidly within days and become established as colonies on solid surfaces, subsequently releasing toxins and/or allergens into the air. The most common indoor molds are *Cladosporium*, *Penicillium*, *Aspergillus* and *Alternaria*. Molds can cause both allergic reactions and chemical toxigenic responses from direct exposure to spores, cell wall components and mycotoxins. Molds and endotoxins can also be found within tobacco smoke (Pauly and Paszkiewicz, 2011).

instance, eight employees in a clinical microbiological laboratory developed acute brucellosis (Staszkiewicz *et al.*, 1991). A survey of the laboratory and the personnel showed that a cryogenically stored clinical isolate of *Brucella* sp. had been thawed and subcultured without the use of a biosafety cabinet. Other than this, the laboratory worker claimed to have used good technique. This example demonstrates the ease with which a bioaerosol can spread within areas where pathogens are handled for research and clinical purposes, and indicates the importance of bioaerosol control methodologies. The following sections describe how bioaerosol formation and spread is actually controlled in the laboratory.

5.8 BIOAEROSOL CONTROL

The control of airborne microorganisms can be handled in a variety of ways. Launching, transport and deposition are all points at which the airborne spread of pathogens can be controlled. The mechanisms used to control bioaerosols include ventilation, filtration, UV treatment, biocidal agents and physical isolation. These are discussed in the following sections.

5.8.1 Ventilation

Ventilation is the method most commonly used to prevent the accumulation of airborne particles. This mechanism involves creating a flow of air through areas where airborne contamination occurs. This can mean simply opening a window and allowing outside air to circulate inward, or use of air-conditioning and heating units that pump outside air into a room. Ventilation is considered one of the least effective methods for controlling airborne pathogens, but is still very important. Ventilation relies on mixing of intramural air with extramural air to reduce the concentration of airborne particles. However, in some cases the addition of extramural air can actually increase airborne particles. For example, one study showed that hospitals in Delhi, India, that relied on ventilation alone contained airborne fungal loads that were higher inside the hospital than those outside. This indicates that ventilation alone may not be sufficient to significantly reduce circulating bioaerosols. Thus, for most public buildings, especially hospitals, other forms of bioaerosol control need to be implemented.

5.8.2 Filtration

Unidirectional airflow filtration is a relatively simple and yet effective method for control of airborne contamination. Some filters, for example, high-efficiency particulate air (HEPA) filters are reported to remove virtually all infectious particles. These types of filters are commonly used in biological safety hoods. However, because of their high cost, they are not often used in building filtration systems. Instead, other filtration systems that rely on baghouse filtration (a baghouse works on the same principle as a vacuum cleaner bag) are used. Typically, air filters (baghouse, HEPA, etc.) are rated using the dust-spot percentage, which is an index of the size of the particles efficiently removed by the filter, with higher percentages representing greater filtration efficiencies. The typical rating for the filters used in most buildings is 30 to 50%. Studies have shown that a 97% dust-spot rating is required to effectively remove virus particles from the air. Other factors that influence filtration efficiency are related to the type of circulation system and how well it mobilizes air within the building, the type of baghouse system used and the filter material chosen (nylon wound, spun fiberglass, etc.) as well as the filter's nominal porosity (1 μm−5 μm). All these factors combine to influence the efficiency of the air filtration and removal of particles including bioaerosols. In spite of the high level of efficiency that can be achieved with filtration, many systems still cannot stop the circulation of airborne microorganisms, especially viruses, and added treatments may be required to ensure that air is safe to breathe.

5.8.3 Biocidal Control

Biocidal control represents an added treatment that can be used to eradicate all airborne microorganisms, ensuring they are no longer viable and capable of causing infection. Many eradication methods are available, for example, superheating, superdehydration, ozonation and UV irradiation. The most commonly used of these methods is UVGI or ultraviolet germicidal radiation. UVGI has been shown to be able to control many types of pathogens, although some microbes show various levels of resistance. The control of contagion using UV irradiation was tested in a tuberculosis (TB) ward of a hospital. Contaminated air was removed from the TB ward through a split ventilation duct and channeled into two animal holding pens that contained guinea pigs. One pen received air that had been treated with UV irradiation; the other received untreated

air. The guinea pigs in the untreated-air compartment developed TB, but none of the animals in the UV-treated compartment became infected. The American Hospital Association (1974) indicated that, properly utilized, UV radiation can kill nearly all infectious agents, although the effect is highly dependent on the UV intensity and exposure time. Thus, major factors that affect survival (temperature, relative humidity, UV radiation, ozone) in the extramural environment can be used to control the spread of contagion in the intramural environment.

5.8.4 Isolation

Isolation is the enclosure of an environment through the use of positive or negative pressurized air gradients and airtight seals. Negative pressure exists when cumulative airflow travels into the isolated region. Examples of this are the isolation chambers of the tuberculosis wards in hospitals used to protect others outside the TB wards from the infectious agent generated within these negative-pressure areas. This type of system is designed to protect other people in the hospital from the pathogens (*Mycobacterium tuberculosis*) present inside the isolation area. Air from these rooms is exhausted into the atmosphere after passing through a HEPA filter and biocidal control chamber.

Positive-pressure isolation chambers work on the opposite principle by forcing air out of the room, thus protecting the occupants of the room from outside contamination. One can reason that the TB ward is a negative-pressure isolation room, while the rest of the hospital, or at least the nearby anterooms, are under positive-pressure isolation. Other examples are the hospitals critical care wards for immunosuppressed patients such as organ transplant, human immunodeficiency virus (HIV)-infected and chemotherapy patients. These areas are protected from exposure to any type of pathogen or opportunistic pathogens. The air circulating into these critical care wards is filtered using HEPA filters, generating purified air essentially free of infectious agents.

5.9 BIOSAFETY IN THE LABORATORY

Many microbiological laboratories work specifically with pathogenic microorganisms, some of which are highly dangerous, especially in association with the aeromicrobiology pathway. Also, many types of equipment, such as centrifuges and vortexes (instruments for mixing of microbial suspensions) that are commonly used in microbiological laboratories can promote the aerosolization of microorganisms. Thus, laboratories and specialized equipment used in these laboratories (e.g., biosafety cabinets) are designed to control the spread of airborne microorganisms. There are essentially four levels of control designed

FIGURE 5.16 Biosafety cabinet Class II. From Telstar Life Science Solutions, photo courtesy J. Bliznick.

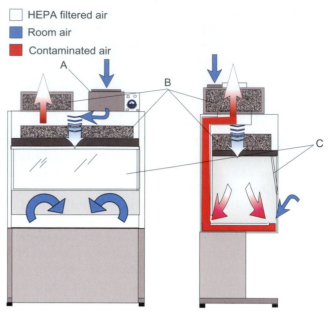

□ HEPA filtered air
■ Room air
■ Contaminated air

FIGURE 5.17 Schematic representation of the airflow paths within a typical Class II biosafety cabinet. Room air is dawn in from the top and from the front as indicated by the blue arrows. The nonpurified atmospheric air (blue) entering from the top of the cabinet is drawn in by an air pump (A) and then is purified by a HEPA filter (B) as it enters the workspace. Nonpurified air from the room (blue) entering from the front of the cabinet passes into the front grill and passes up through the top of the cabinet where it also passes through a HEPA filter before entering the workspace (C). This prevents the experiments in the workspace from being contaminated by airborne room contaminants. As the purified room air is exposed to the work environment and becomes contaminated (red) it is passed through yet another HEPA filter before being exhausted to the atmosphere. This pattern of airflow and purification ensures that the worker and the atmosphere are not exposed to the biohazards contained within the biosafety cabinet.

into laboratories, depending on the type of research being conducted. These levels of control are termed biosafety levels 1—4, with 1 being the lowest level of control and 4 the highest level of control. Within these laboratories, biosafety cabinets are essentially isolation chambers that provide safe environments for the manipulation of pathogenic microorganisms. In this section we will discuss biosafety cabinets and biosafety suits, followed by a short discussion of the actual biosafety levels imposed to achieve specific levels of control.

5.9.1 Biological Safety Cabinets

Biological safety cabinets (BSC) are among the most effective and commonly used biological containment devices in laboratories that work with infectious agents (US Department of Health and Human Services: CDC-NIH, 1993). There are two basic types of biosafety cabinets currently available (Class II and Class III), each of which has specific characteristics and applications that dictate the type of microorganism it is equipped to contain. Properly maintained biosafety cabinets provide safe environments for working with microorganisms. Class II biosafety cabinets are characterized by having considerable

negative-pressure airflow that provides protection from infectious bioaerosols generated within the cabinet (Figures 5.16 and 5.17), and Class III biosafety cabinets are characterized by total containment (Figure 5.18). Class I cabinets are also in existence, but they are no longer produced and are being replaced by Class II cabinets for all applications.

Class II biosafety cabinets, of which there are several types, are suitable for most work with moderate-risk pathogens (Table 5.3). Class II biosafety cabinets operate by drawing airflow past the worker and down through the front grill. This air is then passed upward through conduits and downward to the work area after passing through a HEPA filter. Room air is also drawn into the cabinet through the top of the unit, where it joins the circulating air and passes through the HEPA filter and into the work area. About 70% of the air circulating in the work area is then removed by passing it through the rear grill of the cabinet, where it is discharged into the exhaust system. The remaining 30% is passed through the front grill, essentially recirculating in the cabinet (Figure 5.17).

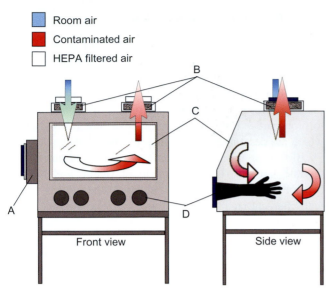

Room air

Contaminated air

HEPA filtered air

FIGURE 5.18 Schematic representation of a Class III biological safety cabinet. This cabinet is completely sealed from the environment. Any materials entering or leaving the cabinet are passed through a chemical dunk tank or autoclave (A) in order to sterilize them and prevent environmental contamination. Air entering or leaving these cabinets is passed through HEPA filters (B). Access to the workspace is by means of rubber gloves (D) and the workspace is visualized through a sealed window (C). These biosafety cabinets are utilized when working with highly pathogenic microorganisms to protect workers and the environment. Class III cabinets can be used to work with all biohazardous agents except those specifically designated for biosafety level 4 containment.

Laboratory personnel require special training in order to properly use Class II cabinets and to ensure proper containment of bioaerosols. One of the major hazards associated with Class II cabinets is the potential for the disruption of the negative airflow. Many mechanical actions can disrupt the protective airflow, such as repeated insertion and withdrawal of arms, opening or closing of doors in the laboratory, or even someone walking past the cabinet while it is in use. Any of these actions can potentially allow the escape of bioaerosols from the cabinet.

The Class III biosafety cabinet (Figure 5.18) is a completely enclosed environment that offers the highest degree of personnel and environmental protection from bioaerosols. Class III cabinets are used for high-risk pathogens (Table 5.3). All operations in the work area of the cabinet are performed through attached rubber gloves. Class III cabinets use complete isolation to protect workers. All air entering the cabinet is filtered using a HEPA filter, and the air leaving the cabinet is filtered by two HEPA filters in series. The exhaust may also include biocidal treatment such as incineration following the HEPA filtration to further ensure complete biological inactivation. In addition to these safeguards, Class III cabinets are connected with airtight seals to all other laboratory equipment (such as incubators, refrigerators and centrifuges)

TABLE 5.3 Examples of Classification of Biological Agents According to Risk

Class	Type of Agent	Agent
Class I	Bacterial Fungal Protozoal Viral	All those which have been assessed for risk and do not belong in higher classes Influenza virus reference strains Newcastle virus
Class II	Bacterial	*Campylobacter* spp. *Clostridium* spp. *E. coli* spp. *Klebsiella* spp. *Mycobacteria* spp. *Shigella* spp. *Vibrio* spp. *Salmonella* spp.
	Fungal	*Penicillium* spp. *Cryptococcus* spp. *Microsporum* spp.
	Protozoal	*Cryptosporidium* spp. *Giardia* spp. *Encephalitozoon* spp. *Enterocytozoon* spp. *Babesia* spp. *Echinococcus* spp. *Entamoeba* spp. *Fasciola* spp. *Leishmania* spp. *Plasmodium* spp. *Schistosoma* spp. *Trypanosoma* spp.
	Viral	Adenoviruses Corona viruses Cowpox virus Coxsackie A and B viruses Echoviruses Hepatitis viruses A, B, C, D and E Epstein–Barr virus Influenza viruses Vaccinia virus Rhinoviruses
Class III	Bacterial	*Brucella* spp. *Mycobacterium bovis* *Mycobacterium tuberculosis* *Rickettsia* spp. *Yersinia pestis*
	Fungal	*Coccidioides immitis* *Histoplasma capsulatum*
	Protozoal	None
	Viral	Dengue virus Monkey pox virus Yellow fever virus
Class IV	Bacterial	None
	Fungal	None
	Protozoal	None
	Viral	Hemorrhagic fever agents Ebola fever virus Marburg virus

Adapted from University of Pennsylvania Biological Safety Manual.

that is needed for working with the pathogens while using the cabinet. The Class III cabinet must also be connected to autoclaves and chemical dunk tanks used to sterilize or disinfect all materials entering or exiting the cabinet.

Another type of containment that typically provides the same level of protection as a Class III biosafety hood is the biological safety suit (Figure 5.19). The biological suit, unlike biosafety cabinets, operates under positive pressure created by an external air supply, thus protecting the wearer. Like the biosafety cabinets, the biosafety suit isolates the laboratory worker wearing it from bioaerosols. Biosafety suits are typically used in airtight complete biocontainment areas, and are decontaminated by means of chemical showers upon exiting the biohazard area. Some biosafety suits are portable and can be used in environments outside the laboratory such as "hot zones" (epidemiological areas that are currently under the influence of epidemic cases of diseases caused by high-risk pathogens) so that microbiologists and physicians working in these areas can minimize their risk of exposure to pathogens. As in biosafety cabinets, the air entering and leaving the biosafety suit passes through two HEPA filters.

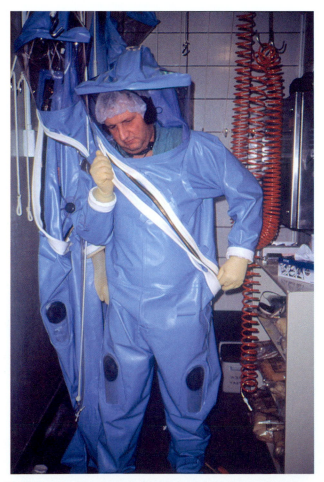

FIGURE 5.19   Biosafety suit. *Source*: Centers for Disease Control.

5.9.2 Biosafety Laboratories

Biosafety laboratories are carefully designed environments where infectious or potentially infectious agents are handled and/or contained for research or educational purposes. The purpose of a biosafety laboratory is to prevent the exposure of workers and the surrounding environment to biohazards. There are four levels of biohazard control, which are designated as biosafety levels 1 through 4.

Biosafety level 1, as defined by the Centers for Disease Control (US Department of Health and Human Services: CDC-NIH, 1993), indicates laboratories where well-characterized agents that are not associated with disease in healthy adult humans are handled. In general, no safety equipment is used other than sinks for hand washing, and only general restrictions are placed on public access to these laboratories. Work with the microorganisms can be done on bench tops using standard microbiological techniques. A good example of a biosafety 1 laboratory is a teaching laboratory used for undergraduate microbiology classes.

Biosafety 2 indicates an area where work is performed using agents that are of moderate hazard to humans and the environment. These laboratories differ from biosafety 1 laboratories in that the personnel have specialized training in the handling of pathogens, and access to the work areas is limited. Many procedures that may cause aerosolization of pathogenic microorganisms are conducted in biological safety level II cabinets or other physical containment equipment, to protect the laboratory workers.

Biosafety 3 indicates laboratories where agents that can cause serious or fatal disease as a result of AMB exposure are handled. As with biosafety 2, all personnel are specifically trained to handle pathogenic microorganisms. All procedures involving these infectious agents are conducted in biological safety level II cabinets or other physical containment devices. These facilities also have permanent locks to control access, negative airflow and filtered ventilation in order to protect the public and the surrounding environments. With certain pathogens used in biosafety 3 laboratories, Class III safety hoods may also be used, and clothes must be changed before leaving the premises.

Biosafety 4 is the highest level of control and is indicated for organisms that have high potential for life-threatening disease in association with aerosolization. To work in these facilities, personnel must have specialized training beyond that required for biosafety levels 2 and 3. Biosafety level 4 laboratories are 100% isolated from other areas of a building and may even be separated from other buildings altogether. Work in these areas is confined exclusively to Class III biological safety cabinets unless one-piece positive-pressure ventilation suits are worn, in which case Class II biosafety cabinets may be used. These laboratories are also specially designed to prevent microorganisms from being disseminated into the environment. The laboratories have complete containment, and require personnel to wear specialized clothing, which is removed and sterilized before leaving the containment areas. Personnel are also required to shower before leaving the facility. In general, all air into and out of these laboratories is sterilized by filtration and germicidal treatment. These facilities represent the ultimate in our ability to control airborne pathogens.

5.9.3 Biological Agent Classification

For any microorganism, defined degrees of risk associated with its use indicate the type of containment needed to ensure the safety of laboratory workers, the public and the environment. There are five classes of organisms. Class I microorganisms are those that pose little or no hazard under ordinary conditions of handling and can be safely handled without special apparatus or equipment. In contrast, Class II are agents of low potential hazard that may cause disease if accidentally inoculated or injected but that can be contained by ordinary laboratory techniques. Class III agents are those that require special containment; they are associated with aerosol disease transmission and special permits are required to import them from outside the country. Class IV agents are those that require extreme containment and are extremely hazardous to laboratory personnel or may cause serious epidemic disease. Finally, Class V agents are restricted foreign pathogens whose importation, possession or use is prohibited by law.

QUESTIONS AND PROBLEMS

1. List the major factors important in the survival of microorganisms in aerosols.
2. What is the major component of biosafety cabinets that remove microorganisms?
3. What is the role of microorganisms in cloud formation?
4. Give an example of a continuous linear source and an example of an instantaneous area source of bioaerosols.
5. Considering a windspeed of 1.5 m/s, an object that is 12 cm tall and normal air viscosity, determine whether conditions around the object would be considered turbulent.
6. Consider an airborne virus and an airborne protozoan with a radius of 30 nm and 1 μm and particle densities of 2.0 and 1.1 g/cm^3, respectively. Under normal gravitational acceleration calculate the terminal velocity for each.
7. Calculate the annual community risk of infection given the following data:
   - aerosolized virus concentration = 7.16/m^3 of air
   - duration of exposure = 8 hours
   - infectivity constant $r = 0.0253$
   - breathing rate = 0.83 m^3 per hour

REFERENCES AND RECOMMENDED READING

American Hospital Association. (1974) "Infection Control in the Hospital," 3rd ed. American Hospital Association, Chicago, pp. 69–117.

Amon, S. S., Schechter, R., Inglesby, T. V., Henderson, D. A., Bartlett, J. G., Ascher, M. S., *et al.* (2001) Botulin toxin as a biological weapon: medical and public health management. *JAMA* **285**, 1059–1070.

Baertsch, C., Paez-Rubio, T., Viau, E., and Peccia, J. (2007) Source tracking aerosols related from land applied Class B biosolids during high-wind events. *Appl. Environ. Microbiol.* **73**, 4522–4531.

Baptista-Rosas, R. C., Hinojosa, A., and Riquelme, M. (2007) Ecological niche modeling of *Coccidioides* spp. in Western North American deserts. *Ann. N.Y. Acad. Sci.* **1111**, 35–46.

Barry, J. M. (2004) The site of origin of the 1918 influenza pandemic and its public health implications. *J. Transl. Med.* **2**, 3.

Baskerville, A., Humphrey, T. J., Fitzgeorge, R. B., Cook, R. W., Chart, H., Rowe, B., and Whitehead, A. (1992) Airborne infection of laying hens with *Salmonella enteritidis* phage type 4. *Vet. Rec.* **130**, 395–398.

Boehm, F., and Leuschner, R. M. (1986) "Advances in Aerobiology," Proceedings of the Third International Conference on Aerobiology, Birkhäuser Verlag, Boston.

Bollin, G. E., Plouffe, J. F., Para, M. F., and Hackman, B. (1985) Aerosols containing *Legionella pneumophila* generated by shower heads and hot-water faucets. *Appl. Environ. Microbiol.* **50**, 1128–1131.

Brankston, G., Gitterman, L., Hirji, Z., Lemieux, C., and Gardam, M. (2007) Transmission of influenza A in human beings. *Lancet Infect. Dis.* **7**, 257–265.

Brooks, J. P., Gerba, C. P., and Pepper, I. L. (2004) Bioaerosol emission, fate, and transport from municipal and animal wastes. *J. Residuals Sci. Technol.* **1**, 13–25.

Brooks, J. P., Tanner, B. D., Gerba, C. P., Haas, C. N., and Pepper, I. L. (2005a) Estimation of bioaerosol risk of infection to residents adjacent to a land applied biosolids site using an empirically derived transport model. *J. Appl. Microbiol.* **98**, 397–405.

Brooks, J. P., Tanner, B. D., Josephson, K. L., Haas, C. N., Gerba, C. P., and Pepper, I. L. (2005b) A national study on the residential impact of biological aerosols from the land application of biosolids. *J. Appl. Microbiol.* **99**, 310–322.

Brooks, J. P., Tanner, B. D., Gerba, C. P., and Pepper, I. L. (2006) The measurement of aerosolized endotoxin from land application of Class B biosolids in Southeast Arizona. *Can. J. Microbiol.* **52**, 150–156.

Brooks, J. P., McLauglin, M. R., Gerba, C. P., and Pepper, I. L. (2012) Land application of manure and Class B biosolids: an occupational and public quantitative microbial risk assessment. *J. Environ. Qual.* **41**, 2009–2023.

Burrow, S. M., Butler, T., Jockel, P., Tust, H., and Kerkweg, A. (2009) Bacteria in the global atmosphere—part-2 modeling of emissions and transport between different ecosystems. *Atm. Chem. Phys.* **9**, 9281–9297.

Chistner, B. C. (2012) Cloudy with a chance of microbes. *Microbe* **7**, 70–75.

Committee on Particulate Control Technology. (1980) "Controlling Airborne Particles," Environmental Studies Board, National Research Council, National Academy of Sciences, Washington, DC.

Cox, C. S., and Heckley, R. J. (1973) Effects of oxygen upon freeze-dried and freeze-thawed bacteria—viability and free-radical studies. *Can. J. Microbiol.* **19**, 189−194.

Deepe, G. S., and Gibbons, R. S. (2008) TNF-(alpha) antagonism generates a population of antigen-specific CD4tCD25t T cells that inhibit protective immunity in murine histoplasmosis. *J. Immunol.* **180**, 1088−1097.

Delort, A. M., Vatilingom, M., Amato, P., Sancelme, M., Prazols, M., Mailot, G., *et al.* (2010) A short overview of the microbial population in clouds: potential in atmospheric chemistry and nucleation process. *Atmos. Res.* **98**, 249−260.

Dimmic, R. L., and Akers, A. B. (1969) "An Introduction to Experimental Aerobiology," Wiley Interscience, New York.

Donaldson, A. I., and Ferris, N. P. (1975) Survival of foot-and-mouth-disease virus in open air conditions. *J. Hyg.* **74**, 409−416.

Fengxiang, C., Qingxuan, H., Lingying, M., and Junbao, L. (1992) Particle diameter of the airborne microorganisms over Beijing and Tianjin area. *Aerobiologia* **8**, 297−300.

Freifelder, D. M. (1987) "Microbial Genetics," Jones and Bartlett, Portolla Valley, CA.

Gentry, T. J., and Pepper, I. L. (2002) Incidence of *Bacillus anthracis* in soil. *Soil Sci.* **167**, 627−635.

Grubman, M. J., and Baxt, B. (2004) Foot and mouth disease. *Clin. Microbiol. Rev.* **17**, 465−493.

Haas, C. N., Rose, J. B., and Gerba, C. P. (2014) "Quantitative Microbial Risk Assessment," 2nd ed. John Wiley and Sons, New York, NY.

Hinton, M., Ali, E. A., Allen, V., and Linton, A. H. (1983) The excretion of *Salmonella typhimurium* in the feces of cows fed milk substitute. *J. Hyg.* **91**, 33−45.

Hugh-Jones, M. E., and Wright, P. B. (1970) Studies on the 1967−8 foot and mouth disease epidemic: the relation of weather to the spread of disease. *J. Hyg.* **68**, 253−271.

Hurst, C. J., Knudsen, G. T., McInerney, M. J., Stetzenbach, L. D., and Walter, M. V. (1997) "Manual of Environmental Microbiology," ASM Press, Washington, DC, pp. 661−663.

Ingold, C. T. (1971) "Fungal Spores—Their Liberation and Dispersal," Clarendon Press, Oxford.

Loosli, C. G., Lemon, H. M., Robertson, O. H., and Appel, E. (1943) Experimental airborne infection. I. Influence of humidity on survival of virus in air. *Proc. Soc. Exp. Biol. Med.* **53**, 205−206.

Meselson, M., Guillemin, J., Hugh-Jones, M., Langmuir, A., Popova, I., Shelokov, A., and Yampolskaya, O. (1994) The Sverdlovsk anthrax outbreak of 1979. *Science* **266**, 1202−1208.

Mohr, A. J. (2001) Fate and transport of microorganisms in air. In "Manual of Environmental Microbiology" (C. J. Hurst, R. L. Crawford, G. R. Knudson, M. J. McInerney, and L. D. Stetzenbach, eds.), Second Ed., ASM Press, Washington, DC, pp. 827−838.

Pauly, J. L., and Paszkiewicz, G. M. (2011) Cigarette smoke, bacteria, mold, microbial toxins and chronic lung inflammation. *J. Oncol.* **2011**, Article id 819129, 13 pages.

Pepper, I. L., Gerba, C. P., and Brusseau, M. L. (2006) "Environmental and Pollution Science," 2nd ed., Elsevier, New York.

Pepper, I. L., Gerba, C. P., Newby, D. T., and Rice, C. W. (2009) Soil: a public health threat or savior? *Crit. Rev. Environ. Sci. Technol.* **39**, 416−432.

Portner, S. M., Hoffman, R. K., and Phillips, C. R. (1965) Microbial control in assembly areas needed for spacecraft. *Air Eng.* **7**, 46−49.

Staszkiewicz, J., Lewis, C. M., Colville, J., Zeros, M., and Band, J. (1991) Outbreak of *Brucella melitensis* among microbiology laboratory workers in a community hospital. *J. Clin. Microbiol.* **29**, 287−290.

Tanner, B. D., Brooks, J. P., Haas, C. N., Gerba, C. P., and Pepper, I. L. (2005) Bioaerosol emission rate and plume characteristics during land application of liquid Class B biosolids. *Environ. Sci. Technol.* **39**, 1584−1590.

Tayler, G. I. (1915) Eddy motion in the atmosphere. *Philos. Trans. R. Soc. Ser. A* **215**, 1−26.

Theunissen, H. J. J., Lemmens-Den Toom, N. A., Burggraaf, A., Stolz, E., and Michel, M. F. (1993) Influence of temperature and relative humidity on the survival of *Chlamydia pneumoniae* in aerosols. *Appl. Environ. Microbiol.* **59**, 2589−2593.

Turnbull, P. C. B., Lindeque, P. M., Le Roux, J., Bennett, A. M., and Parks, S. R. (1998) Airborne movement of anthrax spores from carcass sites in the Etosha National Park, Namibia. *J. Appl. Microbiol.* **84**, 667−676.

U.S. Department of Health and Human Services: CDC-NIH; (1993) "Biosafety in Microbiological and Biomedical Laboratories," 3rd ed., U.S. Government Printing Office, Washington, DC, HHS Publication No. (CDC) 93-8395.

Wathes, C. M., Zaidan, W. A. R., Pearson, G. R., Hinton, M., and Todd, N. (1988) Aerosol infection of calves and mice with *Salmonella typhimurium*. *Vet. Rec.* **123**, 590−594.

Wells, W. F., and Riley, E. C. (1937) An investigation of the bacterial contamination of the air of textile mills with special reference to the influence of artificial humidification. *J. Ind. Hyg. Toxicol.* **19**, 513−561.

Chapter 6

Aquatic Environments

Virginia I. Rich and Raina M. Maier

6.1 INTRODUCTION

The majority of the planet's habitat is aquatic: more than 80% of Earth's surface is aquatic, and the volume of habitat in aquatic systems is vast, spanning a range of environments (Table 6.1; Figure 6.1). These habitats are teeming with microbial life. Microorganisms are key drivers of the planet's biogeochemical cycles (Chapter 16), and this includes large roles for aquatic microbes. While the Amazon Forest has been called the lungs of the planet, roughly 50% of the oxygen that you breathe was actually produced by the photosynthesis of aquatic microbial primary producers. In addition, microbes are the base of aquatic food chains, which supply roughly 15% of the world's protein, and are projected to become an even larger share in the future. The water itself in aquatic environments is a vital resource, supplying water for drinking, agriculture, mining, power generation, semiconductor manufacturing and virtually every other industry. For some of these uses, aquatic microbes may be considered contaminants; as in the case of computer chip manufacturing. For potable water, contamination with pathogens results in approximately 11% of the world's

population still lacking access to safe drinking water. In this chapter, we first define the main aquatic habitat types (planktonic, benthic and their interface), then examine how microbial lifestyles (primary and secondary production) are employed in them. Finally, we describe and provide general microbial characteristics of: (1) marine systems; (2) freshwater systems; and (3) select other aquatic environments.

6.2 MICROBIAL HABITATS IN THE AQUATIC ENVIRONMENT

6.2.1 Physical and Chemical Characteristics

There are a number of typical misconceptions about aquatic habitats, due to how we think about water, for example that it tends to be well mixed. First, aquatic habitats are not homogeneous. Stratification is an important physical structuring of aquatic environments, established due to temperature and salinity differences (see lake example in Figure 6.2). Surface waters are warmed by sunlight, and since warm water is less dense than cold

I.L. Pepper, C.P. Gerba, T.J. Gentry: Environmental Microbiology, Third edition. DOI: http://dx.doi.org/10.1016/B978-0-12-394626-3.00006-5
© 2015 Elsevier Inc. All rights reserved.

water (water is most dense at 4°C), this temperature-driven stratification tends to persist in the absence of mixing (which *does* occur, see below). The thermocline is the layer in aquatic systems where a rapid change in temperature occurs. Salinity differences can also establish stratification, when precipitation or other inputs bring fresher waters over saltier ones, which are denser. (The salinity of aquatic systems can range from freshwater at 0.5‰, to

marine water between 33 and 37‰, to hypersaline systems such as the Dead Sea at 290‰; see Information Box 6.1.) This layering of aquatic environments can act as a barrier

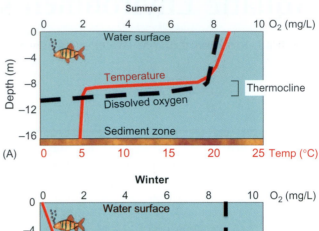
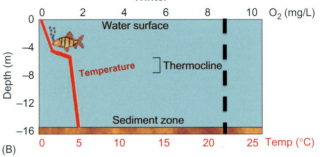

(A)

(B)

FIGURE 6.2 Stratification in a lake. This idealized view of a temperate-region, eutrophic (high-nutrient) lake shows temperature-driven stratification in the summer (A) due to warming of the surface waters. The thermocline, where the temperature drops sharply, acts as a barrier to mixing of deeper waters, thus preventing their oxygenation. In the fall and winter as the surface cools, the thermocline breaks down and mixing occurs, reoxygenating deeper waters (B).

TABLE 6.1 Distribution of Water on Earth

Habitat	Volume km^3	Percent of Total Water
Oceans, seas and bays	1,338,000,000	96.5
Ice caps, glaciers and permanent snow	24,064,000	1.74
Saline groundwater	12,870,000	0.94
Fresh groundwater	10,530	0.76
Ground ice and permafrost	300,000	0.22
Fresh lakes	91,000	0.007
Saline lakes	85,400	0.006
Atmosphere	12,900	0.001
Soil moisture	16,500	0.001
Swamp water	11,470	0.0008
Rivers	2120	0.0002
Biological water	1120	0.0001

From http://en.wikipedia.org/wiki/File:Earth_water_distribution.svg.

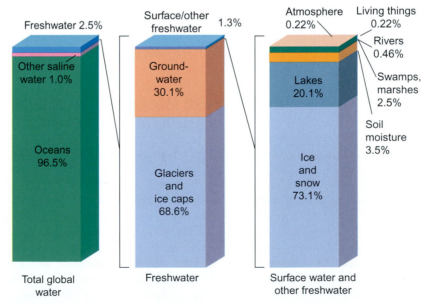

FIGURE 6.1 What are aquatic habitats? A view of the distribution of Earth's water. From http://ga.water.usgs.gov/edu/earthwherewater.html.

to vertical movement of organisms that move passively and/or are very small, including microbes. In addition, the layering can create markedly different chemical conditions in adjacent layers. Surface layers have higher oxygen concentrations due to diffusion and mixing from air, and they experience sunlight, but their typically high primary production can result in nutrient-limiting conditions. Microbes can be adapted for the conditions found in specific layers, and some have developed strategies for staying in certain layers. A hot topic in aquatic systems is thin layers (Sullivan *et al.*, 2010), which consist of layers of distinct chemistry and biology just centimeters to meters thick that can extend horizontally for kilometers. Interest in thin layers is due to the fascinating and very active biology within them, which until recently was simply missed by typically bulk-scale observations. The thermocline itself is often a transition layer of higher cell numbers and heterotrophic activity, because organic matter accumulates there. Likewise, the layer at the very bottom of the water column, directly above the sediment, often supports higher cell numbers, due to the resuspension of nutrients and carbon, as well as cells from the sediment.

A second misconception about aquatic environments is that they are static; that water's high thermal inertia and the vast size of many water bodies create extremely stable environments. In fact, aquatic systems are spatially and temporally highly dynamic. Mixing counteracts stratification, and is caused by the action of winds, currents, tides, upwelling, and temperature and salinity changes. Mixing is critical to bringing oxygen and nutrients to depleted waters. High evaporation rates such as occur in the Tropics can make surface waters saltier, which can help drive mixing as the denser saltier water then sinks. Seasonal temperature changes in surface waters are a major driver of mixing in aquatic systems. Surface waters are warmed by the summer sun and are cooled in the fall and winter. This results in a decrease in the thermocline strength, permitting deeper mixing (Figure 6.2). In addition, fall and winter often bring more storms, which further mix the water column. In some systems with extreme air temperature changes such as the Polar Regions or limited water volumes such as lakes, the thermocline breaks down altogether, allowing mixing throughout the water column. Where air temperatures drop below freezing, ice forms at the surface; in shallow lakes the ice may propagate all the way to the bottom. As an interesting aside, in marine systems, the formation of ice crystals pushes out the "impurities" of salts, creating extremely salty brine channels in sea ice, an extreme habitat where unique microbiology occurs. When ice thaws in springtime in temperate, alpine and polar aquatic systems, mixing through the water column occurs once more, before summer stratification is re-established.

Light is a critical driver of habitat differences in aquatic systems. Light is able to penetrate to a depth of 200 m or more, depending on the turbidity of the water (Figure 6.3). This depth defines the photic zone. In lakes and coastal areas, where the amount of suspended particulate matter in the water is high, light may penetrate less than 1 m. The aphotic zone is the dark water where light does not reach. The presence or absence of light results in very different microbial lifestyles, diversity and activity. It is essential to consider stratification and mixing in tandem with photic zone depth, when thinking about microbes that specialize in the sunlit surface waters. If an aquatic system is highly mixed (for example, in the fall in

---

Information Box 6.1   What Is Salinity?

The average salt concentration in the ocean is approximately 3.5‰. This is more precisely expressed in terms of salinity. Salinity (‰) is defined as the mass in grams of dissolved inorganic matter in 1 kg of seawater after all Br^- and I^- have been replaced by the equivalent quantity of Cl^-, and all HCO_3^- and CO_3^{2-} have been converted to oxide. In terms of salinity, marine waters range from 33 to 37‰, with an average of 35‰.

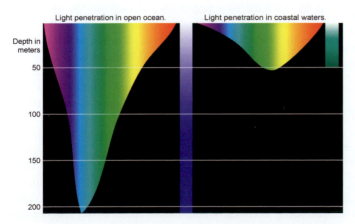

FIGURE 6.3 Light penetration through water, here shown for two ocean habitats, coastal vs. open ocean. Penetration is markedly different due to differences in turbidity resulting from dissolved and particulate matter and microbial cells. In the coastal waters, turbidity is higher, and light penetrates less deeply. Long wavelength light (red) is absorbed by water, while short wavelength light (purple) is scattered, such that the deepest penetrance is for mid-range green and blue wavelengths. Maximum light penetration may be several hundred meters in the open ocean but just tens of meters in coastal waters. From NOAA: Image courtesy Kyle Carothers, NOAA-OE.

many systems), photoautotrophic microbes may not stay in the photic zone long enough to accomplish much primary production. In contrast, stratification allows photoautotrophs to remain in the surface sunlit waters, but may result in nutrient depletion of those waters, which can also limit their primary production and growth.

Another way to consider the structuring of aquatic environments is based on habitats defined with respect to depth below surface. In particular, at the sea surface the air—water interface is referred to as the neuston (Figure 6.4), a habitat with high levels of harsh ultraviolet radiation, biochemicals and nutrients. Due to biomolecule interactions with the air—water interface that cause them to concentrate there, the neuston is comprised of a thin gel-like matrix of biomolecules (mainly lipids, proteins and polysaccharides) where microbes can attach or become trapped. The pelagic zone is a broad term used to describe the water column or planktonic habitat (see Section 6.2.2), and is subdivided on the basis of depth. In marine systems, depth from surface defines the epipelagic, mesopelagic, bathypelagic and abyssopelagic zones. Oceans range in depth up to 11,000 m in the deepest of ocean trenches. In lakes, which can be a few meters to more than 1000 m deep, depth is combined with light penetration to define the surface limnetic zone (where light intensity is at least 1% of sunlight), and deeper profundal zone. Within the pelagic zones, water can be comprised of many microhabitats and be highly structured. Floating or sinking particles create miniature islands of carbon, nutrients and substrates. Particle-associated microbes specialize in living on these islands, in contrast

to the more intuitive free-living lifestyle we imagine in the pelagic habitat. Also, microbes and many macroorganisms produce exopolysaccharides, which cumulatively create an actual mesh structure to broad areas of water. Then, moving below the pelagic zone, the benthos is the sediment habitat underlying the water column (see Section 6.2.3).

Another important and not immediately obvious set of aquatic habitats is defined by microbial associations with macroorganisms. These relationships define two additional habitats: epibiotic, which means attached to the surface of another organism, and endobiotic, which means living within another organism's tissues. Such microbe—macrobe relationships and communication increasingly appear to be the exception rather than the rule in nature (for example, in our own bodies, see Chapter 20), and can result in some particularly innovative and exciting biology. For example, many fish and squid employ bioluminescence generated through diverse microbial relationships. *Vibrio harveyii* is one microbe that uses luminescence in its fascinating endobiotic lifestyle (see Information Box 20.3).

6.2.2 Overview of Planktonic Microbes

Plankton, from Greek word meaning "wanderer" or "drifter," are organisms that live suspended in the water column and drift with the currents, with little or no ability to control their horizontal location. There are three functional groups of plankton, each with microbial members: phytoplankton, bacterioplankton and zooplankton. Pelagic microbial populations can be referred to as bacterioplankton (though notably, despite the name, these include archaeans as well as bacteria), and include photoautotrophs, chemoautotrophs (see Section 6.3.1.1) and heterotrophs. The phytoplankton (Figure 6.5) are the photoautotrophic plankton, which include microbes (cyanobacteria) and eukaryotes (algae, especially the single-celled dinoflagellates and diatoms). The zooplankton are larger heterotrophic plankton, including protozoans such as the intricate foraminiferans and radiolarians and larger organisms such as copepods. Figure 6.6 shows the relationship and interdependence of the various microbial components within a general planktonic food web.

Phytoplankton are the primary producers, which use photosynthesis to fix CO_2 into organic matter. This is a major source of organic carbon and energy, which is transferred to other trophic levels within the web (Figure 6.6). The organic compounds produced by phytoplankton can be divided into two classes, particulate or dissolved, depending on their size. Particulate organic matter (POM) compounds are large macromolecules such as polymers, which make up the structural components of the cells, including cell walls and membranes. Dissolved organic

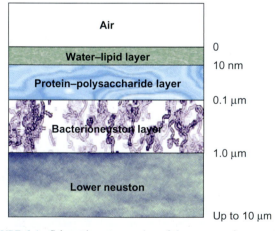

FIGURE 6.4 Schematic representation of the neuston, the upper layer of aquatic environments that can range from 1 to 10 μm in depth. The upper layer that interacts with the atmosphere consists of a water—lipid mixture that has increased surface tension. Below this is a layer of organic matter that accumulates from organic matter rising up the water column. Most scientists consider the neuston an extreme environment (see Chapter 7) because of many factors, including intense solar radiation, large temperature fluctuations, and the natural accumulation of toxic substances including chemicals, organic matter and heavy metals.

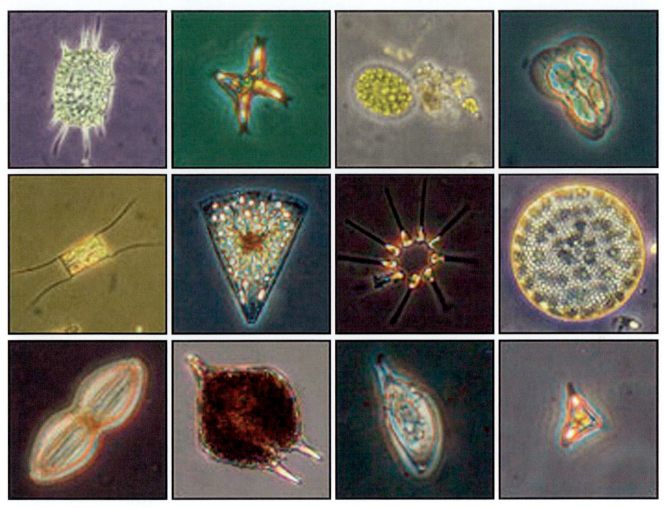

FIGURE 6.5   Examples of phytoplankton.

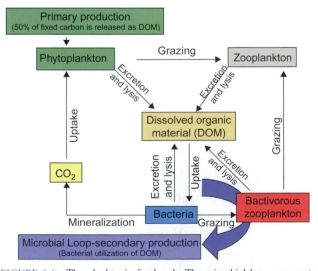

FIGURE 6.6   The planktonic food web. The microbial loop represents a pathway in which the dissolved organic products are efficiently utilized. The role of bacterioplankton is to mineralize important nutrients contained within organic compounds and to convert a portion of the dissolved carbon into biomass. Grazing by bactivorous protozoans provides a link to higher trophic levels. Modified from Fuhrman (1992).

matter (DOM) is composed of smaller soluble material that passes through a filter (pore size 0.7 μm) including amino acids, carbohydrates, organic acids and nucleic acids, which are rapidly taken up by microbes and metabolized (Kirchman, 2004). DOM is an extremely large carbon pool, equal in size to atmospheric CO_2 (Chapter 16).

6.2.3 Overview of Benthic Microbes

The benthos is the transition between the water column and the mineral subsurface. It collects organic material that settles from above, or that is deposited from nearby terrestrial environments (through river inputs, bulk overland flow or groundwater flow). At the surface of the benthos, nutrient and carbon levels are higher than in the directly overlying waters, which often causes dramatically higher microbial numbers (as much as five orders of magnitude) and activity than in the plankton. Since activity is high, oxygen is quickly used up, leading to a steep redox gradient in the sediment; that is, oxygen is replaced by other terminal electron acceptors such as sulfate, nitrate or iron (see Chapter 3).

The benthos supports a physiologically diverse aquatic microbial community. This is because the redox and nutrient gradients described above create numerous microenvironmental niches, of which specific physiological groups of microorganisms are strategically positioned to take advantage. In shallow-water benthic communities (mud flats, river bottoms, etc.), the surface may be dominated by photoautotrophs. Below that, and in aphotic benthic habitats, heterotrophs and/or chemoautotrophs dominate, the latter fixing dissolved CO_2 into biomass using the energy of chemical bonds (see Chapter 3). The cycling of essential nutrients, such as carbon, nitrogen and sulfur, is dependent on a combination of aerobic and anaerobic microbial transformations (Figure 6.7), and can

be viewed as biogeochemical cycling at small habitat scales (see Chapter 16).

In terms of carbon, the incoming organic matter (as POM or DOM) can be degraded aerobically to produce smaller compounds or CO_2, or anaerobically through fermentation into organic acids and CO_2. Organic acids can then act as electron donors for a group of strictly anaerobic bacteria that utilize CO_2 as the final electron acceptor in anaerobic respiration, thus generating methane (CH_4). The methanogenic activity in turn supports the activity of the methane-oxidizing bacteria (methanotrophs), which can use methane and other one-carbon compounds as an energy source, regenerating CO_2. Methanotrophic activity until recently was assumed restricted to the sediment– water interface zone, because it was believed to require oxygen. We now know that anaerobic methane oxidation

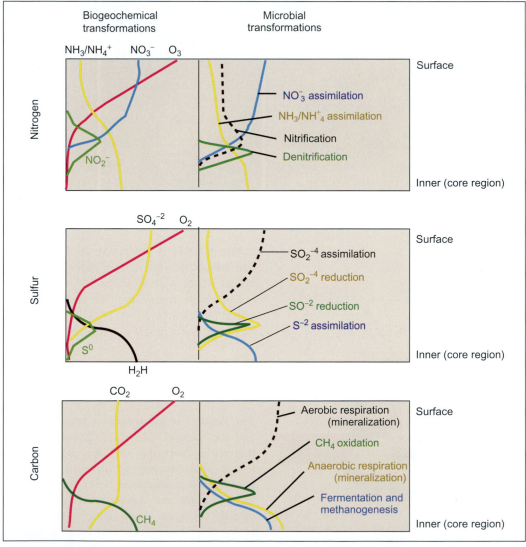

FIGURE 6.7 Biogeochemical profiles and major carbon, nitrogen and sulfur transformations that can be predicted for benthic environments in which oxygen levels are highest at the "surface" layer and are depleted by microbial activity to create anoxic conditions in the "inner" region. Adapted with permission from Pearl and Pinckney (1996).

occurs, performed by archaea and often in close syntrophic (cross-feeding) relationships with bacteria (Orphan *et al.*, 2009), although the details of this lifestyle and relationships continue to be unraveled.

In terms of nitrogen, the decomposition of organic material in the sediment layer generates ammonia from organic debris. Ammonia in the benthos may be used for two purposes: (1) biomass; that is, its assimilation as a source of essential nitrogen by planktonic and sediment microorganisms; and (2) energy; that is, its oxidation as an energy source by chemoautotrophic microorganisms. Ammonia oxidation is often localized at the sediment−water interface, where organisms utilize the release of ammonia by the decomposers and either oxygen (aerobic ammonia oxidation, by bacteria or archaea), or nitrite (anaerobic ammonia oxidation) as terminal electron acceptor (see also Section 16.3.4.2 and Case Study 19.3). The latter reaction, known as anammox, is thus far known only in the ubiquitous bacterial phylum *Planctomycetes* as its terminal electron acceptor (Kuenen, 2008). Nitrification is a two-step process of sequential oxidation of ammonia to nitrite, and then nitrite to nitrate, for example, performed in sequence by the bacterial genera *Nitrosomonas* and *Nitrobacter* (see also Section 16.3). The control of ammonia compounds can be important, especially in alkaline environments, where the undissociated NH_4OH form can be toxic to aquatic animals. The activity of the ammonia-oxidizing or nitrifying microbes can be highly sensitive to the presence of certain DOM, including naturally occurring and industrial chemicals. Therefore, the inhibition of nitrification (ammonia oxidation), which can be detected by an accumulation of ammonia or NO_2, provides a sensitive indicator of the environmental impact of certain toxic pollutants.

6.2.4 At the Interface: Biofilms and Microbial Mats

Interfacial habitats are special, as may have become clear above. They are defined by sharp environmental gradients (for example, of UV and oxygen in the neuston, and of redox conditions in the benthos), and these gradients can create distinct niches, which microbes take advantage of through diverse life history strategies. One strategy is to team up and specialize. While we think of microbial cells in isolation, in fact they are constantly interacting with one another (see Section 19.3.1). A common way is through quorum sensing, a process by which a single cell can "sense" whether a threshold number, or "quorum," of cells is nearby (see also Chapter 20). The coordination of microbial cells through physical, chemical and biological processes can result in the formation of complex,

specialized and diversified structures. Here we discuss the two such structural types, common on surfaces in aquatic habitats.

6.2.4.1 Biofilms

A biofilm is a surface association of microorganisms that are strongly attached through the production of an extracellular polymer matrix. Biofilm-harboring surfaces are usually aquatic or at least moist, and include inert surfaces, such as rocks and the hulls of ships, and living ones, such as a copepod's exoskeleton, or an aquatic plant's submerged leaf. As elaborated below, biofilms have been extensively studied for their role in nutrient cycling and pollution control within the aquatic environment, as well as for their beneficial or detrimental effects on human health.

Biofilm development occurs through microbial attachment to a solid surface (Figure 6.8) in two stages: (1) reversible attachment, which is a transitory physicochemical attraction (including via hydrophobic, electrostatic and van der Waals forces; Marshall, 1985); and (2) irreversible attachment, which is a biologically mediated stabilization reaction (see Section 19.3.2.1). The attached bacteria excrete extracellular polymers, which create a matrix that surrounds the cells and forms a strong chemical bridge to the solid surface. The polymers then provide a matrix for the attachment of additional cells, form internal architecture in the biofilm structure and can create a visibly "slimy" layer on a solid surface (Marshall, 1992). The exopolymer matrix is also an integral component influencing the functioning and survival of biofilms in hostile environments.

Biofilms can be so highly organized that their architecture can rival that of simple macroorganisms. Examination of mature biofilms in their native states (using microscopic techniques, such as confocal laser scanning microscopy) has revealed a complex organization (Costerton *et al.*, 1995). Biofilms can be composed of cone-, mushroom- and column-shaped clusters of cells embedded within the extracellular polymer matrix and surrounded by large void spaces (Wolfaardt *et al.*, 1994; Korber *et al.*, 1995) (Figure 6.9A). The void spaces form channels (Figure 6.9B), which function as a primitive circulatory system by carrying limiting nutrients, such as oxygen, into the exopolymeric matrix. The presence of void spaces increases the biofilm surface area and the efficiency with which nutrients and gases are transferred between the biofilm and the surrounding water. The exact nature of the biofilm architecture depends on numerous factors, including the type of solid surface, the microbial composition of the biofilm and environmental conditions.

Microorganisms benefit from membership in a biofilm community. The extracellular matrix can have several

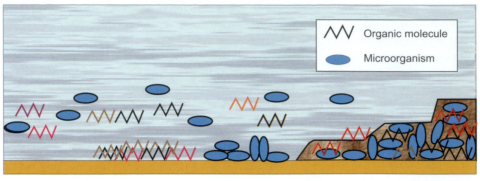

Clean surface   Conditioning   Phase I   Phase II   Mature biofilm
                   film of dissolved   primary microbial   irreversible
                   organic matter   colonization   microbial attachment

FIGURE 6.8   Representation of biofilm formation. Dissolved organic molecules of a hydrophobic nature accumulate at the solid surface—water interface and form a conditioning film. Bacteria approach the solid surface because of water flow and/or active motility. The initial adhesion (phase I) is controlled by various attractive or repulsive physicochemical forces leaking to passive, reversible attachment to the surface. An irreversible attachment is a biological, time-dependent process related to the proliferation of bacterial exopolymers forming a chemical bridge to the solid surface (phase II). By a combination of colonization and bacterial growth, the mature biofilm is formed. It is characterized by cell clusters surrounded by water-filled voids. Adapted from Marshall (1992).

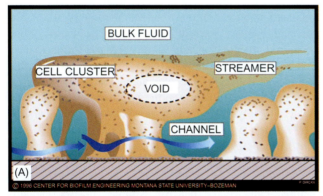

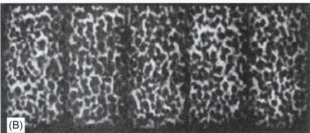

FIGURE 6.9   Biofilm communities are composed of cone-, column- and mushroom-shaped clusters of cells embedded in extracellular polysaccharide and other materials (A). Clusters of cells are surrounded by large void spaces (white networks of channels shown in biofilm in panel (B)) that allow convective flow of fluids through biofilm communities.

functions for the member cells: as an ion-exchange resin to filter and collect essential nutrients; as physicochemical protection against abiotic environmental stressors, such as desiccation or changes in pH and temperature; and against actual grazing by predators. Compared with planktonic cells, biofilm cells are far more resistant to

antibacterial substances, such as antibiotics and disinfectants; this may be due to the barrier of the extracellular matrix, or to an altered physiological state of the attached bacterial cells.

Humans contain biofilms (the most obvious example being dental plaques) and have been using them intentionally for hundreds to thousands of years. Biofilms can act as water purification systems in natural environments, and remove DOM and other contaminants from flowing waters. This property has long been exploited for use in purifying water from municipal (sewage) or industrial sources (Case Study 6.1), with crude management in ancient times giving way to sophisticated "collaboration" with the biofilms in current water purification system designs, including trickling filters in wastewater treatment (Chapter 25) and point-of-use devices (Chapter 19). On the other hand, biofilms are responsible for tooth decay and compromised medical implants. Disease caused by biofilms on medical implants is estimated to account for nearly 60% of all hospital-borne infections, lengthen hospital visits by 2—3 days and increase healthcare costs by $1 billion per year (Davey and O'Toole, 2000). In industry, biofilm control is required for any type of pipeline (Chapter 28), in which biofilms lower the flow capacity, decrease heat-exchange efficiency and catalyze corrosion in the case of metal pipes.

6.2.4.2 Microbial Mats

Microbial mats can be considered a specialized type of biofilm. They are an extreme example of an interfacial aquatic habitat in which many microbial groups are

---

Case Study 6.1   Beneficial Biofilms Remove Cyanide from Gold Mine Effluent and Keep Mines in Business

---

The Homestake Mine in Lead, South Dakota, opened in 1877 during the Black Hills Gold Rush and operated profitably in the gold mining business for decades. As was the case with the operation of many other gold mines, cyanide was used to increase recovery of gold from ore obtained from the mine. As a result, an estimated 4 million gallons of cyanide-laden wastewater were released daily into nearby Whitewood Creek. These extraordinary levels of this toxic waste rendered aquatic life nearly nonexistent in the creek. By 1981, Whitewood Creek was listed as an Environmental Protection Agency (EPA) Superfund site. In 1977, the EPA required Homestake Mine to reduce its discharge of this toxic effluent. Traditional approaches to minimizing discharge of such toxic substances were expensive, and implementation of these approaches would result in closure of the mine.

The Homestake Mine needed an innovative, cost-effective strategy to deal with levels of cyanide in its wastestream. Jim Whitlock, a biochemist and South Dakota native, and Terry Mudder, an environmental engineer, were charged with addressing the problem (Whitlock, 1990). The solution the cross-disciplinary duo devised relied upon a bacterial biofilm, composed primarily of *Pseudomonas*, to remove cyanide and a host of other toxic substances, including ammonia and the metals nickel chromium, from the wastestream. Sets of large discs, called rotating biological contactors (RBCs), served as substrates upon which the pollutant-removing biofilm grew. Each RBC consisted of disks that harbored billions of bacteria across large surface areas (100,000 to 150,000 ft^2). Wastewater passed through a train of five of these RBCs. Each disk rotated at a rate of 1.5 revolutions per minute. Approximately 40% of each disk was submerged in the wastestream at all times. The rotation allowed the biofilm community to contact the wastestream and remove pollutants such as cyanide while meeting some microbial community members' requirements for oxygen. The first two RBCs contained primarily *Pseudomonas* for the removal of cyanide and the metal contaminants, while the remaining RBCs harbored nitrifying bacteria that allowed conversion of ammonia into a less toxic form, nitrate. End products resulting from this treatment were relatively innocuous and included sulfate, carbonate, nitrate and some solids, which were subsequently removed using a clarifier. The treatment facility began operation in 1984 and became more efficient and economical over time. Cyanide removal rates of 99% (from influent levels of 4.1 mg/L to effluent levels of 0.06 mg/L) were obtained. Copper and iron were removed quite efficiently— removal rates of 95–98% were common. Removal of other metals, particularly nickel, chromium and zinc, was less remarkable. Nonetheless, the effluent was free enough of pollutants to allow rainbow trout to reinhabit Whitewood Creek. Thus, this innovative use of biofilms dramatically reduced pollution introduced into the environment by the Homestake Mine, and allowed the mine to continue operations until its closing in 2002. To date, thousands of similar RBCs have been employed worldwide to reduce cyanide levels from industrial wastestreams.

---

laterally tightly compressed into a thin mat of biological activity. While biofilms are typically one to several cell layers thick, microbial mats range from several millimeters to a centimeter thick, and are vertically stratified into distinct layers (Information Box 6.2). Another distinguishing characteristic of microbial mats is that they are based on autotrophy, the fixation of inorganic carbon into biomass, which occurs either photosynthetically or chemosynthetically. Similarly to biofilms, mat microbial groups interact with each other in close spatial and temporal physiological couplings. Microbial mats have been found associated with environments such as the benthic— planktonic interface of hot springs, deep-sea vents, hypersaline lakes and marine estuaries. By supporting most of the major biogeochemical cycles, these mats are largely self-sufficient.

In a photosynthetic microbial mat, the photosynthetic activity of the cyanobacteria creates an oxygenic environment in the upper layer of the mat. Oxygen can become supersaturated during the day, but at night, in the absence of sunlight, microbial respiration rapidly depletes all the available oxygen. Respiration by sulfate-reducing bacteria, considered a strictly anaerobic process, helps decompose the DOM from the cyanobacteria in the upper generally aerobic layers. This apparent contradiction may be resolved temporally, with oxygenic photosynthesis occurring during the day and anaerobic sulfate reduction occurring at night, or spatially, due to the formation of anaerobic microenvironments even in the upper layers, due to the high demand for oxygen by heterotrophic activity.

Microbial mats are unique communities because the interdependent microbial components form clearly stratified and often distinctively colored zones. Mats are often found in extreme environments or in environments where conditions fluctuate rapidly. The cyanobacteria are known to be tolerant of extreme conditions, such as high temperatures or highly saline waters, and thrive in locations where competition from other microbial groups and predation by grazing organisms are limited by the inhospitable environment. Fossilized microbial mats, known as stromatolites, dating back 3.5 billion years were among the first indications of life on Earth (see Information Box 6.2, and Section 16.1.2). At that time, Earth's atmosphere lacked oxygen, and the stromatolites from that era were probably formed with anoxygenic phototrophic bacteria (purple and green sulfur bacteria— see also Section 16.4.3.2).

Information Box 6.2   The Importance of Microbial Mats on Early Earth

The figure (top) shows a cross section of a microbial mat collected by NASA scientists from a hypersaline pond at one of the world's largest salt production facilities in Guerrero Negro, Baja California Sur, Mexico. The smallest gradations in the ruler are in millimeters.

It has been suggested that such mats forced the close proximity of the first aerobic photosynthetic microbes (cyanobacteria) with anaerobic heterotrophs. This proximity was in turn responsible for the adaptation of anaerobic heterotrophic microbes on early Earth to the presence of oxygen, which was extremely toxic to these first heterotrophic forms of life (Hoehler *et al.*, 2001). Today, as shown in the figure below, these mats provide another example of the complexity of biogeochemical cycling in aquatic environments (Des Marais, 2003). In these mats, cyanobacteria photosynthetically generate organic matter (required by heterotrophs) and oxygen (toxic for strict anaerobes). On the other hand, anaerobic heterotrophic activity recycles required nutrients back to the phototrophic community while generating toxic sulfide. Mat microbes have developed strategies to cope with the conundrum posed by these different populations. Note that in this case, the community is driven directly by photosynthesis while activity in the benthic environment is driven indirectly by photosynthesis in the form of DOM.

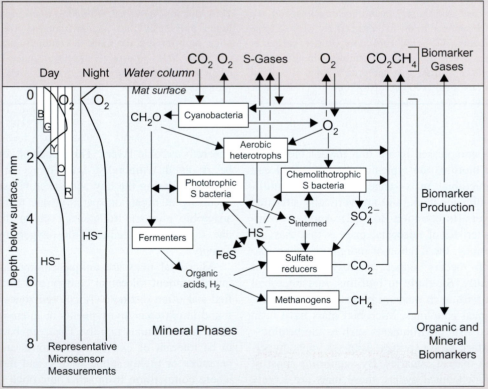

From (top) NASA, 2005 and (bottom) from Fig. 1 from Des Marais, D. 2003. *Biol. Bull.* **204**, 160–167. Reprinted with permission from the Marine Biological Laboratory, Woods Hole, MA.

6.3 MICROBIAL LIFESTYLES IN AQUATIC ENVIRONMENTS

6.3.1 Primary Production

Primary production in the ocean is estimated to be 50—60 petagrams (Pg = 10^{15} g) of carbon per year (De la Rocha, 2006), with freshwater likely accounting for an additional one-to-several Pg (Tranvik *et al.*, 2009). This represents 50% of the total primary production globally. The amount of primary production within a given water column depends on a variety of environmental factors. These factors include: the availability of essential inorganic nutrients, particularly nitrogen and phosphorus; water temperature; the turbidity of the water, which affects the amount of light transmitted through the water column; and the degree of vertical mixing, as described above. The concentration of primary producers in aquatic environments ranges from 10^0 organisms/ml in some benthic habitats, to 10^8 organisms/ml in surface zones.

Open oceans have relatively low primary productivity because of low levels of the essential nutrients nitrogen and phosphorus. The exceptions are areas where currents cause upwelling of deeper waters bringing nutrients from the deep sea. Coastal areas are productive because of the introduction of dissolved and particulate organic material from river outflows and surface runoff from the terrestrial environment. Upwelling can also increase productivity due to wind driven nutrient rich waters, such as off the coast of California, where upwelling-driven productivity supported the large sardine fishery made famous in John Steinbeck's *Cannery Row*. For freshwater environments, smaller and shallower freshwater bodies tend to be nutrient rich or eutrophic, supporting high productivity. Large, deep lakes can be nutrient poor or oligotrophic like the open ocean, with low productivity. However, human activities can significantly increase nutrient loading. Sources of natural nutrient loading include terrestrial runoff, rivers that feed into the lake and plant debris such as leaves. Nutrient loading from human activities includes runoff from animal manures and agricultural runoff, both of which contain high levels of nitrogen and phosphorus, the nutrients most often limiting in aquatic environments.

6.3.1.1 Photoautotrophy vs. Chemoautotrophy

Aquatic primary production is considered, and quantified, almost exclusively as photoautotrophy occurring in sunlit waters. Because photosynthesis is mediated by photopigments with characteristic absorption spectra such as

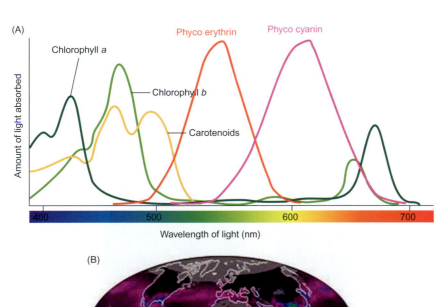

FIGURE 6.10 The absorption spectra of photosynthetic pigments (A). Modeled global primary production (B), from satellite data on chlorophyll abundance, measured by its characteristic spectral signature. *Source*: NASA.

chlorophyll (Figure 6.10A), it can be measured from space through satellite imagery (Figure 6.10B). Chemoautotrophy, the ability to fix carbon using chemical bond energy (i.e., the oxidation of reduced compounds) (see Chapter 3), does not have a similar such marker that is visible from space! Initially, chemoautotropy was believed to occur in extreme aquatic environments such as geothermal hot springs and oceanic hydrothermal vents, but gradually it has been increasingly recognized to occur in many more aquatic habitats. The question has now evolved from "where does chemoautotrophy occur?" to "where does chemoautotrophy not occur?"

Chemoautotrophs are seen throughout the water column, especially in oxygen minimum areas, which occur when high microbial decomposition (usually linked to high overlying primary productivity) strips oxygen from the water. The Gulf of Mexico's "Dead Zone" (Figure 6.11) is an example of a particularly large oxygen minimum area, and received its name because the extreme lack of oxygen in mid-depth waters causes fish die-offs. These areas are expected to increase with global change, due to increased water stratification from warmer temperatures, combined with more extreme precipitation events flushing agricultural fertilizers into aquatic systems. Chemoautotrophs have been discovered to be abundant in oxygen minimum waters, as well as in benthic systems (see Section 6.4.2.2), and are present at varying levels throughout aquatic habitats (Reinthaler et al., 2010; Swan et al., 2011). When inorganic carbon fixation is measured throughout the water column, consistent though low amounts can be seen through the dark subphotic waters (Figure 6.12), cumulatively (due to the large volume involved) representing a large and important though still poorly defined contribution to marine primary production.

Ammonia oxidizing archaeans are one form of ubiquitous pelagic chemoautrophs, virtually unknown until the mid-1990s (Schleper and Nicol, 2010). These marine pelagic Crenarchaota are a minor component of photic communities, but reach 35−40% of open-ocean microbes below 1000 m (Figure 6.13) (Karner et al., 2001). A variety of molecular methods, including gene and transcript surveys, genomics and metagenomics (see Chapter 21), show that chemoautotrophy via aerobic ammonia oxidation is likely the dominant lifestyle for these microbes (Church et al., 2010).

6.3.2 Secondary Production

Although we may think of the general planktonic food web in the aquatic systems as simply involving

HOW A DEAD ZONE FORMS

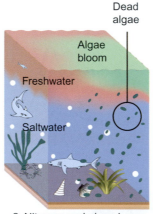 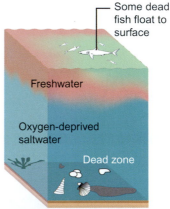

1. During the spring, sun-heated freshwater runoff from the Mississippi River creates a barrier layer in the Gulf, cutting off the saltier water below from contact with oxygen in the air.

2. Nitrogen and phosphorus from fertilizer and sewage in the freshwater layer ignite huge algal blooms. When the algae die, they sink into the saltier water below and decompose, using up oxygen in the deeper water.

3. Starved of oxygen and cut off from resupply, the deeper water becomes a dead zone. Fish avoid the area or die in massive numbers. Tiny organisms that form the vital base of the Gulf food chain also die. Winter brings respite, but spring runoffs start the cycle anew.

FIGURE 6.11   A so-called Dead Zone, where heterotrophic decomposition has stripped the water of oxygen, leading to large regions of anoxia and sometimes resulting in massive fish die-offs.

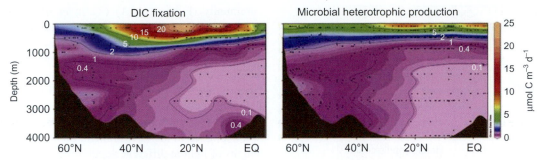

FIGURE 6.12 Chemoautotrophy in the eastern North Atlantic. On the left, dissolved inorganic carbon (DIC) fixation through the water column, measured by ^{14}C-bicarbonate fixation. The strong signal of photosynthesis in surface waters is clear, but low amounts of fixation continue throughout the vast dark waters below. On the right, microbial heterotrophic production as measured by ^{3}H-leucine incorporation showing that the subphotic pelagic chemoautotrophy is on the same order of magnitude as the heterotrophic production at those depths. Image from Reinthaler *et al.* (2010).

bacterioplankton and zooplankton consuming phytoplankton, the primary producers, which in turn are consumed by progressively larger organisms, the actual transfer of carbon and energy between trophic levels is much more complex (Figure 6.6; see also Figure 2.19 for an alternate image of a pelagic food chain). DOM represents a very large pool of carbon, roughly equivalent to the CO_2 in the atmosphere. This is because >50% of the carbon fixed by photosynthesis is released into the water column as DOM, which is rapidly utilized by heterotrophic microbes in a pathway in the aquatic food web referred to as the microbial loop. In this loop, bacterioplankton remineralize a portion of the DOM into CO_2 and nutrients, which in turn fuel new primary production—in fact, microbially recycled nutrients in the ocean's surface waters fuel roughly 80% of marine primary production (Duce *et al.*, 2008). Bacterioplankton also assimilate DOM to produce new biomass of their own, which is referred to as microbial secondary production. Thus, the microbial loop serves to efficiently utilize the DOM released into the water column. Because there is so much DOM, and the microbial loop allows its nutrients, carbon and energy to be retained in the sunlit surface waters to support more growth, the microbial loop is a key concept in aquatic systems.

Why is there so much DOM, and where does it come from? The DOM pool comes primarily from phytoplankton, with contributions from zooplankton and bacterioplankton, as well as from larger organisms through excretion and the lysis of dead cells. Among the phytoplankton, it is known that both "healthy" cells and "stressed" cells (those under some form of environmental stress) release DOM into the water column. In addition, "sloppy" feeding habits of zooplankton and larger organisms that eat phytoplankton may release a portion of their biomass as DOM into the water column. Finally, evidence indicates that as much as 6 to 26% of DOM is released during the lysis of phytoplankton and bacterioplankton by viruses (Ashelford *et al.*, 2003) (see Figure 6.6).

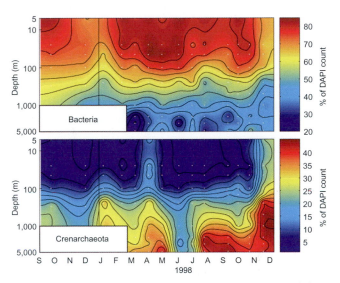

FIGURE 6.13 Marine Crenarchaea are abundant over space and time. This example shows the relative abundance of Bacteria (top) and Crenarchaeota (bottom) at Station ALOHA in the north Pacific off Hawai'i, through the water column over 15 months. Total cell counts were made by epifluorescence using the DAPI nucleic acid stain (see Chapter 13), and the relative proportion of each group was obtained using targeted probes with fluorescence *in situ* hybridization (FISH; see Chapter 13).

Thus, in a real aquatic food web, the heterotrophs (the bacterioplankton and zooplankton) consume each other, DOM, POM and autotrophs (phytoplankton, the main primary producers). The zooplankton in turn are consumed by larger organisms such as fish and other filter feeders. In the open ocean it takes approximately five steps or trophic levels to produce exploitable fish. In coastal zones it takes 1.5 to 3.5 steps to produce fish because primary production levels are higher. There is often a temporal lag between primary and secondary production; Figure 6.14 shows that an increase in phototrophs (as measured via chlorophyll *a*, a photosynthetic pigment) is followed by an increase in heterotrophs in a marine system.

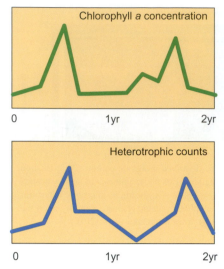

FIGURE 6.14 Diagram of the interrelationship between the concentration of chlorophyll *a*, a photosynthetic pigment and heterotroph density. The concentration of chlorophyll in water is related to the amount of primary production. This in turn influences the amount of secondary production by heterotrophic populations. In this figure, it can be seen that as the chlorophyll *a* concentration increases, it is closely followed by an increase in heterotrophic populations. Thus, secondary production is intimately tied to primary production. Adapted from Rheinheimer (1985).

6.3.2.1 Photoheterotrophy: A Newly Appreciated Microbial Lifestyle

Until recently, microbial life in aquatic systems was imagined to occupy the two main categories of (1) photosynthesizers, living in the surface water and forming the base of the food chain, and (2) the heterotrophs, living solely off that sunlight-fueled primary production. Increasingly, we know this was too simplistic a view. As described above, chemoautotrophy is likely an appreciable additional though poorly understood source of primary production in many aquatic systems. In addition, heterotrophs may not be as exclusively reliant on primary production as we have understood. They may be able to supplement their energy, but not carbon, requirements through the use of sunlight. This photoheterotrophy is not photosynthesis because it does not involve carbon fixation, and, since CO_2 is not being fixed, these organisms are still fundamentally "heterotrophs." There are several different ways that heterotrophs can directly, or indirectly, use sunlight for energy (reviewed in Moran and Miller, 2007), and two of these ways are highlighted briefly here.

First, aerobic anoxygenic phototrophy (AAnP), which uses bacteriochlorophyll to capture sunlight for energy without producing oxygen, appears to be widespread among diverse marine microbial lineages (Moran and Miller, 2007). In the freshwater and marine sites examined so far, the percentage of surface microbial cells carrying the genes for AAnP ranges from <1% to 10% of the

community (Figure 6.15A) (Yutin *et al.*, 2007). Second, a simple type of light-harvesting molecule related to the rhodopsins in our own retinas was discovered in marine microbes (reviewed in DeLong and Béjà, 2010). Since it was initially described in a Proteobacteria, it was termed proteorhodopsin, and was demonstrated in the lab to be a light-activated proton pump that generates proton gradients available for ATP production (Figure 6.15B). In addition, it is expressed in the environment, tuned to different wavelengths of light at different depths, and confers growth benefits in some cultured microbes. The gene encoding proteorhodopsin is present in a remarkable 13−80% of marine surface bacteria and archaeans (DeLong and Béjà, 2010). The current understanding for how both of these strategies works is as a way for heterotrophs to survive lean times, and be faster to respond when new food is again present. This has significant impacts on marine carbon cycling, energy budgets and ideas about the ecology and evolution of these communities. Yet, the prevalence of photoheterotrophy was unknown until very recently, and there is much about it that remains to be understood.

6.4 MARINE ENVIRONMENTS

Marine environments are one of the dominant groups of habitats on the planet. Oceans cover ≈70% of Earth's surface, such that a more accurate name for the planet might be Ocean. Beyond surface area, the ocean's vast volume makes its importance as a habitat even greater. The ocean's average depth is ≈4000 m, and its deepest spots are ≈11,000 m; this ocean habitat is mostly dark (except for bioluminescence) and under high pressure (Figure 6.16 and Table 6.2). It encompasses a remarkable diversity of conditions and life forms, and we are only at the early stages of mapping these.

Marine microbiology has had a major role in propelling environmental microbiology forward (see Information Box 6.3), through the discovery of new physiologies (e.g., photoheterotrophy, Section 6.3.2.1), and the use of cutting-edge methods (e.g., metagenomics; see Chapter 21). These advances were led over the last several decades by a small number of individuals, and have resulted in a now large, robust community of marine microbiologists and microbial oceanographers. The number of global exploration efforts centered on marine microbiology are one indication of the success of this field; these include the International Census of Marine Microorganisms (ICoMM), which has profiled microbial communities from more than 500 global sites using the 16S rRNA gene (see Chapter 21 for how such profiling works); the Global Ocean Survey, which profiled roughly 150 communities using shotgun sequencing (thus capturing a random sampling of microbial genes rather than just the 16S rRNA marker gene; see Chapter 21 for details);

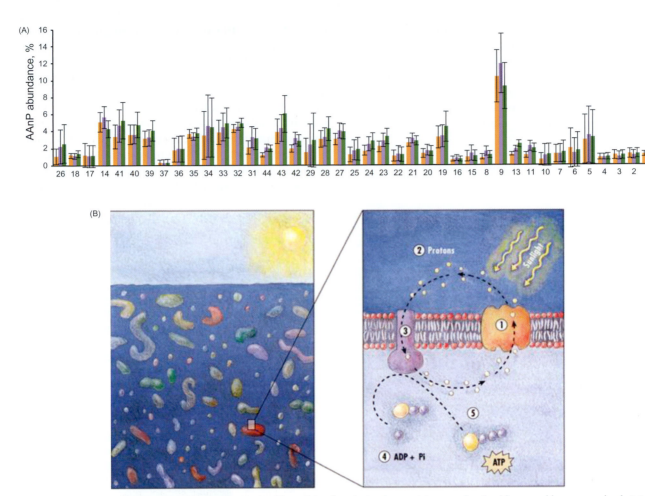

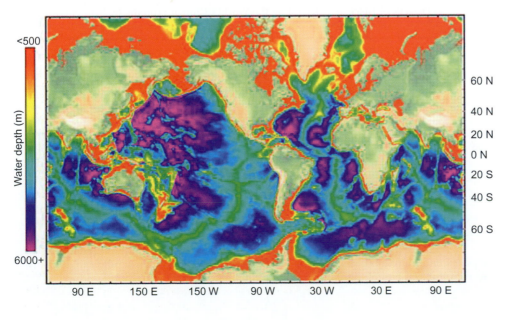

FIGURE 6.15 Photoheterotrophy. (A) The percentage of bacterial cells calculated to contain genes involved in anaerobic anoxygenic phototrophy, in Atlantic and Pacific surface waters, using the metagenomic data from the Global Ocean Survey (numbers on the X-axis refer to sampling station numbers). From Yutin et al. (2007). (B) An artist's rendition of microbial cells in the surface ocean, and how proteorhodopsin works in the cell membrane to capture light energy to pump protons, producing the proton gradient that ATPase can then harness to make ATP. From DeLong and Béjà (2010).

FIGURE 6.16 Global map of ocean water depth. The water depth scale is from less than 500 m (red) to 6,000+ m (purple). From Orcutt et al. (2011).

and the Tara Oceans survey, which is undertaking a trophi-
cally integrated virus-through-zooplankton survey across
more than 150 global sites, using a variety of molecular and
microscopic techniques. The extremely large volumes of
data coming from such efforts continue to fuel discoveries
and novel analyses of marine microorganisms' ecology and
evolution that are not yet possible for microbes in most
other habitats.

6.4.1 Marine Planktonic Communities

The ocean contains diverse microbial habitats, both vertically (neuston to abyssopelagic depths) and horizontally
(coastal upwelling regions versus open ocean gyres, the
Mediterranean versus the Antarctic Ocean). As a general
rule, microbial concentrations are highest within the neuston
and drop markedly below this region (Figure 6.17); surface
waters contain up to $\approx 10^8$ microorganisms/ml and decrease
by more than 10-fold at a depth of 100 m. Coastal oceans
support on average 10-fold higher microbial numbers
than open oceans, due to terrestrial nutrient and carbon
inputs; this is especially true in populated coastal areas
(Rheinheimer, 1985). Oceans have profiles similar to those

TABLE 6.2 Estimated Volumes of Ocean Habitats

Habitat	Vol. (m³)
Water column (<200 m below sea level)	3.0×10^{16}
Water column (200+ m below sea level)	1.3×10^{18}
Hydrothermal plumes[a]	7.2×10^{13} (/yr)
Subsurface ocean	$!10^{16}$
Sediment, all	4.5×10^{17}
Shelf sediment	7.5×10^{16}
Slope sediment	2×10^{17}
Rise sediment	1.5×10^{17}
Abyssal sediment	2.5×10^{16}
0- to 10-cm layer	3.6×10^{13}
Ocean crust[b]	2.3×10^{18}

Adapted from Orcutt et al. (2011).
[a]*The volume of hydrothermal plumes is given as the volume of plume fluid produced per year.*
[b]*The volume of oceanic crust was assumed by multiplying the average thickness of the oceanic crust (7 km by the assumed area of seafloor underlain by crust (65% of Earth's surface, or 3.3×10^{14} m²)).*

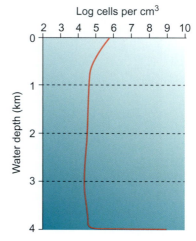

FIGURE 6.17   Cell density through the water column and surface sediments at station BIOTRANS in the north-east Atlantic. Note log scale.
From Jørgensen and Boetius (2007).

Information box 6.3   Marine Microbes as Drivers of Discovery: Doubling the World's Known Proteins by Sequencing the Sargasso Sea

The oceans harbor a staggering number of microorganisms that
mediate biogeochemical cycles which affect the entire planet. In
2003, a team of scientists led by J. Craig Venter (a leader of
sequencing the human genome) embarked on round-the-world
yacht trip to sample surface marine microorganisms and sequence
their genomes. This monumental effort, termed the Global Ocean
Survey, undertook the use of metagenomic analysis (sequence
analysis of all the DNA in a sample, see Chapter 21) to increase
knowledge of ocean microorganisms (Falkowski and de Vargas,
2004; Venter et al., 1994). The nutrient-limited Sargasso Sea, a
two-million square mile portion of the North Atlantic Ocean, was
chosen for the journey's starting place on the assumption that its
oligotrophic status would support a simpler community, more
amenable to sequencing.

Sargasso Sea water samples were passed through filters (from
0.8 μm down to 0.1 μm) to capture microbial members of this
marine environment. Once microorganisms were harvested, the
scientists extracted DNA from the filters. Shotgun sequencing, a
high-throughput technique commonly used in the field of genomics, was then used to characterize the community DNA. Using
this approach, the team generated truly staggering amounts of
data over the next years of their journey. Yooseph et al. (2007)
organized these data into protein clusters of similar sequences, as
a means of analyzing it without relying on reference databases for
matching and identification (an approach which remains more
common as it is easier, but is highly biased by, and limited to, the
composition of the reference database(s) used and typically only
permits analysis of a subset of the data). Using this approach,
Yooseph et al. discovered that the marine microbial protein diversity discovered by their team to date—which was just a portion of
the final total—doubled the world's known diversity of protein
types.

of lakes presented later in the chapter, depending on whether the marine environment is oligotrophic like the open ocean, or eutrophic like coastal waters, especially coastal waters where sewage outpours may be present. Due to numerous efforts such as the International Census of Marine Microorganisms (ICoMM), we now know fairly definitively that pelagic bacterial communities are dominated by surface Cyanobacteria (particularly *Prochlorococcus*, described below), Alphaproteobacteria (driven by SAR11, described below) and Gammaproteobacteria (Zinger *et al.*, 2011) (Figure 6.18A and B).

6.4.1.1 Making Half the Oxygen You Breathe: Marine Phytoplankton

Overall, marine phytoplankton (Figure 6.5) make half of the oxygen you breathe, and this contribution is in turn divided about 50:50 between the cyanobacteria (Figure 6.19) and the eukaryotic algae. Cyanobacteria, dominated by the genera *Prochlorococcus* and *Synechococcus*, account for a quarter of global primary production. As the "coccus" of their names implies, they are small, round cells that are morphologically relatively

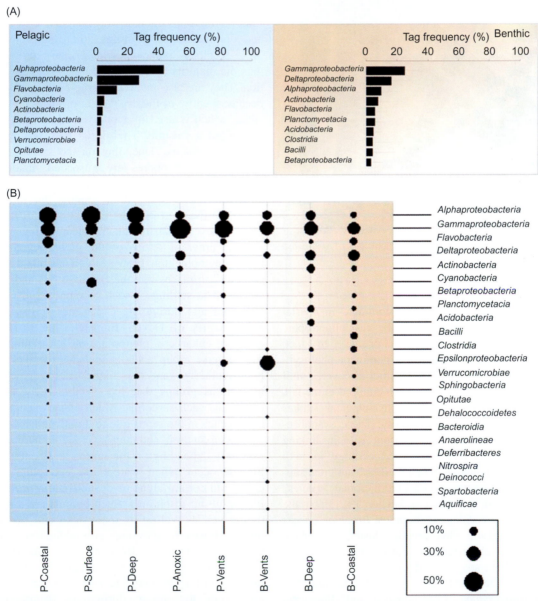

FIGURE 6.18 Bacterial community composition across 509 samples spanning the ocean's pelagic and benthic habitats, from the International Census of Marine Microorganisms. This represents 9.6 million sequences of amplicons of the 16S rRNA gene V6 hypervariable region, sequenced by 454 pyrosequencing (see Chapter 21). (A) The top 10 bacterial classes in each habitat (not including hydrothermal vents and anoxic habitats in the benthic averages). (B) Average abundances of bacterial taxa in various pelagic (P) and benthic (B) ecosystem types. From Zinger *et al.* (2011).

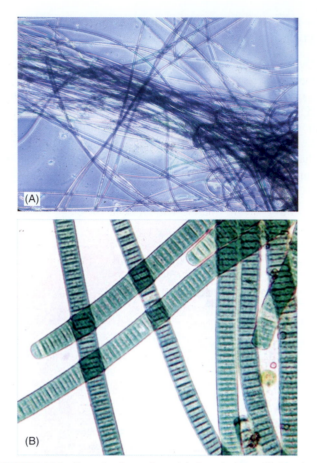

FIGURE 6.19   Example of cyanobacterial phytoplankton. (A) *Lyngbya* and (B) *Oscillatoria*.

details of this process and the nitrogen cycle). Although heterotrophic marine nitrogen fixers exist, cyanobacterial nitrogen fixers are currently considered to be the group performing the vast majority of marine nitrogen fixation.

Eukaryotic phytoplankton are responsible for roughly the other half of marine primary production, thus about a quarter of the primary production on the planet. They are also some of the singularly most beautiful microorganisms known (Figure 6.5). The remarkable morphologies many possess are created by hard outer coverings with intricate designs, the shapes of which can help keep them from sinking out of the photic zone. Eukaryotic phytoplankton are also responsible for much of the periodic bioluminescence seen in surface waters at night. Important groups include the diatoms, coccolithophores and dinoflagellates. Diatoms are responsible for about 20% of global photosynthetic primary production, and thus the majority of eukaryotic phytoplankton's contribution. They are also involved in the cycling of silica in the oceans: 4—50% of the dry weight of their cells is made of silica, which is used in a two-valved extracellular skeleton called a frustule. Their cells may be unicellular or form long chains, depending on the species and conditions. Coccolithophores are a second major group of eukaryotic phytoplankton, whose name comes from the Greek word for "round stone-bearers" due to the calcium carbonate ($CaCO_3$, i.e., chalk) plates that cover their surfaces. They are small but significant primary producers (on the order of $\approx 5\%$ of the global signal), and like many phytoplankton grow to high abundance known as "blooms" in places like the Sargasso Sea and the Gulf of Alaska. The settling of such blooms to old ocean floors millennia ago during the Cretaceous formed the famed White Cliffs of Dover, England. The final important group of eukaryotic phytoplankton is the dinoflagellates. This diverse group actually includes members unable to photosynthesize (and thus living heterotrophic lifestyles), as well as groups that live symbiotically with marine organisms such as corals. Much of the planktonic bioluminescence in the sea is due to dinoflagellates.

Phytoplankton abundance and community composition vary depending on the season and conditions. Algal blooms occur when waters are eutrophic, warm and calm, and certain algae or cyanobacteria proliferate rapidly resulting in blooms. Such blooms are a natural part of the yearly cycling of many lake and ocean ecosystems. However, extreme eutrophication and bloom events can adversely affect the water quality in several ways. As described above, high amounts of primary production settling through the water column can strip oxygen from the water during its decomposition. In addition, water experiencing algal blooms tends to be unpleasant for recreation (fishing, boating and swimming) because of odors and slime. In the worst cases, harmful algal blooms are composed of algae that

nondescript. *Synechococcus* tend to dominate in coastal regions, while *Prochlorococcus* rule the vast oligotrophic surface waters in the center of oceans. In fact, *Prochlorococcus* are the numerically dominant photosynthetic cells on the planet—and yet were not discovered until 1988! These two genera have also been cultured in the lab, and been the focus of extensive physiological studies including the distributions and diversity. These cyanobacterial genera have played an important role in linking microbial diversity to physiology and ecology in both cultures and wild populations. Their studies leant great support to the microbial "ecotype" concept: microbial "species" are extremely difficult to define (see Section 19.2.3), but the rigorous diversity and ecological studies of *Prochlorococcus* clearly revealed related phylogenetic clusters that share defining ecological characteristics (for example, high-light adapted vs. low-light adapted) (Moore *et al.*, 1998). Other important cyanobacteria include lineages that fix nitrogen, including *Trichodesmium*, *Crocosphaera* and *Anabaena*, with *Trichodesmium* believed to account for about half of global marine nitrogen fixation (Bergman *et al.*, 2013). As in all habitats, in the oceans nitrogen gas must be fixed into ammonia to be biologically available (see Chapter 16 for produce potent toxins, sometimes resulting in coastal red tides. These get their name from the most notorious sources, red-pigmented dinoflagellates. The red tides of Florida, known for massive fish-kills washing onto beaches and human respiratory irritation, are caused by the dinoflagellate *Karenia brevis* blooming throughout the Gulf of Mexico, and then accumulating along the Southeast U.S. coast. *K. brevis* is a highly toxic alga that causes human illness, shellfish toxicity, animal and bird mortalities, and reddish water discoloration. As global change conditions exacerbate eutrophication and bloom-supporting conditions (warmer, more stratified waters), harmful algal blooms and red tides are likely to become more frequent and intense.

6.4.1.2 Heterotrophic Marine Microbes

There are several ubiquitous, abundant marine heterotrophic microbes, analogs to the major photoautotrophs *Prochlorococcus* and *Synechococcus*, described above. By far the most important is the alphaproteobacterial SAR11 clade typified by *Pelagibacter ubique*, whose name literally means ubiquitous pelagic bacteria. This clade was discovered in 1990 in 16S rRNA gene surveys and is seen throughout the world's oceans, from the tropics to the poles. Since it could not be grown, it bore the name SAR11, based on its first description as a sequence clade, from the 11th clone sequenced from a clone library made from Sargasso ("SAR") seawater. Finally, in 2002, this clade's "birth announcement" came after *Pelagibacter ubique* was successfully grown in the lab and its physiology began to be studied (Rappé *et al.*, 2002). Remarkably, about a third of all bacterial cells in the ocean are from the SAR11 lineage, which may be its own family in the Alphaproteobacteria. They are equally abundant in shallow and deep waters (Zinger *et al.*, 2011). Why is SAR11 so successful? It has a highly streamlined genome, which allows it to survive and reproduce even in very low-nutrient conditions. It also has diversified into a number of "ecotypes" specialized to different subhabitats, like *Prochlorococcus*. Lastly, its success may be partly due to its ability to supplement its energy needs (though not its carbon needs) by phototrophy, since at least some of the clade's members have the ability to perform photoheterotrophy (see Section 6.3.2.1) (Giovannoni *et al.*, 2005), which may allow them to get through lean times and channel more scarce resources into biomass rather than energy.

There are a number of other important heterotrophic groups in the oceans. They include microbes, such as the Deltaproteobacteria that often dominate deeper water, and Euryarchaeota. Euryarchaoeta are less abundant overall than pelagic Crenarchaeota (which are highlighted in Section 6.3.1.1), but are ubiquitous, and appear to be seasonally important members of surface communities. A recent population genome assembly of one (from a metagenome, see Chapter 21) suggests a photoheterotrophic lifestyle via proteorhodopsin (see Section 6.3.2.1) with their carbon substrates of choice being proteins and lipids (Iverson *et al.*, 2012). Other important marine pelagic heterotrophs include protozoa such as flagellates and viruses (see next section), as well as fungi. Heterotrophs not only consume DOM and POM, but can be active predators; it is estimated that protozoa and viruses are responsible for similar amounts of bacterial mortality (Wommack and Colwell, 2000).

Fungi in the marine habitat occur in endolithic associations with limestone, the shells of sea creatures, sponges and corals (Golubic *et al.*, 2005). They have also been isolated from carbon-rich areas of the water column and benthic habitat. The distribution of fungi in aquatic environments is not well studied; however, there is increasing interest in useful secondary metabolites that marine fungi may produce. Another surprising fact is that fungi have been isolated from sediment samples taken at depths of 5000 m in the Central Indian Basin (Damarem *et al.*, 2006).

6.4.1.3 Marine Viruses

Viruses are important to almost any ecosystem so far studied (see Section 2.4 for a general overview of viruses), and are the most abundant biological entities on the planet (10^{31} on Earth), commonly outnumbering bacteria about 10 to 1. If laid end to end, 10^{31} viruses would stretch 10^8 light years away from Earth, and comprise a biomass equivalent to ≈ 75 million blue whales. Such astronomical numbers have led to speculation that viruses represent the largest unexplored genetic reservoir on Earth.

Ocean viruses were thought nearly nonexistent until 1989, when seawater was concentrated onto electron microscopy grids, allowing their direct observation. Previous culture-based studies had used nonmarine microbial hosts, and thus had unsurprisingly failed to recover marine viruses. Now marine viruses are perhaps the best-studied environmental viruses, and are known to play diverse roles in marine ecosystems (reviewed in Breitbart, 2011). They impact microbes through mortality (cell lysis), horizontal gene transfer and the modulation of host metabolisms. Additionally, they alter global biogeochemistry. In fact, through lysis and the production of POM and DOM, they are responsible for the largest ocean carbon flux (150 Gt/yr) dwarfing all others by >five-fold). Viruses that infect phytoplankton are also important to nutrient cycling, because their lysis of phytoplankton's primary production promotes secondary production, as described in Section 6.3.2. Marine viruses also infect larger marine animals such as fish and crabs, and can thereby cause considerable economic losses for the fishing industry, particularly in aquaculture settings with high host densities.

With new genomics-based windows (Chapter 21) into wild viral communities, the extent of viral impacts on large-scale biogeochemistry, beyond simple lysis, has been increasingly revealed. For example, viruses can contain "host" genes, which are expressed during infection to metabolically reprogram infected cells, in order to maximize viral production. The mechanisms by which this occurs can have profound biogeochemical impacts. The most surprising discovery of this is that cyanobacterial viruses carry "host" photosynthesis genes (reviewed in Breitbart, 2011). These genes are expressed during infection, allowing the infected cells to maintain photosynthesis (and thus energy production) longer, and are hypothesized to power the production of more progeny viruses. A significant portion of marine cyanobacterial photosynthesis in fact appears to be performed by phage-encoded photosynthesis proteins (Sharon *et al.*, 2007). Evolutionarily, these virally encoded photosynthesis genes are a dynamic gene pool, with the diversity generated among phage copies able to recombine back into the host gene pool (Sullivan *et al.*, 2006). This means that viruses appear to help shape global photosystem evolution! Such discoveries demonstrate the central ecological and evolutionary roles of viruses. However, in spite of advances in marine virology, environmental virology is bottlenecked by "unknowns." Fundamental questions such as "who infects whom?" remain open; 90% of each viral genome and metagenome is typically new to science, and cultured model systems insufficiently represent wild diversity. This is an area open for many more discoveries.

6.4.2 Marine Benthic Communities

Benthic environments are habitats of steep redox gradients and abundant microniches, as described in Section 6.2.3 along with the general characteristics of benthic aquatic microbes. The seafloor provides diverse habitats that can be grossly divided into soft-bottom sediments, hard-bottom rocky ocean crusts and hydrothermal vents. The latter is an extreme environment covered in detail in Section 7.4.1. There are a number of other notable but less abundant benthic habitats whose microbial communities have been studied in some detail; these include cold seeps and, though the concept may be surprising, "whale-falls," where the carcasses of whales fuel thriving successional ecosystems (see *Osedax* the "bone-eating" worm, Figure 6.20). Symbioses between microbes and macrobes are a hallmark of many of these systems.

FIGURE 6.20 Two flatfish lurk near the vertebrae of a dead whale about 600 meters below the ocean surface in Monterey Canyon. Researchers at the Monterey Bay Aquarium Institute (MBARI) towed this whale carcass off a beach in Monterey Bay and placed it on the seafloor so that they could study the animals that fed on and colonized the carcass over time. The orange specks on the bones are amphipods, which consume the flesh of the recently sunk whale. MBARI researchers also discovered a new genus of worms (*Osedax*) that burrow into the whale bones after the flesh has been consumed. Instead of legs, these worms have "roots" which infiltrate the whale bone and contain unique endosymbiotic bacteria that can digest collagen within the whale bones and thus provide nutrition for the worms. © 2007 MBARI.

6.4.2.1 Marine Sediments

The seafloor is the final resting place for "marine snow" (sinking organic particles) and larger dead organisms, and this rain of organic matter from above can be an important food source for heterotrophic marine microorganisms. It can also add seasonality to the seafloor, where temperatures and other physicochemical conditions may remain more constant through the year. Rates of decomposition typically decline with depth, as does the amount of organic matter reaching the ocean floor. Concomitantly, the number of cells in the surface sediments typically declines with the seafloor depth.

From the global perspective enabled by ICoMM (one of the few broad efforts to survey diverse benthic habitats), Gammaproteobacteria comprise an average of ≈25% of global benthic bacterial communities surveyed (their analyses so far have focused on bacteria), while Deltaproteobacteria are the next most abundant phylum at ≈16% (Zinger *et al.*, 2011) (Figure 6.18). Overall, these groups, along with the next three most abundant phyla, the Planctomycetales, Actinobacteria and Acidobacteria, contain chemoautotrophs and anaerobic or microaerophilic heterotrophs, making them suited for benthic living. Coastal sediments are distinguished by lineages typically considered terrestrial or indicative of human contamination, the Clostridia and Bacilli (Zinger *et al.*, 2011). In deep-sea sediments, it appears that Acidobacteria in the upper sediments give way to Chloroflexi and the candidate division JS1 in lower sediments (Zinger *et al.*, 2011).

Other research reveals that Archaea and Eukarya are important components of the soft benthos. Crenarchaeota and Euryarchaeota are abundant in sediments, and benthic diatoms, forams and radioladiarans contribute to the benthic food web.

6.4.2.2 Rock-eaters Under the Sea

When we consider seafloor communities, we typically think of soft-bottomed habitats. However, much of the ocean's floor is made of basaltic crust, which extends several kilometers below the seafloor. Due to the resulting massive volume, basaltic crust is actually the largest potential habitat on Earth. But is it actually a habitat— can things live on rocks at, and below, the bottom of the sea? Discoveries just within the last decade suggest that chemolithoautotrophs—"autotrophs" meaning they fix their own carbon, "chemo" meaning they use chemical energy to do it and "lithos" meaning that the chemical energy comes from rock—are abundant, diverse and active in basaltic crusts. In fact, microbial cell numbers on seafloor basalt are typically 3−4 *orders of magnitude* higher than in the overlying waters (Santelli *et al.*, 2008), then decrease in the subseafloor. The physiologies of these abundant cells are just beginning to be investigated, and will likely be diverse since their communities appear more diverse than those in either deep or surface ocean waters (Santelli *et al.*, 2008). However, chemolithoautotrophy appears to be a major lifestyle, through various proposed mechanisms including iron, sulfur and manganese oxidation (Santelli *et al.*, 2008), and as supported by the presence of diagnostic gene sequences, isotopic investigations and incubation experiments (Lever *et al.*, 2013). Even before the last decade's discoveries in this habitat, subseafloor microbes had been estimated to comprise 10−30% of the total living biomass of Earth (Whitman *et al.*, 1998). Understanding planetary carbon and energy cycles is simply not possible without understanding these systems, and their continued exploration is certain to yield new discoveries for years to come.

6.5 FRESHWATER ENVIRONMENTS

Freshwater environments, such as springs, rivers and streams, and lakes, are those not directly influenced by marine waters. The science that focuses on the study of freshwater habitats is called limnology, and the study of freshwater microorganisms is microlimnology. There are two types of freshwater environments: running water, including springs, streams and rivers; and standing water, including lakes, ponds and bogs. These freshwater environments have very different physical and chemical characteristics, and correspondingly different microbial communities and activities. For instance, the microbial community in a lake in Egypt is not the same as the microbial community in one of the Great Lakes in the northeastern United States. In this section we define various freshwater environments and outline the types of microorganisms that inhabit them.

6.5.1 Springs, Streams and Rivers

Springs form wherever subterranean water reaches Earth's surface. Microorganisms, especially bacteria and algae, are often the only inhabitants of springs. In general, photosynthetic bacteria and algae dominate spring environments, with communities ranging from 10^2 to 10^8 organisms/ml. These primary producers are present in the highest concentrations (10^6 to 10^9 organisms/ml) along the shallower edges of the spring and in association with rock surfaces, where light is available and inorganic nutrients are in highest concentrations (Rheinheimer, 1985; Kaplan and Newbold, 1993). Although heterotrophs are also present, numbers are usually low (10^1 to 10^6 organisms/ml) because DOM is low. As they mature and die, photosynthetic populations provide the initial source of organic matter for downstream heterotrophic populations. However, the largest portion of DOM found in surface freshwater originates from surrounding terrestrial sources. This organic input, which originates from sources such as plant exudates, dead plants, animals and microbial biomass, is transported into standing water habitats by mechanisms such as terrestrial runoff, seepage and wind deposition. Thus, we have the image of spring water starting at its source with very low concentrations of DOM and heterotrophs. The DOM and the heterotrophic populations steadily increase as the spring moves away from the source and as inputs of terrestrial organic matter and microbial biomass continue to accumulate (Kaplan and Newbold, 1993).

Springs, as they flow away from their subsurface source, merge with other water sources to form streams and rivers that eventually flow into other bodies of water such as lakes or seas. Streams contain primary producer communities, especially when light can penetrate to the bottom of the stream. Photosynthetic populations range from 10^0 to 10^8 organisms/ml and tend to be present as attached communities associated with biofilms because of the flowing nature of the water column. Phytoplankton (free-living) communities also exist in streams, but because of the constant water movement, they are not spatially stable populations (Rheinheimer, 1985).

As a stream progresses and becomes larger, it tends to accumulate DOM from surface runoff and sediments. The increase in DOM limits the penetration of light and consequently begins to limit photoautotrophic populations.

In turn, heterotrophic populations begin to increase in response to increased DOM. In general, the concentration of heterotrophs in streams and rivers ranges from 10^4 to 10^9 organisms/ml, with microbial numbers increasing as DOM increases. Because of their flow patterns, stream and river waters are for the most part well aerated, meaning that their microbial inhabitants are predominantly aerobic or facultatively aerobic. Although isolated pools that form in rivers act as DOM and POM sinks and support fairly stable heterotrophic planktonic communities, the only truly stable populations in the flowing habitats of streams and rivers are the biofilm and sediment (benthic) communities (Rheinheimer, 1985).

6.5.2 Lakes

Lakes are among the most complex of the freshwater environments. They may range from small ponds to vast lakes that generate their own weather patterns, such as Russia's Lake Baikal, which holds roughly one-fifth of the world's unfrozen surface freshwater, and is the deepest lake on the planet, and the U.S.A.'s Great Lakes. Although often regarded as nonflowing environments, lakes have inflows and outflows. Lakes may have unique chemical composition, and can form extreme environments (see Chapter 7); examples include salt lakes (see Section 6.6.2), bitter lakes that are rich in $MgSO_4$, borax lakes that are high in $Na_2B_4O_7$ and soda lakes that are high in $NaHCO_3$.

Lake microbial communities and their interactions are complex and diverse, reflecting the complexity of the habitat. Lakes contain extensive primary and secondary productive populations that interact dynamically. The primary productivity in the shallow near-shore waters is high, driven predominantly by algae and secondarily by cyanobacteria. The attached communities here are dominated by the presence of filamentous and epiphytic algae. Central lake waters are dominated by phytoplankton, which form distinct community gradients based upon the wavelength and the amount of light that penetrates to a given depth (Figure 6.21). One example of a lake-dwelling microbial phototroph with a specialized niche is *Chlorobium*, a green sulfur bacterium. *Chlorobium* can use longer wavelengths of light than many other phototrophs, meaning they can live deeper. They are also anaerobic organisms, using H_2S rather than H_2O for photosynthesis (see Section 16.4.3.2). Thus, they have a competitive advantage in establishing a niche at depths lower in the water column or even in the surfaces of sediments, where only small amounts of light penetrate, little or no oxygen is present, but hydrogen sulfide is available.

In addition to their phototrophic populations, lakes have extensive heterotrophic communities. Heterotrophic concentrations vary with depth, but there are three areas

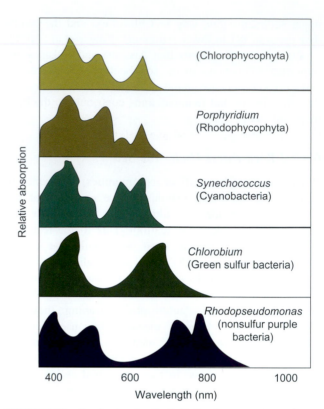

FIGURE 6.21 Graphs showing the light absorbance spectrum of common phytoplanktonic algae and photosynthetic bacteria. It can be seen that each of these groups has a different profile. This enables groups to take advantage of their niche. In general, organisms that are capable of utilizing longer wavelengths are found deeper in the water column. Thus, they do not have to compete with organisms higher in the water column that absorb the shorter wavelengths. Adapted from Atlas and Bartha (1993).

that generally have elevated numbers of heterotrophs (with some parallels to marine systems). These are areas with higher nutrients and organic matter: the neuston layer; the thermocline, where organic debris tends to settle and accumulate; and the upper layer of the benthos, where the heterotrophic populations are mainly anaerobic.

There are some striking differences between oligotrophic and eutrophic lakes; Figure 6.22A and B compare the major bacterial populations typically found in each (Konopka, 1993). Oligotrophic lakes have higher rates (four- to 20-fold) of primary production than eutrophic lakes, due to their deeper light penetration. In oligotrophic lakes, as might be expected, the amount of secondary production is directly coupled to primary production, and secondary production in the photic zone is generally 20 to 30% of primary production. Eutrophic lakes have much higher ($\approx$ three to 80 times) rates of secondary production than oligotrophic lakes, and of a decoupling from primary production than oligotrophic lakes (Atlas and Bartha, 1993).

Apart from their microbial and algal populations, streams, rivers and lakes also contain fungi, protozoa and

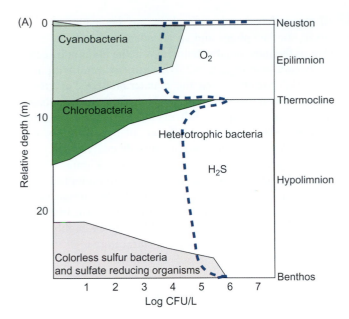

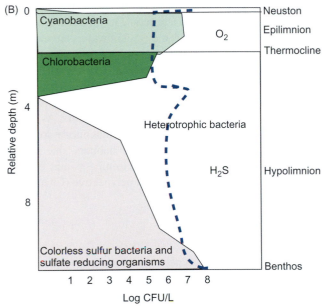

FIGURE 6.22   (A) Schematic representation of bacterial distribution in a typical oligotrophic lake. Notice especially the distribution and concentrations of the photosynthetic populations. Also note the lower concentration of heterotrophs in the upper zone, where cyanobacteria predominate. The large increase in the heterotrophic population between the epilimnion and the hypolimnion is related to the presence of a zone where organic matter accumulates. This area is known as a thermocline and is a zone where the sunlight-warmed surface water (less dense) and the deeper colder water (more dense) meet, forming a density gradient where organic matter accumulates. (B) Schematic representation of a typical eutrophic lake. This figure shows the same groups of organisms as in (A) indicating the localization and relative concentrations throughout the water column. Notice that both the photosynthetic and the heterotrophic populations are considerably higher in a eutrophic lake. Adapted from Rheinheimer (1985).

viruses, which interact and contribute to the functioning of the food web. Fungi serve as parasites of planktonic algae, preventing overpopulation and allowing light to penetrate farther into the water column. Other fungi have simple lifestyles, colonizing surfaces and often forming fungal lawns, whereas still others can have complex predatory lifestyles. A fascinating example is *Zoophagus insidians*, which live attached to filamentous green algae in rivers and lakes. The fungi have long hyphae which trail down into the water column forming fishing lines: when touched by prey such as rotifers, the hyphae rapidly secrete a sticky substance, ensnaring the microscopic animal. The hyphae then grow quickly into the mouth of the rotifer and form a fungal mycelium that absorbs the contents of the animal's body from the inside.

Protozoa and viruses are important predators of aquatic microorganisms. Protozoan populations are typically several orders of magnitude lower than bacterial numbers. They are able to affect the numbers of bacteria and algae because each protozoan is able to consume hundreds of bacteria and algae per day. Protozoa and their food species often exhibit the cyclic, temporally offset population dynamics typical of predator–prey relationships, where the prey is abundant first, then the predators' numbers rise and cull the prey, until food is limiting and predators decrease again. Viral and host populations fluctuate in a similar fashion (Wommack and Colwell, 2000). Viruses in freshwater environments can be very abundant, with viral numbers exceeding bacterial by two orders of magnitude (versus one in most marine habitats). Like marine viruses, freshwater viruses have diverse hosts, including microbes, microalgae, protozoa and larger organisms. As in marine systems, viruses can cause an appreciable amount of bacterial mortality (20 to 60%) (Suttle, 1994; Hennes and Simon, 1995), and the associated release of DOM fuels secondary production (Middledoe *et al.*, 1996), as discussed in Section 6.3.2. Together, protozoan and viral populations help to control the concentrations and biomass of the bacterial and algal communities (Wommack and Colwell, 2000).

6.6 OTHER NOTABLE AQUATIC ENVIRONMENTS

6.6.1 Brackish Waters

Brackish water is a broad term used to describe water whose salinity is between that of fresh and marine water, and these are often transitional areas where such waters mix. An estuary, which is the part of a river that meets the sea, is the best-known example of brackish water.

Estuaries are highly variable environments because the salinity can change drastically over a relatively short distance, ranging from 10‰ to 32‰ (Information Box 6.1), and over time of day due to tidal cycles (for example, high tide bringing saltier marine waters farther up into the estuary). Seasonal increases in freshwater due to rainfall or snowmelt will decrease the salinity at a given point in the estuary. In order to survive here, resident microbes must be adapted to these large fluctuations in salinity. Despite this challenge, estuaries are very productive environments. In general, estuarine primary production is low, due to poor sunlight penetration as a result of high turbidity, which occurs from suspended organic matter brought by river inflow and tidal mixing (Ducklow and Shiah, 1993). However, heterotrophic activity and secondary production are high. Primary and secondary production are decoupled in these systems, because of the large amounts of organic carbon brought by terrestrial runoff and river inflow. In fact, the supply of carbon and nutrients can be so great that in many cases estuaries can actually become anoxic for whole seasons during the year (Ducklow and Shiah, 1993).

6.6.2 Hypersaline Waters

Hypersaline environments include coastal lagoons, salt and soda lakes, salterns (human-made hypersaline ponds for producing salt; Figure 6.23A), deep-sea brine pools (formed from the dissolution of salt during seafloor tectonic activity), brine channels in sea ice, and fermented foods and pickling brines. Hypersaline environments have higher salinities than seawater ($\approx 35‰$) and may even be salt saturated. At room temperature, saturation of freshwater with sodium chloride results in $\approx 270‰$. The saltiest aquatic habitat on Earth may be the hypersaline lakes of McMurdo Dry Valleys in Antarctica, where the salinity can reach $\approx 440‰$—see Section 7.1.1 for a detailed discussion of these lakes. Hypersaline environments are considered extreme because normal cell physiologies cannot withstand the strong salt concentrations: the salinity gradient from inside to outside the cell causes it to rapidly desiccate, losing its cellular water.

Halophiles, a type of extremophile (see Chapter 7 for other examples) adapted to salty habitats, have mechanisms for accumulating or producing nontoxic solutes inside their cells to be isosmotic with the external environment. Their proteins are also specially modified to prevent denaturation at high salt concentrations. For surface-sunlit hypersaline habitats such as salterns, high UV levels are also a challenge, and many of the microbes there have pigments to protect them, as well as efficient DNA protection and repair systems. The microbial pigments, often carotenoids and bacteriorhodopsins, typically lend shallow salt habitats a pink or orange color. While halophiles occur in all three domains of life, there are comparatively few eukaryotic halophiles and an abundance of archaeal ones. One of the most notable halophiles is the ubiquitous genus *Haloquadratum* (Figure 6.23B), a euryarchaean within the class Halobacteria (note the counterintuitive "-*bacteria*" ending despite being within the Archaea), which is shaped just like a flat, square salt crystal! Like other extreme environments, the harsh conditions of hypersaline waters result in lower microbial community diversity, since fewer lineages are able to survive in them (Figure 6.23C). Halophiles are of particular interest to astrobiologists (biologists who study life on Earth that may be similar to life on other planets, and search for that extraterrestrial life), since remnant water on Mars is likely to be highly salty, and also UV levels at the Martian surface are high.

6.6.3 Subterranean Waters

The groundwater environment is in the subsurface and includes shallow and deep aquifers. The characteristics and microbial communities of the groundwater environment have been discussed in Sections 4.2.1.3 and 4.6. Briefly, microorganisms are the sole inhabitants of these environments and bacteria and archaeans are the dominant types of microbe present. In general, levels of microbial activity are low, especially in intermediate and deep aquifers. As shown in Figure 4.18, activity is orders of magnitude lower in these aquifers than in other aquatic habitats, due to low nutrient levels. Many subsurface environments may even be considered extreme from a nutrient perspective (Chapter 7).

6.6.4 Wetlands

Wetlands are a habitat type where soils are seasonally or permanently saturated with water, which can be freshwater, brackish or marine. They contain distinct plant and microbial communities, and represent a diverse and important aquatic habitat. They were discussed in Section 4.2.1.4, and are simply highlighted here. Marine wetlands include mangrove swamps that provide "nursery" habitat for the larvae of many fish species. Brackish wetlands include estuarine marshes that can act as massive filtration systems to decrease total carbon and nutrient loading on coastal waters. Related to this, human-made wetlands are often created as part of wastewater treatment (described in Section 25.6). Freshwater wetlands are the largest natural source of the potent greenhouse gas methane (CH_4) to the atmosphere (Denman *et al.*, IPCC report, 2007). Freshwater wetlands also include peat bogs, which contain a vast amount ($> 30\%$) of the planet's stored soil carbon pool. This stored carbon may be in danger of release to the atmosphere under continued climate change (Denman *et al.*, 2007).

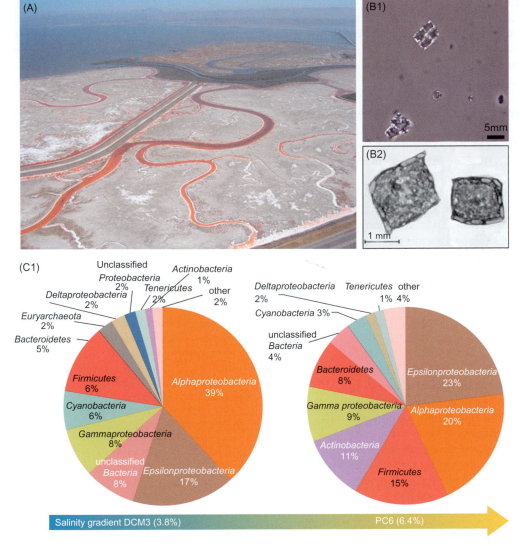

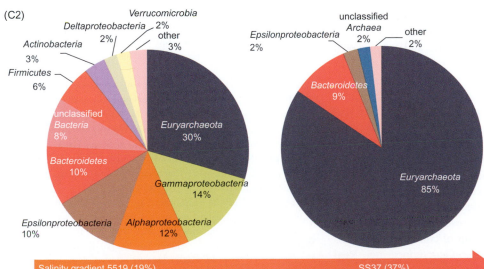

FIGURE 6.23 Hypersaline communities. Diverse hypersaline habitats exist including salt ponds, like those on the shores of California's San Francisco Bay (A). (B) *Haloquadratum* is one of the most memorable halophiles for its box-like shape, reminiscent of a salt crystal. (C) Microbial community composition shifts and becomes less diverse as salinity increases. A study compared Mediterranean warm waters at 35‰ salinity, a hypersaline coastal lagoon in the Galapagos Islands at 6‰, and two Spanish salterns (salt ponds) at 190‰ and 370‰. The research team profiled the communities using metagenomic sequencing (Chapter 21), and these pie charts show the taxonomic breakdown of 16S rRNA gene sequences pulled from those metagenomic datasets (using sequences longer than 100 bp, identity >95% to the taxonomy indicated, total number of 16S rRNA reads per dataset ranging from 1408 to 3333). *Haloquadratum walsbyi* (shown in the inset of (B)), within the Euryarchaeota, was by far the dominant microbe in the two most saline sites, at 15% and 64% of the total reads, respectively. (B) © Mike Dyall-Smith. http://www.nature.com/news/2004/041011/full/news041011-3.html and http://web.aanet.com.au/~aanet/DGBHome/Research/PC_square4_7_opt_L.jpg. (C) From Ghai *et al.* (2011).

QUESTIONS AND PROBLEMS

1. What are the most "important" aquatic habitats by volume and/or by relevance to humans?
2. Describe large-scale and small-scale ways in which aquatic environments are not homogeneous.
3. What is meant by the term microbial loop? Roughly how much primary production is fueled by recycled nutrients?
4. What are biofilms and microbial mats? Where would you expect to find each? Why are they of particular interest to humans?
5. A newly trained environmental engineer is hired to investigate solutions to clogging of water distribution lines by a persistent, thick and gelatinous material. The astute engineer quickly recognizes that this recurring problem may be caused by microorganisms and rushes to isolate and characterize the microorganisms clogging the pipelines. The engineer is successful in culturing several microorganisms in broth cultures (i.e., flasks containing liquid microbiological media) from the material found in the pipelines. In these broth cultures, the engineer determines the amount of an antimicrobial compound necessary to kill these microorganisms. To be certain an adequate amount of this antimicrobial compound is delivered, the engineer adds twice as much as the broth culture-based tests suggested would be necessary. Much to the engineer's surprise and dissatisfaction, the treatment is ineffective in killing the microorganisms found within the pipelines. Why did this treatment fail? What additional measures might the engineer need to take to solve the company's problem with the clogged pipes? Can you devise any novel strategies based on material presented in this chapter?
6. What is photoheterotrophy? Give two specific examples of microbes that live this lifestyle.
7. What is aquatic chemoautotrophy, and why may it be important?
8. What is a thermocline?
9. Describe how marine environments differ from freshwater environments physically, chemically and microbially.
10. What roles do aquatic viruses play in ecosystems?
11. Your first job is as an environmental microbiologist is at a wastewater treatment plant, where you are in charge of the sludge bioreactors. Even before you read the later chapter on wastewater treatment, why do you care about aquatic viruses in your system?
12. You have been hired fresh out of college by a geoengineering consulting firm, Geoengineering Real Solutions. You are put on a team evaluating ocean fertilization with iron as a way to sequester carbon from the atmosphere. The idea is that iron is a limiting micronutrient for a number of phytoplankton in the oligotrophic open oceans. Fertilizing large areas of ocean by dumping iron filings off of tanker ships should cause phytoplankton to bloom, fixing more CO_2 out of the atmosphere, resulting in more carbon getting buried in the deep sea due to sinking particles. How would you use your logic to evaluate whether this solution is likely to be successful, and what additional pieces of information might you need to help your team decide?

REFERENCES AND RECOMMENDED READING

Ashelford, K. E., Day, M. J., and Fry, J. C. (2003) Elevated abundance of bacteriophage infecting bacteria in soil. *Appl. Environ. Microbiol.* **69**, 285–289.

Atlas, R. M., and Bartha, R. (1993) "Microbial Ecology," Benjamin Cummings, Redwood City, CA.

Bergman, B., Sandh, G., Lin, S., Larsson, J., and Carpenter, E. J. (2013) *Trichodesmium*—a widespread marine cyanobacterium with unusual nitrogen fixation properties. *FEMS Microbiol. Rev.* **37**, 286–302.

Breitbart, M. (2011) Marine viruses: truth or dare. *Annu. Rev. Marine Sci.* **4**, 425–448.

Church, M. J., Wai, B., Karl, D. M., and DeLong, E. F. (2010) Abundances of crenarchaeal amoA genes and transcripts in the Pacific Ocean. *Environ. Microbiol.* **12**, 679–688.

Costerton, J. W., Lewandowski, Z., Caldwell, D. E., Korber, D. R., and Lappinscott, H. M. (1995) Microbial biofilms. *Annu. Rev. Microbiol.* **49**, 711–745.

Damarem, S., Raghukumar, C., Muraleedharan, U. D., and Raghukumar, S. (2006) Deep-sea fungi as a source of alkaline and cold-tolerant proteases. *Enzyme Microb. Technol.* **39**, 172–181.

Davey, M. E., and O'Toole, G. A. (2000) Microbial biofilms: from ecology to molecular genetics. *Microbiol. Mol. Biol. Rev.* **64**, 847–867.

De la Rocha, C. L. (2006) The biological pump. In "The Oceans and Maine Geochemistry" (H. Elderfield, ed.), Elsevier, NY, pp. 83–111.

DeLong, E. F., and Béjà, O. (2010) The light-driven proton pump proteorhodopsin enhances bacterial survival during tough times. *PLoS Biol.* **8**, e1000359.

Denman, K. L., Brasseur, G., Chidthaisong, A., Ciais, P., Cox, P. M., Dickinson, R. E., *et al.* (2007) Couplings between changes in the climate system and biogeochemistry. In "Climate Change 2007: The Physical Science Basis Contribution of Working Group I to the Fourth Assessment Report of the Intergovernmental Panel on Climate Change" (S. Solomon, D. Qin, M. Manning, Z. Chen, M. Marquis, K. B. Averyt, *et al.*, eds.) Cambridge University Press, Cambridge, United Kingdom and New York, NY, U.S.A.

Des Marais, D. J. (2003) Biogeochemistry of hypersaline microbial mats illustrates the dynamics of modern microbial ecosystems and the early evolution of the biosphere. *Biol. Bull.* **204**, 160–167.

Duce, R. A., LaRoche, J., Altieri, K., Arrigo, K. R., Baker, A. R., Capone, D. G., *et al.* (2008) Impacts of atmospheric anthropogenic nitrogen on the open ocean. *Science* **320**, 893–897.

Ducklow, H. W., and Shiah, F. (1993) Bacterial production in estuaries. In "Aquatic Microbiology: An Ecological Approach"

(T. E. Ford, ed.), Blackwell Scientific Publications, Cambridge, MA, pp. 261–287.

Falkowski, P. G., and De Vargas, C. (2004) Shotgun sequencing in the sea: a blast from the past? *Science* **304**, 58–60.

Fuhrman, J. (1992) Bacterioplankton roles in cycling of organic matter: the microbial food web. In "Primary Productivity and Biogeochemical Cycles in the Sea" (P. G. Falkowski, and A. D. Woodhead, eds.), Plenum, New York, pp. 361–383.

Ghai, R., Pašić, L., Fernández, A. B., Martin-Cuadrado, A.-B., Mizuno, C. M., McMahon, K. D., *et al.* (2011) New abundant microbial groups in aquatic hypersaline environments. *Sci. Rep.* **1**, 135.

Giovannoni, S. J., Bibbs, L., Cho, J-C., Stapels, M. D., Desiderio, R., Vergin, K. L., *et al.* (2005) Proteorhodopsin in the ubiquitous marine bacterium SAR11. *Nature* **438**, 82–85.

Golubic, S., Radtke, G., and Le Campion-Alsumard, T. (2005) Endolithic fungi in marine ecosystems. *Trends Microbiol.* **13**, 229–235.

Hennes, K. P., and Simon, M. (1995) Significance of bacteriophages for controlling bacterioplankton growth in a mesotrophic lake. *Appl. Environ. Microbiol.* **61**, 333–340.

Hoehler, T. M., Bebout, B. M., and Des Marais, D. J. (2001) The role of microbial mats in the production of reduced gases on the early Earth. *Nature* **19**, 324–327.

Iverson, V., Morris, R. M., Frazar, C. D., Berthiaume, C. T., Morales, R. L., and Armbrust, E. V. (2012) Untangling genomes from metagenomes: revealing an uncultured class of marine Euryarchaeota. *Science* **335**, 587–590.

Jørgensen, B. B., and Boetius, A. (2007) Feast and famine—microbial life in the deep-sea bed. *Nat. Rev. Microbiol.* **5**, 770–781.

Kaplan, L. A., and Newbold, J. D. (1993) Biogeochemistry of dissolved organic carbon entering streams. In "Aquatic Microbiology: An Ecological Approach" (T. E. Ford, ed.), Blackwell Scientific Publications, Cambridge, MA, pp. 139–165.

Karner, M. B., DeLong, E. F., and Karl, D. M. (2001) Archaeal dominance in the mesopelagic zone of the Pacific Ocean. *Nature* **409**, 507–510.

Kirchman, D. L. (2004). DOM and heterotrophic prokaryotes in the oceans. In "The Ocean Carbon Cycle and Climate" (M. Follows, and T. Oguz, eds.), NATO Science Series IV, Earth and Environmental Sciences, vol. 40, Kluwer Academic Publishers, pp. 31–63.

Konopka, A. E. (1993) Distribution and activity of microorganisms in lakes: effects of physical processes. In "Aquatic Microbiology: An Ecological Approach" (T. E. Ford, ed.), Blackwell Scientific Publications, Cambridge, MA, pp. 47–68.

Korber, D. R., Lawrence, J. R., Lappin-Scott, H. M., and Costerton, J. W. (1995) Growth of microorganisms on surfaces. In "Microbial Biofilms" (H. M. Lappin-Scott, and J. W. Costerton, eds.), Cambridge University Press, Cambridge, MA, pp. 15–45.

Kuenen, J. G. (2008) Anammox bacteria: from discovery to application. *Nat. Rev. Microbiol.* **6**, 320–326.

Lever, M. A., Rouxel, O., Alt, J. C., Shimizu, N., Ono, S., Coggon, R. M., *et al.* (2013) Evidence for microbial carbon and sulfur cycling in deeply buried ridge flank basalt. *Science* **339**, 1305–1308.

Marshall, K. C. (1985) Mechanisms of bacterial adhesion at solid–liquid interfaces. In "Bacterial Adhesion: Mechanisms and Physiological Significance" (D. C. Savage, and M. Fletcher, eds.), Plenum, New York, pp. 131–161.

Marshall, K. C. (1992) Biofilms: an overview of bacterial adhesion, activity, and control at surfaces. *Am. Soc. Microbiol. (ASM) News* **58**, 202–207.

Middledoe, M., Jorgensen, N. O. G., and Kroer, N. (1996) Effects of viruses on nutrient turnover and growth efficiency of noninfected marine bacterioplankton. *Appl. Environ. Microbiol.* **62**, 1991–1997.

Moore, L. R., Rocap, G., and Chisholm, S. W. (1998) Physiology and molecular phylogeny of coexisting *Prochlorococcus* ecotypes. *Nature* **393**, 464–467.

Moran, M. A., and Miller, W. L. (2007) Resourceful heterotrophs make the most of light in the coastal ocean. *Nat. Rev. Microbiol.* **5**, 792–800.

Orcutt, B. N., Sylvan, J. B., Knab, N. J., and Edwards, K. J. (2011) Microbial ecology of the dark ocean above, at, and below the seafloor. *Microbiol. Mol. Biol. Rev.* **75**, 361–422.

Orphan, V. J., Turk, K. A., Green, A. M., and House, C. H. (2009) Patterns of 15N assimilation and growth of methanotrophic ANME-2 archaea and sulfate-reducing bacteria within structured syntrophic consortia revealed by FISH-SIMS. *Environ. Microbiol.* **11**, 1777–1791.

Paerl, H. W., and Pickney, J. L. (1996) A mini-review of microbial consortia: their roles in aquatic production and biogeochemical cycling. *Microb. Evol.* **21**, 225–247.

Rappé, M. S., Connon, S. A., Vergin, K. L., and Giovannoni, S. J. (2002) Cultivation of the ubiquitous SAR11 marine bacterioplankton clade. *Nature* **418**, 630–633.

Reinthaler, T., van Aken, H. M., and Herndl, G. J. (2010) Major contribution of autotrophy to microbial carbon cycling in the deep North Atlantic's interior. *Deep Sea Res. Part II: Topical Studies Oceanogr.* **57**, 1572–1580.

Rheinheimer, G. (1985) "Aquatic Microbiology," 3rd ed. John Wiley & Sons, New York, pp. 1–247.

Santelli, C. M., Orcutt, B. N., Banning, E., Bach, W., Moyer, C. L., Sogin, M. L., *et al.* (2008) Abundance and diversity of microbial life in ocean crust. *Nature* **453**, 653–656.

Schleper, C., and Nicol, G. W. (2010) Ammonia-oxidising archaea—physiology, ecology and evolution. In "Advances in Microbial Physiology" (K. P. Robert, ed.), Academic Press, Oxford, pp. 1–41.

Sharon, I., Tzahor, S., Williamson, S., Shmoish, M., Man-Aharonovich, D., Rusch, D. B., *et al.* (2007) Viral photosynthetic reaction center genes and transcripts in the marine environment. *ISME J.* **1**, 492–501.

Sullivan, J. M., McManus, M. A., Cheriton, O. M., Benoit-Bird, K. J., Goodman, L., Wang, Z., *et al.* (2010) Layered organization in the coastal ocean: an introduction to planktonic thin layers and the LOCO project. *Continental Shelf Res.* **30**, 1–6.

Sullivan, M. B., Lindell, D., Lee, J. A., Thompson, L. R., Bielawski, J. P., and Chisholm, S. W. (2006) Prevalence and evolution of core photosystem II genes in marine cyanobacterial viruses and their hosts. *PLoS Biol.* **4**, e234.

Suttle, C. A. (1994) The significance of viruses to mortality in aquatic microbial communities. *Microb. Ecol.* **28**, 237–243.

Swan, B. K., Martinez-Garcia, M., Preston, C. M., Sczyrba, A., Woyke, T., Lamy, D., *et al.* (2011) Potential for chemolithoautotrophy among ubiquitous bacteria lineages in the dark ocean. *Science* **333**, 1296–1300.

Tranvik, L. J., Downing, J. A., Cotner, J. B., Loiselle, S. A., Striegl, R. G., Ballatore, T. J., *et al.* (2009) Lakes and reservoirs as regulators of carbon cycling and climate. *Limnol. Oceanogr.* **54**, 2298–2314.

Venter, J. C., Remington, K., Heidelberg, J. F., Halpern, A. L., Rusch, D., Eisen, J. A., *et al.* (1994) Multicellular organization in a degradative biofilm community. *Appl. Environ. Microbiol.* **60**, 434–446.

Whitman, W. B., Coleman, D. C., and Wiebe, W. J. (1998) Prokaryotes: the unseen majority. *Proc. Nat. Acad. Sci. U.S.A.* **95**, 6578—6583.

Wolfaardt, G. M., Lawrence, J. R., Roberts, R. D., Caldwell, S. J., and Caldwell, D. E. (1994) Multicellular organization in a degradative biofilm community. *Appl. Environ. Microbiol.* **60**, 434—446.

Wommack, K. E., and Colwell, R. R. (2000) Virioplankton: viruses in aquatic ecosystems. *Microbiol. Mol. Biol. Rev.* **64**, 69—114.

Yooseph, S., Sutton, G., Rusch, D. B., Halpern, A. L., Williamson, S. J., Remington, K.*, et al.* (2007) The Sorcerer II Global Ocean Sampling Expedition: expanding the universe of protein families. *PLoS Biol.* **5**, e16.

Yutin, N., Suzuki, M. T., Teeling, H., Weber, M., Venter, J. C., Rusch, D. B., and Béjà, O. (2007) Assessing diversity and biogeography of aerobic anoxygenic phototrophic bacteria in surface waters of the Atlantic and Pacific Oceans using the Global Ocean Sampling expedition metagenomes. *Environ. Microbiol.* **9**, 1464—1475.

Zinger, L., Amaral-Zettler, L. A., Fuhrman, J. A., Horner-Devine, M. C., Huse, S. M., Welch, D. B. M.*, et al.* (2011) Global patterns of bacterial beta-diversity in seafloor and seawater ecosystems. *PLoS ONE* **6**, e24570.

Chapter 7

Extreme Environments

Raina M. Maier and Julia W. Neilson

Extreme environments are important to environmental microbiologists because there is much speculation that such environments harbor unique microorganisms with activities and metabolic strategies that are not only of scientific interest, but also have commercial potential. The interest in extreme environments has engendered support for large research efforts focused on such sites. Scientific agencies such as the European Science Foundation (ESF) and the U.S. National Science Foundation (NSF) are supporting long-term projects that study specific ecological systems which focus on extreme environments. The purpose of these projects is to help provide knowledge that can be used to conserve, protect and manage unique global ecosystems and the biodiversity they sustain, and to help evaluate the potential global significance of poorly understood ecosystems. The ecosystem services (Information Box 7.1) they provide, and the discovery of natural products and processes that can be harnessed for societal benefit in the areas of biotechnology, medicine and remediation, are of immense value. The concept of ecosystem services is becoming more important as the world population grows and we place increasing stress on these fragile environments that cycle critical nutrients and sustain our Earth and aquatic environments.

Microbial communities in extreme environments have adapted to amazing levels of stress. These adaptations are of interest for development of remediation approaches for some contaminated sites including acid mine drainage sites and radioactive waste sites. They also are of interest for applications of novel enzymes adapted to temperature or pH extremes. Finally, they are of interest for understanding evolutionary history and possible impacts of future climate change. Here we describe six different extreme environments: including environments characterized by either low or high temperature; a desert environmental characterized by aridity and UV stress; and three environments where low levels of organic matter and the absence of photosynthesis have resulted in chemoautotrophy playing an important latter role. These are: the marine hydrothermal vents; an acid mine drainage system; and a carbonate cave located in the Sonoran Desert. This is by no means an exhaustive list of extreme environments, but will give an idea of how unique microbial communities develop in such environments, and how novel discoveries are made through the study of such environments.

7.1 LOW TEMPERATURE ENVIRONMENTS

7.1.1 McMurdo Dry Valleys, Antarctica

The McMurdo Dry Valleys in Antarctica represent one of the driest and coldest ecosystems on Earth. The average mean annual surface air temperature is $-27.6°C$ and the average surface soil temperature is $-26.1°C$. This

I.L. Pepper, C.P. Gerba, T.J. Gentry: Environmental Microbiology, Third edition. DOI: http://dx.doi.org/10.1016/B978-0-12-394626-3.00007-7
© 2015 Elsevier Inc. All rights reserved.

FIGURE 7.1   (A) Permanently ice-covered Lake Vanda in the McMurdo Dry Valleys, Antarctica. (B) A sediment core taken by the McMurdo Dry Valley Microbial Observatory researchers from Lake Vanda. Photos courtesy (A) Vladamir Samarkin, (B) A. Chiuchiolo.

Information Box 7.1   Ecosystem Services

Natural ecosystems provide resources, or ecosystem services, that benefit society including provision of clean drinking water and healthy soils for growing crops, regulation of climate, biogeochemical cycling activities, and species diversity. It has been estimated that ecosystems, underpinned by microbial activity, provide at least $33 trillion per year in global services (e.g., climate regulation, nutrient cycling, waste treatment, water supply and regulation) (Costanza et al., 1997). The World Bank 2006 estimate of the gross world product (the value of all final goods and services produced globally) was $48 trillion, a very similar figure. As a global society we have thus far considered ecosystem services as free. However, as Earth's human population continues to grow and place increasing stress on the environment, it is increasingly recognized that we must place a value on ecosystem services that is factored into the gross world products.

---

ecosystem has the only permanently ice-covered lakes on Earth, varying in ice-cover thickness from 3 to 5 meters (Figure 7.1). The permanent ice cover greatly impacts several aspects of normal lake characteristics (see Chapter 6) including:

- reduced wind-driven mixing resulting in vertical transport that is decreased to the level of molecular diffusion
- reduced direct gas exchange between liquid water and the atmosphere
- reduced light penetration
- reduced sediment deposition into the water column

The long mixing times mean that some chemical gradients can exist in the water column for at least 20,000 years before they are dissipated by diffusion. Ecosystem properties in the water columns of the lakes are also controlled by the seasonal uncoupling of photoautotrophic and heterotrophic processes resulting from the unusual solar cycle: 4 months of darkness followed by 4 months of continuous light with twilight in between (MDV, 2007).

The McMurdo Dry Valleys are part of the NSF Long Term Ecological Research network and also serve as an NSF Microbial Observatory. One site that has been studied extensively from a microbial perspective is Lake Fryxell, a freshwater lake that is one of the most productive lakes in the region. Geochemical analysis of the water column shows the presence of oxygen to a depth of 10 meters. There is a sulfide gradient ranging from 0 at the ice−water interface to > 1 mM sulfide (S^{2-}) at the sediment surface (Figure 7.2). A complementary sulfate gradient occurs with low concentrations at the sediment surface building to concentrations > 1.5 mM just below the chemocline (Information Box 7.2). These gradients suggested the existence of sulfur cycling (see Section 16.4). In fact, the researchers studying the site have found a diverse community of phototrophic purple bacteria (Karr et al., 2003), sulfur chemoautotrophs and heterotrophic sulfate reducers. For example, cell numbers of sulfur-oxidizing bacteria were found to peak at 200 cells per ml at a depth of 9.5 m (Sattley and Madigan, 2006). As shown in Figure 7.2, this is precisely where both dissolved oxygen and sulfide coexist in the water column. Three sulfur oxidizers were cultured from lake water samples, all most closely related to *Thiobacillus*

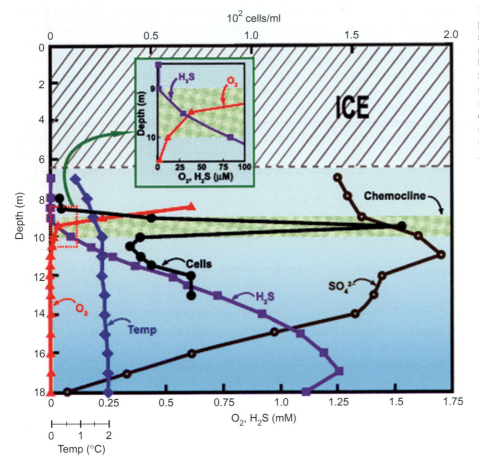

FIGURE 7.2   Diagram showing relevant physicochemical parameters related to the sulfur cycle in Lake Fryxell including cell numbers of sulfur-oxidizing bacteria based on most probable number (see Section 10.3.2) analyses. The expanded scale (inset) shows concentration of sulfide and dissolved oxygen at the chemocline. Diagram from Sattley and Madigan (2006). Reproduced with permission from the American Society for Microbiology Journals Department.

Information Box 7.2   Chemocline in Lake Fryxell

The chemocline is defined as the interface between two different chemistries in a body of water. In Lake Fryxell, the chemocline is the interface between oxygen-rich and anoxic water (Figure 7.2). In general, both nutrients and bacteria tend to accumulate at a chemocline or a thermocline (see Section 6.2.1 and Figure 6.2). Indeed, in Lake Fryxell, the chemocline harbors the largest numbers of sulfur-oxidizing bacteria and also harbors photosynthetic purple sulfur bacteria that consume the hydrogen sulfide, completing a sulfur minicycle in the chemocline.

---

thioparus, a known sulfur oxidizer, but the Lake Fryxell strains were classified as psychrotolerant. The isolates were able to grow in temperatures as low as $-2°C$ with a temperature optimum of $18°C$ and an upper limit of $31°C$. Sulfate-reducing bacteria were also found throughout the water column as determined by analysis for the dissimilatory sulfite reductase gene. Interestingly, several clone groups were highly localized with respect to lake depth suggesting that there are depth-specific population niches (Karr *et al.*, 2005).

In addition to the sulfur gradient, a methane gradient begins at a depth of 12 meters and increases to nearly 1 mM at the sediment surface. This gradient implies the presence of a methane cycle as well as the presence of both methanogenic and methanotrophic microorganisms (Karr *et al.*, 2006). In fact, of 13 archaeal clones found in the site, the majority represent methanogens and anoxic methanotrophs.

What is interesting about cold-adapted microorganisms? Their ability to survive and grow in the cold requires specialized adaptations that have the potential to be harnessed for the benefit of human society. For example, these microorganisms synthesize cold-adapted enzymes which have had to evolve specific structural features that make them highly flexible in comparison to their warm temperature equivalents (Siddiqui and Cavicchioli, 2006). This flexibility, particularly around the active site of the enzyme (the site where the enzyme interacts with its substrate), means that the enzyme can operate efficiently at low temperatures. This also means that at high temperature, the enzyme becomes unstable. In fact, it is these two properties of cold-active enzymes that make them suitable for biotechnological application; their high activity at low temperature and their low stability at elevated temperatures. The ability to carry out a reaction at low temperature can have several advantages. Sometimes, low temperature is more suitable for labile reactants, but also reactions carried out at low temperature require less energy input, which is a direct cost saving. The fact that these enzymes have low stability at elevated temperatures means that the reaction can be stopped quite easily by simply raising the temperature. Low temperature enzymes that have been examined or used in industry include; α-amylase (bread making, textiles, brewing and detergents); cellulase (textiles and the pulp and paper industries); β-galactosidase (removes lactose from milk), lipase (detergents and flavorings), proteases (detergents and meat tenderizers); and xylanase (bread making) (Cavicchioli *et al.*, 2002).

7.2 HIGH TEMPERATURE ENVIRONMENTS

7.2.1 Geothermal Hot Springs

Yellowstone National Park, Montana, U.S.A., has over 10,000 unique geothermal features that contain a wide and varied range of temperature, pH and geochemical profiles (Figure 7.3). This site has been a focus of research interest ever since it was realized that the thermostable DNA polymerase enzyme from *Thermus aquaticus*, which was isolated from Yellowstone, could be used for polymerase chain reaction (PCR) (Saiki *et al.*, 1988) (Chapter 13). *Thermus aquaticus* is an extreme thermophile that is able to grow between 40 and 79°C with an optimum temperature of 70°C (Brock and Freeze, 1969). Hot springs, such as those found in Yellowstone, have

temperatures of up to 100°C. Similarly, deep-sea hydrothermal vents can harbor hyperthermophiles (in fact, one of the record holders for high temperature tolerance is a deep-sea vent organism, *Pyrolobus fumarii*, an Archaea that can tolerate up to 113°C). Genera commonly found in these environments include *Thermus*, *Methanobacterium*, *Sulfolobus*, *Pyrodictium* and *Pyrococcus*.

The study of the thermophilic microorganisms in Yellowstone hot springs has been the subject of another NSF Microbial Observatory. In this case, researchers have examined the microbial community associated with the large amounts of sulfur that are fed into hot springs by geothermal fluids. One study site is called Dragon Spring. The source waters for Dragon Spring are acidic (pH 3.1 at the source); range in temperature from 66 to 73°C; and contain up to 80 μM of dissolved organic carbon. As these fluids reach the surface of the springs, the elemental sulfur in the incoming geothermal fluids forms flocs that are a distinctive feature of the entire outflow channel of the spring (Figure 7.3). This flocculent is the basis for a sulfur-cycling microbial community. Although the flocculent is formed abiotically, it is quickly colonized by two groups of organisms.

The first group is composed of two sulfur-respiring archaeal populations, *Caldisphaera draconis* and *Acidilobus sulfurireducens*. Quantitative PCR analysis (Chapter 13) shows that these two populations represent a major portion, 17 to 37%, of the floc-associated DNA (Boyd *et al.*, 2007) (Information Box 7.3). These isolates face an intriguing problem as sulfur reducers. It is speculated that these microorganisms do not reduce elemental sulfur (S^0) as it exists in the flocculent, but rather, reduce

FIGURE 7.3 The source of Dragon Spring in the Hundred Springs Plain of Norris Geyser Basin, Yellowstone National Park, Wyoming, U.S.A. The whitish-yellow material is flocculent elemental sulfur that becomes quickly colonized by sulfur-respiring Crenarchaea in the domain of Archaea and hydrogen sulfide-oxidizing bacterium *Hydrogenobaculum* spp. Photo courtesy Gill G. Geesey, Montana State University.

Information Box 7.3 Isolation of Novel Microorganisms

Analysis of DNA sequences from the environment can reveal the existence of novel and potentially intriguing microorganisms. But to actually study the physiology of these microbes, they must be cultured. Culturing an unknown microorganism is a major challenge and has been approached in two ways. First, by sequencing the 16S rRNA gene of a novel microbe, one can phylogenetically analyze the organism in relation to its nearest neighbors. This can provide clues to possible metabolic preferences, which can be tested by constructing specialized culture media. The second approach was used by researchers at Dragon Spring in Yellowstone National Park (Boyd *et al.*, 2007). In this case the 16S rRNA genes of the two novel sulfur-respiring microbes recovered from Dragon Spring did not closely match any known sulfur-reducers. Therefore, the scientists analyzed the geochemistry of Dragon Spring and mimicked these conditions in the laboratory. Using this approach they successful cultured the archaeal sulfur-reducers *Caldisphaera draconis* and *Acidilobus sulfurireducens*.

sulfur from polysulfide ($^-$S-S-S-S-S-S-S-S$^-$). Interestingly, polysulfide is not stable at the acidic pH conditions found in Dragon Spring, and it disproportionates almost completely into S^{2-} and S^0 (Rickard and Morse, 2005). Thus these microbes have some, as yet undescribed, mechanism to obtain polysulfide.

The second group that colonizes the flocculent is composed largely of the chemoautotrophic sulfide-oxidizing bacteria from the genus *Hydrogenobaculum*. This thermophile oxidizes the sulfide produced by the sulfur-respiring Archaea back to elemental sulfur (S^0). Thus, a very dynamic and tightly coupled sulfur cycle occurs at the very source of Dragon Spring. Interestingly, as the water flows away from the spring over a distance of a very few meters, there are drastic changes in the biogeochemistry of the system. These include pH increases, temperature decreases, the introduction of oxygen and iron, and arsenic becoming important in the energy flow in the system (Inskeep and McDermott, 2005).

Many mechanisms allow microorganisms to survive at temperatures that would normally denature proteins, cell membranes and even genetic material (Bouzas *et al.*, 2006). The key to enzyme function, whether in cold or hot environments, is the maintenance of an appropriate balance between molecular stability and structural flexibility. One general adaptive mechanism exhibited by thermophilic microorganisms is the production of chaperonins, which are specialized thermostable proteins that help refold and restore other proteins to their functional form following thermal denaturation. In addition, there are microbe-specific adaptations to increase protein stability at high temperature including:

- an increased number of disulfide bridges
- increased interactions among aromatic peptides
- increased hydrogen bonding among peptides

In terms of cell membranes, most hyperthermophiles belong to the Archaea. The archaeal cell membrane differs in structure from the bacterial cell membrane, as shown in Figure 7.4. The archaeal membrane is more thermostable than bacterial membranes (although it should be noted that the membrane structure is the same whether or not an archaean is thermophilic). Finally, in terms of nucleic acids, all hyperthermophiles produce a unique enzyme called DNA gyrase. This gyrase acts to induce positive supercoils in DNA, theoretically providing considerable heat stability (Kikuchi and Asai, 1984; Bouthier de la Tour *et al.*, 1990).

What has been learned about microbial adaptation to life as an extreme thermophile? One example is a recent report of *Thermus thermophilus* HB27, which was isolated from a hot spring in Japan. This organism has a DNA translocator system that allows it to take up DNA very broadly from various members of all three domains of life, Bacteria, Archaea and Eukarya (Schwarzenlander

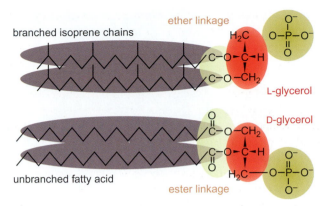

FIGURE 7.4 A comparison of membrane phospholipids from Archaea (top) and Bacteria (bottom). Note that in the Archaea the lipids are linked to glycerol through an ether linkage while for Bacteria an ester linkage is used. Further, archaeal lipids are based on a repeating 5-carbon unit isoprene usually 20 carbons in length. Bacterial lipids are straight chain fatty acids of 16 to 18 carbons in length.

and Averhoff, 2006). Further, this translocator system exhibits extremely high rates of DNA uptake. In fact, this organism has a DNA uptake velocity of 40 kilobase pairs per second (i.e., 40,000 base pairs). This can be compared to some known mesophilic organisms which have uptake velocities that are 2.5 to 10 times lower including *Bacillus subtilis* (4 kbp/sec), *Haemophilus influenzae* (16 kbp/sec) and *Streptococcus pneumoniae* (4 kbp/sec). The authors of this report suggest that the DNA translocator system has been of benefit to this organism by allowing it to adapt quickly to life in extreme environments. Moreover, this DNA translocator might be one of the most powerful tools for interdomain transfer of thermophilic and physiological traits between microorganisms thriving in extreme environments.

What do thermophiles have to offer biotechnology? There are numerous biotechnological applications for enzymes isolated from thermophilic microorganisms and the number of applications is growing rapidly, especially in commercial industry. The prime example is the thermostable DNA polymerase used in PCR. Other examples include proteases, lipases, amylases and xylanases that are used in the agricultural, paper, pharmaceutical, water purification, bioremediation, mining and petroleum recovery industries.

7.3 DESICCATION AND UV STRESS

7.3.1 The Atacama Desert, Chile

The deserts of the world represent both hot and cold semiarid to hyperarid environments where extreme conditions severely limit primary productivity and, thus, the diversity of life. Factors limiting microbial life in the arid deserts include water availability, temperature and the

FIGURE 7.5 The hyperarid core region of the Atacama Desert, Chile. Photo was taken at 974 m elevation (24.073°S, 70.204°W) southeast of Antofagasta. Courtesy Julia W. Neilson.

intensity of UV radiation. Arid and semiarid deserts are characterized by mean annual rainfall levels of 25−200 mm while hyperarid deserts have a mean annual rainfall of <25 mm. Water availability in a desert is not only determined by mean annual rainfall, but also by the combined effects of precipitation (P) and potential evapotranspiration (PET). Hyperarid areas are defined as those with a P/PET ratio less than 0.05 (Houston and Hartley, 2003). These moisture-deficient environments are of global relevance because they occupy 37% of the terrestrial surface (Middleton et al., 1997), and little is known about the arid and hyperarid ecosystems that occupy 12% and 7.5%, respectively, of the global land area.

One of the driest deserts on Earth is the Atacama Desert, Chile, where the hyperarid core of the desert experiences intervals of years to decades with no rain (Betancourt et al., 2000). Due to the lack of available moisture, plants are sparse or completely absent, creating soil conditions with extremely low soil organic carbon levels that further limit the potential diversity of microbial life (Figure 7.5). Examples of total organic carbon levels in a range of desert soil samples include: 0.7% in the Mojave (Eureka Valley, California, U.S.A.); 0.17% in the Sahara (near Abu Simbel, Egypt) (Lester et al., 2007); and 0.02−0.09% in samples taken along an elevational transect through the hyperarid core region of the Atacama Desert (Drees et al., 2006). For comparison, organic carbon in temperate soils ranges from 1 to 5% (see Section 4.2.2.6).

A commonly studied refuge for life in arid environments is the lithobiontic microbial communities that inhabit rock surfaces and subsurface rock pores. These communities contain a surprising diversity of heterotrophic bacteria that are sustained by photoautotrophic nitrogen-fixing microorganisms. These include cyanobacteria and green algae, and are capable of colonizing a

diverse group of minerals including dolomite, granite, gypsum, halite, limestone, quartz and sandstone. These communities have been found in a range of hot and cold deserts including the Atacama, the Mojave, the al-Jafr Basin (Jordan), the deserts of northwestern China and the McMurdo Dry Valleys of Antarctica (Figure 7.6). These hypolithic (inhabit the underside of rock surfaces) and endolithic (inhabit pore spaces within the rocks) communities are believed to exploit the protection offered by mineral surfaces that scatter UV radiation and trap limited water supplies while still allowing enough light penetration for photosynthesis (Dong et al., 2007).

One dominant photoautotroph found in many of these communities is the desiccation and radiation tolerant cyanobacterium Chroococcidiopsis (DiRuggiero et al., 2013). Cyanobacteria, for example Nostoc sp., have been shown to remain in a state of desiccation for months or years at a time (Potts, 1999). This has been measured at a level of 2−5% cell water content, which is one order of magnitude lower than that of eubacterial spores (Gao and Ye, 2007)! Adaptation to desiccation is unique among the extremes experienced by bacteria (i.e., temperature, pH, pressure) because the cells do not grow while desiccated, and the greater portion of their viable lifetime may be spent in the dehydrated state. Thus, cycles of desiccation appear to induce survival strategies for the cells rather than the ability to function under extreme conditions. The survival strategies identified include:

- the ability to protect and repair DNA exposed to UV radiation
- maintenance of protein stability in the dehydrated state
- maintenance of membrane integrity

The primary adaptive mechanism of the Cyanobacteria is the production of an extracellular polysaccharide sheath (EPS). This sheath regulates the uptake and loss of water, serves as a matrix for immobilization of cellular components produced by the cell in response to desiccation, and may protect cell walls during shrinking and swelling (Potts, 1999). Several molecules are produced by the cell in response to desiccation and UV exposure. These are often found in the EPS sheath, and include UV absorbing compounds such as mycosporine-like amino acids and scytonemin, carotenoids and detoxifying enzymes or radical quenchers that provide protection from harmful radicals and oxygen species, and water stress proteins (Gao and Ye, 2007). Water stress proteins are extremely stable and have been found to comprise up to 70% of the soluble proteins in environmental samples of Nostoc commune. In addition, N. commune cells contain trehalose and sucrose, which have the ability to stabilize proteins and protect the integrity of the membrane during desiccation. Electron microscopy indicates that the nucleoplasm of N. commune cells requires 5 days of rehydration before

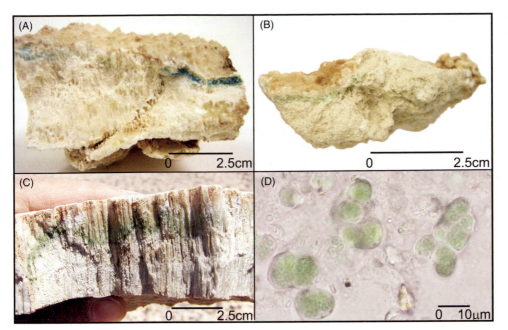

FIGURE 7.6 Samples of soil gypsum with cyanobacteria: (A) Soil gypsic crust sample AT326b from the Atacama Desert; (B) soil gypsic crust sample DG from the Mojave Desert; (C) fibrous gypsum sample JB1 from a secondary vein exposed by erosion at the surface from Al-Jafr Basin, Jordan; (D) light micrograph of cyanobacterium *Chroococcidiopsis* in an enrichment culture from sample AT326b from the Atacama Desert. From Dong *et al.* (2007). Reproduced with permission of AUG.

appearing similar to fully hydrated cells. In addition, a very ordered recovery of gene expression has been documented during rehydration beginning with respiration, followed by photosynthesis and finally nitrogen fixation. Although desiccation and UV tolerance are widespread among cyanobacteria, the interactions are complex and considerable work is still needed to fully understand the mechanisms involved.

Due to the presence of these photoautotrophic nitrogen-fixing primary producers, the lithic communities in arid environments are not dependent on exogenous carbon or nitrogen supplies. This has allowed the investigation of other physical factors that may limit microbial life in hyperarid environments. Two studies have examined such lithic systems to determine the physical factors most limiting to life in hyperarid regions. In the first study, the microbial diversity was characterized in hypolithic crusts on quartz substrates from three hyperarid desert locations in northwestern China (Turpan Depression, Taklimakan Desert and Qaidam Basin). Regression analysis revealed a positive correlation between the availability of liquid water and two diversity indices: the Species Richness ($R^2 = 0.738$) and the Shannon Diversity Index ($R^2 = 0.650$). The availability of liquid water was calculated from the interaction of temperature and rainfall (Pointing *et al.*, 2007). A similar study was conducted by Warren-Rhodes *et al.* (2006) in which translucent quartz and quartzite stones were collected from four locations along an aridity gradient in the Atacama Desert where rainfall declined from 21 to ≤ 2 mm year^{-1}. The percentage of rocks collected that were colonized by hypolithic cyanobacteria declined from 28 to $< 0.1\%$, from the less arid to the most hyperarid region, reinforcing the

conclusion that the absence of available water severely limits microbial biomass in the extreme hyperarid deserts.

Less is known about the energy dynamics of microbial communities that inhabit soils of these expansive hyperarid and arid regions. Intriguing data suggest that the soils of these barren, unvegetated regions contain numerous radiotolerant and halotolerant species, as well as bacteria with phylogenetic associations to chemolithoautotrophic taxa that obtain energy through the oxidation of nitrite, carbon monoxide, iron or sulfur. This suggests a genetic potential for nonphototrophic primary production and geochemical cycling in these arid ecosystems (Neilson *et al.*, 2012).

What knowledge can be gained from the study of organisms capable of survival in extreme hyperarid regions? First, an understanding of where life can occur in the hyperarid regions on Earth helps to narrow and focus the search for life beyond our planet. For example, it has been suggested that life would have been forced into endolithic habitats on Mars as liquid water slowly disappeared from the planet. Second, research suggests that expanded knowledge of microbial diversity in hyperarid regions could be used to evaluate precipitation history. A recent study of soil bacterial diversity in the Atacama Desert was performed along a west/east elevational transect (400−4500 m) through the driest region of the core absolute desert (Drees *et al.*, 2006). Mean annual temperature and precipitation along the transect ranges from 17°C to 7°C and 0.6 to 35.7 mm, respectively. In 2002, soil samples were collected at a depth of 25 cm to obtain bacteria unaffected by surface radiation or eolian (windborne) dispersal. Community DNA was extracted from the samples, amplified by PCR using universal

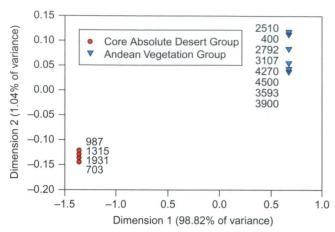

FIGURE 7.7 Kruskal's multidimensional scaling analysis of the DGGE profiles from the Atacama Desert elevational transect analyzed in three dimensions with 99.5% of the variance explained. Data labels represent sample elevations (meters) along transect. The microbial communities in samples along this transect clustered into one of two groups; the core absolute desert group and the Andean group (Drees *et al.*, 2006). Reproduced with permission from the American Society for Microbiology Journals Department.

bacterial 16 S rDNA primers, and the diversity analyzed by denaturing gradient gel electrophoresis (DGGE) (see Section 13.5.2.2). The results revealed the presence of two distinct microbial community types, one found in samples taken from the central hyperarid core and the other occurring in samples taken at elevations above and below the hyperarid central plateau, where sample sites showed evidence of exposure to either onsite precipitation or moisture runoff from higher elevations (Figure 7.7). The two microbial community groups did not correlate exactly with any other specific factors such as temperature, precipitation, TOC or percent plant cover gradients. A second sampling 2 years later, in 2004, indicated that the observed microbial communities were stable over time. Thus, these results suggest that microbial diversity patterns in hyperarid and arid environments can reveal information concerning the frequency of precipitation events or even historical climactic variations. Finally, research groups hope to identify heat and desiccation-tolerant mineral-oxidizing microbes from among the novel populations present in arid deserts that can be exploited for the microbially driven heap-leaching strategy used by mining companies to extract minerals such as copper from low-grade ore (Soto *et al.*, 2013).

7.4 APHOTIC ENVIRONMENTS BASED ON CHEMOLITHOAUTOTROPHY

Photosynthesis is the primary driver of energy dynamics in surface terrestrial and marine ecosystems, where light energy is harnessed for the conversion of atmospheric carbon dioxide to reduced carbon. Despite the abundance of life on the terrestrial surface where photosynthesis dominates, most biomass is actually located beneath the surface and the majority of this biomass is microbial (Whitman *et al.*, 1998). Thus, avenues for understanding alternate nonphotosynthetic primary production strategies can be found in subterranean or deep-sea ecosystems that function in the absence of sunlight. As will be explained in the following sections, a significant amount of research has been devoted to characterizing primary production in subsurface ecosystems to discover and understand the unique metabolic strategies employed for survival (see also Chapter 6).

7.4.1 Deep-sea Hydrothermal Vents

In 1977, geologists first described deep-sea hydrothermal vents (Figure 7.8). These are areas on the ocean floor where, driven by magma-derived hydrothermal convection, hot water laced with minerals flows up through cracks and fissures. The cracks, which are known as hydrothermal vents, often have a buildup of chemical precipitates that resemble chimneys surrounding them (Figure 7.9). Water, reaching temperatures of up to 400°C, is emitted from these vents at rates of 1 to 5 m/sec. In addition, most vent fluids are anoxic, highly reduced, acidic (pH from 2 to 4), and enriched in CO_2, H_2S, CH_4, H_2, Fe^{2+}, Zn^{2+}, Cu^{2+} and other transition metals. As the hot mineral-rich, hydrothermal water emerges from the vent, it quickly mixes with cold seawater forming a dark cloud of mineral precipitates. The appearance of this dark cloud has given the name "black smoker" to these vent chimneys. It was surprising to find whole self-contained ecosystems consisting of microscopic and macroscopic life in this environment, which has no light and extremely high temperature and pressure.

It is the confluence of the superheated hydrothermal vent water, which contains reduced minerals, that can act as electron donor, and the oxidized seawater, which contains a variety of electron acceptors, that serves as the basis for a chemoautotrophic community of microorganisms that sustains the entire heterotrophic component of the vent community ranging from microorganisms to animals (Fisher *et al.*, 2007). Thus, the entire food web in a hydrothermal vent community is based on chemoautotrophy, not photoautotrophy as in surface environments (see Sections 6.3.1.1 and 16.2.2). Examples of electron donor/acceptor pairs supporting autotrophy that have been identified in hydrothermal vents include: H_2 (donor) with O_2, Fe^{3+}, NO_3^-, CO_2, SO_4^{2-}, S^0 (acceptors); H_2S, S^0, $S_2O_3^{2-}$ (donors) with O_2, NO_3^- (acceptors); Fe^{2+} (donor) with O_2 (acceptor). In addition, the presence of CH_4 in vent fluids can support heterotrophic growth of methanotrophic

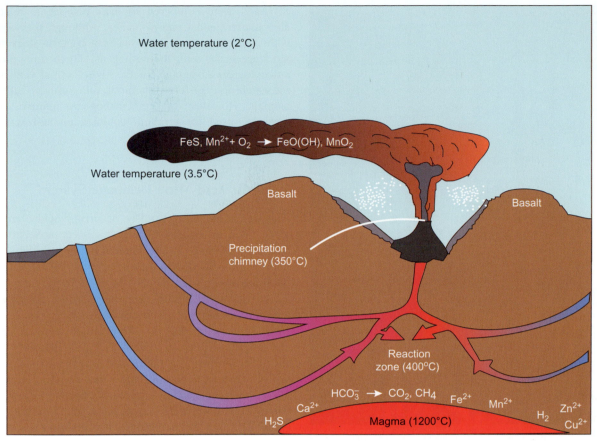

Water temperature (2°C)

FeS, Mn²⁺+ O₂ → FeO(OH), MnO₂

Water temperature (3.5°C)

Basalt

Basalt

Precipitation
chimney (350°C)

Reaction
zone (400°C)

HCO₃⁻ → CO₂, CH₄   Fe²⁺   Mn²⁺   Zn²⁺
Ca²⁺   H₂   Cu²⁺
H₂S   Magma (1200°C)

FIGURE 7.8   Schematic representation of a hydrothermal vent depicting a black smoker rising from the ocean floor creating a plume of chemical-rich superheated water around it.

microorganisms in the presence of either O_2 or SO_4^{2-} as terminal electron acceptors.

The vent communities support macrofauna that rely on the chemoautotrophic bacterial populations as a source of organic carbon. There are at least three major mechanisms for transfer of this bacterial carbon and energy to the next trophic level. The first is an endosymbiotic relationship between vent bacteria and an invertebrate, *Riftia pachyptila*, which has been dubbed "tube worm" (Markert *et al.*, 2007). *Riftia pachyptila* are large tube-shaped creatures that grow from the seafloor (Figure 7.10). These worms have no mouth, gut or any other digestive system, and depend completely on bacteria for their nutrition. Instead of consuming the bacteria, the worms have interior surfaces that are colonized by massive quantities (3×10^{11} bacteria per ounce of tissue) of sulfur-oxidizing chemoautotrophs (Karl, 1995) (see Section 16.4.3.1). Chemoautotrophy is sustained by the presence of H_2S originating from vent fluids and oxygen in the seawater. The worm's body is filled with blood containing large amounts of hemoglobin that binds H_2S. The blood transports the H_2S to the bacteria, which oxidize it and fix CO_2 into organic compounds that nourish the worm. The symbionts benefit from high nutrient

concentrations within the worm's body, which results in high microbial metabolic activity. In turn, the microbially fixed organic carbon is transferred to the host, making *R. pachyptila* one of the fastest growing marine invertebrates that has been studied (Markert *et al.*, 2007).

The second mechanism by which microbially produced carbon and energy is transferred to the next trophic level is termed microbial gardening. In this case, bacterial cultures are maintained by mussels and other invertebrates on specialized appendages such as tentacles and gills. These invertebrates periodically harvest and consume the bacteria, retaining small inocula to initiate the next crop. The third mechanism for carbon transfer to higher trophic levels is direct consumption of free-living bacterial cells, filaments or mats. Crabs, amphipods, predatory fish and even other microorganisms, including bacteria, have been observed to feed directly on the chemoautotrophic or chemoheterotrophic primary producers (Karl, 1995).

What new insights are hydrothermal vent communities offering to environmental microbiologists? It has recently been hypothesized, through comparison of genomic DNA of vent bacteria and closely related pathogenic bacteria, that there are evolutionary links between the

FIGURE 7.9 A photograph of a black smoker vent. This was first published on the cover of *Science* magazine. The vent water is exploding out of the vent at 1−5 meters per second and is 380°C. From Spiess *et al.* (1980). Reprinted with permission from AAAS.

FIGURE 7.10 An adult *Riftia pachyptila* tubeworm community *in situ*. Note the clams surrounding one of the worms in the community. Photo courtesy Andrea D. Nussbaumer, Charles R. Fisher and Monika Bright.

chemolithoautotrophic vent bacteria (which operate as symbionts to vent animals) and important closely related human pathogens (Nakagawa *et al.*, 2007). In this study, the genomes of two deep-sea vent ε-Proteobacteria strains, *Sulfurovum* sp. NBC37-1 and *Nitratiruptor* sp.

SB155-2, were compared to their pathogenic relatives, *Helicobacter* and *Campylobacter*. Although they are not pathogenic, these two deep-sea vent bacteria share many virulence genes with the pathogens. These traits provide an ecological advantage for the hydrothermal vent bacteria which need to form symbiotic relationships with vent animals. However, in their pathogenic relatives, these traits are used for efficient colonization and persistent infection of the host.

7.4.2 An Acid Mine Drainage System

Although there are naturally occurring acidic environments such as Dragon Spring discussed in Section 7.2, acid mine drainage is an excellent example of an anthropogenically caused extreme environment. Mineral mining often focuses on pyrite (FeS_2) deposits that contain metals such as silver (Au), gold (Ag), copper (Cu), zinc (Zn) and lead (Pb), usually as impurities in the pyrite ore. In some cases these metals are part of the sulfide minerals such as chalcopyrite ($CuFeS_2$), chalcocite (CuS), sphalerite (ZnS) and galena (PbS). In either case, mining exposes these sulfide minerals to both air and water, resulting in biologically mediated acid generation (Section 16.4.3). In sites where excess water is generated, acidic leachates form (<pH 2), which are sometimes also high in toxic metals. These leachates are called acid mine drainage (AMD). Despite extremes of acidity and heat and high concentrations of sulfate and toxic metals, a range of specialized microorganisms populate AMD environments.

Given the difficulty of culturing the majority of environmental microorganisms (and especially those found in extreme environments), there has been great interest in the study of the community based on its DNA (Chapter 13). The term metagenome was introduced in 1998 to describe the entire DNA that is represented by the microbial community (see also Chapter 21) (Handelsman *et al.*, 1998). At the time, the focus of metagenomic analysis was on the discovery of new natural products and biochemical pathways. However, the metagenome can also be explored to better understand ecological and evolutionary processes that drive community development (Allen and Banfield, 2005). In fact, this approach was used to study a subsurface AMD community from the Richmond Mine Site, Iron Mountain, California, U.S.A. (Tyson *et al.*, 2004). Like the hydrothermal vent community discussed in Section 7.4, the AMD community, which was taken beneath the surface of this site, is self-sustaining and based on chemoautotrophy rather than photoautotrophy. The pH and temperature of the biofilm sample taken from this site were 0.83 and 42°C, respectively. Solution concentrations of metals in the AMD were 317 mM iron, 14 mM zinc, 4 mM copper and 2 mM arsenic.

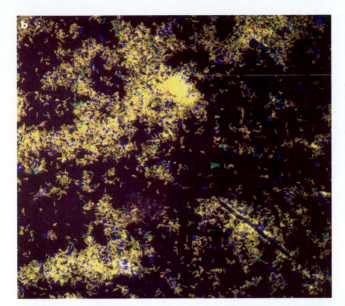

FIGURE 7.11 Fluorescent *in situ* hybridization (FISH) image of the AMD biofilm taken from a subsurface site at the Richmond Mine at Iron Mountain, California, U.S.A. Three different FISH probes were used. The first was specific for bacteria and used a fluorescein isothiocyanate probe which fluoresces green. The second problem was specific for archaeans and used the Cy5 fluor which appears blue. This probe detected the *Ferroplasma* genus. The third probe was one that targeted the *Leptospirillum* genus and used the Cy3 fluor appearing red. Note that overlap of red and green appears yellow and indicates *Leptospirillum* cells. The dominance of yellow in this image shows the dominance of *Leptospirillum*. *From* Tyson *et al.* (2004).

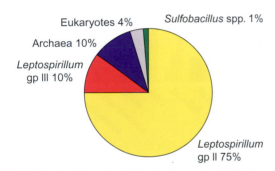

Eukaryotes 4%   *Sulfobacillus* spp. 1%

Archaea 10%

Leptospirillum gp III 10%

Leptospirillum gp II 75%

FIGURE 7.12 Relative microbial abundances in AMD determined using quantitative FISH counts. *From* Tyson *et al.* (2004).

In the metagenomic analysis, a total of 76.2 million base pairs of usable DNA sequence were obtained. These sequences were reconstructed to give nearly complete genomes for two iron-oxidizing bacteria, *Leptospirillum* group II and *Ferroplasma* type II, as well as partial recovery of three other genomes. Fluorescent *in situ* hybridization (Chapter 9) analysis of an AMD sample confirmed the presence of these organisms, and suggested that the *Leptospirillum* group II is dominant (75%) with the Ferroplasma type II representing approximately 10% of the community (Figures 7.11 and 7.12).

What did the researchers learn in terms of evolution and ecology at this site? In terms of evolution, analysis of the *Leptospirillum* group II genome indicated very few nucleotide polymorphisms (changes in DNA sequence), implying that only a single strain of this isolate dominates the community (Tyson *et al.*, 2004). In contrast, for *Ferroplasma* type II, the researchers observed between one and three distinct patterns of nucleotide polymorphisms in the assembled genome. These data suggest that *Ferroplasma* type II has undergone evolutionarily recent homologous recombination resulting in three distinct strains of the organism.

In terms of ecology, analysis of the genomes of these two microbes gave these scientists intriguing insights into how each bacterium processes carbon, fixes nitrogen and generates energy. For example, several CO_2 fixation pathways were identified for *Leptospirillum* group II, indicating that it is definitely chemoautotrophic. In contrast, while the *Ferroplasma* type II has a mechanism to fix CO_2, it also contains a large number of sugar and amino acid transporters, suggesting that it may prefer to metabolize heterotrophically. Both genomes were also examined for nitrogen fixation genes which were found to be absent. However, they were found in one of the partially sequenced genomes for *Leptospirillum* group III indicating an important role for this organism in the AMD biofilm system, although it represents only about 10% of the community as analyzed by fluorescence *in situ* hybridization (Figure 7.12).

The researchers note that the metagenomic analysis of the AMD site was successful in part because the diversity in the biofilm was low, and the frequency of genomic rearrangements and gene insertions/deletions was relatively low (allowing reconstruction of each genome).

7.4.3 A Desert Carbonate Cave

Carbonate caves represent subterranean ecosystems that are largely devoid of phototrophic primary production. Caves are typical features of karst terrains that represent approximately 20% of Earth's dry ice-free surface (Ford and Williams, 2007). Though they are aphotic, caves can receive fixed organic carbon from the surface of Earth from drip water that feeds the growth of the beautiful carbonate formations known as speleothems (Figure 7.13) and from streams that flow through the cave either constantly or ephemerally. However, for desert caves, the allochthonous organic carbon inputs that normally enter with drip water or streams are minimal, resulting in unique highly oligotrophic conditions.

The NSF Kartchner Caverns Microbial Observatory was created to examine the microbial diversity and energy strategies present in a desert cave system (Kartchner, 2008). Kartchner Caverns are located in the Sonoran

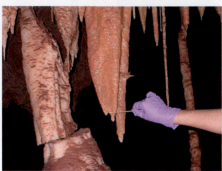

FIGURE 7.13 Speleothems in Kartchner Caverns. (Left) a variety of stalactites with a bacon formation on the left and a soda straw on the right. (Top right) a close-up of small formations including helectites, formations that grow while defying gravity. (Bottom right) sampling a speleothem for metagenome analysis. Photos courtesy Robert Casavant, Arizona State Parks.

Desert and developed within a Lower Carboniferous-age Escabrosa limestone formation in the Whetstone Mountains of southeastern Arizona, U.S.A. The only natural entrance is a very small blow-hole, largely limiting entrance to all but a bat colony, insects, small rodents, amphibians and reptiles. The cave has been developed for tourism but human access is tightly controlled and environmental conditions within the cave have been continuously monitored since 1989 (Tufts and Tenen, 1999).

Unexpectedly high bacterial diversity on Kartchner speleothem surfaces was revealed by a pyrotag analysis or next generation sequencing (Section 13.6.2) with approximately 2000 operational taxonomic units (OTUs) per speleothem. These OTUs were classified into 21 phyla and 12 candidate phyla, and appeared to be dominated by organisms associated with heterotrophic growth (Ortiz et al., 2013a). Evidence for chemolithoautotrophy came from examination of a clone library with full-length 16S rRNA sequences where phylogenetic associations to *Nitrospira* and *Leptospirillum* were revealed, bacteria that obtain energy from nitrite and iron oxidation, respectively.

Metagenome analysis was performed to elucidate the community structure and search for clues to the potential energy dynamics sustaining the diverse speleothem microbial communities (Ortiz et al., 2013b). One key insight discovered in this analysis was the potential contribution of archaeans to the energy dynamics in Kartchner. A recent 16S rRNA pyrotag analysis of soils from around the world found the average abundance of archaeans in soils to be 2%. Archaeal abundance was inversely correlated with C:N ratio, suggesting that archaeans can tolerate or even exploit

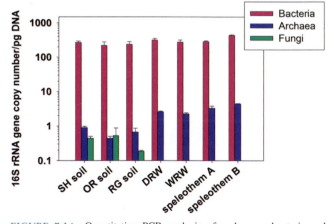

FIGURE 7.14 Quantitative PCR analysis of archaeans, bacteria and fungi in three soil samples taken from above Kartchner Caverns (SH soil, OR soil and RG soil), a dry (DRW) and wet (WRW) rock wall sample taken inside Kartchner, and two speleothem surfaces (A and B) in Kartchner. From Ortiz et al. (2013b)

low nutrient conditions such as those found in Kartchner (Bates et al., 2011). In fact, the taxonomic analysis of the Kartchner speleothem metagenome found that unassembled archaeal sequences comprised 10% of the metagenome. This finding was confirmed using a quantitative PCR analysis which revealed that archaeal abundance in Kartchner speleothem communities is significantly higher than in soil communities from above the cave (Figure 7.14).

Thaumarchaeota dominated the speleothem Archaea and represented the third most abundant phylum in the cave. A significant percentage of the Thaumarchaeota metagenome reads (18%) were associated with chemoautotrophic ammonia-oxidizing archaeans. This led the researchers to examine whether there were Archaea-associated CO_2-fixation genes in the metagenome. In fact, two key genes, including ATP-citrate lyase (required for the rTCA CO_2 fixation pathway), and 4-hydroxybutyryl-CoA dehydratase (required for the HP/HB and the DC/HB CO_2 fixation pathways), showed taxonomic associations to the Thaumarchaeota. Finally, examining a key gene in the ammonia oxidation pathway (*amoA*), the researchers found that half of the speleothem *amoA* genes were classified as archaeal. Taken together, these pieces of evidence suggest that archaeans make a significant contribution to the energy dynamics of this ecosystem by contributing to primary production through ammonia oxidation.

A recent metagenomic analysis of microbial slime in the subterranean aquatic Weebubbie cave below Australia's Nullarbor Plain also found a pattern of aphotic primary production driven by inorganic nitrogen metabolism (Tetu *et al.*, 2013). Like Kartchner, Weebubbie microbial communities included an abundance of ammonia-oxidizing Thaumarchaeota, although the Weebubbie community had a much higher relative abundance of Thaumarchaeota, and the ratio of ammonia oxidizing archaeans to ammonia oxidizing bacteria was more similar to that found in marine environments. Thus, it appears that the Kartchner Caverns ecosystem may represent an oligotrophic terrestrial counterpart to the energy dynamics observed in oligotrophic marine habitats such as Weebubbie cave.

A second aspect of this analysis helps to demonstrate how metagenomes can be used to reveal the functional potential of a community. The Kartchner metagenomes were analyzed in two ways. First, a full analysis of one speleothem was performed by mining multiple databases and employing various analysis strategies to compensate for the limited representation of cave microbes and oligotrophic metagenomes in current databases. Second, four cave speleothem metagenomes were compared to reported metagenomes from other environments including ocean, bulk soil and rhizosphere soil.

This two-pronged analysis was performed to examine potential carbon-fixation strategies in Kartchner. In-depth analysis of one speleothem metagenome detected genes for all six known CO_2-fixation pathways, and suggested a community genetically adapted to low-nutrient conditions. The RuBisCo genes representative of the Calvin—Benson—Bassham cycle were overrepresented in Kartchner speleothem metagenomes relative to bulk soil, rhizosphere soil and deep-ocean communities. The abundances of genes in two other pathways, ATP-citrate lyase (rTCA) and 4-hydroxybutyryl-CoA dehydratase (HP/HB and DC/HB), were comparable to the nutrient-limited ocean environments but greater than both soil and rhizosphere ecosystems. The abundance of HP/HB genes was of particular interest because enzymes in this cycle use

bicarbonate as the active inorganic carbon species, whereas bicarbonate is not a RuBisCo substrate. The HP/HB pathway is hypothesized to be advantageous for chemolithoautotrophic marine archaea because bicarbonate availability under slightly alkaliphilic conditions (e.g., ocean water) is significantly higher than dissolved CO_2 (Berg, 2011), conditions that may also apply to this carbonate cave ecosystem, where drip water pH averages 8.0.

The comparative metagenome analysis also suggested an intriguing survival strategy used by Kartchner speleothem communities. One gene category that was significantly overrepresented in the cave metagenomes was DNA repair enzyme genes belonging to the RAMP (Repair Associated Mysterious Proteins) superfamily. The RAMP family is a group of proteins for which no specific function is known, but that clearly associate with DNA repair mechanisms. The overabundance of DNA repair genes was surprising given the absence of typical DNA-damaging agents such as UV light. It was hypothesized that the exceedingly high calcium concentrations in cave drip water (Legatzki *et al.*, 2012) might be the source of stress. Previous work has suggested that cave bacteria precipitate calcium carbonate as a mechanism to overcome calcium toxicity (Banks *et al.*, 2010). Although little is known for bacteria, in eukaryotic cells calcium toxicity has been linked to DNA strand breakage (Cantoni *et al.*, 1989). Specifically, when cells are under oxidative stress, calcium homeostasis is disrupted, leading to an increase in intracellular calcium concentrations which in turn activates nucleases that damage DNA. Thus, the abundance of DNA repair enzymes may be an adaptation of cave microbes to the presence of extremely high calcium concentrations in the cave ecosystem where the primary mineral found in a speleothem is calcium carbonate.

What new insights does the study of an ecosystem like Kartchner Caverns offer? This oligotrophic subterranean environment supports a unique chemolithoautotrophic microbial community with unusual nutrient cycling strategies. This Microbial Observatory provides insight into the life survival strategies of other ecosystems dominated by oligotrophy including: subsurface soils or aquifers, arid deserts and even extraterrestrial ecosystems. For example, Kartchner provides a template for evaluating the role of archaeans in oligotrophic terrestrial ecosystems; Kartchner can be compared to Movile Cave in Romania (Chen *et al.*, 2009). Like Kartchner, Movile Cave is aphotic and sustained by chemolithoautotrophy; however, the atmosphere in Movile is rich in hydrogen sulfide and methane, and supports high microbial productivity and rich fauna. Microbial mats sustained by sulfur- and ammonia-oxidizers contain a diversity of autotrophic bacterial phylotypes, but no archaeans. The absence of archaeans is in stark contrast to their abundance in the less energy-rich caves discussed, Kartchner and Weebubbie.

QUESTIONS AND PROBLEMS

1. Define the term ecosystem services.
2. Give two examples of extreme environments not discussed in this chapter.
3. Which of the extreme environments discussed in this chapter likely has the slowest growth rates?
4. Of the six extreme environments discussed in this chapter, which are based on chemoautotrophy?
5. If you were a microorganism, what type of extreme environment would you choose to live in?
6. Discuss adaptations for enzymes at low and high temperature.
7. Compare and contrast a thermocline (Chapter 6) and a chemocline.

REFERENCES AND RECOMMENDED READING

Allen, E. E., and Banfield, J. F. (2005) Community genomics in microbial ecology and evolution. *Nat. Rev. Microbiol.* **3**, 489–498.

Banks, E. D., Taylor, N. M., Gulley, J., Lubbers, B. R., Giarrizo, J. G., Bullen, H. A., *et al.* (2010) Bacterial calcium carbonate precipitation in cave environments: a function of calcium homeostasis. *Geomicrobiol. J.* **27**, 444–454.

Bates, S. T., Berg-Lyons, D., Caporaso, J. G., Walters, W. A., Knight, R., and Fierer, N. (2011) Examining the global distribution of dominant archaeal populations in soil. *ISME J.* **5**, 908–917.

Berg, I. A. (2011) Ecological aspects of the distribution of different autotrophic CO_2 fixation pathways. *Appl. Environ. Microbiol.* **77**, 1925–1936.

Betancourt, J. L., Latorre, C., Rech, J. A., Quade, J., and Rylander, K. A. (2000) A 22,000-year record of monsoonal precipitation from northern Chile's Atacama Desert. *Science* **289**, 1542–1546.

Bouthier de la Tour, C., Portemer, C., Nadal, M., Stetter, K. O., Forterre, P., and Duguet, M. (1990) Reverse gyrase, a hallmark of the hyperthermophilic archaebacteria. *J. Bacteriol.* **172**, 6803–6808.

Bouzas, T. D., Barros-Velazquez, J., and Villa, T. G. (2006) Industrial applications of hyperthermophilic enzymes: a review. *Protein Peptide Lett.* **13**, 445–451.

Boyd, E. S., Jackson, R. A., Encarnacion, G., Zahn, J. A., Beard, T., Leavitt, W. D., *et al.* (2007) Isolation, characterization, and ecology of sulfur-respiring Crenarchaea inhabiting acid-sulfate-chloride-containing geothermal springs in Yellowstone National Park. *Appl. Environ. Microbiol.* **73**, 6669–6677.

Brock, T. D., and Freeze, H. (1969) *Thermus aquaticus* gen. n. and sp. n., a nonsporulating extreme thermophile. *J. Bacteriol.* **98**, 289–297.

Cantoni, O., Sestili, P., Cattabeni, F., Bellomo, G., Pou, S., Cohen, M., *et al.* (1989) Calcium chelator quin 2 prevents hydrogen-peroxide-induced DNA breakage and cytotoxicity. *Eur. J. Biochem.* **182**, 209–212.

Cavicchioli, R., Siddiqui, K. S., Andrews, D., and Sowers, K. R. (2002) Low-temperature extremophiles and their applications. *Curr. Opion. Biotechnol.* **13**, 253–261.

Chen, Y., Wu, L., Boden, R., Hillebrand, A., Kumaresan, D., Moussard, H., *et al.* (2009) Life without light: microbial diversity and evidence of sulfur- and ammonium-based chemolithotrophy in Movile Cave. *ISME J.* **3**, 1093–1104.

Costanza, R., D'Arge, R., De Groot, R., Farber, S., Grasso, M., Hannon, B., *et al.* (1997) The value of the world's ecosystem services and natural capital. *Nature* **387**, 253–260.

DiRuggiero, J., Wierzchos, J., Robinson, C. K., Souterre, T., Ravel, J., Artieda, O., *et al.* (2013) Microbial colonization of chasmoendolithic habitats in the hyper-arid zone of the Atcama Desert. *Biogeosciences* **10**, 2439–2450.

Dong, H., Rech, J. A., Jiang, H., Sun, H., and Buck, B. J. (2007) Endolithic cyanobacteria in soil gypsum: occurrences in Atacama (Chilie), Mojave (United States), and Al-Jafr Basin (Jordan) Deserts. *J. Geophys. Res.* **112**, G02030.

Drees, K. P., Neilson, J. W., Betancourt, J. L., Quade, J., Henderson, D. A., Pryor, B. M., *et al.* (2006) Bacterial community structure in the hyper-arid core of the Atacama Desert, Chile. *Appl. Environ. Microbiol.* **72**, 7902–7908.

Fisher, C. R., Takai, K., and Le Bris, N. (2007) Hydrothermal vent ecosystems. *Oceanography* **20**, 14–23.

Ford, D., and Williams, P. (2007) "Karst Hydrogeology and Geomorphology," John Wiley & Sons Ltd, West Sussex.

Gao, K., and Ye, C. (2007) Photosynthetic insensitivity of the terrestrial cyanobacterium *Nostoc flagelliforme* to solar UV radiation while rehydrated or desiccated. *J. Phycol.* **43**, 628–635.

Handelsman, J., Rondon, M. R., Brady, S. F., Clardy, J., and Goodman, R. M. (1998) Molecular biological access to the chemistry of unknown soil microbes: a new frontier for natural products. *Chem. Biol.* **5**, R245–R249.

Houston, J., and Hartley, A. J. (2003) The Central Andean west-slope rainshadow and its potential contribution to the origin of hyperaridity in the Atacama Desert. *Int. J. Climatol.* **23**, 1453–1464.

Inskeep, W. P., and McDermott, T. R. (2005) Geomicrobiology of acid-sulfate-chloride springs in Yellowstone National Park. From: http://www.rcn.montana.edu/pubs/pubview.aspx?nav = 8&pubID = 123.

Karl, D. M. (ed.) (1995) "The Microbiology of Deep-Sea Hydrothermal Vents", CRC Press, Boca Raton, FL, pp. 35–124.

Karr, E. A., Sattley, W. M., Jung, D. O., Madigan, M. T., and Achenbach, L. A. (2003) Remarkable diversity of phototrophic purple bacteria in a permanently frozen Antarctic lake. *Appl. Environ. Microbiol.* **69**, 4910–4914.

Karr, E. A., Sattley, W. M., Rice, M. R., Jung, D. O., Madigan, M. T., and Achenbach, L. A. (2005) Diversity and distribution of sulfate-reducing bacteria in permanently frozen Lake Fryxell, McMurdo Dry Valleys, Antarctica. *Appl. Environ. Microbiol.* **71**, 6353–6359.

Karr, E. A., Ng, J. M., Belchik, S. M., Sattley, W. M., Madigan, M. T., and Achenbach, L. A. (2006) Biodiversity of methanogenic and other Archaea in the permanently frozen Lake Fryxell, Antarctica. *Appl. Environ. Microbiol.* **72**, 1663–1666.

Kartchner Caverns Microbial Observatory website (2008) http://swes.cals.arizona.edu/maier_lab/kartchner/.

Kikuchi, A., and Asai, K. (1984) Reverse gyrase—a topoisomerase which introduces positive superhelical turns into DNA. *Nature* **309**, 677–681.

Legatzki, A., Ortiz, M., Neilson, J. W., Casavant, R. R., Palmer, M. W., Rasmussen, C., *et al.* (2012) Factors influencing the speleothem-specific structure of bacterial communities on the surface of formations in Kartchner Caverns, AZ, U.S.A. *Geomicrobiol. J.* **29**, 422–434.

Lester, E. D., Satomi, M., and Ponce, A. (2007) Microflora of extreme arid Atacama Desert soils. *Soil Biol. Biochem.* **39**, 704–708.

MDV (McMurdo Dry Valley Microbial Observatory website) (2007) http://mcm-vlakesmo.montana.edu/index.htm.

Markert, S., Arndt, C., Felbeck, H., Becher, D., Sievert, S. M., Hugler, M., *et al.* (2007) Physiological proteomics of the uncultured endosymbiont of *Riftia pachyptila*. *Science* **315**, 247−250.

Middleton, N. J., Thomas, D. S. G., (eds), for United Nations Environment Programme (1997) "World Atlas of Desertification," Arnold, London.

Nakagawa, S., Takaki, Y., Shimamura, S., Reysenbach, A. L., Takai, K., and Horikoshi, K. (2007) Deep-sea vent epsilon-proteobacterial genomes provide insights into emergence of pathogens. *Proc. Natl Acad. Sci.* **104**, 12146−12150.

Neilson, J. W., Quade, J., Ortiz, M., Nelson, W. M., Legatzki, A., Tian, F., *et al.* (2012) Life at the hyerarid margin: novel bacterial diversity in arid soils of the Atacama Desert, Chile. *Extremophiles* **16**, 553−566.

Ortiz, M., Neilson, J. W., Nelson, W. M., Legatzki, A., Byrne, A., Yu, Y., *et al.* (2013a) Profiling bacterial diversity and taxonomic composition on speleothem surfaces in Kartchner Caverns, AZ. *Microb. Ecol.* **65**, 371−383.

Ortiz, M., Legatzki, A., Neilson, J. W., Fryslie, B., Nelson, W. M., Wing, R. A., *et al.* (2013b) Making a living while starving in the dark: metagenomic insights into the energy dynamics of a carbonate cave. *ISME J.* In press. Available from: http://dx.doi.org/10.1038/ismej.2013.159.

Pointing, S. B., Warren-Rhodes, K. A., Lacap, D. C., Rhodes, K. L., and McKay, C. P. (2007) Hypolithic community shifts occur as a result of liquid water availability along environmental gradients in China's hot and cold hyperarid deserts. *Environ. Microbiol.* **9**, 414−424.

Potts, M. (1999) Mechanisms of desiccation tolerance in cyanobacteria. *Eur. J. Phycol.* **34**, 319−328.

Rickard, D., and Morse, J. W. (2005) Acid volatile sulfate (AVS). *Marine Chem.* **97**, 141−197.

Saiki, R. K., Gelfand, D. H., Stoffel, S., Scharf, S. J., Higuchi, R., Horn, G. T., *et al.* (1988) Primer-directed enzymatic amplification of DNA with a thermostable DNA polymerase. *Science* **239**, 487−491.

Sattley, W. M., and Madigan, M. T. (2006) Isolation, characterization, and ecology of cold-active, chemolithotrophic, sulfur-oxidizing bacteria from perennially ice-covered Lake Fryxell, Antarctica. *Appl. Environ. Microbiol.* **72**, 5562−5568.

Schwarzenlander, C., and Averhoff, B. (2006) Characterization of DNA transport in the thermophilic bacterium *Thermus thermophilus* HB27. *FEBS J.* **273**, 4210−4218.

Siddiqui, K. S., and Cavicchioli, R. (2006) Cold-adapted enzymes. *Annu. Rev. Biochem.* **75**, 403−433.

Soto, P. E., Galleguillos, P. A., Serón, M. A., Zepeda, V. J., Dmergasso, C. S., and Pimilla, C. (2013) Parameters influencing the microbial oxidation activity in the industrial bioleaching heap at Escondida mine, Chile. *Hydrometallurgy* **133**, 51−57.

Spiess, F. N., MacDonald, K. C., Atwater, T., Ballard, R., Carranza, A., Cordoba, D., *et al.* (1980) East Pacific Rise—hot springs and geophysical experiments. *Science* **207**, 1421−1433.

Tetu, S. G., Breakwell, K., Elbourne, L. D., Holmes, A. J., Gillings, M. R., and Paulsen, I. T. (2013) Life in the dark: metagenomic evidence that a microbial slime community is driven by inorganic nitrogen metabolism. *ISME J.* **7**, 1227−1236.

Tufts, R., and Tenen, G. (1999) Discovery and history of Kartchner Caverns, Arizona. *J. Cave Karst Stud.* **61**, 44−48.

Tyson, G. W., Chapman, J., Hugenholtz, P., Allen, E. E., Ram, R. J., Richardson, P. M., *et al.* (2004) Community structure and metabolism through reconstruction of microbial genomes from the environment. *Nature* **428**, 37−43.

Warren-Rhodes, K. A., Rhodes, K. L., Pointing, S. B., Ewing, S. A., Lacap, D. C., Gómez-Silva, B., *et al.* (2006) Hypolithic cyanobacteria, dry limit of photosynthesis, and microbial ecology in the hyperarid Atacama Desert. *Microbiol. Ecol.* **52**, 389−398.

Whitman, W. B., Coleman, D. C., and Wiebe, W. J. (1998) Prokaryotes: the unseen majority. *Proc. Natl Acad. Sci. U.S.A.* **95**, 6578−6583.

Part III

Detection, Enumeration, and Identification

Environmental Sample Collection and Processing

Ian L. Pepper and Charles P. Gerba

8.1 SOILS AND SEDIMENTS

Soils are discontinuous, heterogeneous environments that contain large numbers of diverse organisms. As described in Chapter 4, soil microbial communities vary with depth and soil type, with surface soil horizons generally having more organisms than subsurface horizons. Communities also vary from site to site, and even within sites because of natural microsite variations that can allow very different microorganisms to coexist side by side. Because of the great variability within communities, it is often necessary to take more than one sample to obtain a representative microbial sample at a particular site. Therefore, the overall sampling strategy will depend on many factors, including the goal of the analyses, the resources available and the site characteristics. The most accurate approach is to take many samples within a given site and perform a separate analysis of each sample. However, in many instances time and effort can be conserved by combining the samples taken to form a composite sample that is analyzed, thereby limiting the number of analyses that need to be performed. Another approach often used is to sample a site sequentially over

time from a small defined location to determine temporal effects on microbes. Because so many choices are available, it is important to delineate a sampling strategy to ensure that quality assurance is addressed. This is done by developing a quality assurance project plan (QAPP) according to the guidelines shown in Information Box 8.1.

8.1.1 Sampling Strategies and Methods for Surface Soils

Bulk soil samples are easily obtained with a shovel or, better yet, a soil auger (Figure 8.1). Soil augers are more precise than simple shovels because they ensure that samples are taken to exactly the same depth on each occasion. This is important, as several soil factors can vary considerably with depth, such as oxygen, moisture content, organic carbon content and soil temperature. A simple hand auger is useful for taking shallow soil samples from areas that are unsaturated. Given the right conditions, a hand auger can be used to take samples to depths of 180 cm in 30 cm increments. However, some soils are

I.L. Pepper, C.P. Gerba, T.J. Gentry: Environmental Microbiology, Third edition. DOI: http://dx.doi.org/10.1016/B978-0-12-394626-3.00008-9
© 2015 Elsevier Inc. All rights reserved.

Information Box 8.1 Collection and Storage Specifications for a Quality Assurance Project Plan (QAPP)

The QAPP involves delineating the details of the sampling strategy, the sampling methods, and the subsequent storage of all samples. The QAPP normally also includes details of the proposed microbial analysis to be conducted on the soil samples.

- Sampling strategies: Number and type of samples, locations, depths, times, intervals
- Sampling methods: Specific techniques and equipment to be used
- Sample storage: Types of containers, preservation methods, maximum holding times

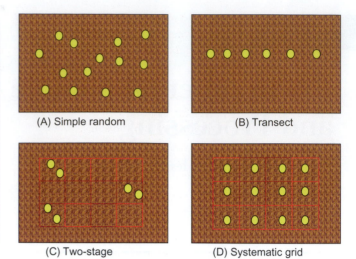

(A) Simple random (B) Transect

(C) Two-stage (D) Systematic grid

FIGURE 8.2 Alternative spatial sampling patterns.

FIGURE 8.1 Hand auger.

simply too compacted or contain too many rocks to allow sampling to this depth. When taking samples for microbial analysis, consideration should be given to contamination that can occur as the auger is pushed into the soil. In this case, microbes that stick to the sides of the auger as it is inserted into the soil and pushed downward may contaminate subsequent cores that are taken. To minimize such contamination, one can use a sterile spatula to scrape away the outer layer of the core and use the inner part of the core for analysis. Contamination can also occur between samples, but this can be avoided by cleaning the auger after each sample is taken. The cleaning procedure involves washing the auger with water, then rinsing it with 75% ethanol or 10% bleach, and finally rinsing with sterile water.

Composite samples can be obtained by collecting equal amounts of soil from samples taken over a wide area and placing them in a bucket or plastic bag. The whole soil mass is then mixed and becomes the composite sample. To reduce the volume of samples to be stored, a portion of the composite sample can be removed, and this becomes the sample for analysis. In all cases, samples should be stored on ice until processed and analyzed.

In some instances, a series of experimental plots or fields need to be sampled to test the effect of a soil amendment, such as fertilizer, pesticide or sewage sludge, on microbial communities. In this case, a soil sample must be taken from each of several plots or fields to compare control nontreated plots with plots that have received an amendment. For example, a researcher might be interested in the influence of inorganic nitrogen fertilizers on soil nitrifying populations. The investigator would then sample an unamended plot (the control) for comparison with a plot that had been treated with inorganic fertilizer. Another example would be the case in which soil amended with sewage sludge is sampled for subsequent viral pathogen analysis. In either example, multiple samples or replicates always give a more refined estimate of the parameters of interest. However, fieldwork can be costly and the number of samples taken must be weighed against the cost of analysis and the funds available. In the examples given, two-dimensional sampling plans can be used to determine the number and location of samples taken. In two-dimensional sampling, each plot is assigned spatial coordinates and set sampling points are chosen according to an established plan. Some typical two-dimensional sampling patterns, including random, transect, two-stage and grid sampling, are illustrated in Figure 8.2.

Random sampling involves choosing random points within the plot of interest, which are then sampled to a defined depth. Transect sampling involves collection of samples in a single direction. For example, transect sampling might be useful in a riparian area, where transects could be chosen adjacent to a streambed and at right angles to the streambed. In this way, the influence of the stream on the microbial community could be evaluated. In two-stage sampling, an area is broken into regular subunits called primary units. Within each primary unit, subsamples can be taken randomly or systematically. This approach might be useful when a site consists of a hillside slope and a level plain, and there is likely to be variability between the primary units. The final example of a sampling pattern is grid sampling, in which samples are taken systematically at regular intervals at a fixed spacing. This type of sampling is useful for mapping an area when little is known about the variability within the soil.

Two-dimensional sampling does not give any information about changes in microbial communities with depth. Therefore, three-dimensional sampling is used when information concerning depth is required. Such depth information is critical when evaluating sites that have been contaminated by improper disposal, or spills of contaminants. Three-dimensional sampling can be as simple as taking samples at 50 cm depth increments to a depth of 200 cm, or can involve drilling several hundred meters into the subsurface vadose zone. For subsurface sampling, specialized equipment is needed, and it is essential to ensure that subsurface samples are not contaminated by surface soil.

Finally, note that there is a specialized zone of soil that is under the influence of plant roots. This is known as the rhizosphere, which is of special interest to soil microbiologists and plant pathologists because of enhanced microbial activity and specific plant–microbe interactions (see Chapter 16). Rhizosphere soil exists as a continuum from the root surface (the rhizoplane) to a point where the root has no influence on microbial properties (generally 2–10 mm). Thus, rhizosphere soil volumes are variable and are difficult to sample. Normally, roots are carefully excavated and shaken gently to remove bulk or nonrhizosphere soil. Soil adhering to the plant roots is then considered to be rhizosphere soil. Although this is a crude sampling mechanism, it remains intact to this day. As a result, the sampling of rhizosphere remains a major experimental limitation, regardless of the sophistication of the microbial analyses that are subsequently performed.

8.1.2 Sampling Strategies and Methods for the Subsurface

Mechanical approaches using drill rigs are necessary for sampling the subsurface environment. This significantly increases the cost of sampling, especially for the deep subsurface. As a result, few cores have been taken in the deep subsurface, and these coring efforts have involved large teams of researchers (see Case Study 4.4). The approach used for sampling either deep or shallow subsurface environments depends on whether the subsurface is saturated or unsaturated. For unsaturated systems, air rotary drilling can be used to obtain samples from depths up to several hundred meters (Chapelle, 1992). In air

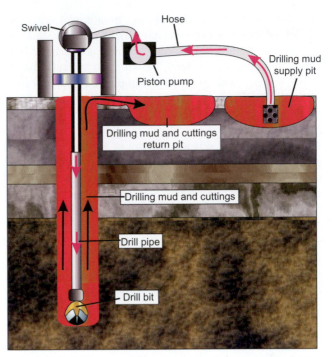

FIGURE 8.3 With rotary drilling the mechanical rotation of a drilling tool is used to create a borehole. Either air (air rotary drilling), or a fluid often called a drilling mud (mud rotary drilling), is forced down the drill stem to displace the borehole cuttings to the outside of the drill and upward to the surface. This figure illustrates mud rotary drilling.

rotary drilling, a large compressor is used to force air down a drill pipe, out the drill bit, and up outside the borehole (Figure 8.3). As the core barrel cuts downward, the air serves to blow the borehole cuttings out of the hole and also to cool the core barrel. This is important, because if the core barrel overheats, microbes within the sample may be effectively sterilized, posing difficulty for subsequent microbial analysis. In normal air drilling, small amounts of water containing a surfactant are injected into the airstream to control dust and help cool the drill bit. However, this increases the possibility of contamination, so cores such as the one drilled in Idaho's Snake River Plain (see Section 4.6.2.1) have been drilled with air alone (Colwell, 1989). To help keep the core barrel cool, the coring was simply done very slowly to avoid overheating. To help maintain sterile conditions and prevent contamination from surface air, all air used in the coring process was prefiltered through a 0.3-μm high-efficiency particulate air (HEPA) filter (see Section 5.8.2). Immediately after the core was collected the surface layer was scraped away with a sterile spatula, and then a subcore was taken using a 60-ml sterile plastic syringe with the end removed. The samples were immediately frozen and shipped to a laboratory, where microbial analyses were initiated within 18 hours of collection.

Saturated subsurface environments are sampled somewhat differently because the sediments are much less cohesive than those found in unsaturated regions. Therefore, the borehole must be held open so that an intact core can be taken and removed at each desired depth. For sampling of depths down to 30 m, hollow-stem auger drilling with push-tube sampling is widely used (Figure 8.4). The auger consists of a hollow tube with a rotating bit at the tip that drills the hole. The outside of the hollow auger casing is reverse threaded so that the cuttings are pushed upward and out of the hole as drilling proceeds. As the borehole is drilled, the casing of the auger is left in place to keep the borehole open. Thus, the casing acts as a sleeve into which a second tube, the core barrel, is inserted to collect the sample when the desired depth has been reached. The core barrel is basically a sterile tube that is placed at the tip of the hollow-stem auger, driven down to collect the sediment sample, and then retrieved. Drilling can then continue to the next desired depth and the coring process repeated. Each core collected is capped, frozen and sent to a laboratory for study. To avoid contamination of samples, the outside of the core is scraped away or the core may be subcored.

For cores that are deeper than 30 m, mud rotary coring is used (Chapelle, 1992). In this case, the hole is again bored using a rotating bit. However, drilling fluids are used to remove the borehole cuttings and to apply pressure to the walls of the borehole to keep it from collapsing. Mud rotary drilling has been used to obtain sediment samples to 1000 m beneath the soil surface. An example of such a core is one taken from the deep subsurface sediments of the Southeast Coastal Plain in South Carolina (see Case Study 4.4). During this coring, samples were retrieved from depths ranging from 400 to 500 m. In order to ensure the integrity of the cores obtained, the drilling fluids were spiked with two tracers, potassium bromide and rhodamine dye. The use of these two tracers allowed researchers to evaluate how far the drilling fluids had penetrated into the cores. Any areas of the cores that are contaminated with tracer must be discarded. The cores were retrieved in plastic liners, frozen, and sent for immediate analysis.

It is important to emphasize that coring either saturated or unsaturated environments is a difficult process

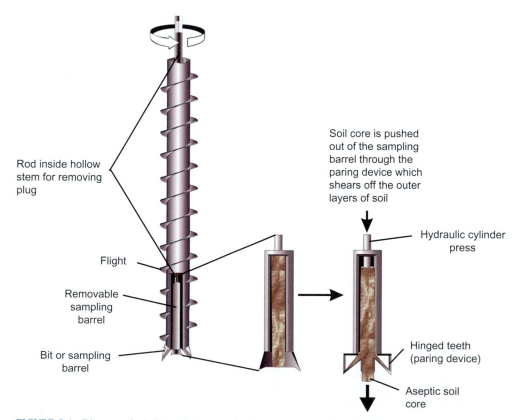

FIGURE 8.4 Diagram of a hollow-stem auger. Note the reverse threading on the outside of the auger. This is used to displace the borehole cuttings upward to the surface. This type of auger was used at Purdue University to collect core samples to a depth of 26 m for microbial and soil analysis as described by Konopka and Turco (1991). A subcore of each core collected is taken using a split spoon sampler or a push tube. In either case, the outside of the core must be regarded as contaminated. Therefore, the outside of the core is shaved off with a sterile spatula or a subcore can be taken using a sterile plastic syringe. Alternatively, as shown in this figure, intact cores are automatically pared to remove the outer contaminated material, leaving an inner sterile core.

for several reasons. First, it may take years to plan and obtain funding to proceed with cores such as those described here for the Snake River Plain and the Southeast Coastal Plain. The actual drilling and recovery of such samples is an engineering problem whose sophistication has only been touched upon in this section. Also keep in mind that the cores obtained are not always truly representative of the sediments from which they are taken. For example, a 1-m core may be compressed considerably in the coring process so that it is difficult to identify exactly the depth from which it was taken. A second difficulty in obtaining representative samples is due to horizontal heterogeneities in the subsurface material. Such heterogeneities can mean that two samples taken a few meters apart may have very different physical, chemical and microbiological characteristics. Finally, for microbial analysis it is not enough merely to retrieve the sample; the logistics of sample storage and analysis must be considered as well.

8.1.3 Sample Processing and Storage

Microbial analyses should be performed as soon as possible after collection of a soil to minimize the effects of storage on microbial communities. Once removed from the field, microbial communities within a sample can and will change regardless of the method of storage. Reductions in microbial numbers and microbial activity have been reported even when soil samples were stored in a field moist condition at 4°C for only 3 months (Stotzky et al., 1962). Interestingly in this study, although the bacterial community changed, the actinomycete community remained unchanged.

The first step in microbial analysis of a surface soil sample usually involves sieving through a 2 mm mesh to remove large stones and debris. However, to do this, samples must often be air dried to facilitate the sieving. This is acceptable as long as the soil moisture content does not become too low, because this can also change the microbial community (Sparkling and Cheshire, 1979). Following sieving, short-term storage should be at 4°C prior to analysis. If samples are stored, care should be taken to ensure that samples do not dry out and that anaerobic conditions do not develop, because this too can alter the microbial community. Storage up to 21 days appears to leave most soil microbial properties unchanged (Wollum, 1994), but again time is of the essence with respect to microbial analysis. Note that routine sampling of surface soils does not require sterile procedure. These soils are continually exposed to the atmosphere, so it is assumed that such exposure during sampling and processing will not affect the results significantly.

More care must be taken with processing subsurface samples for three reasons. First, they have lower cultural counts, which means that an outside microbial contaminant may significantly affect the numbers counted. Second, subsurface sediments are not routinely exposed to the atmosphere, and microbial contaminants in the atmosphere might substantially contribute to microbial types found. Third, it is more expensive to obtain subsurface samples, and often there is no second chance at collection. Subsurface samples obtained by coring are either immediately frozen and sent back to the laboratory as an intact core or processed at the coring site. In either case, the outside of the core is normally scraped off using a sterile spatula or a subcore is taken using a smaller diameter plastic syringe. The sample is then placed in a sterile plastic bag and analyzed immediately or frozen for future analysis.

8.1.3.1 Processing Soil and Sediment Samples for Bacteria

Culture-Based Analysis

Traditional methods of analysis for microbial communities have usually involved either cultural assays utilizing dilution and plating methodology on selective and differential media or direct count assays (see Chapter 10). Direct counts offer information about the total number of bacteria present, but give no information about the number or diversity of populations present within the community. Plate counts allow enumeration of total cultural or selected cultural populations, and hence provide information on the different populations present. However, since less than 1% of soil bacteria is readily culturable (Amann et al., 1995), cultural information offers only a piece of the picture. The actual fraction of the community that can be cultured depends on the medium chosen for cultural counts. Any single medium will select for the populations that are best suited to that particular medium. Thus, the choice of medium is crucial in determining the results obtained. This is illustrated by the data in Table 8.1, which show that whereas direct counts from a series of sediment samples spanning a 5 m depth were similar, the culturable counts varied depending on the type of medium used. A nutritionally rich medium, PTYG, made from peptone, trypticase, yeast extract and glucose, consistently gave counts that were one to three orders of magnitude lower than counts from two different low-nutrient media that were tested. These were a 1:20 dilution of PTYG and a soil extract agar made from a 1:2 suspension of surface soil. These data reflect the fact that most soil microbes exist under nutrient-limited or oligotrophic conditions.

Community DNA Analysis

In recent years, the advantages of studying community DNA extracted from soil samples have become apparent

TABLE 8.1 Total and Viable Cell Counts of Bacteria in Sediment Samples

Depth (m)	Saturation Status	AODC[a] (Cells/g Dry Weight)	Culturable Counts (CFU/g Dry Weight)		
			PTYG[b]	Dilute PTYG[c]	SSA[d]
1.2	Unsaturated	$6.8 \pm 4.9 \times 10^6$	$3.4 \pm 0.9 \times 10^4$	$1.9 \pm 0.4 \times 10^5$	$1.3 \pm 0.2 \times 10^5$
3.1	Interface[e]	$3.4 \pm 2.6 \times 10^6$	$2.0 \pm 0.5 \times 10^4$	$2.6 \pm 0.2 \times 10^6$	$2.9 \pm 0.6 \times 10^6$
4.9	Saturated	$6.8 \pm 4.3 \times 10^6$	$2.6 \pm 0.7 \times 10^3$	$3.5 \pm 0.1 \times 10^6$	$4.1 \pm 0.2 \times 10^6$

Adapted from Balkwill and Ghiorse (1985).
[a]*AODC, acridine orange direct counts. Reproduced by permission of the American Society for Microbiology Journals Department.*
[b]*PTYG, a nutritionally rich medium composed of peptone, trypticase, yeast extract, and glucose.*
[c]*Dilute PTYG, a 1:20 dilution of PTYG medium.*
[d]*SSA, soil extract agar. This medium was made by autoclaving a 1:2 suspension of surface soil in distilled water and then centrifuging and filtering the extract to clarify it.*
[e]*This sample was taken at the interface between the unsaturated and the saturated zone.*

Information Box 8.2   Comparison of Bacterial Fractionation and *In Situ* Lysis Methodologies for the Recovery of DNA from Soil

Issue	Bacterial Fractionation	*In Situ* Lysis
Yield of DNA	$1-5\,\mu g/g$	$1-20\,\mu g/g$
Representative of community	Less representative because of cell sorption	More representative, unaffected cell sorption
Source of DNA recovered	Only bacteria	Mostly bacteria but also fungi and protozoa
Degree of DNA shearing	Less shearing	More shearing
Average size of DNA fragments	50 kb	25 kb
Degree of humic contamination	Less contaminated	More contaminated
Ease of methodology	Slow, laborious	Faster, less labor-intensive

(see Chapter 13). This nonculture-based approach is thought to be more representative of the actual community present than culture-based approaches. In addition to providing information about the types of populations present, this approach can also provide information about their genetic potential. As with any technique, there are limitations to the data that can be obtained with DNA extraction. Therefore, many researchers now use DNA extraction in conjunction with direct and cultural counts to maximize the data obtained from an environmental sample.

Initially, two approaches were developed for isolation of bacterial DNA from soil samples. The first was based on fractionation of bacteria from soil followed by cell lysis and DNA extraction (Holben, 1994). The second method involved *in situ* lysis of bacteria within the soil matrix with subsequent extraction of the DNA released from cells (Information Box 8.2). Subsequent to the development of these two approaches, *in situ* lysis has become the commonly used extraction procedure primarily because it is easier and faster, because it yields more representative DNA, and because commercial kits have made it easier to purify the DNA.

The *in situ* lysis method involves lysing the bacterial cells within the soil and releasing their DNA prior to

extraction of DNA from the sample. Lysis methodology has usually involved a combination of physical and chemical treatments. For bacteria, physical treatments have involved freeze–thaw cycles and/or sonication or bead beating, and chemical treatments have often utilized a detergent such as sodium dodecyl sulfate (SDS) and/or an enzyme such as lysozyme or proteinase (Moré *et al.*, 1994). Following lysis, cell debris and soil particles are removed by precipitation and centrifugation, and the DNA in the supernatant is precipitated with ethanol. The DNA can be further purified by sorption onto homemade or commercial columns packed with ion-exchange resins or gels that can subsequently be rinsed for removal of humic materials that can inhibit DNA analysis. Further purification can be achieved with phenol–chloroform/isoamyl alcohol extractions, followed once more by ethanol precipitations (Xia *et al.*, 1995). Pure samples of DNA are necessary to allow subsequent molecular analyses such as with the polymerase chain reaction (PCR) (see Chapter 13). However, regardless of what purification methodology is employed, each step in the purification process causes loss of DNA. Thus, purified DNA is obtained only at the expense of DNA yield.

Commercial kits, which have optimized the procedures described above, are now available for processing

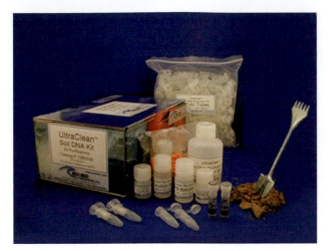

FIGURE 8.5   UltraClean™ Soil DNA Isolation Kit.

Information Box 8.3   Spectroscopic Analysis of DNA

The amount of DNA is estimated from the 260 nm reading. An absorbance reading of 1.0 is equivalent to 50 µg of DNA per ml of solution.

The purity of DNA is estimated from the ratio of the reading at 260 nm to that at 280 nm. A value > 1.7 indicates relatively pure DNA. The maximum theoretical value is 2.0.

soils for community DNA. Examples of such kits are Ultraclean™ Soil DNA Isolation Kit (MoBio) and Fast DNA Spin for Soil (MP Biomedicals) (Figure 8.5). Typically, these kits utilize physical bead beating technology followed by chemical lysis of microbes and subsequent DNA extraction and purification. These kits can even be used to extract DNA from high organic matter environmental samples including compost, sediments, manures and biosolids. But in this case, additional purification of the extracted DNA may be necessary in conjunction with the kit. One common approach is to repeatedly rinse the DNA with guanidine thiocyanate while it is sorbed to the extraction column provided by the kit. Overall these commercial kits have dramatically increased the ease and rapidity of community DNA extractions from soil. Note that there are also kits available to extract community DNA from water samples, e.g., UltraClean™ Water DNA Kit.

Although direct lysis using commercial kits has many advantages, it also has some problems. Sorption of DNA from lysed cells by clay or humic colloids can reduce the yield of extracted DNA (Ogram *et al.*, 1987). Another problem associated with direct lysis is distinguishing free from cellular DNA. Free DNA released from microbes that lysed naturally some time before the DNA extraction can sometimes be protected from degradation by sorption to soil particles (Lorentz and Wackernagel, 1987). This DNA may be extracted along with DNA from the viable cells. In addition, DNA isolated by direct lysis tends to be randomly sheared due to the bead beating procedure associated with most extraction kits. Finally, most kits will lyse all soil microorganisms including fungi and protozoa. Thus, the DNA extracted is not limited to bacterial DNA. Fortunately, fungi ($\approx 10^5$ per gram) and protozoa ($\approx 10^4$ per gram) are present at significantly lower numbers in most soils,

and thus will not contribute significantly to DNA obtained from $10^8 - 10^9$ bacteria per gram, even allowing for the larger genome size of protozoa.

Once a sample of purified DNA is obtained from the soil sample, it can be quantified by ultraviolet (UV) spectroscopy or fluorometry. Normally, UV readings are made at wavelengths of 260 and 280 nm, from which the purity and quantity of DNA can be estimated (Information Box 8.3). One limitation of quantification by UV spectroscopy is that readings will be affected by any compound that absorbs at 260 nm. Quantification by fluorometry is more sensitive and more specific, but does not allow the evaluation of extract purity obtained by comparing readings at 260 and 280 nm. DNA concentrations as low as 1 picogram per µl can be measured with a fluorometer using picogreen dye.

Once the amount of DNA per mass of soil is known, estimates can be made of the microbial community. For bacteria such as *Escherichia coli*, a typical chromosome contains 4−5 million base pairs, equivalent to about 9 fg (9×10^{-15} g) of DNA. However, the amount of DNA per cell varies and other estimates are lower, approximately 4 fg per cell. The amount of DNA per cell can also vary because of chromosome replication occurring faster than cell division, resulting in two or three chromosomes per cell (Krawiec and Riley, 1990). These theoretical DNA estimates can be used to relate total extracted DNA to the number of microbes within a sample. Table 8.2 shows the total DNA extracted from four soils amended with glucose. The amounts of DNA obtained increased with the amount of soil organic matter (silt loam and loam), presumably because of larger sustainable bacterial communities. Extracted DNA also decreased in soils high in clay (clay loam), most likely because of sorption of DNA by soil colloids (Ogram *et al.*, 1987). Overall, the influence of the amendments can be seen over time as microbial communities get larger through growth, resulting in more extractable DNA. The theoretical number of bacterial cells that the extracted DNA represents can be calculated as illustrated below.

For example, at time zero for the clay loam soil:

$$\text{Extracted DNA} = 0.12 \ \mu\text{g/g soil}$$

TABLE 8.2 Total Community DNA Extracted from Four Soils Amended with 1% Glucose and 0.1% Potassium Nitrate[a]

Soil	Extracted DNA (μg/g Soil) with Time (Days)				
	0	2	4	6	8
Clay loam	0.12	0.04	0.21	1.3	0.52
Silt loam	17.80	17.6	16.6	18.4	19.9
Sandy loam	0.63	0.60	1.90	5.50	1.90
Loam	1.30	0.90	1.30	4.20	7.70

[a]*Amendments were added at time zero and DNA was obtained via direct lysis.*

Therefore, if each cell has 4 fg of DNA:

$$\text{Number of cells} = \frac{0.12 \times 10\text{ g}^{-1}\text{ DNA}/g\text{ soil}}{4 \times 10\text{ g}^{-15}\text{ DNA}/\text{cell}}$$
$$= 3.0 \times 10^7\text{ cells}/g\text{ soil}$$

Similar values at time zero for the other soils are 1.6×10^8 cells/g soil (sandy loam), 3.3×10^8 cells/g soil (loam), and 4.5×10^9 cells/g soil (silt loam).

8.1.3.2 Processing Soil Samples for Fungal Hyphae and Spores

As with bacteria, it is also impossible to culture all species of viable fungi from a soil or sediment sample. Cultural methods for fungi are described in Section 10.5. However, several approaches have been developed for direct isolation of fungal hyphae or spores from soil. The first is a soil washing methodology. This involves saturating small volumes of soil with sterile water. Aggregates of soil are gently teased open with a fine jet of water, allowing heavier soil particles to sediment and finer particles to be decanted off. The procedure is repeated several times until only the heavier particles remain. These are then spread in a film of sterile water and examined under a dissecting microscope. Sterile needles or very fine forceps can then be used to obtain any observed fungal hyphae.

For spores, a different approach can be used. Soil samples are placed in separate sterile boxes each of which contains a number of sieves of graded size. The soil samples are washed vigorously in each box, and soil of defined size is retained on each sieve. Spores are determined empirically by plating successive washings (two minutes per wash) from each sieve. Because hyphae are retained by the sieves, any fungal colonies that arise must be due to the presence of spores. Information on fungi present as hyphae can be obtained by plating the washed particles retained by the sieves. However, these washing methods are labor intensive, and rely on trial and error in terms of which size fractions are most relevant with respect to individual spore size.

8.1.3.3 Processing Sludge, Soil and Sediment for Viruses

To assess fully and understand the risks from pathogens in the environment, it is necessary to determine their occurrence in sewage sludges (biosolids), soils to which wastewater or sludge is applied, or marine sediments that may be affected by sewage outfalls or sludge disposal. In fact, the U.S. Environmental Protection Agency currently requires monitoring of sludge for enteroviruses for certain types of land application. To detect viruses on solids it is first necessary to extract them by processes that will cause their desorption from the solid. As with microporous filters (see Section 8.2.2.1), viruses are believed to be bound to these solids by a combination of electrostatic and hydrophobic forces (see Chapter 15). To recover viruses from solids, substances are added that will break down these attractive forces, allowing the virus to be recovered in the eluting fluid (Gerba and Goyal, 1982; Berg, 1987).

The most common procedure for sludges involves collecting 500−1000 ml of sludge and adding $AlCl_3$ and HCl to adjust the pH to 3.5. Under these conditions, the viruses bind to the sludge solids, which are removed by centrifugation, and then resuspended in a beef extract solution at neutral pH to elute the virus. The eluate is then reconcentrated by flocculation of the proteins in the beef extract at pH 3.5, resuspended in 20−50 ml, and neutralized. A major problem with sewage sludge concentrates prepared in this manner is that they often contain substances toxic to cell culture. A diagram of the details of this procedure is shown in Figure 8.6. Similar extraction techniques are used for the recovery of viruses from soils and aquatic sediments (Hurst *et al.*, 2007).

8.2 WATER

8.2.1 Sampling Strategies and Methods for Water

Sampling environmental waters for subsequent microbial analysis is somewhat easier than sampling soils, for a variety of reasons. First, because water tends to be more homogeneous than soils, there is less site-to-site variability between two samples collected within the same vicinity. Second, it is often physically easier to collect water samples because it can be done with pumps and hose lines. Thus, known volumes of water can be collected from known depths with relative ease. Amounts of water

Recovery and concentration of viruses from sewage sludge

Procedure	Purpose
500–2000 ml sludge	
Adjust to pH 3.5 0.005 M AlCl₃	Adsorb viruses to solids
Centrifuge to pellet solids	
Discard supernatant	
Resuspend pellet in 10% beef extract	Elute (desorb) viruses from solids
Centrifuge to pellet solids	
Discard pellet and filter through 0.22 μm filter	Remove bacteria viruses are in supernatant
Assay using cell culture	

FIGURE 8.6   Procedure for recovery and concentration of viruses from sludge.

collected depend on the environmental sample being evaluated, but can vary from 1 ml to 1000 liters. Sampling strategy is also less complicated for water samples. In many cases, because water is mobile, a set number of bulk samples are simply collected from the same point over various time intervals. Such a strategy would be useful, for example, in sampling a river or a drinking water treatment plant. For marine waters, samples are often collected sequentially in time, within the defined area of interest.

Although the collection of the water sample is relatively easy, processing the sample prior to microbial analysis can be more difficult. The volume of the water sample required for detection of microbes can sometimes become unwieldy because the numbers of microbes tend to be lower in water samples than in soil samples (see Chapter 6). Therefore, strategies have been developed to allow concentration of the microbes within a water sample. For larger microbes including bacteria and protozoan parasites, samples are often filtered to trap and concentrate the organisms. For bacteria this often involves filtration using a 0.45-μm membrane filter (see Chapter 10). For protozoan parasites, coarse woven fibrous filters are used. For viruses, water samples are also filtered, but because viral particles are often too small to be physically trapped, collection of the viral particles depends on a combination of electrostatic and hydrophobic interactions of the virus with the filter. The different requirements for processing of water samples for analysis of viruses, bacteria and protozoa are outlined in the next two sections.

8.2.2 Processing Water Samples for Virus Analysis

The detection and analysis of viruses in water samples is often difficult because of the low numbers encountered and the different types that may be present. There are four basic steps in virus analysis: sample collection, elution, reconcentration and virus detection. For sample collection, it is often necessary to pass large volumes of water (100 to 1000 liters) through a filter because of the low numbers of viruses present. The viruses are concentrated from the water by adsorption onto the filter. Recovery of virus from the filter involves elution of the virus from the collection filter, as well as a reconcentration step to reduce the sample volume before assay. Virus detection can be done via cell culture or molecular methods such as PCR. However, both methods can be inhibited by the presence of toxic substances in the water that are concentrated along with the viral particles. Many strategies have been developed to overcome the difficulties associated with analysis of viruses, but they are often time consuming, labor intensive and costly. For example, the cost of enterovirus detection ranges from $600 to $1000 per sample for drinking water. Another problem with analysis of viruses is that the precision and accuracy of the methods used suffer from the large number of steps involved. In particular, the efficiency of viral recovery associated with each step is dependent on the type of virus that is being analyzed. For example, hepatitis A virus may not be concentrated as efficiently as rotavirus by the same process. Variability also results from the extreme sensitivity of these assays. Methods for the detection of viruses in water have been developed that can detect as little as one plaque-forming unit in 1000 liters of water. On a weight-to-weight basis with water, this is a sensitivity of detection of one part in 10^{18}. For comparison, the limit of sensitivity of most analytical methods available for organic compounds is about 1 μg/liter. This corresponds to one part in 10^9.

8.2.2.1 Sample Collection

Virus analysis is performed on a wide variety of water types. Types of water tested include potable water, ground and fresh surface waters, marine waters and sewage. These waters vary greatly in their physical–chemical composition, and contain substances either dissolved or suspended in solution, which may interfere with our ability to employ various concentration methods. The suitability of a virus concentration method depends on the probable virus density, the volume limitations of the concentration method for the type of water and the presence of interfering substances. A sample volume of less than 1 liter may suffice for recovery of viruses from raw and primary sewage. For drinking water and relatively nonpolluted waters, the virus levels are likely to be so low that hundreds or perhaps thousands of liters must be sampled to increase the probability of virus detection. Various methods employed for virus concentration from water are shown in Table 8.3.

Most methods used for virus concentration depend on adsorption of the virus to a surface, such as a filter or mineral precipitate, although hydroextraction and ultrafiltration have been employed (Gerba, 1987). Field systems for virus concentration usually consist of the use of a pump for passing the water through the filter (at rates of 20 to 40 liters per minute), a filter housing and a flowmeter (Figure 8.7A). The entire system can usually be contained in a 20 liter capacity ice chest.

The filters most commonly used for virus collection from large volumes of water are adsorption—elution microporous filters, more commonly known as VIRADEL (for virus adsorption-elution). VIRADEL

TABLE 8.3 Methods Used for Concentrating Viruses from Water

Method	Initial Volume of Water	Applications	Remarks
Filter adsorption—elution			
Negatively charged filters	Large	All but the most turbid waters; best for seawater and sewage	Only system shown useful for concentrating viruses from large volumes of tap water, sewage, seawater, and other natural waters; cationic salt concentration and pH must be adjusted before processing
Positively charged filters	Large	Tap water; does not perform well with seawater because of positive charge	No preconditioning of water necessary at neutral or acidic pH levels
Adsorption to metal salt precipitate, aluminum hydroxide, ferric hydroxide	Small	Tap water, sewage	Has been useful in reconcentration
Charged filter aid	Small	Tap water, sewage	40-liter volumes tested, low cost; used as a sandwich between prefilters
Polyelectrolyte PE60	Large	Tap water, lake water, sewage	Because of its unstable nature and lot-to-lot variation in efficiency for concentration of viruses, the method has not been used in recent years
Bentonite	Small	Tap water, sewage	
Iron oxide	Small	Tap water, sewage	
Glass powder	Large	Tap water, sewage	Columns containing glass powder have been made that are capable of processing 40-liter volumes
Positively charged glass wool	Small to large	Tap water	Positively charged glass wool is inexpensive; used in pipes or columns
Protamine sulfate	Small	Sewage	Very efficient method for concentrating reoviruses and adenoviruses from small volumes of sewage
Hydroextraction	Small	Sewage	Often used as a method for reconcentrating viruses from primary eluates
Ultrafiltration			
Soluble filters	Small	Clean waters	Clogs rapidly even with low turbidity
Flat membranes	Small	Clean waters	Clogs rapidly even with low turbidity
Hollow fiber or capillary	Large	Tap water, lake water, seawater	Up to 100 to 1000 liters may be processed, but water must often be prefiltered; cannot easily be used in the field and requires longer processing time than filters
Reverse osmosis	Small	Clean waters	Also concentrates cytotoxic compounds that adversely affect assay methods; cannot easily be used in the field and requires longer filtering time than filters

FIGURE 8.7   (A) Field VIRADEL system for concentrating viruses from water. (B) Elution of virus from filter with beef extract.

involves passing the water through a filter to which the viruses adsorb. The pore size of the filters is much larger than the viruses, and adsorption takes place by a combination of electrostatic and hydrophobic interactions (Iker et al., 2012). Two general types of filters are available: electronegative (negative surface charge) and electropositive (positive surface charge). Electronegative filters are composed of either cellulose esters or fiberglass with organic resin binders. Because the filters are negatively charged, cationic salts ($MgCl_2$ or $AlCl_3$) must be added in addition to lowering the pH to 3.5. This reduces the net negative charge usually associated with viruses allowing adsorption to be maximized (see Chapter 2). Such pH adjustment can be cumbersome, as it requires modifying the water prior to filtering and the use of additional materials and equipment such as pH meters. The most commonly used electronegative filter is the Filterite. Generally, it is used as a 10-inch (25.4 cm) pleated cartridge with either a 0.22- or 0.45-μm nominal pore size rating. Electronegative filters are ideal when concentrating viruses from seawater and waters with high amounts of organic matter and turbidity (Gerba et al., 1978).

Electropositive filters may be composed of fiberglass or cellulose containing a positively charged organic polymeric resin (1MDS), or nano alumina fibers (NanoCeram), which create a net positive surface charge to enhance adsorption of the negatively charged virus. These filters adsorb viruses efficiently over a wide pH range without a need for polyvalent salts. The 1MDS are less efficient with seawater or water with a pH exceeding 8.0–8.5 (Sobsey and Glass, 1980). The electropositive 1MDS Virozorb is especially manufactured for virus concentration from water.

VIRADEL filter methods suffer from a number of limitations. Suspended matter in water tends to clog the filters, thereby limiting the volume that can be processed and interfering with the elution process. Dissolved and colloidal organic matter in some waters can interfere with virus adsorption to filters, presumably by competing with viruses for adsorption sites. Finally, the concentration efficiency varies depending on the type of virus, presumably because of differences in the isoelectric point of the virus, which influences the net charge of the virus at any pH.

8.2.2.2 Sample Elution and Reconcentration

Adsorbed viruses are usually eluted from the filter surfaces by pressure filtering a small volume (1–2 liters) of an eluting solution through the filter. The eluent is usually a slightly alkaline proteinaceous fluid such as 1.5% beef extract adjusted to pH 9.5 (Figure 8.7B). The elevated pH increases the negative charge on both the virus and filter surfaces, which results in desorption of the virus from the filter. The organic matter in the beef extract also competes with the virus for adsorption on the filter, further aiding desorption. The 1- to 2-liter volume of the elutant is still too large to allow sensitive virus analysis, and therefore a second concentration step (reconcentration) is used to reduce the volume to 20–30 ml before assay. The elution–reconcentration process is shown in detail in Figure 8.8. Overall, these methods can recover enteroviruses with an efficiency of 30–50% from 400- to 1000-liter volumes of water (Gerba et al., 1978; Sobsey and Glass, 1980).

8.2.2.3 Virus Detection

Several options are available for virus detection and are described in detail in Section 10.7. Briefly, virus can be detected by inoculation of a sample into an animal cell

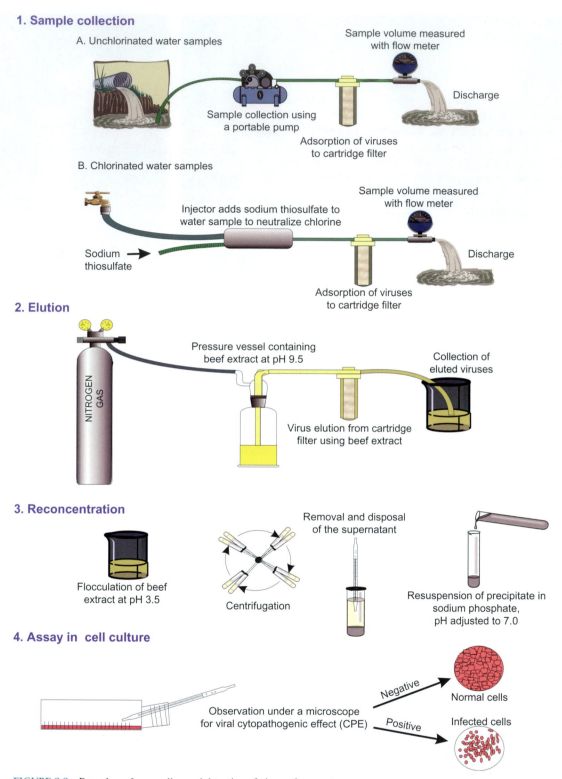

FIGURE 8.8   Procedures for sampling and detection of viruses from water.

culture followed by observation of the cells for cytopathogenic effects (CPE) or by enumeration of clear zones or plaque-forming units (PFU) in cell monolayers stained with vital dyes (i.e., dyes that stain only living nonvirus-infected cells). The PFU method allows more adequate quantitation of viruses because they can easily be enumerated. PCR can also be used to detect viruses directly in either the sample concentrates or the animal cell culture. The overall procedure for sampling and detecting viruses in water is shown in Figure 8.8.

8.2.3 Processing Water Samples for Detection of Bacteria

Processing water samples for bacteria is much simpler than the processing required for viruses. Typically, bacteria are collected and enumerated by one of two different procedures: the membrane filtration and most probable number (MPN) methodologies. Membrane filtration, as the name implies, relies on collection and concentration of bacteria via filtration. In the MPN method, samples are generally not processed prior to the analysis. In both procedures, bacteria are detected via cultural methods using MPN or membrane filtration techniques, described in Chapter 10.

8.2.4 Processing Water Samples for Detection of Protozoan Parasites

As with enteric viruses, it requires ingestion of only a few protozoan parasites to cause infection in humans. As a result, large volumes of tapwater (10 liters or more) or surface water (10 to 100 liters) need to be sampled in order to detect low numbers. The first step usually involves collection of the cysts or oocysts by filtration on pleated cartridge or foam filters (Schaefer, 2007). During filtration, the cysts or oocysts are entrapped on the filters by size exclusion (Figure 8.9). Usually, a pump running at a flow rate of 2 liters per minute is used to collect a sample. The filter is placed in a plastic bag, sealed, stored on ice and sent to the laboratory to be processed within 72 hours.

In the laboratory, the cysts and oocysts are extracted. In the case of the cartridge filter (Envirocheck, Pall Filter, Ann Arbor), an elution buffer (a solution of laureth-12, Tris buffer, EDTA and antifoam) is added to the filter cartridge, and it is placed on a shaker and agitated for five minutes to release the cysts or oocysts. This is followed by pelleting the protozoa by centrifugation and resuspension in a buffer. In the case of the foam filter (Filta-Max, IDEXX, Westbrook, ME), an elution solution of phosphate buffer and 0.01% Tween 20 is added, and the protozoa are squeezed from the flexible filter with a plunger. A great deal of particulate matter is often concentrated along with the cysts and oocysts, and requires further purification by immunomagnetic separation (IMS). In this process, the cysts and oocysts attach to specific antibodies that are associated with magnetic beads (Dynal, Inc., Lake Success, NY), and the beads (with the organisms attached) are removed from solution. After dissociation of the cysts and oocysts from the beads, they are suspended in a small volume of buffer, placed into wells, stained with fluorescent monoclonal antibodies and viewed with an epifluorescence microscope (Figure 8.10). Fluorescent bodies of the correct size and shapes are identified and examined by differential interference contrast microscopy for the

(A)

(B)

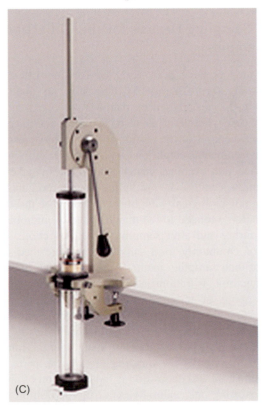

(C)

FIGURE 8.9 Filters used for *Giardia* and *Cryptosporidium* concentration from water. (A) Environcheck® Pleated Cartridge filter; (B) Filtramax® foam filter; (C) plunger system for elution of Filta-max filter. (A) Photo courtesy Pall Filter, Ann Arbor, MI; (C) Photo courstesy IDEXX, Westbrook, ME).

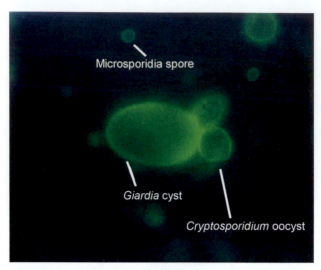

FIGURE 8.10 Immunofluorescence of *Giardia* cysts, *Cryptosporidium* oocysts and *Microsporidia* spores, representative of waterborne pathogenic protozoans.

presence of internal bodies (i.e., trophozites or sporozites). The entire procedure is shown in detail in Figure 8.11.

8.3 AIR

8.3.1 Sampling Devices for the Collection of Air Samples

Many devices have been designed for the collection of bioaerosols (see Chapter 5). Choosing an appropriate sampling device is based on many factors, such as availability, cost, volume of air to be sampled, mobility, sampling efficiency (for the particular type of bioaerosol) and the environmental conditions under which sampling will be conducted. Another factor that must be taken into account, especially when sampling for microorganisms, is the overall biological sampling efficiency of the device. This factor is related to the maintenance of microbial viability during and after sampling. In this section, several types of commonly used samplers are described on the basis of their sampling methods: impingement, impaction, centrifugation, filtration and deposition. Impingement is the trapping of airborne particles in a liquid matrix; impaction is the forced deposition of airborne particles on a solid surface; centrifugation is the mechanically forced deposition of airborne particles using inertial forces of gravity; filtration is the trapping of airborne particles by size exclusion; and deposition is the collection of airborne particles using only naturally occurring deposition forces. The most commonly used devices for microbial air sampling are: the all glass AGI-30 impinger (Ace Glass, Vineland, NJ); the SKC impinger (SKC-West Inc., Fullerton, CA, U.S.A.); and the Anderson six-stage

impaction sampler (6-STG, Andersen Instruments Incorporated, Atlanta, GA).

8.3.1.1 Impingement

The AGI-30 (Figure 8.12) and the SKC glass impingers (Figure 8.13) operate by drawing air through an inlet that is similar in shape to the human nasal passage. The air is transmitted through a liquid medium where the air particles become associated with the fluid and are subsequently trapped. The impingers usually separate at a flow rate of 12.5 L/min at a height of 1.5 m, which is the average breathing height for humans. They are easy to use, inexpensive, portable, reliable, easily sterilized and have high biological sampling efficiency in comparison with many other sampling devices. The impingers tend to be very efficient for particles in the range of 0.8 to 15 μm. The usual volume of collection medium is 20 ml, and the typical sampling duration is approximately 20 minutes, which prevents evaporation during the sampling in warm climates, or the freezing of the liquid medium when sampling at lower temperatures. The SKC biosamplers are more expensive due to the delicate nature of the blown glass which reduces damage to microbes during impingement (Brooks *et al.*, 2005). Another feature of the impingement process is that the liquid and suspended microorganisms can be concentrated or diluted, depending on the requirements for analysis. Liquid impingement media can also be divided into subsamples in order to test for a variety of microorganisms by standard cultural and molecular methods such as those described in Chapters 10 and 13. The impingement medium can also be optimized to increase the relative biological recovery efficiency. This is important, because during sampling the airborne microorganisms, which are already in a stressed state due to various environmental pressures such as ultraviolet (UV) radiation and desiccation, can be further stressed if a suitable medium is not used for recovery. Sampling media range from simple to complex. A simple medium is 0.85% NaCl, which is an osmotically balanced, sampling medium used to prevent osmotic shock of recovered organisms. A more complex medium is peptone (1%), which is used as a resuscitation medium for stressed organisms. Finally, enrichment or defined growth media can be used to sample selectively for certain types of organisms. The major drawback when using these impingers is that there is no particle size discrimination, which prevents accurate characterization of the sizes of the airborne particles that are collected.

8.3.1.2 Impaction

Unlike the impingers, the Andersen six-stage impaction sampler (Andersen 6-STG) provides accurate particle size

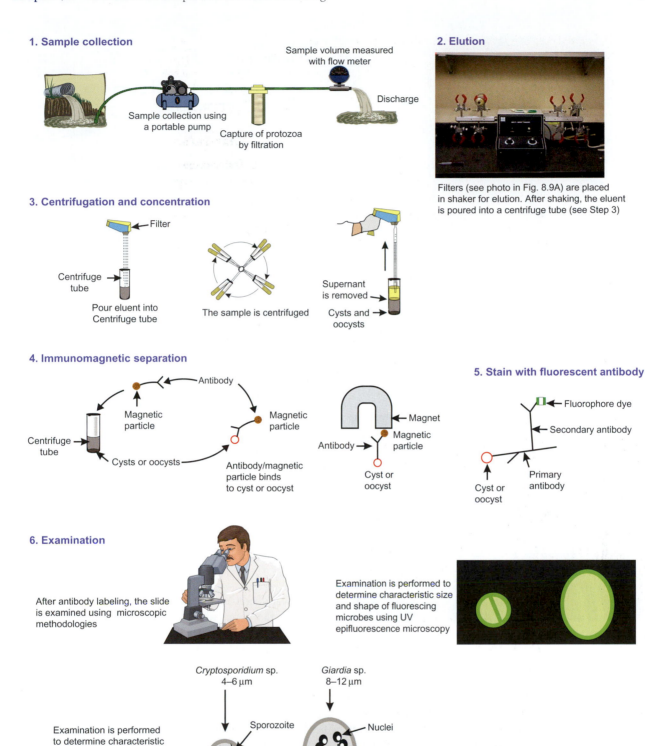

FIGURE 8.11   Procedure for processing and staining water samples for detection of protozoan parasites.

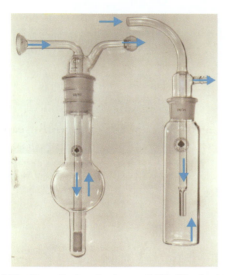

FIGURE 8.12 Two all glass impingers (AGI). The impinger on the right is the classic AGI-30 impinger. Arrows indicate the direction of air flow. The air enters the impinger drawn by suction. As bioaerosols impinge into the liquid medium contained in the bottom of the impinger, the airborne particles are trapped within the liquid matrix.

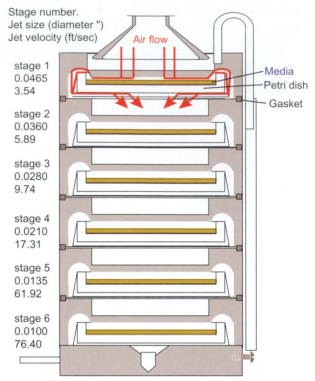

FIGURE 8.14 This is a schematic representation of the Andersen six stage impaction air sampler. Air enters through the top of the sampler and larger particles are impacted upon the surface of the Petri dish on stage 1. Smaller particles, which lack sufficient impaction potential, follow the air stream to the subsequent levels. As the air stream passes through each stage the air velocity increases, thus increasing the impaction potential so that particles are trapped on each level based up on their size. Therefore, larger particles are trapped efficiently on stage 1 and slightly smaller particles on stage 2, and so on, until even very small particles are trapped on stage 6. The Andersen six stage thus separates particles based upon their size.

FIGURE 8.13 An SKC impinger. Photo courtesy John Brooks.

The lower stages correspond to the alveoli and the upper stages to the upper respiratory tract. The Andersen sampler is constructed of stainless steel with glass Petri dishes, allowing sterilization, ease of transport and reliability. It is useful over the same particle size range as for the impingers (0.8 to over 10 μm), corresponding to the respirable range of particles. It is more expensive than the impingers, and the biological sampling efficiency is somewhat lower because of the method of collection, which is impaction on an agar surface. Analysis of viruses collected by impaction is also somewhat difficult, because after impaction, the viruses must be washed off the surface of the impaction medium and collected before assay. In contrast, bacteria or other microorganisms can be grown directly on the agar surface. Alternatively, these microbes can be washed off the surface and assayed using other standard methodologies as described in Chapter 10. The biggest single advantage of the Andersen 6-STG sampler is that particle size determinations can be obtained. Thus, the

discrimination (Figure 8.14). It is described as a multi-level, multiorifice, cascade impactor. The Andersen 6-STG was developed by Ariel A. Andersen in 1958 and operates at a input flow rate of 28.3 L/min. The general operating principle is that air is sucked through the sampling port and strikes agar plates. Larger particles are collected on the first layer, and each successive stage collects smaller and smaller particles by increasing the flow velocity and consequently the impaction potential. The shape of the Andersen sampler does not conform to the shape of the human respiratory tract, but the particle size distribution can be directly related to the particle size distribution that occurs naturally in the lungs of animals.

two reference samplers (impingers and the Andersen 6-STG) complement each other's deficiencies.

8.3.1.3 Centrifugation

Centrifugal samplers use circular flow patterns to increase the gravitational pull within the sampling device in order to deposit particles. The Cyclone, a tangential inlet and return flow sampling device, is the most common type (Figure 8.15). These samplers are able to sample a wide range of air volumes (1–400 L/min), depending on the size of the unit. The unit operates by applying suction to the outlet tube, which causes air to enter the upper chamber of the unit at an angle. The flow of air falls into a characteristic tangential flow pattern, which effectively circulates air around and down along the inner surface of the conical glass housing. As a result of the increased centrifugal forces imposed on particles in the airstream, the particles are sedimented out. The conically shaped

upper chamber opens into a larger bottom chamber, where most of this particle deposition occurs. Although these units are able to capture some respirable-sized particles, in order to trap microorganisms efficiently, the device must be combined with some type of metered fluid flow that acts as a trapping medium. This unit, when used by someone proficient, can be effective for microbiological air sampling. It is relatively inexpensive, easily sterilized and portable, but it lacks high biological sampling efficiency and particle sizing capabilities. Analysis is performed by rising the sampler with an eluent medium, collection of the eluent and subsequent assay by standard methodologies.

8.3.1.4 Filtration and Deposition

Filtration and deposition methods are both widely used for microbial sampling because of cost and portability reasons. Filter sampling requires a vacuum source and involves passage of air through a filter, where the particles are trapped. Membrane filters can have variable pore sizes that tend to restrict flow rates. After collection, the filter is washed to remove the organisms before analysis. Filtration sampling for microorganisms is not highly recommended because it has a low overall sampling efficiency and it is not portable. However, in many cases the low cost makes it an attractive method.

One case where filtration is routinely used is in sampling for airborne lipopolysaccharide (LPS). The sampling and analysis procedure for airborne LPS levels is slightly different from methods used for analysis of airborne microorganisms. The most efficient means of sampling is usually filter collection using polyvinyl chloride or glass fiber membrane filters. Quantification analysis is usually done using a chromogenic Limulus amebocyte lysate assay (Hurst *et al.*, 2007). This system uses a Limulus amebocyte lysate obtained from blood cells of horseshoe crabs (Brooks *et al.*, 2006). The lysate contains an enzyme-linked coagulation system, which is activated by the presence of LPS. With the addition of a substrate, and using luminescence, the system is able to quantitate the amount of environmental LPS by comparison with a standard curve.

Deposition sampling is by far the easiest and most cost-effective method of sampling. Deposition sampling can be accomplished merely by opening an agar plate and exposing it to the wind, which results in direct impaction, gravity settling and other depositional forces. The problems with this method of sampling are: low overall sampling efficiency because it relies on natural deposition, no defined sampling rates or particle sizing and an intrinsic difficulty in testing for multiple microorganisms with varied growth conditions. Analysis of microorganisms collected by depositional sampling is similar to impaction sample analysis.

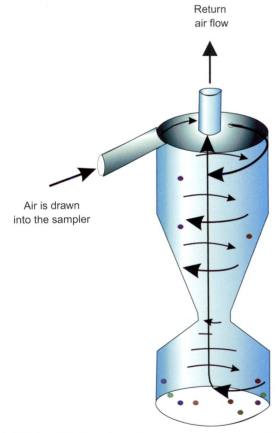

Return air flow

Air is drawn into the sampler

FIGURE 8.15 This is a schematic representation of a tangential inlet and returned flow centrifugal air sampler. Air is drawn into the sampler at an angle (tangential) to the walls of the device so that it circulates around and down the walls. As it circulates the decrease in the diameter of the sampling body causes a dramatic increase in the velocity of the air and subsequently on particle's terminal velocity. This increase in gravitational settling potential causes the particles to be trapped in the lower collection chamber because their "centrifugally increased" mass prevents them from existing with the return air flow.

8.4 DETECTION OF MICROORGANISMS ON FOMITES

Fomites are inanimate objects that may be contaminated with infectious organisms and subsequently serve in their transmission. Clothing, dishes, toys, tabletops and hypodermic needles are examples of common fomites. Fomites can range in size from as small as a particle of household dust, to as large as an entire floor surface. Complexity can vary from a flat tabletop, to a delicate medical instrument. The involvement of fomites in disease transmission was recognized long before the identification of some pathogenic microorganisms. Over 100 years ago, the spread of smallpox among laundry workers was not uncommon. In 1908, outbreaks of smallpox were traced to imported raw cotton contaminated with variola virus in crusts or scabs (England, 1984). Fomites are also believed to be important in the transmission of respiratory viruses, such as rhinovirus. An outbreak of hepatitis B virus, typically a blood-borne virus associated with blood transfusion, was associated with computer cards as the probable agents of transfer. These cards, when handled, inflict small wounds on the fingertips, allowing transmission and entry of the pathogen into a new host (Pattison et al., 1974). Growth of enteric pathogenic bacteria in household sponges and on utensils or surfaces used for food preparation has also been recognized as an important route for transfer of these organisms to other foods or surfaces. Self-inoculation can also occur when the fingers that have handled a sponge or utensil are brought to the mouth.

Fomites may become contaminated with pathogenic microorganisms by direct contact with infectious body secretions or fluids, soiled hands, contaminated foods or settling from the air. For fomites to serve as vehicles of microbial disease, the organisms must be able to survive in association with the fomites, and be successfully transferred to the host. Survival of organisms on a surface is influenced by temperature, humidity, evaporation, desiccation, light, ultraviolet radiation, the physical and chemical properties of the surface and the substance in which the organism is suspended. Enteric and respiratory pathogens may survive from minutes to weeks on fomites, depending on the type of organism and the previously listed factors (Boone and Gerba, 2007).

Sampling of fomites is essential in the food manufacturing industry to assess sanitation practices, and is in common use in the food service and healthcare industries to evaluate cleaning and disinfection efficacy. It is also useful in epidemiological investigation and evaluation of hard surface disinfectants. The approaches most commonly used for detection of bacteria on fomites involve Rodac agar plates and the swab–rinse technique. Rodac dishes are Petri dishes in which the agar fills the entire dish to produce a convex surface, which is then pressed against the surface to be sampled. Selective media can be used for isolation of specific groups of organisms (e.g., m-FC media for fecal coliforms). After incubation, the colonies are counted and reported as colony-forming units (CFU) per cm^2. The swab–rinse method was developed in 1917 for studying bacterial contamination of eating utensils (England, 1984). The method is also suitable for sampling of viruses. A sterile cotton swab is moistened with a buffer or other solution and rubbed over the surface to be sampled. The tip of the swab is then placed aseptically in a container with a sterile collection solution, the container is shaken and the rinse fluid is assayed on an appropriate culture medium, or by a molecular technique such as the PCR method. Other approaches for surface sampling are the use of sponges, vacuum systems using HEPA filters and agar films (Peti-flim, 3M Corporation, Minneapolis, MN), and even laboratory Kimwipes (Yan et al., 2007). Sponges, vacuum systems and wipes allow for sampling of much larger areas than swabs. In practice, usually $100 \, cm^2$ is sampled with a swab or sponge.

QUESTIONS AND PROBLEMS

1. How many samples and what sampling strategy would you use to characterize a portion of land of area $500 \, km^2$? Assume that the land is square in shape with a river running through the middle of it which is contaminating the land with nitrate.

2. Discuss the reasons why deep subsurface sampling is more difficult than surface sampling.

3. A soil is extracted for its community DNA and is found to contain $0.89 \, \mu g$ DNA per g soil. How many bacterial cells does this theoretically involve?

4. When would one utilize electropositive filters for concentrating viruses from environmental samples?

5. When would one utilize electronegative filters for concentrating viruses from environmental samples?

6. In what ways is it easier to sample water for subsequent microbial analysis than it is to sample soil?

7. If you collected a surface soil sample from a desert area in the summertime, when daytime temperatures were in excess of $40°C$, how would you store the soil, and how would you get the soil ready for microbial experiments to be conducted 1 month later? Discuss the pros and cons of various strategies.

8. What sampling strategy would you use to give the most complete picture of all bacteria found in a soil sample?

9. How do electropositive filters concentrate viruses from water? Why are they not effective in concentrating viruses from seawater or from water with a pH above 8.5? What is the principle behind eluting viruses from filter surfaces?

REFERENCES AND RECOMMENDED READING

Amann, R. I., Ludwig, W., and Schleifer, K. H. (1995) Phylogenetic identification and in situ detection of individual microbial cells without cultivation. *Microbiol. Rev.* **59**, 143–169.

Balkwill, D. L., and Ghiorse, W. C. (1985) Characterization of subsurface bacteria associated with two shallow aquifers in Oklahoma. *Appl. Environ. Microbiol.* **50**, 580–588.

Berg, G. (1987) "Methods for Recovering Viruses from the Environment," CRC Press, Boca Raton, FL.

Boone, S. A., and Gerba, C. P. (2007) Significance of fomites in the spread of respiratory and enteric disease. *Appl. Environ. Microbiol.* **73**, 1687–1696.

Brooks, J. P., Tanner, B. D., Josephson, K. L., Haas, C. N., Gerba, C. P., and Pepper, I. L. (2005) A national study on the residential impact of biological aerosols from the land application of biosolids. *J. Appl. Microbiol.* **99**, 310–322.

Brooks, J. P., Tanner, B. D., Gerba, C. P., and Pepper, I. L. (2006) The measurement of aerosolized endotoxin from land application of Class B biosolids in southeast Arizona. *Can. J. Microbiol.* **52**, 150–156.

Chapelle, F. H. (1992) "Ground-Water Microbiology and Geochemistry," John Wiley & Sons, New York.

Colwell, F. S. (1989) Microbiological comparison of surface soil and unsaturated subsurface soil from a semiarid high desert. *Appl. Environ. Microbiol.* **55**, 2420–2423.

England, B. L. (1984) Detection of viruses on fomites. In "Methods in Environmental Virology" (C. P. Gerba, and S. M. Goyal, eds.), Marcel Dekker, New York, pp. 179–220.

Gerba, C. P. (1987) Recovering viruses from sewage, effluents, and water. In "Methods for Recovering Viruses from the Environment" (G. Berg, ed.), CRC Press, Boca Raton, FL, pp. 1–23.

Gerba, C. P., and Goyal, S. M. (1982) "Methods in Environmental Virology," Marcel Dekker, New York.

Gerba, C. P., Farah, S. R., Goyal, S. M., Wallis, C., and Melnick, J. L. (1978) Concentration of enteroviruses from large volumes of tap water, treated sewage, and seawater. *Appl. Environ. Microbiol.* **35**, 540–548.

Holben, W. E. (1994) Isolation and purification of bacterial DNA from soil. In "Methods of Soil Analysis, Part 2, Microbiological and Biochemical Properties," SSSA Book Series No. 5, Soil Science Society of America, Madison, WI, pp. 727–750.

Hurst, C., Crawford, R. L., Garland, J. L., Lipson, D. A., Mills, A. L., and Stetzenbach, L. D. (2007) "Manual of Environmental Microbiology," ASM Press, Washington, DC.

Iker, B. A., Gerba, C. P., and Bright, K. R. (2012) Concentration and recovery of viruses from water: a comprehensive review. *Food Environ. Virol.* **4**, 41–67.

Konopka, A., and Turco, R. (1991) Biodegradation of organic compounds in vadose zone and aquifer sediments. *Appl. Environ. Microbiol.* **57**, 2260–2268.

Krawiec, S., and Riley, M. (1990) Organization of the bacterial chromosome. *Microbiol. Rev.* **54**, 502–539.

Lorentz, M. G., and Wackernagel, W. (1987) Adsorption of DNA to sand, and variable degradation rate of adsorbed DNA. *Appl. Environ. Microbiol.* **53**, 2948–2952.

Moré, M. I., Herrick, J. B., Silva, M. C., Ghiorse, W. C., and Madsen, E. L. (1994) Quantitative cell lysis of indigenous microorganisms and rapid extraction of microbial DNA from sediment. *Appl. Environ. Microbiol.* **60**, 1572–1580.

Ogram, A., Sayler, G. S., and Barkay, T. (1987) The extraction and purification of microbial DNA from sediments. *J. Microbiol. Methods* **7**, 57–66.

Pattison, C. P., Boyer, K. M., Maynard, J. E., and Kelley, P. C. (1974) Epidemic hepatitis in a clinical laboratory: possible association with computer card handling. *J. Am. Med. Assoc.* **230**, 854–857.

Pepper, I. L., Gerba, C. P., and Brusseau, M. L., eds. (1996) "Pollution Science." Academic Press, San Diego, CA.

Schaefer, F. W. (2007) Detection of protozoan parasites in source and finished drinking water. In "Manual of Environmental Microbiology" (C. J. Hurst, R. L. Crawford, J. L. Garland, D. A. Lipson, A. L. Mills, and L. D. Stetzenbach, eds.), ASM Press, Washington, DC, pp. 265–279.

Sobsey, M. D., and Glass, J. S. (1980) Poliovirus concentration from tap water with electropositive adsorbent filters. *Appl. Environ. Microbiol.* **40**, 201–210.

Sparkling, G. P., and Cheshire, M. V. (1979) Effects of soil drying and storage on subsequent microbial growth. *Soil Biol. Biochem.* **11**, 317–319.

Stotzky, G., Goos, R. D., and Timonin, M. I. (1962) Microbial changes occurring in soil as a result of storage. *Plant Soil* **16**, 1–18.

Wollum, A. G. (1994) Soil sampling for microbiological analysis. In "Methods of Soil Analysis, Part 2, Microbiological and Biochemical Properties," SSSA Book Series No. 5, Soil Science Society of America, Madison, WI, pp. 2–13.

Xia, X., Bollinger, J., and Ogram, A. (1995) Molecular genetic analysis of the response of three soil microbial communities to the application of 2,4-D. *Mol. Ecol.* **4**, 17–28.

Yan, Z., Vorst, K. L., Zhang, K. L., and Ryser, E. T. (2007) Use of one-ply composite tissues in an automated optical assay recovery of *Listeria* from food contact surfaces and poultry-processing environments. *J. Food Protect.* **70**, 1263–1266.

Chapter 9

Microscopic Techniques

Timberley M. Roane and Ian L. Pepper

9.1 HISTORY OF MICROSCOPY

Despite the increasing use of advanced genomic and proteomic techniques, microscopy still plays an integral role in the study of microorganisms. Microscopy had its origins in the seventeenth century when the Dutch discovered the ability to magnify objects by combining convex and concave glass lenses. However, who actually invented the first Dutch microscope is not clear. In 1611, Johannes Kepler, a German mathematician and astronomer, found that magnification could be achieved with the use of a convex ocular with a convex objective lens and created the Kepler ocular. Giovanni Faber coined the word "microscope" in 1625 in reference to the ability to see small things. The first use of the microscope to see microorganisms was by the Dutch merchant Antonie van Leeuwenhoek (1632–1723). At the age of 40, van Leeuwenhoek began experimenting with glass lenses, and he eventually made over 400 different microscopes. With these microscopes, van Leeuwenhoek was the first to see "animalcules," which today are known as microorganisms.

The Dutch physicist and astronomer Christiaan Huygens made a crucial discovery in the seventeenth century. The Huygens eyepiece consists of two convex lenses, each with the convex side facing the objective. The lower lens provides a brighter, smaller image from the objective lens, and the upper Huygens lens then focuses the image. The Huygens design is still used today in eyepieces with magnifications of $10\times$ or less.

As the quality of lenses and understanding of resolution and magnification have improved, different types of microscopies have been developed. The first light microscope in the 1600s was followed by the development of the first electron microscope in the 1920s, and the confocal laser scanning microscope and atomic force microscope followed in the 1980s. In this chapter, we describe the basic techniques used in microscopy, introduce advances and show how microscopy is still fundamental to the field of environmental microbiology.

9.2 THEORY OF MICROSCOPY

Regardless of the microscope, microscopy relies heavily on user interpretation. While this makes it a highly subjective tool, microscopy can provide extremely useful information about microorganisms. The human eye alone can resolve about $150\,\mu m$ between two points. The objective of the microscope is to increase the resolution of the human eye. Resolution is the smallest distance between

I.L. Pepper, C.P. Gerba, T.J. Gentry: Environmental Microbiology, Third edition. DOI: http://dx.doi.org/10.1016/B978-0-12-394626-3.00009-0
© 2015 Elsevier Inc. All rights reserved.

two points visible to the eye, aided or unaided by a micro-scope. Similarly, resolving power, a function of the wavelength of light and the aperture of the objective lens used in viewing a specimen (see Information Box 9.1), is the ability to distinguish two points as separate. The resolving power of most light microscopes is 0.2 μm, a 750-fold improvement over what we can actually see.

An aberration in a microscope refers to the inability to image a point in an object as a point. The light micro-scope has five kinds of aberrations: spherical, coma, astigmatism, curvature of field and distortion. Aberrations are functions of the lenses in an optical system, and severe aberrations result in decreased resolution. However, in the light microscope, corrective lenses eliminate aberrations so that the theoretical resolving power can be achieved. For example, spherical aberration is the most common aberration. Spherical aberration results when light rays pass through a lens at different points on the lens, resulting in light rays of different focal lengths (Figure 9.1A). Recall that the wavelength of light determines resolution in the light microscope. Thus, light of varying focal planes or wavelengths results in poor resolution of two points in an object. The use of a light diaphragm corrects spherical aberrations by focusing light rays to a single focal plane (Figure 9.1B). The ability to stack on correcting lenses in the light microscope has eliminated aberrations, allowing the theoretical resolution to be achieved. The electromagnetic lenses in electron microscopes have the same aberrations as the glass lenses of light microscopes. However, there are no glass lenses in the electron microscope and so these aberrations are not as easy to correct. Consequently, although the theoretical resolution of the electron microscope is 0.0002 nm, the actual working resolution is only 0.2 nm.

Perhaps the most important aspect of microscopy is illumination of the sample. Without illumination, the specimen cannot be visualized. In light microscopy, transmitted light or reflected light may be used. The source of illumination can be white light or ultraviolet light. Realize, however, that all microscopic techniques rely on the manipulation of light (or electrons in electron micros-copy) to influence the resolution of a specimen. In Köhler

illumination, for example, a series of condenser lenses and diaphragms are used to focus light rays onto the specimen, increasing not only illumination but also resolution (Figure 9.2).

Magnification is the ability to enlarge the apparent size of an image, and useful magnification is a function of the resolving power of the microscope and the eye, which can be stated as:

$$\frac{\text{Limit of resolution by eye}}{\text{Limit of resolution of microscope}} = \frac{150\ \mu m}{0.2\ \mu m} = 750 \times$$

Note that the ability to magnify an image is almost infinite, but most of this magnification is blurred because

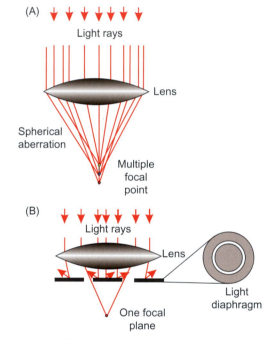

FIGURE 9.1   (A) Spherical aberrations are inherent in any lens, resulting in multiple focal planes of light. (B) The addition of a light diaphragm eliminates spherical aberrations in light microscopes by focusing the light onto one focal plane.

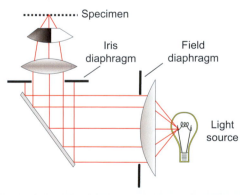

FIGURE 9.2   Schematic of Köhler illumination. Multiple diaphragms inserted between the light source and the condenser focus light directly on the specimen. Adapted from Zieler (1972), McCrone Research Institute, Microscope Publications Division.

Information Box 9.1   Theoretical Resolving Power

The resolving power (RP) is the minimum distance that an optical system can distinguish:

$$RP = \frac{\text{wavelength of light}}{2 \times \text{numerical aperture of objective lens}}$$

For example, for the light microscope

$$RP = \frac{500\ nm}{2 \times 1.25} = 200\ nm = 0.2\ \mu m$$

resolution is limited by the wavelength of the light, and so it becomes empty magnification.

Contrast refers to the ability to distinguish an object from the surrounding medium, and without it, both resolution and magnification become unimportant. More specifically, two points in an object that are resolved separately are not seen separately unless their images are contrasted against their surrounding medium. Because of their small size, bacteria, for example, provide little retardation of the light passing through the cell. The result is low contrast, and the color of the bacterial cell will be similar to its surrounding medium making visualization of the cell difficult. Many of the advances in microscopy have been for the sole purpose of increasing contrast, e.g., phase-contrast microscopy. Dyes and stains, such as methylene blue and safranin, are also used to increase contrast with all types of microscopes.

9.3 VISIBLE LIGHT MICROSCOPY

9.3.1 Types of Light Microscopy

Optical microscopes, also known as light microscopes, have multiple lenses, including ocular, objective and condenser lenses (Figure 9.3). By varying these lenses and light sources, five types of light microscopy can be defined: bright-field, dark-field, phase-contrast, differential interference and fluorescence. Characteristics of each of these types of microscopy are given in Table 9.1.

9.3.1.1 Bright-Field Microscopy

In bright-field microscopy, images are the result of light being transmitted through a specimen. The specimen absorbs some of the light, and the rest of the light is transmitted up through the ocular lens. The specimen will

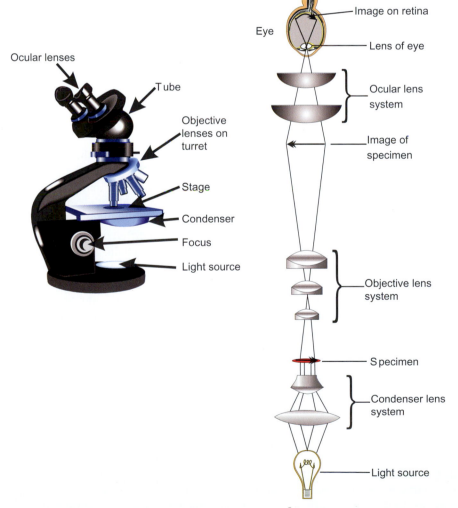

FIGURE 9.3 A typical compound light microscope and its optics.

TABLE 9.1 Comparison of Types of Microscopy

Microscope	Maximum Practical Magnification	Resolution	Important Features
Visible Light as Source of Illumination			
Bright-field	2000×	0.2 μm (200 nm)	Common multipurpose microscope for live and preserved stained specimens; specimen is dark, field is white; provides fair cellular detail
Dark-field	2000×	0.2 μm	Best for observing live, unstained specimens; specimen is bright, field is black; provides outline of specimen with reduced internal cellular detail
Phase-contrast	2000×	0.2 μm	Used for live specimens; specimen is contrasted against gray background; excellent for internal cellular detail
Differential interference	2000×	0.2 μm	Provides brightly colored, highly contrasting, three-dimensional images of live specimens
Ultraviolent Rays as Source of Illumination			
Fluorescence	2000×	0.2 μm	Specimens stained with fluorescent dyes or combined with fluorescent antibodies emit visible light; specificity makes this microscope an excellent diagnostic tool
Electron Beam Forms Image of Specimen			
Transmission electron microscope (TEM)	1,000,000×	0.5 nm	Sections of specimen are viewed under very high magnification; finest detailed internal structure of cells and viruses is shown; used only on preserved material
Scanning electron microscope (SEM)	100,000×	10 nm	Whole specimens are viewed under high magnification; external structures and cellular arrangement are shown; generally used on preserved material
Surface Forces Forms Image of Specimen			
Atomic force microscope (AFM)	1,000,000×	0.5 nm	Can examine live or preserved specimens. Provides surface detail at very high resolution

appear darker than the surrounding brightly illuminated field. Bright-field microscopy is most commonly used to examine morphology; however, due to the small size of microorganisms, in particular the bacteria, bright-field microscopy often requires staining to increase contrast in order to achieve the desired magnification. Other types of light microscopy (described below) use manipulation of light to increase contrast.

Many different types of stains are available, but in general they can be classified as basic dyes, which have a positive charge, or acidic dyes, with a negative charge. Cell components that are negatively charged such as nucleic acids attract basic dyes. In contrast, those that are positively charged, for example some cell-associated proteins, attract acidic dyes. Stains with positive charge attach more readily to the specimen, giving it color, while the background remains unstained. The most important types of positive stains are the simple stains, which involve a single dye such as methylene blue or Rose Bengal. These dyes stain the entire cell so that it takes on the color of the stain used (e.g., pink for Rose Bengal) against an unstained background. Simple stains are useful for size and morphological assessments, as well as for

cell enumeration. Negative stains such as the acidic dye, and India ink, are less common. They are repelled by the negatively charged surface of the cell and so stain the background, which results in the highlighting of the specimen as a silhouette

Differential stains utilize two different dyes designated as the primary dye and the counterstain. The Gram stain, developed by Hans Christian Gram, is the most important differential stain. The Gram stain utilizes crystal violet and safranin dyes to classify bacteria into one of two major categories. Gram-positive bacteria stain purple, whereas Gram-negative bacteria stain red. In both cases, the differential staining is due to differences in cell wall components (see Section 2.2.1). For many environmental isolates, the Gram stain may be inconclusive, and such isolates are designated as Gram variable. In this case, both red and purple cells may be seen, as in the case of *Arthrobacter* spp., which are common soil organisms.

Finally, there are a number of special stains. These special stains are used to identify specific cell components, such as bacterial capsules and spores. One such stain is the acid-fast stain, which was developed to identify difficult-to-stain bacteria. These organisms do not stain with commonly used dyes, such as those used in the Gram stain. Acid-fast bacteria are those that when stained with carbolfuchsin cannot be destained, even with acid. This property is typical of *Mycobacterium* spp., which have mycolic acids on their cell surface. Mycobacteria are of particular interest because they are causative agents of several serious human diseases, including tuberculosis and leprosy. They are also common soil isolates that are slow growing, with many of them having the ability to degrade organic contaminants.

9.3.1.2 Dark-Field Microscopy

Dark-field microscopy can be used to increase the contrast of a transparent specimen. By inserting a central stop before the condenser, some but not all of the light from the condenser is prevented from reaching the objective (Figure 9.4). Only light that is scattered from the edges of the specimen is viewed. Thus, the specimen appears as a bright image against a dark background. Dark-field microscopy is often used to visualize live specimens that have not been fixed or stained. For example, dark-field microscopy has been used to quantify the motility of bacteria and protozoa and to monitor the growth of bacterial microcolonies (Korber *et al.*, 1990). Although gross morphology can be delineated, internal details are not revealed. Murray and Robinow (1994) and Hoppert (2003) describe the nature of dark-field microscopy and its applications.

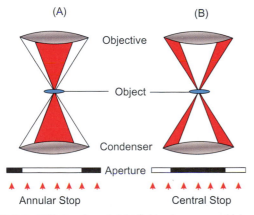

(A)   (B)

Objective

Object

Condenser

Aperture

Annular Stop   Central Stop

FIGURE 9.4 Differing from bright-field microscopy which uses an annular stop (A), phase-contrast microscopy uses a central stop (B), allowing some but not all of the light from the condenser to reach the objective. Adapted from Rochow and Tucker (1994).

9.3.1.3 Phase Contrast Microscopy

Phase-contrast microscopy enhances specimen contrast which aids viewing of high-contrast images of transparent specimens, such as living cells. This technique takes advantage of the fact that although many internal cell components are transparent, they have different densities.

Different densities interact differently with light, thereby creating contrast between internal cellular components and the surrounding medium (Figure 9.5). Phase-contrast microscopy uses a series of diaphragms for separating and recombining direct versus diffracted light rays (Figure 9.6). Köhler illumination is used to focus the light source on one focal plane. Light rays through the Köhler diaphragm are focused as a hollow cone onto the specimen. In the back focal plane of the objective, there is an annular diaphragm or a diffraction plate. The phase of light rays entering the diffraction plate, also called a phase plate, is altered. The degree of retardation of light through the plate results in either lightening or darkening of the specimen.

9.3.1.4 Differential Interference Contrast Microscopy

Differential interference contrast (DIC) microscopy provides brightly colored, highly contrasting three-dimensional images of live specimens. In DIC microscopy, the illuminating beam is split into two separate beams. One beam passes through the specimen, creating a phase difference between the sample beam and the second or reference beam. The two beams are then combined so that they interfere with each other. DIC can allow the detection of small changes in depth or elevation in the sample, thus giving the perception of a three-dimensional image (Figure 9.7).

9.3.1.5 Polarization

Anisotropic light, light that depends on the angle of observation, originates from specimens that have asymmetry in their crystal lattice properties. Anisotropy is observable in liquid and solid crystals; stained glasses; stressed plastic materials; crystallized resins and

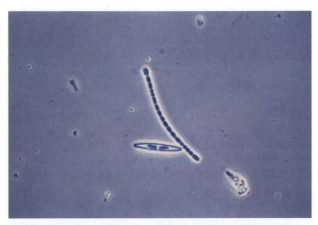

FIGURE 9.5 Phase-contrast image of a free-living nitrogen-fixing cyanobacterium (40 μm length) and the algal cell known as a diatom (12 μm length). Photo courtesy P. Rusin.

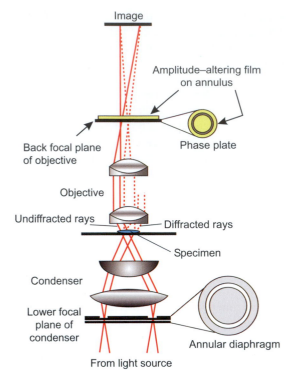

FIGURE 9.8 Polarization microscopy used to view sand grains in a sandy loam soil. The various colors are the result of light interference, which can be used to identify individual minerals. Magnification 400 ×. Photo courtesy T.M. Roane.

FIGURE 9.6 A phase-contrast microscope equipped with an amplitude-altering film on the phase plate to increase specimen contrast. *Handbook of Chemical Microscopy,* Vol. 1, 4th ed. C.W. Mason. Copyright © 1983. Adapted by permission of John Wiley & Sons, Inc.

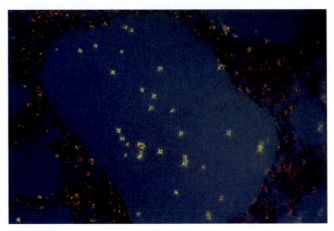

FIGURE 9.9 Polarization microscopy used to view bacteria producing extracellular polymeric substances. EPS producing cells show a cross-hatched illumination due to the resulting diffraction pattern as polarized light passes through the cell-surrounding EPS layer. Magnification 1000 ×. Photo courtesy T.M. Roane.

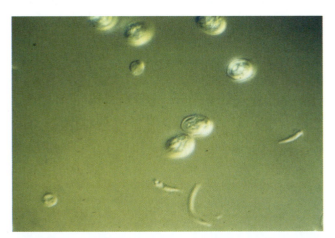

FIGURE 9.7 A differential interference contrast (DIC) image of *Cryptosporidium* with associated sporozoites. Photo courtesy P. Rusin.

polymers; refracting surfaces; synthetic filaments; and biological fibers, cells and tissues. Polarized light is light in one plane that can be used to examine anisotropy in sample materials. Polarization microscopy is traditionally used to determine the optical properties of soil minerals to aid in their identification (Figure 9.8). The optical anisotropy of individual crystals reflects the bonding patterns of units, e.g., molecules or elements, and usually

involves differences in at least two crystallographic directions (at least two directions of polarized light). Multiple anisotropic crystals have optical characteristics above and beyond those of individual crystals. Anisotropy observed in a sample can provide more information about the sample than ordinary unpolarized light.

For example, a result of light polarization is molecular birefringence. Molecular birefringence is manifested by long or flat molecules, especially polymeric macromolecules, and is particularly applicable in the examination of microbially produced extracellular polymers (Figure 9.9). In molecular birefringence, when polarized light encounters a series of atomic dipoles arranged in chains, as in long molecules, the strength of the dipoles causes the light to vibrate lengthwise along the chain, resulting in greater polar anisotropy at the poles. However, side

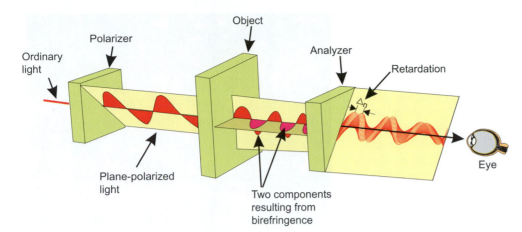

FIGURE 9.10 Schematic showing the function of the polarizer and the analyzer in polarization microscopy. Adapted from Mason (1983).

chains on the molecules tend to reduce the strength of birefringence in the main chain of the molecule, resulting in less polar anisotropy. The patterns and strengths of anisotropy evident in a sample can give indications of the purity and elemental structure of the sample.

The resolution of a specimen's effects on polarized light depends on producing plane polarized light with a polarizer and examining the effects with an analyzer. Two polarizers are used in polarization microscopy: a polarizer and an analyzer (Figure 9.10). For transmitted light, the polarizer is placed between the light source and the substage condenser lens. The analyzer is placed between the objective and ocular lenses. When polarized light from the first polarizer vibrates in a direction that allows it to pass through the analyzer, the field of view in the microscope will be black as the polarizers are crossed with respect to their directions of vibration of light. Thus, in a polarizing microscope, contrast is the result of various interference phenomena throughout the sample. Light interference or retardation at each point in a crystal results in contrast and color on a dark background. In accordance with the Michel-Levy interference spectra based on light retardation through varying sample thicknesses, light interference gives first order gray, high order white and color to the sample image.

9.3.2 Sample Preparation

9.3.2.1 Preparation from Liquid Samples

Sample preparation for microscopy can be as simple as placing a drop on a glass slide, or as complex as the thin sectioning and mounting on a copper grid as with transmission electron microscopy. In general, sample preparation varies with the type of microscopy used and the goal of the microscopic analysis.

Viable microorganisms are generally viewed via wet mounts. Here, cells are suspended in water, saline or some other liquid medium. The liquid maintains viability

and allows locomotion. Wet mounts can be done on a simple glass slide with a coverslip or on specially constructed slides. For example, in the latter case, a drop containing the specimen can be placed on a glass coverslip and then a slide with a concave depression is placed on top of the coverslip. Upon inversion of both coverslip and slide, the drop hangs from the coverslip. The drop is not affected by the glass slide, due to the concave depression, thereby creating a hanging drop. Hanging drop slides are useful in monitoring bacterial motility. Wet mounts are often viewed with phase-contrast or DIC microscopes to maximize specimen contrast.

Although morphology can be determined in a wet mount, this is often difficult because of the lack of contrast and detail between the specimen and the surrounding medium, and also because the microorganisms are moving. Thus, morphology and internal structure are better examined by fixing and staining specimens on glass slides, a process which kills the cells. Fixation of cells involves spreading a thin film of a liquid suspension of cells onto a slide and air drying it, producing a smear. The smear is then fixed on the slide by gently heating it over a flame for a few seconds. The smear is normally stained by the addition of a dye that enables cellular detail to be seen, as in bright-field and fluorescence microscopy.

9.3.2.2 Preparation from Soil Samples

Microorganisms in soil samples can be examined microscopically much like microorganisms from liquid samples, following microbial extraction from the soil. Microorganisms in soil can be ionically bound to soil and soil-associated particulates. To help release bound microorganisms, a combination of physical disruption, e.g., via mechanical mixing, and chemical neutralization can be used. Chemical neutralizers, such as sodium pyrophosphate, help homogenize ionic charges, causing repulsion between soil surfaces and microorganisms. Soil particulates can be removed from the resulting suspension through filtration or centrifugation. The microbial suspension can then be processed similarly to a liquid sample.

Historically, direct examination of microorganisms *in situ*, or within their environment, has been an important tool for microbiologists. Both light and electron microscopies allow direct examination of the form and arrangement of microorganisms in their environments. However, quantitation of actual microbial numbers has been difficult because interfering colloids and soil particles potentially mask large numbers of organisms. An example of direct examination is the buried slide technique.

Rossi *et al.* (1936) first introduced the buried slide. In this technique, a glass microscope slide is embedded in a soil or sediment sample. After a period of incubation, the slide is carefully removed with minimal disturbance, and soil particles with attached microbes can be viewed directly under the light microscope (Figure 4.22 shows an example of a buried slide). Although this method is more than 60 years old, it is still useful in illustrating the abundance of microorganisms in soil and their relationship to each other and to soil particles. Details of this technique can be found in Pepper and Gerba (2004). A variation of the buried slide technique is the pedoscope technique. Here, optically flat capillary tubes (the tubes are square so that all light passing through the tube has the same distance to travel) are buried in soil. Because soil microorganisms grow in pores or within soil aggregates, the relationships seen on the surface of a typical flat glass slide may not be truly representative of the natural state. The pedoscope capillary tubes overcome this by resembling soil pore spaces.

9.4 FLUORESCENCE MICROSCOPY

Fluorescence microscopy is technically a type of light microscopy but it differs in that it utilizes ultraviolet (UV) light sources. This type of microscopy is used in combination with fluorescent dyes, such as acridine orange or fluorescein, which are used to directly stain samples and perform direct counts. More powerfully, fluorescence microscopy can be used to detect specific probes which have been hybridized with the sample to detect the presence of a target molecule such as an antibody (immunolabeling) or a nucleic acid sequence (fluorescence *in situ* hybridization).

9.4.1 Direct Counts

Microbiologists are often interested in determining numbers of microorganisms associated with a given environment or process. There are two main methods for determining microbial numbers. The first involves

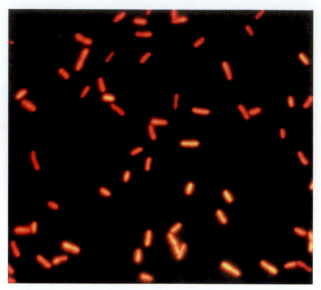

FIGURE 9.11   Acridine orange direct count. Photo courtesy K.L. Josephson.

culture-based assays as discussed in Chapter 10. The second method, known as the direct count, involves direct microscopic observations. Direct count procedures usually provide numbers that are one to two orders of magnitude higher than culturable counts because direct counts include viable, dead and viable but nonculturable (VBNC) organisms (see Section 3.3).

For direct counts, fluorescent stains are used rather than simple stains. A widely used stain for direct microscopy of bacteria is acridine orange (AO), used in obtaining acridine orange direct counts (AODC). Acridine orange intercalates with nucleic acids, and bacteria stained with AO appear either green (high amounts of RNA) or orange (high amounts of DNA) (Figure 9.11). Originally, it was thought that the green or orange color correlated with the viability of the organism; however, this has not been established. Two other important stains used in direct counting are 4,6-diamidino-2-phenylindole (DAPI) and fluorescein isothiocyanate (FITC). More recently, specialty stains have become available such as the LIVE/DEAD® *Bac*Light™ stain from Molecular Probes®. This stain can help differentiate the proportion of live and dead cells in the preparation (Berney *et al.*, 2007) (Figure 9.12). The *Bac*Light™ stain uses a mixture of SYTO® 9 green-fluorescent nucleic acid stain and a red-fluorescent nucleic acid stain, propidium iodide. These stains differ in their ability to penetrate healthy bacterial cells. The SYTO 9 stain will label all bacteria in a population, but the propidium iodide penetrates only bacteria with damaged membranes, causing a reduction in the SYTO 9 stain fluorescence when both dyes are present. Thus, live bacteria appear green (SYTO 9) and dead bacteria appear red (propidium iodide).

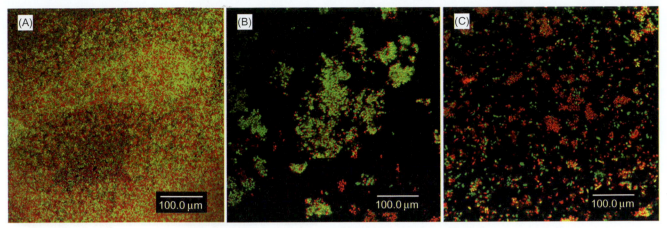

FIGURE 9.12 Bacterial cells stained with LIVE/DEAD BacLight bacterial viability stain to directly visualize the effects of an added antibiotic, vancomycin. These are confocal scanning laser microscope images of *Staphylococcus epidermidis* (SE6) intact biofilms (A), disrupted biofilm (B) and planktonic cells (C) on plastic coverslips after incubation for 24 h with 500 µg/ml of the vancomycin. These images suggest that the antibiotic is more effective on planktonic cells and disrupted biofilms than it is on an intact biofilm. From El-Azizi *et al.* (2005).

Sample preparation for direct stains depends on the type of sample. For pure cultures, samples can be centrifuged and resuspended in a known sample volume, stained and placed on a slide or in a counting chamber that holds a specific volume, such as a hemocytometer or a Petroff−Hauser chamber. Direct microscopy of water samples involves collecting the sample on a Nuclepore filter, staining and counting (Hobbie *et al.*, 1977). Direct microscopy of soil microorganisms usually involves first separating the organisms from soil particles. Soils are treated with a dispersing agent, such as Tween 80 or sodium pyrophosphate, and vortexed or sonicated to remove organisms from soil particles and to disrupt soil aggregates. A known volume of the resulting soil suspension is then stained for counting (Ipsilantis and Sylvia, 2007). In soil suspensions, regardless of the fluorescent stain, there are problems associated with the presence of clay colloids, which can either autofluoresce or nonspecifically bind the fluorescent stain. Colloids also mask the presence of soil microorganisms, a phenomenon known as colloidal interference. At high dilutions, colloidal interference decreases; however, the numbers of organisms are also diluted, which may result in low counts that are not statistically valid. At low dilutions, the numbers of organisms increase, but colloidal interference also increases.

9.4.1.1 Estimating Biomass

Direct counts can be used to estimate the microbial biomass in a sample. Estimates can be calculated in terms of bacterial or fungal biomass as carbon as shown in Table 9.2. To calculate bacterial biomass, some assumptions must be made. Approximate bacterial volumes must be determined using average cell lengths and diameters.

Approximate bacterial numbers are determined using direct count microscopy. For fungi, an estimate of fungal hyphal lengths per gram of sample must be known. In addition, estimates of the solids content for each organism have to be made. It should be noted that biomass estimates can also be made using chemical fumigation methods (see Section 11.4.2.3) or DNA content (see Chapter 13). The obvious limitations associated with estimating biomass are in estimating organism numbers. This problem is exacerbated in soils with high clay content resulting in colloidal interference.

9.4.2 Fluorescent Immunolabeling

A second common application of fluorescence microscopy is the detection of environmental microorganisms using antibodies that are labeled with fluorescent probes (see Chapter 12). The basis of this methodology is that a specific antibody can be designed and used as a probe for almost any target molecule including proteins, nucleic acids, polysaccharides and lipids. Once the target is determined, an antibody is constructed that will bind to the target with both selectivity and sensitivity. The antibody can be directly labeled with a fluorophore (primary detection agent). Fluorescein is most common but there are a number of alternative dyes such as Cy5 or biotin−avidin. Figure 9.13 shows an example of the use of a fluorescent antibody to detect rhizobia in a complex bacterial community.

Although many biomolecules bind selectively to a biological target, the result can be weak fluorescence necessitating the use of a secondary detection reagent, defined as a molecule that can be indirectly linked to the molecule of interest (see Figure 12.10). The labeled secondary detection reagent is designed to bind to the original

TABLE 9.2 Equations for Calculating Biomass

Calculation of Bacterial Numbers in Soil:

$$N_g = N_f \frac{A}{A_m} \frac{V_{sm}}{V_{sa}} D \frac{W_w}{W_d}$$

N_g = number of bacteria per gram dry soil
N_f = bacteria per field
A = area (mm^2) of smear (or filter)
A_m = area (mm^2) of microscope field
V_{sm} = volume (ml) of smear or filter
V_{sa} = volume (ml) of sample
D = dilution
W_w = wet weight soil
W_d = dry weight soil

Calculation of Bacterial Biomass as Carbon:

$$C_b = N_g V_b e \, S_c \frac{\%C}{100} \times 10^{-6}$$

C_b = bacterial biomass carbon (μg/g soil)
N_g = number of bacteria per gram soil
V_b = average volume (μm^3) of bacteria (r^2L; r = bacterial radius, L = length)
e = density (1.1×10^{-3} in liquid culture)
S_c = solids content (0.2 in liquid culture, 0.3 in soil)
$\%C$ = carbon content (45% dry weight)

Calculation of Fungal Biomass Carbon:

$$C_r = \pi r^2 L S_c \%C \times 10^{10}$$

C_r = fungal carbon (μg carbon/g soil)
r = hyphal radius (often 1.13 μm)
L = hyphal length (cm/g soil)
e = density (1.1 in liquid culture, 1.3 in soil)
S_c = solids content (0.2 in liquid culture, 0.25–0.35 in soil)

Adapted from Paul and Clark (1989).

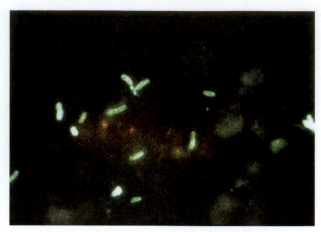

FIGURE 9.13 Use of fluorescent antibodies coupled to fluorescein isothiocyanate to detect antigens. Here rhizobia fluorescence in response to UV irradiation is shown. Photo courtesy I.L. Pepper.

antibody probe thereby linking the fluorescently labeled secondary detection reagent to the target.

9.4.3 Fluorescence *In Situ* Hybridization

Combining identification and visualization of cells in their natural environment is a task of considerable interest to environmental microbiologists. Fluorescence *in situ* hybridization (FISH) has been developed to meet that need. Commonly used in medical applications, it was first developed as a cultivation-independent means of identifying bacterial cells by DeLong *et al.* (1989). FISH involves the use of fluorescently labeled nucleic acid probes to target DNA or RNA sequences within an organism. Due to its high copy number and comparative accessibility, the 16S rRNA gene is by far the most popular target site in bacteria as is the 18S rRNA gene for eukaryotes

(Zwirglmaier, 2005). The probe is designed to selectively target regions of rRNA that consist of evolutionarily conserved or variable nucleotide regions. Thus, by choosing the appropriate rRNA probe sequence, FISH can be used to detect all bacterial cells (a universal probe to a conserved region), or a single population of cells (a strain-specific probe to a variable region). Recent developments have made other nucleic acid sequences feasible targets such as a gene on a high copy number plasmid or stable mRNA transcripts (Amann and Ludwig, 2000).

FISH is a valuable tool for microbiologists interested in detecting otherwise unculturable bacteria, and understanding microbial diversity and complexity of microbial communities. FISH can also provide insight into how microorganisms interact with each other under varying environmental conditions. However, sensitivity can be problematic and is restricted if cells are not actively growing. This results because the number of target rRNA copies within a cell is dependent on metabolic activity. This technique can be useful in analyzing spatial distributions of microorganisms. For example, Maixner *et al.* (2006) used FISH to study the niche differentiation of *Nitrospira* (a nitrifying bacterium) populations in wastewater-associated biofilms. In this study, FISH helped show that the spatial distribution of different populations depends on the nitrite concentration. As a second example, Figure 9.14 shows how FISH can be used to examine bacterial colonization of plant roots grown in mine tailings. In this case, a universal bacterial FISH probe was used to detect all bacteria on the root surfaces. This experiment examined the effect of compost amendment on colonization (Iverson and Maier, 2009). The results show that there was extensive root colonization in the presence of compost (along with good plant growth), while in the absence of compost there was little bacterial colonization of the root and poor plant growth.

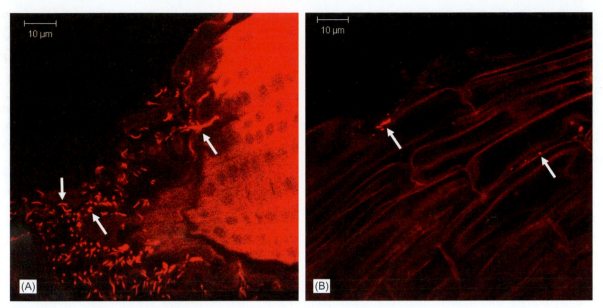

FIGURE 9.14   Use of FISH to compare root colonization of *Buchloe dactyloides* (buffalo grass) grown in mine tailings with either 15 or 0% compost amendment. The FISH probe used was a universal probe EUB338 labeled with the CY3 fluorophore. Samples were visualized on a Zeiss Confocal LSM 510 equipped with a 543 nm laser. Arrows point to bacterial colonies. (A) Heavy bacterial colonization on a root tip grown in mine tailings amended with 15% (w/w) compost. Optical slice is 1.0 μm thick. (B) Minimal colonization of a root grown in unamended mine tailings. Optical slice is 0.6 μm thick. Image courtesy S.L. Iverson.

9.4.4 Confocal Laser Scanning Microscopy

Most commonly used for imaging fluorescent specimens, the confocal laser scanning microscope (CLSM) is often used to document transects through a specimen, such as tissue sections. Computer software can be used to compile the images into a three-dimensional composite. Because nonfocused light is reduced in confocal microscopy, the confocal scanning microscope gives higher resolution, increased contrast and thinner planar views than other forms of light microscopy. Since three-dimensional views can be generated, the CLSM readily lends itself to digital processing, by which images of thin optical sections can be reassembled into a composite, three-dimensional image (Figures 9.12 and 9.14). These images may be viewed as a whole or as individual sections for greater detail. Confocal scanning microscopy is commonly used in bright-field, dark-field and fluorescence microscopies.

In confocal microscopy, a laser beam is used to focus light of a specific wavelength onto the specimen. The confocal scanning microscope has the ability to take optical sections at successive focal planes (known as a Z series). Pinhole apertures are used so that only a small area of the specimen is focused at any given time. Light from the plane of focus enters the detector, eliminating any scattered light, which has the tendency to blur images. The focused light beam moves across the specimen, scanning it, which is required because only a small volume is illuminated at any given time, and a number of these small volumes must be collected for a complete specimen image.

9.4.5 Flow Cytometry

In flow cytometry, microscopic detection of cells or other particles is required as the cells pass through a laser detector. Flow cytometry was first discovered in the 1950s and its uses include the detection of a variety of microorganisms, including bacteria and parasites. As a cell passes through the detector's laser beam, the amount of light scattered in the forward direction and a direction at a 90° angle is measured. These measurements respectively correlate with the size and internal complexity of the particle. The instrument can also measure the fluorescent light emitted by each particle. Data in the flow cytometer are collected as light energy, converted to electrical energy and then plotted on user-defined histograms.

In flow cytometry, particles are separated and flow singly through the detector. Flow cytometer cell sorters have the ability to detect target cells or particles among unwanted ones. The cell sorter vibrates the sample stream, causing it to break into droplets. Information about the particles of interest, such as light scattering and fluorescence criteria, is programmed into the cytometer computer so that when the particle is encountered, the instrument electrically charges the droplet carrying that particle. Oppositely charged deflection plates pull the particles of interest out of the uncharged sample stream toward the charged plate, and ultimately deflect them onto a glass microscope slide or into a collection tube. The droplets containing unwanted particles flow into a waste collection tank.

Flow cytometry is commonly used in environmental microbiology. The FITC method relies on the binding of a fluorescein (FITC)-conjugated antibody to antigens present in the sample. For example, this technique has been used to quantitate *Cryptosporidium* and *Giardia* present in environmental samples (see Figure 8.10 and Section 22.3.1). The FITC-stained sample suspension is aspirated by the flow cytometer and each particle in the sample is examined in the instrument's laser beam. The fluorescein molecule, when excited by the 488 nm laser light, in return emits light at 525 nm. The light energy is detected in the flow cytometer and quantitated. The cysts and oocysts of these organisms are identified by their 90° light scattering and additional FITC fluorescence properties. Collection by the flow cytometer on a glass slide allows additional microscopic analysis for identification.

9.4.6 Developing Methods in Fluorescence Microscopy

Limited resolution is a challenge with most fluorescence microscopy studies. While providing high specimen contrast, cellular and subcellular visualization of structures is limited by the weak resolution. However, techniques have been developed to combine the contrasting ability of fluorescence microscopy with enhanced resolution.

Stimulated emission depletion (STED) fluorescence microscopy increases resolution through the use of lasers to excite very specific locations of specimen-associated fluorophore label. STED microscopy has been used to identify protein complexes in mitochondria (Donnert *et al.*, 2007) and synaptic vesicles in living cells (Westphal *et al.*, 2008). Another increasingly common technique is single-molecule fluorescence imaging. This approach uses a fluorescent microscope with digital detection to track the cellular location of a fluorescence emission (Biteen and Moerner, 2010). Based on the fusion of a protein of interest to a fluorophore label, this type of microscopy allows for investigation of live cells, a major advantage over other types of microscopies. The fluorescent protein fusion allows the cell to remain physiologically active while being microscopically monitored. Continuing developments of fluorescence techniques offer tremendous potential in increasing our understanding of the subcellular structure of microorganisms.

9.5 ELECTRON MICROSCOPY

The electron microscope produces high resolution detail by using electrons instead of light to form images. The extremely short wavelength and focusability of electron beams are responsible for the theoretically high resolving power of electron microscopes. The increased resolution allows a functional magnification of up to 1,000,000 × for the observation of fine structure and detail. Although electron microscopes are conceptually similar to light microscopes, there are some fundamental differences between using light versus electronic illumination (Table 9.3). In the electron microscope, an electron gun aims a beam of electrons at a specimen placed in a vacuum sample chamber. A series of coiled electromagnets are used to focus the beam. As in light microscopy, poor contrast is a problem in electron microscopes, so samples are often stained to increase contrast. Images produced in the electron microscope are in shades of gray, although computerized color may be added in some scopes. The two most common types of electron microscopy are scanning electron microscopy and transmission electron microscopy.

TABLE 9.3 Comparison of Optical Microscopes and Electron Microscopes

Characteristic	Optical	Electron
Illuminating beam	Light beam	Electron beam
Wavelength	7500 Å (visible)	0.086 Å (20 kV)
	2000 Å (ultraviolet)	0.037 Å (100 kV)
Medium	Atmosphere	Vacuum
Lens	Glass lens	Electrostatic lens
Resolving power	2000 Å	3 Å
Magnification	Up to 2000 ×	Up to 1,000,000 ×
Focusing	Mechanical	Electrical
Viable specimen	Yes	No
Specimen requires staining or treatment	Yes/no	Always
Colored image produced	Yes	No

9.5.1 Scanning Electron Microscopy

In the scanning electron microscope (SEM), an image is formed as an electron probe scans the surface of the specimen (Figure 9.15), producing secondary electrons, backscattered electrons, X-rays, Auger electrons and photons of various energies. The SEM uses these signals to produce three-dimensional surface characteristics of specimens (Figure 9.16). There are several advantages of the SEM. These include a large depth of field and the ability to examine bulk samples with low magnification, and lifelike images.

In a typical SEM, an electron gun and multiple condenser lenses produce an electron beam whose rays are aligned through electromagnetic scan coils. Electron-accelerating voltages in the gun range from 60 to 100 kV (kilovolts). A tungsten filament, heated to approximately 2700K, is the illumination source within the gun. Heating the filament causes electrons to be released from the tip of the filament. An image of the surface topography of the specimen is generated by electrons that are reflected (backscattered) or given off (secondary electrons). Contrast in the SEM is enhanced by coating the sample with a thin layer of a conductive metal, e.g., gold or palladium, or even carbon. Image formation itself is the result of rastering the electron beam (from 2 to 200 Å in diameter) back and forth along the specimen surface (Figure 9.17). A visual image corresponding to the signal produced by the interaction

between the beam spot and the specimen at each point along each scan line is simultaneously built up on the face of a cathode ray tube in the same way a television picture is generated. As with all microscopy, interpretation of SEM images is subjective, particularly because SEM images include high resolution, high contrast and varying depths of focus resulting in topography.

Sample preparation for the SEM is relatively straightforward. The general sequence involves: (1) sample fixation with an aldehyde solution; (2) dehydration of the sample (because the sample must be under vacuum in the SEM); (3) mounting of the specimen on a metal stub; and (4) coating of the specimen with a thin layer of electrically conductive material.

9.5.2 Transmission Electron Microscopy

In the SEM, electrons interacting with the surface of the specimen form the image. In transmission electron microscopy (TEM) (Figure 9.18), the image is formed by electrons passing through the specimen. Consequently, the specimens must be thin sectioned to allow the passage

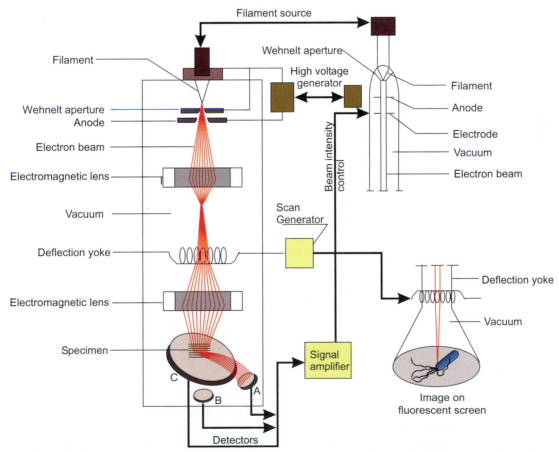

FIGURE 9.15   A typical scanning electron microscope (SEM) and its similarity to a television. They are based on the same principles. Courtesy FEI Company. Reproduced with permission.

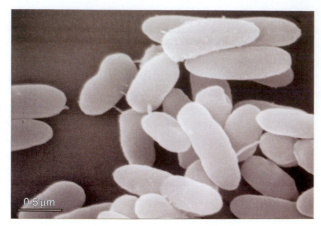

FIGURE 9.16 Scanning electron microscope image of *Pseudomonas aeruginosa* cells. The size bar is 5 μm. Photo courtesy A.A. Bodour.

FIGURE 9.18 A typical transmission electron microscope (TEM). Image courtesy FEI Company.

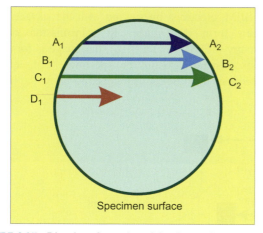

FIGURE 9.17 Direction of scanning of the electron beam on a sample surface in a scanning electron microscope. Diagram courtesy D. Bentley.

of electrons. TEM is often used to view detail of fine structures (Figure 9.19) and internal cell structures (Figure 9.20), due to the selective absorption of electrons by different parts of the specimen.

Sample preparation for the TEM is much more extensive than for the SEM. A sample initially undergoes fixation in glutaraldehyde or formaldehyde to preserve structure. Fixation also protects the sample from damage that may occur during the rest of the preparation. Following fixation, the sample is dehydrated, most commonly by replacing water with ethanol. The ethanol acts as a solvent between the aqueous environment of the cell and the hydrophobic embedding medium. Embedding involves resin infiltration, where the ethanol is replaced by a highly miscible plastic embedding agent, and is later cured at a high temperature ($\approx 70°C$). Curing causes the embedding medium to polymerize and become solid. A microtome equipped with either a glass or diamond knife

is then used to make thin sections approximately 90 nm thick for viewing under the TEM.

A technique called cryo-TEM can be used to directly image small fluid structures in water (Figure 9.19). Cryo-TEM requires extremely rapid cooling (vitrification) of the sample to $-170-185°C$. This technique has been valuable for examining biological molecules and their aggregation behavior in water solutions (Won, 2004). A related technique called electron cryotomography (ECT) provides three-dimensional imaging of intact cells at a resolution of ≈ 4 nm, allowing for subcellular examination of cells in their native state (Tocheva *et al.*, 2010). Increasing in use, ECT has revealed detailed information about septum formation in dividing bacterial cells, bacterial cytoskeletons and subcellular structures associated with motility and chemotaxis. The continued use of ECT will revolutionize our understanding of microbial cell structure.

All lenses are subject to aberrations, and the electromagnetic lenses used in electron microscopy are no exception. Unlike those in light microscopy, however, aberrations in the electron microscope are difficult to resolve because of the inability to add corrective lenses to the optics. Consequently, whereas in the light microscope the theoretical and achievable resolving powers are

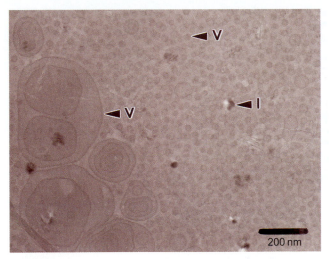

FIGURE 9.19   A cryo-TEM micrograph of the morphology of a rham-nolipid biosurfactant produced by *Pseudomonas aeruginosa* at pH 6.5. This micrograph shows bilayer surfactant vesicles (V) which range in size from 30 nm to several hundred nm. The size bar = 200 nm. Image courtesy J.T. Champion.

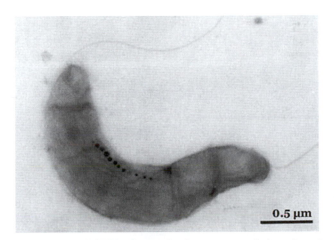

FIGURE 9.20   A TEM micrograph of a magnetotactic bacterium. A single bacterium in cross-section shows a chain of seven electron-dense magnetosomes. Each magnetosome contains a single crystal of a tiny magnet (iron oxide magnetite). Because the motility of this type of bacterium is directed by these magnets, they are referred to as being "magnetotactic." This sample was taken at a depth of 4 meters in the Pettaquamscutt River, Rhode Island. Magnification 116,000×. Taken by Paul Johnson.

similar, in the electron microscope the theoretical resolution is not reached.

9.5.3 Elemental Analysis

One of the advantages of electron microscopy is the ability to perform microanalysis X-ray spectrometry with an energy-dispersive spectrometer (EDS). In EDS, when an electron beam of sufficient energy encounters a surface, X-ray photons of characteristic energies may be emitted via inner-shell ionization (Figure 9.21A and B). The result is a fingerprint of X-ray energies specific for a particular element. By comparison with fingerprints of known elements, an unknown element in a sample can be identified and quantitated. Energy-dispersive spectrometry can be performed with both transmission and scanning electron microscopes. No additional sample preparation is needed beyond that necessary for the electron microscope. EDS applications in environmental microbiology are broad and are discussed in more detail in Chapter 21.

9.6 SCANNING PROBE MICROSCOPY

Scanning probe microscopy deals with imaging surfaces on a very fine scale, even to the level of molecules and groups of atoms. This technique uses an extremely sharp tip (3–50 nm radius of curvature) to scan across the surface of the sample. When the tip moves close to the sample surface, the forces of interaction between the tip and the surface of the sample can be measured. The most common types of scanning probe microscopy are atomic force microscopy, scanning tunneling microscopy and near-field scanning optical microscopy. Such microscopes have the ability to view single atoms with a magnification of 1,000,000×. We discuss the first of these below.

9.6.1 Atomic Force Microscopy

The atomic force microscope (AFM) measures surface contours with a probe or "tip" placed very close to the sample. The image is acquired when the probe is raster-scanned over the sample. Depending on the set-up, this measures either the contour height or the electric potential at any given site. The AFM does not use lenses, so the size of the probe tip rather than diffraction is the limiting factor in image resolution. The fine resolution offered by AFM is allowing scientists to begin to decipher interactions of biological molecules with surfaces (Figure 9.22).

9.7 IMAGING

Micrography, taking an image using a microscope, provides a means of permanently recording an image for both artistic and scientific purposes. Historically, micrographs consisted of images on photographic film, but the use of digital imaging via a digital camera attached to a microscope is now standard. With the development of more sophisticated computers and imaging systems, images can be digitized; this process has higher resolution than traditional photographic methods and creates a more accurate reproduction of a microscopic image. The use of

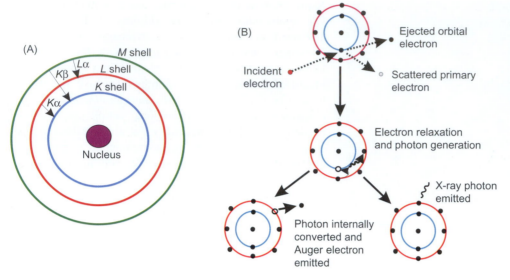

FIGURE 9.21 (A) Electronic transitions in an atom and (B) electron interactions as an electron from the electron beam encounters an atom. X-ray generation is used in electron dispersive spectrometry. Adapted from Goldstein and Yakowitz (1975).

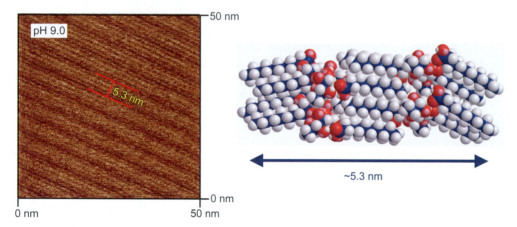

FIGURE 9.22 (Left) An AFM image of monorhamnolipid adsorbed on graphite at pH 4.3. (Right) The information from this image can be used to make a precise molecular model of how rhamnolipid interacts with the surface. Image and model courtesy V. Ochoa and J.E. Pemberton.

computers in the digital processing of microscopic images has allowed automated image processing. Manipulation of images into three dimensions or overlaying images of multiple fluorescent probes is now possible with digital micrographs. The development of increasingly sophisticated imaging is a necessary component of advancing microscopic techniques used in microbiology.

QUESTIONS AND PROBLEMS

1. List four possible applications in environmental microbiology for each of the following microscopic techniques: bright-field, fluorescence, electron and *in situ* microscopy.
2. Direct microscopic counts using acridine orange (AODC) are often two to three orders of magnitude greater than viable counts. Why?
3. If you wanted to determine whether or not a specific membrane protein was being produced by a microorganism, which microscopic techniques might you use and why?
4. List at least four things about a microorganism you can learn from viewing it with bright-field microscopy.
5. Which is most important in microscopy—resolution, contrast or magnification? Why?
6. List two prokaryotic structures that would stain in response to (i) an acidic dye. (ii) A basic dye.

REFERENCES AND RECOMMENDED READING

Recommended Microscopy Texts

Goldstein, J. I., and Yakowitz, H. (1975) "Practical Scanning Electron Microscopy," Plenum, New York.

Hayat, M. A. (2000) "Principles and Techniques of Electron Microscopy Biological Applications," Cambridge University Press, Cambridge, UK.

Hoppert, M. (2003) "Microscopic Techniques in Biotechnology," Wiley-VCH Verlag, Weinheim.

Mason, C. W. (1983) "Handbook of Chemical Microscopy", vol. 1, 4th ed. John Wiley & Sons, New York.

Murphy, D. B. (2001) "Fundamentals of Light Microscopy and Electronic Imaging," Wiley-Liss, New York, NY.

Rochow, T. G., and Tucker, P. A. (1994) "Introduction to Microscopy by Means of Light, Electrons, X-Rays or Acoustics," Plenum, New York.

Chapter References

Amann, R., and Ludwig, W. (2000) Ribosomal RNA-targeted nucleic acid probes for studies in microbial ecology. *FEMS Microbiol. Rev.* **24**, 555–565.

Berney, M., Hammes, F., Bosshard, F., Weilenmann, H. U., and Egli, T. (2007) Assessment and interpretation of bacterial viability by using the LIVE/DEAD BacLight kit in combination with flow cytometry. *Appl. Environ. Microbiol.* **73**, 3283–3290.

Biteen, J. S., and Moerner, W. E. (2010) Single-molecule and superresolution imaging in live bacteria cells. *Cold Spring Harb. Perspect. Biol.* **2**, a000448.

DeLong, E. F., Wickham, G. S., and Pace, N. R. (1989) Phylogenetic stains: ribosomal RNA-based probes for the identification of single cells. *Science* **243**, 1360–1363.

Donnert, G., Keller, J., Wurm, C. A., Rizzoli, S. O., Westphal, V., Schonle, A., *et al.* (2007) Two-color far-field fluorescence nanoscopy. *Biophys. J.* **92**, L67–L69.

El-Azizi, M., Rao, S., Kanchanapoom, T., and Khardori, N. (2005) In vitro activity of vancomycin, quinupristin/dalfopristin, and linezolid against intact and disrupted biofilms of staphylococci. *Ann. Clin. Microbiol. Antimicrob.* **4**, 2.

Goldstein, J. I., and Yakowitz, H. (1975) "Practical Scanning Electron Microscopy," Plenum, New York.

Hobbie, J. E., Daley, R. J., and Jasper, S. (1977) Use of nucleopore filters for counting bacteria by fluorescence microscopy. *Appl. Environ. Microbiol.* **33**, 1225–1228.

Hoppert, M. (2003) "Microscopic Techniques in Biotechnology," Wiley-VCH Verlag, Weinheim.

Ipsilantis, I., and Sylvia, D. M. (2007) Abundance of fungi and bacteria in a nutrient-impacted Florida wetland. *Appl. Soil Ecol.* **35**, 272–280.

Iverson, S. L., and Maier, R. M. (2009) Effects of compost on colonization of roots of plants grown in metalliferous mine tailings, as examined by fluorescence in situ hybridization. *Appl. Environ. Microbiol.* **75**, 842–847.

Korber, D. R., Lawrence, J. R., Zhang, L., and Caldwell, D. E. (1990) Effect of gravity on bacterial deposition and orientation in laminar flow environments. *Biofouling* **2**, 335–350.

Maixner, F., Noguera, D. R., Anneser, K. S., Wegl, G., Wagner, M., and Daims, H. (2006) Nitrite concentration influences the population structure of *Nitrospira*-like bacteria. *Environ. Microbiol.* **8**, 1487–1495.

Mason, C. W. (1983) "Handbook of Chemical Microscopy", vol. 1, 4th ed. John Wiley & Sons, New York.

Murray, R. G. E., and Robinow, C. F. (1994) Light microscopy. In "Methods for General and Molecular Bacteriology" (P. Gerhardt, R. G. E. Murray, W. A. Wood, and N. R. Krieg, eds.), American Society for Microbiology, Washington, DC, pp. 8–20.

Paul, E. A., and Clark, F. E. (1989) "Soil Microbiology and Biochemistry," Academic Press, San Diego.

Pepper, I. L., and Gerba, C. P. (2004) "Environmental Microbiology Laboratory," Academic Press, San Diego.

Rochow, T. G., and Tucker, A. (1994) "Introduction to Microscopy by Means of Light, Electrons, X-Rays or Acoustics," Plenum, New York.

Rossi, G., Riccardo, S., Gesue, G., Stanganelli, M., and Wang, T. K. (1936) Direct microscopic and bacteriological investigations of the soil. *Soil Sci.* **41**, 53–66.

Tocheva, E. I., Li, Z., and Jensen, G. J. (2010) Electron cryotomography. *Cold Spring Harb. Perspect. Biol.* **2**, a003442.

Westphal, V., Rizzoli, S. O., Lauterbach, M. A., Kamin, D., Jahn, R., and Hell, S. W. (2008) Video-rate far-field optical nanoscopy dissects synaptic vesicle movement. *Science* **320**, 246–249.

Won, Y.-Y. (2004) Imaging nanostructured fluids using cryo-TEM. *Korean J. Chem. Eng.* **21**, 296–302.

Zieler, H. W. (1972) "The Optical Performance of the Light Microscope", Part 1. Microscope Publications, Chicago.

Zwirglmaier, K. (2005) Fluorescence in situ hybridisation (FISH)—the next generation. *FEMS Microbiol. Lett.* **246**, 151–158.

Chapter 10

Cultural Methods

Ian L. Pepper and Charles P. Gerba

10.1 INTRODUCTION

Isolation and enumeration of microorganisms by cultural methods is widely used to evaluate the diversity of microbial communities or quantitate a specific organism of interest. However, unlike the assay of pure culture samples, culture techniques involving the diverse microbial communities found in the environment are complex, and the numbers obtained depend on the specific culture techniques.

To determine the appropriate culture method, it is necessary to define which specific microorganism or group of microorganisms is to be enumerated. The methods for enumerating bacteria, fungi, algae and viruses are very different. In addition, within each group, special enumeration techniques may be needed, as in the case of anaerobic or autotrophic bacteria. Precise, standard methods are necessary for assaying indicator organisms such as fecal coliforms, or pathogens such as *Salmonella* spp.

The type of environmental sample under analysis must also be considered when enumerating microorganisms. Different techniques may be necessary for the extraction of the organisms from the sample, and, depending on the number of organisms of interest present in the particular sample, dilution or concentration of the sample may be necessary (see Chapter 8). Selective or differential growth media (see Section 10.4.1.2) may also be necessary to observe a small population of a specific organism when an abundance of other organisms is also present in the sample. For example, the enumeration of *Salmonella* in sewage samples is complicated by the presence of the large numbers of enteric bacteria.

10.2 EXTRACTION AND ISOLATION TECHNIQUES

10.2.1 Extraction of Cells from Soil

To obtain accurate bacterial numbers from soil it is necessary to have efficient recovery of the microorganisms that are attached to soil particles or are present in the pores of soil aggregates. Separation of bacteria from soil particles by physical and chemical dispersion methods is discussed by Lindahl and Bakken (1995). These include hand or mechanical shaking with or without glass beads, mechanical blending and sonication. Different extracting solutions may be used depending on the pH and texture of the soil. A surfactant such as Tween 80 (Difco) may be used, often with the use of a dispersing agent such as sodium pyrophosphate (see also Chapter 8).

Since most soils contain millions of bacteria, the extraction step is followed by serial dilution of the sample

I.L. Pepper, C.P. Gerba, T.J. Gentry: Environmental Microbiology, Third edition. DOI: http://dx.doi.org/10.1016/B978-0-12-394626-3.00010-7
© 2015 Elsevier Inc. All rights reserved.

to separate the microorganisms into individual reproductive units. Frequently, the first step in the dilution process is the addition of 10 g of soil to 95 ml of the extracting solution, which results in a 10^{-1} weight by volume dilution. Sterile water, physiological saline and buffered peptone or phosphate solutions are a few of the solutions commonly used for this step. Although water is convenient and commonly used, it is not preferred because it does not prevent osmotic shock during the dilution process. Additional 10-fold dilutions are normally conducted to allow for individual colonies to form following plating.

10.2.2 Extraction of Cells from Water Samples

Some samples, such as marine water or drinking water, contain low bacterial numbers and require concentration rather than dilution before enumeration (see Section 8.2.1). In this method, a specified volume of water is filtered through a membrane using a vacuum. The bacteria are trapped on the membrane, which is placed on the agar medium or a cellulose pad soaked in medium to allow growth of individual colonies. This is the basis of the membrane filtration technique (see Section 10.4.1.3). Different volumes of water may need to be filtered to obtain the correct concentration of bacteria on the membrane for isolation and counting purposes. In this situation, it is critical to select a type and size of membrane appropriate to the bacteria to be collected. Often a nitrocellulose filter of pore size 0.45 μm is used. Care must be taken during processing of the sample to cause minimal stress to the organism with respect to such factors as processing time, vacuum pressure and desiccation.

10.3 PLATING METHODS

After dilution or concentration, the sample is added to Petri dishes containing a growth medium consisting of agar mixed with selected nutrients. Two different methods are used for application of the diluted sample to the growth medium. In the spread plate method, a 0.1-ml aliquot of selected dilutions of the sample is uniformly spread on top of the solid agar with the aid of a sterile glass rod (Figure 10.1). Alternatively, in the pour plate method, 1-ml aliquots of appropriate sample dilutions are mixed with molten agar (45°C) in a Petri dish and allowed to solidify (Figure 10.2). The spread plate technique is advantageous since it allows colonies to develop on the surface of the agar, making it easier to distinguish different microorganisms on the basis of morphology. It also facilitates further isolation of the colonies. The spread plate method generally gives bacterial counts that

are higher than with the pour plate method (for the same size of inoculant), perhaps because of improved aeration and desegregation of clumps of bacteria that occur with this method.

After plating, the samples are incubated under specified conditions, allowing the bacteria to multiply into macroscopic, isolated colonies known as colony-forming units (CFU). Because it is assumed that each colony-forming unit originates from a single bacterial cell, it is critical for the organisms to be separated into discrete reproductive units in the dilution step prior to plating. In reality, however, colonies may arise from chains or clusters of bacteria, resulting in an underestimation of the true bacterial number. The total number of bacteria of interest is calculated from the number of colonies found on a specific dilution. The range of colonies acceptable for counting is 30−300 on a standard 150-mm-diameter agar plate. Below 30, accuracy is reduced; above 300, accuracy increases only slightly and, in fact, numbers may be reduced by overcrowding and competition between organisms growing on the plate. An example of a typical dilution and plating calculation is shown in Information Box 10.1.

The process of isolating microorganisms from environmental samples often necessitates further isolation of individual colonies arising from the spread or pour plate method, to allow for additional characterization or confirmation. In this technique, a sterile inoculating loop is used to pick colonies from the original agar plate and the loop of bacteria is streaked to dilution on a new agar plate. This process is shown in Figure 10.3.

10.3.1 Unculturable Bacteria

The standard plate count technique for the enumeration of microorganisms is one of the oldest and most widely used techniques in microbiology. Despite its popularity, the dilution and plating technique has been subject to much scrutiny and criticism almost since its inception (Information Box 10.2). One of the main criticisms is that only a small fraction of the total population, which can be observed microscopically, can be cultured on laboratory media. It is well documented that only 1−10% of the number of cells observed with direct microscopic counts can be recovered as viable bacteria using cultural plating techniques (see also Chapter 2).

There are many potential reasons for the "unculturability" of bacteria. For example, strict anaerobic organisms will not be able to be cultured in the presence of oxygen. Another simple reason for the apparent inability of an organism to be cultured is a slow growth rate such that visible colonies do not appear within the normal incubation period of a few days. Other more subtle factors are responsible as shown in Information Box 10.3.

Step 1. Make a 10-fold dilution series

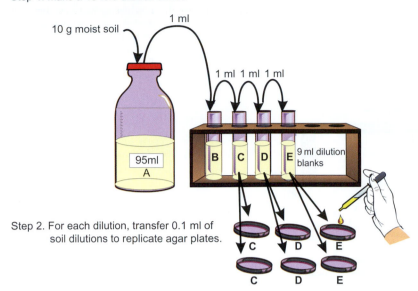

Step 2. For each dilution, transfer 0.1 ml of
        soil dilutions to replicate agar plates.

FIGURE 10.1 Dilution and spread plating technique. Here, soil that initially contains billions of microbes is diluted prior to being spread plated to enable discrete colonies to be seen on each plate. Numbers of colonies on each plate can be related to the original soil microbial population.

Step 3a. A glass spreading rod is flame sterilized.

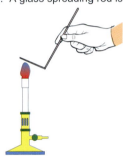

Step 3b. Sample is spread on the surface of
        the agar. This is done by moving the
        spreader in an arc on the surface of the
        agar while rotating the plate.

Step 4. Incubate plates under specified conditions.

Step 5. Count dilutions yielding 30–300 colonies
        per plate. Express counts as CFUs per
        g dry soil.

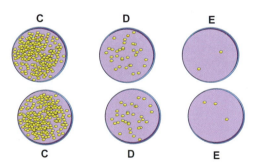

Overall, when isolated on a culture plate, microbial interactions and signals may be absent resulting in a lack of growth in monoculture. Additionally, when faced with this new environment devoid of necessary growth factors, some bacteria may enter into a dormant state of low activity which prevents growth and colony formation. Finally, note that numbers of viable bacteria may appear to be lower due to low efficiencies of extraction from soil, or poor separation of the bacteria during the dilution step. This could result in clumps of bacteria giving rise to a single colony. As a rule, the 1- to 2-log difference between direct and viable counts is referred to as viable but unculturable cells as it is assumed that dead cells would be rapidly degraded (Roszak and Colwell, 1987).

10.3.1.1 New Approaches for Enhanced Cultivation of Soil Bacteria

Scientists have known for decades that the total number of bacteria found in soil samples is far greater than the number that are found using culturable dilution and plating techniques (Joseph et al., 2003). In fact, typically less

Step 1. Make a 10-fold dilution series

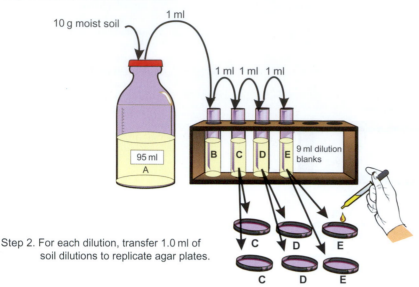

Step 2. For each dilution, transfer 1.0 ml of
        soil dilutions to replicate agar plates.

FIGURE 10.2 Dilution and
pour plating technique. Here, the
diluted soil suspension is incorporated directly in the agar
medium rather than being surface applied as in the case of
spread plating.

Step 3a. Add molten agar cooled to 45°C
          to the dish containing the soil
          suspension.

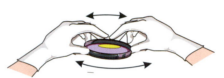

Step 3b. After pouring each plate, replace
          the lid on the dish and gently swirl
          the agar to mix in the inoculum and
          completely cover the bottom of the
          plate.

Step 4. Incubate plates under specified conditions.

Step 5. Count dilutions yielding 30–300 colonies
        per plate. Express counts as CFUs per
        g dry soil.

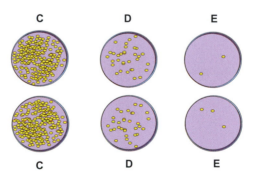

than 1% of the total bacterial population is culturable. The fact that such a small percentage is culturable means that the vast majority of soil organisms have not been obtained in pure culture to allow their characterization. In addition, cultural methods may actually select for less abundant rare microbes (see Section 10.3.1.2). Because of this, there have been several new approaches utilized to enhance the culturability of previously uncultured soil bacteria (Information Box 10.4).

These new approaches can be as simple as using longer incubation periods. Several investigators have seen enhanced culturability utilizing enhanced periods of incubation for colonies to evolve (30 days to 3 months) (Joseph *et al.*, 2003; Stevenson *et al.*, 2004). Other investigators have utilized dilute media with low nutrient content (Janssen *et al.*, 2002), or used hybrid methodologies combining cultural and molecular technologies (Stevenson *et al.*, 2004). Low nutrient media for dilution and plating can involve diluting nutrient broth to 1/100 of its normal concentration, or the use of more traditional soil plating media including soil extract agar derived from the soil under investigation.

Information Box 10.1   Dilution and Plating Calculations

A 10-g sample of soil with a moisture content of 20% on a dry weight basis is analyzed for viable culturable bacteria via dilution and plating techniques. The dilutions were made as follows:

Step		Dilution
10 g soil	→95 ml saline (solution A)	10^{-1} (weight/volume)
1 ml solution A	→9 ml saline (solution B)	10^{-2} (volume/volume)
1 ml solution B	→9 ml saline (solution C)	10^{-3} (volume/volume)
1 ml solution C	→9 ml saline (solution D)	10^{-4} (volume/volume)
1 ml solution D	→9 ml saline (solution E)	10^{-5} (volume/volume)

In plating, 1 ml of solution E is pour plated into an appropriate medium and results in 200 bacterial colonies.

$$\text{Number of CFUs} = \frac{1}{\text{dilution factor}} \times \text{number of colonies}$$

$$= \frac{1}{10^{-5}} \times 200 \text{ CFUs/10g moist soil}$$

$$= 2.00 \times 10^6 \text{ CFUs/g moist soil}$$

But, for 1 g of moist soil,

$$\text{Moisture content} = \frac{\text{moist weight} - \text{dry weight } (D)}{\text{dry weight } (D)}$$

Therefore,

$$0.20 = \frac{1 - D}{D}$$

$$D = 0.833 \text{ g}$$

$$\text{Number of CFUs per g dry soil} = 2.00 \times 10^6 \times \frac{1}{0.833}$$

$$= 2.4 \times 10^6$$

Step 1. Sterilize inoculating loop

Step 2. Obtain culture from an agar plate (A) or from broth (B).

Step 3. Make successive streaks on an agar plate to isolate single colonies

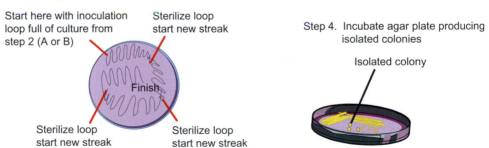

Start here with inoculation loop full of culture from step 2 (A or B)

Sterilize loop start new streak

Finish

Sterilize loop start new streak

Sterilize loop start new streak

Step 4. Incubate agar plate producing isolated colonies

Isolated colony

FIGURE 10.3   Isolation of a bacterial colony using the streak plate technique.

R2A (Difco, Detroit, MI) has also long been utilized for soil dilution and plating. Other new approaches have utilized additions of humic acid or humic-like compounds (anthraquinone disulfonate) to media, or the addition of quorum sensing signaling compounds including acyl homoserine lactones (Stevenson *et al.*, 2004). Recently, a new soil diffusion system was utilized to enrich the growth of novel bacterial taxa (Information Box 10.5). Researchers have also increased the numbers of culturable bacteria from soil by protecting cells from exogenous peroxides. This can be done by adding pyruvate or catalase to growth media, two compounds known to eliminate H_2O_2 (Misunoe *et al.*, 1999). Other researchers have hypothesized that incubation in CO_2-enriched atmospheres with reduced oxygen content may enhance culturability since these are the atmospheric conditions that soil-borne organisms encounter (Stevenson *et al.*, 2004).

Information Box 10.2 Errors and Assumptions Associated with Dilution and Plating Assays

1. Clumping of cells may result in an underestimation of the true count. The assumption of one organism per colony is rarely satisfied as several cells associated with a soil particle may give rise to one colony (sample sonication may alleviate this problem). Microscopic direct counts do not make the assumption.

2. Errors in diluting the soil can arise from either particles not dispersing entirely (less dilution occurs) or from particles settling out of solution prior to the next dilution (more dilution occurs).

3. Only a small fraction of organisms will grow on a given medium. Microscopic direct counts do not make this assumption. Therefore direct counts are often referred to as total counts, which will typically be one to two orders of magnitude greater than the culturable count (see Section 9.4.1).

4. As soil is a heterogeneous medium, biological variability may be high even between adjacent areas of soil.

5. Heavily sporulating organisms are often overemphasized.

6. Slow growing bacteria may not give rise to visible colonies within reasonable time frames (2 weeks).

7. Organisms may become nonviable due to stress imposed during soil extraction.

8. Only aerobic organisms are cultured (unless anaerobic conditions are imposed).

Information Box 10.3 Reasons for the "Unculturability" of Bacteria

- Slow growing bacteria that do not form visible colonies within several days of incubation
- Fastidious growth requirements including specific nutrients, pH regime, incubation temperature or redox conditions (Kopke et al., 2005)
- Competition for nutrients
- Growth inhibition due to bacteriocins or antibacterial substances released by other soil bacteria (Tamaki et al., 2005)
- Need for cross-feeding or metabolic cooperation between species for the provisions of nutrients (Belenguer et al., 2006)
- Requirement for community communication through a network of signals found only in natural environments (Nichols et al., 2008)

Information Box 10.4 New Techniques for Culturing the Unculturable

- Use of low nutrient media (Connon and Giovannoni, 2002)
- Longer incubation periods (Davis et al., 2005)
- Physically reducing the number and diversity of bacteria within mixed samples before cultivation (Song et al., 2009)
- Co-cultivation with helper strains to enhance beneficial interactions (Nichols et al., 2008)
- Addition of spent culture media to fresh media to provide growth stimulants (Kim et al., 2008)
- Addition of signaling molecules (Bruns et al., 2002)
- Encapsulation of individual cells in gel microdroplets (Ben Dov et al., 2009)
- Mimicking natural conditions using microcolony cultivation on polycarbonate membranes with soil extract (Ferrari and Gillings, 2009)
- Use of flow cytometry and cell sorting to isolate individual bacterial cells and cell encapsulation within gel microdroplets (Zengler et al., 2002)
- Use of optical tweezers (infrared laser or Raman tweezers) for micromanipulation of single cells and transfer to growth media (Huang et al., 2009)

Information Box 10.5 Use of a Soil Diffusion System to Enrich the Growth of Novel Bacterial Taxa

The Soil Diffusion System (SDS) has been developed to coax previously uncultivated bacteria into growth. The SDS allows microbes to be cultivated in an environment that more closely mimics the natural soil environment. Specifically, the technology allows bacteria to grow in close association with their native soil, but isolated from the soil by a polycarbonate nano-membrane (0.003 µm). On top of the nano-membrane, a regenerated cellulose filter is placed to support the growing microbial community that arises following inoculation of the cellulose filter with a dilute soil inoculum (10^{-3} dilution). The filter is thought to allow for diffusion of soil-derived organics, signaling molecules and nutrients. This technology has been used to culture previously unculturable bacteria. Thus, it appears that proximity to a living microbial community appears to enhance community growth on the cellulose filter.

Source: Kakumanu and Williams, 2012

More sophisticated approaches to soil dilution and plating have included the use of cell encapsulation wherein individual cells are enclosed within micro gel capsules (Zengler *et al.*, 2002). Researchers are now starting to use molecular characterization of microbial communities to select custom media for previously uncultured bacterial groups (Joseph *et al.*, 2003). The molecular analyses use 16S rRNA gene sequences to identify the major divisions of bacteria present in an environmental sample. This information can aid in the selection of a suite of media to culture the diverse populations present in a given sample. Other recent innovations involve the use of flow cytometry and cell sorting followed by cell encapsulation with gel microdroplets. Finally, optical tweezers for micromanipulation of single cells and subsequent transfer to growth media can be used.

10.3.1.2 Culturing Members of the Soil Rare Biosphere

Microorganisms that are found in low abundance have been referred to as "the rare biosphere" (Sogin et al., 2006). However, little is known of the ecological significance of rare microorganisms within complex microbial communities. In a recent study, agricultural soil samples were analyzed by both conventional cultural techniques and culture-independent pyrosequencing technology (Shade et al., 2012). Specifically, soil samples were diluted and plated on a rhizosphere isolation medium (culture dependent). In addition, the same soil samples were extracted for community DNA (culture independent). Both the community DNA and DNA from the cultured bacteria were subjected to pyrosequencing of the 16 SrRNA genes. Interestingly, the analyses showed that soil bacteria captured by culturing were in very low abundance or absent in the culture-independent community. Thus, molecular culture-independent assays tend to identify the most abundant microbes in an environmental matrix, whereas culture-dependent assays actually give greater access to "the rare biosphere." This raises the question: "Are cultured bacteria actually atypical members of the community?"

10.3.2 Most Probable Number Technique

The most probable number (MPN) technique is sometimes used in place of the standard plate count method to estimate microbial counts in the environment. In this method, the sample to be assayed is dispersed in an extracting solution and successively diluted, as in the plate count. This method relies on the dilution of the population to extinction, followed by inoculation of five to 10 replicate tubes containing a specific liquid medium with each dilution. After incubation, the tubes are scored as plus or minus for growth on the basis of such factors as turbidity, gas production and appearance or disappearance of a substrate (Figure 10.4). Scoring a tube positive for growth means that at least one culturable organism was present in the dilution used for its inoculation. The number of positive and negative tubes at each dilution is used to calculate the number present in the original sample through the use of published statistical MPN tables or computer programs designed to simplify the analysis. MPN tables can be found in American Public Health Association (2012).

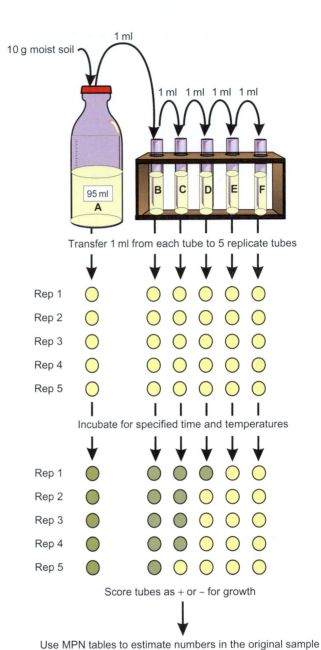

FIGURE 10.4 Most probable number technique. Here, an environmental sample is diluted to extinction. Diluted samples are used to inoculate replicate tubes at each dilution. The presence or absence of the microorganism of interest at a given dilution can be analyzed statistically to estimate the original population in the environment.

MPN methodology is useful because it allows estimation of a functional population of bacteria based on a process-related attribute such as nitrification, nitrogen fixation or sulfur oxidation. It is mandatory to use MPN analysis when enumerating a microorganism requiring broth enrichment prior to culturing. It is also essential when assaying many foods, sediments and soils. However, the MPN technique is very labor intensive, and results are usually less precise than those obtained with direct plating methods.

10.4 CULTURE MEDIA FOR BACTERIA

10.4.1 General Media Used for Culturing Bacteria

Liquid and solid media for the cultivation of microorganisms must contain substances that will support their growth, and the media available are as diverse as the microorganisms themselves. Many media are available from commercial manufacturers such as Difco (Detroit, MI) or BBL (Sparks, MD), and variations and new formulations are published by researchers available through journals and volumes such as the *Handbook of Microbiological Media* (Atlas, 2010). The major nutrient components of microbiological media include: (1) a source of carbon for incorporation in biomass, such as glucose for heterotrophic bacteria or CO_2 for autotrophic bacteria; (2) nitrogen, which is needed for growth and commonly supplied as ammonia, proteins, amino acids, peptones or extracts from plants or meat; (3) buffers to maintain a suitable pH; and (4) growth factors such as defined trace minerals or metals, or undefined factors such as those found in extracts made from the environmental samples themselves. Many media also contain selective components that favor the growth of specific organisms while inhibiting the growth of nontarget organisms (see Section 10.4.1.2).

For solid media used in plating, agar is the most commonly used solidifying agent. It is a polysaccharide of an extract from marine algae. Agar exists as a liquid at high temperatures but solidifies on cooling to 38°C. Although agar ideally does not supply nutritional value to the medium, variations in growth may be observed with different types of agar.

While cultural media are generally used to enumerate or isolate specific microorganisms, they also can be used in metabolic fingerprinting analysis to allow identification of microbial isolates. Such systems include API strips (BioMérieux, Durham, NC) and Enterotube (BBL, Sparks, MD), which are routinely used for clinical identification of Gram-negative microorganisms. For environmental microbiology, the most commonly used of the metabolic fingerprinting systems is Biolog. Biolog is used for identification of single isolates as well as for analysis of community composition. The Biolog Company (Hayward, CA) makes an EcoPlate which contains 31 carbon sources chosen for soil community analysis. The basis for the Biolog system is a 96-well microtiter plate, whereas for the EcoPlate, the 31 carbon sources are each in triplicate wells. Each well is inoculated with the same isolate or community sample. If the substrate is utilized, the well turns purple. The plates can be read either manually or automatically using a plate reader. The EcoPlate can be used to provide a comparison of the metabolic fingerprint of the community before and after a perturbation, or simply to monitor the community for a period of time.

10.4.1.1 Heterotrophic Plate Counts

Heterotrophic bacteria are "consumers" that obtain energy and carbon from organic substances. Heterotrophic plate counts give an indication of the general "health" of the soil as well as an indication of the availability of organic nutrients within the soil. Two basic types of media can be used for this analysis: nutrient-rich and nutrient-poor media. Examples of nutrient-rich media are nutrient agar (Difco), peptone-yeast agar (Atlas, 2010), and soil extract agar amended with glucose and peptone (Atlas, 2010). These media contain high concentrations of peptone, yeast and/or extracts from beef or soil. Nutrient-poor media are often called minimal media, and contain as much as 75% less of these ingredients, often with substitutions such as casein, glucose, glycerol or gelatin. Examples of minimal media are R2A agar (BBL), m-HPC agar (Difco) and soil extract agar with no amendments. In many cases, higher colony counts are obtained with a minimal medium because a large number of oligotrophic organisms in the environment cannot be cultured on a rich medium. This is true for most water and subsurface porous medium samples, for which the nutrient-poor medium, R2A, is often used (Figure 10.5). Growth on a nutrient-poor medium may take longer (5–7 days), and the colony sizes are often smaller than for a rich medium. In addition, the

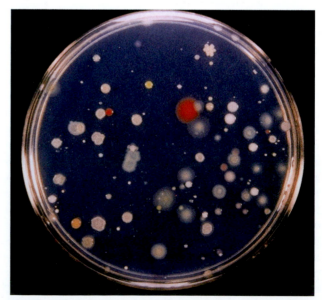

FIGURE 10.5 Heterotrophic colonies on an R2A agar plate. A number of discrete colonies with diverse morphology arise after dilution and plating from soil.

TABLE 10.1 Common Antibiotics Often Used in Selective Media

Name	Spectrum	Mode of Action
Chloramphenicol	Broad spectrum	Inhibits protein synthesis by binding to 50S ribosomal subunit
Erythromycin	Mostly Gram-positive	Inhibits protein synthesis by binding to 50S ribosomal subunit
Tetracycline	Broad spectrum	Inhibits protein synthesis by binding to 30S ribosomal subunit
Streptomycin	Broad spectrum	Inhibits protein synthesis by binding to 30S ribosomal subunit
Polymyxin	Gram-negative bacteria, especially *Pseudomonas*	Disrupts cell membrane
Nalidixic acid	Gram-negative bacteria	Inhibits DNA synthesis
Novobiocin	Gram-negative bacteria	Inhibits DNA synthesis
Trimethoprim	Broad spectrum	Inhibits purine synthesis
Rifampicin	Gram-positive bacteria	Inhibits RNA synthesis
Penicillin	Mostly Gram-positive bacteria	Inhibits cell wall peptidoglycan synthesis

community of isolates from a single sample may be entirely different when plated on the two media types.

10.4.1.2 Culturing Specific Microbial Populations

It is often necessary to detect or enumerate a specific population of microorganisms, or even a very specific bacterial isolate, from the total population of bacteria found in an environmental sample. This may necessitate culturing with one or more specialized media, and often requires that a specified sequence of steps be performed to ensure maximum culturability of the target organism. We will now define the media involved in these steps and explain why they are important.

Pre-enrichment or resuscitation medium is a liquid medium that allows the microorganisms in a sample to begin to actively metabolize and increase in number. There are two reasons for a pre-enrichment medium. First, damaged cells are given time and the necessary nutrients to repair and grow. This step is important because direct inoculation of a sample into a selective enrichment medium (discussed next) may result in the death of some bacteria. This is especially critical for sub-lethally injured organisms, which may not be recoverable under selective conditions. The second purpose of pre-enrichment is to increase the number of cells. This step is crucial when trying to enumerate low numbers of a target organism in environmental samples.

Enrichment medium is a liquid medium that promotes the growth of a particular physiological type of microorganism present in a mixture of organisms, while suppressing the growth of competitive background flora. It aids in the detection of the desired environmental isolate when the specimen contains a high population of normal flora. Enrichment media may be elective or selective. Elective enrichment medium allows growth of a single or limited type of bacteria based on a unique combination of nutritional or physiological attributes. Selective enrichment medium involves the use of inhibitory substances or conditions to suppress or inhibit the growth of most organisms while allowing the growth of the desired organism.

A selective plating medium is a modification of an agar medium to suppress or prevent the growth of one group of organisms while fostering the growth of the desired organism. Antibiotics are among the most widely used and effective selection agents. Cycloheximide is often used in heterotrophic media to inhibit the growth of fungi from soil. Table 10.1 lists some of the more common antibiotics, with their spectrum and mode of action. Other examples of selective agents often added to growth media are metals such as mercury or lead, which are usually used in the presence of a minimal salts medium. Selection may also be accomplished by the use of toxic chemicals such as high salt concentrations or dyes. For example, the dye crystal violet inhibits most Gram-positive bacteria while allowing the Gram-negative bacteria to grow. Adjustments in pH or osmotic conditions are also used for selection. Finally, selectivity may be based on incubation conditions of the inoculated growth medium. These may include temperature or oxygen levels.

A specialized isolation medium contains formulations that meet the nutritional needs of specific groups of organisms, such as *Staphylococcus* or *Corynebacterium*, thereby allowing differentiation and identification.

A differential medium contains ingredients to allow distinction of different microbes growing on the same medium. It includes an indicator, usually for pH, to distinguish certain groups of organisms on the basis of variations in nutritional requirements, and the production of acid or alkali from various carbon sources. This results in a distinguishable morphological characteristic of the colony, usually color. In addition, it may support the growth of a selected group of bacteria while inhibiting the growth of others. Thus, a medium can be selective as well as differential.

Although culturing procedures for pure laboratory strains or clinical samples are fairly straightforward, the same techniques may not be successful for culturing the same organisms found in soil or water. Often it is necessary to follow a series of steps or modifications in procedures along with a combination of media to obtain optimal culturing of the target environmental organism. Furthermore, once colonies are isolated by cultural methods, they may be characterized by physiological (see Chapter 11), immunological (see Chapter 12) or molecular (see Chapter 13) techniques. In the next sections we present five examples of the application of these methods to culture specific microbes from soil and water samples. These examples were chosen to illustrate a wide range of bacteria and problems encountered in the environment. These examples are far from exhaustive, but the selection represents a variety of cultural methodologies, and their application to environmental samples.

10.4.1.3 Fecal Coliforms and the Membrane Filtration Technique

Coliform bacteria are nonpathogenic bacteria that occur in the feces of warm-blooded animals. Their presence in a water sample indicates that harmful pathogenic bacteria may also be present. Coliforms are found in numbers corresponding to the degree of fecal pollution, are relatively easy to detect and are overall hardier than pathogenic bacteria. For these reasons, coliforms are important "indicator organisms" in an environmental sample to assess water quality prior to or in place of culturing other organisms (see also Chapter 23).

The membrane filtration technique is used for the detection of fecal coliforms in water samples (see Figure 23.3). The volume of water required to enumerate fecal coliforms varies, but a volume of 100 ml is normally processed for drinking water, lakes and reservoirs. After filtration, the membrane containing the trapped bacteria is subjected to cultural methods. For fecal coliform analysis, the membrane is applied to a Petri dish containing m-FC agar (Difco) and incubated at 45°C for 24 hours. Fecal coliforms appear as blue colonies and nonfecal coliform colonies are gray to cream colored. Colonies are viewed and counted on a microscope under low magnification. The fecal coliform—membrane filtration procedure uses an enriched lactose medium with 1% rosalic acid to inhibit the growth of noncoliform colonies, and an elevated incubation temperature is a critical component in the selection for fecal coliforms.

This is a published standard method (*Standard Methods for the Examination of Water and Wastewater*, 2012), but there are also variations to help recover injured bacteria. These include an enrichment—temperature acclimation in which the membrane is incubated on nonselective agar at 35°C for 2 hours prior to incubation on the m-FC medium, and deletion of the rosalic acid suppressive agent from the m-FC medium. Individual colonies must also be confirmed to be fecal coliforms.

10.4.1.4 Salmonella and the MPN Technique

Salmonella is an enteric bacterium that is pathogenic to humans and causes a wide range of symptoms, primarily gastroenteritis. It can be transmitted by drinking improperly disinfected water or contaminated recreational water, but infection in the United States is primarily due to foodborne transmission because *Salmonella* infects both beef and poultry. Municipal sewage sludge (biosolids) also contains many microorganisms including *Salmonella*. Some cities now apply biosolids to agricultural soil to improve nutrient quality. Thus, it has become necessary to monitor the biosolids for indicator organisms and pathogens such as *Salmonella*, to ensure the safety of the product. Culture methods for *Salmonella* are varied, complex and time consuming and require final confirmatory tests (Figure 10.6).

The traditional method for detecting *Salmonella* spp. relies on enrichment and a plating technique with subsequent estimation using MPN tables. Selected volumes of biosolids, generally ranging from 10 to 0.01 ml, are added to a pre-enrichment broth and incubated for 18—24 hours at 37°C. Laboratory studies have found peptone and lactose broth to be most effective for the resuscitation of most *Salmonella* spp. Aliquots from the pre-enrichment phase are added to a selective enrichment broth. Numerous media are available for this step, all with different selection techniques. Included are the use of the inhibitory agent sodium selenite in broth and brilliant green and tetrathionate, which are contained in Mueller—Hinton tetrathionate broth (Difco, Detroit, MI). Rappaport—Vassiliadis (RV) broth (Difco) is also widely used; selection is based on resistance of *Salmonella* to malachite green, $MgCl_2$, and ability to grow at a pH of 5.0. Recently, RV broth was modified to include the antibiotic sodium novobiocin with incubation at an elevated temperature of 43°C. This medium, known as NR10 broth, has shown much promise for isolation of *Salmonella* from marine water and possible deletion of the pre-enrichment phase of analysis (Alonso *et al.*, 1992).

Growth from the enrichment phase is plated onto a selective or differential medium such as Hektoen enteric agar (Difco), brilliant green bile agar (Difco), or xylose—lysine—deoxycholate agar (Difco). Characteristics used for differential purposes are production of H_2S, and the inability of *Salmonella* to ferment lactose.

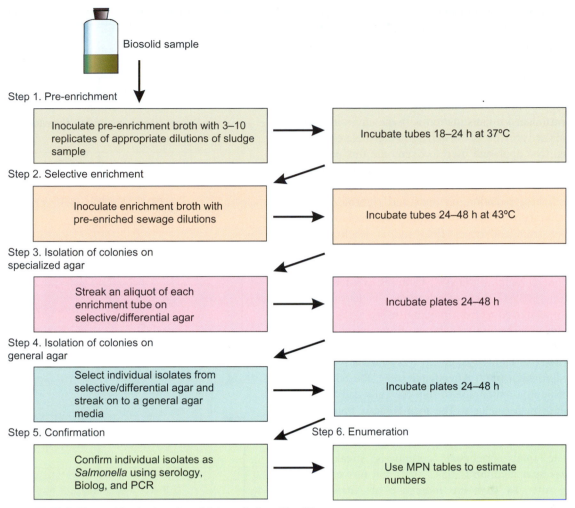

FIGURE 10.6   Protocol for the detection of *Salmonella* from biosolids.

Selective agents include triphenylmethane dyes, antibiotics such as sulfadiazine or sodium deoxycholate salts.

Individual colonies arising from this medium are streaked onto conventional growth media to allow confirmation using biochemical tests such as BIOLOG, serological identification (see Chapter 12), or confirmation using nucleic acid methods such as the polymerase chain reaction (PCR) (see Chapter 13).

10.4.1.5 Isolation of Fluorescent Pseudomonads from Soil

Pseudomonas are Gram-negative aerobic chemoheterotrophs that are commonly found in the environment. They exhibit diverse enzymatic systems and are capable of conducting many biochemical transformations. Fluorescent pseudomonads are characterized by the production of siderophores. Siderophores are compounds that are produced under low levels of iron. They have the ability to chelate iron and transport it into the microbial cell. These microbes also have a yellow–green pigment that diffuses through

agar media during growth and fluoresces under ultraviolet light. Pigment production is enhanced by iron deprivation.

For enumeration of fluorescent pseudomonads, soil is added to an extracting solution and a dilution series is performed as outlined earlier (see Section 10.2). The spread plate technique is used to isolate individual colonies on selective media (see Section 10.4). One medium appropriate for isolation is called S1 medium (Gould *et al.*, 1985). Components of the media include sucrose, glucose, casamino acids, sodium lauroyl sarcosine (SLS), trimethoprim, various salts and agar. The SLS prevents the growth of Gram-positive organisms, and the trimethoprim is an effective inhibitor of facultative Gram-negative organisms. Sucrose and glycerol provide an osmotic stress that selects for the fluorescent pseudomonads. Fluorescent colonies are identified on the medium with the use of ultraviolet light.

10.4.1.6 Isolation of Nitrifying Organisms from Soil

Nitrosomonas and *Nitrobacter* are chemoautotrophic organisms found in soil and water, and are responsible for the oxidation of ammonium to nitrite (*Nitrosomonas*) and nitrite to nitrate (*Nitrobacter*). This process, known as nitrification, is important because it can affect plant growth beneficially, but nitrate also contributes to potable water contamination. Direct plating techniques are difficult, because even with the use of strictly inorganic substrates in the medium, slight amounts of organic material introduced by inoculation with the environmental sample allow growth of faster growing heterotrophic organisms. One approach used to overcome this problem is a lengthy serial enrichment technique that includes soil enrichment, an initial culture enrichment step, a final enrichment step, isolation on agar and rigorous purity checks (Figure 10.7). To begin the isolation of *Nitrobacter* from soil, fresh field soil is leached with sodium nitrite on a daily basis for 2–3 weeks to enrich for *Nitrobacter*. The enriched soil is transferred to a liquid medium containing $NaNO_2$. Cultures are incubated with shaking at 25°C. As soon as turbidity is detected, the culture is transferred to fresh medium using a 1% inoculum. Prompt transfer reduces the development of heterotrophs in the sample. The initial enrichment goes through a series of six to eight passages to fresh medium using a 1% inoculum each time. After these steps, the culture is filtered through a membrane to trap the cells and washed to remove growth products. The membrane is transferred to fresh medium, and again undergoes five or six passages into fresh medium after filtration of cells onto a membrane. The final enrichment culture is streaked on *Nitrobacter* agar medium (Atlas, 2010) and tiny colonies appearing after 14 days are indicative of *Nitrobacter*. Purity of the test culture must be checked by using media with various carbon and nitrogen sources, and osmotic strengths for selection (Schmidt *et al.*, 1973). Numbers of *Nitrobacter* spp. in a soil sample can also be estimated using an MPN technique. After extraction, the soil dilutions are added to replicate tubes of broth medium containing potassium buffers, magnesium salts, trace elements, iron, $NaNO_2$ as the inorganic substrate for oxidation and Bromthymol Blue as the indicator. A drop in pH monitored by a change in color of the tubes from blue–green to yellow over a period of 3–6 weeks indicates the oxidation of NH_4^+ to NO_2^-. Dilutions are scored as + or − for growth, indicated by the presence of NO_2, and the population is estimated using MPN tables.

10.4.1.7 Isolation and Enumeration of 2,4-D Degrading Bacteria in Soil

Researchers often wish to determine whether bacteria that perform a known metabolic function exist in an environmental sample. As an example, the research community is very interested in the fate of pesticides that are introduced into the environment. The herbicide 2,4-D (2,4-dichlorophenoxyacetic acid) is one such pesticide that has

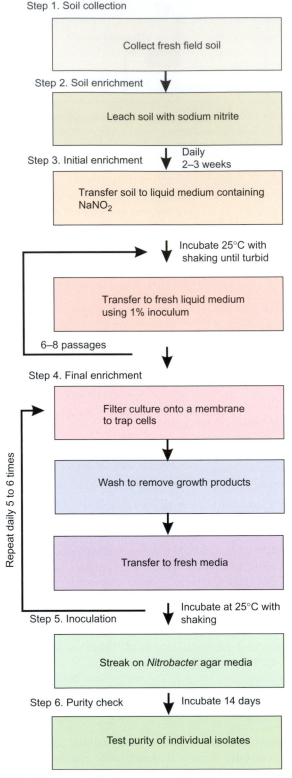

FIGURE 10.7   Protocol for the isolation of nitrifying organisms.

been commonly used in fate and ecology studies (see Case Study 3.1). A cultural medium used for enumeration and isolation of viable 2,4-D degraders is a selective, differential enrichment medium called eosin–methylene blue (EMB)-2,4-D agar (Neilson *et al.*, 1994). This agar contains minimal salts and 2,4-D as the carbon source. Indicators eosin B and methylene blue allow selection for Gram-negative bacteria and differentiation of 2,4-D-degrading colonies, which turn black. In essence, this is using 2,4-D as an elective carbon source, and the indicators as both selective agents and differential agents that distinguish on the basis of the black colonies that arise during 2,4-D degradation.

10.5 CULTURAL METHODS FOR FUNGI

Fungi are ubiquitous in nature and can be found in samples taken from soil, sediments and aquatic environments such as lakes, ponds, rivers, marine water, wastewater and well water. They are heterotrophic organisms, mostly aerobic or microaerophilic in nature. Fungi exist in a variety of morphological and physical states, which makes them difficult to quantitate and identify by cultural techniques. Cultural methods for fungi are similar to those for bacteria but must be modified to restrict bacterial growth. This is normally done by the addition of antibiotics or dyes, such as Rose Bengal, or by lowering the pH of the medium. Cultural methods for fungi are normally used to obtain pure fungal isolates, but these methods are not usually appropriate for quantitative dilution and plating analyses. This is because counts can be highly biased by the presence of spore-forming fungi, where colonies can arise from spores, and the number of culturable colonies obtained will not be a true reflection of the number of colonies in the environmental sample. For cultural assays of bacteria, it is assumed that each colony-forming unit originates from a single bacterial cell. However, in the same enumeration technique for fungi, fungal colonies may develop from a single spore, an aggregate of spores or a mycelial fragment containing more than one viable cell. Keeping in mind this limitation, plating techniques are still a valuable tool for the assessment of fungi in the environment. Techniques for culturing fungi include the pour plate technique (see Section 10.3), the spread plate technique (see Section 10.3) and the membrane filtration technique (see Section 10.4.1.3). These protocols as they relate specifically to the cultivation of fungi can be found in *Standard Methods for the Examination of Water and Wastewater* (2012). Modifications of the dilution and plating technique include directly picking fungal hyphae from samples with subsequent plating on nutrient agar, or washing the soil (or plant roots) and placing a small amount of the washed sample on the agar medium.

Fungi are often isolated on a nonselective agar medium, which allows the isolation of the maximum number of fungal taxa from the sample under study. As with bacteria, a wide selection of nutrient media is available. The most common include potato dextrose agar and malt extract agar (Figure 10.8). In some cases, a bacterial antibiotic such as streptomycin can be added to the agar, or alternatively, the agar can be acidified to pH 4.5 to inhibit bacterial growth.

Often, the scientist is interested in the isolation of a specific physiological group of fungi or a specific taxon such as Basidiomycetes or mycorrhizal fungi. A wide range of selective media have been developed for this purpose (Seifert, 1990). The MPN technique (see Section 10.3.2) is often used for the quantification of mycorrhizal fungi, which do not grow well on plating media. In this technique, soil dilutions are used to inoculate a host plant rather than a nutrient medium. After a specified incubation, plant roots are observed for colonization by the fungi, and calculations are performed using the MPN tables to estimate the numbers present in the original sample. After the initial isolation of fungi from the environmental sample, individual colonies must be selected for further purification and identification.

10.6 CULTURAL METHODS FOR ALGAE AND CYANOBACTERIA

Algae are unicellular or multicellular phototrophic eukaryotic microorganisms that occur in fresh and marine water and moist soil (see Section 2.3). In contrast, the so-called blue−green algae are not true algae—rather they belong to a group of bacteria known as cyanobacteria. The majority of soil algae are obligate photoautotrophs, using light to manufacture organic compounds from inorganic nutrients (phototrophic eukaryotes). Thus, their nutritional requirements include water, light, oxygen, carbon dioxide and inorganic nutrients. A small number of algae are photoheterotrophic, requiring organic compounds for growth. Algae and cyanobacteria can be enumerated by dilution and plating techniques or the MPN assay. Prior to dilution and culturing, because of the filamentous nature of some algal species, it is sometimes necessary to use more forceful methods for extraction than those used for bacteria. These include grinding soil samples with a mortar and pestle prior to extraction, as well as a longer and more vigorous homogenizing step using a blender or glass beads. When enumerating microalgal colony-forming units on solid media, the addition of antibiotics such as penicillin and streptomycin to the algal medium is recommended, to ensure their growth in the presence of the faster growing heterotrophic bacteria and fungi. Alternatively, low doses of UV radiation or exposure to nonspecific bactericides such as formaldehyde or sodium lauryl sulfate may be used to suppress growth of bacteria. An incubation period of 1−3 weeks or even longer may be necessary, requiring special care to ensure that desiccation of the medium does not occur. Recommended air temperatures are 20 to 25°C, with a

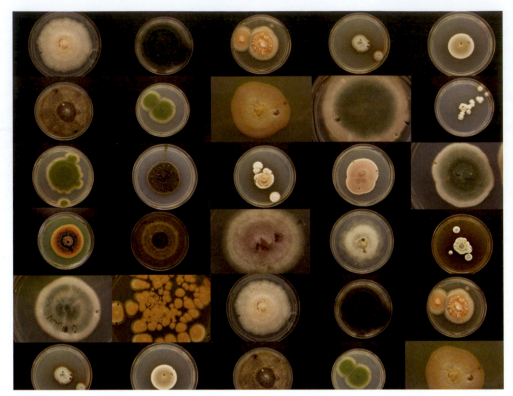

FIGURE 10.8   A collage of pure cultures of fungal isolates from a mine tailings site. Figure courtesy M.O. Mendez.

photoperiod of either 12 hours light/12 hours dark or 16 hours light/8 hours dark.

When isolating algae, the approach may be to select for a specific alga, or to obtain a quantitative index of the entire community. In the latter case, an MPN technique using a growth medium consisting of a soil−water mix is often used. Many media are available for specific enrichment and plating, including Bold's Basa medium for green and yellow−green microalgae (and cyanobacteria) (Figure 10.9).

A good review of culturable methods for algae methods was presented by Castenholz (1988), including several media suitable for freshwater cyanobacteria. The antibiotic cycloheximide is usually added to preclude growth of eukaryotic algae, diatoms and protozoa. The medium should be incubated at 24−26°C under fluorescent light of 2000−3000 lux (cool white light or daylight fluorescent light) (Figure 10.10). For agar plates, samples are usually diluted and applied as a spread plate.

10.7 CELL CULTURE-BASED DETECTION METHODS FOR VIRUSES

Living cells are necessary for virus replication, and originally human volunteers, laboratory animals or embryonated hens' eggs were the only techniques available to study and isolate human viruses. However, the advent of animal cell culture techniques after World War II revolutionized the study of animal viruses, where for the first time viruses could be grown and isolated without the need for animals. Two main types of cell culture are used in virology, the first of which is primary cell culture. In this procedure, cells are removed directly from an animal and can be used to subculture viral cells for a limited number of times (20−50 passages). Thus, a source of animals must be available on a continuous basis to supply the needed cells. Primary monkey kidney cell cultures (commonly from rhesus or green monkeys) are permissive, meaning that they allow the replication of many common human viruses. These are frequently used for the detection of enteric viruses found in the environment. A second type of cell culture is known as continuous cell culture, which is also derived from animals or humans, but may be subcultured indefinitely. Such cell lines can be derived from normal or cancerous tissue.

Cell cultures are initiated by dissociating small pieces of tissue into single cells by treatment with a proteolytic enzyme and a chelation agent (generally trypsin and EDTA). The dispersed cells are then suspended in cell culture medium composed of a balanced salt solution containing glucose, vitamins, amino acids, a buffer system, a pH indicator, serum (usually fetal calf) and antibiotics. The suspended cells are placed in flasks, tubes or Petri dishes, depending on the needs of the laboratory. Plastic

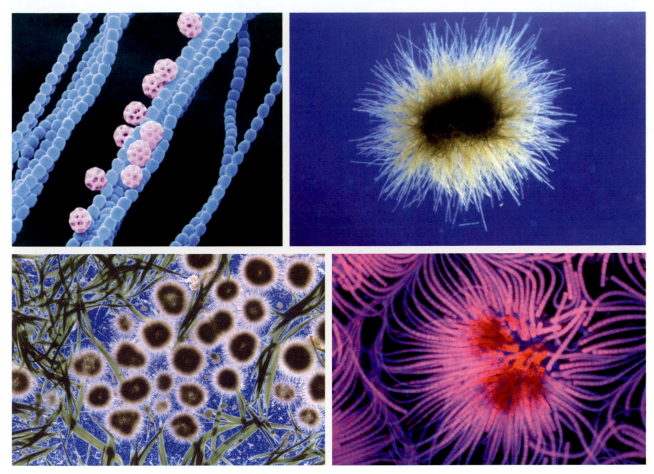

FIGURE 10.9   Various cyanobacteria (blue−green algae).

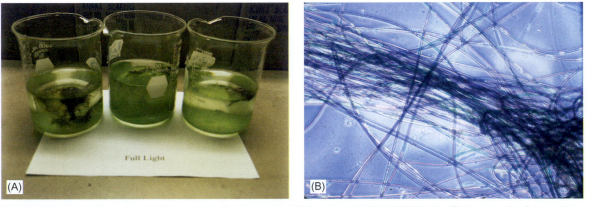

FIGURE 10.10   (A) Cyanobacteria (blue−green algae) grown in full fluorescent light in BG11 broth. (B) An example of a cyanobacterium *Lynabya* seen using phase contrast microscopy.

flasks and multiwell plates are especially prepared for cell culture use, and are currently the most common type of container used (Figure 10.11). The cells attach to the surface of the vessel and begin replicating. Replication ceases when a single layer or monolayer of cells occupies all the available surface of the vessel.

Unfortunately, not all animal viruses will grow in the same cell line, and others have never been successfully grown in cell culture. Viruses require specific receptors for attachment and replication with the host cell, and if these are not present, no replication takes place. For example, rotavirus grows well in the MA-104 cell line, but not the blue green monkey (BGM) cell line. The most commonly used cell line for enterovirus detection in water is the BGM cell line. In recent years, the cell line, CaCO$_2$, originating from a human colon carcinoma, has been found to grow more types of enteric viruses (hepatitis A, astroviruses, adenoviruses, enteroviruses, rotaviruses) than any other cell line, and is seeing increased use in environmental virology. Cell lines commonly used to grow human viruses are shown in Table 10.2.

The two most common methods for detecting and quantifying viruses in cell culture are the cytopathogenic effect (CPE) method and the plaque-forming unit (PFU) method. Cytopathic effects are observable changes that take place in the host cells as a result of virus replication. Such changes may be observed as changes in morphology, including rounding or formation of giant cells, or formation of a hole in the monolayer due to localized lysis of virus-infected cells (Figures 10.12 and 10.13). Different viruses may produce very individual and distinctive CPE. For example, adenovirus causes the formation of grapelike clusters, and enteroviruses cause

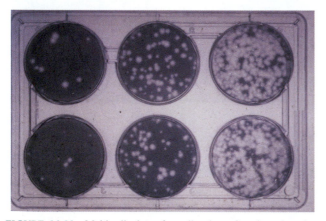

FIGURE 10.11 Multiwell plates for cell culture for virus detection. These wells show increasing dilutions (right to left) of virus on a monolayer. Each clear zone, or plaque, theoretically arises from a single infectious virus particle, i.e. a plaque-forming unit (PFU).

rounding of the cells. The production of CPE may take as little as 1 day to 2−3 weeks, depending on the original concentration and type of virus. Not all viruses may produce CPE during replication in a given cell line. In addition, the virus may grow to high numbers, but no visible CPE is observed. In these instances, other techniques for virus detection must be used. Alternatives include the use of an immunoassay that detects viral antigens, or the use of PCR to detect viral nucleic acid.

Generally, laboratory strains of viruses have been selected for their ability to grow in cell culture easily and rapidly with the production of CPE. In contrast, viruses isolated from environmental samples often come directly from infected animals or humans, and do not grow as rapidly in cell culture. Often, "blind passages" of cell culture showing no CPE are passed on to fresh monolayers of cells before a CPE is observed. Two procedures can be used to quantify viruses utilizing CPE. The first is the serial dilution endpoint or TCID$_{50}$ method, which involves adding serial dilutions of virus suspension to host cells, and subsequently observing the production of CPE over time. The titer or endpoint is the highest viral dilution capable of producing a CPE in 50% of the tissue culture vessels or wells, and is referred to as the medium tissue culture infective dose or TCID$_{50}$. The second method is the most probable number or MPN method, which is similar to that used to quantify bacteria. This can be used with viral cells by observation of CPE in monolayers inoculated with different dilutions of viral suspension, and subsequent use of MPN tables to determine viral numbers.

Some viruses, such as enteroviruses, produce plaques or zones of lysis in cell monolayers overlaid with solidified nutrient medium. These plaques originate from a single infectious virus particle, thus the titer can be quantified by counting the plaques or plaque-forming units (PFU). This can be thought of as analogous to enumeration of bacteria using CFU. There are variations of the basic plaque-forming unit assay, but usually a vital

TABLE 10.2 Commonly Used Continuous Cell Cultures for Isolation and Detection of Enteric Viruses

Cell Culture Line	Virus							
	Adeno	Astro	Coxsackie A	Coxsackie B	Echo	Polio	Rota	Hepatitis A
BGM	−[a]	−	−	+	+	+	−	−
BSC-1	−	*	−	*	*	+	*	*
CaCO$_2$	+	+	*	+	+	+	+	+
Hep-2 (HeLa)	+	*		+	*	*	*	*
RD	*	*	+	−	+	*	*	
RfhK	*	*	*	*	*	*	*	+

[a] +, growth and/or production of cytopathogenic effect (CPE); −, no growth and/or production of CPE; *, no data.

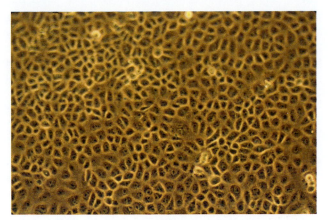

FIGURE 10.12   This figure shows a normal uninfected cell culture monolayer. Compare with Figure 10.13.

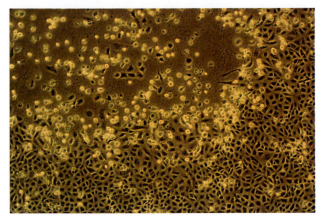

FIGURE 10.13   This figure shows an infected cell culture monolayer exhibiting CPE. Compare with Figure 10.12.

dye, which only stains living cells, is used. As the virus infection in the monolayer spreads and cells lyse, a clear plaque appears in a stained monolayer of living cells (Figure 10.11). A system commonly used for enteroviruses utilizes an agar overlay containing media components, serum and the dye neutral red. A virus kills and lyses infected cells, leaving a clear plaque in a red-stained cell monolayer, because the dye does not stain dead cells.

Cell culture is not 100% effective in detecting all of the viral particles that can be observed through use of an electron microscope. This is because not all the particles are infectious. A noninfectious viral particle may contain no nucleic acid in the capsid, or the nucleic acid may be incomplete. Other reasons for the inefficiency of cell culture may be that there is a lack of cell receptors, or the cell culture incubation period is insufficient for all of the viruses to attach to viral receptors on the host cells. Further, the plaquing efficiency or production of a CPE is influenced by the salts, additives or enzymes added to the medium. For example, good plaque formation for rotaviruses requires the addition of trypsin to the overlay

medium. The susceptibility of cell lines can be enhanced by the addition of 5-iododeoxyuridine for enteric viruses (Benton and Ward, 1982). Generally, for laboratory-grown viruses, the ratio of PFU or infectious virus to total viral particles is 1:100, but it may be much greater for viruses from the host or environment (1:10,000) (Ward *et al.*, 1984). For this reason, techniques such as PCR, which can detect one to 10 copies of a viral genome, are more sensitive than conventional cell culture methods.

Use of cell culture for detection of viruses in environmental samples is also often made difficult by the presence of bacteria, molds and toxic substances. To overcome problems with other microorganisms, antibiotics are added, or the samples are filtered. Unfortunately, this sometimes results in loss of viruses that may be associated with solids. Problems with substances toxic to the cell culture are also difficult to overcome. The easiest solution is to dilute the sample until no more toxicity is observed. Unfortunately, this may require use of an excessive amount of cell culture. Treatment of the sample with Freon or chloroform is another technique that has often been employed.

QUESTIONS AND PROBLEMS

1. Student "A" performs a soil extraction with peptone as the extracting solution. A dilution and plating technique is performed using the spread plate method. The following results are obtained:
   a. 150 colonies are counted on an R2A plate at the 10^5 dilution
   b. 30 colonies are counted on a nutrient agar plate at 10^4 dilution
   Calculate the number of viable bacteria per gram of soil for R2A and nutrient agar. Why are the counts for nutrient agar lower than for the R2A media?
2. Student "B" performs a soil extraction with peptone as the extracting solution. A dilution and plating technique is performed using the pour plate method. The number of bacteria per gram of soil is calculated to be 8×10^8. What dilution plate contained 80 colonies?
3. Student "C" used direct microscopic methods to count bacteria in the same soil as student "A" above. How would student "C's" numbers compare to student "A's" numbers and why?
4. The last student also extracted the same soil in peptone, but let the extraction sit overnight at room temperature before diluting and plating. How would these numbers compare to student "A's" numbers and why?
5. Identify the major errors associated with dilution and plating techniques.
6. What is the difference between a primary cell line and a continuous cell line? Which cell line do you think is better for isolating viruses from the environment? Why?

7. What are the two most common methods used to assay viruses in cell culture?
8. Why are some virus particles unable to be detected by cell culture?
9. Discuss the advantages and disadvantages of cultural methodologies for the detection of microbes in the environment.

REFERENCES AND RECOMMENDED READING

Alonso, J. L., Batella, M. S., Amoros, I., and Rambach, A. (1992) Salmonella detection in marine waters using a short standard method. *Water Res.* **26**, 973–978.

American Public Health Association (2012) "Standard Methods for the Examination of Water and Wastewater," 22nd ed. APHA, Washington, DC.

Atlas, R. M. (2010) "Handbook of Microbiological Media," fourth ed. CRC Press, Boca Raton, FL.

Belenguer, A., Duncan, S. H., Calder, A. G., Holtrop, G., Louis, P., Lobley, G. E., *et al.* (2006) Two routes of metabolic cross feeding between *Bifidobacterium adolescentis* and butyrate producing anaerobes from the human gut. *Appl. Environ. Microbiol.* **72**, 3593–3599.

Ben Dov, E., Kramarsky-Winter, E., and Kushmaro, A. (2009) An in situ method for cultivating microorganisms using a double encapsulation technique. *FEMS Microbiol. Ecol.* **68**, 363–371.

Benton, W. H., and Ward, R. L. (1982) Induction of cytopathogenicity in mammalian cell lines challenged with culturable enteric viruses and its enhancement by 5-iododeoxyuridine. *Appl. Environ. Microbiol.* **43**, 861–868.

Bruns, A., Cypionka, H., and Overmann, J. (2002) Cyclic AMP and acyl homoserine lactones increase the cultivation efficiency of heterotrophic bacteria from the central Baltic Sea. *Appl. Environ. Microbiol.* **68**, 3978–3987.

Castenholz, R. W. (1988) Culturing methods for cyanobacteria. In "Methods in Enzymology" (J. N. Abelson, and M. I. Simon, eds.), vol. 167, Academic Press, San Diego, pp. 68–93.

Connon, S. A., and Giovannoni, S. J. (2002) High throughput methods for culturing microorganisms in very-low-nutrient media yield diverse new marine isolates. *Appl. Environ. Microbiol.* **68**, 3878–3885.

Davis, K. E., Joseph, S. J., and Janssen, P. H. (2005) Effects of growth medium inoculum size, and incubation time on culturability and isolation of soil bacteria. *Appl. Environ. Microbiol.* **71**, 826–834.

Ferrari, B. C., and Gillings, M. (2009) Cultivating fastidious bacteria: viability staining and micromanipulation from a soil substrate membrane system. *Appl. Environ. Microbiol.* **75**, 3352–3354.

Gould, W. D., Hagedorn, C., Bardinelli, T. R., and Zablotowica, R. M. (1985) New selective medium for enumeration and recovery of fluorescent pseudomonads from various habitats. *Appl. Environ. Microbiol.* **49**, 28–32.

Huang, W. E., Ward, A. D., and Whiteley, A. S. (2009) Raman tweezers sorting of single microbial cells. *Environ. Microbiol. Rep.* **1**, 44–49.

Janssen, P. H., Yates, P. S., Grinton, B. E., Taylor, P. M., and Sait., M. (2002) Improved culturability of soil bacteria and isolation in pure culture of novel members of the divisions Acidobacteria, Actinobacteria, Proteobacteria, and Vernacomicrobia. *Appl. Environ. Microbiol.* **68**, 2391–2396.

Joseph, S. J., Hugenholtz, P., Sangwan, P., Osborne, C. A., and Janssen, P. H. (2003) Laboratory cultivation of widespread and previously uncultured soil bacteria. *Appl. Environ. Microbiol.* **69**, 7210–7215.

Kakumanu, M. L., and Williams, M. A. (2012) Soil diffusion system enriches the growth of diverse and previously uncultivated bacterial taxa. *Soil Sci. Soc. Am. J.* **76**, 463–474.

Kim, J. J., Kim, H. N., Masui, R., Kuramitsu, S., Seo, J. H., Kim, K., *et al.* (2008) Isolation of uncultivable anaerobic thermophiles of the family Clostridiaceae requiring growth-supporting factors. *J. Microbiol. Biotechnol.* **18**, 611–615.

Kopke, B., Wilms, R., Engelen, B., Cypionka, H., and Sass, H. (2005) Microbial diversity in coastal subsurface sediments: a cultivation approach using various electron acceptors and substrate gradients. *Appl. Environ. Microbiol.* **71**, 7819–7830.

Lindahl, V., and Bakken, L. R. (1995) Evaluation of methods for extraction of bacteria from soil. *FEMS Microbiol. Ecol.* **16**, 135–142.

Misunoe, Y., Wai, S. N., Takade, A., and Yoshida, S. (1999) Restoration of culturability of starvation-stressed and low-temperature stressed *Escherichia coli* 0157 cells by using H_2O_2-degrading compounds. *Arch. Microbiol.* **172**, 63–67.

Neilson, J. W., Josephson, K. L., Pepper, I. L., Arnold, R. G., DiGiovanni, G. D., and Sinclair, N. A. (1994) Frequency of horizontal gene transfer of a large catabolic plasmid (pJP4) in soil. *Appl. Environ. Microbiol.* **60**, 4053–4058.

Nichols, D., Lewis, K., Orjala, J., Mo, S., Ortenberg, R., O'Connor, P., *et al.* (2008) Short peptide induces an uncultivable microorganism to grow in vitro. *Appl. Environ. Microbiol.* **74**, 4889–4897.

Roszak, D. B., and Colwell, R. R. (1987) Survival strategies of bacteria in the natural environment. *Microbiol. Rev.* **51**, 365–379.

Schmidt, E. L., Molina, J. A. E., and Chiang, C. (1973) Isolation of chemoautrophic nitrifiers from Moroccan soils. *Bull. Ecol. Res. Comm. (Stock.)* **17**, 166–167.

Seifert, K. A. (1990) Isolation of filamentous fungi. In "Isolation of Biotechnological Organisms from Nature" (D. P. Labeda, ed.), McGraw-Hill, New York, pp. 21–51.

Shade, A., Hogan, C. S., Klimowicz, A. K., Linske, M., McManus, P. S., and Handelsman, J. (2012) Culturing captures members of the soil rare biosphere. *Environ. Microbiol.* **14**, 2247–2252.

Sogin, M. L., Morrison, H. G., Huber, J. A., Welch, D. M., Huse, S. M., Neal, P. R., *et al.* (2006) Microbial diversity in the deep sea and the under explored rare biosphere. *Proc. Natl. Acad. Sci. USA* **103**, 12115–12120.

Song, J., Oh, H. M., and Cho, J. C. (2009) Improved culturability of SAR 11 strains in dilution to extinction culturing from the East Sea, West Pacific Ocean. *FEMS Microbiol. Lett.* **295**, 141–147.

Stevenson, B. S., Eichorst, S. A., Wertz, J. T., Schmidt, T. M., and Breznak, J. A. (2004) New strategies for cultivation and detection of previously uncultured microbes. *Appl. Environ. Microbiol.* **70**, 4748–4755.

Tamaki, H., Sekiguchi, Y., Hanada, S., Nakamura, K., Nomura, N., Matsumura, M., *et al.* (2005) Comparative analysis of microbial diversity in freshwater sediment of a shallow eutrophic lake by molecular and improved cultivation-based techniques. *Appl. Environ. Microbiol.* **71**, 2162–2169.

Ward, R. L., Knowlton, D. R., and Pierce, M. J. (1984) Efficiency of human rotavirus propagation in cell culture. *J. Clin. Microbiol.* **19**, 748–753.

Zengler, K., Toledo, G., Rappé, M., Elkins, E. J., Short, J. M., and Keller, M. (2002) Cultivating the uncultured. *Proc. Nat. Acad. Sci. USA* **99**, 15681–15685.

Chapter 11

Physiological Methods

Raina M. Maier and Terry J. Gentry

11.1 INTRODUCTION

A great deal of recent focus in environmental microbiology has been on developing rapid or even real-time techniques for analyzing microbial communities using nonculture-based techniques. Some of these techniques involve analysis of the community DNA extracted from a sample (see Chapter 13) or novel spectroscopic techniques (see Chapter 9). While these techniques have revolutionized the amount of information that we can obtain from a sample, it has also become apparent that these techniques provide information about "who" is in the community but not about "what" activities the community carries out. For this reason, environmental microbiologists have realized that a complete analysis of an environmental sample requires physiological measurements that are made in combination with the newer molecular approaches. This is of particular current interest for assessment of possible impacts that climate change will have on ecosystem function or vice versa.

In an undisturbed environment, the activity measured will reflect a baseline level of activity. This baseline activity level is dependent on a variety of biological, chemical and physical parameters—especially important is the nutrient status of the environment. In an ecological sense, measurement of microbial activity in an undisturbed environment allows one to determine the microbial contribution to nutrient cycling (see Chapter 16) within an ecosystem. For example, in the case of carbon cycling, the focus of microbial activity measurements is to determine accurately the production of new biomass by each microbial component of the community, and estimate the amount of energy that is being stored in a particular group of micro- or macroorganisms. This information allows determination of the transfer of energy between trophic levels in the food chain within a particular ecosystem (see Figure 6.6).

I.L. Pepper, C.P. Gerba, T.J. Gentry: Environmental Microbiology, Third edition. DOI: http://dx.doi.org/10.1016/B978-0-12-394626-3.00011-9
© 2015 Elsevier Inc. All rights reserved.

Measurement of microbial activity can also be used to evaluate the response of a microbial community to a variety of stimuli or disturbances, including naturally variable parameters such as moisture level, and anthropogenically imposed variables such as the addition of fertilizers or contaminant spills. In this case, measurement of microbial activity can provide an indication of the general health of the environment and can be used to evaluate the impact of a disturbance on the microbial community. For example, microbial activity has been used to evaluate the impact of soil quality on certain land use practices such as farming, logging and mining. Microbial activity is also an important indicator for evaluating the process of restoration of disturbed sites. One example of this is natural attenuation, the process by which indigenous microbial populations degrade spilled pollutants within a natural environment (see Section 17.7). Other areas where microbial activity measurements are often used include: engineered systems such as municipal wastewater treatment systems; compost systems; and biological reactor systems; all of which utilize microbial processes for specific purposes.

The goal of this chapter is to examine different types of microbial activity measurements in both pure culture and environmental samples. Although activity measurement in pure culture is relatively straightforward, environmental communities contain diverse microorganisms and physiological types, including aerobic and anaerobic heterotrophs and autotrophs (see Chapter 2). Thus, any measurement of microbial activity in an environmental sample can be designed to be more or less inclusive of different classes and physiological types of microorganisms. As a result, it is important to understand the different activity measurements and their relative specificity. In this chapter, activity measurements in pure culture will be discussed first, including: measurement of substrate disappearance; terminal electron acceptor (TEA) utilization; cell mass increase; and carbon dioxide evolution. Following this, activity measurements in environmental samples will be addressed. Environmental activity measurements have been divided into five broad categories: (1) carbon respiration; (2) incorporation of radiolabeled tracers in cellular macromolecules; (3) adenylate energy charge; (4) enzyme activity assays; and (5) stable isotope probing. Frequently, the nature of the research question being asked, and the resources available, determines which technology is utilized.

11.2 MEASURING MICROBIAL ACTIVITY IN PURE CULTURE

As discussed in Chapter 3, growth in pure culture can be described by the generalized equation representing respiration:

$$\text{Substrate} + \text{nitrogen} + \text{TEA} \rightarrow \text{cell mass} + \text{carbon dioxide} + \text{water} \quad \text{(Eq. 11.1)}$$

In this equation, the TEA (terminal electron acceptor) is oxygen for aerobic conditions or one of several alternate TEAs for anaerobic conditions (see Table 3.3). Examination of this equation shows that there are several ways in which microbial activity can be measured. These include substrate disappearance, TEA utilization, biomass production and carbon dioxide evolution. Each of these approaches is discussed briefly in the following sections.

11.2.1 Substrate Utilization

11.2.1.1 Heterotrophic Substrates

For heterotrophic activity the substrate is carbon based. Such substrates can be measured in many different ways, depending on the chemistry of the substrate molecules, and whether a single component or multiple components are present. For example, in a single-solute system, aromatic compounds such as benzene and all benzene derivatives can be measured easily and sensitively by UV spectrophotometry (Figure 11.1). A UV spectrophotometer measures how much light is absorbed while passing through a sample. The spectrophotometer can be set to a particular wavelength of light that is chosen on the basis of the absorbance spectrum of the compound. For instance, the absorbance spectrum of benzoate, shown in Figure 11.2, indicates that the optimal detection wavelength is 224 nm (note that 224 nm is chosen because there is a great deal of interference by salts such as NO_3^- and SO_4^{2-} below 200 nm). Analogously, a fluorimeter can be used to measure fluorescent molecules such as poly-aromatic hydrocarbons (PAHs) within single-component systems (Figure 11.3). A fluorimeter offers increased sensitivity over UV spectroscopy.

Often, samples from pure cultures and especially environmental samples contain compounds that interfere with

FIGURE 11.1 UV spectrophotometer. Photo courtesy R.M. Maier.

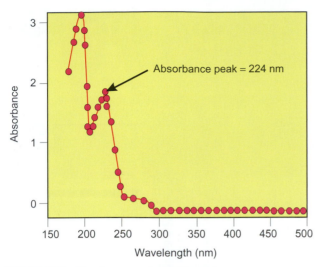

FIGURE 11.2 Absorbance spectrum of benzoate.

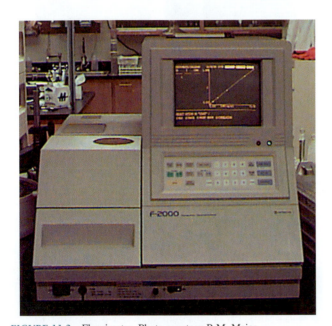

FIGURE 11.3 Fluorimeter. Photo courtesy R.M. Maier.

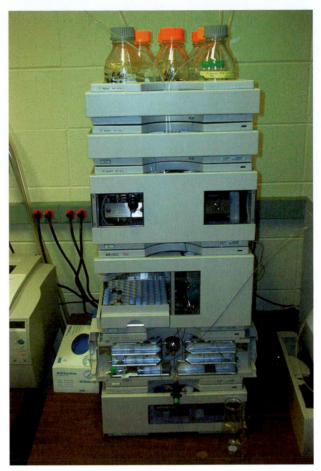

FIGURE 11.4 High performance liquid chromatograph. Photo courtesy T.R. Sandrin.

the quantitation of the target compound. This can be considered a multiple-component system. For example, humic compounds present in soil and water samples contain many aromatic molecules that interfere strongly with measurement of aromatic substrates. Alternatively, there may be multiple components of interest in a sample containing a complex mixture of compounds such as might be found in gasoline or other fuel oils. In this case, a separation or chromatography step is required prior to detection of the target compound. There are different types of chromatography, including liquid chromatography and gas chromatography. One common instrument based on liquid chromatography is the high-performance liquid chromatograph (HPLC) (Figure 11.4). The HPLC forces liquid solvent through a packed column under high pressure to separate the components within a mixture. The separation is dictated by the type of column packing, and is generally based on charge or hydrophobicity. The resulting chromatogram can be analyzed to provide quantitative information about the amount of each component present in the sample (Figure 11.5). For the example chromatogram shown, a UV detector was used for analysis; however, several other types of detectors can be used. These include the photodiode array detector, which is based on measuring light wavelength absorption, but uses a detector array to rapidly measure the entire wavelength spectrum instead of just a single wavelength. Other detectors used in HPLC include refractometers, which are primarily for sugar analysis; evaporative light scattering detectors, used for surfactants and polymers; conductivity meters, used for ionic or charged compounds; and fluorescence detectors, used for analysis of polyaromatic hydrocarbons and fluorescent dyes. A related instrument is the ultra performance liquid chromatograph (UPLC) (Figure 11.6). The UPLC is similar to the HPLC, but uses smaller-sized particles ($<2\,\mu m$) for analyte separation.

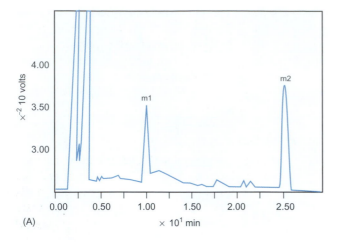

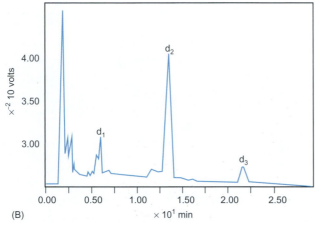

FIGURE 11.5　HPLC chromatogram. These are typical chromatograms of a microbial surfactant, rhamnolipid, produced by *Pseudomonas aeruginosa*. (A) Monorhamnolipid (one rhamnose sugar per molecule) and (B) dirhamnolipid (two rhamnose sugars per molecule). Note that each chromatogram has more than one peak. This is because microbes make a mixture of surfactants that vary in the length of the fatty acid tails. The longer the tail, the more hydrophobic the molecule is, and the longer it is retained on the column. Thus, for monorhamnolipid, the peak labeled m2 has longer lipid tails than the monorhamnolipid in the peak labeled m1. A reverse phase C_{18} column was used for sample separation. Elution was isocratic (meaning the mobile phase composition remained constant during analysis) using a mobile phase of acetonitrile-water (40:60, v/v) at a flow rate of 1 ml/min. A UV detector set at 214 nm was used for detection of the surfactant. Chromatograms courtesy Y. Zhang.

FIGURE 11.6　Ultra performance liquid chromatograph-mass spectrometer (UPLC-MS). Photo courtesy L. Dykes.

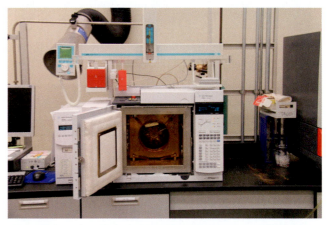

FIGURE 11.7　Gas chromatograph-mass spectrometer (GC-MS). Photo courtesy S. Williams, Office of the Texas State Chemist.

This enables much greater speed, resolution and sensitivity than can be achieved with an HPLC, and is very useful for detection of low-level environmental contaminants, such as antibiotics and pharmaceuticals (Tamtam *et al.*, 2009).

Another instrument based on chromatographic separation of solutes is the gas chromatograph (GC) (Figure 11.7). In gas chromatography, the mobile phase is a mixture of gases instead of liquids. Because the mobile phase is gaseous, compounds amenable to GC analysis must be volatile so that they can move through the GC column with the mobile phase. Therefore, compounds

that can be analyzed by GC are characterized by lower boiling points and higher volatilities. The GC can be equipped with a variety of different detectors. Two of the most commonly used are the flame ionization detector (FID) and the electron capture detector (ECD). The FID is suitable for analysis of hydrocarbons such as BTEX (benzene, toluene, ethylbenzene, xylene), polyaromatic hydrocarbons or aliphatic hydrocarbons. The ECD detector is used for analysis of halogenated materials, primarily chlorinated compounds such as trichloroethylene (TCE). It should be noted that for some of these hydrocarbons both HPLC and GC can be used for analysis but for others, such as the aliphatic hydrocarbons, which do not absorb in the UV range, only GC can be used. In other instances, such as for organic acids that have low volatility, HPLC is the method of choice.

Finally, it should be mentioned that the GC, HPLC and UPLC can all be used in conjunction with a mass spectrometer (MS) to allow further confirmation/identification of the detected compound (Figures 11.6 and 11.7). Mass spectrometers bombard a compound with electrons, causing it to break up and fragment into smaller pieces. The fragmentation pattern can then be interpreted to determine the composition of the original molecule. For routine analyses, most mass spectrometers contain libraries of fragmentation patterns that can be used to provide a best fit match or prediction of the target compound being analyzed. This is especially powerful for detection and characterization of unknown compounds such as metabolites that may be produced during biodegradation, and can be invaluable for elucidation of biodegradation pathways.

11.2.1.2 Chemoautotrophic Substrates

In contrast to chemoheterotrophic activity, for which energy is provided by oxidation of organic or carbon-based substrate, for chemoautotrophic activity the energy providing substrate is an inorganic compound. Important microbial activities that contribute to the cycling of inorganic materials in soil and water environments include chemoautotrophic oxidation of ammonia (nitrification) and sulfur (sulfur oxidation). The following sections discuss the approaches used to measure microbial transformation of these two chemoautotrophic substrates.

Nitrification

Nitrification, the oxidation of ammonia to nitrite and nitrate (see Section 16.3.4), can be measured in several ways. A direct assay involves addition of ^{15}N-labeled ammonia and measurement of the formation of ^{15}N-nitrite and ^{15}N-nitrate as the ^{15}N-ammonia is oxidized (Paerl, 1998). These ^{15}N-labeled products are measured by mass spectroscopy. However, the sensitivity of this measurement is low, and thus large quantities of ammonia that far exceed those found naturally may be required in an experiment. A second less direct approach is to divide the samples being measured into two groups and incubate one group with a nitrification inhibitor such as N-serve (in general, nitrifiers are very sensitive to chemical inhibitors). After incubation, the differences in ammonia and nitrate concentrations between the nitrification inhibited and uninhibited systems can be compared. Finally, a third

approach to the measurement of nitrification is again to divide the samples into two groups, one of which is treated with a nitrification inhibitor. The samples are then incubated with $^{14}CO_2$ to measure CO_2 fixation. The amount of carbon dioxide fixation is then compared in the presence and absence of the nitrification inhibitor to determine nitrification activity.

Sulfur Oxidation

Sulfur oxidation (see Section 16.4.3.1) can be measured by addition of elemental sulfur (S^0) to a sample and measuring the amount of sulfate (SO_4^{2-}) produced during incubation. However, this is a complicated process. One such assay measures sulfate as a precipitate with barium chloride in acid solution (Kelly and Wood, 1998). Following precipitation, the barium in solution is measured by inductively coupled plasma mass spectroscopy (ICP-MS), an instrument that is used to measure metals. While the ICP-MS can distinguish between different metals, it cannot distinguish between the different species of a given metal in solution (see Section 18.8.2 for metal speciation methods). In this assay, as the amount of sulfate increases as a result of sulfur oxidation, the amount of barium in solution decreases through precipitation with the sulfate. A second approach to measurement of sulfur oxidation is to determine O_2 utilization (Padden et al., 1998) or CO_2 uptake in the presence of sulfur compounds. As described above for nitrification, samples can be incubated with $^{14}CO_2$ to measure CO_2 fixation as an estimate of sulfur oxidation.

11.2.2 Terminal Electron Acceptors

A variety of TEAs are available for microbial use and levels of microbial activity are reflected in the disappearance of the TEA and the appearance of the reduced TEA. Measurement of several common TEAs is discussed in the following sections.

11.2.2.1 Oxygen Uptake

The majority of environmental activities measured are under aerobic conditions where oxygen is the required TEA. Oxygen, unlike many other nutrients, is relatively insoluble in water and can become limiting to growth unless it is constantly replenished (Information Box 4.7). Thus, aeration is an important consideration in both pure culture and environmental systems. Normal dissolved oxygen levels in an aqueous sample that is equilibrated with air range up to 10 mg/L, depending on sample temperature and atmospheric pressure. In general, aerobic activity is maintained until the dissolved oxygen concentration in the sample falls below 1 to 2 mg/L. Because oxygen is a limiting nutrient, it is a good measure of

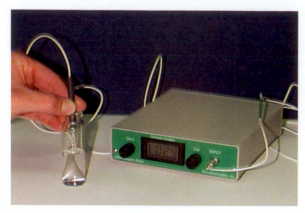

FIGURE 11.8   Oxygen microprobe. Photo courtesy R.M. Maier.

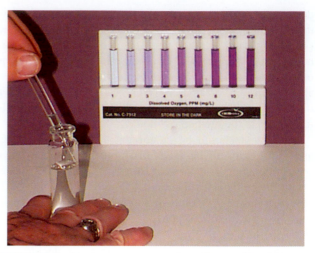

FIGURE 11.9   Oxygen measurement using a Chemets Kit. Photo courtesy R.M. Maier.

growth and activity. Measurement of oxygen in a pure culture is relatively rapid, routine and cheap to perform. This technique is perhaps used most extensively in the wastewater treatment industry to determine the biological oxygen demand (BOD) of wastewater samples (see Section 11.4.2.4) (see also Chapter 25). As discussed in the following sections, oxygen levels in an aqueous sample are generally measured using an oxygen probe or by colorimetric methods.

Oxygen Probes

An oxygen probe is basically an electrode covered by a gas-permeable membrane combined with a meter that converts the electrical signal into an analytical measurement. Oxygen probes do not measure the absolute amount of oxygen in a sample; rather, they measure oxygen amounts relative to an oxygen-saturated solution. A variety of commercially available oxygen probes can be used to measure dissolved oxygen. These include probes that are used for routine BOD measurement, and microprobes that can be used for small sample amounts or can be inserted into a soil column through a port (Figure 11.8). Advances in technology have produced microprobes for the detection of O_2, as well as nitrogen compounds (N_2O, NO_3, NH_4), sulfide, hydrogen and pH (Revsbech and Jørgensen, 1986). The oxygen microelectrodes can be as small as $20\,\mu m$ diameter, allowing the determination of gas flux across critical interfaces.

Colorimetric Assays

Alternatively, a variety of colorimetric assays are available. In this case, a sample is mixed with a series of reagents to produce an oxygen-dependent color. A titration method known as the Winkler method can be used to determine dissolved oxygen in water samples. Winkler titrations are based on the precipitation of dissolved oxygen using manganous sulfate and a potassium hydroxide—potassium iodide mixture (Wetzel and Likens,

1991). The oxygen precipitate, manganic basic oxide, reacts with sulfuric acid to form a manganic sulfate, which in turn reacts with potassium iodide to liberate iodine. The number of moles of iodine liberated is equivalent to the number of moles of oxygen present in the water sample. The liberated iodine can be determined spectrophotometrically in clear water samples, or can be determined by titration with sodium thiosulfate. The basic Winkler method may be affected by the presence of oxidizing or reducing materials that can occur in natural waters and especially in polluted waters. Modifications of the basic method are described in the APHA, AWWA, WEF (2005) manual. There are also several kits available for oxygen measurement that are based on a colorimetric reaction. An example of such an assay is shown in Figure 11.9.

11.2.2.2 Alternative Terminal Electron Acceptors

Although not nearly as commonly measured as oxygen, perhaps because the methodologies are much more complex and time consuming, there are methods available to quantitate alternate TEAs. For example, in denitrification, nitrate (NO_3^-) serves as the terminal electron acceptor and is reduced sequentially to nitrite (NO_2^-), nitric oxide (NO), nitrous oxide (N_2O) and nitrogen (N_2) (see Section 16.3.5.2). One common assay for denitrification involves the use of acetylene (C_2H_2) which blocks the reduction of nitrous oxide to nitrogen (Paerl, 1998). Thus, nitrous oxide accumulates in the system and can be measured using a gas chromatograph with an electron capture detector. A disadvantage of using this method in environmental samples is that acetylene also inhibits nitrification. Since nitrification and denitrification are closely coupled (e.g., the nitrate produced by nitrification is used in denitrification), this inhibition may reduce or eliminate denitrification in the sample. Therefore, in some instances, denitrification is estimated by simply measuring nitrification (see Section 11.2.1.2). A second problem with this technique is that it is not effective at low nitrate concentrations (less than 0.63 mg/L).

Similarly, there are methodologies available for measuring sulfate (SO_4^{2-}) reduction (Tabatabai, 1994a), iron and manganese reduction (Ghiorse, 1994) and methanogenic activity (Zinder, 1998).

11.2.3 Cell Mass

Cell mass in a pure culture is often quantified by performing culturable plate counts or direct microscopic counts (see Chapter 10). This approach is also often used to estimate the cell mass in environmental samples. Other common approaches to estimating cell mass include measurement of the turbidity or the protein content in the culture being studied. Although these measurements are not suitable for environmental samples, they are easy, rapid and reproducible when used for measuring cell mass increases in pure culture.

11.2.3.1 Turbidity

A rapid estimate of cell mass can be obtained by measuring the turbidity of a bacterial suspension. Turbidity measurements are generally made using a colorimeter or spectrophotometer, both of which work by directing a light beam through the sample. The bacteria in the sample scatter the light beam and lower the intensity of the light coming through the suspension. At low bacterial densities, there is a direct linear relationship between the number of bacteria and the amount of light scattered; thus, as the number of bacteria increases, the turbidity of the suspension increases. At high bacterial densities, this relationship becomes nonlinear. Therefore, a standard curve must be constructed to determine the linear range for turbidity measurement for each organism being measured. In theory, a wide range of wavelengths can be used to measure turbidity, although the practical range is between 490 and 550 nm. The sensitivity of the measurement increases at lower wavelengths, but there are often yellow or brown pigments produced during growth that can interfere with turbidity measurements at lower wavelengths.

11.2.3.2 Protein

A variety of methods are available for measurement of protein (Daniels *et al.*, 1994). Currently, the most commonly used assays are based on the Folin reaction (Lowry assay; Lowry *et al.*, 1951); the Coomassie Blue reaction (Bradford assay; Bradford, 1976); and the bicinchoninic acid−copper reaction (Smith *et al.*, 1985). Each of these assays quantifies the amount of protein present by measuring color produced by the reaction of the protein present with the assay reagents. For example, the Lowry assay is based on the development of a blue color resulting from the reaction of the Folin reagent with aromatic amino acids contained in the protein sample. The sample is measured in a spectrophotometer at 550 nm. Similarly, the Bradford assay measures the binding of Coomassie Brilliant Blue G dye to protein amino groups, which causes the development of a blue color that can be measured at 595 nm. The bicinchoninic assay is based on the fact that proteins react with Cu^{2+} to produce Cu^+, which then reacts with bicinchoninic acid to form a purple product that is measured at 562 nm.

The method chosen for a particular application depends on the types of samples and potential for interference by sample components. The basic procedure for protein analysis is first to lyse cells in order to release the protein. This can be done by placing an aliquot of cells into 1 N NaOH, and heating the suspension at 90°C for 10 minutes. It is a good practice to wash the cells prior to lysis to minimize the presence of materials that may interfere with the protein assay. Once the cells are lysed, the protein assay is performed and the resulting sample measured in a spectrophotometer to determine protein quantity.

11.2.4 Carbon Dioxide Evolution

In general, the CO_2 trapping methodologies described here have been used to measure aerobic mineralization. However, the methods could be adapted to allow for measurement of CO_2 evolved under anaerobic conditions as well.

11.2.4.1 CO₂ Trapping

An alkaline trap, usually composed of sodium hydroxide or another strongly basic solution, can be used to trap CO_2 produced during mineralization (Figure 11.10A). The alkaline solution traps carbon dioxide by transforming it into the bicarbonate (HCO_3^-) form:

$$CO_2 + OH^- \rightarrow HCO_3^- \qquad \text{(Eq. 11.2)}$$

The trapped carbonate can be detected by using titrimetric, gravimetric or conductimetric measurements. Perhaps the most common approach to quantifying trapped CO_2 is titration of the alkaline solution that was used to trap CO_2 with a standardized acid solution. The HCO_3^- that was formed in the trapping solution is first precipitated with barium chloride. Then the solution is titrated with hydrochloric acid in the presence of a pH-sensitive dye such as phenolphthalein to indicate the end point of the titration. The amount of CO_2 trapped is then calculated from the reduction of the original hydroxyl ion

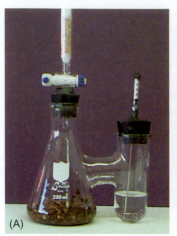

FIGURE 11.11 Liquid scintillation counter. Photo courtesy R.M. Maier.

FIGURE 11.10 (A) Biometer flask for trapping CO_2. At intervals over the incubation period, the CO_2 trapping solution is withdrawn from the sidearm, and then replaced with fresh alkaline solution. As the trapping solution is withdrawn, air is drawn into the flask to replenish the oxygen supply within the sealed flask. The replacement air is drawn through a CO_2 absorbing filter (ascarite). To assay respiration, the amount of CO_2 generated is determined by titration of the trapping solution, and then the cumulative CO_2 production over the incubation period can be calculated. (B) Flask with a test tube trap containing alkali. (C) Stripping chain for $^{14}CO_2$ collection. Photos courtesy R.M. Maier.

concentration in the trapping solution. Alternatively, a simple gravimetric method is to precipitate the carbon dioxide as barium carbonate by the addition of an excess of $BaCl_2$. The precipitate can be collected, dried and weighed. A second gravimetric approach is to trap carbon dioxide in granular soda lime and monitor the gain in mass. Conductimetric quantification of CO_2 trapping is based on the reduction of the number of hydroxyl ions (OH^-), which is directly related to the amount of CO_2 trapped, and can be quantified by a decrease in conductivity of the trapping solution.

11.2.4.2 $^{14}CO_2$ Trapping

The use of ^{14}C-radiolabeled substrates allows for very sensitive and specific measurements of $^{14}CO_2$ evolution. In this case the CO_2 evolved is specific for the substrate

added. This technique is often used to evaluate the feasibility and rates of biodegradation of organic contaminants. The generation of $^{14}CO_2$ offers confirmation that it is actually the organic contaminant in question that is being degraded. In this case, a biometer flask such as shown in Figure 11.10A can be used, and the alkali within the sidearm trap is assayed for radioactivity using liquid scintillation counting (Figure 11.11). Alternatively, a simple alkaline trap can be made by inserting a test tube containing alkali into a flask (Figure 11.10B). A slightly more complex system is the stripping chain shown in Figure 11.10C. Each flask is connected to a series of six traps. The first and fourth are empty to prevent backflow contamination. The second and third traps are filled with a general scintillation cocktail to remove any ^{14}C-labeled volatile organic compounds that may have been produced during biodegradation. The fifth and sixth traps are filled with a phenethylamine-based cocktail that selectively traps CO_2. Although this system is more complex than those shown in Figures 11.10A and B, it offers confirmation that the radioactivity assayed is actually $^{14}CO_2$.

11.3 CHOOSING THE APPROPRIATE ACTIVITY MEASUREMENT FOR ENVIRONMENTAL SAMPLES

Measuring the activity of a diverse population of microbes within an environmental sample is a very different prospect from measuring the microbial activity of a single isolate in the laboratory setting. As discussed in Section 11.2, the microbial activity of a pure culture in a defined medium can easily be determined by measuring cell number (culturable plate counts or direct counts) or biomass (turbidity or protein). However, these measurements are not always practical or even realistic for environmental samples. For example, protein analysis of an environmental sample reflects sources of biomass other than the microbial population, including plant debris or microscopic animals.

Although both culturable and direct counts are commonly used to enumerate bacteria at a particular location, they are less helpful for measuring microbial activity. Culturable plate counts underestimate bacterial numbers in an environmental sample because this process enumerates only the cells compatible with the growth medium and culture conditions (see Chapter 10). Further, although many bacteria are viable in their environmental setting, they are nonculturable in the artificial environment of the laboratory. It should be noted that despite the drawbacks of culturable counts for measurement of total activity within a sample, this technique is still useful for estimation of specific physiological groups, such as sulfate-reducing organisms or hydrocarbon-degrading organisms. Direct counts as a measure of total microbial activity are problematic because they include dormant and moribund cells. Also, it is often a small portion of the total population that is responsible for a given activity, and it is difficult to estimate the activity of this part of the population with direct counts. Finally, both culturable and direct counts fail to enumerate cells that adhere tightly to soil particles or other surfaces. As a result of these problems, the measurement of microbial activity in environmental samples has focused more on the quantification of metabolic activity, such as respiration or synthesis of cellular macromolecules, which provides a more direct reflection of the level of metabolic activity within a sample (Table 11.1). These types of measurements are explained in detail in the following sections of the chapter.

11.4 CARBON RESPIRATION

A primary activity of the microbial community in undisturbed environments is the utilization of naturally occurring organic materials. Heterotrophic activity is a measure of the extent to which microbial populations utilize these organic materials as a carbon and energy source. This process is called respiration, and involves the consumption of oxygen (under aerobic conditions) or an alternate terminal electron acceptor (under anaerobic conditions), with production of carbon dioxide as shown in Eq. 11.1. As discussed in Section 3.4, the carbon from a substrate will be partially evolved as CO_2 and partially utilized to build new cell material. For heterotrophic activity under aerobic conditions, one can use the general assumption that 50% of the carbon goes to CO_2 and 50% goes to cell mass. Under anaerobic conditions a smaller portion of the carbon is used to build cell mass, and a larger portion

goes to CO_2 and methane (see Section 3.4.2). Thus microbial activity within environmental samples can be monitored by measuring TEA consumption or CO_2 evolution. However, for both CO_2 and TEA, the relationship between the measured parameter and the amount of new cell material formed is specific to the organism, the particular substrate and environmental conditions that influence growth rate. Therefore, a conversion efficiency must be estimated to quantify bacterial production absolutely. Even if an absolute value for microbial activity cannot be determined, measurement of the flux of respiratory gases can provide a relative level of microbial activity.

11.4.1 Measurement of Respiratory Gases, CO_2 and O_2, in Laboratory and Field Studies

As described in Sections 11.2.2 and 11.2.4, there are several approaches to measuring CO_2 and in particular O_2 in pure culture. Most of these approaches can also be adapted for use with environmental samples. The presence of respiration gases can also be determined by gas chromatographic analysis, typically using a thermal conductivity detector. Chromatography requires that an air sample is collected in a gastight syringe, and then injected into a stream of inert gas, which carries the sample through a column that has been packed with selective material. The column packing material acts to separate the individual gaseous components of the sample, and allows simultaneous monitoring of O_2 and CO_2 within the same sample. Carbon dioxide in atmospheric samples can also be monitored using an infrared gas analyzer, which detects CO_2 by the absorption of a specific electromagnetic wavelength. Some specific approaches for measurement of respiration gases in terrestrial and aquatic environments are detailed in the following sections.

11.4.1.1 Terrestrial Environments

Microcosm Studies

Microbial activity within soil or subsurface materials can be measured under controlled conditions in a laboratory, or can be measured *in situ* (in the field). In a laboratory study, a sample of the porous medium is typically incubated in a sealed, airtight enclosure, usually referred to as a microcosm. Some examples of microcosms are shown in Figure 11.12. This approach allows one to design complex experiments that can be performed and replicated under relatively controlled conditions. For example, microcosm studies performed in the laboratory allow for the standardization of environmental parameters such as soil moisture content and temperature, for all samples. In the establishment of laboratory microcosms, a common practice is to sieve the soil through a 2-mm sieve to homogenize the sample and to remove large stones or plant debris. The soil

TABLE 11.1 Several Common Methods Used for the Measurement of Microbial Activity in Environmental Samples[a]

Test	Basis of Test	Application	Advantages	Disadvantages
Measurement of respiration gases	Measurement of oxygen utilization or CO_2 production in an environmental sample. The flux of respiration gases provides an indication of overall metabolic activity, which can reflect the level of microbial activity.	Basal respiration measurements reflect microbial metabolism of organic substrates present in the environment. However, respiration by other components, such as plant roots in soil or algae in water, will also be included, depending on the environment.	Field chambers can be built and installed *in situ* to monitor flux of respiration gases in relatively undisturbed samples. The addition of an organic substrate to the sample can be used to indicate the level of potential microbial activity.	Incubation of samples in a closed chamber (microcosms) in order to monitor the flux of respiration gases can create an artificial environment. CO_2 production can be underestimated due to pH-dependent retention of inorganic carbon as bicarbonate.
Respiration of radiolabeled substrates	Metabolism of a radiolabeled substrate is monitored by measuring the evolution of labeled CO_2.	This method is used to determine the potential for metabolism of a foreign substrate, such as an organic pollutant. Also, overall heterotrophic potential can be estimated by determining the turnover of organic substrates that occur naturally in the environment.	The use of radiolabeling results in high sensitivity of the measurement, thus short incubation periods can be used. The use of specific radiolabeled compounds shows the potential for degradation of that specific substrate.	The concentration of substrate added is often greater than the concentration present in the environment, thus the rate of metabolism may be overestimated unless corrective procedures are used.
Microelectrodes	Probes with tips <20 μm in diameter can be inserted into environmental samples to provide a continuous monitoring of activity.	Microelectrodes have been designed to measure specific respiratory activities, including oxygen utilization and nitrate respiration.	Can monitor real-time activity at critical interfaces in biological systems. Especially useful to monitor the interdependence of aerobic and anaerobic processes in the environment.	The instrumentation is delicate and relatively expensive. The small probe size can result in large variations in observed measurements when used on spatially heterogeneous environments such as soil.
Incorporation of radiolabeled thymidine into cellular DNA	Microorganisms will scavenge DNA precursors, such as thymidine, from their environment. By radiolabeling thymidine, the rate of incorporation into DNA can be measured.	The rate of DNA synthesis provides a reasonable estimate of the rate of cell division, thus providing an estimate of microbial biomass production.	When using a short incubation time, thymidine incorporation is thought to measure bacterial DNA production because the rate of bacterial incorporation of thymidine is thought to be much faster than for other organisms which may be present in environmental samples.	Not all bacteria will incorporate exogenously supplied thymidine into DNA. Also, estimating microbial activity requires the development of a conversion factor relating thymidine incorporation to biomass production.
Adenylate energy charge (AEC)	AEC is a weighted ratio of ATP to total adenylates. ATP is quantified using a luciferin–luciferase substrate–enzyme system.	AEC values reflect a continuum between an active microbial community (AEC > 0.8), and a community with a high proportion of dead or moribund cells (AEC <0.4).	AEC can establish the presence of a metabolically active community without the need to incubate the sample or add a surrogate substrate.	AEC is not necessarily a direct measure of microbial activity because adenylates are present in all living organisms and can also be released by decaying cells.
Dehydrogenase assay	This assay measures the rate of oxidation–reduction reactions (electron transport chain activity) by monitoring the reduction of	The rate of reduction of tetrazolium salts reflects the overall activity by all respiring microorganisms.	Actively respiring microorganisms can be visualized microscopically by the deposition of pigmented tetrazolium salts reduction products within	The amount of tetrazolium salts reduced depends on many factors including sample incubation conditions. Direct comparison of activity

(Continued)

TABLE 11.1 (*Continued*)

Test	Basis of Test	Application	Advantages	Disadvantages
	a tetrazolium salt by actively respiring microorganisms.		the cell. The assay has a high sensitivity and can be used to measure activity in low productivity environments.	between samples can be made only when the assay is performed under identical conditions.
Hydrolysis of fluorescein diacetate	Hydrolysis of fluorescein diacetate is performed by a variety of enzymes including esterases, proteases and lipases.	The assay measures enzymatic activity, thus providing an estimate of total microbial activity in an environmental sample.	Fluorescein diacetate hydrolysis produces a highly fluorescent product, fluorescein, which is easily detected.	Enzymes involved in hydrolysis can be intracellular or extracellular.
Stable isotope probing	An isotopically labeled, e.g., ^{13}C, substrate is added to a sample. Analysis of ^{13}C distribution into DNA (DNA-SIP) or cell lipids (PLFA-SIP) provides information about the active populations in the community.	This technique identifies the populations within a community that are actively metabolizing the labeled substrate added.	Links phylogeny with function even if the functional populations cannot be cultured. Can be used to measure carbon flow through trophic levels in the environment.	PLFA-SIP is more sensitive in terms of numbers but gives only very general taxonomic information. DNA-SIP is less sensitive in terms of number but gives precise phylogenetic information. SIP is technically difficult and not yet standardized.
Microarray[b]	A matrix of immobilized nucleic acid probes is used to query the nucleic acid makeup in a sample.	Allows high throughput screening of gene expression in a sample.	Allows direct and rapid inquiry of how microorganisms respond to and interact with their environment.	Cost prohibitive in many cases although technological advances are decreasing costs associated with microarray fabrication. Data interpretation can also be challenging.
Proteomics[b]	Analysis of the proteins expressed by a microbial community under a prescribed set of environmental conditions.	Allows identification of proteins that are differentially expressed and thus likely important in microbial responses to environmental conditions.	Provides the best information available describing the microbial community response to a stimulus.	Analysis is expensive, difficult and time consuming. The technique has not yet been fully developed for use with complex, environmental samples.

[a]*This table is meant to highlight advantages and disadvantages of these techniques, and should not be considered as comprehensive.*
[b]*Microarrays and proteomics are discussed in Chapters 13 and 21, respectively.*

can then be dried (if necessary) and adjusted to a set moisture content. However, the disadvantage of preparing the soil in this manner is that soil structural features are altered, such as soil aggregation and pore size distribution, and this may have an effect on microbial activity.

The use of sealed microcosms allows for the measurement of microbial activity within the sample as determined by the flux of CO_2 and/or O_2 within the headspace atmosphere. Headspace gas samples can be withdrawn using a gastight syringe, and the CO_2 or O_2 concentrations can be determined using gas chromatography. An alternative method is to trap the CO_2 produced in a basic solution using a trap such as that shown in Figure 11.10. One problem that can arise during long-term incubation periods is the depletion of oxygen within the headspace of sealed, airtight flasks. To address this problem, flow-through incubation systems have been devised to allow headspace gases to be replenished, while still allowing quantification of microbial respiration. In this case, CO_2-free air is used as the flow-through gas, and any CO_2 in the air exiting the flask is a direct result of microbial activity. The CO_2 in the effluent air can be trapped in alkali and quantified or fed directly into a CO_2 detection device, such as an infrared detector (Brooks and Paul, 1987). A second problem in some alkaline soils is that CO_2 evolution may be underestimated, because the equilibrium shown in Eq. 11.2 is dependent on pH. In soils with a pH above 6.5, a significant portion of the CO_2

(A)

(B)

FIGURE 11.12 (A) Examples of microcosms used in studies under aerobic conditions. As shown, a microcosm can be any shape or size depending on the size of the sample and the headspace required. In some cases much larger microcosms may be used so that several samples can be removed and analyzed during the experiment. In other cases, volatile substrates are added to the microcosm. When volatile substrates are used it is important that the cap seal be airtight, and also that the cap seal does not absorb the volatile material. To address these problems a Teflon liner can be placed into the cap. (B) Serum vials or bottles are often used as microcosms for studies under anaerobic conditions. The vials or bottles are sealed with a septum and a crimp-top cap. To impose anaerobic conditions, a needle is inserted into the septum in the crimp top and the flask is flushed exhaustively with an inert gas (e.g., N_2) to drive out the oxygen. Photos courtesy (A) R.M. Maier and (B) A. Somenahally.

produced by mineralization will be retained in the soil in the bicarbonate (HCO_3^-) form. Flow-through systems, which provide continuous replenishment of the atmosphere within the incubation flasks, reduce the retention of CO_2 in the soil by maintaining a low concentration of CO_2 in the atmosphere of the microcosm.

Respiration measurements to determine microbial activity in soil should be corrected for possible nonbiological sources of gas exchange. The nonbiological, spontaneous generation of CO_2 can occur in certain soils, especially when moisture is added to a dry soil, or in soils containing free calcium carbonate. This is a particular problem with arid soils that are high in calcium carbonate ($CaCO_3$). Nonbiological oxygen utilization can occur in the spontaneous oxidation of certain chemical elements,

such as iron or copper. The separation of chemical reactions from biological production requires the use of sterile controls. The most common approach to sterilization of soil is to autoclave the soil three times on three consecutive days, or to treat the soil with a chemical agent such as mercuric chloride ($HgCl_2$) to inhibit microbial activity (Rozycki and Bartha, 1981).

Field Studies

Alternatively, field studies can be performed by placing a field chamber over a plot of surface soil and using this set-up to make *in situ* measurements (Figure 11.13). In this way the soil structure is not altered by the sampling procedure and field respiration rates of the indigenous population are more reliably determined. In some cases it may be desirable to measure respiration gases below the soil surface. In this case, a gas sampling probe can be inserted beneath the soil surface and used to withdraw gas samples at discrete depths in the soil profile. However, care must be taken to avoid the creation of preferential flow paths, which could result in the introduction of aboveground atmosphere into the subsurface.

Samples of atmosphere from within the field chamber can be withdrawn using a gastight syringe and then injected into a gas chromatograph for CO_2 and O_2 determination (Figure 11.13A). Alternatively, gas from the chamber can be analyzed in real time in the field. This has most commonly been done for CO_2 with instruments based upon infrared absorption, with at least one company offering a system which can be set up to automatically take in-field measurements (Figure 11.13B; LI-COR; Lincoln, NE, U.S.A.). In addition, newer technologies such as photoacoustic- and Fourier transform infrared spectroscopy (FTIR)-based methods are now being tested which promise the ability not only to measure CO_2 but also simultaneously to measure a large number of other gases in the field (Griffith *et al.*, 2012; Iqbal *et al.*, 2013).

There are several difficulties inherent in field studies. For example, because environmental parameters, such as soil moisture and temperature, cannot be readily controlled, field respiration measurements have more variability than laboratory microcosm measurements. Temperature variations can be especially problematic if the project is so large that it requires several hours (or even days) to collect samples from all of the sites. It is also difficult to perform control studies in the field because sterilization of a soil plot is difficult. Finally, in many environments the respiration of plant roots may contribute significantly to the flux of respiration gases. Because the contributions of heterotrophic bacteria and plant roots cannot be separated in field measurements, CO_2 production has been viewed as a measure of the gross soil metabolic activity. Additional discussion of these issues and suggested gas sampling protocols are

FIGURE 11.13   Examples of field chambers that allow measurement of respiration gases. (A) Gas in the chamber is sampled via a syringe inserted into a septum. (B) A commercial chamber (LI-COR®, Lincoln, NE, U.S.A.) is inserted over a PVC collar (green) in the soil and used to sample and test the gas in the field (via an attached instrument). Figure 11.13A from Pepper *et al.* (2006). Figure 11.13B courtesy J.O. Storlien.

available from the Greenhouse gas Reduction through Agricultural Carbon Enhancement network (GRACEnet) (USDA-ARS, 2013).

11.4.1.2 Aquatic Environments

Oxygen depletion is a common means of determining microbial activity in aquatic environments, in part because of the ease with which dissolved oxygen can be determined. However, many aquatic systems are oligotrophic, especially marine environments, and thus the number and activity of heterotrophic bacteria are limited. Therefore, water samples must be incubated in sealed microcosms for extended period of 12 hours or even longer, in order to detect significant oxygen depletion. Incubation of water samples for the extended periods required to record oxygen demand may create an artificial environment because of alteration of several time-dependent variables in the system. These include: changes in the quantity and the form of organic carbon substrate;

changes in grazing pressure resulting from the presence of bactivorous zooplankton; and the exclusion of larger forms of grazing animals from the microcosm (Pomeroy *et al.*, 1994). With time, these effects can create deviations from the initial respiratory rate. To avoid this type of error, incubation times should be kept as short as possible.

Another factor that must be considered in aquatic systems is the close association between photosynthetic and heterotrophic microorganisms. In order to prevent photosynthetic generation of oxygen, which would interfere with the measurement of heterotrophic activity using oxygen depletion, incubations should be performed in the dark. However, in this case the oxygen demand measured will include both heterotrophic activity and oxygen utilization in the dark respiration cycle of photosynthetic organisms. Because of these inherent difficulties in the measurement of respiration gases in aquatic systems, a better choice of activity measurement is usually one of the tracer methods described in the following sections. These techniques have the advantage of requiring much shorter incubation periods and are more specific for heterotrophic bacteria.

11.4.2 The Application of Respiration Measurements in Environmental Microbiology

11.4.2.1 Basal Rate of Microbial Activity in Soil Samples

Concern about the transport of organic pollutants into subsurface environments and recognition of the potential role of microorganisms in the remediation of pollutants in the subsurface have led to an exploration of the microbial life in subsurface environments. In one study described by Kieft *et al.* (1995), the existence of microorganisms was investigated in sediments ranging in depth from 173.3 to 196.8 m below the surface at a site in south−central Washington State. To determine the existence and distribution of microorganisms, a battery of tests was performed on each subsurface sample, including acridine orange direct counts (AODCs), the mineralization of glucose and basal respiration. Basal respiration was measured as the biological CO_2 production in samples that had received no nutrient amendment. In a parallel experiment, the soil was amended with glucose to determine if nutrient addition was required to detect microbial activity. Basal respiration was determined by transferring 5.0 g of subsurface sediment into 70-ml vials or microcosms, which were then sealed with rubber septa. After 3, 7 and 14 days of incubation (22°C), a sample of headspace gas was withdrawn and analyzed for CO_2 content using gas chromatography. The possibility of nonbiological CO_2 production was investigated using poisoned controls in which $HgCl_2$ (250 μg/ml) was added to preclude microbial activity. Basal respiration in the deep subsurface samples ranged from < 0.001 to 0.664 μg CO_2/g dry soil/h. The highest rates of basal respiration were detected in subsurface samples with the highest total organic carbon content, which were fine-grained sediments originating from lake (lacustrine) deposits. Subsurface deposits originating from buried soil (paleosol) or river (fluvial) sediments were found to have much lower basal respiration. There was strong agreement in the pattern of microbial distribution as revealed by basal respiration, cell number determined by microscopic counts (AODCs) and glucose mineralization. It should be noted that basal respiration is not an *in situ* measurement, and it should be expected that the results obtained will differ from *in situ* respiration rates, which are affected by the physical and environmental parameters of the particular site.

Basal respiration has also been used as an indicator of soil "health" or condition. Dulohery *et al.* (1996) used large (0.5 by 1.0 m) field chambers to determine CO_2 production in forest soils that had been affected by logging activity. Tree harvesting involves the use of heavy equipment, which can alter forest soils in many ways, including removal of topsoil and compaction of the soil profile. A comparison of soil respiration was made between pristine sites and sites where logging activities had taken place. Basal levels of CO_2 production ranged from 52 to 257 mg/m^2/h in undisturbed forest soil sites, depending on the season. CO_2 production was significantly reduced in sites affected by heavy equipment. For example, in areas that had been used as skid trails to haul logs from the site, soil respiration had declined an average of 34%. Attempts to improve soil conditions after logging using fertilizer application and soil tillage were evaluated.

11.4.2.2 Substrate-Induced Microbial Activity

Microbial respiration has been used to evaluate the response of microbial populations to the introduction of organic pollutants, such as petroleum hydrocarbons. In this case, the desired response is that the hydrocarbons be used as a carbon source by microorganisms. This response can be measured as an increase *in situ* respiration above the basal respiration level of the indigenous microbial community. Hinchee and Ong (1992) used subsurface probes to monitor microbial respiration at a number of sites where the subsurface was contaminated with a variety of different hydrocarbon mixtures, including jet fuel and crude oil. The soil probes (Figure 11.14) were installed to monitor atmospheric gases between 1 and 5 m below the ground surface, depending on the site. Probes were installed in areas already contaminated with hydrocarbons and, in order to estimate the basal respiration rate, in areas where there was no contamination. Prior to

measurement, air was injected through the probe into the subsurface for 24 hours to ensure that the subsurface contained an adequate amount of oxygen. The air injection was then turned off, and the microbial utilization of O_2 and production of CO_2 was monitored by withdrawing gas samples at intervals of 2 to 8 hours. Respiration rates were calculated on the basis of O_2 utilization (% O_2/h), and revealed that more than 50% of the available oxygen was utilized within 20 to 80 hours in the hydrocarbon-contaminated sites. This was a significant increase over the approximate 10% reduction in the available oxygen in a comparable but uncontaminated site.

As discussed in Section 11.4, conversion factors are needed to determine the rate of substrate utilization from oxygen uptake. In this study, an estimate of the rate of hydrocarbon biodegradation based on O_2 utilization was made using a hexane as reference hydrocarbon. The stoichiometric relationship for the oxidation of hexane indicated that 9.5 moles of O_2 were required for the complete mineralization of 1 mole of hexane. Based on this relationship, the rate of hexane degradation ranged from 0.4 to 13.0 mg hexane/kg soil/day for sites contaminated with jet fuel, and 3.6 to 19.0 mg hexane/kg soil/day for sites contaminated with crude oil. The results provide an approximation of the rate at which petroleum hydrocarbons are biodegraded within this particular subsurface site if adequate aeration is provided. Such an estimate is important in determining the rate at which the site can be remediated biologically, and whether this rate is high enough to contain the contaminant spill and prevent further migration of the hydrocarbons.

Hydrocarbon biodegradation rates based on CO_2 production were also determined but were less consistent than rates based on O_2 utilization. CO_2 production (percent CO_2/h) was unexpectedly low in contaminated sites

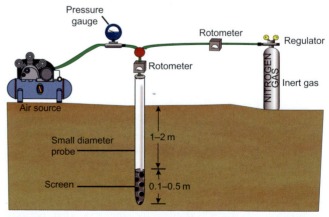

FIGURE 11.14 A subsurface probe used to remove gas samples from various depths within the soil profile. This setup was used to measure *in situ* respiration in a hydrocarbon-contaminated site. Adapted with permission from A.R. and Waste Management Association; Hinchee and Ong (1992).

where the soil pH was above 7.0, indicating the possibility of CO_2 retention in the bicarbonate form by the soil.

11.4.2.3 Microbial Biomass Determination

A measurement of microbial biomass in a soil is commonly performed to give an indication of soil condition and the potential for metabolic activity. This measurement can be made using two different approaches, both based on CO_2 evolution. One approach is the chloroform fumigation—incubation method (see Information Box 11.1; Figure 11.15). A soil sample is first fumigated with chloroform to kill the indigenous microorganisms. Upon reinoculation of the soil, the dead microbial biomass becomes available for microbial consumption. The amount of CO_2 produced by the consumption of this organic material is monitored and used to calculate the initial quantity of microbial biomass. In this procedure, a sample of fumigated soil and a sample of nonfumigated soil, which serves as a control to determine the basal respiration level, are incubated for 10 days in a sealed microcosm. The microorganisms responsible for the respiration will be the very small proportion of microbes that survived the fumigation treatment, or the fumigated soil will be inoculated with a starter culture from the nonfumigated soil. The amount of CO_2 released by mineralization is quantified using an alkaline trap or by GC analysis of headspace gases. The production of CO_2 in the nonfumigated soil represents the basal rate of mineralization, and CO_2 production in the fumigated soil represents primarily the mineralization of microbial biomass. Therefore, the amount of respiration, corrected for the basal rate, is a measure of the amount of microbial biomass present prior to the fumigation step. The amount of carbon held in microbial biomass is calculated as:

$$\text{Biomass C} = \frac{\text{Fc} - \text{Ufc}}{\text{Kc}} \qquad \text{(Eq. 11.3)}$$

where:

> biomass C = the amount of carbon trapped in microbial biomass
> Fc = CO_2 produced by the fumigated soil sample
> Ufc = CO_2 produced by the nonfumigated soil sample
> Kc = fraction of biomass C mineralized to CO_2

The value of Kc is of considerable importance because it expresses the proportion of total metabolized organic carbon that is mineralized. Literature values for Kc range from 0.41 to 0.45. These estimates are usually based on tracer experiments in which the mineralization of radiolabeled substrates, such as glucose or bacterial cells, by microorganisms isolated from the soil is determined in a liquid culture. The proportion of radiolabeled substrate that is mineralized to $^{14}CO_2$ or assimilated into microbial biomass can thus be determined.

Information Box 11.1   Measurement of Microbial Biomass in Soil — The Chloroform Fumigation Method

Step 1. Collect a representative soil sample (see Chapter 8).
Step 2. Perform a fumigation experiment as illustrated in Figure 11.15 as outlined:

To determine Fc:
1. Fumigate a portion of the soil sample with chloroform.
2. Inoculate.
3. Incubate for 10 days.
4. Determine CO_2 as an estimate of biomass mineralization = Fc.

To determine Ufc:
1. Incubate a second portion of the soil sample for 10 days.
2. Determine CO_2 as an estimate of basal mineralization levels = Ufc.

Step 3. Estimate Kc by performing a mineralization experiment using ^{14}C-labeled glucose and a soil inoculum (see Section 11.2.4).

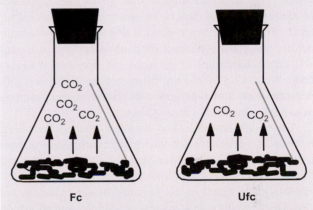

FIGURE 11.15 Microcosms for measurement of microbial biomass in soil. Fc, fumigated with chloroform; Ufc, not fumigated with chloroform.

An alternative method for the estimation of microbial biomass is the substrate-induced respiration method. This method estimates the amount of carbon held in living, nonresting heterotrophic cells by determining the initial respiration response when glucose, a readily metabolized organic carbon source, is added to the soil. The incubation time is limited to several hours, with samples of the mineralized CO_2 taken at regular intervals. The rate of glucose respiration in the soil is expected to increase with time, reflecting both the initial metabolism of the substrate and the fueling of cell division, which results in an increase in the number of metabolizing cells. Thus, the lowest and usually initial rate of glucose respiration is used as an indicator of the microbial biomass present in the soil sample. Calculation of microbial biomass is based on a correlation between the substrate-induced respiration method and the chloroform fumigation-incubation method, described earlier, or other methods for determining microbial biomass in soil. A commonly used relationship was provided by Anderson and Domsch (1978):

$$y = 40.04x + 0.37$$

where:

  y is biomass C (mg/100 g dry weight soil), and
  x is the respiration rate (ml CO_2/100 g dry weight sediment/h)

The main advantage of the substrate induction method, compared with the chloroform fumigation method, is that microbial biomass can be estimated in a much shorter time. To implement this method, preliminary studies are required to determine the appropriate glucose concentration to add to the soil. Increasing glucose concentrations saturate the initial uptake capacity of the microbial population, resulting in a plateau in glucose mineralization. The lowest glucose concentration that yields the maximum respiratory response must be independently determined for each soil type, and then applied to that soil in order to standardize the substrate induction method between different soil types. Examples of estimates of microbial biomass range from 12.8 to 203 mg biomass C in 100 g dry soil, depending on soil type and also on the method used to determine microbial biomass (Martens, 1987).

The substrate-induced respiration method has been modified to determine the individual contribution of bacteria and fungi to total heterotrophic activity within soil by the use of selective antibiotics. Glucose-induced respiration is determined in the presence of either streptomycin, which inhibits prokaryotes, or cycloheximide, which inhibits eukaryotes. Dominance of fungal glucose-induced respiration was evident in a soil sample from a semiarid region (Johnson *et al.*, 1996), and in the early stages of mineralization of plant litter (Beare *et al.*, 1990). In contrast, bacteria are thought to dominate in the rhizosphere and in subsurface sediments. However, it must be recognized that the glucose-responsive, antibiotic-sensitive microbial population may not represent the entire microbial population within an environment, and the use of antibiotics can only yield estimates of the relative contribution of fungi and bacteria to glucose-induced respiration (Johnson *et al.*, 1996). In addition, others have modified the substrate-induced respiration approach to detect microbial activity via heat generation instead of CO_2 production (Information Box 11.2; Figure 11.16).

11.4.2.4 Biological Oxygen Demand (BOD)

The 5-day BOD test was developed as a means of monitoring wastewater quality. The BOD test quantifies the oxygen required to metabolize dissolved organic carbon present after wastewater treatment. Thus, the BOD test can provide an indication of the impact that wastewater treatment plant effluents may have on the receiving waters following discharge. This test has become an industry standard in terms of evaluating water quality and has an established protocol (APHA, AWWA, WEF, 2005). Water samples, usually 250 to 300 ml, are incubated in sealed bottles for 5 days at a constant temperature, $20 \pm 1°C$. The BOD is calculated from the difference in dissolved oxygen concentration measured at the beginning and at the completion of the incubation period. The emphasis of the BOD test is on the determination of the oxygen demand created by the presence of dissolved organic material. Hence, it may be necessary to add inorganic nutrients or even a "seed" solution of heterotrophic bacteria to ensure that the dissolved organic material is in fact degraded.

The oxygen demand within the sediment of freshwater and estuarine environments can have a major impact on oxygen levels in surface waters, especially in shallow lakes and in rivers receiving organic-based waste material. The oxygen demand is created by microbial activity, as well as invertebrate respiration and nonbiological oxidation. Dissolved oxygen standards for the protection of fish and the aquatic ecosystem as a whole have been instituted for lakes and rivers affected by organic waste generated by human activity. Thus, the implementation of water quality standards necessitates quantification of sediment oxygen demand. Bowman and Delfino (1980) described an experimental apparatus for the determination of sediment oxygen demand in which water circulates over a layer of sediment in a sealed container (Figure 11.17). The closed loop of circulating water passes by a dissolved-oxygen probe, which continuously monitors dissolved oxygen levels.

11.4.2.5 Micro-level Measurements of Microbial Activity

The development of microelectrodes has allowed the *in situ* measurement of respiratory gases (and other environmental parameters) at the micro-level in biofilms and environmental samples (Figure 11.18). Mixed community microbial mats, or biofilms, are formed on many different surfaces as bacteria attach to and subsequently proliferate into a densely packed film (see Section 6.2.4). Biofilms can be found on almost any surface imaginable, from rock surfaces in streambeds to filter systems used in wastewater treatment to the surfaces of our teeth. The flux of respiratory gases within and around biofilms has revealed their importance in terms of organic carbon and nutrient cycling.

Use of microelectrodes to measure different respiration gases within biofilms has allowed us to better understand how microorganisms function within biofilms. For example, depth-dependent activity in a microbial mat was studied by Nielsen *et al.* (1990) using a combined O_2 and N_2O

Information Box 11.2   Measurement of Microbial Activity using Infrared Thermography (IRT)

Historically, scientists have primarily measured the substrates and products listed in Eq. 11.1 (e.g., disappearance of substrate and electron acceptors and production of cell mass and CO_2) as indicators of microbial activity. However, depending upon the process and microorganisms involved, there may be other potential indicators of microbial activity, such as temperature. Microbial generation of heat occurs as a by-product of microbial metabolism, and is perhaps best known for its role in the composting process. Advances in temperature detection technology have enabled high-resolution detection of slight changes in temperature with potential application for determining microbial activity. For example, Kluge *et al.* (2013) used infrared thermography (IRT) to monitor the temperature in soil microcosms amended with glucose ($\approx 1\%$ w/w). Within 18 h, there was a significant increase in temperatures of the glucose-amended microcosms as compared to unamended controls, with the temperature increase peaking at 40 h (Fig. 11.16). By 70 h, the temperature of the amended microcosms had decreased to that of the unamended microcosms. The IRT results mirrored CO_2 production (i.e., respiration) in the microcosms, presumably corresponding to metabolism of the added glucose. Potential advantages of using IRT for measuring microbial activity include: (1) it is a non-destructive technique; (2) measurements are easy to conduct; and (3) it allows observation of spatial patterns in activity.

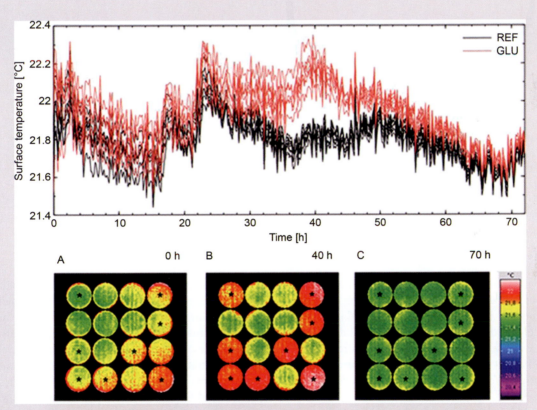

FIGURE 11.16   Surface temperature of 16 soil microcosms 0, 40 and 70 h after either being amended with glucose (indicated by "*") or left unamended (controls). From Kluge *et al.* (2013).

---

microelectrode with a sensor tip that was 20 μm in diameter (Figure 11.18a). In contrast, the biofilms being studied were several centimeters thick. The microelectrode was lowered into the biofilm at intervals as small as 50 μm to record specific respiration activity. Results revealed that oxygen was depleted by microbial activity in the surface layers of the mat, and that beneath the surface, conditions were anoxic. This allowed the development of denitrifying bacterial populations in the interior of the biofilm. These bacteria can respire under anaerobic conditions using NO_3^- as a terminal electron acceptor in place of O_2. Anaerobic respiration of NO_3^-, referred to as denitrification, was estimated using a N_2O microelectrode and acetylene gas. The addition of acetylene gas blocks the reduction of NO_3^- during denitrification so that the intermediate N_2O is formed (see Section 11.2.2.2). Thus, the rate of N_2O production provides an estimate of the rate of anaerobic respiration of NO_3^- (i.e., μmol NO_3^- utilized/cm^3/h).

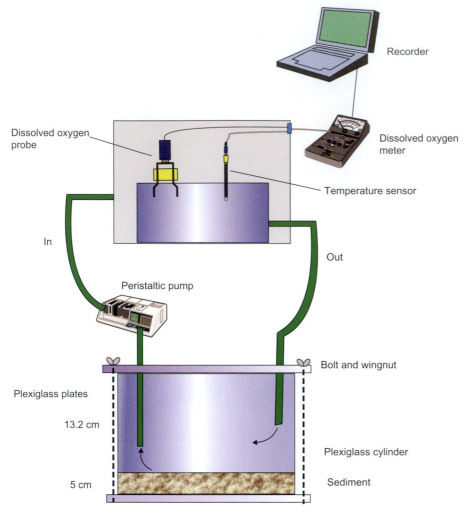

FIGURE 11.17 Apparatus used to determine sediment oxygen demand. Adapted with permission from Bowman and Delfino (1980).

Schramm *et al.* (1996) also revealed the anaerobic nature of microbial activity in the interior of a microbial biofilm using an N_2O microelectrode. In this case, the N_2O microelectrode was coated with immobilized bacteria, which were selected for their ability to reduce NO_3^- metabolically, but only as far as N_2O. Insertion of this microelectrode into the interior of a biofilm recorded the respiration of NO_3^-, a process which could occur only in the absence of oxygen. Using these tools, it has been possible to better understand how the structure of a biofilm affects its function. Specifically, the rapid utilization of oxygen by the abundant bacteria in the upper layers of the biofilm creates an anaerobic environment in the deeper layers that supports the anaerobic respiration of NO_3^-.

11.4.3 Tracer Studies to Determine Heterotrophic Potential

The use of ^{14}C-labeled carbon substrates was discussed in Section 11.2.4.2 as a sensitive way to measure biodegradation of a specific organic compound. This approach can also be used to determine heterotrophic potential in environmental samples. Many radiolabeled substrates are commercially available, including sugars such as glucose, organic acids such as lactic acid, amino acids and even many representative organic pollutants. Mineralization of these substrates can be quantified by measurement of the evolution of $^{14}CO_2$.

The addition of a ^{14}C-labeled substrate to an environmental sample reveals the presence of a degrading population in that sample, and also indicates the rate at which the substrate can be mineralized. However, the rate of substrate mineralization is related to the amount of added substrate up to a saturation limit (see Figure 3.7). Thus, depending on the level of substrate added, this measure may represent basal metabolism, or may represent the potential for a microbial response to higher levels of added substrate. So the dilemma associated with the use of radiolabeled substrates is that although they can provide a very sensitive measurement of substrate utilization, the rate of utilization is usually proportional to the amount of labeled substrate added to the environmental sample. This is especially true for compounds such as

(A)

Epoxy

Ag/AgCl anode

Electrolyte

Soda-lime glass

8533 glass containing
platinum wire

8533 glass tip

1 cm

(B)

FIGURE 11.18   Microelectrodes used to study biofilms and environmental microsites. (A) On the left is the microsensor used for analysis of O_2 and N_2O. There are three cathodes within the outer casing; one is shown behind the plane of the two others. On the right, the tip of the microsensor is enlarged 250 times. (B) Use of microelectrodes to measure porewater constituents in salt marsh sediment. (A) Adapted with permission from Revsbech and Jørgensen (1986). (B) Courtesy B. Sundby.

glucose, which are very rapidly metabolized by microbial populations. If the fundamental concern of the environmental microbiologist is to determine the rate at which naturally occurring organic substrates are utilized, care must be taken not to overestimate basal levels of activity by adding levels of a carbon source that exceed those already present in the system.

To address this problem, a kinetics-based analysis of labeled substrate mineralization has been developed to calculate the indigenous rate of glucose mineralization in environmental samples. A saturation kinetics model was adapted so that the stimulated rates of substrate mineralization can be extrapolated back to the rate at which a substrate is mineralized in an undisturbed sample. This analysis is referred to as the determination of heterotrophic potential (Wright and Burnison, 1979; Ladd *et al.*, 1979). When a radiolabeled substrate, such as ^{14}C-labeled glucose, is added to a soil or water sample, the rate at which $^{14}CO_2$ is evolved is monitored using an alkaline trapping technique. In the case of a water sample, the amount of radiolabeled substrate assimilated into biomass can be monitored by filtering the bacterial cells from the water sample, and then measuring the radioactivity incorporated into the biomass. Combining the mineralization with the assimilation data provides a measure of the total amount of substrate utilized by the cell. If a single concentration of ^{14}C-glucose is considered, then the rate of uptake by a microbial community can be expressed as:

$$v = \frac{f(S_n + A)}{t} \qquad \text{(Eq. 11.4)}$$

where:

v = the rate of uptake (or respiration) by the microorganisms (mass/volume/time)

t = the incubation time

f = the fraction of labeled substrate taken up (or respired) in time t

A = the amount of added substrate (mass/volume)

S_n = the naturally occurring concentration of the substrate (mass/volume)

If S_n is known and A is added such that $A \ll S_n$, then the assumption can be made that the natural rate of uptake is not significantly altered by the presence of A. Using a short incubation time, the fraction of labeled substrate taken up (or respired) can be determined, and can be expressed as:

$$T_n = \frac{S_n + A}{v_n} = \frac{t}{f} \qquad \text{(Eq. 11.5)}$$

where:

v_n is the natural rate of substrate uptake (or respiration), and

T_n is the substrate turnover time due to uptake by the natural population at the natural substrate concentration.

However, in most cases S_n is not known, and A will be added at a concentration much greater than S_n. A situation in which $A \gg S_n$ would result in the stimulation of substrate utilization rates above what would be considered the natural level of microbial activity by the indigenous population. Therefore, a single concentration tracer study would not be appropriate for determining T_n. However, T_n can be calculated by measuring the substrate utilization rate with added substrate, and then extrapolating back to the natural level of microbial activity. The basis of this extrapolation technique is Michaelis—Menten kinetics, which state that as the concentration of a substrate increases, the rate of activity of an enzyme will also increase until a plateau is reached. An analogy can be drawn between the transformation of a substrate by an enzyme and the uptake of an organic substrate by bacteria. Therefore, the rate of uptake of an organic substrate by bacteria increases as the concentration of the added substrate is increased, until the mechanism of substrate uptake has been saturated. Michaelis—Menten kinetics can be adapted for heterotrophic potential studies using the modified Lineweaver—Burk transformation, which states that:

$$\frac{S_n + A}{v_n} = \frac{1}{V_{max}} A + \frac{K + S_n}{V_{max}} \qquad \text{(Eq. 11.6)}$$

where:

V_{max} is the theoretical maximal rate of substrate uptake by the microbial population, and
K is a transport constant defined as the substrate concentration at which $v = \{1/2\} V_{max}$.

Combining Eqs. 11.5 and 11.6 will give:

$$\frac{t}{f} = \frac{1}{V_{max}} A + \frac{K + S_n}{V_{max}} \qquad \text{(Eq. 11.7)}$$

The value of (t/f) can be plotted over several levels of added substrate A, and fitted with a regression line (Figure 11.19). The line is extrapolated back to the x-axis. The slope of the line is $1/V_{max}$, the y-intercept is $(K + S_n)/V_{max}$, and the x-intercept is $K + S_n$.

This kinetic approach generates parameters that describe the *in situ* rate of heterotrophic activity. The natural turnover time, T_n, can be calculated when $A = 0$ using Eqs. 11.4 and 11.6:

$$\frac{t}{f} = \frac{K + S_n}{V_{max}} = \frac{S_n}{v} \qquad \text{(Eq. 11.8)}$$

Other useful parameters include V_{max}, the theoretical maximum rate of substrate uptake, which reflects the abundance and activity of the microbial population. The quantity $(K + S_n)$ may be used as an estimate of the natural substrate concentration, S_n, assuming that K is very small ($K \ll S_n$), which is not necessarily true in all cases.

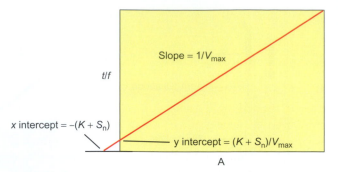

FIGURE 11.19   A Lineweaver—Burke plot.

These parameters can be used to compare the heterotrophic activity of different microbial populations in different environments. However, certain assumptions and limitations must be considered. First, the kinetic approach assumes no significant change in microbial number and no significant depletion of the added substrate during the incubation period. To meet these assumptions, short incubation periods are required. In highly productive environments, incubations of 1 hour may be adequate, but the incubation period must be increased in oligotrophic environments. The temperature of incubation will also affect the measured parameters. All samples should be incubated at the same temperature, preferably at *in situ* temperatures, so that the results are environmentally relevant.

It should also be noted that the heterotrophic potential approach measures only a single substrate and analyzes a diverse microbial population by calculating kinetic parameters as if they were uniform in the population. An assumption is made that the uptake of a substrate by a natural community follows saturation kinetics as if it were a single enzymatic reaction that is not influenced by the presence of other substrates in the environment (Van Es and Meyer-Reil, 1982).

Measurement of heterotrophic activity has been particularly important in understanding how carbon is cycled through the aquatic environment. Heterotrophic bacterial populations have been shown to dominate the utilization of dissolved organic carbon in aquatic environments. Determination of dissolved organic carbon utilization rates revealed that the production of new biomass through heterotrophic processes can equal the amount of biomass formed through primary production by photosynthetic microorganisms. Heterotrophic activity measurements have increased our understanding of the distribution of microbial communities in different ecological zones within the aquatic environment. It has been shown that bacterial abundance and the rate of carbon cycling are greatest in freshwater sediments, as compared with the planktonic, or free-floating, environments. From this, we now understand the ecological importance of sediment-associated bacterial communities in carbon and energy cycling.

11.4.4 Anaerobic Respiration as an Indicator of Microbial Activity

As discussed earlier, soil respiration under aerobic conditions can be monitored using CO_2 production or O_2 utilization as an indicator of carbon utilization. However, many environments are completely anaerobic or contain anaerobic niches, and there are situations in which monitoring the activity of anaerobic respiration is critical. For example, whereas anaerobic environments can support the biotransformation of organic pollutants, in some cases this results in the formation of more toxic metabolites (e.g., the production of vinyl chloride from TCE; Figure 17.2). Anaerobic respiratory pathways include the use of NO_3^-, Fe^{3+} (and other metals), SO_4^- and CO_2 as terminal electron acceptors. For some of these terminal electron acceptors, a gaseous intermediate or end product is formed that can be used as a measure of anaerobic respiration. For example, nitrous oxide (N_2O) is an intermediate in the bacterial reduction of NO_3^- to N_2 (denitrification). Acetylene gas is used to inhibit the complete denitrification of NO_3^-, resulting in the accumulation of N_2O gas (see Section 11.2.2.2). Production of N_2O in samples treated with acetylene gas is proportional to the rate of denitrification. In terms of the soil condition, the presence of N_2O gas suggests a reducing environment in which the indigenous bacteria can utilize NO_3^- as an alternative electron acceptor. The production of N_2O also indicates that nitrification processes at some point resulted in an adequate amount of NO_3^-.

The reduction of CO_2 during anaerobic respiration also produces methane (CH_4) (see Section 3.4). The utilization of CO_2 as a terminal electron acceptor is limited to a group of archaea called the methanogens, and this process requires a strongly reduced environment. The carbon substrates compatible with methanogenesis are simple one- or two-carbon compounds that are produced as a result of the activities of several different groups of microorganisms, including fermentative and acidogenic bacteria. Despite the complexity of the microbial interactions leading to methane production, methanogenic environments are common in soils and are also well characterized for use in the treatment of large volumes of organic waste. Methane production can be a problem in some instances, such as within municipal landfill sites. As landfills age, they often develop highly productive methanogenic populations, and the release of copious amount of CH_4 can create a fire hazard. The most accurate means of determining N_2O and CH_4 is by gas chromatography, although other methods such as photoacoustic-based gas analyzers are also available (see Section 11.4.1.1). Microelectrodes specific for N_2O have also been used to investigate the location of denitrifying activity in inhibited samples, but they are generally less sensitive than gas chromatography (Nielsen *et al.*, 1990).

11.5 INCORPORATION OF RADIOLABELED TRACERS INTO CELLULAR MACROMOLECULES

Quantification of cellular constituents, such as protein or nucleic acids, can be used to monitor the increase in biomass of a microbial population. This approach is often used to monitor the growth of a pure culture on defined media in the laboratory. In the natural environment, numerous sources of these constituents, including plant debris and soil animals, can contribute to the total protein or nucleic acids, and, therefore, these assays would not be specific to microorganisms. A way to make these assays more specific to microorganisms is to measure the incorporation of radiolabeled tracer molecules into cellular macromolecules. This technique is based on the fact that many species of heterotrophic microorganisms can scavenge preformed molecules such as nucleotides or amino acids from their environment, incorporating them directly in cellular constituents. Monitoring the rate at which radiolabeled nucleotides or amino acids are incorporated into essential cellular macromolecules can be used as a means of determining microbial activity. Examples of specific molecules used as tracers are the nucleoside thymidine, labeled with tritium (^3H) which is incorporated into DNA, and the amino acid leucine, labeled with either tritium or carbon-14 (^{14}C), which is incorporated into protein.

There are several advantages of measuring microbial activity using radiolabeled tracer molecules. This technique requires a short incubation period, which is more convenient for field studies and which reduces artifacts created by extended incubation of environmental samples. Further, the use of radiolabeled tracers increases the sensitivity of the measurements, allowing the quantification of very low levels of microbial activity such as those found in extreme environments such as Antarctica (Tibbles and Harris, 1996). There are also several potential sources of error in the tracer method including: (1) the fact that not all bacteria are capable of assimilating tracers into macromolecules; (2) the possible nonspecific incorporation of the label into cellular macromolecules other than the intended target; and (3) the extent of isotope dilution, which may vary both spatially and temporally within a sample site.

11.5.1 Incorporation of Thymidine into DNA

The rate at which a tracer is incorporated into a cellular macromolecule such as DNA provides an indication of the rate of formation of that macromolecule. The measured rate of formation will provide an accurate measure of microbial growth only if the formation rate is directly related to cell division. In a state of growth known as balanced growth, all cell constituents increase at the same rate. Therefore, the doubling time of any cellular macromolecule would equal the doubling time of the whole cell. Although balanced growth can be achieved under controlled laboratory conditions, sustained balanced growth is unlikely under environmental conditions. However, the synthesis of DNA and protein is strongly related to cell division, so that even when growth is unbalanced, their synthesis is expected to provide a reasonable reflection of cell division.

The scavenging of the thymidine nucleoside and its incorporation into DNA by heterotrophic bacteria have been reviewed by Azam and Fuhrman (1984), Moriarty (1986) and Robarts and Zohary (1993). Once transported across the cell membrane, the thymidine is converted into thymine monophosphate by the action of the enzyme thymidine kinase. Further phosphorylation results in incorporation into DNA as the thymine base. In very general terms, the procedure involves adding [3H]thymidine to a water sample containing planktonic bacteria or to a slurry prepared by mixing soil or sediment with water. The incubation time is usually limited to several hours, and the incorporation of labeled thymidine by the cells is terminated by the addition of a chemical inhibitor or by placing the sample in an ice bath. DNA is then extracted from the cells, and the radioactivity quantified to determine the amount of label that has been incorporated. One problem with this technique is that during the labeling and uptake procedure, nonspecific labeling of other macromolecules, such as proteins and RNA, can occur. In this case, purification of the DNA may be required to improve the accuracy of the activity measurement.

One aspect of this assay that makes it useful for measurement of bacterial heterotrophic activity is that the rate of thymidine incorporation into bacterial DNA is much higher than the rate for other organisms tested, such as algae, fungi or protozoa. Therefore, if the uptake study is limited to short incubation periods (several hours), it is generally believed that the incorporation of these precursors into cellular macromolecules will reflect the activity of growing heterotrophic bacteria. However, not all growing bacteria will incorporate exogenously supplied thymidine into DNA. Exceptions have been noted, particularly within the genus *Pseudomonas*, which may limit the usefulness of this technique in certain environments.

A further complication in the interpretation of the results of this assay is that there are both external and internal pools of thymidine or thymidine metabolites that can compete with the added [3H]-thymidine for incorporation into DNA. The internal pool is created by cells in a *de novo* process in which the nucleotides that form DNA are synthesized from cellular components. If this internal nucleotide pool is large, it may compete with scavenged labeled thymidine for incorporation into DNA. There can also be an external pool of thymidine available for uptake. This external pool is composed of extracellular thymidine probably released from dying organisms. This is especially true for sediments, where thymidine becomes sorbed to sediment particles. The existence of the external and internal pools of thymidine metabolites will result in dilution of the added [3H]thymidine, and cause underestimation of the rate of DNA synthesis. Isotope dilution, which refers to the size of the pool into which [3H]thymidine is diluted, can be estimated (Moriarty, 1986; Robarts and Zohary, 1993) and used to correct estimates of DNA synthesis. Further, the addition of sufficient quantities of exogenous labeled thymidine may cause competitive inhibition of the *de novo* synthesis route and thereby limit the size of the internal thymidine pool.

The thymidine incorporation rate can be converted into a measure of microbial activity using a conversion factor that relates the number of new cells formed to a given amount of thymidine incorporated (i.e., cells formed per mole thymidine incorporated). This conversion factor is derived on the basis of measurements, or best possible estimates, of the amount of DNA per cell, the thymidine content of a cell, and the extent of dilution of the labeled thymidine. An alternative way to determine the conversion factor is to relate the [3H]thymidine incorporation rate directly to cell division based on a separate and independent measure of the increase in bacterial numbers. Common estimates of this conversion factor range from 1.3 to 2.0×10^{18} cells/mole [3H]thymidine incorporated.

Based on thymidine incorporation, bacterial growth can range from less than 1×10^6 cells/g soil/day in an oligotrophic aquifer environment to greater than 1×10^9 cells/g soil/day in a marine sediment (Thorn and Ventullo, 1988). The rate of thymidine incorporation can also be used without conversion to estimate cell division. The thymidine incorporation rate provides a relative measure of microbial activity that can be used in controlled experiments to evaluate the impact of a specific factor, such as the toxicity of heavy metals to microbial activity in environmental samples (Diáz-Raviña *et al.*, 1994).

11.5.2 Incorporation of Leucine into Protein

The rate of leucine incorporation into cellular protein has also been used as a measure of microbial activity (Kirchman *et al.*, 1985; Chin-Leo and Kirchman, 1988). Studies have indicated that a majority of bacteria scavenge leucine from their environment and that most of the assimilated radioactive label will be incorporated into proteins. Further, the *de novo* synthesis of leucine is inhibited by sufficient quantities of exogenous leucine.

11.6 ADENYLATE ENERGY CHARGE

Adenosine triphosphate (ATP) is a compound synthesized by actively growing cells as a means of short-term energy storage and transfer. ATP captures metabolic energy in the form of high-energy phosphate bonds, and is transported to sites within the cell where energy is required to drive a biochemical reaction. Adenosine diphosphate (ADP) and adenosine monophosphate (AMP) are precursors of ATP, and together the three forms represent the cellular adenylates. ADP and AMP cycle between the sites where high-energy phosphate bonds are added to form ATP and sites where the phosphate bonds are broken to transfer energy to a metabolic process. The ATP content of the cell varies depending on its level of activity, with rapidly growing cells having a higher ATP content than stressed cells.

ATP has been used to measure microbial biomass in sediments and soils. A fairly constant relationship between ATP and cell biomass ($10-12$ mol ATP/g biomass C) has been measured for soils incubated under specific conditions (Jenkinson, 1988). However, a strong correlation between ATP content and microbial biomass may not hold for microbial populations under conditions of environmental stress, such as in dry or excessively wet soils (Inubushi et al., 1989; Rosacker and Kieft, 1990). Thus, the detection of total adenylates (ATP, ADP and AMP) may provide a better indicator of microbial biomass.

Measurement of the cellular adenylates can also be used to determine microbial activity. The relative abundance of ATP compared with its precursors, ADP and AMP, indicates how rapidly the highest energy state (ATP) is formed. The ATP/ADP/AMP ratio provides a biochemical basis for assessing the physiological and nutritional status of organisms. A measure of the adenylate energy charge (AEC) ratio is the weighted ratio of cellular adenylates:

$$AEC = \frac{ATP + \frac{1}{2}ADP}{ATP + ADP + AMP}$$

High AEC values (>0.8) reflect an active community, intermediate values (0.4 to 0.8) reflect cells in a resting state and low values (<0.4) reflect a high proportion of dead or moribund cells (Kieft and Rosacker, 1991).

Quantification of AEC as a measure of bacterial biomass has some advantages and disadvantages in comparison with approaches already discussed. A major benefit is that AEC provides a measure of microbial activity in environmental samples without the requirement to introduce a substrate and/or the need to incubate the sample for a period of time, as required in respiration or radiolabeled tracer activity measurements. It should be noted, however, that cellular adenylates are present in all living organisms, and its measurement is not selective for microbial populations. Rapid degradation of cellular adenylates from organic debris will minimize nonmicrobial sources. A preincubation of the soil under aerobic conditions may be necessary to ensure that cellular adenylates from sources such as plant root fragments are given an opportunity to be degraded (Sparling et al., 1985).

The quantification of ATP is based on the transfer of energy to a luciferin−luciferase substrate−enzyme system. This system was originally isolated from the abdomen of the common firefly. With the energy supplied by ATP, the enzyme luciferase acts on the substrate luciferin to produce radiant energy that can be detected and quantified and is proportional to the ATP present. Quantification of ATP from environmental samples requires an extraction procedure, followed by the concentration of cellular components into a buffer solution (Martens, 2001). The amount of ATP present is then determined directly using the luciferin−luciferase assay. ADP and AMP in the sample are then converted to ATP by enzymatic reactions and quantified.

AEC has been used to characterize microbial activity in both soil and subsurface environments. AEC analysis in subsurface samples from various depths revealed a wide range of values from relatively high values (0.76), which reflect an active population, to much lower values (0.23), which reflect dead or moribund microbial populations (Kieft and Rosacker, 1991). In surface soils, AEC analysis has indicated a range of microbial activity from inactive (0.5) (Rosacker and Kieft, 1990) to active (0.8) (Brookes et al., 1983; Ciardi et al., 1991). Drying or saturating the soil can cause a decrease in AEC, to below 0.4 in some soils (Inubushi et al., 1989; Ciardi et al., 1991). However, Rosacker and Kieft (1990) reported that air drying caused only a temporary decrease in AEC due to a transient increase in cellular AMP concentration. Increased AMP may have been the result of cellular catabolism of RNA, indicating that stressed cells may utilize endogenous metabolism of cellular macromolecules in order to survive. Such transient increases in AMP by stressed cells may affect the interpretation of AEC results.

The value of AEC measurement is that it can establish the presence of metabolically active microbial populations in environmental samples. For example, AEC has been used to characterize microbial populations in environments contaminated with pollutants, such as aviation fuel (Webster et al., 1992). AEC measurements in polluted environments can reveal the presence of microorganisms that are resistant to the adverse effects of the pollutants. These microorganisms may also be actively metabolizing the pollutants, thus removing them from the environment.

11.7 ENZYME ASSAYS

Enzymes are specialized proteins that combine with a specific substrate and act to catalyze a biochemical reaction. In the soil and sediment environments, enzymatic activity is essential for energy transformation and nutrient cycling reactions. For example, enzymes catalyze the hydrolysis of certain nitrogen-, phosphorus- or sulfur-containing organic compounds, releasing the ammonia, phosphate or sulfate constituents, which are then available for assimilation by other organisms. Also of note are enzymes that catalyze the hydrolysis of plant constituents, such as cellulose, starch and other polysaccharides, releasing the monomeric sugar units, such as glucose, which provide important energy sources for microorganisms. Several reviews have examined enzymatic reactions in the soil environment, and discussed their importance in terms of soil fertility and ecosystem function (Burns, 1978; Tabatabai, 1994b; Morra, 1997; Burns *et al.*, 2013).

One means of monitoring enzymatic reactions is to measure the conversion of a specific substrate into a product. An example is monitoring nitrate reductase activity through the disappearance of nitrate and the formation of elemental nitrogen. More commonly, enzymatic reactions are detected using a bioassay procedure that is specific for a particular class of enzymes. A bioassay utilizes a surrogate substrate that is transformed by a specific class of enzymes, producing a product that has specific properties for detection. An example is the conversion of *p*-nitrophenol phosphate by phosphatase enzymes, producing *p*-nitrophenol and phosphate. The indicator product, *p*-nitrophenol, can then be quantified spectrophotometrically. Alternatively, fluorescent compounds (e.g., 4-methylumbelliferone) can be used to label substrates for a variety of enzymes, including phosphatases. Use of a fluorimeter to detect the fluorescent reaction product allows greater detection sensitivity than can be achieved using spectrophotometric-based approaches (see Section 11.2.1.1).

The enzymatic activity within an environmental sample can originate from a variety of sources. Enzymatic activity can be directly associated with actively growing microorganisms, including bacteria, fungi and actinomycetes. These enzymes may be contained within the cell and are often located within the cell membrane. Enzymes may also be released from the cell, in which case they are called extracellular enzymes. Extracellular enzymes are released from actively growing cells to hydrolyze large polymers, such as the plant polymers, cellulose, hemicellulose and lignin, in order to facilitate their uptake for further metabolism. Alternatively, enzymes may be found outside cells as a result of the decay and disintegration of bacterial, animal or plant cells. Enzymes may be associated for a short time with moribund cells. These enzymes may also become stabilized on clay or humic particles within the soil structure and can remain viable for a period of time (Tabatabai, 1994b). Therefore, any bioassay of enzymatic activity in environmental samples will measure activity from all sources, and provide an indication of total enzymatic potential within that sample.

Characterizing and quantifying enzyme activity in environmental samples can reflect the health of the environment in terms of nutrient cycling and can also reflect soil fertility parameters, such as crop yield. However, there are problems associated with using enzyme assays to directly quantify microbial activity (Nannipieri *et al.*,

1990). First of all, enzymatic activity associated with actively growing microorganisms cannot be easily separated from the activity of extracellular enzymes stabilized in the soil environment or enzymes associated with decaying cells. Another problem is that enzyme assays often require the addition of a surrogate substrate, and as a consequence, the assay determines the potential enzymatic activity and not the actual level of activity in the sample. Enzyme assays are also specific for a particular substrate—enzyme combination, and may not reflect the overall activity of all types of microorganisms. The simultaneous determination of a large number of enzyme assays may be more representative of overall microbial activity, but this approach is more labor intensive.

Many different enzyme assays have been developed to detect either specific or general microbial activity in environmental samples. Some examples are given in Table 11.2. The first several assays in this table measure general microbial activity, and the latter ones are assays for specific activity. One commonly used assay for general microbial activity is the dehydrogenase assay.

11.7.1 Dehydrogenase Assay

Dehydrogenases are intracellular enzymes that catalyze oxidation—reduction reactions required for the respiration of organic compounds. Because dehydrogenases are inactive when outside the cell, this assay is considered a measure of microbial activity. Dehydrogenase reactions can be detected using a water soluble, almost colorless, tetrazolium salt, which when reduced forms a reddish-colored formazan product that can be detected in a variety of ways. Tetrazolium salts compete with other electron acceptors for the reducing power of the electron transport chain. Thus, measurements of the reduction of tetrazolium salts will reflect electron transport chain activity. Hence, this measurement is an index of the general level of activity of a large part of the microbial community, but it is not a direct measure of microbial growth in terms of the production of new biomass. It should also be recognized that all respiring organisms have an electron transport chain, including both aerobic and anaerobic microorganisms. Also included in the measurement are eukaryotic populations, such as algae and fungi. Therefore, the measurement of tetrazolium reduction provides an overall indication of electron transport chain activity.

A commonly utilized tetrazolium salt is 2-(*p*-iodophenyl)-3-(*p*-nitrophenyl)-5-phenyltetrazolium chloride (INT), which is transformed into an intensely colored, water-insoluble formazan (INT-formazan). INT has been used to measure microbial activity in surface waters (Posch *et al.*, 1997), soil and sediment samples (Trevors *et al.*, 1982; Songster-Alpin and Klotz, 1995), subsurface sediments (Beloin *et al.*, 1988) and biofilms (Blenkinsopp

TABLE 11.2 General and Specific Enzyme Assays that can be used to Measure Microbial Activity

Enzyme	Substrate	Description of Assay
Dehydrogenase	Triphenyltetrazolium	Dehydrogenases convert triphenyltetrazolium chloride to triphenylformazan; the triphenylformazan is extracted with methanol and quantitated spectrophotometrically.
Phosphatase	p-Nitrophenol phosphate	Phosphatases convert the p-nitrophenol phosphate to p-nitrophenol, which is extracted in aqueous solution and quantitated spectrophotometrically.
Protease	Gelatin	Gelatin hydrolysis, as an example of proteolytic activity, can be measured by the determination of residual protein.
Amylase	Starch	The amount of residual starch is quantitated spectrophotometrically by the intensity of the blue color resulting from its reaction with iodine.
Chitinase	Chitin	Production of reducing sugars is measured using anthrone reagent.
Cellulase	Cellulose Carboxymethylcellulose	Production of reducing sugars is measured using anthrone reagent. Cellulases alter the viscosity of carboxymethylcellulose, a quantity that can be measured.
Nitrogenase	Acetylene	Nitrogenase, besides reducing dinitrogen gas (N_2) to ammonia (NH_3), is also capable of reducing acetylene (C_2H_2) to ethylene (C_2H_4); the rate of formation of ethylene can be monitored using a gas chromatograph, and the rate of nitrogen fixation can be calculated using an appropriate conversion factor.
Nitrate reductase	Nitrate	Dissimilatory nitrate reductase can be assayed by the disappearance of nitrate or by measuring with a gas chromatograph the evolution of denitrification products, such as nitrogen gas and nitrous oxide, from samples; denitrification can be blocked at the nitrous oxide level by the addition of acetylene, permitting a simpler assay procedure.

From Atlas and Bartha (1993).

and Lock, 1990). Environmental samples are suspended in a solution containing INT and incubated for a matter of hours. Incubations between 1 and 12 hours are usually sufficient with shorter incubation times reducing the chance that the microbial community will undergo significant changes in activity. The production of INT-formazan can be detected by microscopic examination of the red INT-formazan deposits that form within the cells, or by quantifying total INT-formazan production. Microscopic examination gives an indication of the physiological status of the microbial population by determining the percentage of total cells that are actively respiring (Zimmermann *et al.*, 1978). Total cell number can then be determined using a counterstain, such as acridine orange, which stains all cells. The difference between the INT-formazan-containing cells and the total cell count is the proportion of the population that is metabolically active. This method is very sensitive, and can be used to detect electron transport chain activity even when low numbers of microorganisms are present or when samples are incubated at low temperatures (Trevors, 1984). For example, Posch *et al.* (1997) studied planktonic bacteria in a high mountain lake that was covered by a layer of ice at the time of sampling. The lake water was exposed to INT and incubated at the *in situ* temperature, 2°C, in order to determine the physiological status of the planktonic community. They obtained the surprising result that as much as 25% of planktonic bacteria in this oligotrophic lake were actively respiring. However, when using this method, it should be recognized that some of the cells considered inactive may in fact be respiring very slowly, or not utilizing INT as an electron acceptor (Posch *et al.*, 1997).

An alternative to microscopic examination of cells is quantification of the amount of INT-formazan produced. An environmental sample can be exposed to INT for a period of time, and then a solvent, such as methanol, is used to extract the INT-formazan from the cells. INT-formazan can be detected spectrophotometrically, and the total production calculated from a standard curve. In terms of quantifying total INT-formazan production, this measurement provides a relative index by which electron transport chain activities in different samples can be compared. However, INT has a low efficiency as an electron acceptor, and formazan production can be affected by numerous factors including: the concentration of INT used; incubation time; incubation temperature; pH of the sample; and whether the sample was incubated under aerobic or anaerobic conditions. Therefore, a direct comparison of activities of different samples can be made only if identical methods and experimental conditions are used (Trevors, 1984).

Several other forms of tetrazolium salts are available and provide certain advantages. For example, tetrazolium salt sodium 3′-{(1-[(phenylamino)-carbonyl]-3,4-

tetrazolium}-bis(4-methoxy-6-nitro) benzenesulfonic acid hydrate (XTT) has the advantage of producing an orange-colored, water-soluble XTT-formazan product (Roslev and King, 1993). Because XTT-formazan is water soluble, its production can be quantified spectrophotometrically without the need to use solvent extraction. Another tetrazolium salt is 5-cyano-2,3-ditolyl tetrazolium chloride (CTC). The reduced form of CTC is a water-insoluble, red-fluorescent CTC-formazan, which forms deposits within the cell (Rodriguez *et al.*, 1992). The fluorescence of CTC-formazan allows respiring cells to be highly visible when viewed under epifluorescence microscopy. This provides for easier enumeration of respiring cells in environmental samples. Using CTC, Winding *et al.* (1994) determined that actively respiring bacterial cells accounted for only 2 to 6% of the total population in an agricultural soil, and Schuale *et al.* (1993) determined that between 1 and 10% of bacteria

in samples of drinking water were actively respiring. The fluorescent properties of CTC-formazan also provide a means of determining the location of physiologically active cells within attached biofilm communities without having to disrupt the biofilm structure (Yu and McFeters, 1994a). Schuale *et al.* (1993) used CTC to examine the viability of thin biofilms formed by microorganisms present in drinking water (Figure 11.20). Their results revealed that between 5 and 35% of total sessile bacteria were actively respiring. Yu and McFeters (1994b) reported that CTC provided a sensitive measure of the efficacy of biocidal compounds in disinfecting biofilms formed by a potentially pathogenic waterborne bacterium. A disadvantage of the use of CTC is that it is redox sensitive and in a low redox environment can be reduced by an abiotic chemical reaction (Schuale *et al.*, 1993). Therefore, anaerobic pockets within a biofilm formation or within the soil environment may interfere with

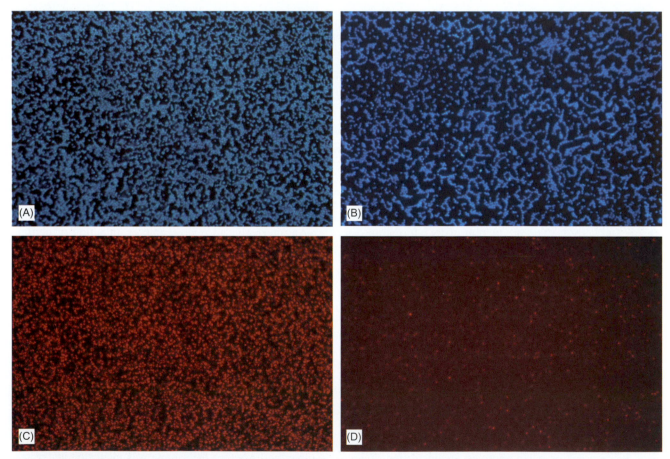

FIGURE 11.20 CTC-staining can be used to detect respiring cells in a biofilm. In this experiment, UV-sterilized, plastic, uncoated, microscope slides were placed in a sterile 1-liter beaker containing 750 ml of 10% R2A medium. The beaker was inoculated with 2 ml of a *Pseudomonas putida* strain 54G cell suspension, and after 24 hours, the slides were taken out and placed into mineral salts medium (starved cells) or R2A medium (unstarved cells). After a further period of incubation, each slide was stained with CTC and then counterstained with DAPI (see Section 9.4). The micrographs that are stained with CTC-DAPI (which stains all cells regardless of physiological state) show that comparable biofilms were formed on both slides (A, unstarved cells and B, starved cells). Examination of the slides stained only with CTC show that nearly the entire biofilm of unstarved cells is respiring (C), while only a few of the starved cells are respiring (D). The results of this experiment were enumerated, and showed that cells that were actively respiring comprised 76.8% of unstarved biofilm cells, but only 9.6% of starved biofilm cells. Data and photos courtesy G. Rodriguez and H. Ridgway, Orange County Water District, Fountain Valley, CA, U.S.A.

the detection of respiring cells, although these are issues for further research.

11.8 STABLE ISOTOPE PROBING

Stable isotope probing (SIP) is a relatively new technique that has great promise for identifying microorganisms involved in specific biogeochemical processes. SIP is similar to the radiolabel-based detection methods discussed previously in this chapter, but it involves the use of a stable (non-radioactive) isotope-labeled substrate to follow the fate of the substrate as it is metabolized by an intact microbial community (Kreuzer-Martin, 2007). The two most commonly used isotopes for SIP are ^{13}C and ^{15}N. As microorganisms metabolize the ^{13}C- or ^{15}N-labeled substrate, they will incorporate the isotope into their biomass. Extraction and analysis of biomarker molecules can then provide information about the fate of these isotopes within the microbial community, and thus insight into which microorganisms were involved in degrading the labeled substrate. In the first demonstration that SIP could work, Radajewski et al. (2000) grew the bacterium *Methylobacterium extorquens* in pure culture on either $^{12}CH_3OH$ ("light" methanol) or $^{13}CH_3OH$ ("heavy" methanol), extracted the DNA from both cultures and subjected the DNA to a density-gradient centrifugation. As can be seen in Figure 11.21, the light and heavy DNA clearly separates into distinct bands.

This technique can also be used to identify which populations within a microbial community are involved in the degradation of a specific compound. In theory, if you provide $^{13}CH_3OH$ to a microbial community instead of a pure culture only DNA from the populations within the community that utilized the $^{13}CH_3OH$ would be labeled with ^{13}C and appear in the heavy isotope band. The remaining populations that did not participate in utilization of the $^{13}CH_3OH$ would appear in the light isotope band. This means that one can separate or target "active" portions of the community from "inactive"

ones. In this example, the active portion of the community includes the populations that utilize $^{13}CH_3OH$, while the inactive portion of the community does not. However, in reality, there is the potential for cross-feeding to occur where metabolites produced by microorganisms, which are degrading the labeled compound, are taken up by other microorganisms that did not directly participate in conversion of the added substrate. This can be particularly challenging if working with recalcitrant substrates that require long incubation times for degradation to occur.

In SIP, the stable isotope most often used is ^{13}C, and the biomarker molecule most often used is DNA, although biomolecules such as 16S rRNA, lipids and proteins are also used (Figure 11.22). The choice of stable isotope and biomarker is important because each alternative has advantages and disadvantages (Neufeld et al., 2006).

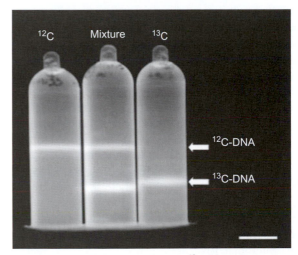

FIGURE 11.21 Centrifugal separation of ^{13}C-labeled from unlabeled DNA for stable isotope probing (SIP) experiments. The three centrifuge tubes were loaded with isotopically distinct DNA. The tube on the left contains only ^{12}C-DNA, the tube on the right contains only ^{13}C-DNA, and the middle tube contains a mixture of ^{12}C-DNA and ^{13}C-DNA. The pure fractions and a mixture of the DNA were extracted from a *M. extorquens* AM155 culture utilizing either ^{12}C- or ^{13}C-methanol as the sole carbon source. Bar, 1 cm. From Radajewski et al. (2000).

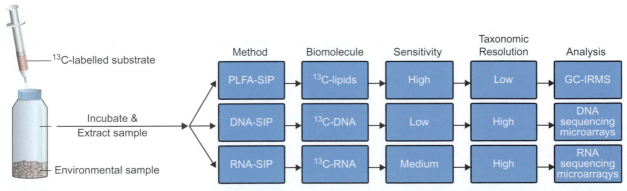

FIGURE 11.22 Example methods and biomolecules used for stable isotope probing of environmental microorganisms. GC-IRMS, gas chromatography-ion ratio mass spectrometry. Adapted from Gutierrez-Zamora and Manefield (2010) and Neufeld et al. (2006).

The use of DNA-SIP or RNA-SIP provides the most complete phylogenetic information but requires the separation of ^{13}C-labeled from unlabeled DNA or RNA (Figure 11.21) before being used for downstream analysis such as DNA sequencing (Chapter 13). This reduces the sensitivity of the technique in terms of microbial numbers that are required and the amount of growth that must occur to incorporate sufficient isotope label into the biomarker. In contrast, the use of lipids, called phospholipid-derived fatty acids or PLFA-SIP, does not require separation of the ^{13}C-labeled from unlabeled fatty acids before being analyzed by gas chromatography-ion ratio mass spectrometry (GC-IRMS). Therefore, it can be a more sensitive technique when microbial numbers and growth rates are low. On the other hand, PLFA-SIP does not give good phylogenetic information, just a gross assignment to a domain (e.g., archaea or bacteria) or a large division within a domain.

The development of new methods that directly combine mass spectrometry with DNA- or RNA-based analyses offers the potential to retain the phylogenetic information provided by nucleic acid-based methods while avoiding the need to first separate the labeled and unlabeled nucleic acids. One such example is the use of microarrays combined with high-resolution secondary ion mass spectrometry imaging (NanoSIMS). Mayali *et al.* (2012) used this approach, which they termed Chip-SIP, to characterize metabolism of labeled substrates by microorganisms in a natural estuarine community. Isotope (^{13}C and ^{15}N)-labeled substrates including amino and fatty acids were added to microcosms containing the microbial communities. After a short incubation ($\leq$24 h), microorganisms were recovered from the microcosm via filtration. Microbial RNA was then extracted and hybridized to a microarray which was designed to work specifically with NanoSIMS, and that targeted organisms found in the estuarine community and related marine environments. Scanning the microarray revealed which phylogenetic probes bound isotopically labeled RNA, and thus identified which microorganisms likely metabolized the added substrate. The ChipSIP analysis found that the labeled substrate was partitioned into a wide variety of microorganisms (81 distinct microbial taxa), likely due to the relatively simple structure of the added compounds (e.g., amino acids) and their ubiquitous presence in natural environments. Researchers are also developing methods that combine NanoSIMS with electron microscopy-based methods (e.g., scanning transmission X-ray microscopy; STXM) that potentially allow detection of individual microorganisms (at the cellular level) which have incorporated an isotope label into their biomass (Behrens *et al.*, 2012). These new approaches that combine physiological methods with other approaches (e.g., nucleic acid-based methods) have the potential, over the next decade,

to revolutionize our understanding of environmental microorganisms and the processes they conduct.

QUESTIONS AND PROBLEMS

1. You are given a pure culture of four different benzoate-degrading microbes and asked to evaluate which has the fastest degradation rate (μ_{max}). Design an experiment to answer this question. In your experimental design be sure to specify the culture medium to be used and how you will analyze benzoate degradation.

2. Next you are given four different soils and asked to evaluate the capacity of each soil to degrade the herbicide 2,4-dichlorophenoxy acetic acid (2,4-D). Design an experiment to answer this question. Be sure to specify how the experiment will be set up and how you will analyze 2,4-D degradation.

3. One of the soils that you tested in problem 2 shows rapid rates of 2,4-D degradation. How would you determine whether there is one or more than one different population degrading the 2,4-D?

4. You are given an assignment to determine the impact of agricultural practices on microbial activity in soil. You collect a series of soil samples from plots that are undisturbed (no crops, no fertilizer), cropped with normal tillage, cropped with reduced tillage and cropped with no tillage. In addition, a high and low fertilizer application rate was used. Thus, there is a combination of seven different treatments. What microbial activity tests would you choose to run and why? How would you set up these tests?

5. You are assigned to restore a site in a national park that has been severely disturbed by overuse. The site currently has no visible plant growth. You would like to add an organic amendment to stimulate microbial activity. You test several amendments for their effect on activity and for their longevity. These include a manure amendment, wood chips and a manure—wood chip mixture. Describe how you would determine the effect of these different amendments on (1) microbial biomass and (2) microbial activity including nitrification and heterotrophic activity.

REFERENCES AND RECOMMENDED READING

Anderson, J. P. E., and Domsch, K. H. (1978) A physiological method for the quantitative measurement of microbial biomass in soils. *Soil Biol. Biochem.* **10**, 215—221.

APHA, AWWA, WEF American Public Health Association, and American Water Works Association and Water Environment Federation (2005) "Standard Methods for the Examination of Water and Wastewater," 21st ed. American Public Health Association, Washington, DC.

Atlas, R. M., and Bartha, R. (1993) "Microbial Ecology," Benjamin/Cummings, Redwood City, CA.

Azam, F., and Fuhrman, J. A. (1984) Measurement of bacterioplankton growth in the sea and its regulation by environmental conditions. In "Heterotrophic Activity in the Sea" (J. E. Hobbie, P. J. Le, and B. Williams, eds.), Plenum, New York, pp. 179–196.

Beare, M. H., Neely, C. L., Coleman, D. C., and Hargrove, W. L. (1990) A substrate-induced respiration (SIR) method for measurement of fungal and bacterial biomass on plant residues. *Soil Biol. Biochem.* **22**, 585–594.

Behrens, S., Kappler, A., and Obst, M. (2012) Linking environmental processes to the in situ functioning of microorganisms by high-resolution secondary ion mass spectrometry (NanoSIMS) and scanning transmission X-ray microscopy (STXM). *Environ. Microbiol.* **14**, 2851–2869.

Beloin, R. M., Sinclair, J. L., and Ghiorse, W. C. (1988) Distribution and activity of microorganisms in subsurface sediments of a pristine study site in Oklahoma. *Microb. Ecol.* **16**, 85–97.

Blenkinsopp, S. A., and Lock, M. A. (1990) The measurement of electron transport system activity in river biofilms. *Water Res.* **24**, 441–445.

Bowman, G. T., and Delfino, J. J. (1980) Sediment oxygen demand techniques: a review and comparison of laboratory and in situ systems. *Water Res.* **14**, 491–499.

Bradford, M. M. (1976) A rapid and sensitive method for the quantitation of microgram quantities of protein utilizing the principle of protein–dye binding. *Anal. Biochem.* **72**, 248–254.

Brookes, P. C., Tate, K. R., and Jenkinson, D. J. (1983) The adenylate energy charge of the soil microbial biomass. *Soil Biol. Biochem.* **15**, 9–16.

Brooks, P. D., and Paul, E. A. (1987) A new automated technique for measuring respiration in soil samples. *Plant Soil* **101**, 183–187.

Burns, R. G. (ed.) (1978) *Soil Enzymes.* Academic Press, New York.

Burns, R. G., DeForest, J. L., Marxsen, J., Sinsabaugh, R. L., Stromberger, M. E., Wallenstein, M. D., *et al.* (2013) Soil enzymes in a changing environment: current knowledge and future directions. *Soil Biol. Biochem.* **58**, 216–234.

Chin-Leo, G., and Kirchman, D. (1988) Estimating bacterial production in marine waters from the simultaneous incorporation of thymidine and leucine. *Appl. Environ. Microbiol.* **54**, 1934–1939.

Ciardi, C., Ceccanti, B., and Nannipieri, P. (1991) Method to determine the adenylate energy charge in soil. *Soil Biol. Biochem.* **23**, 1099–1101.

Daniels, L., Hanson, R. S., and Phillips, J. A. (1994) Chemical analysis. In "Methods for General and Molecular Bacteriology" (P. Gerhardt, R. G. E. Murray, W. A. Wood, and N. R. Krieg, eds.), American Society for Microbiology, Washington, DC, pp. 512–554.

Díaz-Raviña, M., Bååth, E., and Frostegård, Å. (1994) Multiple heavy metal tolerance of soil bacterial communities and its measurement by a thymidine incorporation technique. *Appl. Environ. Microbiol.* **60**, 2238–2247.

Dulohery, C. J., Morris, L. A., and Lowrance, R. (1996) Assessing forest soil disturbances through biogenic gas fluxes. *Soil Sci. Soc. Am. J.* **60**, 291–298.

Ghiorse, W. C. (1994) Iron and manganese oxidation and reduction. In "Methods of Soil Analysis, Part 2, Microbiological and Biochemical Properties," Soil Science Society of America, Madison, WI, pp. 1079–1096.

Griffith, D. W. T., Deutscher, N. M., Caldow, C. G. R., Kettlewell, G., Riggenbach, M., and Hammer, S. (2012) A Fourier transform infrared trace gas analyser for atmospheric applications. *Atmos. Meas. Tech. Discuss.* **5**, 3717–3769.

Gutierrez-Zamora, M-L., and Manefield, M. (2010) An appraisal of methods for linking environmental processes to specific microbial taxa. *Rev. Environ. Sci. Biotechnol.* **9**, 153–185.

Hinchee, R. E., and Ong, S. K. (1992) A rapid in situ respiration test for measuring aerobic biodegradation rates of hydrocarbons in soil. *J. Air Waste Manage. Assoc.* **42**, 1305–1312.

Inubushi, K., Brookes, P. C., and Jenkinson, D. S. (1989) Adenosine 5′-triphosphate and adenylate energy charge in waterlogged soil. *Soil Biol. Biochem.* **21**, 733–739.

Iqbal, J., Castellano, M. J., and Parkin, T. B. (2013) Evaluation of photo-acoustic infrared spectroscopy for simultaneous measurement of N_2O and CO_2 gas concentrations and fluxes at the soil surface. *Global Change Biol.* **19**, 327–336.

Jenkinson, D. S. (1988) Determination of microbial biomass carbon and nitrogen in soil. In "Advances in Nitrogen Cycling in Agricultural Ecosystems" (J. R. Wilson, ed.), C.A.B. International, Wallingford, Oxon, U.K, pp. 368–386.

Johnson, C. K., Vigil, M. F., Doxtader, K. G., and Beard, W. E. (1996) Measuring bacterial and fungal substrate-induced respiration in dry soil. *Soil Biol. Biochem.* **28**, 427–432.

Kelly, D. P., and Wood, A. P. (1998) Microbes of the sulfur cycle. In "Techniques in Microbial Ecology" (R. S. Burlage, R. Atlas, D. Stahl, G. Geesey, and G. Sayler, eds.), Oxford University Press, New York, pp. 31–57.

Kieft, T. L., and Rosacker, L. L. (1991) Application of respiration- and adenylate-based soil microbiological assays to deep subsurface terrestrial sediments. *Soil Biol. Biochem.* **23**, 563–568.

Kieft, T. L., Fredrickson, J. K., McKinley, J. P., Bjornstad, B. N., Rawson, S. A., Phelps, T. J., *et al.* (1995) Microbiological comparisons within and across contiguous lacustrine, paleosol, and fluvial subsurface sediments. *Appl. Environ. Microbiol.* **61**, 749–757.

Kirchman, D., K'ness, E., and Hodson, R. (1985) Leucine incorporation and its potential as a measure of protein synthesis by bacteria in natural aquatic systems. *Appl. Environ. Microbiol.* **49**, 599–607.

Kluge, B., Peters, A., Krüger, J., and Wessolek, G. (2013) Detection of soil microbial activity by infrared thermography (IRT). *Soil Biol. Biochem.* **57**, 383–389.

Kreuzer-Martin, H. W. (2007) Stable isotope probing: linking functional activity to specific members of microbial communities. *Soil Sci. Soc. Am. J.* **71**, 611–619.

Ladd, T. I., Costerton, J. W., and Geesey, G. G. (1979) Determination of the heterotrophic activity of epilithic microbial populations. In "Native Aquatic Bacteria: Enumeration, Activity, and Ecology" (J. W. Costerton, and R. R. Colwell, eds.), ASTM STP 695. American Society for Testing and Materials, Pittsburgh, pp. 180–195.

Lowry, O. H., Rosebrough, N. J., Farr, A. L., and Randall, R. J. (1951) Protein measurement with the Folin phenol reagent. *J. Biol. Chem.* **193**, 265–275.

Martens, R. (1987) Estimation of microbial biomass in soil by the respiration method: importance of soil pH and flushing methods for the measurement of respired carbon dioxide. *Soil Biol. Biochem.* **19**, 77–81.

Martens, R. (2001) Estimation of ATP in soil: extraction and calculation of extraction efficiency. *Soil Biol. Biochem.* **33**, 973–982.

Mayali, X., Weber, P. K., Brodie, E. L., Maberry, S., Hoeprich, P. D., and Pett-Ridge, J. (2012) High-throughput isotopic analysis of RNA microarrays to quantify microbial resource use. *ISME J.* **6**, 1210–1221.

Moriarty, D. J. W. (1986) Measurement of bacterial growth rates in aquatic systems from rates of nucleic acid synthesis. *Adv. Microb. Ecol.* **9**, 245–292.

Morra, M. J. (1997) Assessment of extracellular enzymatic activity in soil. In "Manual of Environmental Microbiology" (C. J. Hurst, G. R. Knudsen, M. J. McInerney, L. D. Stetzenbach, and M. V. Walter, eds.), American Society for Microbiology Press, Washington, DC, pp. 459–465.

Nannipieri, P., Grego, S., and Ceccanti, B. (1990) Ecological significance of the biological activity in soil. In "Soil Biochemistry" (J.-M. Bollag, and G. Stotsky, eds.), vol. 6, Marcel Dekker, New York, pp. 293–355.

Neufeld, J. D., Dumont, M. G., Vohra, J., and Murrell, J. C. (2006) Methodological considerations for the use of stable isotope probing in microbial ecology. *Microb. Ecol.* **53**, 435–442.

Nielsen, L. P., Christensen, P. B., Revsbech, N. P., and Sørensen, J. (1990) Denitrification and oxygen respiration in biofilms: studies with a microsensor for nitrous oxide and oxygen. *Microb. Ecol.* **19**, 63–72.

Padden, A. N., Kelly, D. P., and Wood, A. P. (1998) Chemolithoautotrophy and mixotrophy in the thiophene-2-carboxylic acid-utilizing *Xanthobacter tagetidis*. *Arch. Microbiol.* **169**, 249–256.

Paerl, H. W. (1998) Microbially mediated nitrogen cycling. In "Techniques in Microbial Ecology" (R. S. Burlage, R. Atlas, D. Stahl, G. Geesey, and G. Sayler, eds.), Oxford University Press, New York, pp. 3–30.

Pepper, I. L., Gerba, C. P., and Brusseau, M. L. (2006) "Environmental and Pollution Science," second ed. Academic Press, San Diego, CA.

Pomeroy, L. R., Sheldon, J. E., and Sheldon, W. M., Jr. (1994) Changes in bacterial numbers and leucine assimilation during estimations of microbial respiratory rates in seawater by the precision Winkler method. *Appl. Environ. Microbiol.* **60**, 328–332.

Posch, T., Pernthaler, J., Alfreider, A., and Psenner, R. (1997) Cell-specific respiratory activity of aquatic bacteria studied with the tetrazolium reduction method, cyto-clear slides, and image analysis. *Appl. Environ. Microbiol.* **63**, 867–873.

Radajewski, S., Ineson, P., Parekh, N. R., and Murrell, J. C. (2000) Stable-isotope probing as a tool in microbial ecology. *Nature* **403**, 646–649.

Revsbech, N. P., and Jørgensen, B. B. (1986) Microelectrodes: their use in microbial ecology. *Adv. Microb. Ecol.* **9**, 293–352.

Robarts, R. D., and Zohary, T. (1993) Fact or fiction—bacterial growth rates and production as determined by [methyl-3H]-thymidine? *Adv. Microb. Ecol.* **13**, 371–425.

Rodriguez, G. G., Phipps, D., Ishiguro, K., and Ridgway, H. F. (1992) Use of a fluorescent redox probe for direct visualization of actively respiring bacteria. *Appl. Environ. Microbiol.* **58**, 1801–1808.

Rosacker, L. L., and Kieft, T. L. (1990) Biomass and adenylate energy charge of a grassland soil during drying. *Soil Biol. Biochem.* **22**, 1121–1127.

Roslev, P., and King, G. M. (1993) Application of a tetrazolium salt with a water-soluble formazan as an indicator of viability in respiring bacteria. *Appl. Environ. Microbiol.* **59**, 2891–2896.

Rozycki, M., and Bartha, R. (1981) Problems associated with the use of azide as an inhibitor of microbial activity in soil. *Appl. Environ. Microbiol.* **41**, 833–836.

Schramm, A., Larsen, L. H., Revsbech, N. P., Ramsing, N. B., Amann, R., and Schleifer, K-H. (1996) Structure and function of a nitrifying biofilm as determined by in situ hybridization and the use of microelectrodes. *Appl. Environ. Microbiol.* **62**, 4641–4647.

Schuale, G., Flemming, H-C., and Ridgway, H. F. (1993) Use of 5-cyano-2,3-ditolyl tetrazolium chloride for quantifying planktonic and sessile respiring bacteria in drinking water. *Appl. Environ. Microbiol.* **59**, 3850–3857.

Smith, P. K., Krohn, R. I., Hermanson, G. T., Mallia, A. K., Gartner, R. H., Provenzano, M. D., *et al.* (1985) Measurement of protein using bicinchoninic acid. *Anal. Biochem.* **150**, 76–85.

Songster-Alpin, M. S., and Klotz, R. L. (1995) A comparison of electron transport system activity in stream and beaver pond sediments. *Can. J. Fish. Aquat. Sci.* **52**, 1318–1326.

Sparling, G. P., West, A. W., and Whale, K. N. (1985) Interference from plant roots in the estimation of soil microbial ATP, C, N, and P. *Soil Biol. Biochem.* **17**, 275–278.

Tabatabai, M. A. (1994a) Sulfur oxidation and reduction in soils. In "Methods of Soils Analysis, Part 2, Microbiological and Biochemical Properties," Soil Science Society of America, Madison, WI, pp. 1067–1078.

Tabatabai, M. A. (1994b) Soil enzymes. In "Methods of Soil Analysis, Part 2, Microbiological and Biochemical Properties," Soil Science Society of America, Book Series 5. SSSA, Madison, WI, pp. 775–834.

Tamtam, F., Mercier, F., Eurin, J., Chevreuil, M., and Le Bot, B. (2009) Ultra performance liquid chromatography tandem mass spectrometry performance evaluation for analysis of antibiotics in natural waters. *Anal. Bioanal. Chem.* **393**, 1709–1718.

Thorn, P. M., and Ventullo, R. M. (1988) Measurement of bacterial growth rates in subsurface sediments using the incorporation of tritiated thymidine. *Microb. Ecol.* **16**, 3–16.

Tibbles, B. J., and Harris, J. M. (1996) Use of radiolabelled thymidine and leucine to estimate bacterial production in soils from continental Antarctica. *Appl. Environ. Microbiol.* **62**, 694–701.

Trevors, J. T. (1984) Electron transport system activity in soil, sediment, and pure cultures. *Crit. Rev. Microbiol.* **11**, 83–100.

Trevors, J. T., Mayfield, C. I., and Inniss, W. E. (1982) Measurement of electron transport system (ETS) activity in soil. *Microb. Ecol.* **8**, 163–168.

USDA-ARS (U.S. Department of Agriculture-Agricultural Research Service) (2013) Greenhouse gas Reduction through Agricultural Carbon Enhancement network (GRACEnet). http://www.ars.usda.gov/research/programs/programs.htm?np_code=212&docid=21223.

Van Es, F. B., and Meyer-Reil, L-A. (1982) Biomass and metabolic activity of heterotrophic marine bacteria. *Adv. Microb. Ecol.* **6**, 111–170.

Webster, J. J., Hall, S. M., and Leach, F. R. (1992) ATP and adenylate energy charge determinations on core samples from an aviation fuel spill site at the Travers City, Michigan airport. *Bull. Environ. Contam. Toxicol.* **49**, 232–237.

Wetzel, R. G., and Likens, G. E. (1991) "Limnoligical Analyses," second ed. Springer-Verlag, New York.

Winding, A., Binnerup, S. J., and Sorrensen, J. (1994) Viability of indigenous soil bacteria assayed by respiratory activity and growth. *Appl. Environ. Microbiol.* **60**, 2869–2875.

Wright, R. T., and Burnison, B. K. (1979) Heterotrophic activity measured with radiolabelled organic substrates. In "Native Aquatic Bacteria: Enumeration, Activity, and Ecology" (J. W. Costerton, and R. R. Colwell, eds.), ASTM STP 695. American Society for Testing and Materials, Pittsburgh, pp. 140–155.

Yu, F. P., and McFeters, G. A. (1994a) Rapid in situ assessment of physiological activities in bacterial biofilms using fluorescent probes. *J. Microbiol. Methods* **20**, 1–10.

Yu, F. P., and McFeters, G. A. (1994b) Physiological responses of bacteria in biofilms to disinfection. *Appl. Environ. Microbiol.* **60**, 2462–2466.

Zimmermann, R., Iturriaga, R., and Becker-Birck, J. (1978) Simultaneous determination of the total number of aquatic bacteria and the number thereof involved in respiration. *Appl. Environ. Microbiol.* **36**, 926–935.

Zinder, S. H. (1998) Methanogens. In "Techniques in Microbial Ecology" (R. S. Burlage, R. Atlas, D. Stahl, G. Geesey, and G. Sayler, eds.), Oxford University Press, New York, pp. 113–136.

Chapter 12

Immunological Methods

Hye-Weon Yu, Marilyn J. Halonen and Ian L. Pepper

12.1 INTRODUCTION

There are multiple infectious agents including viruses, bacteria, fungi and parasites that can infect a host and cause disease. However, in many cases, infections within healthy individuals are transient in nature and do not cause permanent damage. The reason for this is due to an individual's immune system which fights off and overcomes the infectious agents. Immunology is the study of the immune systems of higher organisms in relation to disease. In this chapter, we will describe the structure and function of the immune system, and how immunological methods can be used in environmental microbiology via antibody-based laboratory techniques or immunoassays.

The immune system is divided into two functional divisions: the innate and the adaptive immune systems. The innate immunity system acts as a first line of defense against infectious agents or pathogens by recognizing whether or not an entity is "native" to or "foreign" to the body before an infection is established. If determined to be foreign, the adaptive immune system can be triggered into action resulting in the evaluation of the foreign agent. Interestingly, once the body has been exposed to a specific foreign agent, frequently the immune system remembers the agent providing life-long immunity against subsequent infection. The innate and adaptive immune systems involve multiple molecules and cells distributed throughout the body. The important adaptive immune responses are categorized as cell mediated or antibody mediated. The cell-mediated response is produced when a subset of sensitized white blood cells or lymphocytes directly attacks material such as a bacterial cell or virus, which has been determined to be foreign to the body. The antibody-mediated response involves the transformation of a subset of lymphocytes into cells that produces and secretes specific antibodies that target the foreign material, which is termed the antigen. Overall, the two immune responses are triggered when foreign material is introduced into the host as depicted in Figure 12.1.

Here we do not deal with all aspects of immunology or the immune responses *per se*, but instead adapt immunology-based research technologies or immunoassays for the study of microorganisms and chemical contaminants found in the environment. The primary immunological tool used in environmental microbiology is the antibody. In this chapter, an introduction to

I.L. Pepper, C.P. Gerba, T.J. Gentry: Environmental Microbiology, Third edition. DOI: http://dx.doi.org/10.1016/B978-0-12-394626-3.00012-0
© 2015 Elsevier Inc. All rights reserved.

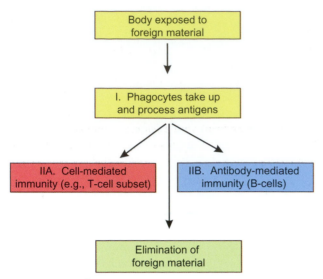

FIGURE 12.1 Flow chart showing the two major immune response systems that comprise the host response to foreign materials. Together these two branches of the immune system work together to create immunity. The cell-mediated response is designed to directly attack and destroy cells determined by the body to be nonself. The antibody-mediated response is the branch of the immune system involved in the formation of antibodies.

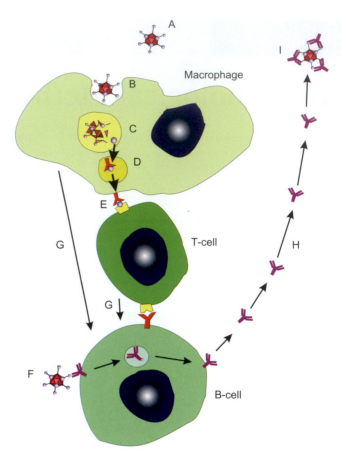

FIGURE 12.2 Schematic representation showing the processes that lead to the formation of antibodies. (A) Foreign material represented here by a virus that has gained access to the body; (B) the virus is phagocytosed by a macrophage; (C) the virus is broken down into subunits by enzymes contained in the phagocytic vacuole; (D) an antigen-presenting molecule escorts antigenic subunits from the virus to the macrophage surface; (E) the antigenic molecule derived from the virus is presented to a T cell; (F) the B cell encounters the virus and expands its antibody production specific to the viral epitope; (G) the macrophage and the T cell release chemicals that stimulate B cells; (H) antibodies are released into circulation; (I) antibodies neutralize the virus in circulation by binding to the viral epitopes inhibiting viral attachment to target cells.

antibodies is given, with respect to the structure of antibodies, the various classes of antibodies and the interaction of antibodies with antigens. Following this introduction, we discuss several of the basic immunological methodologies or immunoassays that are widely used in environmental microbiology. These immunoassays include fluorescent immunolabeling, enzyme-linked immunosorbent assay (ELISA), magnetic bead antigen capture, Western immunoblotting, immunoaffinity chromatography, immunoprecipitation and lateral flow immunoassay. Finally, in order to provide perspective and illustrate how these immunoassays and immunosensors can be used in the field of environmental microbiology, an example of each immunoassay and immunosensor is provided in relation to current research topics such as bioremediation and pathogen detection.

12.2 WHAT IS AN ANTIBODY?

Antibodies, also known as immunoglobulins, are produced by the immune system of higher life forms that help defend the host against foreign invasion. When a host is challenged by an antigen such as bacteria or viruses, the first response of certain host immune cells called macrophages is to engulf these invaders and process them biochemically. This biochemical processing essentially creates a blueprint that is used for the development of an immune response that results in the production of antibodies (Figure 12.2). The unique feature of antibodies produced in response to an antigen is that they are synthesized in such a way that they

are highly specific for that antigen. Thus, they can chemically interact and bind only with that particular antigen, neutralize it, and/or aid in its destruction and removal from the body.

There are five different classes of immunoglobulins (Igs): IgA, IgD, IgE, IgG and IgM. These immunoglobulins differ in many ways including their overall structures (Figure 12.3). The most common type of antibody used for immunoassays is the IgG class of immunoglobulins (Figure 12.4). IgG antibodies are Y-shaped proteins composed of four peptide chains that are joined together by disulfide linkages. There are two major structural fragments or regions of IgGs called the Fc and Fab regions. The antigen-binding fragments (Fabs) of the IgG immunoglobulins are the two identical regions at the top of the molecule, which, as indicated,

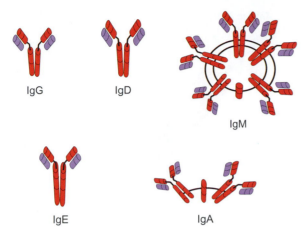

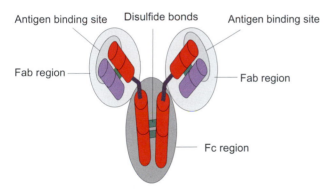

FIGURE 12.3   Schematic representation of the five classes of immunoglobulins. IgG (molecular weight 146,000), IgD (184,000) and IgE (188,000)) are all comprised of one basic subunit. IgG is the major immunoglobulin involved in the humeral response. IgM, which consists of five subunits (970,000), is the second major antibody involved in the humeral response. IgA (385,000) is commonly found in body secretions such as saliva, milk and intestinal fluid. IgD is found in low concentrations in plasma, and IgE is involved in allergic reactions. IgG and IgM are the antibody classes most commonly utilized in immunoassays.

FIGURE 12.4   Schematic representation of an IgG antibody, showing the various regions associated with the antigen−antibody interaction. There are two antigen-binding fragments (Fab), which interact with the antigen. There is also one crystal fragment (Fc), which is the part of the antibody recognized by the host immune system as self. There are two light chains (in purple) joined to two heavy chains (red/orange) by disulfide bonds, and the two heavy chains are in turn joined to each other in a similar fashion.

are the sites of antibody−antigen interaction. The Fc region is the tail of the antibody, and is the fragment that is recognized by the host as "self." Because antibodies are relatively large proteins they can also act as antigens, so the ability to recognize a particular antibody as being "self" prevents the host from responding against its own antibodies. In addition to the major classes of antibodies, minor differences in the protein structure result in subclasses. Variance among different subclasses is less than the variance among different

classes. For example, all rabbits and humans have the same classes of antibodies such as IgG or IgM, but do not have the same subclasses such as IgG1 or IgG2, or the same host recognition sites on the immunoglobulins. This species-specific difference makes the immune system of one species recognize another species' antibodies as being foreign. For example, if you immunize a rabbit (purposefully expose it to a foreign antigen) with a human's antibodies, the rabbit will produce different antibodies against them. Antibodies produced in response to another individual's antibodies are termed antiglobulins.

12.2.1 Antibody Diversity

B cells belong to a group of white blood cells or lymphocytes, and are responsible for antibody diversity through the production of a wide variety of antibodies that can interact with a diverse range of antigens. In fact, the vast populations of B cells have been estimated to have the potential to produce up to 1×10^{10} structurally different IgG antibodies, which in theory could recognize 1×10^{10} different antigens. This enormous diversity has been exploited in the field of environmental microbiology. An animal's immune system can adapt and produce an immune response against many different antigens. Therefore, essentially any bacterium, virus, protein or pollutant that can stimulate an immune response can be used as an antigen to produce specific antibodies. These antibodies can then be used to design immunoassays to aid in the study of that particular bacterium, virus or toxic pollutant.

12.2.2 Antibody Specificity

Specificity for a particular antigen is one characteristic that makes immunology-based methodologies such valuable tools. In essence, once they are produced, antibodies are very precise in their recognition of the particular antigen. This discrimination is based on the molecular structure of the antigen-binding sites located on the Fab portion of the antibody, and on the epitopes or chemically reactive sites of the antigen (Figure 12.5). Antigen−antibody binding is the result of specific chemical interactions (i.e., charge−charge, dipole−dipole, hydrogen bonding and van der Waals) that occur between the antigen and amino acid residues of the antibody that are located in the Fab region. This reaction is so specific that even a small change, such as an alteration of one amino acid in the binding site of the antibody, may weaken or nullify the antigen−antibody binding. It should be noted, however, that even with this specificity, there are still instances in which the antibody may react with more than one antigen. This phenomenon in which an antibody reacts with two unrelated epitopes is termed cross-reactivity.

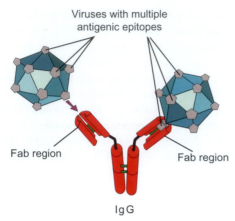

FIGURE 12.5 Schematic representation of the antigen−antibody interaction. This figure shows an IgG antibody binding to an enterovirus. Note that the Fab regions of the antibody are chemically formed to fit perfectly with the antigenic epitopes of the virus.

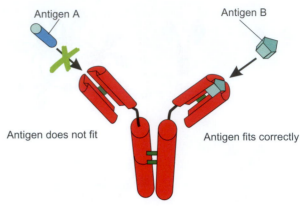

FIGURE 12.6 Schematic representation of an IgG antibody interacting with two different antigens. In the first case, antigen A does not have the correct biochemical conformation so the antibody is unable to interact and bind it. Antigen B has the correct conformation and the antibody is able to interact and bind this antigen tightly.

12.2.3 Antibody Affinity

Affinity is defined as the attraction between an antibody and an antigen. More specifically, affinity is a measure of the strength of this interaction, and is usually expressed as an interaction or association constant. Quantitatively, affinity is the sum of the chemical bonds that form between the antigen and the antibody. These are usually relatively weak interactions such as hydrophobic interactions and hydrogen bonds. Even though individually such chemical bonds are relatively weak, collectively they form very strong and tight interactions. Thus, the strongest binding occurs between epitopes and antibodies only when their shapes are complementary (Figure 12.6). Affinity can be described as the reversible formation of the antigen (Ag)−antibody (Ab) complex by the equation:

$$Ag + Ab \rightleftharpoons AgAb \qquad \text{(Eq. 12.1)}$$

The affinity constant (K) can thus be determined by the mass balance equation, which is expressed in the form:

$$K = [AgAb]/[Ag][Ab] \qquad \text{(Eq. 12.2)}$$

where:

[AgAb] is the concentration of the antigen−antibody complex at equilibrium
[Ag] is the concentration of free antigen-binding sites at equilibrium
[Ab] is the concentration of free antibody binding sites at equilibrium, and
K is the affinity constant or the measure of the strength of the bond formation between Ag and Ab.

Essentially, the higher the K value, the stronger the affinity of an antibody for an antigen. This is a direct consequence of a stronger molecular interaction. This is

important in the development of monoclonal antibodies for immunoassays, especially in the selection of the best hybridoma to use for monoclonal antibody production (see Section 12.2.4). The affinity constant not only gives an indication of which of the antibodies might be best for a certain assay, but it can also indicate the concentration of antibody required for the assay, and the potential for cross-reactivity with other antigens.

12.2.4 Polyclonal and Monoclonal Antibodies

In the past, injecting a laboratory animal with the antigen of interest produced particular antibodies needed for environmental and other research purposes. In response to immunization, the animal produces antibodies that can be collected (in serum) directly from the blood of the animal (Figure 12.7). These blood-derived antibodies are termed polyclonal antibodies because they are not derived from a single B lymphocyte or its progeny, but are instead a product of many different B cells binding in slightly different ways to the same antigen. As a result, this method yields a mixed product that contains a wide variety of antibodies.

To avoid problems associated with use of such antibody mixtures, scientists developed the technology to produce monoclonal antibodies. A monoclonal antibody is an antibody that is the product of a single B cell clone. Production of monoclonal antibodies involves the *in vitro* combination of two types of cells. The first type of cell is a B cell that produces a single, unique antibody. The second type of cell is an immortalized myeloma cell: a cancer cell that is able to thrive and multiply *in vitro*. The specific antibody-producing B cell is fused with the myeloma cell to form a hybrid cell called a hybridoma. This hybridoma

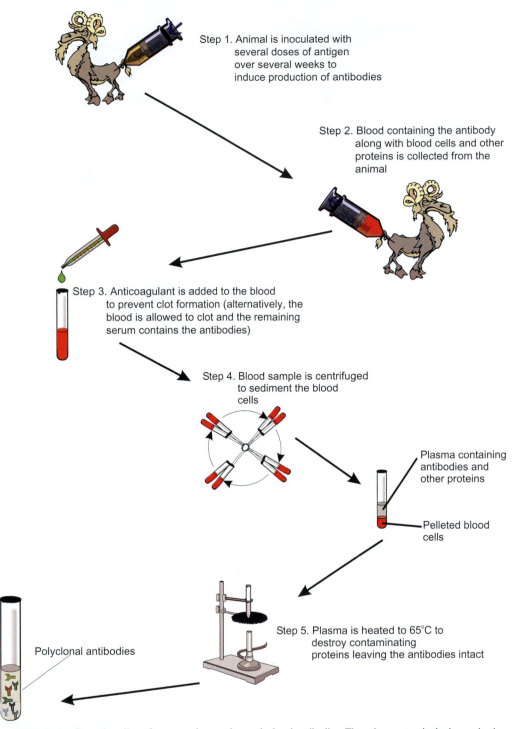

Step 1. Animal is inoculated with
several doses of antigen
over several weeks to
induce production of antibodies

Step 2. Blood containing the antibody
along with blood cells and other
proteins is collected from the
animal

Step 3. Anticoagulant is added to the blood
to prevent clot formation (alternatively, the
blood is allowed to clot and the remaining
serum contains the antibodies)

Step 4. Blood sample is centrifuged
to sediment the blood
cells

Plasma containing
antibodies and
other proteins

Pelleted blood
cells

Step 5. Plasma is heated to 65°C to
destroy contaminating
proteins leaving the antibodies intact

Polyclonal antibodies

FIGURE 12.7   General outline of steps used to produce polyclonal antibodies. The primary step is the immunization of the animal and the collection of the blood containing the antibodies. Following collection, the blood cells are separated by centrifugation and heating denatures other proteins such as complement. This results in a stable suspension of antibodies that can be used in many types of immunoassays.

combines the characteristic of "immortality" with the ability to produce the desired specific antibody in high concentrations and in pure form. The procedure for the production of monoclonal antibodies is schematically diagrammed in Figure 12.8. As a result of the development of monoclonal

antibody technology, scientists are able to produce large amounts of pure and highly specific antibodies.

What are the advantages and disadvantages of using polyclonal or monoclonal antibodies? In general, monoclonal antibodies have higher specificity and lower cross-reactivity

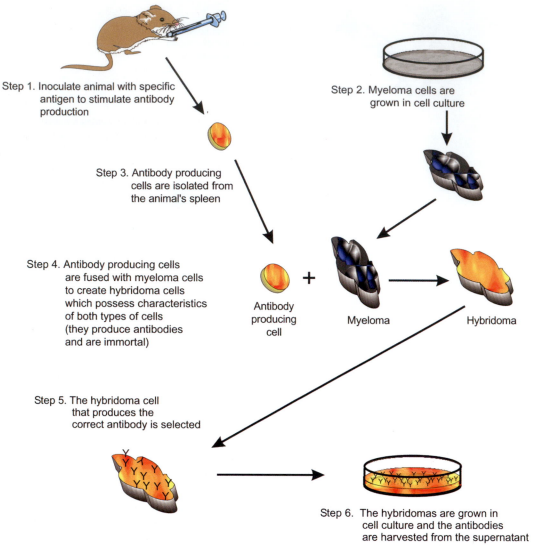

Step 1. Inoculate animal with specific
        antigen to stimulate antibody
        production

Step 2. Myeloma cells are
        grown in cell culture

Step 3. Antibody producing
        cells are isolated from
        the animal's spleen

Step 4. Antibody producing cells
        are fused with myeloma cells
        to create hybridoma cells
        which possess characteristics
        of both types of cells
        (they produce antibodies
        and are immortal)

Antibody
producing
cell   Myeloma   Hybridoma

Step 5. The hybridoma cell
        that produces the
        correct antibody is selected

Step 6. The hybridomas are grown in
         cell culture and the antibodies
         are harvested from the supernatant

FIGURE 12.8   General outline of the steps used to produce monoclonal antibodies. Essentially, the spleen cells from the immunized mouse are removed and combined with myeloma cells using polyethylene glycol to fuse the two. After this the cells that produce the specific antibody needed are selected and cultured in order to produce large quantities of highly purified monoclonal antibody.

than polyclonal antibodies. They can also be produced indefinitely and in relatively large concentrations (>13 mg/ml). Polyclonal antibodies, on the other hand, can be produced more rapidly and much less expensively because they are prepared directly from the serum of immunized animals. Thus, polyclonal and monoclonal antibodies each have benefits and drawbacks that must be considered during the design of any immunoassay.

12.2.5 Antiglobulins

Antiglobulins, as indicated earlier, are antibodies that are specific (usually targeting the Fc portion) for another foreign antibody. Usually, antiglobulins are developed to

recognize a whole antibody class for a specific organism, e.g., a mouse. The antiglobulins can then be attached to a signal molecule, and used as secondary or indirect detection molecules to detect any mouse antibody used in an immunoassay (see Section 12.3). Because antibodies are large proteins with complex structures, they have the potential to be seen as antigens if they do not have the "self" recognition sites common to the host that produced the antibody. For example, an antiglobulin can be raised in a goat by immunizing the goat with mouse IgG antibodies (Figure 12.9). Because the goat does not recognize the mouse antibody as self, it produces its own set of antibodies against the mouse antibodies. Thus, the antibodies produced by the goat are specific for the mouse-derived IgG antibody. Using monoclonal methods for production

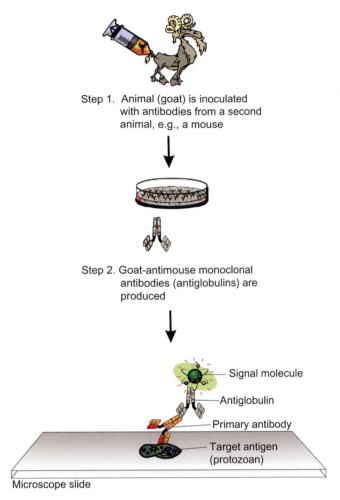

Step 1. Animal (goat) is inoculated
with antibodies from a second
animal, e.g., a mouse

Step 2. Goat-antimouse monoclonal
antibodies (antiglobulins) are
produced

Signal molecule

Antiglobulin

Primary antibody

Target antigen
(protozoan)

Microscope slide

Step 3. Antiglobulin is labeled with a signal molecule and
indirect immunoassay is used to detect the
target antigen

FIGURE 12.9   A general outline of the steps involved in the production of monoclonal antiglobulins. A goat is inoculated with the mouse
antibodies. The goat produces antibodies against the mouse antibody.
These goat anti-mouse monoclonal antibodies are then cloned and
used as secondary labeling antibodies. In this case, a mouse antibody
has been made to detect a protozoal target antigen. The labeled
goat anti-mouse monoclonal antiglobulin is then used to detect the
protozoan—mouse antibody complex.

of specific antibodies, goat antibodies can be derived
that are specific against all mouse IgGs, or even against
specific IgG subclasses, for example IgG2a.

12.3 IMMUNOASSAYS

Immunoassays are analytical methods used for the detection
and/or quantitation of the antigen—antibody interaction. For
the most part, the types of immunoassays used in environmental microbiology are based on quantitation or detection
of antigens as opposed to the characterization of the antigens.

That is, we are usually interested in using immunoassays to
determine how much antigen is in an environmental sample,
and not in characterizing an antibody—antigen interaction, or
the role of the antigen in disease or in the immune response.
However, in order to quantitate or detect the antigen there
must be a way to visualize the antigen—antibody interaction.
This visual signal is produced by the attachment of specific
signal molecules to the antibodies or antiglobulins used to
detect the antigen within an environmental sample.

For almost all types of immunoassays, attachment of
a signal molecule to the antibody and/or antigen is very
important. Many types of signal molecules are used in
immunoassays, including iodine, enzymes, fluorochromes
and radioisotopes. These signal molecules produce a
visual signal that allows quantitation of the specific
antibody—antigen interaction being investigated. The signal is usually indicated by the production of some type of
color change. For example, enzymes such as horseradish
peroxidase and alkaline phosphatase act by enzymatically
cleaving colorless substrates to produce a colored product.
This signal is then detected qualitatively by the naked
eye, or quantitatively using an instrument such as a
spectrophotometer. For fluorochrome signal molecules,
the antigen—antibody interaction can be detected by
exciting the fluor with a particular wavelength of light.
The fluor will emit energy (light) at a second wavelength,
which can then be detected visually or instrumentally.
The fluorescent dye most commonly used as an antibody
label is fluorescein isothiocyanate (FITC). Other examples of fluorescing chemicals used in this type of assay
include R-phycoerythrin, rhodamine and Texas red.
Radioisotopes are quantitated by liquid scintillation
counting (see Figure 11.11) or by exposing the sample to
a photographic emulsion (X-ray film), which produces a
qualitative signal on the film that can be visualized as a
dark spot.

Attachment of an antibody or antigen to a signal
molecule is an almost universal way to allow for visual
detection of the antigen via an immunoassay, which can be
achieved via direct or indirect labeling (Figure 12.10). For
an immunoassay with direct labeling, the primary antibody
(antibody specific for the target) has the signal molecule
attached to it allowing for one-step detection. Indirect
labeling involves two steps. The first is the attachment of a
primary antibody to the target, and the second is the attachment of a secondary (antiglobulin) antibody to the primary
antibody. In indirect labeling, it is the secondary antibody
which has the signal molecule attached. Both methods
work well though both have advantages and disadvantages.
With direct labeling, the binding and signal are usually
more specific because there is a smaller signal-to-
background noise ratio. However, the use of indirect labeling allows for one labeled antibody to be used with many
different primary antibodies provided they are all of the
same type, meaning that each primary antibody does not

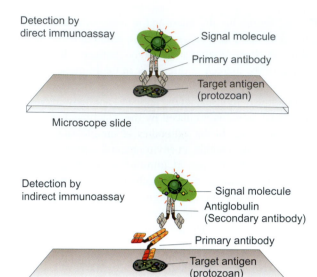

FIGURE 12.10 Direct vs. indirect immunoassay. In the direct assay the primary antibody, also called the detection antibody, is labeled and used to directly detect the antigen. The indirect assay has an extra step. In the first step the primary or detection antibody is bound to the antigen. In the second step a secondary labeled antibody (antiglobulin) is used as the signal molecule. In ELISA the indirect assay is more common because the secondary antibody can be used to detect numerous different primary antibodies and thus the primary antibodies do not have to be individually labeled.

have to be labeled separately. For instance, if you use a mouse to produce monoclonal IgG primary antibodies against four different protozoa (*Microsporidia*, *Giardia*, *Cryptosporidium* and *Entamoeba*), then you can use the same labeled antiglobulin to bind to each of these. Because conjugating signal molecules to antibodies is tedious and often difficult, and because a wide range of antiglobulins conjugated to various signal molecules are available commercially, this is frequently the format of choice. However, if a one-step assay that is slightly more specific is desired, then the use of direct labeling is often preferred.

With this brief introduction to the immunoassays, the following sections provide descriptions of the main types of immunoassays routinely used in environmental microbiology.

12.3.1 Fluorescent Immunolabeling

12.3.1.1 Technique

Fluorescent immunolabeling (immunofluorescence) involves the use of fluorescent signal molecules conjugated to antibodies to interact with and subsequently indicate the presence of a particular antigen by the production of fluorescent light. Detection of the fluorescent signal can be via epifluorescence microscopy as

described in Section 9.4.2. Figures 8.10 (protozoa) and 9.13 (rhizobia) provide examples of immunofluorescent microscopy. The basic procedure for immunofluorescence microscopy is to attach the sample antigen to a microscope slide, add a fluorescent chemical/antibody conjugate specific to the antigen and view the sample under a microscope equipped with a fluorescent light source. When viewing fluorescence-labeled samples under the microscope, the labeled antibody-bound antigen appears bright green against a dark background.

12.3.1.2 Application

One of the main current uses for immunofluorescence is to spatially examine the interaction of an antigen of interest with its environment. For example, Rendón *et al.* (2007) used immunofluorescence labeling to study a pilus common to both commensal and pathogenic *E. coli*, and its interaction with human intestinal epithelial cells. Recall that some *E. coli* are commensal or normal inhabitants of the human gut. However, others, such as the enterohemorrhagic *E. coli* strain O157:H7, are highly pathogenic (Section 22.2.2). Thus, it is important to understand how different *E. coli* strains (both commensal and pathogenic) interact with the intestine. One technique that this group used to study this question was immunofluorescence (Figure 12.11). In this study, the authors showed that there is an *E. coli* common pilus (ECP) that aids all strains in colonization of the epithelial cells. It is suggested that ECP production by the pathogenic strain O157:H7 helps it to mimic commensal *E. coli* strains which gives it a competitive advantage and allows it to avoid detection and elimination during colonization of the human host.

12.3.1.3 Advantages and Disadvantages

One of the primary advantages of fluorescent antibody-based techniques over other detection methodologies is the ease of use. Individuals with a minimal amount of training can perform immunofluorescence assays. Another obvious advantage is that results can be obtained usually within a few hours. In many cases, immunofluorescence is also highly specific and sensitive. On the other hand, there are several problems that must be considered when performing immunofluorescence assays. One serious disadvantage is the potential for cross-reaction between the antibody and nontarget antigens. Such cross-reaction can provide false-positive results using immunofluorescence assays. False-negative results are also possible. This could occur if antigenic epitopes on a target cell were damaged or genetically altered, thus preventing antibody recognition, binding and subsequent detection.

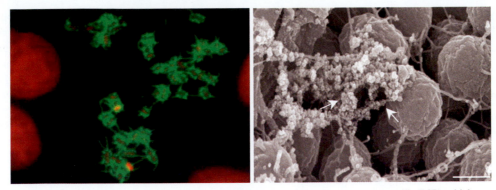

FIGURE 12.11   (Left panel) Immunofluorescent microscopy of the *E. coli* common pili (ECP) which are depicted as green fluorescent structures that mediate the interaction between *E. coli* cells (stained red with propidium iodide). The *E. coli* cells in this sample have adhered to a model epithelial cell. In this case the sample was incubated first with anti-ECP antibody (created in a rabbit) and then with a goat anti-rabbit IgG fluor conjugate. (Right panel) For comparison, we also show an immuno-scanning electron micrograph of the ECP. In this case, the samples were incubated with anti-ECP antibodies followed by incubation with anti-rabbit IgG conjugated to 30 nm gold particles. The white arrows show the antibody—antiglobulin—gold particle complex covering the ECP. (Scale bar: 0.5 μm.) Courtesy Rendón *et al.* (2007).

12.3.2 Enzyme-Linked Immunosorbent Assays

12.3.2.1 Technique

The enzyme-linked immunosorbent assay (ELISA) is a very sensitive method used to detect the presence of antigens or antibodies of interest in a sample. ELISA is typically performed using one of two detection methods: direct or indirect (Figure 12.10), but can also be performed as a competitive assay (see Section 12.3.3). In direct ELISA an enzyme-linked (labeled) antibody is used to directly detect the captured antigen or antibody of interest. In the more common indirect ELISA, a primary antibody is bound to the sample antigen—antibody, and then a secondary labeled antibody (antiglobulin) is used to detect the primary antibody. For any ELISA procedure, the sample antigens—antibodies of interest are concentrated (if necessary) and solubilized in an appropriate buffer. An outline of the steps for an indirect ELISA used to detect an antigen and an antibody is shown in Figure 12.12. For example, for detection of antigen, also called a sandwich ELISA, a capture antibody is attached to a microtiter plate, microcentrifuge tube or other solid support. The antigen is then added and allowed to incubate in order to bind with the antibody (antigen capture). After the antigen is bound, a second detection antibody is added. Usually, the detection antibody recognizes a different antigenic determinant than the capture antibody. Finally, a secondary labeled antibody (antiglobulin) is added which binds to the detection antibody. The secondary antibody is conjugated to a signal molecule, for example, the enzyme alkaline phosphatase. Substrate is then added that causes a color change in response to the presence of the signal molecule. This color change is usually in proportion to the amount of antigen present; thus, the assay becomes quantitative. This makes it possible to quantify the amount of antigen present in a given sample. Once a signal is produced it can be used to visually score the results based on the color change or an automated plate reader can be used. Plate readers provide highly sensitive detection of low-level signals and can determine accurately the strength of a given signal in comparison with a standard curve.

12.3.2.2 Application

Biofilms are specialized environments where microorganisms are firmly attached to surfaces and to one another by exopolymeric substances. Previously, the enumeration of specific organisms within biofilms was done using cultural counts, most probable number (see Chapter 10) or immunofluorescence microscopy. However, the inefficiency of these methods in biofilms due to cells clumping and exopolysaccharide production can lead to significant underestimation of the actual numbers. ELISA has been used as an alternative method to quantifying biomass within biofilms and even protein production in biofilms (Black *et al.*, 2004; Nguyen *et al.*, 2007). In a study by Bouer-Kreisel *et al.* (1996), ELISA was used to quantify populations of *Dehalospirillum multivorans* (*D. multivorans* are bacteria associated with the biodegradation of organic pollutants) in mixed culture biofilms. A standard curve was developed using ELISA and immunofluorescence microscopy to quantify signal from and enumerate serial dilutions of pure cultures of *D. multivorans*. The standard curve related the amount of signal provided by ELISA to the direct counts provided by microscopic enumeration. The two assays were found to be directly proportional. ELISA was then performed on

Antigen Detection
Sandwich Indirect ELISA

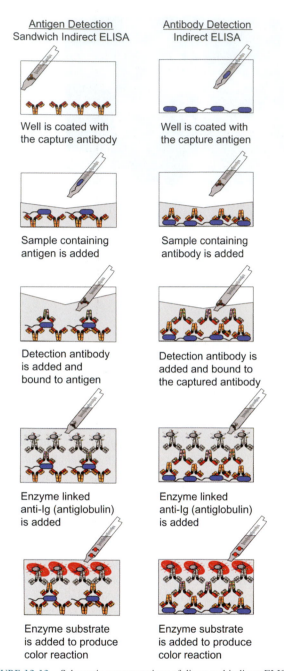

Well is coated with
the capture antibody

Sample containing
antigen is added

Detection antibody
is added and
bound to antigen

Enzyme linked
anti-Ig (antiglobulin)
is added

Enzyme substrate
is added to produce
color reaction

Antibody Detection
Indirect ELISA

Well is coated with
the capture antigen

Sample containing
antibody is added

Detection antibody is
added and bound to
the captured antibody

Enzyme linked
anti-Ig (antiglobulin)
is added

Enzyme substrate
is added to produce
color reaction

FIGURE 12.12 Schematic representations of direct and indirect ELISA. These reactions are usually carried out in a microtiter plate and the color change shown can be detected and quantitated using a plate reader.

actual biofilms, and the results were used to quantify the total biomass of specific organisms responsible for the degradation of organic pollutants in mixed culture biofilms. Thus, the rates of degradation of these pollutants could be directly correlated with biomass using ELISA methodology.

12.3.2.3 Advantages and Disadvantages

There are many advantages of using ELISA over other detection or quantification methods. ELISA is sensitive and

can be quantitative when used in conjunction with standard curves. Disadvantages are similar to all antibody-based methods, and are related to cross-reactivity and nonspecific signal production. ELISAs must also be optimized to provide consistent results especially when using environmental samples. ELISAs, as described above, are also poorly suited for detection of extremely low concentrations of antigens. Other types of ELISAs have been described that are capable of detecting low target concentrations. One of these assays is known as competitive ELISA.

12.3.3 Competitive ELISA

12.3.3.1 Technique

In competitive ELISA, both an enzyme-linked (labeled) control antigen and a sample containing an unknown quantity of unlabeled antigen are added to a sample well coated with antibody (Figure 12.13). In this assay, the sample is added first and the bound antibody captures the unlabeled antigen. The labeled antigen is then added, and a reversible equilibrium is established between the amount of labeled control antigen and sample antigen bound to the antibody depending on their relative concentrations. If the ratio of labeled control antigen to sample antigen is high then the signal will be maximized. If the ratio of labeled control antigen to sample antigen is low, then the signal will be weak. Because the standard curve in a competitive ELISA exhibits the maximum signal at the lowest concentrations of sample antigen, this assay is very sensitive.

12.3.3.2 Application

Various immunoassay test kits based on the competitive ELISA are commercially available for measurement of organic contaminants such as pesticides, petroleum and PCBs in environmental samples (Neilson and Maier, 2001). These kits are approved for use by the U.S. Environmental Protection Agency for detection and monitoring of these contaminants. Several groups have also worked on development of monoclonal antibodies for detection of various metals. Metal detection is problematic due to the small size of the metal atom which does not illicit an immune response and so antibodies are not produced (Neilson and Maier, 2001). This problem has been overcome by using a metal–chelator complex to produce specific monoclonal antibodies (Zhu et al., 2007).

Commercial kits use different approaches to trap the contaminant antigen including antibody-coated tubes, a series of filtering steps or magnetic particles (see Section 12.3.4). Various immunoassay kits have compared favorably in terms of sensitivity with traditional analytical approaches for a variety of organic contaminants including pesticides such as chlorfenapyr (Watanabe et al., 2005) and polychlorinated

Competitive ELISA

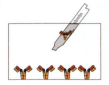

Well is coated
with antibody

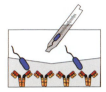

Sample is added and
antigen is captured by
the antibody

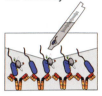

Enzyme linked
antigen is added and
competes for antibody
binding sites

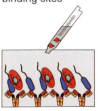

Enzyme substrate
is added to produce
color reaction

FIGURE 12.13 A schematic representation of the procedure used for competitive ELISA. This variation of a typical ELISA shown in Figure 12.12 can be used to detect very low concentrations of antigen.

biphenyls (PCBs) (Schreiber and Pedersen, 1996) with detection limits ranging from high μg per liter (parts per billion) to low mg per liter (parts per million). For example, Schreiber and Pedersen (1996) compared the use of a commercial immunoassay test kit with traditional gas chromatography-mass spectrometer (GC-MS) analysis for monitoring 161 soil, sediment and waste samples in a PCB-contaminated site. Results showed that in comparison to the GC analysis, the immunoassay had 3.1% false positives and 11.2% false negatives out of 161 samples.

12.3.3.3 Advantages and Disadvantages

The advantage of competitive ELISA lies in its ability to detect extremely low antigen concentrations. This is related to the inverse relationship between target concentration and signal strength described previously. In some cases, such as for highly contaminated samples, these assays are so sensitive that extensive dilution may be required, thereby introducing potential for erroneous analysis. Other advantages include ease and rapidity of analysis, portability, the ability to operate at remote locations and a reduced cost compared to conventional measurement. Perhaps the biggest problem associated with immunoassays is related to their selectivity for a contaminant. This is because the antibody can cross-react with molecules that have a structure similar to the contaminant. Immunoassay of polyaromatic hydrocarbons (PAHs) illustrates this point well. A PAH-specific antibody will react to some extent with many different PAH molecules causing potential overestimation of the amount of contaminant present. This was illustrated by Barcelo *et al.* (1998), who showed that immunoassay results overestimated PAH concentration in river water samples by approximately one order of magnitude in comparison to GC-MS analysis. When the contaminant has a more unique structure (as do many of the pesticides), the problem of overestimation is reduced.

12.3.4 Immunomagnetic Separation Assays

12.3.4.1 Technique

Magnetic immunoseparation is an antigen capture methodology that uses antibodies conjugated to paramagnetic beads to attach to, concentrate and purify antigens. Immunomagnetic separation, especially with the development of immunomagnetic nanoparticles, allows for the specific manipulation of microorganisms, proteins and nucleic acids. In its simplest form, an immunomagnetic separation is accomplished with antibody-coated magnetite beads and a magnet. Essentially, the antibodies coated on the magnetic beads bind with antigens in solution, and are then separated from the solution using a magnet (Figure 12.14). The beads used for magnetic separation are small, ranging from 75 nm to 20 μm, and are typically made of iron oxide (magnetite). Such particles react strongly in a magnetic field but do not retain any magnetism when the magnetic field is removed.

12.3.4.2 Application

Immunomagnetic separation was used to recover thermophilic sulfate-reducing bacteria from oil field waters below oil production platforms in the North Sea (Christensen *et al.*, 1992). These bacteria can proliferate in oil field waters and cause considerable problems for oil companies during oil recovery, so the ability to detect their presence is of great value to the petroleum industry. In this example,

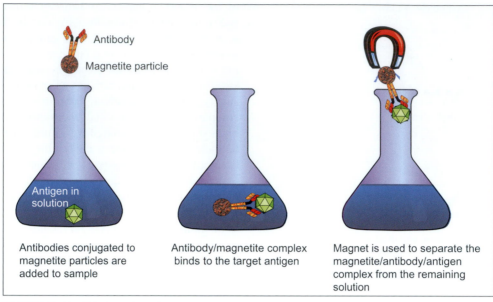

FIGURE 12.14   This is a schematic representation of the principle of immunomagnetic separation. In the first flask is a solution containing the antigen (virus). Antibodies specific to the virus are added to this solution. These antibodies, which have been conjugated (attached chemically) to a paramagnetic particle such as magnetite, then bind the virus. When a magnet is applied to the solution the magnetic particle is attracted to the magnet along with the attached antigen (virus). Immunomagnetic separation is a valuable tool for concentrating and purifying antigens from complex solutions.

immunobeads specific against cell wall antigens of the thermophilic *Thermodesulfobacterium mobile* captured several different isolates from oil-containing strata. Only one of the bacteria isolated by this method was serologically and morphologically identical to the bacterium (*T. mobile*) for which the antibodies were designed. Two other species of bacteria isolated using the immunomagnetic beads were spore forming and similar to *Desulfotomaculum* sp., a sulfate-reducing bacterium that had previously been isolated from oil fields. However, Western blots (see Section 12.3.5) of whole cells showed that the isolates were serologically different from *Desulfotomaculum* sp. This is a good illustration of the fact that an antibody that is designed against one organism can cross-react with other organisms that are serologically different.

Rapid detection of pathogens in food, water and air is one area of current research focus. Yang *et al.* (2007) used immunomagnetic nanoparticles in combination with real-time PCR (Information Box 13.7) to achieve very sensitive detection of the pathogen *Listeria monocytogenes* in milk. With this combined technique, they achieved a sensitivity of 226 CFU per 0.5 ml of milk, which is lower than for most other methods currently available.

12.3.4.3 Advantages and Disadvantages

Advantages of immunomagnetic separation techniques include relatively efficient separation of the target antigen,

low cost and potential for automation. In most cases, immunomagnetic separation is also one of the easiest methods available for specific target isolation. Though cross-reactivity issues are a concern, as with any antibody-oriented methodology, in most cases the problems associated with nonspecific signal production can be overcome. As with any microbiology protocol such immunomagnetic separations require optimization and proper choice of format (i.e., microcentrifuge separation or column matrix separation).

12.3.5 Western Immunoblotting Assays

12.3.5.1 Technique

Western immunoblotting is a three step, binding assay used to identify the presence of target antigens in a complex mixture of many other nontarget antigens such as might be found in environmental samples. This assay can be done with simple dot blot hybridization with a labeled antibody, or an electrophoretic separation followed by hybridization with the labeled antibody. In dot blot hybridization, environmental samples are added directly to an immobilizing nitrocellulose membrane, followed by immunolabeling and signal detection. In the second technique, a sample of antigen is added to a gel and separated by size using electrophoresis. After electrophoretic separation, the sample is transferred to an immobilizing

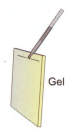

Step 1. Sample is added to an
electrophoresis gel

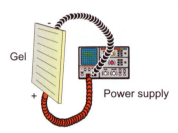

Gel

Power supply

Step 2. Sample is separated by size
as it migrates through the
gel matrix

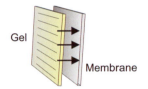

Gel

Membrane

Step 3. Sample is transferred to a
nitrocellulose membrane

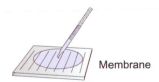

Membrane

Step 4. Labeled antibody is added to the
nitrocellulose membrane and
allowed to bind to the target
antigen

Membrane

Step 5: Target antigen is detected
by color production

FIGURE 12.15 A schematic representation showing the basic steps involved in Western blot immunoassay.

nitrocellulose membrane. This membrane is then incubated with enzyme-labeled or radiolabeled antibodies that specifically bind to the antigen. After incubation, a substrate for enzyme-labeled or photographic film for radiolabeled substrate is used to detect the presence of the

target antigen. Either method (dot blot or electrophoretic separation) indicates the presence and relative quantity of an antigen. If a separation step is used, this also allows molecular size determination of the antigen, which aids in confirming its identity. A schematic diagram of the electrophoretic separation and detection process is shown in Figure 12.15.

12.3.5.2 Application

Methylosinus trichosporium OB3b is a methanotrophic bacterium that has been studied for bioremediation of trichloroethylene (TCE). TCE is a common environmental pollutant (see Section 16.2.3.3). The first step in the degradation pathway of TCE is the enzymatic cleavage of TCE by an enzyme called methane monooxygenase (MMO). Because little was known about the environmental factors that influence the rates of TCE degradation by *M. trichosporium*, studies were undertaken to optimize the cellular expression of the enzyme (MMO). One study used Western blotting to determine the amounts of MMO produced by *M. trichosporium* (Fitch *et al.*, 1993, 1996). The amount of signal produced by the Western blot analysis was compared with a standard curve. The maximum amount of signal was then used as an index of optimized MMO expression conditions, which could then be correlated with optimal rates of TCE degradation. A more recent study used Western blotting to help examine the expression of a cadmium binding protein that was engineered into two TCE degraders to help make the degraders tolerate the presence of toxic levels of cadmium (Lee *et al.*, 2006b). Results showed that the two engineered strains (*Pseudomonas* and *Rhizobium*) recovered the ability to degrade TCE in the presence of cadmium when they expressed the cadmium-binding protein.

12.3.5.3 Advantages and Disadvantages

The obvious advantage of immunoblotting is its ability to specifically detect a particular antigen (target) within a heterogeneous matrix. Further, immunoblotting can detect extremely low levels of target antigen. However, this procedure is not quantitative, and it is subject to the problems inherent to all antibody-based methods.

12.3.6 Immunoaffinity Chromatography Assays

12.3.6.1 Technique

Affinity chromatography is a very powerful method used in purification and concentration of antigens. In affinity chromatography, the antibody is chemically bound to an inert support matrix (usually a glass, latex or plastic bead) in a chromatography column. The sample containing the antigen is eluted through the column, and the antigen is selectively retained within the column while the sample passes through. After the sample is run through the column, the purified antigen is eluted, usually by changing the pH of the column, which causes the antigen to detach from the antibody. The antigen can then pass out of the column and be collected in highly purified form (Figure 12.16). This process provides a very efficient means of both concentration and purification. Immunoaffinity chromatography offers several advantages compared with conventional purification techniques. Not only is the process selective and efficient, but it also enables the processing of large-volume samples with relatively few steps.

12.3.6.2 Application

In terms of application, there is currently a virtual explosion in the development of commercial immunochromatographic assays for various pathogens including protozoa (*Giardia*, *Cryptosporidium*), bacteria (*Yersinia pestis*, *Listeria*) and viruses (parvovirus, rotavirus) that allow rapid screening for these organisms (e.g., Magi *et al.*, 2006; Garcia and Garcia, 2006). These assays are done in cartridges or on test strips.

Another example is the use of immunochromatography for recovery of high value enzymes. Pyranose oxidase is an enzyme made by many types of fungi. It catalyzes the oxidation of D-glucose in the presence of oxygen to form 2-dehydro-D-glucose and hydrogen peroxide (H_2O_2). These types of enzymes are important in many industrial processes to aid in catalysis of reactions that are used to synthesize different carbohydrate products. Schafer *et al.* (1996) described the use of immunoaffinity chromatography for the highly efficient purification of a pyranose oxidase produced by *Phlebiopsus gigantea*. They used antibodies specific for other pyranose oxidases to construct an immunoaffinity column to purify this enzyme from mycelial extracts. The researchers were able to get yields of 71% of highly purified enzyme. Thus, immunoaffinity chromatography is one of the most powerful methodologies for the isolation, purification and concentration of antigens including active enzymes from complex samples.

12.3.6.3 Advantages and Disadvantages

Immunoaffinity chromatography for pathogens is rapid and available at commercial scale. One concern with these assays is that they produce false-positive results. Immunoaffinity chromatography for the purification of high value enzymes is rapid and efficient, but some problems encountered with this technique include a limited degree of sample concentration, and the possibility for column clogging if the sample applied is too turbid.

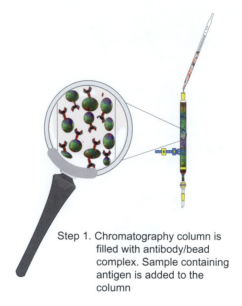

Step 1. Chromatography column is filled with antibody/bead complex. Sample containing antigen is added to the column

Step 2. The antigen is bound and retained within the column by the bead/antibody complex

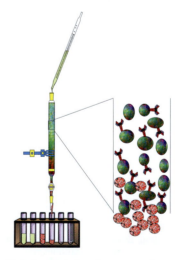

Eluted sample fractions are collected

Step 3. Bound antigen is eluted from the column by adding a high pH solution

FIGURE 12.16 A schematic representation showing the basic steps involved in antibody-mediated chromatographic separation.

Sample concentration is problematic because usually several column volumes are required for target elution after purification. Further, the column matrix can often retain the target even after elution.

12.3.7 Immunocytochemical Assays

12.3.7.1 Technique

An immunocytochemical assay is used for the detection and determination of the cellular localization of target antigens. The purpose of an immunocytochemical assay is to determine where target antigens are localized within a particular cell. For instance, you can determine whether the target antigen (a particular protein, for example) is localized in the cytoplasm or on the cell surface. In many cases, light microscopes are used to determine the location of antigens within a eukaryotic cell. However, immunocytochemical assays more often involve the use of electron microscopes to increase resolution and magnification of the area being studied. Many types of antibody labels can be used for electron microscopy, including heavy metals such as colloidal gold, enzymes such as horseradish peroxidase and proteins such as ferritin (Figure 12.11, left panel). These labels are visualized by the electron microscope as electron dense regions (Bozzola and Russell, 1992). The process involved in the preparation of a sample for immunocytological analysis is complicated. The sample is fixed with a preservative to maintain the original localization and antigenicity of the target. The sample is then embedded in plastic, sectioned, placed on an electron microscope grid, immunolabeled, poststained and finally viewed under the electron microscope.

12.3.7.2 Application

An example of the use of immunocytochemical techniques in environmental microbiology is in the study of the parasitism of certain amoebae by *Legionella pneumophila*. *L. pneumophila* is an intracellular parasite and is the causative agent of Legionnaires' disease (Section 22.2.7). This bacterium is found in aquatic environments and has been shown to parasitize and multiply within some protozoa (Declerck *et al.*, 2009). The sequence of events in the intracellular infection of the amoeba *Hartmannella vermiformis* by *L. pneumophila* was examined by Kwaik (1996) using an immunocytochemical assay. The goal of this study was to compare the intracellular infection of the amoebae with the infection of human alveolar macrophages that occurs during onset of Legionnaires' pneumonia, to aid in understanding the environmental life cycle of this human pathogen. Specifically, these researchers wanted to determine whether the accumulation of

ribosomes and rough endoplasmic reticulum (RER) around amoebic phagosomes containing *L. pneumophila* was similar to that observed in human phagosomes. To accomplish this, monolayers of *H. vermiformis* amoeba cells were infected with *L. pneumophila*. The infected cells were harvested, fixed and embedded in plastic. The samples were then thin sectioned, collected on sample grids and incubated with an antibody specific for Bip, a heat shock protein associated with RER. This was followed by secondary incubation with an antiglobulin labeled with colloidal gold particles (indirect immunoassay), and examination by transmission electron microscopy. The results demonstrated that there is considerable similarity in the ultrastructure of phagosomes containing *L. pneumophila* in both amoeba and humans. Using an immunocytochemical assay, it was also found that similar to the human system, RER-specific protein (Bip) was present in the phagosomal membrane of the amoebae. Further, the localization of the RER and ribosomes in the amoebic phagosome was identical to the localization seen in human macrophage phagosomes. This study helped indicate a possible role of the RER in the protection and growth of the *L. pneumophila* within the phagosomes of both hosts.

12.3.7.3 Advantages and Disadvantages

Immunocytochemical assays used in conjunction with electron microscopy provide high resolution of the spatial interactions in the system being studied. This is not a quantitative technique, but rather one that examines qualitative interactions in the sample being studied. Like all the techniques discussed it is subject to the problems inherent to all antibody-based methods.

12.3.8 Immunoprecipitation Assays

12.3.8.1 Technique

Immunoprecipitation is a methodology that uses the antigen—antibody reaction in solution to semi-quantitatively determine the amount of antigen or antibody in a sample by determining the amount of precipitation or clumping of the antigen—antibody complex. Immunoprecipitation can be used to determine the concentration of low levels of antigen, or can be used to quantify or titer antibodies or antigens. Immunoprecipitation can also be used to determine the optimal concentration ratio for precipitation of an antibody and antigen. Most commonly, a series of reaction tubes are set up, each of which contains a constant titer of antibodies. Antigen is then added in increasing concentration to consecutive tubes. In the initial tubes, where the lowest concentration of antigen has been added, there is no obvious precipitation. As the antigen concentration is increased, the formation of antigen—antibody complexes increases until a visible precipitate is formed. As the antigen concentration is further increased, it will eventually exceed the concentration of antibody present and the amount of precipitate will decrease again (Figure 12.17).

This may seem strange until one entirely understands the nature of antibodies and antigens. IgG antibodies are considered to be bivalent, whereas antigens are most often multivalent; in other words they have multiple antibody reactive sites or epitopes as shown in Figure 12.18A. This means that antibodies are able to bind at least two antigens at once, and antigens can be bound by more than one antibody at a time. When there is excess antibody (Figure 12.18B), multiple antibodies bind to each antigen and no cross-linking beyond this takes place. In this case, there is no visible precipitation and considerable antibody is still found in solution. When antibody and antigen are present in optimal proportions, antigen—antibody cross-linking is much more extensive forming large complexes (Figure 12.18D). This results in precipitation from solution. As the antigen concentration is increased past this optimal antigen—antibody proportion, smaller and smaller complexes form with only one molecule of antigen, no precipitation occurs and excess antigen is found in the supernatant (Figure 12.18C).

12.3.8.2 Application

This technology has been utilized in environmental microbiology to determine the mechanism for the inhibition of certain economically important fungal plant pathogens by another nonpathogenic fungus. It was hypothesized that the fungus *Talaromyces flavus* can control the proliferation of several fungal plant pathogens including *Sclerotinia sclerotionum*, *Rhizoctonia solani* and *Verticillium dahliae* by the production of hydrogen peroxide. *T. flavus* produces hydrogen peroxide as a product of glucose metabolism in the presence of the enzyme glucose oxidase. To determine

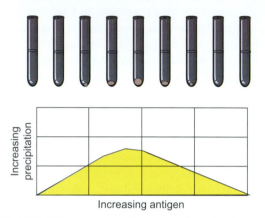

FIGURE 12.17 This is a schematic representation of an immunoprecipitation assay. The test tubes at the top all have constant concentrations of antibody and increasing concentrations of antigen starting from the left. In the middle tubes there is optimum antigen—antibody interaction and precipitation of the antibody and antigen occur. As indicated by the graph on the bottom there is a point where the highest amount of precipitation forms. This type of assay is used to determine optimum antibody to antigen ratios for immunoassays. It is also useful for quantitating either antibody or antigen concentrations in solutions.

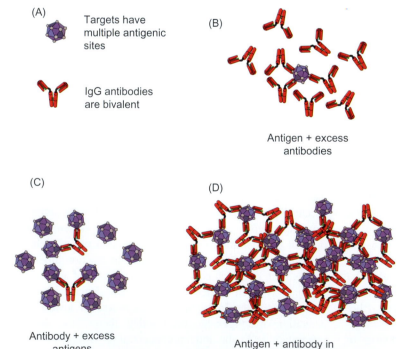

FIGURE 12.18 This is a schematic representation of what occurs in an immunoprecipitation assay. (A) Targets can have multiple antigenic sites. An IgG antibody has two binding sites (bivalent). (B) In this case there is excess antibody so little antibody—antigen binding occurs resulting in no precipitation. (C) In this case there is excess antigen and while antibody—antigen binding occurs, there is little cross-linking and so precipitation does not occur. (D) In this case the amount of antibody and antigen are optimal, there is extensive cross-linking of antibody—antigen complexes and this results in maximum precipitation as shown in Figure 12.17.

whether hydrogen peroxide was responsible for the inhibition, cultures containing the pathogens were incubated with and without glucose oxidase. The cultures without glucose oxidase showed a high percentage of germination, whereas cultures containing glucose oxidase exhibited very low germination (inhibition). To confirm these results, the cultures with glucose oxidase were subjected to immunoprecipitation. Antibody to the glucose oxidase was added to the cultures, effectively removing the glucose oxidase from the culture. After immunoprecipitation of the glucose oxidase, the fungal pathogens recovered and showed high levels of germination. Germination was subsequently halted again when more glucose oxidase was added. This research effectively showed that glucose oxidase is the enzyme produced by *T. flavus* that controls the plant pathogens.

12.3.8.3 Advantages and Disadvantages

In addition to the type of application described, this method is important in the development of immunoassays and the characterization of antigen–antibody interactions. This assay is relatively easy and inexpensive to perform though its use requires careful optimization. Disadvantages vary depending on what type of assay and application is being considered. As with all immunoassays, the possible nonspecific interactions with nontarget antigens are always an issue.

12.3.9 Lateral Flow Immunoassay

12.3.9.1 Technique

Lateral flow immunoassay (LFIA), also known as immunochromatographic assay, is an antibody-based analytical method which uses a prefabricated strip of a carrier membrane containing dry reagents that are activated by applying a fluid sample via capillary action. It can be used as a rapid diagnostic pregnancy test, or to test failure of internal organs (e.g., heart attack or diabetes). Additionally, it can be used to detect contamination of food or water with specific pathogens.

The test strip is composed of a series of capillary zones including a sample pad, conjugate pad, reaction membrane and absorbent pad. These are made of porous polymeric materials, to allow analytes in a sample to travel laterally across the strip and react with reagents on the strip (Figure 12.19A). When a liquid sample is applied onto the sample pad and becomes saturated, the capillary fluid moves to the conjugate pad in which detector regents (target analyte-specific antibody conjugated with label) are stored for immunological recognition of analytes in the sample. Labels are used that generate a visually detectable mark (colored or fluorescent materials), but also allow for an unobstructed flow through the membrane. Labels are

generally fluorescent dyes or nanoparticles, colloidal gold particles, magnetic particles, colored latex beads or dyed liposome. Typically, there are at least two lines in the reaction membrane: the test line and control line. The absence or presence of a colored line in the test and control region indicates a negative (absence of analyte in sample), positive (presence of analyte in sample) and test failure (test strip does not work well) (Figure 12.19A). Two predominant approaches to the test are the sandwich and competitive reaction schemes depending on the molecular weight, and the number of antigenic sites for binding target analytes. For analytes with high molecular weight and more than one epitope (antigenic determinant or antigen-binding site), the sandwich format is applicable using two different antibodies to recognize separate epitopes of target analytes, which are utilized for detector reagents and capture reagents (Figure 12.19B). Once detector reagents in the conjugate pad react with target analytes, the mixture migrates to the reaction membrane to react with capture reagents at the test line that involves capturing a target analyte between two layers of antibodies (i.e., detector reagents and capture reagents). This immunological reaction provides a colored line within the test region, which indicates the presence of target analytes in a sample. In addition, detector reagents bind to secondary antibody (anti-immunoglobulin against detector reagents) that are immobilized at the control line, and generate a colored line for test validation. In the sandwich format, the positive response is directly proportional to the concentration of analyte in the sample. When the analyte is of low molecular weight and has only a single antigenic determinant, the competitive format can be utilized by spraying the antigens as capture reagents at the test line to entrap nonreacted detector reagents with target analytes in the sample (Figure 12.19C). Because detector reagents have limited antibody binding sites against target analytes, in occupied sites detector antibodies can bind to capture reagents at the test line and secondary antibodies at the control line. This generates a colored line at the test and control region during sample migration. In the competitive format, the color of the positive mark is inversely correlated to the analyte concentration. After passing these reaction zones, the fluid moves to the absorbent pad, which wicks the liquid to the end of the strip due to capillary force, thus maintaining the flow and acting as a waste container.

12.3.9.2 Application

A lateral flow immunoassay was developed to detect ricin, a cytotoxin found naturally in the seeds of the castor bean *Ricinus communis*, that has been studied for the application of cancer therapy, but has also been considered for potential use as a bioweapon (Shyu *et al.*, 2002). The immunochromatographic assay was based on the sandwich format using two different monoclonal antibodies to recognize the specific binding sites of the

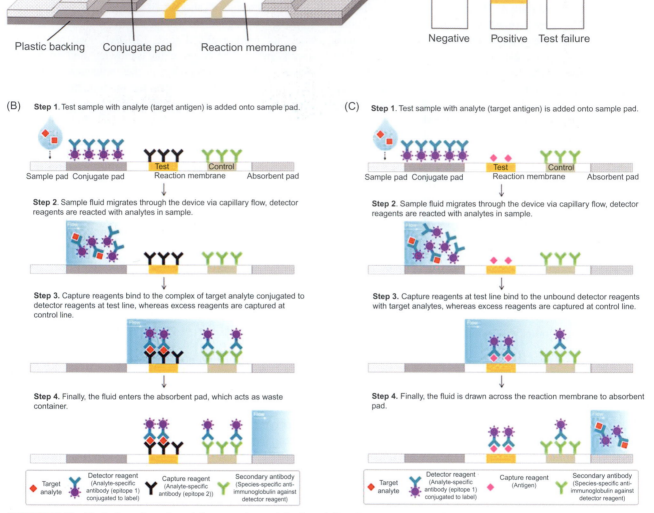

FIGURE 12.19 Typical configuration of the test strip used for lateral flow immunoassay. (A) Sample that is added onto sample pad flows through the strip due to capillary force of fluid, passing each bed and coming into contact with dried reagents to make an immunological reaction sequentially. With color indicators at both the test line and control line, the result can be interpreted as negative (absence of analyte in sample), or positive (presence of analyte in sample). If the test fails, no marks are seen. Depending on immunological properties of analyte, the lateral flow immunoassay can be divided into two types: sandwich format (B); or competitive format (C). Image courtesy H.-W. Yu.

ricin protein. One anti-ricin antibody was immobilized to a defined detection zone on a porous nitrocellulose membrane, while the other anti-ricin antibody was conjugated to colloidal gold particles (25 nm), which served as a detector reagent. Colloidal gold particles at contamination concentrations less than 100 nm develop a red color, such that the test line provides a red color with an intensity proportional to the ricin concentration. With this method, 50 ng/ml of ricin was detected in less than 10 min.

For ensuring the safety of feedstock for animals, a lateral flow dipstick was developed to allow rapid screening of aflatoxin B_1, a secondary metabolite mainly produced by *Aspergillus* spp. This compound, also known as a mycotoxin, is classified as a carcinogenic substance (Delmulle *et al.*, 2005). The test strip was designed using a competitive immunoassay format with aflatoxin B_1–bovine serum albumin conjugate as a capture reagent, and colloidal gold particles (40 nm) coated with anti-aflatoxin B_1 monoclonal antibody as a detector reagent. In the presence of aflatoxin B_1 extracted from the pig feed matrix, the dipstick provides a pink color at the test line, with an inverse intensity relationship against the toxin contents. In this study, the visual detection limit for aflatoxin B_1 was 5 μg/kg within 10 min.

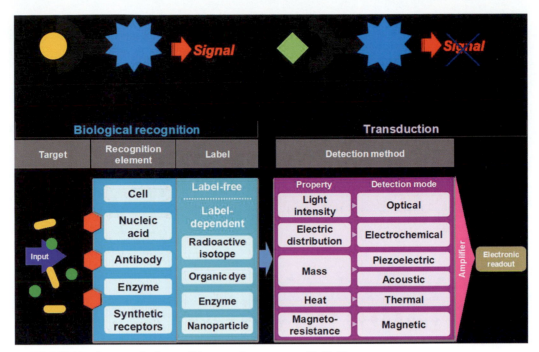

FIGURE 12.20   Conceptual scheme of biosensor. Image courtesy H.-W. Yu.

In addition to toxic compounds, Yan *et al.* (2006) suggested an up-converting phosphor technology (UPT)-based lateral flow immunoassay for quantitative detection of *Yersinia pestis*, which is a pathogenic bacterium that results in plague, a deadly infectious disease. A sandwich immunoassay was employed by using same polyclonal antibodies for detector and capture reagents, in which UPT particles (400 nm) were used as a luminescent label resulting in 10- to 100-fold more sensitivity than assays using conventional reporter systems such as colloidal gold or colored latex beads (Hampl *et al.*, 2001). The signal detection of the strip utilized an infrared laser to excite the UPT particles, then collection of the visible luminescence emitted, and finally its conversion to the voltage as a signal. The detection limit of this assay was 10^4 CFU/ml, taking less than 30 min to perform.

12.3.9.3 Advantages and Disadvantages

Lateral flow immunoassay is a single step assay that requires only the addition of a sample to a prefabricated test strip, with benefits of low cost, commercialization potential, ease of use and rapid detection of various analytes. However, this is not an accurate quantitative technique but rather a presence/absence test for the target analytes in the sample.

12.4 IMMUNOSENSORS

Biosensors are analytical devices that convert a biological response into an electrical signal. They consist of three components including a biological recognition receptor or bioreceptor, which recognizes the target analyte; a transducer, which converts the recognition event into a measurable electrical signal; and some kind of data processing system that allows the data to be displayed in a user-friendly manner. A conceptual depiction of a biosensor is shown in Figure 12.20.

Biosensors can be subdivided into two types based on the kind of biorecognition molecule utilized. Catalytic biosensors employ enzymes and/or microorganisms as the biorecognition molecule to catalyze a reaction with an analyte that results in a product. The other category of biosensors is the affinity biosensors, where the biorecognition molecules can be antibodies, DNA, peptides or lectins. Affinity biosensors are characterized by a binding event between the biorecognition molecule and the analyte (Byrne *et al.*, 2009). A transducer is a device that converts one form of energy to another, in this case the recognition event into a measurable electrical signal (Figure 12.21). The transduction efficiency in turn determines the efficacy of the biosensor including sensitivity, selectivity and signal stability. Generally, the transducer converts the biochemical interactions into measurable electrical signals. Overall, catalytic microbial biosensors are utilized for the assessment of chemical toxicity, whereas affinity biosensors are used for the detection of pathogenic microorganisms and their associated toxins.

Immunosensors are affinity biosensors that utilize antibodies as bioreceptors and rely on the basic Ag—Ab reaction as the recognition event based on the lock and key mechanism of three-dimensional shape fitting and

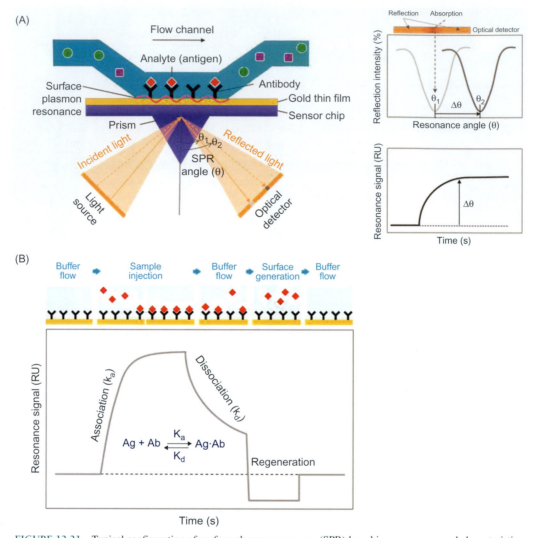

FIGURE 12.21   Typical configuration of surface plasmon resonance (SPR)-based immunosensor and characteristics of a sensorgram. (A) SPR immunosensor can detect the binding of target analyte (antigen) in solution to antibody immobilized on the gold surface of a sensor chip, which alters the refractive index of the medium near the surface. This change can be monitored noninvasively in real time as a plot of resonance signal (proportional to mass change) versus time. (B) Upon injection, binding of the analyte results in an increase in resonance signal (association phase). After equilibrium, the sample is replaced by a running buffer, and the decrease in signal represents the dissociation of analyte from the antigen−antibody complex on the surface. A regeneration solution is used to disrupt binding and regenerate the sensor surface for further measurement. Therefore, the concentration of analyte in the sample and kinetic rate constants (K_a and K_d) can be derived from the sensorgram. Image courtesy H.-W. Yu.

chemical bonds. The transducing agent is the entity that differentiates immunosensors from most standard immunoassays. Immunosensors are becoming increasingly popular for the detection of waterborne pathogens and their associated toxins (Table 12.1) in environmental samples including water.

Aptamers, also known as "artificial antibodies," are single-stranded nucleic acids (RNA or DNA) with defined tertiary structures for selective binding to target molecules (Nutiu and Li, 2003). While aptamers are analogous to antibodies in their range of target recognition and variety of

applications (Stoltenburg *et al.*, 2007), they possess several key advantages (Table 12.2) and have been used widely for constructing a new type of affinity biosensor. Mimicking natural evolution, a selection process called systematic evolution of ligands by exponential enrichment (SELEX) has resulted in many aptamers that can bind to a broad range of biological targets with high affinity and with specificity comparable to that of antibodies, such as small organic molecules, proteins, viruses and cells (Li *et al.*, 2012). There have been attempts to identify aptamers that are specific to target analytes including pathogenic bacteria,

TABLE 12.1 Examples of Immunosensors for the Detection of Pathogens and Associated Toxins

Target Analyte	Label	Transducer	Type of Signal	Limit of Detection	Analysis Time	Matrix	Reference
E. coli	Free	Gold screen-printed electrode	Electrochemical (Impedance)	3.3 CFU/ml	60 min	River, tap water	Escamilla-Gómez *et al.*, 2009
Vibrio cholerae	Alkaline phosphatase	Carbon screen-printed electrode	Electrochemical (Amperometry)	8 CFU/ml	55 min	Ground water	Sharma *et al.*, 2006
Cryptosporidium parvum	Alkaline phosphatase and gold nanoparticle	Indium tin oxide electrode	Electrochemical (Voltammetry)	3 oocysts/ml	70 min	Drinking water	Thiruppathiraja *et al.*, 2011
Cholera toxin	Potassium ferrocyanide-encapsulated liposome	Carbon nanotube	Electrochemical (Voltammetry)	10 pg/L	65 min	Tap water	Viswanathan *et al.*, 2006
Microcystin	Quantum dot	Microbead	Flow cytometry (Fluorescence)	0.5 µg/L	30 min	Tap water	Yu *et al.*, 2012

TABLE 12.2 Comparative Properties of Antibody and Aptamer

Characteristic		Antibody	Aptamer
Target molecule		Antigen (protein or peptide)	Amino acid, peptide, nucleic acid, and microbial cell
Affinity		High (K_d: pM − nM)	High (K_d: pM − nM)
Molecular weight		<150 kD	<50 kD
Size[a]		~139 Å ~122 Å IgG antibody	~25 Å ~21 Å Thrombin aptamer
Chemical stability		Low	High
Reusability		Impossible	Possible
Production	Method	*In vivo* biological process	*In vitro* chemical process
	Development period	<2 months	<6 months
	Cost	High	Low

[a]*Lee* et al. (2006a).

viruses and protozoa. Table 12.3 shows examples of a few publications on the utilization of aptamers as a bioreceptor.

Proof-of-concept experiments have been performed under well-controlled laboratory conditions, but relatively little has been done to test these biosensors in actual environmental water matrices. Environmental water samples pose a number of challenges to biosensor development because of the potential for microbial, particulate and

TABLE 12.3 Select Aptasensors Utilized for Detection of Pathogens and Their Toxins

Target	Aptamer			Biosensor Application				
	DNA/RA	Sequence length (nucleotides)	Binding Site	Label	Detection Mode	Limit of Detection	Analysis Time	Reference
Bacillus thuringiensis	DNA	60	*B. thuringiensis* spore	Quantum dot	Fluorescence	10^3 CFU/ml	>20 min	Ikanovic et al., 2007
E. coli O111	DNA	72	Lipopolysaccharide	Alkaline phosphatase	Voltammetry	112 CFU/ml	>3.5 h	Luo et al., 2012
Salmonella typhimurium	DNA	40	*S. typhimurium* outer membrane Proteins	Nanoparticles	Luminescence	5 CFU/ml	>40 min	Duan et al., 2012
Staphylococcus aureus	DNA	88	N.A.	Nanoparticles	Luminescence	8 CFU/ml	>40 min	Duan et al., 2012

other organic and inorganic contaminants (Connelly and Baeumner, 2012). Moreover, although waterborne pathogens and toxins are typically present at low concentrations even in source water, the development of sampling protocols and techniques to concentrate environmental samples is still in its infancy (Vikesland and Wigginton, 2010).

12.4.1 Surface Plasmon Resonance-Based Immunosensor

12.4.1.1 Technique

Surface plasmon resonance (SPR)-based immunosensors provide a nondestructive optical analysis technique, useful for investigating the interaction of thin-layered biomolecules, especially antigen–antibody reactions, on the surface of sensor chip. As shown in Figure 12.21A, when a light beam is focused onto the gold surface of the sensor chip, evanescent wave photons produced by the incident light interact with the surface plasmons, that is, free oscillating electrons propagating along the surface of the gold film. Resonance occurs at a critical angle (SPR angle) of incident light, and light energy is transferred to the electrons in the metal film surface, causing a reduction of the reflected light from the metal film at the interface between two media of different refractive indices. This phenomenon can be utilized by measuring a change in the angle of light reflection, because the reflected beam covers the same angle range of incident light that depends on the refractive index at the gold surface. At resonance, a minimum in reflected light intensity is observed, and a dark line appears in the band of a photodetector, resulting in a sharp drop (also referred

to as the SPR "dip") in the plot of reflection intensity of this band versus the SPR angle. The SPR angle shifts when biomolecules bind to the surface and change the mass of the surface layer. This change in resonance angle can be monitored in real time as a plot of resonance signal (proportional to mass change) versus time, and subsequently used to calculate the refractive index and concentration of the adhered target analytes.

A typical sensorgram of an SPR-based immunosensor, a plot of resonance signal (expressed as resonance units (RU) versus time, is shown in Figure 12.21. A specific antibody to an analyte of interest is immobilized on the sensor surface with appropriate coupling chemistry. Before injection of the sample, a running buffer flows through a microfluidic flow cell of the SPR instrument. When a sample containing target analyte (antigen) is injected, binding of the analyte to the surface leads to the increase in refractive index, and consequently the resonance signal (association phase). During this association phase, the association rate constant (K_a in $M^{-1}s^{-1}$) of the antigen–antibody interaction (number of binding events per unit of time), which is a measure of antigen–antibody complex formation, can be determined by interpretation of a sensorgram with different concentrations of injected target antigens. At equilibrium, the signal will reach a plateau when the amount of analyte that is associating and dissociating with the antibody is equal. The response level at equilibrium is related to the concentration of target analyte in the sample. When the analyte solution is replaced with buffer, the antigen–antibody complex is allowed to dissociate as a function of the dissociation rate constant (number of dissociation events per unit of time) (K_d in s^{-1}) that describes the antigen–antibody complex stability. After dissociation, the sensor surface can be regenerated for the next measurement using a regeneration solution (for example, high salt or low pH) to remove the remaining bound analyte with a denaturing immobilized antibody. Therefore, this immunosensor can determine the analyte concentration in a sample if an appropriate calibration curve is constructed.

12.4.1.2 Application

An SPR immunosensor was developed for the detection of *Salmonella typhimurium* in contaminated water and food (Oh *et al.*, 2004). When using the SPR immunosensor, the orientation of antibody immobilized on the sensor surface is crucial to increase the activity of the antibody interacting with the analyte of interest. This improves the sensitivity for the detection of very low concentrations of antigens. Since protein G, an immunoglobulin-binding cell wall protein found in *Streptococci* sp., is capable of interacting with the Fc region of the antibody, the self-assembled protein G layer on the gold surface of the sensor using 11-mercaptoundecanoic acid leads to a well-oriented monolayer of antibody molecules for the target analyte. The shift in SPR angle has a linear relationship with the concentration of *S. typhimurium* over the range of 10^2 to 10^9 CFU/ml. Thus, the limit of detection of this sensor was four orders of magnitude more sensitive than a standard ELISA.

Homola *et al.* (2002) improved the sensitivity of a dual-channel SPR immunosensor based on wavelength modulation. They utilized a sandwich-binding format to detect staphylococcal enterotoxin B (SEB) in milk. SEB is an exotoxin produced by *Staphylococcus aureus*. It is one of the toxins responsible for staphylococcal food poisoning in humans, and has been produced by some countries as a biological weapon. The dual-channel (reference and sample channels) SPR sensor can provide compensation for nonspecific sensor response due to temperature-induced changes in sample refractive index. The sandwich binding of a secondary antibody to an analyte–antibody complex on the gold sensor surface leads to an increase in mass change and refractive index, which can amplify the sensor response by a factor of 10. The SPR immunosensor has been shown to be capable of detecting concentrations of SEB as low as 5 and 0.5 ng/ml, using direct and sandwich modes, respectively.

12.4.1.3 Advantages and Disadvantages

The SPR immunosensor is one of the best-known and commercially available optical label-free biosensors, useful for monitoring the specific binding event of an antigen–antibody continuously in real time with high sensitivity. However, mass transport can affect the kinetic analysis and any artifactual change in refractive index other than from the interaction can also give a false-positive signal.

QUESTIONS AND PROBLEMS

1. Describe an immunological assay for the accurate quantitation of *Rhizobium* around root nodules.
2. Describe an immunological assay for detection of *Giardia* in water samples.
3. Immunomagnetic purification methods are becoming popular in environmental microbiology. Sediment slurries can contain paramagnetic particles, which are copurified along with target antigens, making subsequent assay procedures difficult. Describe the problems that would arise with immunomagnetic analysis in this case, and devise potential methods for solving these issues.
4. How could you determine whether an enzyme is intercellular or extracellular using immunoassay techniques?
5. You have two monoclonal antibodies available from two different commercial companies. Both cost the same and both are specific for the rhizobia you are studying in question 1. After ordering both antibodies, how would you determine which antibody was the best for the assay you designed in question 1? How would you perform these evaluations?
6. Rapid detection of biological warfare agents is an emerging area of research. Design a rapid detection method using an immunoassay that can aid in the detection of *Bacillus anthracis*.
7. What immunoassays described offer quantitative results?
8. What immunoassays described allow for purification of antigens from heterogeneous samples?

REFERENCES AND RECOMMENDED READING

Barcelo, D., Oubina, A., Salau, J. S., and Perez, S. (1998) Determination of PAHs in river water samples by ELISA. *Anal. Chim. Acta* **376**, 49–53.

Black, C., Allan, I., Ford, S. K., Wilson, M., and McNab, R. (2004) Biofilm-specific surface properties and protein expression in oral *Streptococcus sanguis. Arch. Oral Biol.* **49**, 295–304.

Bouer-Kreisel, P., Eisenbeis, M., and Scholz-Muramatsu, H. (1996) Quantification of *Dehalospirillum multivorans* in mixed-culture biofilms with an enzyme-linked immunosorbent assay. *Appl. Environ. Microbiol.* **62**, 3050–3052.

Bozzola, J. J., and Russell, L. D. (1992) "Electron Microscopy: Principles and Techniques for Biologists," Jones and Bartlett, Boston.

Byrne, B., Stack, E., Gilmartin, N., and O'Kennedy, R. (2009) Antibody-based sensors: principles, problems and potential for detection of pathogens and associated toxins. *Sensors* **9**, 4407–4445.

Christensen, B., Torsvik, T., and Lien, T. (1992) Immunomagnetically captured thermophilic sulfate-reducing bacteria from North Sea oil field waters. *Appl. Environ. Microbiol.* **58**, 1244–1248.

Connelly, J., and Baeumner, A. (2012) Biosensors for the detection of waterborne pathogens. *Anal. Bioanal. Chem.* **402**, 117–127.

Declerck, P., Behets, J., Margineanu, A., van Hoef, V., De Keersmaecker, B., and Ollevier, F. (2009) Replication of *Legionella pneumophila* in biofilms of water distribution pipes. *Microbiol. Res.* **164**, 593–603.

Delmulle, B. S., De Saeger, S. M., Sibanda, L., Barna-Vetro, I., and Van Peteghem, C. H. (2005) Development of an immunoassay-based lateral flow dipstick for the rapid detection of aflatoxin B1 in pig feed. *J. Agric. Food Chem.* **53**, 3364–3368.

Duan, N., Wu, S., Zhu, C., Ma, X., Wang, Z., Yu, Y., *et al.* (2012) Dual-color upconversion fluorescence and aptamer-functionalized magnetic nanoparticles-based bioassay for the simultaneous detection of *Salmonella typhimurium* and *Staphylococcus aureus*. *Anal. Chim. Acta* **723**, 1–6.

Escamilla-Gómez, V., Campuzano, S., Pedrero, M., and Pingarrón, J. M. (2009) Gold screen-printed-based impedimetric immunobiosensors for direct and sensitive *Escherichia coli* quantisation. *Biosens. Bioelectron.* **24**, 3365–3371.

Fitch, M. W., Graham, D. W., Arnold, R. G., Speitel, G. E., Jr., Agarwal, S. K., Phelps, P., *et al.* (1993) Phenotypic characterization of copper-resistant mutants of *Methylosinus trichosporium* OB3b. *Appl. Environ. Microbiol.* **59**, 2771–2776.

Fitch, M. W., Speitel, G. E., Jr., and Georgiou, G. (1996) Degradation of trichloroethylene by methanol-grown cultures of *Methylosinus trichosporium* OB3b PP358. *Appl. Environ. Microbiol.* **62**, 1124–1128.

Garcia, L. S., and Garcia, J. P. (2006) Detection of *Giardia lamblia* antigens in human fecal specimens by a solid-phase qualitative immunochromatographic assay. *J. Clin. Microbiol.* **44**, 4587–4588.

Hampl, J., Hall, M., Mufti, N. A., Yao, Y. M., MacQueen, D. B., Wright, W. H., *et al.* (2001) Up-converting phosphor reporters in immunochromatographic assays. *Anal. Biochem.* **288**, 176–187.

Homola, J., Dostálek, J., Chen, S., Rasooly, A., Jiang, S., and Yee, S. S. (2002) Spectral surface plasmon resonance biosensor for detection of staphylococcal enterotoxin B in milk. *Int. J. Food Microbiol.* **75**, 61–69.

Ikanovic, M., Rudzinski, W., Bruno, J., Allman, A., Carrillo, M., Dwarakanath, S., *et al.* (2007) Fluorescence assay based on aptamer-quantum dot binding to *Bacillus thuringiensis* spores. *J. Fluoresc.* **17**, 193–199.

Kwaik, Y. A. (1996) The phagosome containing *Legionella pneumophila* within the protozoan *Hartmannella vermiformis* is surrounded by the rough endoplasmic reticulum. *Appl. Environ. Microbiol.* **62**, 2022–2028.

Lee, J. F., Stovall, G. M., and Ellington, A. D. (2006a) Aptamer therapeutics advance. *Curr. Opin. Chem. Biol.* **10**, 282–289.

Lee, W., Wood, T. K., and Chen, W. (2006b) Engineering TCE-degrading rhizobacteria for heavy metal accumulation and enhanced TCE degradation. *Biotechnol. Bioeng.* **95**, 399–403.

Li, L. L., Ge, P., Selvin, P. R., and Lu, Y. (2012) Direct detection of adenosine in undiluted serum using a luminescent aptamer sensor attached to a terbium complex. *Anal. Chem.* **84**, 7852–7856.

Luo, C., Lei, Y., Yan, L., Yu, T., Li, Q., Zhang, D., *et al.* (2012) A rapid and sensitive aptamer-based electrochemical biosensor for direct detection of *Escherichia coli* O111. *Electroanalysis* **24**, 1186–1191.

Magi, B., Canocchi, V., Tordini, G., Cellesi, C., and Barberi, A. (2006) *Cryptosporidium* infection: diagnostic techniques. *Parasitol. Res.* **98**, 150–152.

Neilson, J. W., and Maier, R. M. (2001) Biological techniques for measuring organic and metal contaminants in environmental samples. In "Humic Substances and Chemical Contaminants" (C. E. Clapp, M. H. B. Hayes, N. Senesi, P. R. Bloom, and P. M. Jardine, eds.), Soil Science Society of America, Madison, WI, pp. 255–273.

Nguyen, L. T. T., Cronberg, G., Annadotter, H., and Larsen, J. (2007) Planktic cyanobacteria from freshwater localities in ThuaThien-Hue province, Vietnam. II. Algal biomass and microcystin production. *Nova Hedwigia* **85**, 35–49.

Nutiu, R., and Li, Y. (2003) Structure-switching signaling aptamers. *J. Am. Chem. Soc.* **125**, 4771–4778.

Oh, B. K., Kim, Y. K., Park, K. W., Lee, W. H., and Choi, J. W. (2004) Surface plasmon resonance immunosensor for the detection of *Salmonella typhimurium*. *Biosens. Bioelectron.* **19**, 1497–1504.

Rendón, M. A., Saldaña, Z., Erdem, A. L., Monteiro-Neto, V., Vázquez, A., Kaper, J. B., *et al.* (2007) Commensal and pathogenic *Escherichia coli* use a common pilus adherence factor for epithelial cell colonization. *Proc. Natl Acad. Sci. U.S.A.* **104**, 10637–10642.

Schafer, A., Bieg, S., Huwig, A., Kohring, G., and Giffhorn, F. (1996) Purification by immunoaffinity chromatography, characterization, and structural analysis of a thermostable pyranose oxidase from the white rot fungus *Phlebiopsis gigantea*. *Appl. Environ. Microbiol.* **62**, 2586–2592.

Schreiber, J. A., and Pedersen, T. A. (1996) PCB immunoassay performance evaluation. *Remed. Manage.* July/Aug., 28–31.

Sharma, M. K., Goel, A. K., Singh, L., and Rao, V. K. (2006) Immunological biosensor for detection of *Vibrio cholerae* O1 in environmental water samples. *World J. Microb. Biot.* **22**, 1155–1159.

Shyu, R. H., Shyu, H. F., Liu, H. W., and Tang, S. S. (2002) Colloidal gold-based immunochromatographic assay for detection of ricin. *Toxicon* **40**, 255–258.

Stoltenburg, R., Reinemann, C., and Strehlitz, B. (2007) SELEX—a (r) evolutionary method to generate high-affinity nucleic acid ligands. *Biomol. Eng.* **24**, 381–403.

Thiruppathiraja, C., Saroja, V., Kamatchiammal, S., Adaikkappan, P., and Alagar, M. (2011) Development of electrochemical based sandwich enzyme linked immunosensor for *Cryptosporidium parvum* detection in drinking water. *J. Environ. Monit.* **13**, 2782–2787.

Vikesland, P. J., and Wigginton, K. R. (2010) Nanomaterial enabled biosensors for pathogen monitoring—a review. *Environ. Sci. Tech.* **44**, 3656–3669.

Viswanathan, S., Wu, L. C., Huang, M. R., and Ho, J. A. (2006) Electrochemical immunosensor for cholera toxin using liposomes and poly(3,4-ethylenedioxythiophene)-coated carbon nanotubes. *Anal. Chem.* **78**, 1115–1121.

Watanabe, E., Baba, K., Eun, H., Arao, T., Ishii, Y., Ueji, M., *et al.* (2005) Evaluation of a commercial immunoassay for the detection of chlorfenapyr in agricultural samples by comparison with gas chromatography and mass spectrometric detection. *J. Chromatogr. A* **1074**, 145–153.

Yan, Z., Zhoua, L., Zhao, Y., Wang, J., Huang, L., Hu, K., *et al.* (2006) Rapid quantitative detection of *Yersinia pestis* by lateral-flow immunoassay and up-converting phosphor technology-based biosensor. *Sensor Actuat. B* **119**, 656—663.

Yang, H., Qu, L. W., Wimbrow, A. N., Jiang, X. P., and Sun, Y. P. (2007) Rapid detection of *Listeria monocytogenes* by nanoparticle-based immunomagnetic separation and real-time PCR. *Int. J. Food Microbiol.* **118**, 132—138.

Yu, H. W., Kim, I. S., Niessner, R., and Knopp, D. (2012) Multiplex competitive microbead-based flow cytometric immunoassay using quantum dot fluorescent labels. *Anal. Chim. Acta* **750**, 191—198.

Zhu, X. X., Hu, B. S., Lou, Y., Xu, L. N., Yang, F. L., Yu, H. N., *et al.* (2007) Characterization of monoclonal antibodies for lead-chelate complexes: applications in antibody-based assays. *J. Agric. Food Chem.* **55**, 4993—4998.

Chapter 13

Nucleic Acid-Based Methods of Analysis

Emily B. Hollister, John P. Brooks and Terry J. Gentry

When describing their discovery of the double-stranded structure of DNA, Watson and Crick (1953) famously wrote, "It has not escaped our notice that the specific pairing we have postulated immediately suggests a possible copying mechanism for the genetic material." This understated revelation not only enabled elucidation of the DNA replication process, but also ultimately provided the foundation for the plethora of molecular biology-based methods that have been developed over the past several decades. With these new approaches, microbiologists are now able to use a small sample of microbial nucleic acid to identify microorganisms, track genes, and evaluate genetic expression in the environment. Furthermore, the extraction of nucleic acids directly from environmental samples (soil, water, air), followed by nucleic acid-based analysis, has helped to overcome some of the biases of culture-based assays that require isolation and growth of organisms in the laboratory before they can be studied (see Case Study 4.2). These technologies include a variety of different methods such as: polymerase chain reaction (PCR)-based assays, microarrays, DNA sequencing and metagenomics. Given the breadth and rapid development of nucleic acid-based technologies over the past few decades, it is impossible to cover all of these methods in one textbook let alone one book chapter. Therefore, the goal of this chapter is to present some of the most commonly used nucleic acid-based methodologies, with a focus on their theoretical foundations and applications for detecting and characterizing environmental microorganisms.

13.1 STRUCTURE AND COMPLEMENTARITY OF NUCLEIC ACIDS

Molecular analyses are dependent on the type and sequence of the nucleic acids. Nucleic acids are in one of

I.L. Pepper, C.P. Gerba, T.J. Gentry: Environmental Microbiology, Third edition. DOI: http://dx.doi.org/10.1016/B978-0-12-394626-3.00013-2
© 2015 Elsevier Inc. All rights reserved.

two forms, either deoxyribonucleic acid (DNA) or ribonucleic acid (RNA). Nucleic acids consist of a phosphate sugar backbone, where the 5′ carbon atom of one sugar is covalently linked to the 3′ carbon atom of the adjacent sugar. The difference between the two forms is a single base substitution and the presence or absence of a hydroxyl group in the 3′ carbon of the sugar. DNA is made up of four deoxynucleotide bases: guanine (G); cytosine (C); adenine (A); and thymine (T), while uracil (U) substitutes for thymine in RNA. For convenience, the nucleotide bases are generally referred to by their single letter abbreviations. The bases are divided based on their chemical composition and structure into either purine (G and A) or pyrimidine bases (C, T, U). In DNA, the sugar is deoxyribose, while in RNA the sugar is ribose. As will be discussed shortly, the presence of a hydroxyl group on the 3′ carbon of deoxyribose plays a critical role in replication of DNA.

Structurally, DNA consists of two strands of deoxyribonucleic acids combined together to form a double helix. One strand of DNA is oriented 5′ to 3′, while the complementary strand is oriented 3′ to 5′. These two strands are linked by hydrogen bonds between corresponding pairs of bases. Specifically, G binds only to C, and A binds only to T. For each G-C pairing, there are three hydrogen bonds, whereas for the A-T pairing there are only two hydrogen bonds (Figure 13.1). Because these bases are specific in their ability to bind together to form a base pair (bp), they are said to be complementary to each

other. Single-stranded pieces of DNA (ssDNA) form a double-stranded DNA helix (dsDNA) if it comes in contact with another ssDNA molecule that has a complementary sequence. For example, a DNA strand containing the nine bases 5′-A-T-T-C-G-G-A-A-T-3′ will anneal to its complementary strand 3′-T-A-A-G-C-C-T-T-A-5′ with the resulting dsDNA being nine base pairs in length. Ideally, the two strands of a DNA molecule will be 100% complementary; however, there are cases where mismatches or incorrect base matching occurs which is stabilized by the hydrogen bonds of the surrounding correctly paired bases.

Recall that most bacteria have a single circular DNA chromosome that is in excess of one million base pairs (see Section 2.2.3). This chromosome encodes all of the genetic information that makes each bacterial population unique. For this information to be of use it must be replicated and then translated into RNA and eventually transcribed into structural proteins or enzymes that carry out all the activities of the cell (see Information Box 13.1). In DNA replication, one strand of DNA serves as a template for the synthesis of its complementary strand. The enzyme responsible for replication, DNA polymerase, adds a nucleotide complementary to the template strand to the free hydroxyl at the 3′ carbon of the last sugar on the growing DNA chain, and thus DNA is synthesized only in the 5′ to 3′ direction. The double-stranded, complementary nature of DNA not only forms the basis of its replication within an organism, but is also the foundation

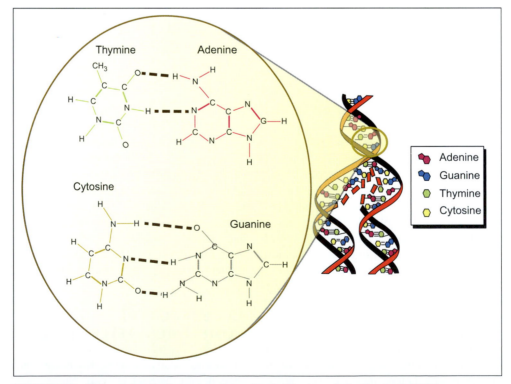

FIGURE 13.1 Hydrogen bonding between purine and pyrimidine bases in double-stranded DNA.

Information Box 13.1   Transcription and Translation

$$DNA \xrightarrow{\text{transcription}} RNA \xrightarrow{\text{translation}} protein$$

How does the cell use the information stored in its chromosome? It must transfer the information into useful activity. This is done in two steps, transcription and translation. In transcription, a DNA sequence is transcribed by the enzyme RNA polymerase to produce an RNA copy that is complementary to the DNA. During translation, the three different types of RNA (mRNA, tRNA, and rRNA) are used to synthesize a protein that is based on an mRNA sequence. In this process, a specific amino acid is inserted into a growing polypeptide chain based on the sequence of the mRNA being translated. Specific tRNA molecules deliver each amino acid to the ribosome, which is constructed of proteins and rRNA and is the site where translation or protein production occurs.

---

of several nucleic acid-based molecular analyses that will be discussed later, including gene probes and the polymerase chain reaction (PCR).

In RNA, G still binds to C, but A binds to U. There are three forms of RNA:

- ribosomal RNA (rRNA), which plays a structural role in creating the ribosome, the site of protein translation
- messenger RNA (mRNA), which is transcribed from a coding region of DNA and is subsequently translated into protein
- transfer RNA (tRNA), which transfers specific amino acids to the ribosome during mRNA translation

Two types of RNA, rRNA and mRNA, are used extensively in molecular genetic analyses to understand phylogenetic relationships, and to evaluate metabolic activity, respectively.

13.2 OBTAINING MICROBIAL NUCLEIC ACIDS FROM THE ENVIRONMENT

The first step in nucleic acid-based methods is often the extraction of nucleic acids from environmental samples. This step is critical for the downstream success and validity of the subsequent analyses. In this process, DNA is concurrently extracted from all of the populations within a sample, generating a mixture of DNA referred to as community DNA. The most common approach to extraction of community DNA from soil is to lyse the bacterial cells *in situ* (direct lysis) (see Section 8.1.3.1). Community DNA is ideally representative of all populations within the sample community; however, in reality, the extraction efficiency of different types of microorganisms can vary widely causing biases in subsequent analyses. Similar

extraction approaches can be used to obtain community RNA for analysis, although there are additional challenges associated with extraction of community RNA from soil, including active RNases and the inherently short half-life of mRNA. Consequently, it is more difficult to assess microbial gene expression in soil. Community RNA extractions generally parallel those of *in situ* lysis of DNA, with the extra step of adding an RNA stabilization agent and/or snap-freezing immediately after sampling. As with soil, microorganisms from either water or air samples can be analyzed via molecular analysis following extraction and purification of DNA/RNA from the collected microbial biomass (see Sections 8.2 and 8.3).

Regardless of the source of biomass or the lysis method, it is often necessary for the nucleic acid extract to undergo one or more purification steps to remove coextracted impurities (e.g., humic acids, metals, carbohydrates), which could interfere with or completely inhibit subsequent analyses. Purification methods include cesium chloride density centrifugation, commercial purification kits employing spin filters and traditional phenol—chloroform extraction followed by ethanol precipitation (see also Chapter 8). It is highly advised that the quantity and quality of extracted nucleic acids be determined (e.g., via UV spectrophotometry) prior to use, and the extract purified as necessary for the desired downstream application. However, the number of purifications performed should be minimized as much as possible since large portions of the extracted nucleic acids may be lost during the purification process. In addition, the presence of humic residues, metals and other inhibitory materials from environmental samples can often still interfere with downstream analyses even after careful purification. Finally, it should be noted that community DNA/RNA extracts may also contain nonmicrobial DNA/RNA originating from larger organisms such as insects and plants in the extracted sample. This may result in misleading concentrations of nucleic acids extracted from samples that contain substantial amounts of nonmicrobial biomass, such as a plant root and its associated microbial community.

Another limitation of nucleic acid extraction methods is that DNA/RNA recovery can vary from sample to sample depending on sample characteristics (e.g., for a soil the clay and salt content) and extraction efficiency. Not surprisingly, this variation can be further magnified when researchers use different methods to extract nucleic acids. This may not be a major issue for individual studies where all of the samples are extracted and processed in the same manner, but it is potentially a major confounding factor for comparison of results across experiments and between research groups, such as in large, collaboration-based projects. In order to address this issue, researchers are attempting to establish international standards for nucleic acid extraction from soils and other matrices (Petric *et al.*, 2011). Once nucleic acids are obtained from either cultured isolates or environmental samples, they can be analyzed with a variety of methods including those discussed in the following sections.

13.3 HYBRIDIZATION-BASED ASSAYS

Gene probe methodology takes advantage of the fact that DNA can be denatured and reannealed (Information Box 13.2). Gene probes consisting of single-stranded DNA can be used to identify the presence of a particular nucleic acid sequence within an environmental sample. Typically, probes are short sequences of DNA known as oligonucleotides, which are complementary to the target

---

Information Box 13.2   Hybridization and Denaturation

When two complementary DNA strands combine together, the process is known as DNA–DNA hybridization, because the resulting dsDNA is a hybrid of the two separate strands. The reverse process, in which dsDNA melts into two single strands, is called denaturation. This can be done chemically or simply by heating the DNA to 94°C. On cooling, the two single strands automatically hybridize back into a double-stranded molecule, a process known as reannealing. Complementarity is an important concept because if one strand sequence is known, the sequence of the other strand is easily deduced. This concept is the basis of many of the nucleic acid–based methodologies discussed in this chapter.

---

sequence of interest. These probes are labeled in some way that facilitates their detection. Probes can be used for a variety of environmental applications including: (1) examination of soil microbial diversity; (2) identification of a particular genotype; and (3) testing for virulence genes of suspected pathogens. Construction of gene probes and specific application methodologies are outlined in detail in manuals such as Green and Sambrook (2012). Here we briefly present the general concepts involved in the construction of gene probes along with some of their practical applications. These concepts are also critical to understanding some of the more sophisticated techniques discussed later, including microarrays.

13.3.1 Marker Selection

Gene probes can be designed to target various functional or phylogenetic genes (Table 13.1) depending upon the objective of the study. The targeted sequence may be unique to a particular microbial species, in which case the gene probe would allow screening of an environmental sample for the presence of that microorganism. Alternatively, the target gene may code for the production of an enzyme unique to a metabolic pathway. In this case, positive gene probe results indicate that the environmental sample contains the genetic potential for that particular activity. This kind of probe can be defined as a functional gene probe. A good example of such a probe is one designed to be complementary to genes coding for enzymes involved in nitrogen fixation. It is important to

---

TABLE 13.1 Commonly Used Marker Genes

Target Type	Examples	Information Obtained
Ribosomal RNA (rRNA) or internal transcribed spacer (ITS)	16S rRNA gene, the small ribosomal subunit (SSU) in Bacteria and Archaea	Taxonomy and phylogeny of Bacteria and Archaea
	18S rRNA gene, the small ribosomal subunit in eukaryotes	Taxonomy and phylogeny of Fungi and other eukaryotes
	ITS, the internal transcribed spacer	Taxonomy of Fungi
Other universal or "housekeeping" genes	*rpoB*, an RNA polymerase subunit	Phylogenetic information, similar to that obtained from rRNA or ITS sequences
	gyrB, a DNA gyrase	
	recA, DNA recombination and repair	
	HSP70, a heat shock protein	
Functional genes	*amoA, nifH, ntcA*	Detection of microorganisms involved in various nitrogen transformations
	dsrAB	Detection of sulfate-reducing Bacteria and Archaea
	phnA, phnI, phnJ	Detection of microorganisms involved in phosphorus transformations

TABLE 13.2 Selected Nucleic Acid Databases

Database and Web Address	Contents and Comments
GenBank® (National Center for Biotechnology Information, NCBI) http://www.ncbi.nlm.nih.gov/	Annotated database of publicly available DNA sequences, various software tools for analyzing genome data, funded by National Institutes of Health
ENA, European Nucleotide Archive http://www.ebi.ac.uk/ena/	Europe's primary resource for nucleotide sequence and annotation
DDBJ, DNA Databank of Japan http://www.ddbj.nig.ac.jp/	Database of nucleic acid sequences generated in Japan
INSDC, International Nucleotide Sequence Database Collaboration http://www.insdc.org/	Collaboration between GenBank, EMBL and DDBJ, which allows for sharing of data between member archives
Ribosomal Database Project (RDP) http://rdp.cme.msu.edu/	Quality-controlled ribosomal sequence data and analysis services, including sequence classification and alignment of bacterial and archaeal 16S rRNA gene and fungal 28S rRNA gene sequences
UNITE http://unite.ut.ee/	Quality-controlled molecular database and tools for the annotation and identification of fungi using ITS sequences
Genomes OnLine Database (GOLD) http://www.genomesonline.org/	Comprehensive listing of completed and ongoing genomes with links to sequence information
JGI, Joint Genome Institute http://www.jgi.doe.gov/	Archive for large-scale sequencing projects generated by the U.S. Department of Energy, plus tools for comparative analysis of genomes and metagenomes
Kyoto Encyclopedia of Genes and Genomes (KEGG) http://www.genome.jp/kegg/	Annotated genes and metabolic pathways within organisms

keep in mind that since DNA probes are being used, only the genetic potential and not expression of the gene (requires mRNA detection) or the actual activity (requires enzyme activity assays) is being detected. Probes designed against specific ribosomal RNA (rRNA) sequences are known as phylogenetic probes. Phylogenetic probes can be specific for groups of bacteria, for example Proteobacteria or even classes of Proteobacteria (e.g., α, β, γ) or can be designed to detect an entire domain (Bacteria, Archaea or Eukarya) in which case they are called universal probes.

13.3.2 Probe Construction and Detection

The basic strategy in the construction of a gene probe, or PCR primer, is to obtain the sequence of a target gene and then select a portion of this sequence for use as a probe. Nucleic acid sequences can be obtained from a variety of databases (Table 13.2), with the U.S. National Institute of Health's genetic sequence database GenBank® and the European EMBL database being two of the largest and most commonly consulted. Several other smaller databases with particular foci are also available. The number of sequences resident within databases has expanded significantly each year (Figure 13.2A). The vast number of sequences currently available facilitates design of target-specific primers and probes. In addition, thousands of whole microbial genomes have been sequenced, serving as the foundation for a new area of research based on comparisons of whole genomes or comparative genomics (Figure 13.2B).

Once a target sequence is obtained, it and other sequences for the same target in other organisms are imported into another program, where the sequences are aligned. This step is critical since it allows identification of conserved and unique regions, which can be subjected to algorithms to design probes that meet appropriate criteria such as length, melting temperature (T_m) and minimal secondary structure. The size of the probe can range from ≈ 20 base pairs to as many as several hundred base pairs. A large number of probes for various organisms are publicly available. One of the best sources for rRNA-targeted probes is the probeBase database which currently contains information for > 2700 probes and > 170 PCR primers (Loy et al., 2007).

Once a probe is selected, it is synthesized and labeled in such a way that it can be detected after hybridization

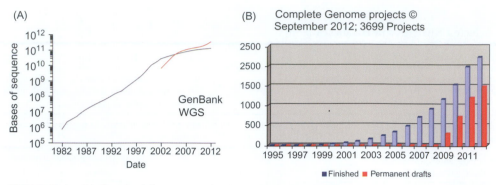

FIGURE 13.2   (A) Growth of the sequence data in GenBank and the Whole Genome Shotgun (WGS) databases. These databases are maintained by the National Center for Biotechnology Information, U.S. National Library of Medicine and are part of the International Nucleotide Sequence Database Collaboration. (B) Sequenced microbial genomes to 2012.

to the target sequence. A researcher has several options when constructing and labeling a probe. In the early days of molecular biology, the only labeling option was radioactivity, but several nonradioactive methods are now available. Radioactive labeling of a probe is typically done by labeling the sequence with a radioactive chemical, such as ^{32}P. Nonradioactive alternatives include probes labeled with digoxigenin (DIG), biotin or fluorescein, which can be incorporated into the sequence by chemical synthesis. The different labels are detected by binding the respective antibody or streptavidin–alkaline phosphate conjugate, which, when reacted with the appropriate substrate, will give a signal (see also Chapter 12).

13.3.3 Dot, Southern and Northern Blots

Dot blotting or dot hybridization is a technique used to evaluate the presence of a specific nucleic acid sequence in a microbial culture. Nucleic acid is extracted from the culture, spotted on a nitrocellulose filter and subsequently probed. Dot blots can be used to indicate the presence or absence of a sequence, or can be used to quantify the target. For quantitation, the relative amount or intensity of hybridization will give an estimate of the quantity of sequences in a sample when compared with similarly spotted standards. Gene probes can also be used to detect a specific gene sequence within bacterial colonies on a Petri plate containing a mixed population of bacteria through use of a process termed colony hybridization or lifts. To perform a colony hybridization, a nylon membrane is lightly pressed onto the Petri plate so that some bacterial cells from each colony adhere to the membrane. Subsequently, in a series of steps, the cells are lysed directly on the membrane, and the DNA is fixed to the filter and then denatured into two single strands. The gene probe is similarly denatured and then added to the membrane. The DNA is allowed to reanneal, and ideally the

single strand of the gene probe will anneal with the complementary target DNA sequence from the bacterial cells. After washing the membrane free of unhybridized probe, the filter undergoes probe detection. After this procedure, only the colonies that contain the specific DNA sequence give a signal. Since the original Petri plate contains the corresponding intact colonies, the viable colony of interest can now be identified, isolated and retained for further study.

In some cases, it may be necessary first to separate the target DNA or RNA into different size fractions before probing. For example, it may be important to know whether a gene is carried on a plasmid or the chromosome. To determine this, all of the plasmids within the microbe being studied can be extracted and separated by gel electrophoresis (Information Box 13.3; Figure 13.3). The plasmid DNA is then transferred onto a nylon membrane by blotting, and the membrane subsequently probed. Only the DNA molecules that contain the target sequence hybridize with the probe, thus allowing detection of those plasmids that contain the target sequence. When the target is DNA, as in the above case, this process is known as Southern blotting or hybridization. Similarly, Northern blotting, which detects RNA, can be used in gene expression studies to detect induction of a specific gene.

13.3.4 Microarrays

Scientists have modified and miniaturized the blotting process to construct "gene chips" or microarrays that enable simultaneous detection of thousands to even millions of different gene targets. Microarrays are basically a collection of oligonucleotides (or gene probes) that have been "arrayed" onto a glass slide or a chip. In contrast to the hybridizations discussed in Section 13.3.3, the probes are bound to the solid matrix, and the target nucleic acids are then hybridized to the bound probes. Depending upon the array design and target nucleic acids, microarrays can

Information Box 13.3   Agarose Gel Electrophoresis

Agarose gel electrophoresis is a fundamental tool in nucleic acid analysis. It is a simple and effective technique for viewing and sizing DNA molecules such as plasmids or DNA fragments, including PCR products. The DNA samples are loaded into wells in an agarose gel medium. Voltage is applied to the gel, causing the DNA to migrate toward the anode because of the negatively charged phosphates along the DNA backbone. The gel is stained with a dye such as ethidium bromide, allowing visualization of the DNA when viewed under ultraviolet (UV) light. The smaller DNA fragments migrate faster through the gel matrix, and the larger fragments migrate more slowly. The molecular weight of the DNA (base pairs [bp]) determines the rate of migration through the gel and is estimated from comparison with standards (i.e., molecular weight markers) of known sizes that are run in parallel on the gel (Figure 13.3).

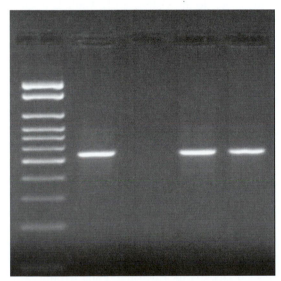

FIGURE 13.3 Agarose gel electrophoresis of PCR-amplified product DNA. Lane M includes a size ladder and lanes " + " and " − " are the positive and negative controls. Lanes 1 and 2 are samples that are positive for the target DNA as shown by presence of the ≈ 550 bp PCR product. Photo courtesy E.C. Martin.

be used to study the presence (DNA), abundance (DNA) and/or activity of microorganisms (mRNA). Although large arrays were initially cost-prohibitive for many laboratories, they are becoming widely used due to technological advances that have decreased the costs associated with probe synthesis, microarray fabrication and template labeling. Furthermore, the rapidly increasing availability of sequence information, including whole genome sequences, makes microarray approaches an enticing option not only for investigating the transcriptome (all genes expressed by an organism; Chapter 21) and subsets

of the transcriptome targeted to specific functions, but also for the screening of environmental samples for expression of particular genes of interest by any member (s) of the microbial community.

Two types of microarray fabrication processes are commonly used: printed (or spotted) arrays and synthesized (*in silico*) arrays. There are two common ways to produce the probes for printed arrays. For printed cDNA arrays, genes of interest ranging in size from ≈ 0.6 kb to 2.4 kb are amplified using the polymerase chain reaction (PCR) to produce cDNA probes (see Section 13.4.5 for explanation of cDNA). For printed oligonucleotide arrays, the sequences of the genes of interest (in some cases, all genes in the genome) are used to design unique 35−70 nucleotide probes. These oligonucleotide probes are then synthesized commercially. Once the probes have been produced, they are robotically deposited onto the microarray slide and subsequently chemically linked to the slide. Some manufacturers synthesize their oligonucleotides directly on the microarray chip in a process called *in silico* synthesis. This involves the use of photolithography and solid-phase DNA synthesis, and has enabled the construction of high-density arrays containing over one million probes.

Microarrays can also vary based upon their targeted organisms and information provided (Table 13.3). The two types of microarrays most commonly used for environmental microbiology are phylogenetic oligonucelotide arrays (POAs) and functional gene arrays (FGAs). The POAs are primarily based upon rRNA (e.g., 16S rRNA) genes, and are used to detect the presence of organisms. This allows a researcher to follow population dynamics and community profile changes across a wide variety of species on the same array. The largest POA currently available is the PhyloChip developed by Gary Andersen's group at Lawrence Berkley National Laboratory. The PhyloChip G2 contains probes representing 9773 clusters of sequences (operational taxonomic units; OTUs) within Bacteria and Archaea, and also probes at higher levels of taxonomic classification (e.g., subfamilies). A newer, commercial version of the PhyloChip (G3) is currently available via Second Genome (San Bruno, CA). The PhyloChip has been used to characterize microbial communities in a variety of samples including urban air samples (Brodie *et al.*, 2007), air in commercial aircraft (Korves *et al.*, 2013), the Gulf of Mexico following the *Deepwater Horizon* oil spill (Hazen *et al.*, 2010), mine tailings (Wakelin *et al.*, 2012) and uranium-contaminated groundwater (Brodie *et al.*, 2006).

In contrast to POAs, FGAs are designed based upon genes directly tied to an environmentally relevant process (e.g., sulfate reduction). This enables researchers not only to use FGAs to detect the presence and potential activity of organisms which may contribute to processes of interest, but also as a measure of microbial activity if mRNA is assayed for the target sequences. The most

TABLE 13.3 Microarrays for Characterizing Environmental Microorganisms

Array	Targeted Microorganisms	Information Provided	Probe Template
Phylogenetic oligonucleotide array	Cultured & uncultured	Phylogenetic	rRNA genes
Functional gene array	Cultured & uncultured	Functional	Functional genes
Community genome array	Cultured	Phylogenetic	Whole genome
Whole-genome open-reading-frame array	Cultured	Phylogenetic & functional	Open reading frames in whole genome
Metagenomic array	Cultured & uncultured	Functional	Environmental DNA

Adapted from *Gentry* et al. *(2006).*

TABLE 13.4 Microbial Genes Represented on the GeoChip 3.0 Functional Gene Microarray

Gene Category	Number of Probes	Covered Sequences
Carbon cycling	5196	10,573
Nitrogen cycling	3763	7839
Phosphorus cycling	599	1220
Sulfur cycling	1504	2042
Organic contaminant degradation	8614	17,441
Metal resistance	4870	10,962
Antibiotic resistance	1594	3944
Energy processes	508	671
Phylogenetic marker (*gyrB*)	1164	2298
Total	27,812	56,990

Adapted from *He* et al. *(2010).*

comprehensive FGA currently available is the GeoChip developed by Jizhong Zhou's group at the University of Oklahoma. GeoChip 3.0 contains 27,812 probes covering 56,990 sequences—primarily representing genes involved in C, N and S biogeochemical cycling, organic chemical degradation and metal resistance (Table 13.4). The GeoChip has been used to characterize microbial communities in a variety of samples including grassland soil (He *et al.*, 2010), reclaimed surface-mine soil (Ng, 2012), the Gulf of Mexico following the *Deepwater Horizon* oil spill (Lu *et al.*, 2012) and remediation of uranium-contaminated groundwater (Van Nostrand *et al.*, 2011).

Other types of arrays include: (1) community genome arrays that use whole genomes of isolated organisms for probe material and can be used to determine the similarity of tested isolates to these strains; (2) whole-genome open-reading-frame arrays that contain probes for all potential genes in a genome and can be used for transcriptomics and comparative genomics; and (3) metagenomic arrays that use cloned environmental DNA as templates for probes and can be used as a high-throughput method to screen these libraries (Section 13.6.2).

Once constructed, microarrays can be employed to characterize microbial communities and determine how microorganisms respond to and interact with their environment. Figure 13.4 shows the steps in a microarray experiment in which the goal is to characterize the microbial community via detection of target sequences (e.g., 16S rRNA or functional genes). The first step is to extract DNA from the sample. If necessary, the DNA can be further purified and/or amplified (Section 13.4). Subsequently, the DNA is labeled, usually with a fluorescent molecule such as Cy3 or Cy5. The labeled DNA is then denatured and hybridized to the probes on the microarray, after which the array is scanned to indicate the extent of hybridization to the various probes. A cutoff value, such as a signal-to-noise ratio of 3.0, is commonly used to differentiate background noise from true target detection. These results are then compared to those for hybridization of other samples to ascertain community diversity, presence of microorganisms or genes of interest, and/or population dynamics.

Additionally, microarrays can be used to determine microbial activity. In this case, mRNA is extracted from two samples: the test or experimental sample which has been exposed to a treatment (e.g., an environmental pollutant) and a reference or control sample that has not been exposed to the treatment. Messenger RNA (mRNA) is extracted from both samples, converted to cDNA via reverse transcription (Section 13.4.5), labeled with two different molecules (e.g., Cy3 and Cy5) and hybridized to the

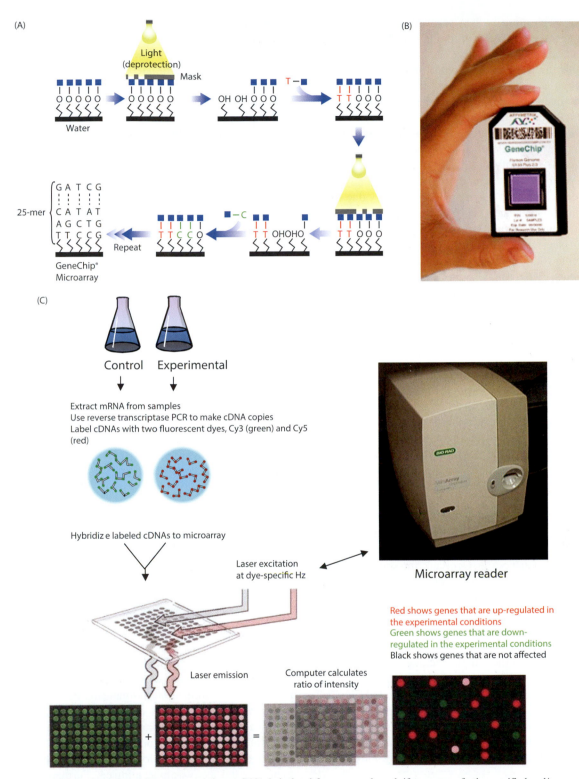

(A)

(B)

(C)

Control   Experimental

Extract mRNA from samples
Use reverse transcriptase PCR to make cDNA copies
Label cDNAs with two fluorescent dyes, Cy3 (green) and Cy5
(red)

Hybridiz e labeled cDNAs to microarray

Laser excitation
at dye-specific Hz

Microarray reader

Red shows genes that are up-regulated in
the experimental conditions
Green shows genes that are down-
regulated in the experimental conditions
Black shows genes that are not affected

Laser emission   Computer calculates
                          ratio of intensity

FIGURE 13.4 Steps in a microarray experiment. DNA is isolated from a sample and, if necessary, further purified and/or amplified. The resulting DNA is then labeled, denatured and hybridized to probes on a microarray. The array is then scanned to detect and quantify the binding of different probes on the array to the target DNA. The availability of different labels allows simultaneous testing of two or more different samples. This dual-labeling approach is commonly used for gene-expression studies where mRNA is extracted from a control and treated sample. The mRNA extracts are converted to cDNA and then labeled with different fluorescent labels, e.g., Cy3 and Cy5. The up- or down-regulation of genes in the treatment sample is then determined by comparison to hybridization in the control sample.

same array. The array is then scanned at the appropriate wavelength for each label, and the up- or down-regulation of each gene in the treatment sample is determined by comparison to hybridization in the control sample.

Although microarrays are powerful analytical tools, they face several challenges when being used to characterize environmental microorganisms (Van Nostrand *et al.*, 2013). First, only targets with high enough similarity or homology to the probes will bind. This often results in a reduced sensitivity of detection. For example, when samples such as soil community DNA are analyzed, a reduced detection limit may result, presumably due to target and probe sequences being diverse and consequently not highly homologous to the designed probes. Due to these issues, microarrays are currently only capable of detecting dominant members ($\geq$5%) of microbial communities. This can also be an issue if only small quantities of DNA or mRNA are available for microarray analysis. Target nucleic acids can be amplified prior to microarray analysis, but this can introduce additional artifacts and further complicate analysis (Wang *et al.*, 2011; Section 13.4).

Although sensitivity can be a challenge, specificity is perhaps the greatest issue with use of microarrays for environmental samples. In contrast to using microarrays with pure cultures where the entire genome sequences of the tested organisms are often known and probes can thus be designed to reduce or eliminate the potential for cross-hybridization, little is known regarding the DNA sequences present in most environmental samples. Therefore, even though microarray probes are usually designed carefully and screened against all relevant sequences in public databases such as GenBank, these sequences may only represent a tiny fraction of the sequences present in a given sample (e.g., while there may be millions of sequences in GenBank for soil microorganisms collected from around the world, they may only represent a small portion of the microorganisms in a forest soil from Huntsville, TX, U.S.A.). Due to this, it is impossible to say that a positive microarray hybridization reflects true detection of that target and not cross-hybridization with an unknown sequence in the sample, although other techniques such as qPCR and DNA sequencing can be used to verify microarray detection of specific targets, and reduce this limitation. Additionally, although some researchers have shown that microarrays can be quantitative, amplification biases and potential cross-hybridization can make the quantitative detection of target sequences difficult when working with environmental samples.

13.3.5 Fluorescence *In Situ* Hybridization (FISH)

Probes with fluorescent labels can be used to investigate cells *in situ*, i.e., in a culture or in an environmental sample in a technique called fluorescence *in situ* hybridization (FISH). In this case, reagents are added that facilitate penetration of the probe through the cell membrane, where the probe ultimately hybridizes to its target sequence. Cells containing the target sequence are visualized under a fluorescence or confocal microscope. This technique has the advantage of allowing visualization of spatial relationships between populations within a community to be elucidated (see Section 9.4.3 and Figure 9.14). Since FISH analyses are conducted *in situ*, an inherent difficulty of this technique is differentiating signal from background noise such as probe nonspecifically sorbed to soil particles or the presence of naturally fluorescing compounds including metals and some soil minerals.

13.4 AMPLIFICATION-BASED ASSAYS

13.4.1 The Polymerase Chain Reaction (PCR)

The primary methodology used for most amplification-based assays is the polymerase chain reaction (PCR). Developed by Kerry Mullis and associates in 1985 (Saiki *et al.*, 1985), PCR has now been available for three decades and, along with its various iterations, has become the "gold standard" assay in microbiology. PCR has become so synonymous with microbiology that the majority of readers of this textbook will not have known a time prior to its use; it is the cell phone of microbiology.

PCR harnesses the biological DNA replication process to produce multiple copies (up to millions in a single reaction) of a DNA sequence *in vitro*. The amplified DNA is then detected using gels or other fluorescence-based techniques and can be used in further applications such as cloning and DNA sequencing. Before PCR, environmental microbiologists relied on culture and physiological assays to unravel the microbiological mysteries of soil, water and other environmental matrices. The two greatest advantages of PCR are that it enables researchers to amplify a DNA target millions of times over and also eliminates the need to isolate and culture microorganisms before they can be detected or studied. Both of these features have greatly enhanced the detection and characterization of environmental microorganisms and revolutionized our understanding of the microbiological realm, far beyond what culture-based assays alone allowed. Furthermore, a wide variety of modified PCR assays have been developed that enable detection of RNA targets, measurement of microbial activity, differentiation of live and dead cells and quantitative detection of microorganisms. In the following sections, we will discuss the basic process of PCR along with several modified PCR assays and their applications to environmental microbiology.

13.4.2 The Steps of PCR

The basic PCR reaction consists of three major steps: (1) DNA melting; (2) primer annealing; and (3) DNA polymerization or elongation. These steps comprise a single PCR amplification cycle and are often repeated 30–40 times in order to amplify a target DNA sequence to produce millions of copies or amplicons (Information Box 13.4; Figure 13.5). The first step relies on adding sufficient heat to melt or separate the antiparallel DNA strands into two distinct molecules. Essentially, this is a process that has been lifted from cell biology; whereas in a cell, helicase is used to separate the strands, heat is used to separate the strands in the amplification tube. This process is usually carried out at 94–95°C for between 10 and 45 seconds. The second step of the amplification cycle involves single-stranded, ≈ 20 nucleotides (nt)-long DNA primers, which ideally would be complementary to the ends of the DNA target. One primer binds to the 3′ end of one of the separated DNA strands, while the other primer binds to the 3′ end of the other, antiparallel strand. The primers bind to the 3′ ends because DNA polymerase requires a free 3′ OH on the primer to polymerize DNA down the template DNA molecule in the opposite direction. This all occurs at a single temperature typically ranging from 45 to 60°C, depending upon the composition of the primers (i.e., the melting point of the primers). Ideally, the temperature is 2–4°C below the melting temperature of the primers. In theory, the higher the temperature at which binding occurs, the more specific the binding will be, though this can vary depending upon the primer and assay. It is important to state that higher temperature binding may also limit assay sensitivity. This step is highly specific and may be the most important step throughout the entire process. Primers are either highly specific or slightly degenerate (Information Box 13.5). When the primers are specific, the primer acts as a mirror opposite, and each nucleotide matches its mirrored partner on the DNA template. Other times, such as with 16S rRNA PCR, the primers are degenerate, meaning they are able to bind varying configurations of the primer binding site. In this case, the primers may be degenerate in up to $\approx 10\%$ of their nt, which enables binding to variations of the primer site. This leads us to the third and final step of the amplification cycle, DNA polymerization. There are many DNA polymerases available, each specializing in specific duties (Information Box 13.6). Here we will focus on the typical DNA polymerase and amplification. Efficient DNA polymerization was not possible until *Taq* polymerase was discovered. *Taq*

Information Box 13.4   The Polymerase Chain Reaction (PCR)

Although advances in reagents and thermocyclers now allow polymerase chain reaction (PCR)-based amplifications to be performed much more rapidly, the basic idea behind PCR has not changed since its advent in the mid-1980s. The general PCR approach is as follows:

Stage 1: DNA melting-hold step. This is the process by which the DNA duplex strands are separated into single-stranded DNA molecules awaiting amplification. The energy added to the process separates the strands by breaking the hydrogen bonds between the individual nucleotides of opposite strands. The step is typically performed at 94–95°C for 5–10 min. Additionally, this step may be used to activate, or "turn on," some DNA polymerases that are designed to require this activation in order to prevent amplification by-products from being formed before the desired cycling conditions are initiated.

Stage 2: Cycling. This is the part of a PCR assay where DNA amplification occurs. This stage is broken into a three-step process (denature, anneal and extension), often repeated between 25 and 45 cycles, depending on the needs of the researcher (Figure 13.5). The denaturing step follows a similar process to the stage 1 hold step, the only difference is that the denaturing cycling step is typically conducted at 95°C for only 15–30 s. As before, this step is used to separate or melt the DNA strands, to prepare for the following step, the annealing step. In the annealing step, primers (small ≈ 20 bp DNA molecules) bind to the separated DNA strands. The temperature at which this is performed is primer dependent, and can vary from 45 to 65°C. Lower annealing temperatures generally result in less specific binding, and thus increase the potential for nontarget amplification. The third and final step in the cycle is the extension step, in which the DNA polymerase synthesizes the new DNA strand beginning from the 3′ end of the bound primer. During this step, DNA polymerase adds new nucleotides to the extending DNA molecule following the guidelines of the original DNA template. Typical conditions for this step are 72°C for approximately 60 s, depending on the size of the expected product and the DNA polymerase.

Stage 3: DNA final extension–hold steps. The final stage consists of a final DNA extension, as well as a final hold step at 4°C. The final extension step consists of DNA polymerase extending unfinished amplification products. This is then followed by a final hold step in which the thermocycler is set to hold a 4°C temperature to prevent DNA degradation until the user can proceed to further analysis of the PCR products.

In the end, the amplification proceeds in an exponential fashion, in which $X_n = 2^n * X_0$: where X_n = the number of amplicons after n cycles, beginning with X_0 DNA template copies.

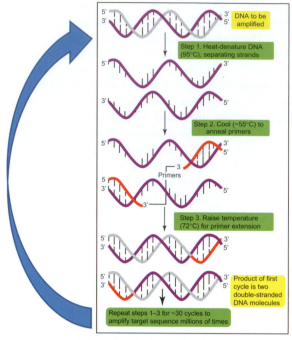

FIGURE 13.5 The PCR cycle. Step 1, denaturation of the DNA template; Step 2, annealing of primers to the single-stranded template; Step 3, extension of the primer to make a complementary copy of the DNA template. These three steps comprise one cycle of PCR.

Information Box 13.5   What Does The 'N' in My PCR Primer Sequence Mean? – Degenerate Primers

If DNA is formed from the nucleotides A, T, G and C, what do the other letters commonly seen in PCR primers and probes (e. g., M, W and N) represent? These symbols are part of the International Union of Pure and Applied Chemistry (IUPAC) standard nomenclature for nucleotide base sequence names and represent degenerate or nonstandard nucleotides that can bind multiple nucleotides. For example, an "M" in a PCR primer would consist of a mixture of A and C at that position, and thus could bind to either T or G in the complementary DNA strand. An "N" is the most extreme example and reflects of mixture of A, T, G and C at that position in the PCR primer, which would thus enable binding to any nucleotide at that position in the complementary DNA strand. These types of nucleotides are commonly used to design degenerate PCR primers or probes that thus enable detection of a wider spectrum of microorganisms by a single primer set or probe.

---

polymerase is a heat-stable DNA polymerase from the thermophilic bacterium *Thermus aquaticus*. Prior to its discovery, the DNA polymerases used in PCR were labile enzymes which required constant addition throughout the PCR process. In addition to being heat stable, *Taq* polymerase can achieve processing rates of anywhere from 35

Information Box 13.6   Not all DNA Polymerases are Created Equal

There are a many different DNA polymerases available from a variety of sources. These enzymes can vary greatly with respect to their sensitivity to reaction contaminants, their optimal reaction temperature, the type and length of template they will amplify, the speed of amplification, and the accuracy of the resulting amplicons. When choosing a DNA polymerase, it is crucial to consider the specific enzyme properties needed for each specific application. For example, many DNA polymerases, including *Taq*, add a single "A" to the 3′ ends of PCR products. This is invaluable for cloning PCR products using the TA cloning process (Section 13.6.1). However, some DNA polymerases produce blunt-ended products. If a researcher wants to clone PCR amplicons produced with these enzymes using a TA cloning system, the amplicons must first be "A"-labeled via a single amplification step with an enzyme such as *Taq* DNA polymerase.

---

to >100 nt per second. This enables assays to occur in as little as 45 min (e.g., real-time PCR), though longer products require longer polymerization. This final step in the PCR cycle normally occurs at 72°C. The three steps in a PCR cycle are then repeated between 25 and 40 times, depending on the end-point of the assay. It is important to note that there are other steps involved in the PCR process. For instance, PCR often begins with a single initial melting step of 95°C for 10 min; this process serves two purposes, to activate a particular type of *Taq* DNA polymerase and to separate the initial DNA templates. In addition, following the final cycle of PCR, the user will often include a final elongation step, which serves to elongate the ends of any PCR product not finished during the previous 40-cycle PCR. Theoretically, a single copy of DNA could be amplified to produce >10^{12} amplicons following 40 PCR cycles (Information Box 13.4); however, this level of amplification is typically not achieved under real-world conditions. For more information about generic PCR processes and steps, the reader is encouraged to view resources such as the DNA Learning Center website (http://www.dnalc.org/resources/animations/pcr.html) hosted by Cold Spring Harbor Laboratory.

Upon completion of the PCR assay, the results obtained are typically plus/minus (presence/absence), meaning that the gene or target of interest has either been detected in the template DNA or it has not; PCR amplicons are typically observed and compared using a gel (Figure 13.3). Here the products are loaded onto an agarose gel to which an electric current is applied allowing the negatively charged DNA molecules to migrate towards the positive end. The amplicons can be compared to positive and negative controls as well as a molecular weight marker. This marker provides a means to compare multiple product sizes, normally from 100 to 1500 bp; the amplicons, of different sizes, migrate along with the electric current and settle out at precise locations, which enables the user to determine sizes in comparison to the marker bands.

In the early days of PCR, thermal cycling occurred in water baths with a user cycling tubes from one water bath to the next water bath; needless to say, this was a time-consuming and labor-intensive effort. The commercialization of PCR brought thermocyclers into the standard laboratory; thermocyclers are now capable of housing hundreds of individual reaction tubes in a single self-contained instrument, capable of rapidly and precisely altering and holding temperatures for exact time points. Newer thermocyclers allow for the use of smaller reagent amounts; often reactions can be completed in a <25 μl-reaction vessel. Many of these apparatuses also enable users to operate multiple temperatures on a single block, also known as gradient cycling, which allows for primer troubleshooting at various temperatures. Gradient cyclers are especially useful in the early stages of PCR assay development, and enable the user to test multiple temperatures corresponding to a single primer set to determine which temperature/primer combination produces the best amplification. Gradient cyclers enable the user to run completely different PCR assays, with completely different PCR primers all functioning at various temperatures. For instance, one row on the block could correspond to 16S rRNA PCR, while another row corresponds to *nosZ* PCR, and, finally, another to *uidA* PCR.

Another specialized thermocycler, one which has quickly become the "gold standard" in most environmental microbiological laboratories, is the real-time PCR thermocycler (aka quantitative PCR). This thermocycler is capable of quickly amplifying and reporting product results. The advantage of the real-time thermocycler is that no gel is needed post-run while simultaneously obtaining quantitative results. The use of this thermocycler and its accompanying PCR assays will be discussed in the section below.

13.4.3 PCR Assays

The basic PCR assay has been modified via changes to primer targets, cycling conditions and template material to greatly expand the utility of PCR-based assays. Quantitative PCR (qPCR), multiplex PCR, reverse transcriptase PCR (RT-PCR), integrated cell culture PCR (ICC-PCR) and enterobacterial repetitive intergenic consensus sequence-PCR (ERIC-PCR) are just a few of the variations of the traditional PCR approach available to the environmental microbiologist.

Perhaps the most important recent addition to the PCR repertoire is qPCR (Information Box 13.7). Quantitative PCR has enabled researchers to harness the sensitivity of PCR, but with the quantitative characteristics of culture-based approaches. The basic premise behind qPCR consists of using fluorescent reporter dyes to detect the amount of newly amplified DNA. Reporter dyes are detected following each reaction cycle, and when the fluorescent signal is plotted against a standard curve, the reaction can be quantified. Until the advent of qPCR, the PCR assay was, for the most part, limited to presence/absence or qualitative (semi-quantitative) assays. In this respect, qPCR has enabled environmental microbiologists to measure and quantify microorganisms that may otherwise be difficult or even impossible to identify by culture methodology.

Quantitative PCR approaches primarily use one of two chemistries for quantification, either TaqMan® or SYBR Green®; the former accommodates a number of fluorescent dyes on an additional labeled internal DNA probe, while the latter is more cost-effective but can only utilize SYBR Green as the fluorophore (reporter signal) (Figure 13.6). An assay with SYBR Green requires an additional step in the form of a melt-curve step. This step applies heat, from 65 to 90°C, to separate the newly amplified DNA strands; a reduction in the fluorescent signal indicates the melting temperature at which DNA denaturation occurs. This final step indicates to the researcher the specificity of the amplified product, provided that it matches with the positive controls. The TaqMan assay relies on the specificity provided by the internal fluorescent probe; this probe binds a small (≈ 15 bp) region of DNA, which is internal to the two primer-binding sites. The inclusion of this internal probe in TaqMan-based assays greatly increases specificity as compared to SYBR Green-based approaches. Both chemistries yield quantitative results, provided a standard curve with known DNA quantities is used (Figure 13.7). More information about qPCR can be found in references including Heid *et al.* (1996) and Dowd and Pepper (2007).

Multiplex PCR is another effective manipulation of the traditional PCR assay. This method enables the detection/measurement of multiple targets in a single reaction assay (Figure 13.8), something that few culture-based approaches can achieve. Multiplex PCR primers can be added to a PCR assay with multiple targets; in some cases gene targets belong to completely different taxonomic groups. In this respect, the user could add PCR primers corresponding to *Campylobacter* spp., 18S rRNA (e.g., eukaryotic DNA) and tetracycline resistance in a single PCR reaction, provided that the primers do not exhibit any cross-reactivity. Typically, though, the reactions address multiple genes of a single pathogen, or varying taxonomic hierarchal genes such as a reaction targeting the family Enterobacteriaceae, the genus *Escherichia* and toxin-specific primers corresponding to *E. coli* O157:H7. The multiplex PCR assay was first adapted to gel approaches where one would read a gel with multiple bands in a single lane, each corresponding to a specific gene of interest. This approach has since been modified to fit with the

---

Information Box 13.7   Quantitative PCR (qPCR)

The most commonly used qPCR chemistries are SYBR Green$^\circledR$ and TaqMan$^\circledR$ (Figure 13.6). SYBR Green involves the use of a fluorescent dye which, when bound to double-stranded DNA can be excited and fluoresce. In this case the fluorescent signal is non-specific, meaning that any dsDNA will fluoresce, including primer-dimers and other non-specific PCR products. SYBR Green is more cost-effective than other qPCR chemistries and in some ways easier to carry out, since only two primers, a polymerase and dye, are needed for the reaction (in addition to basic PCR reagents). TaqMan assays involve the use of *Taq* polymerase and exploits its 5′ nuclease activity. TaqMan assays use an upstream and downstream primer, as in any PCR, with the addition of an internal probe which binds between the two primer-binding sites. This probe is labeled with a fluorescent dye as well as a nonfluorescent quencher; when both are in close proximity, the fluorescent dye is quenched and cannot be seen. TaqMan uses its 5′ nuclease activity, through routine DNA amplification, to cleave this probe, thus liberating the fluorescent dye, and enabling visualization and subsequent quantification. TaqMan assays can be modified to accommodate multiplex PCR, since multiple dyes are available, and each primer/probe combination could accommodate an individual gene of interest. SYBR Green, on the other hand, is not as conducive to multiplex assays due to the difficulty in differentiating multiple products.

The quantitative capability of qPCR is enabled by the generation of a standard curve from different concentrations of target sequences. The standard curve is used for both SYBR Green and TaqMan-based chemistries; the use of the curve enables comparison between samples and known quantities of DNA standards. In Figure 13.7A, DNA standards decrease in quantity from left to right on the figure. In Figure. 13.7B, these DNA standards are regressed against the resulting strength of their fluorescent signals (SYBR Green or TaqMan probes). The curve fit of 0.995 (r^2) indicates a strong correlation between signal strength and DNA quantity; therefore, this curve and subsequent regression equation can be used to determine the amount of genomic units present in an unknown sample.

Another component of qPCR assays is the melt-curve (Figure 13.7C). This is used to determine the specificity of SYBR Green-based qPCR assays. Since SYBR Green yields a signal when bound to dsDNA, the presence of any dsDNA will yield a signal. The use of a melt-curve enables the user to compare the melting temperatures of unknown samples to known positive controls (e. g., the standard curve). During the melt-curve, the temperature is raised (65−90°C) ultimately resulting in denaturation of the double-stranded PCR amplicons, and thus a decrease in the SYBR Green fluorescent signal. Comparison of the PCR amplicon melt-curve to a standard helps to verify the specificity of the amplification. The use of TaqMan-based chemistry reduces the need for melt-curve analyses, since internal DNA probes are designed to bind between the upstream and downstream primer-binding sites thus providing an added layer of specificity.

---

advances of qPCR, where instead of reading multiple bands on a gel, the researcher would read multiple fluorescent signals, each relating to a specific gene target. This approach typically uses TaqMan chemistry, but has been modified to work with SYBR Green by using multiple melting points at the melt-curve step (Figure 13.7C).

The PCR assay can also be modified to provide a form of genotyping using either modified primers or a post-PCR step involving restriction enzymes. The former approach involves the exploitation of repetitive genetic elements within an organism's chromosome, while the latter can be applied to nearly any PCR amplicon. Environmental microbiologists have applied these techniques to a number of treatment-impacted environments to report broadscale changes in a microbial population, or to determine genotypic differences between isolated strains. The application of ERIC-PCR and related PCR fingerprinting approaches enables theoretical discrimination of strains belonging to single species (Section 13.5.1). Restriction digestion of PCR products can also yield fingerprints (e.g., T-RFLP or terminal-restriction fragment length polymorphisms; Section 13.5.2.1) of microbial communities; this method has been commonly used to determine differences in soil bacterial communities, such as in soils under varying

agronomic management systems. Typically, DNA from a microbial community would be extracted, PCR-amplified with 16S rRNA primers, digested with one or two restriction enzymes and then visualized using either a high-resolution gel or capillary-based system. Fluorescently labeled primers facilitate visualization of the digested bands when using the capillary-based approach. As with ERIC-PCR and other PCR-based fingerprinting methods, these approaches carry with them their own caveats. Particularly, inter-lab variability and even inter-operator variability can limit the potential of these assays. Additionally, the presence of similar bands in two different samples which represent different or even multiple organisms. Furthermore, with the decreasing costs of DNA sequencing, many labs have moved away from community fingerprinting methods such as T-RFLP and instead use high-throughput sequencing-based analyses that offer greater resolution and better taxonomic characterization of microbial communities (Chapter 21).

13.4.4 PCR Target Sequences

PCR has become a tool of choice for environmental microbiology, and its assays and uses vary substantially

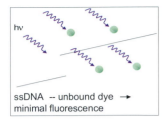

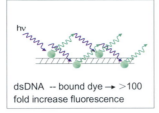

ssDNA -- unbound dye → minimal fluorescence

dsDNA -- bound dye → >100 fold increase fluorescence

SYBR Green I is a fluorescent dye that upon intercalation into double stranded DNA exhibits an increase in fluorescence intensity of greater than 100-fold. Advantages include the relatively low cost of the dye, the fact that it will work with any primer set since it is not sequence specific, and the ability to perform a melt analysis. The main disadvantage is that any nonspecific products formed in the PCR, including primer dimers, are also detected.

TaqMan—Hydrolysis Probe

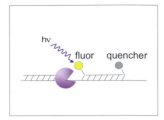

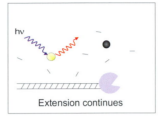

fluor   quencher

Extension continues

A TaqMan probe contains a fluorophore in close proximity to a quencher, such that the presence of the quencher blocks fluorescence. When bound to the target sequence, the probe is cleaved by the polymerase. The necessity of a probe to bind internally to the amplicon results in increased specificity. These assays can be multiplexed, but have the disadvantages of difficult design and cost of the dual labeled probe and, the inability to perform a melt analysis.

Hybridization probes

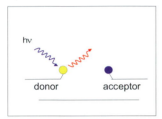

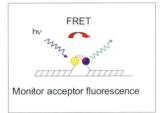

donor   acceptor

FRET

Monitor acceptor fluorescence

Two labeled probes must bind within the amplicon to generate a fluorescence signal as energy is transferred from the donor fluorophore to the acceptor fluorophore, a process known as fluorescence resonance energy transfer (FRET). These assays can be highly specific since two primers and two probes must bind. In addition, detection via hybridization probes is not dependent on a hydrolysis reaction, and, thus, a melt analysis can be conducted. These assays can also be multiplexed.

FIGURE 13.6 Schematic of SYBR Green I, TaqMan and hybridization probe fluorogenic detection approaches for quantitative (real-time) PCR.

given the end product. Products can either be qualitatively measured, such as presence or absence measurements, or quantitatively assessed. Primer design is one factor that drives this process. Target sequences commonly used in environmental microbiology include: 16S and 18S rRNA genes, genes involved in biogeochemical cycling such as nitrogen fixation, antibiotic resistance genes and genes related to pathogenicity (Table 13.1). Additionally, each of these primer targets can be modified for quantitative PCR (qPCR) assays. The 16S rRNA gene is one of the most exploited gene targets in environmental microbial PCR. The 16S rRNA gene is highly conserved due to its importance in cell functionality; thus, any changes to this target are related to phylogenetic similarity (Chen et al., 1989). Some have referred to the gene as a molecular chronometer (Woese, 1987). This gene has proven to be versatile, particularly for bacterial taxonomy, qPCR and library analysis. At its most basic level, the 16S rRNA gene assay can be used in a presence/absence format to identify the taxonomy of a given bacterial isolate. A simple "colony PCR" followed by sequence analysis can be used to classify an isolate. Quantitative PCR involving 16S rRNA primers, on the other hand, provides quantitative data which can be used to enumerate a specific bacterial population in a given environment. There are some

caveats associated with the assay, but for the most part, 16S rRNA qPCR can be used to estimate the total number of bacteria in a sample. The qPCR 16S rRNA assay has been applied to a number of clinical and environmental samples, often replacing traditional plating techniques in many environmental microbiology labs. This method can be used in environments where traditional culture may not yield accurate results, such as environments with fastidious or noncultivable pathogens (Matsuki et al., 2004); qPCR can provide numbers which are comparable to cell counts, typically 10−100-fold greater than culture techniques (Matsuki et al., 2004). The 16S rRNA gene can also be used to determine phylogeny via traditional clone libraries or high-throughput DNA sequencing (Section 13.7; Chapter 21).

Additionally, PCR has many advantages for detecting environmental pathogens and antibiotic resistance (Table 13.5). Various PCR assays have been successfully applied to measure pathogens in environmental matrices such as air, soil, water, food and fomites. The PCR assay can be applied to either isolated microorganisms or complex environmental samples, depending on the purpose of the assay. In the case of isolated microorganisms, a pathogen PCR assay could be used to confirm the pathogen, confirm pathogenicity traits or confirm other functional

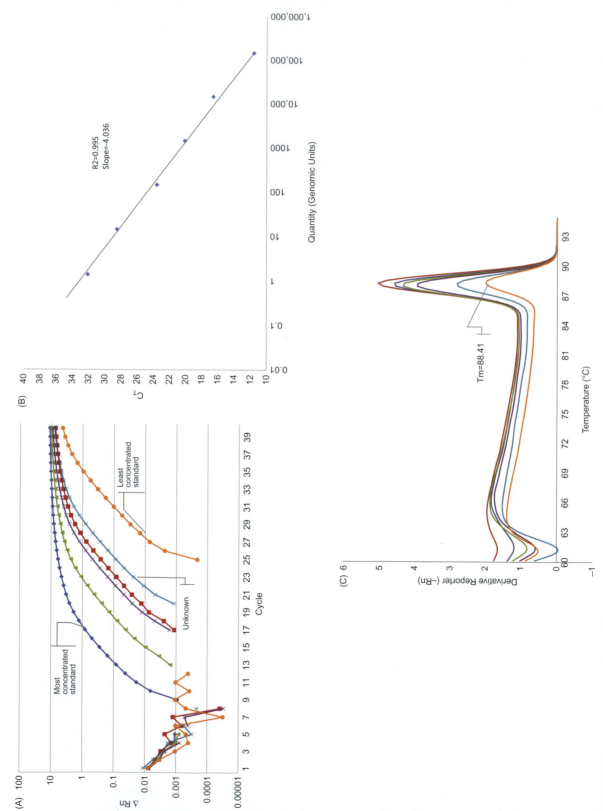

FIGURE 13.7 (A) A typical qPCR amplification plot generated using a 10-fold dilution series of genomic DNA. (B) A standard curve generated from the amplification plot. (C) A melt-curve analysis. Courtesy R.K. Smith.

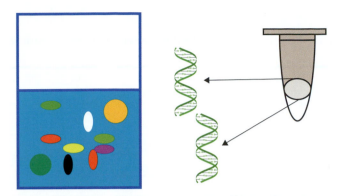

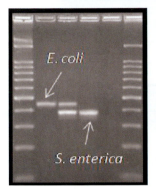

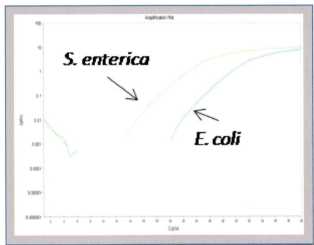

FIGURE 13.8   Schematic representation of multiplex PCR in which multiple sets of primers allow simultaneous amplification of more than one sequence (organism).

genetic traits. In an environmental sample, the PCR assay could be used to confirm the presence of a pathogen as a means to eliminate negative samples before more costly culture techniques are used; for instance this could be applied to reduce the costs associated with virus culture. Common pathogen PCR targets in the environmental lab might include *Escherichia coli* (*uidA*), *Campylobacter jejuni* (*hipO*), *Salmonella* spp. (*invA*), *Norovirus* (capsid), tetracycline resistance (*tetR*) and methicillin-resistant *Staphylococcus aureus* (*mecA*). In these cases, particularly for *C. jejuni* and *Norovirus*, the measurement of difficult-to-culture pathogens can be facilitated by PCR. However, for organisms whose genomic material consists

of only RNA, such as *Norovirus*, this requires additional steps as discussed in the section below.

13.4.5 RNA Amplification Using PCR

Since DNA polymerase is the basis for PCR, all PCR approaches use DNA as their starting template; however, RNA instead of DNA is the genomic material for many viruses, including several pathogens of concern to environmental microbiologists. In addition, since mRNA is the precursor molecule to proteins, mRNA can be used as a reporter molecule for microbial protein production and

TABLE 13.5 Comparison of Various PCR-Based Methods with Traditional, Culture-Based Methods for Pathogen Detection

Advantage/Disadvantage	PCR-Based Method				Culture
	qPCR	RT-PCR	ICC-PCR	Enrichment-PCR	
Reduced time of detection	Yes	Yes	Yes	Yes	No
Increased sensitivity	No	No	Yes	Yes	Yes
Affected by PCR inhibitory substances	Yes	Yes	No	Yes[a]	No
Reduced costs	Yes	Yes	Yes	Yes	Yes[b]
Viable pathogens detected	Yes	Yes	Yes	Yes	Yes
Detects **only** viable/culturable pathogens	No	No[c]	Yes	Yes	Yes
Detects viable but not culturable pathogens	Yes	Yes	No[d]	No[d]	No
Quantitative results	Yes	No	Yes[e]	Yes[e]	Yes

[a]*Some enrichment media contains PCR inhibitory substances.*
[b]*Costs for various pathogens can increase significantly.*
[c]*If mRNA is the starting template, then RT-PCR will detect only viable pathogens.*
[d]*Assumes that a viable-but-not-culturable pathogen would not grow effectively in any of the known culture systems.*
[e]*A most-probable-number approach must be set up during the culture phase.*

an indicator of microbial activity. There are a number of RNA-focused kits which can extract RNA from a host of environments focusing solely on viral RNA or more broad-spectrum extraction of generic microbial RNA (Chapter 8). In considering viral RNA, or any RNA product requiring PCR for detection, the researcher must turn towards a specialized set of enzymes known as reverse transcriptases. In a process known as reverse transcriptase PCR (RT-PCR), these enzymes reverse transcribe RNA to DNA, hence the name, which can then be used in a number of downstream assays, particularly PCR (Figure 13.9). The newly converted cDNA (complementary DNA) serves as the template for PCR. A number of viral detection assays utilize this pretreatment, including assays for enteroviruses and *Norovirus*. In the case of the latter, if not for reverse transcriptase, microbiologists would still be struggling to assay *Norovirus* since no culture-based detection method is available. The RT-PCR process can be modified to accommodate qPCR; for *Norovirus*, RT-qPCR is the only means available for quantifying *Norovirus* levels in the environment (Hill *et al.*, 2010).

Approaches have also been developed to randomly amplify whole-community (environmental) mRNA. This process is relatively straightforward for eukaryotic mRNA due to the presence of a poly(A) tail on the mRNA, which can serve as a primer site. However, this does not work for bacterial mRNA since it does not have a poly(A) tail. Gao *et al.* (2007) developed a method to amplify whole-community mRNA (including bacteria) by using a fusion primer (random primer with an attached T7 RNA

polymerase promoter sequence) to generate cDNA from environmental mRNA. After the cDNA was generated, it was amplified via linear RNA amplification with T7 RNA polymerase prior to conversion to cDNA for microarray analysis. This process resulted in >1000-fold amplification of the RNA template and produced representative and reproducible results, although some amplification biases were detected.

13.4.6 Detection of Live vs. Dead Microorganisms

The DNA-based PCR reactions that we have discussed thus far are capable of many things, the quick identification of a genetic trait (e.g., hours vs. days/weeks for pathogen culture confirmation), quantifying organisms, detection of unculturable microorganisms, etc., but they cannot be used to determine whether the nucleic acid originated from a viable or dead microorganism. Traditionally, this has been the primary argument against the use of PCR as a standard detection method. At least, this is true when considering PCR alone; however, some enterprising researchers have adapted PCR to live, culture-based approaches to solve this issue (Case Study 13.1). ICC-PCR (Blackmer *et al.*, 2000) and broth enrichment-based PCR for bacteria (Thomas *et al.*, 1991) have been used to simultaneously detect genetic traits in a sample while ensuring these traits came from live, theoretically infectious, pathogens. The methods ensure that

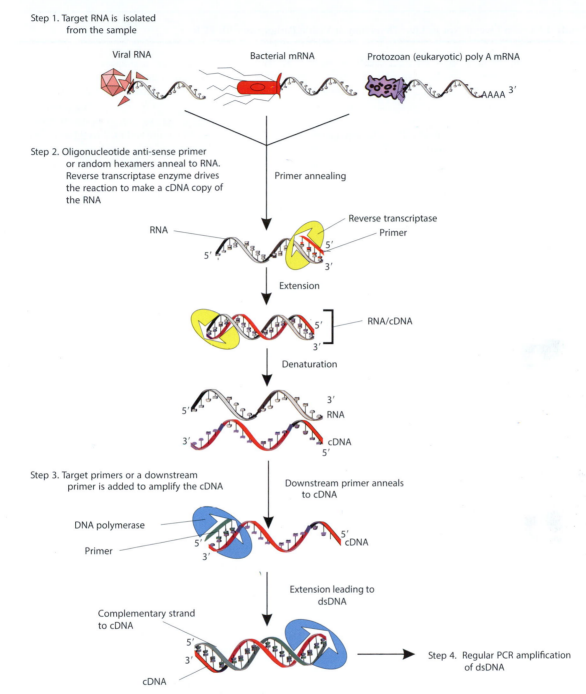

Step 1. Target RNA is isolated
        from the sample

Viral RNA   Bacterial mRNA   Protozoan (eukaryotic) poly A mRNA

Step 2. Oligonucleotide anti-sense primer
        or random hexamers anneal to RNA.
        Reverse transcriptase enzyme drives   Primer annealing
        the reaction to make a cDNA copy of
        the RNA

RNA   Reverse transcriptase
                                                Primer

Extension

                                             RNA/cDNA

Denaturation

                                                RNA

                                                cDNA

Step 3. Target primers or a downstream   Downstream primer anneals
        primer is added to amplify the cDNA   to cDNA

DNA polymerase
                                                cDNA
Primer

Extension leading to
dsDNA

Complementary strand
to cDNA
                                             Step 4. Regular PCR amplification
                                                     of dsDNA
cDNA

FIGURE 13.9   Reverse transcriptase-PCR (RT-PCR) amplification of RNA. RNA is reverse transcribed to synthesize cDNA by random hexamers or a specific antisense primer. PCR can then be performed on the cDNA template.

at least one of the major disadvantages associated with PCR can be overcome (Table 13.5).

Another approach that has been developed to differentiate between PCR detection of live and dead microorganisms is based upon the use of propidium monoazide (PMA; Nocker *et al.*, 2006). PMA only penetrates dead cells. Once inside the cell, it intercalates into the DNA. The PMA can then be photo-induced to covalently cross-link with the DNA, which renders it insoluble and prevents its subsequent extraction. Therefore, the DNA extracted from a treated culture would only originate from live cells, at least theoretically. Like all PCR-based assays, PMA-PCR can have some challenges when being used on environmental samples, but it shows great potential for allowing quantitative, PCR-based detection of viable organisms (van Frankenhuyzen *et al.*, 2011).

Case Study 13.1   To Live or Not to Live: Detection of Viable Pathogens with PCR

Since PCR arrived on the microbiological scene nearly 30 years ago, environmental microbiologists have envisioned a time in which difficult-to-detect pathogens could be easily quantified. Although PCR enabled sensitive pathogen detection from a number of matrices, one caveat has always remained: "... so is the DNA that you detected, from live or dead cells?" The very nature of PCR is based on the *in vitro* amplification of template DNA extracted from cells — either living or dead; thus PCR lacks the discriminating power to only consider DNA from "viable" organisms. Consider the fact that PCR is slowly replacing culture assays and one can see why this issue needs to be addressed. That is not to say that culture is no longer needed; culture techniques are still far more sensitive than PCR, given the sample aliquots allowable ($>$10 g) in a culture assay versus a 25 µl PCR assay with 2 µl of template. There are a few techniques available which enable the researcher to remove "dead" DNA from the equation:

1. The use of intercalating molecules, such as propidium monoazide (Nocker *et al.*, 2006), has enabled researchers to selectively detect DNA from live bacterial cells as these chemicals will only penetrate a compromised cell wall/membrane.

2. The use of mRNA as a starting template in a reverse transcriptase PCR has been shown to be effective as well. In this case, we exploit the cell's natural production of mRNA as an indicator of a live cell; the theory is, if the cell is living it will continue to produce mRNA as it continues with protein synthesis.

3. The combination of culture and PCR methods can yield powerful results; enrichment-based and integrated cell-culture (ICC) PCR have been utilized for the measurement of viable environmental bacteria, viruses, and parasites, and in some ways allows for an extremely sensitive assay (Blackmer *et al.*, 2000).

The theory here relies on the ability for one to culture or enrich these pathogens prior to the PCR assay, thus increasing sensitivity and ensuring that only viable pathogens are detected. However, these enrichment methods are not without their drawbacks; particularly, the fact that quantifiable results can only be obtained through the use of a most-probable-number approach, making the method quite labor intensive. Additionally, the time it takes to enrich may add an extra day to bench work, thus delaying the availability of test results.

13.4.7 Other Nucleic Acid Amplification Methods

In addition to the PCR-based amplification methods discussed in the previous sections, there are also other methods for amplifying DNA. One of the most common uses of the bacteriophage φ29 DNA polymerase is a process often referred to as multiple displacement amplification (MDA), since a single strand of DNA is simultaneously being replicated at different points. Random hexamer oligonucleotides are commonly used as the primers for MDA. In contrast to PCR, the reaction can occur at 30°C and does not require temperature cycling. MDA has several advantages over PCR including having a lower error rate and the ability to generate longer amplicons. Like any amplification process, it may introduce biases (Wang *et al.*, 2011), but it has great potential for whole-genome amplification, especially for single-cell sequencing (Lasken, 2012; Chapter 21).

13.4.8 Challenges for PCR-Based Amplification Methods

The presence of metals, chelating agents and humic acids can limit the effectiveness of a PCR assay, particularly when these traits inhibit the reaction to a point where no amplicons can be obtained. This can be especially troublesome when reporting negative samples; "Did the sample fail to amplify because no genetic target existed, or because of the presence of an inhibitor?" Additionally, the PCR assay can enhance detection limits, making the assay more sensitive than culture; however, PCR is limited by sample aliquot volume. A PCR is typically conducted in volumes of $<$25 µl, thus only a small fraction of the original sample can be accounted for in each tube; therefore, while a theoretical single molecule of DNA can be detected, the fact that such a small amount of sample is accounted for reduces the overall sensitivity of the assay. Assay sensitivity can be increased using broth enrichment approaches, in which the target bacterium can be biologically amplified and detected via PCR assay; this approach is similar to the ICC-PCR virus assay. The assay can be quantified using a most probable number (MPN) approach (McLaughlin *et al.*, 2009). Additionally, sample aliquots can be concentrated prior to PCR, utilizing flocculation, molecular weight filters, membrane filters or centrifugation (Chapter 8). As with any scientific method, PCR is not without its limitations, but it continues to play an important role in the environmental microbiology laboratory, and will continue to do so for the foreseeable future.

13.5 DNA FINGERPRINTING

A variety of DNA fingerprinting approaches have been developed over the past few decades. Although these approaches produce a similar output—a fingerprint or barcode for a microorganism or microbial community—they can vary widely in their approach. The basis for generating the fingerprints has primarily been: (1) digestion of genomic DNA with restriction enzymes; (2) PCR amplification of repeating elements or specific genes; or (3) some combination of PCR amplification and digestion. While the DNA fingerprinting of microbial isolates remains a cornerstone of microbial detection, DNA fingerprinting of microbial communities has been largely replaced by newer technologies such as high-throughput DNA sequencing. In the following sections we will highlight some of the major methods used for DNA fingerprinting and their application to characterization of environmental microorganisms.

13.5.1 Fingerprinting of Microbial Isolates

13.5.1.1 Restriction Fragment Length Polymorphism Analysis

Restriction fragment length polymorphism (RFLP) analysis is frequently used to characterize or differentiate whole genomes or specific PCR products from bacterial isolates. In general, this process involves extracting and purifying genomic DNA from a pure culture of a bacterial isolate, and in some cases PCR amplifying the target sequence. The technique is often used in conjunction with cloning (Section 13.6.1) where primers are used to amplify the target gene from a mixed population of microorganisms. Clones are generated (each containing the target sequence from a single environmental microbe), and then the clones are subjected to RFLP analysis. In all cases, DNA is cut into smaller fragments by the use of restriction enzymes. The fragments of DNA are usually separated by gel electrophoresis. If the restriction enzymes cut at many sites within the genomic DNA, several or even hundreds of DNA fragments can result. The pattern of these fragments following electrophoresis and staining can be used to differentiate genomes or clones

from one another. In addition, if gene probes are used to interrogate the bands (i.e., Southern hybridization), then the location of a particular gene or sequence within fragments can be determined, which may be beneficial for subsequent molecular analyses or for simplification of complex banding patterns. This approach is the basis for the automated RiboPrinter® System (DuPont™, Wilmington, DE, U.S.A.) for typing microbial isolates (Figure 13.10A).

13.5.1.2 PCR of Repetitive Genomic Sequences

PCR fingerprinting techniques employ primers that anneal to sequences that are repeated at multiple locations throughout the genome of an organism. Consequently, amplification results in the generation of multiple products of varying size. Analysis of the resulting amplicons via gel electrophoresis allows differentiation of many environmental isolates. Several PCR fingerprinting techniques have been applied to environmental samples. One PCR fingerprinting method is arbitrarily primed PCR (AP-PCR), also referred to as random amplified polymorphic DNA (RAPD) (Welsh and McClelland, 1990). AP-PCR uses one random primer (10−20 bp), which is annealed at a low temperature, resulting in nonspecific amplification of a genome. Thus, this reaction requires no prior sequence information, and will generate a fingerprint based on the uniqueness of the genome of the isolate or species. Depending on the random primer used, fingerprint patterns can be simple (1−2 bands) or complex (>10 bands).

Other fingerprinting methods use two different primers to PCR amplify repetitive sequence elements in microbial genomes. These elements include REP (repetitive extragenic palindromic), ERIC (enterobacterial repetitive intergenic consensus) (Figure 13.10B; Versalovic et al., 1991) and BOX sequences (Burr et al., 1997). These PCR-based fingerprinting approaches, and their variations, are commonly used to characterize microorganisms. For example, a patented REP-PCR-based approach has been commercialized and is available as the

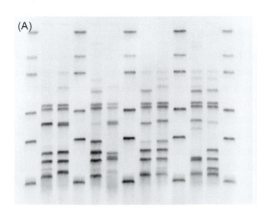

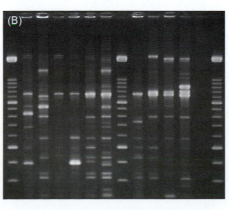

FIGURE 13.10 Examples of DNA fingerprinting approaches for characterizing microbial isolates: (A) RiboPrinter® and (B) enterobacterial repetitive intergenic consensus (ERIC) sequence-PCR fingerprinting of environmental E. coli isolates. RiboPrinter fingerprinting is an automated Southern hybridization-based method; whereas ERIC-PCR fingerprinting is a PCR-based method. Photos courtesy E.C. Martin.

DiversiLab™ System for genotyping of bacteria and fungi (bioMérieux, Marcy l'Etoile, France).

In comparison to RFLP fingerprinting, PCR fingerprinting is simple and rapid. However, it may be too sensitive for some applications, and even slight changes in experimental conditions can result in nonreproducible fingerprints. Such conditions include DNA quality and PCR temperature profiles. Reliability must be taken into consideration when making comparisons between runs, between laboratories and in some cases even between different DNA templates within the same laboratory.

13.5.1.3 Amplified Fragment Length Polymorphism Analysis

In addition to the more traditional fingerprinting techniques described above, some newer fingerprinting approaches rely on fluorescently labeled oligonucleotides. Amplified fragment length polymorphisms (AFLP), originally developed as a nonfluorescent technique, uses selective PCR amplification of genomic restriction fragments to generate fingerprints of individual isolates (Vos *et al.*, 1995). A limited number of primer sets are used for amplification and no prior sequence knowledge of the target is required. In contrast to some of the PCR fingerprinting techniques described earlier, stringent PCR conditions are used in this technique, which results in the generation of robust and reliable DNA fingerprints. The number of fingerprint bands generated is dependent upon the template and primer set used, the latter of which is determined experimentally during an optimization process. There are essentially four steps involved in this technique: (1) restriction digestion of genomic DNA; (2) ligation of oligonucleotide adapters; (3) nonselective and selective amplification of sets of restriction fragments; and (4) gel or capillary electrophoresis analysis of amplified fragments.

AFLP polymorphisms are generally a single nucleotide polymorphism in the restriction sites or in the selective nucleotides adjacent to the restriction sites. The primers used in AFLP target a chimeric DNA that includes the adapter and restriction fragment sequences, with selective nucleotides added to the 3′ ends of some of the primers. In this case, chimeric DNA refers to DNA that is from two sources that have been spliced together. Addition of the selective nucleotides is done so that the primers only prime a subset of the restriction sites, limiting the number of generated fragments. A preselective PCR amplification can be done using nonlabeled primers (with zero or one selective nucleotides) to reduce the number of digestion fragments followed by a "selective" PCR amplification in which the primers have additional selective nucleotides added to their 3′ ends. One of the primers is fluorescently labeled; thus amplification results in the formation of fluorescently labeled amplicons. AFLP primers targeting restriction enzyme sites *Eco*RI

(this is a 6 base pair base cutter that results in rare cuts) and *Mse*I (a 4 base pair base cutter that results in frequent cuts) are the most commonly used, but the approach can be used with other primer pairs designed to target other restriction fragment sites. Typically, the *Eco*RI primer is labeled, and since amplification is accomplished using the *Eco*RI and *Mse*I primers, only *Eco*RI—*Mse*I fragments are detected. A typical AFLP fingerprint contains between 50 and 100 amplified fragments. AFLP technology has been widely used in environmental microbiology and is useful for identifying and typing specific organisms such as plant pathogens (Portier *et al.*, 2006).

13.5.1.4 Pulsed-Field Gel Electrophoresis

Pulsed-field gel electrophoresis (PFGE) was first described by Schwartz and Cantor in 1984. PFGE is similar to the RFLP analyses described above, in that it is another way to detect polymorphisms using restriction enzymes. In addition, PFGE is also used to get good estimations of genome size, by cutting the chromosomal DNA into fragments that can be sized accurately. PFGE is used to detect fragments of higher molecular weight than those in normal RFLP, ranging from 10 to 800 kb. These larger fragments are often generated by the use of rare cutting enzymes, the selection of which are organism dependent (e.g., *Xba*I is often used for typing of *E. coli* O157:H7, *Salmonella* and *Shigella* (Ribot *et al.*, 2006) while *Sfi*I and *Not*I are used for typing *Vibrio cholerae* (Cooper *et al.*, 2006)). Overall cell lysis and restriction digestion occur in plugs made from bacterial broth cultures and molten agarose. The plugs are embedded into wells in the gel prior to electrophoresis. This procedure reduces shearing of the DNA, which is important since large fragments are needed for analysis. The restriction digest is separated by gel electrophoresis of alternately pulsed, perpendicularly oriented electrical fields and stained, resulting in a large-fragment fingerprint. Due to its ability to differentiate very closely related microorganisms, PFGE is used to type microorganisms tied to foodborne outbreaks like hemorrhagic *E. coli*, *Salmonella* and *Shigella* (Information Box 13.8).

13.5.1.5 Plasmid Analysis

Many bacterial isolates contain one or more plasmids of variable size (see Section 2.2.3). In some cases, detection of these plasmids allows identification of specific bacterial isolates. The size and number of plasmids associated with a given bacterium are often unique. Note, however, that plasmids can be transferred to other species or in some cases lost from a bacterium. Plasmids are self-replicating, circular, extrachromosomal DNA molecules that encode genes nonessential for cell survival (e.g., metal and/or antibiotic resistance genes) under nonselective

Information Box 13.8   PulseNet — A System for Tracking Microorganisms Causing Foodborne Disease Outbreaks

The U.S. Centers for Disease Control and Prevention (CDC) sponsors the PulseNet program (http://www.cdc.gov/pulsenet/) for identifying and tracking microorganisms involved in foodborne disease outbreaks. The goals of the program are to enhance early detection of foodborne disease outbreaks, and provide information to assist epidemiologists in identifying the source of the outbreaks. The CDC maintains a database of fingerprints for a variety of microorganisms, such as *E. coli* O157: H7, which were developed using pulsed-field gel electrophoresis (PFGE). PulseNet participants isolate and DNA fingerprint suspected illness-causing microorganisms using standardized PFGE procedures, and the fingerprints are ultimately uploaded to the national database at CDC. Database managers then compare the uploaded fingerprints to other microorganisms in the database, and look for clusters of patterns representing potential outbreaks caused by the same microorganism. Results are reported back to the original lab and local, state and federal agencies as appropriate. As nucleic acid-based methods continue to rapidly advance, PulseNet plans to use new methods such as multilocus sequence typing (MLST) to identify microorganisms in the future.

conditions. Some plasmids are also capable of integrating into the host genome. Curing is the process whereby a plasmid is removed from a bacterium, and it can often be achieved by the application of a stress such as heat shock, or simply if the selective pressure such as a metal or antibiotic is not present in the growth medium. For example, pJP4 is an 80 kb plasmid that encodes some of the enzymes used to degrade 2,4-dichlorophenoxyacetic acid as well as sequences responsible for mercury resistance. This plasmid has been shown to undergo horizontal gene transfer from *Ralstonia eutropha* (now classified as *Cupriavidus necator*) to other soil organisms (see Case Study 13.1) and is not stable within the *Ralstonia* isolate if grown in laboratory culture without selective pressure (Di Giovanni *et al.*, 1996).

Despite the potential for plasmid transfer or loss, plasmids have been used to detect specific bacteria or genes. Specific gene sequences associated with a plasmid or homologous gene sequences among plasmids can be detected using gene probes. Bacteria that are associated with specific plasmids can be identified by plasmid profiles or plasmid fingerprints. Plasmid profiles are prepared by electrophoresis and staining of whole plasmids, whereas plasmid fingerprints are produced by restriction digestion of plasmid DNA, followed by electrophoresis and staining of restriction fragments. For gene probe analysis, bacterial isolates can be lysed and probed via colony

lifts using labeled plasmid DNA sequences. Alternatively, plasmid extracts free of chromosomal DNA can be prepared, electrophoresed and probed, in essence becoming a Southern hybridization.

Regardless of whether plasmid profiles or plasmid fingerprints are generated, this analysis provides a relatively quick method for identification of specific DNA sequences associated with a bacterium. Often these DNA sequences can be correlated with specific phenotypic characteristics such as resistance to antibiotics or metals, or the ability to degrade specific organic contaminants. In addition, plasmids have been used to identify specific bacterial isolates. However, in this case, the greatest disadvantage of plasmid analysis is the potential loss of the plasmid from the bacterium of interest. This can occur through plasmid curing, which could result in false-negative results. On the other hand, transfer of the plasmid to other isolates through horizontal gene transfer could result in false-positive results.

13.5.2 Fingerprinting of Microbial Communities

13.5.2.1 Terminal-Restriction Fragment Length Polymorphism Analysis

A method which has frequently been used to generate molecular fingerprints of microbial communities is terminal-restriction fragment length polymorphism (T-RFLP) analysis. For a T-RFLP analysis, DNA is extracted from a microbial community and then amplified with a primer pair for a specific gene of interest (most commonly the 16S rRNA gene), where one or both of the primers is fluorescently labeled. The amplicon is subsequently digested with one or more restriction enzymes. Fragments are then separated on an automated DNA analyzer, and only fragments containing the fluorescently labeled primer are detected. Primer binding sites are located at the ends (or termini) of the amplicon, and fragments are differentiated based on sequence differences in regions extending from that binding site of the labeled primer; thus, the name T-RFLP. Microbial diversity in a community can be estimated based on the number and peak heights of terminal restriction fragments (T-RF), which are easily visualized on electropherograms. A variety of software programs are available for T-RFLP data analysis, facilitating reproducible results. T-RFLP has been used largely to assess diversity and shifts in microbial communities.

13.5.2.2 Denaturing/Temperature Gradient Gel Electrophoresis

Microbial diversity can also be assessed by subjecting PCR-amplified DNA fragments to denaturing gradient gel electrophoresis (DGGE) or temperature gradient gel electrophoresis (TGGE). Originally developed for the medical field to detect point mutations in genetic linkage studies, it was introduced to microbial ecology by Muyzer *et al.* (1993). In essence, DGGE or TGGE can be used for analysis of PCR-amplified gene sequences obtained from community DNA extractions or mixtures of different bacterial isolates. The DNA amplicons generated are of nearly identical lengths, but with variable sequence composition. These analyses are most frequently applied to 16S rRNA gene amplicons, since regions of sequence variability are bounded by conserved primer sites. However, the techniques can be applied to any PCR where it is expected that generated amplicons will have sufficient sequence variability to allow separation.

Separation is based on changes in the electrophoretic mobility of DNA amplicons as they migrate through a gel containing a linearly increasing gradient of DNA denaturants (urea/formaldehyde or temperature). Changes in amplicon mobility result from partial melting of the double-stranded DNA in discrete regions or domains less resistant to denaturants or temperature. In the case of DGGE, this occurs because the temperature of the gel is held constant, so that the melting of domains varies according to the concentration of the denaturant and therefore position in the gel. When the DNA enters a region of the gel containing sufficient denaturant, a transition from helical to partially melted molecules occurs, with a resultant branching of the molecule that sharply decreases the mobility of the DNA amplicons. Sequence variation within particular domains alters their melting behavior; thus, different PCR amplicons essentially stop migrating at different positions in the denaturing gradient. The use of a GC-rich sequence or GC clamp, which consists of 40−45 bases of a GC-rich sequence, acts as a high-temperature-melting domain and prevents complete melting of the amplicons. The GC clamp is normally attached to the 5′ end of the forward primer. Using these techniques, amplicons with only a few base pair variants can be separated efficiently in a linearly increasing denaturing gradient of urea and formaldehyde at 60°C. In contrast, a linearly increasing temperature gradient in the presence of a high constant concentration of urea and formaldehyde is used for separation of PCR amplicons in TGGE. Theoretically, each band in a DGGE/TGGE profile represents one population within the community. However, in practice, some populations are represented by multiple bands and some bands represent multiple populations. Once separated, specific bands can be excised from the gel and subjected to further analysis such as sequencing to identify dominant members in a community.

13.6 RECOMBINANT DNA TECHNIQUES

13.6.1 Cloning

Recombinant DNA technology or DNA cloning has been widely used to examine the genetics of individual bacteria, archaeans and fungi, as well as mixed communities of microorganisms. Cloning, the process of creating identical copies of a gene, has enabled scientists to find new or closely related genes, as well as characterize and identify unculturable or unknown isolates. Cloning may also be used to examine the activity of specific genes, or in the case of functional metagenomics it may be used to screen for activities within large fragments of DNA isolated directly from the environment.

The process of cloning creates a population of organisms that contains recombinant DNA molecules (Figure 13.11). That is, these organisms maintain and express genes contained within their own genomes while at the same time carrying and utilizing genetic material from other organisms. During the cloning process, a single fragment of "source" DNA is ligated to a cloning vector, and a single vector is introduced into each host. As a result, each transformed host only contains one fragment, or sequence, of source DNA. This is one of the most powerful aspects of cloning—the ability to isolate single sequences of source DNA within a host cell. This host cell can then be propagated, generating many copies of the DNA sequence of interest, which can then be studied further.

A cloning vector is a self-replicating DNA molecule, such as a plasmid or phage, which transfers a DNA fragment between host cells (Figure 13.11). A useful cloning vector has four key properties: (1) it can replicate within the desired host; (2) it has a multiple cloning site (MCS) or basis for insertion of foreign DNA; (3) it contains genes that enable selection of vector-containing host cells; and (4) it provides a quick screening mechanism to allow clones containing a source DNA insert to be identified. The origin of replication impacts both host range and copy number of the vector. Most cloning vectors have a ColE1 (pUC)-derived origin of replication that enables production of high copy numbers of the vector in an *E. coli* host. The multiple cloning site (MCS) contains a concentration of restriction enzyme sites for inserting DNA and/or removing DNA inserts, and the insertion site is flanked by sites complementary to common primer sequences (e.g., M13 and T7) that allow for easy amplification of inserted DNA. For selection, vectors typically encode one or more antibiotic resistance genes. As a result, successfully transformed host cells are selected for based upon their ability to grow on a medium containing the appropriate antibiotic(s). There are a variety of ways

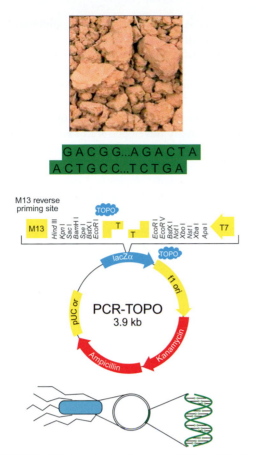

FIGURE 13.11 Cloning of a target gene via PCR amplification. Once the target gene is PCR amplified, the resulting amplicon is cloned into a plasmid vector—the TOPO® TA Cloning System (Life Technologies) in this example. The basis for cloning into this vector is the single "A" added to the 3′ ends of PCR products by many DNA polymerases including *Taq* polymerase. Each terminal "A" overhang can then complement a corresponding "T" in the vector backbone. In the TOPO TA Cloning System, the "ligation" of the insert into the vector backbone is done using a topoisomerase enzyme (TOPO). The illustrated vector contains several common components including: (1) origins of replication (PUC and f1) allowing the vector to multiply in a given host(s); (2) a multiple cloning site (MCS) with a concentration of restriction enzyme sites for inserting DNA and/or removing DNA inserts; (3) antibiotic resistance genes for selection of transformed cells containing the vector; and (4) the *lacZα* gene containing the site for DNA insertion in order to differentiate transformed cells containing DNA inserts from those that do not via the blue–white screening process (Information Box 13.9). The MCS region also contains sites complementary to common primer sequences (e.g., M13 and T7) that allow for easy amplification of inserted DNA. Once the target gene is inserted, the cloning vector can be transformed into a competent bacterial host, which will grow and divide. Subsequently, all progeny bacteria will carry the foreign DNA as long as selection pressure for maintaining the plasmid (i.e., an antibiotic) is included in the growth medium. Plasmid vector illustration adapted from Life Technologies Corporation, Grand Island, NY, U.S.A.

to screen for the presence of a source DNA insert within the cloning vector, but one of the most common is the blue–white selection process that allows recombinant hosts to be identified on the basis of the interruption, or

insertional inactivation, of a reporter gene (Information Box 13.9; Figure 13.12).

PCR has had a dramatic impact on the process of cloning DNA fragments. PCR can be used to rapidly produce large amounts of the source DNA fragment, and the use of appropriately designed primers provides a mechanism of introducing restriction enzyme sites at the ends of the fragments, simplifying the first two steps of the cloning process. A single PCR amplicon can then be directly inserted into a cloning vector and replicated in a host, such as *E. coli*. In addition, cloning vectors have been designed with a "T" on both 3′ ends of a linearized vector backbone that enables binding to the single "A" added to the 3′ ends of PCR products (Figure 13.11) by many DNA polymerases including *Taq* DNA polymerase (Information Box 13.6). Cloning of PCR amplicons has become a very common research tool, and commercial kits for PCR-based cloning are widely available.

13.6.2 Metagenomics

The term metagenomics was first coined by Handelsman *et al.* (1998) in reference to the collective analysis of the gene content of a community of soil microbes. Metagenomics involves the manipulation and analysis of DNA fragments obtained from a mixed community of microorganisms. Two main approaches exist for the characterization of metagenomes: sequence-based analysis and functional screening (Handelsman, 2004) (Figure 13.13). Metagenomics can be used to provide information regarding the structure, function and metabolic potential of a microbial community. Metagenomics studies are currently being performed in a wide variety of environments, ranging from soil and water to bioreactors, built environments and mammalian or insect hosts.

Early metagenomics studies typically used functional screening as a starting point to identify genes of interest for sequencing. This approach relied on cloning large DNA fragments into specialized vectors known as bacterial artificial chromosomes (BACs) or yeast artificial chromosomes (YACs). This process allows multiple genes encoded on a continuous piece of DNA to be cloned into a host together and screened for functional activity. One of the potential pitfalls of this approach, however, is that heterologous expression in the host is required. That is, the host must be able to express the cloned gene content, some of which may require coexpression of other genes. In addition, the codon usage of the host must be compatible with that of the inserted DNA. The major advantage of this approach is that, when heterologous gene expression is achieved, it does not require that the genes of interest be recognized by

Information Box 13.9   Blue−White Screening Process for Identifying Clones Containing Recombinant DNA

Many cloning vectors allow recombinant hosts to be identified on the basis of the interruption, or insertional inactivation, of a reporter gene. One widely used reporter is *lacZ*, a gene which encodes for β-galactosidase, which catalyzes the hydrolysis of lactose into glucose and galactose. Conveniently, it can also cleave the compound X-Gal (5-bromo-4-chloro-3-indolyl-β-D-galactopyranoside) into galactose and 5-bromo-4-chloro-3-hydroxyindole, an insoluble compound with an indigo/blue color. These *lacZ*-based screening systems rely on a phenomenon known as α-complementation. In this process, a mutant β-galactosidase gene that produces a protein lacking several N-terminal residues is included on the host chromosome. Expression of this mutant gene alone produces an inactive β-galactosidase enzyme. However, this mutant enzyme can be made active by complementation of the missing portion of the chromosomal gene via inclusion of the first half of the *lacZ* gene, encoding the N-terminal (α) portion of β-galactosidase, on a cloning vector. When both portions of the gene are expressed, as in a host cell transformed with a cloning vector, an active β-galactosidase enzyme is formed. However, the insertion of source DNA into the cloning vector disrupts the vector-borne portion of the *lacZ* gene, prevents α-complementation, and blocks the formation of a functional β-galactosidase enzyme. As a result, clones containing a vector with inserted DNA can easily be identified when grown on media containing X-Gal, since they will form white colonies (Figure 13.12). In contrast, cells without inserted DNA produce active β-galactosidase, cleave X-Gal and thus produce blue colonies.

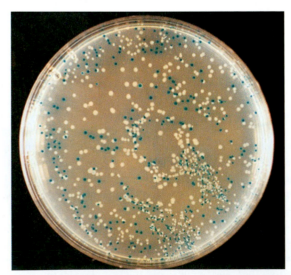

FIGURE 13.12 The blue−white screen for identification of *E. coli* clones containing recombinant DNA. White colonies contain inserted DNA while blue colonies do not.

sequence analysis, and thus allows potentially novel genetic elements to be identified.

The cloning of DNA fragments into BACs or YACs also provides a relatively easy means of targeting genes of interest for sequencing, since BACs and YACs contain known insertion site sequences and can serve as the targets for PCR-based amplification and sequencing primers. BACs were originally developed for cloning of eukaryotic genome fragments, can accommodate DNA inserts >300 kb and are maintained in low copy number in *E. coli*. YACs work similarly, and can incorporate inserts as large as 2 Mb in size; however, YAC transformation efficiencies are 100 times lower than BACs.

Most metagenomic studies today begin at the sequencing stage. Rather than generating DNA fragments and cloning them into BAC or YAC libraries, community DNA is fractionated into small pieces and sequenced directly via high-throughput sequencing (e.g., 454, Illumina and similar platforms). This approach is often described as shotgun sequencing. Using high-throughput sequencing (i.e., pyrosequencing) platforms, DNA fragments are fixed onto small beads and undergo sequencing by synthesis, a process in which thousands to millions of DNA fragments are amplified by PCR and their sequences are determined as each new base is incorporated into their respective PCR products. The advantages of this approach are that it can be used to sequence large amounts of DNA, over a relatively short time span, and at low cost relative to BAC sequencing. The disadvantages are that the length of the sequences that are generated is relatively short (approximately 150−400 bp, depending on the sequencing platform used) compared to what can be generated from BAC fragments (>300 kbp); the function of genes is inferred bioinformatically rather than tested empirically; and the analysis of such large numbers of short sequence fragments can be computationally challenging.

Despite the challenges associated with metagenomic library construction and analysis, metagenomic studies are providing unprecedented insight into the structure, function and genetic potential of microbial communities. Metagenomic sequencing provides access to both culturable and unculturable populations, as well as to underrepresented populations that otherwise go undetected. Because of this, metagenomics is allowing new insights to be gained with respect to how microbial communities function, and it is helping to unlock the vast genetic potential held within microbial communities.

13.6.3 Reporter Gene Systems

Nucleic acid sequences can be utilized not only to detect specific organisms but also to detect the expression of a particular gene. Reporter genes, often simply referred to

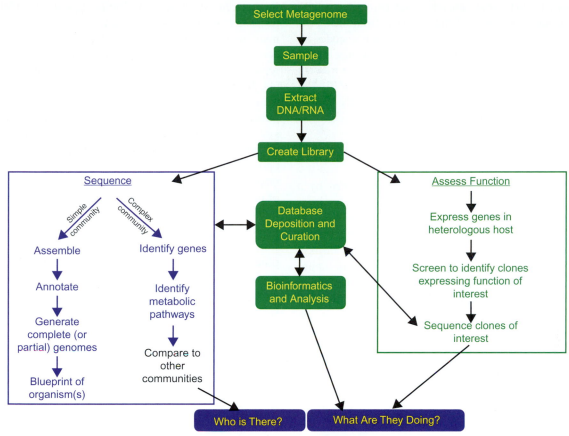

FIGURE 13.13   Metagenomics-based analyses can be used to answer the questions "Who is there?" and "What are they doing?" These projects involve the extraction of nucleic acids from an environmental sample followed by sequence-driven (light blue box) and/or function-driven (light green box) analyses. Techniques such as rRNA gene surveys are often used to determine the degree of diversity in a sample, and thus the appropriate approach for further analysis (e.g., the degree of sequencing or size of clone library necessary). An important point for any metagenomic project is the collection of environmental metadata such as location, vegetation, and soil data. This information is invaluable for environmental interpretation of metagenomic data and comparing results with other projects from around the world. For simpler communities comprised of only a few populations of microorganisms, such as a highly contaminated groundwater sample, it may be possible to assemble metagenomic sequence data to generate complete (or near-complete) genomes and metabolic blueprints of the dominant organisms. For more complex communities, sequence data is usually screened for the presence of genes of interest and compared to data from other communities, without assembly of sequencing reads. Adapted from National Academy of Sciences (2007).

as reporters, are genes that confer an easily detected signal when expressed by the host. The theory behind a reporter gene is that most genes encode products that are not easily detected, whereas the inserted reporter gene produces a product that is easy to detect. Reporters can be used for a variety of purposes including selection, gene expression analyses, promoter activity assessments, determining interactions between proteins such as in two-hybrid screening and determining the presence or bio-availability of pollutants. Antibiotic and metal resistance genes encoded on plasmid vectors can be considered reporters since they provide a selective mechanism for vector-bearing hosts. Reporter genes are also frequently constructed to allow gene expression of a target gene. In this case, a reporter is designed creating a DNA construct where the reporter gene is inserted between the promoter of the gene of interest and the gene of interest (Figure 13.14). Thus, the reporter is turned on (expressed) only when the target gene of interest is expressed. The reporter gene and the gene of interest are thus transcribed together and ultimately translated into a single polypeptide chain (fusion protein). A segment of DNA coding for a flexible polypeptide linker region is generally incorporated into the construct between the reporter and target gene, allowing for proper folding into active conformations of the two fused proteins. The activity of a particular promoter can also be determined using reporter genes. In this case, the reporter gene is placed under control of a target promoter and the activity of the reporter gene product is quantitatively measured.

Several different reporter genes have been used including: *lac*Z, which encodes β-galactosidase (Information Box

13.9); *gusA* (*uidA*), which encodes β-glucuronidase; *xylE*, which encodes catechol 2,3-dioxygenase from the TOL plasmid of *Pseudomonas putida*; *inaZ*, which encodes an ice nucleation protein from *P. syringae*; *lux* genes, which catalyze light production; and genes for various fluorescent proteins such as the green fluorescent protein (GFP). The latter two are among the reporters most commonly used to characterize environmental microorganisms.

13.6.3.1 Luminescent Reporter Genes

Luciferases are light-generating enzymes produced by a wide variety of organisms (e.g., fireflies and biolumines-cent bacteria such as *Aliivibrio fischeri* (Section 20.2.1)). Luciferase and its luciferin substrate are coded by *lux* (*luxCDABE*) genes and are active in the presence of oxy-gen and a source of reducing power such as flavin mono-nucleotide (FMNH$_2$). The *luxAB* genes encode the active form of luciferase, and the *lux*CDE genes encode the syn-thesis of the substrate. Use of the full *luxCDABE* gene cas-sette is therefore advantageous because it does not require the addition of any exogenous substrate. The basis of the *lux* system is that *lux* genes are inserted into the operon being studied, and when that operon is induced,

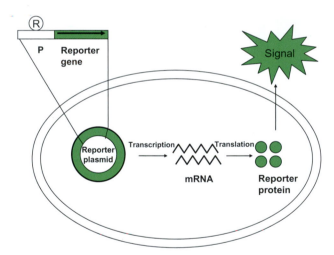

FIGURE 13.14 Regulation of a reporter gene by a regulatory protein. Binding of the regulatory protein R to the promoter P controls transcrip-tion, followed by translation of the mRNA to produce the protein. Both of these steps produce multiple copies of the reporter protein, leading to an increased protein concentration and a measurable signal.

luminescence is given off. The *lux* system is widely used because it is uncommon to find large numbers of indige-nous microorganisms that emit light, and the biolumines-cent signal generated is typically short-lived allowing for repeated sampling, which is required in time course stud-ies. Most *lux* systems require $\approx 10^5$ cells per sample for the detection of light emission, and detection can be done directly in a nondestructive manner. Different *lux*-based systems have been used in numerous experiments. For example, Dorn *et al*. (2005) used a *lux* reporter and a fiber optic detection system to study the real-time *in situ* response of microbes in a soil column system to addition of naphthalene. In this system, optic fibers were emplaced into the soil column and used to collect luminescence over several days to study the response to naphthalene addition. This lux reporter was successfully able to provide real-time information about where naphthalene degradation occurred in the soil column. However, these reporters are typically very sensitive to changes in environmental condi-tions and it is important that the reporter be extensively characterized prior to use (Sørensen *et al*., 2006).

13.6.3.2 Fluorescent Reporter Genes

There are now several fluorescent reporter systems including the green fluorescent protein (GFP) and the red fluorescent protein (DsRed or RFP). The genes for GFP and RFP were isolated from the jellyfish *Aequorea victo-ria* and from *Discosoma* sp. coral, respectively. These bioreporters rely on fluorescent light production instead of luminescence. The difference is that to obtain fluores-cence, the molecule must first be excited by a specific wavelength of light. Following excitation, the fluorescent molecule emits, or fluoresces, at a different wavelength that can be detected and measured. Excitation of GFP with UV light (395 nm) results in a bright green fluores-cence (509 nm), while for RFP, excitation occurs at 558 nm, and emission is at 583 nm. Through mutation, these genes have been modified to produce a broad spec-trum of fluorescent proteins that have different excitation and emission patterns (Figure 13.15).

The fluorescent reporter systems have several advan-tages over *lacZ* and *xylE* fusions. First, as with the *lux* system, because of the abundance of fluorescent proteins

FIGURE 13.15 Examples of the wide variety of fluo-rescent proteins available for use as reporter genes. From Tsien (2010).

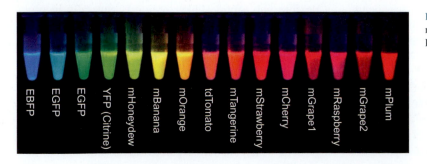

EBFP | EGFP | EGFP | YFP (Citrine) | mHoneydew | mBanana | mOrange | tdTomato | mTangerine | mStrawberry | mCherry | mGrape1 | mRaspberry | mGrape2 | mPlum

---

Case Study 13.2   Use of Reporter Genes to Track Horizontal Gene Transfer in a Microcolony

Seoane *et al.* (2011) used three different fluorescent reporter genes to track transfer of plasmids between *Pseudomonas putida* KT2440 cells in a bacterial microcolony. The cells were inoculated on a nutrient agar slab, and images were taken with confocal laser scanning microscopy (CLSM) every 20 min. Figure 13.16 contains time-lapse images of the microcolony where donor *P. putida* KT2440 cells expressing a red fluorescent protein gene (DsRed) and producing LacIq (to repress GFP expression in donor cells) transferred a plasmid containing the green fluorescent protein (GFP) gene (pWW0 TOL::GFP) to recipient *P. putida* KT2440 cells that also expressed the yellow fluorescent protein (YFP) gene. Therefore, the transconjugants simultaneously expressed GFP and YFP. The GFP signal (transconjugant cells) and the corresponding overlay of all fluorescence signals (all cell types) are displayed on the top and on the bottom rows of Figure 13.16, respectively. Non-dividing, inoculated donors (bottom left of microcolony) have higher red intensities due to previous DsRed maturation. Thick arrows mark the individual cell transferring the plasmid while thin arrows indicate the resulting new transconjugant cell. After 160 min of donor—recipient contact, conjugative transfer was detected (A). It then took < 40 min for this transconjugant to retransfer twice (B, C) and < 20 min for the new transconjugants to retransfer again (D). After 480 min (approximately five division cycles), most of the potential recipient cells in the microcolony contained the plasmid (E).

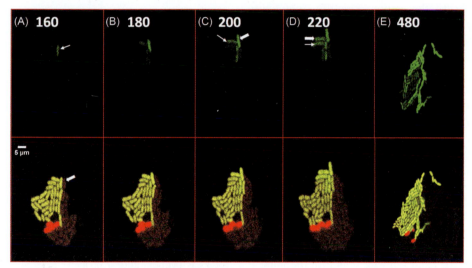

FIGURE 13.16   Use of reporter genes to track horizontal gene transfer in a bacterial microcolony. From Seoane *et al.* (2011).

---

in a single cell, transcription of a gene in individual cells rather than the average transcription of a population of cells can be measured. However, unlike the case of the *lux* system, cellular metabolism is not required for reporter fluorescence, thus allowing *in situ* gene expression by cells that are not actively growing, which is likely to be the case in many environments. Reporter fluorescence is also stable at 37°C, whereas the luciferase may be heat labile. Finally, both fluorescent proteins and *lux* are affected by oxygen levels. Although the fluorescent protein reporter systems function in the absence of O_2, the GFP protein must be oxidized in order to fluoresce. The *lux* system is affected when O_2 levels drop below 5 mg/L; however, luminescence occurs even as low as 0.5 mg O_2/L. Another important factor is that luminescence from a *lux* reporter stops as soon as gene expression stops, while the wild-type GFP protein will continue to fluoresce for as long as it remains intact. New variants of fluorescent proteins have been developed that have decreased stability to allow more accurate detection of real-time gene expression. Additionally, several color variants have been developed with altered excitation and emission spectra that allow for simultaneous tracking of multiple organisms (Case Study 13.2; Figure 13.16). These modifications have greatly increased the utility of fluorescent protein-based reporter systems for applications such as tracking dynamics of protein expression and protein localization within living cells. It should be kept in mind that it is difficult to track low levels of expression with fluorescent proteins and that the intrinsic fluorescence of some bacterial species can preclude their use as a reporter. Thus, the degree of potential background interference along with the desired characteristics of the reporter including ease of detection, sensitivity, ability to quantify and degree of stability should be considered when selecting a reporter gene system. More detailed discussions regarding the advantages and disadvantages of various reporter genes are available in reviews including Burlage (2007) and Leveau *et al.* (2007).

13.7 SEQUENCE ANALYSIS

Advances in DNA sequencing technology have allowed sequence-based analyses to become routine and relatively cost-effective. DNA samples may be sent to commercial laboratories or core facilities for sequencing, but the use of benchtop sequencers in individual laboratories is becoming more common. Multiple databases store and catalog information from sequence-based projects and make it accessible to the entire scientific community (Table 13.2). The sequence data stored in these databases and archives carry accession numbers to facilitate identification, organization and communication about particular sequences or projects. The database entries also typically contain the DNA, RNA or protein coding regions for a given sequence, relevant metadata (e.g., where a sample was collected, when it was collected, the environmental characteristics of the collection site) and citations for associated published research. Software programs, such as BLAST (the Basic Local Alignment Search Tool; Altschul *et al.*, 1997), allow researchers to identify sequences of interest with these databases, translate a DNA sequence into protein and look for homology or relatedness based on sequences. Some of these databases also provide links to gene maps, algorithms that identify restriction enzyme sites and relevant scientific literature.

One of the most common uses of sequence data in the field of environmental microbiology is in the identification of microbes through marker gene sequences. In the case of Bacteria and Archaea, the 16S ribosomal RNA (rRNA) gene sequence is used. For Fungi, the ribosomal internal transcribed spacer (ITS) tends to be used most frequently, but the 18S and 25−28S rRNA genes are also sometimes used. Multiple databases and sequence analysis software programs are available to aid in sequence searches and identification. The National Center for Biotechnology Information (NCBI) provides large, community-generated sequence archives. Other specialized databases, such as the Ribosomal Database Project and the UNITE fungal ITS database, obtain data from NCBI but implement additional quality filtering and annotation.

Researchers typically interact with these databases using BLAST or another similar software package. Programs like BLAST allow researchers to submit a query sequence and compare it to all other sequences in the database. The results of this comparison are then scored for similarity and identity. Searches like this are used when researchers want to find out which sequences in the databases are similar to

---

Information Box 13.10   Phylogenetic Analysis

In the 16S rRNA gene, there are regions that are conserved among all bacteria. These conserved regions flank a sequence region that is variable and that can be used to study the relatedness among different bacteria. These attributes have been exploited to study complex bacterial communities in environmental samples using a combination of PCR and sequencing analysis. Using universal 16S rRNA primers, PCR is performed on community DNA extracted from an environmental sample. The PCR results in amplicons of roughly the same size but varying sequence composition. Once the sequences are obtained, they are aligned and then analyzed using bioinformatics (Chapter 21) to determine the diversity and composition of the community and compare it to communities from other samples.

---

their newly-generated, unknown sequences. In some cases, this information may be used for phylogenetic analyses (Information Box 13.10), allowing researchers to characterize the genetic relationships and ancestry of new sequences relative to those contained in reference databases. For example, a 16S rRNA gene sequence from an unknown soil isolate can be compared with other sequences in the database in order to characterize and identify the organism. By analyzing numerous sequences from a sample, this approach can also be used to determine community composition and diversity (Case Study 13.3; Figure 13.17). In other cases, sequences from multiple, related organisms may be aligned with one another to identify regions of homology (i.e., similarity) as a part of the design process for PCR primers and/or probes.

Molecular databases are built from sequences submitted by researchers all over the world. Sequence data can be found by searching for an accession number assigned to a specific sequence, a functional description such as "hydrolase," a known sequence pattern (e.g., "GGTACCTTGAG") or the name of a gene or organism of interest (e.g., *lacZ*, *E. coli*). A number of search algorithms are available for a wide scope of nucleic acid and protein sequence comparisons. FASTA (Pearson, 1990) and BLAST are well-established algorithms that facilitate effective comparisons of unidentified sequences with those contained in the current databanks. One must take care in choosing database sequences for comparison because quality control of sequence data varies from database to database and misannotation is a common occurrence.

While genomic and metagenomic studies often target the total gene content of an organism or a community, many times it is advantageous to target a single gene or collection of genes instead. The characterization of a target gene (or genes) is known as a marker gene study. As mentioned in Section 13.4.4, the 16S rRNA gene is commonly used as a marker gene for bacterial communities, and the

---

Case Study 13.3   Use of DNA Sequencing to Determine Microbial Diversity in an Antarctic Freshwater Ecosystem

---

The 16S rRNA gene has been used to describe the composition and diversity of a wide variety of biological systems ranging from soil, water, and air to kitchen surfaces, hotel rooms, and insect or mammalian hosts. Prior to the advent of high-throughput sequencing technologies, such as the 454 and Illumina platforms, diversity surveys typically relied on cloning and Sanger sequencing (i.e., dye-terminator sequencing), PCR-fragment-based characterization (e.g., T-RFLP or DGGE), culture-based techniques, or some combination thereof. Although there are advantages and disadvantages to each approach, the general consensus is that high-throughput sequence-based approaches tend to give greater insight into community structure and diversity.

In a recent survey of Lake Tawani, a freshwater lake located in East Antarctica, Huang *et al.* (2013) used a combination of 454 sequencing, cloning and Sanger sequencing, and plate-based cultures to describe community composition and bacterial diversity and compared the outcomes obtained from each of the different approaches. The authors collected mixed water and sediment samples, using one subsample for community DNA extraction and another subsample for the enumeration of bacteria via plate counts. The community DNA was used to generate partial 16S rRNA gene amplicons for 454 sequencing, as well as near-full-length 16S rRNA gene amplicons for cloning and Sanger sequencing. The plate counts yielded $\sim 2 \times 10^3$ colony forming units per

ml, and from this 270 bacterial colonies were selected for identification. The cloning experiment yielded 232 clones with the correct insert size, and the 454 sequencing run yielded 11,235 high-quality sequence reads. The identities of both the culture-derived colonies and 16S clones were confirmed through Sanger sequencing (Figure 13.17).

As in many soil and water-based ecosystems, bacteria belonging to the phylum Proteobacteria were encountered with the greatest frequency, regardless of the technique that was used. The authors found that the three different approaches successfully detected the most dominant taxa in Lake Tawani, but the culture-based, cloning, and 454-based techniques different in their coverage, estimates of relative abundance (i.e., ranking) and ability to detect infrequent, or rare, members of the community. Among the 270 culture-derived colonies, 16 operational taxonomic units (OTUs, a sequence-based approximation of bacterial species) were found. The clone library detected 64 OTUs, and the 454 sequence library detected 498 OTUs. Among these, only 9 OTUs were detected by all three approaches. The authors concluded that high-throughput sequencing provided greater sensitivity and greater coverage than the clone or culture-based techniques, but also noted that the combination of multiple technologies gave a more complete picture than any single approach alone.

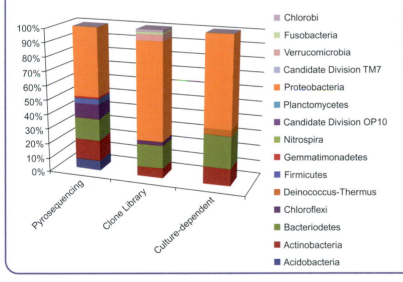

FIGURE 13.17 Relative abundance of bacterial phyla in an East Antarctica freshwater ecosystem. Bacteria were identified through culture-dependent and culture-independent (pyrosequencing and clone library) methodologies. From Huang *et al.* (2013).

---

fungal ITS sequence is often used to characterize the composition of fungal communities. Other common targets include "housekeeping" genes, functional genes and collections of genes sequenced simultaneously in a process known as multi-locus sequencing (Table 13.1).

Marker genes are commonly used to identify individual sequences or isolates, but they can also be used to provide insight into the composition and diversity of entire microbial communities. Informative marker genes harbor several key traits, and these include:

- Ubiquity—the marker should be present in most, if not all, species.
- Genetic conservation—the sequence of the marker should be conserved enough that it can be targeted with PCR primers.
- Variability—in combination with genetic conservation, the marker should also contain variable regions of sequence that allow for the detection of differentiation between species, among lineages and within populations.

For some organisms, single marker genes do not provide enough information to distinguish species or strains from one another. For example, species within the genera *Escherichia* and *Shigella* share so much similarity among their 16S rRNA gene sequences that it is difficult to distinguish them from one another. Likewise, it is often necessary to distinguish pathogenic strains from their less harmful relatives in infection, disease outbreaks or source-tracking scenarios. In cases like this, multilocus sequence typing (MLST) can be used to provide the necessarily additional resolution. MLST involves the sequencing of multiple target genes (often seven or eight genes) from the same organism in order to provide a more complete taxonomic and phylogenetic profile (Pérez-Losada *et al.*, 2013). Once these sequences are generated, they can be compared to reference databases in order to obtain sequence- or strain-type identities (e.g., Aanensen and Spratt, 2005).

13.8 CHOOSING THE APPROPRIATE NUCLEIC ACID-BASED METHOD

With so many nucleic acid-based methods available, you may ask "How can I choose which one to use for my research?" Table 13.6 overviews environmental microbiology applications of several nucleic acid-based methods. Due to the wide variety of methods (and their modifications) available, this table should not be considered to be comprehensive with respect to either the methods listed or their applications. Other reference materials, such as the *Manual of Environmental Microbiology* (Hurst *et al.*, 2007), also provide detailed discussions regarding the applications and advantages and disadvantages of numerous nucleic acid-based methods. Ultimately, the choice of method will be driven by the available research budget and several factors including:

- Research objectives—Detection or quantification of microorganisms? Measurement of microbial activity? Characterization of microbial community composition? Elucidation of environmental function? Identification of genes conferring a specific process?
- Target organisms—Isolated or uncultured microorganisms? Pure culture or mixed community? Simple or complex microbial communities?
- Target material—DNA or RNA?

In many cases, the discovery power of nucleic acid-based methods can be magnified even further when used in combination with physiological methods such as stable isotope probing (Chapter 11). These combined approaches can enable not only the identification of microorganisms but also their connection to a specific environmental process.

The past several decades have seen dramatic advances in the use of nucleic acid-based methods to study environmental microorganisms. These methods have provided insights into the microbial world that were unimaginable just a few years ago. It is exciting to think about the continued methodological advances that will occur over the next few decades. If there is not a current method that will achieve your research objectives, we look forward to you modifying a current method or even developing an entirely novel approach for detecting and characterizing environmental microorganisms.

QUESTIONS AND PROBLEMS

1. What are some of the advantages of using nucleic acid-based methods to detect and characterize environmental microorganisms?
2. What are some of the challenges of using nucleic acid-based methods to detect and characterize environmental microorganisms?
3. What are the major advantages and disadvantages of PCR when it is applied to environmental samples?
4. If you wanted to use PCR to assess microbial activity, what type of PCR would you use and why?
5. There are several features of qPCR that make its use advantageous over conventional PCR, list three and explain why each feature is beneficial.
6. Discuss the differences between the SYBR Green and TaqMan approaches for qPCR. Which qPCR approach is more specific? Why?
7. Compare and contrast ERIC and PFGE methods of DNA fingerprinting.
8. Compare and contrast the *lux* reporter system and the GFP reporter system. Based on these characteristics, which would be more useful for evaluating contaminant transport in a soil column study?
9. Describe what is meant by selection versus screening of a clone library. Explain how the *lacZ* reporter works and incorporate its use into your description.
10. Which nucleic acid-based method(s) would you use to accomplish each of the following objectives? Explain why you chose each method.
    a. Detect *Salmonella* spp. in a soil sample.
    b. Detect both *Salmonella* spp. and *E. coli* in the same sample.
    c. Quantify *Norovirus* in a water sample.
    d. Characterize the presence and diversity of dissimilatory sulfite reductase genes (*dsrAB*) in a groundwater sample.
    e. Determine the bacterial community composition of a fecal sample.
    f. Quantify expression of nitrite reductase genes (*nirK* and *nirS*) in a sediment sample.

TABLE 13.6 Environmental Microbiology Applications of Various Nucleic Acid-based Methods

Technology	Application				Notes
	Microbial Detection	Microbial Quantification	Microbial Activity	Community Characterization	
Polymerase chain reaction (PCR)	Yes	No	No	No	PCR is the most commonly used approach for microbial detection. It is also the basis for many of the other forms of detection and characterization.
Reverse transcriptase PCR (RT-PCR)	No	No	Yes	No	Reverse transcription of mRNA followed by PCR amplification of the resulting cDNA is a primary means of detecting microbial activity.
Quantitative PCR (qPCR)	Yes	Yes	Yes	No	Unlike PCR which is semi-quantitative at best, a variety of qPCR approaches can be used to quantify microorganisms (DNA) and their activity (if RT-PCR is used to convert mRNA to cDNA prior to qPCR).
DNA fingerprinting	Yes	No	No	Yes	DNA fingerprinting approaches such as pulsed-field gel electrophoresis are the basis of microbial typing used for detection of specific strains of microorganisms. These approaches are best used with pure cultures of organisms. Some DNA fingerprinting approaches such as T-RFLP have been used on microbial communities but these approaches have been largely replaced by DNA sequencing.
DNA/RNA sequencing	Yes	No	Yes	Yes	The current standard for microbial community characterization. If RNA template is sequenced (transcriptomics) this can indicate relative activity. Target DNA/RNA is often amplified via PCR methods prior to sequencing.
Fluorescence in situ hybridization (FISH)	Yes	Yes	Yes	No	FISH can be used to detect, quantify and determine activity of microorganisms in situ. However, because it is primarily a microscopy-based approach using labeled probes, only a small number of organisms can be assayed for at a given time.
Microarrays	Yes	No	Yes	Yes	Microarrays can be used to detect microorganisms (DNA) and their activity (usually after RT-PCR of mRNA). Due to the number of organisms that can be tested simultaneously, microarrays can also be used to characterize microbial communities. Although some microarray-based approaches can be quantitative, they are generally not the best approach for microbial quantification.
Reporter genes	No	No	Yes	No	The primary use of reporter genes is to determine activity of organisms engineered to contain the reporter. These genes can also be used to track the survival and movement of a host organism engineered to contain the reporter.

g. Isolate a microbial gene from a soil sample that confers resistance to a newly discovered antibiotic.

h. Determine *in situ* expression of a bacterial gene in a biofilm.

i. Determine if a new pesticide impacts soil microbial community composition.

REFERENCES AND RECOMMENDED READING

Aanensen, D. M., and Spratt, B. G. (2005) The multilocus sequence typing network: mlst.net. *Nucleic Acids Res.* **22**, W728–W733.

Altschul, S. F., Madden, T. L., Schäffer, A. A., Zhang, J., Zhang, Z., Miller, W., *et al.* (1997) Gapped BLAST and PSI-BLAST: a new generation of protein database search programs. *Nucleic Acids Res.* **25**, 3389–3402.

Blackmer, F., Reynolds, K. A., Gerba, C. P., and Pepper, I. L. (2000) Use of integrated cell culture-PCR to evaluate the effectiveness of poliovirus inactivation by chlorine. *Appl. Environ. Microbiol.* **66**, 2267−2268.

Brodie, E. L., DeSantis, T. Z., Joyner, D. C., Baek, S., Larsen, J. T., Andersen, G. L., *et al.* (2006) Application of a high-density oligonucleotide microarray approach to study bacterial population dynamics during uranium reduction and reoxidation. *Appl. Environ. Microbiol.* **72**, 6288−6298.

Brodie, E. L., Desantis, T. Z., Parker, J. P., Zubietta, I. X., Piceno, Y. M., and Andersen, G. L. (2007) Urban aerosols harbor diverse and dynamic bacterial populations. *Proc. Natl. Acad. Sci. U.S.A.* **104**, 299−304.

Burlage, R. S. (2007) Bioreporters, biosensors, and microprobes. In "Manual of Environmental Microbiology" (C. J. Hurst, R. L. Crawford, D. A. Lipson, A. L. Mills, and L. D. Stetzenbach, eds.), 3rd ed., American Society for Microbiology Press, Washington, DC, pp. 169−181.

Burr, M. D., Josephson, K. L., and Pepper, I. L. (1997) An evaluation of DNA based methodologies for subtyping *Salmonella. Crit. Rev. Environ. Sci. Technol.* **28**, 283−323.

Chen, K., Neimark, H., Rumore, P., and Steinman, C. R. (1989) Broad range DNA probes for detecting and amplifying eubacterial nucleic acids. *FEMS Microbiol. Lett.* **48**, 19−24.

Cooper, K. L. F., Luey, C. K. Y., Bird, M., Terajima, J., Nair, G. B., Kam, K. M., *et al.* (2006) Development and validation of a PulseNet standardized pulsed-field gel electrophoresis protocol for subtyping of *Vibrio cholerae. Foodborne Pathog. Dis.* **3**, 51−58.

Di Giovanni, G. D., Neilson, J. W., Pepper, I. L., and Sinclair, N. A. (1996) Gene transfer of *Alcaligenes eutrophus* JMP134 plasmid pJP4 to indigenous soil recipients. *Appl. Environ. Microbiol.* **62**, 2521−2526.

Dorn, J. G., Brusseau, M. L., and Maier, R. M. (2005) Real-time in situ monitoring of bioactive zone dynamics in heterogeneous systems. *Environ. Sci. Technol.* **39**, 8898−8905.

Dowd, S. E., and Pepper, I. L. (2007) PCR: agricultural and environmental applications for soil microbes. In "Manual of Environmental Microbiology" (C. J. Hurst, R. L. Crawford, D. A. Lipson, A. L. Mills, and L. D. Stetzenbach, eds.), 3rd ed., American Society for Microbiology Press, Washington, DC, pp. 676−686.

Gao, H., Yang, Z. K., Gentry, T. J., Wu, L., Schadt, C. W., and Zhou, J. (2007) Microarray-based analysis of microbial community RNAs by whole community RNA amplification. *Appl. Environ. Microbiol.* **73**, 563−571.

Gentry, T. J., Wickham, G. S., Schadt, C. W., He, Z., and Zhou, J. (2006) Microarray applications in microbial ecology research. *Microb. Ecol.* **52**, 159−175.

Green, M. R., and Sambrook, J. (2012) "Molecular Cloning: A Laboratory Manual," 4th ed. Cold Spring Harbor Laboratory Press, Cold Spring Harbor, NY.

Handelsman, J. (2004) Metagenomics: application of genomics to uncultured microorganisms. *Microbiol. Mol. Biol. Rev.* **68**, 669−685.

Handelsman, J., Rondon, M. R., Brady, S. F., Clardy, J., and Goodman, R. M. (1998) Molecular biological access to the chemistry of unknown soil microbes: a new frontier for natural products. *Chem. Biol.* **5**, R248−R249.

Hazen, T. C., Dubinsky, E. A., DeSantis, T. Z., Andersen, G. L., Piceno, Y. M., Singh, N., *et al.* (2010) Deep-sea oil plume enriches indigenous oil-degrading bacteria. *Science* **330**, 204−208.

He, Z., Deng, Y., Van Nostrand, J. D., Tu, Q., Xu, M., Hemme, C. L., *et al.* (2010) GeoChip 3.0 as a high-throughput tool for analyzing microbial community composition, structure, and functional activity. *ISME J.* **4**, 1167−1179.

Heid, C. A., Stevens, J., Livak, K. J., and Williams, P. M. (1996) Real time quantitative PCR. *Genome Res.* **6**, 986−994.

Hill, V. R., Mull, B., Jothikumar, N., Ferdinand, K., and Vinjé, J. (2010) Detection of GI and GII noroviruses in ground water using ultrafiltration and TaqMan real-time RT-PCR. *Food Environ. Virol.* **2**, 218−224.

Hurst, C. J., Crawford, R. L., Lipson, D. A., Mills, A. L., and Stetzenbach, L. D. (eds.) (2007) Manual of Environmental Microbiology. 3rd ed., American Society for Microbiology Press, Washington, DC.

Huang, J. P., Swain, A. K., Thacker, R. W., Ravindra, R., Andersen, D. T., and Bej, A. K. (2013) Bacterial diversity of the rock-water interface in an east antarctic freshwater ecosystem, Lake Tawani(P). *Aquat. Biosyst.* **9**, 4.

Korves, T. M., Piceno, Y. M., Tom, L. M., DeSantis, T. Z., Jones, B. W., Andersen, G. L., *et al.* (2013) Bacterial communities in commercial aircraft high-efficiency particulate air (HEPA) filters assessed by PhyloChip analysis. *Indoor Air* **1**, 50−61.

Lasken, R. S. (2012) Genomic sequencing of uncultured microorganisms from single cells. *Nat. Rev. Microbiol.* **10**, 631−640.

Leveau, J. H. J., Loper, J. E., and Lindow, S. E. (2007) Reporter gene systems useful in evaluation in situ gene expression by soil- and plant-associated bacteria. In "Manual of Environmental Microbiology" (C. J. Hurst, R. L. Crawford, D. A. Lipson, A. L. Mills, and L. D. Stetzenbach, eds.), 3rd ed., American Society for Microbiology Press, Washington, DC, pp. 734−747.

Loy, A., Maixner, F., Wagner, M., and Horn, M. (2007) probeBase—an online resource for rRNA-targeted oligonucleotide probes: new features 2007. *Nucleic Acids Res.* **35**, D800−D804.

Lu, Z., Deng, Y., Van Nostrand, J. D., He, Z., Voordeckers, J., Zhou, A., *et al.* (2012) Microbial gene functions enriched in the deepwater horizon deep-sea oil plume. *ISME J.* **6**, 451−460.

Matsuki, T., Watanabe, K., Fujimoto, J., Takada, T., and Tanaka, R. (2004) Use of 16S rRNA gene-targeted group-specific primers for real-time PCR analysis of predominant bacteria in human feces. *Appl. Environ. Microbiol.* **70**, 7220−7228.

McLaughlin, M. R., Brooks, J. P., and Adeli, A. (2009) Characterization of selected nutrients and bacteria from anaerobic swine manure lagoons on sow, nursery, and finisher farms in the Mid-South USA. *J. Environ. Qual.* **38**, 2422−2430.

Muyzer, G., de Waal, E. C., and Uitterlinden, A. (1993) Profiling of complex microbial populations by denaturing gradient gel electrophoresis analysis of polymerase chain reaction-amplified genes coding for 16S rRNA. *Appl. Environ. Microbiol.* **59**, 695−700.

National Academy of Sciences (2007) "The New Science of Metagenomics: Revealing the Secrets of Our Microbial Planet," The National Academies Press, Washington, DC.

Ng, J. (2012) Recovery of carbon and nitrogen cycling and microbial community functionality in a post-lignite mining rehabilitation chronosequence in east Texas. Ph.D. Dissertation, Texas A&M University, College Station, TX.

Nocker, A., Cheung, C.-Y., and Camper, A. K. (2006) Comparison of propidium monoazide with ethidium monoazide for differentiation of live vs. dead bacteria by selective removal of DNA from dead cells. *Appl. Environ. Microbiol.* **67**, 310−320.

Pearson, W. R. (1990) Rapid and sensitive sequence comparison with FASTP and FASTA. *Methods Enzymol.* **183**, 63−98.

Pérez-Losada, M., Cabezas, P., Castro-Nallar, E., and Crandall, K. A. (2013) Pathogen typing in the genomics era: MLST and the future of molecular epidemiology. *Infect. Gen. Evol.* **16**, 38−53.

Petric, I., Philippot, L., Abbate, C., Bispo, A., Chesnot, T., Hallin, S., *et al.* (2011) Inter-laboratory evaluation of the ISO standard 11063 "Soil quality—Method to directly extract DNA from soil samples." *J. Microbiol. Methods* **84**, 454−460.

Portier, P., Saux, M. F-L., Mougel, C., Lerondelle, C., Chapulliot, D., Thioulouse, J., *et al.* (2006) Identification of genomic species in *Agrobacterium* Biovar 1 by AFLP genomic markers. *Appl. Environ. Microbiol.* **72**, 7123−7131.

Ribot, E. M., Fair, M. A., Gautom, R., Cameron, D. N., Hunter, S. B., Swaminathan, B., *et al.* (2006) Standardization of pulsed-field gel electrophoresis protocols for the subtyping of *Escherichia coli* O157:H7, *Salmonella*, and *Shigella* for PulseNet. *Foodborne Pathog. Dis.* **3**, 59−67.

Saiki, R. K., Scharf, S., Faloona, F., Mullis, K. B., Horn, G. T., Erlich, H. A., *et al.* (1985) Enzymatic amplification of β-globin genomic sequences and restriction site analysis for diagnosis of sickle cell anemia. *Science* **230**, 1350−1354.

Schwartz, D. C., and Cantor, C. R. (1984) Separation of yeast chromosome-sized DNAs by pulsed-field gradient gel electrophoresis. *Cell* **37**, 67−75.

Seoane, J., Yankelevich, T., Dechesne, A., Merkey, B., Sternberg, C., and Smets, B. F. (2011) An individual-based approach to explain plasmid invasion in bacterial populations. *FEMS Microbiol. Ecol.* **75**, 17−27.

Sørensen, S., Burmølle, M., and Hansen, L. H. (2006) Making bio-sense of toxicity: new developments in whole-cell biosensors. *Curr. Opin. Biotechnol.* **17**, 11−16.

Thomas, E. J., King, R. K., Burchak, J., and Gannon, V. P. (1991) Sensitive and specific detection of *Listeria monocytogenes* in milk and ground beef with the polymerase chain reaction. *Appl. Environ. Microbiol.* **57**, 2576−2580.

Tsien, R. Y. (2010) Nobel lecture: constructing and exploiting the fluorescent protein paintbox. *Integr. Biol.* **2**, 77−93.

van Frankenhuyzen, J. K., Trevors, J. T., Lee, H., Flemming, C. A., and Habash, M. B. (2011) Molecular pathogen detection in biosolids with a focus on quantitative PCR using propidium monoazide for viable cell enumeration. *J. Microbiol. Methods* **87**, 263−272.

Van Nostrand, J. D., Wu, L., Wu, W., Huang, Z., Gentry, T. J., Deng, Y., *et al.* (2011) Dynamics of microbial community composition and function during in situ bioremediation of a uranium-contaminated aquifer. *Appl. Environ. Microbiol.* **77**, 3860−3869.

Van Nostrand, J. D., Gentry, T. J., and Zhou, J. (2013) Microarray-based amplification product detection and identification. In "Advanced Techniques in Diagnostic Microbiology" (Y.-W. Tang, and C. W. Stratton, eds.), 2nd ed., Springer, New York, NY, pp. 397−412.

Versalovic, J., Koeuth, T., and Lupski, R. (1991) Distribution of repetitive DNA sequences in eubacteria and application to fingerprinting of bacterial genomes. *Nucleic Acids Res.* **19**, 6823−6831.

Vos, P., Hogers, R., Bleeker, M., Reijans, M., van de Lee, T., Hornes, M., *et al.* (1995) AFLP: a new technique for DNA fingerprinting. *Nucleic Acids Res.* **23**, 4407−4414.

Wakelin, S. A., Anand, R. R., Reith, F., Gregg, A. L., Noble, R. R. P., Goldfarb, K. C., *et al.* (2012) Bacterial communities associated with a mineral weathering profile at a sulphidic mine tailings dump in arid Western Australia. *FEMS Microbiol. Ecol.* **79**, 298−311.

Wang, J., Van Nostrand, J. D., Wu, L., He, Z., Li, G., and Zhou, J. (2011) Microarray-based evaluation of whole-community genome DNA amplification methods. *Appl. Environ. Microbiol.* **77**, 4241−4245.

Watson, J. D., and Crick, F. H. C. (1953) Molecular structure of nucleic acids: a structure for deoxyribose nucleic acid. *Nature* **171**, 737−738.

Welsh, J., and McClelland, M. (1990) Fingerprinting genomes using PCR with arbitrary primers. *Nucleic Acids Res.* **19**, 861−866.

Woese, C. R. (1987) Bacterial evolution. *Microbiol. Rev.* **52**, 221−271.

Part IV

Microbial Communication, Activities, and Interactions with Environment and Nutrient Cycling

Microbial Source Tracking

Channah Rock, Berenise Rivera and Charles P. Gerba

14.1 WATER QUALITY AND FECAL CONTAMINATION

Water quality has been a concern for numerous stake-holders including wastewater utilities, local governments and members of the community, and has been monitored for many decades—in particular since the enactment of the Clean Water Act in 1972. However, more than 30 years after the Clean Water Act was implemented, a significant fraction of U.S. rivers, lakes and estuaries continue to be classified as failing to meet their designated use due to high levels of fecal bacteria (U.S. EPA, 2005). As a consequence, protection from fecal contamination is one of the most important and difficult challenges facing environmental scientists, regulators and communities trying to safeguard public water supplies and waters used for recreation (primary and secondary contact). Traditional water quality monitoring has helped improve water sanitation to protect public health but has also led to economic losses due to closures of recreational beaches, lakes and rivers. Additionally, solutions to contamination are not always readily apparent and easily identifiable. The ability to discriminate between sources of fecal contamination is necessary for a more defined evaluation of human health risks and to make water safe for human use.

The potential sources of fecal contamination causing these impairments can be classified into two groups: point sources that are easily identifiable (e.g., raw and treated sewage and combined sewer overflows) and nonpoint sources that are diffuse in the environment and may be difficult to identify (e.g., agriculture, forestry, wildlife and urban runoff) (Okabe *et al.*, 2007). Understanding the origin of fecal contamination is paramount in assessing associated health risks as well as identifying the actions necessary to remedy the problem (Scott *et al.*, 2002). As a result, numerous methods have been developed to identify fecal contamination as well as differentiate between these sources of pollution. Accurately identifying these sources can help to facilitate the elimination of water-borne microbial disease as a leading threat to public health (Simpson *et al.*, 2002).

14.2 MICROBIAL SOURCE TRACKING METHODS

Microbial source tracking (MST) methods are intended to discriminate between human and nonhuman sources of fecal contamination, and some methods are designed to differentiate between fecal contamination originating from individual animal species (Griffith *et al.*, 2003). MST is an active area of research with the potential to provide important information to effectively manage water resources (Stoeckel *et al.*, 2004).

MST methods are typically divided into two categories (Table 14.1). The first category is called library dependent, relying on isolate-by-isolate identification of bacteria cultured from various fecal sources and water samples and comparing them to a "library" of

I.L. Pepper, C.P. Gerba, T.J. Gentry: Environmental Microbiology, Third edition. DOI: http://dx.doi.org/10.1016/B978-0-12-394626-3.00014-4
© 2015 Elsevier Inc. All rights reserved.

TABLE 14.1 Common Types of Microbial Source Tracking Methods

Library Dependent				Library Independent	
Culture Dependent		Biochemical or Molecular	Culture Independent	
Biochemical	Molecular		Molecular	
– Antibiotic resistance – Carbon utilization	– Rep-PCR – PFGE – Ribotyping	– Bacteriophage – Bacterial culture	– Host-specific bacterial PCR – Host-specific viral PCR – Host-specific quantitative PCR	

Adapted from U.S. EPA (2011).

TABLE 14.2 Commonly Used Terms

Biochemical (phenotypic) methods refer to the ability to physically observe a characteristic of the isolated bacteria that might have been acquired from exposure to different host species or environment. Examples may be the resistance to certain antibiotics or utilization of carbon or nutrient source.

Culture-dependent methods rely on bacteria from water samples being grown or cultured in a lab.

Culture-independent methods isolate and identify DNA directly from a water sample without first having to grow or culture the bacteria from the sample.

Fecal source refers to a human or animal host where a microbe originates in the fecal waste of that host. Depending on the specificity of an MST method, a fecal source might refer to a general group of hosts (e.g., all humans, all animals or a group of animals such as ruminants), or a specific animal host (e.g., cattle, elk, dogs, etc.).

Library-dependent methods identify fecal sources from water samples based on databases of genotypic or phenotypic fingerprints for bacteria strains of known fecal sources.

Library-independent methods identify fecal sources based on known host-specific characteristics of the bacteria without the need of a library.

Microbial source tracking (MST) refers to a group of methods intended to discriminate between human and nonhuman sources of fecal contamination. Some methods are designed to differentiate between fecal contamination originating from individual animal species.

Microbial strain is a genetic variant or subtype of a microorganism (e.g., bacterial species).

Molecular (genotypic) methods utilize variations in the genetic makeup or the DNA of each individual organism or bacterium. This is often referred to as "DNA fingerprinting."

From U.S. EPA (2011).

bacterial strains from known fecal sources. Library-dependent methods require the development of biochemical (phenotypic) or molecular (genotypic) fingerprints for bacterial strains isolated from suspected fecal sources (U.S. EPA, 2005). These fingerprints are then compared to developed libraries for classification (see Table 14.2). The use of fecal bacteria to determine the host animal source of fecal contamination is based on the assumption that certain strains of fecal bacteria are associated with specific host animals, and that strains from different host animals can be differentiated based on phenotypic or genotypic markers (Layton *et al.*, 2006). Library-dependent methods tend to be more expensive and require more time as well as the use of experienced personnel to complete the analysis due to the time it takes to develop a library. Additionally, one of the major disadvantages of library-dependent methods is that libraries tend to be temporally and geographically specific. While this can be useful for a specific location, they are generally not as applicable on a broader watershed scale, or on statewide issues (Information Box 14.1).

The second category is called library independent, and is based on the detection of a specific host-associated genetic marker or gene target identified in the molecular material isolated from a water sample. These methods can help identify sources based on a known host-specific characteristic (genetic marker) of the bacteria without the

Information Box 14.1   Use of Microbial Source Tracking to Identify Pollution Sources in Oak Creek Canyon, Arizona

Federal and state regulations require that a TMDL be established for the impaired waters with oversight by the U.S. Environmental Protection Agency (Simpson *et al.*, 2002). As a result several state departments of environmental quality are looking towards alternative methods to determine the sources of pollution across their states' watersheds. According to the 2010 assessment, the state of Arizona has 21 impaired waters due to *E. coli* levels higher than the set standards (U.S. EPA, 2011). It is anticipated that the number of impaired watersheds will increase by the year 2015. In watersheds where sources are not known or understood, MST techniques can help to identify and also eliminate potential sources of fecal bacteria. In a study conducted by the University of Arizona (Rivera and Rock, 2010), MST methods were chosen within regions across Arizona due to the anticipated source(s) of bacteria not visibly obvious in these watersheds. More specifically, molecular methods were selected to differentiate between human and animal sources of *Bacteroides* present in water samples collected by volunteers across the state using host-specific 16S rDNA (Shanks *et al.*, 2010). Each of the five different watersheds included in this study has unique land-use characterization (urban vs. rural) and potential inputs of pollution within their area. One of the watersheds that has been extensively studied in the state of Arizona is the Oak Creek Watershed near Sedona, Arizona. Oak Creek is specifically known for its frequently visited recreation areas including Slide Rock State Park. Since 1973, *E. coli* bacteria in the water of Oak Creek have been a concern. Southam *et al.* (2000) used DNA fingerprinting to identify the

relative contributions of *E. coli* from source mammals. Human-related sources (from humans, pets, livestock, septic system effluent) accounted for about 33% of all *E. coli* found in Oak Creek, with perhaps a few more percentages attributable to wild animals that are present near the creek foraging on human food waste. The remainder of *E. coli* in Oak Creek was attributed to wildlife including: raccoons (31%), skunks (11%), elk (8%), white-tailed deer (6%), beaver (6%) and other mammals. While the contribution of human influence was significant, such a diverse number of wildlife contributors makes it a challenge to address dispersed nonpoint source pollution with comprehensive and complete measures to reduce *E. coli* loads to acceptable levels. Results of this study indicated both human and bovine inputs across multiple watersheds were causing the water quality impairments. More specifically, of the total 171 surface water samples that were analyzed using molecular methods, 37% were positive for human molecular markers for *Bacteroides*. Because of this research, best management practices or BMPs were implemented to reduce run-off from communities surrounding the creek, and failing septic systems leaching into rivers and lakes were repaired to help reduce contamination. This is one example of how the use of MST methods to identify the sources of fecal pollution can help to empower local regulatory agencies to work with stakeholders within the community to monitor and remediate locations contributing to contamination with the ultimate intent to delist impaired waters.

need of a "library." One of the most widely used library-independent approaches utilizes polymerase chain reaction (PCR) to amplify a gene target that is specifically found in a host population (Shanks *et al.*, 2010). PCR provides the ability to screen genetic material from bacteria (e.g., deoxyribonucleic acid [DNA] or ribonucleic acid [RNA]) isolated from a water sample for a specific sequence or target in a relatively short amount of time. These methods do not depend on the isolation of DNA directly from the original source, although some methods often require a pre-enrichment to increase the sensitivity of the approach (U.S. EPA, 2005).

Recently, there has been an effort to better understand the various types of MST methods available as well as which methods are most useful for the goals of source identification and watershed characterization. According to the U.S. EPA, while there has been significant progress in the past 10 years towards method development, variability among performance measurements and validation approaches in laboratory and field studies has led to a body of literature that is very difficult to interpret (U.S. EPA, 2005). Comparison studies have shown that no single method is clearly superior to the others (U.S. EPA,

2005). Therefore, no single method has emerged as the method of choice for determining sources of fecal contamination in all fecal-impaired water bodies. However, using the appropriate method and appropriate indicator, sources of fecal contamination can be found and characterized as being from either animal or human origin (Simpson *et al.*, 2002). MST based on identification of specific molecular markers can provide a more complete picture of the land uses and environmental health risks associated with fecal pollution loading in a watershed than is currently possible with traditional indicators and methods (Jenkins *et al.*, 2009). MST methods have the ability to identify "who" is contributing to the pollution, whereas traditional culture-based methods only tell you "if" and "when" fecal contamination is present. Table 14.3 describes existing MST methods that are currently being used and the general purposes for each.

There are several detection methods available for library-dependent and library-independent MST. Included in library-dependent MST are methodologies for phenotypic and genotypic analysis. Antibiotic-resistant analysis and carbon source utilization are two commonly used methods for phenotypic analysis. Antibiotic-resistant

TABLE 14.3 Comparison of Molecular Microbial Source Tracking Methods Used for Watershed Experiments

Method	Description	Advantages	Disadvantages	References
Ribotyping	Southern blot of genomic DNA cut with restriction enzymes; probed with ribosomal sequences; discriminates species	Highly reproducible; classifies isolates from multiple sources	Complex; expensive; labor intensive; geographically specific; database required; variations in methodology	Samadpour and Chechowitz, 1995; Farber, 1996; Tynkkynen *et al.*, 1999; Parveen *et al.*, 1999; Farag *et al.*, 2001; Hager, 2001a; Carson *et al.*, 2001; Hartel *et al.*, 2002; Samadpour, 2002; Scott *et al.*, 2003
Pulse-field gel electrophoresis (PFGE)	DNA fingerprinting with rare-cutting restriction enzymes coupled with electrophoretic analysis; discriminates species	Extremely sensitive to minute genetic differences; highly reproducible	Long assay time; limited simultaneous processing; database required	Tynkkynen *et al.*, 1999; Simmons *et al.*, 2000; Hager, 2001b; King and Stansfield, 2002
Denaturing-gradient gel electrophoresis (DGGE)	Electrophoresis analysis of PCR products based on melting properties of the amplified DNA sequences; discriminates species	Works on isolates	Technically demanding; time consuming; limited simultaneous processing; not good on environmental isolates; database required	Farnleitner *et al.*, 2000; Buchan *et al.*, 2001
Repetitive DNA sequences (Rep-PCR)	PCR used to amplify palindromic DNA sequences coupled with electrophoretic analysis; discriminates species	Simple and rapid	Reproducibility a concern; cell culture required; large database required; variability increases as database increases	Dombek *et al.*, 2000; Holloway, 2001
Length heterogeneity PCR (LH-PCR)	Separates PCR products for host-specific genetic markers based on length	Does not require culturing or a database	Expensive equipment; technically demanding	Suzuki *et al.*, 1998; Bernhard and Field, 2000a,b
Terminal restriction fragment length polymorphism analysis (T-RFLP)	Uses restriction enzymes coupled with PCR in which only fragments containing a fluorescent tag are detected	Does not require culturing or a database	Expensive equipment; technically demanding	Bernhard and Field, 2000a,b
Host-specific 16S rDNA	Combine LH-PCR and T-RFLP methods on fecal anaerobes (*Bacteroides* and *Bifidobacterium*); discriminates human and cattle; other markers being developed	Does not require culturing or a database; indicator of recent pollution	Expensive equipment; technically demanding; little known about survival of *Bacteriodes* spp. in environment	Bernhard and Field, 2000a,b

Adapted from Meays et al. (2004).

analysis is a method based on the premise that host animals/humans exposed to antibiotics will release bacteria resistant to those antibiotics, and on the assumption that this selective load would be a mechanism for distinguishing among fecal bacteria from different hosts (U.S. EPA, 2005). This method is labor intensive and time consuming as it requires culturing a large number of antibiotic-resistant isolates, as well as determining which antibiotics are involved with the resistance (Field *et al.*, 2003). Field *et al.* (2003) reported that when sets of isolates from a single type of feces have been evaluated,

rates of correct classification have ranged from approximately 64 to 87%; however, when individual isolates from mixed fecal sources were analyzed, rates of correct classification were lower. As a result, using this method to determine fecal contamination by analyzing resistance to antibiotics has low accuracy. According to Griffith *et al.* (2003), carbon source utilization is similar to antibiotic-resistant analysis, but instead relies on growth patterns created when fecal bacterial isolates are exposed to a number of antibiotics or grown on different carbon sources. While this method can work in the laboratory for

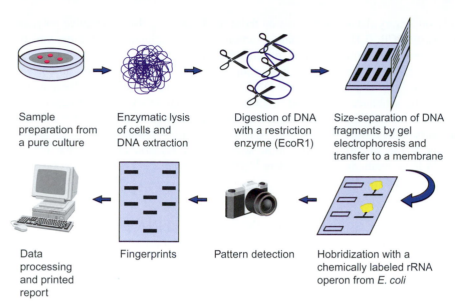

FIGURE 14.1 Ribotyping procedure. From Meays *et al.* (2004).

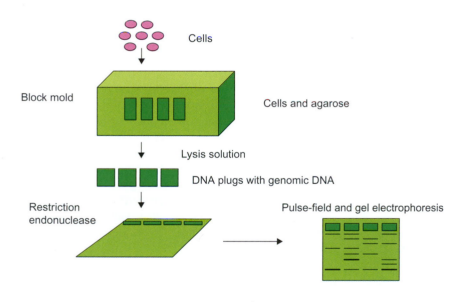

FIGURE 14.2 Pulse-field gel electrophoresis. From Meays *et al.* (2004).

analysis of pure cultures, there are numerous factors in an environmental water body that can influence bacterial nutrient needs that may make this method inconvenient for field determination (Simpson *et al.*, 2002).

Ribotyping, pulsed-field gel electrophoresis (PFGE) and rep-PCR are MST methods that have been used to quantify genotypic characteristics (Griffith *et al.*, 2003). Ribotyping is a technique that involves digestion of restriction enzymes from genomic DNA (Figure 14.1). Fragments are then separated by gel electrophoresis and transferred to a nylon membrane by Southern transfer and subsequent hybridization using a labeled probe of the *E. coli* rRNA genes or the entire operon (U.S. EPA, 2005). According to Field *et al.* (2003), ribotyping is excellent for accurately differentiating between human and animal fecal isolates, but is less successful at pinpointing the animal source and takes approximately 2 weeks to complete. The PFGE method (Figure 14.2) consists of pure culture bacterial cells placed in agarose plugs where the DNA is broken down by using a series of restriction enzymes (Griffith *et al.*, 2003). Both ribotyping and PFGE methods need extensive local collections of strains for comparison. Finally, by using rep-PCR, amplification of a genetic fragment can be visualized by using specific primers that target the region, followed by matching to a library of known sources (Stoeckel *et al.*, 2004). This method requires the establishment of a library containing identified strains. In addition, it is necessary to screen a large geographic region that includes known isolates (Scott *et al.*, 2002).

Other MST methods that do not rely on library databases include amplicon length heterogeneity PCR (LH-PCR), terminal-restriction fragment length polymorphism (T-RFLP) and host-specific PCR. These methods are library independent and distinguish between sources of fecal pollution by recognizing specific genetic sequences distinct to the host fecal bacteria (Griffith *et al.*, 2003). These methods function by looking at the bacterial community as a whole rather than individually. Bernhard and Field (2000a) define LH-PCR and T-RFLP as methods used to explore changes in the sizes of gene fragments due to additions and removal, and to calculate the approximate relative quantity of each gene fragment. T-RFLP analysis is a method of comparative community analysis, and is based on the restriction endonuclease digestion of fluorescently end-labeled PCR products such as the 16S rRNA gene. The digested products are separated by gel electrophoresis and detected on an automated sequence analyzer. The method provides distinct profiles (fingerprints) dependent on the species composition of the communities of the samples. While useful, these methods have only been used against a small number of animal fecal samples and need additional evaluation using other possible sources of contamination (Simpson *et al.*, 2002). In addition, these two methods require expensive equipment and are technically demanding.

There are also MST methods that are an alternative to the library-dependent and library-independent methods. These methods offer a direct measure of viruses found in humans, and aim to detect viruses present in human feces that are not present in other animals, including human enteroviruses, adenoviruses and F$^+$ coliphage (Griffith *et al.*, 2003). While these methods can be useful at determining if human fecal contamination is present, they do not determine other fecal contamination within an impaired water body. Additionally, the isolation and detection of viruses can be cumbersome due to the extensive training, equipment and time needed to accurately determine their presence.

A recent review of the literature has identified an increase in library-independent methods available for watershed characterization. In particular, host-specific bacterial and viral PCR as well as host-specific quantitative PCR methods have recently been developed. In theory, host-specific PCR (library-independent MST) uses genetic marker sequences that are not only specific to fecal bacteria, but are also specific to the host species that produced the feces, allowing discrimination among different potential sources (Field *et al.*, 2003). Host-specific PCR holds promise as an effective method for characterizing a microbial population without first culturing the organisms in question (Scott *et al.*, 2002). Furthermore, these methods are cost effective, rapid and potentially more specific than library-dependent methods. It is anticipated that these host-specific molecular methods will continue to develop with

emphasis on those methods using the quantitative polymerase chain reaction (qPCR) technique that measures the amount of microbial DNA present in the water sample rather than simply detecting a presence or absence of microbial DNA (Santo Domingo *et al.*, 2007). By quantifying the amount of microbial DNA, comparisons can be made regarding the relative impacts of a specific source to a specific location within the watershed. In particular, one of the most widely cited bacteria analyzed for library-independent MST is *Bacteroides*.

14.3 COMMON BACTERIA USED IN SOURCE TRACKING STUDIES: *Bacteroides*

The genus *Bacteroides* contains Gram-negative, non-spore-forming, nonmotile, anaerobic rod bacteria generally isolated from the gastrointestinal tract (GI tract) of humans and animals (Smith *et al.*, 2006). As members of the indigenous flora, they play a variety of roles that contribute to normal intestinal physiology and function. These include beneficial roles such as polysaccharide breakdown or nitrogen cycling (Smith *et al.*, 2006). According to Smith *et al.* (2006) *Bacteroides* generally cause opportunistic infections that can occur any time the integrity of the mucosal wall of the intestine is compromised. Another important aspect of *Bacteroides* biology is their inability to proliferate in the environment as well as their potential to survive in the environment at a rate directly proportional to pathogens of concern. *Bacteroides* survival depends primarily on temperature and presence of predators, and they have been found to survive for up to 6 days under oxygen-stressed conditions (Field and Dick, 2004).

The abundance of this bacterium in human and animal feces has allowed for host-related analysis targeting genes present in the *Bacteroides* genome. Layton *et al.* (2006) suggested that bacteria belonging to the genus *Bacteroides* could be an alternative fecal indicator to *E. coli* or fecal coliform bacteria, because they make up a significant portion of the fecal bacteria population, have little potential for growth in the environment and have a high degree of host specificity that likely reflects differences in host animal digestive systems.

Numerous methodologies have been designed to target specific diagnostic sequences within the *Bacteroides* 16S rRNA gene (which is vital for protein synthesis and therefore present in all bacteria) present in feces from different animals. Field and Dick (2004) developed 16S rRNA gene makers from *Bacteroides* to detect fecal pollution, and to distinguish between human and ruminant (e.g., bovine, goat, sheep, deer and others) sources by PCR. Developing MST methods specific to molecular markers within the target gene will allow differentiation between human and ruminant-associated *Bacteroides*, therefore identifying the possible source of contamination. This approach offers the advantage of circumventing the need for a culturing step, which allows for a more rapid identification of target organism (Scott *et al.*, 2002)

While progress has been made in identifying genetic markers that are useful for MST, few studies have evaluated how these molecular markers used as MST targets vary temporally and spatially following fecal contamination of surface waters (Bower *et al.*, 2005). There are several studies that have used MST methods; in particular, host-associated PCR-based assays targeting *Bacteroides* genetic markers to investigate the sources and levels of fecal pollution in recreational water and watersheds. In a study conducted by Gourmelon *et al.* (2007), three estuaries were compared by PCR using human-specific *Bacteroides* markers in combination with human- and animal-specific targets. In this study PCR was found to be a reliable indicator of fecal contamination. *Bacteroides* was observed in 95% of fecal samples in all sewage treatment plant samples and pig liquid manure. A separate study targeting *Bacteroides* (Shanks *et al.*, 2010) compared seven PCR and qPCR assays targeting *Bacteroides* genes reported to be associated with either ruminant or bovine feces. PCR-indicated prevalence ranged from 54 to 85% for all DNA extracts from 247 individual bovine fecal samples, and specificity (how well the PCR assay detected known bovine fecal samples) ranged from 76 to 100% for the assays studied. A previous study by Griffith *et al.* (2003) using blind samples demonstrated that *Bacteroides* source-specific MST methods identified fecal sources correctly when the sources comprised as little as 1% of the total fecal contamination in the samples. While a wealth of knowledge exists in the literature, there are still many ongoing MST studies targeting the 16S rRNA *Bacteroides* gene to improve detection and watershed characterization.

Although *Bacteroides* MST has been useful for pollution characterization, it is still an emerging science and research is currently being done to validate published methods and better understand the effectiveness of available technologies. Extensive field testing is ongoing to determine the efficacy of published assays and the geographic distributions of presumptively human-specific markers (McLain *et al.*, 2009). Several recent studies have described testing of feces from domestic animals, livestock, bird and mammal wildlife as well as fish and other aquatic species for cross-amplification with human assays and molecular markers previously thought to be human specific (McLain *et al.*, 2009). Therefore, it is critical that MST-based methods be evaluated on a watershed-by-watershed basis to ultimately understand the utility of the methods for accurate pollution characterization.

14.4 APPLICATION OF SOURCE TRACKING

A primary driver of microbial source tracking has been the U.S. Environmental Protection Agency's Total Maximum Daily Load or TMDL program. A TMDL is defined as the maximum amount of a pollutant the water body can receive, and still meet regulated limits for that pollutant. Under the 1972 Clean Water Act, all waters in the United States must be evaluated in the context of applicable water-quality standards, which include water quality criteria designed to protect the water's designated uses (e.g., swimming, fishing). Water bodies that do not meet water quality standards are classified as impaired. A TMDL is defined as the total pollutant (e.g., fecal bacteria, pesticides) load a water body can receive and still meet applicable water quality standards. Waters are designated as having pathogen impairment if fecal indicator bacteria concentrations exceed standards for the water use (e.g., swimming). One of the most common uses of MST is to identify sources of fecal bacterial indicator impairments (e.g., human, livestock, wildlife) for the purpose of prioritizing control. Where impairments are demonstrated as primarily from bacteria sources such as wildlife, which are not practical to control and are thought to pose less risk to human health, TMDL development and implementation may not be warranted.

Overall, while many approaches to source tracking have been developed, only recently has large-scale testing of the different approaches in multiple laboratories been conducted, and even this is limited to only a few contaminant sources. Clearly, there is a need to develop more standardized procedures for source tracking.

QUESTIONS AND PROBLEMS

1. What are the two major approaches used for microbial source tracking? What are the advantages and disadvantages of these methods?
2. What is Total Maximum Daily Load?
3. Can viruses be used in source tracking? How?

REFERENCES AND RECOMMENDED READING

Bernhard, A. E., and Field, K. G. (2000a) Identification of nonpoint sources of fecal pollution in coastal waters by using host-specific 16S ribosomal DNA genetic markers from fecal anaerobes. *Appl. Environ. Microbiol.* **66**, 1587−1594.

Bernhard, A. E., and Field, K. G. (2000b) A PCR assay to discriminate human and ruminant feces on the basis of host differences in *Bacteroides−Prevotella* genes encoding 16S rRNA. *Appl. Environ. Microbiol.* **66**, 4571−4574.

Bower, P. A., Scopel, C. O., Jensen, E. T., Depas, M. M., and McLellan, S. L. (2005) Detection of Genetic markers of fecal indicator bacteria in Lake Michigan and determination of their relationship to *Escherichia coli* densities using standard microbiological methods. *Appl. Environ. Microbiol.* **71**, 8305–8313.

Buchan, A., Alber, M., and Hodson, R. E. (2001) Strain-specific differentiation of environmental *Escherichia coli* isolates via denaturing gradient gel electrophoresis (DGGE) analysis of the 16S-23S intergenetic spacer region. *FEMS Microbiol. Ecol.* **35**, 313–321.

Carson, C. A., Shear, B. L., Ellersieck, M. R., and Asfaw, A. (2001) Identification of fecal *Escherichia coli* from humans and animals by ribotyping. *Appl. Environ. Microbiol.* **67**, 1503–1507.

Dombek, P. E., Johnson, L. K., Zimmerley, S. T., and Sadowsky, M. J. (2000) Use of repetitive DNA sequences and the PCR to differentiate *Escherichia coli* isolates from human and animal sources. *Appl. Environ. Microbiol.* **66**, 2572–2577.

Farag, A. M., Goldstein, J. N., Woodward, D. F., and Samadpour, M. (2001) Water quality in three creeks in the backcountry of Grand Teton National Park, USA. *J. Freshwater Ecol.* **16**, 135–143.

Farber, J. M. (1996) An Introduction to the hows and whys of molecular typing. *J. Food Protect.* **59**, 1091–1101.

Farnleitner, A. H., Kreuzinger, N., Kavka, G. G., Grillenberger, S., Rath, J., and Mach, R. L. (2000) Simultaneous detection and differentiation of *Escherichia coli* populations from environmental freshwaters by means of sequence variations in a fragment of the beta-D-glucuronidase gene. *Appl. Environ. Microbiol.* **66**, 1340–1346.

Field, K. G., and Dick, L. K. (2004) Rapid estimation of numbers of fecal bacteroidetes by use of a quantitative PCR assay for 16S rRNA genes. *Appl. Environ. Microbiol.* **70**, 5695–5697.

Field, K. G., Bernhard, A. E., and Brodeur, T. J. (2003) Molecular approaches to microbiological monitoring: fecal source detection. *Environ. Monitor. Assess.* **81**, 313–326.

Gourmelon, M., Caprais, M. P., Segura, R., Le Mennec, C., Lozach, S., Piriou, J. Y., *et al.* (2007) Evaluation of two library-independent microbial source tracking methods to identify sources of fecal contamination in French estuaries. *Appl. Environ. Microbiol.* **73**, 4857–4866.

Griffith, J. F., Weisberg, S. B., and McGee, C. D. (2003) Evaluation of microbial source tracking methods using mixed fecal sources in aqueous test samples. *J. Water Health* **1**, 141–151.

Hager, M. C. (2001a) Detecting bacteria in coastal waters. Part 1. *J. Surf. Water Qual. Prof.* **2**, 16–25.

Hager, M. C. (2001b) Detecting bacteria in coastal waters. Part 2. *J. Surf. Water Qual. Prof* **2**.

Hartel, P. G., Summer, J. D., Hill, J. L., Collins, J. V., Entry, J. A., and Segars, W. I. (2002) Geographic variability of *Escherichia coli* ribotypes from animals in Idaho and Georgia. *J. Environ. Qual.* **31**, 1273–1278.

Holloway, P. (2001) Tracing the source of E. coli fecal contamination of water using rep-PCR. In "Manitoba Livestock Manure Management Initiative Project: MLMMI 00-02-08," University of Winnipeg.

Jenkins, M. W., Sangam, T., Lorente, M., Gichaba, C. M., and Wuertz, S. (2009) Identifying human and livestock sources of fecal contamination in Kenya with host-specific Bacteroidales assay. *Water Res.* **43**, 4956–4966.

King, R. C., and Stansfield, W. D. (2002) "A Dictionary of Genetics," 6th ed. Oxford University Press, New York, pp. 530.

Layton, A., McKay, L., Williams, D., Garrett, V., Gentry, R., and Sayler, G. (2006) Development of *Bacteroides* 16S rRNA Gene TaqMan-based real-time PCR assays for estimation of total, human, and bovine fecal pollution in water. *Appl. Environ. Microbiol.* **72**, 4214–4224.

McLain, J. E., Ryu, H., Kabiri-Badr, L., Rock, C. M., and Abbaszadegan, M. (2009) Lack of specificity for PCR assays targeting human *Bacteroides* 16S rRNA gene: cross-amplification with fish feces. *FEMS Microbiol. Lett.* **299**, 38–43.

Meays, C. L., Broersma, K., Nordin, R., and Mazumder, A. (2004) Source tracking fecal bacteria in water: a critical review of current methods. *J. Environ. Mgmt.* **73**, 71–79.

Okabe, S., Okayama, N., Savichtcheva, O., and Ito, T. (2007) Quantification of host-specific *Bacteroides–Prevotella* 16SrRNA genetic markers for assessment of fecal pollution in freshwater. *Appl. Microbiol. Biotechnol.* **74**, 890–901.

Parveen, S., Portier, K. M., Robinson, K., Edmiston, L., and Tamplin, M. L. (1999) Discriminant analysis of ribotype profiles of *Escherichia coli* for differentiating human and nonhuman sources of fecal pollution. *Appl. Environ. Microbiol.* **65**, 3142–3147.

Rivera, B., and Rock, C. M. (2010) Microbial Source Tracking in Arizona Watersheds using Host-specific 16S rDNA. Master's thesis, Dept. Soil Water and Environmental Science, The University of Arizona.

Samadpour, M. (2002) Microbial source tracking: principles and practice, US EPA Workshop on Microbial Source Tracking, 5–9. http://www.sccwrp.org/tools/workshops/source_tracking_agenda.html.

Samadpour, M., and Chechowitz, N. (1995) "Little Soos Creek Microbial Source Tracking," Department of Environmental Health, University of Washington.

Santo Domingo, J. W., Bambic, D. G., Edge, T. A., and Wuertz, S. (2007) Quo vadis source tracking? Towards a strategic framework for environmental monitoring of fecal pollution. *Water Res.* **41**, 3539–3552.

Scott, T. M., Rose, J. B., Jenkins, T. M., Farrah, S. R., and Lukasik, J. (2002) Microbial source tracking: current methodology and future directions. *Appl. Environ. Microbiol.* **68**, 5796–5803.

Scott, T. M., Parveen, S., Portier, K. M., Rose, J. B., Tamplin, M. L., Farrah, S. R., *et al.* (2003) Geographical variation in ribotype profiles of *Escherichia coli* isolates from humans, swine, poultry, beef, and dairy cattle in Florida. *Appl. Environ. Microbiol.* **69**, 1089–1092.

Shanks, O. C., White, K., Kelty, C. A., Hayes, S., Sivaganesan, M., Jenkins, M., *et al.* (2010) Performance assessment PCR-based assays targeting Bacteroidales genetic markers of bovine fecal pollution. *Appl. Environ. Microbiol.* **76**, 1359–1366.

Simmons, G. M., Waye, D. F., Herbein, S., Myers, S., and Walker, E. (2000) Estimating nonpoint fecal coliform sources in Northern Virginia's four mile run watershed. In "Proceedings of the Virginia Water Research Symposium. VWRRC Special Report SR-19-2000" (T. Younos, and J. Poff, eds.), pp. 248–267.

Simpson, J. M., Santo Domingo, J. W., and Reasoner, D. J. (2002) Microbial source tracking: state of the science. *Environ. Sci. Technol.* **36**, 5279–5288.

Smith, C. J., Rocha, E. R., and Paster, B. J. (2006) The medically important *Bacteroides* spp. in health and disease. *Prokaryotes* **7**, 381–427.

Stoeckel, D. M., Mathes, M. V., Hyer, K. E., Hagedorn, C., Kator, H., Lukasik, J., *et al.* (2004) Comparison of seven protocols to identify fecal contamination sources using *Escherichia coli*. *Environ. Sci. Technol.* **38**, 6109–6117.

Suzuki, M., Rappe, M. S., and Giovannoni, S. J. (1998) Kinetic bias in estimates of coastal picoplankton community structure obtained by measurements of small-subunit rRNA gene PCR amplicon length heterogeneity. *Appl. Environ. Microbiol.* **64**, 4522−4529.

Tynkkynen, S., Satokari, R., Saarela, M., Mattila-Sandholm, T., and Saxelin, M. (1999) Comparison of ribotyping, randomly amplified polymorphic DNA analysis, and pulsed-field gel electrophoresis in typing of *Lactobacillus rhamnosus* and *L. casei* strains. *Appl. Environ. Microbiol.* **65**, 3908−3914.

U.S. Environmental Protection Agency (2005) Microbial source tracking guide. Document EPA/600/R-05/064. U.S. Environmental Protection Agency, Washington, DC.

U.S. Environmental Protection Agency, Region 10 (2011) Using Microbial Source Tracking to Support TMDL Development and Implementation [Online]. http://www.epa.gov/region10/pdf/tmdl/mst_for_tmdls_guide_04_22_11.pdf.

Chapter 15

Microbial Transport in the Subsurface

Charles P. Gerba, Ian L. Pepper and Deborah T. Newby

Microbial transport through the subsurface is of interest not only from the standpoint of introducing microorganisms for beneficial purposes, but also in the removal of waterborne-disease-causing microorganisms from wastewater. Beneficial applications include introduction of bacteria that may enhance biodegradation of organic contaminants; remediation of metal-contaminated sites; improvement of soil structure; increased crop production; and biological control of plant pathogens. However, to obtain such benefits, the introduced microbe must be able to be transported through the soil and vadose zone. In the case of pathogens, the goal is to limit their transport and use the soil and vadose zone to improve the microbial quality of groundwater. Almost 40 million people in the United States depend on subsurface disposal of sewage wastes via septic tanks, a practice which can potentially contaminate drinking water wells or surface waters. In addition, soils are used to enhance the removal of enteric pathogens from wastewater or contaminated river water (riverbank filtration). Almost half of the waterborne outbreaks reported every year in the United States are due to contaminated groundwater and viral contamination of groundwater is estimated to occur in at least 27% of drinking water wells (Bradbury *et al.*, 2013). Understanding transport of pathogens allows us to determine safe distances from septic tank drainfields and wells used for drinking water.

In this chapter we examine the factors that determine the transport of microorganisms and nucleic acids through soil and the subsurface. These include microbial adhesion to and detachment from solid surfaces, which are governed by the physical–chemical properties of both the porous medium surfaces and the surrounding solution; the surface properties of the microbe; and the impact of water saturation and flow on movement. We also examine microbial survival and activity during transport, approaches to facilitating transport and mathematical models that describe and predict microbial transport.

15.1 FACTORS AFFECTING MICROBIAL TRANSPORT

Transport of microorganisms is governed by a variety of factors including: adhesion processes, filtration effects,

I.L. Pepper, C.P. Gerba, T.J. Gentry: Environmental Microbiology, Third edition. DOI: http://dx.doi.org/10.1016/B978-0-12-394626-3.00015-6
© 2015 Elsevier Inc. All rights reserved.

physiological state of the cells, porous medium characteristics, water flow rates, predation and intrinsic mobility of the cells.

Understanding each of these factors separately is the first step toward a holistic assessment of microbial or nucleic acid transport. With such an understanding it becomes possible to assess exposure that results from the transport of pathogens, or the feasibility of delivering beneficial microbes to a target site.

15.1.1 Microbial Filtration

Transport of microbes and other contaminants occurs within the pore spaces of a soil or subsurface material. One mechanism by which microbial transport is limited is physical straining or filtration of cells by small pores. Filtration of bacterial cells has been shown to be statistically correlated with bacterial size (Gannon *et al.*, 1991). Filtration becomes an important mechanism when the limiting dimension of the microbe is greater than 5% of the mean diameter of the soil particles (Herzig *et al.*, 1970). Thus, for a sandy soil with particle diameters of 0.05 to 2.0 mm, filtration will have a relatively small impact on the retention of bacteria of diameter approximately 0.3 to 2 µm. However, in a soil containing a significant portion of silt or clay particles (particle diameters range from 0.2 to 50 µm), filtration will be a major mechanism of bacterial cell removal. In contrast, studies of factors affecting the movement of particles less than 50 nm in diameter, such as viral particles, have shown that filtration has little effect on movement (Gerba *et al.*, 1991). An example of typical pore sizes found in a sandy loam is shown in Figure 4.8.

Cell shape, defined as the ratio of cell width to cell length, has also been shown to influence bacterial transport through a porous medium. Weiss *et al.* (1995) examined the transport of 14 strains of bacteria suspended in artificial groundwater through columns packed with quartz sand. A comparison of the distributions of size and shape of cells in the effluent with those in the influent suspensions revealed that cells in the effluent were smaller and rounder.

One consequence of microbial filtration is micropore exclusion, which states that bacteria may be excluded from the microporous domain of structured (e.g., aggregated, macroporous) porous media (see Figure 4.4). Most bacteria range from 0.3 to 2 µm in diameter, and micropores or pore throats located in the microporous domain of structured media can be much smaller in size. As a result, bacterial cells are physically excluded from the micropores (Figure 15.1). Thus, the location and rate of microbial activities can vary over a relatively small scale within a porous medium. In other words, microbial activity within the micropores that exclude microbes can be expected to be nonexistent, whereas an immediately

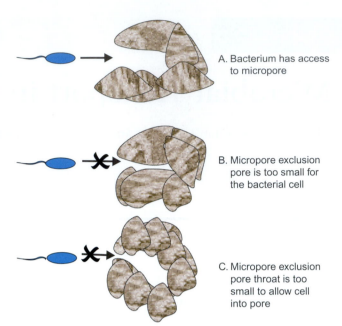

FIGURE 15.1   Exclusion of a bacterial cell from microporous domains in structured porous media.

adjacent site that is colonized may have an extremely high rate of activity.

Micropore exclusion of bacteria can have negative impacts. For example, when contaminants diffuse into micropores that exclude bacteria, they become unavailable for biodegradation. Because diffusion occurs slowly, this is a problem that normally worsens as the contact time between the contaminant and the porous medium increases. This process, known as contaminant aging, results in slower rates of contaminant degradation.

15.1.2 Physiological State

A variety of factors influence the size of a microbe and thus its transport potential. The physiological state of microbial cells is one such factor. When nutrients are not limiting, most cells produce exopolymers that form a capsule on the outer surface of the cell. Exopolymers increase the effective diameter of a cell; help the cell adhere to surfaces; and when released from the cell may modify solid surfaces to promote attachment. All of these, in addition to cell proliferation, may lead to pore clogging. Pore clogging can severely limit transport of bacteria and can lead to poor septic tank performance related to clogging of the drain field, clogging of nutrient injection wellheads used for *in situ* bioremediation and reduced rates of groundwater infiltration in recharge basins.

Under starvation conditions, bacteria tend to decrease in size (0.3 µm or even smaller), round up and shed their exopolymer capsule (Young, 2006). These so called ultramicrobacteria may have increased transport potential,

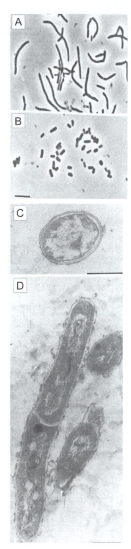

FIGURE 15.2   (A) Phase-contrast micrograph of an isolate grown on *p*-nitrophenol as the sole carbon source. (B) Phase-contrast micrograph of the *p*-nitrophenol degrader after 10 weeks of starvation in phosphate-buffered saline. (C) Electron micrograph of the starved *p*-nitrophenol-degrading isolate. (D) Electron micrograph of the starved *p*-nitrophenol-degrading isolate after resuscitation on *p*-nitrophenol. Modified with permission from Herman and Costerton (1993).

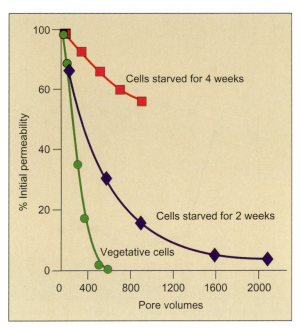

FIGURE 15.3   Permeability reduction profiles of fused glass bead cores injected with *K. pneumoniae* cells either in a vegetative state or starved for a period of time in phosphate-buffered saline. Modified with permission from MacLeod *et al.* (1988).

because both cell size and surface properties are changed. Herman and Costerton (1993) subjected a *p*-nitrophenol degrader isolated from a waste lagoon to starvation by placing it in phosphate-buffered saline for 10 weeks. The difference in cell size and morphology before and after starvation is clearly shown in Figure 15.2A and B. The starved cells were then resuscitated by adding *p*-nitrophenol to the medium (Figure 15.2C and D).

To demonstrate the difference in transport of ultramicrobacteria and normal cells, MacLeod *et al.* (1988) examined the movement of starved and normal *Klebsiella pneumoniae* cells through glass bead columns. The starved cells not only were smaller, but also demonstrated a reduction in capsule production as compared with the vegetative

cells. As expected, the starved cells penetrated further into the column than did the vegetative cells. Figure 15.3 depicts the observed reduction in column permeability as *K. pneumoniae* cells (10^8ml^{-1}) in different metabolic states were injected into the column. The ability of the cells to cause a reduction in permeability within the column was shown to be dependent on the length of starvation prior to inoculation and on the volume of cells at a given concentration injected through the core. Cell distribution within the columns also differed depending on nutrient status. Starved cells were evenly distributed throughout the column, whereas the vegetative cells were found in much higher numbers near the inlet end of the column (Figure 15.4). Upon nutrient stimulation, the starved cells were found to enter a state of growth accompanied by increased exopolymer production. This work demonstrates that inoculation with starved cells followed by nutrient stimulation has potential for increasing bioaugmentation efforts. For example, a starved cell that migrates farther through the terrestrial profile has an increased likelihood of reaching a target contaminated site. Once at the site, the contaminant can serve as a nutrient source, inducing the microbe to enter a metabolically active state.

15.1.3 Microbial Adhesion—The Influence of Cell Surface Properties

The adhesion of microbes to soil particles and vadose zone materials requires an initial interaction between the

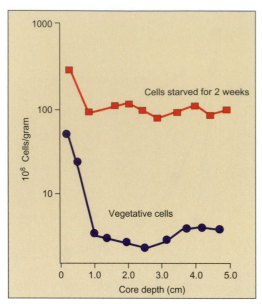

FIGURE 15.4 Differences in the DNA-derived cell distribution in cores injected with either a vegetative cell culture of *K. pneumoniae* or a cell suspension that was starved in phosphate-buffered saline for 2 weeks. Modified with permission from MacLeod *et al.* (1988).

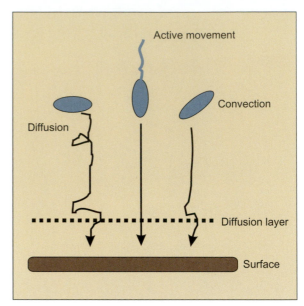

FIGURE 15.5 Different ways in which a cell can approach a solid surface. Modified with permission from van Loosdrecht *et al.* (1990).

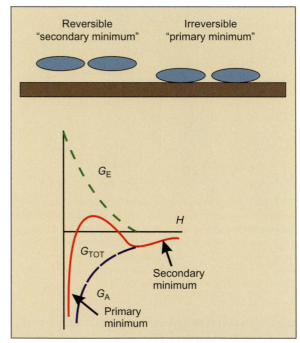

FIGURE 15.6 Gibbs energy of interaction between a sphere (in this case a bacterial cell) and a flat surface having the same charge, according to DLVO theory. G_E, electrostatic interaction; G_A, van der Waals interaction; G_{TOT}, total interaction; H, shortest separation distance between the two surfaces. Modified with permission from van Loosdrecht *et al.* (1990).

cell and a particle surface (van Loosdrecht *et al.*, 1990). Once a cell is in the vicinity of a surface, this initial interaction can occur in one of three ways: diffusion, active movement or advective transport. Diffusion is a result of Brownian motion and allows random interactions of cells with surfaces. The effective rate of diffusion is small; on the order of 40 μm/h. Motile cells may also come in contact with the surface through active movement in response to a chemotactic chemical gradient or in some cases simply by chance. Active movement also occurs on a micrometer scale. Finally, advective transport is due primarily to water movement and can move many orders of magnitude faster than diffusive or active transport (Figure 15.5).

Once contact between the cell and a particle surface has been made, adhesion can take place. Adhesion is a physicochemical process, and depending on the mechanisms involved, can be reversible or irreversible. Reversible adhesion, often thought of as initial adhesion, is controlled primarily by the balance of the following interactions: electrostatic interactions, hydrophobic interactions and van der Waals forces. These are explained in detail in the following sections.

In general, electrostatic interactions are repulsive because both cell and particle surfaces are negatively charged. In contrast, hydrophobic interactions and van der Waals forces tend to be attractive. Initial adhesion occurs when attractive forces overcome repulsive forces. Porous medium properties in conjunction with cell surface properties determine the relative importance of each of these interactions. Figure 15.6 illustrates how the interaction of electrostatic and van der Waals forces governs reversible

adhesion at various distances between the cell and particle surfaces. As can be seen from this figure, when a cell surface is in actual contact with, or very close to, a particle surface, the attractive forces are very strong, creating a primary minimum. The forces governing the primary

FIGURE 15.7 Irreversible attachment is mediated by physical attachment of cells to a surface, which can occur via production of exopolymers or special cell surface structures such as fibrils. Modified with permission from van Loosdrecht *et al.* (1990).

minimum are short-range forces such as hydrogen bonding and ion pair formation. As the two surfaces are separated slightly, e.g., by several nanometers, repulsive forces grow quickly and prohibit adhesion. At slightly longer distances, another shallower minimum exists called the secondary minimum. It is the secondary minimum that is responsible for the initial reversible adhesion of microbes. As shown in Figure 15.6, the cell and particle surfaces are not in actual contact at the secondary minimum. As a result, cells can be removed from the surface easily, for example, by increasing the water flow velocity or by changing the chemistry of the porous medium solution, e.g., ionic strength.

After initial adhesion, cells can become irreversibly attached to a particle surface. Irreversible attachment is a time-dependent process that occurs as a result of the interaction of cell surface structures such as fimbriae with the solid surface or as a result of the production of exopolymers that cement the microbial cell to the surface (Figure 15.7). The role of reversible and irreversible attachment in biofilm formation is discussed further in Chapter 6.

15.1.3.1 Electrostatic Interactions

Electrostatic interactions occur between charged particles. In terms of microbial transport, repulsion is the dominant electrostatic interaction because both the porous medium including organic matter and mineral surfaces are generally negatively charged. As already mentioned, microbial cell surfaces are also generally negatively charged. The negative charge comes primarily from lipoteichoic acids on the surface of Gram-positive bacteria and lipopolysaccharides on the surface of Gram-negative bacteria. Viral protein coats are also predominantly negatively charged. The overall charge on a microbe can be measured by electrostatic interaction chromatography or by electrophoretic mobility. Surface charge varies between types and species of microbes and can be affected by the pH of the matrix solution.

15.1.3.2 van der Waals Forces

Interactions between neutral molecules generally result from van der Waals forces. Van der Waals forces occur because while neutral molecules have no net charge or

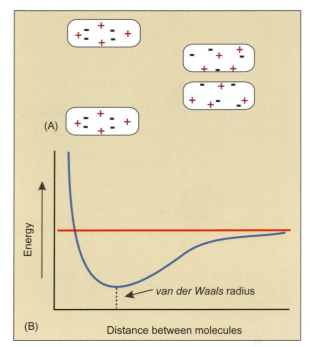

FIGURE 15.8   (A) For a neutral molecule the charge distribution in a molecule can vary to produce a net electrostatic attraction, allowing the molecules to approach very closely. This is a very weak attraction called the van der Waals force. Van der Waals forces can become strong if they are numerous enough. (B) As two molecules approach each other, the van der Waals attractive force increases to a maximum, then decreases and becomes repulsive.

permanent dipole moment, they do have a dynamic distribution of charge. As two molecules approach, this charge distribution can become favorable for interaction between the two molecules (Figure 15.8A). What actually occurs as two molecules approach is that the van der Waals attractive forces increase to a maximum, then decrease and become repulsive (Figure 15.8B). The van der Waals radius is defined as one half the distance between two equivalent atoms at the point of the energy minimum (where attractive forces are at a maximum). Van der Waals radii range from one to several angstroms in length, so these forces are effective only over short distances. Although individual van der Waals interactions are weak, the total attraction between two particles is equal to the sum of all attractive forces between every atom of one particle and every atom of the other particle. Thus, total van der Waals interactions can be quite strong.

15.1.3.3 Hydrophobic Interactions

Hydrophobic interactions refer to the tendency of nonpolar groups to associate in an aqueous environment. Hydrophobicity can be measured in a variety of ways, including contact angle determination, which is done by examining the shape of a drop of water that is placed on a layer of bacterial cells (Figure 15.9). Other methods

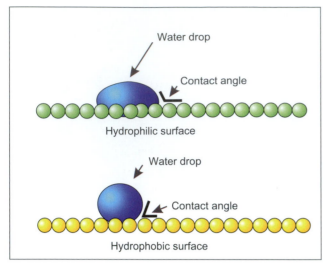

FIGURE 15.9 Water, which is a polar material, spreads out on a hydrophilic or polar surface but forms a round bead on a hydrophobic or nonpolar surface. The angle that describes the interaction of a water droplet with a surface is called the contact angle.

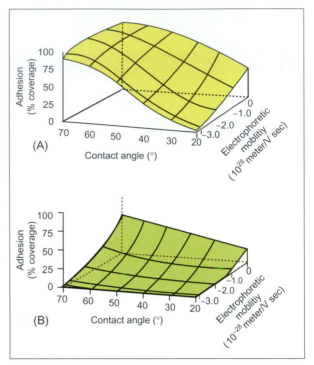

FIGURE 15.10 Relationship between bacterial adhesion to (A) sulfated polystyrene (hydrophobic) and (B) glass (hydrophilic) and bacterial surface characteristics as determined by contact angle measurement and electrophoretic mobility. Modified with permission from van Loosdrecht et al. (1990).

commonly used to assess hydrophobicity include phase partitioning (BATH test) and hydrophobic interaction chromatography. The BATH (bacterial adherence to hydrocarbon) test is a relatively simple test that measures the partitioning of microbial cells between a water phase and an organic phase.

As a result of hydrophobic forces, cells tend to partition from the aqueous phase and accumulate at the solid—water interface, resulting in decreased transport potential. Van Loosdrecht et al. (1990) examined adhesion of a variety of bacteria with different cell surface properties to two surfaces, one hydrophobic (polystyrene) and one hydrophilic (glass). Both cell surface charge and cell surface hydrophobicity were considered in this series of experiments. As shown in Figure 15.10A, cell surface hydrophobicity was the dominant force in determining adhesion to the hydrophobic polystyrene surface. In summary, two trends in cell adhesion can be inferred from this study: (1) adhesion typically decreases with decreasing hydrophobicity of either the solid surface or the cell surface; and (2) adhesion generally increases with decreasing cell surface charge. Knowledge of the combined effects of hydrophobicity and cell surface charge can be used to predict initial microbial adhesion of a particular microbe.

Sanin et al. (2003) examined the effect of starvation on the adhesive properties of three cyanuric acid-degrading bacteria. Microorganisms were independently starved for carbon and nitrogen. Surface hydrophobicities of all three strains remained fairly constant during carbon starvation, but decreased significantly when starved for nitrogen with a concomitant decline in attachment. Understanding starvation responses could have significant impacts on bacterial transport.

Similarly, hydrophobic effects and electrostatic repulsion govern the sorption of viruses depending on viral and subsurface characteristics. Zhuang and Jin (2003) examined the influence of natural organic matter on the retention and transport of two bacteriophages (MS-2 and ΦX174) through sand. MS-2 is an icosahedral phage of diameter 24 to 26 nm and a pI of 3.9, while ΦX174 is 25 to 27 nm in diameter, has a pI of 6.6 and is less hydrophobic than MS-2. In the sand alone the retention was high for MS-2 (99.2%) and much lower for ΦX174 (30%). When the sand was coated with organic matter, the retention of MS-2 decreased to 29% while retention of ΦX174 remained essentially the same (23%). In this case, coating the sand surface with organic matter blocked MS-2 interaction with charged sorption sites. Torkzaban et al. (2006) studied the transport of these two viruses under differing water saturation conditions. Their results show that both viruses are retained more strongly under unsaturated conditions and that retention is further increased at lower pH and higher ionic strength.

In natural systems, dissolved organic matter, most often present in the form of polymers, can also influence microbial adhesion by adsorbing to the microbe and/or solid surfaces (Dexter, 1979). The polymeric coating may affect adhesion by changing the electrostatic, the van der Waals and/or the hydrophobic interactions between the microbe and the solid surface. When polymers adsorb and coat both bacteria and the solid surface completely, adhesion is reduced because of an extra repulsive interaction due to steric hindrance (Fletcher, 1976). However, if only one of the surfaces is covered with polymers, or if both surfaces are partly covered with polymers, one polymer molecule may attach to both surfaces, thus forming a "bridge" between the two surfaces. This reduces the Gibbs free energy of adhesion (Figure 15.6) and results in a strong bond (Dexter, 1979).

15.1.4 Impact of pH on Microbial Transport

The pH of the matrix solution within a porous medium does not seem to have a large effect on bacterial transport. However, viral transport can vary greatly depending on the pH of the porous medium solution. The difference in impact of pH on the transport of bacteria and viruses can be attributed to a variety of factors. Remember that the primary interaction limiting bacterial transport is filtration, not adsorption as it is for viruses. In addition, in contrast to viruses, bacteria have very chemically diverse surfaces, and thus a change in pH would not be expected to alter the net surface charge to the same extent as for the more homogeneous viral surface. Finally, the overall charges on the surfaces of bacteria and viruses differ and will be affected differently by pH changes. This can be expressed in terms of the isoelectric point (pI). The pI is the pH at which the net charge on a particle of interest is zero. For bacteria, the pI usually ranges from 2.5 to 3.5, so the majority of cells are negatively charged at neutral pH. At pH values more acidic than the isoelectric point, a microbe becomes positively charged. This will reduce its transport potential because of increased sorption. This will not happen often with bacterial cells in environmental matrices, given their low pI values; the pH would have to decrease to 2.5 or lower to alter cell surface charge significantly. However, viruses display a wider range of isoelectric points (see Table 15.1) making their net surface charge much more dependent on changes in pH.

A study evaluated the influence of viral isoelectric point and size on virus adsorption and transport, and it was concluded that the isoelectric point was an important factor controlling viral adsorption and transport. However, when virus particles were greater than 60 nm in diameter, viral size became the overriding factor (Dowd *et al.*, 1998).

15.1.5 Impact of Ionic Strength on Transport

It has already been established that frequently the net charge on both particle and cell surfaces is negative, which causes electrostatic repulsion between these two surfaces. The concentration of anions and cations in

TABLE 15.1 Isoelectric Points of Selected Viruses

Virus	pI
Reovirus 3 (Dearing)	3.9
Rhinovirus 2	6.4
Polio 1 (Bruenders)	7.4, 3.8
Polio 1 (Mahoney)	8.2
Polio 1 (Chat)	7.5, 4.5
Polio 1 (Brunhilde)	4.5, 7.0
Polio 1 (LSc)	6.6
Polio 2 (Sabin T2)	6.5, 4.5
Echo 1 (V239)	5.3
Echo 1 (V248)	5.0
Echo 1 (V212)	6.4
Echo 1 (R115)	6.2
Echo 1 (4CH-1)	5.5
Echo 1 (Farouk)	5.1
Coxsackie A21	6.1, 4.8
Hepatitis A	2.8
Parvovirus AA4. X14	2.6
Noro	6.0, 5.0
Smallpox (Harvey)	5.9, 3.3
Influenza A (PRS)	5.3
Encephalomyocarditis (mengo M)	8.4, 4.4
T2 bacteriophage	4.2
T4 bacteriophage	4−5
MS-2 bacteriophage	3.9
PRD-1	3−4
Qβ	5.3
ΦX174	6.6
PM2	7.3

Data compiled from Gerba (1984); Ackermann and Michael (1987); Michen and Graule (2010).

solution is referred to as the ionic strength of the medium. Soil solution ionic strength influences transport primarily through two mechanisms—by altering the size of the diffuse double layer and by influencing soil structure (see Section 4.2).

The negative charges present on mineral particles electrostatically attract cations in the solution. Thus, in the immediate vicinity of the negatively charged surface, there is an excess of cations and a deficit of anions. Further from the particle surface, the cation concentration

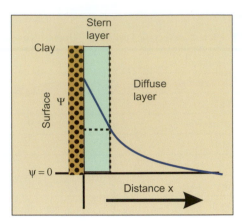

FIGURE 15.11 Illustration of the diffuse double layer. The diffuse double layer is a combination of the charge layer on the surface and the charge in solution. The first monolayer of ions in contact with the surface, the Stern layer, is held tightly to the surface. The second layer, the diffuse layer, responds to the remaining charge on the surface but is held more loosely than the Stern layer ions. This figure shows the energy (ψ) required to bring an ion from the bulk solution to the surface as a function of distance from the surface. Immediately next to the surface the decrease in potential is linearly related to increasing distance from the surface (the Stern layer). As the distance from the surface is increased further, the potential decreases exponentially. Modified with permission from Tan (1993), p. 198, courtesy Marcel Dekker, Inc.

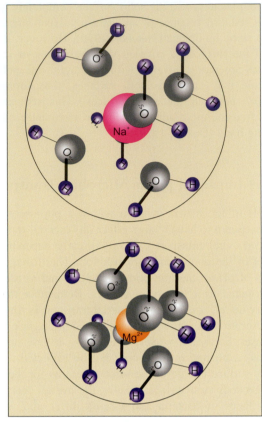

FIGURE 15.12 The radius of hydration of a cation in soil depends on the charge density of the atom. In the example shown, magnesium has a higher charge density than sodium and thus attracts water molecules more strongly resulting in a larger radius of hydration.

decreases until it reaches that of the bulk solution. Thus, the porous medium solution is often thought of as a double layer as depicted in Figure 15.11. The impact of ion concentration on this diffuse double layer will ultimately play a critical role in the transport of microbes. As the ionic strength of the bulk soil solution increases, the difference in cation concentration between the cation-rich layer and the bulk layer is reduced, and thus there is a tendency for cations to diffuse away from particle or cell surfaces. This causes a general compression of the diffuse double layer because the interacting cations and anions neutralize one another. The result is a decreased electric potential, which increases the likelihood of attachment of cells to the surfaces.

In addition to the overall ionic strength, the type of ion contributing to ionic strength is important. This is because the hydration radius of a cation in the soil solution affects the extent of the diffuse double layer and thus the soil structure. The radius of hydration of a particular cation is a function of surface charge density and refers to the radius of the cation and its complexed water molecules (Figure 15.12). In general, monovalent cations have lower surface charge densities and thus larger radii of hydration than divalent cations. Thus, in the presence of high concentrations of monovalent cations such as Na^+, clays tend to be dispersed. Dispersed clays create puddled soils, which are sticky when wet and hard when dry. As a dispersed soil dries, compaction may occur, which reduces pore spaces, inhibiting soil aeration and reducing the capacity for water flow. This adversely affects the

transport potential of microbes. On the other hand, the presence of divalent cations such as Ca^{2+} and Mg^{2+}, with smaller radii of hydration, leads to flocculated soils, which have increased pore space and thus favor transport.

The impact of ionic strength on microbial transport is demonstrated by the following examples. Bai *et al.* (1997) found that fewer cells were recovered in a column study when 2 mM NaCl was used as the percolating solution as compared with the use of artificial groundwater with a lower ionic strength. This observation can be explained in terms of cation concentration and associated electrostatic interactions. Viruses either do not readily adsorb or are released from soil particle surfaces suspended in low-ionic-strength solutions (Landry *et al.*, 1979; Goyal and Gerba, 1979).

How can ionic strength vary in a porous medium? One example is following a rainfall event. When it rains, the added water will generally lower the ionic strength of the soil solution. In addition, a rainfall event results in increased rates of water flow. Both decreased ionic strength and increased flow rate will promote microbial transport. This can result in peaks of microbial contamination in groundwater from the release of bacteria and viruses previously adsorbed.

15.1.6 Cellular Appendages

Bacteria may have a variety of appendages such as pili, flagella or fimbriae. Flagella are responsible for bacterial motility, while fimbriae and pili are involved in attachment. These appendages may all play a role in microbial transport through the terrestrial profile. The influence of bacterial motility on overall transport is generally minimal because extensive continuous water films would be needed to support microbial movement and because motility typically occurs on a micrometer scale. However, in nonflowing systems where no advective transport occurs, motility can increase transport potential over a very small scale. For example, Reynolds *et al.* (1989) evaluated bacteria movement through nutrient-saturated sand-packed cores under static conditions. In this study, transport through the column was four times faster with motile strains of *Escherichia coli* than with nonmotile mutants defective only in flagellar synthesis. Thus, the presence of cellular appendages involved in motility (flagella) can lead to measurable increases in microbial transport under certain circumstances.

Movement caused by flagella is usually a result of chemotaxis. Chemotaxis is the movement of microbes toward beneficial substances or away from inhibitory substances. This type of movement is dependent on the presence of a chemical gradient within continuous films of soil solution. The ability on move in this manner may confer survival advantages on the microbe. For example, chemotaxis is thought to play a role in the movement toward and subsequent infection of legume roots by *Rhizobium*, a nitrogen-fixing bacterium.

In contrast to flagella, the presence of cellular appendages involved in attachment (pili and fimbriae) can reduce microbial transport potential. It is thought that cellular appendages can penetrate the electrostatic barrier thereby facilitating attachment at greater distances from the surface. Functional groups (hydrophobic groups or positive charge sites) on the appendages may facilitate interaction with surfaces leading to increased adsorption. Thus, the presence of appendages may actually decrease microbial transport in some cases.

15.1.7 Hydrogeological Factors

Soil texture and structure, porosity, water content and potential, and water movement through the profile are key hydrogeological factors influencing microbial transport (see Chapter 4). The specific soil and vadose zone layers within a site serve as protective or attenuating zones with regard to contamination of groundwater by microbes (or chemical pollutants) via a variety of mechanisms, including filtration and adhesion (e.g., hydrophobic interactions with the air—water interface). In addition to the site-

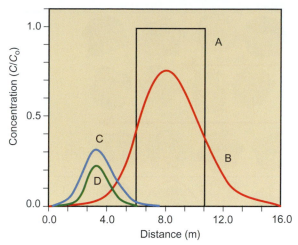

FIGURE 15.13 Effects of various processes on contaminant transport. This figure shows the theoretical distribution of a short pulse of microbes added to a saturated soil column 16 m long. The ordinate represents the relative concentration where C = the microbial concentration in the solution phase at a given point in the column, and C_0 = the influent concentration of microbes. The abscissa represents distance along the column from 0 to 16 m. Pulse A represents microbes that have moved through the column influenced only by advection. Pulse B represents the combined influence of advection and dispersion on microbial distribution. Note that no microbes are lost from the solution phase in either pulse A or B. Pulse C represents addition of adsorption to advective and dispersive processes. In this case microbes are lost to the solid phase and the resulting pulse is smaller and retarded. Finally, pulse D represents the addition of decay to the other three processes, which further removes microbes from the solution phase. Modified with permission from Yates and Yates (1991).

specific makeup of the porous medium, the distance between the soil surface and the vadose—groundwater interface is often a critical factor for determining pollution potential: the greater the distance, the less likely it is that groundwater contamination will occur.

Terms used to describe the flow of water and the transport of dissolved and particulate substances are commonly applied to describe the transport of microbes (Figure 15.13). Advection, the movement of the bulk pore fluid and its dissolved and suspended constituents, is primarily responsible for microbial transport. Dispersion is the combined result of mechanical mixing and molecular diffusion. Mechanical mixing results from the path tortuosity and velocity differences within the pore that depend on pore size and location of the microbe as depicted in Figure 15.14. Spreading due to molecular diffusion, the random movement of very small particles suspended in a fluid, results from the presence of a concentration gradient. It is generally considered negligible with regard to bacterial transport, but can be significant in the transport of smaller particles ($<1\,\mu m$) such as viruses. Finally, adsorption represents the removal of microbes from the bulk solution by reversible and irreversible adhesion.

Because microbes are transported along with the soil solution primarily through advection, the flow rate and degree of saturation of the soil can play significant roles

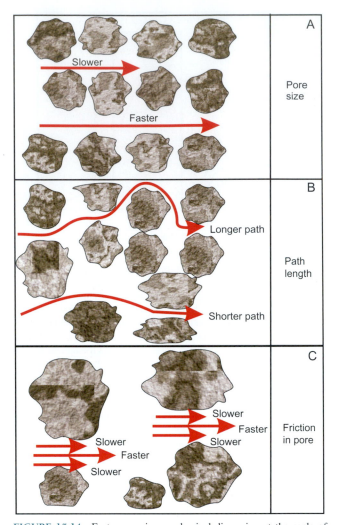

FIGURE 15.14 Factors causing mechanical dispersion at the scale of individual pores. (A) Microbes are transported through small pores more slowly than through large pores; (B) depending on pore sizes and shapes, path lengths can vary considerably; (C) flow rates are slower near the edges of the pore than in the middle. Modified with permission from Fetter (1993), © MacMillan Magazines Limited.

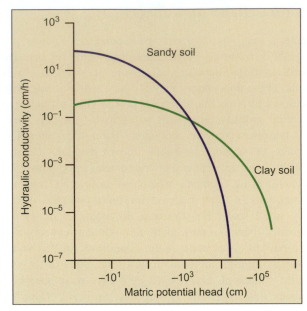

FIGURE 15.15 The hydraulic conductivity of a soil is dependent on the texture and the moisture content of the soil. This figure compares the hydraulic conductivity of a sand and a clay soil as a function of moisture content. Modified with permission from Soil Physics, Jury *et al.* (1991), © John Wiley & Sons, Inc.

in determining transport potential. In general, higher water content and greater flow velocities result in increased transport. For example, virus penetration through columns packed with loamy sand soil under unsaturated flow was 40 cm, compared with a penetration depth of 160 cm during saturated flow (Lance and Gerba, 1984). The reduced penetration is because under unsaturated conditions water is present as a discontinuous film on soil surfaces and, in addition, under unsaturated conditions there is increased interaction of the viruses with soil surfaces thereby increasing the potential for adsorption to soil and the air−water interface.

The flow rate of water through a saturated soil can be calculated using Darcy's law:

$$Q = K \frac{\Delta H \, At}{z} \qquad \text{(Eq. 15.1)}$$

where:

Q is the volume of water moving through the column (m^3)

A is the cross-sectional area of the column (m^2)

t is time (days)

ΔH is the hydraulic head difference between inlet and outlet (m)

K is the hydraulic conductivity constant (m/day), and

z is the length of the column (m).

Hydraulic conductivity can be defined as the ease with which water moves through soil. A hydraulic conductivity greater than 4 cm/hour is considered large, whereas a value less than 0.4 is low. For saturated soils, a coarse-textured material such as sand always has a higher conductivity than a clay soil because it contains larger pores, which hold water less tightly and allow for easier flow.

Darcy's law may also be applied to unsaturated soils; however, in this case, the hydraulic conductivity in Eq. 15.1 is no longer constant. This is because the unsaturated hydraulic conductivity of a soil, K(h), is a nonlinear function of the matric potential, which in turn is related to the water content. Figure 15.15 shows typical K(h) values for a coarse-textured soil (sand) and a fine-textured soil (clay). At saturation (low matric potential), the pores are filled with water. Thus, the coarse-textured soil has a higher conductivity because it contains greater numbers of large pores, where the water is held less tightly. When water is no longer added to the system, these large pores drain first and fairly rapidly, resulting in a pronounced decrease in hydraulic conductivity. As water continues to drain, a point will be reached when the sand and clay soils have similar hydraulic conductivity ($K(h) = -5 \times 10^3$) because the smaller pores in the clay retain water more strongly. From this point on (at higher matric potentials) the clay soil will have higher $K(h)$ because water remains in the smaller pores. As a result, considerably more water is present in clay soils at high matric potential and there is an increased probability of a continuous water film remaining to facilitate microbial transport.

Darcy's law was developed for steady flow, where Q is constant. However, in the subsurface, conditions are dynamic and thus Q is constant only over short periods of time. To account for changing flow, the flow equation is written in differential form to yield the Darcy flux:

$$q = -K\frac{\partial H}{\partial z}$$

(Eq. 15.2)

where $q = Q/At$ (m/day) and $\partial H/\partial z$ is the hydraulic gradient (m/m). By definition, q is the volume of water moving through a 1-m^2 face area per unit time. However, because water moves only through pore space and not through solids, the actual velocity of water moving through soil is considerably higher than q, the Darcy velocity. The pore water velocity is proportional to pore size; however, the average pore water velocity is generally defined as:

$$v = \frac{q}{\theta_w}$$

(Eq. 15.3)

where:

   v is the pore water velocity

   q is the flow rate per unit area determined for Darcy's law, and

   θ_w is the water-filled porosity.

In saturated soils, θ_w is equal to the total porosity, so the pore water velocity is approximated well by the Darcy velocity. However, in unsaturated soils there is a marked increase in pore water velocity over Darcy velocity.

Another factor to consider is hydrologic heterogeneity arising as a function of soil structure. Variations in structure such as cracks, fissures and channels can greatly affect flow rates by creating preferred flow paths, with increased flow velocities. This phenomenon is termed preferential flow. Such structural inconsistencies can greatly increase microbial transport.

15.1.8 Persistence and Activity of Introduced Microbes

Persistence and activity are key transport considerations because, ignoring for the moment the possibility of genetic exchange, it is the movement of live/intact microbes that is of concern to environmental and health considerations. Microbes introduced into a soil environment typically decline rapidly in number through cell death or viral inactivation and the average activity per cell of the surviving microbes is often reduced. Microbial adhesion to particle surfaces tends to provide some degree of protection from adverse factors. Access to and utilization of solid phase-associated nutrients (i.e., electrostatically held cations (NH_4^+, Na^+, K^+, Mg^{2+}) and nutrients released from sorbed organic material) may account, in part, for the increased survival of adsorbed microorganisms. Moisture content plays a multifaceted role in microbial survival. Too little water leads to desiccation, while moisture contents above a certain optimal level may lead to decreased microbial numbers, potentially due to oxygen depletion in saturated pores. In addition, microbial predators such as protozoa tend to be more active at higher soil moisture contents. Microbial sorption to particles and within small pores is thought to provide protection from protozoa, which are typically larger and thus may be excluded from certain pores or bacterial sorption sites. Indeed, in the presence of protozoa, higher percentages of particle-associated bacteria have been observed (Postma et al., 1990).

Immobilization of bacterial cells in a carrier material such as polyurethane or alginate has been investigated as an improved inoculation technique leading to increased survival and degradation capabilities of the inoculum (van Elsas et al., 1992; Hu et al., 1994). For introduction into the soil environment, the carrier material confers on the inoculum some protection against harmful physico-chemical and biological factors.

Although the soil environment is often detrimental to introduced organisms and thus their transport, certain biotic components of the terrestrial profile can increase movement of added microbes. For instance, channels formed by earthworms have been shown to increase transport by creating regions of preferential flow (Thorpe et al., 1996). Similarly, bacterial transport has been shown to be stimulated by root growth (Hekman et al., 1995). Water movement through channels formed by root growth and/or in films along the root surfaces contributes to the increased bacterial dispersion.

15.2 FACTORS AFFECTING TRANSPORT OF DNA

The survival and transport potential of introduced microbes are both issues of concern. However, it is important to realize that a dead or inactivated microbe usually breaks open, releasing its genetic material to the environment. Upon lysis, there is potential for the genetic material to be transported or sorbed to colloids where it can remain protected from degradation (Ogram et al., 1988). Free or desorbed nucleic acids may be reincorporated into other microbes via transformation. This can result in the expression of genes encoded by these nucleic acids or in the potential transport within the intact recipient cell.

Sorption of free DNA depends on several factors, including the mineralogy of the matrix material, ionic strength and pH of the soil solution, and length of the DNA polymer. DNA has a pK_a of approximately 5. At pH values equal to the pK_a the DNA is neutral, and at lower pH values it is positively charged. In either of these states the DNA is subject to adsorption to colloids and to intercalation into certain minerals, such as montmorillonite. This is enhanced by the fact that the pH of the microenvironment surrounding a soil particle may be as much as two or three units below the pH of the bulk solution. However, at higher pH values the DNA is negatively charged and is repelled from the negatively charged surfaces. Ogram *et al.* (1988) determined that the surface pH of some natural soils and sediments may be near the pK_a of DNA, and thus significant amounts of DNA could remain nonsorbed and be present in the aqueous phase. The same group also found that higher molecular weight DNA was sorbed more rapidly and to a greater extent than lower molecular weight DNA. Depending on the specific conditions and soil sample, DNA sorption can be highly variable (Ogram *et al.*, 1988).

15.3 NOVEL APPROACHES TO FACILITATE MICROBIAL TRANSPORT

For a number of applications including bioremediation and oil recovery, delivery of viable microbes is critical to the success of the application. As a result, strategies have been developed to attempt to optimize the "natural conditions" that favor transport. Several novel approaches designed to facilitate microbial transport through the terrestrial profile are being investigated. Formation of ultramicrobacteria and biosurfactants and gene transfer are among those that show potential.

15.3.1 Ultramicrobacteria

Marine bacteria react to starvation by dividing and shrinking to one-third their normal size. Such bacteria are referred to as ultramicrobacteria (UMB). Similarly, isolates obtained from soil can be placed in a nutrient-deprived medium such as phosphate-buffered saline and form UMB. After several weeks of starvation, a distinct morphological change takes place in these cells. As shown in Figure 15.2, the cells shrink to approximately 0.3 μm in size and become rounder. They also lose their capsule layer, thereby becoming less sticky. These bacteria can then be resuscitated by providing a carbon source.

They recover both morphologically and physiologically. Such UMB have been shown to penetrate farther into sandstone cores than their vegetative counterparts. For example, Ross *et al.* (2001) demonstrated in a bench-scale experiment that the permeability of a limestone fracture was reduced by 99% in 22 days following inoculation with an indigenous groundwater community and then flushing with a molasses solution at a carbon loading rate of 1.08×10^{-2} mg carbon per ml/per minute.

Interest in UMB first centered on their potential for use in oil recovery. The ability to control flow through the subsurface has potential use for containment or biotreatment of contaminated sites or to improve oil recovery. For oil recovery, after oil is initially flushed from a geologic formation, removal of further oil residuals becomes more difficult because flow paths have become established. At this point, ultramicrobacteria injected into the formation move relatively easily through established flow paths. They can then be resuscitated by nutrient injection, grow and divide, thereby plugging pores and forcing flow through other regions of the geologic formation. For example, Bossolan *et al.* (2005) examined the response of a *Klebsiella pneumoniae* isolated from an oil well to starvation and resuscitation and determined that this strain was a viable option for transport and growth of microorganisms inside porous media, with possible applications to microbially enhanced oil recovery (MEOR).

15.3.2 Surfactants

Another approach involves the use of a chemical additive, specifically a surfactant, to increase the transport potential of microbes. Bai *et al.* (1997) investigated the influence of an anionic monorhamnolipid biosurfactant on the transport of three *Pseudomonas* strains with various hydrophobicities through soil under saturated conditions. Columns packed with sterile sand were saturated with sterile artificial groundwater, and then three pore volumes of ^{3}H-labeled bacterial suspensions with various rhamnolipid (RL) concentrations were pumped through the column. Four additional pore volumes of rhamnolipid solution were then applied. Rhamnolipid enhanced the transport of all cell types tested but to varying degrees. Recovery of the most hydrophilic strain increased from 22.5 to 56.3%, recovery of the intermediate strain increased from 36.8 to 49.4% and recovery of the most hydrophobic strain increased from 17.7 to 40.5%. Figure 15.16 shows the breakthrough curves for the most hydrophilic strain at different rhamnolipid concentrations.

In this experiment it was found that the surface charge density of the bacteria did not change in the presence of the rhamnolipid, but the negative surface charge density of the porous medium increased. Thus, reduced bacterial sorption may be due to one of several factors including

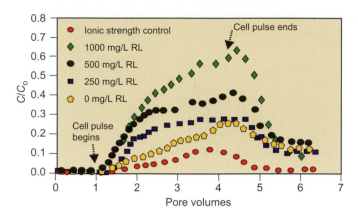

FIGURE 15.16 The effect of a rhamnolipid (RL) biosurfactant on transport of a *Pseudomonas* sp. where C_0 is the CFU/ml in the influent solution, and C is the CFU/ml in the effluent solution. A pore volume is the amount of liquid it takes to fill all of the soil pores in the column. Modified with permission from Bai *et al.* (1997).

an increase in surface charge density caused by rhamnolipid adsorption; solubilization of extracellular polymeric glue; or reduced availability of sorption sites on porous surfaces. The advection–dispersion transport model used to interpret these results suggests that the predominant effect of rhamnolipid was to prevent irreversible adsorption of cells.

Streger *et al.* (2002) investigated the use of surfactants to enhance transport of *Hydrogenophaga flava* ENV735, a bacterium capable of degrading the gasoline oxygenate methyl tert-butyl ether (MTBE), a widespread groundwater contaminant. While several tested surfactants were toxic to this bacterium, one nonionic surfactant, Tween 20, was not toxic and enhanced transport in sand columns. Findings such as this may facilitate delivery of microbial inocula to contaminated sites with no indigenous degraders.

15.3.3 Gene Transfer

Bacteria are intentionally introduced into soil systems in order to manipulate components and/or processes that occur within the soil profile. Typically, enhancement of microbial activities, e.g., organic degradation or metal resistance/immobilization, is the driving force behind their introduction. It may be possible to circumvent some of the factors that limit microbial transport and thus the success of bioremediation in soil via genetic exchange. Gene transfer between organisms may occur through conjugation, transduction or transformation. Transfer events such as these may make it possible to distribute genetic information more readily through the soil. Gene transfer events between an introduced organism and indigenous soil recipients has been shown to occur and in some cases this results in increased degradation of a contaminant (DiGiovanni *et al.*, 1996) (see Case Study 3.1).

Studies have addressed the transport potential of transconjugants, which arise when indigenous bacteria receive a plasmid from an introduced donor through conjugation. In a column study involving a donor inoculum at the column surface, Daane *et al.* (1996) found that transconjugants were limited to the top 5 cm of the column. However, when earthworms were also introduced into the column, not only did the depth of transport of donor and transconjugants increase, depending on the burrowing behavior of the earthworm species, but also the number of transconjugants found increased by approximately two orders of magnitude.

In a separate study, Lovins *et al.* (1993) examined the transport of a genetically engineered *Pseudomonas aeruginosa* strain that contained plasmid pR68.45 and the indigenous recipients of this plasmid in nonsterile, undisturbed soil columns. The surface of the column was inoculated and unsaturated flow conditions were maintained. Transconjugants survived longer in the columns and were found to have moved farther down the column than the donor. The greater survival rate of transconjugants would be expected because these organisms have previously adapted to the particular conditions of the soil. The increased transport could be the result of plasmid transfer to smaller, more mobile bacteria.

In addition, consecutive gene transfer events between indigenous microbes have been suggested as a mode of transfer (DiGiovanni *et al.*, 1996). This would be especially feasible when microbes are present in high densities, such as stationary microbes growing within a biofilm on soil surfaces or in the rhizosphere.

15.4 MICROBIAL TRANSPORT STUDIES

15.4.1 Column Studies

In situ transport experiments involving microorganisms are difficult to conduct. There are many obstacles associated with sampling and manipulating a complex environmental system that make the determination of critical transport factors difficult. As a result, most studies are performed in the laboratory in soil-packed columns where factors that affect transport can be varied individually (Figure 15.17). However, these usually represent homogeneous media of one soil type with no soil structure. Thus,

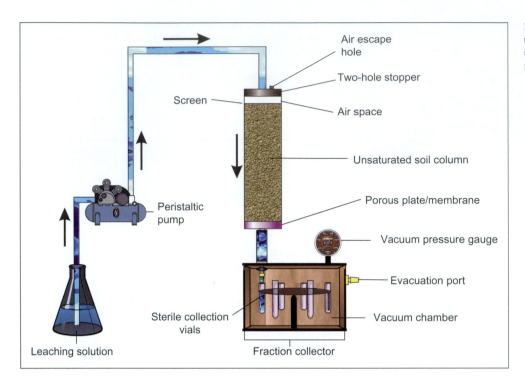

FIGURE 15.17 The setup used to run an unsaturated soil column. Modified with permission from Wierenga (1996).

while they are useful to study individual factors, like infiltration rates, they do not reflect the heterogeneous nature of subsurface environments. To overcome this limitation intact soil cores obtained from a site which retains site heterogeneity rather than using hand-packed more homogeneous columns can be used. For example, Smith *et al.* (1985) observed that under saturated flow conditions packed columns retained at least 93% of the bacterial cells applied, whereas intact cores retained only 21 to 78%, depending on the matrix texture and structure. Increased heterogeneity in soil structure preferential flow and flow velocity are probably key factors accounting for the increased transport within the intact soil cores.

Another limitation of soil cores is that the length of the columns does not reflect full-scale transport over long distances. However, generally, greater removal of microbes occurs in the first meter in most soils, and removal rate tends to decrease with travel distance. Microbial removal determined from laboratory column (usually one meter in length or less) studies can be one to three orders (10 to 1000) of magnitude greater than observed under field conditions (Pang, 2009). Thus, caution should be used in extrapolating results from column studies to distances greater than one meter in the field.

15.4.2 Field Studies

Field studies, in which multiple factors may change simultaneously, provide more relevant information than column studies. Field studies may involve both the use of

tracers (e.g., bacteriophages) and/or the detection of naturally occurring enteric organisms near a source (i.e., septic tank or land treatment of wastewater). This requires the use of sampling wells near the point of injection or source. In an extensive review of field studies on microbial transport, Pang (2009) found that a log function best describes removal for most organisms, with removal rates decreasing with increasing distance from the source, contradicting the conventional transport models and filtration theory. This reflects the hypothesis that heterogeneity among microbial particles themselves (type, size, density, charge, strains, survival characteristics and aggregation with colloids) affects their transport. Approximate microbial removal rates in different subsurface media are shown in Table 15.2.

15.4.3 Tracers

Tracers, chemical or particulate in nature, are often used to estimate microbial transport potential. Tracers are chosen so that their transport will closely mimic that of the microbe of interest. Their use is especially informative with regard to abiotic processes that influence movement of bacteria and viruses through subsurface media. Tracers are advantageous, particularly for field studies, for a variety of reasons. They can be added to a system in high numbers, their transport can be monitored without introducing a risk of infection, and they are typically easy to detect. A number of different tracers have been used in microbial transport studies, including microspheres,

TABLE 15.2 Magnitude of Removal Rates for Different Subsurface Media

Category	Magnitude of Removal Rate Log/Meter	Conditions
Soil	1	Most soil types
	0.1	Clayey soil, clay loam and clayey silt loam
Vadose zone (unsaturated soil)	0.1	Clay and silt, sand, sand-gravels, coarse gravels
Sand aquifers (velocity <2 meters/day)	1	Sand aquifers
Sand and gravel aquifers (velocity < 3 meters/day)	0.1	Less than 17 meters travel distance
	0.1 to 0.01	Less than 177 meters travel distance
	0.0001	210 to 2930 meters travel distance
Coarse gravel aquifers (velocity greater than 50 meters/day)	0.01	
Fractured rock aquifers	1 to 0.1	Clean fractured clay till
Karst limestone aquifers	0.1 to 0.01	Less than 85 meters travel distance
	0.0001	5000 meters travel distance

Modified from Pang (2009).

halides, proteins and dyes. The use of microbe-sized microspheres has the advantage over the use of dissolved tracers since the microspheres should follow the same flow paths as bacteria, even in highly heterogeneous subsurfaces. However, their surfaces may interact with subsurface particles very differently from bacteria. Studies indicate that the tracer choice is often critical in determining relevant estimates of transport but that all tracers have limitations in terms of mimicking microbial transport through the terrestrial profile. Coliphages and bacteria containing a marker (antibiotic resistance, stained) are often used (Harvey, 1997).

Powelson *et al.* (1993) compared the transport of a tracer with that of two phages. This group used potassium bromide, a conservative chemical tracer, and the bacteriophage MS-2 and PRD-1, which were selected because of their low adsorption to soils and their long survival time in the environment. Both the viruses and the conservative chemical tracer arrived at sampling depths in irregular patterns, indicating preferential flow. Virus breakthroughs were later than bromide except when viruses were added after pore clogging had reduced infiltration of the surface-applied sewage effluent. This study demonstrates the variability of relative transport rates that can exist between microbes and a chemical tracer.

Gitis *et al.* (2002) developed fluorescent dye-labeled bacteriophages to provide an additional tool to study virus transport. Advantages of these modified bacteriophages over conventional tracers include the ability to uncouple inactivation and transport phenomena, decreased costs associated with sample preservation, simple quantitation by optical methods and enumeration of individual virus concentrations instead of aggregates.

The ratio of the time it takes for the maximum concentration of a conservative tracer (a tracer that is nonsorbing) to be detected in the column effluent to the corresponding time for the coinjected microorganism defines the retardation factor for the microbe. This factor can be used for comparisons involving the same microbe and different soils or different microbes and the same soil. For example, Bai *et al.* (1997) found that in the absence of rhamnolipid (a surfactant) the retardation factors for three *Pseudomonas* strains through sandy soil ranged from 3.13 to 2.12. This is in comparison to a value of 1, which indicates no retardation. In addition, the retardation factor can be used to assess the occurrence of preferential flow. Preferred flow paths are suggested when the retardation factor is less than 1.0, indicating enhanced microbe transport relative to mean flow velocity (tracer transport). Gerba *et al.* (1991) suggested a worst-case value of 0.5 for the retardation factor when using models to predict microbial transport. One possible explanation for increased microbial transport is pore-size exclusion. Microbes may be excluded from smaller pores, where, on average, water travels more slowly (see Figure 15.14A). Thus, they are forced to travel in the larger pores, with velocities that are higher than that of the soil solution as a whole. Accordingly, transport of microbes can be faster than that of a conservative tracer through the same porous medium.

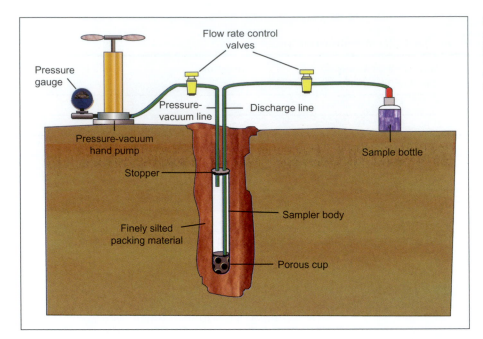

FIGURE 15.18 Example of an experimental setup that allows *in situ* sampling of soil pore water, which can subsequently be examined for microbial or chemical contaminants. Modified with permission from Soil Moisture Equipment Corp., Santa Barbara, CA.

15.5 MODELS FOR MICROBIAL TRANSPORT

Models have been used to predict not only the extent of movement of microbes but also the time required for the microorganisms to arrive at a specific location. Mathematical models designed to predict microbial transport through the subsurface should take into consideration a variety of factors that influence both survival and adsorption. In addition, they should account for changing conditions that the microbe encounters as it migrates through the subsurface profile. Laboratory and field studies should be conducted in an attempt to verify the accuracy of conceptual transport models. Such studies indicate that although models can be useful tools, they can also generate predictions that are off by orders of magnitude (Section 15.4.2). Accordingly, model predictions should be treated with caution.

The heterogeneity of microbes within a population can further complicate the application of models. Each bacterium or virus even of the same species may have slight differences in isoelectric points or factors in their biochemical composition which increase or decrease their transport. Typically, a microbial species that exhibits transport characteristics at one end of the spectrum or the other is chosen. The choice of an organism that does not travel extensively may be useful in assessing problems that may be encountered with bioaugmentation efforts. On the other hand, a microbe known to have high transport potential may be chosen when attempting to determine the minimum distance between a sewage release point and a well for drinking water. Another approach involves modeling the transport of each microbe individually and then combining the results to determine an overall concentration at

a specified time and location. This method may be more accurate but is also much more involved.

A term is generally incorporated in equations to reflect the survival characteristics of the microorganism. This term represents the loss of the microbe, via death or inactivation, resulting from adverse chemical, physical or biological processes. Microbial growth counteracts a portion of this decay, and thus a net rate decay term (i.e., net decay rate = growth rate − death or inactivation rate) is often used (Figure 15.18). Microbial decay rates have been found to vary by several orders of magnitude (Table 15.3), making it necessary to evaluate survival characteristics for each particular microbe. Contaminant transport models, specifically advection−dispersion models and filtration models, are often modified for application to microbial transport. Factors that influence microbial transport can be incorporated into equations governing either of these models. However, the fundamental basis of each model differs, and thus under certain conditions use of a particular model may be preferable.

15.5.1 Advection−Dispersion Models

Unmodified advection−dispersion models assume that the contaminant is in solution and thus has the same average velocity as the matrix solution. Values for average velocity and dispersion of a contaminant are generally obtained from conservative tracer tests. These values may not be appropriate for microbial transport because microbes are not dissolved but are instead suspended particulates in the liquid medium. In order to obtain accurate adsorption data for input into the advection−dispersion

TABLE 15.3 Survival of Microorganisms in Groundwater vs. Temperature

Organism	Temperature ($^{\circ}$C)	Mean Inactivation Rate (Log/Day)	Inactivation Rate Range (Log/Day)
Poliovirus	0−10	0.02	0.005−0.05
	11−15	0.08	0.03−0.2
	16−20	0.1	0.03−0.2
	26−30	0.08	0.006−1.4
Hepatitis A virus	0−10	0.02	0−0.08
	20−30	0.04	0.009−0.1
Echovirus	11−15	0.1	0.05−0.2
	16−20	0.1	0.05−0.2
	21−25	0.2	0.06−0.6
Coxsackievirus	8−20	0.06	0.002−0.2
	25−30	0.1	0.007−0.3
Rotavirus	3−15	0.4	one study
	23.2	0.03	one study
Adenovirus	4	0.0076	one study
	12−22	0.028	0.01−0.047
Coliforms	0−10	0.07	0.03−0.4
	15−20	0.4	0.02−1.5
	21−37	0.3	0.007−2.5
Cryptosporidium	22	0.039	0.025−0.072

Data from John and Rose (2005) and Regnery et al. (2013).

equation, it may be necessary to conduct site- and microbe-specific adsorption studies. Furthermore, both irreversible sorption and reversible adsorption should be considered. The advection−dispersion model often incorporates a decay term along with terms that account for transport with the bulk flow (advection), transport resulting from diffusion and mechanical mixing (dispersion), and adsorption. Advection−dispersion equations can be expanded in order to take into account hydrogeological heterogeneities in addition to the variety of factors that determine microbial survival. Figure 15.13 illustrates the influence of advection, dispersion, adsorption and decay on the transport of a contaminant. A relatively simple advection−dispersion equation is:

$$R_f \partial C / \partial t = -V \partial C / \partial x + D \partial^2 C / \partial x^2 \pm R_x$$

where:

 C is the concentration of microbe (mass/volume)
 x is the distance traveled through the porous medium (length)
 V is the average linear velocity constant (length/time)

R_x is the microbial net decay term (mass/time-volume)
 t is time, and
R_f is the retardation factor, accounting for reversible interaction with the porous medium.

Bai *et al.* (1997) used a one-dimensional advection−dispersion model to assess bacterial transport through a sandy soil in the presence of rhamnolipid (Figure 15.16). They found that three parameters were especially important: R, the retardation factor, which represents the effect of reversible adsorption on cell transport, and two irreversible sticking rate constants, one for instantaneous sorption and the other for rate-limited sorption. They found that all three constants decreased with increasing rhamnolipid concentration; however, the rate-limited sorption sites were affected the most.

15.5.2 Filtration Models

Filtration models, on the other hand, assume that the contaminant is particulate in nature and that its removal is dependent on physical straining and sorption processes. These processes are often combined into a filtration coefficient for the system. Such models take into account mechanisms by which colloids (i.e., microbes) come in contact with particle surfaces and the relative size of the microbe compared with the pores in the medium. The premise of these models is that as the microbial suspension passes through the terrestrial profile, microbes will be removed. Yao *et al.* (1971) demonstrated that filtration models may be applicable in terms of predicting bacterial immobilization during transport through the subsurface. A general filtration equation is:

$$\partial C / \partial x = \lambda C$$

where:

C is the concentration of colloid (mass/volume)
x is the distance traveled through the porous medium (length), and
λ is the filter coefficient (1/length).

According to this equation, microbial removal would be exponential with depth, as is sometimes observed. As with the advection−dispersion equation, terms can be incorporated into the equation to account for net microbial decay.

QUESTIONS AND PROBLEMS

1. Compare and contrast the major factors influencing bacterial versus viral transport through the terrestrial profile. Which type of microbe would you expect to find deeper in the profile following surface application?
2. Choose either a bacterium or virus and design a column experiment to assess its transport potential. Your discussion should include items such as column design, type of matrix material, flow conditions, percolating solution, inoculation and sampling approaches. Support your choices.
3. What are UMB? How can they be used to facilitate bioremediation of contaminated sites?
4. As both soil particles and microbes generally have a net negative surface charge, why is adsorption to matrix material often a factor limiting microbial transport?
5. Why do microbes introduced to a site often die within a few days to weeks? What impact does this have on the transport potential of the introduced microbes?
6. Discuss the advantages and disadvantages of using a soil column to assess microbial transport potential as compared to the use of a column.
7. Why is microbial removal rate greater near the soil surface than at greater distances?
8. Determine the time in days for inactivation of 10^6 adenoviruses in ground water at 4°C.
9. You have packed a column with a sandy soil and are preparing to determine the transport of a particular

bacterial isolate through the porous material. Indicate whether the following changes to your experimental system might be expected to increase or decrease the transport of the bacterium:

a. an increase in cell size
b. a decrease in ionic strength
c. an increase in organic carbon content of the soil
d. an increase in particle size of the sandy soil
e. a bacterium produces copious amounts of exopolymeric material
f. an anionic surfactant is added
g. an anionic surfactant is added
h. the cells are carefully washed with a phosphate buffer three times before the experiment
i. a bacterium covered with fimbriae is used
j. a bacterium with an extremely hydrophobic cell surface is used.

REFERENCES AND RECOMMENDED READING

Ackermann, H. W., and Michael, S. D. (1987) "Viruses of Prokaryotes," CRC Press, Boca Raton, FL, pp. 173−201.

Bai, G., Brusseau, M. L., and Miller, R. M. (1997) Influence of a rhamnolipid biosurfactant on the transport of bacteria through a sandy soil. *Appl. Environ. Microbiol.* **63**, 1866−1873.

Bradbury, K. R., Borchardt, M. A., Gotkowita, S. K., Shu, J., and Hunt, R. J. (2013) Source and transport of human enteric viruses in deep municipal water supply wells. *Environ. Sci. Technol.* **47**, 4096−4104.

Bossolan, N., Godinho, M. J. L., and Volpon, A. G. T. (2005) Growth and starvation of a strain of *Klebsiella pneumoniae* isolated from a Brazilian oil formation. *World J. Microbiol. Biotechnol.* **21**, 1471−1475.

Daane, L. L., Molina, J. A. E., Berry, E. C., and Sadowsky, M. J. (1996) Influence of earthworm activity on gene transfer from *Pseudomonas fluorescens* to indigenous soil bacteria. *Appl. Environ. Microbiol.* **62**, 515−521.

Dexter, S. C. (1979) Influence of substratum critical surface tension on bacterial adhesion—in situ studies. *J. Colloid Interface Sci.* **70**, 346−354.

DiGiovanni, G. D., Neilson, J. W., Pepper, I. L., and Sinclair, N. A. (1996) Gene transfer of *Alcaligenes eutrophus* JMP 134 plasmid pJ4 to indigenous soil recipients. *Appl. Environ. Microbiol.* **62**, 2521−2526.

Dowd, S. E., Pillai, S. D., Wang, S., and Corapcioglu, M. Y. (1998) Delineating the specific influence of virus isoelectric point and size on virus adsorption and transport through sandy soils. *Appl. Environ. Microbiol.* **64**, 405−410.

Fetter, C. W. (1993) Mass transport in saturated media. In "Contaminant Hydrogeology," Macmillan, New York, pp. 43−114.

Fletcher, M. (1976) The effects of proteins on bacterial attachment to polystyrene. *J. Gen. Microbiol.* **94**, 400−404.

Gannon, J. T., Manilal, V. B., and Alexander, M. (1991) Relationship between cell surface properties and transport of bacteria through soil. *Appl. Environ. Microbiol.* **57**, 190−193.

Gerba, C. P., Yates, M. V., and Yates, S. R. (1991) Quantitation of factors controlling viral and bacterial transport in the subsurface. In "Modeling the Environmental Fate of Microorganisms" (C. J. Hurst, ed.), ASM Press, Washington, DC, pp. 77−88.

Gitis, V., Adin, A., Nasser, A., Gun, J., and Lev, O. (2002) Fluorescent dye labeled bacteriophages—a new tracer for the investigation of viral transport in porous media: introduction and characterization. *Water Res.* **36**, 4227–4234.

Goyal, S. M., and Gerba, C. P. (1979) Comparative adsorption of human enteroviruses, simian rotavirus, and selected bacteriophages to soils. *Appl. Environ. Microbiol.* **38**, 241–247.

Harvey, R. W. (1997) Microorganisms as tracers in groundwater injection and recovery experiments: a review. *FEMS Microbiol. Rev.* **20**, 461–472.

Hekman, W. E., Heijnen, C. E., Burgers, S. L. G. E., van Veen, J. A., and van Elsas, J. D. (1995) Transport of bacterial inoculants through intact cores of two different soils as affected by water percolation and the presence of wheat plants. *FEMS Microbiol. Ecol.* **16**, 143–158.

Herman, D. C., and Costerton, J. W. (1993) Starvation-survival of a p-nitrophenol-degrading bacterium. *Appl. Environ. Microbiol.* **59**, 340–343.

Herzig, J. P., Leclerc, D. M., and LeGolf, P. (1970) Flow of suspensions through porous media-application to deep filtration. *Ind. Eng. Chem.* **62**, 8–35.

Hu, Z., Korus, R. A., Levinson, W. E., and Crawford, R. L. (1994) Adsorption and biodegradation of pentachlorophenol by polyurethane-immobilized *Flavobacterium*. *Environ. Sci. Technol.* **28**, 491–496.

John, D. E., and Rose, J. B. (2005) Review of factors affecting microbial survival in groundwater. *Environ. Sci. Technol.* **39**, 7345–7356.

Jury, W. A., Gardner, W. R., and Gardner, W. H. (1991) Water movement in soil. In "Soil Physics," 5th ed., John Wiley & Sons, New York, pp. 73–121.

Lance, J. C., and Gerba, C. P. (1984) Virus movement in soil during saturated and unsaturated flow. *Appl. Environ. Microbiol.* **47**, 335–337.

Landry, E. F., Vaughn, J. M., Thomas, M. Z., and Beckwith, C. A. (1979) Adsorption of enteroviruses to soil cores and their subsequent elution by artificial rainwater. *Appl. Environ. Microbiol.* **38**, 680–687.

Lovins, K. W., Angle, J. S., Wiebers, J. L., and Hill, R. L. (1993) Leaching of *Pseudomonas aeruginosa* and transconjugants containing pR68.45 through unsaturated, intact soil columns. *FEMS Microbiol. Ecol.* **13**, 105–112.

MacLeod, F. A., Lappin-Scott, H. M., and Costerton, J. W. (1988) Plugging of a model rock system by using starved bacteria. *Appl. Environ. Microbiol.* **54**, 365–1372.

Michen, B., and Graule, T. (2010) Isoelectric point of viruses. *J. Appl. Microbiol.* **109**, 388–397.

Ogram, A., Sayler, G. S., Gustin, D., and Lewis, R. J. (1988) DNA adsorption to soils and sediments. *Environ. Sci. Technol.* **22**, 982–984.

Pang, L. (2009) Microbial removal rates in subsurface media estimated from published studies of field experiments and large intact soil cores. *J. Environ. Qual.* **38**, 1531–1559.

Postma, J., Hok-A-Hin, C. H., and van Veen, J. A. (1990) Role of microniches in protecting introduced *Rhizobium leguminosarum* biovar trifolii against competition and predation in soil. *Appl. Environ. Microbiol.* **56**, 495–502.

Powelson, D. K., Gerba, C. P., and Yahya, M. T. (1993) Virus transport and removal in water during aquifer recharge. *Water Res.* **27**, 583–590.

Regnery, J., Mazahirali, A., Wing, A. D., Gerba, C. P., Dickenson, E., Snyder, S., *et al.* (2013) Role of retention time on attenuation of microbial and chemical contaminants during groundwater recharge in indirect potable systems: a review. Submitted for publication.

Reynolds, P. J., Sharma, P., Jenneman, G. E., and McInerney, M. J. (1989) Mechanisms of microbial movement in subsurface materials. *Appl. Environ. Microbiol.* **55**, 2280–2286.

Ross, N., Villemur, R., Deschenes, L., and Samson, R. (2001) Clogging of a limestone fracture by stimulating groundwater microbes. *Water Res.* **35**, 2029–2037.

Sanin, S. L., Sanin, F. D., and Byers, J. D. (2003) Effect of starvation on the adhesive properties of xenobiotic degrading bacteria. *Process Biochem.* **38**, 909–914.

Smith, M. S., Thomas, G. W., White, R. E., and Ritonga, D. (1985) Transport of *Escherichia coli* through intact and disturbed soil columns. *J. Environ. Qual.* **14**, 87–91.

Streger, S. H., Vainberg, S., Dong, H. L., and Hatzinger, P. B. (2002) Enhancing transport of *Hydrogenophaga flava* ENV735 for bioaugmentation of aquifers contaminated with methyl tert-butyl ether. *Appl. Environ. Microbiol.* **68**, 5571–5579.

Tan, K. H. (1993) Colloidal chemistry of inorganic soil constituents. In "Principles of Soil Chemistry," 2nd ed., Marcel Dekker, Inc., New York, pp. 129–205.

Thorpe, I. S., Prosser, J. I., Glover, L. A., and Killham, K. (1996) The role of the earthworm *Lumbricus terrestris* in the transport of bacterial inocula through soil. *Biol. Fertil. Soils* **23**, 132–139.

Torkzaban, S., Hassanizadeh, S. M., Schijven, J. F., de Bruin, H. A. M., and de Roda Husman, A. M. (2006) Virus transport in saturated and unsaturated sand columns. *Vadose Zone J.* **5**, 877–885.

Van Elsas, J. D., Trevors, J. T., Jain, D., Wolters, A. C., Heijnen, C. E., and van Overbeek, L. S. (1992) Survival of, and root colonization by, alginate-encapsulated *Pseudomonas fluorescens* cells following introduction into soil. *Biol. Fertil. Soils* **14**, 14–22.

van Loosdrecht, M. C. M., Norde, W., Lyklema, J., and Zehnder, A. J. B. (1990) Hydrophobic and electrostatic parameters in bacterial adhesion. *Aquat. Sci.* **52**, 103–114.

Weiss, T. H., Mills, A. L., Hornberger, G. M., and Herman, J. S. (1995) Effect of bacterial cell shape on transport of bacteria in porous media. *Environ. Sci. Technol.* **29**, 1737–1740.

Wierenga, P. J. (1996) Physical processes affecting contaminant fate and transport in soil and water. In "Pollution Science" (I. L. Pepper, C. P. Gerba, and M. L. Brusseau, eds.), Academic Press, San Diego, pp. 45–62.

Yao, K. M., Habibian, M. T., and O'Melia, C. R. (1971) Water and waste water filtration: concepts and applications. *Environ. Sci. Technol.* **11**, 1105–1112.

Yates, M. V., and Yates, S. R. (1991) Modeling microbial transport in the subsurface: a mathematical discussion. In "Modeling the Environmental Fate of Microorganisms" (C. J. Hurst, ed.), ASM Press, Washington, DC, pp. 48–76.

Young, K. D. (2006) The selective value of bacterial shape. *Microbiol. Molec. Biol. Rev.* **70**, 660–703.

Zhuang, J., and Jin, Y. (2003) Virus retention and transport as influenced by different forms of soil organic matter. *J. Environ. Qual.* **32**, 816–823.

Chapter 16

Biogeochemical Cycling

Raina M. Maier

16.1 INTRODUCTION

16.1.1 Biogeochemical Cycles

Carbon dioxide concentrations in the atmospheric exceeded 400 parts per million (ppm) for the first time in May 2013, increasing from 315 ppm in 1958 (when the first accurate measurements were made), reports the U.S. National Oceanic and Atmospheric Administration. What has driven this increase? Carbon is cycled between organic forms such as sugar or other cellular building blocks and inorganic forms such as carbon dioxide. Vast amounts of organic matter are photosynthetically produced on Earth each year utilizing carbon dioxide from the atmosphere. Most of this material is consumed and degraded but a part of it, over the millennia, has been stored in permafrost, peat bogs and as fossil fuels. A delicate global balance of organic and inorganic carbon has been maintained largely driven by microbial activity. Human use of stored organic carbon (fossil fuels and peat) and recent warming of permafrost have upset this balance in favor of the release of carbon dioxide to the atmosphere. Most scientists agree that this is having a major impact on global warming, and is an excellent example of a major perturbation of the carbon cycle—one that is occurring during our lifetimes. Cycling between organic and inorganic forms is not limited to carbon. All of the major elements found in biological organisms (see Table 16.1), as well as some of the minor and trace elements, are similarly cycled in predictable and definable ways.

Taken together, the various element cycles are called the biogeochemical cycles. Understanding these cycles allows scientists to understand and predict the development of microbial communities and activities in the environment. There are many activities that can be harnessed in a beneficial way, such as for remediation of organic and metal pollutants, or for recovery of precious metals such as copper or uranium from low-grade ores. There are detrimental aspects of the cycles that can cause global environmental problems, for example the formation of acid rain and acid mine drainage, metal corrosion processes and formation of nitrous oxide, which can deplete Earth's ozone layer (see Chapter 31). As these examples

I.L. Pepper, C.P. Gerba, T.J. Gentry: **Environmental Microbiology, Third edition.** DOI: http://dx.doi.org/10.1016/B978-0-12-394626-3.00016-8
© 2015 Elsevier Inc. All rights reserved.

illustrate, the microbial activities that drive biogeochemical cycles are highly relevant to the field of environmental microbiology. Thus, the knowledge of these cycles is increasingly critical as the human population continues to grow, and the impact of human activity on Earth's environment becomes more significant. In this chapter,

the biogeochemical cycles pertaining to carbon, nitrogen, sulfur and iron are delineated. Although the discussion will be limited to these four cycles, it should be noted that there are a number of other cycles—the phosphorus cycle, the manganese cycle, the calcium cycle and more (Dobrovolsky, 1994).

16.1.2 Gaia Hypothesis

In the early 1970s, James Lovelock theorized that Earth behaves like a superorganism, and this concept developed into what is now known as the Gaia hypothesis. To quote Lovelock (1995), "Living organisms and their material environment are tightly coupled. The coupled system is a superorganism, and as it evolves there emerges a new property, the ability to self-regulate climate and chemistry." The basic tenet of this hypothesis is that Earth's physicochemical properties are self-regulated so that they are maintained in a favorable range for life. As evidence for this, consider that the sun has heated up by 30% during the past 4–5 billion years. Given Earth's original carbon dioxide-rich atmosphere, the average surface temperature of a lifeless Earth today would be approximately 290°C (Table 16.2). In fact, when one compares Earth's present-day atmosphere with the atmospheres found on our nearest neighbors Venus and Mars, one can see that something has drastically affected the development of Earth's atmosphere. According to the Gaia hypothesis, this is the development and continued presence of life. Microbial activity, and later the appearance of plants, have changed the original heat-trapping carbon dioxide-rich atmosphere to the present oxidizing, carbon dioxide-poor atmosphere. This has allowed Earth to maintain an average surface temperature of 13°C, which is favorable to the life that exists on Earth.

How do biogeochemical activities relate to the Gaia hypothesis? These biological activities have driven

TABLE 16.1 Chemical Composition of an *E. coli* Cell

Elemental Breakdown	% Dry Mass of an *E. coli* Cell
Major elements	
Carbon	50
Oxygen	20
Hydrogen	8
Nitrogen	14
Sulfur	1
Phosphorus	3
Minor elements	
Potassium	2
Calcium	0.05
Magnesium	0.05
Chlorine	0.05
Iron	0.2
Trace elements	
Manganese	all trace elements
Molybdenum	combined comprise 0.3%
Cobalt	of dry weight of cell
Copper	
Zinc	

Adapted from Neidhardt et al. (1990).

TABLE 16.2 Atmosphere and Temperatures Found on Venus, Mars and Earth

Gas	Planet			
	Venus	Mars	Earth Without Life	Earth With Life
Carbon dioxide	96.5%	95%	98%	0.03%
Nitrogen	3.5%	2.7%	1.9%	9%
Oxygen	trace	0.13%	0.0	21%
Argon	70 ppm	1.6%	0.1%	1%
Methane	0.0	0.0	0.0	1.7 ppm
Surface temperature (°C)	459	−53	290 ± 50	13

Adapted from Lovelock (1995).

FIGURE 16.1   An example of a living stromatolite (left) and a stromatolite fossil (right). From (left) Reynolds (1999) and (right) Farabee (2008).

the response to the slow warming of the sun resulting in the major atmospheric changes that have occurred over the last 4–5 billion years. When Earth was formed 4–5 billion years ago, a reducing (anaerobic) atmosphere existed. The initial reactions that mediated the formation of organic carbon were abiotic, driven by large influxes of ultraviolet (UV) light. The resulting reservoir of organic matter was utilized by early anaerobic heterotrophic organisms. This was followed by the development of the ability of microbes to fix carbon dioxide photosynthetically. Evidence from stromatolite fossils suggests that the ability to photosynthesize was developed at least 3.5 billion years ago. Stromatolites are fossilized laminated structures that have been found in Africa and Australia (Figure 16.1). Although a hotly debated topic, there is evidence that these structures were formed by photosynthetic microorganisms (first anaerobic, then cyanobacterial) that grew in mats and entrapped or precipitated inorganic material as they grew (Bosak *et al.*, 2007).

The evolution of photosynthetic organisms tapped into an unlimited source of energy, the sun, and provided a mechanism for carbon recycling, i.e., the first carbon cycle (Figure 16.2). This first carbon cycle was maintained for approximately 1.5 billion years. Geologic evidence then suggests that approximately 2 billion years ago, photosynthetic microorganisms developed the ability to produce oxygen. This allowed oxygen to accumulate in the atmosphere, resulting, in time, in a change from reducing to oxidizing conditions. Further, oxygen accumulation in the atmosphere created an ozone layer, which reduced the influx of harmful UV radiation, allowing the development of higher forms of life to begin.

At the same time that the carbon cycle evolved, the nitrogen cycle emerged because nitrogen was a limiting element for microbial growth. Although molecular nitrogen was abundant in the atmosphere, microbial cells

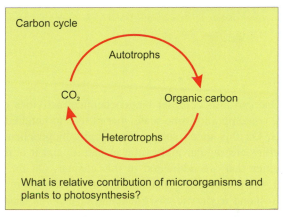

FIGURE 16.2   The carbon cycle is dependent on autotrophic organisms that fix carbon dioxide into organic carbon and heterotrophic organisms that convert organic carbon to carbon dioxide during respiration.

could not directly utilize nitrogen as N_2 gas. Cells require organic nitrogen compounds or reduced inorganic forms of nitrogen for growth. Therefore, under the reducing conditions found on early Earth, some organisms developed a mechanism for fixing nitrogen using the enzyme nitrogenase. Nitrogen fixation remains an important microbiological process, and to this day the majority of nitrogenase enzymes are totally inhibited in the presence of oxygen.

When considered over this geologic time scale of several billion years, it is apparent that biogeochemical activities have been unidirectional. This means that the predominant microbial activities on Earth have evolved over this long period of time to produce changes, and to respond to changes that have occurred in the atmosphere, i.e., the appearance of oxygen and the decrease in carbon dioxide content. Presumably these changes will continue to occur, but they occur so slowly that we do not have the capacity to observe them.

TABLE 16.3 Global Carbon Reservoirs

Carbon Reservoir	Metric Tons Carbon	Actively Cycled
Atmosphere		
CO_2	6.7×10^{11}	Yes
Ocean		
Biomass	4.0×10^{9}	No
Carbonates	3.8×10^{13}	No
Dissolved and particulate organics	2.1×10^{12}	Yes
Land		
Biota	5.0×10^{11}	Yes
Humus	1.2×10^{12}	Yes
Fossil fuel	1.0×10^{13}	Yes
Earth's crust[a]	1.2×10^{17}	No

Data from Dobrovolsky (1994).
[a]This reservoir includes the entire lithosphere found in either terrestrial or ocean environments.

TABLE 16.4 Net Carbon Flux between Selected Carbon Reservoirs

Carbon Source	Flux (Metric Tons Carbon/Year)
Release by fossil fuel combustion	7×10^{9}
Land clearing	3×10^{9}
Forest harvest and decay	6×10^{9}
Forest regrowth	-4×10^{9}
Net uptake by oceans (diffusion)	-3×10^{9}
Annual flux	9×10^{9}

Adapted from Atlas and Bartha (1993).

One can also consider biogeochemical activities on a more contemporary time scale, that of tens to hundreds of years. On this much shorter time scale, biogeochemical activities are regular and cyclic in nature, and it is these activities that are addressed in this chapter. On the one hand, the presumption that Earth is a superorganism that can respond to drastic environmental changes is heartening when one considers that human activity is causing unexpected changes in the atmosphere, such as ozone depletion and buildup of carbon dioxide. However, it is important to point out that the response of a superorganism is necessarily slow (thousands to millions of years), and as residents of Earth we must be sure not to overtax Earth's ability to respond to change by artificially changing the environment in a much shorter time frame.

16.2 CARBON CYCLE

16.2.1 Carbon Reservoirs

A reservoir is a sink or source of an element such as carbon. There are various global reservoirs of carbon, some of which are immense in size and some of which are relatively small (Table 16.3). The largest carbon reservoir is carbonate rock found in Earth's crust. This reservoir is four orders of magnitude larger than the carbonate reservoir found in the ocean, and six orders of magnitude larger than the carbon reservoir found as carbon dioxide in the atmosphere. If one considers these three reservoirs,

it is obvious that the carbon most available for photosynthesis, which requires carbon dioxide, is in the smallest of the reservoirs, the atmosphere. Therefore, it is the smallest reservoir that is most actively cycled. It is small, actively cycled reservoirs such as atmospheric carbon dioxide that are subject to perturbation from human activity. In fact, since global industrialization began in the late 1800s, humans have affected several of the smaller carbon reservoirs. Utilization of fossil fuels (an example of a small, inactive carbon reservoir) and deforestation (an example of a small, active carbon reservoir) are two activities that have reduced the amount of fixed organic carbon in these reservoirs, and added to the atmospheric carbon dioxide reservoir (Table 16.4).

The increase in atmospheric carbon dioxide has not been as great as expected. This is because the reservoir of carbonate found in the ocean acts as a buffer between the atmospheric and sediment carbon reservoirs through the equilibrium equation shown below:

$$H_2CO_3 \rightleftharpoons HCO_3^- \rightleftharpoons CO_2$$

Thus, some of the excess carbon dioxide that has been released has been absorbed by the oceans. However, there has still been a net efflux of carbon dioxide into the atmosphere of approximately 7×10^9 metric tons/year. The problem with this imbalance is that because atmospheric carbon dioxide is a small carbon reservoir, the result of a continued net efflux over the past 100 years or so has been a 28% increase in atmospheric carbon dioxide from 0.026 to 0.033%. A consequence of the increase in atmospheric carbon dioxide is that it contributes to global warming through the greenhouse effect (see also Chapter 31). The greenhouse effect is caused by gases in the atmosphere that trap heat from the sun and cause Earth to warm up. This effect is not solely due to carbon

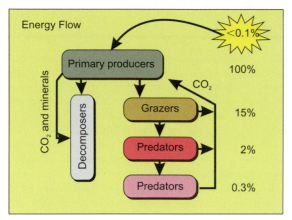

FIGURE 16.3 Diagram of the efficiency of sunlight energy flow from primary producers to consumers.

TABLE 16.5 Net Primary Productivity of Some Natural and Managed Ecosystems

Description of Ecosystem	Net Primary Productivity (g Dry Organic Matter/m²/Year)
Tundra	400
Desert	200
Temperate grassland	Up to 1500
Temperate or deciduous forest	1200–1600
Tropical rain forest	Up to 2800
Cattail swamp	2500
Freshwater pond	950–1500
Open ocean	100
Coastal seawater	200
Upwelling area	600
Coral reef	4900
Corn field	1000–6000
Rice paddy	340–1200
Sugarcane field	Up to 9400

Adapted from Atlas and Bartha (1993).

dioxide; other gases such as methane, chlorofluorocarbons (CFCs) and nitrous oxide all contribute to the problem.

16.2.2 Carbon Fixation and Energy Flow

The ability to photosynthesize allows sunlight energy to be trapped and stored. In this process carbon dioxide is fixed into organic matter (Figure 16.2). Photosynthetic organisms, also called primary producers, include plants and microorganisms such as algae, cyanobacteria, some bacteria and some protozoa. As shown in Figure 16.3, the efficiency of sunlight trapping is very low; less than 0.1% of the sunlight energy that hits Earth is actually utilized. As the fixed sunlight energy moves up each level of the food chain, up to 90% or more of the trapped energy is lost through respiration. Despite this seemingly inefficient trapping, photoautotrophic primary producers support most of the considerable ecosystems found on Earth. Productivity varies widely among different ecosystems depending on the climate, the type of primary producer and whether the system is a managed one (Table 16.5). For example, one of the most productive natural areas is the coral reefs. Managed agricultural systems such as corn and sugarcane systems are also very productive, but it should be remembered that a significant amount of energy is put into these systems in terms of fertilizer addition and care. The open ocean has much lower productivity, but covers the majority of Earth's surface, and so is a major contributor to primary production. In fact, aquatic and terrestrial environments contribute almost equally to global primary production. Plants predominate in terrestrial environments, but with the exception of immediate coastal zones, microorganisms are responsible for most primary production in aquatic environments. It follows that microorganisms are responsible for approximately one-half of all primary production on Earth.

16.2.3 Carbon Respiration

Carbon dioxide that is fixed into organic compounds as a result of photoautotrophic activity is available for consumption or respiration by animals and heterotrophic microorganisms. This is the second half of the carbon cycle shown in Figure 16.2. The end products of respiration are carbon dioxide and new cell mass. An interesting question to consider is the following: if respiration were to stop, how long would it take for photosynthesis to use up all of the carbon dioxide reservoir in the atmosphere? Based on estimates of global photosynthesis, it has been estimated that it would take 30 to 300 years. This illustrates the importance of both legs of the carbon cycle in maintaining a carbon balance (see Information Box 16.1).

The following sections discuss the most common organic compounds found in the environment and the microbial catabolic activities that have evolved in response. These include organic polymers, humus and C_1 compounds such as methane (CH_4). It is important to understand the fate of these naturally occurring organic compounds because degradative activities that have evolved for these compounds form the basis for degradation pathways that may be applicable to organic contaminants that are spilled in the environment (see Chapter 31).

Information Box 16.1   The Role of Soil Microbes in Carbon Sequestration

Currently there is debate about how soil microbial activity may influence global warming (Knorr *et al.*, 2005; Rice, 2006). Depending on the relative rates of microbial respiration versus photosynthetic activity, soils could be either a source or sink for CO_2. The estimated amount of carbon sequestered or stored in world soil organic matter ranges from 1.1 to 1.6×10^{12} metric tons (see Table 16.3). This is more than twice the carbon in living vegetation ($\sim 5.6 \times 10^{11}$ metric tons) or in the atmosphere ($\sim 6.7 \times 10^{11}$ metric tons) (Sundquist, 1993). Hence, even relatively small changes in soil carbon storage could have a significant impact on the global carbon balance. In the last 7800 years, the net carbon reservoir in the soil has decreased by 5.0×10^{10} metric tons largely due to conversion of land to agriculture (Lai, 2004). It is estimated that some of this lost carbon could be recovered through strategic management practices. For example, agricultural

practices that enhance crop productivity (CO_2 uptake) while decreasing microbial decomposition rates (CO_2 release) could be optimized to maximize the sequestration of carbon in the soil reservoir. Such practices, which include conservation set-aside, reduced tillage, and increased crop productivity have been estimated to account for the sequestration of 1.1 to 2.1×10^7 metric tons carbon annually (Lokupitiva and Paustian, 2006). However, global warming could also result in enhanced rates of microbial decomposition of the carbon stored in the soil. For example, Bellamy *et al.* (2005) documented carbon losses from all soils across England and Wales from the recent period 1978–2003 during which global warming has occurred. Overall, the debate is as yet unresolved, and many subtle feedback effects affect the final outcome, including for example the C:N ratio within soil organic matter and the mandatory C:N requirements of soil microbes.

But before looking more closely at the individual carbon compounds, it should be pointed out that the carbon cycle is actually not quite as simple as depicted in Figure 16.2. This simplified figure does not include anaerobic processes, which were predominant on early Earth and remain important in carbon cycling even today. A more complex carbon cycle containing anaerobic activity is shown in Figure 16.4. Under anaerobic conditions, which predominated for the first few billion years on Earth, some cellular components were less degradable than others (Figure 16.5). This is especially true for highly reduced molecules such as cellular lipids. These components were therefore left over and buried with sediments over time, and became the present-day fossil fuel reserves. Another carbon compound produced under anaerobic conditions is methane. Methane is produced in soils as an end product of anaerobic respiration (see Eq. 3.21, Chapter 3). Methane is also produced in significant quantities under the anaerobic conditions found in ruminants such as cows as well as termite guts. It is also produced within landfills (see Chapter 31).

16.2.3.1 Organic Polymers

What are the predominant types of organic carbon found in the environment? They include plant polymers, fungal and bacterial cell wall polymers, and arthropod exoskeletons (Figure 16.6). Because these polymers constitute the majority of organic carbon, they are the basic food supply available to support heterotrophic activity. The three most common polymers are the plant polymers: cellulose; hemicellulose; and lignin (Table 16.6) (Wagner and Wolf, 1998). The various other polymers produced include: starch (plants); chitin (fungi, arthropods); and

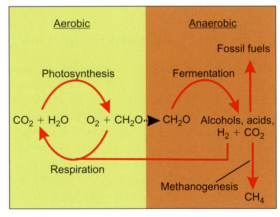

FIGURE 16.4 The carbon cycle, showing both aerobic and anaerobic contributions.

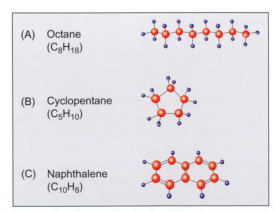

FIGURE 16.5 Examples of petroleum constituents: (A) an alkane; (B) an alicyclic compound; and (C) an aromatic compound. A crude oil contains some of each of these types of compounds but the types and amounts vary in different petroleum reservoirs.

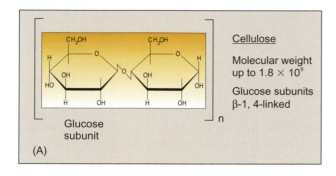

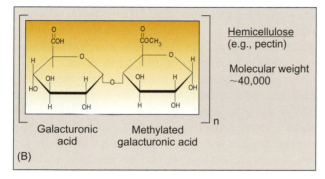

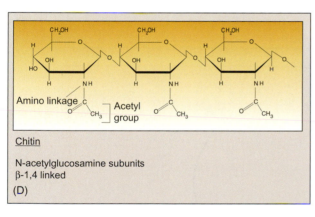

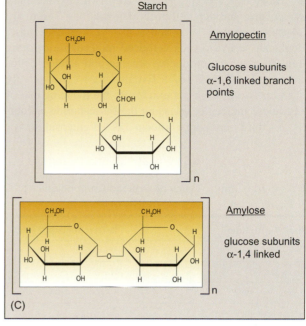

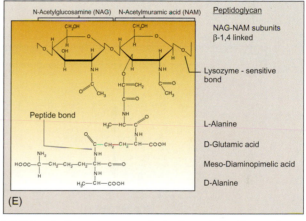

FIGURE 16.6   Common organic polymers found in the environment. (A) Cellulose is the most common plant polymer. It is a linear polymer of β-1,4-linked glucose subunits. Each polymer contains 1000 to 10,000 subunits. (B) Hemicellulose is the second most common polymer. This molecule is more heterogeneous, consisting of hexoses, pentoses and uronic acids. An example of a hemicellulose polymer is pectin. (C) Starch is a polysaccharide synthesized by plants to store energy. Starch is formed from glucose subunits and can be linear (α-1,4 linked), a structure known as amylose, or branched (α-1,4 and α-1,6 linked), known as amylopectin. (D) Chitin is formed from subunits of *N*-acetyl glucosamine linked β-1,4. This polymer is found in fungal cell walls. (E) Bacterial cell walls are composed of polymers of *N*-acetyl glucosamine and *N*-acetylmuramic acid connected by β-1,4 linkages.

peptidoglycan (bacteria). These various polymers can be divided into two groups on the basis of their structures: the carbohydrate-based polymers, which include the majority of the polymers found in the environment, and the phenylpropane-based polymer, lignin.

Carbohydrate-Based Polymers

Cellulose is not only the most abundant of the plant polymers, but it is also the most abundant polymer found on Earth. It is a linear molecule containing β-1,4 linked

glucose subunits (Figure 16.6A). Each molecule contains 1000 to 10,000 subunits with a resulting molecular weight of up to 1.8×10^6. These linear molecules are arranged in microcrystalline fibers that help make up the woody structure of plants. Cellulose is not only a large molecule; it is also insoluble in water. How then do microbial cells get such a large, insoluble molecule across their walls and membranes? The answer is that they have developed an alternative strategy, which is to synthesize and release enzymes, called extracellular enzymes, that can begin the polymer degradation process outside the cell (Deobald and Crawford, 1997). There are two extracellular enzymes that initiate cellulose degradation. These are β-1,4-endoglucanase and β-1,4-exoglucanase. Endoglucanase hydrolyzes cellulose molecules randomly within the polymer, producing smaller and smaller cellulose molecules (Figure 16.7). Exoglucanase consecutively hydrolyzes two glucose subunits from the reducing end of the cellulose molecule, releasing the disaccharide cellobiose. A third enzyme, known as β-glucosidase or cellobiase, then hydrolyzes cellobiose to glucose. Cellobiase can be found as both an extracellular and an intracellular enzyme. Both cellobiose and glucose can be taken up by many bacterial and fungal cells.

Hemicellulose is the second most common plant polymer. This molecule is more heterogeneous than cellulose, consisting of a mixture of several monosaccharides including various hexoses and pentoses as well as uronic acids. In addition, the polymer is branched instead of linear. An example of a hemicellulose polymer is the pectin molecule shown in Figure 16.6B, which contains galacturonic acid and methylated galacturonic acid. Degradation of hemicellulose is similar to the process described for cellulose except that, because the molecule is more heterogeneous, many more extracellular enzymes are involved.

In addition to the two major plant polymers, several other important organic polymers are carbohydrate based. One of these is starch, a polysaccharide synthesized by plants to store energy (Figure 16.6C). Starch is formed from glucose subunits and can be linear (α-1,4 linked), a structure known as amylose, or can be branched (α-1,4 and α-1,6 linked), a structure known as amylopectin. Amylases

TABLE 16.6 Major Types of Organic Components of Plants

Plant Component	% Dry Mass of Plant
Cellulose	15–60
Hemicellulose	10–30
Lignin	5–30
Protein and nucleic acids	2–15

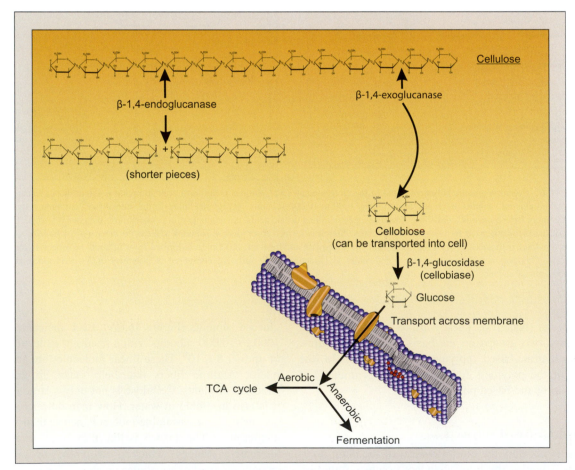

FIGURE 16.7 The degradation of cellulose begins outside the cell with a series of extracellular enzymes called cellulases. The resulting smaller glucose subunit structures can be taken up by the cell and metabolized.

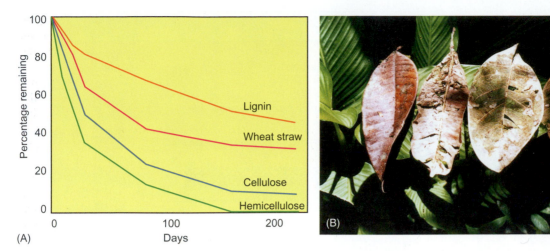

FIGURE 16.8 (A) Decomposition of wheat straw and its major constituents in a silt loam. The initial composition of the wheat straw was 50% cellulose, 25% hemicellulose and 20% lignin. (B) Leaves from a rain forest in different stages of decomposition. Cellulose and hemicellulose are degraded first, leaving the lignin skeleton. (A) Adapted from Wagner and Wolf (1998). Photo taken in Henry Pittier National Park, Venezuela, courtesy C.M. Miller.

(α-1,4-linked exo- and endoglucanases) are extracellular enzymes produced by many bacteria and fungi. Amylases produce the disaccharide maltose, which can be taken up by cells and mineralized. Another common polymer is chitin, which is formed from β-1,4-linked subunits of N-acetylglucosamine (Figure 16.6D). This linear, nitrogen-containing polymer is an important component of fungal cell walls and of the exoskeleton of arthropods. Finally, there is peptidoglycan, a polymer of N-acetylglucosamine and N-acetylmuramic acid, which is an important component of bacterial cell walls (Figure 16.6E).

Lignin

Lignin is the third most common plant polymer, and is strikingly different in structure from all of the carbohydrate-based polymers. The basic building blocks of lignin are the two aromatic amino acids tyrosine and phenylalanine. These are converted to phenylpropene subunits such as coumaryl alcohol, coniferyl alcohol and sinapyl alcohol. Then 500 to 600 phenylpropene subunits are randomly polymerized, resulting in the formation of the amorphous aromatic polymer known as lignin. In plants, lignin surrounds cellulose microfibrils and strengthens the cell wall. Lignin also helps make plants more resistant to pathogens.

Biodegradation of lignin is slower and less complete than degradation of other organic polymers. This is shown experimentally in Figure 16.8A and visually in Figure 16.8B. Lignin degrades slowly because it is constructed as a highly heterogeneous polymer, and in addition contains aromatic residues rather than carbohydrate residues. The great heterogeneity of the molecule precludes the evolution of specific degradative enzymes comparable to those for cellulose. Instead, a nonspecific extracellular enzyme, H_2O_2-dependent lignin peroxidase, is used in conjunction with an extracellular oxidase enzyme that generates H_2O_2. The peroxidase enzyme and H_2O_2 system generate oxygen-based free radicals that react with the lignin polymer to release phenylpropene residues (Morgan *et al.*, 1993). These residues are taken up by microbial cells and degraded as shown in Figure 16.9. Biodegradation of intact lignin polymers occurs only aerobically, which is not surprising because reactive oxygen is needed to release lignin residues. However, once residues are released, they can be degraded under anaerobic conditions.

Phenylpropene residues are aromatic in nature, similar in structure to several types of organic pollutant molecules such as the BTEX (benzene, toluene, ethylbenzene, xylene) and polyaromatic hydrocarbon compounds found in crude oil as well as gasoline and creosote compounds found in wood preservatives (see Chapter 17). These naturally occurring aromatic biodegradation pathways are of considerable importance in the field of bioremediation. In fact, a comparison of the pathway shown in Figure 16.9 with the pathway for degradation of aromatics presented in Chapter 17 shows that they are very similar. Lignin is degraded by a variety of microbes including fungi, actinomycetes and bacteria. The best studied organism with respect to lignin degradation is the white rot fungus *Phanerochaete chrysosporium*. This organism is also capable of degrading several pollutant molecules with structures similar to those of lignin residues (see Information Box 2.6).

16.2.3.2 Humus

Humus was introduced in Chapter 4 and its structure is shown in Figure 4.14. How does humus form? It forms in

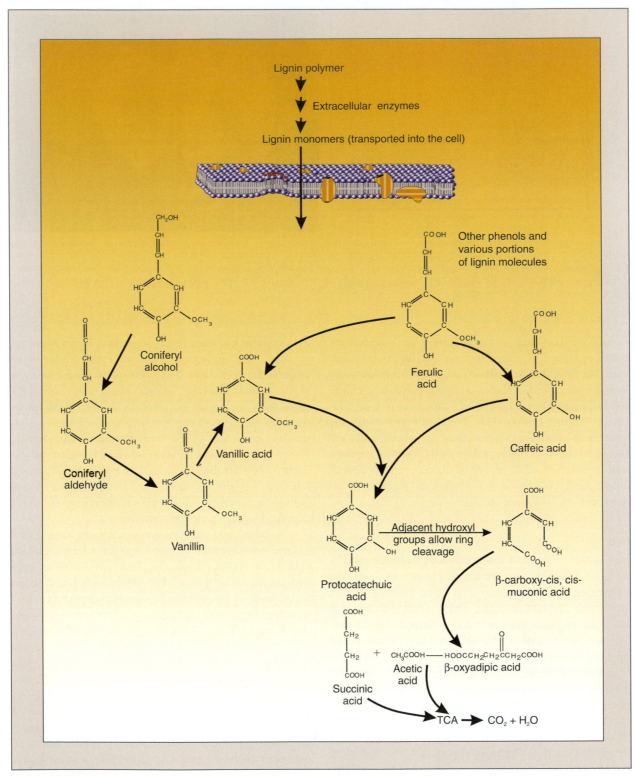

FIGURE 16.9 Lignin degradation. Adapted from Wagner and Wolf (1998).

a two-stage process that involves the formation of reactive monomers during the degradation of organic matter, followed by the spontaneous polymerization of some of these monomers into the humus molecule. Although the majority of organic matter that is released into the environment is respired to form new cell mass and carbon dioxide, a small amount of this carbon becomes available to form humus. To understand this spontaneous process,

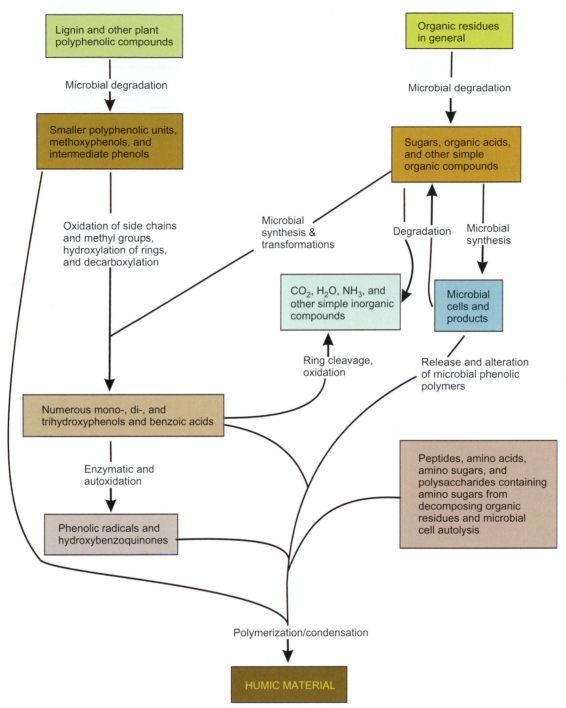

FIGURE 16.10   Possible pathways for the formation of soil humus. Adapted with permission from Wagner and Wolf (1998).

consider the degradation of the common organic polymers found in soil, which were described in the preceding sections. Each of these polymers requires the production of extracellular enzymes that begin the polymer degradation process. In particular, for lignin these extracellular enzymes are nonspecific and produce hydrogen peroxide and oxygen radicals. It is not surprising, then, that some of the reactive residues released during polymer degradation might repolymerize and result in the production of humus. In addition, nucleic acid and protein residues that are released from dying and decaying cells contribute to the pool of molecules available for humus formation. This process is illustrated in Figure 16.10. Considering the wide array of residues that can contribute to humus formation, it is not surprising that humus is even more heterogeneous than lignin. Table 16.7

TABLE 16.7 Chemical Properties of Humus and Lignin

Characteristic	Humic Material	Lignin
Color	Black	Light brown
Methoxyl ($-OCH_3$) content	Low	High
Nitrogen content	3–6%	0%
Carboxyl and phenolic hydroxyl content	High	Low
Total exchangeable acidity (cmol/kg)	≥ 150	≤ 0.5
α-Amino nitrogen	Present	0
Vanillin content	<1%	15–25%

Data from Wagner and Wolf (1998).

TABLE 16.8 Estimates of Methane Released into the Atmosphere

Source	Methane Emission (10^6 Metric Tons/Year)	
Biogenic		
Ruminants	80–100	
Termites	25–150	
Paddy fields	70–120	
Natural wetlands	120–200	
Landfills	5–70	
Oceans and lakes	1–20	
Tundra	1–5	
Abiogenic		
Coal mining	10–35	
Natural gas flaring and venting	10–35	
Industrial and pipeline losses	15–45	
Biomass burning	10–40	
Methane hydrates	2–4	
Volcanoes	0.5	
Automobiles	0.5	
Total	350–820	
Total biogenic	302–665	81–86% of total
Total abiogenic	48–155	13–19% of total

Adapted from Madigan et al. (1997).

compares the different properties of these two complex molecules.

Humus is the most complex organic molecule found in soil, and as a result it is the most stable organic molecule. The turnover rate for humus ranges from 2 to 5% per year, depending on climatic conditions (Wagner and Wolf, 1998). This can be compared with the degradation of lignin shown in Figure 16.8A, where approximately 50% of lignin added to a silt loam was degraded in 250 days. Thus, humus provides a very slowly released source of carbon and energy for indigenous autochthonous microbial populations. The release of humic residues most likely occurs in a manner similar to release of lignin residues. Because the humus content of most soils does not change, the rate of formation of humus must be similar to the rate of turnover. Thus, humus can be thought of as a molecule that is in a state of dynamic equilibrium (Haider, 1992).

16.2.3.3 Methane

Methanogenesis

The formation of methane, methanogenesis, is predominantly a microbial process, although a small amount of methane is generated naturally through volcanic activity (Table 16.8) (Ehrlich, 1996). Methanogenesis is an anaerobic process and occurs extensively in specialized environments including: water-saturated areas such as wetlands and paddy fields; anaerobic niches in the soil; landfills; the rumen; and termite guts. Methane is an end product of anaerobic degradation (see Section 3.4), and as such is associated with petroleum, natural gas and coal deposits.

At present a substantial amount of methane is released to the atmosphere as a result of energy harvesting and utilization. A second way in which methane is released is through landfill gas emissions (see Chapter 31). Although methane makes a relatively minor carbon contribution to the global carbon cycle (compare Table 16.8 with Table 16.2), methane emission is of concern from several environmental aspects. First, like carbon dioxide, methane is a greenhouse gas and contributes to global warming. In fact, it is the second most common greenhouse gas emitted to the atmosphere. Further, it is 22 times more effective than carbon dioxide at trapping heat. Second, localized production of methane in landfills can create safety and health concerns. Methane is an explosive gas at concentrations as small as 5%. Thus, to avoid accidents, the methane generated in a landfill must be managed in some way. If methane is present in concentrations higher than 35%, it can be collected and used for energy. Alternatively, the methane can be burned off

TABLE 16.9 C_1 Compounds of Major Environmental Importance

Compound	Formula	Comments
Carbon dioxide	CO_2	Combustion, respiration and fermentation end product, a major reservoir of carbon on Earth
Carbon monoxide	CO	Combustion product, common pollutant. Product of plant, animal and microbial respiration, highly toxic
Methane	CH_4	End product of anaerobic fermentation or respiration
Methanol	CH_3OH	Generated during breakdown of hemicellulose, fermentation by-product
Formaldehyde	$HCHO$	Combustion product, intermediate metabolite
Formate	$HCOOH$	Found in plant and animal tissues, fermentation product
Formamide	$HCONH_2$	Formed from plant cyanides
Dimethyl ether	CH_3OCH_3	Generated from methane by methanotrophs, industrial pollutant
Cyanide ion	CN^-	Generated by plants, fungi and bacteria. Industrial pollutant, highly toxic
Dimethyl sulfide	$(CH_3)_2S$	Most common organic sulfur compound found in the environment, generated by algae
Dimethyl sulfoxide	$(CH_3)_2SO$	Generated anaerobically from dimethyl sulfide

at concentrations of 15% or higher. However, most commonly, it is simply vented to the atmosphere to prevent it from building up in high enough concentrations to ignite. Although venting landfill gas to the atmosphere does help prevent explosions, it clearly adds to the global warming problem.

The organisms responsible for methanogenesis are a group of obligately anaerobic archaebacteria called the methanogens. The basic metabolic pathway used by the methanogens is:

$$4H_2 + CO_2 \rightarrow CH_4 + 2H_2O \quad \Delta G^{0'} = -130.7 \text{ kJ/mol}$$
$$\text{(Eq. 16.1)}$$

This is an exothermic reaction where CO_2 acts as the TEA and H_2 acts as the electron donor providing energy for the fixation of carbon dioxide. Methanogens that utilize CO_2/H_2 are therefore autotrophic. In addition to the autotrophic reaction shown in Eq. 16.1, methanogens can produce methane during heterotrophic growth on a limited number of other C_1 and C_2 substrates including acetate, methanol and formate. Since there are very few carbon compounds that can be used by methanogens, these organisms are dependent on the production of these compounds by other microbes in the surrounding community. As such, an interdependent community of microbes typically develops in anaerobic environments. In this community, the more complex organic molecules are catabolized by populations that ferment or respire anaerobically, generating C_1 and C_2 carbon substrates as well as CO_2 and H_2 that are then used by methanogens.

Methane Oxidation

Clearly, methane as the end product of anaerobiosis is found extensively in nature. As such, it is an available food source, and a group of bacteria called the methanotrophs have developed the ability to utilize methane as a source of carbon and energy. The methanotrophs are chemoheterotrophic and obligately aerobic. They metabolize methane as shown in Eq. 16.2:

$$\underset{\text{methane}}{CH_4} + O_2 \xrightarrow[\text{monooxygenase}]{\text{methane}} \underset{\text{methanol}}{CH_3OH} \rightarrow \underset{\text{formaldehye}}{HCHO} \quad \text{(Eq. 16.2)}$$
$$\rightarrow \underset{\text{formic acid}}{HCOOH} \rightarrow \underset{\text{carbon dioxide}}{CO_2} + H_2O$$

The first enzyme in the biodegradation pathway is called methane monooxygenase. Oxygenases in general incorporate oxygen into a substrate and are important enzymes in the intial degradation steps for hydrocarbons (see Chapter 17). However, methane monooxygenase is of particular interest because it was the first of a series of enzymes isolated that can cometabolize highly chlorinated solvents such as trichloroethene (TCE) (see Chapter 17). Until this discovery it was believed that biodegradation of highly chlorinated solvents could occur only under anaerobic conditions as an incomplete reaction. The application of methanogens for cometabolic degradation of TCE is a strategy under development for bioremediation of groundwater contaminated with TCE. This is a good illustration of the way in which naturally occurring microbial activities can be harnessed to solve pollution problems.

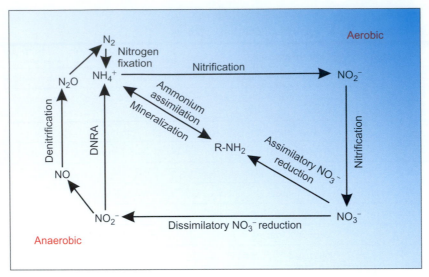

FIGURE 16.11 The nitrogen cycle.

16.2.3.4 Carbon Monoxide and Other C_1 Compounds

Bacteria that can utilize C_1 carbon compounds other than methane are called methylotrophs. There are a number of important C_1 compounds produced from both natural and anthropogenic activities (Table 16.9). One of these is carbon monoxide. The annual global production of carbon monoxide is $3-4 \times 10^9$ metric tons/year (Atlas and Bartha, 1993). The two major carbon monoxide inputs are abiotic. Approximately 1.5×10^9 metric tons/year result from atmospheric photochemical oxidation of carbon compounds such as methane, and 1.6×10^9 metric tons/year results from burning of wood, forests and fossil fuels. A small proportion, 0.2×10^9 metric tons/year, results from biological activity in ocean and soil environments. Carbon monoxide is a highly toxic molecule because it has a strong affinity for cytochromes, and in binding to cytochromes, it can completely inhibit the activity of the respiratory electron transport chain.

Destruction of carbon monoxide can occur abiotically by photochemical reactions in the atmosphere. Microbial processes also contribute significantly to its destruction, even though it is a highly toxic molecule. The destruction of carbon monoxide seems to be quite efficient because the level of carbon monoxide in the atmosphere has not risen significantly since industrialization began, even though CO emissions have increased. The ocean is a net producer of carbon monoxide and releases CO to the atmosphere. In contrast, the terrestrial environment is a net sink for carbon monoxide and absorbs approximately 0.4×10^9 metric tons/year. Key microbes found in terrestrial environments that can metabolize carbon monoxide include both aerobic and anaerobic organisms. Under aerobic conditions *Pseudomonas carboxydoflava* is an example of an organism that oxidizes carbon monoxide to carbon dioxide:

$$CO + H_2O \rightarrow CO_2 + H_2 \qquad \text{(Eq. 16.3)}$$

$$2H_2 + O_2 \rightarrow 2H_2O \qquad \text{(Eq. 16.4)}$$

This organism is a chemoautotroph and fixes the CO_2 generated in Eq. 16.3 into organic carbon. The oxidation of the hydrogen produced provides energy for CO_2 fixation (see Eq. 16.4). Under anaerobic conditions, methanogenic bacteria can reduce carbon monoxide to methane:

$$CO + 3H_2 \rightarrow CH_4 + H_2O \qquad \text{(Eq. 16.5)}$$

A number of other C_1 compounds support the growth of methylotrophic bacteria (Table 16.9). Many, but not all, methylotrophs are also methanotrophic. Both types of bacteria are widespread in the environment because these C_1 compounds are ubiquitous metabolites. In response to their presence, microbes have evolved the capacity to metabolize them under either aerobic or anaerobic conditions.

16.3 NITROGEN CYCLE

In contrast to carbon, elements such as nitrogen, sulfur and iron are taken up in the form of mineral salts and cycle oxidoreductively. For example, nitrogen can exist in numerous oxidation states, from -3 in ammonium (NH_4^+) to $+5$ in nitrate (NO_3^-). These element cycles are referred to as the mineral cycles. The best studied and most complex of the mineral cycles is the nitrogen cycle (Figure 16.11). There is great interest in the nitrogen cycle because nitrogen is the mineral nutrient most in demand by microorganisms and plants. It is the fourth most common element found in cells, making up approximately 12% of cell dry weight, and includes the microbially catalyzed processes of nitrogen fixation, ammonium oxidation (aerobic nitrification and anaerobic anammox),

TABLE 16.10 Global Nitrogen Reservoirs

Nitrogen Reservoir	Metric Tons Nitrogen	Actively Cycled
Atmosphere		
N_2	3.9×10^{15}	No
Ocean		
Biomass	5.2×10^8	Yes
Dissolved and particulate organics	3.0×10^{11}	Yes
Soluble salts (NO_3^-, NO_2^-, NH_4^+)	6.9×10^{11}	Yes
Dissolved N_2	2.0×10^{13}	No
Land		
Biota	2.5×10^{10}	Yes
Organic matter	1.1×10^{11}	Slow
Earth's crust[a]	7.7×10^{14}	No

Adapted from Dobrovolsky (1994).
[a]This reservoir includes the entire lithosphere found in either terrestrial or ocean environments.

TABLE 16.11 Inputs of Reactive Nitrogen from Natural and Anthropogenic Sources

Source	Nitrogen Fixation (Metric Tons/Year)
Natural nitrogen fixation	
Terrestrial	11×10^7
Aquatic	14×10^7
Lightning	1×10^7
Anthropogenic nitrogen fixation	
Terrestrial (managed farming)	4.6×10^7
Fertilizer manufacture	13.6×10^7
Anthropogenic fixed nitrogen mobilization	
Fossil fuel burning	2.5×10^7
Biomass burning	4×10^7
Wetland draining	1×10^7
Land clearing	2×10^7

assimilatory and dissimilatory nitrate reduction, ammonification and ammonium assimilation. Similarly to the carbon cycle, the global nitrogen cycle is currently undergoing major changes due to the ever-increasing demand for nitrogen in both agriculture and industry, fossil fuel burning and land use changes (Galloway *et al.*, 2008). These perturbations also have major consequences for the environment.

16.3.1 Nitrogen Reservoirs

Nitrogen in the form of the inert gas, dinitrogen (N_2), has accumulated in Earth's atmosphere since the planet was formed. Nitrogen gas is continually released into the atmosphere from volcanic and hydrothermal eruptions, and is one of the major global reservoirs of nitrogen (Table 16.10). A second major reservoir is the nitrogen that is found in Earth's crust as bound, nonexchangeable ammonium. Neither of these reservoirs is actively cycled; the nitrogen in Earth's crust is unavailable, and the N_2 in the atmosphere must be fixed before it is available for biological use. Nitrogen fixation is an energy-intensive and relatively slow process carried out by a limited number of microorganisms. Consequently, the amount of fixed nitrogen available is a limiting factor in primary production, and the nitrogen cycle is closely coupled to the carbon cycle. The pool of fixed nitrogen available can be divided into small reservoirs including the organic nitrogen found in living biomass and in dead organic matter and soluble inorganic nitrogen salts (Table 16.10).

These small reservoirs tend to be actively cycled, particularly because nitrogen is often a limiting nutrient in the environment. For example, soluble inorganic nitrogen salts in terrestrial environments can have turnover rates greater than once per day. Nitrogen in plant biomass turns over approximately once a year, and nitrogen in organic matter turns over once in several decades.

16.3.2 Nitrogen Fixation

Ultimately, all fixed forms of nitrogen, NH_4^+, NO_3^- and organic N, come from atmospheric N_2. The relative contributions to nitrogen fixation from undisturbed terrestrial and aquatic environments are compared with those influenced by human activity in Table 16.11 (Canfield *et al.*, 2010). From this table one can see that approximately two-thirds of the N_2 fixed annually is microbial, half from terrestrial environments, including both natural systems and managed agricultural systems, and half from marine ecosystems. The remaining third comes from the manufacture of fertilizers. Recall that nitrogen fixation is energy intensive whether microbial or manufactured, and as a result, fertilizer prices are tied to the price of fossil fuels. As fertilizers are expensive, management alternatives to fertilizer addition have become attractive. These include rotation between nitrogen-fixing crops such as soybeans and nonfixing crops such as corn. Wastewater reuse is another alternative that has become especially popular in the desert southwestern United States for nonfood crops and uses, such as cotton and golf courses, where both water and nitrogen are limiting (see Chapter 27).

TABLE 16.12 Representative Genera of Free-Living Nitrogen Fixers

Status with Respect to Oxygen	Mode of Energy Generation	Genus
Aerobic	Heterotrophic	Azotobacter Beijerinckia Acetobacter Pseudomonas
Facultatively anaerobic	Heterotrophic	Klebsiella Bacillus
Microaerophilic	Heterotrophic	Xanthobacter Azospirillum
Strictly anaerobic	Autotrophic Heterotrophic	Thiobacillus Clostridium Desulfovibrio
Aerobic	Phototrophic (cyanobacteria)	Anabaena Nostoc
Facultatively anaerobic	Phototrophic (bacteria)	Rhodospirillum
Strictly anaerobic	Phototrophic (bacteria)	Chlorobium Chromatium

TABLE 16.13 Rates of Nitrogen Fixation

N_2-Fixing System	Nitrogen Fixation (kg N/hectare/ year)
Rhizobium—legume	200–300
Anabaena—Azolla	100–120
Cyanobacteria—moss	30–40
Rhizosphere associations	2–25
Free-living	1–2

Nitrogen is fixed into ammonia (NH_3) by over 100 different free-living bacteria, both aerobic and anaerobic, as well as some actinomycetes and cyanobacteria (Table 16.12). For example, *Azotobacter* (aerobic), *Beijerinckia* (aerobic), *Azospirillum* (facultative) and *Clostridium* (anaerobic) can all fix N_2. Because fixed nitrogen is required by all biological organisms, nitrogen-fixing organisms occur in most environmental niches. The amount of N_2 fixed in each niche depends on the environment (Table 16.13). Free-living bacterial cells that are not in the vicinity of a plant root fix small amounts of nitrogen (1 to 2 kg N/hectare/year). Bacterial cells associated with the nutrient-rich rhizosphere environment can fix larger amounts of N_2 (2 to 25 kg N/hectare/year). Cyanobacteria are the predominant N_2-fixing organisms in aquatic environments, and because they are photosynthetic, N_2 fixation rates are one to two orders of magnitude higher than for free-living nonphotosynthetic bacteria. An evolutionary strategy developed collaboratively by plants and microbes to increase N_2 fixation efficiency was to enter into a symbiotic or mutualistic relationship to maximize N_2 fixation (see Information Box 16.2). The best studied of these symbioses is the *Rhizobium*—legume relationship, which can increase N_2 fixation to 200 to 300 kg N/hectare/year. This symbiosis irrevocably changes both the plant and the microbe involved but is beneficial to both organisms.

As the various transformations of nitrogen are discussed in this section, the objective is to understand how they are interconnected and controlled. As already mentioned, N_2 fixation is limited to bacteria and is an energy-intensive process (see Information Box 16.3). Therefore, it does not make sense for a microbe to fix N_2 if sufficient amounts are present for growth. Thus, one control on this part of the nitrogen cycle is that ammonia, the end product of N_2 fixation, is an inhibitor for the N_2-fixation reaction. A second control in some situations is the presence of oxygen. Nitrogenase is extremely oxygen sensitive, and some free-living aerobic bacteria fix N_2 only at reduced oxygen tension. Other bacteria such as *Azotobacter* and *Beijerinckia* can fix N_2 at normal oxygen tension because they have developed mechanisms to protect the nitrogenase enzyme.

Summary for Nitrogen Fixation

- N_2 fixation is energy intensive
- End product of N_2 fixation is ammonia
- N_2 fixation is inhibited by ammonia
- Nitrogenase is O_2 sensitive; some free-living N_2 fixers require reduced O_2 tension

16.3.3 Ammonia Assimilation (Immobilization) and Ammonification (Mineralization)

The end product of N_2 fixation is ammonia. In the environment, there exists an equilibrium between ammonia (NH_3) and ammonium (NH_4^+) that is driven by pH. This equilibrium favors ammonium formation at acid or near-neutral pH. Thus, it is generally the ammonium form that is assimilated by cells into amino acids to form proteins, cell wall components such as *N*-acetylmuramic acid and purines and pyrimidines to form nucleic acids. This

Information Box 16.2   The Legume–*Rhizobia* Symbiosis

Gram-negative heterotrophic bacteria classified within the genera *Rhizobium, Bradyrhizobium, Sinorhizobium,* and *Azorhizobium* can form an intriguing symbiosis with leguminous plants. This symbiosis has been well studied because it can meet or nearly meet the nitrogen requirements of economically important legume crops, reducing or eliminating the need for fertilizer application. Grain legumes such as peas, beans, and soybeans can fix about 50% of their total nitrogen requirements, whereas forage legumes such as alfalfa or clover are even more efficient, achieving nitrogen fixation rates as high as 200 to 300 kg per hectare per year. The formation of root nodules on the plant host root system is the result of subtle interactions between the host and the rhizobial endosymbiont (Jones *et al.*, 2007). The plant releases bacteria-specific phenolic compounds called flavonoids that in turn signal rhizobia to produce plant-specific lipo-chitooligosaccharide compounds called Nod factors. The Nod factors activate a series of host plant responses that prepare the plant to form infection threads, which the invading bacteria use to enter the plant. The infection thread is a thin tubule that penetrates into the plant cortex. Plant cells in the cortex receive the invading bacteria, after which they mature into structures known as symbiosomes. The bacteria in the symbiosome then differentiate into bacteroid cells that are basically nitrogen fixation factories. Also contributing to the success of this amazing invasion process is the fact that the rhizobia can evade all host plant immune responses.

As rhizobia are released from the infection thread, cell division occurs, and a visible root nodule begins to be seen (1 to 2 weeks after infection). Within the root nodule, the rhizobia enlarge and

elongate to perhaps five times the normal size of rhizobia and change physiologically to forms known as bacter-oids. The nodule also contains leghemoglobin, which protects the nitrogenase enzyme within the bacteroids from the presence of oxygen. The leghemoglobin imparts a pink color to the interior of the nodule and is indicative of active nitrogen fixation. Thus, examination of the interior of a nodule allows instant determination of whether the nodule is active. Prior to the end of the growing season, nodules begin to break down or senesce, at which point they appear white, green, or brown.

Commercial legume crops are often aided in terms of nitrogen fixation through the application of rhizobial inoculants. This is particularly important when a new legume species is introduced into soils that are free of indigenous rhizobia. In this case, rhizobia introduced into the soil through the use of inocu-lants tend to establish themselves in the available ecological niche and are difficult to displace by any subsequently introduced rhizobia. Therefore, in such situations it is important that the originally introduced rhizobia are appropriate. The desired characteristics include being infective (capable of causing nodule initiation and development), effective (capable of efficient nitrogen fixation), competitive (capable of causing nodule initiation in the presence of other rhizobia), and persistent (capable of surviving in soil between crops in successive years). Usually rhizobia are impregnated into some kind of peat-based carrier with approximately 10^9 rhizobia per gram of peat. Production of commercial rhizobium inoculants is a big business globally.

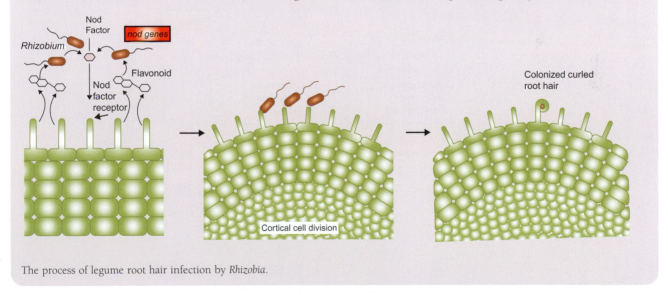

The process of legume root hair infection by *Rhizobia*.

process is known as ammonium assimilation or immobilization. Nitrogen can also be immobilized by the uptake and incorporation of nitrate into organic matter, a process known as assimilatory nitrate reduction (Section 16.3.5.1). Because nitrate must be reduced to ammonium before it is incorporated into organic

molecules, most organisms prefer to take up nitrogen as ammonium if it is available. The process that reverses immobilization, the release of ammonia from dead and decaying cells, is called ammonification or ammonium mineralization. Both immobilization and mineralization of nitrogen occur under aerobic and anaerobic conditions.

Information Box 16.3   The Process of Nitrogen Fixation

Nitrogen gas is a very stable molecule that requires a large amount of energy (946 kJ per mole) to fix into ammonia. The energy for this process arises from the oxidation of carbon sources in the case of heterotrophs or from light in the case of photosynthetic diazotrophs. A diazotroph is a microorganism that can fix nitrogen and so does not require fixed nitrogen for growth. Central to biological nitrogen fixation is the enzyme complex nitrogenase, which is shown below (Zhao et al., 2006). The overall nitrogenise complex consists of two protein components, which in turn consist of multiple subunits. The iron protein termed dinitrogenase reductase is thought to function in the reduction of the molybdenum—iron protein dinitrogenase, which reduces nitrogen gas to ammonia. In the 1980s, it was shown that some nitrogenase enzymes do not contain molybdenum and that vanadium and other metals can substitute for molybdenum (Premakumar et al., 1992). To date, three nitrogenase systems have been identified including Nif, Vnf, and Anf, all of which are sensitive to and inhibited by oxygen. One oxygen-insensitive system has been reported from the actinomycete *Streptomyces thermoautotrophicus*.

The dinitrogenase reductase and dinitrogenase form a complex during which an electron is transferred and the two MgATPs are hydrolyzed to MgADP+ inorganic phosphate (Pi). The two proteins then dissociate and the process repeats. After the dinitrogenase protein has collected sufficient electrons, it binds a molecule of nitrogen gas and reduces it, producing ammonia and hydrogen gas. Thus, during reduction of one N_2 molecule, the two proteins must complex and then dissociate a total of eight times. This is the rate-limiting step of the process and takes considerable time. In fact, it takes 1.25 seconds for a molecule of enzyme to reduce one molecule of N_2. This is why nitrogen-fixing bacteria require a great deal of the nitrogenase enzyme, which can constitute 10—40% of the bacterial cell's proteins.

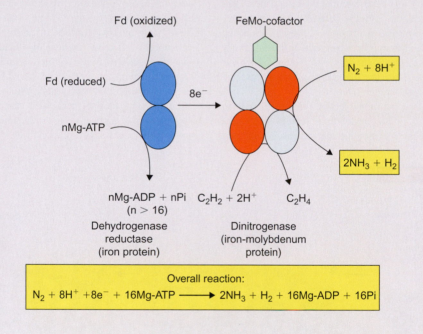

Overall reaction:
$$N_2 + 8H^+ + 8e^- + 16Mg\text{-}ATP \longrightarrow 2NH_3 + H_2 + 16Mg\text{-}ADP + 16Pi$$

16.3.3.1 Ammonia Assimilation (Immobilization)

There are two pathways that microbes use to assimilate ammonium. The first is a reversible reaction that incorporates or removes ammonium from the amino acid glutamate (Figure 16.12A). This reaction is driven by ammonium availability. At high ammonium concentrations (>0.1 mM or >0.5 mg N/kg soil), in the presence of reducing equivalents (reduced nicotinamide adenine dinucleotide phosphate, $NADPH_2$), ammonium is incorporated into α-ketoglutarate to form glutamate using the GOGAT pathway. However, in most soil and many aquatic environments, ammonium is present at low concentrations. Therefore, microbes have a second ammonium uptake pathway that is energy dependent. This reaction is driven by ATP and two enzymes, glutamine synthase and glutamate synthetase (Figure 16.12B). The first step in this reaction adds ammonium to glutamate to form glutamine, and the second step transfers the

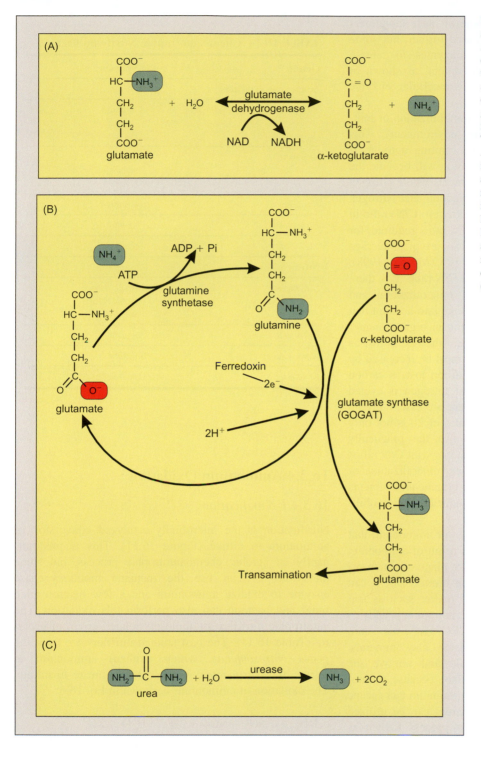

FIGURE 16.12 Pathways of ammonium assimilation and ammonification. Assimilation: (A) The enzyme glutamate dehydrogenase catalyzes a reversible reaction that immobilizes ammonium at high ammonium concentrations. (B) The enzyme system glutamine synthetase— glutamate synthetase (GOGAT), which is induced at low ammonium concentrations. This ammonium uptake system requires ATP energy. Ammonification: Ammonium is released from the amino acid glutamate as shown in (A). (C) Ammonia is also released from urea, a molecule that is found in animal waste and in some fertilizers. Note that the first of these reactions (A and B) occurs within the cell. In contrast, urease enzymes are extracellular enzymes, resulting in release of ammonia to the environment. *Source:* Paul and Clark (1989).

ammonium molecule from glutamine to α-ketoglutarate resulting in the formation of two glutamate molecules.

16.3.3.2 Ammonification (Mineralization)

Ammonium mineralization can occur intracellularly by the reversible reaction shown in Figure 16.12A.

Mineralization reactions can also occur extracellularly. Microorganisms release a variety of extracellular enzymes that initiate degradation of plant polymers. Microorganisms also release a variety of enzymes including proteases, lysozymes, nucleases and ureases that can initiate degradation of nitrogen-containing molecules found outside the cell including proteins, cell walls, nucleic acids and urea. Some of these monomers are taken up by the cell and degraded further, but some are acted upon by extracellular enzymes to release ammonium directly into the environment, as shown in Figure 16.12C for urea and the extracellular enzyme urease.

Which of these two processes, immobilization or mineralization, predominates in the environment? This depends on whether nitrogen is the limiting nutrient. If nitrogen is limiting, then immobilization will become the more important process. For environments where nitrogen is not limiting, mineralization will predominate. Nitrogen limitation is dictated by the carbon/nitrogen (C/N) ratio in the environment. Generally, the C/N ratio required for bacteria is 4 to 5 and for fungi is 10. So, a typical average C/N ratio for soil microbial biomass is 8 (Myrold, 1998). It would then seem logical that at the C/N ratio of 8, there would be a net balance between mineralization and immobilization. However, one must take into account that only approximately 40% of the carbon in organic matter is actually incorporated into cell mass (the rest is lost as carbon dioxide). Thus, the C/N ratio must be increased by a factor of 2.5 to account for the carbon lost as carbon dioxide during respiration. Note that nitrogen is cycled more efficiently than carbon, and there are essentially no losses in its uptake. In fact, a C/N ratio of 20 is not only the theoretical balance point but also the practically observed one. When organic amendments with C/N ratios less than 20 are added to soil, net mineralization of ammonium occurs. In contrast, when organic amendments with C/N ratios greater than 20 are added, net immobilization occurs.

There are numerous possible fates for ammonium that is released into the environment as a result of ammonium mineralization. It can be taken up by plants or microorganisms and incorporated into living biomass, or it can become bound to nonliving organic matter such as soil colloids or humus. In this capacity, ammonium adds to the cation-exchange capacity (CEC) of the soil. Ammonium can become fixed inside clay minerals, which essentially trap the molecule and remove the ammonium from active cycling. Also, because ammonium is volatile, some mineralized ammonium can escape into the atmosphere. Finally, ammonium can be utilized by chemoautotrophic microbes in a process known as nitrification.

Summary for Ammonia Assimilation and Mineralization

- Assimilation and mineralization cycles ammonia between its organic and inorganic forms
- Assimilation predominates at C:N ratios > 20; mineralization predominates at C:N ratios < 20

TABLE 16.14 Chemoautotrophic Nitrifying Bacteria

Genus	Species
Ammonium Oxidizers	
Nitrosomonas	europaea
	eutrophus
	marina
Nitrosococcus	nitrosus
	mobilis
	oceanus
Nitrosospira	briensis
Nitrosolobus	multiformis
Nitrosovibrio	tenuis
Nitrite oxidizers	
Nitrobacter	winogradskyi
	hamburgensis
	vulgaris
Nitrospina	gracilis
Nitrococcus	mobilis
Nitrospira	marina

16.3.4 Ammonium Oxidation

16.3.4.1 Nitrification

Nitrification is the microbially catalyzed conversion of ammonium to nitrate (Figure 16.11). This is predominantly an aerobic chemoautotrophic process, but some methylotrophs can use the methane monooxygenase enzyme to oxidize ammonium and a few heterotrophic fungi and bacteria can also perform this oxidation. The autotrophic nitrifiers are a closely related group of bacteria (Table 16.14). The best studied nitrifiers are from the genus *Nitrosomonas*, which oxidizes ammonium to nitrite, and *Nitrobacter*, which oxidizes nitrite to nitrate. The oxidation of ammonium is shown in Eq. 16.6:

$$NH_4^+ + 1.5O_2 \rightarrow NO_2^- + 2H^+ + H_2O \quad \Delta G^{0'} = -267.5 \text{ kJ/mol} \quad \text{(Eq. 16.6)}$$

This is a two-step energy-producing reaction, and the energy produced is used to fix carbon dioxide. There are two things to note about this reaction. First, it is an inefficient reaction requiring 34 moles of ammonium to fix 1 mole of carbon dioxide. Second, the first step of this reaction is catalyzed by the enzyme ammonium monooxygenase. This first step is analogous to Eq. 16.2, where the enzyme methane monooxygenase initiates oxidation of methane. Similarly to methane monooxygenase, ammonium monooxygenase has broad substrate specificity, and can be used to oxidize pollutant molecules such as TCE cometabolically (see Section 17.6.2.1).

The second step in nitrification, shown in Eq. 16.7, is even less efficient than the first step, requiring approximately 100 moles of nitrite to fix 1 mole of carbon dioxide:

$$NO_2^- + 0.5O_2 \rightarrow NO_3^- \quad \Delta G^{0'} = -87 \text{ kJ/mol} \quad \text{(Eq. 16.7)}$$

These two types of nitrifiers, i.e., those that carry out the reactions shown in Eqs. 16.6 and 16.7, are generally found together in the environment. As a result, nitrite does not normally accumulate in the environment. Nitrifiers are sensitive populations. The optimum pH for nitrification is 6.6 to 8.0. In environments with pH <6.0, nitrification rates are slowed, and below pH 4.5, nitrification seems to be completely inhibited.

Heterotrophic microbes that oxidize ammonium include some fungi and some bacteria. These organisms gain no energy from nitrification, so it is unclear why they carry out the reaction. The relative importance of autotrophic and heterotrophic nitrification in the environment has not yet been clearly determined. Although the measured rates of autotrophic nitrification in the laboratory are an order of magnitude higher than those of heterotrophic nitrification, some data for acidic forest soils have indicated that heterotrophic nitrification may be more important in such environments (Myrold, 1998).

Nitrate does not normally accumulate in natural, undisturbed ecosystems. There are several reasons for this. One is that nitrifiers are sensitive to many environmental stresses. But perhaps the most important reason is that natural ecosystems do not have much excess ammonium. However, in agricultural systems that have large inputs of fertilizer, nitrification can become an important process resulting in the production of large amounts of nitrate. Other examples of managed systems that result in increased nitrogen inputs into the environment are feedlots, septic tanks and landfills. The nitrogen released from these systems also becomes subject to nitrification processes. Because nitrate is an anion (negatively charged), it is very mobile in soil systems, which also have an overall net negative charge. Therefore, nitrate moves easily with water and this results in nitrate leaching into groundwater and surface waters. There are several health concerns related to high levels of nitrate in groundwater, including methemoglobinemia and the formation of nitrosamines. High levels of nitrate in surface waters can also lead to eutrophication and the degradation of surface aquatic systems.

Summary for Nitrification

- Nitrification is an aerobic process that produces nitrite and nitrate
- Nitrification is sensitive to a variety of chemical inhibitors and is inhibited at low pH
- Nitrification in managed systems can result in nitrate leaching and groundwater contamination

16.3.4.2 Anammox

A relative recent discovery is that ammonium oxidation can also occur under anaerobic conditions using nitrite as the terminal electron acceptor (Kuenen, 2008). This process, known as anammox, is the bacterial oxidation of ammonium using nitrite as the electron acceptor resulting in the production of nitrogen (Eq. 16.8):

$$NH_4^+ + NO_2^- \rightarrow N_2 + NO_3^- \quad \Delta G^{0'} = -356 \text{ kJ/mol}$$
$$\text{(Eq. 16.8)}$$

Anammox was first discovered in a bench-scale wastewater treatment reactor. Since then it has been observed in a wide variety of both aquatic and terrestrial ecosystems with naturally low levels of oxygen. These include marine and freshwater sediments and anoxic water columns, anoxic terrestrial environments, as well as a number of managed systems including wastewater treatment plants, aquaculture and landfill leachate treatment systems (Thamdrup, 2012). All anammox bacteria identified are associated with the phylum Planctomycetes. This phylum is characterized by extremely slow growth rates and in addition has internal membrane-bound structures. In anammox bacteria one such structure is the "anammoxosome," the organelle where the ammonium oxidation and energy generation reactions take place. Anammox bacteria have not yet been isolated in pure culture (see also Chapter 2).

Anammox bacteria require habitats where both ammonium and nitrite are present. This occurs in the vicinity of aerobic–anaerobic interfaces. At these interfaces, ammonium is released both from the degradation of organic matter (either aerobically or anaerobically) and from dissimilatory nitrate reduction to ammonium (DNRA), an anaerobic process. Nitrite can be produced from nitrate reduction (anaerobic) and also from ammonium that has diffused to an oxic region and then undergone nitrification (Figure 16.11). Thus, competition for anammox substrates occurs at the aerobic–anaerobic interface making this an extremely complex activity.

An immediate and useful application for annamox that was recognized early is the removal of excess ammonium in wastewater treatment. Traditionally, this was done in a two-step process using nitrifying bacteria to oxidize the ammonium to nitrate under aerobic conditions, and then switching to anoxic conditions to allow denitrification to reduce the nitrate to N_2 (Figure 16.11). This is an energy intensive process and can be replaced by an anammox process. Anammox requires that the first step of nitrification take place (using ammonia oxidizing bacteria) converting some of the ammonium to nitrite (Eq. 16.6). After the first step of nitrification occurs, anammox substrates, nitrite and ammonium are all present. Anammox bacteria can then proceed (at approximately a 1:1 substrate ratio) resulting in the production of N_2. Research has shown that these steps can take place together in a reactor at one-third to one-half the cost of traditional ammonium removal.

A second discovery was made after anammox was first described. It is now thought that a significant portion (30 to 50%) of the nitrogen loss in ocean environments that had been attributed to denitrification is actually due to anammox. In particular, in ocean zones where oxygen is less than 0.64 mg/L, regions termed oxygen minimum zones, anammox is now thought to dominate as the main cause of fixed nitrogen loss. This is of concern because scientists have suggested that ocean oxygen minimum zones are expanding as a result of climate change. This may in turn cause increased losses of fixed nitrogen through anammox, and a reduction in this very important nutrient in ocean environments.

Summary for Anammox

- Anammox is an anaerobic chemolithoautotrophic process
- Anammox results in the loss of fixed nitrogen to dinitrogen gas
- Anammox is important in ammonium removal in wastewater treatment, and in the loss of fixed nitrogen from oxygen minimum zones in the ocean

16.3.5 Nitrate Reduction

What are the possible fates of nitrate in the environment? Nitrate leaching into groundwater and surface waters is one possible fate. In addition, nitrate can be taken up and incorporated into living biomass by plants and microorganisms. The uptake of nitrate is followed by its reduction to ammonium, which is then incorporated into biomass. This process is called assimilatory nitrate reduction or nitrate immobilization. Finally, microorganisms can utilize nitrate as a terminal electron acceptor in anaerobic respiration to drive the oxidation of organic compounds. There are two separate pathways for this dissimilatory process, one called dissimilatory nitrate reduction to ammonium, where ammonium is the end product, and one called denitrification, where a mixture of gaseous products including N_2 and N_2O is formed.

16.3.5.1 Assimilatory Nitrate Reduction

Assimilatory nitrate reduction refers to the uptake of nitrate, its reduction to ammonium and its incorporation into biomass (see Figure 16.12A and B). Most microbes utilize ammonium preferentially, when it is present, to avoid having to reduce nitrate to ammonium, a process requiring energy. So, if ammonium is present in the environment, assimilatory nitrate reduction is suppressed. Oxygen does not inhibit this activity. In contrast to microbes, for plants that are actively photosynthesizing and producing energy, the uptake of nitrate for assimilation is less problematic in terms of energy. In fact, because nitrate is much more mobile than ammonium, it is possible that in the vicinity of the plant roots, nitrification of ammonium to nitrate makes nitrogen more available for plant uptake. Because this process incorporates nitrate into biomass, it is also known as nitrate immobilization (see Section 16.3.3).

Summary for Assimilatory Nitrate Reduction

- Nitrate taken up must be reduced to ammonia before it is assimilated
- Ammonia inhibits this process
- O_2 does not inhibit this process

16.3.5.2 Dissimilatory Nitrate Reduction

Dissimilatory Nitrate Reduction to Ammonium

There are two separate dissimilatory nitrate reduction processes both of which are used by facultative chemoheterotrophic organisms under microaerophilic or anaerobic conditions (Figure 16.11). The first process, called dissimilatory nitrate reduction to ammonium (DNRA), uses nitrate as a terminal electron acceptor to produce energy to drive the oxidation of organic compounds. The end product of DNRA is ammonium:

$$NO_3^- + 4H_2 + 2H^+ \rightarrow NH_4^+ + 3H_2O$$
$$\Delta G^{0'} = -603 \text{ kJ/8e}^- \text{ transfer}$$

(Eq. 16.9)

The first step in this reaction, the reduction of nitrate to nitrite, is the energy-producing step. The further reduction of nitrite to ammonium is catalyzed by an NADH-dependent reductase. This second step provides no additional energy, but it does conserve fixed nitrogen and also regenerates reducing equivalents through the reoxidation of $NADH_2$ to NAD. These reducing equivalents are then used to help in the oxidation of carbon substrates. In fact, it has been demonstrated that under carbon-limiting conditions, nitrite accumulates (denitrification predominates), while under carbon-rich conditions, ammonium is the major product (DNRA predominates). A second environmental factor that selects for DNRA is low levels of available electron acceptors. It is therefore not surprising that this process is found predominantly in saturated, carbon-rich environments such as stagnant water, sewage sludge, some high-organic-matter sediments and the rumen. Table 16.15 lists a variety of bacteria that perform DNRA. It is interesting to note that most of the bacteria on this list have fermentative rather than oxidative metabolisms.

Summary for Dissimilatory Reduction of Nitrate to Ammonia

- Anaerobic respiration using nitrate as TEA
- Inhibited by O_2
- Not inhibited by ammonia
- Found in a limited number of carbon-rich, TEA-poor environments
- Fermentative bacteria predominate

Denitrification

The second type of dissimilatory nitrate reduction is known as denitrification. Denitrification refers to the microbial reduction of nitrate, through various gaseous inorganic forms, to N_2. This is the primary type of dissimilatory nitrate reduction found in soil, and as such is of concern because it cycles fixed nitrogen back into N_2. This process removes a limiting nutrient from the environment. Further, some of the gaseous intermediates formed during denitrification, e.g., nitrous oxide (N_2O), can cause depletion of the ozone layer and can also act as a greenhouse gas contributing to global warming (see Chapter 31). The overall reaction for denitrification is:

$$NO_3^- + 5H_2 + 2H^+ \rightarrow N_2 + 6H_2O$$
$$\Delta G^{0'} = -888 \text{ kJ/8e}^- \text{ transfer} \qquad \text{(Eq. 16.10)}$$

Denitrification, when calculated in terms of energy produced for every eight-electron transfer, provides more energy per mole of nitrate reduced than DNRA. Thus, in a carbon-limited, electron acceptor-rich environment, denitrification will be the preferred process because it provides more energy than DNRA. The relationship between denitrification and DNRA is summarized in Figure 16.13.

The four steps involved in denitrification are shown in more detail in Figure 16.14. The first step, reduction of nitrate to nitrite, is catalyzed by the enzyme nitrate reductase. This is a membrane-bound molybdenum–iron–sulfur protein that is found not only in denitrifiers but also in

TABLE 16.15 Bacteria that Utilize Dissimilatory Nitrate or Nitrite to Ammonium (DNRA)

Genus	Typical Habitat
Obligate Anaerobes	
Clostridium	Soil, sediment
Desulfovibrio	Sediment
Selenomonas	Rumen
Veillonella	Intestinal tract
Wolinella	Rumen
Facultative Anaerobes	
Citrobacter	Soil, wastewater
Enterobacter	Soil, wastewater
Erwinia	Soil
Escherichia	Soil, wastewater
Klebsiella	Soil, wastewater
Photobacterium	Seawater
Salmonella	Sewage
Serratia	Intestinal tract
Vibrio	Sediment
Microaerophiles	
Campylobacter	Oral cavity
Aerobes	
Bacillus	Soil, food
Neisseria	Mucous membranes
Pseudomonas	Soil, water

Adapted from Tiedje (1988).

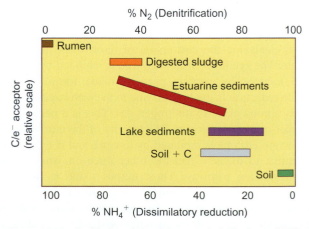

FIGURE 16.13 Partitioning of nitrate between denitrification and DNRA as a function of available carbon/electron (C/e⁻) acceptor ratio. Adapted from J.M. Tiedje in *Biology of Anaerobic Microorganisms*, A.J.B. Zehnder, ed. © 1988. Reprinted by permission of John Wiley & Sons, Inc.

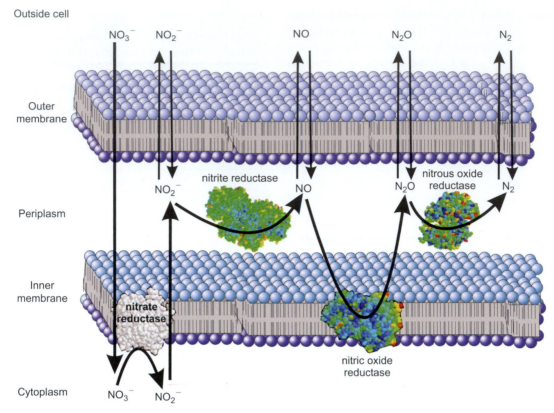

FIGURE 16.14   The denitrification pathway. Adapted from Myrold (1998).

DNRA organisms. Both the synthesis and the activity of nitrate reductase are inhibited by oxygen. Thus, both denitrification and DNRA are inhibited by oxygen. The second enzyme in this pathway is nitrite reductase, which catalyzes the conversion of nitrite to nitric oxide. Nitrite reductase is unique to denitrifying organisms and is not present in the DNRA process. It is found in the periplasm and exists in two forms, a copper-containing form and a heme form, both of which are distributed widely in the environment. Synthesis of nitrite reductase is inhibited by oxygen and induced by nitrate. Nitric oxide reductase, a membrane-bound protein, is the third enzyme in the pathway, catalyzing the conversion of nitric oxide to nitrous oxide. The synthesis of this enzyme is inhibited by oxygen and induced by various nitrogen oxide forms. Nitrous oxide reductase is the last enzyme in the pathway, and converts nitrous oxide to dinitrogen gas. This is a periplasmic copper-containing protein. The activity of the nitrous oxide reductase enzyme is inhibited by low pH, and is even more sensitive to oxygen than the other three enzymes in the denitrification pathway. Thus, nitrous oxide is the final product of denitrification under conditions of high oxygen (in a relative sense, given a microaerophilic niche) and low pH. In summary, both the synthesis and activity of denitrification enzymes are controlled by oxygen. Enzyme activity is more sensitive to oxygen than enzyme synthesis as shown in Figure 16.15. The amount of dissolved oxygen in equilibrium with water at $20°C$ and 1 atm pressure is 9.3 mg/L. However, as little as 0.5 mg/L or less inhibits the activity of denitrification enzymes. Therefore, nitrous oxide reductase is the most sensitive denitrification enzyme, and it is inhibited by dissolved oxygen concentrations of less than 0.2 mg/L.

Whereas the denitrification pathway is very sensitive to oxygen, neither it nor the DNRA pathway is inhibited by ammonium as is the assimilatory nitrate reduction pathway. However, the initial nitrate level in an environmental system can help determine the extent of the denitrification pathway. Low nitrate levels tend to favor production of nitrous oxide as the end product. High nitrate levels favor production of N_2 gas, a much more desirable end product.

Organisms that denitrify are found widely in the environment and display a variety of different characteristics in terms of metabolism and activities. In contrast to DNRA organisms, which are predominantly heterotrophic using fermentative metabolism, the majority of denitrifiers are also heterotrophic, but use respiratory pathways of metabolism. However, as shown in Table 16.16, some denitrifiers are autotrophic, some are fermentative and some are associated with other aspects of the nitrogen cycle; for example, they can fix N_2.

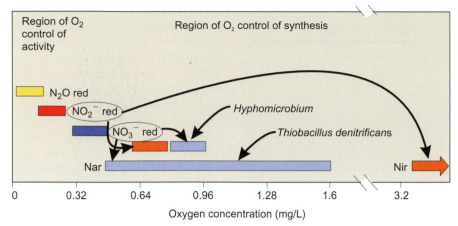

FIGURE 16.15   Approximate regions of oxygen concentration that inhibit the enzyme activity and synthesis for three steps in the denitrification pathway. Adapted from J.M. Tiedje in *Biology of Anaerobic Microorganisms*, A.J.B. Zehnder, ed. © 1998. Reprinted by permission of John Wiley & Sons, Inc.

Summary for Denitrification

- Anaerobic respiration using nitrate as TEA
- Inhibited by O_2
- Not inhibited by ammonia
- Produces a mix of N_2 and N_2O
- Many heterotrophic bacteria are denitrifiers

16.4 SULFUR CYCLE

Sulfur is the tenth most abundant element in Earth's crust. It is an essential element for biological organisms, making up approximately 1% of the dry weight of a bacterial cell (Table 16.1). Sulfur is not generally considered a limiting nutrient in the environment except in some intensive agricultural systems with high crop yields. Sulfur is cycled between oxidation states of +6 for sulfate (SO_4^{2-}) and −2 for sulfide (S^{2-}) (Figure 16.16). In cells, sulfur is required for synthesis of the amino acids cysteine and methionine, and is also required for some vitamins, hormones and coenzymes. In proteins, the sulfur-containing amino acid cysteine is especially important because the formation of disulfide bridges between cysteine residues helps govern protein folding, and hence activity. All of these compounds contain sulfur in the reduced or sulfide form. Cells also contain organic sulfur compounds in which the sulfur is in the oxidized state. Examples of such compounds are glucose sulfate, choline sulfate, phenolic sulfate and two ATP-sulfate compounds that are required in sulfate assimilation, and can also serve to store sulfur for the cell. Although the sulfur cycle is not as complex as the nitrogen cycle, the global impacts of the sulfur cycle are extremely important, including the formation of acid rain, acid mine drainage and corrosion of concrete and metal.

16.4.1 Sulfur Reservoirs

Sulfur is outgassed from Earth's core through volcanic activity. The sulfur gases released, primarily sulfur dioxide (SO_2) and hydrogen sulfide (H_2S), become dissolved in the ocean and aquifers. Here, the hydrogen sulfide forms sparingly soluble metal sulfides, mainly iron sulfide (pyrite), and sulfur dioxide forms metal sulfates with calcium, barium and strontium as shown in Eqs. 16.11 and 16.12:

$$2S^{2-} + Fe^{2+} \rightarrow FeS_2 \ (pyrite) \qquad (Eq.\ 16.11)$$

$$SO_2^{2-} + Ca^{2+} \rightarrow CaSO_4 \ (gypsum) \qquad (Eq.\ 16.12)$$

This results in a substantial portion of the outgassed sulfur being converted to rock. Some of the gaseous sulfur compounds find their way into the upper reaches of the ocean and the soil. In these environments, microbes take up and cycle the sulfur. Finally, the small portions of these gases that remain after precipitating and cycling find their way into the atmosphere. Here, they are oxidized to the water-soluble sulfate form, which is washed out of the atmosphere by rain. Thus, the atmosphere is a relatively small reservoir of sulfur (Table 16.17). Of the sulfur found in the atmosphere, the majority is found as sulfur dioxide. Currently, one-third to one-half of the sulfur dioxide emitted to the atmosphere is from industrial and automobile emissions, due to the burning of fossil fuels. A smaller portion of the sulfur in the atmosphere is present as hydrogen sulfide, and is biological in origin.

TABLE 16.16 Genera of Denitrifying Bacteria

Genus	Interesting Characteristics
Organotrophs	
Alcaligenes	Common soil bacterium
Agrobacterium	Some species are plant pathogens
Aquaspirillum	Some are magnetotactic, oligotrophic
Azospirillum	Associative N_2 fixer, fermentative
Bacillus	Spore former, fermentative, some species thermophilic
Blastobacter	Budding bacterium, phylogenetically related to *Rhizobium*
Bradyrhizobium	Symbiotic N_2 fixer with legumes
Branhamella	Animal pathogen
Chromobacterium	Purple pigmentation
Cytophaga	Gliding bacterium; cellulose degrader
Flavobacterium	Common soil bacterium
Flexibacter	Gliding bacterium
Halobacterium	Halophilic
Hyphomicrobium	Grows on one-C substrates, oligotrophic
Kingella	Animal pathogen
Neisseria	Animal pathogen
Paracoccus	Halophilic, also lithotrophic
Propionibacterium	Fermentative
Pseudomonas	Commonly isolated from soil, very diverse genus
Rhizobium	Symbiotic N_2 fixer with legumes
Wolinella	Animal pathogen
Phototrophs	
Rhodopseudomonas	Anaerobic, sulfate reducer
Lithotrophs	
Alcaligenes	Uses H_2, also heterotrophic, common soil isolate
Bradyrhizobium	Uses H_2, also heterotrophic, symbiotic N_2 fixer with legumes
Nitrosomonas	NH_3 oxidizer
Paracoccus	Uses H_2, also heterotrophic, halophilic
Pseudomonas	Uses H_2, also heterotrophic, common soil isolate
Thiobacillus	S-oxidizer
Thiomicrospira	S-oxidizer
Thiosphaera	S-oxidizer, heterotrophic nitrifier, aerobic denitrification

From Myrold (1998).

The largest reservoir of sulfur is found in Earth's crust and is composed of inert elemental sulfur deposits, sulfur—metal precipitates such as pyrite (FeS_2) and gypsum ($CaSO_4$), as well as sulfur associated with buried fossil fuels. A second large reservoir that is slowly cycled is the sulfate found in the ocean, where it is the second most common anion (Dobrovolsky, 1994). Smaller and more actively cycled reservoirs of sulfur include sulfur found in biomass and organic matter, in the terrestrial and ocean environments. Two recent practices have caused a disturbance in the global sulfur reservoirs. The first is strip mining, which has exposed large areas of metal sulfide ores to the atmosphere, resulting in the formation of acid mine drainage. The second is the burning of fossil fuels, a sulfur reservoir that was quite inert until recently. This has resulted in sulfur dioxide emissions into the atmosphere with the resultant formation of acid rain.

16.4.2 Assimilatory Sulfate Reduction and Sulfur Mineralization

The primary soluble form of inorganic sulfur found in soil is sulfate. Whereas plants and most microorganisms incorporate reduced sulfur (sulfide) into amino acids or other sulfur-requiring molecules, they take up sulfur in the oxidized sulfate form and then reduce it internally (Widdel, 1988). This is called assimilatory sulfate reduction. Cells assimilate sulfur in the form of sulfate because it is the most available sulfur form, and also because sulfide is toxic. Sulfide toxicity occurs inside the cell when sulfide reacts with metals in cytochromes to form metal sulfide precipitates, destroying cytochrome activity. However, under the controlled conditions of sulfate reduction inside the cell, the sulfide can be removed immediately and incorporated into an organic form (see Information Box 16.4). Although this process does protect the cell from harmful effects of the sulfide, it is an energy-consuming reaction. After sulfate is transported inside the cell, ATP is used to convert the sulfate into the energy-rich molecule adenosine 5'-phosphosulfate (APS) (Eq. 16.14). A second ATP molecule is used to transform APS to 3'-phosphoadenosine-5'-phosphosulfate (PAPS) (Eq. 16.15). This allows the sulfate to be reduced to sulfite and then sulfide in two steps (Eqs. 16.16 and 16.17). Most commonly, the amino acid serine is used to remove sulfide as it is reduced, forming the sulfur-containing amino acid cysteine (see Eq. 16.18).

The release of sulfur from organic forms is called sulfur mineralization. The release of sulfur from organic molecules occurs under both aerobic and anaerobic conditions. The enzyme serine sulfhydrylase can remove sulfide from cysteine in the reverse of the reaction shown in Eq. 16.18, or a second enzyme, cysteine sulfhydrylase,

FIGURE 16.16   The sulfur cycle.

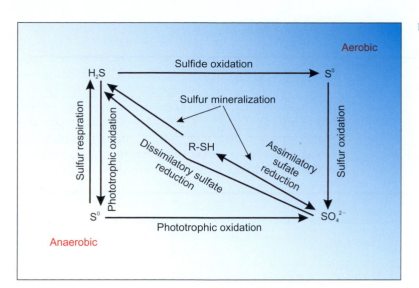

TABLE 16.17   Global Sulfur Reservoirs

Sulfur Reservoir	Metric Tons Sulfur	Actively Cycled
Atmosphere		
SO_2/H_2S	1.4×10^6	Yes
Ocean		
Biomass	1.5×10^8	Yes
Soluble inorganic ions (primarily SO_4^{2-})	1.2×10^{15}	Slow
Land		
Living biomass	8.5×10^9	Yes
Organic matter	1.6×10^{10}	Yes
Earth's crust[a]	1.8×10^{16}	No

Adapted from Dobrovolsky (1994).
[a]*This reservoir includes the entire lithosphere found in either terrestrial or ocean environments.*

Information Box 16.4   Processing of Sulfate for Uptake into Bacteria

$$SO_4^{2-} \text{(outside cell)} \xrightarrow{\text{active transport}} SO_4^{2-} \text{(inside cell)} \quad \text{(Eq. 16.13)}$$
sulfate

$$ATP + SO_4^{2-} \xrightarrow{\text{ATP sulfurylase}} APS + PP_i \quad \text{(Eq. 16.14)}$$
adenosine 5'-phosphosulfate

$$ATP + APS \xrightarrow{\text{APS phosphokinase}} PAPS$$
3'-phosphoadenosine-5'-phosphosulfate
$$\text{(Eq. 16.15)}$$

$$2(R-SH) + PAPS \xrightarrow{\text{PAPS reductase}}$$
thioredoxin (reduced)
$$\text{(Eq. 16.16)}$$
$$\text{sulfite} + PAP + RSSR$$
AP-3'-phosphate   thioredoxin (oxidized)

$$SO_3^{2-} + 3NADPH \rightarrow S^{2-} + 3NADP^+ \quad \text{(Eq. 16.17)}$$
sulfite   sulfide

$$O\text{-acetyl-L-serine} + S^{2-} \xrightarrow{O\text{-acetylserine sulfhydrylase}} \quad \text{(Eq. 16.18)}$$
$$\text{L-cysteine} + \text{acetate} + H_2O$$

can remove both sulfide and ammonia as shown in Eq. 16.19:

$$\text{cysteine} \xrightarrow{\text{cysteine sulfhydrylase}} \text{serine} + H_2S \quad \text{(Eq. 16.19)}$$

In marine environments, one of the major products of algal metabolism is the compound dimethylsulfoniopropionate (DMSP), which is used in osmoregulation of the cell. The major degradation product of DMSP is dimethylsulfide (DMS). Both H_2S and DMS are volatile compounds and therefore can be released to the atmosphere. Once in the atmosphere, these compounds are photooxidized to sulfate (Eq. 16.20):

$$H_2S/DMS \xrightarrow{\text{UV light}} SO_4^{2-} \xrightarrow{H_2O} H_2SO_4 \quad \text{(Eq. 16.20)}$$
sulfuric acid

Normal biological release of reduced volatile sulfur compounds results in the formation of approximately 1 kg SO_4^{2-}/hectare/year. The use of fossil fuels, which all contain organic sulfur compounds, increases the amount of sulfur released to the atmosphere to up to 100 kg SO_4^{2-}/hectare/year in some urban areas. Exacerbating this

TABLE 16.18 Sulfur-Oxidizing Bacteria

Group	Sulfur Conversion	Habitat Requirements	Habitat	Genera
Obligate or facultative chemoautotrophs	$H_2S \rightarrow S^0$ $S^0 \rightarrow SO_4^{2-}$ $S_2O_3^{2-} \rightarrow SO_4^{2-}$	H_2S—O_2 interface	Mud, hot springs, mining surfaces, acid mine drainage, soil	*Acidothiobacillus, Sulfobacillus, Thiomicrospira, Achromatium, Beggiatoa, Thermothrix*
Anaerobic phototrophs	$H_2S \rightarrow S^0$ $S^0 \rightarrow SO_4^{2-}$	Anaerobic, H_2S, light	Shallow water, anaerobic sediments meta- or hypolimnion, anaerobic water	*Chlorobium, Chromatium, Ectothiorhodospira, Thiopedia, Rhodopseudomonas*

Adapted from Germida (1998).

problem is the fact that reserves of fossil fuels that are low in sulfur are shrinking, forcing the use of reserves with higher sulfur content. Burning of fossil fuels produces sulfite as shown in Eq. 16.21:

$$\text{Fossil fuel combustion} \rightarrow SO_2 \xrightarrow{+H_2O} \underset{\text{sulfurous acid}}{H_2SO_3}$$

(Eq. 16.21)

Thus, increased emission of sulfur compounds to the atmosphere results in the formation of sulfur acid compounds. These acidic compounds dissolve in rainwater and can decrease the rainwater pH from neutral to as low as pH 3.5, a process also known as the formation of acid rain. Acid rain damages plant foliage, causes corrosion of stone and concrete building surfaces, and can affect weakly buffered soils and lakes.

Summary for Sulfur Assimilation and Mineralization

- Sulfur is taken up as sulfate rather than sulfide
- Assimilation and mineralization cycles sulfur between its organic and inorganic forms
- Mineralization predominates at C:S ratio <200; assimilation predominates at C:S >400

16.4.3 Sulfur Oxidation

In the presence of oxygen, reduced sulfur compounds can support the growth of a group of chemoautotrophic bacteria under strictly aerobic conditions and a group of photoautrophic bacteria under strictly anaerobic conditions (Table 16.18). In addition, a number of aerobic heterotrophic microbes, including both bacteria and fungi, oxidize sulfur to thiosulfate or to sulfate. The heterotrophic sulfur oxidation pathway is still unclear, but apparently no energy is obtained in this process. Chemoautotrophs are considered the predominant sulfur oxidizers in most environments. However, because many

chemoautotrophic sulfur oxidizers require a low pH for optimal activity, heterotrophs may be more important in some aerobic, neutral to alkaline soils. Further, heterotrophs may initiate sulfur oxidation, resulting in a lowered pH that is more amenable for chemoautotrophic activity.

16.4.3.1 Chemoautotrophic Sulfur Oxidation

Of the chemoautotrophs, most oxidize sulfide to elemental sulfur, which is then deposited inside the cell as characteristic granules (Eq. 16.22):

$$H_2S + 1/2O_2 \rightarrow S^0 + H_2O \quad \Delta G^{0'} = -218 \text{ kJ/mol}$$

(Eq. 16.22)

The energy provided by this oxidation is used to fix CO_2 for cell growth. In examining Eq. 16.22, it is apparent that these organisms require both oxygen and sulfide. However, reduced compounds are generally abundant in areas that contain little or no oxygen. So these microbes are microaerophilic; they grow best under conditions of low oxygen tension (Figure 16.17). Characteristics of marsh sediments that contain these organisms are their black color due to sulfur deposits, and their "rotten egg" smell due to the presence of H_2S. Most of these organisms are filamentous and can easily be observed by examining a marsh sediment under the microscope and looking for small white filaments.

Some chemoautotrophs, most notably *Acidothiobacillus thiooxidans*, can oxidize elemental sulfur as shown in Eq. 16.23:

$$S^0 + 1.5O_2 + H_2O \rightarrow SO_4^{2-} + 2H^+ \quad \Delta G^{0'} = -532 \text{ kJ/mol}$$

(Eq. 16.23)

This reaction produces acid, and as a result, *A. thiooxidans* is extremely acid tolerant with an optimal growth pH of 2. It should be noted that there are various *Acidothiobacillus* species, and these vary widely in

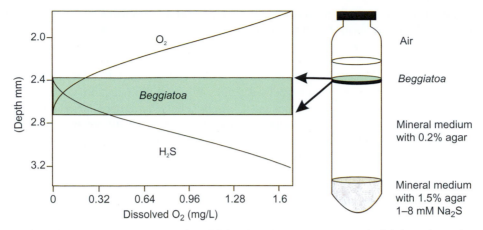

FIGURE 16.17   Cultivation of the sulfur oxidizing chemolithotroph *Beggiatoa*. At right is a culture tube with sulfide agar overlaid with initially sulfide-free soft mineral agar. The air space in the closed tube is the source of oxygen. Stab-inoculated *Beggiatoa* grows in a narrowly defined gradient of H_2S and oxygen as shown. Adapted with permission from *Microbial Ecology* by R.M. Atlas and R. Bartha. © 1993, by Benjamin Cummings.

their acid tolerance (Baker and Banfield, 2003). However, the activity of *A. thiooxidans* in conjunction with the iron-oxidizing, acid-tolerant, chemoautotroph *Acidothiobacillus ferrooxidans* is responsible for the formation of acid mine drainage, an undesirable consequence of sulfur cycle activity. It should be noted that the same organisms can be harnessed for the acid leaching and recovery of precious metals from low-grade ore, also known as biometallurgy. Thus, depending on one's perspective, these organisms can be very harmful or very helpful.

Although most of the sulfur-oxidizing chemoautotrophs are obligate aerobes, there is one exception, *Acidothiobacillus denitrificans*, a facultative anaerobic organism that can substitute nitrate as a terminal electron acceptor for oxygen as shown in Eq. 16.24:

$$S^0 + NO_3^- + CaCO_3 \rightarrow CaSO_4 + N_2 \qquad \text{(Eq. 16.24)}$$

In the above equation, the sulfate formed is shown as precipitating with calcium to form gypsum. *A. denitrificans* is not acid tolerant, and has an optimal pH for growth of 7.0.

Summary for Chemoautotrophic Sulfur Oxidation

- It is an aerobic process, but most sulfur oxidizers are microaerophilic
- Some heterotrophs oxidize sulfur but obtain no energy
- This process can result in the formation of acid mine drainage
- This process can be used in metal recovery

16.4.3.2 Photoautotrophic Sulfur Oxidation

Photoautotrophic oxidation of sulfur is limited to green and purple sulfur bacteria (Table 16.17). This group of bacteria evolved on early Earth when the atmosphere contained no oxygen. These microbes fix carbon using light energy, but instead of oxidizing water to oxygen, they use an analogous oxidization of sulfide to sulfur:

$$CO_2 + H_2S \rightarrow S^0 + \text{fixed carbon} \qquad \text{(Eq. 16.25)}$$

These organisms are found in mud and stagnant water, sulfur springs and saline lakes. In each of these environments, both sulfide and light must be present. Although the contribution to primary productivity is small in comparison with aerobic photosynthesis, these organisms are important in the sulfur cycle. They serve to remove sulfide from the surrounding environment, effectively preventing its movement into the atmosphere and its precipitation as metal sulfide.

Summary for Photoautotrophic Sulfur Oxidation

- It is an anaerobic process that is limited to the purple and green sulfur bacteria
- It is responsible for a very small portion of total photosynthetic activity

16.4.4 Sulfur Reduction

There are three types of sulfur reduction. The first, already discussed in Section 16.4.2, is performed to assimilate sulfur into cell components (Widdel, 1988). Assimilatory sulfate reduction occurs under either aerobic or anaerobic conditions. In contrast, there are two dissimilatory pathways, both of which use an inorganic form of sulfur as a terminal electron acceptor. In this case, sulfur reduction occurs only under anaerobic conditions. The two types of sulfur that can be used as terminal electron acceptors are elemental sulfur and sulfate. These two types of metabolism are differentiated as sulfur respiration and dissimilatory sulfate reduction. *Desulfuromonas acetooxidans* is an example of a bacterium that grows on small carbon compounds such as acetate, ethanol and propanol, and uses elemental sulfur as the terminal electron acceptor as shown in Eq. 16.26:

$$CH_3COOH + 2H_2O + 4S^0 \rightarrow 2CO_2 + 4S^{2-} + 8H^+$$
acetate

(Eq. 16.26)

However, the use of sulfate as a terminal electron acceptor seems to be the more important environmental process. The following genera, all of which utilize sulfate as a terminal electron acceptor, are found widely distributed in the environment, especially in anaerobic sediments of aquatic environments, water-saturated soils and animal intestines: *Desulfobacter, Desulfobulbus, Desulfococcus, Desulfonema, Desulfosarcina, Desulfotomaculum* and *Desulfovibrio*. Together these organisms are known as the sulfate-reducing bacteria (SRB). They can utilize H_2 as an electron donor to drive the reduction of sulfate as shown in Eq. 16.27:

$$4H_2 + SO_4^{2-} \rightarrow S^{2-} + 4H_2O$$ (Eq. 16.27)

Thus, SRB compete for available H_2 in the environment, as H_2 is also the electron donor required by methanogens. It should be noted that this is not usually a chemoautotrophic process because most SRB cannot fix carbon dioxide. Instead, they obtain carbon from low-molecular-weight compounds such as acetate or methanol. The overall reaction for utilization of methanol is shown in Eq. 16.28:

$$4CH_3OH + 3SO_4^{-2} \rightarrow 4CO_2 + 3S^{2-} + 8H_2O$$ (Eq. 16.28)
methanol

Both sulfur and sulfate reducers are strict anaerobic chemoheterotrophic organisms that prefer small carbon substrates such as acetate, lactate, pyruvate and low-molecular-weight alcohols. Where do these small carbon compounds come from in the environment? They are by-products of fermentation of plant and microbial biomass that occurs in anaerobic regions. Thus, the sulfate reducers are part of an anaerobic consortium of bacteria including fermenters, sulfate reducers and methanogens, which together act to completely mineralize organic compounds to carbon dioxide and methane (see Chapter 3). More recently, it has been found that some SRB can also metabolize more complex carbon compounds including some aromatic compounds and some longer chain fatty

acids. These organisms are being looked at closely to determine whether they can be used in remediation of contaminated sites that are highly anaerobic, and that would be difficult to oxygenate.

The end product of sulfate reduction is hydrogen sulfide. What are the fates of this compound? It can be taken up by chemoautotrophs or photoautotrophs and reoxidized; it can be volatilized into the atmosphere; or it can react with metals to form metal sulfides. In fact, the activity of sulfate reducers and the production of hydrogen sulfide are responsible for the corrosion of underground metal pipes. In this process, the hydrogen sulfide produced reacts with ferrous iron metal to make more iron sulfide.

Summary for Sulfur Reduction

- Anaerobic respiration using SO_4^{2-} or S^0 as a TEA
- Completely inhibited by O_2
- Produces H_2S which can cause metal corrosion

16.5 IRON CYCLE

16.5.1 Iron Reservoirs

Iron is the fourth most abundant element in Earth's crust. Iron generally exists in three oxidation states: $0, +2$ and $+3$ corresponding to metallic iron (Fe^0), ferrous iron (Fe^{2+}) and ferric iron (Fe^{3+}). In the environment, iron is actively cycled between the $+2$ and $+3$ forms (Figure 16.18). Under aerobic conditions iron is usually found in its most oxidized form (Fe^{3+}), which has low aqueous solubility. Under reducing or anaerobic conditions Fe^{3+} is reduced to the ferrous form, Fe^{2+}, which has higher solubility. Iron is an essential but minor element for biological organisms, making up approximately 0.2% of the dry weight of a bacterial cell (Table 16.1). Although the amount of iron in a cell is low, it has a very important function as a part of enzymes that are used in respiration and photosynthesis, both processes that require electron transfer.

16.5.2 Iron in Soils and Sediments

Iron is generally not a limiting nutrient in soil due to its high abundance in Earth's crust. However, even though iron abundance is high, the bioavailability of most iron minerals is quite limited. Thus, microorganisms have developed strategies to obtain iron from its mineral form, usually from iron oxides or iron oxyhydroxides. The best studied strategy is the use of iron chelators known as siderophores (Figure 16.19). Siderophores are synthesized and released from the cell, where they bind Fe^{3+}, which helps keep this low solubility form of iron in solution.

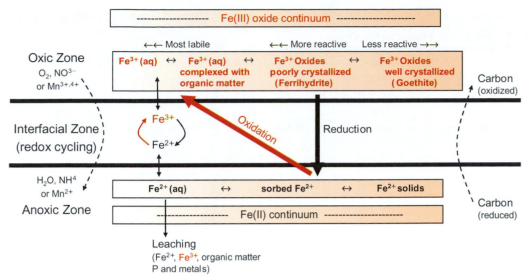

FIGURE 16.18 Conceptual diagram of iron redox cycling in soils and sediments. Iron atoms can cycle many times through oxidized and reduced states before being lost from the soil profile. Consequently, electron fluxes associated with iron may greatly exceed the oxidation capacity represented by the mass of iron oxides present at any given time. Courtesy Aaron Thompson.

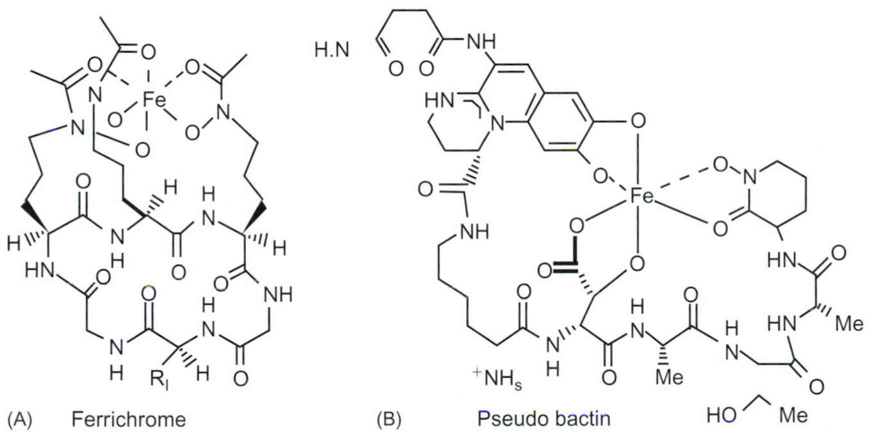

FIGURE 16.19 Two siderophores showing the coordination of the siderophores with iron. Ferrichrome is produced by fungi including *Aspergillus*, *Ustilago* and *Penicillium*. Pseudobactin is a bacterial siderophore made by *Pseudomonas* sp.

(A) Ferrichrome   (B) Pseudo bactin

The soluble iron—siderophore complex is recognized by and binds to siderophore-specific receptors on the cell surface. The iron is released from the complex, reduced and taken up into the cell as Fe^{2+}, its more soluble form.

16.5.3 Iron in Marine Environments

Unlike terrestrial and sediment environments, iron is considered a limiting nutrient in the modern marine environment (Raiswell, 2006). In fact, the extent of this limitation has been the focus of a vigorous debate in recent years. Recent data seem to indicate that for one-third of Earth's oceans, those that are more nutrient rich and support higher numbers of phytoplankton, iron is the limiting nutrient to growth (Boyd *et al.*, 2007). How does iron enter the ocean environment? As shown in Table 16.19, the largest flux of iron entering the oceans is

fluvial in nature, meaning that it enters as suspended particles or dissolved iron carried by rivers, or is associated with glacial sediments (Jickells *et al.*, 2005). For the most part, this iron is deposited in the sediments of near-coastal areas, and does not reach the open ocean. Therefore, in the open ocean, the major pathway of iron entry is eolian, or as dust that is carried through the atmosphere mainly from desert and other arid environment land surfaces.

16.5.4 Iron Oxidation

16.5.4.1 Chemoautotrophs

Under aerobic conditions, ferrous iron tends to oxidize to the ferric form. Ferrous iron will autoxidize or spontaneously oxidize under aerobic conditions at pH >5. Iron

TABLE 16.19 Global Iron Fluxes to the Ocean

Source	Flux Teragrams (1×10^9 kg) Per Year
Fluvial particulate total iron	625–962
Fluvial dissolved iron	1.5
Glacial sediments	34–211
Atmospheric	16
Coastal erosion	8
Hydrothermal	14
Authigenic (release from deep-sea sediments during diagenesis)	5

Adapted from Jickells et al. (2005).

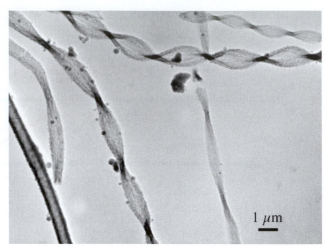

FIGURE 16.20 Iron hydroxide- and iron oxyhydroxide-coated stalks of *Gallionella* and sheaths of *Leptothrix*. In this transmission electron micrograph, dark areas are the metal deposits. From University of California Berkeley Geomicrobiology Program, 2008.

oxidation also occurs biotically. In fact, reduced iron is an important source of energy for several specialized genera of chemoautotrophic bacteria, the iron-oxidizers (Ehrlich, 1996):

$$Fe^{2+} + H^+ + 1/4O_2 \rightarrow Fe^{3+} + 1/2H_2O \quad \Delta G^{0'}$$
$$= -40 \text{ kJ/mol}$$

(Eq. 16.29)

Note that compared to ammonia (Eq. 16.6) or sulfur (Eqs. 16.22, 16.23) oxidation, the yield of energy in iron oxidation is quite low. Nevertheless, this is an exploitable niche and an important biological reaction, which can have considerable environmental consequences.

Iron oxidation is most often associated with acidic environments. For example, the acidophilic thermophile *Sulfolobus* is an archaean that was isolated from acidic hot springs. But iron oxidation is perhaps best studied in association with acid mine drainage and the acidophilic bacterium *Acidothiobacillus ferrooxidans*. Interestingly, although *A. ferrooxidans* is best studied (because it has been most easily cultured), nonculture-based analysis suggests that other iron-oxidizers may actually play more important roles in the creation of acid conditions in mine tailings and the formation of acid mine drainage (Baker and Banfield, 2003). These include Bacteria such as *Leptospirillum ferrooxidans* and *L. ferriphilum*, *Sulfobacillus acidophilus*, *S. thermosulfooxidans* and *Acidimicrobium ferrooxidans*, as well as some Archaea, e.g., *Ferroplasma acidiphilum*.

Iron oxidation has also been described in several marine genera at neutral pH. This process is problematic at neutral pH because: (1) the energy yield is very low; and (2) at neutral pH, spontaneous oxidation of Fe^{2+} to Fe^{3+} (in the form of iron oxyhydroxides) occurs rapidly in the presence of oxygen and competes with the

biological reaction (Figure 16.18). Neutriphilic iron-oxidizers overcome these problems with very specific niche requirements. They position themselves in regions with low O_2 tension and high and constant Fe^{2+} concentrations. One marine environment that meets these requirements occurs in hydrothermal vents on the sea floor (see Chapter 7). Here, anoxic vent fluids charged with Fe^{2+} rapidly come in contact with the cold, oxygenated ocean water. Similar conditions can occur in municipal and industrial water pipelines.

The best-studied neutrophilic iron-oxidizers are *Gallionella* and *Leptothrix*. *Gallionella* is capable of chemoautotrophic growth using Fe^{2+} as sole energy source under microaerobic conditions. *Leptothrix* is a sheath-forming chemoheterotrophic organism that oxidizes both Fe^{2+} and Mn^{2+}, depositing an iron–manganese encrusted coating on its sheath. As can be seen in Figure 16.20, these microbes can deposit copious amounts of iron (and manganese) minerals on their surfaces. This can have serious economic consequences in some instances. For example, this process can cause extensive biofouling and corrosion of water pipelines.

Summary for Iron Oxidation

- Iron oxidation is a chemoautotrophic, aerobic process usually found under acidic conditions
- This process participates in the formation of acid mine drainage (and can be used for metal recovery)
- Neutriphilic iron oxidation occurs in a more limited number of environments where, due to iron mineral deposition, this process can result in biofouling and corrosion

16.5.4.2 Photoautotrophs

As shown below, some members of the purple and green bacteria can use Fe^{2+} as an electron donor to carry out anaerobic photosynthesis coupled to photoautotrophic growth (i.e., growth involving photosynthesis and oxidation):

$$4Fe^{2+} + CO_2 + 11H_2O \xrightarrow{light} C(H_2O) + 4Fe(OH)_3 + 8H^+$$

(Eq. 16.30)

It has been proposed that iron-based photosynthesis represents the transition between anaerobic photosynthesis that developed on early Earth and aerobic photosynthesis that began approximately 2 billion years ago (Jiao and Newman, 2007). In fact, it is thought that Fe^{2+} was the most widespread source of reducing power from 1.6 to 3.8 billion years ago, where under reducing conditions, Fe^{2+} was favored over Fe^{3+}. Under these conditions, the amount of iron in the marine environment was considerably higher (0.1 to 1 millimoles/L) than it is today (0.03 to 1 nanomoles/L) (Jickells *et al.*, 2005; Bosak *et al.*, 2007).

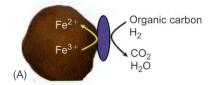
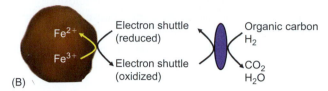

FIGURE 16.21 Two primary strategies are used in iron respiration to facilitate electron transfer between microbial cells and iron oxide surfaces. (A) Cells can come in direct contact with the oxide surface. (B) An electron shuttle can be used to mediate the electron transfer between the cell and iron oxide surfaces.

Summary for Photoautotrophs

- Phototrophic iron oxidation is a photoautotrophic, anaerobic process that is limited to purple and green sulfur bacteria
- This process is thought to be a key transition step between anaerobic photosynthesis found on early Earth and modern aerobic photosynthesis

16.5.5 Iron Reduction

Iron is microbially reduced for two purposes, assimilation and energy generation. Assimilatory iron reduction is the reduction of Fe^{3+} for uptake and incorporation into cell constituents. As described in Section 16.5.2, this usually involves the release of siderophores, which complex Fe^{3+} in the environment exterior to the cell. The iron–siderophore complex then delivers the Fe^{3+} to the cell, which is reduced to Fe^{2+} as it is taken up.

Dissimilatory iron reduction or iron respiration is the use of Fe^{3+} as a terminal electron acceptor for the purpose of energy generation during anaerobic respiration. Due to its abundance in Earth's crust, Fe^{3+} found in iron oxides and oxyhydroxides serves as an important terminal electron acceptor for anaerobic heterotrophic bacteria. Because iron respiration has been an important activity during the evolution of Earth, there is wide diversity among the bacteria and archaeans capable of carrying out this activity (Weber *et al.*, 2006). The problem for the iron-reducers is that most of the iron in environment is relatively unavailable (Figure 16.18). So, microorganisms have developed some very interesting strategies to solve this problem (Lovley, 2000). As shown in Figure 16.21,

the first is to make direct contact with an iron oxide surface. In this case, the iron reductase is a membrane-bound enzyme allowing direct access of the enzyme with the substrate. A second strategy is to use an electron shuttle that can act as an intermediate in transferring electrons from the cell to the iron oxide surface. Possible electron shuttles in the environment include humic acids or other molecules that contain quinone-like structures that are reduced to a corresponding hydroquinone form. Humic acids are considered an exogenous electron shuttle source, or one that is obtained from the environment outside the cell. Alternatively, some bacteria can make and release endogenous electron shuttles. For example, *Geothrix fermentans* releases a quinone-like electron shuttle during growth on lactate (electron donor) and poorly crystalline iron oxides (electron acceptor) (Nevin and Lovley, 2002).

Summary for Iron Reduction

- Assimilatory iron reduction is generally mediated by iron-chelating molecules called siderophores
- Dissimilatory iron reduction is the use of ferric iron as a terminal electron acceptor during anaerobic respiration. Due to the abundance of iron in soils and sediments, it is thought that this is an important process in anaerobic environments

QUESTIONS AND PROBLEMS

1. Describe both microbial and non-microbial contributions to the carbon cycle.
2. Give an example of
   a. a small actively cycled reservoir
   b. a large actively cycled reservoir
   c. a large inactively cycled reservoir

3. Describe how the ocean has reduced the expected rate of increase of CO_2 in the atmosphere since industrialization began.

4. What is the concept behind fertilization of the ocean with iron?

5. What strategy is used by microbes to initiate degradation of large plant polymers such as cellulose?

6. Define what is meant by a greenhouse gas and give two examples. For each example describe how microorganisms mediate generation of the gas, and then describe how human activity influences generation of the gas.

7. Both autotrophic and heterotrophic activities are important in element cycling. For each cycle discussed in this chapter (carbon, nitrogen, sulfur and iron), name the most important heterotrophic and autotrophic activity. Justify your answer.

8. What would happen if microbial nitrogen fixation suddenly ceased?

REFERENCES AND RECOMMENDED READING

Atlas, R. M., and Bartha, R. (1993) "Microbial Ecology," Benjamin Cummings, New York.

Baker, B. J., and Banfield, J. F. (2003) Microbial communities in acid mine drainage. *FEMS Microbiol. Ecol.* **44**, 139–152.

Bellamy, P. H., Loveland, P. J., Bradley, R. I., Lark, R. M., and Kirk, G. J. (2005) Carbon losses from all soils across England and Wales 1978–2003. *Nature* **437**, 245–248.

Bosak, T., Greene, S. E., and Newman, D. K. (2007) A likely role for anoxygenic photosynthetic microbes in the formation of ancient stromatolites. *Geobiology* **5**, 119–126.

Boyd, P. W., Jickells, T., Law, C. S., Blain, S., Boyle, E. A., Buesseler, K. O., *et al.* (2007) Mesoscale iron enrichment experiments 1993–2005: synthesis and future directions. *Science* **315**, 612–617.

Canfield, D. E., Glazer, A. N., and Falkowski, P. G. (2010) The evolution and future of earth's nitrogen cycle. *Science* **330**, 192–196.

Deobald, L. A., and Crawford, D. L. (1997) Lignocellulose biodegradation. In "Manual of Environmental Microbiology" (C. J. Hurst, G. R. Knudsen, M. J. McInerney, L. D. Stetzenbach, and M. V. Walter, eds.), American Society for Microbiology, Washington, DC, pp. 730–737.

Dobrovolsky, V. V. (1994) "Biogeochemistry of the World's Land," CRC Press, Boca Raton, FL.

Ehrlich, H. L. (1996) "Geomicrobiology," Marcel Dekker, New York.

Farabee, M. (2008) Paleobiology: the precambrian: life's genesis and spread. In "Online Biology Book", Estrella Mountain Community College, Avondale, Arizona. http://www.emc.maricopa.edu/faculty/farabee/BIOBK/BioBookPaleo2.html.

Galloway, J. N., Townsend, A. R., Erisman, J. W., Bekunda, M., Cai, Z., Feney, J. R., *et al.* (2008) Transformation of the nitrogen cycle: recent trends, questions, and potential solutions. *Science* **320**, 889–892.

Germida, J. J. (1998) Transformations of sulfur. In "Principles and Applications of Soil Microbiology" (D. M. Sylvia, J. J. Fuhrmann,

P. G. Hartel, and D. A. Zuberer, eds.), Prentice-Hall, Upper Saddle River, NJ, pp. 346–368.

Haider, K. (1992) Problems related to the humification processes in soils of temperate climates. In "Soil Biochemistry" (G. Stotzky, and J.-M. Bollag, eds.), vol. 7, Marcel Dekker, New York, pp. 55–94.

Jiao, Y., and Newman, D. K. (2007) The pio operon is essential for phototrophic Fe(II) oxidation in *Rhodopseudomonas palustris* TIE-1. *J. Bacteriol.* **189**, 1765–1773.

Jickells, T. D., An, Z. S., Andersen, K. K., Baker, A. R., Bergametti, G., Brooks, N., *et al.* (2005) Global iron connections between desert dust, ocean biogeochemistry, and climate. *Science* **308**, 67–71.

Jones, K. M., Kobayashi, H., Davies, B. W., Taga, M. E., and Walker, G. C. (2007) How rhizobial symbionts invade plants: the *Sinorhizobium-Medicago* model. *Nat. Rev. Microbiol.* **5**, 619–633.

Knorr, W., Prentice, I. C., House, J. I., and Holland, E. A. (2005) Long-term sensitivity of soil carbon turnover to global warming. *Nature* **433**, 298–301.

Kuenen, J. G. (2008) Anammox bacteria: from to discovery to application. *Nat. Rev. Microbiol.* **6**, 320–326.

Lai, R. (2004) Soil carbon sequestration impacts on global climate change and food security. *Science* **304**, 1623–1627.

Lokupitiya, E., and Paustian, K. (2006) Agricultural soil greenhouse gas emissions: a review of national inventory methods. *J. Environ. Qual.* **35**, 1413–1427.

Lovelock, J. (1995) "The Ages of Gaia," W. W. Norton, New York.

Lovley, D. R. (2000) Fe(III) and Mn(IV) reduction. In "Environmental Microbe–Metal Interactions" (D. R. Lovley, ed.), ASM Press, Washington, DC, pp. 3–30.

Madigan, M. T., Martinko, J. M., and Parker, J. (1997) "Brock Biology of Microorganisms," 8th ed. Prentice Hall, Upper Saddle River, NJ.

Morgan, P., Lee, S. A., Lewis, S. T., Sheppard, A. N., and Watkinson, R. J. (1993) Growth and biodegradation by white-rot fungi inoculated into soil. *Soil Biol. Biochem.* **25**, 279–287.

Myrold, D. D. (1998) Transformations of nitrogen. In "Principles and Applications of Soil Microbiology" (D. M. Sylvia, J. J. Fuhrmann, P. G. Hartel, and D. A. Zuberer, eds.), Prentice-Hall, Upper Saddle River, NJ, pp. 218–258.

Neidhardt, F. C., Ingraham, J. L., and Schaechter, M. (1990) "Physiology of the Bacterial Cell: A Molecular Approach," Sinauer Associates, Sunderland, MA.

Nevin, K. P., and Lovley, D. R. (2002) Mechanisms for accessing insoluble Fe(III) oxide during dissimilatory Fe(III) reduction by *Geothrix fermentans*. *Appl. Environ. Microbiol.* **68**, 2294–2299.

Paul, E. A., and Clark, F. E. (1989) "Soil Microbiology and Biochemistry," Academic Press, New York.

Premakumar, R., Jacobson, M.R., Loveless, T.M., and Bishop, P.E. (1992) Characterization of transcripts expressed from nitrogenase-3 structural genes of *Azotobacter vinelandii*. *Can J Microbiol.* **38**, 929–36.

Raiswell, R. (2006) Towards a global highly reactive iron cycle. *J. Geochem. Explor.* **88**, 436–439.

Reynolds, S. (1999) Stromatolites. In "MLSSA Journal," The Marine Life Society of South Australia, Inc., Adelaide, Australia, http://www.mlssa.asn.au/.

Rice, C. W. (2006) Introduction to special section on greenhouse gases and carbon sequestration in agriculture and forest. *J. Environ. Qual.* **35**, 1338–1340.

Sundquist, E. T. (1993) The global carbon dioxide budget. *Science* **259**, 934–941.

Thamdrup, B. (2012) New pathways and processes in the global nitrogen cycle. *Annu. Rev. Col. Evol. Syst.* **43**, 407–428.

Tiedje, J. M. (1988) Ecology of denitrification and dissimilatory nitrate reduction to ammonium. In "Biology of Anaerobic Microorganisms" (A. J. B. Zehnder, ed.), John Wiley & Sons, New York, pp. 179–244.

Wagner, G. H., and Wolf, D. C. (1998) Carbon transformations and soil organic matter formation. In "Principles and Applications of Soil Microbiology" (D. M. Sylvia, J. J. Fuhrmann, P. G. Hartel, and D.

A. Zuberer, eds.), Prentice-Hall, Upper Saddle River, NJ, pp. 259–294.

Weber, K. A., Achenbach, L. A., and Coates, J. D. (2006) Microorganisms pumping iron: anaerobic microbial iron oxidation and reduction. *Nat. Rev. Microbiol.* **4**, 752–764.

Widdel, F. (1988) Microbiology and ecology of sulfate- and sulfur-reducing bacteria. In "Biology of Anaerobic Microorganisms" (A. J. B. Zehnder, ed.), John Wiley & Sons, New York, pp. 469–585.

Zhao, Y., Bian, S. M., Zhou, H. N., and Huang, J. F. (2006) Diversity of nitrogenase systems in diazotrophs. *J. Integr. Plant Biol.* **48**, 745–755.

Part V

Remediation of Organic and Metal Pollutants

Chapter 17

Microorganisms and Organic Pollutants

Raina M. Maier and Terry J. Gentry

17.1 INTRODUCTION

Global release of industrial and agricultural chemicals has resulted in widespread environmental pollution. The energy production industry alone generates large amounts of waste during the processing of coal and oil, and also nuclear energy production. As just one example, the United States Environmental Protection Agency (USEPA) estimates that more than 1 million underground storage tanks (USTs), predominantly employed for gasoline storage, have been in service in the United States. As of 2013, there have been over 510,000 confirmed accidental releases from these USTs (USEPA, 2013). This type of contamination is known as a point source. In contrast, the application of pesticides and fertilizers over vast land areas can result in what is called nonpoint source contamination. It has been reported that >90% of the monitored streams and >55% of shallow groundwater sites on agricultural and urban lands are contaminated with pesticides (Gilliom et al., 2006).

Groundwater contamination is a critical issue from two perspectives. First, groundwater constitutes approximately 97% of all available fresh water on Earth. In the USA, groundwater sources account for around 37% of the water used for public potable water supplies, and around 98% of that used for private water supplies (e.g., rural residents) (Hutson et al., 2004). Second, there is a hydrologic interchange between surface and subsurface water systems, such that there is a conduit for moving groundwater contamination into surface waters (see Information Box 17.1). As a result of accidental or intentional releases of hazardous waste in the United States, over 1.4 million acres of chemical plumes in groundwater require remediation (NRC, 2003). Furthermore, there is little doubt that these chemical plumes are contributing to the contamination of surface waters including streams and lakes.

The objective of this chapter is to examine microbial interactions with organic pollutants that can be harnessed

I.L. Pepper, C.P. Gerba, T.J. Gentry: Environmental Microbiology, Third edition. DOI: http://dx.doi.org/10.1016/B978-0-12-394626-3.00017-X
© 2015 Elsevier Inc. All rights reserved.

Information Box 17.1   An Endangered River—Groundwater Connections

The Colorado River is currently impacted by two different groundwater contaminant plumes. The source of the first plume is a plant near Henderson, Nevada, where, in the 1950s the U.S. Navy commissioned a manufacturing plant to produce perchlorate, a rocket fuel component. Perchlorate is a health concern because it is a competitive inhibitor of iodide transport to the thyroid (Greer *et al.*, 2002). As a result of this operation, the groundwater under the plant now contains approximately 9.3 million kg of perchlorate dissolved in 34 billion liters of water (Hogue, 2003). This groundwater feeds into the Las Vegas Wash, a tributary of the Colorado River that flows into Lake Mead. The groundwater plume contributed between 226 and 453 kg/day of perchlorate into the wash in the 1990s, resulting in perchlorate levels as high as 24 parts per billion in Lake Mead (Hogue, 2003).

The second plume involves an abandoned uranium tailings site once operated by Atlas Corporation near Moab, Utah. The site is unlined and contains approximately 9.5 million metric tons of uranium wastes, including 1.6 billion liters of contaminated liquid. As a result, the uranium content in the groundwater near the Moab site exceeds the EPA groundwater standards at uranium tailings piles by 530-fold (ORNL, 1998). A study conducted by the Oak Ridge National Laboratory estimates that 36,520 liters/day or 25.4 liters/min of uranium-contaminated water enters the Colorado River at the Moab site (ORNL, 1998). Furthermore, the tailings at Moab contain high levels of ammonia, a marker of mill contamination, which is also leaching into the groundwater and eventually entering the Colorado River.

to help prevent contamination, and also to remediate contaminated sites. The United States has passed a series of environmental laws mandating the cleanup of such sites, but the cost of cleanup has been estimated to be in the trillions of dollars. Therefore, as a society we are examining the cleanup issue from several perspectives. The first is related to the cleanup target and the question "How clean is clean?" As you can imagine, the stricter the cleanup provision such as lower contaminant concentrations, the greater the attendant cleanup costs. It may require tens to hundreds of millions of dollars for complete cleanup of a large, complex, hazardous waste site. In fact, it may be impossible to clean many sites completely. It is very important, therefore, that the physical feasibility of cleanup and the degree of potential risk posed by the contamination be weighed against the economic impact, and the future use of the site. The second perspective being considered is whether natural microbial activities in the environment can aid in the cleanup of contaminated sites, and whether these activities will occur rapidly enough naturally, or whether they can and should be enhanced. These two perspectives are closely tied together because although microbial activities can reduce contamination significantly, they often do not completely remove it.

17.2 ENVIRONMENTAL LAW

Society began responding to environmental concerns long ago, beginning with the recognition that our environment is fragile, and that human activities can have a great impact on it. This led to the creation of Yellowstone National Park in 1872, and the assignment of the care of forested public domain lands to the U.S. Forest Service in 1897. After World War II, as the pollution impacts of industrialization began to be apparent, Congress began to

legislate in the area of pollution control. This legislation culminated in the major federal pollution control statutes of the 1970s that now constitute a large body of law called environmental law. Federal environmental law consists of laws in the conservation and pollution control areas, along with key planning and coordination statutes.

Environmental law is constantly changing and evolving as we try to respond to shifting priorities and pressures on resources. Sometimes changes are made to the law to allow further contamination or risk of contamination to occur for the "good of society." An example is oil exploration in environmentally sensitive areas such as Alaska and the Atlantic and Pacific coasts. On the other hand, some laws can be made more stringent. Again, using an example from the oil industry, whereas refinery wastes are heavily regulated for disposal, the same types of wastes generated in an oil field were not regulated, and were routinely buried without treatment. When attention was drawn to this practice, laws were enacted to require the oil industry to implement proper oil field disposal practices. As these examples suggest, environmental law comprises a complex body of laws, regulations and decisions now established in the United States. This body of law, which evolved quite quickly in comparison with labor, tax, banking and communications laws, already ranks with these other areas in size and complexity (Arbuckle *et al.*, 1987).

The term "environmental law" came into being with the enactment of the National Environmental Policy Act (NEPA) in early 1970 (Information Box 17.2). Although environmental law can vary considerably from state to state and even from city to city, a series of major federal environmental protection laws have been enacted that pertain to the generation, use and disposition of hazardous waste. NEPA requires each agency, government or industry that proposes a major action which may have a significant

---

Information Box 17.2   U.S. Title I Congressional Declaration of National Environmental Policy Sec. 101[42 USC § 4331]

(a) The Congress, recognizing the profound impact of man's activity on the interrelations of all components of the natural environment, particularly the profound influences of population growth, high-density urbanization, industrial expansion, resource exploitation, and new and expanding technological advances and recognizing further the critical importance of restoring and maintaining environmental quality to the overall welfare and development of man, declares that it is the continuing policy of the Federal Government, in cooperation with State and local governments, and other concerned public and private organizations, to use all practicable means and measures, including financial and technical assistance, in a manner calculated to foster and promote the general welfare, to create and maintain conditions under which man and nature can exist in productive harmony, and fulfill the social, economic, and other requirements of present and future generations of Americans.

(b) In order to carry out the policy set forth in this Act, it is the continuing responsibility of the Federal Government to use all practicable means, consistent with other essential considerations of national policy, to improve and coordinate Federal plans, functions, programs, and resources to the end that the Nation may—

1. fulfill the responsibilities of each generation as trustee of the environment for succeeding generations;
2. assure for all Americans safe, healthful, productive, and aesthetically and culturally pleasing surroundings;
3. attain the widest range of beneficial uses of the environment without degradation, risk to health or safety, or other undesirable and unintended consequences;
4. preserve important historic, cultural, and natural aspects of our national heritage, and maintain, wherever possible, an environment which supports diversity, and variety of individual choice;
5. achieve a balance between population and resource use which will permit high standards of living and a wide sharing of life's amenities; and
6. enhance the quality of renewable resources and approach the maximum attainable recycling of depletable resources.

(c) The Congress recognizes that each person should enjoy a healthful environment and that each person has a responsibility to contribute to the preservation and enhancement of the environment.

---

effect on the human environment to prepare an environmental impact statement (EIS). The EIS must address the environmental impact of the proposed action and any reasonable alternatives that may exist. The types of projects that NEPA covers are landfills, roads, dams, building complexes, research projects and any private endeavor requiring a federal license that may affect the environment. *NEPA does not mandate particular results, and it does not require a federal agency to adopt the least environmentally damaging alternative.* Because of this, NEPA can be thought of as an "environmental full disclosure law," which requires the applicant to take a "hard look" at the environmental consequences of its action. Thus, an EIS allows environmental concerns and planning to be integrated into the early stages of project planning. Unfortunately, an EIS is often done as an afterthought and becomes a rationale for a project that may be a poor alternative.

A series of environmental laws have been passed since NEPA to protect our natural resources (Table 17.1). These include laws such as the Clean Air Act and the Clean Water Act, which protect air and water resources. There are also laws that govern the permitting of the sale of hazardous chemicals and laws that mandate specific action to be taken in the cleanup of contaminated sites. When the Comprehensive Environmental Response, Compensation and Liability Act (CERCLA), more commonly known as

Superfund, was enacted, it became clear that technology for cleaning up hazardous waste sites was needed. Early remedial actions for contaminated sites consisted of excavation and removal of the contaminated soil to a landfill. Very soon, it became apparent that this was simply moving the problem around, not solving it. As a result, the Superfund Amendments and Reauthorization Act (SARA) was passed in 1986. This act added several new dimensions to CERCLA. SARA stipulates cleanup standards, and mandates the use of the National Contingency Plan to determine the most appropriate action to take for site cleanup.

Two types of responses are available within Superfund: (1) removal actions in response to immediate threats, e.g., removing leaking drums, and (2) remedial actions, which involve cleanup of hazardous sites. The Superfund provisions can be used when a hazardous substance is released or there is a substantial threat of a release that poses imminent and substantial endangerment to public health and welfare. The first step is to place the potential site into the Superfund Site Inventory. After a preliminary assessment and site inspection, the decision is made as to whether or not the site will be placed on the National Priority List (NPL). Sites placed on this list are those deemed to require a remedial action. Currently, there are more than 1200 sites on the NPL. The National Contingency Plan (NCP) is the next component. The purpose of the NCP is to characterize

TABLE 17.1 History of Environmental Law

Law	Year Passed	Goals
Clean Air Act (CAA)	1970	Sets nationwide ambient air quality standards for conventional air pollutants. Sets standards for emissions from both stationary and mobile sources (e.g., motor vehicles).
Clean Water Act (CWA)	1972	Mandates "fishable/swimmable" waters wherever attainable. Provides for (1) a construction grants program for publicly owned water treatment plants and requires plants to achieve the equivalent of secondary treatment; (2) a permit system to regulate point sources of pollution; (3) areawide water quality management to reduce nonpoint sources of pollution; (4) wetlands protection, sludge disposal and ocean discharges; (5) regulation of cleanup of oil spills.
Surface Mining Control and Reclamation Act (SMCRA)	1977	Regulates coal surface mining on private lands and strip mining on public lands. Prohibits surface mining in environmentally sensitive areas.
Resource Conservation and Recovery Act (RCRA)	1976	Provides a comprehensive management scheme for hazardous waste disposal. This includes a system to track the transportation of wastes and federal performance standards for hazardous waste treatment, storage and disposal facilities. Open dumps are prohibited.
Toxic Substances Control Act (TOSCA)	1976	Requires premarket notification of EPA by the manufacturer of a new chemical. Based on testing information submitted by the manufacturer or premarket test ordered by EPA (including biodegradability and toxicity), a court injunction can be obtained barring the chemical from distribution or sale. EPA can also seek a recall of chemicals already on the market. It is this Act that prohibits all but closed-circuit uses of PCBs.
Comprehensive Environmental Response, Compensation and Liability Act (CERCLA)	1980	Commonly known as Superfund, this Act covers the cleanup of hazardous substance spills, from vessels, active or inactive facilities. Establishes a Hazardous Substances Response Trust Fund, financed by a tax on the sale of hazardous chemicals, to be used for removal and cleanup of hazardous waste releases. Cleanup costs must be shared by the affected state. Within certain limits and subject to a few defenses, anyone associated with the release is strictly liable to reimburse the fund for cleanup costs, including damage to natural resources.
Superfund Amendments and Reauthorization Act (SARA)	1986	SARA provides cleanup standards and stipulates rules through the National Contingency Plan for the selection and review of remedial actions. It strongly recommends that remedial actions use on-site treatments that "permanently and significantly reduce the volume, toxicity, or mobility of hazardous substances" and requires remedial action that is "protective of human health and the environment, that is cost-effective, and that utilizes permanent solutions and alternative treatment technologies or resource recovery technologies to the maximum extent practicable."
National Contingency Plan (NCP)	1988	A five-step process to use in evaluation of contaminated sites and suggest the best plan for remediation.

the nature and extent of risk posed by contamination, and to evaluate potential remedial options. The investigation and feasibility study components are normally conducted concurrently and with a "phased" approach. This allows feedback between the two components. The selection of the specific remedial action to be used at a particular site is a very complex process. The goals of the remedial action are that it be protective of human health and the environment, that it maintains protection over time and that it maximizes waste treatment.

How does microbiology fit into remedial action strategies? Biological remediation approaches have been found to be more economical and often have better public acceptance than traditional physical/chemical approaches. The remainder of this chapter deals with biodegradation of organic contaminants and ways in which these processes can be harnessed to remediate contaminated sites.

17.3 THE OVERALL PROCESS OF BIODEGRADATION

Biodegradation is the breakdown of organic contaminants that occurs due to microbial activity. As such, these organic contaminants can be considered as a microbial food source or substrate. Biodegradation of any organic compound can be thought of as a series of biological degradation steps or a pathway that ultimately results in the oxidation of the parent compound. Often, the degradation of these compounds results in the generation of energy as described in Chapter 3.

Complete biodegradation or mineralization involves oxidation of the parent compound to form carbon dioxide and water, a process that provides both carbon and energy for growth and reproduction of cells. Figure 17.1 illustrates the mineralization of an organic compound under either aerobic or anaerobic conditions. The series of degradation

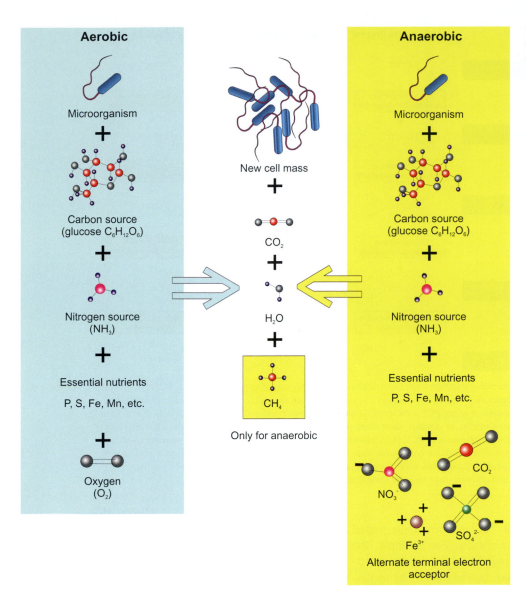

FIGURE 17.1 Aerobic (blue) or anaerobic (yellow) mineralization of an organic compound.

steps constituting mineralization is similar whether the carbon source is a simple sugar such as glucose, a plant polymer such as cellulose or a pollutant molecule. Each degradation step in the pathway is catalyzed by a specific enzyme made by the degrading cell. Enzymes are most often found within a cell, but are also made and released from the cell to help initiate degradation reactions. Enzymes found external to the cell are known as extracellular enzymes. Extracellular enzymes are important in the degradation of macromolecules such as the plant polymer cellulose (see Section 16.2.3.1). Macromolecules must be broken down into smaller subunits outside the cell to allow transport of the smaller subunits into the cell. Biotransformation may stop at any step in the biodegradation pathway if the appropriate enzyme, either internal or extracellular, is not present (Figure 17.2). In fact, lack of appropriate biodegrading enzymes is one common reason for persistence of organic contaminants, particularly those

with unusual chemical structures that existing enzymes do not recognize. Thus, contaminant compounds that have structures similar to those of natural substrates are normally easily degraded. Those that are quite dissimilar to natural substrates are often degraded slowly or not at all.

Some organic contaminants are only partially degraded by environmental microorganisms. This can result from absence of the appropriate degrading enzyme as mentioned earlier. A second type of incomplete degradation is cometabolism, in which a partial oxidation of the substrate occurs but the energy derived from the oxidation is not used to support microbial growth. The process occurs when organisms possess one or more enzymes that coincidentally can degrade a particular contaminant in addition to its target substrate. Thus, such enzymes are nonspecific. Cometabolism can occur during periods of active growth, or can result from the interaction of resting (nongrowing) cells with an organic compound. Cometabolism is difficult

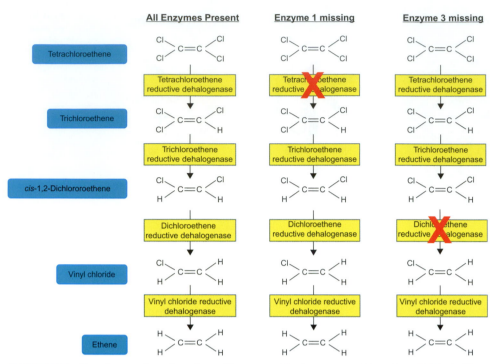

FIGURE 17.2 Stepwise transformation of tetrachloroethene (also known as perchloroethene; PCE). A different enzyme catalyzes each step of the pathway. Some microorganisms such as *Dehalococcoides ethenogenes* strain 195 are capable of using PCE as a terminal electron acceptor and completely dechlorinating it to the non-toxic compound ethene. In contrast, most other organisms either lack the ability to produce the enzymes needed to initiate transformation of PCE (e.g., missing enzyme 1) or can only transform it partially resulting in the accumulation of intermediate compounds such as *cis*-1,2-dichloroethene or vinyl chloride (e.g., missing enzyme 3). The potential accumulation of biodegradation intermediates is a concern since some of these chemicals, such as vinyl chloride, may be more toxic than the parent compound.

to measure in the environment, but has been demonstrated for some environmental contaminants. For example, the industrial solvent trichloroethene (TCE; also known as trichloroethylene) can be oxidized cometabolically by methanotrophic bacteria that grow on methane as a sole carbon source (Suttinun *et al.*, 2013). TCE is of great interest for several reasons. It is one of the most frequently reported contaminants at hazardous waste sites, it is a suspected carcinogen and it is generally resistant to biodegradation. As shown in Figure 17.3, the first step in the oxidation of methane by methanotrophic bacteria is catalyzed by the enzyme methane monooxygenase. This enzyme is so nonspecific that it can also cometabolically catalyze the first step in the oxidation of TCE when both methane and TCE are present. The bacteria receive no energy benefit from this cometabolic degradation step. The subsequent degradation steps shown in Figure 17.3 may be catalyzed spontaneously, by other bacteria, or in some cases by the methanotroph. This is an example of a cometabolic reaction that may have great significance in remediation. Research is currently investigating the application of these methanotrophs, as well as cometabolizing microorganisms that grow on toluene, ethylene, propylene, propane, butane and even ammonia, to TCE-contaminated sites.

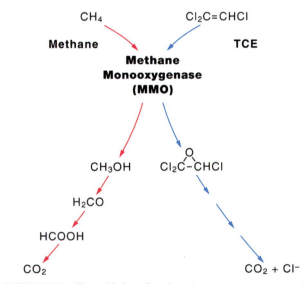

FIGURE 17.3 The oxidation of methane by methanotrophic bacteria is catalyzed by the enzyme methane monooxygenase. The same enzyme can act nonspecifically on trichloroethene (TCE). Subsequent TCE degradation steps may be catalyzed spontaneously, by other bacteria, or in some cases by the same methanotroph. From Pepper *et al.* (2006).

Partial or incomplete degradation can also result in polymerization or synthesis of compounds that are more complex and stable than the parent compound. This occurs when

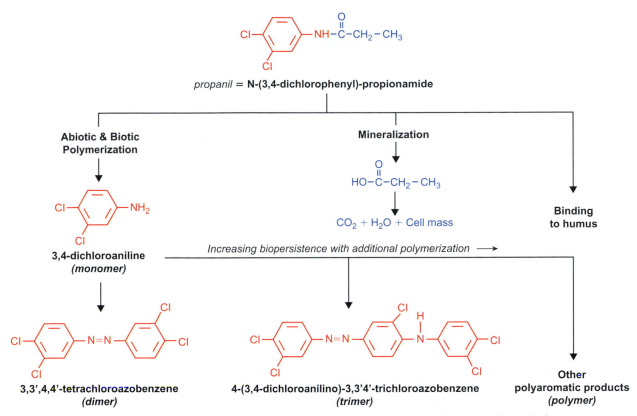

FIGURE 17.4 Polymerization reactions that occur with the herbicide propanil during biodegradation. Propanil is a selective post-emergence herbicide used in growing rice. It is toxic to many annual and perennial weeds. The environmental fate of propanil is of concern because it, like many other pesticides, is toxic to most crops except for cereal grains. It is also toxic to fish. Care is used in propanil application to avoid contamination of nearby lakes and streams. From Pepper *et al.* (2006).

initial degradation steps, often catalyzed by extracellular enzymes, create reactive intermediate compounds. These highly reactive intermediate compounds can then combine with each other or with other organic matter present in the environment. This is illustrated in Figure 17.4, which shows some possible polymerization reactions that occur with the herbicide propanil during biodegradation. These include formation of dimers or larger polymers, which are quite stable in the environment. Stability is due to low bioavailability (high sorption and low solubility), lack of degrading enzymes and the fact that some of these residues become chemically bound to the soil organic matter fraction.

17.4 CONTAMINANT STRUCTURE, TOXICITY AND BIODEGRADABILITY

The vast majority of the organic carbon available to microorganisms in the environment is material that has been photosynthetically fixed (plant material). Of concern are environments that receive large additional inputs of carbon from agriculture or industry (petroleum products, organic solvents, pesticides). Although many of these chemicals can be readily degraded because of their structural similarity to naturally occurring organic carbon, the amounts added may exceed the existing carrying capacity of the environment. Carrying capacity is defined here as the maximum level of microbial activity that can be expected under existing environmental conditions. Microbial activity may be limited by both biological and physical–chemical factors. These factors include low numbers of microbes, insufficient oxygen or nutrient availability, and suboptimal temperature or water availability. These factors are discussed further in Section 17.5. Microbial activity, whether degradation occurs and the rate of degradation, also depend on several factors related to the structure and physical–chemical properties of the contaminant (Miller and Herman, 1997) (see Information Box 17.3).

17.4.1 Genetic Potential

The onset of contaminant biodegradation generally follows a period of adaptation or acclimation of indigenous microbes, the length of which depends on the contaminant structure. The efficient cycling of plant-based organic matter by soil microorganisms can promote the rapid degradation of organic contaminants that have a chemical

---

Information Box 17.3   Genetic and Contaminant Structure Factors that Impact Biodegradation

- **Phenotypic genetic potential.** The presence and expression of appropriate degrading genes by the indigenous microbial community.
- **Toxicity.** The inhibitory effect of the contaminant on cellular metabolism.
- **Bioavailability.** The effect of limited water solubility and sorption on the rate at which a contaminant is taken up by a microbial cell.

- **Contaminant structure.** Includes both steric and electronic effects. Steric effects involve the extent to which substituent groups on a contaminant molecule sterically hinder recognition by the active site of the degrading enzyme. Electronic effects involve the extent to which substituent groups electronically interfere with the interaction between the active site of the enzyme and the contaminant. Electronic effects can also alter the energy required to break critical bonds in the molecule.

Information Box 17.4   Enhanced Degradation of Atrazine in Soil—An Example of Microbial Adaptation Through Gene Transfer

The herbicide atrazine has been used extensively to control broadleaf and grass weeds in several major crops including corn. Due partly to its environmental persistence, contamination of surface water and groundwater with atrazine has been a major concern over the past few decades. However, recent studies have found that atrazine is degraded rapidly in many fields where it has been applied repeatedly over several years. In many of these locations, the half-life of atrazine has been reduced to <5 days in comparison to the 26 to 142 day half-lives found in sites where atrazine

has not previously been applied (Krutz *et al.*, 2010). This enhanced degradation appears to be due to adaptation of the soil microbial communities, largely through acquisition of plasmid-encoded catabolic genes via horizontal gene transfer processes. While this enhanced degradation may be beneficial for reducing the potential environmental impacts of atrazine, it can greatly diminish atrazine's residual activity against weeds thereby necessitating the use of other herbicides or alternative weed control strategies, in order to maintain current levels of crop production.

---

structure similar to those of natural soil organic compounds (Section 3.3.1). Previous exposure to a contaminant through repeated pesticide applications or through frequent spills will create an environment in which a biodegradation pathway is maintained within an adapted community. Adaptation of microbial populations most commonly occurs by induction of enzymes necessary for biodegradation followed by an increase in the population of biodegrading organisms (Leahy and Colwell, 1990).

Naturally occurring analogues of certain contaminants may not exist, and previous exposure may not have occurred. Degradation of these contaminants requires a second type of adaptation that involves a genetic change such as a mutation or a gene transfer (Information Box 17.4). This results in the development of new metabolic capabilities. The time needed for an adaptation requiring a genetic change, or for the selection and development of an adapted community, is not yet predictable, but it may require weeks to years or may not occur at all (van der Meer, 2006).

17.4.2 Toxicity

Chemical spills and engineered remediation projects, such as landfarming of petroleum refinery sludges, can involve extremely high contaminant concentrations. In these cases,

toxicity of the contaminant to microbial populations can slow the remediation process. One common type of toxicity is that associated with nonionic organic contaminants such as petroleum hydrocarbons or organic solvents. This toxicity is mainly due to a nonspecific narcotic-type mode of action, which is based on the partitioning of a dissolved contaminant into the lipophilic layer of the cell membrane, which causes a disruption of membrane integrity (Sikkema *et al.*, 1995). This effect is important because, due to hydrophobic interactions, the cell membrane is a major site of organic contaminant accumulation in microorganisms. In addition, functional groups such as halogens and even the molecular weight of a compound influence its toxicity to microbial cells (Kenawy *et al.*, 2007).

Models have been developed that relate bioconcentration (the accumulation of a hydrophobic contaminant by a cell or organism) and toxicity to the physicochemical attributes or descriptors of the organic contaminant. These models are referred to as quantitative structure–activity relationship (QSAR) models. A number of models have been developed based on different attributes such as structure, functional groups and metabolic pathways of degradation (Pavan and Worth, 2006). Specific descriptors can include hydrophobicity and molecular connectivity (which represents the surface topography of a compound). As might be expected, no one QSAR works for all compounds, in fact some studies show that such models correctly predict biodegradability 68 to 91% of the time depending on the model used and the set of contaminants used to test the models (Tunkel *et al.*, 2005; Pavan and Worth, 2006). Refinement of these models is currently an international effort because the use of QSAR is expected to increase in the future due to: (1) the high costs associated with experimental determination of contaminant persistence, bioconcentration and toxicity, and (2) international pressure to reduce the use of animal testing (Pavan and Worth, 2006).

17.4.3 Bioavailability

For a long time biodegradation was thought to occur if the appropriate microbial enzymes were present. As a result, most research focused on the actual biodegradative process, specifically the isolation and characterization of biodegradative enzymes and genes. There are, however, two steps in the biodegradative process. The first is the uptake of the substrate by the cell, and the second is the metabolism or degradation of the substrate. Assuming the presence of an appropriate metabolic pathway, degradation of a contaminant can proceed rapidly if the contaminant is available in a water-soluble form. However, degradation of contaminants with limited water solubility or those that are strongly sorbed to soil or sediments can be limited due to their low bioavailability (Maier, 2000).

Growth on an organic compound with limited water solubility poses a unique problem for microorganisms. Most microorganisms obtain substrate from the surrounding aqueous phase, but the opportunity for contact between the degrading organism and an organic compound with low water solubility is limited. Such a compound may be present in a liquid or solid state, both of which can form a two-phase system with water. Liquid hydrocarbons can be less or more dense than water, forming a separate phase above or below the water surface. For example, polychlorinated biphenyls (PCBs) and chlorinated solvents such as TCE are denser than water, and form a separate phase below the water surface. Solvents less dense than water, such as benzene and other petroleum constituents, form a separate phase above the water surface. There are three possible modes of microbial uptake of a liquid organic (Figure 17.5):

1. Utilization of the solubilized organic compound (Figure 17.5A).
2. Direct contact of cells with the organic compound. This can be mediated by cell modifications, such as fimbriae (Rosenberg *et al.*, 1982), or cell surface hydrophobicity (Zhang and Miller, 1994), which increase attachment of the cell to the organic compound (Figure 17.5B).

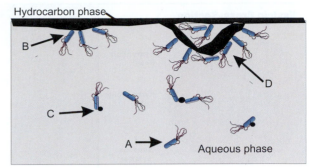

FIGURE 17.5 A water environment with a petroleum oil phase floating on the surface. This is typical of what might occur when oil is spilled in the ocean. There are several ways in which microbes reach the oil phase in this type of situation. (A) Microbes taking up hydrocarbons dissolved in the aqueous phase surrounding degrading cells. This uptake mode becomes limiting as the aqueous solubility of the hydrocarbon decreases. (B) Uptake via direct contact of degrading cells at the aqueous–hydrocarbon interface of large oil drops in water. This uptake mode is limited by the interfacial area between the water and hydrocarbon phase. (C) Uptake through direct contact of degrading cells with fine or submicrometer-size oil droplets dispersed in the aqueous phase. This uptake mode is limited by the formation of such droplets. In the ocean environment, wave action can create substantial dispersion of oil. In a soil environment, such dispersion is more limited. (D) Enhanced uptake as a result of production of biosurfactants or emulsifiers that effectively increase the apparent aqueous solubility of the hydrocarbon, or allow better attachment of cells to the hydrocarbon.

3. Direct contact with fine or submicrometer size substrate droplets dispersed in the aqueous phase (Figure 17.5C).

The mode that predominates depends largely on the water solubility of the organic compound. In general, direct contact with the organic compound plays a more important role (modes 2 and 3) as water solubility decreases.

Some microbes can enhance the rate of uptake and biodegradation as a result of production of biosurfactants or emulsifiers (Figure 17.5D). There are two effects of biosurfactants. First, they can effectively increase the aqueous solubility of the hydrocarbon through formation of micelles or vesicles that associate with hydrocarbons (Figure 17.6). Second, they can facilitate attachment of cells to the hydrocarbon by making the cell surface more hydrophobic, and thus better able to stick to a separate oil phase (Figure 17.7). This makes it possible to achieve greatly enhanced biodegradation rates in the presence of biosurfactants (Herman *et al.*, 1997).

For organic compounds in the solid phase, e.g., waxes, plastics or polyaromatic hydrocarbons (PAHs), there are only two modes by which a cell can take up the substrate:

1. Direct contact with the substrate
2. Utilization of solubilized substrate

Available evidence suggests that for solid-phase organic compounds, utilization of solubilized substrate is

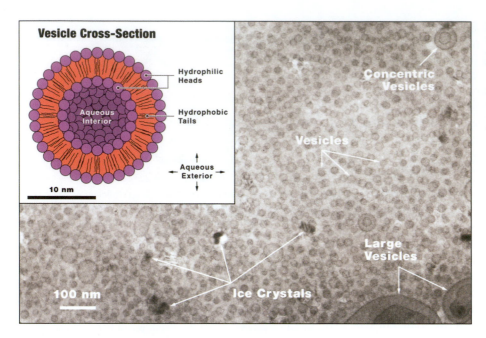

FIGURE 17.6 Cryo-transmission electron micrograph of a microbially produced surfactant, rhamnolipid. In water, this compound spontaneously forms aggregates such as the vesicles shown here. Hydrocarbons like to associate with the lipid-like layer formed by the hydrophobic tails of the surfactant vesicles. These tiny surfactant—hydrocarbon structures are soluble in the aqueous phase. From Pepper *et al.* (2006).

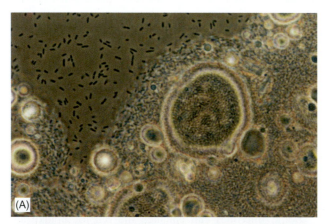

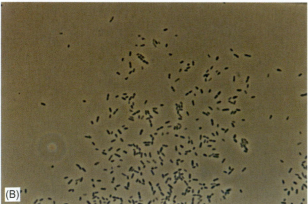

FIGURE 17.7 Phase-contrast micrographs showing the effect of a biosurfactant on the ability of *Pseudomonas aeruginosa* ATCC 15524 to stick to hexadecane droplets (magnification ×1000). (A) Addition of rhamnolipid biosurfactant (0.1 mM) causes cells to clump and to stick to oil droplets. (B) No biosurfactant is present and individual cells do not clump, and do not stick to oil droplets in the solution. Photos courtesy D.C. Herman.

most important. Thus, low water solubility has a greater impact on degradation of solid-phase organic compounds than on liquid-phase organics.

Another factor that affects bioavailability of an organic compound is sorption of the compound by soil or sediment (Novak *et al.*, 1995). Depending on the sorption mechanism, organic compounds can be weakly (hydrogen bonding, van der Waals forces, hydrophobic interactions) or strongly (covalent bonding) bound to soil. Sorption of weakly bound or labile residues is reversible, and when a sorbed residue is released back into solution it becomes available for microbial utilization (Scow, 1993). Bioavailability can also be reduced by the diffusion of contaminants into soil matrix microsites that are inaccessible to bacteria because of pore-size exclusion (Alexander, 1995). There is evidence that the proportion of labile residues made available by desorption decreases with the length of time the residues are in the soil. Thus, as contaminants age and become sequestered more deeply within inaccessible microsites (Figure 4.7), bioavailability, and therefore biodegradation, can be expected to decrease.

Finally, some contaminants may be incorporated into soil organic matter by the catalytic activity of a wide variety of oxidative enzymes that are present in the soil matrix. The incorporation of contaminants into soil organic matter is called humification, a process that is usually irreversible and that may be considered as one factor in the aging process (Bollag, 1992). These bound or humified residues are released and degraded only very slowly as part of the normal turnover of humic material in soil (see Section 16.2.3.2).

17.4.4 Contaminant Structure

17.4.4.1 Steric Effects

Some types of contaminant structures can lead to low degradation rates even if the contaminant structure is similar to naturally occurring molecules. The presence of branching or functional groups can slow degradation by changing the chemistry of the degradation reaction site. The reaction site is the contact area between a degradative enzyme and the contaminant substrate where a transformation step occurs. When the reaction site is blocked by branching or a functional group, contact between the contaminant and enzyme at the reaction site is hindered. This is known as a steric effect and is illustrated in Figure 17.8, which compares two structures, an eight-carbon *n*-alkane (A) and the same eight-carbon backbone with four methyl branches (B). Whereas octane is readily degradable by the pathway shown in Figure 17.12, the four methyl substituents in 3,3,6,6-tetramethyl octane inhibit degradation at both ends of the molecule. Branching or functional groups can also affect transport of the substrate across the cell membrane, especially if the transport is enzyme assisted. Steric effects usually increase as the size of the functional group increases (Pitter and Chudoba, 1990).

17.4.4.2 Electronic Effects

Functional groups may also contribute electronic effects that hinder biodegradation by affecting the interaction between the contaminant and the enzyme. Functional groups can be electron donating (e.g., CH_3) or electron withdrawing (e.g., Cl), and therefore can change the electron density of the reaction site. In general, functional groups which add to the electron density of the reaction site increase biodegradation rates, and functional groups that decrease the electron density of the reaction site decrease biodegradation rates. To illustrate the relationship between functional group electronegativities and rate of biodegradation, Pitter and Chudoba (1990) compared the electronegativity of a series of ortho-substituted phenols with their biodegradation rates. Five different functional groups were tested, and it was found that as the electronegativity of the substituents increased, biodegradation rates decreased (Figure 17.9).

17.5 ENVIRONMENTAL FACTORS AFFECTING BIODEGRADATION

A number of parameters influence the survival and activity of microorganisms in any given environment. One factor that has great influence on microbial activity is organic matter, the primary source of carbon for heterotrophic microorganisms in most environments. Surface soils have a relatively high and variable organic matter content, and therefore are characterized by high microbial numbers and diverse metabolic activity (see Chapter 4). In contrast, the subsurface unsaturated (vadose) zone and saturated zone usually have a much lower content and diversity of organic matter, resulting in lower microbial numbers and activity. Exceptions to this rule are some areas of the saturated zone that have high flow or recharge rates, which can lead to numbers and activities of microorganisms similar to those found in surface soils.

Occurrence and abundance of microorganisms in an environment are determined not only by available carbon but also by various physical and chemical factors. These include oxygen availability, nutrient availability, temperature, pH, salinity and water activity. Inhibition of biodegradation can be caused by a limitation imposed by any one of these factors, but the cause of the persistence of a contaminant is sometimes difficult to determine. Perhaps the most important factors controlling contaminant biodegradation in the environment are oxygen availability, organic matter content, nitrogen availability and contaminant bioavailability. Interestingly, the first three of these factors can change considerably depending on the location of the contaminant. Figure 17.10 shows the relationship between organic carbon, oxygen and microbial activity in a profile of the terrestrial ecosystem including surface soils, the vadose zone and the saturated zone.

17.5.1 Redox Conditions

Redox conditions are very important in determining the extent and rate of contaminant biodegradation. For most contaminants, aerobic biodegradation rates are much higher than anaerobic biodegradation rates. For example, petroleum-based hydrocarbons entering the aerobic zones of freshwater lakes and rivers are generally susceptible to microbial degradation, but oil accumulated in anaerobic

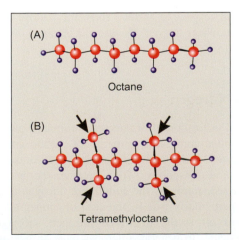

FIGURE 17.8 The structure of (A) octane, which is readily degradable, and (B) a tetramethyl-substituted octane, which is not degradable because the methyl groups block the enzyme—substrate catalysis site.

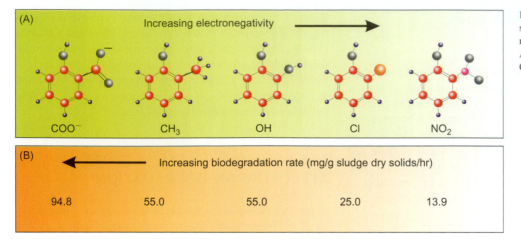

(A)

Increasing electronegativity →

COO⁻   CH₃   OH   Cl   NO₂

(B)

← Increasing biodegradation rate (mg/g sludge dry solids/hr)

94.8   55.0   55.0   25.0   13.9

FIGURE 17.9 Various ortho-substituted phenols and their respective biodegradation rates. Adapted from Pitter and Chudoba (1990).

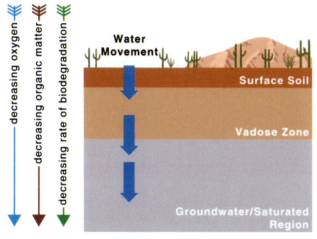

FIGURE 17.10 There are three major locations where contamination can occur in terrestrial ecosystems: surface soils, the vadose zone and the saturated zone. The availability of both oxygen and organic matter varies considerably in these zones. As indicated, oxygen and organic matter both decrease with depth, resulting in a decrease in biodegradation activity with depth. From Pepper *et al.* (2006).

sediments can be highly persistent. Oxygen is especially important for degradation of highly reduced hydrocarbons such as the alkanes. For example, low molecular weight alkanes such as methane do not degrade anaerobically. Higher molecular weight alkanes, such as hexadecane ($C_{16}H_{34}$), can occur, but degradation is very limited, and usually is only found in historically petroleum-contaminated sites. In contrast, there are some highly chlorinated compounds (e.g., perchloroethene (PCE)) that are recalcitrant under aerobic conditions, but amenable to biotransformation under anaerobic conditions.

17.5.2 Organic Matter Content

Surface soils have large numbers of microorganisms. Bacterial numbers commonly range from 10^7 to 10^{10} per

gram of soil with somewhat lower fungal numbers, 10^5 to 10^6 per gram of soil. In contrast, microbial populations in deeper regions such as the deep vadose zone and groundwater region are often lower by two orders of magnitude or more (see Chapter 4). This large decrease in microbial numbers with depth is due primarily to differences in organic matter content. Both the vadose zone and the groundwater region have low amounts of organic matter. One result of low total numbers of microorganisms is that a low population of contaminant degraders may be present initially. Thus, biodegradation of a particular contaminant may be slow until a sufficient biodegrading population has been built up. A second reason for slow biodegradation in the vadose zone and groundwater region is that because a low amount of organic matter is present, the organisms in this region are often dormant. This can cause their response to an added carbon source to be slow, especially if the carbon source is a contaminant molecule that has low bioavailability, or to which the organisms have not had prior exposure.

Because of these trends in oxygen availability and organic matter content, several generalizations can be made with respect to surface soils, the vadose zone and the groundwater region (Figure 17.10 and Information Box 17.5).

17.5.3 Nitrogen

Microbial utilization of organic contaminants, particularly hydrocarbons composed primarily of carbon and hydrogen, creates a demand for essential nutrients such as nitrogen and phosphorus. Thus, biodegradation can often be improved simply by the addition of nitrogen fertilizers. This is particularly true in the case of biodegradation of petroleum oil spills, in which nitrogen shortages can become acute. In general, microbes have an average C:N ratio within their biomass of about 5:1 to 10:1 depending on the type of microorganism. Therefore, a ratio of

Information Box 17.5   Biodegradation in Surface and Subsurface Environments

- Biodegradation in surface soils is primarily aerobic and rapid. Zones of anaerobic activity develop in areas of high microbial activity.
- Biodegradation in the vadose zone is also primarily aerobic. Low organic matter content results in low microbial numbers leading to significant acclimation times to allow biodegrading populations to build up.

- Biodegradation in the deep groundwater region is also initially slow because of low microbial numbers. Conditions can rapidly become anaerobic because of lack of available dissolved oxygen. Biodegradation in shallow groundwater regions is initially more rapid because of higher microbial numbers but is similarly slowed by low dissolved oxygen availability.

approximately 100:10:1 (C:N:P) is often used in such sites. However, in some instances, quite different ratios have been used. For example, Wang and Bartha (1990) found that effective remediation of hydrocarbons in soil required the addition of nitrogen and phosphorus to maintain a C:N ratio of 200:1 and a C:P ratio of 1000:1. Why were the C:N and C:P ratios maintained at levels so much higher than the cell C:N and C:P ratios? As discussed in Section 16.3.3, it is because much of the hydrocarbon that is metabolized is released as carbon dioxide, so that much of the carbon is lost from the system. In contrast, almost all of the nitrogen and phosphorus metabolized is incorporated into microbial biomass, and thus is conserved in the system.

17.5.4 Other Environmental Factors

17.5.4.1 Temperature

Hydrocarbon degradation has been reported to occur at a range of temperatures from close to freezing to more than 30°C. Bacteria can adapt to temperature extremes in order to maintain metabolic activity; however, seasonal temperature fluctuations in the natural environment have been shown to affect the rate at which degradation occurs (Palmisano et al., 1991). For example, the degradation rates of hexadecane and naphthalene in a river sediment were reduced approximately 4.5-fold and 40-fold, respectively, in winter (0−4°C) compared with summer (8−21°C) samples (Wyndham and Costerton, 1981).

17.5.4.2 pH

In soils, the rate of hydrocarbon degradation is influenced by pH with the highest rates generally observed at neutral pH. However, microorganisms have been isolated from historically contaminated sites that have adapted to growth on hydrocarbons even at very acidic pH levels (pH 2−3). It has been observed that the diversity of these microorganisms is lower than for their counterparts that grow at neutral pH (Uyttebroek et al., 2007).

17.5.4.3 Salinity

In typical terrestrial or freshwater ecosystems, co-contamination with moderate-to-high levels of salinity tends to slow hydrocarbon degradation (Ulrich et al., 2009). In marine ecosystems, hydrocarbons are frequently introduced naturally from oil seeps and natural gas deposits as well as anthropogenically from oil tanker spills and discharges; therefore, marine environments tend to contain microbial populations adapted for degradation of hydrocarbons under the salinity levels typically found in these ecosystems (see also Chapter 31). However, elevated salinity levels may also slow degradation even in marine ecosystems (Mille et al., 1991). Also, in contrast to hydrocarbon degraders found in soil which exhibit great diversity, marine hydrocarbon degraders seem to be a small group of specialized obligate hydrocarbon utilizers that have been named marine obligate hydrocarbonoclastic bacteria (OHCB). The OHCB respond rapidly to the addition of hydrocarbons, and have been shown to transiently increase up to 90% of the total microbial community in response to added hydrocarbon (Yakimov et al., 2007). Recognized OHCB genera include Alcanivorax, Cycloclasticus, Marinobacter, Thallassolituus and Oleispira. Related bacteria were found to dominate the bacterial community within a "cloud" of dispersed oil in the Gulf of Mexico following the BP Deepwater Horizon oil leak in 2010 (see Case Study 31.1; Hazen et al., 2010). Which genera predominate in any given environment depends on environmental conditions. For example, Oleispira are psychrophiles that dominate in cold waters.

17.5.4.4 Water Activity

Optimal conditions for activity of aerobic soil microorganisms occur when between 38% and 81% of the soil pore space is filled with water (also referred to as percent saturation). In this range of water content, water and oxygen availability are maximized. At higher water contents, the slow rate of oxygen diffusion through water limits oxygen replenishment, thereby limiting aerobic activity. At lower water contents, water availability becomes limiting. Why is the optimal percent saturation range so

broad? It is because optimal activity really depends upon a combination of factors including water content and available pore space. Available pore space is measured as bulk density, which is defined as the mass of soil per unit volume (g/cm^3). This means that in any given soil, increasing bulk density indicates increasing compaction of the soil. In a soil that is loosely compacted (lower bulk density), a water saturation of 70% represents more water (more filled small pores and pore throats) than in a highly compacted soil. Therefore, in a soil with low bulk density oxygen, diffusion constraints become important at lower water saturation than for highly compacted soils.

17.6 BIODEGRADATION OF ORGANIC POLLUTANTS

17.6.1 Pollutant Sources and Types

In 2011, the United States used more than 6.8 billion barrels of oil for heating, generation of electricity and transportation (USEIA, 2012). Other sources of energy are coal, natural gas and nuclear energy. Large amounts of waste, including solvents, acids, bases and metals, are also produced by the paper, transportation, electronics, defense and metals industries. The EPA estimates that the global market for pesticides in 2007 was over $39 billion of which the United States accounted for 32% of the market (Grube *et al.*, 2011). As these figures demonstrate, both industry and agriculture produce large amounts of chemicals. Inevitably, some of these find their way into the environment as a result of normal handling procedures and accidental spills.

Figure 17.11 shows various contaminant molecules that are added to the environment in significant quantities by anthropogenic activities (see Information Box 17.6). The structure of most contaminant molecules is based on one of the first three structures shown in Figure 17.11: aliphatic, alicyclic or aromatic. By combining or adding to these structures, a variety of complex molecules can be formed that have unique properties useful in industry and agriculture. The objective of this section is to become familiar with these structures and their biodegradation pathways so that given a structure, a reasonable biodegradation pathway can be predicted. An excellent resource for obtaining biodegradation information for a variety of compounds under both aerobic and anaerobic conditions is the University of Minnesota Biocatalysis/Biodegradation Database (http://umbbd.msi.umn.edu/).

17.6.2 Aliphatics

There are several common sources of aliphatic hydrocarbons that enter the environment as contaminants. These include: straight-chain and branched-chain structures found in petroleum hydrocarbons; the linear alkyl benzenesulfonate (LAS) detergents; and the one- and two-carbon halogenated compounds such as chloroform and TCE that are commonly used as industrial solvents. Some general rules for aliphatic biodegradation are presented in Information Box 17.7, and specific biodegradation pathways are summarized in the following sections for alkanes/alkenes and chlorinated aliphatics.

17.6.2.1 Alkanes

Aerobic Conditions

Because of their structural similarity to fatty acids and plant paraffins, which are ubiquitous in nature, many microorganisms in the environment can utilize *n*-alkanes (straight-chain alkanes) as a sole source of carbon and energy. In fact, it is easy to isolate alkane-degrading microbes from any environmental sample. As a result, alkanes are usually considered to be the most readily biodegradable type of hydrocarbon. Biodegradation of alkanes occurs with a high biological oxygen demand (BOD) using one of the two pathways shown in Figure 17.12. The more common pathway is the direct incorporation of one atom of oxygen onto one of the end carbons of the alkane by a monooxygenase enzyme resulting in the formation of a primary alcohol. Alternatively, a dioxygenase enzyme can incorporate both oxygen atoms into the alkane to form a hydroperoxide. The end result of both pathways is the production of a primary fatty acid. There are also examples in the literature of diterminal oxidation, with both ends of the alkane oxidized, and of subterminal oxidation, with an interior carbon oxidized (Britton, 1984).

Fatty acids are common metabolites found in all cells. They are used in the synthesis of membrane phospholipids and lipid storage materials. The common pathway used to catabolize fatty acids is known as β-oxidation, a pathway that cleaves off consecutive two-carbon fragments (Figure 17.12). Each two-carbon fragment is removed by coenzyme A as acetyl-CoA, which then enters the tricarboxylic acid (TCA) cycle for complete mineralization to CO_2 and H_2O. If you think about this process, it becomes apparent that if one starts with an alkane that has an even number of carbons, the two-carbon fragment acetyl-CoA will be the last residue. If one starts with an alkane with an odd number of carbons, the three-carbon fragment propionyl-CoA will be the last residue. Propionyl-CoA is then converted to succinyl-CoA, a four-carbon molecule that is an intermediate of the TCA cycle.

What types of alkanes do microbes most prefer? In general, midsize straight-chain aliphatics (*n*-alkanes C_{10} to C_{18} in length) are utilized more readily than *n*-alkanes with either shorter or longer chains. Long-chain *n*-alkanes are utilized more slowly because of low bioavailability

Hydrocarbon Type	Structure	Name	Physical state at room temp.	Source and uses
Aliphatics		propane n=1	gas	Petroleum contains both linear and branched aliphatics. The gasoline fraction of crude oil is 30–70% aliphatic depending on the source of the crude oil.
		hexane n=4	liquid	
		hexatriacontane n=34	solid	
Alicyclics		cyclopentane	liquid	Petroleum contains both unsubstituted and alkyl substituted alicyclics. The gasoline fraction of crude oil is 20–70% alicyclic depending on the source of the crude oil.
	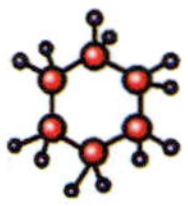	cyclohexane	liquid	
Aromatics	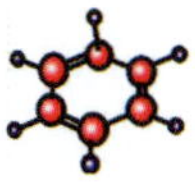	benzene	liquid	Petroleum contains both unsubstituted and alkyl substituted aromatics. The gasoline fraction of crude oil is 10–15% depending on the source of the crude oil.
	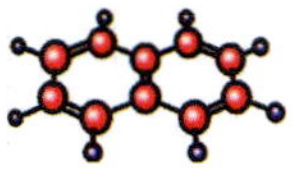	naphthalene	solid	
	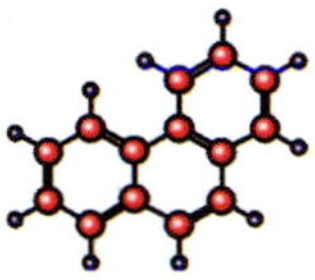	phenanthrene	solid	
Substituted aliphatics	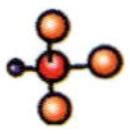	chloroform	liquid	Anthropogenically manufactured and used as solvents and degreasing agents, and in organic syntheses.
		trichloroethene (TCE)	liquid	
Substituted aromatics	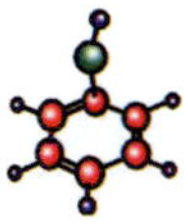	phenol	liquid	Found in coal tar or manufactured and used as a disinfectant; and in manufacture of resins, dyes and industrial chemicals.
		pentachlorophenol	liquid	Manufactured and used as an insecticide, defoliant, and wood preservative.
	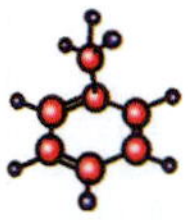	toluene	liquid	Found in tar oil; used in manufacture of organics, explosives, and dyes. Also used as a solvent.
	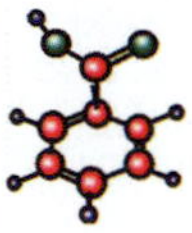	benzoate	liquid	Found in plants and animals and manufactured for use as a food preservative and dye component, and in curing tobacco.

FIGURE 17.11   Representative pollutant structures.   (Continued)

Biaryl hydrocarbons	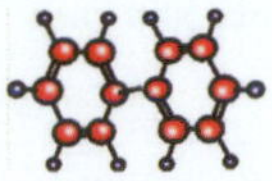	biphenyl	solid	Biphenyl is the parent compound of variously chlorinated biphenyl mixtures known as the PCBs. PCBs are used as transformer oils and plasticizers.
	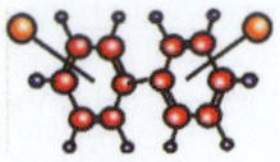	polychlorinated biphenyls (PCBs)	liquid	
Heterocyclics	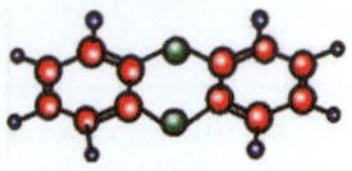	dibenzodioxin	solid	Dioxins are created during incineration processes and are contaminants associated with the manufacture of herbicides including 2,4-D and 2,4,5-T.
	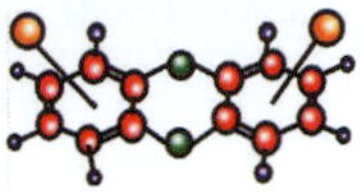	chlorinated dioxins	solid	
		pyridine	liquid	Found in coal tar. Used as a solvent and synthetic intermediate.
	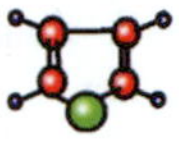	thiophene	liquid	Found in coal tar, coal gas and crude oil. Used as a solvent and in manufacture of resins, dyes, and pharmaceuticals.
Pesticides Organic acids		2,4-dichlorophenoxy acetic acid	solid	Broadleaf herbicide
Organophosphates	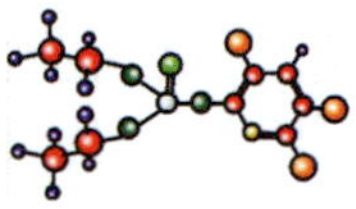	chlorpyrifos	solid	Used as an insecticide and an acaricide
Triazenes	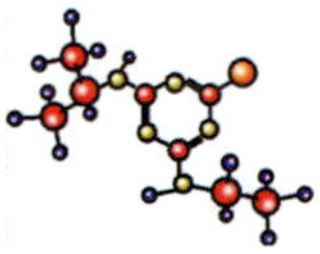	atrazine	solid	Selective herbicide
Carbamates	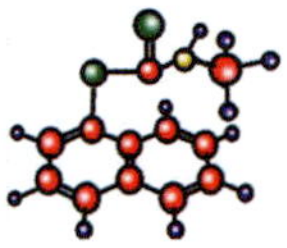	carbaryl	solid	Contact insecticide
Chlorinated hydrocarbons	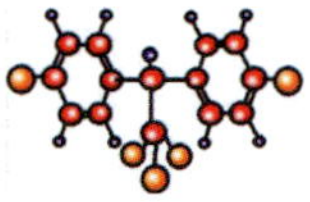	1,1,1-trichloro-2,2-bis-(4-chlorophenyl)-ethane (DDT)	solid	Contact insecticide
		methyl bromide	gas	Used to degrease wool, extract oil from nuts, seeds and flowers, used as an insect and soil fumigant.

Carbon
Hydrogen
Oxygen
Chlorine
Nitrogen
Sulfur
Phosphorus
Bromine

FIGURE 17.11 (Continued).

Information Box 17.6   The 2011 Agency for Toxic Substances Disease Registry (ATSDR) Top Twenty Pollutants

By U.S. law, the ATSDR and EPA are required to prepare a list, in order of priority, of substances that are most commonly found at facilities on the National Priorities List (NPL) and which are determined to pose the most significant potential threat to human health due to their known or suspected toxicity, and potential for human exposure at these NPL sites. This list is revised every two years as additional information becomes available.

1 Arsenic
2 Lead
3 Mercury
4 Vinyl Chloride
5 Polychlorinated Biphenyls
6 Benzene
7 Cadmium

8 Benzo(a)pyrene
9 Polycyclic Aromatic Hydrocarbons
10 Benzo(b)fluoranthene
11 Chloroform
12 Aroclor 1260
13 DDT
14 Aroclor 1254
15 Dibenzo(a,h)anthracene
16 Trichloroethene
17 Chromium, Hexavalent
18 Dieldrin
19 Phosphorus, White
20 Hexachlorobutadiene

From: the Agency for Toxic Substances & Disease Registry, 2011.

Information Box 17.7   General Rules for Degradation of Non-halogenated Aliphatic Compounds

- Aerobic conditions:
  - aliphatics are readily degraded
  - midsize straight-chain aliphatics (*n*-alkanes C_{10} to C_{18} in length) are utilized more readily than *n*-alkanes with either shorter or longer chains
  - saturated aliphatics and alkenes are degraded similarly
  - hydrocarbon branching decreases biodegradability

- Anaerobic conditions:
  - degradation is very limited for low molecular weight alkanes; higher molecular weight alkanes, e.g., hexane, are biodegraded following "activation" through the addition of fumarate
  - "activated" aliphatics including alkenes, alchohols, or acids are readily degraded

---

resulting from extremely low water solubilities. For example, the water solubility of decane (C_{10}) is 0.052 mg/L, and the solubility of octadecane (C_{18}) is almost 10-fold less (0.006 mg/L). Solubility continues to decrease with increasing chain length. In contrast, short-chain *n*-alkanes have higher aqueous solubility, e.g., the water solubility of butane (C_4) is 61.4 mg/L, but they are toxic to cells. Short-chain alkanes are toxic to microorganisms because their increased water solubility results in increased uptake of the alkanes, which are then dissolved in the cell membrane. The presence of these short alkanes within the cell membrane can alter the fluidity and integrity of the cell membrane.

The toxicity of short-chain *n*-alkanes can be mediated in some cases by the presence of free-phase oil droplets. Protection occurs because the short-chain alkanes partition into the oil droplets. This results in reduced bioavailability because the aqueous phase concentration is decreased. Thus, *n*-alkane degradation rates will differ depending on whether the substrate is present as a pure compound or in a mixture of compounds.

Biodegradability of aliphatics is also negatively influenced by branching in the hydrocarbon chain. The degree of resistance to biodegradation depends on both the number of branches and the positions of methyl groups in the molecule. Compounds with a quaternary carbon atom (four carbon—carbon bonds) such as that shown in Figure 17.8B are extremely stable because of steric effects, as discussed in Section 17.4.4.1.

Alkenes are hydrocarbons that contain one or more double bonds. The majority of alkene biodegradability studies have used 1-alkenes as model compounds (Britton, 1984). These studies have shown that alkenes and alkanes have comparable biodegradation rates. The initial step in 1-alkene degradation can involve attack at the terminal or a subterminal methyl group as described for alkanes. Alternatively, the initial step can be attack at the double bond, which can yield a primary or secondary alcohol or an epoxide. Each of these initial degradation products is further oxidized to a primary fatty acid, which is degraded by β-oxidation as shown in Figure 17.12 for alkanes.

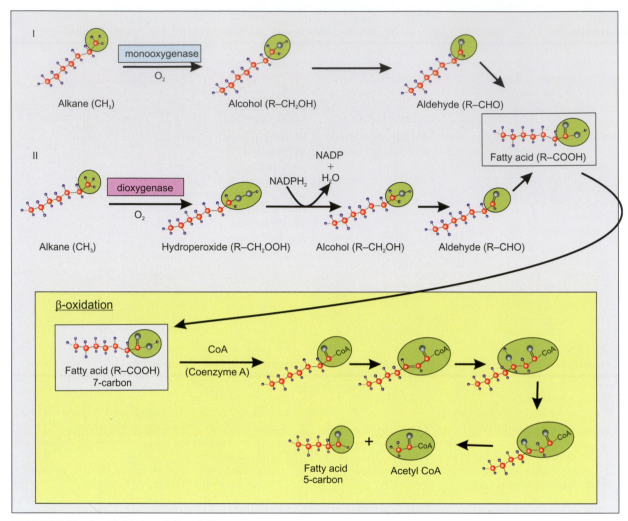

FIGURE 17.12 Aerobic biodegradation of alkanes.

Anaerobic Conditions

In comparison to aerobic conditions, aliphatic hydrocarbons, which are highly reduced molecules, are degraded slowly, if at all, anaerobically (Figure 17.13). This is supported by the fact that hydrocarbons in natural underground reservoirs of oil (which are under anaerobic conditions) are not degraded despite the presence of microorganisms. The current view is that low molecular weight alkanes (e.g., methane) do not energetically support anaerobic degradation. This is because they must be activated or functionalized prior to degradation. In contrast, higher molecular weight aliphatics have been shown to undergo degradation using a unique pathway, where the first step in biodegradation involves activation through the addition of a four-carbon oxygen-containing moiety, fumarate, into the alkane (Figure 17.14). Once oxygen has been introduced into the molecule through the addition of

fumarate, it is mineralized, although the exact pathway has not been elucidated (Widdel and Rabus, 2001). Aliphatics that are already activated, including both alkenes and aliphatics containing oxygen (aliphatic alcohols and ketones), are readily biodegraded anaerobically.

17.6.2.2 Halogenated Aliphatics

Chlorinated solvents such as trichloroethene (TCE, $Cl_2=CHCl$) and perchloroethene (PCE, $Cl_2C=CCl_2$) have been extensively used as industrial solvents. As a result of improper use and disposal, these solvents are among the most frequently detected types of organic contaminants in groundwater. The need for efficient and cost-effective remediation of solvent-contaminated sites has stimulated interest in the biodegradation of these C_1 and C_2 halogenated aliphatics. General rules for biodegradation of chlorinated aliphatics are provided in Information Box 17.8.

Substrate	Structure	Aerobic Biodegradation	Anaerobic Biodegradation
Lower MW alkanes		+++	−/+, only denitrifying conditions
Higher MW alkanes		+++	+/−, only in historically petroleum contaminated sites
Alkenes		+++	+/−, only in historically petroleum contaminated sites
Phenols	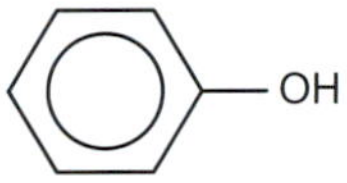 ⬡—OH	+++	++
Aromatic Acids	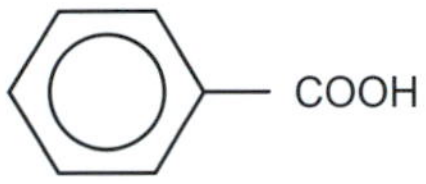 ⬡—COOH	+++	++
Alkylphenols	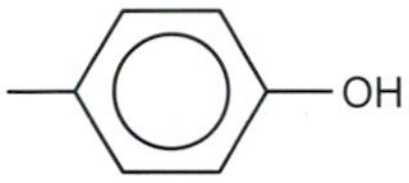 —⬡—OH	+++	++
BTEX (benzene, toluene, xylenes)	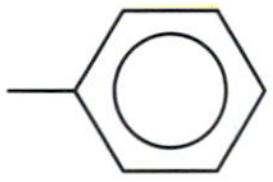 ⬡	+++	+/−, only in historically petroleum contaminated sites
PAH (polyaromatic hydrocarbons) up to three rings	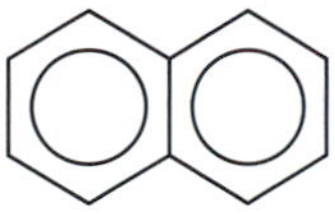 ⬡⬡	+++	+/−, only in historically petroleum contaminated sites
Lignin/Humus		++	

FIGURE 17.13   A comparison of biodegradability of aliphatic and aromatic hydrocarbons under aerobic and anaerobic conditions. A (−) means no biodegradation; (+) means increasing ease of biodegradation. An increasing number of (+) means increasing ease of biodegradation. Courtesy James A. Field.

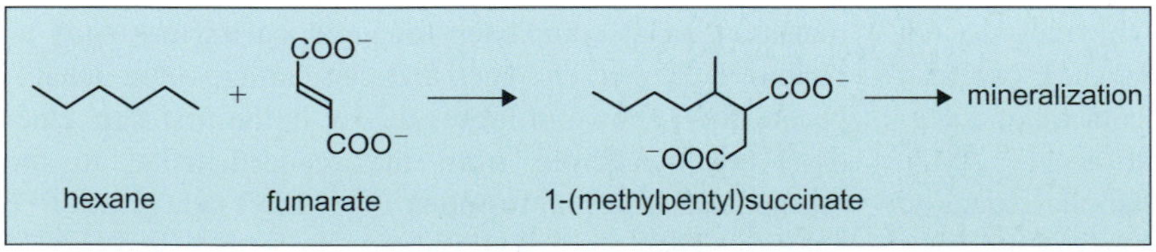

FIGURE 17.14   Anaerobic biodegradation of alkanes.

Aerobic Conditions

Under aerobic conditions, halogenated aliphatics are generally degraded more slowly than aliphatics without halogen substitution. For example, although 1-chloroalkanes ranging from C_1 to C_{12} are degraded as a sole source of carbon and energy in pure culture, they are degraded more slowly than their nonchlorinated counterparts. The presence of two or three chlorines bound to the same carbon atom inhibits aerobic degradation (Janssen et al., 1990). For example, while TCE is degraded under aerobic conditions, PCE is not. These results can be explained by the decreasing electronic effects of the chlorine atom on the enzyme−carbon reaction center as the alkane chain length increases (see Section 17.4.4.2).

Biodegradation of halogenated aliphatics occurs by one of two basic types of reactions (Figure 17.15). Substitution

Information Box 17.8   General Rules for Degradation of Chlorinated Aliphatics

- Aerobic conditions
  - aliphatics with one or two chlorines are generally mineralized
  - as the number of halogens increases, biodegradation rates decrease and reactions tend to become cometabolic
  - highly chlorinated aliphatics, e.g. PCE, are recalcitrant

- Anaerobic conditions
  - most chlorinated aliphatics undergo partial transformation, not mineralization
  - aliphatics with two or more chlorines are transformed either through cometabolism (slow) or halorespiration (rapid)

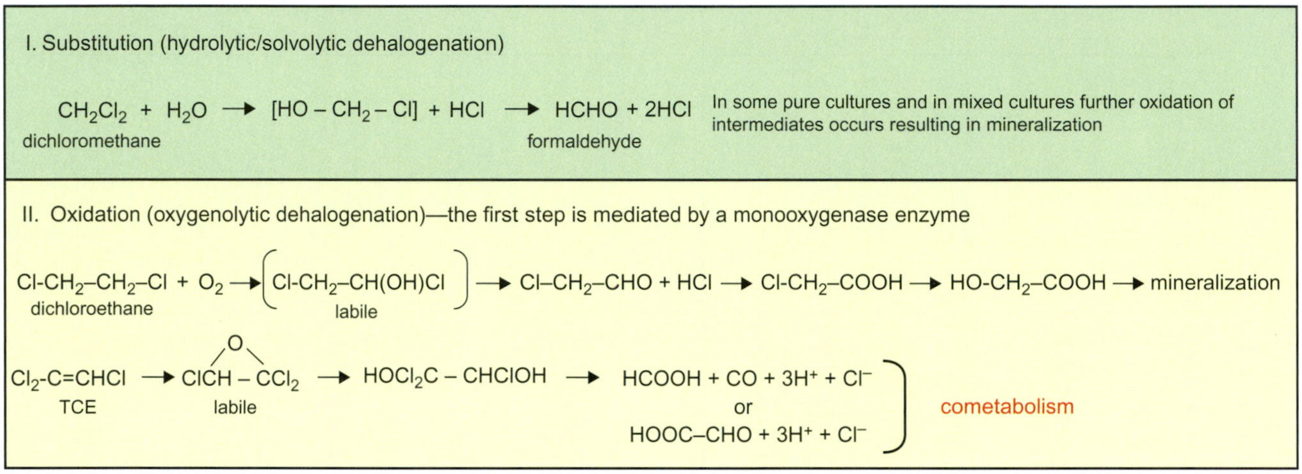

I. Substitution (hydrolytic/solvolytic dehalogenation)

$$CH_2Cl_2 + H_2O \rightarrow [HO-CH_2-Cl] + HCl \rightarrow HCHO + 2HCl$$

dichloromethane   formaldehyde

In some pure cultures and in mixed cultures further oxidation of intermediates occurs resulting in mineralization

II. Oxidation (oxygenolytic dehalogenation)—the first step is mediated by a monooxygenase enzyme

$$Cl\text{-}CH_2\text{--}CH_2\text{-}Cl + O_2 \rightarrow \left(Cl\text{-}CH_2\text{--}CH(OH)Cl \right) \rightarrow Cl\text{--}CH_2\text{--}CHO + HCl \rightarrow Cl\text{--}CH_2\text{--}COOH \rightarrow HO\text{-}CH_2\text{--}COOH \rightarrow \text{mineralization}$$

dichloroethane   labile

$$Cl_2\text{-}C{=}CHCl \rightarrow Cl\overset{O}{CH}-CCl_2 \rightarrow HOCl_2C-CHClOH \rightarrow \begin{array}{c} HCOOH + CO + 3H^+ + Cl^- \\ \text{or} \\ HOOC\text{--}CHO + 3H^+ + Cl^- \end{array}$$

TCE   labile   cometabolism

FIGURE 17.15   Aerobic degradation pathways for chlorinated aliphatics via (I) substitution or (II) oxidation.

is a nucleophilic reaction (the reacting species brings an electron pair) in which the halogens on a mono- or dihalogenated compound are substituted by a hydroxy group. Oxidation reactions are catalyzed by a select group of monooxygenase and dioxygenase enzymes that have been reported to oxidize highly chlorinated C_1 and C_2 compounds (e.g., TCE). These monooxygenase and dioxygenase enzymes are produced by bacteria and oxidize a variety of nonchlorinated compounds including methane, ammonia, toluene and propane. These enzymes do not have exact substrate specificity, and thus they can also participate in either the metabolic or the cometabolic degradation of chlorinated aliphatics (Bhatt *et al.*, 2007). Figure 17.15 shows an example of a metabolic oxidation that supports growth (dichloroethane) and an example of a cometabolic oxidation (TCE). For cometabolic reactions, usually, a large ratio of enzyme substrate to chlorinated aliphatic is required to achieve cometabolic degradation of the chlorinated aliphatic (Section 17.3, Figure 17.3).

Anaerobic Conditions

In some very limited instances, C_1 chlorinated aliphatics such as chloromethane and dichloromethane can serve as a source of carbon and energy to support growth; however,

chlorinated aliphatics are generally metabolized under anaerobic conditions primarily through two processes: (1) cometabolism; and (2) use as a terminal electron acceptor to support growth, a process called halorespiration. Both of these processes usually result in partial transformation of the substrate rather than complete mineralization. In general, the process of removing chlorines under anaerobic conditions is referred to as reductive dehalogenation. Reductive dehalogenation can be mediated by reduced transition metal complexes found in coenzymes such as vitamin B12 or in enzymes that can participate in dehalogenation. As shown in Figure 17.16, in the first step, electrons are transferred from the reduced metal to the halogenated aliphatic, resulting in an alkyl radical and free halogen. Then the alkyl radical can either scavenge a hydrogen atom (1), or lose a second halogen to form an alkene (2). Conversely, a small group of bacteria, specifically *Dehalococcoides* spp. including *D. ethenogenes* strain 195, can use even highly chlorinated compounds such as PCE and TCE directly as terminal electron acceptors and sequentially dechlorinate them completely to form ethene (Fennel *et al.*, 2004; Figure 17.2).

It is now clearly recognized that reductive dehalogenation is a very important process in contaminated environments for highly chlorinated aliphatics and chlorinated

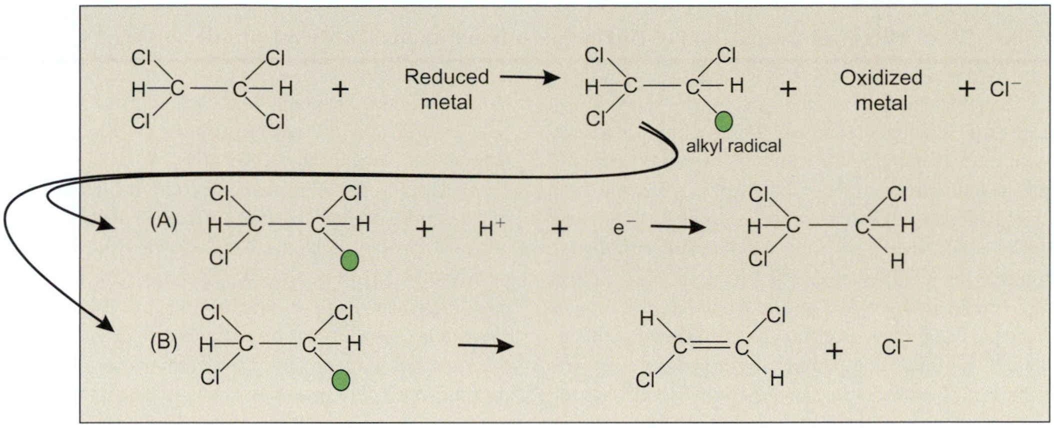

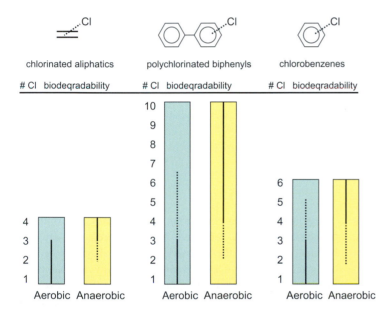

FIGURE 17.16 Reductive dehalogenation of tetrachloroethane to trichloroethane (A) or dichloroethene (B).

chlorinated aliphatics

polychlorinated biphenyls

chlorobenzenes

Cl biodegradability

Cl biodegradability

Cl biodegradability

Aerobic Anaerobic

Aerobic Anaerobic

Aerobic Anaerobic

FIGURE 17.17 The effect of increasing numbers of chlorines on biodegradability under aerobic and anaerobic conditions. Solid lines indicate ready biodegradation, dashed lines indicate biodegradation has been observed on some occasions, no line indicates no biodegradation. Courtesy James A. Field.

compounds in general. This is because anaerobic conditions favor the degradation of highly chlorinated compounds, whereas aerobic conditions favor the degradation of mono- and di-substituted halogenated compounds (Figure 17.17). Recall that TCE and PCE are among the most common groundwater contaminants. Under aerobic conditions, PCE is inert, and while TCE can be cometabolized aerobically, this process requires optimization of the electron donor to TCE ratio. However, under anaerobic conditions, halorespiration of both TCE and PCE can occur quite rapidly resulting in lesser chlorinated metabolites that become amenable for aerobic biodegradation. If appropriate populations of microorganisms (e.g., *Dehalococcoides* spp.) are present, TCE and PCE can even be completely dechlorinated to ethene through use of the chlorinated organic as a terminal electron acceptor during metabolism of a corresponding electron donor (e.g., H_2 or a C_1 or C_2 organic compound such as ethanol). Interestingly, the chlorinated, inorganic

compound perchlorate (ClO_4^-) can also be degraded by a similar mechanism (Information Box 17.9).

17.6.3 Alicyclics

Alicyclic hydrocarbons (Figure 17.11) are major components of crude oil, 20 to 70% by volume. They are commonly found elsewhere in nature as components of plant oils and paraffins, microbial lipids and pesticides (Trudgill, 1984). The various components can be simple, such as cyclopentane and cyclohexane, or complex, such as trimethylcyclopentane and various cycloparaffins. The use of alicyclic compounds in the chemical industry, and the release of alicyclics to the environment through industrial processes, other than oil processing and utilization, is more limited than for aliphatics and aromatics. Consequently, the issue of health risks associated with human exposure to

---

Information Box 17.9   Microbial Degradation of Perchlorate—Implications for Microbial Life on Mars?

Perchlorate is a contaminant of emerging concern—especially in groundwater. It exists as an anion consisting of a chlorine atom surrounded by four oxygen atoms (ClO_4^-), and has been used in a variety of products including rocket fuel and fireworks. It has now been detected as a contaminant in water supplies in over 20 U.S. States (Information Box 17.1). Since perchlorate can interfere with thyroid activity in humans, the USEPA has decided to regulate perchlorate under the Safe Drinking Water Act, and is developing a proposed national primary drinking water regulation for perchlorate. Although perchlorate is an inorganic compound, its degradation pathway shares some similarities with the degradation of the chlorinated solvents TCE and PCE. For example, a wide array of bacteria are now known to be capable of using perchlorate as a terminal electron acceptor, under anaerobic conditions, during the metabolism of

electron donors including hydrogen, organic acids, alcohols and reduced iron, and ultimately converting the perchlorate into chloride (Coates and Achenbach, 2004). The knowledge of these microbial transformations of perchlorate along with the discovery of high levels of perchlorate on Mars have recently led some to speculate about the potential for the perchlorate to enable the existence of microbial life on Mars (McKay *et al.*, 2013). Obviously, one way that perchlorate could potentially support microbial life on Mars is by serving as an electron acceptor during the microbial reduction of ferrous iron or other electron donors. In addition, perchlorate could indirectly support microbial life through its ability to depress the freezing-point of water thus extending the temperature range under which life may be possible.

Information Box 17.10   General Rules for Degradation of Alicyclics

- Aerobic conditions
  - alicyclics are readily degradable
  - unsubstituted alicyclics typically form a lactone intermediate
  - a consortium is usually required for mineralization of unsubstituted alicyclics

- Anaerobic conditions
  - alicyclics are readily degradable under sulfate-reducing conditions
  - degradation is more limited under methanogenic conditions

---

alicyclics has not reached the same level of importance as for the other classes of compounds, especially the aromatics. As a result, far less research has focused on the study of alicyclic biodegradation (see Information Box 17.10).

Aerobic Conditions

It is difficult to isolate pure cultures that degrade alicyclic hydrocarbons using enrichment techniques. Although microorganisms with complete degradation pathways have been isolated (Trower *et al.*, 1985), alicyclic hydrocarbon degradation is thought to occur primarily by commensalistic and cometabolic reactions as shown for cyclohexane in Figure 17.18. In this series of reactions, one organism converts cyclohexane to cyclohexanol (step 1) cometabolically during the oxidation of propane, but is unable to further transform the compound. A second organism that is unable to oxidize cyclohexane to cyclohexanol can perform the subsequent transformations (step 2 onward) including lactonization, ring opening and mineralization of the remaining aliphatic compound (Perry, 1984).

Cyclopentane and cyclohexane derivatives that contain one or two OH, C=O or COOH groups are readily metabolized, and such degraders are easily isolated from environmental samples. In contrast, degradation of

alicyclic derivatives containing one or more CH_3 groups is inhibited. This is reflected in the decreasing rate of biodegradation for the following series of alkyl derivatives of cyclohexanol: cyclohexanol > methylcyclohexanol > dimethylcyclohexanol (Pitter and Chudoba, 1990).

Anaerobic Conditions

Anaerobic biodegradation of complex mixtures of alicyclic compounds in gas condensate has been demonstrated under methanogenic and sulfate-reducing conditions. Gas condensate is the mixture of hydrocarbons in raw natural gas. It is a mixture containing primarily aliphatics (C_2–C_{12}), cyclopentanes, cyclohexanes and aromatics (BTEX = benzene, toluene, ethylbenzene, xylene). In examining the fate of the alicyclic components of gas condensate, sulfate-reducing conditions were found to support the anaerobic biodegradation of unsubstituted cyclopentanes and cyclohexanes as well as those with one methyl or ethyl substitution. In contrast, biodegradation under methanogenic conditions was much less extensive (Townsend *et al.*, 2004). Some alicyclics showed recalcitrance under both methanogenic and sulfate-reducing conditions including dimethyl-substituted cyclopentanes and cyclohexanes.

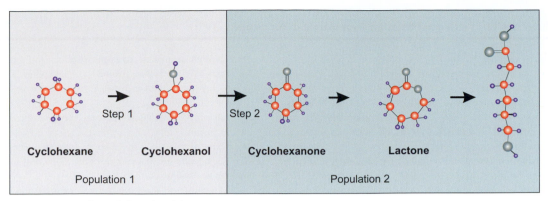

FIGURE 17.18 Degradation of cyclohexane.

17.6.4 Aromatics

Aromatic compounds contain at least one unsaturated ring system with the general structure C_6R_6, where R is any functional group (Figure 17.11). Benzene (C_6H_6) is the parent hydrocarbon of this family of unsaturated cyclic compounds. Compounds containing two or more fused benzene rings are called polyaromatic hydrocarbons (PAHs; also known as polycyclic aromatic hydrocarbons). Aromatic hydrocarbons are natural products; they are part of lignin and are formed as organic materials are burned, for example, in forest fires. However, the addition of aromatic compounds to the environment has increased dramatically through activities such as fossil fuel processing, and utilization and burning of wood and coal.

The quantity and composition of the aromatic hydrocarbons are of major concern when evaluating a contaminated site because several components of the aromatic fraction have been shown to be carcinogenic to humans. Aromatic compounds also have demonstrated toxic effects toward microorganisms.

17.6.4.1 Unsubstituted Aromatics

Aerobic Conditions

A wide variety of bacteria and fungi can carry out aromatic transformations, both partial and complete, under a variety of environmental conditions (see Information Box 17.11; Johnsen et al., 2005). Under aerobic conditions, the most common initial transformation is a hydroxylation that involves the incorporation of molecular oxygen. The enzymes involved in these initial transformations are either monooxygenases or dioxygenases. In general, prokaryotic microorganisms transform aromatics by an initial dioxygenase attack to cis-dihydrodiols. The cis-dihydrodiol is rearomatized to form a dihydroxylated intermediate, catechol. The catechol ring is cleaved by a second dioxygenase either between the two hydroxyl groups (ortho pathway) or next to one of the hydroxyl groups (meta pathway) and further degraded to completion (Figure 17.19).

Most eukaryotic microorganisms do not mineralize aromatics; rather, they are processed for detoxification and excretion. This is done by an initial oxidation with a cytochrome P-450 monooxygenase, which incorporates one atom of molecular oxygen into the aromatic compound, and reduces the second to water, resulting in the formation of an arene oxide. This is followed by the enzymatic addition of water to yield a trans-dihydrodiol (Figure 17.20). Alternatively, the arene oxide can be isomerized to form phenols, which can be conjugated with sulfate, glucuronic acid and glutathione. These conjugates are similar to those formed in higher organisms, which are used in the elimination of aromatic compounds.

A small group of eukaryotes, the lignolytic fungi, can completely mineralize aromatic compounds in a process known as lignolytic degradation. The lignin structure is based on two aromatic amino acids, tyrosine and phenylalanine. In order to degrade an amorphous aromatic-based structure such as lignin, the white rot fungi release nonspecific extracellular enzymes such as laccase or H_2O_2-dependent lignin peroxidase. These enzymes generate oxygen-based free radicals that react with the lignin polymer to release residues that are taken up by the cell and degraded. Since the lignin structure is based on an aromatic structure and the initial enzymes used to degrade lignin are nonspecific, the white rot fungi are able to use the same activity to degrade a variety of aromatic contaminants. The most famous of the white rot fungi is *Phanerochaete chrysosporium*, which has been demonstrated to degrade a variety of aromatic compounds (see Information Box 2.6)

Often the capacity for aromatic degradation is plasmid mediated (Ghosal et al., 1985). Plasmids can carry both individual genes and operons encoding partial or complete biodegradation of an aromatic compound. An example of the latter is the NAH7 plasmid, which codes for the entire naphthalene degradation pathway. The NAH7 plasmid was obtained from *Pseudomonas putida*, and contains genes that encode the enzymes for the first 11 steps of naphthalene oxidation. This plasmid or closely related plasmids are frequently found in sites that are contaminated with

Information Box 17.11   General Rules for Degradation of Aromatics

- Aerobic conditions
  - aromatics are degraded by a wide variety of bacteria, fungi, and algae
  - bacteria use an initial dioxygenase attack to form cis-dihydrodiols and then a catechol intermediate. In this case the aromatic is mineralized and used as a source of carbon and energy
  - eukaryotes use an initial monooxygenase attack to form trans-dihydrodiols, via an arene oxide. In this case the aromatic is being processed for detoxification and excretion. Alternatively, some fungi use lignolytic enzymes and have been shown to mineralize aromatics but not as a sole source of carbon and energy

- low MW aromatics are degraded much more rapidly than high MW aromatics. Aromatics > three rings are resistant to biodegradation and may not serve as a sole source of carbon and energy
- microbial adaptation occurs from chronic aromatic exposure
- many of the genes involved in aromatic, especially PAH, degradation are plasmid encoded.
- Anaerobic conditions
  - mineralization generally requires a consortium
  - benzoyl-CoA is the common degradation intermediate

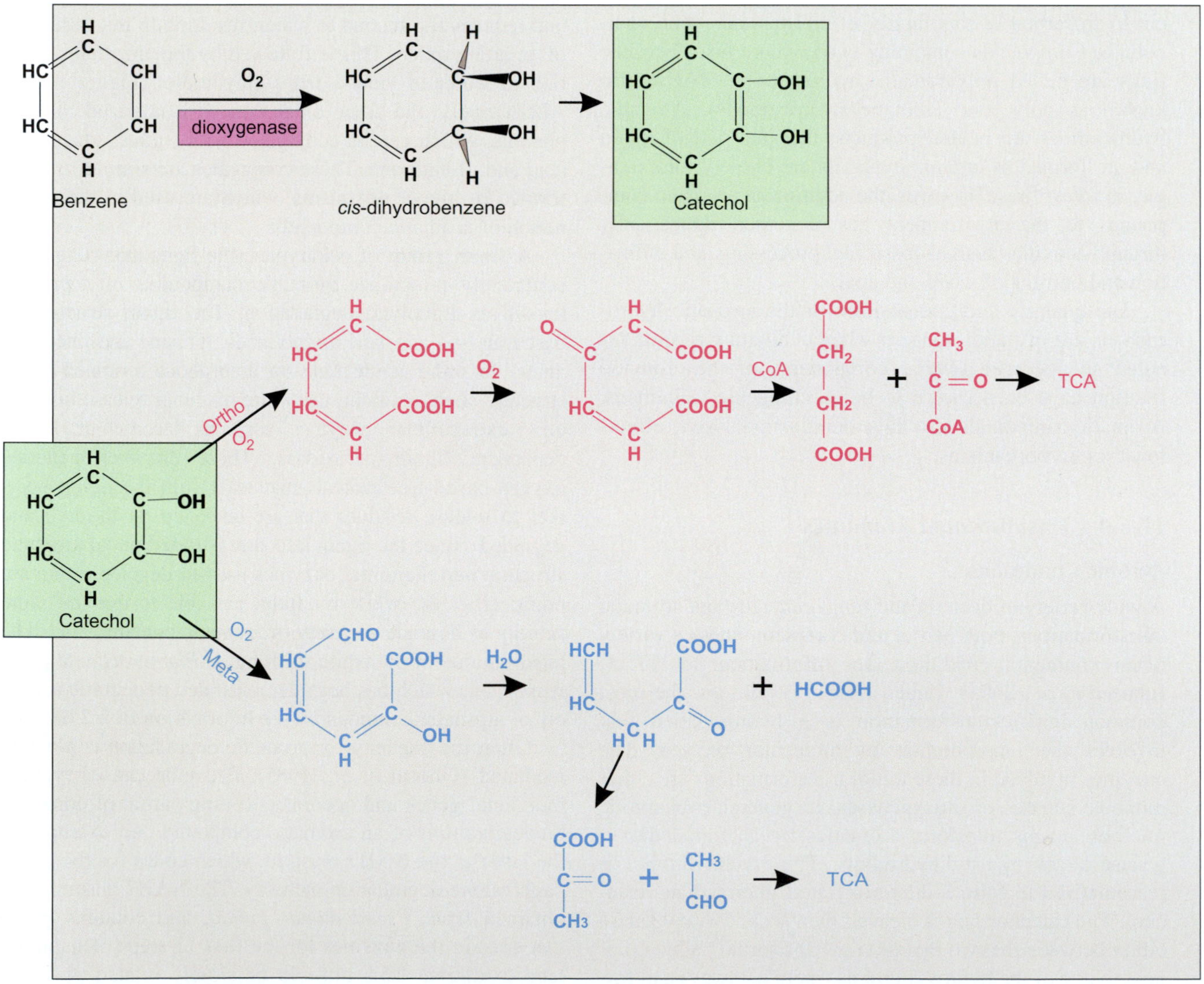

FIGURE 17.19   Incorporation of oxygen into the aromatic ring by the dioxygenase enzyme, followed by meta or ortho ring cleavage.

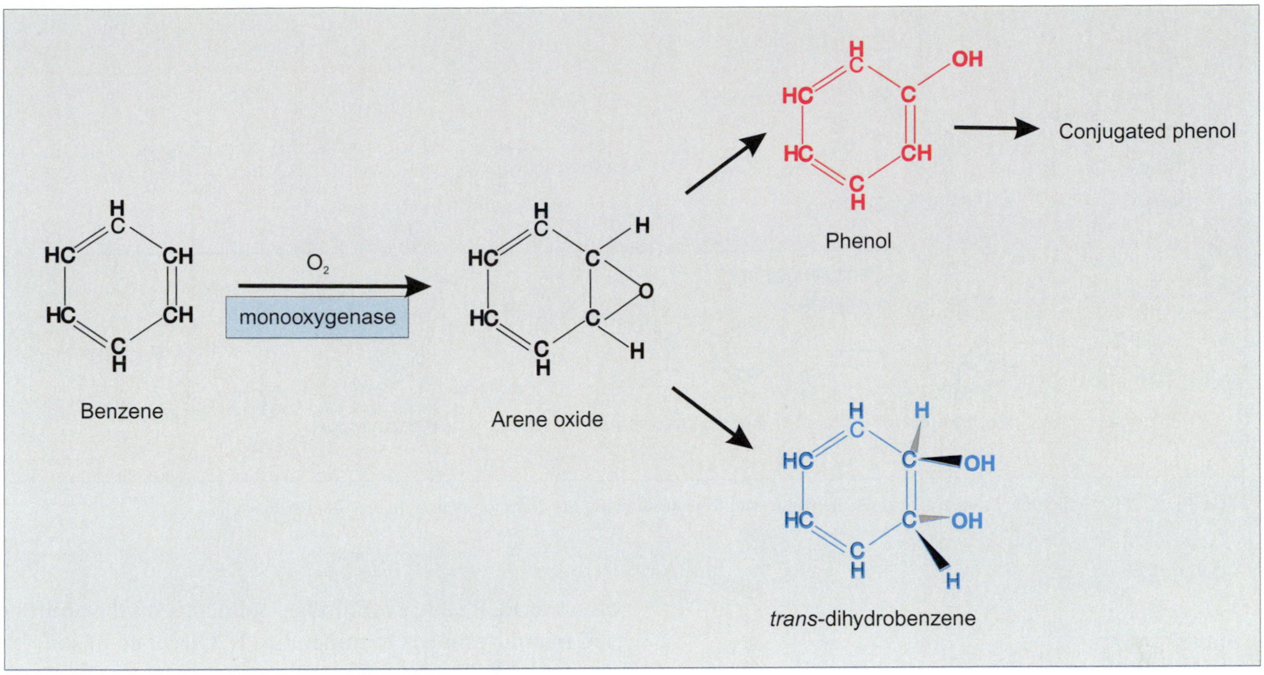

FIGURE 17.20 Fungal monooxygenase incorporation of oxygen into the aromatic ring.

PAHs (Ahn *et al.*, 1999). This plasmid has also been used to construct a luminescent bioreporter gene system. Here the lux genes that cause luminescence have been inserted into the *nah* operon in the NAH plasmid. When the *nah* operon is induced by the presence of naphthalene, both naphthalene-degrading genes and the lux gene are expressed. As a result, naphthalene is degraded and the reporter organism luminesces. Such reporter organisms are currently being used as sensors to study the temporal and spatial activity of the reporter in soil systems.

In general, aromatics composed of one, two or three condensed rings are transformed rapidly and often completely mineralized, whereas aromatics containing four or more condensed rings are transformed much more slowly, often as a result of a cometabolic attack. This is due to the limited bioavailability of these high-molecular-weight aromatics. Such PAHs have very limited aqueous solubility, and sorb strongly to particle surfaces in soil and sediments. However, it has been demonstrated that chronic exposure to aromatic compounds will result in increased transformation rates because of adaptation of an indigenous population to growth on aromatic compounds.

Anaerobic Conditions

Like aliphatic hydrocarbons, aromatic compounds can be completely degraded under anaerobic conditions. Anaerobic mineralization of aromatics produces benzoyl CoA as the common degradation intermediate (Figure 17.21). If the aromatic is oxygenated such as for benzoate, biodegradation occurs rapidly and even at rates comparable to aerobic

conditions (Figure 17.13). However, under anaerobic conditions, a mixed microbial community works together even though each of the microbial components requires a different redox potential. For example, mineralization of benzoate can be achieved by growing an anaerobic benzoate degrader in co-culture with a methanogen and sulfate reducer. The initial transformations in such a system are often carried out fermentatively, and this results in the formation of aromatic acids, which in turn are transformed to methanogenic precursors such as acetate, carbon dioxide and formate. These small molecules can then be utilized by methanogens (Figure 17.22). Such a mixed community is called a **consortium**. It is not known how this consortium solves the problem of requiring different redox potentials in the same vicinity in a soil system. Clearly, higher redox potentials are required for degradation of the more complex substrates such as benzoate, leaving smaller organic acid or alcohol molecules that are degraded at lower redox potentials. To ultimately achieve degradation may require that the organic acids and alcohols formed at higher redox potential be transported by diffusion or by movement with water (advection), to a region of lower redox potential. On the other hand, it may be that biofilms form on the soil surface, and that redox gradients are formed within the biofilm allowing complete degradation to take place.

17.6.4.2 Substituted Aromatics

One group of aromatics of special interest is the **chlorinated aromatics**. These compounds have been used

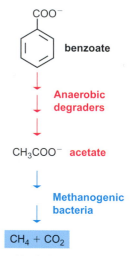

FIGURE 17.21 Anaerobic biodegradation of benzoate and toluene showing the common benzoyl-CoA intermediate.

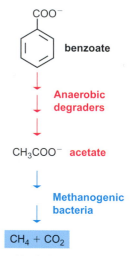

FIGURE 17.22 Anaerobic biodegradation of aromatic compounds by a consortium of anaerobic bacteria. From Pepper *et al.* (2006).

extensively as solvents and fumigants (e.g., dichlorobenzene), and wood preservatives (e.g., pentachlorophenol (PCP)), and are parent compounds for pesticides such as 2,4-dichlorophenoxyacetic acid (2,4-D) and DDT. The difficulty for aerobic microbes in the degradation of chlorinated aromatics is that the common intermediate in aromatic degradation is catechol (see Figure 17.19). Catechol formation requires two adjacent unsubstituted carbons so that hydroxyl groups can be added. Chlorine substituents can block these sites. Some bacteria solve this problem by removing a chlorine using a dehalogenase or monooxygenase enzyme.

Chlorinated phenols are particularly toxic to microorganisms. In fact, phenol itself is very toxic and is used as a disinfectant. Chlorination adds to toxicity, which increases with the degree of chlorination. For example, van Beelen and Fleuren-Kemilä (1993) quantified the

effect of PCP and several other pollutants on the ability of soil microorganisms to mineralize [^{14}C]acetate in soil. The amount of PCP required to reduce the initial rate of acetate mineralization by 10% ranged between 0.3 and 50 mg/kg dry soil, depending on the soil type. High concentrations of PCP also have inhibitory effects on PCP-degrading microorganisms. For example, Alleman *et al.* (1992) investigated the effect of PCP on six species of PCP-degrading fungi. They showed that increasing the PCP concentration from 5 to 40 mg/L decreased fungal growth, and decreased the ability of the fungi to degrade PCP.

Methylated aromatic derivatives, such as toluene, constitute another common group of substituted aromatics. These are major components of gasoline and are commonly used as solvents. These compounds can initially be attacked either on the methyl group or directly on the ring as shown in Figure 17.23. This can be compared to anaerobic degradation of toluene shown in Figure 17.21. Alkyl derivatives of aromatics are attacked first at the alkyl chain, which is shortened by β-oxidation to the corresponding benzoic acid or phenylacetic acid, depending on the number of carbon atoms. This is followed by ring hydroxylation and cleavage.

17.6.5 Dioxins and PCBs

Dioxins and dibenzofurans are created during waste incineration and are part of the released smoke stack effluent. Once thought to be one of the most potent carcinogens known, 2,3,7,8-tetrachlorodibenzo-*p*-dioxin (TCDD) is associated with the manufacture of 2,4-D and 2,4,5-trichlorophenoxy acetic acid (2,4,5-T), hexachlorophene and other pesticides that have 2,4,5-T as a precursor. Current thinking is that TCDD is less dangerous in terms of carcinogenicity and teratogenicity than once thought, but that noncancer risks including diabetes, reduced IQ

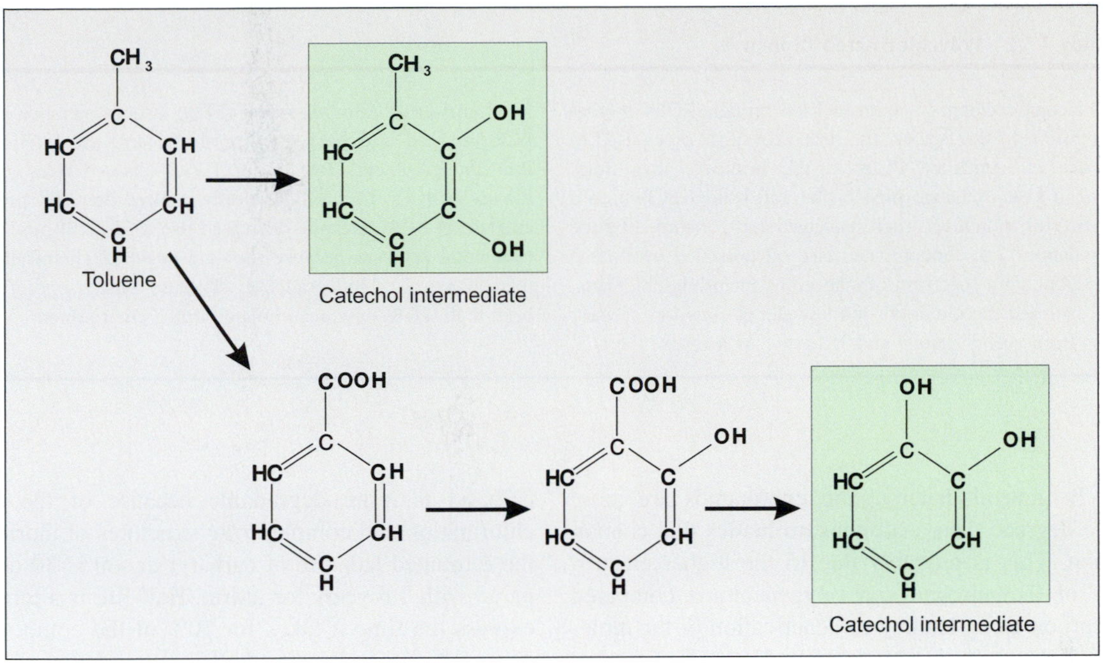

FIGURE 17.23 Aerobic biodegradation of toluene.

and behavioral impacts may be more important. The structure of TCDD (Figure 17.11) and its low water solubility, 0.002 mg/L, result in great stability of this molecule in the environment.

Although bacterial and fungal biodegradation of TCDD has been demonstrated, the extent of biodegradation is very minimal. For example, a mixture of six bacterial strains isolated from TCDD-contaminated soil obtained from Seveso, Italy, was able to produce a metabolite presumed to be 1-hydroxy-TCDD (Phillippi et al., 1982). However, less than 1% of the original TCDD was degraded in 12 weeks. Other work has focused on the anaerobic reductive dechlorination of TCDD and more highly chlorinated isomers (Kao et al., 2001).

Biphenyl is the unchlorinated analogue or parent compound of the polychlorinated biphenyls (PCBs), which were first described in 1881 (Waid, 1986). The PCBs consist of different chlorine-substituted biphenyls of which, in theory, there are 209 possible isomers. Only approximately 100 actually exist in commercial formulations. The aqueous solubility of biphenyl is 7.5 mg/L, and any chlorine substituent decreases the water solubility. In general, the water solubility of monochlorobiphenyls ranges from 1 to 6 mg/L, compared to 0.08 to 5 mg/L for dichlorobiphenyls. In contrast, for hexachlorobiphenyl, the aqueous solubility is just 0.00095 mg/L.

By 1930, because of their unusual stability, PCBs were widely used as nonflammable heat-resistant oils in heat transfer systems, as hydraulic fluids and lubricants, as transformer fluids, in capacitors, as plasticizers in food packaging materials, and as petroleum additives. PCBs were used as mixtures of variously chlorinated isomers and marketed under various trade names, e.g., Aroclor (U.S.A.), Clophen (Germany), Phenoclor (Italy), Kanechlor (Japan), Pyralene (France), and Soval (Russia). The U.S. domestic sales of Aroclors 1221−1268 (where the last two numbers indicated the percent chlorination) went from 32 million pounds in 1957 to 80 million pounds in 1970, with the most popular blend being 1242. PCB use in transformers and capacitors accounted for about 50% of all Aroclors used.

Past use of PCBs has resulted in their accumulation in the environment from waste dumps and spills and as a result of PCB manufacturing processes (see Case Study 17.1). Although some PCB degradation occurs, it is limited by low bioavailability, by the recalcitrance of highly chlorinated PCB congeners under aerobic conditions, and by incomplete degradation under anaerobic conditions. The extensive research that has been performed to understand PCB degradation has suggested several strategies for promoting biodegradation. Of these, the most promising is the use of a sequential anaerobic−aerobic process first to allow removal of chlorines using halorespiration, and then allow mineralization of the less chlorinated congeners.

17.6.6 Heterocyclic Compounds

Heterocyclics are cyclic compounds containing one or more heteroatoms (nitrogen, sulfur or oxygen) in addition to carbon atoms. The dioxins already discussed as well as other compounds shown in Figure 17.11 fall into this

---

Case Study 17.1   Polychlorinated Biphenyls

A rice oil factory accident in Japan in 1968 brought PCBs international attention. In the factory, the heat exchanger pipes used to process rice oil contained PCBs as the heat exchange fluid. Unnoticed, a heat exchange pipe broke and leaked PCBs into a batch of rice oil, which was then packaged and consumed by the local population. The contaminated rice oil poisoned over 1000 people, producing a spectrum of symptoms including chloracne, gum and nail bed discoloration, joint swelling, emission of waxy secretions from eyelid glands, and lethargy. As a result, the U.S. Food and Drug Administration (FDA) issued tolerance levels for PCBs in food and packaging products, and the Environmental Protection Agency (EPA) under TOSCA, issued rules governing the use of PCBs. This has drastically reduced domestic production and use of PCBs. Despite decreased use, PCBs still pose an environmental problem because they are not only chemically stable, they also resist biodegradation. Because past use of PCBs has been high, PCBs have accumulated in the environment.

---

category. In general, heterocyclic compounds are more difficult to degrade than analogous aromatics that contain only carbon. This is probably due to the higher electronegativity of the nitrogen and oxygen atoms compared with the carbon atom, leading to deactivation of the molecule toward electrophilic substitution. Heterocyclic compounds with five-membered rings and one heteroatom are readily biodegradable, probably because five-membered ring compounds exhibit higher reactivity toward electrophilic agents, and hence are more readily biologically hydroxylated. The susceptibility of heterocyclic compounds to biodegradation decreases with increasing number of heteroatoms in the molecules.

17.6.7 Pesticides

Pesticides are the biggest nonpoint source of chemicals added to the environment. The majority of the currently used organic pesticides are subject to extensive mineralization within the time of one growing season or less. Synthetic pesticides show a bewildering variety of chemical structures, but most can be traced to relatively simple aliphatic, alicyclic and aromatic base structures already discussed. These base structures bear a variety of halogen, amino, nitro, hydroxyl, carboxyl and phosphorus substituents. For example, the chlorophenoxyacetates, such as 2,4-D and 2,4,5-T, have been released into the environment as herbicides over the past 50 years. Both of these structures are biodegradable and aerobic pathways are presented in Figure 17.24.

As an exercise, examine the pesticide structures presented in Figure 17.25. For each set of pesticides, predict which is more easily degraded under aerobic conditions. You are correct if you predicted 2,4-D for the first set. It is rapidly degraded by soil microorganisms and although 2,4,5-T is also degraded, the degradation is much slower. For the second set of pesticides, propham is more degradable. In propachlor, the extensive branching so close to the ring structure blocks biodegradation. In the third set,

carbaryl is more degradable because of the extensive chlorination and complex ring structures of aldrin. In fact, the estimated half-life of carbaryl in soil is 30 days, compared with 1.6 years for aldrin. Half-life is a term used to express the time it takes for 50% of the compound to be degraded. Generally, five half-lives are believed sufficient for the compound to be completely degraded. Finally, in the fourth set, methoxychlor is more degradable than DDT. In this case, the half-lives are even longer, 1 year for methoxychlor and 15.6 years for DDT.

17.7 BIOREMEDIATION

The objective of bioremediation is to exploit naturally occurring biodegradative processes to clean up contaminated sites (NRC, 1993). There are several types of bioremediation. *In situ* bioremediation is the in-place treatment of a contaminated site. *Ex situ* bioremediation may be implemented to treat contaminated soil or water that is removed from a contaminated site. Biostimulation, which is the modification of environmental conditions (e.g., addition of oxygen, nitrogen) to enhance the biodegradation activity of indigenous microorganisms, is often used to increase the speed and effectiveness of bioremediation. In contrast, intrinsic bioremediation or natural attenuation is the indigenous level of contaminant biodegradation that occurs without any stimulation or treatment. All of these types of bioremediation continue to receive increasing attention as viable remediation alternatives for several reasons. These include generally good public acceptance and support, good success rates for some applications and a comparatively low cost of bioremediation when it is successful. As with any technology, there are also drawbacks. Success can be unpredictable because a biological system is being used. A second consideration is that bioremediation rarely restores an environment completely. Often the residual contamination left after treatment is strongly sorbed and not available to microorganisms for degradation. Over a long period of time (years), these residuals

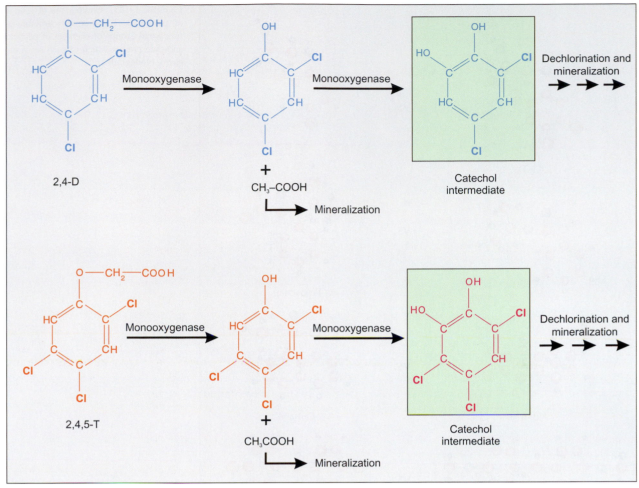

FIGURE 17.24   Aerobic biodegradation of 2,4-dichlorophenoxyacetic acid (2,4-D) and 2,4,5-trichlorophenoxyacetic acid (2,4,5-T).

can be slowly released. There is little research concerning the fate and potential toxicity of such released residuals, and therefore there is both public and regulatory concern about the importance of residual contamination.

Although it is often not thought of as bioremediation, domestic sewage waste has been treated biologically for many years with good success. Interestingly, even sewage treatment is undergoing reexamination in the light of detection of trace levels of endocrine disrupting compounds (EDCs) in treated wastewater. These compounds mimic hormone activities in mammalian endocrine systems and arise from pharmaceuticals and personal care products (PPCPs) that are in sewage but are not completely removed by conventional drinking and wastewater treatment plants (Snyder *et al.*, 2003). Removal of EDCs and PPCPs is incomplete for two reasons; diverse chemical structures that require acclimation, and low concentrations which may fail to induce biodegradation pathways (Information Box 17.12).

In application of bioremediation to problems other than sewage treatment, it must be kept in mind that biodegradation is dependent on the pollutant structure and bioavailability. Therefore, bioremediation success will depend on the type of pollutant or pollutant mixture present, and the type of microorganisms present. The first successful application of bioremediation outside sewage treatment was the cleanup of oil spills, and success in this area is now well documented (see Chapter 31). In the past few years, many new bioremediation technologies have emerged that are being used to address other types of pollutants including (USEPA, 2001):

- Volatile organic compounds (including chlorinated VOCs)
- Polyaromatic hydrocarbons
- Pesticides, herbicides
- Explosives

Several key factors are critical to successful application of bioremediation: environmental conditions, contaminant and nutrient availability, and the presence of degrading microorganisms. If biodegradation does not occur, the first thing that must be done is to isolate the

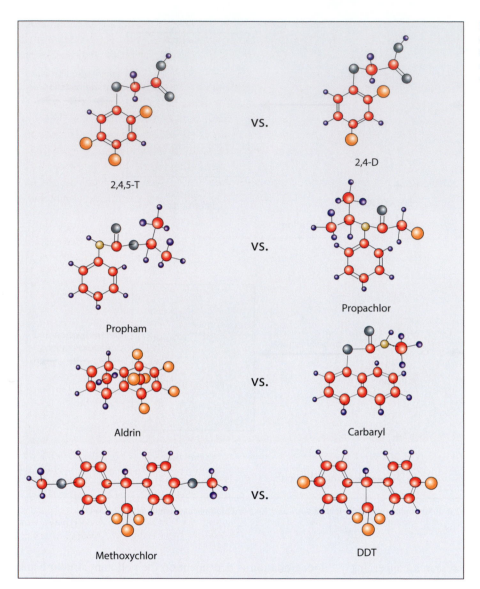

FIGURE 17.25 Comparison of four sets of pesticides. Examine the structure and functional groups of each set of pesticides and then predict which of each set is more easily biodegraded.

Information Box 17.12   Contaminant Concentration, Biodegradation and Growth

Assuming a given contaminant is biodegradable, the contaminant concentration will dictate whether biodegradation pathways are induced and at what rate biodegradation occurs. Four cases can be defined:

Contaminant concentration	Biodegradation behavior
very low	no induction, no biodegradation, no growth
low	induction, biodegradation does not support growth, energy used for maintenance, biodegradation rate is steady state
medium	induction, biodegradation, growth with increasing biodegradation rates until some limiting condition is reached
high	toxicity, no biodegradation, no growth

factor limiting bioremediation, and sometimes this can be a very difficult task. Initial laboratory tests using soil or water from a polluted site can usually determine whether degrading microorganisms are present and whether there is an obvious environmental factor that limits biodegradation, such as extremely low or high pH or lack of nitrogen and/or phosphorus. However, sometimes the limiting factor is not easy to identify. Often pollutants are present as mixtures, and one component of the pollutant mixture can have toxic effects on the growth and activity of degrading microorganisms. Low bioavailability due to sorption and aging is another factor that can limit bioremediation and can be difficult to evaluate in the environment.

Bioremediation was not considered as an option for cleanup of contaminated sites until the 1980s, but has since become an established alternative for remediation of many sites worldwide including numerous Superfund sites in the United States (USEPA, 2001). Overall, from

1982 to 2008, bioremediation has been used as a remedy for control of source contamination zones at 13% of Superfund sites and has been used for treatment of groundwater at 38% of Superfund sites (USEPA, 2010).

17.7.1 Addition of Oxygen or Other Gases

One of the most common limiting factors in bioremediation is availability of oxygen. Oxygen is an element required for aerobic biodegradation. In addition, oxygen has low solubility in water, and a low rate of diffusion (movement) through both air and water. The combination of these three factors makes it easy to understand that inadequate oxygen supplies will slow bioremediation. Several technologies have been developed to overcome a lack of oxygen. A typical bioremediation system used to treat a contaminated aquifer as well as the contaminated zone above the water table is shown in Figure 17.26A. This system contains a series of injection wells or galleries, and a series of recovery wells that comprise a two-pronged approach to bioremediation. First, the recovery wells remove contaminated groundwater, which is treated above ground, in this case using a bioreactor containing

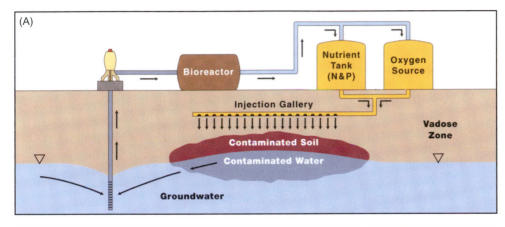

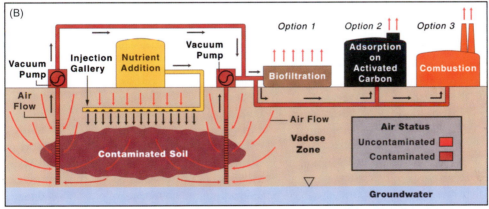

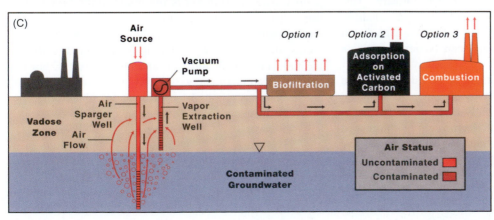

FIGURE 17.26 Bioremediation approaches: (A) *In situ* bioremediation in the vadose zone and groundwater. Nutrient and oxygen are pumped into the contaminated area to promote *in situ* processes. This figure also shows *ex situ* treatment. *Ex situ* treatment is for water pumped to the surface and uses an aboveground bioreactor, as shown, to biodegrade contaminants. Alternatively, other nonbiological methods can be used to remove contaminants from the water, e.g., air stripping, activated carbon, oil–water separation or oxidation. Following treatment, an injection well returns the contaminant-free water to the aquifer. (B) Bioventing and biofiltration in the vadose zone. Air is slowly drawn through the contaminated site (bioventing), which stimulates *in situ* aerobic degradation. Volatile contaminants removed with the air can be treated biologically using a biofilter as shown or by adsorption on activated carbon, or by combustion. (C) Bioremediation in groundwater by air sparging. Air is pumped into the contaminated site to stimulate aerobic biodegradation. Volatile contaminants brought to the surfaced are treated by biofiltration, activated carbon or combustion.

microorganisms that are acclimated to the contaminant. This would be considered *ex situ* treatment. Following bioreactor treatment, the clean water is supplied with oxygen and nutrients, and then it is reinjected into the site. The reinjected water provides oxygen and nutrients to stimulate *in situ* biodegradation. In addition, the reinjected water flushes the vadose zone to aid in removal of the contaminant for aboveground bioreactor treatment. This remediation scheme is a very good example of the use of a combination of physical, chemical and biological treatments to maximize the effectiveness of the remediation treatment.

Bioventing is a technique used to add oxygen directly to a site of contamination in the vadose zone (unsaturated zone). In bioventing alone, air is injected at very low flow rates into the contaminated zone to promote biodegradation. Alternatively in some cases, flow rates can be increased to combine soil vapor extraction technology and bioremediation. In this case, extracted vapor-phase contaminants are treated aboveground either biologically or chemically, and in addition, *in situ* bioremediation is stimulated. The bioventing zone is highlighted by red arrows in Figure 17.26B and includes the vadose zone and contaminated regions just below the water table. As shown in this figure, a series of wells have been constructed around the zone of contamination. To initiate bioventing, a vacuum is drawn on these wells to force accelerated air movement through the contamination zone. This effectively increases the supply of oxygen throughout the site, and thus the rate of contaminant biodegradation. If the rate of air movement is increased further, contaminants are volatilized and removed as air is forced through this system. This contaminated air can be treated biologically by passing the air through aboveground soil beds in a process called biofiltration, as shown in Figure 17.27 (Jutras *et al.*, 1997).

In contrast, air sparging is used to add oxygen to the saturated zone (Figure 17.26C). In this process, an air sparger well is used to inject air under pressure below the water table. The injected air displaces water in the soil matrix, creating a temporary air-filled porosity. This causes oxygen levels to increase, resulting in enhanced biodegradation rates. In addition, volatile organics will volatilize into the airstream, and can be removed by a vapor extraction well.

Methane is another gas that can be added with oxygen in extracted groundwater and reinjected into the saturated zone. Methane is used specifically to stimulate methanotrophic activity and cometabolic degradation of chlorinated solvents. As described in Chapter 16, methanotrophic organisms produce the enzyme methane monooxygenase to degrade methane, and this enzyme also cometabolically degrades several chlorinated solvents. Cometabolic degradation of chlorinated solvents is presently being tested in field trials to determine the usefulness of this technology.

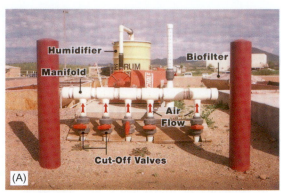

Soil vapor extraction unit

Side view of a biofilter

FIGURE 17.27 Bioremediation of gasoline vapors from a leaking underground storage tank. (A) Red arrows denote the direction of airflow out of the ground and through the manifold, which controls the airflow. The air then passes through the humidifier, a tank containing water, and into the biofilter (the large red vertical cylinders serve as protective barriers to prevent damage to the manifold). (B) Air flows from the manifold into the soil in the biofilter through a perforated pipe running lengthwise along the bottom of the biofilter.

17.7.2 Nutrient Addition

A common bioremediation treatment is the addition of nutrients, in particular nitrogen and phosphorus. Many contaminated sites contain organic wastes that are rich in carbon but contain minimal amounts of nitrogen and phosphorus. Nutrient addition is illustrated in the bioremediation schemes shown in Figure 17.26A and B. Injection of nutrient solutions takes place from an aboveground batch-feed system. The goal of nutrient injection is to optimize the ratio of carbon, nitrogen and phosphorus (C:N:P) in the site to approximately 100:10:1. However, sorption of added nutrients can make it difficult to achieve the optimal ratio accurately.

17.7.3 Sequential Anaerobic—Aerobic Degradation

The rapid biodegradation of many priority pollutants requires both anaerobic and aerobic stages. As already discussed, aerobic conditions favor the biodegradation of compounds with fewer halogen substituents, and anaerobic conditions favor the biodegradation of compounds with a high number of halogen substituents. However, complete biodegradation of highly halogenated aliphatics under anaerobic conditions often does not take place. Therefore, some researchers have proposed the use of a sequential anaerobic and aerobic treatment. Initial incubation under anaerobic conditions is used to decrease the halogen content, and subsequent addition of oxygen creates aerobic conditions to allow complete degradation to proceed aerobically. This approach was used successfully to treat a groundwater plume containing TCE, DCE, vinyl chloride and petroleum hydrocarbons (Morkin *et al.*, 2000).

17.7.4 Addition of Surfactants

Surfactant addition has been proposed as a technique for increasing the bioavailability and hence biodegradation of contaminants (see Section 17.4.3). Surfactants can be synthesized chemically and are also produced by many microorganisms, in which case they are called biosurfactants. Surfactants work similarly to industrial and household detergents that effectively remove oily residues from machinery, clothing or dishes. As shown in the inset in Figure 17.6, individual contaminant molecules can be "solubilized" inside surfactant micelles. These micelles range from 5 to 10 nm in diameter. Alternatively, surfactant molecules can coat oil droplets and emulsify them into solution—a property that makes them useful in dispersants for "breaking up" oil spills (Information Box 17.13). In addition, surfactants can enhance the ability of microbes to stick to oil droplets. There have been extensive laboratory and field tests performed with both synthetic- and bio-surfactants. While these

materials definitely increase the bioavailability of organic contaminants, they do not always stimulate biodegradation. However, enough successful tests have been performed to indicate that, if chosen carefully, surfactants can be used to enhance the remediation process (Maier and Soberon-Chavez, 2000; Martienssen and Schirmer, 2007).

17.7.5 Addition of Microorganisms or DNA

If appropriate biodegrading microorganisms are not present in soil or if microbial populations have been reduced because of contaminant toxicity, specific microorganisms can be added as "introduced organisms" to enhance the existing populations. This process is known as bioaugmentation. Scientists are also capable of creating "superbugs," organisms that can degrade pollutants at extremely rapid rates. Such organisms can be developed through successive adaptations under laboratory conditions, or can be genetically engineered.

Although bioaugmentation, with naturally occurring or engineered organisms, has been demonstrated to increase contaminant degradation in numerous lab-based studies, it generally has not been successful for remediation in actual field sites (Gentry *et al.*, 2004; Stroo *et al.*, 2013). The problem is that introduction of a microorganism to a contaminated site may fail for at least two reasons. First, the introduced microbe often cannot establish a niche in the environment. In fact, these introduced organisms often do not survive in a new environment beyond a few weeks. Second, there are difficulties in delivering the introduced organisms to the site of contamination, because microorganisms, like contaminants, can be strongly sorbed by solid surfaces.

Information Box 17.13   Use of Dispersants to Reduce Environmental Impacts from the BP *Deepwater Horizon* Oil Leak

In 2010, the BP *Deepwater Horizon* oil leak released an estimated 780 million liters of oil into the Gulf of Mexico. In order to help prevent the generation of large oil slicks and reduce the amount of oil reaching the shoreline, dispersants, including Corexit 9500A®, were applied to the water above the leak. Dispersants generally contain a mixture of chemicals, including surfactants and solvents, that are used to emulsify oil into droplets which are more likely to remain suspended in the water column. The application of dispersants to the BP *Deepwater Horizon* oil leak succeeded in reducing the amount of oil that reached the surface, and likely helped to reduce the impact of oil slicks on the shoreline ecosystem. In addition, the increased surface area of the dispersed oil likely increased its potential for biodegradation. This is supported by studies which found greatly increased microbial numbers in the cloud of dispersed oil (Atlas and Hazen, 2011).

However, in the process of containing the oil leak, millions of liters of dispersants were ultimately applied to the Gulf waters thus raising potential concerns about the impacts of the dispersants themselves on marine ecosystems. For example, one study reported that although the toxicity of the dispersant Corexit 9500A® to a marine rotifer was similar to that of the oil being treated, the toxicity of the oil mixed with the dispersant was increased over 52-fold in comparison to the toxicity of the oil alone (Rico-Martínez *et al.*, 2013). It remains to be seen whether the benefits of using such dispersants to reduce the impacts of oil spills on shoreline ecosystems is being offset by negative impacts on other marine ecosystems. In addition, research is ongoing to cost-effectively produce less toxic dispersants for use in such oil spills. There is great interest in the use of biosurfactants for this purpose (see also Chapter 31).

Despite these obstacles, one case in which bioaugmentation has been successful at the field scale is the remediation of soil and groundwater contaminated with chlorinated solvents such as TCE and PCE (Case Study 17.2; Stroo *et al.*, 2013). Why has bioaugmentation succeeded for remediation of these compounds when it so frequently fails for other applications? Two possible reasons are: (1) the unique nature of the chemicals and organisms involved, and (2) the alteration of site conditions to create a niche for the added microorganisms. Unlike petroleum hydrocarbons and many other contaminants, PCE and TCE are used only by a very select group of microorganisms as electron acceptors to support growth under anaerobic conditions (e.g., reductive dehalogenation, see Section 17.6.2.2). Thus, there is less competition for the chemical substrate between the added and indigenous organisms than occurs for many other compounds, e.g., petroleum. In addition, these sites are usually converted from aerobic to anaerobic conditions, through addition of electron donors, prior to addition of the bioaugmentation culture. This creates an immediate niche for the added microorganisms to establish and begin reductive dehalogenation. Since anaerobic conditions generally result in lower microbial numbers, diversity and activity than that found in aerobic conditions, this likely further decreases competition from indigenous microorganisms.

Currently, very little is known about microbial transport and establishment of environmental niches. These are areas of active research, and in the next few years scientists may gain a further understanding of microbial behavior in soil ecosystems. However, until we discover how to successfully deliver and establish introduced microorganisms, bioaugmentation will not be a viable bioremediation option for the majority of contaminated sites.

If bioaugmentation does not work, one way to take advantage of the superbugs that have been developed is to use them in bioreactor systems under controlled conditions. Extremely efficient biodegradation rates can be achieved in bioreactors that are used in aboveground treatment systems.

Another bioaugmentation strategy is to add specific genes that can confer a specific degradation capability to indigenous microbial populations. The addition of degradative genes relies on the delivery and uptake of the genetic material by indigenous microbes. There are two approaches that can be taken in delivery of genes. The first is to use microbial cells to deliver the DNA via conjugation. The second is to add "naked" DNA to the soil to allow uptake via transformation. This second approach may reduce the difficulty of delivery since DNA alone is much smaller than a whole cell. However, little is known as yet about these two approaches. As discussed in Case Study 3.1, Di Giovanni *et al.* (1996) demonstrated that

---

Case Study 17.2 Bioaugmentation-based Remediation of Groundwater Contaminated with Chlorinated Ethenes

Although bioaugmentation has been shown to increase the degradation of many compounds in lab-bases studies, it often fails when applied to field sites under environmental conditions. However, one area in which bioaugmentation has been successfully demonstrated at the field scale is in the remediation of sites contaminated with chlorinated ethenes such as trichloroethene (TCE) and tetrachloroethene (PCE). A classic example of this is the remediation of a site at Kelly Air Force Base in San Antonio, TX, U.S.A. This site was contaminated with up to 1 mg of PCE per liter of groundwater and also lower amounts of TCE and cis-1,2-dichloroethene (cDCE) (Major *et al.*, 2002). The scientists first conducted laboratory microcosm tests using soil and groundwater from the site to determine the potential for biostimulation of the indigenous microbial community to dechlorinate the PCE and TCE. Amendment of the microcosms with various electron donors (e.g., methanol and lactate) expedited dehalogenation to cDCE, but did not further dehalogenate it to vinyl chloride or ethene (see Figure 17.2 for details on the degradation pathway). Since biostimulation alone was not effective, microcosms were also bioaugmented with a dechlorinating, enrichment culture (KB-1). This culture, containing populations of *Dehalococcoides* spp., was originally obtained from a TCE-contaminated site in southern Ontario, Canada. In contrast to the biostimulated microcosms, the bioaugmented microcosms demonstrated complete dechlorination of TCE to ethene. Based upon these results, the scientists then conducted a field-scale test of bioaugmentation to remediate the contaminated soil and groundwater *in situ*. For the first 89 days of the experiment, water was recirculated through the site without electron donor addition in order to equilibrate and verify the system hydraulics. After day 89, methanol and acetate were added as electron donors for generation of anaerobic conditions. On day 176, bioaugmentation began with addition of the KB-1 culture. Similar to the laboratory microcosm tests, concentrations of PCE began to decrease following biostimulation with a corresponding accumulation of cDCE, but little conversion to vinyl chloride or ethene. However, once the site was bioaugmented with the KB-1 culture, the cDCE began also to disappear with a corresponding accumulation of ethene. Concentrations of PCE, TCE and cDCE all ultimately decreased to <5 μg/L. A combination of PCR- and DNA sequencing-based approaches were used to verify that the bioaugmentation culture was spread throughout the treatment zone where the dechlorination occurred, thus providing further evidence that dechlorination of the PCE and TCE was due to the presence and activity of the added microbial culture.

gene transfer can occur in soil, resulting in 2,4-D degradation activity. However, whether such transfer is common, and whether conditions are conducive or inhibitory to such transfer, is not known.

QUESTIONS AND PROBLEMS

1. Why is there concern about the presence of organic contaminants in the environment?
2. Describe the different factors that can limit biodegradation of organic contaminants in the environment.
3. Draw and name an aliphatic, alicyclic and aromatic structure, each with six carbons.
4. Outline the biodegradation pathway for each of the structures that you just drew under aerobic conditions.
5. Why are aerobic conditions usually preferred for biodegradation of organic contaminants? Under what conditions might anaerobic biodegradation be preferred?
6. Compare the advantages and disadvantages of intrinsic, *ex situ* and *in situ* bioremediation.
7. You have been hired to bioremediate a site in which the groundwater is contaminated with petroleum. Groundwater samples have a strong sulfide smell and gas chromatographic analysis of the samples shows negligible biodegradation of the petroleum has occurred. What is your recommendation?
8. Kleen Co. is in charge of a site in Nevada that was used for pesticide preparation. As a result of years of operation, the groundwater below this site has elevated levels of pesticides (up to 20 mg/L). Your initial investigation shows that: (1) the pesticide-containing plume is neither growing nor shrinking in size; (2) there are pesticide degraders in the plume; (3) and the dissolved oxygen levels in the plume range from 2 to 4 mg/L. This site is not being used presently, and the groundwater is not used for drinking water purposes. What is your best recommendation based on these site characteristics and on your knowledge of cost of remediation?

REFERENCES AND RECOMMENDED READING

Ahn, Y., Sanseverino, J., and Sayler, G. (1999) Analyses of polycyclic aromatic hydrocarbon degrading bacteria isolated from contaminated soils. *Biodegradation* 10, 149–157.

Alexander, M. (1995) How toxic are toxic chemicals in soil? *Environ. Sci. Technol.* 29, 2713–2717.

Alleman, B. C., Logan, B. E., and Gilbertson, R. L. (1992) Toxicity of pentachlorophenol to six species of white rot fungi as a function of chemical dose. *Appl. Environ. Microbiol.* 58, 4048–4050.

Arbuckle, J. G., Bryson, N. S., Case, D. R., Cherney, C. T., Hall, R. M., Jr., Martin, H. C., *et al.* (1987) "Environmental Law Handbook," 9th ed. Government Institutes, Rockville, MD.

Atlas, R. M., and Hazen, T. C. (2011) Oil biodegradation and bioremediation: a tale of the two worst spills in U.S. history. *Environ. Sci. Technol.* 45, 6709–6715.

Bhatt, P., Kumar, M. S., Mudliar, S., and Chakrabarti, T. (2007) Biodegradation of chlorinated compounds—a review. *Crit. Rev. Environ. Sci. Technol.* 37, 165–198.

Bollag, J.-M. (1992) Decontaminating soil with enzymes. *Environ. Sci. Technol.* 26, 1876–1881.

Britton, L. N. (1984) Microbial degradation of aliphatic hydrocarbons. In "Microbial Degradation of Organic Compounds" (D. T. Gibson, ed.), Marcel Dekker Inc., New York, NY, pp. 89–129.

Coates, J. D., and Achenbach, L. A. (2004) Microbial perchlorate reduction: rocket-fuelled metabolism. *Nat. Rev. Microbiol.* 2, 569–580.

Di Giovanni, G. D., Neilson, J. W., Pepper, I. L., and Sinclair, N. A. (1996) Gene transfer of *Alcaligenes eutrophus* JMP134 plasmid pJP4 to indigenous soil recipients. *Appl. Environ. Microbiol.* 62, 2521–2526.

Fennel, D. E., Nijenhuis, I., Wilson, S. F., Zinder, S. H., and Häggblom, M. M. (2004) *Dehalococcoides ethenogenes* strain 195 reductively dechlorinates diverse chlorinated aromatic pollutants. *Environ. Sci. Technol.* 38, 2075–2081.

Gentry, T. J., Rensing, C., and Pepper, I. L. (2004) New approaches for bioaugmentation as a remediation technology. *Crit. Rev. Environ. Sci. Technol.* 34, 447–494.

Ghosal, D., You, I.-S., Chatterjee, D. K., and Chakrabarty, A. M. (1985) Plasmids in the degradation of chlorinated aromatic compounds. In "Plasmids in Bacteria" (D. R. Helinski, S. N. Cohen, D. B. Clewell, D. A. Jackson, and A. Hollaender, eds.), Plenum Press, New York, NY, pp. 667–686.

Gilliom, R. J., Barbash, J. E., Crawford, C. G., Hamilton, P. A., Martin, J. D., Nakagaki, N., *et al.* (2006) "Pesticides in the Nation's Streams and Ground Water, 1992–2001", U.S. Geological Survey Circular 1291.

Greer, M. A., Goodman, G., Pleus, R. C., and Greer, S. E. (2002) Health effects assessment for environmental perchlorate contamination: the dose response for inhibition of thyroidal radioiodine uptake in humans. *Environ. Health Perspect.* 110, 927–937.

Grube, A., Donaldson, D., Kiely, T., and Wu, L. (2011) "Pesticides Industry Sales and Usage. 2006 and 2007 Market Estimates," USEPA, http://www.epa.gov/pesticides/pestsales/07pestsales/market_estimates2007.pdf.

Hazen, T. C., Dubinsky, E. A., DeSantis, T. Z., Andersen, G. L., Piceno, Y. M., Singh, N., *et al.* (2010) Deep-sea oil plume enriches indigenous oil-degrading bacteria. *Science* 330, 204–208.

Herman, D. C., Lenhard, R. J., and Miller, R. M. (1997) Formation and removal of hydrocarbon residual in porous media: effects of bacterial biomass and biosurfactants. *Environ. Sci. Technol.* 31, 1290–1294.

Hogue, C. (2003) Rocket-fueled river. *Chem. Eng. News* 81, 37–46.

Hutson, S. S., Barber, N. L., Kenny, J. F., Linsey, K. S., Lumia, D. S., and Maupin, M. A. (2004) "Estimated Use of Water in the United States in 2000, USGS Circular 1268," U.S. Geological Survey.

Janssen, D. B., Oldenhuis, R., and van den Wijngaard, A. J. (1990) Hydrolytic and oxidative degradation of chlorinated aliphatic compounds by aerobic microorganisms. In "Biotechnology and Biodegradation" (D. Kamely, A. Chakrabarty, and G. S. Omenn, eds.), Gulf Publishing Company, Houston, TX.

Johnsen, A. R., Wick, L. Y., and Harms, H. (2005) Principles of microbial PAH-degradation in soil. *Environ. Pollut.* 133, 71–84.

Jutras, E. M., Smart, C. M., Rupert, R., Pepper, I. L., and Miller, R. M. (1997) Field scale biofiltration of gasoline vapors extracted from beneath a leaking underground storage tank. *Biodegradation* **8**, 31–42.

Kao, C. M., Chen, S. C., Liu, J. K., and Wu, M. J. (2001) Evaluation of TCDD biodegradability under different redox conditions. *Chemosphere* **44**, 1447–1454.

Kenawy, E. R., Worley, S. D., and Broughton, R. (2007) The chemistry and applications of antimicrobial polymers: a state-of-the-art review. *Biomacromolecules* **8**, 1359–1384.

Krutz, L. J., Shaner, D. L., Weaver, M. A., Webb, R. M. T., Zablotowicz, R. M., Reddy, K. N., *et al.* (2010) Agronomic and environmental implications of enhanced s-triazine degradation. *Pest Manag. Sci.* **66**, 461–481.

Leahy, J. G., and Colwell, R. R. (1990) Microbial degradation of hydrocarbons in the environment. *Microbiol. Rev.* **54**, 305–315.

Maier, R. M. (2000) Bioavailability and its importance to bioremediation. In "Bioremediation" (J. J. Valdes, ed.), Kluwer Academic Publishers, the Netherlands, pp. 59–78.

Maier, R. M., and Soberon-Chavez, G. (2000) *Pseudomonas aeruginosa* rhamnolipids: biosynthesis and potential environmental applications. *Appl. Microbiol. Biotechnol.* **54**, 625–633.

Major, D. W., McMaster, M. L., Cox, E. E., Edwards, E. A., Dworatzek, S. M., Hendrickson, E. R., *et al.* (2002) Field demonstration of successful bioaugmentation to achieve dechlorination of tetrachloroethene to ethene. *Environ. Sci. Technol.* **36**, 5106–5116.

Martienssen, M., and Schirmer, M. (2007) Use of surfactants to improve the biological degradation of petroleum hydrocarbons in a field site study. *Environ. Technol.* **28**, 573–582.

McKay, C. P., Stoker, C. R., Glass, B. J., Davé, A. I., Davila, A. F., Heldmann, J. L., *et al.* (2013) The Icebreaker Life mission to Mars: a search for biomolecular evidence for life. *Astrobiology* **13**, 334–353.

Mille, G., Almallah, M., Bianchi, M., van Wambeke, F., and Bertrand, J. C. (1991) Effect of salinity on petroleum biodegradation. *Fresenius J. Anal. Chem.* **339**, 788–791.

Miller, R. M., and Herman, D. H. (1997) Biotransformation of organic compounds—remediation and ecotoxicological implications. In "Soil Ecotoxicology." (J. Tarradellas, G. Bitton, and D. Rossel, eds.), Lewis Publishers, Boca Raton, FL, pp. 53–84.

Morkin, M., Devlin, J. F., Barker, J. F., and Butler, B. J. (2000) *In situ* sequential treatment of a mixed contaminant plume. *J. Contam. Hydrol.* **45**, 283–302.

NRC (National Research Council) (1993) "*In Situ* Bioremediation, When Does it Work?", National Academy Press, Washington, DC.

NRC (National Research Council) (2003) "Environmental Cleanup at Navy Facilities: Adaptive Site Management," National Academy Press, Washington, DC.

Novak, J. M., Jayachandran, K., Moorman, T. B., and Weber, J. B. (1995) Sorption and binding of organic compounds in soils and their relation to bioavailability. In "Bioremediation—Science & Applications" (H. Skipper, and R. F. Turco, eds.), Soil Science Society of America Special Publication Number 43, Soil Science Society of America, Madison, WI, pp. 13–32.

ORNL (Oak Ridge National Laboratory) (1998) "Limited Groundwater Investigation of the Atlas Corporation Moab Mill, Moab, Utah," U.S. Department of Energy, Washington, DC.

Palmisano, A. C., Schwab, B. S., Maruscik, D. A., and Ventullo, R. M. (1991) Seasonal changes in mineralization of xenobiotics by stream microbial communities. *Can. J. Microbiol.* **37**, 939–948.

Pavan, M., and Worth, A. P. (2006) Review of QSAR models for ready biodegradation. EUR Scientific and Technical Research Series Report EUR 22355, EN-DG Joint Research Centre, Institute for Health and Consumer Protection.

Pepper, I. L., Gerba, C. P., and Brusseau, M. L. (2006) "Environmental and Pollution Science," 2nd ed. Academic Press, San Diego, CA.

Perry, J. J. (1984) Microbial metabolism of cyclic alkanes. In "Petroleum Microbiology" (R. Atlas, ed.), Macmillan, New York, NY, pp. 61–97.

Phillippi, M., Schmid, J., Wipf, H. K., and Hütter, R. (1982) A microbial metabolite of TCDD. *Experientia* **38**, 659–661.

Pitter, P., and Chudoba, J. (1990) "Biodegradability of Organic Substances in the Aquatic Environment," CRC Press, Ann Arbor, MI.

Rico-Martínez, R., Snell, T. W., and Shearer, T. L. (2013) Synergistic toxicity of Macondo crude oil and dispersant Corexit 9500A® to the *Barchionus plicatilis* species complex (Rotifera). *Environ. Pollut.* **173**, 5–10.

Rosenberg, E., Bayer, E. A., Delarea, J., and Rosenberg, E. (1982) Role of thin fimbriae in adherence and growth of *Acinetobacter calcoaceticus* RAG-1 on hexadecane. *Appl. Environ. Microbiol.* **44**, 929–937.

Scow, K. M. (1993) Effect of sorption-desorption and diffusion processes on the kinetics of biodegradation of organic chemicals in soil. In "Sorption and Degradation of Pesticides and Organic Chemicals in Soil" (D. M. Linn, T. H. Carski, M. L. Brusseau, and F.-H. Chang, eds.), Soil Science Society of America Special Publication, Madison, WI, pp. 73–114.

Sikkema, J., de Bont, J. A. M., and Poolman, B. (1995) Mechanisms of membrane toxicity of hydrocarbons. *Microbiol. Rev.* **59**, 201–222.

Snyder, S. A., Westerhoff, P., Yoon, Y., and Sedlak, D. L. (2003) Pharmaceuticals, personal care products, and endocrine disruptors in water: implications for the water industry. *Environ. Eng. Sci.* **20**, 449–469.

Stroo, H. F., Leeson, A., and Ward, C. H. (eds.) (2013) "SERDP ESTCP Environmental Remediation Technology," vol. 5, Springer, New York.

Suttinun, O., Luepromchai, E., and Müller, R. (2013) Cometabolism of trichloroethylene: concepts, limitations, and available strategies for sustained biodegradation. *Rev. Environ. Sci. Biotechnol.* **12**, 99–114.

Townsend, G. T., Prince, R. C., and Suflita, J. M. (2004) Anaerobic biodegradation of alicyclic constituents of gasoline and natural gas condensate by bacteria from an anoxic aquifer. *FEMS Microbiol. Ecol.* **49**, 129–135.

Trower, M. K., Buckland, R. M., Higgins, R., and Griffin, M. (1985) Isolation and characterization of a cyclohexane-metabolizing *Xanthobacter* sp. *Appl. Environ. Microbiol.* **49**, 1282–1289.

Trudgill, P. W. (1984) Microbial degradation of the alicyclic ring. In "Microbial Degradation of Organic Compounds" (D. T. Gibson, ed.), Marcel Dekker, Inc., New York, NY, pp. 131–180.

Tunkel, J., Mayo, K., Austin, C., Hickerson, A., and Howard, P. (2005) Practical considerations on the use of predictive models for regulatory purposes. *Environ. Sci. Technol.* **39**, 2188–2199.

Ulrich, A. C., Guigard, S. E., Foght, J. M., Semple, K. M., Pooley, K., Armstrong, J. E., *et al.* (2009) Effect of salt on aerobic biodegradation of petroleum hydrocarbons in contaminated groundwater. *Biodegradation* **20**, 27–38.

USEIA (U.S. Energy Information Administration) (2012) Annual energy review 2011. DOE/EIA–0384(201). Washington, DC. http://www.eia.gov/totalenergy/data/annual/pdf/aer.pdf.

USEPA (U.S. Environmental Protection Agency) (2001) Use of bioremediation at Superfund sites. EPA 542-R01-019, Washington, DC.

USEPA (U.S. Environmental Protection Agency) (2010) Superfund remedy report, 13th ed. EPA 542-R-10-004, Washington, DC.

USEPA (U.S. Environmental Protection Agency) (2013) UST program facts. Washington, DC. http://www.epa.gov/oust/pubs/ustfacts.pdf.

Uyttebroek, M., Vermeir, S., Wattiau, P., Ryngaert, A., and Springael, D. (2007) Characterization of cultures enriched from acidic polycyclic aromatic hydrocarbon-contaminated soil for growth on pyrene at low pH. *Appl. Environ. Microbiol.* **73**, 3159−3164.

van Beelen, P., and Fleuren-Kemilä, A. K. (1993) Toxic effects of pentachlorophenol and other pollutants on the mineralization of acetate in several soils. *Ecotox. Environ. Saf.* **26**, 10−17.

van der Meer, J. R. (2006) Environmental pollution promotes selection of microbial degradation pathways. *Front. Ecol. Environ.* **4**, 35−42.

Waid, J. S. (ed.) (1986) *PCBs and the Environment*, vols. I & II, CRC Press, Boca Raton, FL.

Wang, X., and Bartha, R. (1990) Effects of bioremediation on residues, activity and toxicity in soil contaminated by fuel spills. *Soil Biol. Biochem.* **22**, 501−505.

Widdel, F., and Rabus, R. (2001) Anaerobic biodegradation of saturated and aromatic hydrocarbons. *Curr. Opin. Biotechnol.* **12**, 259−276.

Wyndham, R. C., and Costerton, J. W. (1981) Heterotrophic potentials and hydrocarbon biodegradation potentials of sediment microorganisms within the Athabasca oils sands deposit. *Appl. Environ. Microbiol.* **41**, 783−790.

Yakimov, M. M., Timmis, K. N., and Golyshin, P. N. (2007) Obligate oil-degrading marine bacteria. *Curr. Opin. Biotechnol.* **18**, 257−266.

Zhang, Y., and Miller, R. M. (1994) Effect of a *Pseudomonas* rhamnolipid biosurfactant on cell hydrophobicity and biodegradation of octadecane. *Appl. Environ. Microbiol.* **60**, 2101−2106.

Chapter 18

Microorganisms and Metal Pollutants

Timberley M. Roane, Ian L. Pepper and Terry J. Gentry

18.1 METALS IN THE ENVIRONMENT

Metals pose a very different pollution problem than organics (see Chapter 17). Metals cannot be degraded through biological, chemical or physical means to an innocuous by-product. More specifically, while the chemical nature of a metal can be changed through oxidation or reduction, the elemental nature of a metal remains the same. Consequently, metals are persistent and more difficult to remove from the environment.

An important concept to define with respect to metals is bioavailability (see Information Box 18.1). A bioavailable metal is one that can be taken up by a microorganism, plant or animal. Bioavailable metal usually consists of the ionic species that can be readily transformed into free ionic species in solution. Given this definition, metals can clearly exist in both bioavailable and unavailable forms in the environment, and it is only the bioavailable portion that can exert toxicity on microbes, plants or animals. As a result, the total metal in a sample does not necessarily reflect the degree of biological metal toxicity, making it difficult to accurately assess the extent of risk posed by metal contamination. Only recently have investigators begun to try to elucidate the ecological significance of bioavailable metal concentrations.

Because of the toxicity and the ubiquity of metals in the environment, microorganisms have developed multiple ways of dealing with both essential and unwanted toxic metals. Levels of essential metals have to be carefully regulated to ensure sufficient supply while avoiding toxicity. This process is often referred to as metal homeostasis, as opposed to resistance. All organisms need to maintain homeostasis of different essential metals such as copper, iron, manganese and zinc to maintain cell functioning. The most common way microorganisms deal with excess metal is to pump the metal ions out of their cells while

I.L. Pepper, C.P. Gerba, T.J. Gentry: Environmental Microbiology, Third edition. DOI: http://dx.doi.org/10.1016/B978-0-12-394626-3.00018-1
© 2015 Elsevier Inc. All rights reserved.

Information Box 18.1   Total versus Soluble Metal

Normally in environmental samples the soluble metal is a small fraction of the total metal. For example, Kong (1998) found that the soluble metal concentration in sediment slurries initially amended with 20 mg/L cadmium, copper or chromium were below detection limits of 0.03−0.04 mg/L. Furthermore, at 100 mg/L added metal, only 1 mg/L cadmium and 0.12 mg/L copper and chromium were found in the aqueous phase. An even more dramatic example is a site that was contaminated with lead-based paint in Camp Navajo, Colorado. The total amount of lead in the soil was 17,520 mg/kg. This can be compared to background levels of 2−200 mg/kg! Extraction of the soil with a 10 mM KNO_3 solution yielded only 7 mg/kg lead. Extraction with a metal chelator (DTPA) yielded 1289 mg/kg, and finally, extraction with mild acid yielded 13,209 mg/kg. This example illustrates first the large difference between total and soluble metal concentrations and, second, how important it is to define the extraction procedure used to determine soluble metal.

simultaneously restricting metal uptake. In addition, some microorganisms have mechanisms to sequester and immobilize metals, whereas others actually enhance metal solubility in the environment. The goal of this chapter is to demonstrate how the presence of metals influences both the degree and type of metal resistance mechanisms expressed, and how microbial resistance, in turn, can influence the fate of metals in the environment. The first part of the chapter introduces metals and their interaction with the physicochemical components of the environment. The remainder of the chapter then focuses on specific metal−microbe interactions including mechanisms of metal resistance, positive and negative effects of metal−microbe interactions, and applications of microorganisms in metal mining and remediation of metal-contaminated sites.

18.2 CAUSE FOR CONCERN

Concern over metal pollution was once primarily related to mining activity and industrial waste. Now reports of metal-related contamination can be found almost daily in the news, including reports on mercury in fish and arsenic in drinking water. The environmental levels of metals in many locations around the world continue to increase, in some cases to toxic levels, due to contributions from a wide variety of industrial and domestic sources. For example, anthropogenic emissions of lead, cadmium, vanadium and zinc exceed those from natural sources by up to 100-fold.

Metal-contaminated environments pose serious health and ecological risks. Metals, such as aluminum, antimony, arsenic, cadmium, lead, mercury and silver, cause adverse effects including heart disease, liver damage, cancer, neurological and cardiovascular disease, central nervous system damage, encephalopathy, hypophosphatemia and sensory disturbances. The problem of mercury pollution came into focus in Minamata Bay, Japan, after the discovery of high levels of methylmercury in fish and shellfish that resulted in thousands of poisonings and hundreds of deaths (Kudo *et al.*, 1998). The mercury contamination originated from a chemical factory that generated small amounts of highly toxic and bioavailable methylmercury during its manufacturing process, which was disposed of into Minamata Bay, and ultimately accumulated in fish. It is also likely that microbial activity in the sediment converted elemental mercury that was disposed of into the bay into methylmercury. Assignment of responsibility and compensation for this tragedy continues to this day.

Lead is a second metal of concern because lead poisoning of children is common and leads to behavioral problems resulting from impaired mental function and even semipermanent brain damage. The Centers for Disease Control (CDC) and the Agency for Toxic Substances and Disease Registry (ATSDR) estimate that 10% of children in the United States have blood lead levels greater than 10 μg/dl, a potentially toxic level. Historians have speculated that the decline of the Roman Empire may have been due in part a decrease in the mental skills of the ruling class as a result of lead poisoning from wine stored in pottery lined with lead and from lead water pipes. Although contamination of drinking water supplies and concentration of metals in edible fish are of particular concern, soils and sediments are the major sinks for metal-containing wastes as the production of domestic and industrial wastes increases.

18.3 METALS DEFINED

There are three classes of metals: metals; metalloids; and heavy metals. Metals, in general, are a class of chemical elements that form lustrous solids that are good conductors of heat and electricity. However, not all metals fit this definition; for example, mercury is a liquid. Elements such as arsenic, boron, germanium and tellurium are generally considered metalloids or semimetals because their properties are intermediate between those of metals and those of nonmetals. Heavy metals are defined in a number of ways based on cationic-hydroxide formation, a specific gravity greater than 5 g/ml, complex formation, hard−soft acids and bases, and, more recently, association with biological and environmental toxicity. This chapter will focus on the metals most commonly associated with metal pollution: arsenic (As); cadmium (Cd); copper (Cu); chromium (Cr); mercury (Hg); lead (Pb); and zinc (Zn). The metals most commonly associated with severe pollution at

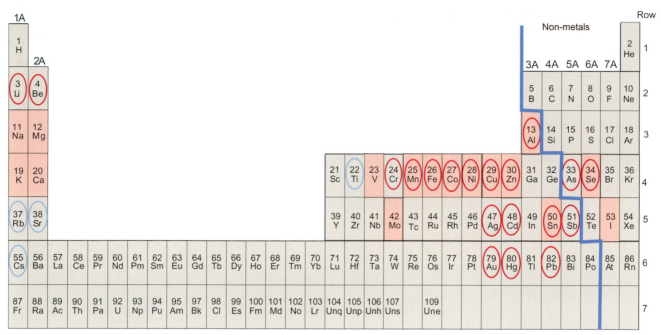

FIGURE 18.1 Periodic table showing the essential metals (pink squares), the toxic metals (circled in red) and the nonessential, nontoxic metals (circled in blue). Essential metals at high concentrations can be toxic; while some limited biological function has been associated with some toxic metals, e.g., chromium and cadmium. Many of the radionuclides, such as uranium (U) and radium (Ra), are thought to be toxic but the associated mechanisms are not well understood.

Superfund sites throughout the United States include arsenic, barium, cadmium, nickel, lead and zinc.

As a result of the complexity of the chemical definitions, metals have also been classified into three additional classes on the basis of their biological functions and effects: (1) the essential metals with known biological functions; (2) the toxic metals and metalloids; and (3) the nonessential metals with no known biological effects.

18.3.1 The Essential Metals

Metals currently known to have essential functions in microorganisms include Al, Ca, Co, Cu, Fe, K, Mg, Mn, Mo, Na, Ni, Se, Sn, V, W and Zn (Figure 18.1). Chromium is also thought to be essential, although this is still in dispute. Metals such as Na, K, Mg and Ca are required by all organisms. Tungsten (W), on the other hand, appears to be essential only in hyperthermophilic bacteria, such as *Pyrococcus furiosus*, found in hydrothermal vents (Holden and Adams, 2003). Tungstate is thought to replace molybdate in these environments. In general, essential metals are required for enzyme catalysis, molecule transport, protein structure, charge neutralization and the control of osmotic pressure. Required metals are usually transported into the cell via membrane transport systems. It is important to note that although these metals are essential for microbial growth

and metabolism, at high concentrations even essential metals can become toxic, e.g., Cu and Se.

18.3.2 The Toxic Metals

Although the toxic metals and metalloids, such as Ag, Al, As, Au, Ge, Cd, Cr, Hg, Pb, Sb and Sn, have been thought to have no known biological function, new data suggest some limited physiological need. For example, while Cd is considered nonessential, marine phytoplankton can functionally substitute cadmium, cobalt and zinc for one another to maintain the activity of enzymes when the needed element is limiting.

Metals are predominantly present as cationic species, and metalloids are predominantly present as anionic species. Radionuclide metals like uranium are also toxic, and are of increasing concern as a result of their enrichment and careless disposal in the past. Metal toxicity can occur in a number of ways including the displacement of essential metals from their normal binding sites on biological molecules (e.g., arsenic and cadmium compete with phosphate and zinc, respectively), inhibition of enzymatic functioning and disruption of nucleic acid structure. Intuitively, one would expect the toxicity of these metals to increase as concentration increases; however, for microorganisms, recent studies have found that, in some cases, higher metal concentrations activate aggressive resistance mechanisms that increase microbial tolerance to these metals. Overall, the toxicity of a metal depends to a large extent on its speciation, which in turn influences metal bioavailability.

18.3.3 The Nontoxic Nonessential Metals

The nontoxic nonessential metals include Rb, Cs, Sr and Ti. These metals sometimes accumulate in cells as a result of nonspecific sequestration and transport. Cation replacement is the general biological effect; for example, Cs^+ replacement of K^+, but there seems to be no apparent effect on the cell (Avery, 1995). In general, the appearance of nontoxic metals in the cell results from elevated environmental levels of the metals.

18.4 METAL SOURCES

18.4.1 Anthropogenic Sources

Metal pollution results when human activity disrupts normal biogeochemical activities, or results in disposal of concentrated metal wastes. Sometimes a single metal is involved, but more often mixtures of metals are present. Mining; ore refinement; nuclear processing; and the industrial manufacture of batteries, metal alloys, electrical components, paints, preservatives and insecticides are examples of processes that produce metal by-products. Examples of specific metal contaminants include arsenic, copper and zinc salts that have been used extensively as pesticides in agricultural settings; silver salts that are used to treat skin burns; and lead, which is utilized in the production of batteries, cable sheathing, pigments and alloys. Other examples include mercury compounds that are used in electrical equipment, paints, thermometers, and fungicides and as preservatives in pharmaceuticals and cosmetics. Triorganotin compounds, such as tributyltin chloride and triphenyltin chloride, can be used as antifouling agents in marine paints because of their toxicity to plankton and bacteria. The extent of metal pollution becomes even more obvious when one considers the amount of waste generated in metal processing. For example, for every kilogram of copper produced in the United States, 198 kg of copper-laden waste is produced (Debus, 1990).

Thus, while metals are ubiquitous in nature (Table 18.1), human activities have caused metals to accumulate in soil. Such contaminated soils provide a metal sink from which surface waters, groundwaters and the vadose zone can become contaminated. Metal contamination has occurred for centuries since metals have been mined and used extensively throughout human history. Archeological evidence unearthed in Timma, Israel, indicates that mining and smelting of ores has been carried out in Western civilization since at least 4500 B.C.E. (Debus, 1990). Roman lead and zinc mines in Wales are still a source of contamination nearly 2000 years

TABLE 18.1 Typical Background Levels of Metals in Soil and Aquatic Systems

Metal	Fresh water[a] (μm)	Seawater[b] (μm)	Soil[c] (μm)
Gold (Au)	ND[d]	0.0028	ND
Aluminum (Al)	Trace[e]	0.37	2.63×10^7
Arsenic (As)	Trace	0.040	660
Barium (Ba)	ND	0.22	31,623
Cadmium (Cd)	0.00053	0.00098	5.37
Cobalt (Co)	0.012	0.0068	1349
Chromium (Cr)	Trace	0.00096	19,054
Cesium (Cs)	Trace	0.0023	447
Copper (Cu)	0.010	0.047	4.667
Mercury (Hg)	Trace	0.0010	1.48
Manganese (Mn)	0.18	0.036	1.10×10^5
Nickel (Ni)	ND	0.12	6.761
Lead (Pb)	0.00029	0.00014	478
Tin (Sn)	Trace	0.0067	851
Zinc (Zn)	0.30	0.153	7.585

[a]From Goldman and Horne (1983), Leppard (1981) and Sigg (1985).
[b]From Bidwell and Spotte (1985).
[c]From Lindsay (1979).
[d]ND, no data reported.
[e]Trace, levels below detection.

after they were first used. Atmospheric metal concentrations have also increased. Contaminated soil contributes to high metal concentrations in the air through metal volatilization and creation of windborne dust particles. In addition, industrial emissions and smelting activities cause release of substantial amounts of metals into the atmosphere. For example, in 1973, a lead smelter in northern Idaho in the United States released an estimated 27,215 kg of lead per 1.6 km^2 within a 6-month period (Keely *et al.*, 1976).

18.4.2 Natural Sources

Naturally occurring high metal concentrations can also be found as a result of the weathering of parent materials that contain high levels of metals. For example, Stone and Timmer (1975) found a natural copper concentration as high as 10% in surface peat that was filtering copper-rich spring water in New Brunswick, Canada; Forgeron (1971) described a natural surface soil with up to 3% lead and zinc at a site on Baffin Island, Canada; and Warren *et al.*

Information Box 18.2   Arsenic in Drinking Water

Arsenic is naturally widely distributed throughout Earth's crust and arsenic poisoning, particularly from groundwater, affects millions of people worldwide. Arsenic in drinking water can cause bladder, lung and kidney cancer, as well as skin lesions, hyperkeratosis (skin discoloration and thickening) and weakening of the blood vessels leading to gangrene. Concern over arsenic toxicity in the United States resulted in January 2006 in a change in the maximum concentration limit (MCL) from 50 to 10 parts per billion in drinking water. There are severe drinking water crises in some parts of the world including Bangladesh, India, Mexico and several countries in Southeast Asia. In Bangladesh alone, it is estimated that 28−62% of the 125 million inhabitants are exposed to toxic levels of arsenic ranging up to 300 parts per billion or higher (Smith *et al.*, 2000). These exposures are due in large part to the installation of tube wells that were constructed in the 1970s to replace surface water supplies that were often contaminated with pathogens. These wells are up to 200 m in depth and underlie parent media that are naturally high in arsenic. Unfortunately, the water was not tested for arsenic and thus the problem went unnoticed for many years until arsenic toxicity symptoms became increasingly observed in the population. While arsenic-containing water can be safely used for washing and bathing, it is not suitable for drinking or food preparation and thus, for the latter, safe water sources must be identified. This arsenic problem has elevated global interest in processes controlling the fate, mobility and ecotoxicology of arsenic in soil and water.

(1966) reported a mercury concentration of 1−10 mg/kg in soil overlying a cinnabar (HgS) deposit in British Columbia. One metalloid currently receiving attention in countries around the world is arsenic. The concern is contamination of groundwater which serves as the source of drinking water and, in some cases, irrigation water. This contamination is most often from naturally occurring arsenic in the parent minerals that make up the soils and subsurface in these areas (Information Box 18.2). Regardless of the source, metals are of concern because they cannot be degraded and therefore accumulate in the environment, which results in the potential for increased exposure and toxicity over time.

18.5 METAL SOLUBILITY, BIOAVAILABILITY AND SPECIATION

The aqueous phase or soluble metal in a soil is usually a small fraction of the total metal present. Most metal is found:

- An inorganic precipitates (e.g., oxides, carbonates) that are either part of or form surface coatings on the solid mineral phase

- Sorbed to inorganic/organic soil colloids
- Sorbed to or complexed with organic matter that is bound to mineral surfaces

For example, soluble cadmium was measured in two soils that were amended with cadmium nitrate. The first was Brazito sandy loam with 7% total organic carbon (TOC), and a cation exchange capacity (CEC) of 6.8 milliequivalents/100 g. The second was a Gila loam soil with 0.2% TOC and CEC of 15 milliequivalents/100 g. In the Brazito soil, 650 mg/kg of cadmium were required to obtain a soluble concentration of 10 mg/L (1.5% of the metal was soluble). For the Gila soil, 1050 mg/kg were required to obtain a similar soluble concentration (in this case 1% of the metal was soluble) (Maslin and Maier, 2000). In both cases, small amounts of metal were soluble, but the actual solubility was dependent on soil properties such as CEC and TOC.

In addition to total and soluble metal, the term bioavailable metal has been defined. Bioavailable metal is the concentration that can be taken up by plants or microbes. Although the soluble metal in a system can provide an approximation of bioavailable metal, this approximation is not entirely accurate. In some cases, metals that are loosely associated with organic matter, colloidal particles or even mineral surfaces can be accessed by plants or microbes, and thus, while not soluble, are considered bioavailable. Even so, the bioavailable concentration of a metal (like the soluble concentration) is generally very low compared to the total metal present. However, approaches have been developed to measure bioavailable metal (Section 18.8.2). For example, one can measure metal bioavailability in a soil or water sample using a microbial biosensor (Chapter 13)—an organism that responds in a measureable way to the amount of metal it can take up from a given system.

Much of the research on metal solubility and bioavailability has been conducted in soil systems because understanding the fate of metals in soils and sediments is crucial to determining metal effects on biota, metal leaching to groundwater and metal transfer up the food chain (Figure 18.2). The environmental hazards posed by metals are directly linked to their concentrations in the soil solution. High bioavailable metal concentrations in the soil solution result in greater plant uptake and/or leaching of metals. In contrast, metals that are precipitated or complexed in the soil solid phase pose a greatly reduced environmental hazard.

The speciation of a metal involves the identification of its specific forms. This is important because each specific form of a metal may have different solubility, bioavailability and toxicity. A good example of this pertains to the Minamata Bay mercury poisonings described in Section 18.2. The same amount of methylmercury (CH_3Hg^+), dimethylmercury (($CH_3)_2Hg$) and inorganic

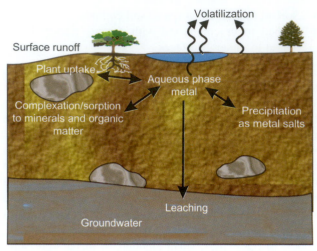

FIGURE 18.2 Potential fates and transformations of metals in the soil environment. The ultimate fate of metals in soil may to be dissolved in groundwater. Metal transformations may also occur at the surface of a soil particle or colloid or in the soil solution.

mercury (Hg^0) in fish tissue exerts very different toxicities to a human consuming the fish since the toxicity of $(CH_3)_2Hg > CH_3Hg^+ > Hg^0$ (Figure 18.3). In fact, a tragic event unfolded in 1997, when an internationally respected professor of chemistry at Dartmouth College (New Hampshire, U.S.) died following exposure to one drop of dimethylmercury that fell on her gloved hand.

Several abiotic and biotic factors can affect the chemical speciation and bioavailability of metals in the environment. These factors (discussed below) include metal chemistry, sorption to clay minerals and organic matter, pH, redox potential and the microorganisms present, e.g., some solubilize metals while others precipitate metals. All of these factors interact to influence metal speciation, solubility, bioavailability and the overall toxicity of metals in the environment. Thus, it must be emphasized that determination of the total concentration of a metal in a soil is not enough to predict toxicity in biological systems. It is the bioavailable amount that is most important. As such, the biosensors described above as well as environmental biological indicators or biomarkers for metal toxicity are being developed and studied. Indicators or biomarkers are populations that can be either microbial or arthropod, whose presence or absence provides information on the toxicity of metals in an environmental sample. It is hoped that such biomarkers will be of use in predicting the effects of metal bioavailability on environmental quality.

18.5.1 Metal Chemistry

Whether a metal is cationic or anionic in nature helps determine its fate and bioavailability in an environment. Most metals are cationic which means they exhibit a positive charge when in their free ionic state, and are most reactive with negatively charged surfaces. Thus, in soil,

cationic metals such as Pb^{2+} or Ca^{2+} strongly interact with the negative charges on clay minerals and with anionic salts, such as phosphates and sulfates. Positively charged metals are additionally attracted to negatively charged functional groups such as hydroxides and thiols on humic residues. Unfortunately, cationic metals are also attracted to negatively charged cell surfaces where they can be taken up and cause toxicity. Cationic metals sorb to both soil particles and cell surfaces with varying strengths or adsorption affinities. For example, of the common soil cations, aluminum binds more strongly than calcium or magnesium:

$$Al^{3+} > Ca^{2+} = Mg^{2+} > K^+ > Na^+$$

As this affinity series shows, the size and charge of the cationic metal helps to determine the strength of adsorption. When in excess, metal cations will compete for the limited number of cation exchange sites present in a soil with the larger multivalent cations replacing smaller monovalent cations such as Na^+ (see Figure 4.11). For example, Al^{3+} has such strong affinity for clay surfaces that it is primarily found as $Al(OH)_3$ which has extremely low bioavailability. Since many of the toxic metals are large, divalent cations, they have high adsorption affinities, and thus are not readily exchanged for common soil cations such as Na^+, K^+ or Ca^{2+}. Components in the soil solution also affect metal solubility. Specifically, phosphates, sulfates and carbonates in the soil solution form sparingly soluble metal-salt compounds.

Negatively charged or anionic metals, such as AsO_4^{3-} (arsenate), are attracted to positively charged surfaces. In soils, anions can be sorbed to negatively charged clays by divalent cation bridging, using cations such as Ca^{2+} or Mg^{2+}. In summary, the mobility of a metal in the environment is strongly influenced by the nature and intensity of its charge.

18.5.2 Cation Exchange Capacity

One of the most important factors affecting metal bioavailability is the soil cation exchange capacity (CEC), which is dependent on both the organic matter and clay content of the soil (see Section 4.2.2.4). Cation exchange reflects the capacity for a soil to sorb metals. Thus, the toxicity of metals in soils with high CEC (organic and clay soils) is often low even at high total metal concentrations. In contrast, sandy soils with low CEC, and therefore low metal binding capacity, show decreased microbial activity at comparatively low total metal concentrations, indicative of higher metal bioavailability.

Clay minerals provide a wide range of negative adsorption sites for cationic metals. These include planar sites associated with isomorphous replacements, edge sites derived from partly dissociated $Si-OH$ groups at the

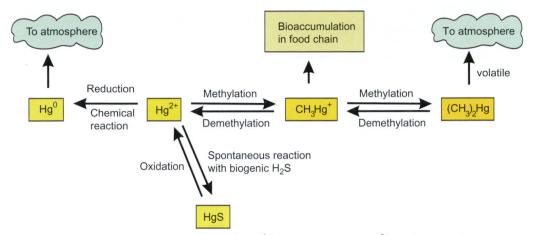

FIGURE 18.3 Microbially-mediated reactions with Hg^{2+} in the environment. Hg^{2+} can be reduced to elemental Hg^0 by chemical reaction with humic acids or by microbially-mediated reactions which are believed to be a detoxification mechanism. Hg^{2+} can be precipitated by reaction with H_2S produced under sulfate-reducing conditions but can also be released by microbial oxidation of HgS. Methylation of Hg^{2+} produces organometals, which can accumulate in the tissue of living organisms. The production of organometals may to some extent be balanced by demethylation reactions occurring in both aerobic and anaerobic environments. Based on Gadd (1993) and Ehrlich (1996).

edges of clay minerals, and interlayer sites located between clay platelets. Permanently charged sites on clay minerals interact with metallic ions by means of nonspecific electrostatic forces. Metallic oxides and hydrated metal oxides offer surface sites for the sorption of metals. Iron, aluminum and manganese oxides are an important group of minerals that form colloidal size particles, which, in the presence of water, assume various hydrated forms able to strongly retain most metals in the soil. Soil organic matter contains both humic and nonhumic substances such as carbohydrates, proteins and nucleic acids, which are normally quickly degraded, as well as less readily degraded substances including lignin, cellulose and hemicellulose. The nonhumic organics in soil are relatively short-lived, and have little influence on the long-term fate of metals.

Humic substances are relatively stable and patchily coat the particle surfaces of natural soils. Humic substances contain a variety of organic functional groups that are able to interact with metals. These functional groups include carboxyl, carbonyl, phenyl, hydroxyl, amino, imidazole, sulfhydryl and sulfonic groups. Metals complexed with humic substances are generally not bioavailable, and therefore less toxic to biological systems.

18.5.3 Redox Potential

Metal bioavailability changes in response to changing redox conditions. Under oxidizing or aerobic conditions (+800 to 0 mV), many metals are often found as soluble cationic forms, e.g., Cu^{2+}, Cd^{2+}, Fe^{3+}, Pb^{2+} and Ca^{2+}. In contrast, reduced or anaerobic conditions (0 to −400 mV), such as those commonly found in sediments

or saturated soils, often result in metal precipitation. For example, in areas rich in sulfur and sulfate-reducing bacteria, the sulfide that is generated is available to form nontoxic, insoluble sulfide deposits, e.g., CuS and PbS. As another example, in soils rich in carbonates, metals are precipitated as metal carbonates, e.g., $CdCO_3$. Conversely, reducing conditions generally increase the bioavailability of arsenic by stimulating microbial reduction and dissolution of iron oxides (with which arsenate is often associated) thus releasing previously bound arsenate. Under reducing conditions, the arsenate can also be microbially reduced to arsenite, which is more soluble and bioavailable (Information Box 18.3).

18.5.4 pH

For cationic metals, the pH of a system can have an appreciable effect on metal solubility, and, hence, metal bioavailability. At high pH, metals are predominantly found as insoluble metal mineral phosphates and carbonates, while at low pH they are more commonly found as free ionic species or as soluble organometals. The pH of a system also affects metal sorption to soil surfaces. The effect of pH on metal sorption is principally the result of changes in the net charge on soil and organic particles. As the pH increases, the electrostatic attraction between a metal and soil constituents is enhanced by increased pH-dependent CEC. This is in addition to the decrease in metal solubility that occurs as pH is increased. Accordingly, the net effect of increased soil pH is to decrease metal bioavailability. In contrast, as soil pH decreases, metal solubility increases while pH-dependent charge decreases, and metal bioavailability is increased.

Information Box 18.3   Impact of Soil Microorganisms on Arsenic in Rice

When grown in soil containing arsenic, rice tends to accumulate higher concentrations of arsenic in comparison to most other food crops. This is due to the continuous flooding practices commonly used in rice cultivation. Under aerobic conditions, most of the arsenic in soil occurs as arsenate (bound to iron oxides) and has limited bioavailability. Once flooded, the soil becomes anaerobic with iron-reducing bacteria reducing and dissolving the iron oxides, thus releasing the arsenate which is also microbially reduced to arsenite. In addition, other soil microorganisms can convert the arsenic to methylated arsenic species. The reduced and methylated arsenic species are much more bioavailable and are accumulated by rice. One management practice that has promise for decreasing rice uptake of arsenic is the use of intermittent flooding, and other less water-intensive management systems, that maintain the soil at a higher redox potential, thus decreasing arsenic bioavailability. In fact, studies have shown that these practices can decrease rice grain arsenic concentrations by over 40% compared to traditional, continuous flooding practices (Somenahally *et al.*, 2011). These systems also have the potential benefits of requiring less water for rice production and generating lower amounts of greenhouse gases such as methane. Research is ongoing to determine the optimal water management systems for maintaining high rice yields yet minimizing these negative aspects of production.

Metalloids such as chromium, arsenic and selenium are usually found in the oxyanion form, e.g., chromate (CrO_4^{2-}), arsenate (AsO_4^{3-}) and selenate (SeO_4^{2-}). In a soil with low pH, these anionic species may become increasingly sorbed as the overall positive charge on soil particles increases. As the pH increases, the fate of these anionic species becomes highly dependent on other environmental factors including redox and metal speciation.

The influence of pH, redox and organic and inorganic minerals on the chemical form and nature of metals demonstrates how important and difficult it is to clearly define metal speciation and bioavailability. Metal bioavailability is further complicated by the complexity of microbiological interactions with metals.

18.6 METAL TOXICITY EFFECTS ON THE MICROBIAL CELL

Due to ionic interactions, metals bind to many cellular ligands and displace essential metals from their normal binding sites (Figure 18.4). For example, cadmium can replace zinc in a cell. Metals also disrupt proteins by binding to sulfhydryl groups, and nucleic acids by binding to phosphate or hydroxyl groups. As a result, protein and DNA conformation are changed and function is disrupted. For example, cadmium often outcompetes catalytic zinc within enzymes, rendering the enzyme inactive, and also nonspecifically binds to DNA, inducing single-strand breaks. Metals may also affect oxidative phosphorylation and membrane permeability, as seen with vanadate and mercury. Metals, such as copper and iron, can exist in different redox states and may generate reactive oxygen species that damage proteins and nucleic acids. Microorganisms often use specific transport pathways to bring essential metals across the cell membrane into the cytoplasm. Unfortunately, toxic metals can also cross membranes, via diffusion, nonspecific uptake systems, or pathways designed for other metals. For instance, Cd^{2+} uptake occurs via the active uptake system for Mn^{2+} in many bacteria. Figure 18.5 illustrates various mechanisms used to transport metals into the cell.

These metal—microbe interactions result in decreased growth, abnormal morphological changes and inhibition of biochemical processes in individual cells. The toxic effects of metals can also be seen at the community level. In response to metal toxicity, overall community numbers and diversity can both decrease. However, few studies have addressed community resistance. While individual microbial populations may be quite metal resistant, how do microbial populations interact with each other when toxic concentrations of metals are present? Further, is it possible for metal-resistant populations to interact in such a way as to confer resistance on a consortium of organisms? Likewise, are there symbiotic relationships between metal-resistant and metal-sensitive populations such that the metal-sensitive organism receives protection from metal toxicity while providing the metal-resistant organism with some essential nutrient or carbon source? The answers to these questions are currently being sought through research of microbial metal resistance.

18.7 MECHANISMS OF MICROBIAL METAL RESISTANCE AND DETOXIFICATION

Microorganisms are believed to have evolved metal resistance because of their exposure to toxic metals on early Earth (Information Box 18.4). Additional development of metal resistance in response to recent exposure to metal pollution over the past 50 years has also been observed. Anthropogenic contamination of the environment with metals has motivated the need for research concerning microbial metal toxicity and resistance, to better understand the fate of metals in the environment, and to develop new remediation techniques for metal-contaminated sites. Microorganisms directly influence the fate of metals in the environment, and so may provide the key to decreasing current contamination.

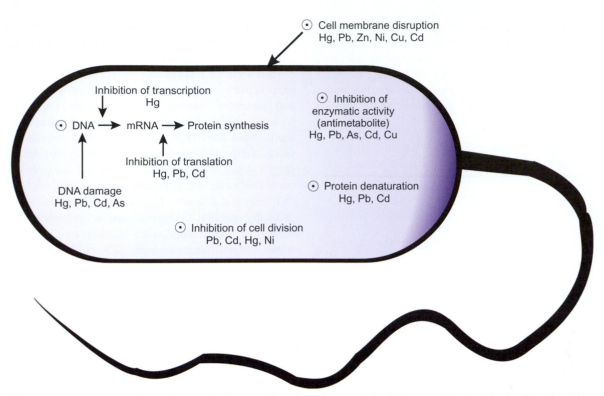

FIGURE 18.4   Summary of the various toxic influences of metals on the microbial cell, demonstrating the ubiquity of metal toxicity. Metal toxicity generally inhibits cell division and metabolism. As a result of this ubiquity, microorganisms have to develop "global" mechanisms of resistance that protect the entire cell from metal toxicity.

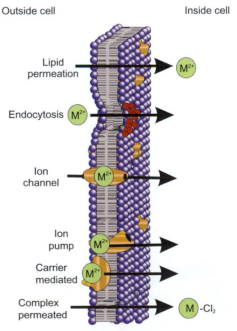

FIGURE 18.5   Mechanisms of metal (M) flux across the microbial cell membrane. Adapted from Simkiss and Taylor (1989).

Microorganisms have evolved ingenious mechanisms of metal resistance and detoxification in response to metals in the environment (Figure 18.6). Many of the resistance determinants are encoded on the chromosome, but some are encoded on mobile genetic elements such as plasmids and transposons. Microbial metal resistance may be divided into three categories. These include:

- General resistance mechanisms that do not require metal stress
- General resistance mechanisms that are activated by metal stress
- Resistance mechanisms that are dependent on a specific metal for activation

General mechanisms of metal resistance often serve other functions. For example, slime layer production, while effectively providing a barrier against metal entry into the cell, also serves in surface adhesion and protection against desiccation and predation. The sole purpose of metal-dependent mechanisms, both specific and general, is cell protection from metal toxicity.

18.7.1 General Mechanisms of Metal Resistance

Binding of metals to extracellular materials immobilizes the metal and prevents its entry into the cell. Metal binding to anionic functional groups on cell surfaces occurs with a large number of cationic metals, including cadmium, lead, zinc and iron. For example, sulfhydryl, carboxyl, hydroxyl, sulfonate, amine, amide and phosphonate groups are examples of functional groups that strongly bind to metals. Binding of metals by microbial cells is important ecologically since the binding of metals by cell surfaces plays a dominant role in the distribution of metals, especially in aquatic environments. In practical terms, the ability of cells to sorb metals has been developed into a technology used to remove metals from contaminated waste streams.

Extracellular binding usually occurs on slime layers or exopolymers composed of carbohydrates, polysaccharides and sometimes nucleic and fatty acids (Schiewer and Volesky, 2000). These exopolymers or extracellular polymeric substances (EPS) are common in natural environments and provide microbial protection against desiccation, phagocytosis and parasitism. Microbial exopolymers are particularly efficient in binding heavy metals, such as lead, cadmium and uranium. Exopolymer functional groups are generally negatively charged, and consequently, the efficiency of metal:exopolymer binding is pH dependent. Metal detoxification through EPS production results in metal immobilization and prevention of metal entry into the cell. For example, the immobilization of lead by exopolymers has been demonstrated in several bacterial genera, including *Staphylococcus aureus*, *Micrococcus luteus* and *Azotobacter* spp. In fact, extracellular polymeric metal binding is a common resistance mechanism against lead.

A second extracellular molecule produced microbially that complexes metals is the siderophore. Siderophores are iron-complexing, low-molecular-weight organic compounds. Their biological function is to harvest iron in environments where concentration is low, and to facilitate its transport into the cell. Siderophores may interact with other metals that have chemistry similar to that of iron, such as aluminum, gallium and chromium (which form trivalent ions similar in size to iron). By binding metals, siderophores can reduce metal bioavailability and thereby metal toxicity. For example, siderophore complexation reduces copper toxicity in cyanobacteria.

Information Box 18.4   Evolution of Metal Resistance

A question plaguing environmental microbiologists is how do microorganisms become metal resistant? Earth is estimated to be 4.7 billion years old and microbial life is thought to have appeared approximately 4 billion years ago. Since microorganisms were the first life forms on Earth, early resistance mechanisms developed in response to the toxic metals that existed early on in Earth when life began. Genetic sequencing of hundreds of microbial genomes provides evidence supporting the early development of metal resistance. However, another possible scenario is that microorganisms have recently developed metal resistance in response to increasing anthropogenic pollution with various metals. Metal concentrations in the atmosphere, in surface soils and in both surface and groundwaters have increased with industrialization. On average, a mutation occurs once in every 10^6 base pairs, so microorganisms have the capacity to rapidly respond to metal-contaminated environments and evolve metal resistance. The answer to the origins of microbial metal resistance is probably a combination of both early and recent exposure to toxic metals. Future study of the physiological and the genetic diversity in microbial metal resistance mechanisms will provide intriguing insights into how microorganisms adapt to and maintain responses to environmental pressures.

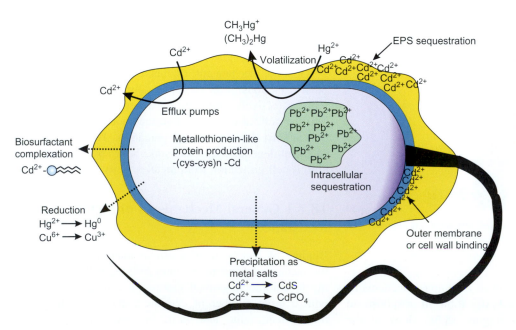

FIGURE 18.6 In response to metal toxicity, microorganisms use a variety of mechanisms to resist and detoxify harmful metals. These mechanisms of resistance may be intracellular or extracellular and may be specific to a particular metal, or a general mechanism able to interact with a variety of metals.

Biosurfactants are a class of compounds produced by many microorganisms that in many cases are excreted from the cell. Biosurfactants have been investigated for their ability to complex metals such as cadmium, lead and zinc (Maier and Soberon-Chavez, 2000). Biosurfactant complexation can actually increase the apparent solubility of metals; however, the biosurfactant-complexed metal is not toxic to cells. It is not yet clear whether biosurfactants are produced specifically to reduce metal toxicity. However, evidence shows that biosurfactant-producing microorganisms can be isolated in greater diversity from metal-contaminated environments than from uncontaminated ones. More information on biosurfactants is available in Chapter 17.

Finally, metal bioavailability can be influenced by common metabolic by-products that result in metal reduction. In this case, soluble metals are reduced to less-soluble metal salts, including sulfide and phosphate precipitates. For example, under aerobic conditions *Citrobacter* spp. can enzymatically produce phosphate, which results in the precipitation of lead and copper. Under anaerobic conditions, high H_2S concentrations from sulfate-reducing bacteria, such as *Desulfovibrio* spp., readily cause metal precipitation as metal sulfides.

18.7.2 Metal-Dependent Mechanisms of Resistance

Free-living, nonsymbiotic microorganisms use energy-dependent metal efflux systems to remove metals from the cell. These mechanisms effectively pump toxic ions that have entered the cell back out of the cell via ATPase pumps or chemiosmotic ion/proton pumps. Arsenic resistance in a number of bacteria involves the enzymatic reduction of arsenate to arsenite followed by arsenite efflux, either by a transporter protein called ArsB or by an unrelated ACR3 transporter. Arsenate reduction and subsequent arsenite efflux are common in archaea, bacteria and fungi.

There are also some plasmid-encoded genes that can confer even higher (and more complex) levels of resistance (Figure 18.7). For example, plasmid R773 contains genes encoding the regulator ArsR, the transporter ArsB, the arsenate reductase ArsC, the ATPase ArsA and an arsenite chaperone ArsD. Arsenate enters the cell through proteins involved in phosphate-specific transport (Pst) with initial binding by the phosphate binding protein PhoS. The reduction of arsenate to arsenite is then mediated by the ArsC enzyme, an NADPH-dependent cytoplasmic protein (Mukhopadhyay *et al.*, 2002). Finally, together, the ArsA and ArsB proteins form a complex that uses ATP for the active efflux of arsenite from the cell. This complex can confer much higher levels of resistance than ArsB alone.

Another well-studied example of efflux-based resistance is cadmium resistance (Figure 18.8). The toxicity of

Cd is primarily due to its binding to sulfhydryl groups of proteins and causing single-stranded DNA breaks. Several cadmium-resistant organisms have been studied, including what was then known as *Alcaligenes eutrophus*, and *Bacillus subtilis*, *Escherichia coli*, *Listeria* spp., *Pseudomonas putida*, *Staphylococcus aureus*, some cyanobacteria, fungi and algae. Most microorganisms possess a chromosomally encoded P-type ATPase termed CadA or ZntA, which pumps Cd^{2+} out of the cytoplasm using ATP hydrolysis as an energy source. Most of these pumps can also transport other cations such as Pb^{2+} and Zn^{2+}. In addition to these P-type ATPases, there are systems in Gram-negative bacteria responsible for transport of Cd^{2+} and other metals from the periplasm across the outer membrane. The best-studied system includes the *czc* genes from *Cupriavidus metallidurans* CH34, where *czc* stands for cobalt, zinc and cadmium resistance (Nies, 2003).

Either in place of or in addition to metal efflux, bacteria can use intracellular metal resistance mechanisms. Possibly the best-known mechanism involves metal binding or sequestration by metallothioneins or similar proteins. Primarily documented in higher microorganisms, plants, algae, yeast and some fungi, metallothioneins are low-molecular-weight, cysteine-rich proteins with a high affinity for cadmium, zinc, copper, silver and mercury metals. Their production is induced by the presence of metals, and their primary function is metal detoxification. Metallothioneins are being found in an increasing number of microorganisms, including bacteria. Metal binding by metallothioneins can result in cellular accumulations visible as electron dense areas within the cell matrix. Suspected deposits are confirmed using electron dispersive spectroscopy that can identify the metal.

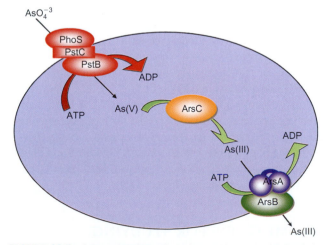

FIGURE 18.7 Schematic of arsenic resistance demonstrating both the influx and efflux systems for arsenic. Arsenate enters the cell via a phosphate-specific transport pathway. Once in the cell, arsenate is reduced to arsenite, and an efflux mechanism then pumps arsenite out of the cell via an anion pump that is fueled by ATP. Note that arsenic is not detoxified by this mechanism because arsenite can still be toxic.

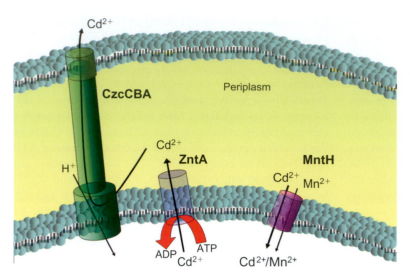

FIGURE 18.8 Cadmium influx and efflux in a Gram-negative bacterial cell. Cadmium crosses the cytoplasmic membrane via a manganese transport pathway (MntH), which relies on membrane potential (see purple structure). Cadmium can also enter the cell via the zinc transporter ZntA (see red arrow). The cadmium efflux system (CzcCBA) excretes cadmium via a $Cd^{2+}/2H^+$ antiport protein (see green structure).

The methylation of metals is considered to be a metal-dependent mechanism of resistance because only some metals are methylated. Methylation involving the addition of methyl or ethyl groups (e.g., conversion of Hg^{2+} to CH_3Hg^+) increases metal volatility, and can increase metal toxicity as a result of increased lipophilicity, thus increasing permeation across cell membranes. However, methylation of some metals, such as selenium, decreases their toxicity. Methylation has also been observed with arsenic, lead and tin. Methylation facilitates metal diffusion away from the cell, and in this way effectively decreases overall metal toxicity. In this manner, methylation has been known to remove significant amounts of metal from contaminated surface waters, sewage and soils.

Mercury is unusual in that it can additionally be volatilized through reduction. Mercury resistance may involve the enzymatic reduction of Hg^{2+} to elemental mercury (Hg^0) in both Gram-positive and Gram-negative bacteria. Often plasmid mediated, two additional pathways of mercury resistance involve the detoxification of organomercurial compounds via cleavage of C−Hg bonds by an organomercurial lyase (MerB), followed by reduction of Hg^{2+} to Hg^0 by a flavin adenine dinucleotide (FAD)-containing, NADPH-dependent mercuric reductase (MerA). Specific to inorganic mercury, the MerP protein in the periplasmic space shuttles Hg^{2+} to the membrane-bound MerT protein, which releases Hg^{2+} to the cytoplasm. Once in the cytoplasm, Hg^{2+} is reduced to Hg^0 by mercuric reductase (Figure 18.9).

18.8 METHODS FOR STUDYING METAL−MICROBIAL INTERACTIONS

Unique considerations apply when studying metal−microorganism interactions. Metal concentrations on a macroscale poorly reflect the toxic influences of metals at the microscale. Recall that total metal concentrations do not accurately assess the biologically toxic concentration. Because metals are not biodegradable, it is difficult to determine whether and how a metal is being detoxified when the total metal concentration does not change. However, new and exciting technical and analytical developments in metal chemistry, microbiology and molecular biology are now making it possible to expand our understanding of how microorganisms influence metal fates in the environment. A few of these approaches are highlighted here.

18.8.1 Culture Medium

The culturing of metal-resistant microorganisms in the laboratory often occurs in either nutrient-rich or chemically defined media, which may contain yeast extract, phosphate buffers and amino acids that bind metal ions. A neutral medium pH is an additional factor that may increase metal binding in culture media. The presence and amount of these reagents strongly influence metal bioavailability, thereby influencing metal toxicity to microorganisms. Thus, depending on the growth medium, metal toxicity will vary. For example, it has been shown that the alga *Chlamydomonas reinhardtii* accumulates more metal (Cd^{2+} or Cu^{2+}) and shows less metal toxicity when grown on a medium containing high levels of phosphate (Wang and Dei, 2006). Similarly, for the bacterium *Comamonas testosteroni*, cadmium toxicity followed a dose-dependent pattern in minimal chemically defined media, but was not dose dependent in an organically rich medium (Hoffman *et al.*, 2005). Consequently, several factors need to be taken into consideration when choosing a culture medium to assess microbial metal resistance. Most importantly, medium components must be defined and chosen in such a way as to minimize metal binding. This applies to both

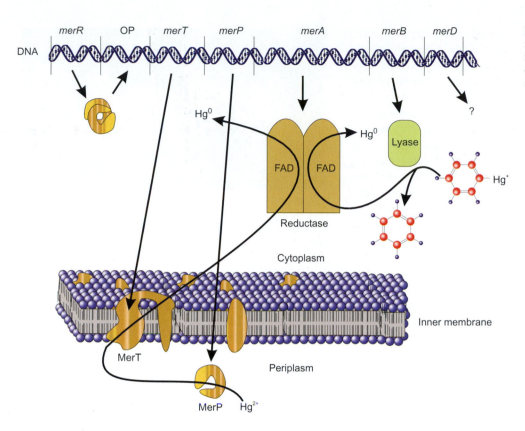

FIGURE 18.9 Proposed model for bacterial mercury resistance encoded by the *mer* operon. Adapted with permission from Silver *et al.* (1986).

carbon substrates and to buffers. For example, phosphate buffers strongly precipitate metals. Recall that phosphate production in some microorganisms confers protection from certain metals. Nonmetal-binding buffers, including the sulfonic acids such as MES [2-(*N*-morpholino)ethanesulfonic acid; $C_6H_{13}NO_4SH_2O$], $pKa = 6.15$, and PIPES (1,4-piperazinediethanesulfonic acid; $C_8H_{18}N_2O_6S_2$), $pKa = 6.80$, optimize metal bioavailability in culture media (although even these buffers can alter the toxicity of metals). Finally, pH strongly influences metal bioavailability. Metals readily precipitate as carbonic salts at pH >7.0. Therefore, the pH should be kept slightly acidic ($\approx$pH 6.0) to maintain metal solubility.

18.8.2 Measurement of Total, Soluble and Bioavailable Metal

To experimentally determine the relationship between metal bioavailability and toxicity, and to determine the rate and extent that microorganisms sequester metals, one must be able to determine both total and soluble metal concentrations. Total metal concentration is determined by digestion of the sample with acids such as nitric or perchloric acid. This process dissolves soil particles and releases even tightly bound metals. Soluble metal determination often involves extraction with the weak acid DTPA (diethylenetriamine pentaacetic acid), or extraction with deionized

water to release loosely bound, readily exchangeable metals. In either case, metal concentrations in the extract are determined using atomic absorption spectroscopy or inductively coupled plasma atomic emission spectroscopy.

When studying metal resistance, it is assumed that a decrease in the soluble metal concentration corresponds to the amount of metal sequestered by the cell. However, care must be taken in interpreting these data since metals often precipitate with culture medium components. In addition, metals may bind to the walls of flasks and test tubes. Controls with no inoculum are therefore crucial in distinguishing between biological and chemical metal removal. It should be noted that while it is relatively easy to determine a macroscale estimate of bioavailability in the environment, such an estimate does not necessarily reflect microscale metal concentrations. So even if very low soluble levels of metal are measured, it is likely that in some micropores (where most soil microorganisms live) substantial levels of bioavailable metal may be encountered. This would explain why microorganisms in metal-contaminated environments with no detectable soluble metal may exhibit extreme resistance.

Flame or flameless atomic absorption spectroscopy (AAS), inductively coupled plasma atomic emission spectroscopy (ICPAES) and inductively coupled plasma mass spectroscopy (ICPMS) are efficient techniques for the determination of metal ions in solution. For AAS, metal is determined by aspirating a metal solution into an air−acetylene flame to atomize the metal. A metal-specific lamp is placed into the AAS and is used to determine the difference in light absorbance between a reference source and the metal solution. This difference reflects the amount of metal present. ICPAES determination is based on light emitted from metal atom electrons in the excited state. An argon plasma is used to produce the excited state atoms. Matrix interference can be a significant problem with these techniques, and some samples need to be acid digested before analysis. Spectroscopy can be used for any metal; however, detection limits vary for each metal. For example, for AAS, the limit of detection is $1\ \mu g/L$ $(8.9 \times 10^{-3}\ \mu M)$ for cadmium and $700\ \mu g/L$ $(2.9\ \mu M)$ for uranium. For comparison, the detection limits for ICPAES are $1\ \mu g/L$ and $75\ \mu g/L$ $(0.32\ \mu M)$ for cadmium and uranium, respectively.

A newer technique is ICPMS, which can speciate as well as quantify metals. The ability to speciate metals yields valuable information regarding metal transformations and ratios of toxic to nontoxic metals important in risk management and environmental assessment. ICPMS detects an element's unique mass-to-charge ratio through ionization in an argon plasma. The use of ICPMS is increasing due to its greater sensitivity (parts per trillion level) than either AAS or ICPAES.

Ion-selective electrodes (ISEs), which are available for some metals such as cadmium, lead and arsenic, provide a quick way to determine metal concentrations in solution to approximately 10^{-6} M. The electrodes are easy to use, relatively inexpensive and are not strongly influenced by sample color or turbidity. They are, however, influenced by the presence of other ions in solution. ISEs are different from AAS and ICPAES in that they allow the determination of extracellular free metal ion in solution following cellular interactions. In this case, the ISE will only measure free metal ions, and does not measure complexed metal even if the complexed metal is soluble. However, this can be considered an advantage since the ISE measures only metal that would likely be bioavailable.

One of the only methods available to truly measure metal bioavailability is through the use of microbial biosensors (see also Chapter 13). These biosensors have been developed to report bioavailable metal concentration in environmental samples. The biosensor is created using recombinant DNA technology (see Chapter 13) to construct a plasmid in which a strictly regulated promoter is connected to a sensitive reporter gene. The best studied example is the mercury resistance (*mer*) operon which causes the reduction of Hg^{2+} to Hg^0 (by the *merA* gene product, the mercuric reductase) and degradation of methylmercury (by the *merB* gene product, the organomercurial lyase). This is beneficial because it reduces the toxicity of mercury to the bacterial cell (Figure 18.9). The *mer* promoter is activated when Hg^{2+} binds to the regulatory protein MerR. Indicator bacteria that contain gene fusions between the promoter of the *mer* operon and

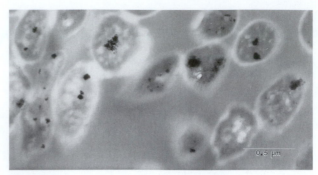

FIGURE 18.10 Transmission electron micrograph of a bacillus exhibiting an intracellular accumulation (dark material) of lead in response to the production of a metallothionein-like protein. Courtesy T.M. Roane.

a reporter gene (such as luminescence) are able to detect Hg^{2+} (Selifnova *et al.*, 1993). This promoter–reporter gene concept has also been used with other metals including arsenic, cadmium, zinc, lead, lead ions and also xenobiotic compounds (Rensing and Maier, 2003).

Whereas the methods discussed so far can detect the presence of a metal, transmission electron microscopy (TEM) provides an effective means of locating and visualizing suspected metal deposits associated with microorganisms (Figure 18.10). This technique is particularly useful for determining whether microorganisms sequester metals inside or outside of the cell. When TEM is coupled with energy-dispersive X-ray spectroscopy (EDS), the metal element can be identified. Metals emit characteristic X-rays as the electron beam interacts with deposits. Elements can be identified by signature spectral lines (Figure 18.11). High- and low-magnification TEM micrographs can help distinguish between intracellular and extracellular metal interactions, depending on the location of metal deposition.

Newer X-ray-based techniques, such as X-ray absorption spectroscopy (XAS) and X-ray fluorescence (XRF) are gaining popularity due to their ability to detect and even speciate metals *in situ* within samples. One of the most used absorption-based approaches is X-ray absorption near-edge structure (XANES) spectroscopy. This method uses an X-ray source, such as a synchrotron, to excite and shift the target's photoabsorption cross-section thus revealing elemental composition and valence states. In contrast, the fluorescence-based approaches, such as micro X-ray fluorescence (μXRF) and related imaging and spectroscopy approaches, instead use the emission following X-ray excitation to spatially observe, quantify and even speciate metals within various types of samples (Figures 18.12 and 18.13). Although access to these approaches is currently limited due to the specialized equipment needed, they have the major analytical advantage of enabling *in situ* determination of elemental composition and speciation within environmental samples (Ginder-Vogel and Sparks, 2010).

18.9 MICROBIAL METAL TRANSFORMATIONS

18.9.1 Oxidation-Reduction

Many microbial transformations of metals occur due to their use as terminal electron acceptors in anaerobic respiration (metal reduction), or their use as an energy substrate in which the metal is oxidized. A number of metals and metalloids are subject to redox cycling in the environment including iron, manganese, selenium and arsenic. For example, the oxidation of arsenite (As(III)) to arsenate (As(V)) can be described by the following chemical reaction:

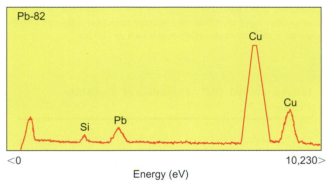

FIGURE 18.11   X-ray analysis of the suspected lead (Pb) deposit in Figure 18.10. Copper and silica were present from the copper grid and embedding medium, respectively, used in mounting the sample for analysis. Courtesy T.M. Roane.

$$2H_3AsO_3 + O_2 \rightarrow HAsO_4^{2-} + 3H^+$$
$$\Delta G^{o'} = -256 \text{ kJ/mole}$$

Arsenite oxidation can occur as an abiotic process, but microorganisms play an important role in arsenite oxidation in natural systems. As(III) oxidation can also serve as a detoxification reaction since As(III) is up to 50 times more toxic to bacterial cells than As(V) in most biological systems (Silver *et al.*, 2002). A variety of arsenite oxidizing bacteria have been identified including both chemoautotrophs and heterotrophs (Ehrlich, 2002).

Reduction of arsenate to arsenite occurs by one of two mechanisms—dissimilatory reduction or detoxification. In detoxification, reduction of arsenate is not coupled to respiration and does not provide energy for the bacterium—it is simply reduced to arsenite and then exported out of the cell. In dissimilatory reduction, As(V) is used as the terminal electron acceptor during anaerobic respiration. An example of such a reaction is the growth of *Bacillus arsenicoselenatis* using lactate as the electron donor and As(V) as the TEA. The reaction can be described by the following equation (Oremland *et al.*, 2002):

$$Lactate + 2HAsO_4^- + H^+ \rightarrow acetate^- + 2H_2AsO_3^- + HCO_3^-$$
$$\Delta G_f^o = -23.4 \text{ kJ/mole}$$

Studies of the rates of arsenate reduction by this mechanism show half-lives averaging around 30 hours (Inskeep *et al.*, 2002). For dissimilatory reduction of As(V) to occur, however, there must be high enough arsenic concentrations to support growth, and strict anaerobic

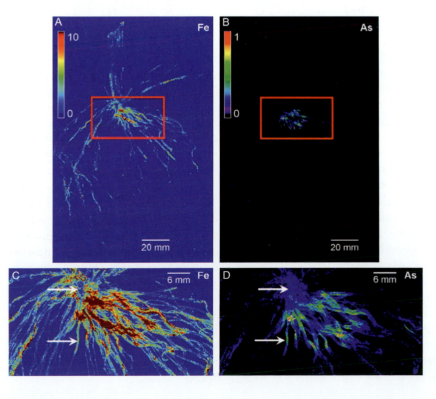

FIGURE 18.12   X-ray fluorescence images of iron (A, C) and arsenic (B, D) distributed within the root system of rice grown in contaminated soil. Images C and D represent magnified images of boxes in A and B, and the arrows represent areas where arsenic is abundant while iron is not. Note the spatial variability in iron and arsenic with highest concentrations occurring on the older roots near the shoot base. From Seyfferth *et al.* (2010).

Highly pigmented

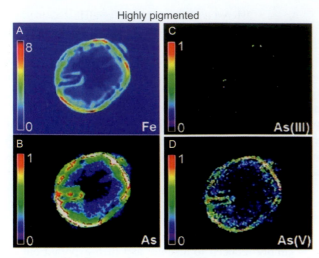

FIGURE 18.13   Cross-sectional computed X-ray tomography of a rice root grown in contaminated soil showing the spatial distribution of total iron (A), total arsenic (B), arsenite (C) and arsenate (D). Note the colocalization of iron and arsenic and that the majority of arsenic on and in the root is oxidized [As(V)]. From Seyfferth *et al.* (2010).

conditions may be required. Environments such as sediments, hot springs and freshwater and marine systems can support such conditions.

In addition to some metals being directly reduced during anaerobic respiration, metals can also be reduced indirectly through reaction with other reduced products such as sulfides. In fact, for some metals, such as uranium, this may actually be a major mechanism for their reduction in the environment (Figure 18.14).

The oxidation and reduction of metals can have profound practical implications. For example, in subsurface geological formations, metals are often found in a reduced state as, for example, pyrite (FeS_2). Pyrite is often associated with metal ore deposits. Pyrite is stable until it is exposed to oxygen by mining activities, e.g., strip mining. Upon the introduction of oxygen, a combination of autoxidation and chemoautotrophic microbial oxidation of iron and sulfur results in the production of large amounts of acid. Acid, in turn, facilitates metal solubilization, resulting in a metal-rich acidic leachate called acid mine drainage (Information Box 18.5). Contaminating groundwater and over 10,000 miles of rivers in the United States alone, acid mine drainage is highly toxic to plants and animals, often resulting in widespread fish kills. Acid mine drainage is a problem associated with many types of mining activity including subsurface mining, where metal deposits become exposed to atmospheric oxygen; strip mining, where large expanses of land are exposed to oxygen; and mine tailing wastes, which are large deposits of processed or spent ore.

Microbially induced corrosion of metal pipes and fuel and storage tanks is a second significant problem of concern. Corrosion occurs due to cooperation between two groups of bacteria, the anaerobic chemoheterotrophic sulfate-reducing bacteria (SRB) and the aerobic

A. Direct reduction

B. Indirect reduction

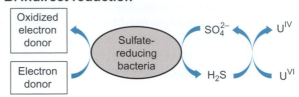

FIGURE 18.14   Examples of mechanisms for bacterial reduction of metals: (A) direct reduction through use as a terminal electron acceptor in anaerobic respiration and (B) indirect reduction following abiotic reactions with reduced chemical species such as sulfides.

Information Box 18.5   Formation of Acid Mine Drainage

The formation of acid from pyrite ore is a complex mechanism that involves the oxidation of both iron and sulfur. The initial reaction leading to the formation of acid mine drainage (AMD) is the spontaneous chemical oxidation of pyrite (FeS_2):

$$4FeS_2 + 14O_2 + 4H_2O \rightarrow 4Fe_2 + (OH)_3^- + 8SO_4^{2-} + 8H^+$$

As the local pH decreases due to the formation of acid, the sulfur- and iron-oxidizing bacterium, *Acidithiobacillus ferrooxidans*, further acidifies the environment. An acidophilic chemoautotroph, *A. ferrooxidans*, derives energy for carbon fixation, and growth from the oxidation of inorganic sulfur- and iron-containing compounds, such as pyrite. Because of the acidophilic nature of *A. ferrooxidans*, as the pH decreases, the microbially-facilitated oxidation of iron increases. Once oxidized, the iron can contribute to the formation of more acid:

$$Fe_3^+ + 3H_2O \rightarrow Fe(OH)_{3ppt} + 3H^+$$

Or the ferric iron can aid in the further chemical oxidation of pyrite:

$$FeS_2 + 14Fe_3^+ + 8H_2O \rightarrow 15Fe_2^+ + 2SO_4^{2-} + 16H^+$$

Note that this reaction produces acid and regenerates reduced or ferrous iron, which can then be reoxidized by *A. ferrooxidans*. Acid mine drainage can have a pH as low as 2. The produced leachate dissolves metal-containing ore, resulting in high concentrations of soluble and bioavailable toxic metals.

chemoautotrophic iron-oxidizing bacteria. These two groups of bacteria work together to create an environmental niche on pipe surfaces that is favorable for their simultaneous activity, even though one group requires oxygen and the other does not.

18.9.2 Methylation

The microbial methylation of metals not only results in increased metal mobility because some organometals are volatile, but also because in some cases it can change the toxicity of the metal. Methylation involves the transfer of methyl groups (CH_3) to metals and metalloids, e.g., lead, mercury, arsenic and selenium. The resulting organometal is more lipophilic than the metal species. This results in the potential of bioaccumulation and biomagnification in food webs (Figure 18.3).

Methylation of mercury occurs in the sediments of lakes, rivers and estuaries, where organic matter concentrations are high and redox conditions are favorable for the activity of sulfate-reducing bacteria, the primary generators of methylmercury (Drott et al., 2007). The most important intracellular agent of mercury methylation is believed to be methylcobalamine (CH_3CoB_{12}), a derivative of vitamin B_{12}. Methylation reactions can be summarized as follows:

$$CH_3CoB_{12} + Hg^{2+} + H_2O \rightarrow CH_3Hg^+ + H_2OCoB_{12}^+$$

methylcobalamine   methylmercury

$$CH_3CoB_{12} + CH_3Hg^+ + H_2O \rightarrow (CH3)_2Hg + H_2OCoB_{12}^+$$

methylcobalamine   dimethylmercury

The dominant product formed is the salt of the methylmercuric ion, CH_3Hg^+ (methylmercury), because the volatile dimethylmercury (($CH_3)_2Hg$) forms at a much slower rate. Microbially mediated reactions affecting the fate of Hg^{2+} are shown in Figure 18.3. Since methylation actually increases the toxicity of mercury, methylation of mercury may facilitate diffusion of both methylmercury and dimethylmercury from the cell more easily than Hg^{2+}. In contrast, methylation of selenium directly decreases its toxicity of selenium (see Case Study 18.1).

Mercury is used extensively in the electrical industry, instrument manufacturing, electrolytic processes and chemical catalysis. Mercury salts and phenylmercury compounds are also used as fungicides and disinfectants. Approximately 10,000 metric tons of mercury are produced worldwide annually. Fossil fuel burning releases an additional 3000 metric tons. Methylmercury compounds are highly lipophilic and neurotoxic. Several outbreaks of mercury poisoning have occurred throughout history. In Minimata Bay, Japan, release of mercury-containing effluents by a chemical processing plant resulted in serious illness in people who consumed fish with elevated levels of mercury. Another example is the Great Lakes in the United States, which until the 1970s had relatively uncontrolled releases of polychlorinated biphenyls (PCBs), dioxins and mercury. For a period of time, parts of the lakes were closed to fishing, but the problem has improved due to restricted use and release of the organic and metal pollutants. At this time, health advisories are in effect that make recommendations about the type and amounts of fish that can be safely consumed.

Arsenic is another example of a metal that is methylated as a resistance mechanism. It is methylated by some bacteria, such as *Rhodopseudomonas palustris*, and many fungi, such as *Scopulariopsis brevicaulis* to mono-, di- and trimethylarsine, volatile forms of arsenic (Information Box 18.6). Arsenic poisonings have occurred in the past when fungi growing on damp wallpaper converted and volatilized arsenate (used as a coloring agent) in the wallpaper. Illness occurred upon inhalation of the resulting methylated arsenic species. There is also growing evidence that microbial methylation of soil arsenic has led to elevated levels of methylated arsenic in rice from some parts of the world (Somenahally et al., 2011).

18.10 PHYSICOCHEMICAL METHODS OF METAL REMEDIATION

The remedial methods used to treat contaminated soil or sediments may be broadly divided into two main categories:

- Methods aimed at preventing movement of metals to the immediate surroundings, also called immobilization
- Methods aimed at metal removal

The goal in metal immobilization is to reduce metal solubility. Two immobilization strategies include pH alteration and addition of organic matter. Since metal solubility decreases with increasing pH, metal solubility should be reduced when site pH is raised. Liming is sometimes used to increase soil pH causing precipitation of contaminating metals as calcic and phosphoric metal-containing minerals. Amendment with organic matter can also aid in metal immobilization as a result of the electrostatic attraction between metals and organic particles. The addition of organic matter may involve the addition of highly organic waste material, such as biosolids. Often sites containing high levels of toxic metals have little or no vegetation. Revegetation of such sites, while sometimes difficult to achieve, is a good way to increase organic matter content.

Metal removal from soils or sediments can be achieved by excavation (which simply moves the problem to another location) or by using soil washing techniques. Soil washing methods rely on chemicals to facilitate metal removal. Washing with acidic solutions or chelating agents, e.g., ethylenediaminetetraacetic acid (EDTA) or nitrilotriacetic acid (NTA), solubilizes metals, enhancing removal from the system. One problem with the use of these chemical agents is the residual toxicity left by the washing agent after treatment. Newer biodegradable chelating agents show promise for metal removal. For example, 84% of nickel in a spent catalyst was removed with the biodegradable chelating agent [S,S]-ethylenediaminedisuccinic acid (EDDS) (Chauhan et al., 2012), while studies show the degree of metal removal by chelating agents can be metal specific (Tandy et al., 2004). Researchers are also looking at biological alternatives to these chemicals. For example,

Case Study 18.1   Selenium Bioremediation in San Joaquin Valley, California

Selenium is known to bioaccumulate and can cause death and deformities in waterfowl. Agriculture is the primary cause of selenium contamination in the San Joaquin Valley, but other anthropogenic sources of selenium include petroleum refining, mining and fossil fuel combustion. Once in soil, selenium exists as selenate (SeO_4^{2-}, Se_6^+), selenite (SeO_3^{2-} Se_4^+), elemental selenium (Se^0), dimethylselenide [DMSe; $(CH_3)_2Se$], and/or dimethyldiselenide [DMDSe; $(CH_3)_2Se_2$]. The most toxic forms of selenium are selenate and selenite; elemental selenium is considered insoluble and is least toxic of the selenium species. For selenium, the methylated forms of selenium are 500−700 times less toxic than the inorganic forms.

As a result of agricultural practices, largely irrigation, soils found in the San Joaquin Valley have elevated selenium levels (400 to 1000 mg/kg soil). Drainage waters in the valley can contain up to 4200 mg/L. Consequently, in some areas, selenium has concentrated to hazardous levels in evaporation ponds and in soils. The Kesterson Reservoir, located within the valley, is one such area where selenium contamination has resulted in extensive bird kills.

The bioremediation of the San Joaquin Valley and specifically the Kesterson Reservoir is based on the ability of a large number of microorganisms to reduce selenium oxyanions to the insoluble, Se^0 or to the volatile, methylated forms, e.g., DMSe. Under anaerobic conditions, organisms such as *Wolinella succinogenes* and *Desulfovibrio desulfuricans* can reduce SeO_4^{2-} and SeO_3^{2-} to elemental selenium. However, reduction to insoluble forms of selenium is not sufficient to stabilize the selenium pool within the soil matrix. Both microbial reoxidation and resolubilization are facilitated by organisms including *Acidithiobacillus ferrooxidans* and *Bacillus megaterium*. Consequently, long-term remediation of

contaminated soils requires selenium removal, hence the role of selenium -volatilizing microorganisms becomes important.

Laboratory experiments initially conducted with Kesterson sediments found that certain environmental conditions influence microbial selenium volatilization. Researchers found that increased soil moisture (− 33 kPa), soil mixing, increased temperature (35°C) and application of an organic amendment increased selenium volatilization. In greenhouse experiments, sediment samples containing 60.7 mg/kg sediment were treated with various carbon sources to enhance microbial selenium methylation (Karlson and Frankenberger, 1990). The highest amount of selenium volatilization was seen upon the addition of citrus peel ($\approx 44\%$), compared with manure (19.5%), pectin (16.4%) and straw plus nitrogen (8.8%). Without amendments, selenium volatilization was 6.1%.

On the basis of promising laboratory and greenhouse results, field plots ($3.7 \times 3.7\ m^2$) were set up at the Kesterson Reservoir. Plots were treated with different carbon amendments, including manure, gluten, citrus and casein, in an attempt to stimulate microbial activity and selenium methylation. With periodic tilling and irrigation, approximately 68−88% of the total amount of selenium was removed from the top 15 cm of soil within 100 months (Flury *et al.*, 1997). The highest rates of selenium removal were in soils amended with casein.

The remediation of selenium-contaminated soils is a successful example of how metal-transforming microorganisms can be used to detoxify and remove metals from affected systems. Microbial selenium remediation is also an excellent example of how laboratory experimentation has led to a viable approach for remediating selenium-contaminated environments. The effective microbial remediation of other metals will need development of strategies similar to those used in this case study.

---

biosurfactants have been explored as alternative "green" soil washing agents for metal-contaminated soils (Maier and Soberon-Chavez, 2000).

While metal removal by excavation may be appropriate when the area of contamination is small or there is immediate risk to human health, increasing cost and shrinking landfill space emphasize the need for cheaper, environmentally friendly alternatives. Following excavation, contaminated soils must be stored in a hazardous waste containment facility or incinerated. In some situations, however, excavation can exacerbate the problem. For example, excavation of sediments, called dredging, can actually result in increased metal toxicity. Metal sediments are often anaerobic and the metals within the sediment exist in an immobile, reduced state. Exposure to oxidizing conditions results in metal oxidation and increased metal solubility, increasing both bioavailability and transport. There is current discussion about whether physically removing metal-containing

sediments is more detrimental than leaving them in place.

Incineration of soils can be used to remove metals from soils. However, incineration is not only expensive and impractical for large volumes of soil, but also releases metals to the atmosphere only to be deposited elsewhere. In addition, such thermal treatment of soil also destroys important soil properties, destroying soil structure and soil biota.

The nonbiological remediation of aquatic systems, including surface water, groundwater and wastewater, is fairly straightforward, albeit costly. Metals are removed and concentrated from contaminated waters through flocculation, complexation and/or precipitation. Lime addition precipitates metals as metal hydroxides. Chelating agents complex metals and can be recovered with a change in pH. Electroreclamation methods include ion exchange, reverse osmosis and electrochemical recovery of metals. At an estimated operating cost of $1.1 million per year, chemical treatment of the acid mine water discharging from the Argo

Tunnel in the Clear Creek Superfund site in Colorado involves precipitation of metals using sodium hydroxide to increase the pH to 10 or above. The metal precipitate is then earmarked for landfill disposal. Plans are under way to replace the sodium hydroxide neutralization with lime neutralization due to expected cost savings. This can be compared to a biological approach used to treat mine waste effluents from the Homestake Mine in Lead, South Dakota, in the United States (see Case Study 6.1).

A developing approach for the removal of organic and metal contaminants from water is the use of permeable

reactive barriers (PRB) containing materials such as zero valent iron. In the case of metal contaminants, the goal is to convert the metal into a less-toxic form and/or immobilize it within the PRB as the contaminated water passes through. For example, a zero valent- or oxidized-iron barrier would promote the sorption and precipitation of arsenic within the barrier, resulting in the water exiting the PRB having decreased levels of arsenic. This approach has tremendous potential for containment and/or remediation; however, in practice, it has faced major challenges including microbial and chemical fouling of the PRBs that prematurely shorten their effective lifetime. Continued advances in PRB materials and coatings may improve their longevity and applications in the future.

Information Box 18.6   Discovery of Bacterial Genes for Arsenic Methylation

It has been recognized for years that some bacteria could methylate arsenic; however, it was not known what mechanism was used to do this and how widespread this was in the environment. In 2006, Qin *et al.* took advantage of the explosion in the number of sequenced bacterial genomes and scoured these data to find homologues to eukaryotic genes known to encode for methylation of arsenic. They found a subset of these genes that appeared to be under the control of an arsenic regulatory gene. In order to prove that these genes encoded the ability to methylate arsenic, Qin and colleagues cloned the putative gene, dubbed *arsM*, from *Rhodopseudomonas palustris*, into an arsenic hypersensitive *E. coli* strain. This enabled the *E. coli* to convert arsenite into various methylated species with trimethylarsine as the end product. Additional research indicates that this gene occurs in a large variety of microorganisms, and is widespread in nature, thus likely having a major impact on the global arsenic cycle.

18.11 MICROBIAL APPROACHES IN THE REMEDIATION OF METAL-CONTAMINATED SOILS AND SEDIMENTS

The goals of microbial remediation of metal-contaminated soils and sediments are to:

- Immobilize the metal *in situ* to reduce metal bioavailability and mobility
- Remove the metal from the soil (Figure 18.15)

There are several proposed methods for microbial remediation of metal-contaminated soils including: microbial leaching; microbial surfactants; microbially induced metal volatilization; and microbial immobilization and complexation.

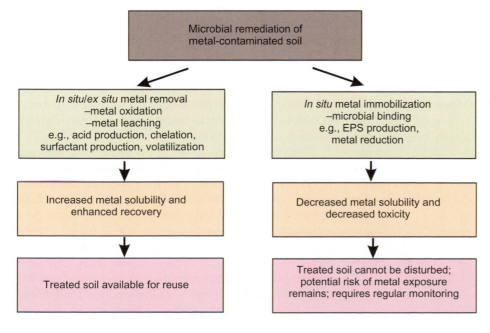

FIGURE 18.15 Microbial metal remediation in metal-contaminated soils relies on either metal removal or, more commonly, metal immobilization. Metal removal is generally more expensive but is ideal because following treatment the soil is available for reuse. In metal immobilization, soil reuse may be limited because of the continued potential risk of exposure.

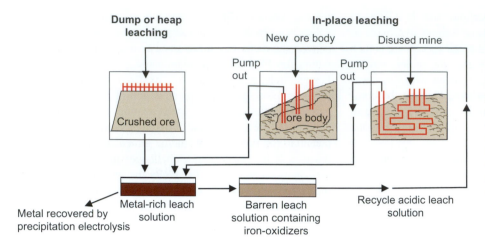

FIGURE 18.16 Various approaches to bioleaching. Metals can be recovered from ores that are in place in the ground if the hydrological conditions permit, or in dumps or heaps on the ground. Some of these heaps can be hundreds of feet high. In each case, an acidic leach solution created and maintained by iron-oxidizers is flushed through the ore, dissolving the metals. The metal-laden leachate is subjected to a precipitation or electrolysis process to remove the metal and then the spent leach solution is recycled back onto the ore body.

Certain microorganisms, such as *Acidithiobacillus ferrooxidans*, can facilitate the removal of metals from soil through metal solubilization or leaching via the same acidification process as seen with the formation of acid mine drainage. Generally used in the recovery of economically valuable metals from ores, bioleaching has also been used to recover copper, lead, zinc and uranium from tailings. The process uses acidophilic iron- and sulfur-oxidizers (e.g., *Acidithiobacillus*, *Leptospirillum*) and is considered to be environmentally friendly (Rawlings, 2002). These microorganisms can participate in both direct bioleaching and indirect bioleaching of metals from a variety of ores. Copper is the major metal recovered using bioleaching.

There are two commercial-scale approaches for bioleaching. The first is used primarily for copper and involves recycling leach liquor through a copper sulfide ore body. As shown in Figure 18.16, this can be done *in situ*, or on ore heaps placed on pads on the ground. *In situ* bioleaching can occur either in a spent mine or can be applied to a new unmined ore body. However, *in situ* bioleaching requires suitable hydrologic conditions to allow efficient collection of the leachates, and also to ensure that leachates do not go off-site. Heap bioleaching or dump bioleaching usually involves mining the ore, crushing it and then placing it in piles on an irrigation pad. The leach liquor is applied to the top of the heap and percolates through the ore, collecting metals. The metal-laden leach liquor is collected from the bottom of the pile, processed to remove the metals, and then recycled onto the top of the pile.

The second commercial-scale approach for bioleaching involves the use of a series of continuous-flow bioreactors, a much more costly process. This process is usually used for high value metals such as gold. However, the principle is the same—the bioreactors are filled with ore and leach liquor is cycled through the bioreactors to remove the metals from the ore (Rawlings, 2002). Metals recovered by leaching can be concentrated by complexation with chelating agents or precipitation with lime. Bioleaching also has potential in the removal of metals from contaminated soils and metal-containing sludges. Unfortunately, this aspect of microbial leaching has received little attention.

Microorganisms can also increase metal solubility for recovery through the production of surfactants. Because of their small size, biosurfactants are a potentially powerful tool in metal remediation. Bacterial surfactants are water-soluble, low-molecular-weight molecules (<1500) that can move relatively freely through soil pores. In addition to their small size, biosurfactants have a high affinity for metals so that, once complexed, contaminating metals can be removed from the soil by soil flushing. Some surfactants, such as the rhamnolipid produced by *Pseudomonas aeruginosa*, show specificity for certain metals, such as cadmium and lead (Ochoa-Loza *et al.*, 2001). Biosurfactant specificity allows the optimization of removal of a particular metal. Related to biosurfactants, the higher molecular weight ($\approx 10^6$) bioemulsifiers such as emulsan, produced by *Acinetobacter calcoaceticus*, can also aid in metal removal and are increasingly being looked at as a potential application for metal recovery (Gutnick and Bach, 2000).

Like leaching, methylation of metals can increase metal bioavailability and toxicity. Some methylated metals are more lipophilic than their nonmethylated counterparts. In spite of the possible increased toxicity, many microorganisms still volatilize metals to facilitate their removal from the immediate environment. Because methylation enhances metal removal, methylation of certain metals has been used as a remediation strategy. The most famous example is the removal of selenium from contaminated soil in San Joaquin, California (see Case Study 18.1) by selenium volatilizing microorganisms. Mercury is another metal commonly methylated by microorganisms. However, mercury is susceptible to bioaccumulation in the food chain, posing serious health risks to the human population, and therefore removal of mercury by volatilization would not be an acceptable approach.

Immobilization strategies include metal sequestration which takes advantage of the ability of some microorganisms to produce metal-complexing polymers (both extracellular and intracellular), or to convert metals to a less-soluble form. Recall that exopolymers have high affinities for various metals. The overall approach in microbial metal sequestration is to introduce the polymer-producing microorganism into the contaminated soil and allow the organism to grow and replicate, thereby increasing the amount of polymer present in the soil, and increasing the number of organisms producing the polymer. Microbial metal sequestration has been shown to be effective in laboratory studies but has yet to be proven effective in the field. A second immobilization strategy is to create reducing or anaerobic conditions which results in the reduction and precipitation of metals. For example, the reduction of sulfate to sulfide under anaerobic conditions can lead to the formation of metal sulfide precipitates that are immobile. Likewise, the reduction of uaramium (uranium(VI)) to a less-soluble form (uranium (IV)) has been demonstrated to dramatically lower concentrations of dissolved uranium in groundwater in field-scale studies (Case Study 18.2). It should be noted that this approach is very metal specific, since reduced forms of some metals (e.g., arsenic) are actually more soluble than their oxidized counterparts. Also, this approach would require that reducing conditions be maintained at the site in order to prevent reoxidation of the sequestered metals. Although immobilization strategies are generally more economical than removal strategies and appear to have tremendous potential for many sites, there is not yet sufficient evidence to confirm the long-term effectiveness of immobilization.

18.12 MICROBIAL APPROACHES IN THE REMEDIATION OF METAL-CONTAMINATED AQUATIC SYSTEMS

Microbially facilitated removal of metals from water is based on the ability of microorganisms to complex and precipitate metals, resulting in both detoxification and removal from the water column. Specific interactions for metal removal include metal binding to microbial cell surfaces and exopolymer layers, intracellular uptake, metal volatilization and metal precipitation via microbially facilitated metal redox reactions (Figure 18.17). Although these microbial mechanisms can effectively remove metals from contaminated aquatic systems, it is important to note that the metals are not destroyed and still have to be disposed of properly.

Wetland treatment is a cost-effective and efficient method for removal of metals from contaminated waters, such as acid mine drainage. Metal reductions are often greater than 90% (Scholz and Xu, 2002). Wetland remediation is based on microbial adsorption of metals, metal bioaccumulation, bacterial metal oxidation and sulfate reduction. The high organic matter content of wetlands provided by high plant and algal growth encourages both the growth of sulfate-reducing microorganisms and metal sorption to the organic material. Although these various processes contribute to the removal of toxic metals from the water column, the metals are not destroyed. Consequently, wetlands are constantly monitored for any environmental change that may adversely affect metal removal. For example, a decrease in pH may solubilize precipitated metals, or a disturbance of the wetland sediment may change the redox conditions and oxidize reduced metals. Wetlands are resilient systems, and as long as new vegetative growth and organic inputs occur, wetlands can effectively remove metals for an indefinite period of time.

The most common treatment for metal-contaminated waters is with microbial biofilms. Many microorganisms, including *Pseudomonas*, *Arthrobacter*, *Bacillus*, *Citrobacter*, *Streptomyces* and the yeasts *Saccharomyces* and *Candida*, produce exopolymers as part of their growth regime. Metals have high affinities for these anionic exopolymers. Microbial biofilms may be viable or nonviable when used in remediation. In general, the biofilm is immobilized on a support as contaminated water is passed through the support (Figure 18.18). Often, a mixture of biofilm-producing organisms grows on these supports, providing a constant supply of fresh biofilm. For example, live *Citrobacter* spp. biofilms are used to remove uranium from contaminated water. Both *Arthrobacter* spp. biofilms and biomass (nonliving) are used in recovery of cadmium, chromium, copper, lead and zinc from wastewaters. Nonliving *Bacillus* spp. biomass preparations effectively bind cadmium, chromium, copper, mercury and nickel, among other metals. The success of microbial biomass in metal recovery from contaminated waters has led to the commercial sale of several biomass products. For example, AMT-BIOCLAIM (*Bacillus* biomass) and AlgaSORB (*Chlorella vulgaris*) are commercially available immobilized, nonliving preparations for treating metal-contaminated water. Interestingly, microbial biofilms are also used in the treatment of metal-contaminated marine waters; however, marine bacteria such as *Deleya venustas* and *Moraxella* sp. are used. Microbial biofilms are likewise used in the removal of metals from domestic wastewater. In domestic waste treatment, the important biofilm-producing organisms include *Zoogloea*, *Klebsiella* and *Pseudomonas* spp. Complexed metals are removed from the wastewater via sedimentation before release from the sewage treatment plant.

Case Study 18.2 Bioremediation of Uranium in a Contaminated Aquifer

Many U.S. Department of Energy (DOE) facilities are highly contaminated as a result of post-WWII nuclear weapons programs. One of these sites is located at the Y-12 National Security Complex in Oak Ridge, TN. This site has extensive subsurface contamination due to the disposal of millions of liters of wastes containing uranium, other metals, organic solvents and acids in unlined disposal ponds from 1951 to 1983.

These wastes subsequently leached from the ponds, contaminating the underlying vadose zone and groundwater with a variety of contaminants including metals, radionuclides and organic solvents. In addition, this resulted in the site having a very low pH (≈ 4) and extremely high levels of nitrate and sulfate due to the acids (e.g., nitric and sulfuric acids) used in processing.

A research group consisting primarily of scientists from Oak Ridge National Laboratory and Stanford University conducted a series of experiments to investigate the potential for *in situ* bioremediation of the site (Wu *et al.*, 2006). Their goal was to decrease the potential for uranium to migrate offsite, and pollute nearby public waterways. Given the extent and depth of the contamination, the group decided that the best remediation option was to biologically reduce the metals thus decreasing their aqueous solubility and sequestering them *in situ*. However, in order to do this, the site had to be significantly modified. The pH was raised to ≈ 6 to facilitate greater biological activity. Also, in order to achieve stable reduction of uranium, the exceedingly high levels of nitrate (>2 g/L) first needed to be decreased. The logical process for decreasing nitrate levels was denitrification, but this could not be lone *in situ* since large amounts of cell biomass and gas produced during denitrification could potentially complicate the remediation process by clogging and/or changing flow paths in the aquifer. Also, denitrification by-products could possibly reoxidize reduced uranium. It was therefore decided to perform the denitrification aboveground and use a combined aboveground and belowground approach for the remediation. Groundwater was pumped aboveground, where volatiles such as TCE were stripped, and the water was chemically neutralized to precipitate out dissolved metals. The water was then pumped into a fluidized bed reactor, where ethanol was added as an electron donor for denitrification. The treated water was then pumped back into the aquifer, where ethanol was again added to serve as an electron donor to stimulate microbial reduction of uranium from the more soluble and mobile uranium (VI) form to the insoluble uranium (IV) form.

After approximately 70 days, aqueous uranium concentrations decreased $\sim 80\%$ from their initial levels (~ 50 mg/L), largely coinciding with the removal of residual nitrate via denitrification and a transition to sulfate-reducing conditions. After approximately 2 years of bioremediation, the groundwater uranium levels were further reduced to below 30 μg/L (the U.S. EPA Maximum Contaminant Level for uranium in drinking water). In other words, the water was improved so dramatically that it was considered safe for human consumption (at least in terms of uranium concentration)! X-ray absorption near-edge structure (XANES) spectroscopy confirmed that the uranium was reduced *in situ* with as much as 82% being U(IV) in sediment samples. The microorganisms and exact mechanisms responsible for the uranium reduction were not known, but microarray and DNA-sequencing results revealed that uranium reduction corresponded with large increases in the populations of iron-reducers, including *Geobacter* spp., and sulfate-reducers, including *Desulfovibrio* spp., related to organisms that have been reported to reduce uranium. Research is ongoing to determine the long-term stability of the immobilized uranium at the site.

S-3 Disposal Ponds

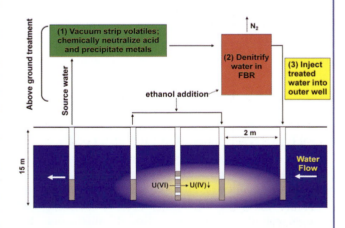

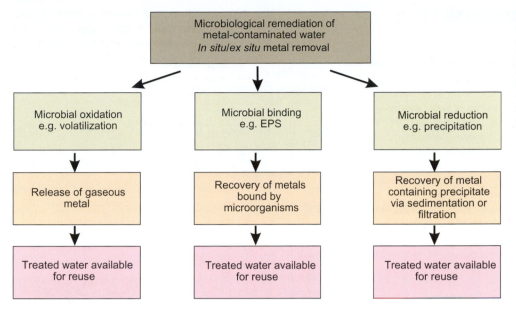

FIGURE 18.17 Microbial metal remediation approaches for metal-contaminated waters. In each method, the treated water is safe to release into the environment. Both metals and microorganisms can easily be recovered during treatment for proper disposal.

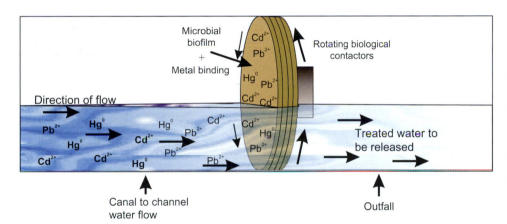

FIGURE 18.18 Schematic demonstrating how microbial biofilms are used in removing metals from contaminated waste streams. The biofilm located on the rotating drum accumulates metals as the water passes through the drum. The treated water can be safely released. The biofilm may either be viable or nonviable. When viable, the biofilm rarely needs to be replaced; however, nonliving biofilms need to be replaced periodically because their metal removal efficiency will decrease with time.

QUESTIONS AND PROBLEMS

1. Address the structural differences between prokaryotic and eukaryotic cells and how these differences influence cell resistance or sensitivity to metal toxicity.

2. In a metal contaminated lake, discuss metal bioavailability throughout the water column including the sediment.

3. Is dredging a viable option for removal of metal contamination in an aquatic system? What are the potential problems?

4. What factors need to be considered when bioaugmenting a metal-contaminated site with a metal-resistant microorganism?

5. Which chemical groups in proteins are most reactive with metals? In nucleic acids? In membranes?

6. Metal-resistant microorganisms are often isolated from noncontaminated environments (with no prior metal exposure). Discuss possible reasons why.

7. A metal-contaminated soil has been remediated using a metal-complexing microorganism. What factors need to be considered to ensure that the metal does not become "reavailable"?

8. Summarize the possible mechanisms of metal resistance in microorganisms and discuss which mechanism would be most effective in remediating a metal-contaminated surface soil, metal-contaminated soil in the vadose zone, a metal-contaminated stream and metal-contaminated groundwater?

9. Discuss the fate (both chemical and biological) of lead in (a) an acid soil, pH 4.0; (b) a neutral soil; (c) a basic soil, pH 8.5; (d) an anaerobic soil; and (e) an aerobic–anaerobic soil interface.

REFERENCES AND RECOMMENDED READING

Recommended Readings on Soil, Microbial and Metal Interactions

Alloway, B. J. (ed.) (2013) Heavy metals in soils: trace metals and metalloids in soils and their bioavailability, 3rd ed. In "Environmental Pollution, vol. 22, Springer, Dordrecht.

Lovley, D. R. (2000) "Environmental Microbe—Metal Interactions," ASM Press, Washington, DC.

Nies, D. H., and Silver, S. (eds.) (2007) "Microbiology Monographs", vol. 6, Springer-Verlag, Berlin, Heidelberg.

Stolz, J. F., and Oremland, R. S. (eds.) (2011) "Microbial Metal and Metalloid Metabolism: Advances and Applications". ASM Press, Washington, DC.

Chapter References

Avery, S. V. (1995) Cesium accumulation by microorganisms—uptake mechanisms, cation competition, compartmentalization and toxicity. *J. Ind. Microbiol.* **14**, 76–84.

Bidwell, J. P., and Spotte, S. (1985) "Artificial Seawaters: Formulas and Methods," Jones and Barlett Publishers, Boston, MA.

Chauhan, G., Pant, K. K., and Nigam, K. D. P. (2012) Extraction of nickel from spent catalyst using biodegradable chelating agent EDDS. *Ind. Eng. Chem. Res.* **51**, 10354–10363.

Debus, K. H. (1990) Mining with microbes. *Technol. Rev.* **93**, 50–57.

Drott, A., Lambertsson, L., Bjorn, E., and Skyllberg, U. (2007) Importance of dissolved neutral mercury sulfides for methyl mercury production in contaminated sediments. *Environ. Sci. Technol.* **41**, 2270–2276.

Ehrlich, H. L. (1996) "Geomicrobiology," 3 ed. Marcel Dekker, New York.

Ehrlich, H. L. (2002) Bacterial oxidation of As(III) compounds. In "Environmental Chemistry of Arsenic" (W. T. Frankenberger, Jr., ed.), Marcel Dekker, Inc., New York, pp. 313–327.

Flury, M., Frankenberger, W. T., Jr., and Jury, W. A. (1997) Long-term depletion of selenium from Kesterson dewatered sediments. *Sci. Total Environ.* **198**, 259–270.

Forgeron, F. D. (1971) Soil geochemistry in the Canadian Shield. *Can. Min. Metall.* **64**, 37–42.

Gadd, G. M. (1993) Microbial formation and transformation of organometallic and organometalloid compounds. *FEMS Microbiol. Rev.* **11**, 297–316.

Ginder-Vogel, M., and Sparks, D. L. (2010) Chapter 1—the impacts of X-ray absorption spectroscopy on understanding soil processes and reaction mechanisms. In "Developments in Soil Science, vol. 34, Synchrotron-Based Techniques in Soils and Sediments" (B. Singh, and M. Gräfe, eds.), Elsevier, pp. 1–26.

Goldman, C. R., and Horne, A. J. (1983) "Limnology," McGraw-Hill, New York.

Gutnick, D. L., and Bach, H. (2000) Engineering bacterial biopolymers for the biosorption of heavy metals; new products and novel formulations. *Appl. Microbiol. Biotechnol.* **54**, 451–460.

Hoffman, D. R., Okon, J. L., and Sandrin, T. R. (2005) Medium composition affects the degree and pattern of cadmium inhibition of naphthalene biodegradation. *Chemosphere* **59**, 919–927.

Holden, J. F., and Adams, M. W. W. (2003) Microbe—metal interactions in marine hydrothermal environments. *Curr. Opin. Chem. Biol.* **7**, 160–165.

Inskeep, W. P., McDermott, T. R., and Fendorf, S. (2002) Arsenic (V)/(III) cycling in soils and natural waters: chemical and microbiological processes. In "Environmental Chemistry of Arsenic" (W. T. Frankenberger, Jr., ed.), Marcel Dekker, New York, pp. 183–215.

Karlson, U., and Frankenberger, W. T., Jr. (1990) Volatilization of selenium from agricultural evaporation pond sediments. *Sci. Total Environ.* **92**, 41–54.

Keely, J. F., Hutchinson, F. I., Sholley, M. G., and Wai, C. M. (1976) Heavy metal pollution in the Coeur d'Alene mining district. In "Project Technical Report to National Science Foundation," University of Idaho, Moscow.

Kong, I. C. (1998) Metal toxicity on the dechlorination of monochlorophenols in fresh and acclimated anaerobic sediment slurries. *Water Sci. Technol.* **38**, 143–150.

Kudo, A., Fujikawa, Y., Miyahara, S., Zheng, J., Takigami, H., Sugahara, M., *et al.* (1998) Lessons from Minamata mercury pollution. Japan—after a continuous 22 years of observation. *Water Sci. Technol.* **38**, 187–193.

Leppard, G. G. (1981) "Trace Element Speciation in Surface Waters," Plenum Press, New York.

Lindsay, W. L. (1979) "Chemical Equilibria in Soils," John Wiley and Sons Inc., New York.

Maier, R. M., and Soberon-Chavez, G. (2000) *Pseudomonas aeruginosa* rhamnolipids: biosynthesis and potential applications. *Appl. Microbiol. Biotechnol.* **54**, 625–633.

Maslin, P., and Maier, R. M. (2000) Rhamnolipid-enhanced mineralization of phenanthrene in organic-metal co-contaminated soils. *Biorem. J.* **4**, 295–308.

Mukhopadhyay, R., Rosen, B. P., Phung, L. T., and Silver, S. (2002) Microbial arsenic: from geocycles to genes and enzymes. *FEMS Microbiol. Rev.* **26**, 311–325.

Nies, D. H. (2003) Efflux-mediated heavy metal resistance in prokaryotes. *FEMS Microbiol. Rev.* **27**, 313–339.

Ochoa-Loza, F. J., Artiola, J. F., and Maier, R. M. (2001) Stability constants for the complexation of various metals with a rhamnolipid biosurfactant. *J. Environ. Qual.* **30**, 479–485.

Oremland, R. S., Newman, D. K., Kail, B. W., and Stolz, J. F. (2002) Bacterial respiration of arsenate and its significance in the environment. In "Environmental Chemistry of Arsenic" (W. T. Frankenberger, Jr., ed.), Marcel Dekker, Inc., New York, pp. 273–295.

Qin, J., Rosen, B. P., Zhang, Y., Wang, G., Franke, S., and Rensing, C. (2006) Arsenic detoxification and evolution of trimethylarsine gas by a microbial arsenite S-adenosylmethionine methyltransferase. *Proc. Nat. Acad. Sci U.S.A.* **103**, 2075–2080.

Rawlings, D. E. (2002) Heavy metal mining using microbes. *Annu. Rev. Microbiol.* **56**, 65–91.

Rensing, C., and Maier, R. M. (2003) Issues underlying use of biosensors to measure metal bioavailability. *Ecotoxicol. Environ. Saf.* **56**, 140–147.

Schiewer, S., and Volesky, B. (2000) Biosorption processes for heavy metal removal. In "Environmental Microbe—Metal Interactions" (D. R. Lovley, ed.), ASM Press, Washington, DC, pp. 329–362.

Scholz, M., and Xu, J. (2002) Performance comparison of experimental constructed wetlands with different filter media and macrophytes treating industrial wastewater contaminated with lead and copper. *Bioresour. Technol.* **83**, 71–79.

Selifnova, O., Burlage, R., and Barkay, T. (1993) Bioluminescent sensors for detection of bioavailable Hg(II) in the environment. *Appl. Environ. Microbiol.* **59**, 3083–3090.

Seyfferth, A. L., Webb, S. M., Andrews, J. C., and Fendorf, S. (2010) Arsenic localization, speciation, and co-occurrence with iron on rice (*Oryza sativa* L.) roots having variable Fe coatings. *Environ. Sci. Technol.* **44**, 8108−8113.

Sigg, L. (1985) Metal transfer mechanisms in lakes; the role of settling particles. In "Chemical Processes in Lakes" (W. Stumm, ed.), John Wiley, New York.

Silver, S., Rosen, B. P., and Misra, T. K. (1986) In "Fifth International Symposium on the Genetics of Industrial Micro-Organisms" (M. Alacevic, D. Hranueli, and Z. Toman, eds.), Karlovac, Yugoslavia, pp. 357−376.

Silver, S., Phung, L. T., and Rosen, B. P. (2002) Arsenic metabolism: resistance, reduction, and oxidation. In "Environmental Chemistry of Arsenic" (W. T. Frankenberger, Jr., ed.), Marcel Dekker, New York, pp. 247−272.

Simkiss, K., and Taylor, M. G. (1989) Metal fluxes across the membranes of aquatic organisms. *Rev. Aquat. Sci.* **1**, 173.

Smith, A. H., Lingas, E. O., and Rahman, M. (2000) Contamination of drinking-water by arsenic in Bangladesh: a public health emergency. *Bull. World Health Organ.* **78**, 1093−1103.

Somenahally, A. S., Hollister, E. B., Yan, W., Gentry, T. J., and Loeppert, R. H. (2011) Water management impacts on arsenic speciation and iron-reducing bacteria in contrasting rice-rhizosphere compartments. *Environ. Sci. Technol.* **45**, 8328−8335.

Stone, E. L., and Timmer, V. R. (1975) Copper content of some northern conifers. *Can. J. Bot.* **53**, 1453−1456.

Tandy, S., Bossart, K., Mueller, R., Ritschel, J., Hauser, L., Schulin, R., *et al.* (2004) Extraction of heavy metals from soils using biodegradable chelating agents. *Environ. Sci. Technol.* **38**, 937−944.

Wang, W.-X., and Dei, R. C. H. (2006) Metal stoichiometry in predicting Cd and Cu toxicity to a freshwater green alga *Chlamydomonas reinhardtii*. *Environ. Pollut.* **142**, 303−312.

Warren, H. V., Delevault, R. E., and Barakso, J. (1966) Some observations on the geochemistry of mercury as applied to prospecting. *Econ. Geol. Ser. Can.* **61**, 1010−1028.

Wu, W.-M., Carley, J., Fienen, M., Mehlhorn, T., Lowe, K., Nyman, J., *et al.* (2006a) Pilot-scale in situ bioremediation of uranium in a highly contaminated aquifer. 1. Conditioning of a treatment zone. *Environ. Sci. Technol.* **40**, 3978−3985.

Chapter 19

Microbial Diversity and Interactions in Natural Ecosystems

Terry J. Gentry, Ian L. Pepper and Leland S. Pierson III

19.1 MICROBIAL COMMUNITIES

Adapting a commonly used phrase, "no microorganism is an island, entire of itself," most environmental microorganisms exist as part of a community. This community may be as simple as two populations coexisting on a fomite (Chapter 30); slightly more complex such as cyanobacteria, sulfate-reducing bacteria and other microorganisms in a stratified, microbial mat (see Information Box 4.4 and Chapter 6); or very complex such as soil, which commonly contains thousands of species per gram. In addition to soils, complex microbial communities are found in essentially all natural ecosystems including plants and surface- and groundwaters. Specialized communities such as rhizospheres or biofilms can have particularly large and diverse community membership (megacommunities). These communities exhibit great diversity as well as great redundancy in terms of potential activity. Both abiotic and biotic pressures drive the evolution of these complex microbial communities so that their composition is highly dynamic, and reflects the rate of change being imposed on the ecosystem—whether it is

through natural successional events, climate change or anthropogenic impacts. Abiotic pressures include pH, redox potential, water availability, temperature, salinity and organic matter, all of which can vary from the micro- to the macrosite level. Biotic pressures are imposed by different populations competing for similar nutrient resources, such as organic carbon or nitrogen, which are normally limiting. Although there is great debate over exactly how many different populations there are in a given ecosystem, there is agreement that it is a large number. Soil typically contains 10^8 to 10^{10} bacteria per gram based on direct counts (see Section 4.4.1). If every population was present at 10^6 per gram, this would mean that there are 100 to 10,000 different populations in every gram of soil! These populations exist in close proximity to each other, and either compete or work synergistically for resources.

How do these different populations respond to these pressures? They tend to acquire genes that allow for previously unavailable activities, or enhanced rates of activities that already exist (Pál *et al.*, 2005). Gene acquisition can be through point mutation events that alter regulation or

I.L. Pepper, C.P. Gerba, T.J. Gentry: Environmental Microbiology, Third edition. DOI: http://dx.doi.org/10.1016/B978-0-12-394626-3.00019-3
© 2015 Elsevier Inc. All rights reserved.

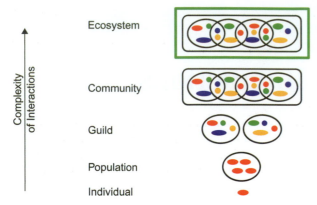

FIGURE 19.1 Ecological organization within a microbial community. Adapted from Atlas and Bartha (1998).

kinetics of enzymes, or alternatively, genes can be acquired through lateral or horizontal gene transfer (Section 2.2.7). The large number of populations present, and their rapid ability to respond to environmental pressures, results in an ever-changing and evolving microbial community within any given ecosystem. In fact, even closely related microorganisms can be quite different. For example, an analysis of four completely sequenced *E. coli* strains showed up to a 29.25% difference in the gene content of the four genomes (Coenye *et al.*, 2005). Thus, diversity in normal healthy soils is necessarily very large to take advantage of numerous niches that can develop. In contrast, diversity in stressed or extreme environments that have very specific, strong selective pressures (e.g., high temperature, low pH) tends to be much lower (Chapter 7).

The importance of these microbial communities is unquestionable. In fact, natural ecosystems provide a myriad of resources that benefit society by helping to maintain Earth's biosphere. These include climate regulation, biogeochemical (nutrient) cycling, waste treatment, water supply and regulation, as well as healthy soils for growing crops. It has been estimated that ecosystems, underpinned by microbial activity, provide at least $33 trillion per year in global services (Costanza *et al.*, 1997) (Information Box 7.1). In addition to this, microbial communities are responsible for a wide array of natural products including antibiotics and anticancer agents that are being explored for the benefit of society. Clearly, then, natural microbial communities are important, and fundamental to the success of these communities in natural ecosystems is their vast diversity, which enables them to adapt to changing conditions. This adaptation response can result in transient community changes in response to new environmental conditions that may be imposed naturally or by anthropogenic activity, or can result in the evolution of new communities in response to long-term ecosystem changes. In this chapter, we will discuss the diversity of environmental microorganisms, how they

interact with other organisms and the environment, and how their diversity and interactions impact microbial adaptation, ecosystem function and the discovery of natural products such as antibiotics.

19.2 MICROBIAL DIVERSITY IN NATURAL SYSTEMS

19.2.1 What is a Microbial Community?

Ecologists use a hierarchical classification system to describe organisms and their communities (Figure 19.1). Individual microorganisms that are genetically related and perform the same function in a proximate location (i.e., that occupy the same niche) are referred to collectively as a population. Populations that compete for the same resources are grouped into a guild. All of the guilds (and thus populations) present in a specific environment constitute the microbial community. The community in a defined sample, such as a plant root, is a species assemblage (this is what is commonly referred to as the "microbial community" in research papers since only a portion of the microbial community has typically been characterized). The microbial community along with the other biotic and abiotic components of its environment constitutes an ecosystem.

19.2.2 What is Microbial Diversity?

From a community perspective, microbial diversity is defined as the amount of variation (i.e., genetic, morphological and functional differences) in microbial populations occupying a given environment. A variety of different diversity measures are commonly used for characterizing microbial communities (Information Box 19.1). The basis for these determinations is the number (richness), equitability of distribution (evenness) and identity (community composition) of different organisms (i.e., species) in a sample(s). The most commonly reported microbial diversity data include measures of species richness and/or evenness (often using diversity indices such as Shannon–Wiener; α-diversity) and community composition (β-diversity). It is important to point out that any diversity estimate, including those based upon rRNA genes, are subject to many limitations (Chapter 21). Furthermore, unless the entire microbial community in a sample has been sequenced, the sequence data represents only a portion (sometimes very small) of the entire community. The diversity data from this sample is then extrapolated to estimate the diversity of the community. This is subject to many biases, such as the size of the DNA sequence library used; therefore, diversity estimates should be used carefully and interpreted with caution (Gihring *et al.*, 2012).

Information Box 19.1   Measures of Microbial Diversity

Measure	Description
Alpha (α) diversity	Species diversity in a sample unit. This is often determined using species richness or a diversity index.
Beta (β) diversity	Amount of compositional variation in a sample or set of sample units.
Gamma (γ) diversity	Overall diversity in a collection of sample units.
Species richness	Number of species in a sample unit.
Species evenness	Equitability of species abundance in a sample unit.
Diversity index	Quantitative measure of diversity based upon factors such as species richness and evenness.

Two of the most commonly used indices are Shannon—Wiener and Simpson's.

Information Box 19.2   Criteria for Distinguishing Soil Bacterial Species

Technique	Cutoff for Species Distinction
DNA—DNA hybridization (i.e., the degree of association between the total genomic DNA of two species)	<70% DNA—DNA re-association
16S rRNA gene sequence identity	<97 to 99%

19.2.3 What is a Species?

The discussion of microbial communities and diversity is critically dependent on the concept of what constitutes a bacterial species and how two different species can be distinguished. Thus, before one can address diversity in a given system, the fundamental unit that determines diversity must be defined. This has been the subject of debate for many years. This is especially problematic for bacteria because we cannot easily observe them in their environment, and they can change rapidly through mutation, genome alteration or reduction, and by the acquisition of genes from other, often distantly related, organisms. Therefore, scientists have not been able to develop a classification system for bacteria that is based on evolutionary and ecological processes as they have for higher forms of life. Rather, microbiologists began by classifying microorganisms on the basis of morphology and selected physiological traits, especially those important for human health. The advent of molecular techniques and the increasing availability of rapid and inexpensive DNA sequencing technologies have allowed reexamination of microbial classification. Although there is still not complete consensus, we seem to be slowly moving closer toward a definition of bacterial species.

In the 1970s, before DNA sequencing was available, DNA—DNA hybridization was used to examine whether two organisms were the same or different. In this technique, both differences in gene content and differences in nucleotide sequence in shared genes contribute to the amount of hybridization that occurs. The standard used to differentiate species has been 70% DNA—DNA hybridization. Above this level, the two organisms are considered to be the same species and below this level, they are considered to be different. Although this technique was one of the first developed, results are consistent with newer techniques. The disadvantage of DNA—DNA hybridization is that it is time-consuming because only two organisms are compared at a time.

Currently, because of the ease of the technique, sequence-based techniques are being employed to examine diversity. Here, the criterion for similarity is based on sequence divergence of homologous genes (Cohan and Perry, 2007). The most commonly used target has been the 16S rRNA gene, which is universally found in all bacteria; millions of 16S rRNA sequences are now publicly available from a number of different databases (Table 13.1). The criterion for similarity based upon comparison of 16S rRNA sequences is at least 97% sequence similarly, although more recently it is been suggested that even a 1% difference may indicate different species. Each unique group of sequences (i.e., those with ≥97% matching sequence identity) is classified into an operational taxonomic unit (OTU; also known as a phylotype) which serves as a "species" classification for the purpose of diversity determinations. However, this approach is not without its shortcomings. For example, it has been suggested that 16S rRNA sequences do not discriminate beyond the genus level. When comparing DNA—DNA hybridization and 16S rRNA sequence similarity, it has been shown that less than 97% sequence homology always yields <70% DNA—DNA hybridization. But in some cases, greater than 97% sequence homology may also yield <70% DNA—DNA hybridization suggesting that the hybridization technique is more discriminating (Information Box 19.2). Also, the <97—99% sequence identity for species determination was largely based upon the full-length 16S rRNA genes. However, most current microbial diversity studies using newer sequencing technologies are based upon partial (<500 bp) sequences of 16S rRNA genes. This can greatly impact the taxonomic resolution of the sequence data depending upon the length and region of the 16S rRNA gene that is sequenced (Mizrahi-Man *et al.*, 2013). Furthermore, even the comparison of complete 16S rRNA gene sequences may fail to reflect subtle but functionally important differences, such as the presence of pathogenicity genes, prophages, etc.

The lack of discrimination provided by 16S rRNA gene analysis has engendered the development of the multilocus sequence typing (MLST) approach to examine diversity among closely related species. MLST involves the sequencing of three to eight genes (each gene is a locus), and comparison of these sequences (Cohan and Perry, 2007). Thus, while the 16S rRNA gene may allow definition to the genus level, MLST can provide further definition to the species level. Furthermore, as DNA sequencing continues to become more affordable, whole-genome sequencing will likely play a greater role in microbial typing in the future. The final step to take is to define a systematic approach for bacteria that is based on ecology and evolution rather than simply on genetic similarity. Additional information on currently used molecular methods for determining microbial diversity and community composition is provided in Chapter 21.

19.2.4 Microbial Diversity in Soil and Ocean Environments

In Section 19.1 we estimated, simply using a numbers game, that the number of different bacterial populations or species in soil could range from 100 to 10,000. However, microbial diversity has actually been measured in soils, and other environments, using a variety of methods. One of the first approaches examined diversity by using DNA-reassociation kinetics of pooled genomic DNA from an environmental sample. This was first applied to a bacterial community from a forest soil sample collected in Norway. This approach yielded an estimate of 4000 different genomes per gram of dry soil (Torsvik *et al.*, 1990). In this case, total counts were 1.5×10^{10} bacteria per gram. A second study used the same approach to compare diversity in an uncontaminated soil from an agricultural field site in Germany, with adjacent soils that were amended with low or high amounts of metal-contaminated sludge. This study estimated 16,000, 6400 and 2000 genomes per gram wet soil for the uncontaminated, low metal and high metal soils, respectively (Sandaa *et al.*, 1999). More recently, the results from the latter study were reexamined using a modified computational analysis of the DNA-reassociation data. This study estimated that there were actually 8,300,000, 64,000 and 7900 genomes in the uncontaminated, low-metal and high-metal soils, respectively (Gans *et al.*, 2005).

At the moment, most diversity estimates are made using rRNA genes (e.g., 16S rRNA) from community DNA samples. Using this approach, diversity estimates are generally somewhat lower than when DNA-reassociation kinetics are used. A statistical analysis is

TABLE 19.1 Estimates of Bacterial Diversity (Based Upon 16S rRNA Sequences) in Different Environments

Environment	Diversity Estimate (Species Richness)	Source
Polar desert soil	2935	Fierer *et al.*, 2012
Agricultural soil	3409	Hollister *et al.*, 2013
Diesel-contaminated soil	3259	Sutton *et al.*, 2013
Hypersaline soil	5285	Hollister *et al.*, 2010
Arctic tundra soil	6965	Fierer *et al.*, 2012
North Atlantic Ocean	6997	Sogin *et al.*, 2006
Tropical forest soil	8772	Fierer *et al.*, 2012
Temperate grassland soil	10,253	Fierer *et al.*, 2012
Temperate forest soil	12,150	Fierer *et al.*, 2012

then performed on the number of unique OTUs recovered in comparison to the total number of OTUs sequenced. This approach also involves comparing the total number of OTUs in the community relative to the abundance of the most prevalent OTUs in the community (Curtis *et al.*, 2002). These statistical approaches have provided estimates of bacterial diversity in several different natural soil environments ranging from a few thousand to >10,000 OTUs or phylotypes (Table 19.1). Although soil fungi are also diverse, they are usually less diverse than bacterial populations within the same environment. For example, Hollister *et al.* (2013) estimated that an agricultural soil contained ≈ 300 species (phylotypes) of fungi in comparison to >3000 species of bacteria.

Natural bacterial communities in marine waters are generally less abundant and diverse than in soils. For example, a survey of paired samples taken from the Juan de Fuca Ridge in the northeast Pacific Ocean showed that microbial counts ranged from 3.9×10^4 to 1.8×10^5 cells per ml water (Sogin *et al.*, 2006). (This can be compared to 10^8 to 10^{10} cells per gram of soil.) This study examined three paired samples taken at the same site but at different depths. In addition, two samples were taken from deep-sea thermal vents. Each sample was 1−2 liters in size and microbes were collected on a filter. Analysis of diversity was based on the 16S rRNA gene and resulted in an estimate that

TABLE 19.2 Impact of Major Environmental Factors on Microbial Diversity

Factor	Impact on Microbial Diversity
pH	Maximum diversity at neutral pH (6 to 8). Extreme pHs result in reduced diversity.
Vegetation	Different plants may stimulate and/or inhibit different microbial populations.
Water content	Greater diversity with moderate water content. Water-saturated conditions decrease diversity due to less spatial isolation of organisms and also the generation of anaerobic conditions.
O_2 concentration	Greater diversity under aerobic conditions.
Temperature	Extremely high or low temperatures reduce diversity.
Organic matter content	Higher organic matter content results in higher diversity.
Soil depth	Decreasing diversity with increasing depth from surface.
Addition of organic substrates	Addition of a single, organic substrate often results in a reduction in diversity due to the stimulation of a subset of the microbial community.
Soil tillage	Decreased diversity due to soil homogenization and reduction in microsite variation.
Addition of organic pollutants	Similar to organic substrates in general, often a reduction in diversity due to stimulation of specific populations but also potentially toxicity of the xenobiotic to other populations.
Addition of metal pollutants	Reduction in diversity due to toxicity to some populations.

between 1184 and 3290 OTUs were present in the samples. A study of the Sargasso Sea, which is located in the North Atlantic Ocean, estimated the presence of 1800 species using a multilocus sequencing approach. Based on rRNA genes alone, this was reduced to 1164 unique sequences (Venter *et al.*, 2004). In this case, several hundred liters of water were collected via filtration.

These two studies give very similar estimates of diversity in the marine environment. However, the samples studied were much larger in volume than 1 gram of soil. One theoretical effort to directly compare bacterial diversity in soil and marine samples is provided by Curtis *et al.* (2002). This analysis was based on relating the total number of bacteria in the sample to the number of bacteria in the least abundant species. This approach estimated that oceans have 160 species per mL while soils have 6400–38,000 species per gram. These scientists further extrapolated these numbers to estimate that the entire bacterial diversity of the ocean is 2×10^6 species. This can be contrasted to the diversity in one ton of soil which was estimated to be 4×10^6 species.

19.2.5 Environmental Factors that Impact Microbial Diversity

Many factors can impact soil microbial diversity including pH (Lauber *et al.*, 2009), presence and type of vegetation (Lauber *et al.*, 2009), contamination (Hemme *et al.*, 2010),

amendment with organic substrates (Hollister *et al.*, 2013), temperature (Castro *et al.*, 2010), depth (Hansel *et al.*, 2008), water content (Zhou *et al.*, 2002), oxygen content (Somenahally *et al.*, 2011) and level of CO_2 (Dunbar *et al.*, 2012) (Table 19.2). In general, properties that increase the heterogeneous nature of soils (Chapter 4) tend to result in increased microbial diversity. For example, greater microbial diversity is typically found in soils that have well-established structure (i.e., soil aggregates), and that are not saturated with water, since these conditions result in more spatial isolation within the soil communities thus encouraging higher microbial diversity (Torsvik and Øvreås, 2002). Even seemingly benign events, such as sheep urinating in a pasture, may impact microbial communities (Nunan *et al.*, 2006). As discussed in Case Study 4.1, Lauber *et al.* (2009) used 16S rRNA sequencing to investigate the bacterial communities in 88 soils from North and South America, and found that that soil pH was one of the main drivers of bacterial diversity, and was more important than other factors such as vegetation type and soil carbon at a continental scale. However, the authors suggested that other factors may be more important at local or regional scales. Fierer *et al.* (2012) expanded upon this by assaying 16 soils from different biomes using both 16S rRNA-based and shotgun metagenomics-based sequencing (Chapter 21). They found a strong correlation between richness of the 16S rRNA-based phylotypes and the metagenomics-based functional genes. Moreover, the grassland and forest environments

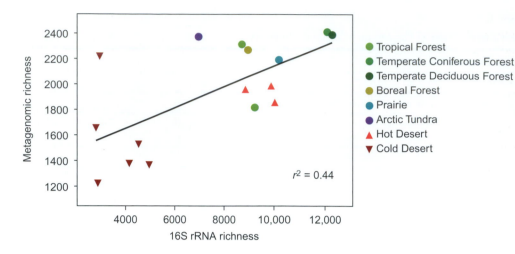

FIGURE 19.2 Diversity of microbial communities in 16 soils from a variety of ecosystems. The y-axis shows the taxonomic richness (number of phylotypes) of the bacterial community. The x-axis shows the functional gene richness. From Fierer *et al.* (2012).

generally contained more diverse microbial communities than either of the more extreme (hot or cold desert) environments (Figure 19.2).

Although dramatic changes such as imposing anaerobic conditions on a previously aerobic soil by flooding it (as commonly used in rice cultivation) can greatly impact microbial communities, the effects of subtle and/or gradual changes, such as may result from climate change, may be more difficult to determine. This is especially true for situations where multiple factors may change simultaneously and even interact. For example, Castro *et al.* (2010) investigated the impact of multiple drivers of climate change (increased atmospheric CO_2, increased temperature and varying precipitation). They found that the level of precipitation had a greater effect than increased CO_2 or temperature on the diversity of the soil bacterial and fungal communities. However, precipitation also had the greatest effect on the abundance and diversity of the plant community, so the observed changes in the microbial community may have been an indirect response to the plant changes, and not directly as a result of the environmental changes. There have been some reported impacts of elevated CO_2 on soil microbial communities for individual ecosystems. For example, one study looked at the effects of 10 years of elevated CO_2 in six ecosystems and found no systematic effect on bacterial biomass, richness or community composition, although ecosystem-specific responses and some trends across sites (e.g., decreased populations of *Acidobacteria* Group 1) were found in response to elevated CO_2 (Dunbar *et al.*, 2012). Additionally, it should be noted that most studies still only characterize the most abundant organisms in a sample. Therefore, even though a study does not detect a response of the microbial community to an environmental disturbance, the impact on less abundant members of the community may be missed unless these populations are specifically targeted (Case Study 4.2).

19.2.6 Functional Diversity and the Resilience of Microbial Communities

Whereas great progress has been made in terms of exploring soil microbial diversity, most of this information has been discovered based on phylogenetic genes (e.g., rRNA genes), which do not provide direct indication of the functional role that these organisms play in the environment. Scientists usually infer potential functional roles for these organisms based upon their relatedness (i.e., similarity of their rRNA genes) to other organisms whose function is known. However, this is complicated by the diversity and widespread distribution of many functional genes. Sequence diversity among functional genes (e.g., nitrite reductase and nitrous oxide reductase) is normally greater than that of ribosomal genes. Further, microbes from all three domains of life (Archaea, Bacteria and Eukarya) have been shown to be able to participate in some of the same processes (e.g., denitrification; Ward, 2002). Therefore, linking phylogeny and function remains a challenging task. However, emerging approaches such as environmental functional gene arrays and metagenomic sequencing are making this more possible (He *et al.*, 2010; Hemme *et al.*, 2010). Additionally, the integration of activity- and community composition-based measurements, such as microradioautography with specific radiolabeled substrates in combination with fluorescent *in situ* hybridization (FISH) of microbial cells, is helping to provide direct linkage between environmental organisms and the processes they facilitate *in situ* (Torsvik and Øvreås, 2002).

Although the relationship between soil microbial diversity and functional diversity remains largely unknown, diverse soil communities are believed to enhance ecosystem stability, productivity and resilience towards stress and disturbance (Torsvik and Øvreås, 2002). Redundancy with respect to functional diversity may enable soil microbial communities to be active even with environmental parameters that constantly change, including temperature, soil moisture content and nutrient availability. Once a certain level of diversity is reached in a community, all of the functions necessary for ecosystem function exist within members of the microbial community. Beyond this point, additional diversity does not provide additional functions but does instead provide functional diversity and ostensibly ecosystem stability (Figure 19.3).

Following a change or disturbance, a microbial community can have at least four responses: (1) the composition stays the same—known as resistance; (2) the composition is altered but later returns to its original composition—known as resilience; (3) the composition is altered but still performs like the original community due to functional redundancy of the community; or (4) the composition is altered and performs differently than the original community (Figure 19.4).

Allison and Martiny (2008) reviewed over 70 studies that experimentally exposed microbial communities to different disturbances. They found that in the vast majority of cases, the microbial communities were sensitive to the disturbance: 60% to increased CO_2 levels; 84% to nitrogen/phosphorus/potassium fertilization; 82% to temperature; and 83% to carbon amendments. In another example, a study evaluating seasonal and environmental changes on soil microbial community composition found that, although bacterial biomass did not change significantly during the seasons, culturable and molecular techniques did demonstrate significant variations in community composition (Smit *et al.*, 2001). Interestingly, in this study and others, culture-dependent and molecular techniques identify very different microbial populations in soil.

This indicates that microbial communities are very fluid, at least taxonomically, and are not resistant to perturbations. If a change occurs, it is possible for the community to be resilient and later return to the original community composition. However, due to the complexities of microbial communities and interactions, this does not seem likely and data are lacking to support the idea that this occurs regularly. The more likely responses appears to be an impacted microbial community with an altered composition that either does or does not perform like the original community (Case Study 19.1; Table 19.3; Figure 19.5).

Interestingly, anthropogenic activities targeting ecosystem restoration of severely impacted environments can create microbial communities that function similarly to natural ecosystems. For example, mining of ores for copper is a significant industry throughout many regions of the world. Following the extraction of the copper-containing minerals, there is a need to deposit the processed ore. These so-called mine tailings are often piped into desert

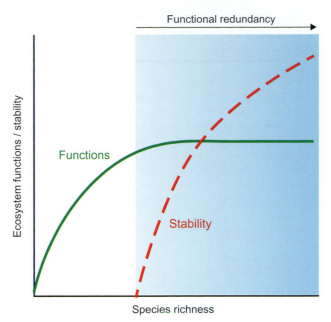

FIGURE 19.3 Relationship between functional redundancy and ecosystem stability. As microbial species are added to an ecosystem, this increases the functional capability of the microbial community. Beyond a point, additional species do not add new functions, but they do serve to increase the functional redundancy and thus stability of the ecosystem. Adapted from Konopka (2009).

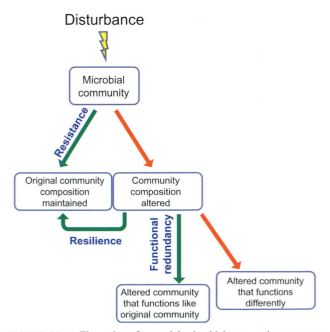

FIGURE 19.4 Illustration of potential microbial community responses to disturbance and the resulting impacts on ecosystem processes. Note that multiple mechanisms (e.g., resistance, resilience and functional redundancy) can help maintain microbial function(s) following a perturbation. Adapted from Allison and Martiny (2008).

Case Study 19.1　Recovery of Soil Microbial Processes, Populations and Communities Following Reclamation of a Lignite Surface Mine

Surface mining for lignite (coal) results in the destruction of the original soil profile characteristics, and therefore alters the physical, chemical and biological conditions. After mining, these sites are reclaimed by backfilling with the previously removed soil (i.e., overburden) and reconstructing the site to its original slopes and contours. In addition, the sites are revegetated with either native or improved (e.g., commercial timber) plant species.

Numerous studies have investigated the recovery of soil chemical and physical properties following reclamation. However, there is very limited information on the recovery of soil microbial communities following reclamation, and even less that correlates this with ecosystem function. To address this knowledge-gap, Ng (2012) conducted a study to determine the amount of time required for recovery of soil physical, chemical and microbial characteristics in a 40-year chronosequence of reclaimed mine soils at the Big Brown lignite mine in East Texas. A similarly vegetated site nearby was used as the unmined reference site.

Following reclamation, many of the soil physical and chemical properties were immediately returned to conditions that met or exceeded those of the soil of the unmined reference site. Nutrient distribution throughout the soil profile required at least 5 years before any stratification was observed. Carbon and nitrogen sustained premined levels through the profile after 15 years. Soil microbial biomass levels and carbon and nitrogen mineralization required 15 to 20 years before returning to unmined levels (Table 19.3). Additionally, numbers of bacteria and fungi (as determined with qPCR) recovered within 20 years. However, the microbial communities (as determined using 16S rRNA gene sequencing and functional gene microarray analysis) did not return to the same composition even after 40 years (Figure 19.5). Interestingly, the 10- and 15-year reclamation soils were more similar to the unmined reference site than the 30- and 40-year post-reclamation sites were. This suggests that the bacterial communities initially became more similar to the reference soil, up to around 15 years, and then deviated into a different bacterial community. Since this corresponded with the recovery of major soil processes (e.g., carbon and nitrogen mineralization), it appears that the functional redundancy of the microbial community contributed to the recovery of soil ecosystem functions, even though the microbial community did not return to its original composition.

Source: Ng (2012).

TABLE 19.3 Recovery of Soil Microbial Processes, Populations and Communities following Reclamation of a Lignite Surface Mine

Years Since Reclamation	Microbial Biomass		Mineralization		Microbial Numbers		Microbial Community Composition
	C	N	C	N	Bacteria	Fungi	
0	X	X	X	X	X	X	?
5	X	X	X	X	X	X	?
10	X	X	X	X	X	X	?
15	√	√	X	X	X	X	?
20	√	√	√	√	√	√	?
30	√	√	√	√	√	√	?
40	√	√	√	√	√	√	?

Adapted from Ng (2012).
X = *Soil quality parameter lower than for unmined conditions.*
√ = *Soil quality parameter equals (or exceeds) unmined conditions.*
? = *Soil quality parameter is different from unmined conditions, but it is unclear if this is better or worse.*

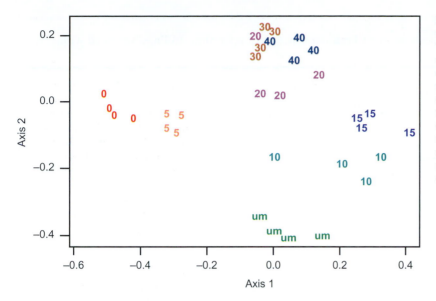

FIGURE 19.5 Changes in soil bacterial communities following reclamation of a lignite surface mine. The graph represents a non-metric multidimensional scaling analysis of 16S rRNA gene pyrosequencing data. Data include four replicate samples of soil from each site along a reclamation chronosequence of 0 to 40 years (0, 5, 10, etc.) along with soil of an unmined (um) reference site. Note that the bacterial community becomes more similar to the original (reference) community up to around 10 years post-reclamation, but then begins to diverge into a relatively stable community that is distinct from the original community. Adapted from Ng (2012).

Case Study 19.2   Community Diversity Dynamics of Mine Tailings Amended with Class A Biosolids

The bacterial community characteristics of mine tailings amended with Class A biosolids were monitored over a 10-year period. Specifically, samples were taken: 3 weeks; 3.5 years; 8 years; and 10 years after biosolid amendment, and subjected to community DNA extraction followed by cloning and sequencing. The 16S rRNA gene sequence analysis showed that the most persistent bacterial populations were members of major soil bacterial phyla including *Proteobacteria*, *Actinobacteria*, *Firmicutes*, *Bacteroidetes* and *Acidobacteria* (Table 19.4). This is consistent with most studies of soil bacterial communities (Kuske *et al.*, 1997; Sun *et al.*, 2004; Janssen 2006). In this study, both the reclaimed mine tailings and the desert soil were comprised of bacteria similar to those others have found in desert soils from the southwestern United States (Kuske *et al.*, 1997) and in most typical soils. It is also of interest that the percentage of unclassified bacteria from the desert soil was in fact slightly higher than the percentage from the other sites. This illustrates the fact that our knowledge of desert soil bacteria is limited and warrants further research.

Initially, sequences affiliated with the phyla *Bacteroidetes* and *Firmicutes* seemed to dominate at the early stages of the study (3 weeks to 3.5 years), but eventually their prevalence diminished with time. This might be due to the fact that *Bacteroidetes* and *Firmicutes*, along with *Actinobacteria*, are known to dominate 80% of identified bacteria in the human gut (Mariat *et al.*, 2009). Thus it could be expected that after the initial application of biosolids, biosolid-associated bacteria (not soil bacteria) would dominate microbial populations. This confirms the hypothesis of this study in which amendment of Class A biosolids into nutrient-poor mine tailings would ultimately lead to the establishment of a functionally redundant soil bacterial population followed by subsequent revegetation. Overall, the results indicated that biosolid-treated mine tailings had eventually acquired diversity levels approaching that of the desert soil.

Source: Pepper et al. (2012).

---

areas to a depth of 35 m. Mine tailings are essentially crushed rock and resemble soil. However, tailings have very low cation exchange capacities, minimal microbial populations and almost zero organic matter content. They support scant revegetative growth and are subject to wind erosion and dust storms. But, large additions of organic matter supplied as "Class A biosolids" can result in a functional soil with respect to microbial characteristics, which is sustainable over a 10-year period resulting in extensive revegetation of the tailings (Case Study 19.2, see also Case Study 26.1).

Functional diversity can also be an important mechanism that allows for soil microbial communities to successfully respond to anthropogenic-induced changes to the soil environment, as in the case of metal and/or herbicide additions to soil, or other soil amendments. In the case of co-contaminant additions to soil (metal + organic), a metal-resistant bacterium with the appropriate catabolic genes is necessary for effective biodegradation of the herbicide (Roane and Pepper, 2000). Populations of organisms with these twin properties can arise via one of two mechanisms.

TABLE 19.4 Bacterial Community Composition in a Chronosequence of Copper Mine Tailings Amended with Class A Biosolids

Phylum	Mine Tailings				Desert Soil
	Time Since Biosolids Applied				
	3 Weeks	3.5 Years	8 Years	10 Years	
	% of bacterial community				
Proteobacteria	23.9	25.2	30.7	31.4	29.0
Alpha*	21.1	49.3	63.9	57.6	48.1
Beta	59.2	13.4	10.8	12.9	26.6
Delta	1.4	1.5	7.2	1.2	8.9
Gamma	16.9	31.3	12.0	28.2	12.7
Unclassified	1.4	4.5	6.0	0.0	3.8
Actinobacteria	23.9	9.3	31.4	41.6	25.7
Firmicutes	16.2	52.6	5.8	2.2	1.1
Acidobacteria	0.0	1.1	14.4	3.6	17.3
Bacteroidetes	26.6	2.6	3.6	9.9	12.1
Chloroflexi	3.0	3.0	3.6	1.5	0.7
TM7	0.3	0.7	1.1	3.6	0.7
Unclassified bacteria	5.7	3.0	9.0	2.6	11.0

Adapted from Pepper et al. (2012).
Communities were characterized using 16S rRNA gene clone libraries. A natural desert soil is presented for comparison.
*Values listed for the classes of Proteobacteria (alpha, beta, etc.) represent the percentage of the total number of Proteobacteria belonging to each respective class.

First, preexisting cells with the desired attributes may already exist within the soil but at low population numbers. In this case, metal-resistant cells with the ability to degrade the herbicide are at a competitive advantage, and their cell density will increase over time. In another scenario, horizontal gene transfer of genes encoding for metal resistance may occur via plasmid transfer to a cell which already has the necessary degradative genes, but previously lacked metal resistance. Following gene transfer, cell proliferation of the newly formed transconjugants can occur (Newby and Pepper, 2002). In either case, this process is termed "adaptation", which can occur within months, weeks or even days. Adaptation explains why repeated amendments of herbicide such as 2,4-dichlorophenoxyacetic acid (2,4-D) to soil result in enhanced areas of soil microbial degradation, relative to the first application. The time required for adaptation can also vary depending on the amount of metal added to soil, as shown in Figure 19.6. Here, as the amount of cadmium added to the soil increases, the adaptive time taken for degradation of 2,4-D increases. Without cadmium addition, no degradation of 2,4-D occurred during the first 7 days, whereas at the highest cadmium addition the adaptive time period was 21 days. Also, without cadmium, the 2,4-D degraded in 21 days, whereas with a cadmium amendment of 240 µg/g soil, degradation was only complete after 35 days.

In soil impacted by human activity, bacterial communities can be affected either adversely or beneficially due to the selective pressures imposed following anthropogenic inputs into soil. For example, Zerzghi et al. (2010) documented enhanced soil bacterial diversity following 20 years of continuous land application of biosolids. In contrast, metal additions to soil can reduce soil bacterial diversity (Kelly et al., 1999).

19.3 MICROBIAL INTERACTIONS

19.3.1 Microbe–Microbe Interactions

Since environmental microorganisms often exist in close proximity to each other (e.g., as microcolonies or biofilms on soil particles), this increases the likelihood of microbial interactions occurring. These interactions may be positive, such as in commensalism and synergism, or negative, such as in competition, amensalism and predation (Table 19.5). In a commensal interaction, one population is benefited but another population(s) is not directly affected. This may include one population metabolizing a metabolic by-

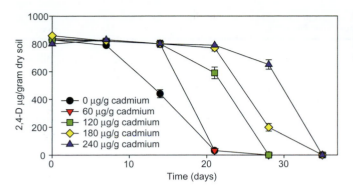

FIGURE 19.6 Biodegradation of 800 μg/g of 2,4-dichlorophenoxya-cetic acid (2,4-D) in the presence of varying amounts of cadmium. Note that the onset of 2,4-D biodegradation is delayed as the cadmium concentration increases. Unpublished data, Pepper (2002).

TABLE 19.5 Types of Environmental Microbial Interactions

Interaction	Effect of Interaction		Example
	Population 1	Population 2	
Mutualism (symbiosis)	↑	↑	Relationship of leguminous plants with nitrogen-fixing rhizobia in which the plant provides the bacteria with organic carbon substrates and a growth habitat in exchange for fixed nitrogen.
Synergism	↑	↑	Syntrophic associations between fermentative bacteria and methanogenic archaeans in which bacteria hydrolyze organic compounds and ferment the products, resulting in generation of hydrogen, acetate, and formate. These materials are then consumed by archaeans, thus maintaining their concentrations at low levels and keeping the overall fermentation reactions energetically favorable.
Commensalism	↔	↑	Cometabolism of cyclohexane to cyclohexanol by one microbial population (which gains no energy from the process) followed by metabolism of the cyclohexanol by a second population.
Neutralism	↔	↔	Two spatially separated microbial populations that have no interaction.
Predation (parasitism)	↑	↓	Protozoa grazing on bacterial populations.
Amensalism	↔ or ↑	↓	Production of an antibiotic by *Streptomyces* spp. that inhibits the growth of fungi.
Competition (antagonism)	↓	↓	Two populations of heterotrophic bacteria (e.g., *Streptomyces* and *Pseudomonas* spp.) attempting to metabolize the same organic substrate.

Adapted from Atlas and Bartha (1998) and Bottomley (2005).
↑ = positive effect; ↔ = no effect; ↓ = negative effect.

product of another population. A classic example of this is the conversion of cyclohexane to cyclohexanol by one microbial population which gains no energy from the process. This conversion is an example of cometabolism or "fortuitous metabolism" reflecting the lack of benefit to the population initiating the transformation (Chapter 17). Other microbial populations (which cannot metabolize cyclohexane) can then metabolize the cyclohexanol. In the example above, if both microbial populations benefited

from the interaction, these organisms would be classified as a consortium (pl. consortia) that worked together to metabolize the cyclohexane.

In synergistic interactions that benefit both (or all) microbial populations involved, the association may either be obligatory or optional, depending upon the organisms. An example of a generally optional synergistic interaction is syntrophy. Syntrophism, also known as cross-feeding, consists of multiple microbial populations

---

Information Box 19.3   Afla-Guard®—An Example of Competitive Exclusion

Production of toxins (e.g., aflatoxin) by environmental fungi such as *Aspergillus flavus* is a major problem for many crops. Recently, a biological control product has been developed and marketed for control of these fungi in crops such as corn. The basis for this control is use of a fungal strain that does not produce toxins. Spores of these atoxigenic fungi are added to fields on a carrier material, and following germination, the fungi colonize the

developing ears of corn. The presence of these organisms serves to competitively prevent naturally occurring, toxin-producing fungal strains from colonizing the corn. Although this approach is subject to a variety of challenges, like any biological control mechanism, it has been shown to be capable of reducing aflatoxin levels by 85% or more!

Information Box 19.4   Fungistasis—An Example of Microbial Competition and Antagonism

It has long been recognized that adding organic amendments such as compost to soil can suppress many soil-borne plant pathogens. When this suppression involves the inhibition of fungal spore germination and hyphal growth, it is referred to as fungistasis. One possible explanation for this suppression is the presence of an active and diverse soil microbial community that competes with and inhibits proliferation of the plant pathogens. Although the

exact mechanisms responsible for fungistasis are not clear, it appears that the process is at least partially due to decreased nutrient availability (due to competition from other microorganisms), and microbial production of inhibitory chemicals such as volatile organic compounds (VOCs) (Garbeva *et al.*, 2011). No matter what mechanism(s) is involved, this provides another example of the benefits of a diverse soil microbial community.

---

being involved in meeting each other's metabolic needs. One example of this is the association between fermentative bacteria and methanogenic archaea. The fermentative bacteria hydrolyze organic compounds and ferment the products resulting in generation of hydrogen, acetate and formate. These fermentation products are then consumed by the archaea, thus maintaining their concentrations at low levels and keeping the overall fermentation reactions energetically favorable. Some syntrophic interactions appear to involve the direct transfer of electrons between different populations through exchange of hydrogen, formate, cysteine or even through conductive nanowires (Sieber *et al.*, 2012).

If the synergistic interactions involve specific microorganisms (i.e., are species specific) or are obligatory, this is referred to as mutualism. Mutualism is commonly used synonymously with symbiosis. There are numerous examples of symbiotic relationships between different microorganisms (e.g., lichens), and between microorganisms and other organisms (e.g., *Rhizobium* spp.—legume interactions and mycorrhizal plants (Section 19.3.2.2)).

If two or more microorganisms are competing for the same resources or space, this will result in competition that will negatively affect one or more of the involved microorganisms. A good example of this is the use of atoxigenic strains of *Aspergillus flavus* to prevent the colonization of crops, such as corn, by strains of environmental fungi that produce mycotoxins (Information Box 19.3). Additionally, amensalism (i.e., antagonism) can

occur due to the production of compounds that are toxic to other organisms. This includes antibiotics (Section 19.4), metabolic by-products such as alcohols and interference with cell–cell communication within a population (Chapter 20). Predation, or even parasitism, can also occur as microorganisms feed on other microorganisms. Examples of this include protozoa that graze on bacteria, fungi that trap nematodes and viruses that infect other types of microorganisms. In many cases, even though a negative interaction between microbial populations can be observed, it can be difficult to ascertain exactly which process (or processes) is responsible for inhibition of the other microbial groups (Information Box 19.4). If different microbial populations have no effect, either positive or negative, on each other, this is referred to as neutralism. This is easy to envision as a possibility for microorganisms that are spatially separated; however, in most cases, microorganisms that are spatially close will have some direct or indirect impact on each other as they metabolize and replicate.

19.3.2 Microbial Megacommunities

Microbes, particularly bacteria, like to adhere to surfaces, and surface-associated microbial communities normally have higher concentrations of microbes than microbial communities suspended in the water column, which are known as the plankton (see Chapter 6). Two such microbial megacommunities are biofilm and rhizosphere communities. Biofilm communities range from 10^8 to 10^{10} CFU/cm^2 (Sjollema *et al.*, 2011). Similarly, rhizosphere populations can be 10^8 to 10^9 CFU per gram of soil (Duineveld and Van Veen, 1999).

19.3.2.1 Biofilm Communities

Biofilm communities are complex microbial mega-populations consisting mostly of bacteria, but also other microbes including algae, protozoa and viruses (as bacteriophages). Biofilms form wherever there are water-associated surfaces, with moisture being a prerequisite for biofilm formation. Biofilms develop naturally on diverse surfaces including: water distribution pipes (see Chapter 28); rocks within rivers or lakes; and even our teeth. Sometimes biofilm formations are encouraged as in the case of the zooleal film which develops within the trickling filters of wastewater treatment plants (see Chapter 25). Microbial mats are specialized biofilms dominated by phototrophic prokaryotes, the cyanobacteria or blue−green algae. Lower layers of the mat can contain anaerobic sulfate-reducing bacteria (Risatti *et al.*, 1994).

The formation and structures of biofilms are described in Chapter 6. Depth-dependent activities of biofilms are discussed in Chapter 7. Here we focus on biofilm communities and diversity. Biofilm communities are typically embedded within a complex mixture of macromolecules including both proteins and exopolysaccharides (EPS). EPS have been implicated as essential for biofilm architecture including the aggregation of bacterial cells, cell−cell recognition and communication, and gene transfer (Fleming and Wingender, 2010). Some bacteria including *Pseudomonas aeruginosa* also produce substantial amounts of extracellular DNA, or eDNA, and such eDNA may actually be a requirement for biofilm formation (Whitechurch *et al.*, 2002). Overall, biofilm communities tend to be highly structured, with *Pseudomonas* spp. being particularly dominant members, and vital to the success of the biofilm community (Case Study 19.3)

19.3.2.2 Rhizosphere Communities

As we have seen, most normal soils do not contain abundant microbial substrates or nutrients, because microbial communities quickly utilize any substrates that are available. In contrast, the rhizosphere is a unique soil environment found in close proximity to plant roots, where substrates are more abundant because of the influence of the plant itself. Increased substrate availability in turn results in enhanced microbial activity, numbers and diversity. Thus, the rhizosphere exists because of complex soil−plant−microorganism interactions. Ultimately, microbial gene expression and diversity in the rhizosphere is controlled by these interactions, which in turn are influenced by environmental factors.

The term rhizosphere was coined by Hiltner in 1904 to describe the part of the soil that is influenced by plant roots. Originally, the rhizosphere was thought to extend 2 mm outward from the root surface. Now it is recognized that the rhizosphere can extend 5 mm or more as a series of gradients of organic substrate, pH, O_2, CO_2 and H_2O. Essentially two regions of the rhizosphere are now recognized: (1) the rhizosphere soil; and (2) the soil in direct contact with the plant root, which is the rhizoplane. Microorganisms also inhabit the root itself and are known as endophytes. The rhizosphere effect is caused by the release of organic and inorganic compounds from the plant roots, which can include root exudates, secretions, lysates or plant mucilages. It has also been shown that many families of plants release living root border cells through a programmed development process. These released cells synthesize novel compounds not produced while attached to the root, and can influence microbial behaviors adjacent to the root (Hawes *et al.*, 2000).

Overall, a vast number of different kinds of microorganisms are found in the rhizosphere, and their numbers generally decrease from the rhizoplane outward toward bulk soil. The rhizosphere effect is often evaluated in terms of R/S ratios, where R = the number of microbes in the rhizosphere and S = the number of microbes in bulk soil. Thus, the greater the R/S ratio, the more pronounced the rhizosphere effect. R/S ratios vary for specific bacteria and fungi but numbers in the rhizosphere can be two to three orders of magnitude higher than in bulk soil.

Bacteria are the most numerous inhabitants of the rhizosphere and R/S ratios can typically be 20:1. In addition to increases in the overall bacterial population within the rhizosphere, specific soil−plant−microbe interactions have evolved that can either benefit or harm the plant. For example, *Agrobacterium* spp. are soil-borne bacteria which cause crown gall diseases (Section 20.2.3). In other cases, the abundant, diverse and active microbial populations in the rhizosphere can function as a "microbial buffer zone" that helps to protect the plant from soil-borne pathogens.

There are two examples of well-studied beneficial rhizosphere microorganisms. The first is bacteria capable of biological dinitrogen fixation, which is the process of converting atmospheric dinitrogen gas into ammonia. Free-living bacteria including species of *Azotobacter* and *Azospirillum* can be found within rhizosphere populations and provide associated plants with a source of fixed nitrogen (NH_3). In contrast to the free-living nitrogen-fixing organisms, the legume−rhizobia association involves a formal symbiosis in which both partners benefit (Information Boxes 16.2 and 16.3). Here, Gram-negative, heterotrophic bacteria originally classified within the genus *Rhizobium* interact with leguminous plants using an

Case Study 19.3   Biofilm Diversity and Community Structure

A recent study evaluated a new point-of-use (POU) device that relies on pathogen inactivation via biofilms developed within a newly manufactured 7-mm thick foam material termed "biofoam." Biofilms within the POUs were developed at three different locations in the U.S. (Montana, Michigan and North Carolina) using three different types of surface waters but under identical conditions. Following harvest, biofilms were analyzed utilizing 454 pyrosequencing of the 16S rRNA genes. Unexpectedly, it was found that there was a remarkable degree of shared community membership between the biofilms from the three locations. The similarity of class-level taxonomy of major biofilm operational taxonomic units (OTUs; ≥97% sequence identity cutoff) for three replicate POU biofilms from each location is shown in Figure 19.7. The top 100 shared OTUs represented 240,000 of the 306,000 raw sequences, and 25% of the shared sequences were classified within the genus *Pseudomonas* (class Gammaproteobacteria). In addition, members of the bacterial community found within the core microbiome of biofoam were closely associated with organisms commonly found in activated sludge, drinking water biofilms, rhizosphere, phyllosphere and soil ecosystems (Table 19.6).

The bacterial communities from each site were strikingly similar despite the fact that they were developed using different source waters. The OTUs classified to the genus level belong to many taxa that are well characterized for functions like the production of EPS, eDNA, quorum sensing molecules, proteases and chitinases, as well as numerous biochemical transformations such as biopolymer and PAH degradation. In addition, planctomycetes were prevalent, capable of the anaerobic oxidation of ammonium to nitrogen gas in the Annamox reaction, perhaps as a detoxifying mechanism.

Overall, this study suggests that unique microbial communities may self-assemble and that key members of the community are necessary for successful biofilm formation. Thus, similar to megacities, although all biofilms are unique, every biofilm may require fundamental structure and requirements in order to function.

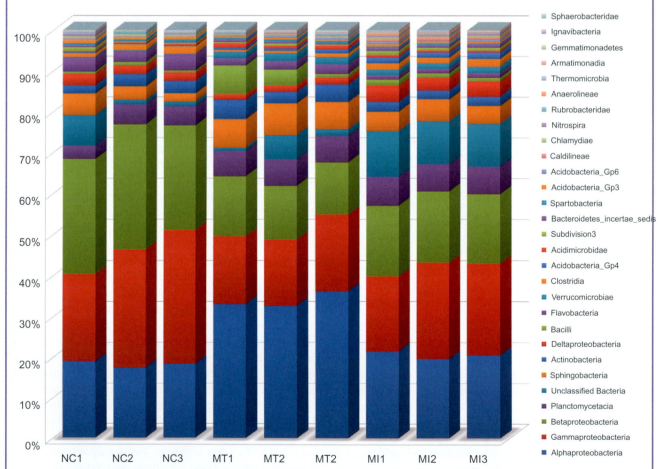

FIGURE 19.7   Bacterial composition (class level) of biofilms in point-of-use (POU) devices used to treat drinking water at three locations in the U.S. NC = North Carolina; MT = Montana; and MI = Michigan. Three replicates were analyzed at each site. From Iker *et al.*, (2013).

TABLE 19.6 Summary of Shared Bacteria in Point-of-Use (POU) Devices Used to Treat Drinking Water at Three Locations in the U.S.

OTU	# of Sequences	Classification	Environment(s) With Similar Organisms
001	66281	*Pseudomonas*	Biofilms (plants, environment, engineered water)
002	20032	Unclassified	Soil, water
003	18065	*Sphingomonas*	Soil, water
004	12951	*Sporosarcina*	Soil, water
005	12596	*Arenimonas*	Drinking water
006	11078	*Rhodobacterium*	Soil, water
007	10126	*Janthinobacterium*	Water biofilms
008	8073	*Rhizobiales*	Rhizosphere
009	7426	*Flavobacterium*	Soil, water
010	6104	*Massilia*	Rhizosphere
011	5934	*Bacteroidetes*	Water
012	5776	*Phyllobacteriaceae*	Phyllosphere

Adapted from Iker et al., (2013).
Table includes the number of community sequences belonging to each OTU (operational taxonomic unit; ≥ 97% similarity cutoff), their closest taxonomic identification, and other environments that have been found to harbor similar organisms.

elaborate cell–cell signaling process (Chapter 20) that results in profound physiological changes in both organisms. These bacteria are known colloquially as rhizobia and are characterized as fixing nitrogen for the plant host in return for carbon sources supplied by the plant as photosynthates. The symbiosis occurs within newly formed root organs called root nodules that develop in response to the presence of specific soil-borne rhizobia. During nodulation the bacteria and the plant contribute to the production of leghemoglobin, which maintains low internal O_2 levels so nitrogenase does not become inactivated. The rhizobia themselves undergo physiological changes, are known as bacteroids, and actually conduct the process of nitrogen fixation. As the plant host matures, ultimately the root nodules senesce as the bacteria become less active and new nodules are formed. Another group of bacteria, *Frankia* spp., form symbiotic relationships with over 200 species of woody plants and shrubs. Since *Frankia* spp. are members of Actinomycetes, plants in these associations are often referred to as actinorhizal plants.

A second important group of rhizosphere microorganisms is the mycorrhizal fungi, which also form symbioses with plants. These fungi act as an extension of the plant root system which aids in the uptake of almost all plant nutrients, but in particular phosphorus, which typically has low solubility and therefore availability in the soil solution. Such fungi assist in the plant uptake of nutrients from dilute solutions by scavenging soil nutrients, utilizing active transport mechanisms to concentrate nutrients against steep concentration gradients. They appear to increase the bioavailability of these compounds for the plant. When released from fungal hyphae, such nutrients can be taken up by plant roots. In addition, when nutrients are stored within the fungus, the fungus can act as a reservoir of nutrients for future plant utilization. The mechanisms that cause the fungus to release its nutrients are not well understood. The plant supplying the fungus with carbon compounds, mostly as hexose sugars, completes the mutualistic association. Thus, each symbiont aids the other in terms of required nutrients.

Mycorrizal fungi become endemic in most soils and form extensive networks of fungal hyphae that can connect different plant species. In addition, on larger root systems, different fungi can infect the same root system. Mycorrhizal fungi naturally infect most plants, but in some commercial cropping systems such as the establishment of pine seedlings in pots, plants can be infected with known highly effective strains of fungi. There are several different types of mycorrhizal fungi. The vesicular–arbuscular mycorrhizae (VAM) are also known as endomycorrhizal fungi, which as the name implies are found mostly within the internal tissues of the root. This type of fungus is frequently found in fertile soils, and is characterized by the presence of smooth vesicles and branched arbuscules that are involved in the storage and transfer of nutrients between the fungus and the plant (Figure 19.8). About 90% of all vascular plants are

(A)

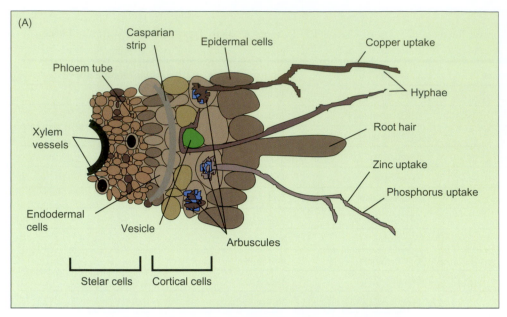

(B)

FIGURE 19.8 (A) A typical endomycorrhiza showing hyphae extending beyond the root epidermis into the rhizosphere. Intracellular arbuscules are also visible. (B) Highly magnified picture of a *Glomus* sp. arbuscule. Photo courtesy Mark Brundrett.

associated with such fungal symbionts. The VAMs are classified into several genera, notably *Glomus* and *Gigaspora* spp., within the phylum *Glomeromycota*. The orchidaceous mycorrhizae fungi (Basidiomycetes) are much more specific than other VAMs, and infect only plants of the orchid family, which contains thousands of species, most of which are tropical. The physiological relationship between the orchid and these fungi is different because in this association, it is the fungus that supplies the plant with a source of carbon. This is the only type of mycorrhizal association in which the carbon flow is into the plant from the fungus. In some cases, mature orchids can therefore live without conducting photosynthesis. It is also of interest that many orchids are associated with *Rhizoctonia* spp., including *R. solani*, which are common plant pathogens.

The ericaceous mycorrhizae are fungi characterized by association with a specific group of plants known as the Ericaceae, which form important plant communities in moors, swamps and peat. The plants involved include heathers, rhododendrons and azaleas, which are often found on nutrient-poor, acidic soil at high altitudes and at colder latitudes. The fungi involved are typical of the endomycorrhizal fungi in that they have intracellular hyphae, but they do not form arbuscules. In this association, the fungus supplies the plant with nutrients, and the plant supplies the fungus with carbon substrate. The fungi also seem to be able to make the plants more tolerant of heavy metals and other soil contaminants. Most of the fungi involved seem to be members of the Ascomycetes.

The Ectomycorrhizae form associations that are characterized by intercellular (between cell) hyphae as opposed to the intracellular (within cell) penetration of the VAMs. These mycorrhizas are formed on the roots of woody plants, with a thick fungal sheath developing around the terminal lateral branches of roots (Figure 19.9). This is also known as the mantle and is connected to the network of intercellular hyphae found in the root cortex

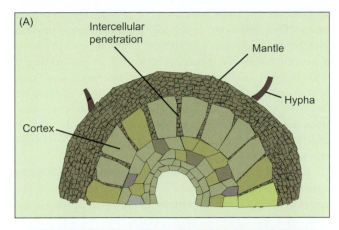

FIGURE 19.9 (A) Cross-section of ectomycorrhizal rootlet showing the exterior fungal sheath or mantle and intercellular penetration. (B) White ectomycorrhizal mantle on a tree root, courtesy Terry W. Henkel.

known as the Hartig net. The plants involved with these mycorrhizas are all trees or shrubs, whereas the fungi involved are often Basidiomycetes or Ascomycetes. Carbon substrate is supplied by the plant to the fungus, and minerals, in particular phosphates, are supplied by the fungus to the plant. Ectomycorrhizal fungi are commonly used as inoculants for pine seedlings, and other plants, in containers prior to planting in reforestation projects.

19.3.3 Microbial Interactions Impacting Animal and Human Pathogens

Many environments, such as soil and aquatic systems, contain microbial pathogens of animals and humans. With the exception of geo-indigenous pathogens (naturally occurring and capable of metabolism, growth and reproduction in soil), such as *Coccidioides immitis* and *Naegleria fowleri*, these pathogens generally do not regrow in the environment, but instead exist transiently while in between hosts. However, while in the environment, these organisms are

negatively affected by microbial interactions such as competition and predation. These effects from indigenous, antagonistic microorganisms are commonly referred to as "biological factors" when discussing the environmental fate of pathogens (Chapter 22), and play critical roles in water and wastewater treatment processes. For example, biofilms of complex microbial communities, known as zooleal film (Chapter 25), are vital to the secondary treatment of wastewater and function to degrade organic materials and also decrease levels of pathogens. In slow sand-filtration point-of-use devices, biofilms referred to as schmutzdecke are known to inactivate pathogens, including viruses (Bauer *et al.*, 2011). Even if the pathogenic organisms are "native" to the soil, they are subject to the same plethora of inhibitory microbial interactions that affect introduced organisms.

While most of these microbe−microbe interactions are considered to be negative with respective to population levels of pathogens, in some case these interactions can actually result in increased pathogen levels. For example, although *Bacillus anthracis* (the causative agent of anthrax) is widely found in soil, there has been some debate about whether it actively grows in soil, or just exists in the soil as spores while in between host organisms (Pepper and Gentry, 2002). However, recent evidence (Dey *et al.*, 2012) has demonstrated that after *B. anthracis* spores are ingested by soil-dwelling amoeba they can germinate and multiply (under environmental conditions) within the amoeba, ultimately resulting in the death of the amoeba and the release of the additional *B. anthracis* cells into the environment.

19.4 MICROBIAL DIVERSITY AND NATURAL PRODUCTS

Terrestrial and aquatic environments are home for billions of microorganisms including bacteria and fungi. In this chapter we have focused primarily on bacterial diversity, and as we have seen, the diversity of environmental bacteria is enormous. Although not as diverse as bacteria, fungal populations are also extremely diverse with one million different species estimated to exist (Gunatilaka, 2006). Overall, from the less than 1% of bacterial species and 5% of fungal species that have been identified, a treasure chest of natural products critical to maintaining human health and welfare have been discovered (Table 19.7). These compounds represent a small portion of the elaborate compounds that environmental microorganisms produce in order to communicate with, stimulate and/or inhibit other microorganisms.

Actinomycetes and fungi are particularly rich sources of metabolites with novel biological activities including antibiotics. Antibiotics are compounds produced by microorganisms that kill or inhibit other microorganisms. Thus, they are a class of chemicotherapeutic agents that

TABLE 19.7 Products Derived from Environmental Microorganisms and Other Natural Sources

Item	Extent %	Reference
Prescription drugs	40	Strobel and Daisy, 2003
New chemical products registered by U.S. Food and Drug Administration	49	Brewer, 2000
Approved drugs between 1989 and 1995	60	Grabley and Thiericke, 1999
Approved cancer drugs between 1983 and 1994	60	Concepcion et al., 2001
Approved antibacterial agents between 1983 and 1994	78	Concepcion et al., 2001

Source: Pepper et al. (2009).

can be used to control infectious disease. Since antibiotics are natural products obtained largely from environmental microorganisms, soils and similarly diverse environments are the ultimate source of antibiotics. The first and perhaps most effective antibiotic discovered was penicillin, isolated by Sir Alexander Fleming in 1929 from the soil-borne fungus *Penicillium*. This antibiotic has proved to be highly effective in treating many bacterial infections, including staphylococcal and pneumonococcal infections. Later in 1943, another potent antibiotic was discovered by Selman Waksman, a feat for which he later received the Nobel Prize. This antibiotic, streptomycin, was isolated from the actinomycete *Streptomyces griseus*. Since then, soil actinomycetes have been shown to be the source of numerous antibiotics. In fact, over 50% of all known antibiotics are derived from the genus *Streptomyces* (Kieser et al., 2000).

However, bacteria are prokaryotic organisms with the ability to metabolize and replicate quickly. They are also very adaptable genetically. Hence, when confronted with an antibiotic, a genetic or mutational change may confer resistance to the antibiotic. Thus, the more that antibiotics are used, the more likely it is that antibiotic-resistant strains will develop. This is of great concern since several human pathogenic bacteria are becoming resistant to popular antibiotics (see Section 31.4).

More recently, interest has centered on rhizosphere bacteria and endophytic microbes as a new source of natural products including antibiotics. Endophytes are bacteria or fungi that live within plants (e.g., in roots) without pathogenic effects. In contrast, rhizosphere organisms reside in soil adjacent to, and under the influence of, plant roots. In both cases, the microbes receive plant metabolites or exudates as a source of nutrition. In return, many of the microbes, especially the endophytes, provide metabolites that protect the plants. They are proving to be a source of natural products effective in controlling a wide variety of human pathogenic microbes, and new source of antibiotics. For example, the endophytic *Streptomyces* sp. strain NRRL 30562 produces wide spectrum antibiotics known as munumbicins capable of controlling multi drug-resistant strains of *M. tuberculosis* (Castillo et al., 2002).

Endophytes are also proving to be useful as anticancer agents. Paclitaxel, the world's first billion-dollar anticancer drug, is produced by many endophytic fungi associated with the yew (*Taxus*) species (Strobel and Daisy, 2003). Other beneficial endophytic natural products include: pestacin with antioxidant activity (Harper et al., 2003), bioinsecticides (Findlay et al., 1997) and insect repellents (Daisy et al., 2002). Other public health benefits are derived from antidiabetic agents that act as an insulin mimetic (Zhang et al., 1999) and immunosuppressive drugs (Lee et al., 1995). New cultural techniques (see Chapter 10) and cloning of community DNA extracted from soil (see Section 13.6) are further enhancing the discovery of beneficial natural products from the vast diversity of environmental microorganisms (Daniel, 2004).

QUESTIONS AND PROBLEMS

1. Calculate the number of soil bacteria that surround a typical homeowner on a quarter acre lot. Assume that an acre furrow slice (one acre to a depth of one foot) weighs 2 million pounds and contains the vast majority of the soil-borne bacteria.
2. What is the bacterial diversity in the quarter acre lot from question 1, assuming it is situated in Minnesota?
3. How could you increase microbial diversity in the soil from questions 1 and 2?
4. Is it always beneficial for an ecosystem to have a diverse microbial community? Explain your reasoning. Compare your answer to the way that most crops are currently grown.
5. Why does the cloning of community DNA extracted from soil potentially increase the availability of beneficial natural products that were previously unknown?
6. What is the largest impact of soil microorganisms on human health?

REFERENCES AND RECOMMENDED READING

Allison, S. D., and Martiny, J. B. H. (2008) Resistance, resilience, and redundancy in microbial communities. *Proc. Natl Acad. Sci. U.S.A.* **105**, 11512–11519.

Atlas, R. M., and Bartha, R. (1998) "Microbial Ecology: Fundamentals and Applications," Benjamin/Cummings, Menlo Park, CA.

Bauer, R., Dizer, H., Graeber, I., Rosenwinkel, K-H., and Lopez-Pila, J. M. (2011) Removal of bacterial fecal indicators, coliphages and enteric adenovirues from waters with high fecal pollution by slow sand filtration. *Wat. Res.* **45**, 439−452.

Bottomley, P. J. (2005) Microbial ecology. In "Principles and Applications of Soil Microbiology" (D. M. Sylvia, J. J. Fuhrmann, P. G. Hartel, and D. A. Zuberer, eds.), 2nd ed., Pearson Prentice-Hall, Upper Saddle River, NJ, pp. 222−241.

Brewer, S. (2000) The relationship between natural products and synthetic chemistry in the discovery process. In "Biodiversity: New Leads for Pharmaceutical and Agrochemical Industries" (S. K. Wrigley, M. A. Hayes, R. Thomas, E. J. T. Crystal, and N. Nicholson, eds.), The Royal Society of Cambridge, Cambridge, United Kingdom, pp. 59−65.

Castillo, U. F., Strobel, G. A., Ford, E. J., Hess, W. M., Jensen, J. B., Albert, H., *et al.* (2002) Munumbicins, wide-spectrum antibiotics produced by *Streptomyces* NRRL 30562, endophytic on Kennedia nigriscans. *Microbiology* **148**, 2675−2685.

Castro, H. F., Classen, A. T., Austin, E. E., Norby, R. J., and Schadt, C. W. (2010) Soil microbial responses to multiple experimental climate change drivers. *Appl. Environ. Microbiol.* **76**, 999−1007.

Coenye, T., Gevers, D., Van de Peer, Y., Vandamme, P., and Swings, J. (2005) Towards a prokaryotic genomic taxonomy. *FEMS Microbiol. Rev.* **29**, 147−167.

Cohan, F. M., and Perry, E. B. (2007) A systematics for discovering the fundamental units of bacterial diversity. *Curr. Biol.* **17**, R373−R386.

Concepcion, G. P., Lazuro, J. E., and Hyde, K. D. (2001) Screening for bioactive novel compounds. In "Bio-Exploitation of Filamentous Fungi" (S. B. Pointing, and K. D. Hyde, eds.), Fungal Diversity Press, Hong Kong, pp. 93−130.

Costanza, R., D'Arge, R., De Groot, R., Farber, S., Grasso, M., Hannon, B., *et al.* (1997) The value of the world's ecosystem services and natural capital. *Nature* **387**, 253−260.

Curtis, T. P., Sloan, W. T., and Scannell, J. W. (2002) Estimating prokaryotic diversity and its limits. *Proc. Natl Acad. Sci. U.S.A.* **99**, 10494−10499.

Daisy, B. H., Strobel, G. A., Castillo, U., Erza, D., Sears, J., Weaver, D., *et al.* (2002) Naphthalene and insect repellent is produced by *Muscodor vitigenus*, a novel endophytic fungus. *Microbiology* **148**, 3737.

Daniel, R. (2004) The soil metagenome—a rich resource for the discovery of novel natural products. *Cur. Opinion Biotechnol.* **15**, 199−204.

Dey, R., Hoffman, P. S., and Glomski, I. J. (2012) Germination and amplification of anthrax spores by soil-dwelling amoebas. *Appl. Environ. Microbiol.* **78**, 8075−8081.

Duineveld, B. M., and Van Veen, J. A. (1999) The number of bacteria in the rhizosphere during plant development: relating colony forming units to different reference units. *Biol. Fert. Soils* **28**, 285−291.

Dunbar, J., Eichorst, S. A., Gallegos-Graves, L. V., Silva, S., Xie, G., Hengartner, N. W., *et al.* (2012) Common bacterial responses in six ecosystems exposed to 10 years of elevated atmospheric carbon dioxide. *Environ. Microbiol.* **14**, 1145−1158.

Fierer, N., Leff, J. W., Adams, B. J., Nielsen, U. N., Bates, S. T., Lauber, C. L., *et al.* (2012) Cross-biome metagenomic analyses of soil microbial communities and their functional attributes. *Proc. Natl Acad. Sci. U.S.A.* **109**, 21390−21395.

Findlay, J. A., Bethelezi, S., Li, G., and Sevek, M. (1997) Insect toxins from an endophyte fungus from wintergreen. *J. Nat. Prod.* **60**, 1214.

Fleming, H-C., and Wingender, J. (2010) The biofilm matrix. *Nat. Rev. Microbiol.* **8**, 623−633.

Gans, J., Wolinsky, M., and Dunbar, J. (2005) Computational improvements reveal great bacterial diversity and high metal toxicity in soil. *Science* **309**, 1387−1390.

Garbeva, P., Gera Hol, W. H., Termorshuizen, A. J., Kowalchuk, G. A., and de Boer, W. (2011) Fungistasis and general soil biostasis—a new synthesis. *Soil Biol. Biochem.* **43**, 469−477.

Gihring, T. M., Green, S. J., and Schadt, C. W. (2012) Massively parallel rRNA gene sequencing exacerbates the potential for biased community diversity comparisons due to variable library sizes. *Environ. Microbiol* **14**, 285−290.

Grabley, S., and Thiericke, R. (eds.) (1999) "Drug Discovery from Nature". Springer-Verlag, Berlin, Germany.

Gunatilaka, A. A. L. (2006) Natural products from plant associated microorganisms: distribution, structural diversity, bioactivity and implications for their occurrence. *J. Nat. Prod.* **69**, 509−526.

Hansel, C. M., Fendorf, S., Jardine, P. M., and Francis, C. A. (2008) Changes in bacterial and archaeal community structure and functional diversity along a geochemically variable soil profile. *Appl. Environ. Microbiol.* **74**, 1620−1633.

Harper, J. K., Ford, E. J., Strobel, G. A., Arif, A., Grant, D. M., Porco, J., *et al.* (2003) Pestacin: a 1,3-dihydro isobenzofuran from *Pestalotipsis* microspora possessing antioxidant and antimycotic activities. *Tetrachedron* **59**, 2471.

Hawes, M. C., Gunawardena, U., Miyasaka, S., and Zhao, X. (2000) The role of root border cells in plant defense. *Trends Plant Sci.* **5**, 128−133.

He, Z., Deng, Y., Van Nostrand, J. D., Tu, Q., Xu, M., Hemme, C. L., *et al.* (2010) GeoChip 3.0 as a high-throughput tool for analyzing microbial community composition, structure, and functional activity. *ISME J.* **4**, 1167−1179.

Hemme, C. L., Deng, Y., Gentry, T. J., Fields, M. W., Wu, L., Barua, S., *et al.* (2010) Metagenomic insights into evolution of a heavy metal-contaminated groundwater microbial community. *ISME J.* **4**, 660−672.

Hollister, E. B., Engledow, A. S., Hammett, A. J., Provin, T. L., Wilkinson, H. H., and Gentry, T. J. (2010) Shifts in microbial community structure along an ecological gradient of hypersaline soils and sediments. *ISME J.* **4**, 829−838.

Hollister, E. B., Hu, P., Wang, A. S., Hons, F. M., and Gentry, T. J. (2013) Differential impacts of brassicaceous and non-brassicaceous oilseed meals on soil bacterial and fungal communities. *FEMS Microbiol. Ecol.* **83**, 632−641.

Iker, B. C., Camper, A. K., Sobsey, M. D., Rose, J. B., and Pepper, I. L., (2013) Biofoam: a new medium for smart biofilms. In review.

Janssen, P. H. (2006) Identifying the dominant soil bacterial taxa in libraries of 16S rRNA and 16S rRNA genes. *Appl. Environ. Microbiol* **72**, 1719−1728.

Kelly, J. J., Häggblom, M., and Tate, R. L., III (1999) Changes in soil microbial communities over time resulting from one time application of zinc: a laboratory microcosm study. *Soil Biol. Biochem* **31**, 1455−1465.

Kieser, T., Bibb, M. J., Buttner, M. J., Chater, K. F., and Hopwood, D. A. (2000) "Practical Streptomyces Genetics," John Innes Foundation, Norwich, UK.

Konopka, A. (2009) What is microbial community ecology? *ISME J.* **3**, 1223–1230.

Kuske, C. R., Barns, S. M., and Busch, J. D. (1997) Diverse uncultivated bacterial groups from soils of the arid Southwestern United States that are present in many geographic regions. *Appl. Environ. Microbiol.* **63**, 3614–3621.

Lauber, C. L., Hamady, M., Knight, R., and Fierer, N. (2009) Pyrosequencing-based assessment of soil pH as a predictor of soil bacterial community structure at the continental scale. *Appl. Environ. Microbiol.* **75**, 5111–5120.

Lee, J., Lobkovsky, E., Pliam, N. B., Strobel, G. A., and Clardy, J. (1995) Subglutinals A and B: immunosuppressive compounds from the endophytic fungus *Fusarium subglutinams*. *J. Org. Chem.* **60**, 7076–7077.

Mariat, D., Firmesse, O., Levenz, F., Guimaraes, V. D., Sokol, H., Dores, J., *et al.* (2009) The *Firmicutes/Bacteroidetes* ratio of the human microbiota changes with age. *BMC Microbiol.* **9**, 123.

Mizrahi-Man, O., Davenport, E. R., and Gilad, Y. (2013) Taxonomic classification of bacterial 16S rRNA genes using short sequencing reads: evaluation of effective study designs. *PLoS ONE* **8**, e53608.

Newby, D. T., and Pepper, I. L. (2002) Dispersal of plasmid pJP4 in unsaturated and saturated 2,4-dichlorophenoxyacetic acid contaminated soil. *FEMS Microbiol. Ecol.* **39**, 157–164.

Ng, J. (2012) Recovery of carbon and nitrogen cycling and microbial community functionality in a post-lignite mining rehabilitation chronosequence in east Texas. Ph.D. dissertation, Texas A&M University, College Station, TX.

Nunan, N., Singh, B., Reid, E., Ord, B., Papert, A., Squires, J., *et al.* (2006) Sheep-urine-induced changes in soil microbial community structure. *FEMS Microbiol. Ecol.* **56**, 310–320.

Pál, C., Papp, B., and Lercher, M. J. (2005) Adaptive evolution of bacterial metabolic networks by horizontal gene transfer. *Nat. Genet.* **37**, 1372–1375.

Pepper, I. L., and Gentry, T. J. (2002) Incidence of *Bacillus anthracis* in soil. *Soil Sci.* **167**, 627–635.

Pepper, I. L., Gerba, C. P., Newby, D. T., and Rice, C. W. (2009) Soil: a public health threat or savior? *Crit. Rev. Environ. Sci. Technol.* **39**, 416–432.

Pepper, I. L., Zerzghi, H. G., Bengson, S. A., Iker, B. C., Banerjee, M. J., and Brooks, J. P. (2012) Bacterial populations within copper mine tailings: long-term effects of amendment with Class A biosolids. *J. Appl. Microbiol.* **13**, 569–577.

Risatti, J. B., Capman, W. C., and Stahl, D. A. (1994) Community structure of a microbial mat: the phylogenetic dimension. *Proc. Natl Acad. Sci. U.S.A.* **91**, 10173–10177.

Roane, T. M., and Pepper, I. L. (2000) Microbial responses to environmentally toxic cadmium. *Microbiol. Ecol.* **38**, 358–364.

Sandaa, R. A., Torsvik, V., Enger, O., Daae, F. L., Castberg, T., and Hahn, D. (1999) Analysis of bacterial communities in heavy metal-contaminated soils at different levels of resolution. *FEMS Microbiol. Ecol.* **30**, 237–251.

Sieber, J. R., McInerney, M. J., and Gunsalus, R. P. (2012) Genomic insights into syntrophy: the paradigm for anaerobic metabolic cooperation. *Ann. Rev. Microbiol.* **66**, 429–452.

Sjollema, J., Rustema-Abbing, M., van der Mei, H. C., and Busscher, H. J. (2011) Generalized relationship between numbers of bacteria and their viability in biofilms. *Appl. Environ. Microbiol.* **77**, 5027–5029.

Smit, E., Leeflang, P., Gommans, S., Van den Broek, J., Vans, M. S., and Wernars, K. (2001) Diversity and seasonal fluctuations of the dominant members of the bacterial soil community in a wheat field as determined by cultivation and molecular methods. *Appl. Environ. Microbiol.* **67**, 2284–2291.

Sogin, M. L., Morrison, H. G., Huber, J. A., Welch, D. M., Huse, S. M., Neal, P. R., *et al.* (2006) Microbial diversity in the deep sea and the underexplored "rare biosphere." *Proc. Natl Acad. Sci. U.S.A.* **103**, 12115–12120.

Somenahally, A. S., Hollister, E. B., Loeppert, R. H., Yan, W., and Gentry, T. J. (2011) Microbial communities in rice rhizosphere altered by intermittent and continuous flooding in fields with long-term arsenic application. *Soil Biol. Biochem.* **43**, 1220–1228.

Strobel, G., and Daisy, B. (2003) Bioprospecting for microbial endophytes and their natural products. *Microbiol. Mol. Biol. Rev.* **67**, 491–502.

Sun, H. Y., Deng, S. P., and Raun, W. R. (2004) Bacterial community structure and diversity in a century-old manure-treated agroecosystem. *Appl. Environ. Microbiol.* **70**, 5868–5874.

Sutton, N. B., Maphosa, F., Morillo, J. A., Al-Soud, W. A., Langenhoff, A. A. M., Grotenhuis, T., *et al.* (2013) Impact of long-term diesel contamination on soil microbial community structure. *Appl. Environ. Microbiol.* **79**, 619–630.

Torsvik, V., and Øvreås, L. (2002) Microbial diversity and function in soil: from genes to ecosystems. *Curr. Opin. Microbiol.* **5**, 240–245.

Torsvik, V., Goksoyr, J., and Daae, F. L. (1990) High diversity in DNA of soil bacteria. *Appl. Environ. Microbiol.* **56**, 782–787.

Venter, J. C., Remington, K., Heidelberg, J. F., Halpern, A. L., Rusch, D., Eisen, J. A., *et al.* (2004) Environmental genome shotgun sequencing of the Sargasso Sea. *Science* **304**, 66–74.

Ward, B. B. (2002) How many species of prokaryotes are there? *Proc. Natl Acad. Sci. U.S.A.* **99**, 10234–10236.

Whitechurch, C. B., Tolker-Nielsen, T., Ragas, P. C., and Mattick, J. S. (2002) Extracellular DNA required for bacterial biofilm formation. *Science* **295**, 1487.

Zerzghi, H., Brooks, J. P., Gerba, C. P., and Pepper, I. L. (2010) Influence of long-term land application of Class B biosolids on soil bacterial diversity. *J. Appl. Microbiol.* **109**, 698–706.

Zhang, B., Salituro, G., Szalkowski, D., Li, Z., Zhang, Y., Royo, L., *et al.* (1999) Discovery of small molecule insulin mimetic with antidiabetic activity in mice. *Science* **284**, 974–977.

Zhou, J., Xia, B., Treves, D. S., Wu, L-Y., Marsh, T. L., O'Neill, R. V., *et al.* (2002) Spatial and resource factors influencing high microbial diversity in soil. *Appl. Environ. Microbiol.* **68**, 326–334.

Chapter 20

Microbial Communication: Bacteria/Bacteria and Bacteria/Host

Leland S. Pierson III, Raina M. Maier and Ian L. Pepper

20.1 INTRODUCTION

Bacteria were the first organisms to evolve on Earth, and were present approximately 2 billion years before the first eukaryotes appeared. Thus, bacteria were critical to the development of the biosphere that enabled the evolution of higher life forms. Since all eukaryotic organisms evolved in the presence of bacteria, they are intimately associated with bacteria. The range of these associations can vary from: (1) bacteria living on the surfaces of the host (saprophytes); (2) those that benefit the host; or (3) those that are detrimental or pathogenic to the host. Since both bacteria and hosts need to sense their environment and the presence of the other, bacteria and their hosts have evolved complex mechanisms of signaling between and among themselves and each other, as a means of communication.

Our current knowledge regarding bacterial physiology and development is based primarily on *in vitro* experiments using pure, single microbial cultures. However, this fails to accurately represent the complexity that bacteria face in their natural environments since they rarely exist in isolation. In nature, most bacteria are members of complex micro, macro- or even megacommunities that predominantly exist as surface-associated biofilms composed of cells embedded in a complex matrix composed of self-synthesized extracellular polysaccharides and DNA known as a biofilm (see Chapters 6 and 19). Often, the survival of a given bacterial species is dependent on the ability of the individual bacterial cells within that population to communicate among themselves, and/or between themselves and other organisms. These other organisms include unrelated bacteria, organisms that share the same ecological niche or eukaryotic hosts, including plants, nematodes, insects, animals and humans.

The environments that bacteria inhabit are complex and subject to rapid change. In order to be successful, bacteria must be able to sense and respond rapidly to these changes by altering the expression of specific genes and metabolic pathways. This ultimately affects the behavior of the bacteria. It is now well recognized that most bacteria produce signals that allow communication between cells. Fundamentally, cells communicate by emitting specific chemical signals, into a particular

I.L. Pepper, C.P. Gerba, T.J. Gentry: Environmental Microbiology, Third edition. DOI: http://dx.doi.org/10.1016/B978-0-12-394626-3.00020-X
© 2015 Elsevier Inc. All rights reserved.

environment inhabited by other organisms. When, or if, the concentration of the signal reaches a level where other cells are able to perceive it, known as the threshold concentration, the gene expression of all the organisms present becomes modified. This cell—cell communication is important for coordinating gene expression within a single population of bacteria (intraspecies signaling), between bacterial populations (interspecies signaling) and between bacteria and other organisms (interkingdom signaling) (Figure 20.1).

Communication signals consist of a wide variety of chemical structures. The primary requirements for these signals are that they are small, they can be released from cells either by passive diffusion or active transport, and that other cells possess the ability to recognize them and alter behavioral patterns in response to their presence. Because these signals alter bacterial behaviors, they have been referred to as bacterial pheromones. Scientists are now beginning to appreciate the world of bacteria—bacteria and bacteria—host signaling. This chapter discusses the current understanding of bacterial signaling using examples of communication systems, including: signaling in Gram-negative bacteria via quorum sensing with N-acyl homoserine lactones; signaling in Gram-positive bacteria via γ-butyryl lactones and small peptide signals; and signaling via autoinducer-2 (AI-2), autoinducer-3 (AI-3) and bacterial muropeptides. In addition, this chapter will touch upon several additional areas of communication such as bacterial eavesdropping,

bacterial signal interference (quorum quenching) and interkingdom signaling.

20.2 SIGNALING VIA QUORUM SENSING IN GRAM-NEGATIVE BACTERIA

Quorum sensing is the regulation of gene expression in response to levels of diffusible signal molecules, which usually correlate with population density, i.e., a sufficient number of cells or a quorum must be present in order for gene expression to occur. Specifically, quorum sensing bacteria produce and release chemical signal molecules or autoinducers that control gene expression of the whole bacterial population. Both Gram-positive and Gram-negative bacteria use quorum sensing systems, but each tends to utilize different chemical signals to control target gene expression.

20.2.1 N-acyl Homoserine Lactones (AHLs)

In Gram-negative bacteria, the best-studied diffusible signals are the N-acyl homoserine lactones (AHL) (Table 20.1). To date, over 50 bacterial species have been shown to produce AHL signals (Scott et al., 2006). This signal class consists of a conserved homoserine lactone ring moiety connected to a fatty acyl side chain (Dong and Zhang, 2005). The specificity of AHL signals is

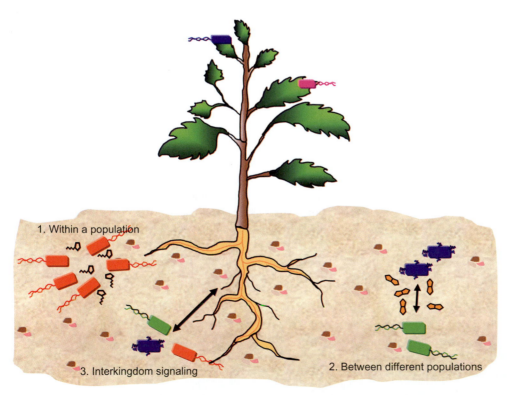

1. Within a population

3. Interkingdom signaling

2. Between different populations

FIGURE 20.1 Examples of cell—cell signaling interactions in nature. (1) Signaling within a single bacterial population. Signals such as AHLs (Gram-negative) and γ-butyrolactones (Gram-positive) are commonly used in bacterial communication within a population. (2) Signaling also occurs between unrelated bacteria. The AHLs have been shown also to participate in communication between different bacteria, as has the universal Autoinducer AI-2. (3) It is increasingly being recognized that signaling occurs between bacteria and eukaryotic hosts such as plants.

TABLE 20.1 Structures of Bacterial Communication Signals

N-Acyl homoserine lactone (AHL) signals, common names, and representative organisms that produce them

Organism	Common name	Structure
Pseudomonas	C4-HSL	
Vibrio harveyi	4-OH-C4-HSL	
Erwinia carotovora, Pantoea stewartii	3-Oxo-C6-HSL	
Pseudomonas	C6-HSL	
Ralstonia solanacearum	C8-HSL	
Agrobacterium	3-Oxo-C8-HSL	
Pseudomonas	3-Oxo-C10-HSL	
Pseudomonas	C12-HSL	

Other types of communication signals

An A1-2 autoinducer	
p-Coumaroyl-HSL	
γ-butyrolactone (produced by *Streptomyces* spp.)	
Halogenated furanone	
Epinephrine	

determined by the length of the fatty acyl side chain as well as the types and number of side chain modifications. AHL-mediated regulation of gene expression is one of the best known examples of quorum sensing.

An excellent example of a quorum sensing system is that of the bioluminescent marine bacterium *Vibrio fischeri* (now classified as *Aliivibrio fischeri* by some). This bacterium exists in a symbiotic association with the marine squid *Euprymna scolopes* (Information Box 20.1). This tiny nocturnal squid contains specialized organs called light organelles that are colonized only by *V. fischeri*, which is ubiquitous in the ocean at low cell densities. Immature *E. scolopes* have cilated "arms" that collect sea water and pass it over the empty light organelles. When *V. fischeri* comes in contact with the light organelles, it colonizes the organs and is supplied nutrients by the host. Colonization by *V. fischeri* induces the loss of the squid's ciliated "arms" by apoptosis (programmed cell death), and also causes the bacteria to lose their flagella and reduce their cell size, indicating a true symbiosis. The eukaryotic squid host provides the prokaryotic bacterium with a nutrient-rich environment in which to live. In return the bacterium produces bioluminescence, or light. The benefit of bioluminescence for the squid may be several-fold. As one example, it may serve as an anti-predation strategy in which light production enables the squid to counter-illuminate itself using the light from *V. fischeri*. This counter-illumination is aimed downwards and enables the squid to avoid casting a shadow beneath it on nights when light from the stars and moon penetrates the seawater, thus allowing the squid to be invisible to predators beneath it. Alternatively, it may enable the squid to locate each other in the darkness of the oceans depths.

The occurrence of bioluminescence is correlated with the cell-population density of the bacteria in the host. As the population of bacterial cells increases it produces and releases an AHL signal into its extracellular environment, which is the eukaryotic squid's light organ. Due to the physical boundaries of the organ, the concentration of the AHL increases, and hence acts as a signal which communicates to the bacteria that they are inside the host as opposed to outside in the seawater. The AHL also initiates a signaling cascade that results in the emission of light. The squid can flush the light organelles until the bacterial population size and signal concentration is below the threshold required for bioluminescence. Hence, the squid controls the level of bioluminescence (Information Box 20.1)

The simplest molecular model for quorum sensing regulation involves two proteins. The first is an AHL synthase (I protein) encoded by a gene commonly referred to as an *I* gene (*luxI, phzI, traI, lasI,* etc.), which

converts cellular precursors into one or more AHL signals. The second is an AHL-responsive regulatory protein (R protein), encoded by a gene referred to as an *R* gene (*luxR, phzR, traR, lasR,* etc.), required for the activation (or in some cases, the repression) of specific genes. At low cell densities, the AHL signal either diffuses out of the cell following a concentration gradient, or is actively transported out of the cell. As cell density increases, the concentration of AHL signal accumulates within the cell. Upon reaching a threshold concentration, the AHL interacts with the R protein resulting in dimerization of the R protein. This causes the R protein dimer to bind to a specific sequence in the promoter of the quorum sensing-regulated gene(s). This binding of the R protein results in enhanced recruitment of RNA polymerase that activates gene expression (Information Box 20.1).

Many Gram-negative bacteria have been shown to utilize quorum sensing to regulate the expression of diverse traits. In all cases, increasing cell numbers result in increased AHL signal concentration. This in turn results in interaction with the R protein that alters the binding affinity of the R protein for a specific sequence located within the promoter regions for genes under quorum sensing control (Dunlap, 1999; Zhu and Winans, 1999; Qin *et al.*, 2000). Evidence that production of AHL signals is required for quorum sensing expression has been shown for many Gram-negative bacteria. An early example was the demonstration that inactivation of the *V. fischeri luxI* gene results in no light production *in vitro* unless exogenous AHL is supplied. Many Gram-negative plant-associated soil bacteria also contain quorum sensing regulatory systems. The first example that AHLs were required for bacterial gene expression on plant roots was demonstrated when a *phzI* AHL mutant of *Pseudomonas chlororaphis* had a 1000-fold reduction in expression of the quorum sensing regulated phenazine genes on wheat roots (Wood and Pierson, 1996, see below). Other evidence that AHLs are important includes the discovery that concentrations of C_4-HSL and 3-oxo-C_{12}-HSL, two AHL signals produced by the opportunistic human pathogen *Pseudomonas aeruginosa*, can be detected in sputum samples of infected patients (Erickson *et al.*, 2002).

This gene regulation mechanism was originally named quorum sensing because it was believed that it enabled a bacterium to determine its own population size or "quorum" (Fuqua *et al.*, 2001). It is now recognized that a single bacterial cell will activate quorum sensing-regulated genes in the presence of sufficient AHL signal. Thus, it is the concentration of AHL, not the number of bacteria *per se*, that determines gene expression patterns (Dulla and Lindow, 2008). This has important implications regarding the effect of AHL signaling on bacterial behavior in single or mixed species populations.

Information Box 20.1   Quorum Sensing in a Marine Squid

Quorum sensing was first discovered in the late 1960s during studies on a light-producing marine squid, *Euprymna scolopes* (see top left figure). This tiny nocturnal squid contains specialized organs called light organelles that are colonized by a single luminescent bacterium, *Vibrio fischeri*. The ability of *Vibrio fischeri* to luminesce is contained on an operon (the *lux* operon) that encodes enzymatic machinery that results in the release of photons of light (see signaling pathway figure below right). The first gene in the operon, *luxI*, encodes for an AHL synthase (LuxI) that converts cellular precursors into the AHL signal C6-HSL. Upstream of the *lux* operon is *luxR*, which encodes the transcriptional protein (LuxR) required for activation of high levels of expression of the *lux* operon. In the absence of AHL signal, LuxR is inactive.

At low cell densities, the AHL signals generated by LuxI diffuse passively out of the cell following a concentration gradient. Thus, the *lux* operon is not expressed.

As the bacterial population size increases, the number of C6-HSL signals accumulates in the light organelle and thus within each bacterial cell. When a sufficient concentration of C6-HSL is reached within the bacterial cell, it interacts with the LuxR protein, causing LuxR to dimerize. Dimerization of LuxR allows it to bind to the *lux* operon promoter region and increase expression of the *lux* operon. Note that expression of the operon results in increased levels of LuxI, resulting in even more C6-HSL signal production and ensuring a rapid onset of light production.

Interestingly, the squid can control the amount of light produced by *V. fischeri* in the light organelles either by covering the light organelle with its black ink sac or by reducing the *V. fischeri* population in the light organelle by flushing out excess bacteria with seawater.

Euprymna scolopes, a bioluminescent squid. From the National Science Foundation, 2005a.

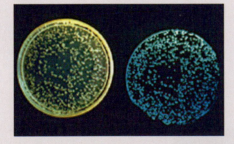

Colonies of *Vibrio fischeri*. (Left) Photo taken under light source. (Right) Photo taken in the dark showing the bacteria luminescing.

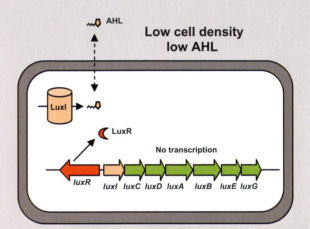

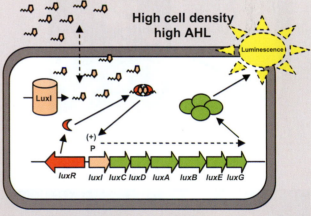

From the National Science Foundation, 2005b.

20.2.2 Interspecies Cross-Signaling

Quorum sensing was originally termed "autoinduction" as it was first identified and studied in single species bacterial communities. More recently, it is recognized that communication via AHL signals occurs between related and unrelated bacterial populations, as well as between bacteria and their eukaryotic hosts. The ability of AHLs to serve as communication signals between species of bacteria (referred to as interspecies signaling, cross-talk or cross-communication) is now widely recognized. One of the first demonstrations of cross-communication utilized the beneficial root-colonizing "rhizosphere" bacterium *Pseudomonas chlororaphis* strain 30-84 (Pierson *et al.*, 1998). *P. chlororaphis* produces three pigmented antibiotics called phenazines. Phenazines are nitrogen-containing broad-spectrum compounds synthesized by the products of the phenazine operon (*phzXYFABCDO*) (Mavrodi *et al.*, 1998). One of the phenazines produced by *P. chlororaphis* is colored bright orange. Phenazine production is regulated, in part, by the PhzR/PhzI quorum sensing system. PhzI is an AHL synthase that produces the AHL C_6-HSL (Table 20.1), and PhzR is the transcriptional regulator that responds to the AHL signal.

To test the hypothesis that *P. chlororaphis* could cross-communicate with other members of the wheat rhizosphere community, 800 culturable bacterial strains from the rhizosphere (the zone surrounding the plant root; Chapter 19) of wheat plants from different U.S. geographic regions were utilized. These were spotted individually onto a lawn of a *phzI* mutant of strain 30-84 that did not produce orange phenazines because it could not produce the C_6-HSL quorum sensing signal. Hence, the *phzI* mutant lawn appeared white. Approximately 8% of the library strains restored phenazine production to the *phzI* mutant as indicated by restoration of orange pigmentation in the lawn, a phenomenon termed positive cross-communication (Figure 20.2).

Of even greater importance, cross-communication between various rhizosphere strains was demonstrated *in situ* on wheat roots using a *phzI⁻*, *phzB::inaZ* reporter of strain 30-84 (Pierson *et al.*, 1998). This nomenclature indicates that the reporter does not produce phenazine as it is defective in the PhzI AHL synthase, and that it has a reporter gene encoding ice nucleation activity inserted within the genomic *phzB* biosynthetic gene (*phzB::inaZ*). For further explanation of reporter genes, see Section 13.6.3. Thus, this reporter expresses ice nucleation activity only when the phenazine biosynthesis operon has been induced. However, since the reporter's *phzI* gene is defective, it does not express the phenazine operon, and so has a 1000-fold decrease in ice nucleation

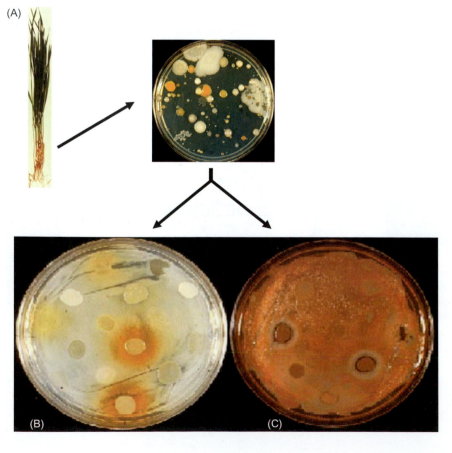

(A)

(B)

(C)

FIGURE 20.2 Positive and negative cross-talk by rhizosphere bacteria and *Pseudomonas chloroaphis*. (A) Collection of wheat rhizosphere strains. (B) Positive cross-talk on a lawn of a *phzI* mutant of *P. chloroaphis* unable to produce phenazines. The spots are rhizosphere isolates that restore various levels of phenazine production via AHL signals. (C) Negative cross-talk on a lawn of wild-type *P. chloroaphis*. The spots are rhizosphere isolates that inhibit phenazine production by producing diffusible signals (Morello *et al.*, 2004). Photos courtesy L.S. Pierson III.

activity as compared to the wild-type $phzI^+$ strain. When the $phzI^-$, $phzB::inaZ$ reporter was grown with several of the 800 unrelated wheat rhizosphere strains, ice nucleation activity by the reporter strain was restored to wild-type levels on roots in soil, demonstrating that quorum sensing was required for phenazine production on roots, and that communication occurred between different bacterial populations via AHL signals! Thus, different bacteria can communicate via AHL signals in a natural environment (the wheat rhizosphere).

A large number of diverse Gram-negative bacteria have been shown to utilize quorum sensing as a key regulatory mechanism. Additionally, many currently noncultatable bacteria also appear to produce quorum sensing signals or related compounds as detected by several different quorum sensing reporter bacteria. This widespread occurrence of quorum sensing in bacteria indicates that quorum sensing plays important roles in bacterial ecology. It is difficult to obtain direct evidence for many of the ecological roles of quorum sensing. Some good examples include the demonstration that quorum sensing defective mutants of bacterial pathogens have reduced ability to persist in the host and to cause disease, and that mutants of beneficial bacteria such as *P. chlororaphis* defective in quorum sensing are impaired in their ability to persist and prevent disease. More recently, quorum sensing has been shown to be critical to the ability of several bacteria to form biofilms on surfaces and tissues (see Lazar, 2011 for a review).

Some of the many possible ecological roles of quorum sensing include: (1) coordination of gene expression within a single bacterial population; (2) coordination of gene expression and bacterial behavior among multiple populations; (3) avoidance of host defense responses; and (4) direct signaling between the bacterium and the host organism (Section 20.4.5).

20.2.2.1 Additional Quorum Sensing Signals

Another class of bacterial signal molecules was identified from the photosynthetic soil bacterium *Rhodopseudomonas palustris* (Schaefer *et al.*, 2008). Similar to the better studied AHL signals, this signal molecule contains a homoserine lactone (HSL) ring. However, in contrast to the fatty acid lipid side chain joined to the HSL in AHLs, this signal utilizes p-coumaroyl, a major monomer component of plant lignin. This signal was named pC-HSL to distinguish it from the classical AHL signals (Table 20.1). Since bacteria do not synthesize p-coumaroyl, it must come from an exogenous source, presumably a lignin-containing plant host. When pC-HSL reaches a threshold concentration in *R. palustris*, several genes involved in bacterial chemotaxis become activated as well as many others for which the function is currently unknown. A fascinating interaction dependent on the levels of pC-HSL quorum sensing

signal occurs between *Emiliania huxleyi*, an environmentally important marine plant-like microalga involved in algal blooms, and *Phaeobacter gallaeciensis*, an α-proteobacterium that colonizes the alga and gains nutrients and a surface to colonize (Seyedsayamdost *et al.*, 2011). As the algal population lives in relatively nutrient-poor marine seawater, it benefits from this association with *P. gallaeciensis*, which produces the growth hormone phenylacetate acid that stimulates algal growth. In addition, *P. gallaeciensis* also produces a broad-spectrum antibiotic called tropodithietic acid that prevents parasitic pathogens from attacking *E. huxleyi*. However, as the algal cells age, lignin in the algal cell wall breaks down, resulting in accumulation of p-coumaric acid, which is in turn converted by *P. gallaeciensis* into pC-HSL. When the level of pC-HSL reaches a threshold concentration, *P. gallaeciensis* activates quorum sensing regulated genes that produce novel algaecides known as roseobacticides. The production of roseobacticides converts *P. gallaeciensis* from a beneficial to an opportunistic pathogen of *E. huxleyi*, causing algal death and release of *P. gallaeciensis* to colonize younger algal cells.

The discovery of this new class of pC-HSL signals has expanded greatly our ideas of potential HSL-based bacterial signals—from molecules synthesized completely from bacterial components to molecules synthesized from bacteria- and plant-derived compounds. This class of signal could integrate the need for a cell density-based quorum of bacteria with a requirement for the presence of a plant host. The identification of this new signal opens up the possibility that there are probably many more novel types of signals yet to be discovered.

20.2.3 Quorum Sensing in *Agrobacterium tumefaciens*, a Ubiquitous Plant Pathogen

Agrobacterium tumefaciens is commonly found in soil and is a plant pathogenic bacterium with an extremely wide host range (>140 plant genera). *A. tumefaciens* causes crown gall disease, so named because the symptoms usually occur at the soil surface or "crown" of the plant (Figure 20.3). Typical disease symptoms include the development of galls, tumor-like growths due to excessive plant cell division at the site of infection. The disease is often easily identified in a variety of dicotyledonous plants, particularly stone fruits, roses and grapes. The ability of *A. tumefaciens* to cause crown gall disease depends on genes necessary for tumor induction that are found within a large 180-kb plasmid called the Ti plasmid (Figure 20.4). This plasmid contains virulence (*vir*) genes required for the processing and transfer of a specific region of the Ti plasmid, known as T-DNA, to the plant. The *vir* genes themselves consist of about 35-kb

FIGURE 20.3 The symptoms of crown gall disease on grapevines caused by *Agrobacterium tumefaciens*. © Queen's Printer for Ontario, 2003. Reproduced with permission.

of DNA, and are essential for tumor formation although they themselves are not transferred into the plant. The induction of the *vir* genes occurs following exposure to signal molecules synthesized by the plant in response to wounding. This explains why crops that rely on root cuttings are particularly susceptible to crown gall disease. One of the signal molecules has been identified as the phenolic compound acetosyringone. This molecule plus sugar monomers, which are precursors of the plant cell wall, are sensed by *Agrobacterium* through the *virA* and *virG* genes, which control the expression of all other *vir* genes. The *virA* gene produces a protein located in the cell wall that appears to sense the phenolic compound directly. This protein has a cytoplasmic domain that becomes activated and in turn activates the cytoplasmic VirG protein, which subsequently activates all other *vir* genes.

Following transfer of the T-DNA from the bacterium to the plant, the T-DNA is targeted to the plant nucleus where it is integrated into a plant chromosome where it codes for the synthesis of two plant growth regulators, auxin and cytokinin, as well as for a group of amino acid derivatives known as "opines." It is fascinating that the promoters that drive the expression of these genes are closely related to eukaryotic promoters. The constitutive synthesis of these plant growth hormones gives rise to the symptoms of crown gall disease. The Ti plasmid in the

A. tumefaciens on the roots also contains genes that allow the bacterium to utilize the unusual opine amino acids now being produced by the plant cells as a food source. In essence, the bacterium "engineers" the plant to produce a novel food source and to undergo multiple rounds of cell division to increase the plant surface area for bacterial colonization and opine production.

So how does quorum sensing fit into *A. tumefaciens* infection? During the initial rapid growth of the bacterium inside wounded plant tissues, some bacterial cells inadvertently lose the Ti plasmid and their ability to utilize opines. The Ti plasmid contains a quorum sensing system comprised of *traR* and *traI*. This quorum sensing-based strategy is used to ensure that the plasmid is maintained in the population. This is based on maintaining rapid rates of plasmid transfer via conjugation to any cells that may have lost the plasmid (Figure 20.5). Thus, activation of plasmid conjugation by opine-induced quorum sensing control serves to ensure that all members of the *A. tumefaciens* community contain a copy of the Ti plasmid and are able to utilize opines for growth.

20.3 SIGNALING IN GRAM-POSITIVE BACTERIA

Gram-positive bacteria do not utilize the AHL-mediated quorum sensing communication systems found in Gram-negative bacteria. One possible reason is that Gram-positive bacteria lack a porous outer membrane, and instead contain a thick peptidoglycan layer which may restrict diffusion of AHL signals through the cell wall. Instead, some Gram-positive bacteria utilize γ-butyrolactones, molecules that have some structurally similarity to AHLs, to regulate specific gene expression in a cell density-dependent manner. However, the majority of Gram-positive bacteria utilize small peptides as their primary communication molecules.

20.3.1 Gamma-Butyrolactones

The first bacterial communication signals discovered in the 1960s were the gamma-butyrolactones (γ-butyrolactones) produced by *Streptomyces* spp. These organisms are Gram-positive soil bacteria (actinomycetes) that undergo cellular differentiation and are known to produce many secondary metabolites. In fact, many antibiotics in use today are derived from *Streptomyces* spp. (see Section 19.4). A handful of γ-butyrolactones have been purified from different *Streptomyces* species (Table 20.1). These signaling compounds superficially resemble AHLs, and, analogously to AHLs, they differ in their stereochemistry, the length of their fatty acid side chains and side branch number. Both γ-butyrolactones and AHLs are biologically active at

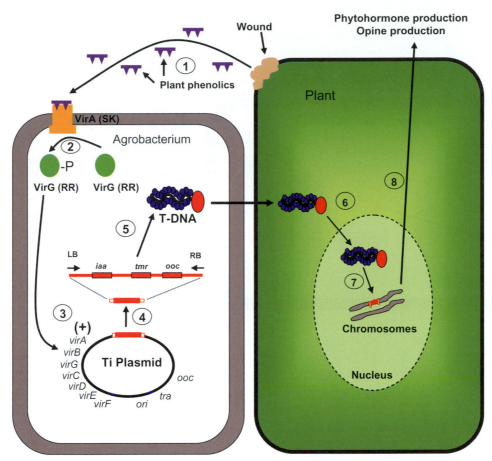

FIGURE 20.4 The plant—*A. tumefaciens* interaction that results in crown gall formation. In soil, *A. tumefaciens* is attracted to plant roots due to their release of root exudates. Wounded plant roots release additional phenolic compounds such as acetosyringone (step 1). Acetosyringone is recognized by a two-component regulatory system (VirA/VirG) encoded by the Ti plasmid as a signal that a plant wound is present (step 2). This recognition induces the expression of a complex region of the Ti plasmid called the *vir* region (for virulence) (step 3). The *vir* region encodes several proteins that interact with a 25 kb T-DNA region on the Ti plasmid. Some of the *vir* gene products are responsible for excising a copy of the T-DNA (step 4) while others encode a type four secretory system (T4SS) that is involved in transferring the copy of the T-DNA across the bacterial and plant cell walls and into the plant cell cytoplasm (steps 5 and 6). The end result is the random insertion of the T-DNA into one of the plant chromosomes (step 7). The integrated T-DNA contains genes that encode for the production of plant hormones and genes that encode enzymes for the production of opines, unusual amino acid derivatives, by the plant cells (step 8). The T-DNA-directed production of opines provides a unique carbon and nitrogen food source for the growth of *A. tumefaciens*. The T-DNA-directed production of growth hormones results in uncontrolled cell division, and the development of the symptoms typical of *A. tumefaciens* infection, plant galls.

extremely low concentrations at the nano- to micromolar level. However, despite these similarities, Gram-negative AHL receptors do not respond to γ-butyrolactones and vice versa, indicating that each sensory system is specific to its own signal type. Hence, AHLs and γ-butyrolactones represent different "languages" used by bacteria. Recall that most AHLs activate target gene expression by altering the affinity of a transcriptional regulatory protein that binds to a promoter region and recruits RNA polymerase to stimulate gene expression. In contrast, γ-butyrolactones usually act by alleviating repression of gene expression.

In other words, they cause a repressor protein to dissociate from the promoter region of the target gene(s), which results in subsequent gene expression.

The compound A-factor (2-isocapryloyl-3*R*-hydroxy-methyl-γ-butyrolactone) made by the soil bacterium *Streptomyces griseus* was the first γ-butyrolactone identified, and is the best known example of this type of cell density-dependent signaling. The A-factor stimulates aerial mycelium formation and production of the antibiotic streptomycin by regulating the expression of the transcriptional activator AdpA (Figure 20.6).

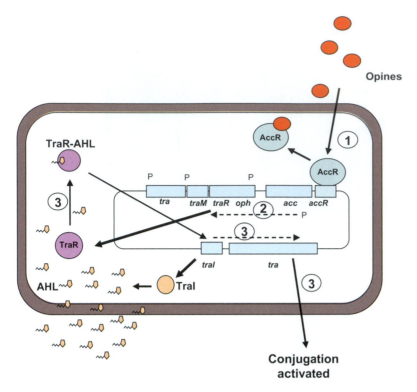

FIGURE 20.5 Quorum sensing control of *A. tumefaciens* infection. *A. tumefaciens* needs to maintain the Ti plasmid during the plant infection process. Therefore, the Ti plasmid contains genes (*tra* genes for transfer) involved in conjugation and transfer of the Ti plasmid back into *A. tumefaciens* cells that may have lost the plasmid. Conjugation is regulated by the TraR/TraI quorum sensing system, a classic quorum sensing regulatory system in which TraR is stimulated by the AHL C8-HSL signal produced by TraI. Uniquely, the TraR/TraI system is only active in the presence of plant-provided opines. In the absence of opines, a repressor protein, AccR, binds within the opine promoter regions blocking the expression of the genes required for uptake and catabolism of opines. When opines are present, however, they bind to AccR (step 1), causing it to dissociate from the promoters and allowing expression of the opine uptake and catabolic regions (step 2). The *traR* promoter region also contains an AccR-binding sequence (step 2). The consequence of this is that TraR is produced only when opines are present. When present, TraR recognizes the *A. tumefaciens* AHL signal and activates bacterial conjugation (step 3).

20.3.2 Peptide Signaling

Most Gram-positive bacteria utilize a variation of a quorum sensing system that incorporates a two-component regulatory system (see Information Box 20.2) to form a combined system that has been termed a three-component quorum sensing system. This combined system consists of a cell membrane-localized histidine kinase (HSK) sensor protein and a cytoplasmic response-regulator protein (RR), coupled to an autoinducing peptide (AIP) secreted by the producing cell (Lyon and Novick, 2004).

The ubiquity of this type of signaling is exemplified by the Gram-positive foodborne pathogen *Staphylococcus aureus*. This microbe's genome contains approximately 17 putative two-component systems (Rasmussen *et al.*, 2000), all of which are believed to be involved in bacteria—bacteria or bacteria—environment signaling! Perhaps the best-studied three-component quorum sensing system is the regulation of exotoxin production by *S. aureus*. These heat-stable exotoxins, including toxic shock syndrome toxin 1 [TSST-1], cause illness in animals and humans (Diggle *et al.*, 2003). As shown in Figure 20.7, *S. aureus* utilizes a cell-density sensing mechanism to activate virulence gene (exotoxin) expression, while simultaneously repressing surface factors to avoid host detection.

A second example of Gram-positive cell—cell signaling includes a group of bacteria known as probiotics.

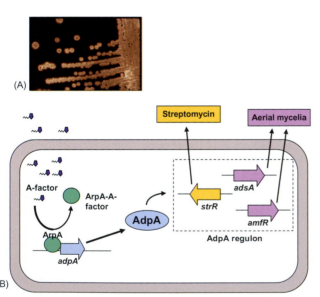

FIGURE 20.6 Quorum sensing in *Streptomyces griseus*. (A) *S. griseus* colonies on an agar plate. (B) Mode of action of A-factor. When A-factor levels are low, the ArpA repressor protein is bound to the promoter region of *adpA* blocking its expression. Once A-factor has accumulated to a threshold concentration within the cell, usually at the mid-exponential phase of growth, it binds to the promoter-bound ArpA. This alters the conformation of ArpA resulting in dissociation from the *adpA* promoter and AdpA production. AdpA subsequently induces the expression of a number of genes (the Adp regulon) including *strR*, which encodes an activator of streptomycin production, and *amfR* and *adsA*, which encode activators for aerial mycelium formation.

Information Box 20.2   Two-component Regulatory Systems

Two-component regulatory systems comprise a sensor protein and a response-regulator (RR) protein. The sensor protein is normally located within the cell's outer membrane and can detect changes in the external environment surrounding the cell. The sensor protein then communicates these changes to the

response-regulator protein inside the cell. The response-regulator protein in turn regulates the expression of key genes to allow an appropriate response to the external stimulus. Communication between the sensor protein and the response-regulator protein is via phosphorylation–dephosphorylation reactions.

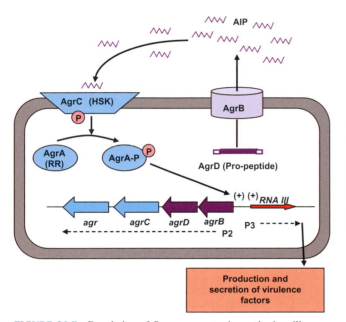

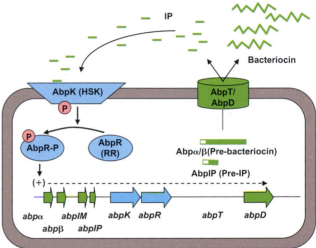

FIGURE 20.8   Probiotic production is regulated by an auto-inducing peptide in *L. salivarius*. Each *L. salivarius* cell produces a small amount of a pre-inducing peptide (AbpIP) that is processed and secreted into the environment by the ABC transporter complex AbpT/AbpD as inducing peptide (IP). As the numbers of *L. salivarius* increase, the level of IP increases until it is recognized by the HSK AbpK that, in turn, phosphorylates AbpR. Phosphorylated AbpR up-regulates production of the pre-bacteriocin genes (*abp118α* and *abp118β*). These encode enzymes that synthesize the pre-bacteriocin that is subsequently processed and secreted by the AbpT/AbpD complex and that then inhibits harmful bacteria.

FIGURE 20.7   Regulation of *S. aureus* exotoxin synthesis utilizes an auto-inducing peptide (AIP) signal. The *agr* locus encodes four proteins (AgrA, AgrB, AgrC and AgrD). The AgrD protein is an AIP that is processed and secreted by AgrB, a membrane-associated protease. As the bacterial cell density increases, the AIP concentration accumulates. When AIP reaches a critical level, it binds to the AgrC/AgrA two-component regulatory system. AgrC is a cell membrane-localized histidine sensor kinase (HSK) that, when bound to AIP, transfers a phosphate group to AgrA, a cytoplasmically-localized response regulator (RR). Phosphorylated AgrA (AgrA-PO$_4$) activates transcription from a promoter (P3) that expresses a small noncoding regulatory RNA (RNA III). RNA III is involved in the activation of a number of virulence genes, including those involved in production and secretion of several exoproteins, enterotoxins, exfoliatins, hemolysins, leukocidins and lipases. Additionally, AgrA-PO$_4$ serves to repress the expression of several bacterial surface proteins. Because cell surface components are often the triggers for host defense responses, the repression of expression of these surface proteins might assist the bacterium in evading recognition by the host.

studied probiotic that colonizes the human intestine and produces a broad-spectrum bacteriocin effective against a number of foodborne and medically important bacterial pathogens (Flynn *et al.*, 2002). *L. salivarius* utilizes a three-component regulatory system to control bacteriocin production (Figure 20.8).

20.4 OTHER TYPES OF SIGNALING

20.4.1 Universal Signals Autoinducer-2 and Autoinducer-3

In 1997 a novel type of universal bacterial signal was reported that is quite different from AHL (Gram-negative) and peptide (Gram-positive) signals (Xavier and Bassler, 2003). This new class of signal, termed

Probiotics are intestinal bacteria that exert positive effects on the health of the human or animal host by interfering with the ability of deleterious bacteria to colonize (Guarner and Schaafsma, 1998). Probiotic bacteria inhibit colonization via the production of extracellular peptides known as bacteriocins (Riley and Wertz, 2002). *Lactobacillus salivarius* UCC118 is an example of a well-

Autoinducer 2 (AI-2), is a family of related furanosyl-borate diester molecules that is produced by over 55 Gram-positive and Gram-negative bacteria (Table 20.1). All of these bacteria contain a synthase gene named *luxS* that if inactivated results in the loss of AI-2 production. Scientists are just beginning to understand the AI-2 signal. So far, AI-2 has clearly been shown to be involved in signaling in two species, *Salmonella enterica* serovar Typhimurium and *Vibrio harveyi* (see Information Box 20.3) (reviewed in Vendeville *et al.*, 2005). A second potential universal bacterial signal, Autoinducer-3 (AI-3), was reported in 2003 (Sperandio

---

Information Box 20.3   *Vibrio harveyi* **Bioluminescence**

The bioluminescent bacterium *Vibrio harveyi* has a well-characterized AI-2 regulatory system. Similar to *V. fischeri*, *V. harveyi* can colonize a number of marine organisms and also produces light via a bioluminescence operon (*lux* operon). *V. harveyi* also can exist in high numbers in a free-living state. Fascinatingly, *V. harveyi* has been implicated as the causative agent of milky seas, a phenomenon in which *V. harveyi*-generated bioluminescence can cover areas of the ocean the size of Connecticut. These milky seas have been observed by merchant vessels and are visible from space (see figure on left, from Miller *et al.*, 2005). However, *V. harveyi* regulates light production much differently than *V. fischeri* (Information Box 20.1). *V. harveyi* contains a LuxR regulatory protein required for *lux* operon activation, but the LuxR protein does not require an AHL signal. Instead, in

V. harveyi, there is a reversible phosphorylation cascade involving four proteins, LuxP, LuxQ, LuxU, and LuxO. In the absence of AI-2, there is an induction of a signal cascade mechanism from the periplasmic protein LuxP to the cytoplasmic proteins LuxQ, LuxU, and LuxO. Phosphorylated LuxO (LuxO-PO$_4$), in conjunction with the sigma factor RpoN (σ^{54}), results in activation of a small noncoding regulatory RNA (sRNA) that results in the degradation of the *luxR* mRNA and therefore no luminescence. Alternatively, when sufficient AI-2 is present, it causes the phosphorylation cascade to go from LuxO to LuxQ, resulting in dephosphorylation of LuxO. In this instance, LuxO is inactive and the luxR mRNA transcript is protected, resulting in production of luminescence.

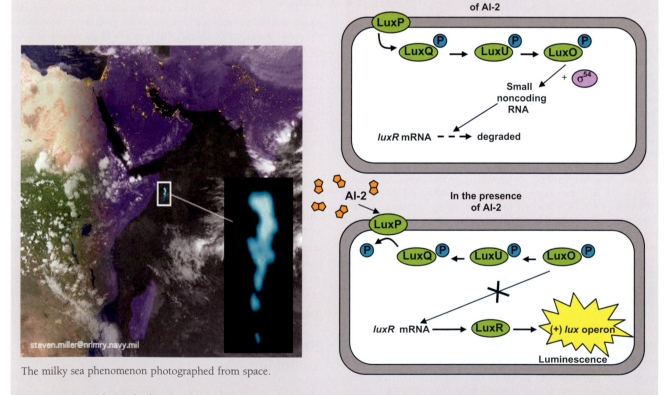

The milky sea phenomenon photographed from space.

From: http://www.lifesci.ucsb.edu/~biolum/organism/milkysea.html. Steve Miller (2005)· Naval Research Lab. Used with permission.

et al., 2003). AI-3 is chemically distinct from AI-2, although its exact structure is not yet known.

20.4.2 Bacterial Muropeptides

Although the exact structure of bacterial cell walls can vary among different organisms (i.e., Gram-negative versus Gram-positive bacteria; Figure 2.3), all bacterial cell walls contain peptidoglycan as a major component. This peptidoglycan layer is composed of repeating disaccharide subunits of β(1-4)-*N*-acetylglucosamine–β(1-4)-*N*-acetylmuramic acid. While the basic disaccharide subunits are conserved, there can be large variations in the specific structures and chain lengths of these peptidoglycan components. Enzymatic digestion of the peptidoglycan layer results in the generation of fragments referred to as muropeptides (Boudreau *et al.*, 2012). Because the integrity of the cell wall is of paramount importance for survival, bacterial cell walls have characteristics that make them more recalcitrant to degradation. For example, bacterial cell walls contain unique components, such as the presence of D-amino acid stereoisomers (e.g., D-alanine), which are rarely found in other organisms. This is why many antibiotics used to treat bacterial infections target the cell wall, which makes them specific to prokaryotes. Recently, muropeptides have been recognized as signals utilized by bacteria and their hosts. For example, bacteria use the detection of muropeptides to signal that cell wall-degrading antibiotics are present; bacteria that form long-term survival structures such as spores use muropeptides as germinants to trigger vegetative growth; and eukaryotic organisms can use muropeptides to sense that a bacterium is present within host tissues.

20.4.2.1 Gram-Negative Bacterial Muropeptide Sensing

Growth and cell division of Gram-negative bacteria requires the constant synthesis and turnover of peptidoglycan. Peptidoglycan synthesis involves the formation of a cytoplasmic lipid precursor known as Lipid II that is transported through the cytoplasmic membrane into the periplasmic space, where it interacts with a group of proteins collectively known as penicillin-binding proteins (PBPs). Existing peptidoglycan is rapidly turned over during cell growth and division. One product of the breakdown of peptidoglycan is the formation of anhydromuropeptides. These can leave the cell, but are transported across the membrane into the cytoplasm by the enzyme AmpG permease. Inside the cell, the muropeptide is further broken down and recycled for peptidoglycan synthesis. Although peptidoglycan recycling occurs in Gram-negative bacteria, loss of this pathway does not result in reduced cell growth or any obvious phenotypic

deficiency, suggesting that this recycling pathway plays another role for the cell. It was observed that reductions in available muropeptide levels, similar to the result of the addition of β-lactam antibiotics that degrade the peptidoglycan layer, results in the induction of resistance pathways to this class of antibiotics. The current hypothesis is that the relative concentrations of muropeptides synthesized *de novo*, versus transported into the cytoplasm by the AmpG permease, control the expression of a regulatory protein known as AmpC. Reductions in *de novo* muropeptide synthesis increase *ampC* transcription that results in the activation of a series of β-lactamases that convey high level resistance to β-lactam antibiotics.

20.4.2.2 Gram-Positive Muropeptide Sensing

Gram-positive bacteria such as *Staphylococcus aureus* have cell walls that lack an outer membrane. Instead, they have a single cytoplasmic membrane surrounded by a thick exposed peptidoglycan layer. Antibiotics effective against Gram-negative bacteria are often ineffective against Gram-positive bacteria, attributed partly to the presence of this thicker peptidoglycan layer. One exception to this is the glycopeptide vancomycin, which is effective against Gram-positive bacteria. Vancomycin binds to the peptidoglycan biosynthetic precursor Lipid II at the −D-Ala-D-Ala- stem. Resistance to vancomycin occurs by two mechanisms. The first, identified in enterococcal Gram-positive lactic acid bacteria, has the −D-Ala-D-Ala- moiety of Lipid II replaced with −D-Ala-D-Lac-, which has a lower affinity for vancomycin. The multidrug-resistant pathogen known as methicillin-resistant *S. aureus* (MRSA) contains a thicker and more cross-linked peptidoglycan believed to confer increased vancomycin resistance as compared to *S. aureus*. MRSA is often acquired in hospitals, where β-lactam antibiotics are commonly used to control Gram-negative bacteria (which usually co-occur with *S. aureus*). However, the combination of β-lactam antibiotics in conjunction with vancomycin can result in MRSA strains with high levels of vancomycin resistance. This is due to the fact that β-lactam antibiotics act by mimicking the −D-Ala-D-Ala- moiety of Lipid II, and by binding irreversibly to the PBPs involved in peptidoglycan biosynthesis. The resulting high-level vancomycin resistance is hypothesized to be the combined result of the replacement of the susceptible PBP by a spontaneous variant that does not bind to β-lactam antibiotics, and the replacement of −D-Ala-D-Ala- with −D-Ala-D-Lac- in Lipid II, which does not bind vancomycin efficiently. These β-lactam-/vancomycin-resistant MRSA derivatives are known as BIVR-MRSA (β-lactam-induced vancomycin-resistant-MRSA). Interestingly, BIVR-MRSA cells release more muropeptides into the surrounding medium than *S. aureus* or MRSA cells. Specifically, analysis of culture filtrates has identified high levels of a specific muropeptide (GlcNAc-MurNAc-L-Ala-D-*i*Gln-L-Lys-(e-Gly$_4$)-D-Ala-Gly$_2$). When this purified muropeptide was added to cultures of BIVR-MRSA cells in the absence of β-lactam antibiotics, the cells grew much more rapidly than vancomycin-only treated cells. Currently, the basis for this enhanced growth rate is unclear, but is problematic to the control of MRSA infections.

Gram-positive bacteria were originally thought to lack a mechanism for peptidoglycan recycling. This was thought to be due to the lack of an enclosed periplasmic space, since Gram-positive bacteria lack an outer membrane. Additionally, up to half of the peptidoglycan layer appeared to be released during cell growth (Mauck *et al.*, 1971). However, several homologues of genes for muropeptide-recycling have been identified in *Bacillus subtilis* (Litzinger *et al.*, 2010). Hence, it is possible that Gram-positive bacteria could also alter transcriptional patterns resulting in drug resistance in response to perturbations in peptidoglycan recycling.

One unique aspect of several Gram-positive bacteria is the ability to form long-term survival structures known as spores, in response to adverse environmental conditions, such as limited nutrient availability. The regulation of the development of these resting structures is highly ordered and temporally complex. We know even less about how these metabolically minimized structures sense their surroundings, and the molecular mechanisms involved in initiation of germination. During spore formation, the peptidoglycan undergoes multiple structural alterations that result in enhanced structural integrity and resistance to degradation. Germination of dormant spores involves multiple sensory pathways. One of these sensory pathways involves muropeptides. These muropeptides have been classified as spore germinants, and are believed to bind to cytoplasmic membrane-bound, eukaryotic-like serine/threonine kinases (STKs). Synthetic muropeptides have been identified that also bind to spore STKs. It has been shown that *Bacillus subtilis* spores contain a PrkC STK that responds to one muropeptide, while spores of *S. aureus* contain a similar PrkC STK, but respond to a different muropeptide. Replacement of the *B. subtilis* PrkC with the *S. aureus* PrkC STC resulted in *B. subtilis* spores that only germinated in the presence of the *S. aureus* muropeptide. Recognition of the specific muropeptide occurs via a penicillin-binding-associated and serine/threonine kinase-associated (PASTA) domain in the PrkC STK. Of additional interest is that the PrkC STKs respond preferentially to muropeptides produced by growing rather than non-growing cultures, indicating that the structures of muropeptides change dependent on the growth stage of the cells.

20.4.2.3 Recognition of Bacterial Muropeptides as Part of Eukaryotic Host Immunity

Humans, animals and plants contain evolutionarily conserved innate immunity systems. This innate immunity utilizes pattern-recognition receptors (PRRs) that become activated upon binding to highly conserved bacterial structures known as microbial-associated molecular patterns (MAMPs). These receptors help the host recognize "nonself" molecules. Examples of bacterial structures recognized by PRRs include flagella, lipopolysaccharides (LPS) and muropeptides. These have also been referred to as PAMPs (pathogen-associated molecular patterns) when used in reference to pathogenic bacteria. There are three basic classes of PRR receptors: the extracellular Toll-like receptors; the nucleotide-binding domain/leucine-rich repeat receptors (NLRs); and the retinoic acid-inducible gene receptors (RLRs). In addition to MAMP recognition by plant and animal hosts (termed pathogen-triggered immunity, or PTI), plants recognize strain-specific pathogen protein effectors that are transported into the host cell during the interaction (termed effector-triggered immunity, or ETI). There is considerable cross-talk among PRRs, probably to ensure that an immune response is generated only when the combined inputs of multiple PRRs reach a threshold value. This may increase the ability of the host to distinguish between the presence of the normal beneficial microbiome and the presence of pathogens.

In animal cells, the two NLR receptors (NOD1 and NOD2) recognize different specific muropeptide structures generated from the enzymatic degradation of peptidoglycan. Interestingly, one muropeptide called muramyl dipeptide (MDP) is produced from enzymatic degradation of both Gram-negative and Gram-positive bacterial peptidoglycan, and has been used as an immune adjuvant for over 40 years. Animal cells also contain a second class of muropeptide receptors called peptidoglycan-recognition proteins (PGRPs), which differ from NOD receptors in that they recognize polymeric peptidoglycan.

20.4.3 Eavesdropping on the Party Line

Although many bacteria produce signals, many other bacteria do not produce signals (as far as we know) but still have the ability to "listen in" on the conversations. This eavesdropping may allow an incoming bacterium to detect the metabolic state of the community to determine whether expression of specific traits is appropriate. For example, there is evidence that *Pseudomonas aeruginosa* listens in on the indigenous microbial community during the infection process. *P. aeruginosa* is an opportunistic pathogen that is problematic for immunocompromised patients, and it is a primary cause of morbidity and mortality in patients suffering from cystic fibrosis, a hereditary life-threatening childhood disease. This ubiquitous Gram-negative bacterium primarily colonizes the lungs in cystic fibrosis patients, where it exacerbates mucus formation.

Enteric bacteria, which include the genera *Escherichia*, *Salmonella*, *Klebsiella*, *Enterobacter* and *Citrobacter*, all contain a single LuxR homologue named SdiA, but lack a corresponding LuxI homologue (Soares and Ahmer, 2011). Genes such as SdiA are considered orphan quorum sensing regulators. Although these bacteria cannot produce AHLs, they respond to a broad range of exogenous AHL signals produced by other bacteria. For example, *Salmonella enterica* serovar Typhimurium, a major cattle and poultry pathogen responsible for salmonellosis, uses a SdiA regulator that recognizes exogenous AHL signals to activate genes in its chromosome (*srgE*), and an operon (*rck*) contained on a virulence plasmid that encodes for resistance to the host immune response and for production of factors such as pili involved in pathogenesis. In some cases, these orphan LuxR genes encode proteins that form dimers with the known LuxR protein making them inaccessible to AHL binding. This may represent a mechanism to prevent premature expression of pathogenicity traits that would induce a host defense response. However, we are only beginning to touch the tip of this fascinating "iceberg."

20.4.4 Quorum Quenching and Quorum Sensing Inhibition

If the accumulation of AHL signals confers a competitive advantage on a microorganism, then other organisms might develop mechanisms to thwart this advantage by breaking down the AHL signal or otherwise interfering with the signaling system. This phenomenon was first termed "quorum quenching" (Dong *et al.*, 2001). More than 20 genera of bacteria are known to degrade AHL signals (Uroz *et al.*, 2009). The ability to breakdown AHL signals occurs primarily by two enzymatic activities—that of AHL lactonases that open the homoserine lactone (HSL) ring, and AHL acylases that cleave the fatty acid sidechain from the HSL. Although *P. aeruginosa* uses AHL signaling to regulate the expression of genes required for successful pathogenesis, it also has three acylase enzymes that degrade AHL signals (Huang *et al.*, 2003). How the concomitant synthesis and turnover of AHL signals influences pathogenicity is currently under study. Some nitrogen-fixing *Sinorhizobium* spp. that colonize and nodulate legume roots (e.g., alfalfa, pea) (Chapter 19) contain at least six AHL-degrading enzymes (Krysciak *et al.*, 2011). Some isolates of *Bacillus cereus*, a Gram-positive soil bacterium that does not itself produce AHL signals, encode a lactonase, reducing the activity of AHL signals 1000-fold. The *Bacillus* spp. gene *aiiA*, which encodes the lactonase, was cloned into potato and shown to reduce virulence by the AHL-dependent soft rot pathogen *Erwinia caratovora* (Dong *et al.*, 2001). In comparison, a plant pathogenic *Ralstonia* spp. was

shown to produce a bacterial AHL acylase (Lin *et al.*, 2003). The gene encoding this enzyme, *aiiD*, was cloned and shown in *E. coli* to inactivate several AHL signals by cleaving the homoserine lactone ring from the fatty acid side chain.

Degradation of AHL signals probably is important for more reasons than reducing a competitor's advantage. It is possible that without AHL turnover, the accumulation of AHLs would render them poor signaling molecules (Diggle *et al.*, 2006). Additionally, we know that many pathogenic bacteria utilize quorum sensing via AHL signals to control the production of pathogenicity proteins required for successful infection and disease. Premature expression of these genes would trigger strong host responses that would interfere with the ability of the pathogen to invade successfully. Therefore, degradation of AHL signals would prevent this early expression or "giving one's position away," until sufficient bacterial numbers had accumulated to successfully infect the host.

There is quite a range of variation among hosts with respect to the ability to degrade AHL signals. For example, the legume alfalfa degrades AHL signals by hydrolysis (Götz *et al.*, 2007), while other plants such as *Arabidopsis* and barley cannot degrade AHLs. AHLs are quite stable on *Arabidopsis* roots, and are taken up by the plant and transported to upper plant parts (von Rad *et al.*, 2008). In addition to AHL degradation, some plants also produce compounds that interfere with bacterial quorum sensing. L-Canavanine, an arginine analogue produced by alfalfa, blocks activation of *Sinorhizobium* spp. quorum sensing, which results in loss of production of an exopolysaccharide (EPSII) on the surface of the bacterial cell that is required for colonization of alfalfa roots (Keshavan *et al.*, 2005). However, why a plant host would want to block colonization by a potentially beneficial nitrogen fixing root symbiont is unclear.

Some organisms utilize quorum quenching to block colonization by bacteria. A well-studied example is the marine red alga *Delisea pulchra* (Rasmussen *et al.*, 2000). This alga produces a halogenated furanone [(5Z)-4-bromo-5-bromomethylene-3-butyl-2(5H)-furanone] (Table 20.1) that interferes with AHL quorum sensing by binding to the AHL receptor protein LuxR which results in its accelerated turnover in the cell. This halogenated furanone has broad effects on quorum sensing as it was also shown to inhibit the AI-2 quorum sensing system in *E. coli* (Ren *et al.*, 2001).

In other cases, hosts may encourage colonization by beneficial bacterial species that degrade AHL signals produced by pathogenic bacteria. An example is the bacterium *Bacillus* isolate QSI-1 isolated from the intestine of Prussian carp (*Carassius auratus gibelio*), the wild form of goldfish (Chu *et al.*, 2010). Strain QSI-1 produces an enzyme that degrades AHLs. When strain QSI-1 was fed to fish, it significantly reduced the ability of the

FIGURE 20.9 Production of the purple pigment violacein by an AHL reporter derivative of strain *Chromobacterium violaceum* that lacks its endogenous LuxI AHL synthase. *C. violaceum* strain CV026 (Teplitski *et al.*, 2000) was streaked in a V-pattern on the right side of the plate, and an *E. coli* strain that contains a plasmid with an AHL synthase gene was streaked in a V-pattern on the left. Recognition of sufficient AHL produced by the *E. coli* strain is visible by purple violacein production by CV026. Photo courtesy L.S. Pierson III.

freshwater fish pathogen *Aeromonas hydrophila* to cause infection, consistent with the hypothesis that degradation of the AHL signal required by the pathogen resulted in the protection of the fish. The authors proposed that strain QSI-1 is a probiotic for aquaculture.

Several plants also produce AHL signal mimics that activate or inhibit bacterial quorum sensing (Teplitski *et al.*, 2000). The AHL reporter strain *Chromobacterium violaceum* only produces a purple pigment (violacein) if supplied with exogenous AHL signal (Figure 20.9). Pea seedling root exudates were identified that blocked activation of violacein production by added AHL in *C. violaceum*. In contrast, these pea root exudates activated several other bacterial AHL reporters. The model legume *Medicago truncatula* produces a collection of root exudate compounds that also differ in their effect on quorum sensing depending on the reporter system used (Gao *et al.*, 2003). The unicellular alga *Chlamydomonas reinhardtii*, which is found in soil and fresh water, produces a number of compounds that also affect quorum sensing systems differentially (Teplitski *et al.*, 2004). Multiple biosensor strains are facilitating the identification of quorum sensing-inhibiting compounds, and several of these compounds have promise as treatments for reducing bacterial pathogenicity.

20.4.5 Interkingdom Signaling: Sociobiology and the Concept of Holobionts

Eukaryotic hosts and their associated bacteria have evolved complex mechanisms of both signaling and signal perception in order to monitor each other's status. As we saw earlier in this chapter, bacteria utilize a wide range of small signaling molecules to control the expression of traits important for their optimal growth and survival under different environmental conditions. Known

bacterial signals include AHLs, AI-2, AI-3, oligopeptides and muropeptides (cell wall components). The role of quorum quenching in modulating the effects of quorum sensing on the expression of key traits has also been discussed.

The discovery of bacterial signaling has led to a paradigm shift in the way we view the microbial world. It was first believed that quorum sensing allowed a single population of bacteria to coordinate group-specific behaviors such as colonization, light production and pathogenicity by controlling gene expression in response to specific signals (intraspecies communication). This concept alone was revolutionary as it meant that bacteria did not react or behave as single, isolated cells, but instead were capable of organizing their behaviors coordinately, analogous to a multicellular organism. Next it was shown that unrelated bacteria could communicate (cross-talk) with each other via signals, and that some bacteria evolved the ability to eavesdrop on these conversations (interspecies communication). It is now recognized that this signaling communication can occur between bacteria and their eukaryotic hosts, a phenomenon known as interkingdom signaling (Rumbaugh, 2007). Interkingdom signaling is defined as the exploitation of signal transduction pathways by the signaling compounds of one organism to alter the behavior, through changes in gene transcription, of an organism from a different kingdom. Interkingdom communication can result in: (1) the ability of bacteria to eavesdrop on their host by responding to host signals; (2) interference of bacterial signaling via host-produced interfering signals; or (3) recognition of bacterial signals by the host resulting in altered host gene expression, such as that required for an immune response.

This section discusses signaling between prokaryotic bacteria and their eukaryotic hosts, i.e., interkingdom signaling. As we will see, signaling occurring between hosts and their associated microorganisms influences the outcomes of the interactions, whether beneficial or detrimental, to both members. Evidence is accumulating that bacterial signals, such as AHLs, can serve as bacterial hormones that modify patterns of gene expression in the eukaryotic host, and serve to alter the host's fitness in its current environment.

20.4.5.1 Signaling Between Bacteria and Fungi

In nature, the majority of bacteria live in mixed communities with other bacteria and fungi, and many bacteria are known to actively colonize fungal hyphae. Not surprisingly, communication among bacteria and fungi can affect their behaviors. For example, an AHL produced by the opportunistic human pathogen *P. aeruginosa* induces morphological changes in the human-associated yeast *Candida albicans* (Hogan *et al.*, 2004). Specifically, McAlester *et al.* (2008) showed that *C. albicans* responded to the bacterial AHL 3-oxo-C_{12}HSL by growing only in the yeast form that is resistant to killing by *P. aeruginosa*. In return, the fungal-produced metabolite farnesol affected *P. aeruginosa* by causing reduced levels of the *Pseudomonas* quinolone signal (PQS) and pyocyanin (Cugini *et al.*, 2007).

20.4.5.2 Signaling Between Bacteria and Plants

Since microorganisms were the first life form on Earth, plants evolved in their presence, and it is logical that these groups of organisms evolved mechanisms of communication that are only now being recognized. Research has shown that plants change patterns of gene expression, and alter developmental pathways in response to bacterial AHL signals.

Communication between plants and bacteria was first studied extensively in the beneficial *Rhizobium* spp.–legume symbiotic association (reviewed in Brencic and Winans, 2005). Studies have documented the exquisite signal communication that occurs by both the bacteria and plant, resulting in colonization, infection, nodule development and nitrogen fixation (Information Boxes 16.2 and 16.3; Section 19.3.2.2). Communication between the plant pathogen *Agrobacterium tumefaciens* and its host has also been studied extensively, and was discussed in Section 20.2.3. Both of these plant-associated bacteria interact with their plant hosts by invading plant tissues, the former beneficially while the latter as a pathogen. However, the majority of plant-associated bacteria exist by colonizing the plant surface. Recent work suggests that interkingdom communication between hosts and microbes is both widespread and occurs bi-directionally.

In other examples, AHL signals can alter the expression of plant defense genes, a phenomenon known as induced systemic resistance (ISR) in plants. For example, AHL produced during root colonization by *Serratia liquefaciens* MG1 induced increased resistance against the fungal leaf pathogen *Alternaria alternata* in tomato plants (Schuhegger *et al.*, 2006). The AHL signal oxo-C_{14}-HSL, when added to *Arabidopsis thaliana*, increased its ability to resist infection by *Pseudomonas syringae* pv. tomato DC3000 (Schikora *et al.*, 2011). Pretreatment of plants with this AHL also blocked infection by the biotrophic fungal pathogens *Golovinomyces orontii* (on *Arabidopsis*) and *Blumeria graminis* (on barley). AHLs produced by *Serratia plymuthica* protected cucumber seedlings from the fungal damping-off disease caused by *Pythium aphanidermatum*, and tomato and bean from *Botrytis cinerea*.

However, not all AHL impacts on plants are beneficial. For example, treatment of *Nicotiana attenuata* by C_6-HSL resulted in reduced production of a protease inhibitor in leaves, resulting in increased herbivory of the leaves by the tobacco hornworm *Manduca sexta* (Heidel *et al.*, 2010). Recently, it was shown that the length and

modification of the AHL side chain impacts how the plant recognizes the AHL and its effect on host defense responses. Because unrestricted expression of plant defense pathways have detrimental effects on plant growth and development, plants only activate defense pathways upon pathogen recognition. For example, in *Arabidopsis thaliana*, recognition of bacterial pathogens requires the response of two mitogen-activated defense protein kinases (AtMPK3, AtMPK6) to conserved bacterial elicitors called MAMPs (microbial-associated molecular patterns) such as flagella. Treatment of plants with a purified subflagellar peptide induces a strong but transient activation of MPK3 and MPK6. However, pretreatment with AHL resulted in a prolonged activation of MPK3 and MPK6, and a stronger plant defense response against bacterial MAMPs (Schikora *et al.*, 2011).

AHL signals can also play important roles in modifying plant development. Addition of AHLs can alter approximately 33% of plant protein patterns (Mathesius *et al.*, 2003). The structure of the AHL is critical. For example, short chain AHLs (C_4 and C_6) promoted root growth (von Rad *et al.*, 2008), long chain AHLs (C_{10}) caused roots to shorten and thicken, while C_{12} AHL caused root hair induction (Ortíz-Castro *et al.*, 2008). These data indicate that colonization of plants by different bacteria that produce different AHL signals can clearly alter plant root architecture, and hence plant growth and development.

The bacterial rice pathogen *Xanthomonas campestris* pv. oryzae (Xoo) contains a LuxR homologue (OryR), but lacks a LuxI-type AHL synthase and produces no AHL signals. Ferluga *et al.* (2007) showed that OryR binds and activates 1,4-β-cellobiosidase, an enzyme required for increased virulence in response to macerated rice, but not to any known AHL signal. Thus, Xoo appears to control virulence by a LuxR receptor protein by interacting directly with a rice plant component.

20.4.5.3 Signaling Between Bacteria and Human Cells

Humans are very interested in the coexistence of bacteria and their hosts since there is a need to understand our human signaling with our human microbial flora. An average human is estimated to be composed of 10^{13} mammalian cells and 10^{14} bacterial cells (Hughes *et al.*, 2009). This means that humans are outnumbered 10 to 1 by their bacterial partners. Unfortunately, to date the best understood interkingdom signaling between bacteria and human cells is based on studies with pathogenic bacteria.

As mentioned earlier, the opportunistic pathogen *P. aeruginosa* can colonize the lung tissues of immunocompromised patients such as those with cystic fibrosis or severe burns, where it is a significant cause of mortality (Antunes *et al.*, 2010). The bacterium contains two quorum sensing systems (LasR/LasI and RhlR/RhlI). The gene *lasI* encodes for the synthesis of 3-oxo-C_{12}-HSL while *rhlI* encodes for C_4-HSL. The RhlR/RhlI system is hierarchically under the control of the LasR/LasI system. There is an additional orphan LuxR receptor QscR which recognizes 3-oxo-C_{12}-HSL. These genes control the expression of multiple genes required for *P. aeruginosa* to be a successful pathogen, including the production of elastase, alkaline protease, pyocyanin, rhamnolipids and exotoxin A. Besides the effects of these AHLs on bacterial gene expression, 3-oxo-C_{12}-HSL has been shown to influence host defensive responses. For example, 3-oxo-C_{12}-HSL induces host interleukin-8 (IL-8) and cyclooxygenase COX-2 secretion from human bronchial cells, while causing the simultaneous down-regulation of lymphocyte proliferation and production of tumor necrosis factor TNF-α, IL-2 and reduced apoptosis of macrophages and neutrophils.

In other cases, signaling between bacteria and human cells involves the ability to recognize signals from other bacteria and the host. Bacterial fatty acid-based AHL signals and mammalian lipid-based hormones have significant similarities both structurally and in their mechanisms of action. The human pathogen EHEC (enterohemorrhagic *E. coli*) O157:H7 colonizes intestinal epithelial cells in which it must interact with the resident gastrointestinal microflora. Mammalian cells, including intestinal epithelial cells, produce the catecholamine stress hormones epinephrine and norepinephrine which participate in coordinating host adaptive responses to stress (Table 20.1). Strain O157:H7 senses the quorum sensing signal AI-3 produced by the resident gastrointestinal microflora, and the presence of epinephrine and norepinephrine through its histidine kinase QseC, and activates the expression of multiple pathogenicity genes in response to this signal cocktail. Expression of these pathogenicity genes results in hemorrhagic colitis and hemolytic uremic syndrome. Thus, in this bacterium, signals from both the host and bacterium are interlinked. Although not directed involved in AI-3 production, a *luxS* mutant of strain O157:H7 unable to synthesize the universal signal AI-2 (discussed earlier) produces little AI-3 signal. Interestingly, this *luxS* mutant can be restored to pathogenicity by the addition of epinephrine and norepinephrine (Sperandio *et al.*, 2003; Hughes *et al.*, 2009). Thus, this pathogenic bacterium utilizes host-derived stress hormones to cue when to express its pathogenicity mechanisms.

20.4.5.4 The Holobiont and Hologenome Hypotheses

Chemical signaling between cells is a basic tenet of multicellularity or multicellular organisms. The ability of bacteria to communicate with, and respond to, their hosts via production and recognition of signals suggests that they be included in the concept of host multicellularity. Thus, if we consider the intracellular microflora (endosymbionts) and the extracellular microflora (exosymbionts) to be essential components of a healthy host, then they become part of its innate multicellularity. The term "holobiont" was coined to describe the collection of all cellular components (microbial and host) that comprise a complete organism such as a plant, animal and human (Zilber-Rosenberg and Rosenberg, 2008). Equally fascinating is the concept of the "hologenome," referring to the hypothesis that the hologenome is the sum of the genetic information contained in the host plus all of its microbiota. The recent completion of the sequencing of several human genomes indicates that humans contain approximately 20,500 genes. Work on the human microbiota (microbiome) estimates that there are approximately 2000 species of bacteria inhabiting the human body. If the average size of a bacterial genome is conservatively estimated to be 2500 genes, this indicates that bacteria potentially contribute 5 million additional genes. Of this total amount, if only 250,000 bacterial genes are considered unique, then the final numbers are 250,000 bacterial genes to 20,500 human genes, indicating that our microflora could be contributing 12-fold more genetic information than we contain. Hence, it could be argued that we humans are technically more "microbial" than we are "human." Continued investigation into human—microbial (and similar) interactions will undoubtedly continue to reveal fascinating details about the diverse and complex communication mechanisms that microorganisms have evolved, and the resulting interconnectedness of life on Earth.

20.5 SUMMARY AND CORE CONCEPTS

Originally, bacteria were thought to be single-celled organisms that sensed and responded to environmental inputs individually. However, bacteria are now believed to be able to communicate among themselves both within a single population and between unrelated populations to essentially behave analogously to multicellular organisms. This communication is dependent on a combination of characterized and uncharacterized signals. Communication between microbes and their hosts is known to directly affect the expression of bacterial genes that encode functions critical to all aspects of bacterial survival and bacterial—bacterial and bacterial—host interactions. These include: (1) coordination of gene expression within a single population; (2) coordination of gene expression among unrelated populations; (3) avoidance of host defense responses; (4) coordination of virulence gene expression; (5) inhibition of a competitor's gene expression; and (6) inhibition or stimulation of host colonization.

Besides allowing communication between bacterial populations, quorum sensing signals have been shown to facilitate communication between bacteria and eukaryotic hosts such as fungi, algae, plants and animals. This signal-dependent communication appears to be a two-way street in that the eukaryotic host can produce quorum sensing signal mimics that directly influence the expression of bacterial genes involved in host−microbe interactions, and bacterial quorum sensing signals can also influence eukaryotic host gene expression patterns. In most higher organisms, many critical processes are dependent on the endogenous microbial flora (microbiome) associated with the host. Thus, these communication networks represent key ecological control points that directly determine the outcome of host−microbe interactions. Understanding these communication networks may facilitate large-scale improvements in bacterial−host interactions, pathogen suppression, host health, bioremediation and treatment of water and wastewater.

QUESTIONS AND PROBLEMS

1. Compare the benefits and limitations of AHL-mediated quorum sensing to peptide-mediated quorum sensing.

2. Design an experiment to determine the effect of bypassing quorum sensing control (i.e., make target gene expression constitutive) on the ability of a human pathogenic bacterium to infect its host.

3. Since higher organisms evolved in the presence of bacteria, it makes inherent sense that these diverse organisms communicate with each other. What processes can you think of that might require bacterial−host cooperation?

4. We have seen examples of both positive cross-talk and negative cross-talk (signal interference) between bacteria and between bacteria and hosts. Are these forms of cross-talk community-wide, or could selected subpopulations be differentially affected? Can you devise an experiment to test this idea?

5. It has become clear that we as humans depend on many of the microorganisms that reside within us to carry out many important activities. Can you identify some of these processes? How would you verify that microbial associations are involved?

REFERENCES AND RECOMMENDED READING

Antunes, L. C. M., Ferreira, R. B. R., Buckner, M. M. C., and Finlay, B. B. (2010) Quorum sensing in bacterial virulence. *Microbiology* **156**, 2271−2282.

Boudreau, M. A., Fisher, J. F., and Mobashery, S. (2012) Messenger functions of the bacterial cell wall-derived muropeptides. *Biochemistry* **51**, 2974−2990.

Brencic, A., and Winans, S. C. (2005) Detection of and response to signals involved in host-microbe interactions by plant-associated bacteria. *Microbiol. Mol. Biol. Rev.* **69**, 155−194.

Chu, W., Lu, F., Zhu, W., and Kang, C. (2010) Isolation and characterization of new potential probiotic bacteria based on quorum-sensing system. *J. Appl. Microbiol.* **110**, 202−208.

Cugini, C., Calfee, M. W., Farrow, J. M., III, Morales, D. K., Pesci, E. C., and Hogan, D. A. (2007) Farnesol, a common sesquiterpene, inhibits PQS production in Pseudomonas aeruginosa. *Mol. Microbiol.* **65**, 896−906.

Diggle, S. P., Winzer, K., Chhabra, S. R., Worrall, K. E., Camara, M., and Williams, P. (2003) The *Pseudomonas aeruginosa* quinolone signal molecule overcomes the cell density-dependency of the quorum sensing hierarchy, regulates rhl-dependent genes at the onset of stationary phase and can be produced in the absence of LasR. *Mol. Microbiol.* **50**, 29−43.

Diggle, S. P., Cornelis, P., Williams, P., and Camara, M. (2006) 4-quinolone signalling in *Pseudomonas aeruginosa*: old molecules, new perspectives. *Int. J. Med. Microbiol.* **296**, 83−91.

Dong, Y. H., and Zhang, L. H. (2005) Quorum sensing and quorum-quenching enzymes. *J. Microbiol.* **43**, 101−109.

Dong, Y. H., Wang, L. H., Xu, J. L., Zhang, H. B., Zhang, X. F., and Zhang, L. H. (2001) Quenching quorum-sensing-dependent bacterial infection by an N-acyl homoserine lactonase. *Nature* **411**, 813−817.

Dulla, G., and Lindow, S. E. (2008) Quorum size of *Pseudomonas syringae* is small and dictated by water availability on the leaf surface. *Proc. Natl Acad. Sci. U.S.A.* **105**, 3082−3087.

Dunlap, P. V. (1999) Quorum regulation of luminescence in *Vibrio fischeri. J. Mol. Microbiol. Biotechnol.* **1**, 5−12.

Erickson, D. L., Endersby, R., Kirkham, A., Stuber, K., Vollman, D. D., Rabin, H. R., et al. (2002) *Pseudomonas aeruginosa* quorum-sensing systems may control virulence factor expression in the lungs of patients with cystic fibrosis. *Infect. Immunol.* **70**, 1783−1790.

Ferluga, S., Bigirimana, J., Höfte, M., and Venturi, V. (2007) A LuxR homologue of *Xanthomonas oryzae* pv. *oryzae* is required for optimal rice virulence. *Mol. Plant Pathol.* **8**, 529−538.

Flynn, S., van Sinderen, D., Thornton, G. M., Holo, H., Nes, I. F., and Collins, J. K. (2002) Characterization of the genetic locus responsible for the production of ABP-118, a novel bacteriocin produced by the probiotic bacterium *Lactobacillus salivarius* subsp. *salivarius* UCC118. *Microbiology* **148**, 973−984.

Fuqua, C., Parsek, M. R., and Greenberg, E. P. (2001) Regulation of gene expression by cell-to-cell communications: acyl-homoserine lactone quorum sensing. *Annu. Rev. Genet.* **35**, 439−468.

Gao, M., Teplitski, M., Robinson, J. B., and Bauer, W. D. (2003) Production of substances by *Medicago truncatula* that affect bacterial quorum sensing. *Mol. Plant Microb. Interact.* **16**, 827−834.

Götz, C., Fekete, A., Gebefuegi, I., Forczek, S. T., Fuksová, K., Li, X., et al. (2007) Uptake, degradation and chiral discrimination of N-acyl-D/L-homoserine lactones by barley (*Hordeum vulgare*) and yam bean (*Pachyrhizus erosus*) plants. *Anal. Bioanal. Chem.* **389**, 1447−1457.

Guarner, F., and Schaafsma, G. J. (1998) Probiotics. *Int. J. Food Microbiol.* **39**, 237−238.

Heidel, A. J., Oz, B., and Baldwin, I. T. (2010) Interaction between herbivore defense and microbial signaling: bacterial quorum-sensing compounds weaken JA-mediated herbivore resistance in *Nicotiana attenuata. Chemoecology* **20**, 149−154.

Hogan, D. A., Vik, A., and Kolter, R. (2004) A *Pseudomonas aeruginosa* quorum-sensing molecule influences *Candida albicans* morphology. *Mol. Microbiol.* **54**, 1212−1223.

Huang, J. J., Han, J. I., Zhang, L. H., and Leadbetter, J. R. (2003) Utilization of AHL quorum signals for growth by a soil pseudomonad and *P. aeruginosa* PAO1. *Appl. Environ. Microbiol.* **69**, 5941–5949.

Hughes, D. T., Clarke, M. B., Yamamoto, K., Rasko, D. A., and Sperandio, V. (2009) The QseC adrenergic signaling cascade in enterohemorrhagic *E. coli* (EHEC). *PLoS Pathog.* **5**, 1–13.

Keshavan, N. D., Chowdhary, P. K., Haines, D. C., and González, J. E. (2005) L-Canavanine made by *Medicago sativa* interferes with quorum sensing in *Sinorhizobium meliloti*. *J. Bacteriol.* **187**, 8427–8436.

Krysciak, D., Schmeisser, C., Preuss, S., Riethausen, J., Quitschau, M., Grond, S., *et al.* (2011) Involvement of multiple loci in quorum quenching of autoinducer I molecules in the nitrogen-fixing symbiont *Rhizobium* (*Sinorhizobium*) sp. strain NGR234. *Appl. Environ. Microbiol.* **77**, 5089–5099.

Lazar, V. (2011) Quorum sensing in biofilms: how to destroy the bacterial citadels or their cohesion/power? *Anaerobe* **17**, 280–285.

Lin, Y. H., Xu, J. L., Wang, L. H., Ong, S. L., Leadbetter, J. R., and Zhang, L. H. (2003) Acyl-homoserine lactone acylase from *Ralstonia* strain XJ12B represents a novel and potent class of quorum-quenching enzymes. *Mol. Microbiol.* **47**, 849–860.

Litzinger, S., Fischer, S., Polzer, P., Diederichs, K., Welte, W., and Mayer, C. (2010) Structural and kinetic analysis of *Bacillus subtilis* N-acetylglucosamine reveals a unique Asp-His dyad mechanism. *J. Biol. Chem.* **285**, 35675–35684.

Lyon, G. J., and Novick, R. P. (2004) Peptide signaling in *Staphylococcus aureus* and other Gram-positive bacteria. *Peptides* **25**, 1389–1403.

Mathesius, U., Mulders, S., Gao, M., Teplitski, M., Caetano-Anollés, G., Rolfe, B. G., *et al.* (2003) Extensive and specific responses of a eukaryote to bacterial quorum-sensing signals. *Proc. Natl Acad. Sci. U.S.A.* **100**, 1444–1449.

Mauck, J., Chan, L., and Glaser, L. (1971) Turnover of the cell wall of Gram-positive bacteria. *J. Biol. Chem.* **246**, 1820–1827.

Mavrodi, D. V., Ksenzenko, V. N., Bonsall, R. F., Cook, R. J., Boronin, A. M., and Thomashow, L. S. (1998) A seven-gene locus for synthesis of phenazine-1-carboxylic acid by *Pseudomonas fluorescens* 2-79. *J. Bacteriol.* **180**, 2541–2548.

McAlester, G., O'Gara, F., and Morrissey, J. P. J. (2008) Signal-mediated interactions between *Pseudomonas aeruginosa* and *Candida albicans*. *Med. Microbiol.* **57**, 563–569.

Miller, S. D., Haddock, S. H. D., Elvidge, C. D., and Lee, T. F. (2005) Detection of a bioluminescent milky sea from space. *Proc. Natl Acad. Sci. U.S.A.* **102**, 14181–14184.

Morello, J., Pierson, E. A., and Pierson, L. S. (2004) Negative cross-communication among wheat rhizosphere bacteria: effect on antibiotic production by the biological control bacterium *P. aureofaciens* 30-84. *Appl. Environ. Microbiol.* **70**, 3103–3109.

National Science Foundation website (2005a) An adult *Euprymna scolopes*, a species of bioluminescent sepiolid squid. Image courtesy McFall-Ngai, M. J. and Ruby, E. G., University of Hawaii, http://www.nsf.gov/news/mmg/mmg_disp.cfm?med_id=51886&from=search_list.

National Science Foundation website (2005b) Colonies of the bioluminescent marine bacterium *Vibrio fischeri*. Image courtesy J. W. Hastings, Harvard University, through E. G. Ruby, University of Hawaii, http://www.nsf.gov/news/mmg/mmg_disp.cfm?med_id=51885&from=search_list.

Ortíz-Castro, R., Martínez-Trujillo, M., and López-Bucio, J. (2008) N-acyl-L-homoserine lactones: a class of bacterial quorum-sensing signals alter post-embryonic root development in *Arabidopsis thaliana*. *Plant Cell Environ.* **31**, 1497–1509.

Pierson, E. A., Wood, D. W., Cannon, J. A., Blachere, F. M., and Pierson, L. S., III (1998) Interpopulation signaling via N-acyl-homoserine lactones among bacteria in the wheat rhizosphere. *Mol. Plant Microb. Interact.* **11**, 1078–1084.

Qin, Y., Luo, Z. Q., Smyth, A. J., Gao, P., von Bodman, S. B., and Farrand, S. K. (2000) Quorum-sensing signal binding results in dimerization of TraR and its release from membranes into the cytoplasm. *EMBO J.* **19**, 5212–5221.

Rasmussen, T., Manefield, M., Andersen, J. B., Eberl, L., Anthoni, U., Christophersen, C., *et al.* (2000) How *D. pulchra* furanones affect quorum sensing and swarming motility in *S. liquefaciens* MG. *Microbiology* **146**, 3237–3244.

Ren, D., Sims, J. J., and Wood, T. K. (2001) Inhibition of biofilm formation and swarming of *Escherichia coli* by (5Z)-4-bromo-5-(bromomethylene)-3-butyl-2(5H)-furanone. *Environ. Microbiol.* **3**, 731–736.

Riley, M. A., and Wertz, J. E. (2002) Bacteriocins: evolution, ecology, and application. *Ann. Rev. Microbiol.* **56**, 117–137.

Rumbaugh, K. P. (2007) Convergence of hormones and autoinducers at the host/pathogen interface. *Anal. Bioanal. Chem.* **387**, 425–435.

Schaefer, A. L., Greenberg, E. P., Oliver, C. M., Oda, Y., Huang, J. J., Bittan-Banin, G., *et al.* (2008) A new class of homoserine lactone quorum-sensing signals. *Nature* **454**, 595–600.

Schikora, A., Schenk, S. T., Stein, E., Molitor, A., Zuccaro, A., and Kogel, K. H. (2011) N-acyl-homoserine lactone confers resistance toward biotrophic and hemibiotrophic pathogens via altered activation of AtMPK6. *Plant Physiol.* **157**, 1407–1418.

Schuhegger, R., Ihring, A., Gantner, S., Bahnweg, G., Knappe, C., Vogg, G., *et al.* (2006) Induction of systemic resistance in tomato by N-acyl-L-homoserine lactone-producing rhizosphere bacteria. *Plant Cell Environ.* **29**, 909–918.

Scott, R. A., Weil, J., Le, P. T., Williams, P., Fray, R. G., von Bodman, S. B., *et al.* (2006) Long- and short-chain plant-produced bacterial N-acyl homoserine lactones become components of phyllosphere, rhizosphere, and soil. *Mol. Plant-Microbe Interact.* **19**, 227–239.

Seyedsayamdost, M. R., Case, R. J., Kolter, R., and Clardy, J. (2011) The Jekyll-and-Hyde chemistry of *Phaeobacter gallaeciensis*. *Nat. Chem.* **3**, 331–335.

Soares, J. A., and Ahmer, B. M. M. (2011) Detection of acyl-homoserine lactones by *Escherichia* and *Salmonella*. *Curr. Opin. Microbiol.* **14**, 188–193.

Sperandio, V., Torres, A. G., Jarvis, B., Nataro, J. P., and Kaper, J. B. (2003) Bacteria–host communication: the language of hormones. *Proc. Natl Acad. Sci. U.S.A.* **100**, 8951–8956.

Teplitski, M., Robinson, J. B., and Bauer, W. D. (2000) Plants secrete substances that mimic bacterial AHL signal activities & affect population density-dependent behaviors in associated bacteria. *Mol. Plant Microb. Interact.* **13**, 637–648.

Teplitski, M., Chen, H., Rajamani, S., Gao, M., Merighi, M., Sayre, R. T., *et al.* (2004) *Chlamydomonas reinhardtii* secretes compounds that mimic bacterial signals and interfere with QS regulation in bacteria. *Plant Physiol.* **134**, 137–146.

Uroz, S., Dessaux, Y., and Oger, P. (2009) Quorum sensing and quorum quenching: the yin and yang of bacterial communication. *ChemBioChem* **10**, 205–216.

Vendeville, A., Winzer, K., Heurlier, K., Tang, C. M., and Hardie, K. R. (2005) Making sense of metabolism: autoinducer-2, LuxS and pathogenic bacteria. *Nat. Rev.* **3**, 383–396.

von Rad, U., Klein, I., Dobrev, P. I., Kottova, J., Zazimalova, E., Fekete, A., *et al.* (2008) Response of *Arabidopsis thaliana* to N-hexanoyl-DL-homoserine-lactone, a bacterial quorum sensing molecule produced in the rhizosphere. *Planta* **229**, 73–85.

Wood, D. W., and Pierson, L. S., III (1996) The phzI gene of *Pseudomonas aureofaciens* 30-84 is responsible for the production of a diffusible signal required for phenazine antibiotic production. *Gene* **168**, 49–53.

Xavier, K. B., and Bassler, B. L. (2003) LuxS quorum sensing: more than just a numbers game. *Curr. Opin. Microbiol.* **6**, 191–197.

Zhu, J., and Winans, S. C. (1999) Autoinducer binding by the quorum-sensing regulator TraR increases affinity for target promoters in vitro and decreases TraR turnover rates in whole cells. *Proc. Natl Acad. Sci. U.S.A.* **96**, 4832–4837.

Zilber-Rosenberg, I., and Rosenberg, E. (2008) Role of microorganisms in the evolution of animals and plants: the hologenome theory of evolution. *FEMS Microbiol. Rev.* **32**, 723–735.

Chapter 21

Bioinformation and 'Omic Approaches for Characterization of Environmental Microorganisms

Emily B. Hollister, John P. Brooks and Terry J. Gentry

21.1 INTRODUCTION

In biology, the term 'ome generally refers to the entirety or totality of a collection of specific things. For example, a biome is a collection of living organisms, and a genome refers to the collection of genes within a single organism. 'Omics, then, are fields of study that deal with these collections and involve the characterization and consideration of multiple molecules simultaneously. When botanist Hans Winkler proposed the term "genome" to describe a collection of chromosomes in the 1920s, he probably had no idea how widely the 'ome suffix would come to be used. We commonly study genomes of individual organisms or the metagenomes of communities in order to: (1) understand functional potential; (2) discern phylogenetic relationships; and (3) evaluate heredity (e.g., horizontal gene transfer) at the DNA level. The 'omics concept extends well beyond DNA, however, and can include RNA transcripts, proteins and metabolites, and these are often referred to as the 'omics cascade (Figure 21.1). In this cascade:

- The genome (or metagenome) contains information about *what can happen* (i.e., functional potential);
- The transcriptome (or metatranscriptome) contains information about *what appears to be happening* (i.e., which genes are being expressed);
- The proteome (or metaproteome) contains information about the molecules that *make things happen*; and
- The metabolome contains information about *what has happened recently or is currently happening*.

Although the 'omics cascade captures many of the major 'omics disciplines under study today, a variety of other 'omics have emerged in recent years. Some are subdisciplines of the major 'omics fields mentioned above (e.g., glycomics, lipidomics, interactomics), while others remain emerging concepts, and have yet to be embraced as standalone disciplines in mainstream science.

In addition, the field of bioinformatics has developed to provide the statistical and computational approaches necessary for evaluating the increasingly large and complex datasets that 'omics technologies are producing. In fact, with the rapid expansion in technologies such as DNA sequencing, the analysis and interpretation of 'omics datasets are often the most challenging parts of 'omics -based experiments. In this chapter, we will discuss the primary

I.L. Pepper, C.P. Gerba, T.J. Gentry: Environmental Microbiology, Third edition. DOI: http://dx.doi.org/10.1016/B978-0-12-394626-3.00021-1
© 2015 Elsevier Inc. All rights reserved.

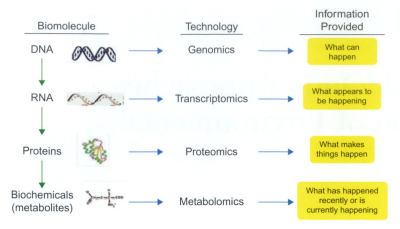

'omics-based methods currently being used to characterize environmental microorganisms, and also approaches for analyzing and interpreting the "bioinformation" that these studies generate.

21.2 GENOMICS AND COMPARATIVE GENOMICS

The term genome describes the total collection of an organism's hereditary information. Genomes are often encoded as DNA and stored in chromosomes, mitochondria, plasmids and/or chloroplasts. However, for many viruses, the genome is composed of RNA only. Advances in DNA sequencing technologies have resulted in the ability to produce vast amounts of sequence information. Where sequencing was once limited to specific gene targets or relatively short DNA fragments, it is now routinely applied to whole genomes. The first whole genome sequence of a free-living organism, *Haemophilus influenzae*, was completed in 1995 (Fleischmann *et al.*, 1995). According to the Genomes Online Database (GOLD, see Table 13.2) as of July 2013, nearly 7000 genomes had been sequenced (in complete or draft stage), and thousands more were listed as ongoing projects. The availability of such large quantities of genome sequence information has spawned a field of study known as comparative genomics. Comparative genomics studies seek to identify similarities and differences in the genes and gene content of various organisms, and a variety of data management systems and analysis platforms have evolved aid in these efforts. The Joint Genome Institute (JGI) provides such a platform in their Integrated Microbial Genomes (IMG) system (Markowitz *et al.*, 2010).

By examining the similarities and differences among genomes, comparative genomics attempts to draw inferences with respect to the function of particular genes, identify regulatory regions and find evidence of evolution and/or genetic exchange, by providing insights into the mobility of chromosomal sections and lateral gene transfer. For

example, bacterial and archaeal thermophiles often share the same habitats, and there is abundant evidence from genomic analysis that lateral gene transfer is common in the group. Specifically, the *Thermotoga maritima* genome has been estimated to have approximately 20% of genes that have primary homology to hyperthermophilic Archaea, principally *Pyrococcus* spp. (Nelson *et al.*, 1999). When comparative genomic approaches were used to study the thermophilic carboxydotroph, *Carboxydothermus hydrogenoformans*, a variety of interesting features, including conserved genes involved in sporulation and a *Rhodosporillum rubrum*-like carbon monoxide dehydrogenase operon, were discovered (Wu *et al.*, 2005). In addition, it was revealed that approximately 30% of the open reading frames in the genome have high similarity to genes in methanogenic Archaea. This observed sequence similarity has led researchers to hypothesize that extensive lateral genetic exchange has occurred between *C. hydrogenoformans* and methanogens (González and Robb, 2000). The close association of methanogens and carboxydotrophic bacteria in the environment suggests that at the very least there is a high potential for exchange of metabolites between the two groups. These examples illustrate the power of comparative genomics in taking nucleic acid sequences and inferring functionality of individual genes as well as potential interactions and genetic exchanges between members of a particular microbial community.

An emerging area of comparative genomics is single-cell genomics (Laskin, 2012). One of the major benefits of nucleic acid-based methods is the ability to circumvent the need to culture microorganisms before they can be characterized, thus enabling the characterization of difficult-(or impossible)-to-culture microorganisms. However, when applied to environmental samples containing diverse communities of microorganisms, these approaches can usually only provide information for a handful of genes (e.g., 16S rRNA), or at best partially assembled genomes for the most dominant organisms in the samples. However, new techniques such as microfluidics and microencapsulation are allowing researchers to isolate and grow individual microorganisms (Zengler *et al.*, 2005; Wessel *et al.*, 2013). When combined with whole-genome amplification methods, these approaches are now enabling researchers to obtain sufficient DNA from one initial microbial cell to determine its entire genome, and thus get a better understanding of its potential environmental function—without ever isolating it on traditional laboratory media (Figure 21.2)! This is particularly powerful when used in combination with other methods such as FISH (Section 13.3.5) to target and select for specific groups of microorganisms that may be less abundant and thus would largely be missed with shotgun sequencing-based metagenomics approaches (Podar *et al.*, 2007).

21.3 METAGENOMICS

As discussed in Section 13.6.2, the term metagenomics was first coined by Handelsman *et al.* (1998) in reference to the collective gene content of a community of microorganisms (e.g., those in a soil sample). The definition of metagenomics has since been expanded by the scientific community to generally include any technique that is based upon analysis of DNA extracted from environmental samples. This broader definition of metagenomics would include 16S rRNA sequencing and related phylogenetic fingerprinting techniques; however, it should be noted that some researchers do not consider these methods (e.g., 16S rRNA sequencing) to be true "metagenomic" techniques.

Over the past two decades, metagenomics-based assays have become the standard for characterizing microbial communities, and have been used in countless studies to determine the structure, function and metabolic potential of microbial communities in a wide variety of environments (Table 19.1). The largest application has been 16S rRNA gene sequencing for determining bacterial diversity and community composition, although a variety of other marker genes have been used, and an increasing number of studies are randomly sequencing environmental DNA.

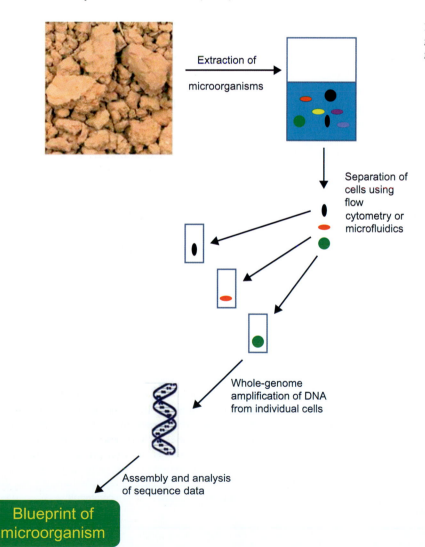

FIGURE 21.2 Overview of a single-cell genomics-based approach for characterizing the genomes of environmental microorganisms.

Extraction of microorganisms

Separation of cells using flow cytometry or microfluidics

Whole-genome amplification of DNA from individual cells

Assembly and analysis of sequence data

Blueprint of microorganism

The ability of metagenomics-based methods to characterize environmental microorganisms without having to first isolate and culture them has allowed the discovery of many previously unknown microorganisms and elucidation of their environmental functions, such as the major contributions of Archaea to ammonia oxidation in a variety of ecosystems (Section 4.4.3).

Although earlier metagenomics studies began with cloning environmental DNA into vectors prior to functional analysis or DNA sequencing, most metagenomics studies today go directly from DNA extraction to sequencing (Figure 13.13). If specific genes are targeted (e.g., 16S rRNA), they can be amplified prior to sequencing (see Section 13.4). Alternatively, the extracted DNA can be sequenced without amplification of any specific genes. This approach is often described as shotgun sequencing. In this process, community DNA is extracted and fractionated into small pieces (if necessary) and sequenced directly via high-throughput sequencing (e.g., 454, Illumina and similar platforms). Following sequencing and processing for quality control (see Section 21.7.1), the reads are either: (1) directly compared to databases for taxonomic and/or functional annotation; or (2) assembled together into longer stretches of DNA which can provide better information since they then represent larger portions of the genome(s) (see Section 21.7.2). Commonly used databases include those available from the National Center for Biotechnological Information (NCBI) and the Metagenomics Analysis Server (MG-RAST) (Section 21.7.2.3; Table 21.1). If higher-order functional identification is required, genes can be categorized using a database such as the Kyoto Encyclopedia of Genes and Genomes (Kanehisa et al., 2004); such databases facilitate identification of specific functional and enzymatic pathways. At the moment, the assembly of metagenomics data from environmental samples is extremely challenging due to the complexity of microbial communities in these environments, and the lack of a good set of reference sequences from a diverse microbial community to serve as a scaffold for assembling the sequences (Thomas et al., 2012). In general, assembly of metagenomics data is limited to only extremely dominant members of simple communities such as those in acid mine drainage (Case Study 21.1 and Figure 21.3; Tyson et al., 2004) or contaminated groundwater (Hemme et al., 2010). Another challenge for assembly is the relatively short read-lengths (<500 bp) of many currently used sequencing methods. This not only makes assembly more difficult, but it also makes direct annotation of the reads more difficult since they often contain only partial gene sequences. However, the development of newer sequencing technologies, such as that of Pacific Biosciences, promise the ability to provide longer reads (>3000 bp) that will encompass entire genes, and possibly even operons, and will thus allow for better taxonomic classification and/or functional prediction. Additionally, the large sequence datasets produced can be computationally challenging to analyze. However, a variety of analysis pipelines and software programs have been developed, and are continually being updated, that facilitate and are standardizing the processing and analysis of these types of datasets (see Section 21.7).

Case Study 21.1   Metagenomics-based Characterization of Dominant Microorganisms in an Acid Mine Drainage Biofilm

One of the first studies to reconstruct putative genomes of environmental microorganisms solely from metagenomic sequence data was the work by Tyson et al. (2004) on an acid mine drainage community in California, U.S.A. Although the site was extremely acidic (pH 0.83), an extensive biofilm existed on the surface of water from the mine. Using fluorescence in situ hybridization (FISH) and 16S rRNA sequencing, the scientists determined that the biofilm community was relatively simple, and was dominated ($\approx 75\%$ of community) by a single group of related bacteria, Leptospirillum group II. The scientists then cloned and sequenced the extracted DNA followed by assembly of the reads. Due to the simplicity of the biofilm community, the sequences were successfully assembled into near-complete genomes for two groups of Bacteria and Archaea: Leptospirillum group II and Ferroplasma type II (Figure 21.3), as well as partial assembly of three other genomes. Both of the near-complete genomes contained putative genes commonly found in microorganisms living in similar, extreme sites including genes for efflux of heavy metals and various other detoxification mechanisms. A number of novel cytochrome genes, which were potentially involved in iron oxidation, were also detected. Since the site was in the deep subsurface, it received little-to-no inputs of carbon and nitrogen from the surface, and therefore would require at least some of the members of the microbial community to fix both carbon and nitrogen. Genes for carbon fixation were found in the Leptospirillum group II genome, but Ferroplasma type II appeared to require external sources of carbon. Interestingly, neither Leptospirillum group II nor Ferroplasma type II contained genes for nitrogen fixation, suggesting that other members of the community most likely fulfilled this vital role for the community. The metagenomic sequencing data from this study provided some initial insights into the metabolism of dominant microorganisms in the biofilm community. In addition, the biofilm is an ideal model community since it is: (1) a relatively simple community dominated by a few microbial populations; and (2) contains a large amount of biomass per unit volume. This enabled a variety of other 'omics methods including transcriptomics and proteomics to be used to validate and expand insights into the ecology of the acid mine drainage biofilm community.

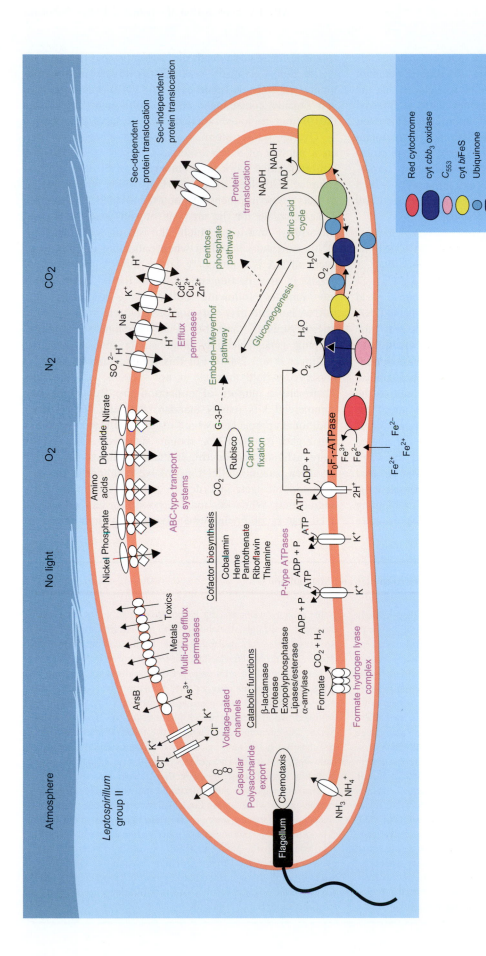

FIGURE 21.3 Metagenomic reconstruction of microbial metabolism in an acid mine drainage community. Constructed from the annotation of 2180 ORFs identified in the assembled *Leptospirillum* group II genome (63% with putative assigned function). The cell diagram is shown within a biofilm that is attached to the surface of an acid mine drainage stream (viewed in cross-section). From Tyson *et al.* (2004).

Despite the unprecedented insight that metagenomics is allowing into the diversity, structure and genetic potential of microbial communities, it should be recognized that the function of genes, from metagenomics data, is inferred bioinformatically rather than tested empirically. However, this initial characterization and prediction of a microbial community's genomic capabilities can serve as the platform for further characterization using other additional 'omics-based assays such as transcriptomics and proteomics, which can verify whether these putative genes are expressed and produce the predicted proteins (Case Study 21.1).

21.4 TRANSCRIPTOMICS

Modern genomic techniques such as metagenomics yield vast amounts of data; however, this data represents the DNA potential of a biological system, not necessarily the expressed phenotype. To unlock the expressed fraction of genomics, one must turn to RNA or protein expression, transcriptomics (a.k.a. metatranscriptomics) and proteomics, respectively. Since RNA, specifically mRNA, represents the product of DNA transcription, it is a logical target for transcriptomics-based analyses. Many metatranscriptomics analyses are less hypothesis driven and may be considered more exploratory in nature. Conversely, some transcriptomics studies focus investigation on expression of targeted genes, and additionally rely on other 'omics to complete the picture (see Case Study 21.2). A number of studies applying transcriptomics to various environmental matrices are

available for more in-depth discussion beyond the scope of this section: Carvalhais et al. (2012) (review of transcriptomics and soil); de Menezes et al. (2012) (transcriptomics and organic contaminant degradation); and Kyle et al. (2010) (transcriptomics applied to E. coli survival on food).

Overall, transcriptomics analyses have been conducted on a number of sample matrices. Much of the original transcriptomics work was conducted with clinical fecal samples (Gosalbes et al., 2011), which given similar caveats as environmental samples, provided for an applicable template for the analysis of soil, water and plant rhizosphere matrices. Much like sample collection for DNA, care must be taken when collecting mRNA; however, mRNA is notoriously labile. mRNA will typically persist in an environmental sample for no more than a few minutes following collection. Additionally, the mRNA half-life may vary for different environments and microorganisms, and by gene function, with housekeeping genes yielding more stable mRNA products (Selinger et al., 2003). For this reason, samples must be preserved within minutes, if not seconds, of collection. There are a number of collection protocols, including commercial kits (easily standardized) and "homemade" traditional approaches, which often yield larger quantities and higher quality RNA, though standardization may be more difficult if conducting latitudinal studies.

Often, sample collection involves immediate freezing in liquid nitrogen in order to prevent enzymatic RNA degradation. While this may be possible when working in a laboratory or greenhouse environment, it may not be

---

Case Study 21.2   Combining 'Omics: Metatranscriptomics and Metabolomics

Combining 'omic analyses yields more useful data than a single analysis in many cases. For example, the application of transcriptomics- and metabolomics-based analyses can reveal the relationships between genes and their final functional activity. At the most basic level, one analysis may provide useful insight while the other may not; a more complex analysis may reveal intricate relationships between transcriptional control and metabolic function. A study by Ishii et al. (2007) aimed to marry the two analyses in the study of common environmental (substrate abundance and reduction) and genetic (missing enzymatic pathways) pressures imposed on Escherichia coli K-12. Global responses were measured using a combination of qRT-PCR (quantitative real-time PCR) to measure targeted mRNA transcripts, and liquid chromatography and time-of-flight mass spectrometry to measure metabolome response. Additionally, DNA microarrays and 2D-differential gel electrophoresis were used to measure relative gene and protein expression, respectively. From these data, the scientists generated an expression index, which took data, separately, from each analysis type and scaled the responses to permit comparisons across all analyses. The analyses revealed gradual increases in mRNA and protein levels using both targeted and

global analyses when placing E. coli under high growth rate conditions. Interestingly, metabolites did not significantly increase. Reducing substrate availability additionally demonstrated few changes in metabolites compared to the control. Finally, the authors disrupted the enzymatic network by disrupting individual genes; but only subtle changes were noted in mRNA and protein expression of central carbon enzymatic pathways. The study demonstrated two approaches which allow E. coli to quickly react to genetic and environmental changes. The results of the study suggest that E. coli has built-in structural redundancy (in enzymatic pathways), which absorbs sudden changes in available substrate as well as loss of single gene function. Results also suggest that E. coli maintained the same metabolic rate (as demonstrated by metabolomics) while up-regulating enzyme expression (as demonstrated by targeted and global transcriptomics). This study demonstrates the stability that E. coli's enzymatic pathways provide along with the ability to rapidly respond to environmental pressures. Discovery of this information was only made possible through use of the multiple 'omics approach, in which one assay demonstrated changes in the system, while the other assay was incapable of detecting responses.

feasible for environmental work. These situations may necessitate the use of RNA stabilizing buffers such as the MO BIO LifeGuard™ Soil Preservation Solution (MO BIO Laboratories, Inc., Carlsbad, CA). These buffers facilitate the collection of mRNA from environmental samples with immediate preservation while in the greenhouse or field. While this step preserves the total RNA in a biological system, the extraction of RNA from intact cells is still necessary prior to analysis. As with DNA extraction procedures, most RNA extraction from commercial kits involves bead-beating technology and the capture of RNA in a stable buffer which can be frozen and subsequently analyzed. However, copurifying soil and fecal humic acids and contaminating organic molecules and metals can affect the quality of the final RNA products (see Chapter 8).

Once mRNA is safely collected and preserved, it needs to be converted to cDNA (complementary DNA; Section 13.4.5). However, mRNA is often present as a small fraction of the total RNA (mostly rRNA and tRNA). Therefore, mRNA is often enriched or selectively isolated from total RNA. As with sample collection and RNA extraction, there are a number of commercial approaches available, including the use of exonuclease treatment (targeting rRNA), and subtractive hybridization using magnetic beads coupled with oligos specific for rRNA and tRNA, which are subsequently removed from the solution. However, in environmental and clinical samples, eukaryotic mRNA may be present at high levels; in these cases, eukaryotic mRNA can be removed by targeting mRNA containing $3'$ poly-A tails (Bailly *et al.*, 2007). Following mRNA enrichment, cDNA is most often the template of choice for most downstream applications. In these cases, reverse transcriptase and either specific primers or random oligos are applied, as in most other methods requiring cDNA synthesis (see Section 13.4.5).

As with DNA metagenomics work, the choice of the sequencing system depends on the length of the intended sequence product and anticipated coverage needed for a specific biological system. Currently, most metatranscriptomics work is conducted using 454 or Illumina systems, the former producing larger sequence products (≈ 500 bp), while the latter provides for smaller sequences (≈ 150 bp), but a larger number of products (<1 Gb vs. 600 Gb). Each system satisfies different study objectives as longer reads are used to map repetitive sequence regions, while some studies require deeper coverage depth. As sequencing methods continue to develop, other platforms will likely be adopted for use in metatranscriptomics.

Following sequencing, bioinformatic analysis removes poor quality and short read sequences. Sequence ends are also trimmed and data analyzed for the presence of rRNA sequences (which can still be present, despite mRNA enrichment), which are promptly removed from the library. Typically, sequences are compared to available databases which assign gene function and identification. However, most metatranscriptomic projects include comparisons of gene relative frequency, and whether a gene is up- or down-regulated. In this case, gene frequencies are normalized to gene abundances from a control metagenome, preferably from the same environmental matrix. Similarly, control metatranscriptomes allow for comparison to treated samples or to various time points, depending on the study objectives. As with metagenomic work, assembly may also be necessary, though the complexity of environmental samples may prohibit this. Various assemblers are available and consist of programs commonly employed in metagenomic work such as Genovo (Laserson *et al.*, 2011) and Newbler (454 Life Sciences, Branford, CT, U.S.A.). A transcriptomic specific assembler such as Velvet (Zerbino and Birney, 2008) can also be used (see Section 21.7 for additional details on bioinformatics).

21.5 PROTEOMICS

Although DNA- and RNA-based methods can provide tremendous insights into the environmental roles of microorganisms, proteins, not genes, are directly responsible for the majority of microbial processes. Therefore, measurement of these proteins (i.e., enzymes) can provide a more direct measurement of microbial activity. The proteins produced by a given microorganism under a given set of conditions are collectively referred to as the proteome. In contrast to the genome, the proteome is much more variable (like the transcriptome) with different proteins being produced depending upon the stage of cell metabolism and the environmental stimuli present.

Studying the proteome has the potential to provide unique information about cell function, and the mechanisms behind cell responses to different stimuli. Specifically, proteomics-based approaches allow identification of proteins that are differentially expressed and, thus, likely to be important in the microbial response to environmental conditions. Proteomics-based studies of environmental effects on microorganisms typically involve the following:

- Exposure of microorganisms to a condition of interest
- Isolation of proteins from each population
- Separation of proteins
- Protein identification

The first two steps in proteomics-based studies are relatively easy. There are a host of effective methods available to isolate and purify the heterogeneous protein mixtures made by microorganisms. However, separating the proteins contained within these complex mixtures represents one of the most challenging aspects of proteomics. Two strategies to separate proteins are commonly used: two-dimensional polyacrylamide gel electrophoresis (2D-PAGE) and liquid chromatography-mass spectrometry (LC-MS). In 2D-PAGE, proteins are first separated according to their isoelectric points (pI), the pH at which the protein has no net charge. The second dimension of 2D-PAGE separates proteins based on their masses using a polyacrylamide gel. The resulting gel contains many spots, each ideally containing a single protein that can be identified using mass spectrometry-based methods described below.

Alternatively, LC can be used for protein separation. In this approach, proteins from a given population are pooled and digested enzymatically into their constituent peptides. These peptides are separated by LC (see Section 11.2.1.1) which allows for the separation of molecules based on charge or hydrophobicity. Proteins in the original population of cells are identified on the basis of these peptides as described below. LC-based separation of proteins can be more readily automated, and may be more reproducible than 2D-PAGE.

Once separated, proteins must be identified to gain insight into mechanisms by which microorganisms interact with the environment. Mass spectrometry is currently the tool of choice for this task. Intact proteins are broken down enzymatically (i.e., digested) into smaller peptides and analyzed by mass spectrometry. Once accurate masses of the peptides are obtained, the protein from which the peptides originated can be identified. This approach to protein identification is known as peptide mass fingerprinting (PMF). When PMF fails, other types of mass spectrometry can be used to obtain direct amino acid sequence data that can be useful for protein identification. As differentially expressed proteins are identified, the investigator gains insight into mechanisms by which the microorganism responds to a particular environmental condition (Westermeier and Naven, 2002).

Studies have demonstrated that the comprehensive, high-throughput nature of proteomics-based approaches is also well suited to elucidating biodegradative pathways. For example, Kim et al. (2004) examined biodegradation pathways of an aromatic-degrading pseudomonad (Pseudomonas sp. K82) using 2D-PAGE followed by mass spectrometric identification of proteins. Using this approach, the investigators discovered three metabolic pathways, each of which was induced to a different degree by three different aromatic compounds.

As with recent research in metagenomics, applications of proteomics to microbial ecosystems are emerging and offer promise to link microbial species within complex communities to function (Hettich et al., 2013). Termed metaproteomics or community proteomics, these approaches are designed to isolate as many proteins as possible from a microbial community to learn more about which microorganisms perform what tasks within a community (Figure 21.4). For example, Ram et al. (2005) used metaproteomics to investigate and characterize an acid mine drainage biofilm community similar to the one described in Case Study 21.1 and Figure 21.3. As with most proteomics-based approaches, this approach was facilitated by genomic sequence data (Figure 21.5). Specifically, the authors constructed a database of 12,148 predicted protein sequences from the similar biofilm community previously characterized using metagenomics (Tyson et al., 2004). Using this database and an LC-mass spectrometry approach to protein identification, the authors identified 2033 individual proteins. Most were produced by members of the genus Leptospirillum and were involved with adaptation to this extremely acidic (pH ≈ 0.8), metal-laden environment. Many proteins could not be assigned a function, yet were highly prevalent. One of these, which was previously identified by the metagenomics approach as possibly playing a role in iron oxidation, was confirmed to be a novel cytochrome involved in iron oxidation and acid mine drainage formation. A subsequent study found that the proteome changed during development of the biofilm (Mueller et al., 2011). For example, the dominant organism, Leptospirillum group II, produced more enzymes for metabolism of 1- and 2-carbon compounds and protein synthesis during early biofilm development, and more stress-related and iron oxidation proteins, likely related to acid mine drainage formation, as the biofilm developed and resources likely became more limiting (Figure 21.6).

Despite the promise of metaproteomics, many impediments to its broader use exist. The need for a universal method to exhaustively extract proteins from complex communities, particularly those indigenous to soil, is of paramount importance. In addition, the sensitivity of detection of existing methods is limited, and approaches are only capable of identifying proteins from microbial populations that comprise >1% of a community. Furthermore, additional metagenomics data are needed in order to better predict the suite of proteins produced by environmental microbial communities and accurately interpret metaproteomics data (Figure 21.5). Nevertheless, metaproteomics is a developing and promising area of research, and will likely be increasingly used over the next decade to study the activity and functions of environmental microorganisms.

21.6 METABOLOMICS

Metabolomics consists of the study of low molecular weight metabolites. Environmental metabolomics consists of metabolites produced by interactions between microorganisms, small eukaryotes, plants, animals, predators and the presence of abiotic pressures and stimulants.

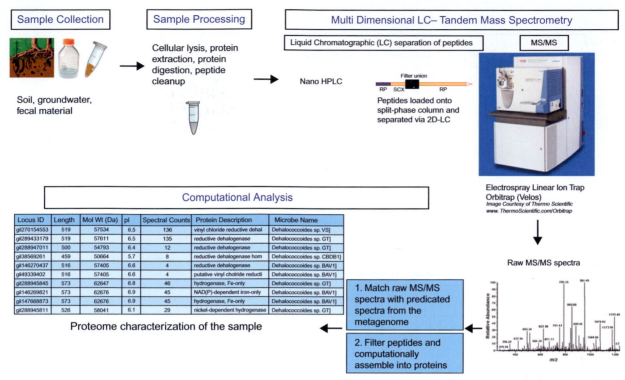

FIGURE 21.4 Experimental flowchart for sample preparation and measurement in a metaproteomics experiment. Sample collection and processing steps must be optimized to match the nature of the material to be analyzed, in terms of biomass amount and complexity, matrix composition, sample heterogeneity, etc. The resulting proteome sample is digested with trypsin and loaded onto a biphasic HPLC column for concomitant 2D-separation and MS analysis via nanoelectrospray-based ionization of eluting peptides. Acquisition of parent peptide ion (MS1) mass and fragmentation (MS/MS or MS2) information provides an experimental dataset containing hundreds of thousands of spectra that can be computationally matched to the predicted proteome obtained from metagenomics information. From Hettich *et al.* (2013).

Common metabolites (≤1500 Da) consist of organic acids (e.g., glycolytic intermediates), amino acids (e.g., protein intermediates) and various saccharides (e.g., monosaccharides and cleaved sugars).

As with genomics, transcriptomics and proteomics studies, the goal of metabolomics is often to elucidate the function of a microorganism or microbial community; however, proteomics and metabolomics reveal information related to the "final" genome product. Similarly, metabolomics characterizes the interactions between microbial constituents and their environment, or between microbial and other higher-order ecological organisms such as plants and animals. Metabolomics has been used as an exploratory tool (Dunn, 2008), to uncover the functional status of microbial populations and single cells in their environment, revealing community and ecological structure. Targeted metabolomics enables the user to focus upon a specific metabolite, for instance when a treatment may dictate the up- or down-regulation of a product, while global metabolomics views the biological system and its metabolites as a whole. A number of studies or reviews describing metabolomics and various environmental matrices are listed for further information beyond the scope of this section: Zhang *et al.* (2010) (review); Ito *et al.* (2013) (contaminated feedstock);

Liebeke *et al.* (2009) (benchtop single culture study); and Bundy *et al.* (2009) (review).

Metabolites are broken down into two groups: the endometabolome and exometabolome, which are metabolites contained intracellularly and extracellularly, respectively. Like transcriptomics, the study of intracellular metabolites can be more difficult, as these molecules are more fleeting and in a constant state of flux. Metabolome complexity and study objectives involving intra- or extra-cellular metabolites determine the type of extraction and processing. Once metabolites are extracted, they are subjected to identification with a number of instruments such as gas and liquid chromatography-mass spectrometry, Raman spectroscopy and nuclear magnetic resonance (NMR). In many instances, depending on the complexity of the biological system, the study will call for a combination of two or more of these instruments (Dunn, 2008; Case Study 21.2).

Regardless of platform, a large amount of metabolic data is typically generated. In many instances, the metabolites under investigation are unknown and global in perspective; therefore, query databases are required to deduce the function and purpose of the metabolite. Metabolites are often identified as products or intermediates of environmental

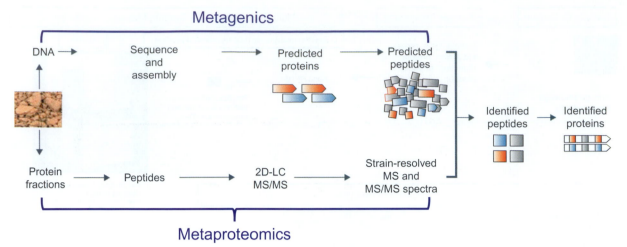

FIGURE 21.5   Integrated use of metagenomics and metaproteomics for characterizing microbial communities. DNA is extracted from biological samples, fragmented and sequenced. The resulting sequence reads are then assembled and/or binned. After gene annotation, the protein-sequence database is constructed and an *in silico* trypsin digest is performed on the predicted proteins, resulting in a peptide database (top). From the same or similar biological samples, total community protein is extracted and then digested using trypsin. Peptide separation by two-dimensional (2D) nano-liquid chromatography (LC) and tandem mass spectrometry (MS/MS) is performed (see Figure 21.4). The spectra are matched to peptides in the database, and after filtering, a list of identified peptides is obtained. Based on their unique occurrence in one protein in the whole database, certain peptides (unique peptides, colored red and blue) can be tracked back to their corresponding proteins and thus permit reliable protein identification. Nonunique peptides (gray) cannot be used to uniquely identify a protein, but these data are used in the calculation of protein coverage and abundance measures. The identified proteins are placed back into the genomic context of the organisms they are derived from to allow for the biological mining of the data. Adapted from VerBerkmoes *et al.* (2009).

populations under stress due to the overall health of a system. Given the relatively novel nature of metabolomics, particularly in environmental sciences, very few databases exist to facilitate identification of environmental metabolites. Common databases consist of the Human Metabolome Database and Kyoto Encyclopedia of Genes and Genomes; commonly used databases can be found at http://www.metabolomicssociety.org/databases.

21.7 BIOINFORMATION

21.7.1 Bioinformatics and Analysis of Marker Gene Data

21.7.1.1 16S rRNA and Other Marker Genes

As discussed in Chapter 13, marker genes, such as ribosomal RNA (rRNA) genes or the internal transcribed spacer (ITS), are frequently used to characterize the composition of bacterial, archaeal and fungal communities. Marker genes are useful because they allow for the relatively rapid characterization of the composition and diversity of microbial communities. The 16S rRNA gene is the most commonly used marker gene for the characterization of Bacteria and Archaea, while the ITS tends to be favored among microbiologists for the characterization of fungi. That notwithstanding, the

18S rRNA and 28S rRNA genes are also commonly used for the characterization of fungal communities, and are frequently employed as an alternative to the ITS region when detailed phylogenetic information is needed.

Recall, good marker genes share the characteristics of:

- Ubiquity—the marker should be present in most, if not all, target species
- Genetic conservation—the sequence of the marker should be conserved sufficiently that it can be targeted with PCR primers
- Variability—in combination with genetic conservation, the marker should also contain regions of sequence that are variable and allow for differentiation between species, among lineages and within populations.

Given these characteristics, marker genes are well suited to serve as targets for sequence-based community surveys. Using high-throughput sequencing platforms, such as 454, Ion Torrent or Illumina, researchers are now able to generate large quantities of sequence information allowing them to describe the structure and diversity of microbial communities of interest.

21.7.1.2 Platforms for Sequence Analysis

Due to the generation of large quantities of marker gene sequences, there is a subsequent need to analyze and

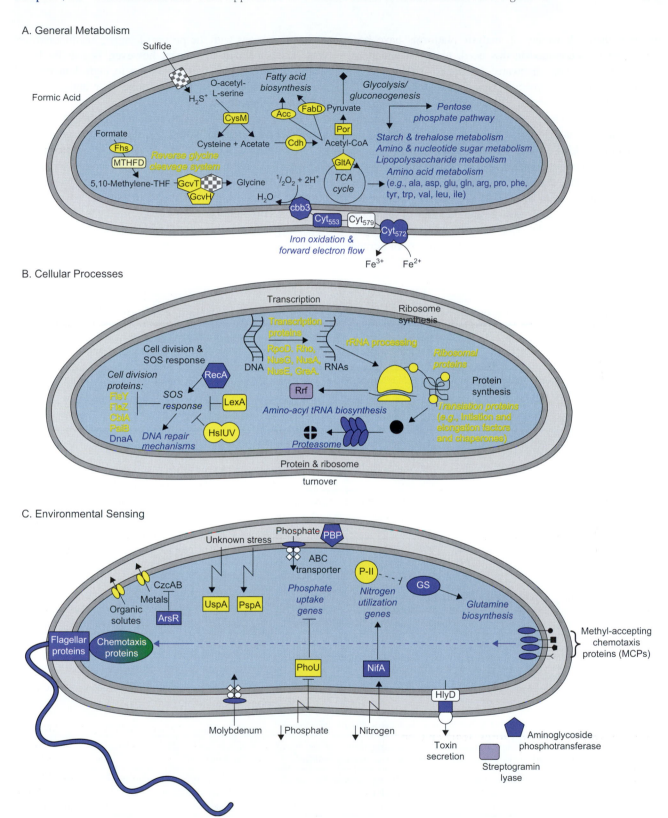

FIGURE 21.6 Physiological changes of the dominant bacteria, *Leptospirillum* group II, in an acid mine drainage biofilm as the biofilm matures. Figure depicts significant changes in *Leptospirillum* group II proteins involved in (A) general metabolism, (B) cellular processes and (C) environmental sensing. Proteins with yellow fill and pathway headings in yellow font (e.g., "Fhs" and "Reverse glycine cleavage system") were significantly more abundant in early and intermediate growth stages, and proteins with blue fill and pathway headings in blue font (e.g., "Cyt$_{572}$" and "Pentose phosphate pathway") were significantly more abundant in late growth stage samples. Proteins labeled in white were detected by proteomics, but did not demonstrate a biologically relevant abundance pattern. Proteins filled with a gray-checked pattern were not detected or are unknown. From Mueller *et al.* (2011).

interpret them. A variety of analysis platforms have been developed to accommodate this need, many of which are open-source and/or freeware packages (Table 21.1). Examples of these include standalone tool sets like MOTHUR (Schloss *et al.*, 2009) and QIIME (Caporaso *et al.*, 2010). Others are web-based portals like the Ribosomal Database Project Pyrosequencing Pipeline (Cole *et al.*, 2009), VAMPS (http://vamps.mbl.edu/), the Genboree Microbiome Toolset (Riehle *et al.*, 2012) and PlutoF (Abarenkov *et al.*, 2010). Many of the web-based portals feature the functions of MOTHUR and QIIME, some utilize custom algorithms, and most feature additional platform-specific analysis modules. One of the biggest advantages of web-based platforms is that they link the features of popular

analysis packages with the power of larger, institutional servers. Their main disadvantage, however, is that by being shared resources, they can be subject to high demand, and one may sometimes have to wait longer than anticipated for results to be processed.

Marker gene analysis platforms tend to revolve around a core set of functions. These include: (1) the conversion of raw sequence data (i.e., sff or fastq files) into FASTA format; (2) quality filtering of sequences; (3) separation of pooled sequences into their originating samples on the basis of barcode tags; (4) data "reduction" to allow for increased computational efficiency; and (5) detection of potentially chimeric reads. Beyond these features, many platforms also offer algorithms that: (1) attempt to

TABLE 21.1 Common Platforms for Sequence Analysis and Their Capabilities

Platform/Package	Website	Features	Data Types Analyzed	Reference
Marker gene analysis				
QIIME	http://qiime.org	Quality filtering, separation of sequence by barcode, OTUs, taxonomic identities, diversity analyses, between community comparisons	Marker gene sequences; developed for 16S but can be used with 18S or ITS sequence	Caporaso *et al.*, 2010
MOTHUR	http://www.mothur.org	Quality filtering, separation of sequence by barcode, OTUs, taxonomic identities, diversity analyses, between community comparisons	Marker gene sequences; developed for 16S but can be used with ITS or other marker gene sequences	Schloss *et al.*, 2009
Ribosomal Database Project	http://rdp.cme.msu.edu	Archive submission portal; quality filtering; taxonomic identities; calculation of some diversity indices	Largely developed to support 16S analysis; includes 28S database for fungi	Cole *et al.*, 2009
VAMPS	http://vamps.mbl.edu	Wraps features of QIIME and MOTHUR; includes links to data from large projects like the Human Microbiome Project and the Microbiome of the Built Environment	16S rRNA gene sequences	Huse *et al.*, 2010
Genboree Microbiome Toolset	http://genboree.org	Web-based platform for QIIME; offers additional custom analysis modules	16S rRNA gene sequences	Riehle *et al.*, 2012
PlutoF	http://unite.ut.ee/workbench.php	Quality filtering, separation of sequence by barcode, OTUs, taxonomic identities	ITS sequences	Abarenkov *et al.*, 2010
(Meta)genome analysis				
IMG and IMG/M	http://img.jgi.doe.gov	Quality filtering; genome and metagenome assembly and annotation; comparative analysis of genomes or metagenomes	Shotgun genomes and metagenomes	Markowitz *et al.*, 2010
MG-RAST	http://metagenomics.anl.gov	Quality filtering; taxonomic and functional annotation; no assembly provided	Shotgun metagenomes, marker gene surveys	Meyer *et al.*, 2008
CAMERA	http://camera.calit2.net	Quality filtering; metagenome assembly and annotation; viral diversity analyses	Shotgun metagenomes, marker gene surveys for Bacteria, Archaea and viruses	Sun *et al.*, 2011
EBI Metagenomics	https://www.ebi.ac.uk/metagenomics	Sequence archiving; quality filtering; taxonomic analysis of 16S reads; functional annotation	Shotgun genomes, metagenomes, marker gene surveys	Hunter *et al.*, 2011

minimize errors as a result of sequencing "noise"; (2) cluster sequences into operational taxonomic units (OTUs) on the basis of similarity; (3) assign identities to each sequence through comparison to reference databases; and (4) perform additional analyses including the calculation of diversity indices, evaluation of sample-to-sample similarities and differences, and detection of features that distinguish one community from another.

21.7.1.3 Quality Criteria

The sequencing process is inherently prone to error (i.e., the incorporation of incorrect base calls during sequencing). Such errors include substitutions made by DNA polymerases, chimeric sequence formation and the difficulties entailed in reliably reproducing homopolymeric regions of sequence (Schloss et al., 2011). Although these error rates vary among sequencing platforms and tend to be relatively low, their cumulative effects on marker gene survey data can alter our perception of microbial community diversity. As a result, it is common to employ a series of quality filters to the sequence data prior to analysis. These include:

- **The removal of low quality sequences**
  As each base is incorporated during a sequencing reaction, a score indicating the quality of each base call is also generated and recorded into the sequencing record. The greater the number of errors in a stretch of sequence, the lower the quality score tends to be. Sequences can be trimmed according to quality scores, and this can be done in one of two ways. The first involves trimming away low-scoring regions of sequence from each read and retaining what remains. The second removes entire sequences from a data set on the basis of average read quality. Typically, sequences with an average quality score lower than 20 are removed.

  From time to time, a base position cannot be called with certainty. These are known as ambiguous base calls, and they are indicated in a stretch of sequence by the letter N (e.g., ATCCN). Sequences containing ambiguous base calls are indicators of poor sequence quality (Huse et al., 2007), and are typically removed from analysis.

- **The removal of sequences that are too long or too short**
  Sequences that are very short or very long, relative to the expected sequence length for a given sequencing platform, tend to be of lower quality and contain large numbers of errors (Huse et al., 2007). As a result, users typically filter out these sequences. For example, it is common to remove sequences that are shorter than 200 bp or longer than 1000 bp from sequence runs generated on the Roche 454 platform, which average 450 bp in length.

- **The removal of sequences containing exceptionally long homopolymers**
  Homopolymeric runs are regions of sequence in which the same base call is incorporated multiple times in a row. The sequence ACGGGGGGGGTC, for example, contains a homopolymer of seven guanine residues. Although homopolymers do exist in nature, they can occur erroneously during the sequencing process (Huse et al., 2007). Some sequencing platforms (the Roche 454 platform, in particular) have difficulty reproducing homopolymeric sequences correctly. As a precaution against spurious homopolymers, most analysis platforms allow users to define an acceptable homopolymer length (e.g., a homopolymer limit of 6 is commonly utilized), and filter out sequences containing longer homopolymeric spans.

- **Barcode and primer trimming and the removal of sequences containing mismatches to their barcode or primer sequences**
  High-throughput sequencing platforms offer the ability to multiplex samples for sequencing. Multiplexing allows pools of DNA amplicons originating from multiple samples to be mixed together and sequenced simultaneously. The incorporation of barcodes into the amplicon sequences permits them to be sorted bioinformatically and attributed back to their sample of origin. Barcodes, also known as tags, are typically short (i.e., 8—12 bp in length) sequences that can be ligated onto PCR products after they are produced or incorporated into the sequencing primer.

  Although barcodes provide a means for assigning reads to their sample of origin, they also represent an additional opportunity for quality control. Typically, sequences that contain errors (i.e., incorrect base calls) in their barcode sequence are considered to be of low quality and are removed from analysis, although some protocols will accept one or two mismatches (Caporaso et al., 2010; Schloss et al., 2011). This is also true of primer sequences. Once sequences have been evaluated for barcode and primer mismatches and pooled by sample of origin, the barcode and primer sequences are trimmed away.

21.7.1.4 Removal of Chimeras

In Greek mythology, the chimera was described as a monster that was part lion, part goat and part snake. During the PCR process, it is possible for DNA polymerase to begin copying one target, become disrupted and finish its amplification cycle by picking up copying a second target. The resulting product is a hybrid of the two original templates and is commonly referred to as a chimera, or chimeric sequence (Figure 21.7). It is estimated that chimeric reads may account for 5% or more of sequence libraries (Ashelford et al., 2005), and the risk for chimera production is potentially problematic when one is trying to characterize the composition and diversity of a mixed microbial community.

The detection of chimeras typically involves the comparison of each individual read to all others within a sequence library or a reference database. Those that appear to have strong similarities to two different and divergent "parent" sequences are typically flagged as potential chimeras. Multiple software packages for the detection of chimeras are available. The earliest ones were developed for the analysis of small sequence libraries and are generally not capable of analyzing large, high-throughput sequence libraries [e.g. Pintail (Ashelford *et al.*, 2005), Chimera Check (Cole *et al.*, 2007), Bellerophon (Huber *et al.*, 2004)]. Newer packages like ChimeraSlayer (Haas *et al.*, 2011), UChime (Edgar *et al.*, 2011) and B2C2 (Gontcharova *et al.*, 2010) are more frequently used for this purpose. Regardless of the chimera detection package that one chooses, users are cautioned to consider that the output generated only identifies *potential* chimeras. The results should be reviewed in greater detail, when possible, as "true" (i.e., nonchimeric) sequences can be flagged incorrectly.

21.7.1.5 The Operational Taxonomic Unit (OTU) Concept

The concept of a bacterial species can be difficult to define. Revisions to existing taxonomies are published on a regular basis with phylogenetic relationships constantly being redefined on the basis of new molecular information (Information Box 21.1). Horizontal gene transfer between individual bacteria obscures relationships that are defined on the basis of function, and it is widely acknowledged by microbiologists that we have only just begun to characterize and classify the extensive diversity of microbial species.

With all of this as a background, sequence-based surveys emerged as a means of characterizing individual bacteria and microbial communities at large. As a means of grappling with the questions of how to quickly distinguish one species from another when many species are present

in a given sample, the concept of the operational taxonomic unit (OTU) emerged. DNA—DNA hybridization studies have long been a gold standard for defining species similarity, but scientists noticed that bacteria that share high levels of similarity via DNA—DNA hybridization also shared a high degree of similarity between their 16S rRNA gene sequences (Stackenbrandt and Goebel, 1994). This concept has also been applied to fungi (O'Brien *et al.*, 2005; Amend *et al.*, 2010), although the ITS region is typically utilized instead of the small ribosomal subunit.

The OTU is a computational construct that is used to represent species, and it is heavily utilized in the field of microbial ecology. OTUs are defined on the basis of sequence similarity, and typically a 97% sequence similarity cutoff is employed. That is, if two sequences have 97% of their base calls in common over the entire length of both sequences, they are considered to belong to the same OTU. OTUs are convenient in that they represent an entity that can be counted and used as the basis for diversity estimates (Schloss and Handelsman, 2005), and they are not tied to known biological diversity (i.e., they

---

Information Box 21.1   The Evolving Taxonomy of Microorganisms

One of the challenges for phylogenetic classification of microbial communities and interpretation of these data is the dramatic evolution of microbial taxonomy, especially over the past few decades. For example, one of the most-studied 2,4-D-degrading bacteria was originally named *Alcaligenes eutrophus* JMP134 after its isolation from soil (Don and Pemberton, 1981); however, a search of the literature will find that this bacterium has been referred to by multiple names over the past three decades including:

Alcaligenes eutrophus JMP134
↓
Ralstonia eutropha JMP134
↓
Wautersia eutropha JMP134
↓
Cupriavidus necator JMP134
↓
Cupriavidus pinatubonensis JMP134

These changes have occurred as the bacterium and related organisms have been reclassified in light of new information for a variety of properties including: lipid; composition; 16S rRNA gene sequence, DNA—DNA hybridization; and phenotype. Although these continual changes are improving the taxonomic classification of microorganisms, they can make it even more difficult to draw functional inferences for environmental microorganisms based solely upon comparison of sequence data for phylogenetic marker genes (e.g., 16S rRNA) to previously classified organisms, whose characterization may have been published under a different name(s) in the literature.

FIGURE 21.7 A chimeric sequence may be generated during the PCR process when DNA polymerase begins replicating one strand of DNA and finishes on another. The resulting chimera contains sequence from one parent template at the 5′ end and the other parent template at the 3′ end. The detection of chimeric sequences often involves BLAST-like searches of reference databases or the other reads produced in the same sequence library, in an attempt to identify reads that share a high degree of similarity with multiple "parent" sequences.

can be used to quantify previously undescribed or uncharacterized organisms). They also allow large, complex collections of sequence data to be summarized quickly in text format. However, one of the major downfalls of OTUs is that without additional characterization, they lack the ability to convey information about phylogenetic relationships, or the degree of similarity, shared with other OTUs. Although all of the sequences that belong to an OTU are, by definition, closely related to one another (i.e., 97% sequence similarity is often used as the cutoff for all sequences within an OTU), the ability to discern whether "OTU A" and "OTU B" are similar to one another can quickly become lost.

21.7.1.6 Diversity Analyses

What is diversity? Biological and ecological diversity are concepts that deal with richness, variability and variety within the context of an environmental system (i.e., a defined unit) (see also Chapter 19). This may be genetic diversity, organism diversity or ecological diversity (Magurran, 2004). In the context of microbial communities, we typically consider aspects of all three. Genetic diversity, often in the form of marker gene sequences, is used as a proxy to describe organism diversity (i.e., OTUs or species), and communities of microorganisms are compared with one another in an attempt to describe the richness and variation that exists within and between communities.

Two key concepts contribute to our understanding of diversity. For the sake of discussion, we will use the terms "species" and "communities" here, but other entities (e.g., genes, taxonomic families) could be used in their place. The first of these concepts is richness, or the number of different types of species that exist within a community. This is a relatively easy concept to define, but in practice it is often difficult to quantify with 100% certainty, because it is extremely difficult to sample microbial communities exhaustively.

The second concept that contributes to our understanding of diversity is evenness, a term which describes the variability of species abundances. An extremely "even" community is one in which all species are present in similar proportions. As an example of an even community, consider an assemblage that contains four species, each of which accounts for 25% of the individuals (or biomass) in the community. In contrast, an "uneven" community is one in which large disparities exist with respect to the relative abundances of its members. Like the example provided above, an uneven community could also contain four species, but in this case, one species accounts for 60% of the community, the second accounts for 30% of the community, the third accounts for 7% of the community and the last accounts for the remaining 3%.

As a means of communicating information about diversity, the concepts of richness and evenness are often communicated as a single value, known as a diversity index. Multiple diversity indices have been developed (Information Box 21.2), and each has strengths, weaknesses and biases (Magurran, 2004). A full discussion of these is beyond the scope of this chapter, but some of the most commonly utilized indices and their applications will be described here.

Alpha diversity refers to the diversity of a defined unit, sample, assemblage or habitat (Rosenzweig, 1995), and it is often described in terms of species or OTU richness, the Shannon (or Shannon–Weiner) index and/or the Simpson index. Because indices like Shannon and Simpson can be biased by disparities in sampling effort or sample size (Magurran, 2004), it is common to subsample sequence (or OTU) libraries to an even depth before calculating diversity index values in order to facilitate head-to-head comparisons between one's samples. Typically, this is accomplished by randomly selecting an equal number of sequences from each sample in one's study prior to the calculation of diversity values.

The Shannon index (H') (Shannon and Weaver, 1949) is based on information theory and attempts to quantify the uncertainty surrounding one's ability to predict, in advance, the identity of an organism sampled at random from a dataset (or community). It is based on the idea that both the number of species in a community and their relative abundances contribute to the "complexity" of a community, and thus the likelihood of being able to correctly predict the identity of an organism randomly sampled from the community. The Shannon index is calculated as:

$$H' = -\Sigma\rho_i \ln \rho_i$$

where ρ_i is the proportion of the ith species in the community. This could be species "A," "B," "C," etc. The value of the proportion of each species in the community multiplied by the log of that value is calculated for every species in the community, and then summed to generate the Shannon index score. Natural log, log 2 or log 10 can be used, but natural log is commonly employed. Although larger Shannon index values are generally considered to represent greater levels of diversity, the means by which the index is calculated make it difficult to interpret whether changes to the statistic are a result of changes to richness, evenness or both. Despite this, the Shannon index is commonly utilized to describe microbial diversity.

Like the Shannon index, Simpson's index (Simpson, 1949) deals with probabilities. More specifically, it attempts to define the probability of any two organisms being drawn from the same community belonging to the same species. Although defining these probabilities inherently deals with defining the number of species in a community (i.e., its richness), the Simpson index tends to have a greater focus on species dominance (i.e., evenness)

Information Box 21.2   Diversity Indices

Although diversity can be characterized at multiple levels (Chapter 19), diversity indices are most frequently used to describe alpha-diversity and beta-diversity. An introduction to commonly used diversity indices is provided below, but many others have been developed and may be encountered in the literature.

Commonly used alpha-diversity indices include:

Species richness (observed species)—a count of the number of unique species that occur in a sample or community

Shannon (H′)—the Shannon (or Shannon–Wiener) index considers both the number of unique species and their relative abundances within a sample (Shannon and Weaver, 1949). Larger values reflect communities with greater species richness and evenness, while lower numbers reflect communities with fewer species and/or a very uneven distribution among them (e.g., one species may account for a very large percentage of the community).

Simpson (D)—the Simpson index evaluates the relative abundances of all species in a community, and attempts to define the probability that any two organisms drawn from the same community will be of the same species (Simpson, 1949; Magurran, 2004). Small values of the Simpson index tend to reflect communities with high richness and low dominance, and high values reflect communities with (potentially) lower richness and high dominance (i.e., most of the community belongs to one or a few species). The Simpson index is often presented in inverse form (1/D or 1 − D) so that large numbers represent increasing evenness and smaller numbers represent increasing dominance.

Chao I—the Chao I index is a correction factor for observed richness. It evaluates the number of species that occur once (singletons) versus those that occur twice (doubletons), and attempts to estimate the number of species that would be captured if the entire community could be sampled exhaustively (Chao, 1984).

Rarefaction—Rarefaction is not an index, but rather a technique used to assess species richness. It involves plotting the number of unique species detected versus the number of organisms sampled. The shape of the resulting curve is used to indicate the "completeness" of a survey. A curve that flattens and reaches a clear asymptote suggests that the majority of the diversity in a community has been captured, while one that maintains a steep slope indicates that more sampling is needed.

Phylogenetic diversity—also known as Faith's diversity (Faith, 1992), this index quantifies the total length of the branches needed to account for a set of taxa on a phylogenetic tree. Increasing values of the index reflect increasing levels of diversity within the community being described.

Commonly used beta-diversity indices include:

Sørensen index—evaluates the degree of similarity between two communities by quantifying the number of species shared in common, relative to the total number of species held in both communities (Sørensen, 1948). This metric can be used with presence/absence (i.e., binary) data.

Jaccard index—evaluates the degree of similarity between two communities by quantifying the number of species shared in common relative to the sum of the number of species uniquely held by each community (Jaccard, 1908). This metric can be used with presence/absence data.

Bray–Curtis dissimilarity—Bray–Curtis dissimilarity (Bray and Curtis, 1957) is an extension of the Sørensen index. Calculated the same way, it is allows for quantitative values (i.e., counts or relative abundances) to be used instead of binary data.

Unifrac distance—a measurement that reflects the amount of branch length shared by two or more communities when their members are placed on a common phylogenetic tree (Lozupone and Knight, 2005). The Unifrac distance is equivalent to "1 minus the fraction of shared branch length."

---

than it does on species richness. The Simpson index (D) is calculated as:

$$D = \Sigma \rho_i^2$$

where dominance (D) is calculated as the sum of the squared proportions of all species in a given community. Large values of D are typically interpreted to represent high dominance and low diversity, whereas small values of D tend to represent lower dominance, higher diversity communities. Because the interpretation of these values is not necessarily intuitive, ecologists commonly calculate inverse Simpson (1/D) or subtract Simpson from 1 (1 − D) to obtain a value that is more easily interpreted.

Once one has described the diversity within a community, it is common to want to compare diversity among communities along gradients (Whittaker, 1960) or separated by space and time. This is also known as beta diversity. A typical first step in describing beta diversity is to calculate the degree of similarity shared between communities in terms of their species composition, species distribution or both. The terms similarity and distance are frequently used to describe the degree to which two communities resemble one another, and they are the inverse of one another (i.e., a similarity of 30% is equivalent to a dissimilarity or distance of 70%).

Multiple approaches to calculating community similarity exist. As mentioned above, some indices, like the

Sörensen or Jaccard indices, consider only the presence or absence of species between two samples. Others, like the Bray—Curtis distance, Spearman distance, Hellinger distance, consider species' presence/absence and relative abundances. The Unifrac distance (Lozupone and Knight, 2005), a third type of measure, attempts to place community similarities and distances in a phylogenetic context, and quantifies the amount of phylogeny (i.e., branch length on a common phylogenetic tree) shared between two communities. Once similarity or dissimilarity values have been calculated among a set of communities, they are commonly communicated using ordination plots. Nonmetric multidimensional scaling (NMDS) and principal coordinates analysis (PCoA) plots are frequently used for this purpose (Figure 19.5).

21.7.1.7 Phylogenetic Analyses

Phylogenetic analysis is another route for analyzing marker gene sequences, especially in the case of 16S rRNA gene data. The use of the Unifrac metric helps to place communities in a phylogenetic context by first building a large phylogenetic tree, and then calculating the amount of the tree that is shared between two or more communities. The amount of phylogenetic diversity within a single sample can also be calculated this way.

From the perspective of single sequences or OTUs, phylogenetic analysis more typically involves trying to place a sequence or OTU of interest into phylogenetic context by comparing it with sequences of known origin. This process is similar to that which is used to generate OTUs: (1) a collection of sequences is gathered; (2) all sequences are compared with one another to determine the amount of sequence similarity that they share with one another; (3) these distances are interpreted and used to identify "nearest neighbors" (i.e., closest relatives), and may be used to construct a phylogenetic tree. This approach is commonly used to help describe the identity of a sequence or OTU whose best match in a public database is an uncultured or unclassified bacterium (or archaeon or fungus). It has also been used to identify highly novel organisms and provide justification for the addition of new phyla (Hugenholtz et al., 1998), and potentially even taxonomic domains (Wu et al., 2011).

21.7.2 Bioinformatics and Analysis of Genomic/Metagenomic Data

Common first steps in the analysis of genomic or metagenomic data are an assessment of sequence quality and the removal of low-quality reads. Depending on the downstream analyses that will be performed, this can be a very important step. As each base is incorporated during a sequencing reaction, a score indicating the quality of each

base call is also generated and recorded into the sequencing record. Often, base calls at the 5' and 3' ends of a sequence read are of lower quality than those that are incorporated in the middle. Likewise, overly long or extremely short reads also tend to be of lower quality, especially those produced on the 454 platform (Huse et al., 2007). Quality scores can be used by some assembly algorithms, but many commonly used assemblers do not take them into consideration (Mende et al., 2012). As a result, trimming and quality filtering of raw sequence reads is often advised, and tends to lead to more accurate genome and metagenome assemblies (DiGuistini et al., 2009; Mende et al., 2012).

21.7.2.1 Assembly-Based Approaches

Once a genome or metagenome has been sequenced, it is much like a jigsaw puzzle (or a collection of many jigsaw puzzles). It represents a large collection of pieces, some of which are informative on their own, and others of which yield better information and a more complete picture once they are assembled and placed in context with other fragments. Also like a jigsaw puzzle, genome and metagenomic sequence data may contain pieces (i.e., sequence fragments) that are duplicated, misshapen (i.e., contain errors) or missing. These add to the challenge of sequence assembly and interpretation, but they do not preclude it completely.

During the assembly process, fragments of sequence that originated from the same parent sequence are identified, and ordered relative to one another to build a larger, contiguous strand of sequence, also known as a contig. Contigs are typically constructed by identifying regions of common, overlapping sequence that are shared between the two smaller sequence fragments. Depending on the sequencing approach used, spatial information (i.e., known distances between fragments) may also be available to aid in the assembly process, and provide a degree of quality control. For example, if it is known that the ends of two different fragments should be oriented 1000 bp apart from one another, the distance can be used as a placeholder, which helps to constrain (i.e., control) the addition of new sequences and contigs. As multiple contigs are joined into longer and longer sequences, scaffolds are formed. Scaffolds are not necessarily contiguous runs of sequence, but can include gaps of known length. Depending on the complexity of the sample and the depth to which it is sequenced, assembly from metagenomic sequencing can yield high-quality draft, or even complete genome sequences.

Multiple approaches and software packages have been developed for the purpose of sequence assembly. The earliest assemblers were designed to piece together single genomes with fragments of relatively long read length. As the high-throughput sequencing of shorter gene fragments and the sequencing of mixed communities (i.e., metagenomes) became more common, newer assemblers designed to handle greater levels of complexity were developed [e.g., Velvet (Zerbino and Birney, 2008), SOAP (Li *et al.*, 2010)].

Regardless of the sequence data or assembly algorithm used, the assembly process can be quite computationally intensive. As a result, preprocessing algorithms have emerged to help reduce the complexity and redundancy of the input data, and reduce the computational load required to complete the assembly (Pell *et al.*, 2012). This is particularly important in complex and highly diverse communities, such as those found in soil, where large amounts of sequence data must be generated in order to provide adequate coverage of the community.

21.7.2.2 Mapping to Reference Genomes

A reference genome, also known as a reference assembly, is a collection of nucleic acid sequence and annotation information describing the gene content of an organism. The nucleic acid sequences may be assembled (i.e., pieced together from smaller sequence fragments) into contigs, scaffolds or complete chromosomes. Often, open reading frames (ORFs) for individual genes will be identified, and attempts will be made to annotate or assign an identity and/or function to each of the ORFs.

Reference genomes play an important role in shaping our interpretation of new genomes and metagenomic data. Just as the "reference" portion of their name implies, reference genomes can serve as a framework for describing the gene content—both in terms of taxonomic origin and potential function—of new genomes and metagenomes. A common step in analyzing shotgun sequence datasets, like metagenomes, is to "map" the unassembled reads to a collection of reference genomes. This can be done using BLAST searches, but fast, memory-efficient alignment algorithms, such as bowtie (Langmead *et al.*, 2009) or BWA (Li and Durbin, 2010), are more commonly used for this purpose. The mapping algorithms search for regions of homology (i.e., similarity) between the reference genome and the sequence of interest. The amount of similarity that they share, the degree of coverage (i.e., amount of the genome that generates matches within your pool of shotgun sequences) and the depth of coverage (i.e., the number of copies of each gene or genome found within the shotgun sequence pool) influence the quality of the amount of information that can be derived from the "map."

The mapping of reads to reference genomes can be used to identify and remove host-derived reads if the community of interest has come from a plant, animal or insect host. By mapping metagenomic reads from various body sites sampled during the Human Microbiome Project to a reference (human) genome, it was discovered that human-derived reads accounted for approximately 1% of the sequences generated from stool, but 80% of more of the sequences generated from samples of saliva, the anterior nares (nostril) and vagina (Human Microbiome Project Consortium, 2012). While the identification and removal of host "contamination" represents an important application of reference genome mapping, the technique can also be used to evaluate the potential origins of your reads. For example, by mapping metagenomic sequence reads to reference genomes, researchers studying a mixed-community, cellulosic bioreactor system were able to determine that their reactor harbored a variety of cellulose- and xylose-degrading bacteria, including *Clostridium thermocellum*, *Thermoanaerobacterium thermosaccharolyticum* and *Moorella thermoacetica* (Hollister *et al.*, 2012). They also learned that their reactor-housed bacteria shared some similarity with previously sequenced *Bacillus* spp.; however, the degree of similarity was low enough and the maps sparse enough to suggest that they had encountered novel, or at least unsequenced, species.

Historically, collections of reference genomes have been biased toward the inclusion of model organisms, pathogens and other organisms of economic or biotechnological importance, but in recent years, large-scale sequencing projects like the Human Microbiome Project (HMP) (Nelson *et al.*, 2010) and the Genomic Encyclopedia of Bacteria and Archaea (GEBA) (Wu *et al.*, 2009) have increased the scope and size of reference genome collections, systematically generating new genome sequences in the attempt to fill out the underrepresented portions of the microbial tree of life. They have utilized innovative isolation and culture techniques (Pope *et al.*, 2011), single cell sequencing (Rinke *et al.*, 2013) and in some cases assembly from metagenome sequences (Hess *et al.*, 2011).

Since the first bacterial genome (*Haemophilus influenza*) was sequenced in the mid-1990s (Fleischmann *et al.*, 1995), the collection of publicly available reference genomes has grown to include > 6000 high-quality draft or completed bacterial and archaeal reference genomes and > 300 eukaryotic reference genomes (Genomes Online Database, http://www.genomesonline.org, July 2013). A recent evaluation of the publically available reference genome collection found that the addition of new genomes as a result of the Human Microbiome Project reference genome sequencing initiative resulted in a 20−40% improvement in read recruitment from human metagenome samples than would have been possible previously (Nelson *et al.*, 2010). Likewise, the recent release of > 200 genomes generated by single cell sequencing is estimated to have increased the phylogenetic coverage of publically available reference genomes by $>11\%$ (Rinke *et al.*, 2013). Although this growth is impressive, much more work remains to be done (Fodor *et al.*, 2012). Our understanding and appreciation of the microbial world is fundamentally linked to the information contained in reference genome collections, and it is anticipated that continued efforts to expand these collections will provide new insight into microbial structure, function and evolution.

21.7.2.3 Databases

As the ability to generate genome and metagenome sequence data has grown, so too has the need to analyze, store and share it. Even as improved algorithms for sequence assembly and annotation are developed, the archiving, analysis and dispersal of genome and metagenomic sequence data is no trivial task. Powerful servers with large storage capacity are typically required to handle the data associated with these large and ever-growing projects. These resource requirements are often greater than individual academic laboratories can support, but centralized databases and similar repositories also serve a valuable purpose in their ability to facilitate the sharing of data within the scientific community.

A variety of databases have been developed with these needs in mind. Some, like the Sequence Read Archive (SRA) at the National Center for Biotechnology Information (NCBI) or the European Nucleotide Archive (ENA) at the European Molecular Biology Laboratory, house compressed versions of the raw (or sometimes assembled) sequence data and associated metadata from genome and metagenomic sequencing projects. Others, like the Integrated Microbial Genomes and Metagenomes (IMG) system (Markowitz et al., 2010), MG-RAST (Meyer et al., 2008), CAMERA (Sun et al., 2011) and the EBI Metagenomics service (https://www.ebi.ac.uk/metagenomics/), will house data, but also allow users to upload genome or metagenomic sequence data and analyze it. Common options offered by these platforms often include sequence assembly, annotation and the ability to carry out comparative analyses.

In addition to the archiving and analysis of genome and metagenome data, the rapid growth of genomic and metagenomic sequencing projects had led to the need to track and catalogue them. Despite the fact that the costs associated with generating sequence have declined, storage and dissemination of data still remain a challenge, and preventing the duplication of projects can help to reduce these burdens. The Genomes OnLine Database (GOLD) (Pagani et al., 2012), first established in 1997, has emerged to fill this need. GOLD serves as a central repository for information about sequencing projects, including genomes and metagenomes, as well as genome resequencing projects, single cell sequencing projects and (meta)transcriptomes. Information catalogued in GOLD includes project type, sequencing status (e.g., targeted, in progress, complete), project metadata, organism phylogeny and contact information for the scientist or research group leading the project efforts.

21.7.3 Integration of 'Omics Data

The ability to generate multiple 'omics datasets from the same system, at the same point in time, has the potential to provide a highly detailed picture of the system's biology and ecology. Efforts to integrate multiple 'omics technologies with one another are still relatively few, especially in mixed microbial communities, and they often rely on the layering of 'omics data onto reference pathways or the correlation of one 'omics data set with another.

21.7.3.1 Layering of -Omics Data Using Reference Databases

Reference databases, such as the Kyoto Encyclopedia of Genes and Genomes (KEGG) (Kanehisa et al., 2012), BioCyc and MetaCyc (Caspi et al., 2012), provide curated, and often experimentally verified, information regarding metabolic pathways, and the enzymes, reactions, compounds and genes that allow them to function. Although many metabolic pathways occur commonly across the Tree of Life, these reference databases also include information regarding the (known) taxonomic distribution of particular genes and pathways.

Reference databases like KEGG and MetaCyc can serve as a platform to support the exploration of 'omics data, and new software tools are emerging to allow these databases to serve as a platform for 'omics integration. The KEGG database can be accessed directly through a web-based interface (http://www.genome.jp/kegg/), allowing users to explore genes, compounds and pathways, or to input information about differentially detected genes and compounds to assess which functional pathways may be affected. Likewise, the interactive Pathways Explorer (iPath, http://pathways.embl.de/) is a web-based tool that allows for the visualization, analysis and customization of pathway information, and the Pathview package (Luo and Brouwer, 2013) is a standalone package for multi-'omics integration.

The value of multiple-'omics datasets is often realized in the ability of one data type to confirm or refute the results of another. For example, following the *Deepwater Horizon* oil spill in the Gulf of Mexico in 2010, Mason et al. (2012) used a combination of metagenomic, metatranscriptomic and single-cell genomic sequencing to characterize microbial community responses. The shotgun metagenomic sequencing results revealed that, relative to the microbial communities inhabiting uncontaminated seawater, the hydrocarbon-exposed communities were significantly enriched in genes related to motility, chemotaxis and aliphatic hydrocarbon degradation, as well as the degradation of more recalcitrant compounds, including benzene, toluene and polycyclic aromatic hydrocarbons (see also Chapter 31). Analysis of the transcriptome of these communities confirmed that the expression of the chemotaxis, motility and aliphatic hydrocarbon degradation genes was significantly enhanced. Surprisingly, though, the transcriptomic analysis found that the expression of genes related to the degradation of the more recalcitrant compounds had not changed, demonstrating that although differences in gene abundance profiles can provide strong clues about how a system works, the addition of transcriptomic and/or metabolomic data can be important to pinpoint which genes, compounds and metabolites are actually being used.

21.7.3.2 Correlation and Network-Based Approaches

The layering of 'omics data onto reference pathways allows for the exploration of these datasets in the context of known, well-characterized reactions and pathways. In the case of multi-'omics studies, it allows functionally related data types (e.g., genes and compounds involved in the same reaction) to be considered in the context of known biology and biochemical reactions. While layering approaches can be very useful, they are not necessarily designed to convey global patterns and relationships within and between 'omics datasets.

In contrast, correlation and network-based approaches can be employed to identify co-occurrence and/or co-abundance patterns within and between 'omics datasets. Correlation and network-based approaches are relatively simple, and are often naïve to the errors inherent in 'omics measurements, and the biology that they are being used to describe. These approaches often employ Pearson or Spearman correlations and can be as simple as asking, "Does gene 'A' occur with similar abundances as genes 'B' and 'C' or metabolite 'D' across all of my samples, or do some bacteria always (or never) occur together in my samples?" While these may seem like simple questions, the potential does exist for correlations to identify artifacts of the data rather than true biological relationships (Friedman and Alm, 2012). As such, more sophisticated methods for examining correlations have been proposed. These include partial least squares regression (Pir et al., 2006), sparse correlations for compositional data (Friedman and Alm, 2012) and generalized boosted linear models (Faust et al., 2012).

Despite the potential pitfalls of correlation-based analyses, they have been used with success in many studies, and have the potential to reveal new insights into the biology of a system of interest. For example, within-'omic (i.e., analyses using a single-'omic technology) correlations have been used to assign functional context to genes of unknown identity (Wang et al., 2012; Buttigieg et al., 2013), identify genes, metabolites or

bacteria that are associated with, or characteristic of, ecological or environmental subtypes (Bhavnani et al., 2011; Barberan et al., 2012; Greenblum et al., 2012), and provide insight into the culture of previously uncultivable organisms (Duran-Pinedo et al., 2011). Examples of cross-'omic (i.e., multi-'omic) correlations are fewer in number, but recent attempts to integrate transcriptome and metabolite datasets from laboratory chemostats (Pir et al., 2006) and metabolites and community composition in the human gut (McHardy et al., 2013) have been described in the literature. The integration of mixed 'omics datasets is considered to be at the forefront of science and has tremendous potential for characterizing environmental microorganisms; however, it represents a technological challenge that remains to be fully resolved, especially for the study of complex microbial communities.

QUESTIONS AND PROBLEMS

1. Discuss the potential advantages and disadvantages of the various 'omics approaches for characterizing environmental microorganisms.
2. What is the value of reference genomes?
3. What kind(s) of information can be learned from 16S rRNA gene sequences? From genomic or metagenomic sequencing?
4. Which 'omics approach provides the most direct indication of microbial activity.
5. Discuss the major quality criteria used for processing DNA sequence data.
6. What is microbial diversity? How can it be determined using 'omics -based approaches?
7. Discuss the major limitations of 'omics approaches for studying microbial community diversity.

REFERENCES AND RECOMMENDED READING

Abarenkov, K., Tedersoo, L., Nilsson, R. H., Vellak, K., Saar, I., Veldre, V., et al. (2010) PlutoF—a web based workbench for ecological and taxonomic research, with an online implementation for fungal ITS sequences. Evol. Bioinform. Online 6, 189—196.
Amend, A. S., Seifert, K. A., Samson, R., and Bruns, T. D. (2010) Indoor fungal composition is geographically patterned and more diverse in temperate zones than in the tropics. Proc. Natl. Acad. Sci. U.S.A. 107, 13748—13753.
Ashelford, K. E., Chuzhanova, N. A., Fry, J. C., Jones, A. J., and Weightman, A. J. (2005) At least 1 in 20 16S rRNA sequence records currently held in public repositories is estimated to contain substantial anomalies. Appl. Environ. Microbiol. 71, 7724—7736.
Bailly, J., Fraissinet-Tachet, L., Verner, M. C., Debaud, J. C., Lemaire, M., Wesolowski-Louvel, M., et al. (2007) Soil eukaryotic functional diversity, a metatranscriptomic approach. ISME J. 1, 632—642.

Barberan, A., Bates, S. T., Casamayor, E. O., and Fierer, N. (2012) Using network analysis to explore co-occurrence patterns in soil microbial communities. *ISME J.* **6**, 343–351.

Bhavnani, S. K., Victor, S., Calhoun, W. J., Busse, W. W., Bleecker, E., Castro, M., *et al.* (2011) How cytokines co-occur across asthma patients: from bipartite network analysis to a molecular-based classification. *J. Biomed. Inform.* **44**, S24–S30.

Bray, J. R., and Curtis, J. T. (1957) An ordination of the upland forest communities in southern Wisconsin. *Ecol. Monographs* **27**, 325–349.

Bundy, J. G., Davey, M. P., and Viant, M. R. (2009) Environmental metabolomics: a critical review and future perspectives. *Metabolomics* **5**, 3–21.

Buttigieg, P. L., Hankeln, W., Kostadinov, I., Kottmann, R., Yilmaz, P., Duhaime, M. B., *et al.* (2013) Ecogenomic perspectives on domains of unknown function: correlation-based exploration of marine metagenomes. *PLoS One* **8**, e50869.

Caporaso, J. G., Kuczynski, J., Stombaugh, J., Bittinger, K., Bushman, F. D., Costello, E. K., *et al.* (2010) QIIME allows analysis of high-throughput community sequencing data. *Nat. Methods* **7**, 335–336.

Carvalhais, L. C., Dennis, P. G., Tyson, G. W., and Schenk, P. M. (2012) Application of metatranscriptomics to soil environments. *J. Microbiol. Meth.* **91**, 246–251.

Caspi, R., Altman, T., Dreher, K., Fulcher, C. A., Subhraveti, P., Keseler, I. M., *et al.* (2012) The MetaCyc database of metabolic pathways and enzymes and the BioCyc collection of pathway/ genome databases. *Nucleic Acids Res.* **40**, D742–D753.

Chao, A. (1984) Nonparametric estimation of the number of classes in a population. *Scand. J. Statist.* **11**, 265–270.

Cole, J. R., Chai, B., Farris, R. J., Wang, Q., Kulam-Syed-Mohideen, A. S., McGarrell, D. M., *et al.* (2007) The Ribosomal Database Project (RDP-II): introducing myRDP space and quality controlled public data. *Nucleic Acids Res.* **35**, D169–D172.

Cole, J. R., Wang, Q., Cardenas, E., Fish, J., Chai, B., Farris, R. J., *et al.* (2009) The Ribosomal Database Project: improved alignments and new tools for rRNA analysis. *Nucleic Acids Res.* **37**, D141–D145.

de Menezes, A., Clipson, N., and Doyle, E. (2012) Comparative metatranscriptomics reveals widespread community responses during phenanthrene degradation in soil. *Environ. Microbiol.* **14**, 2577–2588.

DiGuistini, S., Liao, N., Platt, D., Robertson, G., Seidel, M., Chan, S. K., *et al.* (2009) De novo genome sequence assembly of a filamentous fungus using Sanger, 454 and Illumina sequence data. *Genome Biol.* **10**, R94.

Don, R. H., and Pemberton, J. M. (1981) Properties of six pesticide degradation plasmids isolated from *Alcaligenes paradoxus* and *Alcaligenes eutrophus. Appl. Environ. Microbiol.* **145**, 681–686.

Dunn, W. B. (2008) Current trends and future requirements for the mass spectrometric investigation of microbial, mammalian, and plant metabolomes. *Phys. Biol.* **5**, 1–24.

Duran-Pinedo, A. E., Paster, B., Teles, R., and Frias-Lopez, J. (2011) Correlation network analysis applied to complex biofilm communities. *PLoS One* **6**, e28438.

Edgar, R. C., Haas, B. J., Clemente, J. C., Quince, C., and Knight, R. (2011) UCHIME improves sensitivity and speed of chimera detection. *Bioinformatics* **27**, 2194–2200.

Faith, D. P. (1992) Conservation evaluation and phylogenetic diversity. *Biol. Conserv.* **61**, 1–10.

Faust, K., Sathirapongsasuti, J. F., Izard, J., Segata, N., Gevers, D., Raes, J., *et al.* (2012) Microbial co-occurrence relationships in the human microbiome. *PLoS Comput. Biol.* **8**, e1002606.

Fleischmann, R. D., Adams, M. D., White, O., Clayton, R. A., Kirkness, E. F., Kerlavage, A. R., *et al.* (1995) Whole-genome random sequencing and assembly of *Haemophilus influenzae* Rd. *Science* **269**, 496–512.

Fodor, A. A., DeSantis, T. Z., Wylie, K. M., Badger, J. H., Ye, Y., Hepburn, T., *et al.* (2012) The "Most Wanted" taxa from the human microbiome for whole genome sequencing. *PLoS One* **7**, e41294.

Friedman, J., and Alm, E. J. (2012) Inferring correlation networks from genomic survey data. *PLoS Comput. Biol.* **8**, e1002687.

Gontcharova, V., Youn, E., Wolcott, R. D., Hollister, E. B., Gentry, T. J., and Dowd, S. E. (2010) Black Box Chimera Check (B2C2): a windows-based software for batch depletion of chimeras from bacterial 16S rRNA gene datasets. *Open Microbiol. J.* **4**, 47–52.

González, J. M., and Robb, F. T. (2000) Genetic analysis of *Carboxydothermus hydrogenoformans* carbon monoxide dehydrogenase genes cooF and cooS. *FEMS Microbiol. Lett.* **191**, 243–247.

Gosalbes, M. J., Durban, A., Pignatelli, M., Abellan, J. J., Jimenez-Hernandez, N., Perez-Cobas, A. E., *et al.* (2011) Metatranscriptomic approach to analyze the functional human gut microbiota. *PLoS One* **6**, 1–9.

Greenblum, S., Turnbaugh, P. J., and Borenstein, E. (2012) Metagenomic systems biology of the human gut microbiome reveals topological shifts associated with obesity and inflammatory bowel disease. *Proc. Natl. Acad. Sci. U.S.A.* **109**, 594–599.

Haas, B. J., Gevers, D., Earl, A. M., Feldgarden, M., Ward, D. V., Giannoukos, G., *et al.* (2011) Chimeric 16S rRNA sequence formation and detection in Sanger and 454-pyrosequenced PCR amplicons. *Genome Res.* **21**, 494–504.

Handelsman, J., Rondon, M. R., Brady, S. F., Clardy, J., and Goodman, R. M. (1998) Molecular biological access to the chemistry of unknown soil microbes: a new frontier for natural products. *Chem. Biol.* **5**, R248–R249.

Hemme, C. L., Deng, Y., Gentry, T. J., Fields, M. W., Wu, L., Barua, S., *et al.* (2010) Metagenomic insights into evolution of a heavy metal-contaminated groundwater microbial community. *ISME J.* **4**, 660–672.

Hess, M., Sczyrba, A., Egan, R., Kim, T. W., Chokhawala, H., Schroth, G., *et al.* (2011) Metagenomic discovery of biomass-degrading genes and genomes from cow rumen. *Science* **331**, 463–467.

Hettich, R. L., Pan, C., Chorney, K., and Giannone, R. J. (2013) Metaproteomics: harnessing the power of high performance mass spectrometry to identify the suite of proteins that control metabolic activities in microbial communites. *Anal. Chem.* **85**, 4203–4214.

Hollister, E. B., Forrest, A. K., Wilkinson, H. H., Ebbole, D. J., Tringe, S. G., Malfatti, S. A., *et al.* (2012) Mesophilic and thermophilic conditions select for unique but highly parallel microbial communities to perform carboxylate platform biomass conversion. *PLoS One* **7**, e39689.

Huber, T., Faulkner, G., and Hugenholtz, P. (2004) Bellerophon: a program to detect chimeric sequences in multiple sequence alignments. *Bioinformatics* **20**, 2317–2319.

Hugenholtz, P., Goebel, B. M., and Pace, N. R. (1998) Impact of culture-independent studies on the emerging phylogenetic view of bacterial diversity. *J. Bacteriol.* **180**, 4765–4774.

Human Microbiome Project Consortium (2012) A framework for human microbiome research. *Nature* **486**, 215–221.

Hunter, C., Cochrane, G., Apweiler, R., and Hunter, S. (2011) Chapter 38: the EBI Metagenomics Archive, integration and analysis resource. In "Handbook of Molecular Microbial Ecology, Volume I: Metagenomics and Complementary Approaches" (F. J. de Bruijn, ed.), Wiley-Blackwell, Hoboken, New Jersey, U.S.A, pp. 333–340.

Huse, S. M., Huber, J. A., Morrison, H. G., Sogin, M. L., and Welch, D. M. (2007) Accuracy and quality of massively parallel DNA pyrosequencing. *Genome Biol.* **8**, R143.

Huse, S. M., Welch, D. M., Morrison, H. G., and Sogin, M. L. (2010) Ironing out the wrinkles in the rare biosphere through improved OTU clustering. *Environ. Microbiol.* **7**, 1889–1898.

Ishii, N., Nakahigashi, K., Baba, T., Robert, M., Soga, T., Kanai, A., *et al.* (2007) Multiple high-throughput analyses monitor the response of E. coli to perturbations. *Science* **316**, 593–597.

Ito, T., Tanaka, M., Shinkawa, H., Nakada, T., Ano, Y., Kurano, N., *et al.* (2013) Metabolic and morphological changes of an oil accumulating trebouxiophycean alga in nitrogen-deficient conditions. *Metabolomics* **9**, S178–S187.

Jaccard, P. (1908) Nouvelles recherches sur la distribution florale. *Bulletin de la Société Vaudoise des Sciences Naturelles* **44**, 223–270.

Kanehisa, M., Goto, S., Kawashima, S., Okuno, Y., and Hattori, M. (2004) The KEGG resource for deciphering the genome. *Nucleic Acids Res.* **32**, D277–D280.

Kanehisa, M., Goto, S., Sato, Y., Furumichi, M., and Tanabe, M. (2012) KEGG for integration and interpretation of large-scale molecular data sets. *Nucleic Acids Res.* **40**, D109–D114.

Kim, S. I., Kim, J. Y., Yun, S. H., Kim, J. H., Lee, S. H., and Lee, C. (2004) Proteome analysis of *Pseudomonas* sp. K82 biodegradation pathways. *Proteomics* **4**, 3610–3621.

Kyle, J. L., Parker, C. T., Goudeau, D., and Brandl, M. T. (2010) Transcriptome analysis of *Escherichia coli* O157:H7 exposed to lysastes of lettuce leaves. *Appl. Environ. Microbiol.* **76**, 1375–1387.

Langmead, B., Trapnell, C., Pop, M., and Salzberg, S. (2009) Ultrafast and memory-efficient alignment of short DNA sequences to the human genome. *Genome Biol.* **10**, R25.

Laserson, J., Jojic, V., and Koller, D. (2011) Genovo: de novo assembly for metagenomes. *J. Comput. Biol.* **18**, 429–443.

Laskin, R. S. (2012) Genomic sequencing of uncultured microorganisms from single cells. *Nat. Rev. Microbiol.* **10**, 631–640.

Li, H., and Durbin, R. (2010) Fast and accurate long-read alignment with Burrows-Wheeler transform. *Bioinformatics* **26**, 589–595.

Li, R., Zhu, H., Ruan, J., Qian, W., Fang, X., Shi, Z., *et al.* (2010) De novo assembly of human genomes with massively parallel short read sequencing. *Genome Res.* **20**, 265–272.

Liebeke, M., Brozel, V. S., Hecker, M., and Lalk, M. (2009) Chemical characterization of soil extract as growth media for the ecophysiological study of bacteria. *Appl. Microbiol. Biotechnol.* **83**, 161–173.

Lozupone, C., and Knight, R. (2005) UniFrac: a new phylogenetic method for comparing microbial communities. *Appl. Environ. Microbiol.* **71**, 8228–8235.

Luo, W., and Brouwer, C. (2013) Pathview: an R/bioconductor package for pathway-based data integration and visualization. *Bioinformatics* **29**, 1830–1831.

Magurran, A. E. (2004) "Measuring Biological Diversity," Blackwell Publishing, Maldan, MA.

Markowitz, V. M., Chen, I. M, Palaniappan, K., Chu, K., Szeto, E., Grechkin, Y., *et al.* (2010) The integrated microbial genomes system: an expanding comparative analysis resource. *Nucleic Acids Res.* **38**, D382–D390.

Mason, O. U., Hazen, T. C., Borglin, S., Chain, P. S. G., Dubinsky, E. A., Fortney, J. L., *et al.* (2012) Metagenome, metatranscriptome and single-cell sequencing reveal microbial response to Deepwater Horizon oil spill. *ISME J.* **6**, 1715–1727.

McHardy, I., Goudarzi, M., Tong, M., Ruegger, P., Schwager, E., Weger, J., *et al.* (2013) Integrative analysis of the microbiome and metabolome of the human intestinal mucosal surface reveals exquisite inter-relationships. *Microbiome* **1**, 17.

Mende, D. R., Waller, A. S., Sunagawa, S., Järvelin, A. I., Chan, M. M., Arumugam, M., *et al.* (2012) Assessment of metagenomic assembly using simulated next generation sequencing data. *PLoS One* **7**, e31386.

Meyer, F., Paarmann, D., D'Souza, M., Olson, R., Glass, E., Kubal, M., *et al.* (2008) The metagenomics RAST server—a public resource for the automatic phylogenetic and functional analysis of metagenomes. *BMC Bioinformatics* **9**, 386.

Mueller, R. S., Dill, B. D., Pan, C., Belnap, C. P., Thomas, B. C., VerBerkmoes, N. C., *et al.* (2011) Proteome changes in the initial bacterial colonist during ecological succession in an acid mine drainage biofilm community. *Environ. Microbiol.* **13**, 2279–2292.

Nelson, K. E., Clayton, R. A., Gill, S. R., Gwinn, M. L., Dodson, R. J., Haft, D. H., *et al.* (1999) Evidence for lateral gene transfer between Archaea and bacteria from genome sequence of *Thermotoga maritima*. *Nature* **399**, 323–329.

Nelson, K. E., Weinstock, G. M., Highlander, S. K., Worley, K. C., Creasy, H. H., Wortman, J. R., *et al.* (2010) A catalog of reference genomes from the human microbiome. *Science* **328**, 994–999.

O'Brien, H. E., Parrent, J. L., Jackson, J. A., Moncalvo, J. M., and Vilgalys, R. (2005) Fungal community analysis by large-scale sequencing of environmental samples. *Appl. Environ. Microbiol.* **71**, 5544–5550.

Pagani, I., Liolios, K., Jansson, J., Chen, I. M., Smirnova, T., Nosrat, B., *et al.* (2012) The Genomes OnLine Database (GOLD) v.4: status of genomic and metagenomic projects and their associated metadata. *Nucleic Acids Res.* **40**, D571–D579.

Pell, J., Hintze, A., Canino-Koning, R., Howe, A., Tiedje, J. M., and Brown, C. T. (2012) Scaling metagenome sequence assembly with probabilistic de Bruijn graphs. *Proc. Natl. Acad. Sci. U.S.A.* **109**, 13272–13277.

Pir, P., Kirdar, B., Hayes, A., Onsan, Z. I., Ulgen, K., and Oliver, S. (2006) Integrative investigation of metabolic and transcriptomic data. *BMC Bioinf.* **7**, 203.

Podar, M., Abulencia, C. B., Walcher, M., Hutchison, D., Zengler, K., Garcia, J. A., *et al.* (2007) Targeted access to the genomes of low-abundance organisms in complex microbial communities. *Appl. Environ. Microbiol.* **73**, 3205–3214.

Pope, P. B., Smith, W., Denman, S. E., Tringe, S. G., Barry, K., Hugenholtz, P., *et al.* (2011) Isolation of Succinivibrionaceae implicated in low methane emissions from Tammar wallabies. *Science* **333**, 646–648.

Ram, R. J., VerBerkmoes, N. C., Thelen, M. P., Tyson, G. W., Baker, B. J., Blake, R. C., *et al.* (2005) Community proteomics of a natural microbial biofilm. *Science* **308**, 1915–1920.

Riehle, K., Coarfa, C., Jackson, A., Ma, J., Tandon, A., Paithankar, S., *et al.* (2012) The Genboree Microbiome Toolset and the analysis of 16S rRNA microbial sequences. *BMC Bioinformatics* **13**, S11.

Rinke, C., Schwientek, P., Sczyrba, A., Ivanova, N. N., Anderson, I. J., Cheng, J-F., *et al.* (2013) Insights into the phylogeny and coding potential of microbial dark matter. *Nature* **499**, 431−437.

Rosenzweig, M. L. (1995) "Species Diversity in Space and Time," Cambridge University Press, Cambridge, UK.

Schloss, P. D., and Handelsman, J. (2005) Introducing DOTUR, a computer program for defining operational taxonomic units and estimating species richness. *Appl. Environ. Microbiol.* **71**, 1501−1506.

Schloss, P. D., Westcott, S. L., Ryabin, T., Hall, J. R., Hartmann, M., Hollister, E. B., *et al.* (2009) Introducing Mothur: open-source, platform-independent, community-supported software for describing and comparing microbial communities. *Appl. Environ. Microbiol.* **75**, 7537−7541.

Schloss, P. D., Gevers, D., and Westcott, S. L. (2011) Reducing the effects of PCR amplification and sequencing artifacts on 16S rRNA-based studies. *PLoS One* **6**, e27310.

Selinger, D. W., Saxena, R. M., Cheung, K. J., Church, G. M., and Rosenow, C. (2003) Global RNA half-life analysis in *Escherichia coli* reveals positional patterns of transcript degradation. *Genome Res.* **13**, 216−223.

Shannon, C. E., and Weaver, W. (1949) "The Mathematical Theory of Communication," University of Illinois Press, Urbana, IL.

Simpson, E. H. (1949) Measurement of diversity. *Nature* **163**, 688.

Stackenbrandt, E., and Goebel, B. M. (1994) Taxonomic note: a place for DNA−DNA reassociation and 16S rRNA sequence analysis in the present species definition in bacteriology. *Int. J. Syst. Bacteriol.* **44**, 846−849.

Sun, S., Chen, J., Li, W., Altintas, I., Lin, A., Peltier, S., *et al.* (2011) Community cyberinfrastructure for Advanced Microbial Ecology Research and Analysis: the CAMERA resource. *Nucleic Acids Res.* **39**, D546−D551.

Sørensen, T. (1948) A method of establishing groups of equal amplitude in plant sociology based on similarity of species and its application to analyses of the vegetation on Danish commons. *Kongelige Danske Videnskabernes Selskab* **5**, 1−34.

Thomas, T., Gilbert, J., and Meyer, F. (2012) Metagenomics—a guide from sampling to data analysis. *Microb. Inform. Exp.* **2**, 3.

Tyson, G. W., Chapman, J., Hugenholtz, P., Allen, E. E., Ram, R. J., Richardson, P. M., *et al.* (2004) Community structure and metabolism through reconstruction of microbial genomes from the environment. *Nature* **428**, 37−43.

VerBerkmoes, N. C., Denef, V. J., Hettich, R. L., and Banfield, J. F. (2009) Functional analysis of natural microbial consortia using community proteomics. *Nat. Rev. Microbiol.* **7**, 196−205.

Wang, P. I., Hwang, S., Kincaid, R. P., Sullivan, C. S., Lee, I., and Marcotte, E. M. (2012) RIDDLE: reflective diffusion and local extension reveal functional associations for unannotated gene sets via proximity in a gene network. *Genome Biol.* **13**, R125.

Wessel, A. K., Hmelo, L., Parsek, M. R., and Whiteley, M. (2013) Going local: technologies for exploring bacterial microenvironments. *Nat. Rev. Microbiol.* **11**, 337−348.

Westermeier, R., and Naven, T. (2002) Proteomics technology. In "Proteomics in Practice: A Laboratory Manual of Proteome Analysis" (J. Adams, ed.), Wiley-VCH, Darmstadt, Germany, pp. 1−160.

Whittaker, R. H. (1960) Vegetation of the Siskiyou Mountains, Oregon and California. *Ecol. Monographs* **30**, 279−338.

Wu, D., Hugenholtz, P., Mavromatis, K., Pukall, R., Dalin, E., Ivanova, N. N., *et al.* (2009) A phylogeny-driven genomic encyclopaedia of Bacteria and Archaea. *Nature* **462**, 1056−1060.

Wu, D., Wu, M., Halpern, A., Rusch, D. B., Yooseph, S., Frazier, M., *et al.* (2011) Stalking the fourth domain in metagenomic data: searching for, discovering, and interpreting novel, deep branches in marker gene phylogenetic trees. *PLoS One* **6**, e18011.

Wu, M., Ren, Q., Durkin, A. S., Daugherty, S. C., Brinkac, L. M., Dodson, R. J., *et al.* (2005) Life in hot carbon monoxide: the complete genome sequence of *Carboxydothermus hydrogenoformans* Z-2901. *PLoS Genet.* **1**, 563−574.

Zengler, K., Walcher, M., Clark, G., Haller, I., Toledo, G., Holland, T., *et al.* (2005) High-throughput cultivation of microorganisms using microcapsules. *Methods Enzymol.* **397**, 124−130.

Zerbino, D. R., and Birney, E. (2008) Velvet: algorithms for de novo short read assembly using de Bruijn graphs. *Genome Res.* **18**, 821−829.

Zhang, W., Li, F., and Nie, L. (2010) Integrating multiple 'omics analysis for microbial biology: applications and methodologies. *Microbiology* **156**, 287−301.

Part VI

Water- and Foodborne Pathogens

Environmentally Transmitted Pathogens

Charles P. Gerba

22.1 ENVIRONMENTALLY TRANSMITTED PATHOGENS

Although humans are continually exposed to a vast array of microorganisms in the environment, only a small proportion of these microbes are capable of interacting with the host in such a manner that infection and disease will result. Disease-causing microorganisms are called pathogens. Infection is the process in which the microorganism multiplies or grows in or on the host. Infection does not necessarily result in disease, since it is possible for the organism to grow in or on the host without producing an illness (see Chapter 23). In the case of enteric infections (i.e., diarrhea) caused by *Salmonella*, only half of the individuals infected develop clinical signs of illness. A frank pathogen is a microorganism capable of producing disease in both normal healthy and immunocompromised persons. Opportunistic pathogens are usually capable of causing infections only in immunocompromised individuals such as burn patients, patients taking antibiotics, those with impaired immune systems or elderly patients with diabetes. Opportunistic pathogens are common in the environment, and may be present in the human gut or skin without causing disease.

To cause illness, the pathogen must usually first grow within or on the host. The time between infection and the appearance of clinical signs and symptoms (diarrhea, fever, rash, etc.) is the incubation time (Table 22.1). This may range from as short as 6 to 12 hours in the case of norovirus diarrhea, or up to 30 to 60 days for hepatitis A virus, which causes liver disease. At any time during infection the pathogen may be released into the environment by the host in feces, urine or respiratory secretions. Although the maximum release may occur at the height of the disease, it can also precede the first signs of clinical illness. In the case of hepatitis A virus, the maximum excretion in the feces occurs before the onset of signs of clinical illness. The concentration of organisms released into the environment varies with the type of organism and the route of transmission (Table 22.2). The concentration of enteric viruses during gastroenteritis may be as high as $10^{10}-10^{12}$ per gram of feces.

Pathogenic microorganisms usually originate from an infected host (either human or other animal), or directly from the environment. Many human pathogens can only be transmitted by direct or close contact with an infected person or animal. Examples include the herpes virus, *Neisseria gonorrhoeae* (gonorrhea) and *Treponema pallidum*

I.L. Pepper, C.P. Gerba, T.J. Gentry: Environmental Microbiology, Third edition. DOI: http://dx.doi.org/10.1016/B978-0-12-394626-3.00022-3
© 2015 Elsevier Inc. All rights reserved.

TABLE 22.1 Incubation Time for Common Enteric Pathogens

Agent	Incubation Period	Modes of Transmission	Duration of Illness
Adenovirus	8–10 days	Fecal–oral–respiratory	8 days
Campylobader jejuni	3–5 days	Food ingestion, direct contact	2–10 days
Cryptosporidium	2–14 days	Food or water ingestion, direct and indirect contact	Weeks to months
Escherichia coli ETEC	16–72 h	Food or water ingestion	3–5 days
EPEC	16–48 h	Food or water ingestion, direct and indirect contact	5–12 days
EHEC	72–120 h	Food/ingestion, direct or indirect contact	2–15 days
Giardia lamblia	7–14 days	Food or water ingestion, direct and indirect contact	Weeks to months
Norovirus	24–48 h	Food or water ingestion, direct and indirect contact, aerosol?	1–2 days
Rotavirus	24–72 h	Direct and indirect contact	4–6 days
Hepatitis A	30–60 days	Fecal–oral, fomites	2–4 weeks
Salmonella	16–72 h	Food ingestion, direct and indirect contact	2–7 days
Shigella	16–72 h	Food or water ingestion, direct and indirect contact	2–7 days
Yersinia enterocolitica	3–7 days	Food ingestion, direct contact	1–3 weeks

TABLE 22.2 Concentration of Enteric Pathogens in Feces

Organism	Per Gram of Feces
Protozoan parasites	10^6–10^7
Helminths	
Ascaris	10^4–10^5
Enteric viruses	
Enteroviruses	10^3–10^7
Rotavirus	10^{10}
Adenovirus/Norovirus	10^{11}
Enteric bacteria	
Salmonella spp.	10^4–10^{10}
Shigella	10^5–10^9
Indicator bacteria	
Coliforms	10^7–10^9
Fecal coliforms	10^6–10^9

(syphilis). This is because their survival time outside the host is very brief. Pathogens transmitted through the environment may survive from hours to years outside the host, depending on the organism and the environment. Pathogens may exit a host in respiratory secretions from the nose and mouth, or be shed on dead skin or in feces, urine, saliva or tears. Thus, they may contaminate the air, water, food, or inanimate objects (fomites) (see Chapter 30). When contaminated air is inhaled or food consumed, the organisms are effectively transmitted to another host, where the infection process begins again. Airborne transmission can occur via release from the host in droplets (i.e., coughing) or through natural (surf at a beach) or human activities (cooling towers, showers) (see Chapter 5). Some organisms may be carried great distances, hundreds of meters or miles (e.g., Legionnaires' disease and foot-and-mouth disease). Virus transmission by the airborne route may be both direct and indirect. Infection of a host may be by direct inhalation of infectious droplets, or through contact with fomites on which the airborne droplets have settled. Hand or mouth contact with the organism on the surface of a fomite results in the transfer of the organism to the portal of entry, i.e., nose, mouth or eye.

Microorganisms transmitted by the fecal–oral route are usually referred to as enteric pathogens because they infect the gastrointestinal tract. They are characteristically stable in water and food, and, in the case of enteric bacteria, are capable of growth outside the host under the right environmental conditions.

Waterborne diseases (Table 22.3) are those transmitted through the ingestion of contaminated water that serves as the passive carrier of the infectious agent. The classic waterborne diseases, cholera and typhoid fever, which have frequently ravaged densely populated areas throughout human history, have been effectively

TABLE 22.3 Classification of Water-related Illnesses Associated with Microorganisms

Class	Cause	Example
Waterborne	Pathogens that originate in fecal material and are transmitted by ingestion	Cholera, typhoid fever
Water-washed	Organisms that originate in feces are transmitted through contact because of inadequate sanitation or hygiene	Trachoma
Water-based	Organisms that originate in the water or spend part of their life cycle in aquatic animals and come in direct contact with humans in water or by inhalation	Schistosomiasis, Legionellosis
Water-related	Microorganisms with life cycles associated with insects that live or breed in water	Yellow fever

controlled by the protection of water sources and by treatment of contaminated water supplies. In fact, the control of these classic diseases illustrates the importance of water supply treatment which played an important role in the reduction of infectious diseases. Other diseases caused by bacteria, viruses, protozoa and helminths may also be transmitted by contaminated drinking water. However, it is important to remember that waterborne diseases are transmitted by the fecal–oral route, from human to human or animal to human, so that drinking water is only one of several possible sources of infection.

Water-washed diseases are those closely related to poor hygiene and improper sanitation. In this case, the availability of a sufficient quantity of water is generally considered more important than the quality of the water. The lack of water for washing and bathing contributes to diseases that affect the eye and skin, including infectious conjunctivitis and trachoma, as well as to diarrhea illnesses, which are a major cause of infant mortality and morbidity in developing countries. Diarrheal diseases may be directly transmitted through person-to-person contact, or indirectly through contact with contaminated foods and utensils used by persons whose hands are fecally contaminated. When enough water is available for hand washing, the incidence of diarrheal diseases has been shown to decrease dramatically, as has the prevalence of enteric pathogens such as *Shigella*.

Water-based diseases are caused by pathogens that either spend all (or essential parts) of their lives in water or depend on aquatic organisms for the completion of their life cycles. Examples of such organisms are the parasitic helminth *Schistosoma* and the bacterium *Legionella*, which cause schistosomiasis and Legionnaires' disease, respectively.

Water-related diseases, such as yellow fever, dengue, filariasis, malaria, onchocerciasis and sleeping sickness, are transmitted by insects that breed in water (e.g., mosquitoes that carry malaria) or live near water (e.g., the flies that transmit the filarial infection onchocerciasis). Such insects are known as vectors.

22.2 BACTERIA

22.2.1 *Salmonella*

In the United States, the concept of waterborne and foodborne disease was poorly understood until the late 19th century. During the Civil War (1861–1865), encamped soldiers often disposed of the waste upriver, but drew drinking water from downriver. This practice resulted in widespread dysentery. In fact, dysentery, together with its sister disease, typhoid fever (*Salmonella typhi*), was the leading cause of death among soldiers of all armies until the 20th century. It was not until the end of the 19th century that this state of affairs began to change. At that time, the germ theory was generally accepted, and steps were taken to treat wastes properly, and protect drinking water and food supplies.

In 1890, more than 30 people out of every 100,000 in the United States died of typhoid. But by 1907, water filtration was becoming common in most U.S. cities, and in 1914 chlorination was introduced. Because of these new practices, the national typhoid death rate in the United States between 1900 and 1928 dropped from 36 to five cases per 100,000 people. The lower death toll was largely the result of a reduced number of outbreaks of waterborne diseases. In Cincinnati, for instance, the yearly typhoid rate of 379 per 100,000 people in the years 1905–1907 decreased to 60 per 100,000 people between 1908 and 1910, following the inception of sedimentation and filtration treatment of drinking water. The introduction of chlorination after 1910 further decreased this rate.

Salmonella is a very large group of rod-shaped, Gram-negative, bacteria comprising more than 2000 known serotypes that are members of the family Enterobacteriaceae. All these serotypes are pathogenic to humans, and can cause a range of symptoms from mild gastroenteritis to severe illness or death. *Salmonella* are capable of infecting a large variety of both cold- and warm-blooded animals. Typhoid fever, caused by *S. typhi*, and paratyphoid fever, caused by *S. paratyphi*, are normally found only in humans, although *S. paratyphi* is found in domestic animals on rare occasions. In the United States, salmonellosis is due primarily to foodborne transmission, due to bacteria that infect beef and poultry and are subsequently capable of growing in these foods. Salmonellosis is the leading cause of foodborne illness in the United States. Since the route of transmission is fecal–oral, any food or water contaminated with feces may transmit the organism to a new host.

An estimated 94 million cases of gastroenteritis caused by *Salmonella* species occur globally each year; and of these, nearly 80 million cases are foodborne (Majowicz *et al.*, 2010). In the United States, a little over 1 million cases are estimated to occur annually. All age groups are susceptible, but symptoms are most severe in the elderly, infants and the infirm. The onset time is 12 and 36 hours after ingestion of contaminated food or water. Intestinal disease occurs with the penetration of *Salmonella* organisms from the gut lumen into the lining of the intestines, where inflammation occurs and an enterotoxin is produced. The immediate symptoms include nausea, vomiting, abdominal cramps, diarrhea, fever and headache. Acute symptoms may last for 1 to 2 days or may be prolonged, again depending on host factors and individual strain characteristics.

S. typhi and *S. paratyphi* A, B and C produce typhoid and typhoid-like (paratyphoid) fever in humans. Any of the internal organs may be infected. The fatality rate of untreated cases of typhoid fever is 10%, compared to less than 1% for most forms of salmonellosis. *Salmonella* septicemia (bacteria multiplying in the blood) has been associated with the subsequent infection of virtually every organ system.

Typhoid fever presents a very different clinical picture than salmonellosis. The onset of typhoid fever (1 to 3 weeks) is usually insidious, with fever, headache, anorexia, enlarged spleen, and coughing, and with constipation being more common than diarrhea in adults. Paratyphoid presents a similar clinical picture but tends to be milder. A carrier state may follow infection (<1.0 to 3.9% of the population). A carrier is a person who is permanently infected and may transmit the organism, but does not demonstrate any signs or symptoms of disease. The organism is usually carried in the gallbladder and secreted in the bile.

22.2.2 *Escherichia coli* and *Shigella*

Escherichia coli is a Gram-negative rod found in the gastrointestinal tract of all warm-blooded animals, and is usually considered a harmless organism. However, several strains are capable of causing gastroenteritis; among these are the enterotoxigenic (ETEC), enteropathogenic (EPEC), enteroaggregative (EAEC), enteroinvasive (EIEC) and enterohemorrhagic (EHEC) *E. coli*. They are grouped by their mechanisms of pathogenesis (Figure 22.1). All of the *E. coli* are spread by the fecal–oral route of transmission.

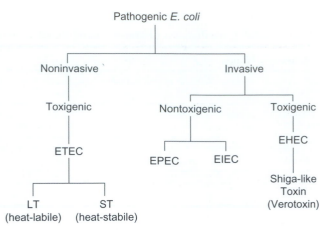

FIGURE 22.1　Mechanisms of pathogenic *Escherichia coli* strains.

The enterotoxigenic *E. coli* are a major cause of traveler's diarrhea in persons from industrialized countries who visit less developed countries, and it is also an important cause of diarrhea in infants and children in less developed countries (Table 22.4). Following an incubation period of 10–72 hours, symptoms including cramping, vomiting, diarrhea (may be profuse), prostration and dehydration occur. The illness usually lasts less than 3 to 5 days. Disease is caused by two toxins, one the heat-labile toxin and the other the heat-stable toxin. The ETEC *E. coli* are usually species specific; that is, humans are the reservoir for the strains causing diarrhea in humans. Only a few outbreaks have been documented in the United States. Conservative estimates suggest that each year, approximately 210 million cases and 380,000 deaths occur (Wenneras and Erling, 2004). EPEC is the oldest recognized category of diarrhea-causing *E. coli*. Diarrheal disease caused by this group of *E. coli* is virtually confined to infants less than 1 year of age. It is associated with infant summer diarrhea, outbreaks of diarrhea in nurseries and community epidemics of infant diarrhea. Symptoms include watery diarrhea with mucus, fever and dehydration. The diarrhea can be severe and prolonged with a high fatality rate (as high as 50%). Since the 1960s, EPEC has largely disappeared as an important cause of infant diarrhea in North America and Europe. However, it remains a major agent of infant diarrhea in South America, Africa and Asia. Humans, cattle and pigs can be infected by this organism. Thus, foods implicated in outbreaks are raw beef and chicken.

The disease caused by EIEC closely resembles that caused by *Shigella*. Illness begins with severe abdominal cramps, watery stools and fever. Disease is usually self-limiting with no known complications. Any age group is susceptible. This type of *E. coli* carries a plasmid that enables it to invade cells lining the gastrointestinal tract, resulting in a mild form of dysentery. EIEC infections are endemic in less developed countries, although occasional

TABLE 22.4 Summary of Pathogenic *E. coli* Incidence and Epidemiology

Pathogenic E. Coli	Site of Infection	Associated Disease	Incidence	Target Population	Significant Transmission Route
ETEC	Small intestine	Travelers' diarrhea, chronic childhood diarrhea (in developing countries)	16 U.S. outbreaks (1996–2003); prevalence 1.4% in patients with diarrhea; 79,420 cases of travelers' diarrhea each year (in the U.S.)	International travelers and children in developing countries	Food (raw produce, street vendors) and water
EPEC	Small intestine	Infant diarrhea	Hundreds of thousands of deaths worldwide	Children in developing countries	Water, infant formula
EHEC	Large intestine	Hemorrhagic colitis, HUS	110,000 cases and 61 deaths annually in the U.S.	All ages	Food (beef, produce), person-to-person, water, animals
EIEC	Large intestine	Dysentery	Low in developed countries	Children in developing countries	Water (rare), person-to-person
EAEC	Intestine	Watery diarrhea with or without blood in the stool, acute and chronic	Developed and developing countries	Children and adults, travelers	Food, water, person-to-person

cases and outbreaks are reported in industrialized countries. Outbreaks also have been associated with hamburgers and unpasteurized milk.

The EHEC *E. coli* (also referred to as *Shigella* toxin producing *E. coli* or STEC) were described in 1982 when a multistate epidemic of hemorrhagic colitis occurred in the United States, and was shown to be due to a specific serotype known as *E. coli* O157: H7. This organism produces two toxins called verotoxins I and II. These toxins are closely related to or identical to the toxin produced by *Shigella dysenteriae*. The toxin production depends on the presence of a prophage. A prophage is a bacterial virus that inserts its own DNA into a bacterial chromosome. A plasmid codes for a novel type of fimbria (hair-like projection) that enabled the organism to adhere to the intestinal lining and initiate disease.

Enteroaggregative *Escherichia coli* (EAEC), an increasingly recognized cause of diarrhea in children in developing countries, has been particularly associated with persistent diarrhea (more than 14 days), a major cause of illness and death. Recent outbreaks implicate EAEC as a cause of foodborne illness in industrialized countries.

EHEC infections are now recognized to be an important problem in North America, Europe and some areas of South America. The illness usually includes severe cramping and diarrhea, which is initially watery but becomes grossly bloody. The illness is usually self-limiting and lasts for an average of 8 days. However, some victims,

particularly the very young, develop hemolytic–uremic syndrome (HUS) resulting in renal failure and hemolytic anemia (Figure 22.2). This disease can result in permanent loss of kidney function. In older individuals, HUS plus two other symptoms, fever and neurologic symptoms, constitutes thrombotic thrombocytopenic purpura (TTP). This illness can have a mortality rate in the elderly as high as 50%. Most outbreaks are associated with undercooked or raw hamburger. Raw milk, unpasteurized fruit juices and vegetables contaminated with cow dung have also been implicated. Infected cattle are believed to be a major environmental source of this organism. Waterborne outbreaks involving both nondisinfected groundwater and recreational waters have also occurred. Various studies have shown that *E. coli* O157:H7 is commonly present in domestic sewage at concentrations from 10 to 100 colony forming units (CFU) per 100 ml, and in wastewater from slaughterhouses from 100 to 1000 CFU/100 ml (Nwachuku and Gerba, 2008).

Shigella is closely related to *E. coli*. Four species have been described: *S. dysenteriae*, *S. flexneri*, *S. boydii* and *S. sonnei*. *S. dysenteriae* causes the most severe disease, and *S. sonnei* causes the mildest symptoms. *S. sonnei* is the serotype most often found in the United States (Lee *et al.*, 1991). *Shigella* very rarely occurs in animals. It is principally a disease of humans and occasionally of other primates such as monkeys and chimpanzees. The organism is often found in water polluted with human

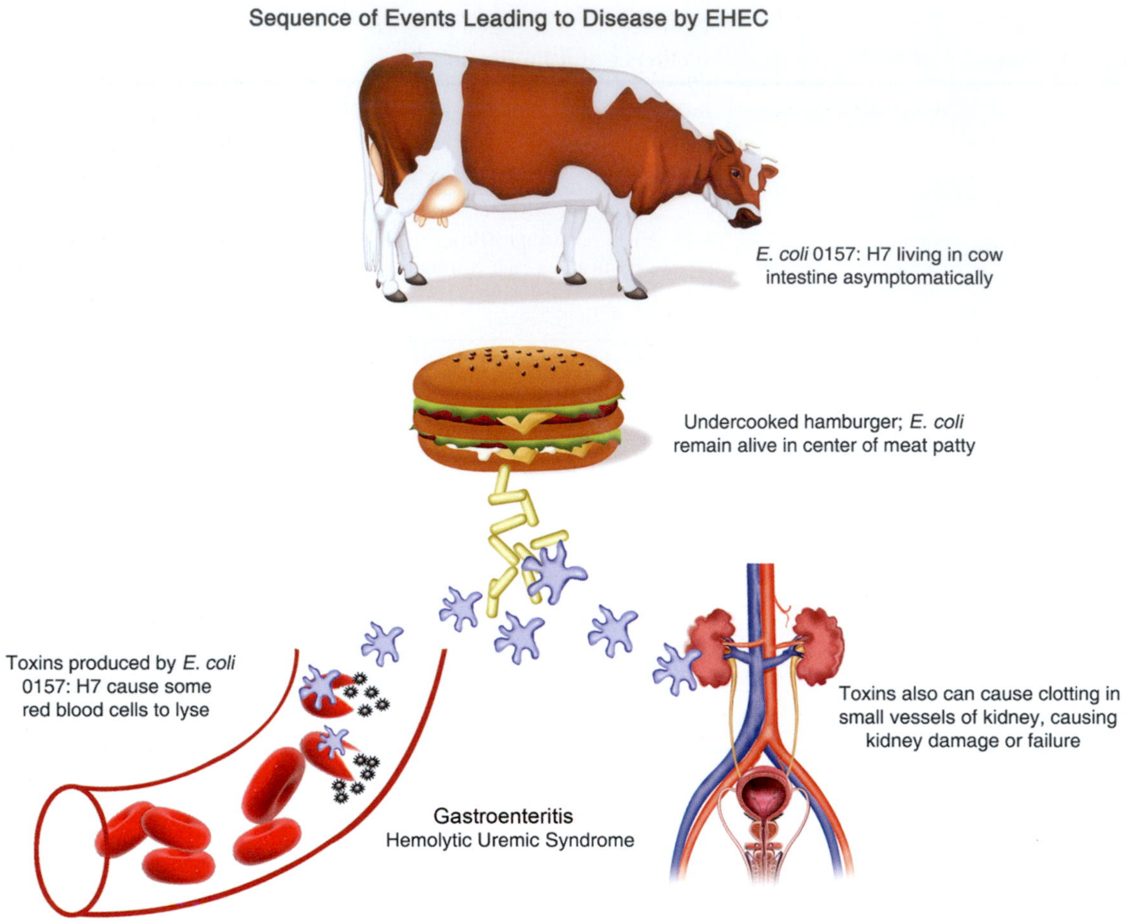

Sequence of Events Leading to Disease by EHEC

E. coli 0157: H7 living in cow intestine asymptomatically

Undercooked hamburger; E. coli remain alive in center of meat patty

Toxins produced by E. coli 0157: H7 cause some red blood cells to lyse

Toxins also can cause clotting in small vessels of kidney, causing kidney damage or failure

Gastroenteritis
Hemolytic Uremic Syndrome

FIGURE 22.2   Sequence of events leading to disease by EHEC.

sewage, and is transmitted by the fecal–oral route. *Shigella* has been a common cause of waterborne recreational outbreaks in the United States. However, most cases of shigellosis are the result of person-to-erson transmission through the fecal–oral route. Secondary attack rates are 20–40% of household contacts. An estimated 440,000 cases of shigellosis occur annually in the United States (Gupta *et al.*, 2004). *Shigella* is associated with certain foods such as salads, raw vegetables, milk and dairy products, and poultry. After an onset time of 12 hours to 1 week, symptoms of abdominal pain, cramps, diarrhea and vomiting may occur. The organisms multiply in the cells of the gastrointestinal tract and spread to neighboring cells, resulting in tissue destruction. Some infections are associated with ulceration, rectal bleeding and drastic dehydration, with fatality rates as high as 10–15%. Infants, the elderly and the infirm are most susceptible.

22.2.3 *Campylobacter*

Campylobacter jejuni is a Gram-negative curved rod. It is relatively fragile, and sensitive to environmental stress (including an oxygen content of 21%, drying, heating, contact with disinfectants or acidic conditions). Before 1972, when methods were developed for its isolation from feces, it was believed to be primarily an animal pathogen causing abortion and enteritis in sheep and cattle. Recent surveys have shown that *C. jejuni* is a leading cause of bacterial diarrheal illness in the United States, with an estimated 2 million cases per year (Samuel *et al.*, 2004). Although *C. jejuni* is not carried by healthy individuals in the United States or Europe, it is often isolated from healthy cattle, chickens, birds and even flies. It is sometimes present in nonchlorinated water sources such as streams and ponds.

C. jejuni infections cause diarrhea with fever, abdominal pain, nausea, headache and muscle pain. The illness usually occurs 2–5 days after the ingestion of the contaminated food or water. Illness generally lasts 7–10 days, and relapses are not uncommon. Although anyone can be infected with *C. jejuni*, children under 5 years of age and young adults (15–29 years old) are more frequently afflicted than other age groups.

Surveys show that 20 to 100% of retail chickens are contaminated with *C. jejuni*. This is not surprising because many healthy chickens carry these bacteria in their intestinal tracts. Nonchlorinated water may also be a source of infection. However, properly cooking chicken, pasteurizing milk and chlorinating water will kill the bacteria (Blaser *et al.*, 1986).

C. jejuni has been isolated from 22% of coastal and estuary water samples in concentrations ranging from 10 to 230 per 100 ml, and from 28% of river samples in concentrations of 10 to 36 cells per 100 ml. Carter *et al.* (1987) found *Campylobacter* in 10−44% of pond water samples. However, it is thought that virtually all surface waters contain *Campylobacter*. Recovery rates from surface waters are highest in the fall and winter months, and lowest during the spring and summer months. *Campylobacter* density does not show a significant correlation with the isolation of indicator bacteria such as fecal or total coliforms (Carter *et al.*, 1987).

This bacterium dies off rapidly in stream water at 37°C, showing a 9-log decrease within 3−12 days (Rollins and Colwell, 1986). However, the organism can remain viable for extended periods at cooler temperatures, surviving over 120 days at 4°C in stream water.

22.2.4 Yersinia

Yersinia enterocolitica and *Y. pseudotuberculosis* are small, rod-shaped, Gram-negative bacteria. *Y. enterocolitica* causes diarrhea often with a bloody stool, while *Y. pseudotuberculosis* causes fever and abdominal pain, usually on the right side. *Yersinia enterocolitica* is a ubiquitous microorganism, with the majority of isolates being recovered from asymptomatic carriers, infected animals, contaminated food and untreated water. Most are nonpathogenic having no clinical importance (Rahman *et al.*, 2011). Only a few serotypes have been associated with illness. Pigs have been shown to be a major reservoir of pathogenic *Y. enterocolitica* involved in human infections, particularly for strains of serotype 4/O:3. *Y. enterocoltica* has been largely associated with foodborne outbreaks, but waterborne outbreaks have also occurred. *Yersinia pseudotuberculosis* also appears to occur in water; animal- and food-associated infections occur primarily in the northern hemisphere, and carrots and iceberg lettuce have been implicated in outbreaks.

Symptoms usually begin 24 to 48 hours after ingestion of contaminated food or drink. Yersiniosis is frequently characterized by such symptoms as gastroenteritis with diarrhea and/or vomiting. However, fever and abdominal pain are the hallmark symptoms. *Yersinia* infections mimic appendicitis, but the bacteria may also cause infections of other sites such as wounds, joints and the urinary tract. Overall, yersiniosis is more common in Northern Europe, Scandinavia and Japan than in the United States. The greatest incidence of disease is seen during cold seasons.

22.2.5 Vibrio

Dr. John Snow (1813−1858) of London was one of the first to make a connection between certain infectious diseases and drinking water contaminated with sewage. In his famous study of London's Broad Street pump, published in 1854, he noted that people afflicted with cholera were clustered in a single area around the Broad Street pump, which he determined was the source of the infection. When, at his insistence, city officials removed the handle of the pump, Broad Street residents were forced to obtain their water elsewhere. Subsequently, the cholera epidemic in that area subsided.

The Gram-negative genus *Vibrio* contains more than one member that is pathogenic to humans. The most famous member of the genus is still *V. cholerae*. Cholera is transmitted through the ingestion of fecally contaminated food and water. Cholera remains prevalent in many parts of Central America, South America, Asia and Africa.

V. cholerae serogroup Ol includes two biovars, cholerae (classical) and El Tor, each of which includes organisms of the Inaba and Ogawa serotypes. A similar enterotoxin is elaborated by each of these organisms, so the clinical pictures are similar. Asymptomatic infection is much more common than disease, but mild cases of diarrhea are also common. In severe untreated cases, death may occur within a few hours and the fatality rate without treatment may exceed 50%. This is due to a profuse watery diarrhea referred to as rice-water stools. The rice-water appearance is due to the shedding of intestinal mucosa and epithelial cells. With proper treatment, the fatality rate is below 1%. Humans are the only known natural host. Thus, the reservoir for *V. cholerae* is human, although environmental reservoirs may exist, apparently in association with copepods or other phytoplankton.

Vibrios that are biochemically indistinguishable, but do not agglutinate in *V. cholerae* serogroup Ol antiserum, were formerly known as nonagglutinable vibrios (NAGs) or noncholera vibrios (NCVs). They are now included in the species *V. cholerae*. Some of these strains produce enterotoxin but most do not. Thus reporting of non-Ol *V. cholerae* infections as cholera is inaccurate and leads to confusion. *V. mimicus* is a closely related species that can cause diarrhea, some strains of which produce an enterotoxin indistinguishable from that produced by *V. cholerae*.

The following example illustrates the possible devastating impacts of this disease. A cholera epidemic caused by *V. cholerae* Ol began in January 1991 in Peru, and spread to Central America and South America. A total of 1,041,422 cases occurred with 9642 deaths (MMWR, 1995). The epidemic was believed to have been initiated by inadequate chlorination of drinking water. A second example is an outbreak in southern Asia, an epidemic caused by the newly recognized strain *V. cholerae* O139 which began in late 1992 and continued to spread to at least 11 countries. This latter strain of *V. cholerae* also produces severe watery diarrhea and dehydration that is indistinguishable from the illness caused by *V. cholerae* OI.

Vibrio cholerae is a native marine organism, and its potential for transmission to humans is related to a complex ecology that controls its occurrence and concentration in the marine food chain (Lipp *et al.*, 2002). Warmer temperatures in combination with elevated pH and plankton blooms can influence *V. cholerae* attachment, growth and multiplication in the aquatic environment, particularly in association with copepods (Figure 22.3). Thus, factors such as climate change, climate variability (El Niño-Southern Oscillation) and monsoons can influence the *V. cholerae* concentration in marine waters and exposure via the marine food chain to humans (see also Chapter 31).

Another species, *V. parahaemolyticus*, is usually transmitted by contaminated food. The ingestion of inadequately cooked seafood, any food cross-contaminated by handling raw seafood or food rinsed with contaminated seawater may transmit this disease. A period of time at room temperature is generally necessary to allow multiplication of the organisms. Disease symptoms include watery diarrhea and abdominal cramps in the majority of cases, sometimes with nausea, vomiting, fever and headache. Usually, the disease is self-limiting and lasts 1 to 7 days. Most cases are reported during the warmer months, and marine coastal environments are a natural habitat. The organisms have been found in marine silt, in coastal waters and in fish and shellfish.

Of all foodborne infectious diseases, infection with *V. vulnificus* is one of the most severe; the fatality rate for *V. vulnificus* septicemia exceeds 50% (Tacket *et al.*, 1984). Cases are most commonly reported during warm-weather months (April—November), and are often associated with eating raw oysters. Many patients are found to have had a preexisting liver disease, usually associated with alcohol use or viral hepatitis. All of these latter patients had eaten raw oysters 1—2 days before the onset of symptoms. These symptoms include thrombocytopenia, bulbous skin lesions, hypotension and shock. Cases have also been reported that arose through the contamination of a wound by seawater or seafood drippings.

22.2.6 *Helicobacter*

In 1982, a physician in Australia cultured a Gram-negative, spiral-shaped bacterium observed in biopsied tissue of a stomach ulcer. Initially called *Campylobacter pylori* based on biochemical and morphological characteristics, the organism is now named *Helicobacter pylori* (Figure 22.4). The stomach mucosa contains cells that secrete proteolytic enzymes and hydrochloric acid. Other specialized cells produce a layer of mucus that protects the stomach itself from digestion. If this mucous layer is disrupted, the ensuing inflammation leads to an ulcer. *H. pylori* has been shown to bind to O^- blood group antigens on the gastric epithelial cells. People of this blood group are twice as likely to develop gastric ulcers.

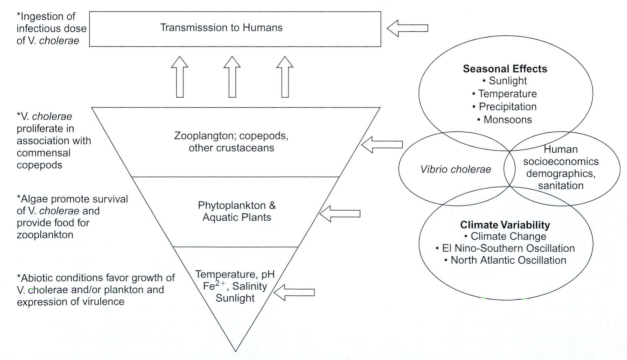

FIGURE 22.3 Hierarchical model for environmental cholera transmission (Lipp *et al.*, 2002).

H. pylori produces large amounts of urease, an enzyme that converts urea to the alkaline product ammonia. This results in a high pH in the local region.

Humans are thought to be a natural host for *H. pylori*. Person-to-person spread is probably the dominant source of transmission, although controversy exists over whether the fecal—oral or oral—oral route of transmission predominates. *H. pylori* has been isolated from feces (Thomas *et al.*, 1992) and detected in saliva (Nguyen *et al.*, 1993). Infection may also be spread by ingesting contaminated food or water. Several studies have shown a greater risk of acquiring *H. pylori* infections from drinking untreated well water (Klein *et al.*, 1991; Rolle-Kampczyk *et al.*, 2004; Reavis, 2005). A recent study found a possible outbreak associated with using a public drinking water supply (Melius *et al.*, 2012). Nonculturable *H. pylori* organisms have been detected in river water and groundwater in the United States (Hegarty *et al.*, 1999).

22.2.7 *Legionella*

Legionella is the causative agent of Legionnaires' disease and Pontiac fever. Both syndromes are characterized initially by anorexia, malaise, myalgia and headache. Within 24 hours, a fever ensues with chills. A nonproductive cough may occur and abdominal pain and diarrhea are seen in many patients. In Legionnaires' disease, chest radiographs show areas of consolidation indicative of pneumonia and, indeed, respiratory failure may occur. This disease has a 15% fatality rate in hospitalized cases. Pontiac fever is not associated with pneumonia or death. Patients recover spontaneously in 2—5 days.

Legionella are poorly staining Gram-negative bacilli that require cysteine and other specific nutrients when grown on artificial laboratory media. At least 40 species of *Legionella* have been shown to cause disease, but the most prominent pathogenic species is *L. pneumophila*, which first received extensive attention after an outbreak in 1976 in Philadelphia. The disease is found worldwide,

with most sporadic cases occurring during the summer and fall months. The reservoir is primarily aquatic. Hot water systems, air-conditioning cooling towers and evaporative condensers have all been implicated in outbreaks, as have decorative fountain and retail store misters (Figure 22.5). The organism has also been found in creeks and ponds, and the soil from their banks. The bacterium survives for months in tap and distilled water. The primary route of transmission is thought to be through the inhalation of aerosols. Exposure is not uncommon, as reflected in serologic assays that show that 1—20% of the general population have antibodies to *L. pneumophila*.

Concentrations of 1.4×10^4 to 1.7×10^5 cells per liter of *Legionella* spp. have been detected in raw drinking water sources using direct fluorescent antibody (DFA) techniques (Tison and Seidler, 1983). The concentrations of *Legionella* found in distribution water samples by DFA were as follows (CFU/L): chlorinated water, $< 8 \times 10^3$ to 1.4×10^4; water treated by slow sand filtration and chlorination, $< 5.4 \times 10^3$ to 4.6×10^4; water treated by flocculation, filtration and chlorination, $< 8 \times 10^3$ to 2.2×10^4. Zacheus and Martikainen (1994) found that 30% of hot water systems in apartment buildings contained *Legionella*. The mean number of *L. pneumophila* was 2.7×10^3 CFU/L with a range of < 50 to 3.2×10^5 CFU/L. For all positive hot water systems, *Legionella* was also isolated from the hot water tap and shower head. *Legionella pneumophila* was isolated from 12.5, 29.0 and 37.5% of the hot water distribution systems receiving chlorinated groundwater, unchlorinated groundwater and chlorinated surface water, respectively.

A great deal of work has been carried out on the survival and growth of legionellae in potable water distribution systems and plumbing in hospitals and homes. Legionellae appear to be more resistant to chlorine than *E. coli*. For example, *Legionella* has been shown to exist in potable water systems even when exposed to 0.75 to 1.5 ppm free chlorine residual. Such protection is afforded by intracellular growth in protozoa such as *Tetrahymena pyriformis* and *Acanthamoeba castellani* (Fields *et al.*, 1984; Moffat and Tompkins, 1992).

Legionella survives well at 50°C, and environmental isolates are able to grow in tap water at temperatures as high as 42°C (Figure 22.6). The enhanced survival and growth in these systems have been linked to stagnation stimulated by rubber fittings in the plumbing system (Colbourne *et al.*, 1988), and trace concentrations of metals such as iron, zinc and potassium (States *et al.*, 1989). Sediment promotes the growth of *Legionella*, as does stagnation in the hot water tank. *Legionella* can be removed from hot water heaters by raising the temperature to over 60°C near the heating element and to over 50°C at outlets, combined with regular flushing (Meenhorst *et al.*, 1985).

The overall attack rate of pneumonia in the United States is 12—15 cases per 1000 persons per year, resulting

FIGURE 22.4 *Helicobacter pylori* from Goai Nobel, 2005.

FIGURE 22.5 (A) Sources of *Legionella* in the environment. (B) Cooling towers. Outbreaks of Legionnaires' disease have been commonly traced to cooling towers.

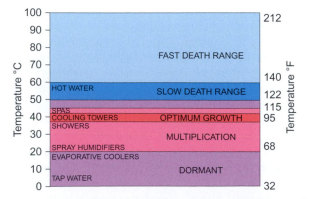

FIGURE 22.6 The effect of temperature on the growth of *Legionella pneumophila*.

in approximately 3,957,000 cases annually. Pneumonia is the sixth leading cause of death in the United States, with an estimated annual cost of $23 billion. *L. pneumophila* causes 4.1 to 20.1% of community-acquired cases, many of which result in hospitalization (Marrie, 1994). These data suggest that *L. pneumophila* is a major cause of serious cases of pneumonia.

22.2.8 Opportunistic Bacterial Pathogens

An opportunistic pathogen is one that usually causes disease only in those whose immune system is compromised. The weakened immune system may be due to very young or old age, pregnancy, cancer therapy, immunosuppressive drugs, human immunodeficiency virus (HIV) and other causes. Opportunistic pathogens are numerous in the environment. Many opportunistic bacterial pathogens are found in surface and drinking waters.

Concern has been generated in the drinking water industry regarding the health effects of heterotrophic bacteria that are found in tap water, bottled water and other sources of potable water. Heterotrophic bacteria are those that require organic carbon rather than carbon dioxide as a carbon source. All human bacterial pathogens are heterotrophic.

Most of the heterotrophic bacteria in drinking water are not human pathogens. However, some of the genera, including *Legionella*, *Mycobacterium*, *Pseudomonas*, *Acinetobacter*, *Stenotrophomonas* and *Aeromonas*, include species that are opportunistic pathogens. See Section 22.2.7 for a discussion of the genus *Legionella*.

The most important opportunistic pathogen in the genus *Pseudomonas* is *P. aeruginosa*, which is primarily a nosocomial (hospital-acquired) pathogen responsible for 10% of nosocomial infections. *Pseudomonas* spp. infects the lungs, urinary tract, eye, burn patients and wounds. Human disease is often associated with water-related reservoirs such as swimming pools, whirl-pools, hot tubs, taps (faucets) and contact lens solutions. In hospitals, growth of *P. aeruginosa* in the drinking water taps appears to be the major source infecting patients (Asghari *et al.*, 2013). The source of community-acquired infections of cystic fibrosis patients appears to be largely from environmental exposure (Ranganathan *et al.*, 2013). It is important as the ultimate cause of death in these patients is often lung infections by *P. aeruginosa*. Although *P. aeruginosa* is found in drinking water, it is not ubiquitous. Results of surveys showed that 2% of bottled water and 2–3% of tap water samples contain *P. aeruginosa* at concentrations between 1 and 2300 organisms/ml (Allen and Geldreich, 1975).

Acinetobacter is the second most frequently isolated, nonfermentative, Gram-negative rod in the clinical laboratory. However, it is generally considered to be of low virulence. Up to 25% of healthy adults carry this organism in the respiratory tract. The human skin is the likely source for most outbreaks of hospital infections, but sinks and taps have also been implicated in some outbreaks. As an opportunistic pathogen, *Acinetobacter* is involved in nosocomial urinary tract infections, bacteremia, wound infections and pneumonia. *Acinetobacter* can also be a cause of community-acquired pneumonia and urinary tract infections. *Acinetobacter* has been isolated from 97% of natural surface water samples with a concentration of 0.1–100 cells/ml (Baumann, 1968). Bifulco *et al.* (1989) isolated *Acinetobacter* from 38% of groundwater supplies at an arithmetic mean density of 8 CFU/100 ml in the positive samples.

Acinetobacter has been isolated from 5 to 92% of distribution water samples (LeChevallier *et al.*, 1980;

Bifulco *et al.*, 1989). It comprised 1.0 to 5.5% of the heterotrophic plate count (HPC) flora in drinking water samples (Payment *et al.*, 1988) with concentrations of 6–21 CFU/ml. It has also been found in 5 to 35% of bottled water samples, at concentrations of 2–30 CFU/ml.

Pseudomonas maltophilia has been reclassified as *Stenotrophomonas maltophilia* and is the third most commonly isolated nonfermentative Gram-negative rod in clinical laboratories. This organism can colonize the body and cause disease. Risk factors for infection include stays in intensive care units, mechanical ventilation, antibiotic treatment and cancer. Diseases it can cause include septicemia, pneumonia, wound infections and more rarely meningitis and endocarditis.

S. maltophilia constituted 5.7% of the HPC found in raw surface water samples and 0–1.2% of the flora in distribution water samples (LeChevallier *et al.*, 1980). Using PCR, *S. maltophilia* was detected in 3.7 to 36.4% of unchlorinated drinking water samples, derived from surface waters in the Netherlands (Van der Wielen and van der Kooij, 2013). The bacterium has also been found in 2% of bottled water samples at concentrations of 2–22 CFU/ml and in < 5% of cooler water samples at 34 CFU/ml.

Aeromonas hydrophila, *A. sobria* and *A. caviae* are very similar and were all referred to as *A. hydrophila* until 1976, when they were first split into separate species. Papers written before 1985 may use the term *A. hydrophila* including all three species and biochemical variants. The three Gram-negative species are biochemically very similar, and have all been implicated as diarrheal agents in humans. The exact mechanism for diarrhea has not been elucidated.

A strong association has been found between drinking untreated water and the occurrence of diarrhea with the isolation of *Aeromonas* spp. Higher counts of *A. hydrophila* in distribution water correlate with greater frequency of diarrheal isolates; however, no waterborne outbreaks have been documented.

Children and the elderly tend to be affected most often. As with many diarrheal agents, outbreaks of diarrhea, in which *Aeromonas* has been implicated, have been associated with day care centers. *Aeromonas* is frequently found in environmental water samples. It has been recovered from 0.6 to 18.2% of natural freshwater samples at a concentration range of 0.1–3600 CFU/ml. It has also been recovered from 0.9 to 27% of distribution water samples at an average concentration of 0.022 CFU/ml. However, even at large oral doses of up to 10^{10} CFU, *A. hydrophila* failed to produce diarrhea in human volunteers (Morgan *et al.*, 1985).

Mycobacteria are rod-shaped bacteria that contain high levels of lipid (waxy) material. They are among the most resistant nonspore-forming bacteria to chlorine and other common disinfectants. The *Mycobacterium avium* complex (MAC) consists of at least 28 serovars of two distinct species, *M. avium* and *M. intracellulare*. It also included three serovars of *M. scrofulaceum* in the past, but the inclusion is no longer appropriate due to recent advances in mycobacterial systematics. Pulmonary disease caused by MAC has dramatically increased in the United States over the last three decades (Kendall and Winthrop, 2013). The disease is most common among individuals over 60 years of age. Predisposing factors include age, chronic lung disease, bronchogenic carcinoma, previous gastrectomy and AIDS. MAC can also cause pulmonary disease, osteomyelitis and septic arthritis in people with no known predisposing factors (Jones *et al.*, 1995). However, disease by MAC can be lethal and is difficult to treat because of resistance to many antimycobacterial agents.

It is believed that municipal drinking water systems are an important reservoir for MAC since it is among the most common bacteria identified in biofilms in chlorinated drinking water distribution systems and fixtures (i.e., showerheads). Epidemiological investigations have associated water sources with infections by atypical mycobacteria (Burns *et al.*, 1991). These bacteria can multiply in water that is oligotrophic (Falcao *et al.*, 1993) or essentially free of nutrients, including dialysis water, and they are relatively resistant to disinfection by chlorination and chloramines (Collins *et al.*, 1984). Atypical mycobacteria are widespread in water environments. Mycobacteria have been isolated from 11 to 38% of raw water samples at concentrations of <0.1−48 CFU/ml. MAC has also been found in up to 50% of municipal and private drinking water samples at concentrations of 0.01−5.2 CFU/ml.

Hospital water systems often harbor MAC and may be a source of nosocomial infection. *M. avium* has primarily been associated with hot water systems, and in some cases, a hot water system may be persistently colonized by the same strain of *M. avium* (Whiley *et al.*, 2012).

22.2.9 Blue−Green Algae

Blue−green algae or cyanobacteria occur in an enormous diversity of habitats, freshwater and marine, as plankton (free floating), mats and periphyton (attached to surfaces). Hot spring mats of some *Oscillatoria* develop up to temperatures of 62°C (Figure 22.7). They have many beneficial functions such as nitrogen fixation and cycling of nutrients in the food chain.

Despite their beneficial roles in the environment, cyanobacteria sometimes become problematic. Occasionally, they increase rapidly resulting in cyanobacterial blooms (Figure 22.8). Blooms are associated with eutrophic water, especially with levels of total phosphorus >0.01 mg/L and levels of ammonia- or nitrate-nitrogen >0.1 mg/L. Optimal temperatures for blooms are 15−30°C, and optimal pH is 6−9. Calm or mild wind

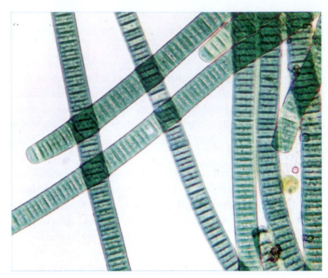

FIGURE 22.7 *Oscillatoria*—a blue−green algae. Courtesy Michael Clayton (2006).

FIGURE 22.8 Cyanobacterial bloom. Courtesy C.P. Gerba.

conditions sometimes allow blooms to cover the water surface, but the highest concentrations of cyanobacteria may occur at depths ranging from 2 to 9 m, which will not be visible from the shore. The offending bacteria may also grow in the sediment. These blooms can impart an off-taste and odor to the water, and/or result in the production of toxins.

The most common complaints related to such blooms are of taste and odor. Geosmin and 2-methylisoborneol (MIB) can produce odors at levels as low as 1.3−10 and 6.3−29 ng/L, respectively (Young *et al.*, 1996). The odor

TABLE 22.5 Cyanobacterial Compounds Producing Off-tastes and Odors

Compound	Odor	Taste
Geosmin	Earthy, musty, grassy	Musty, earthy, stale
MIB	Musty, earthy, peaty	Musty, earthy, stale
Isobutylmethoxypyrazine	Woody, stale, musty	Creosote, stale, dusty
Isopropylmethoxypyrazine	Sooty, dusty, cabbage	Musty, vegetable water
Octa-1,3-diene	Musty	
Hexanal	Green apple-like	
Octan-1-ol	Rancid	
β-Cyclocitral	Tobacco	

TABLE 22.6 Cyanobacterial and Types of Toxins Produced

Genus	Toxins Produced
Anabaena	Anatoxin-a, hepatotoxins
Aphanizomenon	Saxitoxin, neosaxitoxin, hepatotoxins
Alexandrium	Saxitoxin
Cylindrospermopsis	Hepatotoxin
Nodularia	Nodularins
Oscillatoria	Neurotoxins, hepatotoxins
Microcystis	Microcystins

TABLE 22.7 Characterization of Cyanobacterial Toxins

Toxin	Characterization of Toxin
Anatoxins	Neurotoxins
Microcystins	Hepatotoxins
Nodularins	Hepatotoxins
Saxitoxins	Neurotoxins

produced by geosmin is described as earthy, and that of MIB as musty or camphorous smelling. Concentrations of MIB and geosmin are usually highest in summer and fall. Several compounds produced by cyanobacteria can cause off-tastes and odors as shown in Table 22.5.

Geosmin is produced by several cyanobacteria including *Oscillatoria, Anabaena, Lyngbya, Phormidium, Symploca* (Narayan and Nunez, 1974), *Aphanizominon* and *Fischerella* (Wu and Juttner, 1988). *Lyngbya, Oscillatoria* and *Phormidium* (Izaguirre, 1992) are the most common genera producing MIB. Some strains of *Diplocystis* and *Schizothrix* can also cause off-tastes and odors. *Microcystis* release some odorous sulfur compounds, especially when they decay.

Many cyanobacteria found in algal blooms can produce toxins that cause liver damage, neural damage and gastrointestinal (GI) disturbances. This has been well documented in many wild animal and livestock cases and implicated in human cases as well. *Microcystis* is the number one offender worldwide. Other toxin-producing genera include *Anabaena, Aphanizomenon, Alexandrium, Cylindrospermopsis, Nodularia, Nostoc* and *Oscillatoria* (Turner *et al.*, 1990). Different types of toxins produced are shown in Tables 22.6 and 22.7. Most toxic species are associated with temperate rather than tropical climates.

In livestock and wild animals, the hepatotoxins cause weakness, anorexia and liver damage. They can be lethal within minutes to a few days. Neurotoxins can cause twitching, muscle contraction, convulsions and death. Signs and symptoms in humans associated with the ingestion of water with algal blooms are dizziness, headaches, muscle cramps, nausea, vomiting, gastroenteritis and pneumonia (Phillip *et al.*, 1992). Long-term exposure to toxins is associated with liver cancer as well (Carmichael, 1994).

22.3 PARASITOLOGY

The study of parasitology embodies a large diversity of eukaryotic organisms. This group includes organisms that are unicellular, multicellular and multinucleate; aerobic and anaerobic; motile and nonmotile; sexual and asexual. For this chapter parasites will be grouped into two categories: protozoa and helminthes (Table 22.8). Protozoa are unicellular microorganisms in the Kingdom Protista, and are classified according to their means of locomotion: flagella, cilia, pseudopodia or no locomotion. Some are parasitic, although the majority are nonparasitic. Some undergo a sexual stage, whereas others reproduce by asexual means: fission, budding or schizogony. Helminths belong to Kingdom Animalia, and include roundworms, flatworms, tapeworms and flukes. They are multicellular complex organisms containing organs and tissue. They usually develop in soil or in intermediate hosts to complete their life cycle and have elaborate life cycles, including larva, eggs and adult stages. Parasitic protozoa and helminths can be acquired from the environment—water, soil and contaminated food—and have had great health and economic

TABLE 22.8 Classification of some Environmentally Transmitted Protozoa and Helminthes

Protozoa
Phylum Apicomplexa
Cyclospora cayetanensis *Cryptosporidium parvum* *Toxoplasma gondii*
Phylum Microspora
Enterocytozoon bieneusi *Encephalitozoon cuniculi* *Encephalitozoon hellem* *Encephalitozoon intestinalis*
Phylum Sarcomastigophora
Entamoeba histolytica *Giardia lamblia* *Naegleria fowleri*
Helminthes
Phylum Nematoda
Ascaris lumbricoides *Necator americanus* *Trichuris trichiura*
Phylum Platyhelminthes
Class Cestoidea *Taenia saginata* Class Trematoda *Schistosoma mansoni*

impacts in many developing and developed countries. In the United States, there are two protozoan genera that are especially of concern: *Giardia* and *Cryptosporidium*. They form hardy cysts and oocysts that can survive water treatment disinfection and are one of the biggest concerns of water utilities today. Some of the characteristics of environmentally transmitted parasites are summarized in Table 22.9.

22.3.1 Protozoa

22.3.1.1 Giardia lamblia

Giardia lamblia was first described in 1681 by Antonie van Leeuwenhoek, who found them in his own feces. He called the trophozoites "animalcules." In 1859, Vilern Lambl rediscovered *Giardia* by finding the trophozoites in stools of young children with diarrhea. It was not until

the early 20th century that physicians began associating diarrhea with the presence of *Giardia* in stools. In 1954, Robert Rendtorff confirmed infectivity in human volunteers with oral administration of *Giardia* cysts.

Giardia is the most frequently identified intestinal parasite in the United States (Schaefer, 2006). Worldwide it is estimated that 13% of adults and up to 50% of children are infected with *Giardia* (Schaefer, 2006). In the United States, it is one of the most frequent identified causes of waterborne disease (Craun *et al.*, 2010).

Humans become infected with *G. lamblia* by ingesting the environmentally resistant stage, the cyst (Figure 22.9). Once ingested, it passes through the stomach and into the upper intestine. The increase in acidity via passage through the stomach stimulates the cyst to excyst, which releases two trophozoites into the upper intestine. The trophozoites (Figure 22.10) attach to the epithelial cells of the small intestine. It is believed that the trophozoites use their sucking disks to adhere to epithelial cells. The adherence to the cells flattens the villi, causing malabsorption and diarrhea by not allowing adsorption of water and nutrients across the intestine. This can cause both acute and chronic diarrhea within 1−4 weeks of ingestion of cysts resulting in foul-smelling, loose and greasy stools. Once the trophozoites detach from the epithelial cells and travel down the intestine, cholesterol starvation is believed to stimulate the trophozoites to encyst and pass back into the environment as a cyst. Giardiasis can be treated with metronidazole (Flagyl) or nitazoxanide. Although in some cases the symptoms will spontaneously disappear without treatment, in most cases without treatment the symptoms will wax and wane for many months. In symptomatic patients, more trophozoites than cysts are excreted into the feces and cannot withstand the harshness outside the human body. In asymptomatic humans, mostly cysts are passed in stools; therefore *Giardia* carriers can serve as a source of cysts in the environment.

When *Giardia* cysts enter the environment they can survive for prolonged periods. *G. lamblia* cysts have been documented to survive for up to 77 days at 8°C and 4 days at 37°C in distilled water (Bingham *et al.*, 1979). In another study, cysts of *Giardia muris* (a species that infects mice but is often used as a model for *G. lamblia*) were suspended in lake and river water, and found to survive 28 days in lake water at a depth of 15 feet at 19.2°C. At a 30-foot depth at a temperature of 6.6°C, the cysts remained viable for 56 days.

It is still controversial whether animals such as beavers produce a *G. lamblia* strain that can infect humans. Studies have shown that beavers can be infected with *G. lamblia* from humans, but the reverse has not been demonstrated (Erlandsen *et al.*, 1988). In one study, 40 to 45% of beavers in Colorado were found to be infected with *Giardia* and shedding up to 1×10^8 cysts/animal/

TABLE 22.9 Some Characteristics of Environmentally Transmitted Parasites

Organism	Infective Form (Size)	Mechanism of Transmission	Distribution	Reservoirs
Giardia lamblia	Cyst (8—16 µm)	Person—person waterborne, foodborne	Worldwide	Humans, beavers, muskrats, voles
Cryptosporidium parvum	Oocyst (3—6 µm)	Person—person waterborne, foodborne	Worldwide	Many vertebrates, especially cattle
Entamoeba histolytica	Cyst (10—16 m)	Person—person waterborne, foodborne	Areas of poor sanitation	Usually humans (potentially pigs, primates, and dogs)
Naegleria fowleri	Trophozoite (10—15 µm)	Trophozoite swims up nasal cavity	Worldwide	None: free-living in aquatic or soil environment
Cylospora cayetanensis	Sporulated oocyst (8—10 µm)	Waterborne, foodborne	Asia, Caribbean, Mexico, and Peru	None known
Enterocytozoon bieneusi	Spore 1—1.6 µm	Fecal—oral	Not known	None known
Encephalitozoon hellem	Spore 1—2.5 µm	Urine—oral	Not known	None known
Encephalitozoon cuniculi	Spore 1—2.5 µm	Fecal/urine—oral	Not known	Laboratory rabbits, rodents, dogs
Encephalitozoon intestinalis	Spore 1—2 µm	Fecal/urine—oral	Not known	None known
Toxoplasma gondii	Sporulated (11—12 µm)	Oral ingestion from soil or litterbox (oocyst), undercooked meats (oocyst), undercooked meats (tissue cysts), Waterborne	Worldwide	Cats (definitive host), humans, sheep, goats, pigs, cattle, birds (intermediate hosts)
Ascaris lumbricoides	Embryonated egg	Oral from soil contact	Worldwide	Humans
Trichuris trichiura	Embryonated egg	Waterborne, foodborne	Worldwide especially tropics	Humans
Necator americanus	Filariform larva	Skin penetration	Tropical Africa, Asia, Central and South America and Caribbean	Humans
Ancylostoma duodenale	Filariform larva	Skin penetration, oral from soil contact	Tropical Africa, Asia, Central and South America and Caribbean	Humans
Taenia saginata	Cysticercus	Ingestion of undercooked beef containing cysticerci, waterborne	Worldwide	Cattle (intermediate host)
Schistosoma mansoni	Cercaria	Penetrate skin	Arabia, Africa, South America, and Caribbean	Humans, snail (intermediate host)

day, making them a major source of *Giardia* in the environment (Hibler and Hancock, 1990). Other animals that may contribute to *Giardia* numbers in the environment are muskrats, where 95% of the population is infected. Various other animals found to be infected are cattle, goats, sheep, pigs, cats and dogs (Erlandsen, 1995). To date, no infections with *G. lamblia* in humans have been directly linked to an animal host, but there is evidence that suggests that animal-source *Giardia* could potentially infect humans. Studies based upon isoenzyme analysis

and pulsed field gel electrophoresis (PFGE) banding patterns did not find a difference between cysts from beaver hosts and human hosts (Isaac-Renton *et al.*, 1993).

22.3.1.2 Cryptosporidium

Cryptosporidium was first described by Tyzzer in 1907, when he identified the organism in the intestinal epithelium of a mouse. It was not identified as a human pathogen until 1976, when it was described in the stools of

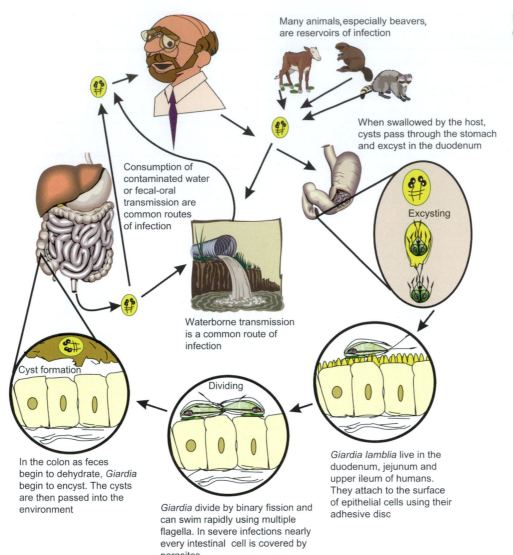

FIGURE 22.9 Life cycle of *Giardia lamblia*.

Many animals, especially beavers, are reservoirs of infection

When swallowed by the host, cysts pass through the stomach and excyst in the duodenum

Excysting

Consumption of contaminated water or fecal-oral transmission are common routes of infection

Waterborne transmission is a common route of infection

Cyst formation

Dividing

In the colon as feces begin to dehydrate, *Giardia* begin to encyst. The cysts are then passed into the environment

Giardia divide by binary fission and can swim rapidly using multiple flagella. In severe infections nearly every intestinal cell is covered by parasites

Giardia lamblia live in the duodenum, jejunum and upper ileum of humans. They attach to the surface of epithelial cells using their adhesive disc

Giardia Lamblia
Infectious dose: approx. 200   Size: 8 to 10 microns in length

FIGURE 22.10 Trophozite of *Giardia lamblia*, the reproductive stage of the waterborne parasite. Photo courtesy E.A. Myer.

immunocompromised hosts. Since that time, there have been several waterborne outbreaks, the most notable being the Milwaukee outbreak in April 1993, which infected over 400,000 people (MacKenzie *et al.*, 1994) and killed more than 50 (Case Study 22.1).

It is now recognized that there are several species of *Cryptosporidium* capable of infecting humans. The most common species infecting humans is *C. hominis*, which infects only humans, and *C. parvum*, which primarily infects cattle, but also humans. *Cryptosporidium hominis* has a complex life cycle involving both sexual and asexual stages (Figure 22.11). The host ingests sporulated oocysts (ranging from 3 to 6 μm in diameter) from contaminated water, food or direct contact (Figure 22.12). In the small intestine, the oocyst excysts, releasing four sporozoites, which attach to the epithelial cells of the mucosa. The sporozoite becomes enveloped by the microvilli, which fuse and elongate to cover the sporozoite. The sporozoite then

Case Study 22.1   Cryptosporidiosis in Milwaukee

Early in the spring of 1993, heavy rains flooded the rich agricultural plains of Wisconsin. These rains produced an abnormal runoff into a river that drains into Lake Michigan, from which the city of Milwaukee obtains its drinking water. The city's water treatment plant seemed able to handle the extra load: it had never failed before, and all existing water quality standards for drinking water were properly met. Nevertheless, by April 1, thousands of Milwaukee residents came down with acute watery diarrhea, often accompanied by abdominal cramping, nausea, vomiting, and fever. In a short period of time, more than 400,000 people developed gastroenteritis, and more than 100—mostly elderly and infirm individuals—ultimately died, despite the best efforts of modern medical care. Finally, after much testing, it was discovered that *Cryptosporidium* oocysts were present in the finished drinking water after treatment. These findings pointed to the water supply as the likely source of infection; and on the evening of April 7, the city put out an urgent advisory for residents to boil

their water. This measure effectively ended the outbreak. All told, direct costs and costs associated with loss of life are believed to have exceeded $150 million.

The Milwaukee episode was the largest waterborne outbreak of disease ever documented in the United States. But what happened? How could such a massive outbreak occur in a modern U.S. city in the 1990s? And how could so many people die? Apparently, high concentrations of suspended matter and oocysts in the raw water resulted in failure of the water treatment process—a failure in which *Cryptosporidium* oocysts passed right through the filtration system in one of the city's water treatment plants, thereby affecting a large segment of the population. And among this general population were some whose systems could not withstand the resulting illness. In immunocompetent people, *Cryptosporidiosis* is a self-limiting illness; it is very uncomfortable, but it goes away of its own accord. However, in the immunocompromised, Cryptosporidiosis can be unrelenting and fatal.

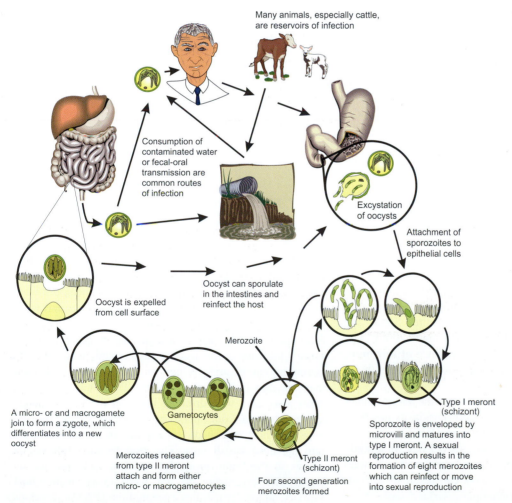

FIGURE 22.11 Life cycle of *Cryptosporidium hominis*.

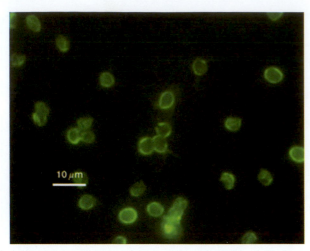

FIGURE 22.12 *Cryptosporidium parvum* oocysts stained with fluorescein isothiocyanate-labeled antibodies. Photo courtesy H.D.A. Lindquest, U.S. Environmental Protection Agency.

matures into a trophozoite and into a schizont. The schizont, an asexual reproductive form in which multiple mitosis occurs followed by cytokinesis, results in eight first-generation merozoites. The cell ruptures, releasing the merozoites, which then infect neighboring epithelial cells, and schizogony occurs again, but forming only four second-generation merozoites. When the cell ruptures, the merozoites attach to uninfected epithelial cells and form either macrogametocytes or microgametocytes. The macrogametocytes and microgametocytes further divide, and form macrogametes and microgametes, respectively. They join, forming a zygote, which differentiates to form the unsporulated oocyst, which is then expelled from the cell surface, sporulates, and is shed in the host's feces.

Within 3–10 days after ingestion of oocysts, nonbloody, voluminous, watery diarrhea begins and lasts for 10–14 days in most immunocompetent hosts. In normal healthy individuals the disease is usually self-limiting. Nitazoxanide has been approved for treatment of diarrhea caused by *Cryptosporidium* in people with healthy immune systems. In the immunocompromised host (e.g., AIDS patients), the disease can be life threatening. The prevalence of cryptosporidiosis in the United States and Europe is 1 to 3% (Sterling and Marshall, 2006). The oocysts are very infectious and the presence of low numbers of oocysts in water or food poses a health threat (Messner *et al.*, 2001).

Cryptosporidium oocysts can enter the environment via human and animal wastes. Cryptosporidiosis has been reported in many domestic animals, especially cattle. An infected calf can excrete 10^{10} oocysts per day. In a study by Kemp *et al.* (1995), farm drains were found to contain 0.06 to 19.4 oocysts per liter. This can result in agricultural land runoff that can contaminate surface water.

Cryptosporidium forms an extremely hardy oocyst that survives chlorine disinfection as commonly practiced at conventional water treatment plants. It is now the most commonly identified cause of recreational outbreaks in the United States, largely associated with chlorinated swimming pools (Hlavsa *et al.*, 2011). It has also been found to survive for weeks in surface waters (Johnson *et al.*, 1997).

22.3.1.3 Entamoeba histolytica

Entamoeba histolytica was discovered in 1873 by D.F. Lösch in St. Petersburg, Russia, although its life cycle was not determined until Dobell did so in 1928. It causes amoebic dysentery (bloody diarrhea), and is the third most common cause of parasitic death in the world. The world prevalence exceeds 500 million infections with more than 40,000 deaths each year. The protozoan only infects humans. There are two sizes of cysts, small (5–9 μm) and large (10–20 μm), with each cyst producing eight trophozoites in the host. Only the larger cyst has been associated with disease; the smaller cyst tends to be associated with a commensal lifestyle (the organism benefits from the host, while the host is unaffected). About 2–8% of people infected develop invasive amoebic dysentery in which the trophozoites actively invade the intestinal wall, bloodstream and liver. It is unknown why this occurs. This organism is generally a problem in developing countries where sanitation is substandard, and is transmitted via contaminated food and water. Humans are the main reservoir, although pigs, monkeys and dogs have also been found to serve as reservoirs. No waterborne outbreaks have occurred in the United States since 1984. *Entamoeba* is not as resistant to disinfectants as *Giardia* and *Cryptosporidium*, and is easily removed during water treatment.

22.3.1.4 Naegleria fowleri

Naegleria fowleri is an amoeboflagellate, changing between a cyst, amoeba and flagellate with the amoeba stage dominant. The free-living protozoa are ubiquitous and found throughout the world in fresh waters (John, 1982). Cysts are usually present in low numbers, but when the water temperature exceeds 35°C (hot springs and warm stagnant waters), the amoeba transforms to the flagellated form quite rapidly, which enables the microorganism to swim. Infections are usually associated with children swimming in natural springs or warm waters, although they are rare in the United States. There have been four cases associated with drinking water in the United States: in Arizona and Louisiana. The flagellate swims into the nose of a host and sheds its flagella (or it may be forced into the nose via diving). The amoeba then follows the nerves to the brain, producing a toxin that liquefies the brain. The organisms do not form cysts in the host. Primary amoebic meningoencephalitis (PAM) develops,

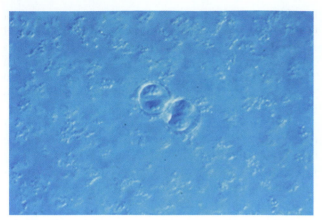

FIGURE 22.13   Oocyst of *Cyclospora cayotanensis*; 100 × under DIC. Photo courtesy H. Smith.

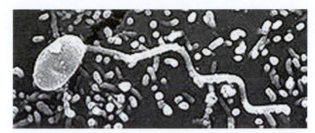

FIGURE 22.14   *Microsporidium* spore with tube by which it penetrates host cells. Courtesy Centers for Disease Control.

causing severe, massive headaches. Death usually follows 4−6 days later. Diagnosis is most frequently postmortem upon brain examination. Treatment (amphotericin B) is available if the diagnosis is made quickly enough, although permanent brain damage may already have occurred.

22.3.1.5 *Cyclospora* sp.

Cyclospora cayetanensis is an emerging waterborne and foodborne protozoan pathogen. It was first identified in the intestine of a mole in 1870 by Eimer. It was not recognized as a human pathogen until the early 1980s, and then it was believed to be a cyanobacterium (Soave and Johnson, 1995). It is a coccidian protozoan in the phylum Apicomplexa. It produces a round spherical oocyst (Figure 22.13) measuring 8×10 μm which contains two sporocysts, each containing two sporozoites. Its sporulation life cycle is typical of a coccidian protozoan parasite. Desiccation of the organism kills it, and therefore it must be in an aquatic environment during maturation.

Cyclosporiasis causes voluminous, explosive, watery, nonbloody diarrhea in addition to abdominal cramps, nausea and fatigue. Illness associated with *Cyclospora* averages 43 days in the immunocompetent host, which can be compared to cryptosporidiosis, which has a duration of only 10 to 14 days (Ortega *et al.*, 1993). It can be treated with Bactrim (trimethoprim−sulfamethoxazole), which appears to eliminate the parasite (Knight, 1995).

Acquisition of *Cyclospora* is not completely understood, as the host range and reservoirs are not known. However, water and food are believed to be a major route of infection, because for the organism to be infectious it must mature (sporulate) in the environment for 2 weeks (Ortega *et al.*, 1993). Water can potentially become contaminated with sporulated oocysts, but most infections have been associated with contaminated produce. Outbreaks related to imported contaminated raspberries, snow peas, basil and mesclun lettuce have been implicated

in various outbreaks of cyclosporiasis in the United States (CDC, 2004). Several suspected waterborne outbreaks have been documented (Rebold *et al.*, 1994; Huang *et al.*, 1995; Baldursson and Karanis, 2011). Infections have also been associated with individuals who live in or visit the Caribbean Islands, Central America and South America, Southeast Asia and Eastern Europe (Knight, 1995).

22.3.1.6 Microsporidia

Microsporidia, the nontaxonomic name to describe organisms belonging to the phylum Microspora, were first described in 1857, when Nägeli identified *Nosema bombycis*, a microsporidian responsible for destruction of the silkworm industry. To date, over 1000 species of microsporidia infecting insects, invertebrates and all five phyla of vertebrate hosts have been described. Microsporidia are for the most part considered to be opportunistic pathogens in humans. There were only a handful of documented cases before the advent of the AIDS epidemic. Since then there have been hundreds of documented cases in immunocompromised patients. However, there have also been cases documented among the immunocompetent. Five genera have been associated with the majority of human infections: *Enterocytozoon bieneusi*, *Encephalitozoon hellem*, *Encephalitozoon cuniculi*, *Encephalitozoon intestinalis*, *Pleistophora* spp. and *Nosema corneum*. The first four have the potential to be waterborne because they are shed in feces and urine. *E. bieneusi*, *E. hellem* and *E. intestinalis* are the most common cause of microsporidian infections in patients with AIDS (Curry and Canning, 1993). In addition, they are much smaller (0.8×1.5 μm depending on species) than other parasites, and potentially more difficult to remove by water treatment filtration.

The microsporidian spore has the potential of being transmitted by water. The life cycle of microsporidia contains three stages: the environmentally resistant spore, merogony and sporogony. The spore is ingested by a host or possibly inhaled in some cases. Once in the body, it infects cells and goes through merogony followed by sporogony, which results in production of resistant infective spores (Figure 22.14). The spores are then shed via bodily fluids such as urine and excreta. Once in the environment they have a strong potential to enter water sources. *E. intestinalis* spores have been identified in sewage, surface and ground waters, supporting the notion of environmental transmission. The spores are highly resistant to heat inactivation and drying. Waller (1979) found that *E. cuniculi* survived 98 days at 4°C and 6 days at 22°C.

Enterocytozoon bieneusi, *Encephalitozoon hellem*, *Encephalitozoon cuniculi* and *Encephalitozoon intestinalis* cause a variety of illnesses. *E. bieneusi* causes diarrhea and wasting disease. It is the most important cause of microsporidiosis in AIDS patients. Several surveys have determined that 7 to 30% of AIDS patients who have unexplained chronic diarrhea are infected with *E. bieneusi* (Weber *et al.*, 1994). *E. intestinalis* is similar to *E. bieneusi* in that it infects the intestines and causes diarrhea, but it can also infect kidneys and bronchial and nasal cells. It can infect macrophages, which allows it to disseminate throughout the body. It is secreted in feces and urine, which supports the notion of water transmission. *E. cuniculi* is not an intestinal parasite but it can be shed in urine (Zeman and Baskin, 1985), and therefore environmental transmission is a possibility. It has also been described infecting many different mammals, which means that there could be many animal reservoirs that can contaminate the environment. *E. hellem* has been recognized and shown to cause eye infections (keratoconjunctivitis) and disseminated infections such as ureteritis and pneumonia. It does not invade the intestine, but it can be shed in the urine (Schwartz *et al.*, 1994). Both water- and foodborne outbreaks have been documented (Cotte *et al.*, 1999; Decraene *et al.*, 2012).

22.3.1.7 *Toxoplasma gondii*

Toxoplasma gondii, the causative agent of toxoplasmosis, is an intestinal coccidium of felines with a very wide range of intermediate hosts. *T. gondii* causes a wide range of clinical conditions including brain damage in children, lymphadenopathy, ocular disease and encephalitis. In the immunocompetent adult it causes symptoms that could be mistaken for influenza. It is estimated that 13% of the world's population is infected with *T. gondii* (Hughes, 1985). Acquisition of the organism can be through contact with infected undercooked meat containing a bradyzoite (tissue cyst), and through contact with the environmental stage, the oocysts. The oocyst is excreted only in cat feces (domestic and wild) (Figure 22.15). Congenital infection occurs when a pregnant woman becomes infected for the first time. Once infected, tachyzoites can cross the placenta and infect the fetus. Of the fetuses that become infected, 15% have severe complications. The oocysts can persist in soil and water, which can serve as a route of transmission. One study in England found that 100% of wild cats had antibodies to *T. gondii* (McOrist *et al.*, 1991). Various studies of cats in the United States have

found a seropositive prevalence of approximately 40% (Dubey and Beattie, 1988). The high prevalence of *T. gondii* in the cat population demonstrates that felines can be a significant source of oocysts in the environment. The oocyst can survive 18 months in the soil (-20 to 33°C) in Kansas (Frenkel *et al.*, 1975) and over 410 days in water (Yilmaz and Hopkins, 1972).

There have been several drinking water outbreaks of toxoplasmosis, and water may be a more important route than previously thought (Baldursson and Karanis, 2011; Karanis *et al.*, 2013). Several epidemiological studies have associated increase risk of infection with the consumption of unfiltered drinking water (Jones *et al.*, 1995). *T. gondii* caused a waterborne outbreak in British Colombia, Canada, where more than 110 people including 12 newborns were infected after exposure to an unfiltered water supply (Mullens, 1996). In a later study of the area, four of seven domestic cats found in the watershed had antibodies to *T. gondii*, potentially being the initial source of contamination (Stephen *et al.*, 1996).

The oocysts are very resistant to inactivation by iodine, chlorine and ozone (Wainwright *et al.*, 2007). The cysts can be inactivated in meat by either freezing at -12°C or cooking to an internal temperature of 67°C (Dubey, 1996).

22.3.2 Nematodes

Nematodes are nonsegmented roundworms belonging to the phylum Nematoda. The majority of roundworms are free living in soil and fresh and salt water. The typical nematode has a flexible outer cuticle that protects the worm. Nematodes move via a muscular system and most lay eggs.

22.3.2.1 *Ascaris lumbricoides*

Ascaris lumbricoides is probably the most prevalent parasitic infection, with over 1 billion humans affected or about 22% of the world's population (CDC, 2013). The prevalence is quite high in some regions of the world. Infection percentages range from 40 to 98% in Africa, to 73% in Southeast Asia, and to 45% in Central America and from South America, and 1 to 2% in the United States (Freedman, 1992).

Infection in humans occurs by ingestion of embryonated eggs (Figures 22.16 and 22.17). There are no known animal reservoirs. The eggs are swallowed and the larvae hatch in the small intestine. The larvae develop into second-stage larvae, which penetrate the lumen and enter into the bloodstream and capillaries. They travel via pulmonary circulation to the liver and heart, where the larvae develop into third-stage organisms that can lodge in the alveolar space (Figure 22.18). This migration through the lungs can cause pneumonitis (Loeffler's syndrome). The immature worms leave the lungs and travel through the bronchi, trachea and epiglottis. They are then swallowed and arrive in the small intestine. There they undergo two molts, mate and produce eggs. There is medical treatment against the adult intestinal worm: mebendazole or pyrantel pamoate. Symptoms usually correspond to the worm load, and a heavy worm load can lead to intestinal blockage. The adult worm can reach more than 30 cm in size. Although most infections are mild,

more than 20,000 people die annually with complications caused by intestinal blockage (Freedman, 1992).

An adult worm can produce more than 200,000 eggs per day. In 2–4 weeks after deposition in soil, they embryonate if the soil conditions are suitable (humid and warm) and are infectious. The eggs can survive months before embryonation if soil conditions are not appropriate. The eggs can survive freezing, chemicals, disinfectants and sewage treatment. Because of their large size ($35 \times 55\ \mu m$), they accumulate in sewage sludge. This is

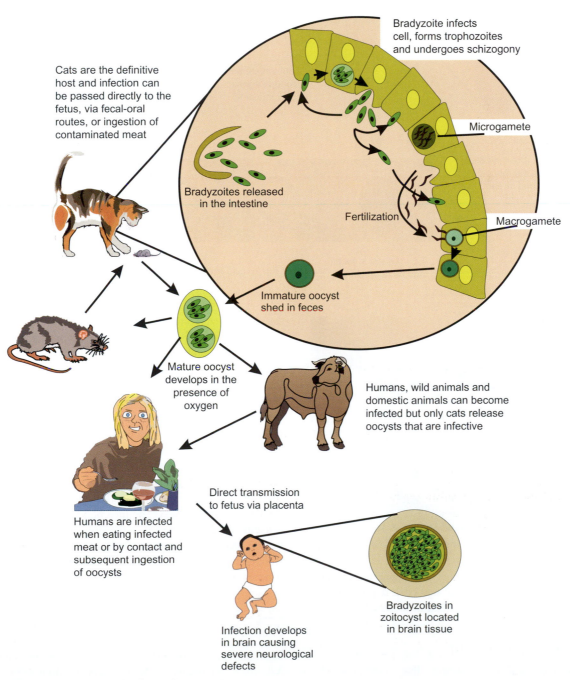

FIGURE 22.15   Life cycle of *Toxoplasma gondii*.

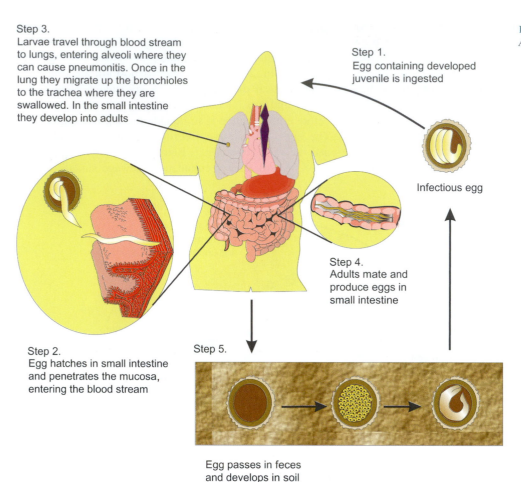

Step 3.
Larvae travel through blood stream to lungs, entering alveoli where they can cause pneumonitis. Once in the lung they migrate up the bronchioles to the trachea where they are swallowed. In the small intestine they develop into adults

Step 1.
Egg containing developed juvenile is ingested

FIGURE 22.16 Life cycle of *Ascaris lumbricoides*.

Infectious egg

Step 4.
Adults mate and produce eggs in small intestine

Step 2.
Egg hatches in small intestine and penetrates the mucosa, entering the blood stream

Step 5.

Egg passes in feces and develops in soil

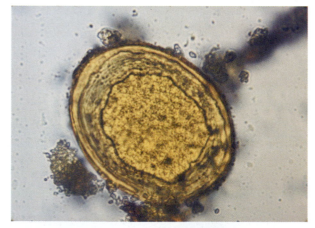

FIGURE 22.17 *Ascaris lumbricoides* ovum. Photo courtesy P. Watt.

especially a concern with sewage sludge-amended soils. In one study, 31−53% of eggs that had been deposited in soil were still viable 10 years later (Brudastov *et al.*, 1971).

22.3.2.2 *Trichuris trichiura*

Trichuris trichiura is a worm that measures about 30−50 mm in length and is referred to as a whipworm, as

the worm's shape resembles a whip. It has been estimated that there are over 1 billion infections worldwide (de Silva *et al.*, 2003), and it is the third most common nematode infection. It is common in the southeastern United States as the weather conditions are ideal for egg survival in the soil (Figure 22.19). The egg must be deposited in the soil and requires 21 days in moist, shady, warm soil to embryonate. In one study, 20% of ova deposited in soil were viable after 18 months (Burden *et al.*, 1976). Infection occurs in humans via ingestion of contaminated water or soil. The worms can survive for years in a host, causing disease symptoms of diarrhea or constipation, anemia, inflamed appendix, vomiting, flatulence and insomnia. The infection is diagnosed by identification of worms or eggs in the stool. The infection can be treated successfully with mebendazole. To prevent transmission, education on hand washing and sanitary feces disposal is necessary.

22.3.2.3 *Necator americanus* and *Ancylostoma duodenale*

There are two major hookworm species that infect humans: *Necator americanus* (New World hookworm) and *Ancylostoma duodenale* (Old World hookworm). The species

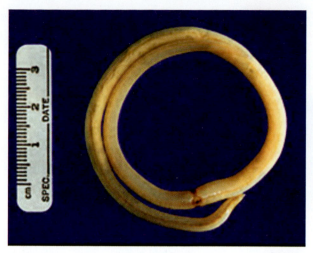

FIGURE 22.18 *Ascaris lumbricoides* adult. Photo courtesy Centers for Disease Control and Prevention.

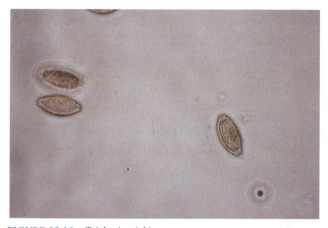

FIGURE 22.19 *Trichuris trichiura* eggs.

are differentiated by mouth parts of adults and body size. There are no known reservoirs. They inhabit the small intestine and feed on intestinal mucosa and blood. They secrete an anticoagulant, causing great blood loss and anemia. They are the leading cause of iron deficiency in the tropics.

N. americanus have round cutting teeth and are 7−10 mm in length. They lay an average of 10,000 eggs a day, and enter humans via skin penetration. Each worm consumes 0.03 ml of blood per day. *A. duodenale* are larger (10−12 mm in length) and cause even greater blood loss. Each worm consumes 0.26 ml of blood per day. *A. duodenale* also lays more eggs (28,000), and are orally infective as well as able to penetrate the skin. In addition, *A. duodenale* also have sharp cutting teeth.

The eggs embryonate once they are passed into the small intestine (Figure 22.20). The eggs further develop into the rhabditiform larvae within 48 hours in warm,

moist sandy or loamy soil. The larvae feed, grow and molt twice and then transform to filariform larvae. The filariforms do not eat. They seek out the highest point in the surroundings (e.g., top of grass blade) waiting for a host. Upon contact with skin, they penetrate the tissue and pass through a hair follicle or cut. They burrow through the subcutaneous tissue, and then through the capillaries to the lungs. In the lungs, they break out of the alveolar capillaries, and migrate up the bronchi and trachea, where they are swallowed and enter the stomach and small intestine. They can live an average of 5 years but have been found to survive up to 15 years. The larvae can survive up to 6 weeks in moist, shady sandy or loamy soil. They do not survive well in clay soil, dry conditions or at temperatures below freezing or greater than 45°C.

22.3.3 Cestodes (*Taenia saginata*)

Cestodes are tapeworms consisting of a flat segmented body and a scolex (head) containing hooks and/or suckers and grooves for attachment. The segments are called proglottids and pregnant segments are called gravids. The adults are parasitic and live in the intestinal lumen of many vertebrates.

Taenia saginata is transmitted by infected beef products and is the most common tapeworm found in humans. It is present in every country where beef is consumed (Figure 22.21). Cattle become infected from eating grass or soil contaminated with human waste containing gravid proglottids. This occurs in areas where night soil (human waste) is used as fertilizer. The proglottids can survive in the environment for weeks. One study found that they could survive 71 days in liquid manure, 16 days in untreated sewage and up to 159 days on grass (Jepsen and Roth, 1952). The eggs hatch in the duodenum, and hexacanths (tapeworm embryos containing six pairs of hooklets) are released. The hexacanths penetrate the mucosa, enter the intestine and travel throughout the body. They then enter muscle and form a cysticercus (larval tapeworm enclosed in a cyst), which becomes infective in a few months. Humans become infected when they eat undercooked beef containing such cysticerci. The cysticerci can be inactivated at 56°C or by freezing at −5°C for 1 week (Schmidt and Roberts, 1989). Within 2−12 weeks the worm begins shedding gravid proglottids. The average worm length is 10−15 feet. Symptoms of infection are abdominal pain, headache, nausea, diarrhea, intestinal blockage and loss of appetite (contrary to the belief that tape worms infections cause an increase in appetite). The infection can be diagnosed by examining a gravid proglottid or scolex. The best methods for prevention are sanitation (proper disposal of human wastes) and thorough cooking of beef.

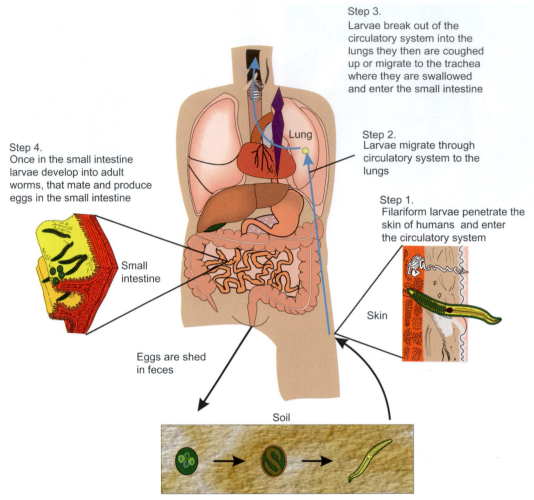

Step 3.
Larvae break out of the circulatory system into the lungs they then are coughed up or migrate to the trachea where they are swallowed and enter the small intestine

Step 4.
Once in the small intestine larvae develop into adult worms, that mate and produce eggs in the small intestine

Lung

Step 2.
Larvae migrate through circulatory system to the lungs

Step 1.
Filariform larvae penetrate the skin of humans and enter the circulatory system

Small intestine

Skin

Eggs are shed in feces

Soil

In soil the egg develops into embryo then hatches into larva

FIGURE 22.20 Life cycle of *Necator americanus*.

22.3.4 Trematodes (*Schistosoma mansoni*)

Trematodes, or flukes, are bilaterally symmetric worms that have two deep suckers and flame cells (for excretion). The suckers are used for both attachment and locomotion. The life cycles are complex, with trematodes being either hermaphrodites (adults have both female and male gonads) or schistosomes (separate sexes). Trematodes require an intermediate host (snail) to complete their life cycle; the human is the definitive host, excreting eggs in the feces.

Three species of *Schistosoma* are medically important. In the past, *S. japonicum* and *S. haematobium* were the main causes of schistosomiasis, but now *S. mansoni* is recognized to be the most widespread of the three. The genus *Schistosoma* is responsible for more than 200 million infections worldwide and causes up to 200,000 deaths annually (Hopkins, 1992). More than 400,000 of those infected live in the United States (West and Olds,

1992). However, all these cases are imported as the intermediate host, the freshwater snail (*Biomphalaria* sp.), is not present in the United States. *S. mansoni* is distributed through most of Africa and the Middle East but is also found in parts of Central America and South America as well as some Caribbean Islands (Savioli *et al.*, 1997).

The larvae of schistosomes of bird and mammals can penetrate human skin, causing what is known as "swimmer's itch." These schistosomes do not mature in humans. This occurs in the Great Lakes region and along the coast of California.

The life cycle of *S. mansoni* is complicated, and human infection begins with the cercariae penetrating human skin. In addition, each parasitic stage involves different organs and thus different medical symptoms. The cercariae from infected snails are found in freshwater bodies. Once the cercariae penetrate the skin, they transform into the schistosomula, the first parasitic stage. This causes the disease cercarial dermatitis, which

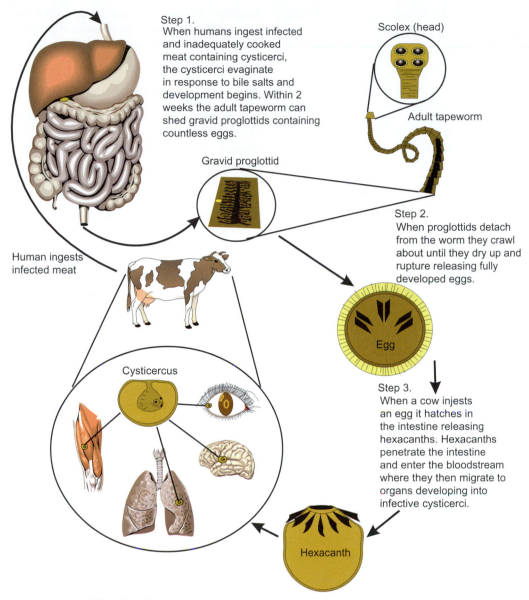

Step 1.
When humans ingest infected and inadequately cooked meat containing cysticerci, the cysticerci evaginate in response to bile salts and development begins. Within 2 weeks the adult tapeworm can shed gravid proglottids containing countless eggs.

Scolex (head)

Adult tapeworm

Gravid proglottid

Human ingests infected meat

Step 2.
When proglottids detach from the worm they crawl about until they dry up and rupture releasing fully developed eggs.

Egg

Cysticercus

Step 3.
When a cow injests an egg it hatches in the intestine releasing hexacanths. Hexacanths penetrate the intestine and enter the bloodstream where they then migrate to organs developing into infective cysticerci.

Hexacanth

FIGURE 22.21   Life cycle of *Taenia saginata*.

consists in immediate or delayed hypersensitivity to the penetration. After 1−2 days of living in the subcutaneous tissues, the schistomsomula migrate through the lungs to the liver. The organisms mature through this migration, developing into male or female adult worms. The adult worms then migrate to the circulatory system and to other organs. Each pair of adult worms live an average of 3−10 years and potentially up to 30 years (Lucey and Maguire, 1993). They also produce 300 eggs a day, only a percentage of which reach the lumen to be passed in the feces. Many of the eggs become trapped in the small blood vessels and tissues, causing granulomas and intestinal, visceral, liver and fibroobstructive diseases. The acute form of disease, Katayama fever, can occur in previously uninfected persons 2−6 weeks after

exposure to cercariae-contaminated water. Chemotherapy is available using praziquantel, although no vaccine is available.

No decrease of infection has been noted, and in fact the number of cases has increased over the past 50 years. The spread of this parasite has occurred as water development projects and population movements have introduced it into previously uninfected regions. However, it has been controlled in some areas: the Caribbean, excluding Puerto Rico, and Brazil. Strategies that have been successful are chemotherapy, health education, water supply treatment and sanitation. Very little success has been achieved by eradicating the intermediate host, the snail, with either mulluscicides or predator fish.

TABLE 22.10 Human Enteric Viruses

Enterovirus
Hepatitis A virus
Reoviruses
Rotaviruses
Adenoviruses
Astroviruses
Torovirus
Human caliciviruses (norovirus, sapporovirus)
Hepatitis E virus
Picobirnaviruses
Bocaviruses
Coronaviruses

22.4 VIRUSES

22.4.1 Enteric Viruses

Viruses are a leading cause of gastroenteritis, in particular in infants and young children, in which they are a major cause of mortality worldwide. Four major groups of human gastroenteritis viruses have been identified: rotavirus; enteric adenovirus; caliciviruses (norovirus and sapporovirus); and astrovirus (Table 22.10). Of these, norovirus is of note because it has become the enteric virus most commonly associated with water- and foodborne illness worldwide. Although endemic viral gastroenteritis can be transmitted person to person by the oral–fecal route, outbreaks of viral gastroenteritis may be triggered by contamination of a common water or food source.

Many human viruses can infect the gastrointestinal tract and be excreted in the feces into the environment. It has been estimated that an individual with an enteric viral infection may excrete 10^{11} viral particles per gram of feces. Once in the environment, viruses can reach water supplies, recreational waters, crops and shellfish, through contact with sewage, land runoff, solid waste landfills and septic tanks.

Diseases caused by enteric viruses range from trivial to severe or even fatal. Waterborne outbreaks caused by enteric viruses are difficult to document because many infections by these agents are subclinical; i.e., the virus may replicate in an individual, resulting in virus shedding but without signs of overt disease. Therefore, an individual with waterborne infection but without overt disease may infect others, who in turn may become ill, spreading the infection throughout the community. In addition, epidemiological techniques lack the sensitivity to detect low-level transmission of viruses through water. Recreational activities in swimming pools have sometimes resulted in waterborne outbreaks caused by norovirus,

hepatitis A virus, coxsackievirus, echovirus and adenoviruses. Enteric viruses from infected individuals may contaminate recreational waters by direct contact or by fecal release.

22.4.1.1 Astroviruses

Astroviruses were first observed by electron microscopy in diarrheal stools in 1975. These agents are icosahedral viruses with a starlike appearance and with a diameter of approximately 28 nm. Astroviruses have a single-stranded RNA (ssRNA) genome. The sequencing of the astrovirus genome allowed the establishment of its own virus family the Astroviridae, with *Mamastrovirus* which infects mammals and *Avastrovirus* which infects birds. Serological assays and sequence analysis have resulted in the identification of eight distinct serotypes which infect humans.

Astrovirus type 1 seems to be the most prevalent strain in children. Type 4 has been associated with severe gastroenteritis in young adults. Astrovirus-like particles have been found in the feces of a number of animals suffering from a mild self-limiting diarrheal infection, but no antigenic cross-reactivity has been found between these agents and human astroviruses. Astrovirus infections occur throughout the year, with a peak during the winter–spring seasons in temperate zones.

Astroviruses cause a mild gastroenteritis after an incubation period of 3 to 4 days. Overt disease is common in 1- to 3-year-old children. However, adults and young children are also affected. Outbreaks of astrovirus infection have been associated with oysters and drinking water (Kurtz and Lee, 1987; Maunula *et al.*, 2004).

22.4.1.2 Adenoviruses

Adenoviruses are double-stranded DNA (dsDNA) icosahedral viruses approximately 70 nm in diameter with protruding spikes called pentons (Figure 22.22). At least 57 human adenovirus types have been identified. Although there are many avian and mammalian adenovirus types, they are species specific. Adenoviruses can replicate in the respiratory tract, the eye mucosa, the intestinal tract, the urinary bladder and the liver.

Most adenovirus human illness is associated only with one-third of adenovirus types. Although many adenovirus infections are subclinical, these viruses may cause: acute respiratory disease (types 1–7, 14 and 21); conjunctivitis (types 3, 7, 8, 11, 14, 19 and 37); acute hemorrhagic cystitis (11 and 21); acute respiratory disease (ARD) of military recruits (types 3, 4, 7, 14 and 21); and gastroenteritis (types 31, 40, 41 and 52) (Table 22.11). Adenovirus type 36 has been associated with obesity in humans and animals (Atkinson *et al.*, 2005).

Adenoviruses gain access to susceptible individuals through the mouth, the nasopharynx or the conjunctiva. Although initial infection may occur via the respiratory route, fecal—oral transmission accounts for most adenovirus infections in young children because of the prolonged shedding of viruses in feces (Horwitz, 1996).

Nose, throat and eye infections caused by adenoviruses have been associated with improperly disinfected swimming pool water. The enteric adenoviruses 40 and 41 have been recognized as the second most important etiological agents of viral gastroenteritis in children. These viruses, in contrast to other adenoviruses, are not shed in respiratory secretions; thus, their transmission is limited to the oral—fecal route. They have been associated with several outbreaks associated with drinking water (Divizia *et al.*, 2004).

Adenoviruses are among the most common viruses in untreated sewage, where they have been found in concentrations 10 times greater than that of the enteroviruses. Contaminated inanimate surfaces may play a significant role in adenovirus transmission because of its stability to drying. At room temperature, adenovirus 2 survives for 8 and 12 weeks at low (7%) and high (96%) relative humidity, respectively, and is more resistant than poliovirus 2, vaccinia virus, coxsackievirus B3 and herpes virus (Boone and Gerba, 2007). Longer survival of adenoviruses in

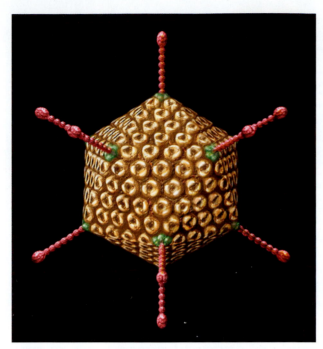

FIGURE 22.22   Adenovirus.

TABLE 22.11 Classification of Human Adenoviruses

Subgenus	Serotype	Human Illness
A	12	Meningoencephalitis
	18, 31	Diarrhea
B	3	Acute febrile pharyngitis; adenopharyngo-conjunctival fever; pneumonia; follicular conjunctivitis; fatal infection in neonates
	7	Acute febrile pharyngitis; adenopharyngo-conjunctival fever; acute respiratory disease with pneumonia; fatal infection in neonates; meningoencephalitis
	11	Follicular conjunctivitis; hemorrhagic cystitis in children
	21	Hemorrhagic cystitis in children; fatal infection in neonates
	14, 16, 50, 55	Acute respiratory disease with pneumonia
	34, 35	Acute and chronic infection in patients with immunosuppression and AIDS
C	1, 2, 6	Acute febrile pharyngitis; pneumonia in children
	5	Acute febrile pharyngitis; pertussis-like syndrome; acute and chronic infection in patients with immunosuppression and AIDS
D	8, 19, 37 9, 10, 13, 15, 17, 42 19, 20, 22—29	Epidemic keratoconjunctivitis
	30	Fatal infection in neonates
	32, 33, 36, 38	Asymptomatic (36 obesity)
	39, 42—51, 53, 54, 56	Acute and chronic infection in patients with immunosuppression and AIDS
E	4	Respiratory infection
F	40, 41	Diarrhea

TABLE 22.12 Some Human Enteroviruses and Paraechoviruses and Clinical Illness

Virus	Serotypes	Clinical Illness
Poliovirus	3 types	Paralysis, aseptic meningitis, febrile illness
Enterovirus A Coxsackievirus	12 types A 2–7 A 8–16	Paralysis, aseptic meningitis, hand, foot, and mouth disease, encephalitis
Enterovirus	71 76	Herpangina, exanthema, diarrhea
Enterovirus B Coxsackievirus	37 types B1-B6, A9	Aseptic meningitis, paralysis, exanthema, respiratory diseases
Echovirus	1–9, 11–21, 24–33	Diarrhea
Enterovirus	69, 73–91	Pericarditis, myocarditis, febrile illness
Enterovirus C	11 types	Paralysis, aseptic meningitis
Coxsackievirus A	1, 11, 13, 15, 17–22, 24	Myocarditis, encephalitis
Enterovirus D Enterovirus	68, 70	Pneumonia, acute hemorrhagic conjunctivitis
Paraechoviruses	1–6	Pericarditis, herpangina, respiratory disease

water has also been observed. Adenovirus type 5 survives longer in tap water, at either 4 or 18°C, than either poliovirus 1 or echovirus 7 (Bagdasar'yan and Abieva, 1971), and enteric adenoviruses 40 and 41 survive longer than poliovirus 1 and HAV in tapwater and seawater (Enriquez *et al.*, 1995b). The increased survival of the enteric adenoviruses in tapwater and seawater, and the faster inactivation in sewage, may indicate that these viruses are inactivated by different mechanisms than those affecting the enteroviruses. The enteric adenoviruses are more thermally stable than poliovirus 1, which is inactivated faster at temperatures above 50°C (Enriquez *et al.*, 1995a). In addition, enteric adenoviruses 40 and 41 are more resistant to ultraviolet (UV) light disinfection than poliovirus type 1 and coliphage MS-2, which has been suggested as a model for enteric virus disinfection (Meng and Gerba, 1996).

The increased resistance showed by enteric adenoviruses, compared with other enteric viruses, may be associated with the double-stranded nature of their DNA, which, if damaged, may be repaired by the host cell DNA repair mechanisms. This mechanism would not be effective with ssRNA genome viruses such as poliovirus 1 or HAV. It has been suggested that the longer survival of the enteric adenoviruses in tapwater and seawater may be associated with DNA damage, and the faster inactivation in sewage may result from protein capsid damage.

22.4.1.3 Enteroviruses and Paraechoviruses

The enteroviruses are members of the family Picornaviridae, which are among the smallest ribonucleic acid (RNA) viruses. "Picornavirus" means small RNA virus. Enteroviruses and Paraechoviruses are icosahedral viruses approximately 27 to 32 nm in diameter. Enteroviruses are divided into five groups (Table 22.12). The nucleic acid of enteroviruses consists of ssRNA. These are the viruses most often detected in sewage polluted water. However, their apparent higher prevalence may be associated, in part, with available cell lines for their propagation, because many pathogenic enteric viruses such as HAV, enteric adenoviruses, rotavirus, norovirus and other small round viruses are difficult to grow in conventional cell lines.

Although viruses belonging to the *Enterovirus* and *Paraechovirus* genera are capable of causing a wide variety of clinical conditions, from asymptomatic to disabling, or even fatal infections, they often do not cause overt disease (Table 22.12).

Despite the fact that there are bovine, porcine, simian and murine enteroviruses, it is believed that humans are the only natural host of human enteroviruses. There are currently more than 100 human enterovirus types known (Piralla *et al.*, 2013). Enteroviruses replicate primarily in the gastrointestinal tract, and may be shed in large numbers (approximately 10^{10}/g feces). The most common forms of transmission include the fecal−oral and respiratory routes (Figure 22.23). Waterborne transmission may be considered a form of fecal−oral transmission in which the responsible vehicle is water instead of hands or fomites. Although enteroviruses are readily found in fecally contaminated drinking or recreational waters,

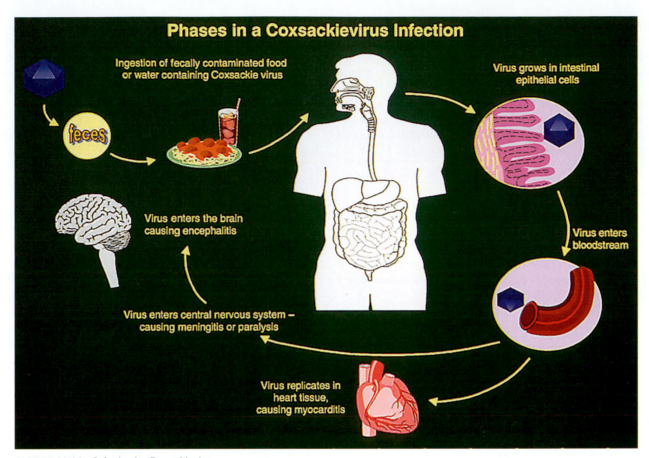

FIGURE 22.23   Infection by Coxsackieviruses.

waterborne enterovirus infection has been only occasionally documented. Waterborne outbreaks related to enteroviruses are difficult to document because many infections by these agents are subclinical. Therefore, an individual with waterborne infection without overt disease may infect others, who in turn may become ill, spreading the infection further. Attack rates of enteroviruses vary depending on the virus and the host age. Asymptomatic infections by poliovirus outnumber symptomatic disease (10:1), whereas symptomatic infection by echovirus 9 is relatively high (10:7) (Morens and Pallansch, 1995).

Both disease and infection caused by polioviruses are age related. Generally, infection is more common in infants, but adults and older children are more severely affected. However, some exceptions exist. Coxsackievirus B virus infection is usually more severe in newborns than in older children and adults, often causing fulminant myocarditis, encephalitis, hepatitis and death. Coxsackieviruses are the most prevalent nonpolio enteroviruses (Morens and Pallansch, 1995). Coxsackieviruses are also the most common nonpolio enteroviruses isolated from water and wastewater. These viruses have been associated with several serious illnesses (Table 22.12). Coxsackievirus B5 infection has been associated with recreational water. In an outbreak

at a boys' summer camp, this virus was isolated from lake water; however, person-to-person contact appeared to be the main form of transmission (Hawley *et al.*, 1973).

It is believed that almost all enteroviruses can be transmitted by the fecal−oral route; however, it is not clear if all of them can additionally be transmitted by the respiratory route. Airborne transmission of enterovirus might include aerosol spread or direct exposure to respiratory secretions. Fecal−oral transmission may predominate in areas with poor sanitary conditions, whereas respiratory transmission may occur more often with better sanitation. In temperate climates, enteroviruses are more common during the summer season. In the United States, most enterovirus isolations (82%) occur from June to October. However, vaccine strains of poliovirus are isolated year-round because of routine vaccination of children. In contrast, in tropical and semitropical areas, enteroviruses do not show seasonality. Transmission of enteroviruses within a household is usually started by young children; then the infection spreads quickly to other family members, especially in larger families living under crowded conditions with poor hygiene. Paralytic poliomyelitis and perhaps some nonpolio enteroviral diseases are more often observed in developing countries, where sanitary conditions are improving. With poor hygiene, most individuals are infected at a very early age, when infection rarely results in overt disease and maternal immunity limits infection. This early exposure to the virus elicits a protective immune response on reexposure to these viruses later in life. In contrast, when early exposure is prevented or delayed as a result of better sanitation, an initial poliovirus infection is likely to occur at an older age, when maternal immunity has waned and the possibility of developing a more severe clinical condition is greater.

22.4.1.4 Hepatitis A Virus (HAV)

HAV is a picornavirus morphologically indistinguishable from other members of the same family. This agent was formerly classified as a member of the *Enterovirus* genus (enterovirus 72). However, differences in nucleotide and amino acid sequences resulted in its classification as the only member of the hepatovirus group.

The average HAV incubation is approximately 30 days, but it may vary from 10 to 50 days, a variation associated with the dose. Infection with very few particles results in longer incubation periods and vice versa. The period of communicability extends from early in the incubation period to about a week after the development of jaundice. The greatest danger of spreading the disease to others occurs during the middle of the incubation period, well before the first presentation of symptoms. During this period the patient remains asymptomatic; however, active shedding of the virus is the norm. Therefore, it is during the incubation time that the infected individual has the highest potential for spreading HAV.

Hepatitis A is usually a mild illness, which almost always results in complete recovery. Severity and disease manifestation are age related. An estimated 80 to 95% of infected children younger than 5 years of age do not develop overt disease, whereas clinical manifestations are observed in approximately 75 to 90% of infected adults. The mortality rate in children 14 years of age or younger is 0.1%; this rate rises to 0.3% in individuals between the ages of 15 and 39 years and 2.1% in those older than 40 years.

Hepatitis A is characterized by sudden onset of fever, malaise, nausea, anorexia and abdominal discomfort, followed in several days by jaundice. In contrast to hepatitis B, HAV infection is not chronic. HAV is excreted in feces of infected people, and can produce clinical disease when susceptible individuals consume contaminated water or foods. Water, shellfish and produce are the most frequent sources. Contamination of foods by infected workers in food processing plants and restaurants is not uncommon.

HAV is very stable in the environment. It is more resistant to high temperatures than poliovirus. HAV can survive at 60°C for 1 hour, and temperatures up to 85 to 95°C are needed to inactivate it in shellfish. Poor sanitation and crowding facilitate HAV transmission.

Therefore, outbreaks of hepatitis A are common in institutions, crowded house projects, prisons and military forces. In developing countries, the incidence of disease in adults is relatively low because of exposure to the virus in childhood. Most individuals 18 years and older possess an immunity that provides lifelong protection against reinfection.

The survival of HAV on hands and its transfer to hands or inanimate surfaces were studied by Mbithi *et al.* (1992). They found that approximately 20% of the initial HAV inoculated on hands remained infectious for at least 4 hours, and that HAV inoculated onto a stainless steel surface survived for 2 hours. They also determined that exerting higher pressure and friction between HAV-contaminated hands or fomites resulted in more efficient transfer of this virus to clean hands.

With better sanitation, HAV infection shifts to older individuals, and the incidence of overt disease increases. In more developed countries, low levels of HAV transmission occur. However, disease outbreaks are relatively common in most of these countries as a significant segment of the population is susceptible to HAV infection. The introduction of a vaccine for HAV has resulted in a steady decline in cases in the United States (see Case Study 22.2).

22.4.1.5 Hepatitis E Virus (HEV)

The hepatitis E virus is the leading cause of acute viral hepatitis among young and middle-aged adults in developing countries. HEV has a diameter of 32−34 nm and an ssRNA genome. Currently, HEV is placed in a sole genus *Hepevirus* within a new family Hepeviridae. It has been suggested that the higher prevalence of HEV in adults may be due to a silent infection early in life, with subsequent waning of immunity after 10 to 20 years, when they again become susceptible to infection by HEV (Bradley, 1992). Important epidemiological features of HEV infection are the frequent occurrence of outbreaks associated with consumption of sewage-polluted water and its severity, particularly among pregnant women, in whom the case-fatality rate may be as high as 25% (Balayan, 1993). HEV virus also infects swine and other animals, and has been transmitted by the consumption of undercooked meat (Tamada *et al.*, 2004). To date, no outbreak has occurred in the United States.

Hepatitis E is clinically indistinguishable from hepatitis A. It is characterized by jaundice, malaise, anorexia, abdominal pain, arthralgia and fever. The incubation period for hepatitis E varies from 2 to 9 weeks. The disease is most often seen in young to middle-aged adults (15−40 years old). The disease is usually mild and resolves in 2 weeks, leaving no long-term effects, with a relatively low fatality rate (0.1−1%). There is no evidence of immunity against HEV in a population that has been exposed to this virus (Margolis *et al.*, 1997).

Case Study 22.2   A Massive Waterborne Outbreak of Infectious Hepatitis

Approximately 35,000 young adults in New Delhi, India came down with hepatitis in the fall of 1955. There were 73 deaths recorded among those affected, one of the largest viral waterborne outbreaks ever documented. New Delhi obtains its drinking water from the Jamuna River. Because of flooding following the monsoon season, water drawn from the river for treatment and distribution became heavily contaminated with sewage from a nearby creek. From November 11 to November 16, raw wastewater freely entered the drinking water treatment plant. The staff at the plant, aware of the contamination, increased the dose of alum coagulant and chlorine in an effort to eliminate contaminating pathogens. About three weeks later, the first hepatitis cases were identified. The observed incubation period ranged from 22 to 60 days, with an average of 40 days. Surprisingly, no increase in other types of viral, bacterial, or parasitic infections was observed during the hepatitis epidemic, possibly because of the partial success of emergency disinfection using high levels of chlorine.

In an attempt to isolate the epidemic's causative agent, researchers at the Virus Research Center in Poona inoculated eight different cell lines with a variety of clinical specimens obtained from hepatitis patients. Unfortunately, no viruses were isolated from any of the samples tested. Although the etiologic agent was not identified, it was believed that the epidemic was caused by the hepatitis A virus (HAV). A retrospective study conducted in 1980 showed that the outbreak in New Delhi was not caused by HAV or hepatitis B virus (HBV) but by the so-called enterically transmitted non-A, non-B (ENANB) hepatitis virus. This agent was later renamed hepatitis E virus (HEV). Although waterborne outbreaks of hepatitis E have occurred in many regions of the world, outbreaks have almost exclusively been in developing countries. HEV waterborne outbreaks primarily affect young adults, and fatalities among pregnant women are not uncommon.

The raw sewage contamination of drinking water that led to the large HEV outbreak in New Delhi was thought to be controlled by the use of high doses of alum coagulant and chlorine. Although this measure prevented an outbreak of bacterial, parasitic, and other enteric viral diseases, it failed to eliminate HEV. Therefore, prevention of contamination, rather than extensive treatment, offers the best alternative for safe drinking water.

22.4.1.6 Rotavirus

Rotaviruses are classified within the family Reoviridae. These viruses have a characteristic genome consisting of 11 dsRNA segments surrounded by a distinctive two-layered protein capsid. Particles are approximately 70 nm in diameter (Figure 22.24). Six serological groups (A–F) have been identified; groups A, B and C infect both humans and animals, and D, E and F have been detected only in animals. Group A rotaviruses have been associated with the majority of infantile acute gastroenteritis cases, group B with severe diarrhea epidemics in adults in China and group C with sporadic cases of diarrhea in children, but their clinical importance has not been determined. Within each group, rotaviruses are classified into serotypes.

Rotaviruses are the most important agents of infantile gastroenteritis around the world. Group A rotavirus is endemic worldwide. It is the leading cause of severe diarrhea among infants and children, and accounts for about half of the cases requiring hospitalization. In temperate areas, it occurs primarily in the winter, but in the tropics it occurs throughout the year. Group B rotavirus, also called adult diarrhea rotavirus, has caused major epidemics of severe diarrhea affecting thousands of persons of all ages in China. Group C rotavirus has been associated with rare and sporadic cases of diarrhea in children in many countries.

Rotaviruses are shed in large numbers, up to 10^{10} viral particles per gram of feces (White and Fenner, 1994). Names applied to the infection caused by the most common and widespread group A rotavirus include: acute gastroenteritis; infantile diarrhea; winter diarrhea; acute nonbacterial infectious gastroenteritis; and acute viral gastroenteritis.

Rotaviruses are transmitted by the fecal–oral route. Exposure by person-to-person spread via contaminated hands is probably one of the most important routes by

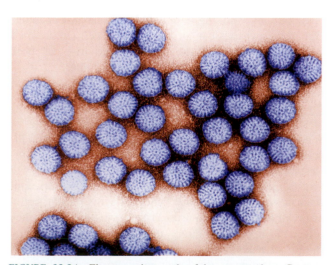

FIGURE 22.24 Electron micrograph of human rotavirus. Courtesy Centers for Disease Control.

which rotaviruses are transmitted. Institutions or close communities such as pediatric wards, day care centers and family homes are usually most affected by outbreaks of gastroenteritis caused by rotaviruses. Because of high infectivity, rotavirus-infected food handlers may contaminate foods that require handling but do not require further cooking, such as cakes, salads and fruits. Among adults, multiple foods served at banquets were implicated in two outbreaks, and an outbreak related to contaminated municipal water occurred in Colorado in 1981. Several large outbreaks of group B rotavirus involving millions of persons as a result of sewage contamination of drinking water supplies have occurred in China since 1982. Although to date, outbreaks caused by group B rotavirus have been confined to mainland China, seroepidemiological surveys have indicated a lack of immunity in the United States to this group of viruses. Rotaviruses are quite stable in the environment and have been found in estuary samples at levels as high as one to five infectious particles per gallon (Anonymous, 1992). Sanitary measures adequate for bacteria and parasites seem to be ineffective in endemic control of rotavirus, as similar incidences of rotavirus infection are observed in countries with both high and low health standards. Rotaviruses have caused numerous water and foodborne outbreaks in both adults and children (Gerba et al., 1996; Greening, 2006).

Rotavirus gastroenteritis is a self-limiting disease, which can be mild to severe. It is characterized by vomiting, watery diarrhea and mild fever. Asymptomatic rotavirus excretion has been well documented and may play a role in perpetuating endemic disease. The incubation period ranges from 1 to 3 days. Symptoms often start with vomiting followed by 4−8 days of diarrhea. As with other viral gastroenteritis, rotavirus gastroenteritis treatment consists of fluid and electrolyte replacement; if it is untreated, severe diarrhea with dehydration and death may occur. Individuals of all ages are susceptible to rotavirus infection. Premature infants, children 6 months to 2 years of age, the elderly and the immunocompromised are particularly prone to develop more severe symptoms. In adults, rotaviral infection is usually subclinical. Each year rotavirus causes millions of cases of diarrhea in developing countries, almost 2 million resulting in hospitalization, and an estimated 453,000 resulting in the death of a child younger than 5 years, 85% of whom live in developing countries (Tate et al., 2012). The introduction of a vaccine for children has been reducing the number of serious cases of rotavirus in the United States. Rotavirus infection does not result in an efficient or long-lasting immunity. Therefore, rotavirus infection in the same child often occurs up to six times during childhood.

Some animal rotaviruses, such as SA-11, are readily propagated in cell culture; the human rotaviruses, however, are rather fastidious. These viruses have not been grown efficiently in any conventional tissue culture system.

22.4.1.7 Human Caliciviruses

The use of electron microscopy (EM) since the early 1970s for the examination of fecal specimens from individuals suffering from nonbacterial gastroenteritis has shown many previously unknown viruses. Among these, viruses such as adenoviruses and rotaviruses are larger and well defined; thus, they are relatively easy to identify by EM. However, many smaller (20−40 nm) viruses without a distinctive morphology are often present in fecal samples of patients suffering from gastroenteritis (Figure 22.25). Norwalk virus was the first to be described (Kapikian et al., 1972). Subsequently, other small round viruses were observed in diarrheal stools, namely Montgomery County, Hawaii, Wollan, Ditchling, the Parramatta, the Cockle agents and minirotavirus. Norwalk-like viruses are in the calicivirus family, and are now divided into two genera which infect humans; Norovirus and Sapovirus. There are also several genera of caliciviruses that only infect animals. Norovirus are major causes of both food- and waterborne disease. It is estimated that 67% of the foodborne illness in the United States are caused by noroviruses (Gerba and Kayed, 2003). Caliciviruses are nonenveloped viruses with a diameter of approximately 26 to 35 nm and a positive-sense ssRNA genome.

Infectivity studies with volunteers have shown that individual susceptibility is more important than acquired immunity. It has been suggested that genetically determined factors are the primary determinants of resistance to Norovirus infection, perhaps at the level of cellular receptor sites (i.e., blood group antigens expressed in the gut). In developing countries, antibodies to Norovirus are acquired early in life, and the peak incidence of illness may also occur among younger age groups than in developed nations. Noroviruses and related viruses usually produce a mild and brief illness, lasting 1 to 2 days. It is characterized by nausea and abdominal cramps, followed commonly by vomiting in children and diarrhea in adults. Human noroviruses cannot be grown in cell culture on a continuous basis, and the murine Norovirus has been often used as a model for environmental studies.

22.4.2 Respiratory Viruses

Worldwide, respiratory illnesses are the most common illnesses in humans and most have a viral etiology. Respiratory disease is associated with a large number of viruses, including: rhinoviruses; coronaviruses; parainfluenza viruses; respiratory syncytial virus (RSV); influenza virus; and adenovirus. These viruses, when they infect the upper respiratory tract, can cause acute viral rhinitis or pharyngitis (common cold); when the primary site of infection is the lower respiratory tract, they can cause laryngotracheitis (croup), bronchitis or pneumonia.

FIGURE 22.25   Norwalk virus (norovirus) capsid.

Mortality related to acute respiratory disease may be especially significant in children and in the elderly. In adults, temporary disability results in important economic loss. Respiratory infection often results from self-inoculation, when virus-contaminated hands or fingers rub the eyes or when viruses are introduced into the mouth or nose. Another important route of transmission of respiratory viruses is inhalation of contaminated aerosols.

22.4.2.1 Rhinoviruses

Rhinoviruses (Latin *rhino*, nose) belong to the family Picornaviridae. Two important characteristics differentiate rhinoviruses from enteroviruses: stability at low pH values and temperature. Whereas the rhinoviruses are inactivated at pH values below 6, the enteroviruses are stable at low pH values. In contrast, rhinoviruses are stable at temperatures (50°C) that would inactivate most enteroviruses (Couch, 1996). Human rhinoviruses have an icosahedral morphology, containing a single-stranded RNA genome. The diameter of these viruses is approximately 25 to 30 nm. Although no etiologic agent is identified in half of the acute upper respiratory illnesses, it has been estimated that 30 to 50% are caused by rhinoviruses. In fact, rhinovirus infection is probably the most common type of human acute infection (Gwaltney, 1997). The optimal temperature for rhinovirus growth is 33 to 35°C, which corresponds to the normal temperature of the nasal mucosa. This may explain why these viruses propagate most efficiently in the upper respiratory tract. These viruses can be propagated in monkey and human cell lines. There are more than 100 different human rhinovirus serotypes. Although these viruses are species specific, an equine rhinovirus can infect other species, including humans. Experimentally, chimpanzees and gibbons have been infected with human rhinoviruses, but these animals do not develop disease. Volunteer studies have shown that infection can be started with less than one tissue culture median infectious dose ($TCID_{50}$).

Rhinovirus infection, after 1 to 4 days of incubation, is characterized by nasal obstruction and discharge, sneezing, scratchy throat, mild cough and malaise. The relatively high rate of rhinovirus infection may be associated with the large number of rhinovirus serotypes, and the fact that the same serotype can infect an individual more than once. In addition, immunity to rhinovirus infection is short lived.

Although rhinovirus infection occurs throughout the year, in temperate climates its frequency increases during colder months, and in the tropics the peak incidence occurs during the rainy season. Experiments with volunteers have failed to associate exposure to low

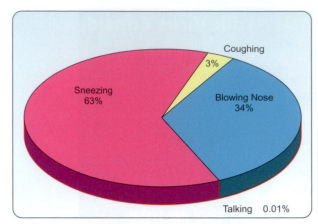

FIGURE 22.26 Source of rhinovirus dispersion. Data from Buckland and Tyrrell (1964).

temperatures with an increased susceptibility to rhinovirus infection. Therefore, the cause of this seasonality remains unclear. Nevertheless, it has been suggested that colder temperatures or rain may increase the survivability of these viruses and/or promote crowding conditions, in which the virus may propagate more efficiently.

During rhinovirus infection, the highest concentration of viruses is found in the nasal and pharyngeal mucosa. In contrast, in saliva, rhinoviruses are found in low numbers, and inconsistently. This suggests that aerosols may not play an important role in the transmission of these agents, as sneezing and coughing generate aerosols composed mainly of saliva (Figure 22.26). In fact, studies with volunteers have shown that rhinovirus transmission by hand contact and self-inoculation of the eye or nasal mucosa is much more efficient than through aerosols (Hendley et al., 1973).

The high concentration of rhinoviruses in nasal secretions easily leads to hand contamination. Rhinoviruses have been isolated from 40 to 90% of hands of ill persons, and from 6 to 15% of environmental objects such as doorknobs, dolls, coffee cups and glasses. It is believed that contamination of objects in the environment may play a significant role in the transmission of rhinoviruses. This, and the ability of rhinoviruses to survive on human skin and on inanimate objects, may explain the more prevalent self-inoculation form of transmission. Both finger-to-eye and finger-to-nose contact are part of normal human behavior. Therefore, experimental rhinovirus transmission has been effectively interrupted by hand disinfection. Rhinoviruses are relatively stable on inanimate surfaces. They can survive at least 3 hours on human skin, nonporous materials such as plastic surfaces, Formica, stainless steel and hard synthetic fabrics such as nylon and dacron. In porous materials such as paper tissue and cotton fabric, rhinoviruses can survive for 1 hour (Hendley et al., 1973).

22.4.2.2 Paramyxovirus

Parainfluenza viruses and respiratory syncytial viruses (RSV) belong to the family Paramyxoviridae. These viruses are spherical enveloped particles of heterogeneous size, ranging from 125 to 250 nm. The paramyxovirus viral particle has a lipid envelope covered with spikes of about 10 nm. The nucleic acid of these agents consists of a single-stranded RNA molecule (Ginsberg, 1990).

RSV and parainfluenza viruses do not survive well in the environment. If suspended in a protein-free medium at 4°C, 90 to 99% of their infectivity is lost within 4 hours. Organic solvents and detergents rapidly inactivate these viruses by dissolving their lipid envelopes. Both RSV and parainfluenza viruses cause a variety of illnesses, primarily in infants and young children. The most important are common cold, bronchitis, bronchiolitis, laryngotracheobronchitis (croup) and pneumonia.

22.4.2.3 Parainfluenza Viruses

The parainfluenza viruses constitute two of the four genera of the family Paramyxoviridae: *Paramyxovirus* (parainfluenza virus types 1 and 3) and *Rubulavirus* (parainfluenza virus types 2 and 4). Parainfluenza viruses infect most people during childhood. Types 1 and 2 are often associated with croup in infants and type 3 with bronchiolitis and pneumonia; type 4 seldom causes illness. Although parainfluenza infections in children and infants can result in serious disease, most infections are subclinical. In adults, these viruses can cause a mild cold, but, as in children, most infections are also subclinical. Survival studies have shown that parainfluenza virus type 3 remains infectious in aerosol particles for at least 1 hour (Miller and Artenstein, 1967).

Disease caused by parainfluenza viruses is observed throughout the year. Infection by parainfluenza virus types 1 and 2 occurs endemically, but small epidemics caused by these agents are observed every 2 years. Parainfluenza virus type 3 infects approximately 60% of infants during the first 2 years of life, reaching 80% by 4 years of age, whereas infection by types 1 and 2 does not reach 80% until 10 years of age. In children, an incubation period of 2 to 4 days has been estimated for parainfluenza virus illness. Studies have shown that a child infected with parainfluenza type 3 may shed viruses over an average of 8 days, but shedding viruses for up to 4 weeks has been documented (Frank et al., 1981). During the fall, parainfluenza has been found on 30% of common fomites in office buildings in the United States (Boone and Gerba, 2010).

22.4.2.4 Respiratory Syncytial Virus

RSV belongs to the family Paramyxoviridae, as the only member of the genus *Pneumovirus*. The characteristics of

RSV have already been described. This virus is the most important respiratory pathogen during infancy and early childhood. It causes approximately half of the cases of bronchiolitis and 25% of cases of pneumonia in infants. Approximately 90,000 hospital admissions and 4500 deaths are associated each year with RSV in the United States in both infants and young children. In approximately half of infants younger than 8 months, the infection spreads to the lower respiratory tract, resulting in life-threatening bronchitis, bronchiolitis, bronchopneumonia and croup (Ginsberg, 1990). Asymptomatic first infection with RSV is rare; almost 100% of infected children develop disease. Because of poor protective immunity, reinfection is very common, but the disease is not as severe. Inoculation of adult volunteers with RSV has shown that the nose and eye mucosae are the most important portals of entry of this virus. Efficient transmission of RSV by large droplets or by touching occurs when susceptible individuals are in close contact with children shedding the virus, but not when they are exposed to small-particle aerosol (McIntosh, 1997).

The incubation period of RSV respiratory disease is 4 to 5 days. Replication of RSV in the upper respiratory tract can reach concentrations of 10^4 to 10^6 TCID$_{50}$/ml of secretion, with higher titers in infants. Infected individuals may shed RSV for up to 3 weeks. In temperate climates, RSV infection may occur throughout the year, but it peaks during winter months with few occurrences in the summer; in tropical areas, outbreaks often occur during the rainy season.

22.4.2.5 Influenza Viruses

The influenza viruses belong to three genera of the family Orthomyxoviridae: influenza virus A, influenza virus B and influenza virus C. These viruses, approximately 80 to 120 nm in size, possess a lipid envelope with a genome consisting of eight segments of single-stranded RNA. The influenza virus type A is further classified into many subtypes according to host of origin, year and geographic location of first isolation.

It is not known why several recent influenza pandemics have started in China. It has been suggested that the large pig, duck and human population in the Canton area may facilitate coinfection of animals with influenza viruses that originated from different species, leading to genetic reassortment and to the generation of viruses with novel antigenic and virulence characteristics. There is strong evidence that aquatic birds are the main reservoir of all influenza viruses in other species. For example, the catastrophic influenza pandemic of 1918, in which over 20 million people died, is believed to have been caused by an influenza A virus derived from a bird.

Influenza A virus is a highly contagious agent, causing epidemics of an acute respiratory infection known as influenza, with high mortality in the elderly. Usually, about 75% of all influenza deaths occur in individuals over 55 years of age. Influenza A virus infection may be asymptomatic, but more often it may be manifested by a wide variety of clinical conditions, ranging from an annoying flu to a fatal pneumonia. For instance, a subtype of influenza A virus that was responsible for epidemics of severe disease at the beginning of the century caused minor illness in older people during widespread epidemics in the 1980s. The reason for this wide variation is poorly understood, but it has been speculated that age, underlying illnesses, previous exposure to a similar influenza A virus subtype and virus virulence may be associated with the type of disease presentation. Mortality due to influenza is significant and varies from season to season. During nine of 20 influenza seasons in the United States (1972−1992), more than 20,000 people died each season; during four seasons, more than 40,000 deaths were recorded (Centers for Disease Control, 1997).

Influenza A virus can be transmitted most efficiently through aerosols. It has been demonstrated that influenza A virus remains infectious for at least 1 hour in aerosols at room temperature (Murphy and Webster, 1996). Clinical manifestations of influenza start rather suddenly, about 1 to 4 days after infection. A disabling syndrome characterized by high fever, together with muscle pain, sore throat, nasal congestion, conjunctivitis, cough and headache, is usually the norm.

Influenza epidemics occur intermittently. In temperate climates they usually occur from early fall to late spring, but in tropical regions epidemics are observed throughout the year. Influenza epidemics spread rapidly and tend to occur worldwide. Although variable, epidemics caused by influenza A virus are observed every 2 to 4 years, whereas influenza B virus epidemics normally occur every 3 to 6 years. Immunity to influenza is long lived; however, it is virus subtype specific. Epidemiological studies have shown that individuals previously infected with influenza A subtype H1N1 during the 1957 pandemic were resistant to infection when the same subtype reappeared in the 1977 pandemic, but they were fully susceptible when exposed to other influenza A subtypes.

22.4.2.6 SARS

Severe acute respiratory virus, or SARS, is a serious respiratory virus that resulted in a worldwide outbreak in 2003 resulting in about 8000 cases and nearly 800 deaths (Figure 22.27). It is caused by a coronavirus believed to have originated from bats in Southeast Asia, and later transmitted to other animals. It causes a fever with chills and coughing, with 10 to 20% of persons infected developing diarrhea. It is believed to be spread by contact droplets or from fomite contact. The airborne route does not appear to be the major route of spread as close contact appears necessary for effective spread under most conditions. It is present in both the respiratory droplets and feces. It appears capable of surviving for a few days in liquids and for 24 hours on fomites (Rabenau *et al.*, 2005). Current evidence suggests that the SARS coronavirus originated in bats, and spread to humans either directly or through animals held in Chinese markets.

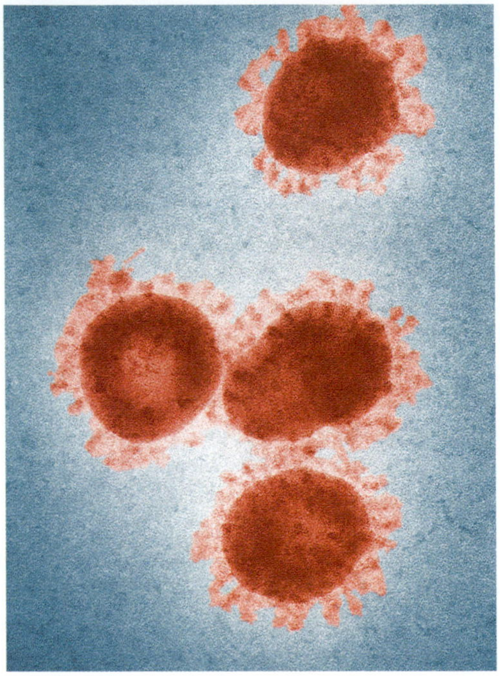

22.5 FATE AND TRANSPORT OF ENTERIC PATHOGENS IN THE ENVIRONMENT

There are many potential routes for the transmission of excreted enteric pathogens. The ability of an enteric pathogen to be transmitted by any of these routes depends largely on its resistance to environmental factors, which control its survival and its capacity to be carried by water or air, as it moves through the environment. Some routes can be considered "natural" routes for the transmission of waterborne disease, but others—such as the use of domestic wastewater for groundwater recharge, large-scale aquaculture projects or land disposal of disposable diapers—are actually new routes created by modern human activities.

Human and animal excreta are sources of pathogens. Humans become infected by pathogens through consumption of contaminated foods, such as shellfish from contaminated waters or crops irrigated with wastewater; from drinking contaminated water; and through exposure to contaminated surface waters, as may occur during bathing or at recreational sites. Furthermore, those individuals infected by the above processes become sources of infection through their excrement, thereby completing the cycle.

In general, viral and protozoan pathogens survive longer in the environment than enteric bacterial pathogens (Figure 22.28). How long a pathogen survives in a

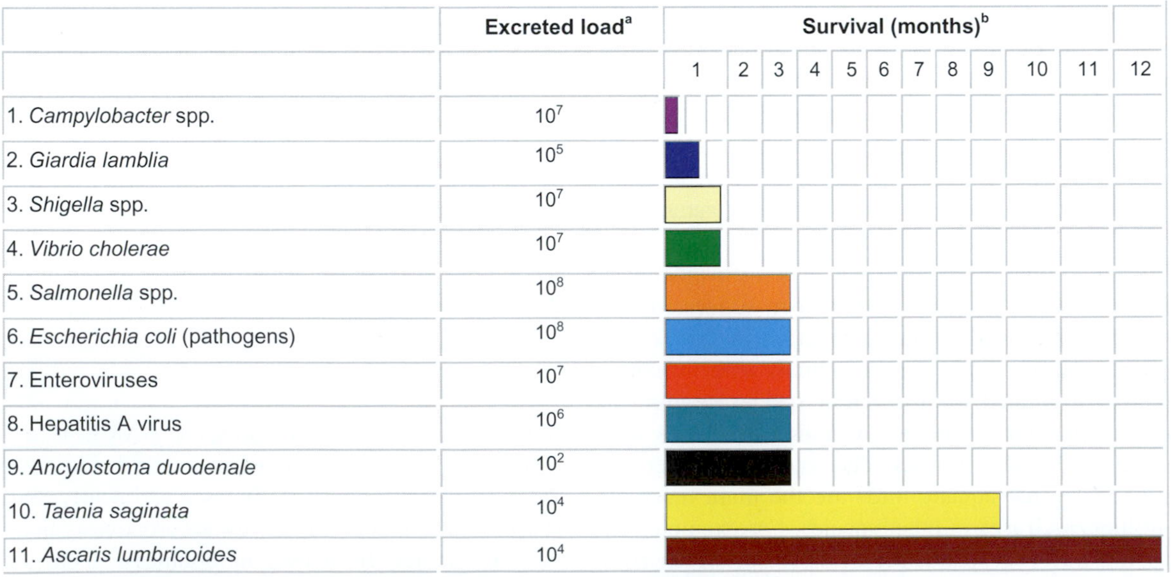

FIGURE 22.27 SARS virus. Courtesy Centers for Disease Control.

	Excreted load[a]	Survival (months)[b]											
		1	2	3	4	5	6	7	8	9	10	11	12
1. *Campylobacter* spp.	10^7												
2. *Giardia lamblia*	10^5												
3. *Shigella* spp.	10^7												
4. *Vibrio cholerae*	10^7												
5. *Salmonella* spp.	10^8												
6. *Escherichia coli* (pathogens)	10^8												
7. Enteroviruses	10^7												
8. Hepatitis A virus	10^6												
9. *Ancylostoma duodenale*	10^2												
10. *Taenia saginata*	10^4												
11. *Ascaris lumbricoides*	10^4												

[a] Typical average number of organisms/g feces
[b] Estimated average life of infective stage at 20–30°C.
(Modified from Feachem *et al.*, 1983).

FIGURE 22.28 Survival times of enteric pathogens in water, wastewater, and soil and on crops.

TABLE 22.13 Environmental Factors Affecting Enteric Pathogen Survival in Natural Waters

Factor	Remarks
Temperature	Probably the most important factor; longer survival at lower temperatures; freezing kills bacteria and protozoan parasites, but prolongs virus survival.
Moisture	Low moisture content in soil can reduce bacterial populations.
Light	UV in sunlight is harmful.
pH	Most are stable at pH values of natural waters. Enteric bacteria are less stable at pH >9 and pH <6
Salts	Some viruses are protected against heat inactivation by the presence of certain cations.
Organic matter	The presence of sewage usually results in longer survival.
Suspended solids or sediments	Association with solids prolongs survival of enteric bacteria and virus.
Biological factors	Native microflora is usually antagonistic.

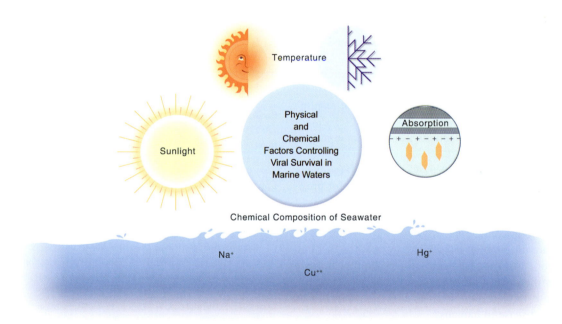

FIGURE 22.29   Factors affecting the survival of enteric bacteria and viral pathogens in seawater.

particular environment depends on a number of complex factors, which are listed in Table 22.13. Of all of the factors, temperature is probably the most important. Temperature is a well-defined factor with a consistently predictable effect on enteric pathogen survival in the environment. Usually, the lower the temperature, the longer the survival time. However, freezing temperatures generally result in the death of enteric bacteria and protozoan parasites. Enteric viruses, however, can remain infectious for months or years at freezing temperatures. Moisture—or lack thereof—can cause decreased survival, and UV light from the sun is a major factor in the inactivation of indicator bacteria in surface waters; thus, die-off

in marine waters can be predicted by amount of exposure to daylight. Viruses are much more resistant to inactivation by UV light.

Many laboratory studies have shown that the microflorae of natural waters and sewage are antagonistic to the survival of enteric pathogens. It has been shown, for example, that enteric pathogens survive longer in sterile water than in water from lakes, rivers and oceans. Bacteria in natural waters can feed upon indicator bacteria. Suspended matter (clays, organic debris and the like) and fresh or marine sediments have been shown to prolong their survival time (Figure 22.29).

QUESTIONS AND PROBLEMS

1. What are pathogens? What is an enteric pathogen?
2. What is the difference between a waterborne and a water-based pathogen? List three of each.
3. Which group of enteric pathogens survives longest in the environment and why?
4. What are some of the niches in which *Legionella* can grow to high numbers?
5. Why are *Cryptosporidium* and *Giardia* major causes of waterborne disease in the United States today?
6. What are the names of the environmentally resistant forms of waterborne protozoan parasites?
7. What group of animals are the reservoirs of *Toxoplasma gondii*?
8. What is the difference between a frank pathogen and an opportunistic pathogen? Give examples of each.
9. Which groups of pathogens cannot grow in the environment outside a host? Which ones can?
10. Which virus is the leading cause of childhood gastroenteritis worldwide?
11. Which virus is most commonly associated with waterborne disease outbreaks?
12. Which bacterium is the leading cause of gastroenteritis in the United States?
13. Fomites are important in the spread of what respiratory viruses?
14. Why are respiratory infections more common at certain times of the year?
15. What virus can be transmitted by contact with the eyes?
16. What type of hepatitis has a high mortality for pregnant women?
17. What type of virus causes eye infections in persons swimming in contaminated waters?
18. What are nosocomial infections?
19. What organism is responsible for typhoid? Cholera? Winter diarrhea in infants?
20. Why are only certain strains of *E. coli* capable of causing disease in humans?
21. What type of pathogenic *E. coli* is transmitted primarily by cattle feces?
22. What are the factors determining the survival of enteric pathogens in the environment? Which is the most important factor?
23. The SARS virus is thought to originate from what group of animals?
24. What group of animals is a major source of influenza in humans?

REFERENCES AND RECOMMENDED READING

Allen, M. J., and Geldreich, E. E. (1975) Bacteriological criteria for groundwater quality. *Groundwater* **13**, 45–52.

Anonymous (1992) "Foodborne Pathogenic Microorganisms and Natural Toxins," Federal Drug Administration, Washington, DC.

Asghari, F. B., Nikaeen, M., and Mirhendi, H. (2013) Rapid monitoring of *Pseudomonas* in hospital water systems: a key priority in prevention of nosocomial infection. *FEMS Microbiol. Lett.* **343**, 77–81.

Atkinson, R. L., Dhurandhar, N. V., Allison, D. B., Bowen, R. L., Israel, B. A., Albu, J. B., *et al.* (2005) Human adenovirus-36 is associated with increased body weight and paradoxical reduction of serum lipids. *Int. J. Obes.* **29**, 281–286.

Bagdasar'yan, G. A., and Abieva, R. M. (1971) Survival of enteroviruses and adenoviruses in water. *Hyg. Sanit.* **36**, 333–337.

Balayan, M. S. (1993) Hepatitis E virus infection in Europe: regional situation regarding laboratory diagnosis and epidemiology. *Clin. Diagn. Virol.* **1**, 1–9.

Baldursson, S., and Karanis, P. (2011) Waterborne transmission of protozoan parasites: review of worldwide outbreak—an update 2004–2010. *Wat. Res.* **45**, 6603–6614.

Baumann, P. (1968) Isolation of *Acinetobacter* from soil and water. *J. Bacteriol.* **96**, 39–42.

Bifulco, J. M., Shirey, J. J., and Bissonnette, G. K. (1989) Detection of *Acinetobacter* spp. in rural drinking water supplies. *Appl. Environ. Microbiol.* **55**, 2214–2219.

Bingham, A. K., Jarroll, E., and Meyer, E. (1979) *Giardia* spp.: physical factors of excystation in vitro, and excystation vs. eosin exclusion as determinants of viability. *Exp. Parasitol.* **47**, 284–291.

Blaser, M. J., Smith, P. F., Wang, W. L. L., and Hoff, J. C. (1986) Inactivation of *Campylobacter* by chlorine and monochloramine. *Appl. Environ. Microbiol.* **51**, 307–311.

Boone, A. A., and Gerba, C. P. (2010) The prevalence of human parainfluenza virus 1 on indoor office fomites. *Food Environ. Virol.* **2**, 41–46.

Boone, S. A., and Gerba, C. P. (2007) Significance of fomites in the spread of respiratory and enteric viral disease. *Appl. Environ. Microbiol.* **73**, 1687–1696.

Bradley, D. W. (1992) Hepatitis E: epidemiology, aetiology and molecular biology. *Rev. Med. Virol.* **2**, 19–28.

Brudastov, A. N., Lemelev, V. R., Kholnukhanedov, S. K., and Krasnos, L. N. (1971) The clinical picture of the migration phase of ascaris in self-infection. *Medskaya Parazitol.* **40**, 165–168.

Buckland, F. E., and Tyrrell, D. A. (1964) Experiments on the spread of colds. 1. Laboratory studies on the dispersal of nasal secretion. *J. Hyg.* **62**, 365–377.

Burden, D. J., Whitehead, A., Green, E. A., McFadzean, J. A., and Beer, R. (1976) The treatment of soil infected with human whipworm. *Trichuris trichiura. J. Hyg.* **77**, 377–382.

Burns, D., Wallace, R. J., Jr., Schultz, M. E., Zhang, Y., Zubairi, S. Q., Pang, Y., *et al.* (1991) Nosocomial outbreak of respiratory tract colonization with *Mycobacterium fortuitum*: demonstration of the usefulness of pulsed-field gel electrophoresis in an epidemiological investigation. *Am. Rev. Respir. Dis.* **144**, 1153–1159.

Carmichael, W. W. (1994) The toxins of cyanobacteria. *Sci. Am.* **273**, 64–72.

Carter, A. M., Pacha, R. E., Clark, G. W., and Williams, E. A. (1987) Seasonal occurrence of *Campylobacter* spp. in surface waters and their correlation with standard indicator bacteria. *Appl. Environ. Microbiol.* **53**, 523–526.

Centers for Disease Control (2004) Outbreak of cyclosporiasis associated with snow peas—Pennsylvania. *MMWR* **53**, 876–878.

Centers for Disease Control (CDC) (1997) Influenza surveillance—United States, 1992–93 and 1993–94. *CDC Surveill. Summ.* **46**, SS-1.

Centers for Disease Control (CDC) (2013) Ascariasis. http://www.cdc.gov/parasites/ascariasis, accessed July 10, 2013.

Colbourne, J. S., Dennis, P. J., Trew, R. M., Beery, C., and Vesey, G. (1988) *Legionella* and public water supplies. Proceedings for the International Conference on Water and Wastewater Microbiology, Newport Beach, CA, February 8–11, 1988, Vol. 1.

Collins, C. H., Grange, J. M., and Yates, M. D. (1984) Mycobacteria in water. *J. Bacteriol.* **57**, 193–211.

Cotte, L., Rabodonirina, M., Chapuis, F., Bailly, F., Bissuel, F., Raynal, C., *et al.* (1999) Waterborne outbreak of intestinal microsporidiosis in persons with and without human immunodeficiency virus infection. *J. Infect. Dis.* **180**, 2003–2008.

Couch, R. B. (1996) Rhinoviruses. In "Virology" (B. N. Fields, D. M. Knipe, P. M. Howley, *et al.*), Raven Press, New York, pp. 713–734.

Craun, G. F., Brunkard, J. M., Yoder, J. S., Roberts, V. A., Carpenter, J., Wade, T., *et al.* (2010) Causes of outbreaks associated with drinking water in the United States from 1971 to 2006. *Clin. Microbiol. Rev.* **23**, 507–528.

Curry, A., and Canning, E. (1993) Human microsporidiosis. *J. Infect.* **27**, 229–236.

de Silva, D. R., Booker, S., Hotez, P. J., Montresor, A., Engles, D., and Savioli, L. (2003) Soil-transmitted helminth infections: updating the global picture. *Trends Parasitol.* **19**, 547–551.

Decraene, V., Lebbad, M., Botero-Kleiven, S., Gustavsson, A. M., and Lofdahl, M. (2012) First reported foodborne outbreak associated with microsporidia, Sweden, October 2009. *Epidemiol. Infect.* **140**, 519–527.

Divizia, M., Gabrieli, R., Donia, D., Macaluso, A., Bosch, A., Guix, S., *et al.* (2004) Waterborne gastroenteritis outbreak in Albania. *Wat. Sci. Technol.* **50**, 57–61.

Dubey, J. P. (1996) Strategies to reduce transmission of *Toxoplasma gondii* to animals and humans. *Vet. Parasitol.* **64**, 65–70.

Dubey, J. P., and Beattie, C. P. (1988) "Toxoplasmosis of Animals and Man," CRC Press, Boca Raton, FL.

Enriquez, C. E., Garzon-Sandoval, J., and Gerba, C. P. (1995a) Survival, detection and resistance to disinfection of enteric adenoviruses. Proceedings of the 1995 Water Quality Technology Conference, American Water Works Association, New Orleans, pp. 2059–2086.

Enriquez, C. E., Hurst, C. J., and Gerba, C. P. (1995b) Survival of the enteric adenoviruses 40 and 41 in tap, sea, and wastewater. *Wat. Res.* **29**, 2548–2553.

Erlandsen, S., Sherlock, L., Januschka, M., Schupp, D., Schaefer, F., Jakubowski, W., *et al.* (1988) Cross-species transmission of *Giardia* spp.: inoculation of beavers and muskrats with cysts of human, beaver, mouse, and muskrat origin. *Appl. Environ. Microbiol.* **54**, 2777–2785.

Erlandsen, S. L. (1995) Biotic transmission—giardiasis a zoonosis? In "*Giardia*: From Molecules to Disease" (R. Thompson, J. Reynoldson, and A. Lymbery, eds.), University Press, Cambridge, pp. 83–97.

Falcao, D. P., Valentini, S. R., and Leite, C. Q. F. (1993) Pathogenic or potentially pathogenic bacteria as contaminants of fresh water from different sources in Araraquara, Brazil. *Wat. Res.* **27**, 1737–1741.

Feachem, R. G., Bradley, D. J., Garelick, H., and Mara, D. D. (1983) "Sanitation and Disease: Health Aspects of Excreta and Wastewater Management," John Wiley and Sons, Chichester.

Fields, B. S., Shotts, E. B., Jr., Feeley, J. C., Gorman, G. W., and Martin, W. T. (1984) Proliferation of *Legionella pneumophila* as an intracellular parasite of the ciliated protozoan *Tetrahymena pyriformis*. *Appl. Environ. Microbiol.* **47**, 467–471.

Frank, A. L., Taber, L. H., Wells, C. R., Wells, J. M., Glezen, W. P., and Paredes, A. (1981) Patterns of shedding of myxoviruses and paramyxoviruses in children. *J. Infect. Dis.* **144**, 433–441.

Freedman, D. O. (1992) Intestinal nematodes. In "Infectious Diseases" (S. L. Gorbach, J. G. Bartlett, and N. Blacklow, eds.), W. B. Saunders, Philadelphia, pp. 2003–2008.

Frenkel, J. K., Ruiz, A., and Chinchilla, M. (1975) Soil survival of *Toxoplasma* oocysts in KS and Costa Rica. *Am. J. Trop. Med. Hyg.* **24**, 439–443.

Gerba, C. P., and Kayed, D. (2003) A major cause of foodborne illness. *J. Food Sci.* **68**, 1136–1142.

Gerba, C. P., Rose, J. B., Haas, C. N., and Crabtree, K. D. (1996) Waterborne rotavirus: a risk assessment. *Wat. Res.* **30**, 2929–2940.

Ginsberg, H. S. (1990) Paramyxoviruses. In "Microbiology" (B. D. Davis, R. Dulbecco, H. N. Eisen, and H. S. Ginsberg, eds.), 4th ed., J. B. Lippincott, Philadelphia, pp. 947–959.

Greening, G. E. (2006) Human and animal viruses in food. In "Viruses in Foods" (S. Goyal, ed.), Springer, New York, pp. 5–42.

Gupta, A., Polyak, C. S., Bishop, R. D., Sobel, J., and Mintz, E. D. (2004) Laboratory-confirmed shigellosis in the United States, 1989–2002: epidemiologic trends and patterns. *Clin. Infect. Dis.* **38**, 1372–1377.

Gwaltney, J. M. (1997) Rhinoviruses. In "Viral Infections of Humans, Epidemiology and Control" (A. S. Evans, and R. A. Kaslow, eds.), 4th ed., Plenum, New York, pp. 815–838.

Hawley, H. B., Morin, D. P., Geraghty, M. E., Tomkow, J., and Phillips, A. (1973) Coxsackievirus B epidemic at a boys' summer camp. *JAMA* **226**, 33–36.

Hegarty, J. P., Dowd, M. T., and Baker, K. H. (1999) Occurrence of *Helicobacter pylori* in surface water in the United States. *J. Appl. Microbiol.* **87**, 697–701.

Hendley, J. O., Wenzel, R. P., and Gwaltney, J. M. (1973) Transmission of rhinovirus colds by self-inoculation. *N. Engl. J. Med.* **288**, 1361–1363.

Herwaldt, B. L., Craun, G. F, Stokes, S. L., and Juranek, D. D. (1992) Outbreaks of waterborne disease in the United States: 1989–90. *J. Am. Wat. Works Assoc.* **84**, 129–135.

Hibler, C., and Hancock, C. (1990) Waterborne giardiasis. In "Drinking Water Microbiology" (G. McFeters, ed.), Springer-Verlag, New York, pp. 271–293.

Hlavsa, M. C., Roberts, V. A., Hill, V. R., Kahler, A. M., Orr, M., Garrison, L. E., *et al.* (2011) Surveillance for waterborne disease outbreaks and other health events associated with recreational water—United States, 2007–2008. *MMWR Surveill. Summ.* **23**, 1–32.

Hopkins, D. R. (1992) Homing in on helminths. *Am. J. Trop. Med. Hyg.* **46**, 626.

Horwitz, M. S. (1996) Adenoviruses. In "Fields Virology" (B. N. Fields, D. M. Knipe, and P. M. Howley, eds.), 3rd ed., Lippincott-Raven, Philadelphia, PA, pp. 2149–2171.

Huang, P., Wever, W., Sosin, D., Griffin, P., Long, E., Murphy, J., *et al.* (1995) The first reported outbreak of diarrheal illness associated with *Cyclospora* in the United States. *Ann. Intern. Med.* **123**, 409–414.

Hughes, H. P. A. (1985) How important is toxoplasmosis? *Parasitol. Today* **1**, 41–44.

Isaac-Renton, J., Corderio, C., Sarafis, K., and Shahriar, H. (1993) Characterization of *Giardia duodenalis* isolates from a waterborne outbreak. *J. Infect. Dis.* **167**, 431–440.

Izaguirre, G. (1992) A copper-tolerant *Phormidium* species from Lake Mathews, CA, that produces 2-methylisoborneol and geosmin. *Wat. Sci. Technol.* **25**, 217−223.

Jepsen, A., and Roth, H. (1952) Epizootiology of *Cysticercus bovis*-resistance of the eggs of *Taenia saginata*. Report 14. *Vet. Cong.* **22**, 43−50.

John, D. T. (1982) Primary amebic meningoencephalitis and the biology of *Naegleria fowleri*. *Annu. Rev. Microbiol.* **36**, 101−103.

Johnson, D. C., Enriquez, C. E., Pepper, I. L., Davis, T. L., Gerba, C. P., and Rose, J. B. (1997) Survival of *Giardia, Cryptosporidium*, poliovirus and *Salmonella* in marine waters. *Wat. Sci. Technol.* **35**, 261−268.

Jones, A. R., Bartlett, J., and McCormack, J. G. (1995) *Mycobacterium avium* complex (MAC) osteomyelitis and septic arthritis in an immunocompetent host. *J. Infect.* **30**, 59−62.

Kapikian, A. Z., Wyatt, R. G., Dolin, R., Thornhill, T. S., Kalica, A. R., and Chanock, R. M. (1972) Visualization by immune electron microscopy of 27 nm particle associated with acute infectious nonbacterial gastroenteritis. *J. Virol.* **10**, 1075−1081.

Karanis, P., Aldeyarbi, H. M., Mirhashemi, M. E., and Khalil, K. M. (2013) The impact of the waterborne transmission of *Toxoplasma gondii* and analysis for water detection: an overview and update. *Environ. Sci. Pollut. Res. Int.* **20**, 86−99.

Kemp, J. S., Wright, S. E., and Bukhari, Z. (1995) On farm detection of *Cryptosporidium parvum* in cattle, calves and environmental samples. In "Protozoan Parasites and Water" (W. B. Betts, D. Casemore, C. Fricker, H. Smith, and J. Watkins, eds.), The Royal Society of Chemistry, Cambridge, UK.

Kendall, B. A., and Winthrop, K. L. (2013) Update on the epidemiology of pulmonary nontuberculus mycobacteria infections. *Semin. Res. Crit. Care Med.* **34**, 87−94.

Klein, P. D., Graham, D. Y, Gaillour, A., Opekun, A. R., and Smith, E. O. (1991) Water source as risk factor for *Helicobacter pylori* infection in Peruvian children. *Lancet* **337**, 1503−1506.

Knight, P. (1995) Once misidentified human parasite is a cyclosporan. *ASM News* **61**, 520−522.

Kurtz, J. B., and Lee, T. W. (1987) Astroviruses: human and animal. In "Novel Diarrhea Viruses" (G. Bock, and J. Whelan, eds.), John Wiley & Sons, Chichester, UK, pp. 92−107.

LeChevallier, M. W, Seidler, R. J., and Evans, T. M. (1980) Enumeration and characterization of standard plate count bacteria in chlorinated and raw water supplies. *Appl. Environ. Microbiol.* **40**, 922−930.

Lee, L. A., Shapiro, C. N., Hargrett-Bean, N., and Tauxe, R. V. (1991) Hyperendemic shigellosis in the United States: a review of surveillance data for 1967−1988. *J. Infect. Dis.* **164**, 894−900.

Lipp, E. K., Hug, A., and Colwell, R. R. (2002) Effects of global climate on infectious disease: the cholera model. *Clin. Microbiol. Rev.* **15**, 757−770.

Lucey, D. R., and Maguire, J. H. (1993) Schistosomiasis. *Infect. Dis. Clin. North Am.* **7**, 635−653.

MMWR (1995) Update, *Vibrio cholerae* O1 Western hemisphere, 1991−1994, and V. cholerae O139—Asia, 1994. *MMWR* **44**, 215−219.

MacKenzie, W, Hoxie, N., Proctor, M., Gradus, M., Blair, K., Peterson, D., *et al.* (1994) A massive outbreak in Milwaukee of *Cryptosporidium* infection transmitted through the public water supply. *N. Engl. J. Med.* **331**, 161−167.

Majowicz, S. E., Musto, J., Scallan, E., Angulo, F. J., O'Brien, S. J., Jones, T. F., *et al.* (2010) The global burden of nontyphoidal *Salmonella* gastroenteritis. *Clin. Infect. Dis.* **15**, 882−889.

Margolis, H. S., Alter, M. J., and Hadler, S. C. (1997) Viral hepatitis. In "Viral Infections of Humans, Epidemiology and Control" (A. S. Evans, and R. A. Kaslow, eds.), 4th ed., Plenum, New York, pp. 363−418.

Marrie, T. J. (1994) Community-acquired pneumonia. *Clin. Infect. Dis.* **18**, 501−515.

Maunula, L., Kalso, S., Von Bonsdorf, C. H., and Ponka, A. (2004) Wading pool water contaminated with both noroviruses and astroviruses as the source of gastroenteritis. *Epidemiol. Infect.* **132**, 737−743.

Mbithi, J. N., Springthorpe, V. S., and Sattar, S. A. (1992) Survival of hepatitis A virus on human hands and its transfer on contact with animate and inanimate surfaces. *J. Clin. Microbiol.* **30**, 757−763.

McIntosh, K. (1997) Respiratory syncytial virus. In "Human Enterovirus Infections" (H. A. Robart, ed.), ASM Press, Washington, DC.

McOrist, S., Boid, R., Jones, T, Easterbee, N., Hubbard, A., and Jarrett, O. (1991) Some viral and protozool diseases in the European wildcat (*Felis silvestris*). *J. Wildlife Dis.* **27**, 693−696.

Meenhorst, P. L., Reingold, A. L., Groothuis, D. G., Gorman, G. W, Wilkinson, H. W, McKinney, R. M., *et al.* (1985) Water-related nosocomial pneumonia caused by *Legionella pneumophila* serogroup 1 and 10. *J. Infect. Dis.* **152**, 356−363.

Melius, E. J., Davis, S. I., Redd, J. T., Lewin, M., Herlihy, R., Henderson, A., *et al.* (2012) Estimating the prevalence of active *Helicobacter pylori* infection in a rural community with global positioning system technology-assisted sampling. *Epidemiol. Infect.* **17**, 1−9.

Meng, Q. S., and Gerba, C. P. (1996) Comparative inactivation of enteric adenovirus, polio virus, and coliphages by ultraviolet irradiation. *Wat. Res.* **30**, 2665−2668.

Messner, M. J., Chappell, C. L., and Okhuysen, P. C. (2001) Risk assessment for *Cryptosporidium*: a hierarchical Bayesian analysis of human dose response data. *Wat. Res.* **35**, 3934−3940.

Miller, W. S., and Artenstein, M. S. (1967) Aerosol stability of three acute disease viruses. *Proc. Soc. Exp. Biol. Med.* **125**, 222−227.

Moffat, J. E., and Tompkins, L. S. (1992) A quantitative model of intracellular growth of *Legionella pneumophila* in *Acanthamoeba castellanii*. *Infect. Immun.* **60**, 292−301.

Morens, D. M., and Pallansch, M. A. (1995) Epidemiology. In "Human Enterovirus Infections" (H. A. Rotbart, ed.), ASM Press, Washington, DC, pp. 3−24.

Morgan, D. R., Johnson, P. C., DuPont, H. L., Satterwhite, T. K., and Wood, L. V. (1985) Lack of correlation between known virulence properties of *Aeromonas hydrophila* and enteropathogenicity for humans. *Infect. Immun.* **50**, 62−65.

Mullens, A. (1996) I think we have a problem in Victoria: MD's respond quickly to toxoplasmosis outbreak in BC. *Can. Med. Assoc. J.* **154**, 1721−1724.

Murphy, B. R., and Webster, R. G. (1996) Orthomyxoviruses. In "Fields Virology" (B. N. Fields, D. M. Knipe, P. M. Howley, *et al.*), 3rd ed., Lippincott-Raven, Philadelphia, pp. 1397−1445.

Narayan, L. V, and Nunez, W. J., III (1974) Biological control: isolation and bacterial oxidation of the taste-and-odor compound geosmin. *J. Am. Wat. Works Assoc.* **66**, 532−536.

Nguyen, A. M., Engstrand, L., Genta, R. M., Graham, D. Y., and el-Zaatari, F. A. (1993) Detection of *Helicobacter pylori* in dental plaque by reverse-transcriptase–polymerase chain reaction. *J. Clin. Microbiol.* **31**, 783–787.

Nwachuku, N., and Gerba, C. P. (2008) Occurrence and persistence of *Escherichia coli* O157: H7 in water. *Rev. Environ. Sci. Biotechnol.* **7**, 267–273.

Ortega, Y., Sterling, C., Gilman, R., Cama, V, and Diaz, F. (1993) *Cyclospora* species—a new protozoan pathogen of humans. *N. Engl. J. Med.* **328**, 1308–1312.

Payment, P., Gamache, F., and Paquette, G. (1988) Microbiological and virological analysis of water from two water filtration plants and their distribution systems. *Can. J. Microbiol.* **34**, 1304–1309.

Phillip, R., Brown, M., Bell, R., and Francis, F. (1992) Health risks associated with recreational exposure to blue–green algae (cyanobacteria) when windsurfing and fishing. *Health Hyg.* **13**, 115–119.

Piralla, A., Daleno, C., Scala, A., Greenberg, D., Usonis, V., Principi, N., *et al.* (2013) Genome characterization of enteroviruses 117 and 118: a new group with human enterovirus species C. *PLoS One* **8**, e60641.

Rabenau, H. F., Cinatl, J., Morgenstern, B., Bauer, G., Preiser, W., and Doerr, H. W. (2005) Stability and inactivation of SARS coronavirus. *Med. Microbiol. Immunol.* **194**, 1–6.

Rahman, A., Bonny, T. S., Stonesaovapak, S., and Ananchaipattans, C. (2011) *Yersinia enterocolitica* studies and outbreaks. *J. Pathogens.* doi: 10.2061/2011/239391 Article ID 239391, 11 pages.

Ranganathan, S. C., Skoric, B., Ramsay, K. A., Carzino, R., Gibson, A. M., Hart, E., *et al.* (2013) Geographical differences in first acquisition of *Pseudomonas aeruginosa* in cystic fibrosis. *Ann. Am. Thorac. Soc.* **10**, 108–114.

Reavis, C. (2005) Rural health alert: *Helicobacter pylori* in well water. *J. Am. Acad. Nurse Pract.* **17**, 283–289.

Rebold, J., Hoge, C., and Shlim, D. (1994) *Cyclospora* outbreak associated with chlorinated drinking water. *Lancet* **344**, 1360–1361.

Rolle-Kampczyk, U. E., Fritz, G. J., Diez, U., Lehmann, I., Richter, M., and Herbarth, O. (2004) Well water—one source of *Helicobacter pylori* colonization. *Int. J. Hyg. Environ. Health* **207**, 363–368.

Rollins, D. M., and Colwell, R. R. (1986) Viable but nonculturable stage of *Campylobacter jejuni* and its role in survival in the natural aquatic environment. *Appl. Environ. Microbiol.* **52**, 531–538.

Samuel, M. C., Vugia, D. J., Shallow, S., Marcus, R., Seglar, S., McGivern, T., *et al.* (2004) Epidemiology of sporadic *Campylobacter* infection in the United States and declining trend in incidence. *Clin. Infect. Dis.* **38**, S165–S174.

Savioli, L., Renganathan, E., Montresor, A., Davis, A., and Behbehani, K. (1997) Control of schistosomiasis—a global picture. *Parasitol. Today* **13**, 444–448.

Schaefer, F. W. (2006) *Giardia lamblia.* In "Waterborne Pathogens," 2nd ed. American Water Works Association, Denver, CO, pp. 209–215.

Schwartz, D. R., Bryan, R. T, Weber, R., and Visvesvara, G. (1994) Microsporidiosis in HIV positive patients: current methods for diagnosis using biopsy, cytologic, ultrastructural, immunological, and tissue culture techniques. *Folia Parasitol.* **41**, 101–109.

Schmidt, G., and Roberts, L. (eds.) (1989) "Foundations of Parasitology", 4th ed., Times Mirror/Mosby, St. Louis.

Soave, R. N., and Johnson, W. (1995) *Cyclospora*: conquest of an emerging pathogen. *Lancet* **345**, 667–668.

States, S. J., Kuchta, J. M., Conley, L. F, Wolford, R. S., Wadowsky, R. M., and Yee, R. B. (1989) Factors affecting the occurrence of legionnaires' disease bacterium in public water supplies. In "Biohazards of Drinking Water Treatment" (R. A. Larson, ed.), Lewis, Chelsea, MI, pp. 67–83.

Stephen, C., Naines, D., Bellinger, T., Atkinson, K., and Schwantje, H. (1996) Serological evidence of *Toxoplasma* infection in cougars on Vancouver Island, British Columbia. *Can. Vet. J.* **37**, 241.

Sterling, C. R., and Marshall, M. M. (2006) *Cryptosporidium parvum* and *Cryptosporidium hominis.* In "Waterborne Pathogens," 2nd ed. American Water Works Association, Denver, CO, pp. 193–197.

Tacket, C. O., Brenner, F., and Blake, P. A. (1984) Clinical features and an epidemiological study of *Vibrio vulnificus* infections. *J. Infect. Dis.* **149**, 558–561.

Tamada, Y., Yano, K., Yatsuhashi, H., Inoue, O., Mawatari, F., and Ishibashi, H. (2004) Consumption of wild boar linked to cases of hepatitis E. *J. Hepatol.* **40**, 869–870.

Tate, J. E., Burton, A. H., Boschi-Pinto, C., Steele, A. D., Duque, J., and Parashar, U. D. (2012) 2008 estimate of worldwide rotavirus-associated mortality in children younger than 5 years before the introduction of universal rotavirus vaccination programmes: a systematic review and meta-analysis. *Lancet Infect. Dis.* **12**, 136–141.

Thomas, J. E., Gibson, G. R., Darboe, M. K., Dale, A., and Weaver, L. T. (1992) Isolation of *Helicobacter pylori* from human feces. *Lancet* **340**, 1194–1195.

Tison, D. L., and Seidler, J. (1983) *Legionella* incidence and density in potable drinking water supplies. *Appl. Environ. Microbiol.* **45**, 337–339.

Turner, P. C., Gammie, A. J., Hollinrake, K., and Codd, G. A. (1990) Pneumonia associated with contact with cyanobacteria. *Br. Med. J.* **300**, 1440–1441.

Van der Wielen, P. W., and van der Kooij, D. (2013) Nontuberculous mycobacteria, fungi, and opportunistic pathogens in unchlorinated drinking water in the Netherlands. *Appl. Environ. Microbiol.* **79**, 825–834.

Wainwright, K. E., Miller, M. A., Barr, B. C., Gardner, I. A., Melli, A. C., Essert, T., *et al.* (2007) Chemical inactivation of *Toxoplasma gondii* oocysts in water. *J. Parasitol.* **93**, 925–931.

Waller, T. (1979) Sensitivity of *Encephalitozoon cuniculi* to various temperatures, disinfectants and drugs. *Lab. Anim.* **13**, 227–230.

Weber, R., and Bryan, R. (1994) Microsporidial infections in immunodeficient and immunocompetent patients. *Clin. Infect. Dis.* **19**, 517–521.

Wenneras, C., and Erling, V. (2004) Prevalence of enterotoxigenic *Escherichia coli*-associated diarrhea and carrier state in the developing world. *J. Health Popul. Nutr.* **22**, 370–382.

West, P. M., and Olds, F. R. (1992) Clinical schistosomiasis. *R. I. Med. J.* **75**, 179.

Whiley, H., Keegan, A., Giglo, S., and Bentham, R. (2012) *Mycobacterium avium* complex—the role of potable water in disease transmission. *J. Appl. Microbiol.* **113**, 223–232.

White, D. O., and Fenner, F. J. (1994) "Medical Virology," 4th ed. Academic Press, San Diego.

Wu, J. T, and Juttner, F. (1988) Effect of environmental factors on geosmin production by *Fischerella muscicola. Wat. Sci. Technol.* **20**, 143–148.

Yilmaz, S. M., and Hopkins, S. H. (1972) Effects of different conditions on duration of infectivity of *Toxoplasma gondii* oocysts. *J. Parasitol.* **58**, 938–939.

Young, W. E., Horth, H., Crane, R., Ogden, T., and Arnott, M. (1996) Taste and odor threshold concentrations of potential potable water contaminants. *Wat. Res.* **30**, 331–340.

Zacheus, O. M., and Martikainen, P. J. (1994) Occurrence of legionellae in hot water distribution systems of Finnish apartment buildings. *Can. J. Microbiol.* **40**, 993–999.

Zeman, D., and Baskin, G. (1985) Encephalitozoonosis in squirrel monkeys (*Saimiri sciureus*). *Vet. Pathol.* **22**, 24–31.

Chapter 23

Indicator Microorganisms

Charles P. Gerba

23.1 THE CONCEPT OF INDICATOR ORGANISMS

The routine examination of environmental samples for the presence of intestinal pathogens is often a tedious, difficult and time-consuming task. Thus, it has been customary to tackle this issue by looking for indicator microorganisms, whose presence indicates that pathogenic microorganisms may also be present. Developed for the assessment of fecal contamination, the indicator concept depends on the fact that certain nonpathogenic bacteria occur in the feces of all warm-blooded animals. These bacteria can often be isolated and quantified by simple bacteriological methods more easily than pathogenic microbes. Detection of these bacteria in water can mean that fecal contamination has occurred, and suggests that enteric pathogens may also be present.

For example, coliform bacteria, which normally occur in the intestines of all warm-blooded animals, are excreted in great numbers in feces. In polluted water, coliform bacteria are found in densities roughly proportional to the degree of fecal pollution. Because coliform bacteria are generally hardier than disease-causing bacteria, their absence from water is an indication that the water is bacteriologically safe for human consumption.

Conversely, the presence of the coliform group of bacteria is indicative that other kinds of microorganisms capable of causing disease may also be present, and that the water is potentially unsafe to drink.

In 1914, the U.S. Public Health Service adopted the coliform group as an indicator of fecal contamination of drinking water. Many countries have adopted coliforms and other groups of bacteria as official standards for drinking water, recreational bathing waters, wastewater discharges and various foods. Indicator microorganisms have also been used to assess the efficacy of food processing and water and wastewater treatment processes. As an ideal assessor of fecal contamination, it has been suggested that they meet the criteria listed in Table 23.1. Unfortunately, no single indicator meets all of these criteria. Thus, various groups of microorganisms have been suggested and used as indicator organisms. Concentrations of indicator bacteria found in wastewater and feces are shown in Tables 23.2 and 23.3.

Indicators have traditionally been used to suggest the presence of enteric pathogens; however, today we recognize that there is rarely a direct correlation between bacterial indicators and human pathogens (Ashbolt *et al.*, 2001). As such, the use of indicators is better defined by their intended purpose (Table 23.4). Thus, process indicators are used to assess the efficacy of a treatment process

I.L. Pepper, C.P. Gerba, T.J. Gentry: Environmental Microbiology, Third edition. DOI: http://dx.doi.org/10.1016/B978-0-12-394626-3.00023-5
© 2015 Elsevier Inc. All rights reserved.

TABLE 23.1　Criteria for an Ideal Indicator Organism

- The organism should be useful for all types of water.
- The organism should be present whenever enteric pathogens are present.
- The organism should have a reasonably longer survival time than the hardiest enteric pathogen.
- The organism should not grow in water.
- The testing method should be easy to perform.
- The density of the indicator organism should have some direct relationship to the degree of fecal pollution.
- The organism should be a member of the intestinal microflora of warm-blooded animals.

TABLE 23.2　Estimated Levels of Indicator Organisms in Raw Sewage

Organism	CFU* per 100 ml
Coliforms	$10^7 - 10^9$
Fecal coliforms	$10^6 - 10^7$
Fecal streptococci	$10^5 - 10^6$
Enterococci	$10^4 - 10^5$
Escherichia coli	$10^6 - 10^7$
Clostridium perfringens	10^4
Staphylococcus (coagulase positive)	10^3
Pseudomonas aeruginosa	10^5
Acid-fast bacteria	10^2
Coliphages	$10^2 - 10^3$
Bacteroides	$10^7 - 10^{10}$

*CFU = colony forming units

TABLE 23.3　Microbial Flora of Animal Feces

Animal Group	Average Density Per Gram		
	Fecal Coliforms	Fecal Streptococci	Clostridium perfringens
Farm animals			
Cow	230,000	1,300,000	200
Pig	3,300,000	84,000,000	3980
Sheep	16,000,000	38,000,000	199,000
Horse	12,600	6,300,000	<1
Duck	33,000,000	54,000,000	–
Chicken	1,300,000	3,400,000	250
Turkey	290,000	2,800,000	–
Animal pets			
Cat	7,900,000	27,000,000	25,100,000
Dog	23,000,000	–	–
Wild animals			
Mouse	330,000	7,700,000	<1
Rabbit	20	47,000	<1
Chipmunk	148,000	6,000,000	–
Human	13,000,000	3,000,000	1580

Modified from Geldreich (1978).

TABLE 23.4　Definitions and Examples of Indicator Microorganisms

Group	Definition and Examples
Process indicator	A group of organisms that demonstrate the efficacy of a process, such as total heterotrophic bacteria or total coliforms for chlorine disinfection
Fecal indicator	A group of organisms that indicate the presence of fecal contamination, such as the fecal coliforms or Escherichia coli
Index and model organisms	A group or species indicative of pathogen presence and behavior, respectively, such as E. coli as index for Salmonella and male-specific coliphages as models for human enteric viruses

Modified from Ashbolt et al. (2001).

(e.g., drinking water treatment), while fecal indicators indicate the presence of fecal contamination. An index (or model) organism represents the presence and behavior of a pathogen in a given environment.

23.2 TOTAL COLIFORMS

The coliform group, which includes *Escherichia*, *Citrobacter*, *Enterobacter* and *Klebsiella* species, is relatively easy to detect (Figure 23.1). Specifically, this group includes all aerobic and facultatively anaerobic, Gram-negative, nonspore-forming, rod-shaped bacteria that produce gas upon lactose fermentation in prescribed culture media within 48 hours at 35°C.

The coliform group has been used as the standard for assessing fecal contamination of recreational and drinking waters for almost a century. Through experience, it has been learned that absence of this organism in 100 ml of drinking water ensures the prevention of bacterial water-borne disease outbreaks. However, it has also been

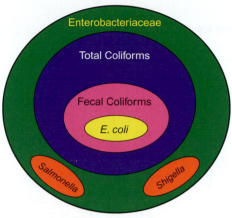

FIGURE 23.1 Relationships between indicators in three Enterobacteriacea.

TABLE 23.5 Deficiencies with the Use of Coliform Bacteria as Indicators of Water Quality

- Regrowth in aquatic environments
- Regrowth in distribution systems
- Suppression by high background bacterial growth
- Not indicative of a health threat
- No relationship between enteric protozoan and viral concentration

Modified from Gleeson and Gray (1997).

---

learned that a number of deficiencies in the use of this indicator exist (Table 23.5).

All members of the coliform group have been observed to regrow in natural surface and drinking water distribution systems (Gleeson and Gray, 1997). The die-off rate of coliform bacteria depends on the amount and type of organic matter in the water, and its temperature. If the water contains significant concentrations of organic matter and is at an elevated temperature, the bacteria may increase in numbers. This phenomenon has been observed in eutrophic tropical waters, waters receiving pulp and paper mill effluents, wastewater, aquatic sediments and organically enriched soil (i.e., soil amended with biosolids) after periods of heavy rainfall. Of greatest concern is the growth of recovery of injured coliform bacteria in a drinking water distribution system, because this may give a false indication of fecal contamination. Coliforms may colonize and grow in biofilms found on distribution system pipes, even in the presence of free chlorine. *Escherichia coli* is 2400 times more resistant to free chlorine when attached to a surface than when it is suspended as free cells in water (LeChevallier *et al*., 1988).

Because large numbers of heterotrophic bacteria in the water may mask the growth of coliform bacteria on selective media used for their isolation, true numbers of coliforms may be underestimated. This often becomes a problem when aerobic heterotrophic bacterial numbers exceed 500 per ml. Finally, the longer survival time and greater resistance to disinfectants of pathogenic enteric viruses and protozoan parasites limits the use of coliform bacteria as an indicator for these organisms. However, the coliform group of bacteria has proved its merit in the assessment of the bacterial quality of water. Three methods are commonly used to identify coliforms in water. These are the most probable number (MPN), the membrane filter (MF), and the presence−absence (P−A) tests.

23.2.1 The Most Probable Number (MPN) Test

The MPN test allows detection of the presence of coliforms in a sample and estimation of their numbers (see also Section 10.3.2). This test consists of three steps: a presumptive test, a confirmed test and a completed test. In the presumptive test (Figure 23.2A), lauryl sulfate−tryptose−lactose broth is placed in a set of test tubes with different dilutions of the water to be tested. Usually, three to five test tubes are prepared per dilution. These test tubes are incubated at 35°C for 24 to 48 hours, and subsequently examined for the presence of coliforms, which is indicated by gas and acid production. Once the positive tubes have been identified and recorded, it is possible to estimate the total number of coliforms in the original sample by using an MPN table that gives numbers of coliforms per 100 ml. In the confirming test (Figure 23.2B), the presence of coliforms is verified by inoculating selective bacteriological agars such as Levine's eosin−methylene blue (EMB) agar or Endo agar with a small amount of culture from the positive tubes. Lactose-fermenting bacteria are indicated on the medium by the production of colonies with a green sheen or colonies with a dark center. In some cases, a completed test (not shown in Figure 23.2) is performed in which colonies from the agar are inoculated back into lauryl sulfate−tryptose−lactose broth to demonstrate the production of acid and gas.

23.2.2 The Membrane Filter Test

The membrane filter (MF) test also allows scientists to determine the number of coliforms in a sample, but it is easier to perform than the MPN test because it requires fewer test tubes and less labor (Figure 23.3) (see also Chapter 8). In this technique, a measured amount of water (usually 100 ml for drinking water) is passed through a membrane filter (pore size 0.45 μm) that traps bacteria on its surface. This membrane is then placed on a thin absorbent pad that has been saturated with a specific medium

(A) Presumptive test

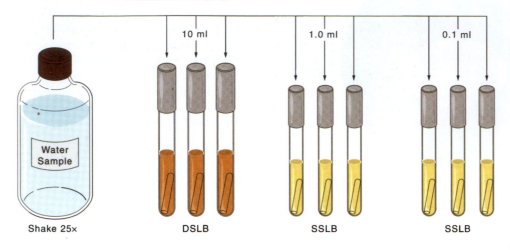

Transfer the specified volumes of sample to each tube.
Incubate 24 h at 35°C.

Tubes that have 10% gas or more are considered positive. The number of
positive tubes in each dilution is used to calculate the MPN of bacteria.

(B) Confirming test

One of the positive tubes is selected, as indicated by the presence of gas trapped
in the inner tube, and used to inoculate a streak plate of Levine's EMB agar and
Endo agar. The plates are incubated 24 h at 35°C and observed for typical
coliform colonies.

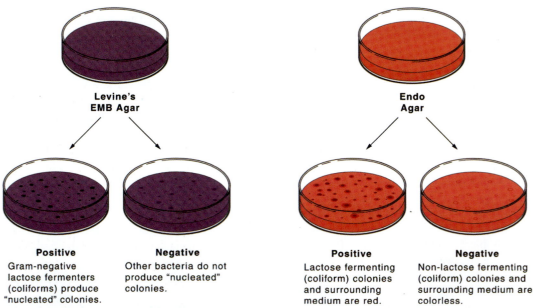

Levine's EMB Agar		Endo Agar	
Positive	**Negative**	**Positive**	**Negative**
Gram-negative lactose fermenters (coliforms) produce "nucleated" colonies.	Other bacteria do not produce "nucleated" colonies.	Lactose fermenting (coliform) colonies and surrounding medium are red.	Non-lactose fermenting (coliform) colonies and surrounding medium are colorless.

FIGURE 23.2   Procedure for performing an MPN test for coliforms on water samples: (A) presumptive test and (B) confirming
test. DSLB = double strength lauryl sulfate broth, SSLB = single strength lauryl sulfate broth.

designed to permit growth and differentiation of the
organisms being sought. For example, if total coliform
organisms are sought, a modified Endo medium
(m-Endo) is used.

For coliform bacteria, the filter is incubated at 35°C
for 18−24 hours. The success of the method depends on
using effective differential or selective media that can
facilitate identification of the bacterial colonies growing
on the membrane filter surface (see Figure 23.3). To
determine the number of coliform bacteria in a water
sample, the colonies having a green sheen with m-Endo
media are enumerated.

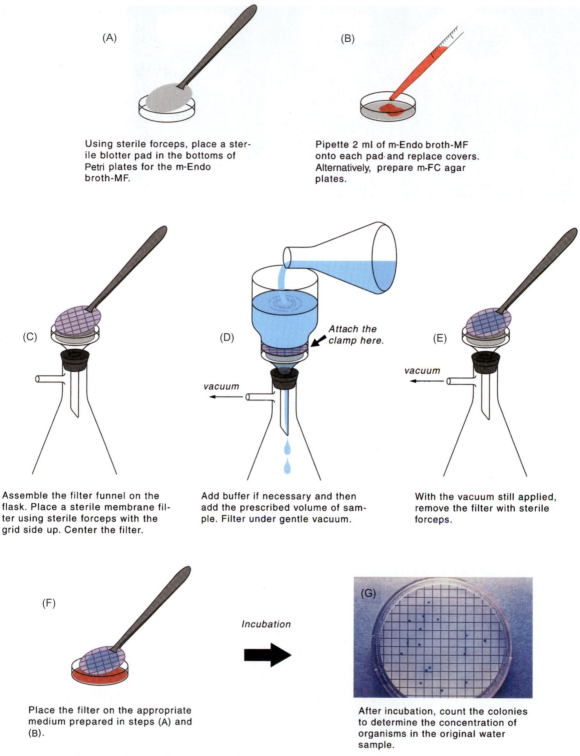

(A)

Using sterile forceps, place a sterile blotter pad in the bottoms of Petri plates for the m-Endo broth-MF.

(B)

Pipette 2 ml of m-Endo broth-MF onto each pad and replace covers. Alternatively, prepare m-FC agar plates.

(C)

Assemble the filter funnel on the flask. Place a sterile membrane filter using sterile forceps with the grid side up. Center the filter.

(D)

Attach the clamp here.

vacuum

Add buffer if necessary and then add the prescribed volume of sample. Filter under gentle vacuum.

(E)

vacuum

With the vacuum still applied, remove the filter with sterile forceps.

(F)

Incubation

Place the filter on the appropriate medium prepared in steps (A) and (B).

(G)

After incubation, count the colonies to determine the concentration of organisms in the original water sample.

FIGURE 23.3   Membrane filtration for determining the coliform count in a water sample using vacuum filtration.

23.2.3 The Presence—Absence (P—A) Test

Presence—absence tests (P—A tests) are not quantitative tests; instead, they answer the question of whether the target organism is present in a sample or not. A single tube of lauryl sulfate—tryptose—lactose broth as used in the MPN test, but without dilutions, would be used in a P—A test. In recent years, enzymatic assays have been developed that allow the simultaneous detection of total coliform bacteria and *E. coli* in drinking water. The assay can be a simple

FIGURE 23.4 Detection of indicator bacteria with Colilert. (A) Addition of salts and enzyme substrates to water sample; (B) yellow color indicating the presence of coliform bacteria; (C) fluorescence under long-wave ultraviolet light indicating the presence of *E. coli*.

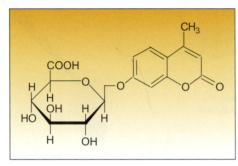

FIGURE 23.5 The structure of 4-methylumbelliferyl-β-*D*-glucuronide (MUG).

P−A test or an MPN assay. The Colilert system (Figure 23.4) is one such assay: it is based on the fact that total coliform bacteria produce the enzyme β-galactosidase, which hydrolyzes the substrate *o*-nitrophenyl-β-*D*-galactopyranoside (ONPG) to yellow nitrophenol. *E. coli* can be detected at the same time by incorporation of a fluorogenic substrate, 4-methylumbelliferone glucuronide (MUG) (Figure 23.5), which produces a fluorescent end product after interaction with the enzyme β-glucuronidase found in *E. coli*, but not in other coliforms. The end product is detected with a long-wave ultraviolet (UV) lamp. The Colilert test is performed by adding the sample to a single bottle (P−A test) or MPN tubes that contain powdered ingredients consisting of salts or specific enzyme substrates that serve as the only carbon source for the organisms (Figure 23.4A). After 24 hours of incubation, samples positive for total coliforms turn yellow (Figure 23.4B), whereas *E. coli*-positive samples fluoresce (blue color) under long-wave UV illumination in the dark (Figure 23.4C).

23.3 FECAL COLIFORMS AND *ESCHERICHIA COLI*

Although the total coliform group has served as the main indicator of water pollution for many years, several of the organisms in this group are not limited to fecal sources.

Thus, methods have been developed to restrict the enumeration to coliforms that are more clearly of fecal origin—that is, the fecal coliforms (Figure 23.1). These organisms, which include the genera *Escherichia* and *Klebsiella*, are differentiated in the laboratory by their ability to ferment lactose with the production of acid and gas at 44.5°C within 24 hours. In general, this test indicates fecal coliforms; it does not, however, distinguish between human and animal contamination. The frequent occurrence of coliform and fecal coliform bacteria in unpolluted tropical waters, and their ability to survive for considerable periods of time in these waters outside the intestine, have suggested that these organisms occur naturally in tropical waters (Toranzos, 1991), and that new indicators for these waters need to be developed.

Some have suggested the use of *E. coli* as an indicator, because it can easily be distinguished from other members of the fecal coliform group (e.g., absence of urease and presence of β-glucuronidase), and is more likely to indicate fecal pollution. Fecal coliforms also have some of the same limitations in use as the coliform bacteria, i.e., regrowth and less resistant to water treatment than viruses and protozoa.

Fecal coliforms may be detected by methods similar to those used for coliform bacteria. For the MPN method, EC broth is used, and for the membrane filter method, m-FC agar is used for water analysis. A medium known as m-T7 agar has been proposed for use in the recovery of injured fecal coliforms from water (LeChevallier *et al.*, 1983), and results in greater recovery from water. The Colilert test has the advantage of detecting coliforms and *E. coli*, the principal fecal coliform, simultaneously within 24 hours.

23.4 FECAL STREPTOCOCCI (*ENTEROCOCCI*)

The fecal streptococci are a group of Gram-positive Lancefield group D streptococci (Figure 23.6). The fecal

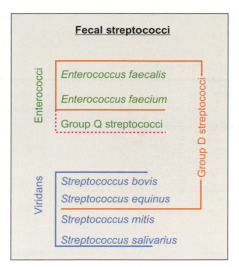

Fecal streptococci

FIGURE 23.6 Definition of the terms "enterococci," "group D streptococci" and "fecal streptococci" based on *Streptococcus* species belonging to each group.

TABLE 23.6 The FC/FS Ratio

FC/FS Ratio	Source of Pollution
>4.0	Strong evidence that pollution is of human origin
2.0–4.0	Good evidence of the predominance of human wastes in mixed pollution
0.7–2.0	Good evidence of the predominance of domestic animal wastes in mixed pollution
<0.7	Strong evidence that pollution is of animal origin

streptococci belong to the genera *Enterococcus* and *Streptococcus* (Sadowsky and Whitman, 2011). The genus *Enterococcus* includes all streptococci that share certain biochemical properties, and have a wide range of tolerance of adverse growth conditions. The enterococci can be found in soil, water, dairy products, food and plants. They are differentiated from other streptococci by their ability to grow in 6.5% sodium chloride, pH 9.6 and 45°C, and include *Ent. avium*, *Ent. faecium*, *Ent. durans*, *Ent. faecalis* and *Ent. gallinarium*. In the water industry, the genus is often given as *Streptococcus* for this group. Of the genus *Streptococcus*, only *S. bovis* and *S. equinus* are considered to be true fecal streptococci. These two species of *Streptococcus* are predominantly found in animals; *Ent. faecalis* and *Ent. faecium* are more specific to the human gut. It has been suggested that a fecal coliform/fecal streptococci (FC/FS) ratio of 4 or more indicates a contamination of human origin, whereas a ratio below 0.7 is indicative of animal pollution (Geldreich and Kenner, 1969) (Table 23.6). However, the validity of the FC/FS ratio has been questioned. Further, this ratio is valid only for recent (24 hours) fecal pollution.

Both the membrane filtration method and MPN method may also be used for the isolation of fecal streptococci. The membrane filter method uses fecal *Streptococcus* agar with incubation at 37°C for 24 hours. All red, maroon and pink colonies (due to reduction of 2,4,5-triphenyltetrazolium chloride to formazan, a red dye) are counted as presumptive fecal streptococci. Confirmation of fecal streptococci is by subculture on bile aesculin agar and incubation for 18 hours at 44°C. Fecal streptococci form discrete colonies surrounded by a brown or black halo due to aesculin hydrolysis, and *Ent. faecalis* organisms are considered to be more specific to

the human gut. Enterococci are considered to have certain advantages over the coliform and fecal coliform bacteria as indicators:

- They rarely grow in water.
- They are more resistant to environmental stress and chlorination than are coliforms.
- They generally persist longer in the environment (Gleeson and Gray, 1997).

The concentration of enterococci in surfaces waters has been shown to be related to the risk of gastroenteritis among recreational bathers, and standards have been developed for acceptable levels of enterococci (Cabelli, 1989).

23.5 CLOSTRIDIUM PERFRINGENS

Clostridium perfringens is a sulfite-reducing anaerobic spore former; it is Gram positive, rod shaped and exclusively of fecal origin. The spores are very heat resistant (75°C for 15 minutes), persist for long periods in the environment and are very resistant to disinfectants. The hardy spores of this organism limit its usefulness as an indicator because it is often found in soils and sediments. However, it has been suggested that it could be an indicator of past pollution, a tracer of less hardy indicators and an indicator of removal of protozoan parasites or viruses during drinking water and wastewater treatment (Payment and Franco, 1993). It is used as an indicator for drinking water in Europe (Bitton, 2011).

23.6 BACTEROIDES AND BIFIDOBACTERIUM

Bifidobacterium and *Bacteroides* are anaerobic bacteria that have also been suggested as potential indicators. *Bacteriodes* are Gram-positive rods found in the gut of humans and animals. Because they are strict anaerobes they do not survive long in the environment, and have been suggested as indicators of recent fecal pollution. They are more common in the human gut than *E. coli* and represent 30% of the total number of fecal isolates (Sadowsky and Whitman, 2011). Because some of the *Bifidobacterium* are primarily associated with humans, it has been suggested that they can be used to help distinguish between human and animal contamination. Until recently the isolation of this strict anaerobe has been difficult; however, use of polymerase chain reaction has made it more feasible.

23.7 HETEROTROPHIC PLATE COUNT

An assessment of the numbers of aerobic and facultatively anaerobic bacteria in water that derive their carbon and energy from organic compounds is conducted via the heterotrophic plate count or HPC. This group includes Gram-negative bacteria belonging to the following genera: *Pseudomonas*, *Aeromonas*, *Klebsiella*, *Flavobacterium*, *Enterobacter*, *Citrobacter*, *Serratia*, *Acinetobacter*, *Proteus*, *Alcaligenes*, *Enterobacter* and *Moraxella*. The heterotrophic plate counts of microorganisms found in untreated drinking water and chlorinated distribution water are shown in Table 23.7 (LeChevallier *et al.*, 1980). These bacteria are commonly isolated from surface waters and groundwater, and are widespread in soil and vegetation (including many vegetables eaten raw). Some members of this group are opportunistic pathogens (e.g., *Aeromonas*, *Pseudomonas*), but no conclusive evidence is available to demonstrate their transmission by ingestion of drinking water. In drinking water, the number of HPC bacteria may vary from less than 1 to more than 10^4 CFU per ml, and members are influenced mainly by temperature, presence of residual chlorine and level of assimilable organic matter. In reality, these counts themselves have no or little health significance. However, there has been concern because the HPC can grow to large numbers in bottled water and charcoal filters on household taps. In response to this concern, studies have been performed to evaluate the impact of HPC on illness. These studies have not demonstrated a conclusive impact on illness in persons who consume water with high HPC. Although the HPC is not a direct indicator of fecal contamination, it does indicate variation in water quality, and the potential for pathogen survival and regrowth. These bacteria may also interfere with coliform and fecal coliform detection when present in high numbers. It has been recommended that the HPC should not exceed 500 per ml in tap water (LeChevallier *et al.*, 1980).

Heterotrophic plate counts are normally done via the spread plate method using yeast extract agar incubated at 35°C for 48 hours. A low-nutrient medium, R$_2$A (Reasoner and Geldreich, 1985), has been widely used for

TABLE 23.7 Identification of HPC Bacteria in Untreated Drinking Water and in a Chlorinated Distribution System

Organism	Distribution Water	Untreated Drinking Water
	% of the total number of organisms identified	% of the total number of organisms identified
Actinomycetes	10.7	0
Arthrobacter spp.	2.3	1.3
Bacillus spp.	4.9	0.6
Corynebacterium spp.	8.9	1.9
Micrococcus luteus	3.5	3.2
Staphylococcus aureus	0.6	0
S. epidermidis	5.2	5.1
Acinetobacter spp.	5.5	10.8
Alcaligenes spp.	3.7	0.6
Flavobacterium meningosepticum	2.0	0
Moraxella spp.	0.3	0.6
Pseudomonas alcaligenes	6.9	2.5
P. cepacia	1.2	0
P. fluorescens	0.6	0
P. mallei	1.4	0
P. maltophilia	1.2	5.7
Pseudomonas spp.	2.9	0
Aeromonas spp.	9.5	15.9
Citrobacter freundii	1.7	5.1
Enterobacter agglomerans	1.2	11.5
Escherichia coli	0.3	0
Yersinia enterocolitica	0.9	6.4
Hafnia alvei	0	5.7
Enterobacter aerogenes	0	0.6
Enterobacter clonane	0	0.6
Klebsiella pneumoniae	0	0
Serratia liquefaciens	0	0.6
Unidentified	18.7	17.8

Modified from LeChevallier et al. *(1980).*

(A) Preparation of the Top Agar

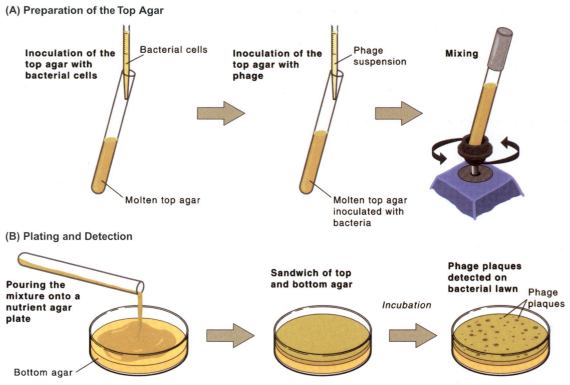

Inoculation of the top agar with bacterial cells — Bacterial cells — Molten top agar

Inoculation of the top agar with phage — Phage suspension — Molten top agar inoculated with bacteria

Mixing

(B) Plating and Detection

Pouring the mixture onto a nutrient agar plate — Bottom agar

Sandwich of top and bottom agar

Incubation

Phage plaques detected on bacterial lawn — Phage plaques

FIGURE 23.7   Technique for performing a bacteriophage assay.

disinfectant-damaged bacteria. This medium is recommended for use with an incubation period of 5–7 days at 28°C. HPC numbers can vary greatly depending on the incubation temperature, growth medium and length of incubation.

23.8 BACTERIOPHAGES

Because of their constant presence in sewage and polluted waters, the use of bacteriophages (or bacterial viruses) as appropriate indicators of fecal pollution has been proposed. These organisms have also been suggested as indicators of viral pollution. This is because the structure, morphology and size, as well as the behavior in the aquatic environment of many bacteriophages, closely resemble those of enteric viruses. For these reasons, they have also been used extensively to evaluate virus resistance to disinfectants, the fate of viruses during water and wastewater treatment, and as indicators for viruses in surface and groundwater. The use of bacteriophages as indicators of fecal pollution is based on the assumption that their presence in water samples denotes the presence of bacteria capable of supporting the replication of the phage. Two groups of phages in particular have been studied: the somatic coliphages, which infect *E. coli* host strains through cell wall receptors, and the F-specific RNA coliphages, which infect strains of *E. coli* and related bacteria through the F^+ or sex pili. A significant

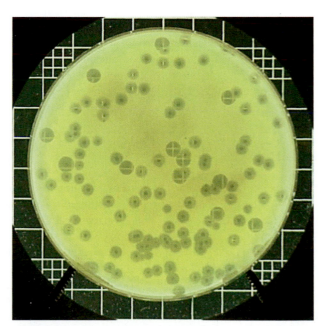

FIGURE 23.8   Bacteriophage plaques on a bacterial lawn.

advantage of using coliphages is that they can be detected by simple and inexpensive techniques that yield results in 8–18 hours. Both a plating method (the agar overlay method) and the MPN method can be used to detect coliphages (Figures 23.7 and 23.8) in volumes ranging from 1 to 100 ml. The F-specific coliphages (male-specific phage) have received the greatest amount of attention because they are similar in size and shape to many of the pathogenic human enteric viruses. Coliphage f2, φX174, MS-2, Qβ and PRD-1 are the phages that have most commonly been used as tracers and for evaluation of disinfectants. Because F-specific phages are infrequently detected in human fecal matter and show no direct relationship to the fecal pollution level, they cannot be considered indicators of fecal pollution (Havelaar *et al.*, 1990). However, their presence in high numbers in wastewaters, and their relatively high resistance to chlorination, contributes to their consideration as an index of wastewater contamination and as potential indicators of enteric viruses.

Bacteriophages of *Bacteroides fragilis* have also been suggested as potential indicators of human viruses in the environment (Tartera and Jofre, 1987). *Bacteroides* spp. are strict anaerobes and are a major component of human feces, so bacteriophages active against these organisms have the potential to be suitable indicators of viral contamination.

Bacteriophages that infect *B. fragilis* appear to be exclusively human in origin (Tartera and Jofre, 1987), and appear to be present only in environmental samples contaminated with human fecal pollution. This may help to differentiate human from animal contamination. They are absent from natural habitats, which is a considerable advantage over coliphages, which are found in habitats other than the human gut. They are unable to multiply in the environment (Tartera *et al.*, 1989), and their decay rate in the environment appears to be similar to that of human enteric viruses. However, their host is an anaerobic bacterium that involves a complicated and tedious methodology, which limits their suitability as a routine indicator organism.

23.9 OTHER POTENTIAL INDICATOR ORGANISMS

A number of other organisms have also been considered to have potential as alternative indicator organisms or for use in certain applications (e.g., recreational waters). These include *Pseudomonas* spp., yeasts, acid-fast mycobacteria (*Mycobacterium fortuitum* and *M. phlei*), *Aeromonas* and *Staphylococcus*.

Within the genus *Pseudomonas*, the species of significant public health concern is *P. aeruginosa*, a Gram-negative, nonsporulating, rod-shaped bacterium. The most common diseases associated with this organism are eye, ear, nose and throat infections. It is also the most common opportunistic pathogen causing life-threatening infections in burn patients and immunocompromised individuals. A characteristic of the pseudomonad is that it can produce the blue−green pigment pyocyanin, or the fluorescent pigment fluorescein, or both. Numerous cases of folliculitis, dermatitis, ear (swimmer's ear) and urinary tract infections are due to *P. aeruginosa* associated with

swimming in contaminated water, or poorly maintained swimming pools and hot tubs. Because of this association and its consistent presence in high numbers in sewage, *P. aeruginosa* has been suggested as a potential indicator for water in swimming pools, hot tubs and other recreational waters (Cabelli, 1978). However, as this organism is known to be ubiquitous in nature and can multiply under natural conditions (it can even grow in distilled water), it is believed to be of little value for fecal contamination studies.

Coliforms have been used for many years to assess the safety of swimming pool water, yet contamination is often not of fecal origin with infections associated primarily with the respiratory tract, skin and eyes. For this reason, *Staphylococcus aureus* and *Candida albicans*, a Gram-positive bacterium and a yeast, respectively, have been proposed as better indicators of this type of infection associated with swimming. Recreational waters may serve as a vehicle for skin infections caused by *S. aureus*, and some observers have recommended that this organism be used as an additional indicator of the sanitary quality of recreational waters, because its presence is associated with human activity in recreational waters (Charoenca and Fujioka, 1993).

The genus *Aeromonas* includes straight facultatively anaerobic gram-negative rods that are included in the family Vibrionaceae. Only *Aeromonas hydrophila* has received attention as an organism of potential sanitary significance. *Aeromonas* occurs in uncontaminated waters as well as in sewage and sewage-contaminated waters. The organism can be pathogenic for humans, other warm-blooded animals and cold-blooded animals including fish. Foodborne outbreaks associated with *A. hydrophila* have been documented, and it is considered an opportunistic pathogen in humans. Because of its association with nutrient-rich conditions, it has been suggested as an indicator of the nutrient status of natural waters.

23.10 STANDARDS AND CRITERIA FOR INDICATORS

Bacterial indicators such as coliforms have been used for the development of water quality standards. For example, the U.S. Environmental Protection Agency (U.S. EPA) has set a standard of no detectable coliforms per 100 ml of drinking water. A drinking water standard is legally enforceable in the United States. If these standards are violated by water suppliers, they are required to take corrective action or they may be fined by the state or federal government. Authority for setting drinking water standards was given to the U.S. EPA in 1974 when Congress passed the Safe Drinking Water Act. Similarly, authority for setting standards for domestic wastewater discharges is given under the Clean Water Act (see Table 17.1). In contrast, standards for recreational waters and wastewater reuse are determined by the individual states. Microbial standards set by various government bodies in the United States are shown in Table 23.8. Standards used by the European Union are given in Table 23.9.

Criteria and guidelines are terms used to describe recommendations for acceptable levels of indicator microorganisms. They are not legally enforceable, but serve as guidance indicating that a potential water quality problem may exist. Ideally, all standards would indicate that an unacceptable public health threat exists, or that some relationship exists between the amount of illness and the level of indicator organisms. Such information is difficult to acquire because of the involvement of costly epidemiological studies that are often difficult to interpret because of confounding factors (see Chapter 24). An area where epidemiology has been used to develop criteria is that of recreational swimming. Epidemiological studies in the United States have demonstrated a relationship between swimming-associated gastroenteritis and the densities of enterococci (Figure 23.9) and *E. coli*. No

TABLE 23.8 U.S. Federal and State Standards for Microorganisms

Authority	Standards
U.S. EPA	
Safe Drinking Water Act	0 coliforms/100 ml
Clean Water Act	
Wastewater discharges	200 fecal coliforms/100 ml
Sewage sludge	<1000 fecal coliforms/4 g
	<3 *Salmonella*/4 g
	<1 enteric virus/4 g
	<1 helminth ovum/4 g
California	
Wastewater reclamation for irrigation	≤ 2.2 MPN/100 ml coliforms
Food and Drug Administration	
Shellfish growing areas[a]	14 MPN/100 ml fecal coliforms

[a]FDA (2005).

TABLE 23.9 Drinking Water Criteria of the European Union

Tap water	
Escherichia coli	0/100 ml
Fecal streptococci	0/100 ml
Sulfite-reducing clostridia	0/20 ml
Bottled Water	
Escherichia coli	0/250 ml
Fecal streptococci	0/250 ml
Sulfite-reducing clostridia	0/50 ml
Pseudomonas aeruginosa	0/250 ml

From European Union (1995).

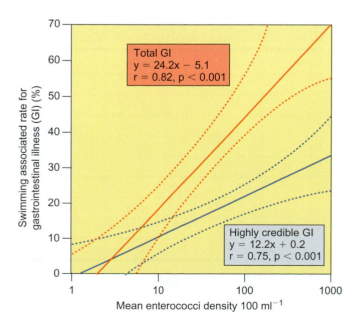

FIGURE 23.9 Dose−response relationships produced by the work of Cabelli *et al.* (1982).

relationship was found for coliform bacteria (Cabelli, 1989). It was suggested that a standard geometric average of 35 enterococci per 100 ml be used for marine bathing waters. This would mean accepting a risk of 1.9% of the bathers developing gastroenteritis (Kay and Wyer, 1992). Numerous other epidemiological studies of bathing-acquired illness have been conducted. These studies have shown slightly different relationships to illness and that other bacterial indicators were more predictive of illness rates (Kay and Wyer, 1992). These differences probably arise for a variety of reasons including: the different sources of contamination (raw versus disinfected waste-water); types of recreational water (marine versus fresh); types of illness (gastroenteritis, eye infections, skin complaints); immune status of the population; and the length of observation. Various guidelines for acceptable numbers of indicator organisms have been in use (Table 23.10), but there is no general agreement on standards.

The use of microbial standards also requires the development of standard methods and quality assurance or quality control plans for the laboratories that conduct the monitoring. Knowledge of how to sample and how often

to sample is also important. All of this information is usually defined in the regulations when a standard is set. For example, frequency of sampling may be determined by the size (number of customers) of the utility providing the water. Sampling must proceed in some systematic fashion so that the entire system is characterized. For drinking water, no detectable coliforms are allowed in the United States (Table 23.8). However, in other countries a certain level of coliform bacteria is allowed. Because of the wide variability in numbers of indicators in water, some positive samples may be allowed or tolerance levels or averages may be allowed. Usually, geometric averages are used in standard settings because the distribution of bacterial numbers is often skewed. This prevents one or two high values from overestimating of high levels of contamination, which would appear to be the case of arithmetic averages (see Table 23.11).

Geometric averages are determined as follows:

$$\log \bar{x} = \frac{\sum (\log x)}{N} \qquad \text{(Eq. 23.1)}$$

$$\bar{x} = \text{anti} \log (\log \bar{x}) \qquad \text{(Eq. 23.2)}$$

where:

N is the number of samples,
$\bar{x}$ is the geometric average, and
x is the number of organisms per sample volume.

As can be seen, standard setting and the development of criteria is a difficult process and there is no ideal standard. A great deal of judgment by scientists, public health officials and the regulating agency is required.

TABLE 23.10 Guidelines for Recreational Water Quality Standards

Country or Agency	Regime (samples/time)	Criteria or Standard[a]
U.S. EPA	5/30 days	200 fecal coliforms/100 ml <10% to exceed 400 per ml
		Fresh water[b] 33 enterococci/100 ml 126 E. coli/100 ml
		Marine waters[b] 35 enterococci/100 ml
European Economic Community	2/30 days[c]	500 coliforms/100 ml
		100 fecal coliforms/100 ml
		100 fecal streptococci/100 ml
		0 Salmonella/liter
		0 Enteroviruses/10 liters
Ontario, Canada	10/30 days	≤ 1000 coliforms/100 ml
		≤ 100 fecal coliforms/100 ml

From Saliba (1993); U.S. EPA (1986).
[a]*All bacterial numbers in geometric means.*
[b]*Proposed, 1986.*
[c]*Coliforms and fecal coliforms only.*

TABLE 23.11 A Comparison of Arithmetic and Geometric Averages of Bacterial Numbers in Water

MPN[a]	Log
2	0.30
110	2.04
4	0.60
150	2.18
1100	3.04
10	1.00
12	1.08
198 = arithmetic average	1.46 = log $\bar{x}$ antilog $\bar{x}$ = 29
	29 = geometric average

[a]*MPN, most probable number.*

QUESTIONS AND PROBLEMS

1. What are some of the criteria for indicator bacteria?
2. What is the difference between standards and criteria?
3. Why are geometric means used to report average concentrations of indicator organisms?
4. Calculate the arithmetic and geometric averages for the following data set: fecal coliforms/100 ml on different days on a bathing beach were reported as 2, 3, 1000, 15, 150 and 4000.
5. Define coliform and fecal coliform bacteria. Why are they not ideal indicators?
6. Why have coliphage been suggested as indicator organisms?
7. What are two methods that can be used to detect indicator bacteria in water?
8. Calculate the most probable number (MPN) for the following dataset:

Volume added to each tube	Number of positive tubes
10 ml	3
1.0 ml	1
0.1 ml	0

9. What is the difference between a fecal indicator organism and a process indicator? Give an example of each.
10. How are bacteriophages used as indicators? What is a coliphage?
11. How many bathers would you expect to develop gastroenteritis: (1) if 35 enterococci/100 ml were detected in the water; or (2) if 100 enterococci/100 ml were present? (See Figure 23.8.)
12. Which indicator would be best for indication of long-term sewage pollution? Which one would be best as a short-term indicator of recent sewage pollution?

REFERENCES AND RECOMMENDED READING

Ashbolt, N. J., Grawbow, W. O. K., and Snozzi, M. (2001) Indicators of microbial water quality. In "Water Quality: Guidelines, Standards and Health" (L. Fewtrell, and J. Bartram, eds.), IWA Publishing, London, pp. 289–315.

Bitton, G. (2011) "Wastewater Microbiology," 4th ed. Wiley-Blackwell, Hoboken, NJ.

Cabelli, V. (1978) New standards for enteric bacteria. In "Water Pollution Microbiology" (R. Mitchell, ed.), vol. 2, Wiley-Interscience, New York, pp. 233–273.

Cabelli, V. J. (1989) Swimming-associated illness and recreational water quality criteria. Water Sci. Technol. 21, 13–21.

Cabelli, V. J., Dufour, A. P., McCable, L. J., and Levin, M. (1982) Swimming associated gastroenteric illness and water quality. Am. J. Epidemiol. 115, 606–616.

Charoenca, N., and Fujioka, R. S. (1993) Assessment of Staphylococcus bacteria in Hawaii recreational waters. Water Sci. Technol. 27, 283–289.

European Union (EU) (1995) Proposed for a Council Directive concerning the quality of water intended for human consumption. Com (94) 612 Final. Offic. J. Eur. Union 131, 5–24.

Food and Drug Administration (2005) National Shellfish Sanitation Program. Guide for the Control of Molluscan Shellfish. Washington, DC.

Geldreich, E. E. (1978) Bacterial populations and indicator concepts in feces, sewage, storm water and solid wastes. In "Indicators of Viruses in Water and Food" (G. Berg, ed.), Ann Arbor Science, Ann Arbor, MI, pp. 51–97.

Geldreich, E. E., and Kenner, B. A. (1969) Comments on fecal streptococci in stream pollution. J. Water Pollut. Control Fed. 41, R336–R341.

Gleeson, C., and Gray, N. (1997) "The Coliform Index and Waterborne Disease," E and FN Spon, London.

Havelaar, A. H., Hogeboon, W. M., Furuse, K., Pot, R., and Horman, M. P. (1990) F-specific RNA bacteriophages and sensitive host strains in faeces and wastewater of human and animal origin. J. Appl. Bacteriol. 69, 30–37.

Kay, D., and Wyer, M. (1992) Recent epidemiological research leading to standards. In "Recreational Water Quality Management, vol. 1, Coastal Waters" (D. Kay, ed.), Ellis Horwood, Chichester, UK, pp. 129–156.

LeChevallier, M. W., Cameron, S. C., and McFeters, G. A. (1983) New medium for improved recovery of coliform bacteria from drinking water. Appl. Environ. Microbiol. 45, 484–492.

LeChevallier, M. W., Cawthen, C. P., and Lee, R. G. (1988) Factors promoting survival of bacteria in chlorinated water supplies. Appl. Environ. Microbiol. 54, 649–654.

LeChevallier, M. W., Seidler, R. J., and Evans, T. M. (1980) Enumeration and characterization of standard plate count bacteria in chlorinated and raw water supplies. Appl. Environ. Microbiol. 40, 922–930.

Payment, P., and Franco, E. (1993) Clostridium perfringens and somatic coliphages as indicators of the efficiency of drinking water treatment for viruses and protozoan cysts. Appl. Environ. Microbiol. 59, 2418–2424.

Reasoner, D. J., and Geldreich, E. E. (1985) A new medium for enumeration and subculture of bacteria from potable water. Appl. Environ. Microbiol. 49, 1–7.

Sadowsky, J., and Whitman, R. L. (2011) "The Fecal Bacteria," ASM Press, Washington, DC.

Saliba, L. (1993) Legal and economic implication in developing criteria and standards. In "Recreational Water Quality Management" (D. Kay, and R. Hanbury, eds.), Ellis Horwood, Chichester, UK, pp. 57–73.

Tartera, C., and Jofre, J. (1987) Bacteriophage active against *Bacteroides fragilis* bacteriophage as indicators of the virological quality of water. *Water Sci. Technol.* **18**, 1623–1637.

Tartera, C., Lucena, F., and Jofre, J. (1989) Human origin of *Bacteroides fragilis* bacteriophage present in the environment. *Appl. Environ. Microbiol.* **55**, 2696–2701.

Toranzos, G. A. (1991) Current and possible alternative indicators of fecal contamination in tropical waters: a short review. *Environ. Toxicol. Water Qual.* **6**, 121–130.

U.S. EPA. United States Environmental Protection Agency (1986) Ambient water quality. Criteria—1986. EPA440/5-84-002. Washington, DC.

Risk Assessment

Charles P. Gerba

The task of interpreting data on the occurrence of pathogens in the environment is often critical in making decisions about potential health risks and corrective actions. Quantitative risk assessment (QRA) is an approach that allows the quantitative expression of risk in terms infection, illness or mortality from microbial pathogens. In this format, such information can be utilized by decision makers to determine the magnitude of such risks, and weigh the costs and benefits of corrective action. The purpose of this chapter is to provide a general background to the topic of risk analysis, and how it can be used in the problem-solving processes.

24.1 THE CONCEPT OF RISK ASSESSMENT

Risk, which is common to all life, is an inherent property of everyday human existence. It is therefore a key factor in all decision making. Risk assessment or analysis, however, means different things to different people: Wall Street analysts assess financial risks, and insurance companies calculate actuarial risks, while regulatory agencies estimate the risks of fatalities from nuclear plant accidents, the incidence of cancer resulting from industrial emissions and habitat loss associated with increases in human populations. All these seemingly disparate activities have in common the concept of a measurable phenomenon called risk that can be expressed in terms of probability. Thus, we can define risk assessment as the process of estimating both the probability that an event will occur, and the probable magnitude of its adverse effects—economic, health or safety related, or ecological—over a specified time period. For example, one might estimate the probable incidence of cancer in the community where a chemical was spilled over a period of years. Or one might calculate the health risks associated with the presence of pathogens in drinking water or food.

There are, of course, several varieties of risk assessment. Risk assessment as a formal discipline emerged in the 1940s and 1950s, paralleling the rise of the nuclear industry. Safety hazard analyses have been used since at least the 1950s in the nuclear, petroleum refining and chemical processing industries, as well as in aerospace. Health risk assessments, however, had their beginnings in 1986 with the publication of the "Guidelines for carcinogenic risk assessment" by the Environmental Protection Agency (EPA). Microbial risk assessment is relatively new, beginning in the mid-1980s, but has already been used in the development of government regulations (Regli et al., 1991).

24.2 ELEMENTS OF RISK ANALYSIS

Risk analysis framework has three basic components: risk assessment, risk management and risk communication (Figure 24.1). Chemical risk assessment has been used to judge the safety of our food and water supply. Such assessments are important in setting standards for chemical contaminants in the environment. Whether chemical or microbial, risk assessment consists of four basic steps:

- Hazard identification—Defining the hazard and nature of the harm: for example, identifying a contaminant

I.L. Pepper, C.P. Gerba, T.J. Gentry: Environmental Microbiology, Third edition. DOI: http://dx.doi.org/10.1016/B978-0-12-394626-3.00024-7
© 2015 Elsevier Inc. All rights reserved.

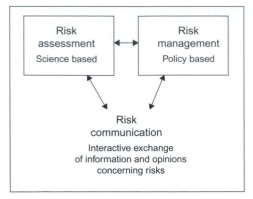

FIGURE 24.1   Risk analysis framework.

(e.g., a chemical, such as lead or carbon tetrachloride, or a microbial pathogen, such as *Legionella*), and documenting its toxic effects on humans.

- **Exposure assessment**—Determining the concentration of a contaminant in the environment and estimating its rate of intake. For example, determining the concentration of *Salmonella* in a meat product and the average dose a person would ingest.
- **Dose—response assessment**—Quantitating the adverse effects arising from exposure to a hazardous agent based on the degree of exposure. This assessment is usually expressed mathematically as a plot showing the response in living organisms to increasing doses of the agent (e.g., rotavirus).
- **Risk characterization**—Estimating the potential impact (e.g., human illness or death) of a microorganism or chemical based on the severity of its effects and the amount of exposure.

Once the risks are characterized, various regulatory options are evaluated in a process called risk management, which includes consideration of social, political and economic issues, as well as the engineering problems inherent in a proposed solution. One important component of risk management is risk communication, which is the interactive process of information and opinion exchange among individuals, groups and institutions. Risk communication includes the transfer of risk information from expert to nonexpert audiences. In order to be effective, risk communication must provide a forum for balanced discussions of the nature of the risk, lending a perspective that allows the benefits of reducing the risk to be weighed against the costs.

In the United States, the passage of federal and state laws to protect public health and the environment has expanded the application of risk assessment. Major federal agencies that routinely use risk analysis include the Food and Drug Administration (FDA), the Environmental Protection Agency (EPA) and the Occupational Safety and Health Administration (OSHA). Together with state agencies, these regulatory agencies use risk assessment in a variety of situations:

- Setting standards for concentrations of toxic chemicals or pathogenic microorganisms in water or food
- Assessing the risk from the release of genetically altered organisms
- Conducting baseline analyses of contaminated sites or facilities to determine the need for remediation and the extent of cleanup required
- Performing cost—benefit analyses of contaminated-site cleanup or treatment options (including treatment processes to reduce exposure to pathogens)
- Developing cleanup goals for contaminants for which no federal or state authorities have promulgated numerical standards: evaluating acceptable variance from promulgated standards and guidelines (e.g., approving alternative concentration limits)
- Constructing "what if" scenarios to compare the potential impact of remedial or treatment alternatives, and to set priorities for corrective action
- Evaluating existing and new technologies for effective prevention, control or mitigation of hazards and risks (e.g., new drinking water treatment technologies)
- Articulating community public health concerns and developing consistent public health expectations among different localities

Risk assessment provides an effective framework for determining the relative urgency of problems and the allocation of resources to reduce risks. Using the results of risk analyses, we can target prevention, remediation or control efforts toward areas, sources or situations in which the greatest risk reductions can be achieved with the resources available. However, risk assessment is not an absolute procedure carried out in a vacuum; rather, it is an evaluative, multifaceted, comparative process. Thus, to evaluate risk, we must inevitably compare one risk with a host of others. In fact, the comparison of potential risks associated with several problems or issues has developed into a subset of risk assessment called comparative risk assessment. Some commonplace risks are shown in Table 24.1. Here we see, for example, that risks from chemical exposure are fairly small compared with those associated with driving a car or smoking cigarettes.

Comparing different risks allows us to comprehend the uncommon magnitudes involved and to understand the level, or magnitude, of risk associated with a particular hazard. But comparison with other risks cannot itself establish the acceptability of a risk. Thus, the fact that the chance of death from a previously unknown risk is about the same as that from a known risk does not necessarily imply that the two risks are equally acceptable. Generally, comparing risks along a single dimension is not helpful when the risks are widely perceived as qualitatively different. Rather, we must take account of certain

TABLE 24.1 Examples of Some Commonplace Risks in the United States

Risk	Lifetime Risk of Mortality
Cancer from cigarette smoking (one pack per day)	1:4
Death in a motor vehicle accident	2:100
Homicide	1:100
Home accident deaths	1:100
Cancer from exposure to radon in homes	3:1000
Death from hepatitis A	3:1000
Exposure to the pesticide aflatoxin in peanut butter	6:10,000
Diarrhea from rotavirus	1:10,000
Exposure to typical EPA maximum chemical contaminant levels	1:10,000–1:10,000,000

Based on data in Wilson and Crouch (1987) and Gerba and Rose (1993).

qualitative factors that affect risk perception and evaluation when selecting risks to be compared. Some of these qualifying factors are listed in Table 24.2. We must also understand the underlying premise that voluntary risk is always more acceptable than involuntary risk. For example, the same people who cheerfully drive their cars every day—thus incurring a 2:100 lifetime risk of death by automobile—are quite capable of refusing to accept the 6:10,000 involuntary risk of eating peanut butter contaminated with aflatoxin.

In considering risk, then, we must also understand another principle—the *de minimis* principle, which means that there are some levels of risk so trivial that they are not worth bothering about. However attractive it is, this concept is hard to define, especially if we are trying to find a *de minimis* level acceptable to an entire society. Understandably, regulatory authorities are reluctant to be explicit about an "acceptable" risk. (How much aflatoxin would you consider acceptable in your peanut butter and jelly sandwich? How many dead insect parts?) Some prefer the term "tolerable risk," i.e., the level of risks we can accept given the economic costs, and social and scientific constraints. But it is generally agreed that a lifetime risk

TABLE 24.2 Factors Affecting Risk Perception and Risk Analysis

Factor	Conditions Associated with Increased Public Concern	Conditions Associated with Decreased Public Concern
Catastrophic potential	Fatalities and injuries grouped in time and space	Fatalities and injuries scattered and random
Familiarity	Unfamiliar	Familiar
Understanding	Mechanisms or process not understood	Mechanisms or process understood
Controllability (personal)	Uncontrollable	Controllable
Voluntariness of exposure	Involuntary	Voluntary
Effects on children	Children specifically at risk	Children not specifically at risk
Effects manifestation	Delayed effects	Immediate effects
Effects on future generations	Risk to future generations	No risk to future generations
Victim identity	Identifiable victims	Statistical victims
Dread	Effects dreaded	Effects not dreaded
Trust in institutions	Lack of trust in responsible institutions	Trust in responsible institutions
Media attention	Much media attention	Little media attention
Accident history	Major and sometimes minor accidents	No major or minor accidents
Equity	Inequitable distribution of risks and benefits	Equitable distribution of risks and benefits
Benefits	Unclear benefits	Clear benefits
Reversibility	Effects irreversible	Effects reversible
Origin	Caused by human actions or failures	Caused by acts of nature

From Covello et al. (1988).

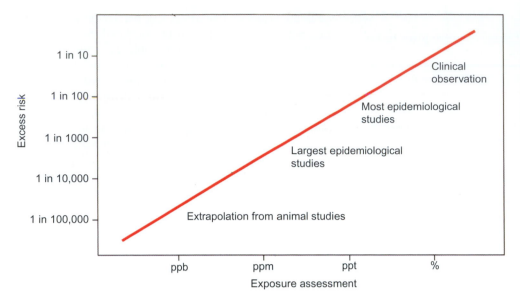

FIGURE 24.2 Sensitivity of epidemiology in detecting risks of regulatory concern. The generalized units of ppt (parts per thousand), ppm (parts per million), ppb (parts per billion), and % (parts per hundred) are used here for comparative purposes. Modified from National Research Council (1993).

on the order of one in a million (or 10^{-6}) is trivial enough to be acceptable for the general public. Although the origins and precise meaning of a one-in-a-million acceptable risk remain obscure, its impact on product choices, operations and costs is very real, for example, hundreds of billions of dollars in hazardous waste site cleanup decisions alone. The levels of acceptable risk can vary within this range. Levels of risk at the higher end of the range (10^{-4} rather than 10^{-6}) may be acceptable if only a few people are exposed rather than the entire populace. For example, workers dealing with production of solvents can often tolerate higher levels of risk than can the public at large. These higher levels are justified because workers tend to be a relatively homogeneous, healthy group, and because employment is voluntary; however, the level of risks would not be acceptable for the same solvents for the general population.

24.3 THE PROCESS OF RISK ASSESSMENT

24.3.1 Hazard Identification

The first step in risk assessment is to determine the nature of the hazard. For pollution-related problems, the hazard in question is usually a specific chemical, a physical agent (such as irradiation), or a microorganism identified with a specific illness or disease. Thus, the hazard identification component of a pollution risk assessment consists of a review of all relevant biological and chemical information bearing on whether or not an agent poses a specific threat.

As Figure 24.2 shows, clinical studies of disease can be used to identify high risks (between 1:10 and 1:100),

whereas most epidemiological studies can detect risks down to 1:1000, and very large epidemiological studies can examine risks in the 1:10,000 range. However, risks lower than 1:10,000 cannot be studied with much certainty using epidemiological approaches. Because regulatory policy objectives generally strive to limit risks below 1:100,000 for life-threatening diseases such as cancer, these lower risks are often estimated by extrapolating from the effects of high doses given to animals.

24.3.2 Exposure Assessment

Exposure assessment is the process of measuring or estimating the intensity, frequency and duration of human exposures to an environmental agent. Exposure to contaminants can occur via inhalation, ingestion of water or food, or the skin. Contaminant sources, release mechanisms, transport and transformation characteristics are all important aspects of exposure assessment, as are the nature, location and activity patterns of the exposed population. (This explains why it is critical to understand the factors and processes influencing the transport and fate of a contaminant.)

An exposure pathway is the course that a hazardous agent takes from a source to a receptor (e.g., human or animal) via environmental carriers or media—generally, air (volatile compounds, particles) or water (soluble or colloidal compounds). An exception is electromagnetic radiation, which needs no medium. The exposure route, or intake pathway, is the mechanism by which the transfer occurs—usually by inhalation, ingestion and/or dermal contact (Figure 24.3). Direct contact can result in a local effect at the point of entry, and/or in a systemic effect.

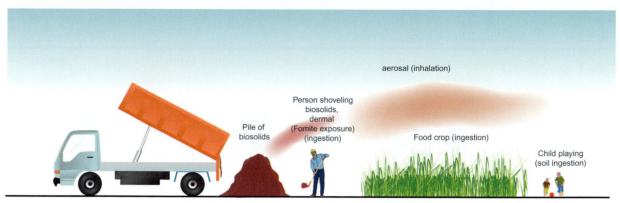

FIGURE 24.3   Potential routes of pathogen exposure from land application of biosolids.

The quantitation of exposure, intake or potential dose can involve equations with three sets of variables:

- Concentrations of chemicals or microbes in the media
- Exposure rates (magnitude, frequency, duration)
- Quantified biological characteristics of receptors (e.g., body weight, absorption capacity for chemicals, level of immunity to microbial pathogens)

Exposure concentrations are derived from measured, monitored and/or modeled data. Ideally, exposure concentrations should be measured at the points of contact between the environmental media and current or potential receptors. It is usually possible to identify potential receptors and exposure points from field observations and other information. However, it is seldom possible to anticipate all potential exposure points, and measure all environmental concentrations under all conditions. In practice, a combination of monitoring and modeling data, together with a great deal of professional judgment, is required to estimate exposure concentrations.

In order to assess exposure rates via different pathways, one has to consider and weigh many factors. For example, in estimating exposure to a substance via drinking water, one first has to determine the average daily consumption of that water. But this is not as easy as it sounds. Studies have shown that daily fluid intake varies greatly from individual to individual. Moreover, total water intake depends on how much fluid is consumed as tap water, and how much is ingested in the form of soft drinks and other nontap water sources. Tap water intake also changes significantly with age, body weight, diet and climate. Because these factors are so variable, the EPA has suggested a number of very conservative "default" exposure values that can be used when assessing contaminants in tap water, vegetables, soil and the like (see Table 24.3).

One also has to consider how much of the population to include in the exposure. For example, the average consumption of tap water is 1.5 liters/day; some persons drink less than this and some significantly more. Approximately 95% of the population consumes 2 liters or less. If we assume the exposure to be 1.5 liters/day, then we are excluding half the population, which consumes more than this amount. Using the 2 liter/day value allows inclusion of 95% of the population.

Event trees have been used to estimate exposures (see Example Calculation 24.1). Event trees are useful to estimate exposure by analysis of pathogen loads upstream, when the actual concentration of the harmful agent that reaches the target is too low to measure.

24.3.3 Dose−Response Assessment

All chemical and microbial contaminants are not equal in their capacity to cause adverse effects. To determine the capacity of agents to cause harm, we need quantitative toxicity or infectivity data. These data can sometimes be derived from occupational, clinical and epidemiological studies. Most toxicity data, however, come from animal experiments in which researchers expose laboratory animals, mostly mice and rats, to increasingly higher concentrations or doses, and observe their corresponding effects. The result of these experiments is the dose−response relationship—a quantitative relationship that indicates the agent's degree of toxicity to exposed species. Dose is normalized as milligrams of substance, or number of organisms ingested, inhaled or absorbed (in the case of chemicals) through the skin per kilogram of body weight per day (mg/kg/day). Responses or effects can vary widely—from no observable effect, to temporary and reversible effects (e.g., enzyme depression caused by some pesticides or diarrhea caused by viruses), to permanent organ injury (e.g., liver and kidney damage caused by chlorinated solvents, heavy metals or viruses), to chronic functional impairment (e.g., bronchitis or emphysema arising from smoke damage), to death.

TABLE 24.3　EPA Standard Default Exposure Factors

Land Use	Exposure Pathway	Daily Intake	Exposure Frequency (days/year)	Exposure Duration (years)
Residential	Ingestion of potable water	2 liters/day	350	30
	Ingestion of soil and dust	200 mg (child)	350	6
		100 mg (adult)		24
	Inhalation of contaminants	20 m^3 (total)	350	30
		15 m^3 (indoor)		
Industrial and commercial	Ingestion of potable water	1 liter	250	25
	Ingestion of soil and dust	50 mg	250	25
	Inhalation of contaminants	20 m^3 (workday)	250	25
Agricultural	Consumption of homegrown produce	42 g (vegetable) (fruit 80 g)	350	30
Recreational	Consumption of locally caught fish	54 g	350	30
Swimming		10−100 ml[a]	1−10	−

Modified from Kolluru (1993).
[a]*Per event.*

Example Calculation 24.1　Estimating Exposure: Application of Event Trees

Event trees can be used to simplify the process of modeling the various pathways and visualizing how the infectivity of a pathogen changes through various processes and routes of exposure (Gale, 2003; Stine *et al.*, 2005). Often times the concentration of the pathogen is unknown in the media to which an individual or population may be exposed. In addition, the exposed concentration may be at levels that cannot be measured (e.g., below the detection limit of the method employed). Through the use of an event tree, an estimate can be made of the amount of pathogen in the media to which an individual may be exposed, and the

uncertainties defined. Event trees are also useful in determining where the greatest amount of uncertainty may exist (die-off fate in the soil, transfer from the soil to plants, etc.).

The following is an example of an event tree used to estimate the exposure of *Salmonella* from biosolids applied to a food crop.

Event tree for estimating exposure from *Salmonella* when biosolids are used on farmland to grow lettuce (numbers above arrows show percent reductions in pathogen load for a given event):

Raw sewage 2.89 × 10^7 CFU/ton —0.83→ Raw sewage sludge 2.4 × 10^7 CFU/ton —0.01→ Anaerobic digestion 2.4 × 10^5 CFU/ton —0.01→ Dilution after incorporation into soil 2.4 × 10^3 CFU/ton ↓

Salmonella ingested per person per year or 2.1 × 10^{-4} CFU/g ←446.5/g— g/lettuce consumed in a year per person 4.8 × 10^{-7} CFU/g = Amount transferred to lettuce 4.8 × 10^{-4} CFU/ton ←0.02— Decay in the soil after 5 months 2.4 × 10^{-2} CFU/ton

---

The goal of a dose−response assessment is to obtain a mathematical relationship between the amount (concentration) of a toxicant or microorganism to which a human is exposed and the risk of an adverse outcome from that dose. The data resulting from experimental studies are presented

as a dose−response curve, as shown in Figure 24.4. The ordinate represents the dose and the abscissa represents the risk that some adverse health effect will occur. In the case of a pathogen, for instance, the ordinate may represent the risk of infection and not necessarily illness.

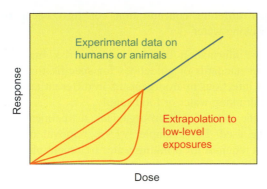

FIGURE 24.4   Extrapolation of dose–response curves. Adapted from U.S. EPA (1990).

TABLE 24.4 Primary Models Used for Assessment of Nonthreshold Effects[a]

Model	Comments
One hit	Assumes (1) single stage for cancer, (2) malignant change induced by one molecular or radiation interaction. *Very conservative.*
Linear multistage	Assumes multiple stages for cancer. *Fits curve to the experimental data.*
Multihit	Assumes several interactions needed before cells become transformed. *Least conservative model.*
Probit	Assumes probit (lognormal) distribution for tolerances of exposed population. *Appropriate for acute toxicity; questionable for cancer.*

Modified from Cockerham and Shane (1994).
[a]*All the models assume that exposure to the pollutant will always produce an effect, regardless of dose.*

TABLE 24.5 Lifetime Risks of Cancer Derived from Different Extrapolation Models[a]

Model Applied	Lifetime Risk (mg/kg/day) from Toxic Chemical
One hit	6.0×10^{-5} (1 in 17,000)
Multistage	6.0×10^{-6} (1 in 167,000)
Multihit	4.4×10^{-7} (1 in 2.3 million)
Probit	1.9×10^{-10} (1 in 5.3 billion)

From U.S. EPA (1990).
[a]*All risks are for a full lifetime of daily exposure. The lifetime is used as the unit of risk measurement because the experimental data reflect the risk experienced by animals over their full lifetimes. The values shown are upper confidence limits on risks.*

However, dose–response curves derived from animal studies must be interpreted with care. The data for these curves are necessarily obtained by examining the effects of large doses on test animals. This is because the large costs involved in testing limits the numbers of animals that can be used—it is both impractical and cost prohibitive to use thousands (even millions) of animals to observe just a few individuals that show adverse effects at low doses (e.g., risks of 1:1000 or 1:10,000). Researchers must therefore extrapolate low-dose responses from their high-dose data. And therein lies the rub: dose–response curves are subject to controversy because their results change depending on the method chosen to extrapolate from the high doses actually administered to laboratory test subjects, to the low doses humans are likely to receive in the course of everyday living.

This controversy revolves around the choice of mathematical models that have been proposed for extrapolation to low doses. Unfortunately, since there are no data available to validate these models, they cannot be proved or disproved, and so there is no way to know which model is the most accurate. The choice of models is therefore strictly a policy decision, which is usually based on understandably conservative assumptions. Thus, for non-carcinogenic chemical responses, the assumption is that some threshold exists below which there is no toxic response; that is, no adverse effects will occur below some very low dose (say, one in a million). Carcinogens, however, are considered nonthreshold—that is, the conservative assumption is that exposure to any amount of carcinogen creates some likelihood of cancer. This means that the only "safe" amount of carcinogen is zero, so the dose–response plot is required to go through the origin (0), as shown in Figure 24.4.

In the microbiological literature the term "minimum infectious dose" is used frequently, implying that a threshold dose exists for microorganisms. In reality, the term used usually refers to the ID_{50} or the dose at which 50% of the animals or humans exposed became infected or exhibit any symptoms of an illness. Existing infectious dose data are compatible with nonthreshold responses, and the term "infectivity" is probably more appropriate when referring to differences in the likelihood of an organism causing an infection. For example, the probability of a given number of ingested rotaviruses causing diarrhea is greater than that for *Salmonella*. Thus, the infectivity of rotavirus is greater than that of *Salmonella*.

There are many mathematical models to choose from for modeling risk. These include the one-hit model, the multistage model, the multihit model and the probit model. The characteristics of these models for nonthreshold effects are listed in Tables 24.4 and 24.5.

24.3.4 Risk Characterization

24.3.4.1 Uncertainty Analysis

Uncertainty is inherent in every step of the risk assessment process. Thus, before we can begin to characterize any risk, we need some idea of the nature and magnitude of uncertainty in the risk estimate. Sources of uncertainty include:

- Extrapolation from high to low doses
- Extrapolation from animal to human responses
- Extrapolation from one route of exposure to another
- Limitations of analytical methods to measure the organism
- Estimates of exposure

Although the uncertainties are generally much larger in estimates of exposure and the relationships between dose and response (e.g., the percent mortality), it is important to include the uncertainties originating from all steps in a risk assessment as part of risk characterization.

Two approaches commonly used to characterize uncertainty are sensitivity analyses and Monte Carlo simulations. In sensitivity analyses, the uncertain quantities of each parameter (e.g., average values, high and low estimates) are varied, usually one at a time, to find out how changes in these quantities affect the final risk estimate. This procedure gives a range of possible values for the overall risk and provides information on which parameters are most crucial in determining the size of the risk. In a Monte Carlo simulation, however, it is assumed that all parameters are random or uncertain. Thus, instead of varying one parameter at a time, a computer program is used to select distributions randomly every time the model equations are solved, the procedure being repeated many times. The resulting output can be used to identify values of exposure or risk corresponding to a specified probability, say the 50th percentile or 95th percentile. Details of these methods of dealing with uncertainty can be found in the text *Quantitative Microbial Risk Assessment* (Haas *et al.*, 2014).

24.3.4.2 Risk Projection and Management

The final phase of the risk assessment process is risk characterization. In this phase, exposure and dose–response assessments are integrated to yield probabilities of effects occurring in humans under specific exposure conditions. Quantitative risks are calculated for appropriate media (air, water, food) and pathways. For example, the risks of lead in water are estimated over a lifetime assuming:

1. that the exposure is 2 liters of water ingested over a 70-year lifetime; and

2. that different concentrations of lead occur in the drinking water

This information can be used by risk managers to develop standards or guidelines for specific toxic chemicals or infectious microorganisms in different media, such as the drinking water or food supply.

In the case of a microorganism, a treatment strategy may be used. For example, 99.9% removal of *Giardia* cysts is required for drinking water treatment plants in the United States to ensure that the yearly risk of infection in a community is no greater than 1:10,000. The assumption is made that this amount of removal will guarantee a concentration of *Giardia* cysts in the finish water that would not result in a risk greater than 1:10,000 (Gerba *et al.*, 1997; Teunis *et al.*, 1997)

24.4 MICROBIAL RISK ASSESSMENT

Outbreaks of waterborne disease caused by microorganisms usually occur when the water supply has been obviously and significantly contaminated. In such high-level cases, the exposure is manifest, and cause and effect are relatively easy to determine. However, exposure to low-level microbial contamination is difficult or impossible to determine epidemiologically. We know, for example, that long-term exposure to microbes can have a significant impact on the health of individuals within a community, but we need a way to measure that impact.

For some time, methods have been available to detect the presence of low levels (one organism per 1000 liters) of pathogenic organisms in water, including enteric viruses and protozoan parasites. The trouble is that the risks posed to the community by these low levels of pathogens in a water supply over time are not like those posed by low levels of chemical toxins or carcinogens. For example, it takes just one amoeba in the wrong place at the wrong time to infect one individual, whereas the same individual would have to consume some quantity of a toxic chemical to be comparably harmed. Microbial risk assessment is, therefore, a process that allows us to estimate responses in terms of the *risk of infection* in a quantitative fashion. Microbial risk generally follows the steps used in other health-based risk assessments—hazard identification, exposure assessment, dose–response and risk characterization. The differences are in the specific assumptions, models and extrapolation methods used. The United States Environmental Protection Agency and the United States Department of Agriculture have developed a guidance document for the conduct of microbial risk assessments for water and food (USEPA, 2012).

Hazard identification in the case of pathogens is complicated because several outcomes—from asymptomatic infection to death (Figure 24.5)—are possible, and these

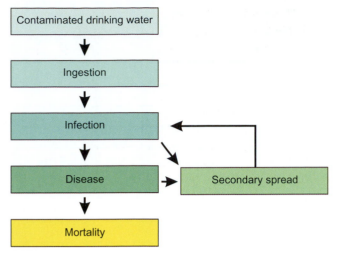

FIGURE 24.5 Outcomes of enteric viral exposure.

TABLE 24.6 Ratio of Subclinical Infections with Enteric Viruses

Virus	Frequency of Clinical Illness[a] (%)
Poliovirus 1	0.1−1
Coxsackie	
A16	50
B2	11−50
B3	29−96
B4	30−70
B5	5−40
Echovirus	
Overall	50
9	15−60
18	Rare−20
20	33
25	30
30	50
Hepatitis A virus (adults)	75
Rotavirus	
Adults	56−60
Children	28
Astrovirus (adults)	12.5

From Gerba and Rose (1993).
[a]The percentage of the individuals infected who develop clinical illness.

outcomes depend on the complex interaction between the pathogenic agent (the "infector") and the host (the "infectee"). This interaction, in turn, depends on the characteristics of the host as well as the nature of the pathogen. Host factors, for example, include: preexisting immunity; age; nutrition; ability to mount an immune response; and other nonspecific host factors. Agent factors include type and strain of the organism, as well as its capacity to elicit an immune response.

Among the various outcomes of infection is the possibility of subclinical infection. Subclinical (asymptomatic) infections are those in which the infection (growth of the microorganism within the human body) results in no obvious illness such as fever, headache or diarrhea. That is, individuals can host a pathogen microorganism—and transmit it to others—without ever getting sick themselves. The ratio of clinical to subclinical infection varies from pathogen to pathogen as shown in Table 24.6. Poliovirus infections, for instance, seldom result in obvious clinical symptoms; in fact, the proportion of individuals developing clinical illness may be less than 1%. However, other enteroviruses, such as the Coxsackieviruses, may exhibit a greater proportion. In many cases, such as for rotaviruses, the probability of developing clinical illness appears to be completely unrelated to the dose an individual receives via ingestion (Ward *et al.*, 1986). Rather, the likelihood of developing clinical illness depends on the type and strain of the virus as well as host age, nonspecific host factors and possibly preexisting immunity. The incidence of clinical infection can also vary from year to year for the same virus, depending on the emergence of new strains or genotypes.

Another outcome of infection is the development of clinical illness. Several host factors play a major role in this outcome. The age of the host is often a determining factor. In the case of hepatitis A, for example, clinical illness can vary from about 5% in children younger than 5 years of age to 75% in adults. In contrast, children are more likely to develop rotaviral gastroenteritis than are adults. Immunity is also an important factor, albeit a variable one. That is, immunity may or may not provide long-term protection from reinfection, depending on the enteric pathogen. It does not, for example, provide long-term protection against the development of clinical illness in the case of the Norovirus or *Giardia*. However, for most enteroviruses and for the hepatitis A virus, immunity to reinfection is believed to be lifelong. Other undefined host factors may also control the odds of developing illness. For example, in experiments with Norovirus, human volunteers who did not become infected upon an initial exposure to the virus also did not respond to a second exposure. In contrast, volunteers who developed gastroenteritis upon the first exposure also developed illness after the second exposure.

The ultimate outcome of infection—mortality—can be caused by nearly all enteric organisms. The factors that control the prospect of mortality are largely the same factors that control the development of clinical illness. Host age, for example, is significant. Thus, mortality for hepatitis A and poliovirus is greater in adults than in children. In general, however, one can say that the very young, the

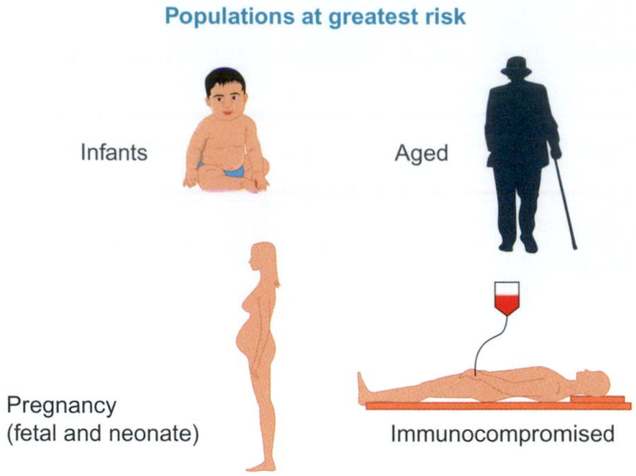

Populations at greatest risk

FIGURE 24.6 Populations at greatest risk of serious illness and mortality from pathogens.

TABLE 24.7 Case–Fatality Rates Observed for Enteric Pathogens in Nursing Home versus General Population

Organism	Case–fatality Rate (%) in General Population	Case–fatality Rate (%) in Nursing Homes
Campylobacter jejuni	0.1	1.1
Escherichia coli O157:H7	0.2	11.8
Salmonella	0.1	3.8
Rotavirus	0.01	1.0

Modified from Gerba et al. *(1996).*

TABLE 24.8 Case–Fatality Rates for Enteric Viruses and Bacteria

Organism	Case–fatality Rate (%)
Viruses	
Poliovirus 1	0.90
Coxsackie	
A2	0.50
A4	0.50
A9	0.26
A16	0.12
Coxsackie B	0.59–0.94
Echovirus	
6	0.29
9	0.27
Hepatitis A	0.30
Rotavirus	
Total	0.01
Hospitalized	0.12
Bacteria	
Shigella	0.2
Salmonella	0.1
Escherichia coli O157:H7	0.2
Campylobacter jejuni	0.1

From Gerba and Rose (1993) and Gerba et al. *(1995).*

elderly and the immunocompromised are at the greatest risk of a fatal outcome of most illnesses (Figure 24.6) (Gerba *et al.*, 1996). For example, the case–fatality rate (%) for *Salmonella* in the general population is 0.1%, but it has been observed to be as high as 3.8% in nursing homes (Table 24.7). In North America and Europe, the reported case–fatality rates (i.e., the ratio of cases to fatalities reported as a percentage of persons who die) for enterovirus infections range from less than 0.1 to 0.94%, as shown in Table 24.8. The case–fatality rate for common enteric bacteria ranges from 0.1 to 0.2% in the general population. Enteric bacterial diseases can be treated with antibiotics, but no treatment is available for enteric viruses.

Recognizing that microbial risk involves a myriad of pathogenic organisms capable of producing a variety of outcomes that depend on a number of factors—many

of which are undefined—one must now face the problem of exposure assessment, which has complications of its own. Unlike chemical-contaminated water, microorganism-contaminated water does not have to be consumed to cause harm. That is, individuals who do not actually drink, or even touch, contaminated water also risk infection because pathogens—particularly viruses—may be spread by person-to-person contact or subsequent contact with contaminated inanimate objects (such as toys). This phenomenon is described as the secondary attack rate, which is reported as a percentage. For example, one person infected with poliovirus can transmit it to 90% of the persons with whom they associate. This secondary spread of viruses has been well documented for waterborne outbreaks of several diseases, including that caused by Norovirus, whose secondary attack rate is about 30%.

The question of dose is another problem in exposure assessment. How does one define "dose" in this context? To answer this question, researchers have conducted a number of studies to determine the infectious dose of enteric microorganisms in human volunteers. Such human experimentation is necessary because determination of the infectious dose in animals and extrapolation to humans is often impossible. In some cases, for example, humans are

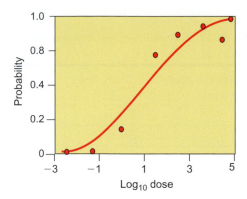

FIGURE 24.7 Dose—response for human rotavirus by oral ingestion.

TABLE 24.9 Best Fit Dose—Response Parameters for Enteric Pathogens Ingestion Studies

Agent	Best Model	Model Parameters
Echovirus 12	Beta—Poisson	$\alpha = 0.374$ $\beta = 186.69$
Rotavirus	Beta—Poisson	$\alpha = 0.26$ $\beta = 0.42$
Poliovirus 1	Exponential	$r = 0.009102$
Poliovirus 1	Beta—Poisson	$\alpha = 0.1097$ $\beta = 1524$
Poliovirus 3	Beta—Poisson	$\alpha = 0.409$ $\beta = 0.788$
Cryptosporidium	Exponential	$r = 0.004191$
Giardia lamblia	Exponential	$r = 0.02$
Salmonella	Exponential	$r = 0.00752$
Escherichia coli	Beta—Poisson	$\alpha = 0.1705$ $\beta = 1.61 \times 10^6$

Modified from Regli et al. (1991).

the primary or only known host. In other cases, such as that of *Shigella* or Norovirus, infection can be induced in laboratory-held primates, but it is not known whether the infectious dose data can be extrapolated to humans. Much of the existing data on infectious doses of viruses has been obtained with attenuated vaccine viruses, or with avirulent laboratory-grown strains, so that the likelihood of serious illness is minimized. An example of a dose—response curve for a human feeding study with rotavirus is shown in Figure 24.7. Information can also be used from food or waterborne outbreaks if the number of organisms ingested in the food or water is known. Use of such data suggests that the doses needed for infection and disease are lower than indicted from controlled human feeding experiments, probably reflecting differences in virulence of the organisms, and heterogeneity in outcome among a diverse population (i.e., varying degrees of inherent susceptibility) (Teunis *et al.*, 2004).

Next, one must choose a dose—response model, whose ordinate is the dose and whose abscissa is the risk of infection (see Figure 24.7). The choice of model is critical so that risks are not greatly over- or underestimated. A modified exponential (beta—Poisson distribution) or a log-probit (simple lognormal, or exponential, distribution) model is commonly used to describe the probability of infection in human subjects for many enteric microorganisms (Haas, 1983). These models have been found to best fit the experimental data. Both are nonthreshold models. For the beta—Poisson model the probability of infection from a single exposure, P, can be described as follows:

$$P = 1 - (1 + N/\beta)^{-\alpha} \qquad \text{(Eq. 24.1)}$$

where N is the number of organisms ingested per exposure and α and β represent parameters characterizing the host—virus interaction (dose—response curve). Some values for α and β for several enteric waterborne pathogens are shown in Table 24.9; these values were determined from human studies. For some microorganisms, an exponential model may better represent the probability of infection.

$$P = 1 - \exp(-rN) \qquad \text{(Eq. 24.2)}$$

In this equation, r is the fraction of the ingested microorganisms that survive to initiate infections (host—microorganism interaction probability). Table 24.9 shows examples of results of both models for several organisms.

These models define the probability of the microorganism overcoming the host defenses (stomach pH, finding a susceptible cell, nonspecific immunity, etc.) to establish an infection in the host. When one uses these models, one estimates the probability of becoming infected after ingestion of various concentrations of pathogen. For example, Case Study 24.1 shows how to calculate the risk of acquiring a viral infection from consumption of contaminated drinking water containing echovirus 12 using Eq. 24.1.

Annual and lifetime risks can also be determined, again assuming a Poisson distribution of the virus in the water consumed (assuming daily exposure to a constant concentration of viral contamination), as follows:

$$P_A = 1 - (1 - P)^{365} \qquad \text{(Eq. 24.3)}$$

where P_A is the annual risk (365 days) of contracting one or more infections; and

$$P_L = 1 - (1 - P)^{25,550} \qquad \text{(Eq. 24.4)}$$

where P_L is the lifetime risk (assuming a lifetime of 70 years = 25,550 days) of contracting one or more infections.

Case Study 24.1 Application of a Virus Risk Model to Characterize Risks from Consuming Shellfish

It is well known that infectious hepatitis and viral gastroenteritis are caused by consumption of raw or, in some cases, cooked clams and oysters. The concentration of echovirus 12 was found to be 8 plaque-forming units (PFU) per 100 g in oysters collected from coastal New England waters. What are the risks of becoming infected and ill from echovirus 12 if the oysters are consumed? Assume that a person usually consumes 60 g of oyster meat in a single serving:

If there are 8 PFU per 100 g of oyster, then for 60 g of oyster, $N = 4.8$ PFU consumed

From Table 24.9, $\alpha = 0.374$, $\beta = 186.69$. The probability of infection from Eq. 24.1 is then

$$P = 1(1 + 4.8/186.69)^{-0.374} = 9.4 \times 10^{-3}$$

If the percentage of infections that result in risk of clinical illness is 50%, then from Eq. 24.5 one can calculate the risk of clinical illness:

Risk of clinical illness $= (9.4 \times 10^{-3})(0.50) = 4.7 \times 10^{-3}$

If the case-fatality rate is 0.001%, then from Eq. 24.6:

Risk of mortality $= (9.4 \times 10^{-3})(0.50)(0.001) = 4.7 \times 10^{-6}$

If a person consumes oysters 10 times a year with 4.8 PFU per serving, then one can calculate the risk of infection in 1 year from Eq. 24.3:

Annual risk $= P_A = 1 - (1 - 9.4 \times 10^{-3})^{365} = 9.7 \times 10^{-1}$

TABLE 24.10 Risk of Infection, Disease and Mortality for Rotavirus

Virus Concentration per 100 liters	Daily Risk	Annual Risk
	Infection	
100	9.6×10^{-2}	1.0
1	1.2×10^{-3}	3.6×10^{-1}
0.1	1.2×10^{-4}	4.4×10^{-2}
	Disease	
100	5.3×10^{-2}	5.3×10^{-1}
1	6.6×10^{-4}	2.0×10^{-1}
0.1	6.6×10^{-5}	2.5×10^{-2}
	Mortality	
100	5.3×10^{-6}	5.3×10^{-5}
1	6.6×10^{-8}	2.0×10^{-5}
0.1	6.6×10^{-9}	2.5×10^{-6}

Modified from Gerba and Rose (1993).

Risks of clinical illness and mortality can then be determined by incorporating terms for the percentage of clinical illness and mortality associated with each particular virus:

$$\text{Risk of clinical illness} = PI \qquad (\text{Eq. 24.5})$$
$$\text{Risk of mortality} = PIM \qquad (\text{Eq. 24.6})$$

where I is the percentage of infections that result in clinical illness and M is the percentage of clinical cases that result in mortality.

Application of this model allows estimation of the risks of infection, development of clinical illness and mortality for different levels of exposure. As shown in Table 24.10, for example, the estimated risk of infection from one rotavirus in 100 liters of drinking water (assuming ingestion of 2 liters per day) is 1.2×10^{-3}, or almost one in a thousand (1:1000) for a single-day exposure. This risk would increase to 3.6×10^{-1}, or approximately one in three, on an annual basis. As can be seen from this table, the risk of developing of clinical illness also appears to be significant for exposure to low levels of rotavirus in drinking water. Example Calculation 24.2

Example Calculation 24.2 Risk Assessment for Rotavirus in Drinking Water

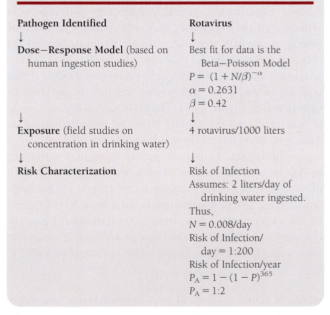

Pathogen Identified → Rotavirus

↓

Dose−Response Model (based on human ingestion studies) → Best fit for data is the Beta−Poisson Model
$P = (1 + N/\beta)^{-\alpha}$
$\alpha = 0.2631$
$\beta = 0.42$

↓

Exposure (field studies on concentration in drinking water) → 4 rotavirus/1000 liters

↓

Risk Characterization → Risk of Infection
Assumes: 2 liters/day of drinking water ingested.
Thus,
$N = 0.008$/day
Risk of Infection/day $= 1:200$
Risk of Infection/year
$P_A = 1 - (1 - P)^{365}$
$P_A = 1:2$

illustrates an example of a calculation for a risk assessment for rotavirus in drinking water.

The EPA has recommended that any drinking water treatment process should be designed to ensure that

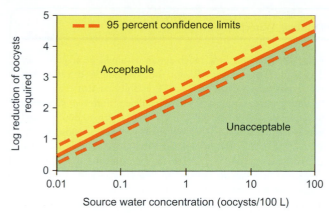

FIGURE 24.8   Relationship of influent *Cryptosporidium* concentration and log reduction by treatment necessary to produce acceptable water. Data from Haas *et al.* (1996).

TABLE 24.11   **Comparison of Outbreak Data to Model Predictions for Assessment of Risks Associated with Exposure to *Salmonella***

Food	Dose (CFU)	Amount Consumed	Attack Rate (%)	Predicted P (%)
Water	17	1 liter	12	12
Pancretin	200	7 doses	100	77
Ice cream	102	1 portion	52	54
Cheese	100–500	28 g	28–36	53–98
Cheese	10^5	100 g	100	>99.99
Ham	10^6	50–100 g	100	>99.99

Modified from Rose et al. *(1995).*

human populations are not subjected to risk of infection greater than 1:10,000 for a yearly exposure. To achieve this goal, it would appear from the data shown in Table 24.10 that the virus concentration in drinking water would have to be less than one per 1000 liters. Thus, if the average concentration of enteric viruses in untreated water is 1400/1000 liters, treatment plants should be designed to remove at least 99.99% of the virus present in the raw water. A further application of this approach is to define the required treatment of a water source in terms of the concentration of a disease-causing organism in that supply. Thus, the more contaminated the raw water source, the more treatment is required to reduce the risk to an acceptable level. An example of this application is shown in Figure 24.8. The plausibility of validation of microbial risk assessment models has been examined by using data from foodborne outbreaks in which information has been available on exposure and outcomes (Rose *et al.*, 1995; Crockett *et al.*, 1996). These studies suggest that microbial risk assessment can give reasonable estimates of illness from exposure to contaminated foods (Table 24.11).

In summary, risk assessment is a major tool for decision making in the regulatory arena. This approach is used to explain chemical and microbial risks as well as ecosystem impacts. The results of such assessments can be used to inform risk managers of the probability and extent of environmental impacts resulting from exposure to different levels of stress (contaminants). Moreover, this process, which allows the quantitation and comparison of diverse risks, lets risk managers utilize the maximum amount of complex information in the decision-making process. This information can also be used to weight the cost and benefits of control options and to develop standards or treatment options (see Case Study 24.2).

QUESTIONS AND PROBLEMS

1. What are some differences between the risks posed by chemicals and those posed by microorganisms?
2. What are some of the potential applications of risk assessment?
3. What is the difference between risk assessment and risk management?
4. Why is the selection of the dose–response curve so important in risk assessment?
5. What is meant by a threshold dose–response curve? Give arguments for a threshold and a nonthreshold dose–response curve for microorganisms.
6. What is the difference between a voluntary and an involuntary risk? Give examples of both.
7. List the four steps in a formal health risk assessment.
8. Does infection always lead to illness with enteric pathogens? What are the factors that determine morbidity and mortality outcomes with microbial infections?
9. What types of dose–response curve best reflect pathogen exposure?
10. What are some potential applications of microbial risk assessment?
11. Calculate the risk of infection from rotavirus during swimming in polluted water. Assume that 50 ml of water is ingested during swimming and the concentration of virus was one per 10 liters. What would the risk be in a year if a person went swimming 10 times in the same water with the same concentration of rotavirus?

Case Study 24.2   How Do We Set Standards for Pathogens in Drinking Water?

In 1974 the U.S. Congress passed the Safe Drinking Water Act giving the U.S. Environmental Protection Agency the authority to establish standards for contaminants in drinking water. Through a risk analysis approach, standards have been set for many chemical contaminants in drinking water. Setting standards for microbial contaminants proved more difficult because (1) methods for the detection of many pathogens are not available, (2) days to weeks are sometimes required to obtain results, and (3) costly and time-consuming methods are required. To overcome these difficulties, coliform bacteria had been used historically to assess the microbial quality of drinking water. However, by the 1980s it had become quite clear that coliform bacteria did not indicate the presence of pathogenic waterborne *Giardia* or enteric viruses. Numerous outbreaks had occurred in which coliform standards were met, because of the greater resistance of viruses and *Giardia* to disinfection. A new approach was needed to ensure the microbial safety of drinking water.

To achieve this goal a new treatment approach was developed called the Surface Treatment Rule (STR). As part of the STR, all water utilities that used surface waters as their source of potable water would be required to provide filtration to remove *Giardia* and enough disinfection to kill viruses. The problem facing the EPA was how much removal should be required. To deal with this issue, the EPA for the first time used a microbial risk assessment approach. The STR established that the goal of treatment was to ensure that microbial illness from *Giardia lamblia* infection should not be any greater than 1 per 10,000 exposed persons annually (10^{-4} per year). This value is close to the annual risk of infection from waterborne disease outbreaks in the United States (4×10^{-3}). Based on the estimated concentration of *Giardia* and enteric viruses in surface waters in the United States from the data available at the time, it was required that all drinking water treatment plants be capable of removing 99.9% of the *Giardia* and 99.99% of the viruses. In this manner it was hoped that the risk of infection of 10^{-4} per year would be achieved. The STR went into effect in 1991.

To better assess whether the degree of treatment required is adequate, the EPA developed the Information Collection Rule, which requires major drinking water utilities that serve surface waters to the public to analyze these surface waters for the presence of *Giardia, Cryptosporidium*, and enteric viruses for a period of almost 2 years. Utilities that have heavily contaminated source water may require greater levels of treatment in the future (see Fig. 24.5).

---

12. Using the exposure data in Example Calculation 24.1 determine the risk of infection and disease from *Salmonella* from a yearly exposure from lettuce.

13. If the concentration of *Cryptosporidium* oocysts in a stream is one oocyst per liter, how many $\log_{10}$ removals would you need to reduce the risk to 1:10,000, of becoming infected in a year, assuming two liters of drinking water are ingested per day? What treatment methods in series would you need to achieve this level of removal (see Chapter 27)?

14. What has a lower greater infectivity rotavirus or *Salmonella*?

REFERENCES AND RECOMMENDED READING

Cockerham, L. G., and Shane, B. S. (1994) "Basic Environmental Toxicology," CRC Press, Boca Raton, FL.

Covello, V., von Winterfieldt, D., and Slovic, P. (1988) Risk communication: a review of the literature. *Risk Anal.* **3**, 71–182.

Crockett, C. S., Haas, C. N., Fazil, A., Rose, J. B., and Gerba, C. P. (1996) Prevalence of shigellosis: consistency with dose–response information. *Int. J. Food. Prot.* **30**, 87–99.

Gale, P. (2003) Developing risk assessments of waterborne microbial contaminates. In "The Handbook of Water and Wastewater Microbiology" (D. Mara, and N. Horm, eds.), Academic Press, London, pp. 263–280.

Gerba, C. P., and Rose, J. B. (1993) Estimating viral disease risk from drinking water. In "Comparative Environmental Risks" (R. Cothern, ed.), Lewis, Boca Raton, FL.

Gerba, C. P., Rose, J. B., and Haas, C. N. (1995) Waterborne disease—who is at risk? In "Water Quality Technology Proceedings," American Water Works Association, Denver, CO, pp. 231–254.

Gerba, C. P., Rose, J. B., and Haas, C. N. (1996) Sensitive populations: who is at the greatest risk? *Int. J. Food Microbiol.* **30**, 113–123.

Gerba, C. P., Rose, J. B., Haas, C. N., and Crabtree, K. D. (1997) Waterborne rotavirus: a risk assessment. *Water Res.* **12**, 2929–2940.

Haas, C. N. (1983) Estimation of risk due to low levels of microorganisms: a comparison of alternative methodologies. *Am. J. Epidemiol.* **118**, 573–582.

Haas, C. N., Crockett, C. S., Rose, J. B., Gerba, C. P., and Fazil, A. M. (1996) Assessing the risk posed by oocysts in drinking water. *J. Am. Water Works Assoc.* **88**, 113–123.

Haas, C. N., Rose, J. B., and Gerba, C. P. (2014) "Quantitative Microbial Risk Assessment," 2nd ed. John Wiley & Sons, New York.

Kolluru, R. V. (1993) "Environmental Strategies Handbook," McGraw-Hill, New York.

National Research Council (NRC) (1993) "Managing Wastewater Coastal Urban Areas," National Academy Press, Washington, DC.

Regli, S., Rose, J. B., Haas, C. N., and Gerba, C. P. (1991) Modeling the risk from *Giardia* and viruses in drinking water. *J. Am. Water Works Assoc.* **83**, 76–84.

Rose, J. B., Haas, C. N., and Gerba, C. P. (1995) Linking microbiological criteria for foods with quantitative risk assessment. *J. Food Saf.* **15**, 11–132.

Stine, S. C., Song, I., Choi, C. Y., and Gerba, C. P. (2005) Application of microbial risk assessment to the development of standards for enteric pathogens in water used to irrigate fresh produce. *J. Food Protect.* **68**, 913–918.

Teunis, P. F. M., Medema, G. J., Kruidenier, L., and Havelaar, A. H. (1997) Assessment of the risk of infection of *Cryptosporidium* and *Giardia* in drinking water from a surface water source. *Water Res.* **31**, 333–1346.

Teunis, P. F. M., Takumi, K., and Shinagawa, K. (2004) Dose response for infection by *Escherichia coli* O157:H7 from outbreak data. *Risk Anal.* **24**, 401–417.

U.S. Environmental Protection Agency (USEPA) (1990) Risk Assessment, Management and Communication of Drinking Water Contamination. EPA/625/4–89/024, Washington, DC.

U.S. Environmental Protection Agency (USEPA) (2012) Microbial Risk Assessment Guideline for Pathogenic Microorganisms with Focus on Food and Water. EPA/100/J–12/001, Washington, DC.

Ward, R. L., Berstein, D. I., and Young, E. C. (1986) Human rotavirus studies in volunteers of infectious dose and serological response to infection. *J. Infect. Dis.* **154**, 871–877.

Wilson, R., and Crouch, E. A. C. (1987) Risk assessment and comparisons: an introduction. *Science* **236**, 267–270.

Part VII

Wastewater Treatment and Disinfection

Chapter 25

Municipal Wastewater Treatment

Charles P. Gerba and Ian L. Pepper

25.1 THE NATURE OF WASTEWATER (SEWAGE)

The cloaca maxima, the "biggest sewer" in Rome, had at one time enough capacity to serve a city of one million people. This sewer, and others like it, simply collected wastes and discharged them into the nearest lake, river or ocean. This expedient made cities more habitable, but its success depended on transferring the pollution problem from one place to another. Although this worked reasonably well for the Romans, it does not work well today. Current population densities are too high to permit a simple dependence on transference. Thus, modern-day sewage is treated before it is discharged into the environment. In the latter part of the nineteenth century, the design of sewage systems allowed collection with treatment to lessen the impact on natural waters. Today, more than 15,000 wastewater treatment plants treat approximately 150 billion liters of wastewater per day in the United States alone. In addition, septic tanks, which were also introduced at the end of the nineteenth century, serve approximately 25% of the U.S. population, largely in rural areas.

Domestic wastewater is primarily a combination of human feces, urine and "graywater." Graywater results from washing, bathing and meal preparation. Water from various industries and businesses may also enter the system. People excrete 100−500 grams wet weight of feces and 1−1.3 liters of urine per person per day (Bitton, 2011). Major organic and inorganic constituents of untreated domestic sewage are shown in Table 25.1.

The amount of organic matter in domestic wastes determines the degree of biological treatment required. Three tests are used to assess the amount of organic matter: biochemical oxygen demand (BOD); chemical oxygen demand (COD); and total organic carbon (TOC).

The major objective of domestic waste treatment is the reduction of BOD, which may be either in the form of solids (suspended matter) or soluble. BOD is the amount of dissolved oxygen consumed by microorganisms during the biochemical oxidation of organic (carbonaceous BOD) and inorganic (ammonia) matter. The methodology for measuring BOD has changed little since it was developed in the 1930s.

The 5-day BOD test (written BOD_5) is a measure of the amount of oxygen consumed by a mixed population of heterotrophic bacteria in the dark at 20°C over a period of 5 days. In this test, aliquots of wastewater are placed in a 300-ml BOD bottle (Figure 25.1) and diluted in phosphate buffer (pH 7.2) containing other inorganic elements

I.L. Pepper, C.P. Gerba, T.J. Gentry: Environmental Microbiology, Third edition. DOI: http://dx.doi.org/10.1016/B978-0-12-394626-3.00025-9
© 2015 Elsevier Inc. All rights reserved.

TABLE 25.1 Typical Composition of Untreated Domestic Wastewater

Contaminants	Concentration (mg/L)		
	Low	Moderate	High
Solids, total	350	720	1200
Dissolved, total	250	500	850
Volatile	105	200	325
Suspended solids	100	220	350
Volatile	80	164	275
Settleable solids	5	10	20
Biochemical oxygen demand[a]	110	220	400
Total organic carbon	80	160	290
Chemical oxygen demand	250	500	1000
Nitrogen (total as N)	20	40	85
Organic	8	15	35
Free ammonia	12	25	50
Nitrites	0	0	0
Nitrates	0	0	0
Phosphorus (total as P)	4	8	15
Organic	1	3	5
Inorganic	3	5	10

From Pepper et al. (2006b).
[a]*Five-day test, (BOD_5, 20°C).*

FIGURE 25.1 BOD bottle.

where:

D_1 = initial dissolved oxygen (DO), D_5 = DO at day 5, and

P = decimal volumetric fraction of wastewater utilized.

If the dilution water is seeded:

$$BOD(mg/L) = \frac{(D_1 - D_5) - (B_1 - B_5)f}{P} \qquad (Eq.\ 25.2)$$

where:

D_1 = initial DO of the sample dilution (mg/L)
D_5 = final DO of the sample dilution (mg/L)
P = decimal volumetric fraction of sample used
B_1 = initial DO of seed control (mg/L)
B_5 = final DO of seed control (mg/L), and
f = ratio of seed in sample to seed in control
= (% seed in D_1)/(% seed in B_1).

Because of depletion of the carbon source, the carbonaceous BOD reaches a plateau called the ultimate carbonaceous BOD (Figure 25.2). The BOD_5 test is commonly used for several reasons:

- To determine the amount of oxygen that will be required for biological treatment of the organic matter present in a wastewater
- To determine the size of the waste treatment facility needed
- To assess the efficiency of treatment processes
- To determine compliance with wastewater discharge permits

The typical BOD_5 of raw sewage ranges from 110 to 440 mg/L (see Example Calculation 25.1). Conventional sewage treatment will reduce this by 95%.

(N, Ca, Mg, Fe) and saturated with oxygen. Sometimes acclimated microorganisms or dehydrated cultures of microorganisms, sold in capsule form, are added to municipal and industrial wastewaters, which may not have a sufficient microflora to enable the BOD test to be carried out. In some cases, a nitrification inhibitor is added to the sample to determine only the carbonaceous BOD.

Dissolved oxygen concentration is determined at time 0, and, after a 5-day incubation, by means of an oxygen electrode, chemical procedures (e.g., Winkler test) or a manometric BOD apparatus. The BOD test is carried out on a series of dilutions of the sample, the dilution depending on the source of the sample. When dilution water is not seeded, the BOD value is expressed in milligrams per liter, according to the following equation (APHA, 1998).

$$BOD(mg/L) = \frac{D_1 - D_5}{P} \qquad (Eq.\ 25.1)$$

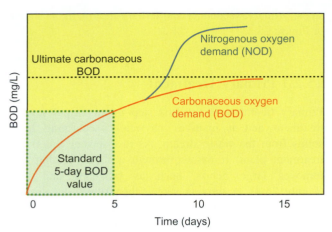

FIGURE 25.2   Carbonaceous and nitrogenous BOD.

TABLE 25.2 Types and Numbers of Microorganisms Typically Found in Untreated Domestic Wastewater

Organism	Concentration (per ml)
Total coliforms	$10^5–10^6$
Fecal coliforms	$10^4–10^5$
Fecal streptococci	$10^3–10^4$
Enterococci	$10^2–10^3$
Shigella	Present
Salmonella	$10^0–10^2$
Clostridium perfringens	$10^1–10^3$
Giardia cysts	$10^{-1}–10^2$
Cryptosporidium cysts	$10^{-1}–10^1$
Helminth ova	$10^{-2}–10^1$
Enteric virus	$10^1–10^2$

From Pepper et al. (2006b).

Example Calculation 25.1   Calculation of BOD

Determine the 5-day BOD (BOD_5) for a wastewater sample when a 15-ml sample of the wastewater is added to a BOD bottle containing 300 ml of dilution water, and the dissolved oxygen is 8 mg/L. Five days later the dissolved oxygen concentration is 2 mg/L.

Using Eq. 25.1:

$$BOD\ (mg/L) = \frac{D_1 - D_5}{P}$$

$$D_1 = 8\ mg/L$$
$$D_5 = 2\ mg/L$$
$$P = \frac{15\ ml}{300\ ml} = 5\% = 0.05$$

$$BOD_5 = \frac{8 - 2}{0.05} = 120\ mg/L$$

Chemical oxygen demand (COD) is the amount of oxygen necessary to oxidize all of the organic carbon completely to CO_2 and H_2O. COD is measured by oxidation with potassium dichromate ($K_2Cr_2O_7$) in the presence of sulfuric acid and silver, and is expressed in milligrams per liter. In general, 1 g of carbohydrate or 1 g of protein is approximately equivalent to 1 g of COD. Normally, the ratio BOD/COD is approximately 0.5. When this ratio falls below 0.3, it means that the sample contains large amounts of organic compounds that are not easily biodegraded.

Another method of measuring organic matter in water is the TOC or total organic carbon test. TOC is determined by oxidation of the organic matter with heat and oxygen, followed by measurement of the CO_2 liberated with an infrared analyzer. Both TOC and COD represent the concentration of both biodegradable and nonbiodegradable organics in water.

Pathogenic microorganisms are almost always present in domestic wastewater (Table 25.2). This is because large numbers of pathogenic microorganisms may be excreted by infected individuals. Both symptomatic and asymptomatic individuals may excrete pathogens. For example, the concentration of rotavirus may be as high as 10^{10} virions per gram of stool, or 10^{12} in 100 g of stool (Table 25.3). Infected individuals may excrete enteric pathogens for several days or as long as a few months. The concentration of enteric pathogens in raw wastewater varies depending on the following:

- The incidence of the infection in the community
- The socioeconomic status of the population
- The time of year
- The per-capita water consumption

The peak incidence of many enteric infections is seasonal in temperate climates. Thus, the highest incidence of enterovirus infection is during the late summer and early fall. Rotavirus infections tend to peak in the early winter, and *Cryptosporidium* infections peak in the early spring and fall. The reason for the seasonality of enteric infections is not completely understood, but several factors may play a role. It may be associated with the survival of different agents in the environment during the different seasons. *Giardia*, for example, can survive winter temperatures very well. Alternatively, excretion differences among animal reservoirs may be involved, as is the case with

TABLE 25.3 Incidence and Concentration of Enteric Viruses and Protozoa in Feces in the United States

Pathogen	Incidence (%)	Concentration in Stool (per gram)
Enteroviruses	10−40	10^3-10^8
Hepatitis A virus	0.1	10^8
Rotavirus	10−29	$10^{10}-10^{12}$
Giardia	3.8	10^6
	18−54[a]	10^6
Cryptosporidium	0.6−20	10^6-10^7
	27−50[a]	10^6-10^7

[a]Children in day care centers.

TABLE 25.4 Estimated Levels of Enteric Organisms in Sewage and Polluted Surface Water in the United States

Organism	Concentration (per 100 ml)	
	Raw Sewage	Polluted Stream Water
Coliforms	10^9	10^5
Enteric viruses	10^2	1−10
Giardia	$10-10^2$	0.1−1
Cryptosporidium	1−10	$0.1-10^2$

From U.S. EPA (1998).

Cryptosporidium. Finally, it may well be that greater exposure to contaminated water, as in swimming, is the explanation for increased incidence in the summer months.

Concentrations of enteric pathogens are much greater in sewage in the developing world than in the industrialized world. For example, the average concentration of enteric viruses in sewage in the United States has been estimated to be 10^3 per liter (Table 25.4), whereas concentrations as high as 10^5 per liter have been observed in Africa and Asia.

25.2 CONVENTIONAL WASTEWATER TREATMENT

The primary goal of wastewater treatment is the removal and degradation of organic matter under controlled conditions. Complete sewage treatment comprises three major

steps: primary, secondary and tertiary treatment, as shown in Figure 25.3.

25.2.1 Primary Treatment

Primary treatment is the first step in municipal sewage treatment and it involves physically separating large solids from the waste stream. As raw sewage enters the treatment plant, it passes through a metal grating that removes large debris, such as branches and tires (Figure 25.4). A moving screen then filters out smaller items such as diapers and bottles (Figure 25.5), after which a brief residence in a grit tank allows sand and gravel to settle out. The waste stream is then pumped into the primary settling tank (also known as a sedimentation tank or clarifier), where about half the suspended organic solids settle to the bottom as sludge (Figure 25.6). The resulting sludge is referred to as primary sludge. Microbial pathogens are not effectively removed from the effluent in the primary process, although some removal occurs.

Dissolved air flotation (DAF) is a more recent innovation for removing suspended solids from sewage, which is now being introduced into new wastewater treatment plants as an alternative to conventional primary sedimentation processes. DAF clarification is achieved by dissolving air in the wastewater under pressure, and then releasing the air at atmospheric pressure in a flotation tank or basin. This occurs in the front end of the DAF tank known as the "contact zone." The resulting air bubbles that form attach to floc particles and suspended solids. Frequently, coagulants are added to the wastewater prior to the DAF tank to produce the flocs. The floc−bubble aggregates are then carried by water into the second DAF zone known as the "separation zone." Here, free bubbles and floc−bubble aggregates rise to the surface of the tank forming a concentrated sludge blanket that can be removed by skimming devices (Edzwald, 2010). DAF clarifiers remove suspended solids more rapidly than does conventional primary sedimentation and are cost effective from an engineering standpoint.

25.2.2 Secondary Treatment

Secondary treatment consists of biological degradation, in which the remaining suspended solids are decomposed by microorganisms, and the number of pathogens is reduced. In this stage, the effluent from primary treatment usually undergoes biological treatment in a trickling filter bed (Figure 25.7), an aeration tank (Figure 25.8) or a sewage lagoon (see Section 25.3). A disinfection step is generally included at the end of the treatment.

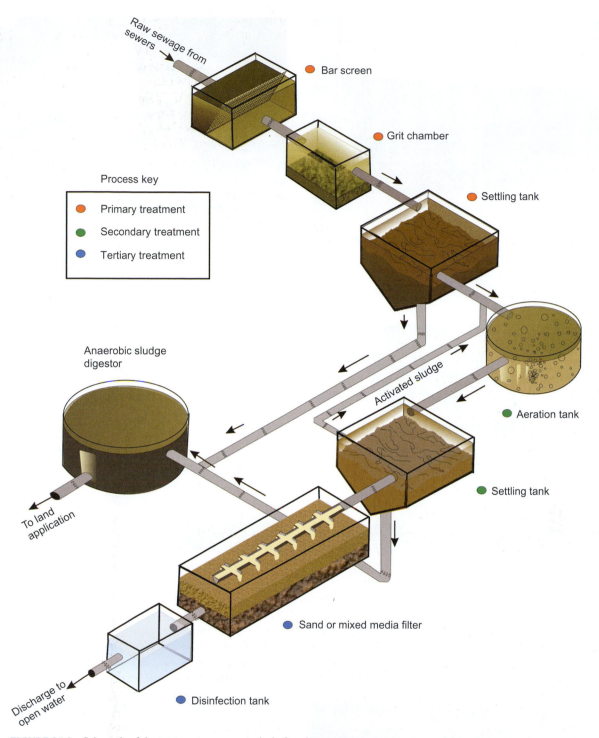

Process key

🟠 Primary treatment

🟢 Secondary treatment

🔵 Tertiary treatment

Raw sewage from sewers

🟠 Bar screen

🟠 Grit chamber

🟠 Settling tank

🟢 Aeration tank

Activated sludge

🟢 Settling tank

Anaerobic sludge digestor

To land application

🔵 Sand or mixed media filter

Discharge to open water

🔵 Disinfection tank

FIGURE 25.3   Schematic of the treatment processes typical of modern wastewater treatment.

25.2.2.1 Trickling Filters

In modern wastewater treatment plants, the trickling filter is composed of plastic units (Figures 25.9 and 25.10). In older plants, or developing countries, the filter is simply a bed of stones or corrugated plastic sheets through which wastewater drips (see Figure 25.7). This is one of the earliest systems introduced for biological waste treatment. The effluent is pumped through an overhead sprayer onto the filter bed, where bacteria and other microorganisms have formed a biofilm on the filter surfaces. These microorganisms intercept the organic material as it trickles past and decompose it aerobically.

FIGURE 25.4 Removal of large debris from sewage via a "bar screen."

FIGURE 25.7 A trickling filter bed. Here, rocks provide a matrix supporting the growth of a microbial biofilm that actively degrades the organic material in the wastewater under aerobic conditions.

FIGURE 25.5 Removal of small debris via a "moving screen."

FIGURE 25.8 Secondary treatment: an aeration basin.

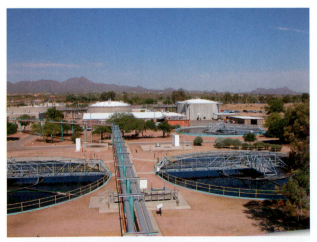

FIGURE 25.6 Three clarifiers (foreground—blue) where suspended organic solids settle out as primary sludge. Also see the two anaerobic sludge digestors in the background (white).

The media used in trickling filters may be stones, ceramic material, hard coal or plastic media. Plastic media of polyvinyl chloride (PVC) or polypropylene are used today in high-rate trickling filters. As the organic matter passes through the trickling filter, it is converted to microbial biomass, which forms a thick biofilm on the filter medium. The biofilm that forms on the surface of the filter medium is called a zooleal film. It is composed of bacteria, fungi, algae and protozoa. Over time, the increase in biofilm thickness leads to limited oxygen diffusion to the deeper layers of the biofilm, creating an anaerobic environment near the filter medium surface. As a result, the organisms eventually slough from the surface and a new biofilm is formed. BOD removal by trickling filters is approximately 85% for low-rate filters (U.S. EPA, 1977). Effluent from the trickling filter usually passes into a final clarifier to further separate solids from effluent.

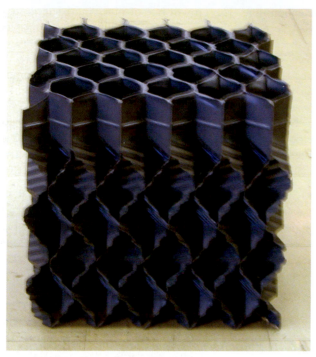

FIGURE 25.9   A unit of plastic material used to create a biofilter (trickling filter). The diameter of each hold is approximately 5 cm. From Pepper *et al.* (2006a).

FIGURE 25.10   A trickling biofilter or biotower. This is composed of many plastic units stacked upon each other. Dimensions of the biofilter may be 20 m diameter by 10–30 m depth. From Pepper *et al.* (2006a).

25.2.2.2 Conventional Activated Sludge

Aeration-tank digestion is also known as the activated sludge process. In the United States, wastewater is most commonly treated by this process. Effluent from primary treatment is pumped into a tank and mixed with a bacteria-rich slurry known as activated sludge. Air or pure oxygen pumped through the mixture encourages bacterial growth and decomposition of the organic material. It then goes to a secondary settling tank, where water is

siphoned off the top of the tank and sludge is removed from the bottom. Some of the sludge is used as an inoculum for primary effluent. The remainder of the sludge, known as secondary sludge, is removed. This secondary sludge is added to primary sludge from primary treatment, and is subsequently anaerobically digested to produce biosolids (Chapter 26). The concentration of pathogens is reduced in the activated sludge process by antagonistic microorganisms as well as adsorption to or incorporation in the secondary sludge.

An important characteristic of the activated sludge process is the recycling of a large proportion of the biomass. This results in a large number of microorganisms that oxidize organic matter in a relatively short time (Bitton, 2011). The detention time in the aeration basin varies from 4 to 8 hours. The content of the aeration tank is referred to as the mixed-liquor suspended solids (MLSS). The organic part of the MLSS is called the mixed-liquor volatile suspended solids (MLVSS), which is the nonmicrobial organic matter as well as dead and living microorganisms and cell debris. The activated sludge process must be controlled to maintain a proper ratio of substrate (organic load) to microorganisms or food-to-microorganism ratio (F/M) (Bitton, 2011). This is expressed as BOD per kilogram per day. It is expressed as:

$$\frac{F}{M} = \frac{Q \cdot BOD_5}{MLSS \cdot V} \qquad (Eq.\ 25.3)$$

where:

Q = flow rate of sewage in million gallons per day (MGD)
BOD_5 = 5-day biochemical oxygen demand (mg/L)
MLSS = mixed-liquor suspended solids (mg/L)
V = volume of aeration tank (gallons)

F/M is controlled by the rate of activated sludge wasting. The higher the wasting rate, the higher the F/M ratio. For conventional aeration tanks the F/M ratio is 0.2–0.5 lb BOD_5/day/lb MLSS, but it can be higher (up to 1.5) for activated sludge when high-purity oxygen is used. A low F/M ratio means that the microorganisms in the aeration tank are starved, leading to more efficient wastewater treatment.

The important parameters controlling the operation of an activated sludge process are: organic loading rates; oxygen supply; and control and operation of the final settling tank. This tank has two functions: clarification and thickening. For routine operation, sludge settleability is determined by use of the sludge volume index (SVI) (Bitton, 2011).

SVI is determined by measuring the sludge volume index, which is given by the following formula:

$$SVI = \frac{V \cdot 1000}{MLSS} \qquad (Eq.\ 25.4)$$

where V = volume of settled sludge after 30 minutes (ml/L).

The microbial biomass produced in the aeration tank must settle properly from suspension so that it may be wasted or returned to the aeration tank. Good settling occurs when the sludge microorganisms are in the endogenous phase, which occurs when carbon and energy sources are limited, and the microbial specific growth rate is local (Bitton, 2011). A mean cell residence time of 3–4 days is necessary for effective settling (Metcalf and Eddy, 2003). Poor settling may also be caused by sudden changes in temperature and pH, absence of nutrients and presence of toxic metals and organics. A common problem in the activated sludge process is filamentous bulking, which consists of slow settling and poor compaction of solids in the clarifier. Filamentous bulking is usually caused by the excessive growth of filamentous microorganisms. The filaments produced by these bacteria interfere with sludge settling and compaction. A high SVI (> 150 ml/g) indicates bulking conditions. Filamentous bacteria are able to predominate under conditions of low dissolved oxygen, low F/M, low nutrient and high sulfide levels. Filamentous bacteria can be controlled by treating the return sludge with chlorine or hydrogen peroxide to kill filamentous microorganisms selectively.

25.2.2.3 Nitrogen Removal by the Activated Sludge Process

Activated sludge processes can be modified for nitrogen removal to encourage nitrification followed by denitrification. The establishment of a nitrifying population in activated sludge depends on the wastage rate of the sludge, and therefore on the BOD load, MLSS and retention time. The growth rate of nitrifying bacteria (μ_n) must be higher than the growth rate (μ_h) of heterotrophs in the system. In reality, the growth rate of nitrifiers is lower than that of heterotrophs in sewage; therefore, a long sludge age is necessary for the conversion of ammonia to nitrate. Nitrification is expected at a sludge age greater than 4 days (Bitton, 2011).

Nitrification must be followed by denitrification to remove nitrogen from wastewater. The conventional activated sludge system can be modified to encourage denitrification. Two such processes are:

- Single sludge system (Figure 25.11A). This system comprises a series of aerobic and anaerobic tanks in lieu of a single aeration tank.
- Multisludge system (Figure 25.11B). Carbonaceous oxidation, nitrification and denitrification are carried out in three separate systems. Added methanol or settled sewage serves as the source of carbon for denitrifiers.

25.2.2.4 Phosphorus Removal by the Activated Sludge Process

Phosphorus can also be reduced by the activity of microorganisms in modified activated sludge processes. The process depends on the uptake of phosphorus by the microbes during the aerobic stage and subsequent release during the anaerobic stage. One of several systems in use is the A/O (anaerobic/oxic) process. The A/O process consists of a modified activated sludge system that includes an anaerobic zone (detention time 0.5–1 hour) upstream of the conventional aeration tank (detention

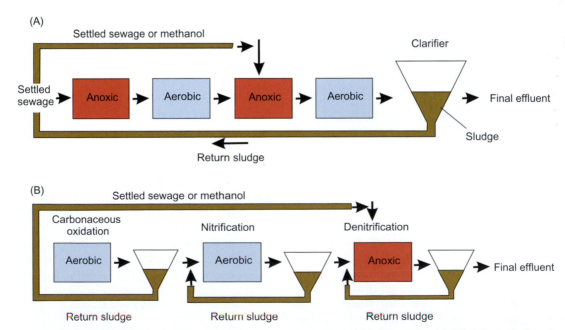

FIGURE 25.11 Denitrification systems: (A) single-sludge system; (B) multisludge system. Modified from Curds and Hawkes (1983).

time 1−3 hours). Figure 25.12 illustrates the microbiology of the A/O process. During the anaerobic phase, inorganic phosphorus is released from the cells as a result of polyphosphate hydrolysis. The energy liberated is used for the uptake of BOD from wastewater. Removal efficiency is high when the BOD/phosphorus ratio exceeds 10 (Metcalf and Eddy, 2003). During the aerobic phase, soluble phosphorus is taken up by bacteria, which synthesize polyphosphates, using the energy released from BOD oxidation.

25.2.2.5 The Bardenpho Process

The Bardenpho process is an advanced modification of the activated sludge process, which results in nutrient removal of nitrogen and phosphorus via microbial processes that occur in a multistage biological reactor (Figure 25.13). This reactor removes high levels of BOD, suspended solids, nitrogen and phosphorus.

- Fermentation stage: Activated sludge is returned from the clarifier and undergoes microbial fermentation and phosphate is released.

- First anoxic stage: Mixed liquor containing nitrates from the third stage is recycled here, and mixed with conditioned sludge from the fermentation stage, in the absence of oxygen. Heterotrophic denitrifying bacteria reduce BOD by utilizing carbonaceous substrate while using nitrate as a terminal electron acceptor, which is reduced to gaseous nitrogen.
- Nitrification stage: Oxygen is introduced allowing for heterotrophic aerobic respiration, which further oxidizes BOD. At the same time ammonia is aerobically nitrified to nitrate, and phosphate is taken up and utilized by microbes. Mixed liquor containing the nitrates is recycled back to the first anoxic stage.
- Second anoxic stage: The remaining liquor from the nitrification stage is passed into this second anoxic stage, where nitrate (in the absence of oxygen) is again reduced to nitrogen gas. This results in low effluent nitrate concentrations.
- Re-aeration stage: This is an aerobic environment that ensures that phosphate taken up microbially is not released in the final clarifier.

Overall the Bardenpho process results in an effluent that is low in nitrates and phosphates. Bardenpho pro-

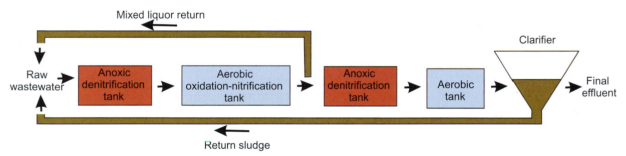

FIGURE 25.12   Microbiology of the A/O process.

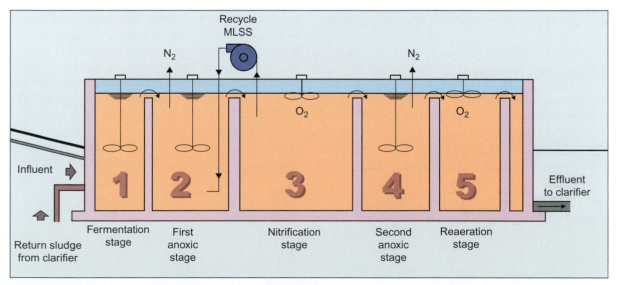

FIGURE 25.13   The five-stage Bardenpho process for microbial nutrient removal.

cesses utilizing autotrophic denitrification have also been evaluated utilizing "spent caustic" as an electron donor. Spent caustic is produced in petrochemical processes and contains adsorbed hydrogen sulfide which acts as substrate for autotrophic denitrifiers (Park *et al.*, 2010).

25.2.2.6 Membrane Bioreactors

Membrane bioreactors (MBR) are a combination of biological treatment with membrane separation by microporous or ultrafiltration membranes. The process consists of a tank and a membrane unit either located external to the bioreactor or submerged directly within it (Figure 25.14). The membranes act to retain suspended solids and maintain a high biomass concentration within the bioreactor thereby functioning as a replacement for sedimentation. Membranes come with various pore sizes and can be dense or porous. Separation by dense membranes relies on physicochemical interactions between the permeating components and the membrane material and is known as reverse osmosis or nanofiltration. Porous membranes have larger pore size, separate particles mechanically and are referred to as ultrafiltration or microfiltration. The microbial bioreactor is normally maintained aerobically, but can be operated anaerobically, or with alternating aerobic/anaerobic phases to enhance microbial nitrification followed by denitrification.

Membrane bioreactors have several advantages including a much smaller area needed than conventional activated sludge and a high quality effluent. A membrane bioreactor effectively displaces three individual process steps in a conventional treatment plant (primary settling, activated sludge and reduces the need for disinfection). The major advantages of MBRs include: good quality effluent, reduced reactor volume and net sludge production. The major disadvantages include high operating costs and membrane fouling (Chang *et al.*, 2002). Operating costs can be reduced by integrating microbial fuel cells and membrane bioreactors (Wang *et al.*, 2012).

25.2.3 Tertiary Treatment

Tertiary treatment of effluent involves a series of additional steps after secondary treatment to further reduce

organics, turbidity, nitrogen, phosphorus, metals and pathogens. Most processes involve some type of physicochemical treatment such as coagulation, filtration, activated carbon adsorption of organics, reverse osmosis and additional disinfection. Tertiary treatment of wastewater is practiced for additional protection of wildlife after discharge into rivers or lakes. Even more commonly, it is performed when the wastewater is to be reused for irrigation (e.g., food crops, golf courses), for recreational purposes (e.g., lakes, estuaries) or for drinking water.

25.2.4 Removal of Pathogens by Sewage Treatment Processes

There have been a number of reviews on the removal of pathogenic microorganisms by activated sludge and other wastewater treatment processes (Leong, 1983). This information suggests that significant removal especially of enteric bacterial pathogens can be achieved by these processes (Table 25.5). However, disinfection and/or advanced tertiary treatment are necessary for many reuse applications to ensure pathogen reduction. Current issues related to pathogen reduction are: treatment plant reliability; removal of new and emerging enteric pathogens of concern; and the ability of new technologies to effect pathogen reduction. Wide variation in pathogen removal can result in significant numbers of pathogens passing through a process for various time periods. The issue of reliability is of major importance if the reclaimed water is intended for recreational or potable reuse, where short-term exposures to high levels of pathogens could result in significant risk to the exposed population.

Compared with other biological treatment methods (i.e., trickling filters), activated sludge is relatively efficient in reducing the numbers of pathogens in raw wastewater. Both sedimentation and aeration play a role in pathogen reduction. Primary sedimentation is more effective for the removal of the larger pathogens such as helminth eggs, but solid-associated bacteria and even viruses are also removed. During aeration, pathogens are inactivated by antagonistic microorganisms and by environmental factors such as temperature. The greatest removal probably occurs by adsorption or entrapment of the

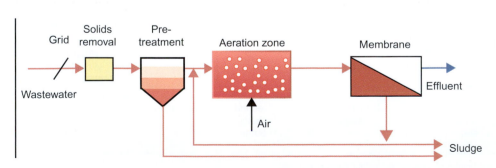

FIGURE 25.14 Membrane bioreactor treatment train showing an external membrane unit.

TABLE 25.5 Pathogen Removal during Sewage Treatment

	Enteric Viruses	*Salmonella*	*Giardia*	*Cryptosporidium*
Concentration in raw sewage (per liter)	10^5–10^6	5000–80,000	9000–200,000	1–3960
Primary treatment[a] % removal	50–98.3	95.8–99.8	27–64	0.7
Number remaining (per liter)	1700–500,000	160–3360	72,000–146,000	
Secondary treatment[b] % removal	53–99.92	98.65–99.996	45–96.7	
Number remaining (per liter)	80–470,000	3–1075	6480–109,500	
Secondary treatment[c] % removal	99.983–99.9999998	99.99–99.999999995	98.5–99.99995	2.7[d]
Number remaining (per liter)	0.007–170	0.000004–7	0.099–2951	

[a]*Primary sedimentation and disinfection.*
[b]*Primary sedimentation, trickling filter or activated sludge, and disinfection.*
[c]*Primary sedimentation, trickling filter or activated sludge, disinfection, coagulation, filtration, and disinfection.*
[d]*Filtration only.*

TABLE 25.6 Average Removal of Pathogen and Indicator Microorganisms in a Wastewater Treatment Plant, St. Petersburg, Florida

	Raw Wastewater to Secondary Wastewater		Secondary Wastewater to Postfiltration		Postfiltration to Postdisinfection		Postdisinfection to Poststorage		Raw Wastewater to Poststorage	
	Percentage	$\log_{10}$	Percentage	$\log_{10}$	Percentage	$\log_{10}$	Percentage	$\log_{10}$	Percentage	$\log_{10}$
Total coliforms	98.3	1.75	69.3	0.51	99.99	4.23	75.4	0.61	99.999992	7.1
Fecal coliforms	99.1	2.06	10.5	0.05	99.998	4.95	56.8	0.36	99.999996	7.4
Coliphage[a]	82.1	0.75	99.98	3.81	90.05	1.03	90.3	1.03	99.999997	6.6
Enterovirus	98.0	1.71	84.0	0.81	96.5	1.45	90.9	1.04	99.999	5.0
Giardia	93.0	1.19	99.0	2.00	78.0	0.65	49.5	0.30	99.993	4.1
Cryptosporidium	92.8	1.14	97.9	1.68	61.1	0.41	8.5	0.04	99.95	3.2

[a]*Escherichia coli host ATCC 15597.*

organisms within the biological floc that forms. The ability of activated sludge to remove viruses is related to the ability to remove solids. This is because viruses tend to be solid associated, and are removal along with the floc. Activated sludge typically removes 90% of the enteric bacteria and 80–99% of the enteroviruses and rotaviruses (Rao *et al.*, 1986). Ninety percent of *Giardia* and *Cryptosporidium* can also be removed (Rose and Carnahan, 1992), being largely concentrated in the sludge. Because of their large size, helminth eggs are effectively removed by sedimentation, and are rarely found in sewage effluent in the United States, although they may be detected in the sludge. However, although

the removal of the enteric pathogens may seem large, it is important to remember that initial concentrations are also large (i.e., the concentration of all enteric viruses in 1 liter of raw sewage may be as high as 100,000 in some parts of the world).

Tertiary treatment processes involving physicochemical processes can be effective in further reducing the concentration of pathogens and enhancing the effectiveness of disinfection processes by the removal of soluble and particulate organic matter (Table 25.6). Filtration is probably the most common tertiary treatment process. Mixed-media (sand, gravel, coal) filtration is most effective in the reduction of protozoan parasites. Usually, greater removal of *Giardia* cysts occurs than of *Cryptosporidium* oocysts because of the larger size of the cysts (Rose and Carnahan, 1992). Removal of enteroviruses and indicator bacteria is usually 90% or less. Addition of coagulant can increase the removal of poliovirus to 99% (U.S. EPA, 1992a).

Coagulation, particularly with lime, can result in significant reductions of pathogens. The high pH conditions (pH $11-12$) that can be achieved with lime can result in significant inactivation of enteric viruses. To achieve removals of 90% or greater, the pH should be maintained above 11 for at least an hour (Leong, 1983). Inactivation of the viruses occurs by denaturation of the viral protein coat. The use of iron and aluminum salts for coagulation can also result in 90% or greater reductions in enteric viruses. The degree of effectiveness of these processes, as in other solids separating processes, is highly dependent on the hydraulic design and, in particular, coagulation and flocculation. The degree of removal observed in bench-scale tests may not approach those seen in full-scale plants, where the process is more dynamic.

Reverse osmosis and ultrafiltration are also believed to result in significant reductions in enteric pathogens. Removal occurs by size exclusion. Removal of enteric viruses in excess of 99.9% can be achieved (Leong, 1983).

25.2.5 Removal of Organics and Inorganics by Sewage Treatment Processes

In addition to nutrients such as nitrogen and phosphorus, and microbial pathogens, there are other constituents within sewage that need to be kept at low concentrations. These include inorganics exemplified by metals, and organic priority pollutants. Metals and organics are normally associated with the solid fraction of sewage, and neither is significantly removed by sewage treatment. However, when point source control mechanisms are implemented to prevent industrial discharges, the concentration of metals and organics within sewage can be significantly reduced. In particular, over the past 15 years in the United States this has resulted in decreased metal concentrations. More recently, there has been concern over the presence of pharmaceuticals such as endocrine disruptors in sewage.

25.3 OXIDATION PONDS

The next two sections discuss several alternatives to large-scale modern wastewater treatment process discussed in Section 25.2. The first of these are sewage lagoons and are often referred to as oxidation or stabilization ponds. These are the oldest of the wastewater

FIGURE 25.15 An oxidation pond. Typically these are only $1-2$ meters deep, and small in area.

treatment systems. Usually no more than a hectare in area and just a few meters deep, oxidation ponds are natural "stewpots," where wastewater is detained while organic matter is degraded (Figure 25.15). A period of time ranging from 1 to 4 weeks (and sometimes longer) is necessary to complete the decomposition of organic matter. Light, heat and settling of the solids can also effectively reduce the number of pathogens present in the wastewater.

The following four categories of oxidation ponds are often used in series:

- Aerobic ponds (Figure 25.16A), which are naturally mixed, must be shallow (up to 1.5 m) because they depend on penetration of light to stimulate algal growth that promotes subsequent oxygen generation. The detention time of wastewater is generally 3 to 5 days.
- Anaerobic ponds (Figure 25.16B) may be 1 to 10 m deep, and require a relatively long detention time of 20 to 50 days. These ponds, which do not require expensive mechanical aeration, generate small amounts of sludge. Often, anaerobic ponds serve as a pretreatment step for high-BOD organic wastes rich in protein and fat (e.g., meat wastes) with a heavy concentration of suspended solids.
- Facultative ponds (Figure 25.17) are most common for domestic waste treatment. Waste treatment is provided by both aerobic and anaerobic processes. These ponds range in depth from 1 to 2.5 m and are subdivided into three layers: an upper aerated zone; a middle facultative zone; and a lower anaerobic zone. The detention time varies between 5 and 30 days.
- Aerated lagoons or ponds (Figure 25.18), which are mechanically aerated, may be $1-2$ m deep and have a detention time of less than 10 days. In general, treatment depends on the aeration time and temperature, as well as the type of wastewater. For example, at 20°C an aeration period of 5 days results in 85% BOD removal.

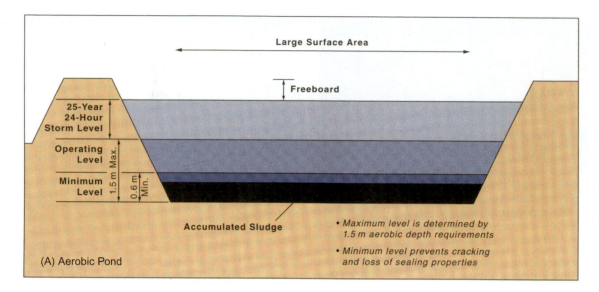

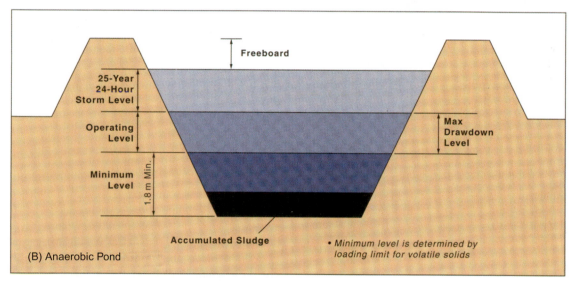

FIGURE 25.16   Pond profiles: (A) aerobic waste pond profile, and (B) anaerobic waste pond profile.

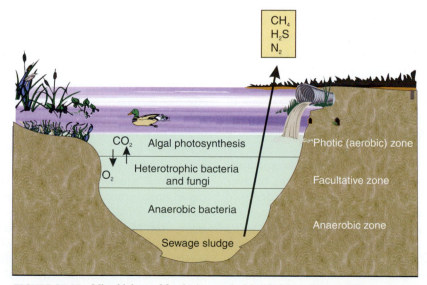

FIGURE 25.17   Microbiology of facultative ponds. Modified from Bitton (1980).

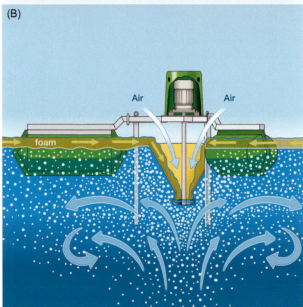

FIGURE 25.18   (A) Aerated lagoon, and (B) floating aeration device.

Because sewage lagoons require a minimum of technology and are relatively low in cost, they are most common in developing countries. However, biodegradable organic matter and turbidity are not as effectively reduced as during activated sludge treatment.

Given sufficient retention times, oxidation ponds can cause significant reductions in the concentrations of enteric pathogens, especially helminth eggs. For this reason, they have been promoted widely in the developing world as a low-cost method of pathogen reduction for wastewater reuse for irrigation. However, a major drawback of ponds is the potential for short-circuiting because of thermal gradients even in multi-pond systems designed for long retention times (i.e., 90 days). Even though the amount of short-circuiting may be small, detectable levels of pathogens can often be found in the effluent from oxidation ponds.

Inactivation and/or removal of pathogens in oxidation ponds is controlled by a number of factors including: temperature; sunlight; pH; bacteriophages; predation by other microorganisms; and adsorption to or entrapment by settleable solids. Indicator bacteria and pathogenic bacteria may be reduced by 90–99% or more, depending on retention times.

25.4 SEPTIC TANKS

Until the middle of the twentieth century in the United States, many rural families and quite a few residents of towns and small cities depended on pit toilets or "outhouses" for waste disposal. In rural areas of developing countries these are still used. These pit toilets, however, often allowed untreated wastes to seep into the groundwater, allowing pathogens to contaminate drinking water supplies. This risk to public health led to the development of septic tanks and properly constructed drain fields. Primarily, septic tanks serve as repositories where solids are separated from incoming wastewater, and biological digestion of the waste organic matter can take place under anaerobic conditions. In 2007, 20% (26.1 million) of the homes in the United States depended on septic tanks. Approximately 20% of all new homes constructed use septic tanks. Most septic tanks are located in the eastern United States (Figure 25.19). In a typical septic tank system (Figure 25.20), the wastewater and sewage enter a tank made of concrete, metal or fiberglass. There, grease and oils rise to the top as scum, and solids settle to the bottom. The wastewater and sewage then undergo anaerobic bacterial decomposition, resulting in the production of a sludge. The wastewater usually remains in the septic tank for just 24–72 hours, after which it is channeled out to a drain field. This drain field or leach field is composed of small perforated pipes that are embedded in gravel below the surface of the soil. Periodically, the residual sludge in the septic tank known as septage is pumped out into a tank truck, and taken to a treatment plant for disposal.

Although the concentration of contaminants in septic tank separate is typically much greater than that found in domestic wastewater (Table 25.7), septic tanks can be an effective method of waste disposal where land is available and population densities are not too high. Thus, they are widely used in rural and suburban areas. But as suburban population densities increase, groundwater and surface water pollution may arise, indicating a need to shift to a commercial municipal sewage system. (In fact, private septic systems are sometimes banned in many suburban areas.) Moreover, septic tanks are not appropriate for every area of the country. They do not work well, for example, in cold, rainy climates, where the drain field may be too wet for proper evaporation, or in areas where the water table is shallow. High densities of septic tanks can also be responsible for nitrate contamination of

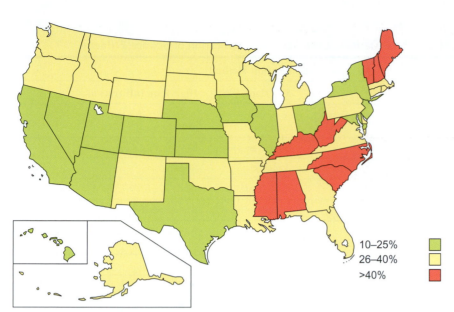

FIGURE 25.19 Percentage of U.S. residents utilizing septic tanks for onsite wastewater treatment. *Source:* U.S. Census Bureau, 1990.

10–25%
26–40%
>40%

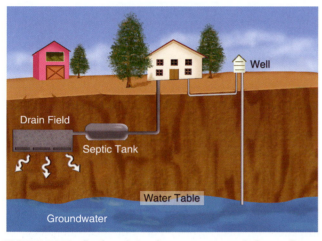

FIGURE 25.20 Septic tank (on-site treatment system). *Source:* Pepper *et al.* (2006).

TABLE 25.7 Typical Characteristics of Septage

Constituent	Concentration (mg/L)	
	Range	Typical Value
Total solids	5000–100,000	40,000
Suspended solids	4000–100,000	15,000
Volatile suspended solids	1200–14,000	7000
BOD_5, 20°C	2000–30,000	6000
Chemical oxygen demand	5000–80,000	30,000
Total Kjeldahl nitrogen (as N)	100–1600	700
Ammonia, NH_3 (as N)	100–800	400
Total phosphorus (as P)	50–800	250
Heavy metals[a]	100–1000	300

From Pepper et al. (2006b).
[a]*Primarily iron (Fe), zinc (Zn), and aluminum (Al).*

groundwater. Finally, most of the waterborne disease outbreaks associated with groundwater in the United States are thought to result from contamination by septic tanks.

25.5 LAND APPLICATION OF WASTEWATER

Although treated domestic wastewater is usually discharged into bodies of water, it may also be disposed of via land application for crop irrigation, or as a means of additional treatment and disposal. The three basic methods used in the application of sewage effluents to land include: low-rate irrigation; overland flow; and high-rate infiltration. Characteristics of each of these are listed in Table 25.8. The choice of a given method depends on the

conditions prevailing at the site under consideration (loading rates, methods of irrigation, crops and expected treatment).

With low-rate irrigation (Figure 25.21A), sewage effluents are applied by sprinkling or by surface application at a rate of 1.5 to 10 cm per week. Two-thirds of the water is taken up by crops or lost by evaporation, and the remainder percolates through the soil matrix. The system must be designed to maximize denitrification in order to avoid pollution of groundwater by nitrates. Phosphorus is immobilized within the soil matrix by fixation or

TABLE 25.8 General Characteristics of the Three Methods Used for Land Application of Sewage Effluent

Factor	Application Method		
	Low-Rate Irrigation	Overland Flow	High-Rate Infiltration
Main objectives	Reuse of nutrients and water, wastewater treatment	Wastewater treatment	Wastewater treatment, groundwater recharge
Soil permeability	Moderate (sandy to clay soils)	Slow (clay soils)	Rapid (sandy soils)
Need for vegetation	Required	Required	Optional
Loading rate	1.5–10 cm/week	5–14 cm/week	>50 cm/week
Application technique	Spray, surface	Usually spray	Surface flooding
Land required for flow of 10^6 liters/day	8–66 hectares	5–16 hectares	0.25–7 hectares
Needed depth to groundwater	About 2 cm	Undetermined	5m or more
BOD and suspended solid removal	90–99%	90–99%	90–99%
N removal	85–90%	70–90%	0–80%
P removal	80–90%	50–60%	75–90%

From Pepper et al. (2006b).

precipitation. The irrigation method is used primarily by small communities and requires large areas, generally on the order of 5–6 hectares per 1000 people.

In the overland flow method (Figure 25.21B), wastewater effluents are allowed to flow for a distance of 50–100 m along a 2–8% vegetated slope and are collected in a ditch. The loading rate of wastewater ranges from 5 to 14 cm a week. Only about 10% of the water percolates through the soil, compared with 60% that runs off into the ditch. The remainder is lost as evapotranspiration. This system requires clay soils with low permeability and infiltration.

High-rate infiltration treatment is also known as soil aquifer treatment (SAT) or rapid infiltration extraction (RIX) (Figure 25.21C). The primary objective of SAT is the treatment of wastewater at loading rates exceeding 50 cm per week. The treated water, most of which has percolated through coarse-textured soil, is used for groundwater recharge, or may be recovered for irrigation. This system requires less land than irrigation or overland flow methods. Drying periods are often necessary to aerate the soil system and avoid problems due to clogging. The selection of a site for land application is based on many factors including: soil types; drainable and depth; distance to groundwater; groundwater movement; slope; underground formations; and degree of isolation of the site from the public.

Inherent in land application of wastewater are the risks of transmission of enteric waterborne pathogens. The degree of risk is associated with the concentration of

pathogens in the wastewater and the degree of contact with humans. Land application of wastewater is usually considered an intentional form of reuse, and is regulated by most states. Because of limited water resources in the western United States, reuse is considered essential. Usually, stricter treatment and microbial standards must be met before land application. The highest degree of treatment is required when wastewater will be used for food crop irrigation, with lesser treatment for landscape irrigation or fiber crops. For example, the State of California requires no disinfection of wastewater for irrigation and no limits on coliform bacteria. However, if the reclaimed wastewater is used for surface irrigation of food crops and open landscaped areas, chemical coagulation (to precipitate suspended matter), followed by filtration and disinfection to reduce the coliform concentration to 2.2/100 ml, is required. In some cities excess effluent is disposed of in river beds that are normally dry. Such disposal can create riparian areas (Figure 25.22).

Because high-rate infiltration may be practiced to recharge aquifers, additional treatments of secondary wastewater may be required. However, as some removal of pathogens can be expected, the treatment requirement may be less. The degree of treatment needed may be influenced by the amount or time it takes the reclaimed water to travel from the infiltration site to the point of extraction, and the depth of the unsaturated zone. The greatest concern has been with the transport of viruses, which, because of their small size, have the greatest chance of traveling large distances within the subsurface.

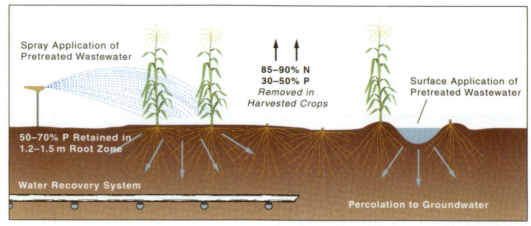

(A) Low-Rate Irrigation

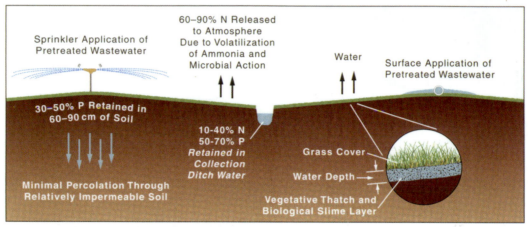

(B) Overland Flow

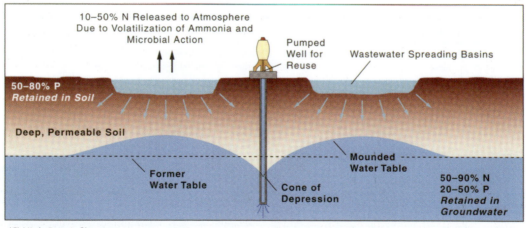

(C) High-Rate Infiltration

FIGURE 25.21   Three basic methods of land application of wastewater.

Factors that influence the transport of viruses are discussed in Chapter 15. Generally, several meters of moderately fine-textured, continuous soil layer are necessary for virus reductions of 99.9% or more (Yates, 1994).

25.6 WETLANDS SYSTEMS

Wetlands, which are typically less than 1 m in depth, are areas that support aquatic vegetation and foster the growth of emergent plants such as cattails, bulrushes, reeds, sedges and trees. They also provide important wetland habitat for many animal species. Recently, wetland areas have been receiving increasing attention as a means of additional treatment for secondary effluents. The vegetation provides surfaces for the attachment of bacteria, and aids in the filtration and removal of such wastewater contaminants as biological oxygen and excess carbon. Factors involved in the reduction of wastewater

FIGURE 25.22 Effluent outfall of the Roger Road Wastewater Treatment Plant in Tucson, Arizona. Here, extensive growth of vegetation due to the effluent produces a riparian habitat.

contaminants are shown in Table 25.9. Although both natural and constructed wetlands have been used for wastewater treatment, recent work has focused on constructed wetlands because of regulatory requirements. Two types of constructed wetland systems are in general use: (1) free water surface (FWS) systems; and (2) subsurface flow systems (SFS). An FWS wetland is similar to a natural marsh because the water surface is exposed to the atmosphere. Floating and submerged plants, such as those shown in Figure 25.23A, may be present. SFS consist of channels or trenches with relatively impermeable bottoms filled with sand or rock media to support emergent vegetation.

During wetland treatment, the wastewater is usable. It can, for instance, be used to grow aquatic plants such as water hyacinths (Figure 25.23B) and/or to raise fish for human consumption. The growth of such aquatic plants provides not only additional treatment for the water but also a food source for fish and other animals. Such aquaculture systems, however, tend to require a great deal of land area. Moreover, the health risk associated with the production of aquatic animals for human consumption in this manner must be better defined.

There has been increasing interest in the use of natural systems for the treatment of municipal wastewater as a form of tertiary treatment (Kadlec and Wallace, 2008). Artificial or constructed wetlands have a higher degree of biological activity than most ecosystems; thus transformation of pollutants into harmless by-products or

TABLE 25.9 Principal Removal and Transformation Mechanisms in Constructed Wetlands Involved in Contaminant Reduction

Constituent	Free Water System	Subsurface Flow	Floating Aquatics
Biodegradable organics	Bioconversion by aerobic, facultative, and anaerobic bacteria on plant and debris surfaces of soluble BOD, adsorption, filtration	Bioconversion by facultative and anaerobic bacteria on plant and debris surfaces	Bioconversion by aerobic, facultative, and anaerobic bacteria on plant and debris surfaces
Suspended solids	Sedimentation, filtration	Filtration, sedimentation	Sedimentation, filtration
Nitrogen	Nitrification/denitrification, plant uptake, volatilization	Nitrification/denitrification, plant uptake, volatilization	Nitrification/denitrification, plant uptake, volatilization
Phosphorus	Sedimentation, plant uptake	Filtration, sedimentation, plant uptake	Sedimentation, plant uptake
Heavy metals	Adsorption to plant and debris surfaces	Adsorption to plant roots and debris surfaces, sedimentation	Absorption by plants, sedimentation
Trace organics	Volatilization, adsorption, biodegradation	Adsorption, biodegradation	Volatilization, adsorption biodegradation
Pathogens	Natural decay, predation, UV irradiation, sedimentation, excretion of antimicrobials from roots of plants	Natural decay, predation, sedimentation, excretion of antimicrobials from roots of plants	Natural decay, predation, sedimentation

From Pepper et al. (2006b).

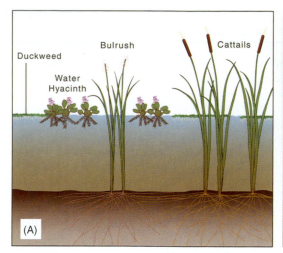

FIGURE 25.23 (A) Common aquatic plants used in constructed wetlands. (B) An artificial wetland system in San Diego, California, utilizing water hyacinths.

essential nutrients for plant growth can take place at a rate that is useful for the treatment of municipal wastewater (Case Study 25.1). Most artificial wetlands in the United States use reeds or bull rushes, although floating aquatic plants such as water hyacinths and duckweed have also been used. To reduce potential problems with flying insects, subsurface flow wetlands have also been built (Figure 25.25). In these types of wetlands all of the flow of the wastewater is below the surface of a gravel bed containing plants tolerant of water-saturated soils. Most of the existing information on the performance of these wetlands concerns coliform and fecal coliform bacteria. Kadlec and Wallace (2008) have summarized the existing literature on this topic. They point out that natural sources of indicators in treatment wetlands never reach zero because wetlands are open to wildlife. Reductions in fecal coliforms are generally greater than 99%, but there is a great deal of variation, probably depending on the season, type of wetland, numbers and type of wildlife and retention time in the wetland. Volume-based and area-based bacterial die-off models have been used to estimate bacterial die-off in surface flow wetlands (Kadlec and Wallace, 2008).

In one study of a mixed-species surface flow wetland with a detention time of approximately 4 days, several other types of microorganisms were examined. Results showed that *Cryptosporidium* was reduced by 53%, *Giardia* by 58% and enteric viruses by 98% (Karpiscak *et al.*, 1996).

25.7 SLUDGE PROCESSING

Primary, secondary and even tertiary sludges generated during wastewater treatment are a major by-product of the treatment process. These sludges, in turn, are usually subjected to a variety of treatments. Raw sludge is sometimes subjected to screening to remove coarse materials including grit that cannot be broken down biologically. Thickening is usually done to increase the solids content of the sludge. This can be achieved via centrifugation which increases the solids content to approximately 12%. Dewatering can further concentrate the solids content to 20−40%. This is normally achieved via filtration, or by the use of drying beds. Conditioning enhances the separation of solids from the liquid phase. This is usually accomplished by the addition of inorganic salts such as: alum; lime; ferrous or ferric salts; or synthetic organic polymers known as polyelectrolytes. All of these processes reduce the water content of the sludge, which ultimately reduces transportation costs to the final disposal and/or utilization site.

Finally, stabilization technologies are available, reducing both the solids content of the sludge and inactivating pathogenic microbes present in the sludge.

25.7.1 Stabilization Technologies

25.7.1.1 Aerobic Digestion

This consists of adding air or oxygen to sludge in a 4- to 8-foot-deep open tank. The oxygen concentration within the tank must be maintained above 1 mg/L to avoid the production of foul odors. The mean residence time in the tank is 12−60 days depending on the tank temperature. During this process, microbes aerobically degrade organic substrate, reducing the volatilize solids content of the sludge by 40−50% (U.S. EPA, 1992b). Digestion temperatures are frequently moderate or mesophilic (30−40°C). By increasing the oxygen content, thermophilic digestion can be induced (>60°C). By increasing

Case Study 25.1   Sweetwater Wetlands Infiltration—Extraction Facility in Tucson, Arizona

Tucson, Arizona, is located in the Sonoran Desert in the southwestern United States. Because of limited water supplies reclamation of wastewater is critical to meet water needs in the region. To meet these needs a system was built to provide tertiary effluents derived from an activated sludge/trickling filter system of sufficient quality to be used for landscape irrigation. The system is composed of several components that allows for various treatments and storage of tertiary effluent (Figures 25.22 and 25.23). A tertiary treatment plant filters the secondary effluent (to reduce turbidity and microorganisms) and then provides additional disinfection. The backwash from the filters is then discharged into an artificial wetland for treatment. When the water exists the wetland it is discharged into infiltration basins where it is further treated. In times of low reclaimed water demand (winter) the tertiary effluent may be discharged into the infiltration basins.

The subsurface aquifer is then used as a storage facility, the water then being pumped to the surface (extraction) when needed during periods of peak demand.

The multiple barriers of conventional and natural technologies are design to enhance the removal of chemical and microbial contaminates. Filtration of the secondary wastewater during tertiary treatment allows for reduction of the larger protozoan parasites (which are more resistant to disinfection than enteric bacteria and viruses) and more effective disinfection. In the wetlands protozoan parasites settle out and bacteria and viruses are reduced by inactivation by sunlight (UV light) and microbial antagonism. Infiltration of the water through the soil results in further removal of pathogens by filtration and adsorption to soil particle (especially viruses) (see Figure 25.24).

FIGURE 25.24 Aerial view of Sweetwater Reclamation Facility. Numbered blue areas are infiltration basins. Photo courtesy Water Reuse Association.

the temperature and the retention time, the degree of pathogen inactivation can be enhanced. Pathogen concentrations ultimately determine the treatment level of the product. Class B biosolids can contain many human pathogens (see Chapter 26). Class A biosolids, which result from more stringent and enhanced treatment, contain very low or nondetectable levels of pathogens. The degree of treatment, for Class A versus Class B, has important implications on the reuse potential of the material for land application (see Chapter 26). Aerobic digestion generally results in the production of Class B biosolids.

25.7.1.2 Anaerobic Digestion

This type of microbial digestion occurs under low redox conditions, with low oxygen concentrations. Carbon dioxide is the major terminal electron acceptor used (see Chapter 3), and results in the conversion of organic substrate to methane and carbon dioxide. This process reduces the volatile solids by 35—60% (Bitton, 2011), and results in the production of Class B biosolids. The advantages and disadvantages of anaerobic digestion relative to aerobic digestion are shown in Information Box 25.1.

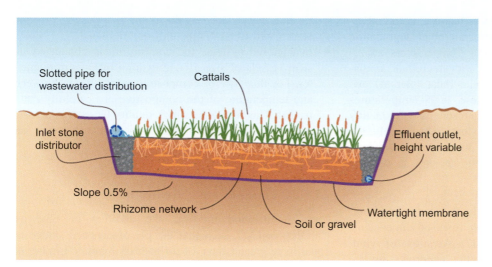

FIGURE 25.25 Cross-section of a subsurface wetland.

Information Box 25.1   Advantages and Disadvantages of Anaerobic Digestion

Advantages
- No oxygen requirement, which reduces cost
- Reduced mass of biosolids due to low energy yields of anaerobic metabolism (see also Chapter 3)
- Methane produced, which can be used to generate electricity
- Enhanced degradation of xenobiotic compounds

Disadvantages
- Slower than aerobic digestion
- More sensitive to toxics

Adapted from Bitton (2011).

FIGURE 25.26   Wood bulking agent for composting. The wood is shredded to increase the surface area of bulking agent for composting.

Information Box 25.2   Factors Affecting Efficient Composting

Temperature.Adequate aeration and moisture must be maintained to ensure temperatures reach 60°C, to inactivate microbial pathogens.
Aeration.Air must be provided via blowers or by turning.
Moisture.Conditions must be neither too moist, which promotes anaerobic activity, nor too dry, which limits microbial activity.
C:N ratio.The C:N ratio of the substrate should be maintained around 25:1, to ensure adequate but not excessive amounts of nitrogen for the microbes.
Surface area of bulking agent.Shredded material should be used to increase substrate surface area for microbial metabolism.

Source: Pepper et al. (2006a).

25.7.2 Sludge Processing to Produce Class A Biosolids

Class B biosolids that arise following digestion can be further treated to Class A levels prior to land application. The three most important technologies to achieve this goal are: composting; lime treatment; and heat treatment.

25.7.2.1 Composting

Composting consists of mixing sludge with a bulking agent that normally has a high C:N ratio (Figure 25.26). This is necessary because of the low C:N ratio of the sludge. The mixtures are normally kept moist but aerobic. These conditions result in very high microbial activity, and the generation of heat that increases the temperature of the composting material. Factors affecting the composting process are shown in Information Box 25.2. There are three main types of composting systems:

- The aerated static pile process typically consists of mixing dewatered digested sludge with wood chips. Aeration of the pile is normally provided by blowers during a 21-day composting period. During this active composting period, temperatures increase to the meso-philic range (20−40°C) where microbial degradation occurs via bacteria and fungi. Temperatures subsequently increase to 40−80°C, with microbial populations dominated by thermophilic (heat tolerant) and spore-forming organisms. These high temperatures inactivate pathogenic microorganisms, and frequently result in a Class A biosolid product. Subsequently, the compost is cured for at least 30 days, during which time temperatures within the pile decrease to ambient levels.

- The windrow process is similar to the static pile process except that instead of a pile, the sludge and bulking agent are laid out in long rows of dimensions: 2 m × 3 m × 80 m (Figure 25.27). Aeration for windrows is provided by turning the windrows several

FIGURE 25.27 Biosolid composting via the windrow process. Here three windrows are illustrated.

times a week. Once again, if the composting process is efficient, Class A biosolids are produced.

- In enclosed systems the composting is conducted in steel vessels of size 10−15 m high by 3−4 m diameter. For this type of composting, aeration via blowers and temperature of the composting are carefully controlled. This results in a high quality Class A compost, with little or no odor problems. However, costs of enclosed systems are higher.

25.7.2.2 Lime and Heat Treatment

Lime stabilization involves the addition of lime as $Ca(OH)_2$ or CaO, such that the pH of digested sludge is equal to or greater than 12 for at least 2 hours. Liming is very effective at inactivating bacterial and viral pathogens, but less so for parasites (Bitton, 2011). Lime stabilization also reduces odors, and can result in a Class A biosolid product.

Heat treatment involves heating sludge under pressure to temperatures up to 260°C for 30 minutes. This process kills microbial pathogens and parasites, and also further dewaters the sludge.

25.7.2.3 The Cambi Thermal Hydrolysis Process

The Cambi process utilizes thermal hydrolysis as a pretreatment to anaerobic digestion. This increases the

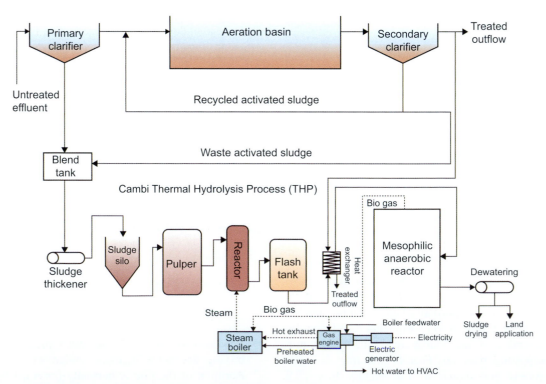

FIGURE 25.28 Cambi thermal hydrolysis process (THP).

FIGURE 25.29 Cambi thermal hydrolysis—preferred technology for Cotton Valley and Whitlingham. © Mott MacDonald Group Limited 2013.

microbial degradation of organic volatile solids and increases the amount of biogas obtained. This process also facilitates a higher degree of separation of solid and liquid phase after digestion. The process is depicted in Figure 25.28. Sludge generated during primary and secondary treatment is dewatered to approximately 15–20% dry solids content, preheated to 100% in the pulper tank, then heated to 150 to 170°C under 8–9 bar pressure in the reactor. In the flash tank the sludge cools to about 100°C and the released steam is recirculated. Further cooling to 35°C occurs via the heat exchangers where more energy is recycled via the production of hot water. Following this, the sludge is subjected to mesophilic anaerobic digestion, leading to the production of biogas and Class A biosolids, since no pathogens can survive the steam treatment. A typical plant is shown in Figure 25.29. Cambi is now well established in several countries within Europe, particularly the United Kingdom and Norway. The first thermal hydrolysis plant to be built in the U.S. is the Blue Plains treatment plant in Washington DC, with startup in 2014. This plant will treat up to 450 dry tons of sludge per day, and the biogas produced will cover the entire steam needs of the plant saving $20 million a year from the energy saved.

Overall, 50% of all biosolids is land applied in the U.S. with most of it currently being Class B (see Chapter 26).

QUESTIONS AND PROBLEMS

1. What are the three major steps in modern wastewater treatment?
2. Why is it important to reduce the amount of biodegradable organic matter and nutrients during sewage treatment?
3. When would tertiary treatment of wastewater be necessary?
4. What are some types of tertiary treatment?
5. What are the processes involved in the removal of heavy metals from wastewater during treatment by artificial wetlands.
6. What are the three types of land application of wastewater? Which one is most likely to contaminate the groundwater with enteric viruses? Why? What factors determine how far viruses will be transported in groundwater? How does nitrogen removal occur? Phosphorus removal?
7. What is the major contaminant in groundwater associated with the use of on-site treatment systems?
8. What factors may determine the concentration of enteric pathogens in domestic raw sewage?
9. Five milliliters of a wastewater sample is added to dilution water in a 300-ml BOD bottle. If the following results are obtained, what is the BOD after 3 days and 5 days?

Time (days)	Dissolved oxygen (mg/L)
0	9.55
1	4.57
2	4.00
3	3.20
4	2.60
5	2.40
6	2.10

10. List some advantages and disadvantages of the wetland treatment of sewage.
11. What is the major mechanism of pathogen removal during activated sludge treatment?
12. What treatment process would you need to obtain an 8-log_{10} reduction of (i) enteric viruses from raw sewage, and (ii) *Giardia*?
13. How effective do you think sunlight is in killing *Cryptosporidium*? Enteric viruses?

REFERENCES AND RECOMMENDED READING

APHA (1998) "Standard Methods for Water and Wastewater," American Public Health Association, Washington, DC.

Bitton, G. (2011) "Wastewater Microbiology," 4th ed. Wiley-Liss, New York.

Chang, I. S., LeClech, P., Jefferson, B., and Judd, S. (2002) Membrane fouling in membrane bioreactors for wastewater treatment. *J. Environ. Eng.* **128**, 1018–1029.

Curds, C. R., and Hawkes, H. A. (1983) Ecological aspects of used water treatment, In "The Processes and their Ecology, **vol 3. Academic Press, London, 1-113.**

Edzwald, J. K. (2010) Dissolved air flotation and me. *Wat. Res.* **44**, 2077–2106.

Kadlec, R. H., and Wallace, S. (2008) "Treatment Wetlands," CRC Press, Boca Raton, FL.

Karpiscak, M. M., Gerba, C. P., Watt, P. M., Foster, K. E., and Falabi, J. A. (1996) Multi-species plant systems for wastewater quality improvements and habitat enhancement. *Wat. Sci. Technol.* **33**, 231–236.

Leong, L. Y. C. (1983) Removal and inactivation of viruses by treatment processes for portable water and wastewater—a review. *Wat. Sci. Technol.* **15**, 91–114.

Metcalf and Eddy, Inc. (2003) "Wastewater Engineering," McGraw-Hill, New York.

Neilson, J. W., Josephson, K. L., Pepper, I. L., Arnold, R. B., DiGiovanni, G. D., and Sinclair, N. A. (1994) Frequency of horizontal gene transfer of a large catabolic plasmid (pJP4) in soil. *Appl. Environ. Microbiol.* **60**, 4053–4058.

Newby, D. T., Gentry, T. J., and Pepper, I. L. (2000b) Comparison of 2,4-dichlorophenoxyacetic acid degradation and plasmid transfer in soil resulting from bioaugmentation with two different pJP4 donors. *Appl. Environ. Microbiol.* **66**, 3399–3407.

Park, S., Seon, J., Byun, I., Cho, S., Park, T., and Lee, T. (2010) Comparison of nitrogen removal and microbial distribution in wastewater treatment process under different electron donor conditions. *Bioresour. Technol.* **101**, 2988–2995.

Pepper, I. L., Brooks, J. P., and Gerba, C. P. (2006a) Pathogens in Biosolids. *Adv. Agron.* **90**, 1–41.

Pepper, I. L., Gerba, C. P., and Brusseau, M. L. (2006b) "Environmental and Pollution Science," 2nd ed. Academic Press, San Diego, CA.

Rao, V. C., Metcalf, T. G., and Melnick, J. L. (1986) Removal of pathogens during wastewater treatment. In "Biotechnology" (H. J. Rehm, and G. Reed, eds.), vol. 8, VCH, Berlin, pp. 531–554.

Rose, J. B., and Carnahan, R. P. (1992) "Pathogen removal by full scale wastewater treatment," Report to Florida Department of Environmental Regulation, Tallahassee, Fl.

Rose, J. B., and Gerba, C. P. (1991) Use of risk assessment for development of microbial standards. *Wat. Sci. Technol.* **24**, 29–34.

Rusin, P., Enriquez, C. E., Johnson, D., and Gerba, C. P. (2000) Environmentally transmitted pathogens. In "Environmental Microbiology" (R. M. Maier, I. L. Pepper, and C. P. Gerba, eds.), Academic Press, San Diego, pp. 447–489.

U.S. Census Bureau, 1990. Historical Census of Housing Tables: Sewage Disposal. Available from <http://www.census.gov/hhes/www/housing/census/historic/sewage.html>.

U.S. EPA (U.S. Environmental Protection Agency) (1977) Wastewater treatment facilities for sewered small communities. EPA−62511−77−009, Washington, DC.

U.S. EPA (U.S. Environmental Protection Agency) (1992a) Guidelines for water reuse. EPA/625/R−92/004, Washington, DC.

U.S. EPA (U.S. Environmental Protection Agency) (1992b) Technical support document for land application of sewage sludge, vol. I EPA 822/R−93−001A.

U.S. EPA (U.S. Environmental Protection Agency) (1994) A plain English guide to the EPA part 503 biosolids rule. EPA 832/R−93/003 U.S. Environmental Protection Agency.

U.S. EPA (U.S. Environmental Protection Agency) (1996) Technical support document for the round two sewage sludge pollutants. EPA−822−R−96−003.

U.S. EPA (U.S. Environmental Protection Agency) (1997) Exposure factors handbook, EPA/600/P−95/002, August 1997.

U.S. EPA (U.S. Environmental Protection Agency) (1998) Volume I— General factors, exposure factors handbook. Update to exposure factors handbook, EPA/600/8−89/043, May 1989.

U.S. EPA (U.S. Environmental Protection Agency) (1999a) Environmental regulations and technology: control of pathogens and vector attraction in sewage sludge. EPA/625/R−92/013. Available online: <http://www.epa.gov/ttbnrmrl/625/R-92/013.htm>.

U.S. EPA (U.S. Environmental Protection Agency) (1999b) "Environmental regulation and technology. Control of pathogens and vector attraction in sewage sludge EPA/652/R−92/013. Revised 1999," Office of Research and Development, U.S. Environmental Protection Agency, Washington, DC, 177 pp.

Wang, Y-P., Liu, X-W., Li, W-W., Wang, Y-K., Shery, G. P., Zeng, R. J., and Yu, H-Q. (2012) A microbial fuel cell-membrane bioreactor integrated system for cost-effective wastewater treatment. *Appl. Energy* **98**, 230–235.

Yates, M. V. (1994) Monitoring concerns and procedures for human health effects. In "Wastewater Reuse for Golf Course Irrigation," CRC Press, Boca Raton, FL, 143–171.

Chapter 26

Land Application of Organic Residuals: Municipal Biosolids and Animal Manures

John P. Brooks, Charles P. Gerba and Ian L. Pepper

26.1 INTRODUCTION TO ORGANIC RESIDUALS

The term "organic residuals" includes several different waste categories. Among them are the organic fraction of municipal solid waste, animal wastes or manure, and municipal biosolids that comprise the organic solids remaining after sewage treatment. In the United States, approximately 450,000 animal feeding operations (AFOs), some of which are concentrated animal feeding operations (CAFOs), collectively produce over 100 million dry tons of manure per year (Burkholder *et al.*, 2007). In contrast, approximately 16,500 municipal wastewater treatment plants operating in the U.S. produce a relatively small 7.2 million dry tons annually (NEBRA, 2007). Of these, the largest $\cong 3300$ generate more than 92% of the total quantity of biosolids in the U.S. (NEBRA, 2007). Table 26.1 shows the amount of waste produced for each respective animal production industry. Conversely, an average 68 kg human produces approximately 37 kg of waste per year and 6.5 million

dry tons per year for all municipalities combined. Both types of residuals are used beneficially for crop production through land application. Overall, animal manures are applied to about 10% of available agricultural land with greater than 90% of the total available animal manures being land applied. In contrast, only 0.1% of available agricultural land is spread with biosolids, accounting for 55% of the available biosolids (NEBRA, 2007; Brooks *et al.*, 2011). Though biosolids represent only a small fraction of total organic residuals produced, they are the most processed, most regulated, most studied and most controversial, with respect to disposal and beneficial reuse. In contrast, raw animal manures are not treated and are not regulated. In fact, certified organic farmers can utilize animal manures as a fertilizer and soil amendment, provided crops grown for human consumption are harvested at least 90 days after the last application (Organic Trade Association, 2012). The objective of this chapter is to compare and contrast microbial aspects of land application of municipal biosolids and animal manures.

I.L. Pepper, C.P. Gerba, T.J. Gentry: Environmental Microbiology, Third edition. DOI: http://dx.doi.org/10.1016/B978-0-12-394626-3.00026-0
© 2015 Elsevier Inc. All rights reserved.

TABLE 26.1 Annual Amount of Waste Residual Produced per Industry (USEPA, 2004)

	Per Animal (1000 pounds live weight) Kg yr^{-1}	Total for an Individual CAFO (1000 animal units) Ton yr^{-1}	Total for the Industry Ton yr^{-1}
Cattle	9525	10,500	8.1×10^8
Dairy	13,607	15,000	1.9×10^8
Poultry	30,000	33,000	1.2×10^8
Swine	13,200	14,500	1.8×10^8

26.2 LAND APPLICATION OF BIOSOLIDS AND ANIMAL WASTES: A HISTORICAL PERSPECTIVE AND CURRENT OUTLOOK

Use of animal wastes and manures as a fertilizer source for agricultural crop production has been practiced since the days of the Roman Empire. During the twentieth century in both the United States and Europe, operations on small agricultural farms frequently consisted in both crop and animal production. Consequently, animal wastes were naturally land applied to enhance crop production. Although fossil fuel-based fertilizers replaced much of the use of manures following World War II, the practice continues today worldwide. The rise of locally grown, "organic" fresh produce and the increased costs associated with fuel-based fertilizers have renewed interest in manure as a fertilizer. Ten years ago, manure was considered to be a waste by-product of the animal production industry; however, it is now considered a commodity.

In the United States, land application of municipal wastewater and biosolids has been practiced for its beneficial effects and for disposal purposes since the advent of modern wastewater treatment about 160 years ago. In England in the 1850s, "sewage farms" were established to dispose of untreated sewage. By 1875, about 50 farms were utilizing land treatment in England, and many others close to other major cities in Europe. In the United States, sewage farms were established by about 1900. At this same time, primary sedimentation and secondary biological treatment was introduced as a rudimentary form of wastewater treatment, and land application of "sludges" began. It is interesting to note that prior to wastewater treatment, "sludge" *per se* did not exist. Municipal sludge in Ohio was used as a fertilizer as early as 1907.

Since the early 1970s, more emphasis has been placed on applying sludge to cropland at rates to supply adequate nutrients for crop growth (Hinesly *et al.*, 1972). In the 1970s and 1980s, many studies were undertaken to investigate the potential benefits and hazards of land application, in both the U.S. and Europe. Ultimately in 1993, U.S. federal regulations were established via the

Information Box 26.1   Definitions of Sewage Sludge and Biosolids

Sewage sludge. The solid, semisolid, or liquid residue generated during the treatment of domestic sewage in a treatment works.
Biosolids. Two different definitions have been developed:

EPA: The primarily organic solid product yielded by municipal wastewater treatment processes that can be beneficially recycled (whether or not they are currently being recycled).
National Research Council (2002): Sewage sludge that has been treated to meet the land-application standards in the Part 503 rule or any other equivalent land-application standards or practices.

"Part 503 Sludge Rule." This document—"The Standards for the Use and Disposal of Sewage Sludge" (U.S. EPA, 1993) was designed to "adequately protect human health and the environment from any reasonably anticipated adverse effect of pollutants." As part of these regulations, two classes of treatment were defined as "Class A and Class B" biosolids, with different restrictions for land applications, based on the level of treatment. The term "biosolids" was coined in the 1990s by a University of Arizona faculty member. The distinction between sewage sludge and biosolids is described in Information Box 26.1, and it is important to note that the term biosolids implies treatment to defined levels. The requirements for Class A versus Class B biosolids are defined in Information Box 26.2.

Land application increased when restrictions were placed on "ocean dumping." By the year 2000, 60% of all biosolids were land applied in the U.S. Currently, most land application of biosolids in the U.S. utilizes Class B biosolids. However, due to public concern over potential hazards, in some areas of the U.S., land application of Class B biosolids has been banned. Thus, by 2004, only 55% of all biosolids was applied to soil for agronomic, silvicultural and/or land restoration processes. The remaining 45% was disposed of in municipal landfills or incinerated (NEBRA, 2007), with about two-thirds of the non-land-applied material being landfilled. Of the total applied to soils, 74% was on farmland for agricultural purposes (NEBRA, 2007). A recent report indicates that approximately 200 million farmers worldwide grow crops in fields fertilized with human waste (IWMI, 2010).

Contrary to municipal wastewater treatment sludge, CAFOs and their manures are a relatively new advancement in egg/dairy/meat production systems. Though manure has been around since the "dawn of time," the idea of a CAFO has not. An AFO is defined as a feedlot or a facility where animals are kept for greater than 45 days; cattle grazing on pasture are exempt (U.S. EPA, 2004). CAFOs are then designated based on numerical criteria such as greater than 300 cattle or 9000 broiler chickens (U.S. EPA, 2004). Thus, all CAFOs are AFOs, but not all AFOs are CAFOs. Since the 1960s, the vast majority of animals raised for food and their products have been produced in CAFOs. This movement has led to the concentration of most food animals into less than 20% of all AFOs. For all AFO and CAFO food animal production, it has always been the responsibility of the owner to dispose of the manure, with reliance on disposal to nearby fields, thereby keeping costs low. The vast majority of AFO owners apply manure to owned lands, or rely on the sale or "giving away" of manure to other landowners. Manure land application has not been governed by any specific law or federal regulation; however, guidelines exist for suggested rates of land application manuring based on nutrient requirements, typically nitrogen or phosphorus, of the crop to be grown. Most states require nutrient management plans to be established prior to the establishment of a new CAFO. These plans essentially establish how the CAFO owner will dispose of the manure in both a quantitative (i.e., how much) and qualitative (i.e., which crop) manner.

Thus far, the level of scrutiny reserved for biosolids land application has not been similarly applied to manure. Anecdotally, the public has regarded manure as a "natural" material, and thus it has escaped intense criticism despite knowledge of pathogens, antibiotic resistant bacteria and nutrient runoff concerns that are all associated with land application of manures. In 1972, the Clean Water Act identified AFOs as potential pollutant sources, resulting in CAFO regulations being set in place in 1976. Increase in CAFO size necessitated revision of the regulations in 2008. Thus, the current U.S. EPA CAFO rule requires that CAFOs that discharge or propose to discharge waste need to apply for a permit, along with the establishment of a nutrient management plan. The rule was challenged, and now an AFO can apply for an exception provided that the manure can be appropriately stored to prevent accidental release (i.e., runoff contaminated with manure) during a 24-hour, 100-year storm event. These limits and rules are specifically designed to reduce discharge to surface water, and do not govern the land application of manure to soil, unless there is a threat of effluent runoff to surface water. Guidelines for manure land application also suggest harvest delays when using manure on "organic" marketed food crops. These guidelines suggest a delay of 90 to 120 days between land application and harvest, depending on the level of interaction between the manure/soil matrix and the food crop edible parts. Apart from these rules and guidelines, no other governing document regulates land application of manure.

Biosolids and manure are applied to agricultural and nonagricultural lands as a soil amendment because they improve the chemical and physical properties of soils, and because they contain nutrients for plant growth. Land application on agricultural land is utilized to grow food crops such as corn or wheat, and nonfood crops such as cotton. Nonagricultural land application includes forests, rangelands, public parks, golf courses and cemeteries. Biosolids and manure are also used to revegetate severely disturbed lands such as mine tailings or strip mine areas (Case Study 26.1).

26.2.1 Class A Versus Class B Biosolids

Biosolids are divided into two classes on the basis of pathogen content: Class A and Class B (Information Box 26.2). Class A biosolids are treated to reduce the presence of pathogens to below detectable levels, and can be used without any pathogen-related restrictions at the application site. Class A biosolids can also be bagged and sold to the public. Class B biosolids are also treated to reduce pathogens, but still contain detectable levels of them. Class B biosolids have site restrictions to minimize the potential for human exposure, until environmental factors such as heat, sunlight or desiccation have further reduced pathogen numbers. Class B biosolids cannot be sold or given away in bags or other containers or used at sites used by the public.

Information Box 26.2 Part 503 Pathogen Density Limits for Class A and B Biosolids

Standard Density Limits (Dry Weight)
Pathogen or Indicator
Class A
  Salmonella <3 MPN/4 g total solids or
  Fecal coliforms <1000 MPN/g and
  Enteric viruses <1 PFU/4 g total solids and
  Viable helminth ova <1/4 g total solids
Class B
  Fecal coliform density <2,000,000 MPN/g total
  solids

Adapted from U.S. EPA (2000).

Case Study 26.1   Reclamation and Revegetation of Mine Tailings using Biosolid Amendment

Mine tailings are formed in two ways: by the initial removal of vegetation, soil, and bedrock to expose the valuable copper containing ores, and then by the disposal of the crushed rock after the ore has been removed. Typically these tailings are 30−40 meters deep. Mine tailings are not the ideal medium on which to grow plants. The crushed rock consists of large and small fragments with large spaces in between them. In addition, there is no organic material; the cation exchange capacity (CEC) is very low; the water holding capacity of the material is poor to nonexistent; and there are few macronutrients (NPK) available for the plants. Soil biota, in the form of bacteria and fungi, are only present in low numbers.

The overall objective of this study was to evaluate the efficacy of dried biosolids as a mine tailing amendment to enhance site stabilization and revegetation. Mine tailing sites were established at ASARCO Mission Mine close to Sahuarita Arizona. Site 1 (December 1998) was amended with 248 tons ha^{-1} of Class A biosolids. Site 2 (December 2000) and Site 3 (April 2006) were amended with 371 tons ha^{-1} and 270 tons ha^{-1} respectively. Site D, a neighboring native desert soil acted as a control for evaluation of soil microbial characteristics. Surface amendment of Class A biosolids showed a 4 log$_{10}$ increase in HPCs compared to unamended tailings, with the increase being maintained for a >10-year period. Microbial activities such as nitrification, sulfur oxidation and dehydrogenase activity were also sustained throughout the study period. Finally, note that extensive revegetation of the sites occurred (Figures 26.1−26.3). 16S rRNA clone libraries obtained from community DNA suggest that mine tailings amended with biosolids achieve diversity and bacterial populations similar to native soil bacterial phyla, ten years post-application (see Case Study 19.2). Thus addition of Class A biosolids to copper mine tailings in the desert southwest increased soil microbial numbers, activity and diversity relative to unamended mine tailings. Overall, the addition of biosolids resulted in a functional soil with respect to microbial characteristics which were sustainable over a ten year period.

FIGURE 26.1   Mine tailings prior to biosolids amendment.

FIGURE 26.2   Mine tailings 2 years after biosolids application.

26.2.2 Methods of Land Application of Organic Residuals

26.2.2.1 Land Application of Biosolids

The method of land application of biosolids essentially depends on the percent solids contained within them, which determines whether the biosolids are liquid in nature, or a "cake" (Information Box 26.3). Figures 5.12−5.14 all illustrate methods of land application of biosolids and can be summarized as:

- Injection. Liquid biosolids are injected to a soil depth of 30 cm. Injection vehicles simultaneously disc the

field. Injection processes reduce odors and bioaerosols, as well as the risk of runoff to surface waters.
- Surface application. Liquid or cake biosolids are surface applied and subsequently tilled into the soil.
- Slingers are also utilized to throw the material through the air as a means of land application.

26.2.2.2 Land Application of Animal Manures

Land application techniques for manure are far more varied than for land application of municipal biosolids. This is due, in part, to the variability associated with the various AFOs, which can include manure from cattle,

FIGURE 26.3   Mine tailings 3 years after biosolids application.

FIGURE 26.4   Liquid manure injection: coulters cut a path in the pasture with injectors applying swine liquid manure effluent below the surface in a furrow. Photo courtesy J.P. Brooks.

Information Box 26.3   Land Application Methods

% solids	Nature of biosolids	Method of application
8	Liquid	Spray application (Figure 5.12)
2	Liquid	Sprinkler system (Figure 5.13)
>20	Cake	Spreaders or slingers (Figure 5.14)

FIGURE 26.5   Subsurface banding (dark section in soil probe at 2.5" mark) with banding applicator (inset). Photo courtesy H. Tewolde.

dairy, poultry and swine industries. In addition, there are smaller operations specializing in niche foods such as ostrich, lamb and bison, which contribute to AFO manure burden. Even within a large industry such as poultry, variability is considerable, given the specific subindustries such as egg production and turkey farms. Egg layer farms typically produce liquid manure, whereas turkey producers produce a litter/fecal matter solid mixture. Each AFO produces a different kind of manure, making the standardization afforded by the USEPA Part 503 Rule a milestone which is difficult to reach. Typically prior to land application, most AFOs store manure in a shed or lagoon for a period of time, depending on season and demand. Storage type depends on solid content, with shed storage reserved for solid manure, while lagoons are utilized for liquid or slurry manure. Composting or "unofficial composting" (typically consists of long-term manure storage, without temperature monitoring or any other handling) are common pretreatments prior to land application. Likewise, in lieu of land application, some poultry producers opt to combust or burn their litter for energy production.

Manure can vary in physical characteristics from a slurry (<2% solids) to a cake (>50% solids); thus, land application can be quite varied. The microbiology

associated with each manure type can also vary dramatically based on production, storage and animal management. Pasture, cotton, food crops and forage are the major crops utilizing manure. Typically, manure land application is conducted as close as possible to the producing AFO. However, given the increasing costs of traditional fertilizers, it is now easier to justify transport of manure across greater distances. As with biosolids, manure handling is based on solid content. Figures 26.4–26.6 all illustrate methods of land application of animal manures, which can be summarized as:

- Injection. Liquid manure can be injected into soil, particularly useful for row crops.
- Subsurface banding. Dried or caked poultry litter can be banded into soil. The band slowly disperses nutrients over time, which is particularly useful for row crops.

- **Slinging.** Dry or solid manure can be surface spread onto pasture, hay or row crops.
- **Center pivot or reel-gun irrigation.** Low solids liquid manure can be surface spread to pasture or hay lands.
- **Surface deposition.** Manure/feces are deposited on pasture lands during typical grazing periods.

26.3 POTENTIAL MICROBIAL HAZARDS ASSOCIATED WITH CLASS B BIOSOLIDS, ANIMAL MANURES AND LAND APPLICATION

Both Class B biosolids and animal manures are known to contain pathogens including bacteria and protozoan parasites. Biosolids (but typically not animal manures) also contain human pathogenic viruses. Over the past 10–15 years

FIGURE 26.6 Liquid manure application using a center pivot system and a reel gun (inset). Photo courtesy M.R. McLaughlin.

a variety of potential microbial hazards associated with Class B biosolids and to a lesser extent animal manures have been identified (Case Study 26.2). Many of these issues involved the potential for infection from pathogens associated with organic residues. These pathogens of concern are described in Section 26.4 and the risks associated with such pathogens are described in Section 26.5. However, in addition to pathogens, other potential hazards have centered on biological but nonpathogenic issues such as antibiotic resistant bacteria, endotoxin and prions.

Although manure is known to contain bacterial and parasitic pathogens, there have only been a few instances where human viral pathogens have been found in manure, including hepatitis E virus in pigs and norovirus in cattle. Normally, viral pathogens are exclusive to municipal wastes. Given that manure is generally not treated, bacterial counts tend to be greater in manures than in municipal-treated biosolids. In addition to pathogens, manure is known to contain high levels of antibiotic-resistant pathogenic and commensal (i.e., normal, nonpathogenic) bacteria.

26.3.1 Antibiotic-resistant Bacteria

A major area of concern with the general public has focused on the potential for antibiotic-resistant bacteria that reside in both animal manures and biosolids, due to the potential for subsequent transfer of the resistance to pathogens. Bacteria are prokaryotic organisms with the ability to metabolize and replicate very quickly. They are also very adaptable genetically. When confronted with an antibiotic, if there is even one bacterial cell with a genetic or mutational change that confers resistance to that antibiotic, it will subsequently allow for the proliferation of antibiotic-resistant bacteria. Thus, the more that antibiotics are used,

---

Case Study 26.2   The University of Arizona, National Science Foundation Studies on Biosolids and Land Application

During the period 1999 through 2014 the University of Arizona was funded by the National Science Foundation to conduct studies on "Water Quality" and "Water and Environmental Technology." One of the focal areas of research included studies on potential hazards associated with Class B biosolids and land application of such material. Table 26.2 highlights some of the studies undertaken. Many of the topics evaluated were highly controversial with the general public because of the concern for potential microbial infections and/or illness due to pathogens found within Class B biosoilds. Particularly worrisome for the public was the potential for infections to occur within communities offsite, following transport of pathogens as bioaerosols, or transport via leaching into underground aquifers. However, research studies on both issues showed very limited transport by

either route. Hence, exposure via such mechanisms was low, and subsequently, risks to human health were also low. Additional information on the bioaerosol studies are in Section 5.6.5. Other biological concerns included the potential for disease from endotoxin, *S. aureus* and infectious proteins known as prions. However, it was shown that neither *S. aureus* nor prions survived wastewater treatment. Finally, note that regrowth of *Salmonella* did not occur following land application, indicating that site restrictions following land application would allow for inactivation of biosolid-associated pathogens without secondary regrowth. Overall, the studies indicate that the risk of adverse health effects from biosolid amended soil is low. However, diligence is necessary and the fate and transport of all emerging microbial contaminants still need to be evaluated.

TABLE 26.2 Microbial Issues Associated with Land Application of Biosolids

Issue	Concern	Outcome of Study	References
Occurrence of *Staphylococcus aureus* in biosolids	Community infections from *S. aureus* associated with biosolids amended soil	*S. aureus* does not survive wastewater treatment	Rusin *et al.*, 2003
Aerosolized bacteria and virus	Offsite community infections from bioaerosols	Limited microbial transport via bioaerosols and negligible risk to offsite communities	Brooks *et al.*, 2005a, b; Tanner *et al.*, 2005
Endotoxin	Community exposure to aerosolized endotoxin due to endotoxin associated with biosolids	Most aerosolized endotoxin derived from soil	Brooks *et al.*, 2006; Brooks *et al.*, 2007
Groundwater contamination	Groundwater contamination with viruses following transport through soil and vadose zone	Very limited transport of viruses because virus sorb to biosolids	Chetochine *et al.*, 2006
Antibiotic-resistant bacteria (ARB)	Presence of antibiotics in biosolids will increase the numbers of ARB in soil subsequent transfer of resistance to pathogens	No increase in soil ARB	Brooks *et al.*, 2007
Salmonella	*Salmonella* regrowth in biosolids and soil following land application	*Salmonella* only regrows in Class A biosolids under saturated conditions. No regrowth in amended soil	Zaleski *et al.*, 2005a
Prions	Prion infection of animals and humans following land application of prions	Prions do not survive wastewater treatment	Miles *et al.*, 2013

TABLE 26.3 Antibiotic-resistant Bacteria, as a Percentage of Total Cultured Heterotrophic Plate Count Bacteria in Environmental and Food Samples

Sample	Antibiotic Resistant (%)			
	Ampicillin[a]	Cephalothin[a]	Ciprofloxacin[a]	Tetracycline[a]
Biosolids	4.3	21.2	1.8	1.9
Composted manure	0.0	0.3	0.0	0.3
Compost	9.7	21.8	3.4	1.2
Fresh manure	0.2	0.7	1.1	0.3
Pristine soil	8.1	10.1	3.1	2.4
Dust	4.9	7.8	8.3	11.2
Ground water	60.3	41.2	22.9	21.0
Raw chicken	47.1	60.3	0.0	0.0
Raw ground beef	16.3	8.7	2.0	3.9
Head lettuce	29.9	35.8	1.5	4.5
Shredded lettuce	14.9	10.5	0.0	0.3
Tomato	0.6	20.6	0.2	0.3

Modified from Brooks et al. *(2007).*
[a]*Ampicillin (32 μg ml^{-1}), cephalothin (32 μg ml^{-1}), ciprofloxacin (4 μg ml^{-1}) and tetracycline (16 μg ml^{-1}).*

the greater the likelihood of antibiotic-resistant strains developing. The greatest concern with antibiotic resistance is the potential for human pathogenic strains to become resistant to overused antibiotics, such that the antibiotic can no longer contain the infectious agent. As is typical in most niches, commensal bacteria tend to dominate the pathogenic bacteria at levels which are orders of magnitude greater than those pertaining to the pathogens. This creates a haven for antibiotic-resistant genes, which all have the potential to transfer to true or opportunistic pathogens. The widespread,

sometimes indiscriminate, use of antibiotics has raised the questions: (1) Can antibiotic resistant genes be transferred from nonpathogenic bacteria to human pathogenic strains in the environment? (2) Can antibiotic resistance in the environment, via residual land application, be transferred to the public?

Brooks *et al.* (2006) evaluated the incidence of antibiotic-resistant bacteria (ARBs) in biosolids and a variety of other environmental samples and foodstuffs. Table 26.3 shows that Class B biosolids did not contain unusually high numbers of ARBs, and that in fact, the relative incidence was less than that found in pristine soil. Interestingly, ARB concentrations were also lower than those found in common foodstuffs such as lettuce. Therefore, food itself could be an important route of exposure to ARBs. Rates of gene transfer in soil are thought to be a relatively infrequent event without selective pressure (Neilson *et al.*, 1994), which reduces the risk of antibiotic-resistant gene transfer to human pathogenic bacteria. Finally, note that soil itself is the original source of human antibiotics.

Antibiotic use in the livestock and poultry industries has gradually increased over the past three decades in direct relation to the increasing number of CAFOs in operation. Throughout this gradual cultural shift in livestock production, the need for antibiotics has increased as stocking densities and production cycles have increased. The Union of Concerned Scientists predicted the amount of antibiotics used in the industries at up to 50 million pounds annually (Chee-Sanford *et al.*, 2009), with nearly half being used as a means to increase production. The Animal Health Institute refutes this total, stating that approximately 20.5 million pounds of antibiotic are used annually with approximately one-tenth thereof used to increase production (Chee-Sanford *et al.*, 2009). These discrepancies highlight how little is known regarding this topic, and how contentious these issues truly are, particularly with news cycles reporting increasing antibiotic resistance in our food supply or higher incidences of nosocomial infections. Regardless, livestock industries account for a large amount of antibiotic use in the United States. Antibiotics are used: (1) to treat infections and to prevent diseases; and (2) as a prophylactic, thus increasing production. It is with the latter that most concern or blame is associated.

In either case, as opposed to human antibiotic use, treating livestock with antibiotics is conducted in a manner that promotes the treatment of nondiseased animals. Typically, CAFO animals are not individually treated for a disease. If there is an outbreak of a disease-causing pathogen, farm managers typically react by not treating just the diseased individuals (perhaps only 100 of 20,000), but by treating the entire flock or herd. This increases the likelihood for antibiotic resistance, as resistance genes can be promoted in healthy as well as diseased members of the host population.

Brooks and McLaughlin (2009) and Brooks *et al.* (2010) described the presence of antibiotic-resistant bacteria in swine and poultry CAFOs. The presence of antibiotic-resistant bacteria in swine CAFOs appeared to be influenced by the type of management employed by the producer; specifically, the presence of younger piglets increased the amount of resistance in commensal *E. coli*. In general, younger piglets led to resistance to an extra class of antibiotics (Brooks and McLaughlin, 2009). In

some instances, regulatory and media pressures have forced industries to reduce antibiotic use, as has been noted in the poultry industry. Brooks *et al.* (2010) noted the overall lack of antibiotic resistance in poultry CAFO manure, and an overall decrease among staphylococci, enterococci and *E. coli* in comparison to previous studies (Brooks *et al.*, 2009).

Ultimately, the concern is for the potential movement of antibiotic-resistant bacteria and genes from the "farm to the plate." Movement from the farm to the product and ultimately the consumer remains a poorly understood area (Marshall and Levy, 2011). Three potential routes exist for the transfer to occur: (1) via consumption of undercooked food; (2) via clonal spread from the occupationally exposed; (3) or from indirect manure contamination onto fresh food crops (e.g., environmental spread). Sufficient evidence exists to support clonal spread from the occupationally exposed (Marshall and Levy, 2011), while the other two routes are poorly understood. Contamination of fresh food crops via runoff, land application of manure/biosolids or feral animals has been hypothesized as indicating potential sources of contamination (Brooks *et al.*, 2012a). Antibiotic resistance phenotypes have been demonstrated to move via aerosols or runoff, though in very small amounts and over small distances from the CAFO (Brooks *et al.*, 2009, 2012b; Chinivasagam *et al.*, 2009). Brooks *et al.* (2009) demonstrated that runoff from plots receiving litter was more concentrated with antibiotic-resistant enterococci, which was characteristic of the litter and thus demonstrated that antibiotic resistant bacteria are transported as readily as any other bacteria.

26.3.2 Endotoxin

Another issue associated with biosolids and manure is the presence of endotoxin. Endotoxin, or lipopolysaccharide (LPS) derived from the cell wall of Gram-negative bacteria, is a highly immunogenic molecule present ubiquitously in the environment (see also Section 5.6.6) (Michel, 2003). Biosolids contain large populations of bacteria, and therefore are another potential source of endotoxin. Although most surfaces contain some traces of dust-associated endotoxin, it is primarily of concern as an aerosol, since most human endotoxin ailments are pulmonary associated (Sharif El *et al.*, 2004). Exposures to aerosolized endotoxin have been studied due to occupational exposures to cotton dust, composting plants and feed houses (Castellan *et al.*, 1987). Exposures to levels of endotoxin as little as 0.2 endotoxin units (EUs) per m^3 derived from poultry dust have been found to cause acute pulmonary ailments such as decreases in forced expiratory volume (Donham *et al.*, 2000). Chronic effects such as asthma and chronic bronchitis have been found to be due to exposures to endotoxin from cotton dust of as little as 10 EU per m³ on a daily basis (Olenchock, 2001).

Endotoxin concentrations in a variety of environmental samples have been investigated, and data show that the endotoxin level in Class B biosolids is similar in magnitude to that of other wastes including animal manures and compost. Since the relevance of this to human health is via inhalation, the potential for aerosolization of endotoxin during land application of biosolids and manure has also caused concern. Brooks *et al.* (2006) showed that endotoxin values measured during biosolids application were comparable to those found in untreated agricultural soils. Therefore, aerosolization of soil particles can result in endotoxin aerosolization, regardless of whether biosolids are involved. This is not surprising since bacterial concentrations in soil routinely exceed 10⁸ per gram, with a majority of bacteria being Gram negative. Soil particles containing sorbed microbes can be aerosolized and hence act as a source of endotoxin (see also Section 5.6.6).

A number of studies have investigated endotoxin in CAFOs, be it cattle, poultry or swine (Dungan and Leytem 2009; Brooks *et al.*, 2010). The majority of studies report endotoxin levels greater than those recommended for farms (Dungan, 2010). However, the majority of endotoxin associated with CAFOs has been confined to open cattle/dairy farms (Dungan, 2010), and swine and poultry interior housing (Brooks *et al.*, 2010; Dungan, 2010). For example, swine barns were found to have mean concentrations of endotoxin of 4385 EU per m³ (Duchaine *et al.*, 2001), while composting plants ranged from 10 to 400 EU per m³ (Clark *et al.*, 1983). Endotoxin release from open lot CAFOs (Dungan, 2010) and building exhaust fans (Brooks *et al.*, 2010) has been shown to be at levels of ≈ 800 and 100 EU m⁻³, respectively, with rapid decreases to near background levels just beyond the point source. It can be assumed that, as with municipal biosolids land application, the majority of aerosolized endotoxin will most likely arise from the dry soil surrounding the site; however, some manure, such as dry poultry litter, will be very prone to endotoxin release. Litter endotoxin levels are approximately one order of magnitude greater than those of typical Class B biosolids (Brooks *et al.*, 2007). In all cases with endotoxin, the severity of the exposure is unknown since not all endotoxin is bioactive, and thus not all exposures are equal. Overall, land application of residuals and aerosolized endotoxin remains an area that is poorly understood by environmental microbiologists.

26.3.3 Prions

Prions are infectious proteins that can result in animal or human disease (see also Section 2.5.2.2). Transmissible spongiform encephalopathies (TSE) are a group of neurological prion diseases of mammals, which in humans include Kuru, Creutzfeldt—Jakob disease (CJD), sporadic Creutzfeldt—Jakob disease (sp CJD) and variant Creutzfeldt—Jakob disease (VCJD) (Prusiner, 2004; Miles *et al.*, 2013). Animal diseases such as bovine spongiform encephalopathy (BSE) are of particular concern. Prions have been detected in the environment at low concentrations (Nichols *et al.*, 2009), and could originate from slaughterhouse wastes. Such wastes could reach wastewater treatment plants, and therefore interest has focused on whether or not prions survive wastewater treatment. If prions survived treatment, then they could end up within biosolids, with subsequent potential exposure of animals following land application. Adding to this concern is the fact that prions are reported to be very resistant to extreme physical conditions including irradiation and heat, and chemical treatment including acids, bases and oxidizing agents (Taylor, 2000).

Within the last few years it has been reported that prions are capable of surviving a very common wastewater treatment, namely mesophilic anaerobic digestion (Kirchmayr *et al.*, 2006; Hinckley *et al.*, 2008). However, these studies utilized an immunoblot method of detection of the prions, which did not distinguish between infectious and noninfectious prions. More recently, Miles *et al.* (2013) developed an assay that only detected infectious prions. This assay utilized a standard scrapie cell assay linked to an enzyme linked immuno-spot reaction (ELISPOT) for infectious prion detection. Using this assay and miniature anaerobic digestors (Figure 26.7) the influence of various wastewater treatments on infectious prion inactivation was evaluated (Table 26.4).

These data show a quantifiable reduction of infectious prions in wastewater during the normal period of anaerobic digestion (21 days), at both mesophilic and thermophilic

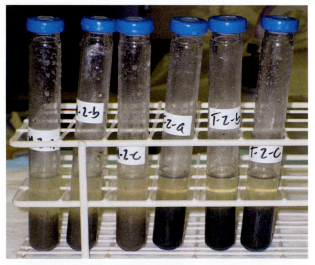

FIGURE 26.7 Sealed test tubes utilized as anaerobic digestor microcosms. Photo courtesy Syreeta Miles, the University of Arizona.

TABLE 26.4 Influence of Wastewater Treatment on Infectious Prion Inactivation

Treatment	Incubation Period	Decrease in Infectious Prions
Mesophilic anaerobic digestion	21 days	4.2 $\log_{10}$
Thermophilic anaerobic digestion	21 days	4.7 $\log_{10}$
Lime treatment of Class B biosolids	2 hours	2.9 $\log_{10}$

Adapted from Miles et al. (2013).

TABLE 26.5 Pathogens and Levels (Geometric Mean) Commonly Found in Waste Residuals

	Bovine	Poultry	Swine	Raw sludge	Class B biosolids
			CFU, PFU, MPN g^{-1}		
Campylobacter jejuni	≈150	≈340	≈460	≈3100	≈2.0
E. coli	≈170				
Listeria monocytogenes	≈600	≈180	≈210	≈2400	≈25
Salmonella	≈630	≈60	≈50	≈2400	≈25
Adenovirus				≈130	≈40
Enterovirus				≈40	≈4.0
Norovirus				≈2.7 × 10^5	≈1700
Cryptosporidium spp.	≈7.0			≈30	0.7

Table modified from Brooks et al. (2012a).

temperatures. In addition, lime treatment of Class B biosolids was shown to be particularly effective in inactivating infectious prions. Overall, the data suggest that prions do not survive wastewater treatment, and that land application of biosolids is not a viable route of human or animal exposure to prions.

26.4 PATHOGENS OF CONCERN IN ORGANIC RESIDUALS

Pathogenic bacteria and protozoa are known to reside within both Class B biosolids and animal manures. Pathogenic viruses can also be found in biosolids, but not in animal manures. Note also that by definition, Class A biosolids do not contain detectable pathogens. Pathogens routinely associated with either organic residual are shown in Table 26.5.

Whereas human pathogenic viruses are found exclusively in Class B biosolids, concentrations of the bacterial pathogens are normally found in higher concentrations in animal manures than in biosolids, most likely due to the fact that manures do not undergo treatment (Brooks *et al.*, 2012a).

Manure is known to contain a wide and varied array of bacterial and parasitic pathogens, and depending on its origin, can be a source of *Campylobacter jejuni*, *Escherichia coli* O157:H7, *Salmonella* spp., *Listeria monocytogenes*, *Cryptosporidium parvum* and *Giardia lamblia* (Guan and Holley, 2003; Hutchison *et al.*, 2004; McLaughlin *et al.*, 2009). Two issues associated with pathogens found in residuals are regrowth and reactivation.

26.4.1 Regrowth and Reactivation of Pathogens within Organic Residuals

26.4.1.1 Class B Biosolids

Regrowth and reactivation have both been documented as occurring in biosolids, but the two terms are not synonymous (Chen *et al.*, 2011). Reactivation is defined as a large increase in fecal coliform or *E. coli* in biosolids collected immediately after centrifugation or other dewatering processes, when compared with the feed into the dewatering equipment (WERF, 2006). Regrowth refers to an additional increase in the density of fecal indicators or *E. coli* upon storage of the biosolids over a period of hours or days.

Reactivation is of concern since studies have documented a large increase in fecal coliforms of several orders of magnitude in a short period of time that would preclude increases due to normal growth that could occur due to binary fission. The phenomenon was first observed by Donald Hendrickson in 2001 (Hendrickson *et al.*, 2004). Because it could not rationally be explained, reactivation was immediately controversial, and had implications for the designation of biosolids as Class A or B. This resulted in numerous studies on the process of reactivation, which indicated that reactivation did occur following dewatering by centrifugation, but not following dewatering with the use of a belt filter press (Erdal *et al.*, 2003). Different hypotheses have been developed to explain the phenomenon of reactivation (Information Box 26.4).

Many studies have also evaluated the potential for growth and/or regrowth of indicators and pathogens in land amended biosolids with either Class A or B material. These studies have resulted in a number of terms being coined to explain the increase in numbers (Information Box 26.5). Studies evaluating the growth and/or regrowth

---

Information Box 26.4   Main Hypotheses Developed to Explain Reactivation

- Clumping of bacteria when the biosolids were originally assayed, followed by desegregation of clumps into single cells following dewatering
- Formation of viable but nonculturable bacteria (VBNC) during wastewater treatment, and subsequent reactivation of the VBNC due to a signaling substance released into the centrate during centrifugation (WERF, 2006)

To date, the VBNC hypothesis is the most likely explanation for reactivation. Use of quantitative polymerase chain reaction (qPCR) to enumerate *E. coli* showed that copy numbers were not significantly different before and after dewatering, which supports the VBNC concept (Higgins *et al.*, 2007). Reactivation not only potentially affects the designation of biosolids as Class A or B, but also raises the possibility of reactivation of pathogens. Increased numbers of fecal coliforms in Class A biosolids has also been reported (Jolis, 2006).

Information Box 26.5   Terms and Definitions Utilized for Increased Numbers of Pathogens and Indicators

Growth:	Increase in detectable numbers of a known microbial population over time.
Regrowth:	Increase in numbers after a period of decline in numbers.
Recolonization:	Reintroduction of bacteria into biosolids followed by growth.
Reactivation:	Large rapid increase in numbers that cannot be ascribed to growth by binary fission.

---

of *Salmonella* and fecal indicators have produced mixed results, some showing increased numbers following land application, and some showing no such increase in numbers. This is most likely due to the different ways in which studies have been conducted, including laboratory studies versus field studies. In addition, some studies have monitored numbers of organisms that survived wastewater treatment and are subsequently introduced into soil via biosolids, as compared to other studies where laboratory strains of organisms have been inoculated into biosolids and the numbers monitored (Zaleski *et al.*, 2005b). Normally, it is a thought that monitoring the organisms that survive treatment is the preferred option, with field studies being more "real world" than laboratory studies. Whereas the growth and regrowth of fecal indicators in biosolid amended soil has frequently been noted, corresponding studies showing growth of *Salmonella* are far less frequent (Zaleski *et al.*, 2005b). Also, regrowth of fecal indicators is frequently associated with increased moisture following rainfall events (Pepper *et al.*, 1993).

Regrowth of *Salmonella* in Class A biosolids was observed after rainfall produced saturated conditions (Zaleski *et al.*, 2005a). Subsequently, this was shown to be recolonization following contamination with bird feces, since the *Salmonella* serotypes identified prior to the increase in numbers were different from those identified after the rainfall event. This has implications for the storage of Class A biosolids, which should be covered during storage for two reasons: (1) to prevent saturated conditions during rainfall events; and (2) to prevent recolonization by bird or animal feces.

26.4.1.2 Manure

Very few studies have demonstrated the regrowth of either indicator or pathogenic bacteria in manure. Composted manure has been demonstrated to support regrowth of *E. coli* O157:H7, particularly in compost with high moisture levels, above 30%, and low background bacterial counts. In other instances, regrowth of enterococci and *E. coli* have been demonstrated in poultry litter applied land following simulated rainfall (Brooks *et al.*, 2009, 2012b), and cow pats in fields (Sinton *et al.*, 2007). As in the situations with compost and biosolids, the driving factor behind regrowth was the presence of readily available organic nutrients and substrates and moisture. Finally, note that when pathogens are introduced into soil, some may adapt and be capable of survival within the soil, but only at the cost of the loss of pathogenicity (Ishii *et al.*, 2006). Similarly, *E. coli* has been shown to lose virulence during manure storage (Duriez *et al.*, 2008).

However, regardless of how low the incidence of infections from pathogens in soil is, people want to know how likely it is that *they* will get infected. To answer this question we can utilize the process of quantitative microbial risk assessment.

26.5 QUANTITATIVE MICROBIAL RISK ASSESSMENT OF PATHOGENS IN ORGANIC RESIDUALS

Quantitative microbial risk assessment (QMRA) can be used to compare pathogen risks from manure to municipal biosolids. Based on recent and historical data, pathogen levels can be estimated for a variety of manure sources as well as for Class B biosolids. Following the risk paradigm (see Chapter 24) the risk for a specific exposure scenario can be determined, utilizing the data from Table 26.4 and transport and inactivation decay models. A comprehensive study recently determined the risks inherent from various exposures to bovine, swine and poultry manures, raw sludge and Class B biosolids (Information Box 26.6).

These comparisons highlighted the importance of waste treatment, time and dilution in attenuating pathogen levels. When risks from biosolids and manure were directly compared for *Campylobacter*, *Cryptosporidium*, *Listeria* and *Salmonella*, manure invariably had greater risks under any of the exposure scenarios (e.g., fomites, fresh food crops, aerosol). However, biosolids pose an additional risk due to the incidence of human viruses within biosolids. The differences in risk can be explained by a large number of bacterial pathogens in manure with low infectivity, while biosolids risk can be attributed to a low level of highly infectious viral pathogens. The study also demonstrated the conservative nature of the EPA Part 503 rule, which dictates long delays between land application and fresh food crop harvest. Overall, a 4-month delay was more than adequate to reduce risk for nearly all microbial pathogens' acceptable levels, except viral pathogens and *Cryptosporidium*. Occupational risk can also be considered from exposures inherent to the land application and handling of waste residuals. Public

---

Information Box 26.6   Application of Risk Paradigm to Pathogens in Land Applied Manures and Biosolids

QMRA approaches can be applied to various manure or biosolids land application scenarios including: fomite contact; soil ingestion (both intentional and accidental); crop ingestion; runoff contamination; and aerosol exposures. Modeling these scenarios can be difficult, often requiring knowledge of pathogen survival in soil amended with residuals, or runoff water transport characteristics. As an example of calculating predicted risk, the following risk paradigm is applied to fomite and fresh food crops contaminated with pathogens from biosolids or manures.

1. Define *Salmonella* level (i.e., define hazard and hazard level)—for this simulation, we assume *Salmonella* at 160 and 5 CFU g^{-1} in bovine manure and Class B biosolids, respectively.
2. Define the route of exposure; in this case we define exposure via: (a) fomite contact; and (b) fresh food crop ingestion. For this simulation we can refer to the exposure models established by Brooks *et al.* (2012a):
   a. Fomite contamination (*fc*),
      $fc = rc \times ft$;
      where:
      rc = residual pathogen level per g defined in step 1,
      ft = the amount of residual transferred to the fomite, 0.1 g, and
      fc = the pathogen level on the fomite. We will assume no decay or decrease in pathogen concentration.
   b. Fresh food crop contamination (*cc*),
      $cc = rc \times dr \times (1/10^{sr}) \times rr \times wr \times 1000$;
      where:
      dr = soil dilution rate of 1.75×10^{-3}
      sr = pathogen decay rate in soil, equivalent to 0.422 $log_{10} 7 \ d^{-1}$

rr = the percentage of soil particles remaining on a crop following harvest, equivalent to 0.02 or 2%
wr = the percentage of soil particles remaining on a crop following washing, equivalent to 0.10 or 10%, and
cc = the crop pathogen level per kg.

3. Model the dose—response using dose exposures and dose response models:
   a. Dose exposure (*d*), $d = ec \times ds$;
      where:
      ec = pathogen level in each separate exposure scenario define in 2(a) or 2(b), *fc* and *cc*, respectively, and
      ds = the dose amount.
      i. In the fomite situation, a fomite to hand (43%) and hand to mouth (36%) transfer rate are used, which modify the previous equation to $d = ec \times (0.43 \times 0.36)$.
      ii. *ds* for the food crop scenario is equivalent to 0.292 kg for a 68 kg adult. We will assume the 0.292 kg dose as a one-time exposure.
   b. The beta—Poisson dose—response model for *Salmonella* consists of $P_i = 1 - [1 + (d/N_{50}) \times (2^{1/\alpha} - 1)]^{-\alpha}$;
      where:
      d = the dose exposure from step 3a
      $N_{50} = 2.4 \times 10^4$, $\alpha = 3.1 \times 10^{-1}$, and
      P_i is the probability of infection
4. The risk is then calculated and characterized.
   a. P_i for *fomite* exposures to *Salmonella* = 3×10^{-4} chance of infection from exposures to bovine manure on a fomite with no decay.
      Exposure to biosolids contaminated fomites is equivalent to 8×10^{-6}.
   b. The *crop* exposure yielded risks of 9×10^{-6} for bovine exposures, while biosolids exposures yielded risks of 2×10^{-7}.

risks arise from exposures either during land application (e.g., aerosol drift), or following land application (e.g., fresh food crops). Differences in these risks can be substantial as occupational exposures occur when the waste residuals are "fresh" with high concentrations of pathogens, while public exposure occurs following multiple intervening steps such as environmental inactivation, dilution and disinfection steps. The application of QMRA to this field is still new and exciting, and reminds us that much is unknown regarding pathogen behavior in soil, water, and air or on crops, and that many of these interactions are dynamic and unpredictable.

QUESTIONS AND PROBLEMS

1. Define the differences between sewage, sewage sludge, Class A biosolids and Class B biosolids.
2. What are the major hazard differences between land application of Class B biosolids and cow manure?
3. Discuss the difference in the potential for infections from land application of Class B biosolids versus animal manures.
4. Which gives the greatest risk of infection during land application of organic residuals: direct contact with the land applied residual, or indirect effects to communities due to offsite transport of pathogens as aerosols or via runoff to crops?
5. Using the equations located in Information Box 24.6, calculate the risk of infection for a *Salmonella* level of 3000 CFU g^{-1} and a soil pathogen decay rate (sr) of 1.422 log$_{10}$ 7 d^{-1} on a fresh food crop contaminated with poultry manure. Assume 7 days of decay, a soil dilution rate of 1.75×10^{-3}, and 2 and 10% of soil particles remaining on crops following harvest and washing, respectively. Assume a 68 kg adult and a one-time exposure.
6. How could you increase the pathogen level on a fomite or crop without increasing the starting pathogen level? What type of scenario(s) would increase the pathogen level in residual waste? Is pathogen level (rc) the most important variable in these QMRAs? Why or why not?
7. Would wearing gloves, in the fomite simulation, always reduce risk?

REFERENCES AND RECOMMENDED READING

Brooks, J. P., Tanner, B. D., Gerba, C. P., Haas, C. N., and Pepper, I. L. (2005a) Estimation of bioaerosol risk of infection to residents adjacent to a land applied biosolids site using an empirically derived transport model. *J. Appl. Microbiol.* **98**, 397−405.

Brooks, J. P., Tanner, B. D., Josephson, K. L., Haas, C. N., Gerba, C. P., and Pepper, I. L. (2005b) A national study on the residential impact of biological aerosols from the land application of biosolids. *J. Appl. Microbiol.* **99**, 310−322.

Brooks, J. P., Tanner, B. D., Gerba, C. P., and Pepper, I. L. (2006) The measurement of aerosolized endotoxin from land application of Class Biosolids in Southeast Arizona. *Can. J. Microbiol.* **52**, 150−156.

Brooks, J. P., Rusin, P. A., Maxwell, S. L., Rensing, C., Gerba, C. P., and Pepper, I. L. (2007) Occurrence of antibiotic-resistant bacteria and endotoxin associated with the land application of biosolids. *Can. J. Microbiol.* **53**, 1−7.

Brooks, J. P., and McLaughlin, M. R. (2009) Antibiotic resistant bacterial profiles of anaerobic swine lagoon effluent. *J. Environ. Qual.* **38**, 2431−2437.

Brooks, J. P., Adeli, A., Read, J. J., and McLaughlin, M. R. (2009) Rainfall simulation in greenhouse microcosms to assess bacterial-associated runoff from land-applied poultry litter. *J. Environ. Qual.* **38**, 218−229.

Brooks, J. P., McLaughlin, M. R., Scheffler, B., and Miles, D. M. (2010) Microbial and antibiotic resistant constituents associated with biological poultry litter within a commercial poultry house. *Sci. Total Environ.* **408**, 4770−4777.

Brooks, J. P., Brown, S., Gerba, C. P., King, G. M., O'Connor, G. A., and Pepper, I. L. (2011) Land application of organic residuals: Public health threat or environmental benefit? American Society of Microbiologists. Washington, DC.

Brooks, J. P., McLaughlin, M. R., Gerba, C. P., and Pepper, I. L. (2012a) Land application of manure and Class B biosolids: an occupational and public quantitative microbial risk assessment. *J. Environ. Qual.* **41**, 2009−2023.

Brooks, J. P., McLaughlin, M. R., Adeli, A., and Miles, D. M. (2012b) Runoff release of fecal bacterial indicators as influenced by two poultry manure application rates and AlCl$_3$ treatment. *J. Water Health* **10**, 619−628.

Burkholder, J., Libra, B., Weyer, P., Heathcote, S., Kolpin, D., Thorne, P. S., and Wichman, M. (2007) Impacts of waste from concentrated animal feeding operations on water quality. *Environ. Health Perspect.* **115**, 308−312.

Castellan, R. M., Olenchock, S. A., Kinsley, K. B., and Hankinson, J. L. (1987) Inhaled endotoxin and decreased spirometric values: an exposure-response relation for cotton dust. *New Engl. J. Med* **317**, 605−610.

Chee-Sanford, J. C., Mackie, R. I., Krapac, I. G., Lin, Y. F., Yannarell, A. C., Maxwell, S., and Aminov, R. I. (2009) Fate and transport of antibiotic residues and antibiotic resistance genes following land application of manure waste. *J. Environ. Qual.* **38**, 1086−1108.

Chen, Y. C., Murthy, S. N., Hendrickson, D., Araujo, G., and Higgins, M. J. (2011) The effect of digestion and dewatering on sudden increases and regrowth of indicator bacteria after dewatering. *Wat. Environ. Res.* **83**, 773−783.

Chetochine, A., Brusseau, M. L., Gerba, C. P., and Pepper, I. L. (2006) Leaching of phage from Class B biosolids and potential transport through soil. *Appl. Environ. Microbiol.* **72**, 665−671.

Chinivasagam, H. N., Tran, T., Maddock, L., Gale, A., and Blackall, P. J. (2009) Mechanically ventilated broiler sheds: a possible source of aerosolized *Salmonella, Campylobacter*, and *Escherichia coli*. *Appl. Environ. Microbiol.* **75**, 7417−7425.

Clark, C. S., Rylander, R., and Larsson, L. (1983) Levels of gram-negative bacteria, *Aspergillus fumigatus*, dust and endotoxin at compost plants. *Appl. Environ. Microbiol.* **45**, 1501−1505.

Donham, K. J., Cumro, D., Reynolds, S. J., and Merchant, J. A. (2000) Dose-response relationships between occupational aerosol exposures and cross-shift declines of lung function in poultry workers: recommendations for exposure limits. *J. Occup. Environ. Med.* **42**, 260—269.

Duchaine, C., Thorne, P. S., Merizux, A., Grimard, Y., Whitten, P., and Cormier, Y. (2001) Comparison of endotoxin exposure assessment by bioaerosol impinger and filter-sampling methods. *Appl. Environ. Microbiol.* **67**, 2775—2780.

Dungan, R. S. (2010) Fate and transport of bioaerosols associated with livestock operations and manures. *J. Anim. Sci* **88**, 3693—3706.

Dungan, R. S., and Leytem, A. B. (2009) Airborne endotoxin concentrations at a large open-lot dairy in Southern Idaho. *J. Environ. Qual.* **38**, 1919—1923.

Duriez, P., Zhang, Y., Lu, Z., Scott, A., and Top, E. (2008) Loss of virulence genes in *Escherichia coli* populations during manure storage on a commercial swine farm. *Appl. Environ. Microbiol.* **74**, 3935—3942.

Erdal, Z. K., Mendenhall, T. C., Neely, S. K., Wagoner, D. L., and Quigly, C. (2003) Implementing improvements in a North Caroliner Residuals Management Program. Proc. WEF/AWWA/CWEA Joint Residuals and Biosolids Management Conference; Baltimore, MD.

Guan, T. Y., and Holley, R. A. (2003) Pathogen survival in swine manure environments and transmission of human enteric illness. A review. *J. Environ. Qual.* **32**, 383—392.

Hendrickson, D., Denard, D., Farrell, J., Higgins, M., and Murthy, S. (2004) Reactivation of fecal coliforms after anaerobic digestion and dewatering. Proc. Water Environment Federation Annual Biosolids and Residuals Conference; Salt Lake City, Utah. Water Environment Federation: Alexandria, VA.

Higgins, M. J., Chen, Y. C., Murthy, S. N., Hendrickson, D., Farrell, J., and Shafer, P. (2007) Reactivation and growth of non-culturable *E. coli* in anaerobically digested biosolids after dewatering. *Wat. Res.* **44**, 665—673.

Hinckley, G. T., Johnson, C. J., Jacobson, K. H., Bartholomay, C., McMahon, K. D., McKenzie, D., *et al.* (2008) Persistence of pathogenic prion protein during simulated wastewater treatment processes. *Environ. Sci. Technol.* **42**, 5254—5259.

Hinesly, T. D., Jones, R. L., and Ziegler, E. L. (1972) Effects on corn by application of heated anaerobically digested sludge. *Comp. Sci.* **13**, 26—30.

Hutchison, M. L., Walters, L. D., Avery, S. M., Synge, B. A., and Moore, A. (2004) Levels of zoonotic agents in British livestock manures. *Lett. Appl. Microbiol.* **39**, 207—214.

Ishii, S., Ksoll, W. B., Hicks, R. E., and Sadowski, M. J. (2006) Presence and growth of naturalized *Escherichia coli* in temperate soils from Lake Superior watersheds. *Appl. Environ. Microbiol.* **72**, 612—621.

IWMI (International Water Management Institute) (2010) In "Wastewater irrigation and health: assessing and mitigating risk in low-income countries" (P. Dreschel, C. A. Scott, L. Raschid-Sally, M. Redwood, and A. Bahri, eds.), Earthscan, London/Sterling, VA.

Jolis, D. (2006) Regrowth of fecal coliforms in Class A biosolids. *Wat. Environ. Res.* **78**, 442—445.

Kirchmayr, R., Reichl, H. E., Schildorfer, H., Braun, R., and Somerville, R. A. (2006) Prion protein: detection in "spiked" anaerobic sludge and degradation experiments under anaerobic conditions. *Wat. Sci. Technol.* **53**, 91—98.

Marshall, B. M., and Levy, S. B. (2011) Food animals and antimicrobials: impacts on human health. *Clin. Microbiol. Rev.* **24**, 718—733.

McLaughlin, M. R., Brooks, J. P., and Adeli, A. (2009) Characterization of selected nutrients and bacteria from anaerobic swine manure lagoons on sow, nursery, and finisher farms in the Mid-South USA. *J. Environ. Qual.* **38**, 2422—2430.

Michel, O. (2003) Role of lipopolysaccaride (LPS) in asthma and other pulmonary conditions. *J. Endotoxin Res.* **9**, 293—300.

Miles, S. L., Sun, W., Field, J. A., Gerba, C. P., and Pepper, I. L. (2013) Survival of infectious prions during anaerobic digestion of municipal sewage sludge and lime stabilization of Class B biosolids. *J. Res. Sci. Technol.* **10**, 69—75.

Miles, S. L., Takizawa, K., Pepper, I. L., and Gerba, C. P. (2011) Survival of infectious prions in water. *J. Environ. Sci. Health* **A46**, 938—943.

NEBRA (North East Biosolids and Residuals Association) (2007) A National Biosolids Regulation, Quality, End Use, and Disposal Survey. Final Report, July 20, 2007.

Neilson, J. W., Josephson, K. L., Pepper, I. L., Arnold, R. B., DiGiovanni, G. D., and Sinclair, N. A. (1994) Frequency of horizontal gene transfer of a large catabolic plasmid (pJP4) in soil. *Appl. Environ. Microbiol.* **60**, 4053—4058.

Nichols, T. A., Pulford, B., Wyckoff, A. C., Meyerett, C., Michel, B., Gertig, K., *et al.* (2009) Detection of protease-resistant cervid prion protein in water from a CWD-endemic area. *Prion* **3**, 171—183.

NRC (National Research Council) (2002) "Biosolids Applied to Land: Advancing Standards and Practices," National Academy Press, Washington, DC.

Olenchock, S. A. (2001) Airborne endotoxin. In "Manual of Environmental Microbiology" (C. J. Hurst, R. L. Crawford, G. R. Knudsen, M. J. McInerney, and L. D. Stetzenbach, eds.), 2nd ed., ASM Press, Washington, DC, pp. 814—826.

Organic Trade Association (2012) Manure Facts. Washington, DC.

Pepper, I. L., Josephson, K. L., Bailey, R. L., Burr, M. D., and Gerba, C. P. (1993) Survival of indicator organisms in Sonoran desert soil amended with sewage sludge. *J. Environ. Sci. Health* **A28(6)**, 1287—1302.

Prusiner, S. B. (ed.) (2004) "Prion Biology and Diseases," Cold Spring Harbor Laboratory Press, Woodbury, New York.

Rusin, P., Maxwell, S., Brooks, J., Gerba, C., and Pepper, I. (2003) Evidence for the absence of *Staphylococcus aureus* in land applied biosolids. *Environ. Sci. Technol.* **37**, 4027—4030.

Sharif El, N., Douwes, J., Hoe, P. H. M., Doekes, G., and Nemery, B. (2004) Concentrations of domestic mite and pet allergens and endotoxin in Palestine. *Allergy* **59**, 623—631.

Sinton, L. W., Braithwaite, R. R., Hall, C. H., and Mackenzie, M. L. (2007) Survival of indicator and pathogenic bacteria in bovine feces on pasture. *Appl. Environ. Microbiol.* **73**, 7917—17925.

Tanner, B. D., Brooks, J. P., Haas, C. N., Gerba, C. P., and Pepper, I. L. (2005) Bioaerosol emission rate and plume characteristics during land application of liquid Class B biosolids. *Environ. Sci. Technol.* **39**, 1584—1590.

Taylor, D. M. (2000) Inactivation of transmissible degenerative encephalopathy agents: a review. *Vet. J.* **159**, 10—17.

U.S. EPA (U.S. Environmental Protection Agency) (1993) The standards for the use or disposal of sewage sludge. In "Final 40 CFR Part 503 Rules. EPA 822/Z—93/001," U.S. Environmental Protection Agency.

U.S. EPA (U.S. Environmental Protection Agency) (2000) A guide to field storage of biosolids and the organic by-products used in agriculture and for soil resource management. EPA/832—B—00—007, Washington, DC.

U.S. EPA (U.S. Environmental Protection Agency) (2004) Risk assessment evaluation for concentrated animal feeding operations. EPA 600/R−04/042.

Water Environment Federation (WERF) (2006) Reactivation and regrowth of fecal coliforms in anaerobically digested biosolids. Technical Practice Update Report: Alexandria, VA.

Zaleski, K. J., Josephson, K. L., Gerba, C. P., and Pepper, I. L. (2005a) Potential regrowth and recolonization of *Salmonella* and indicators in biosolids and biosolids amended soil. *Appl. Environ. Microbiol.* **71**, 3701−3708.

Zaleski, K. J., Josephson, K. L., Gerba, C. P., and Pepper, I. L. (2005b) Survival, growth, and regrowth of enteric indicator and pathogenic bacteria in biosolids, compost, soil, and land applied biosolids. *J. Residuals Sci. Technol.* **2**, 49−63.

Chapter 27

Recycled Water Treatment and Reuse

Channah Rock, Charles P. Gerba and Ian L. Pepper

27.1 RECYCLED WATER REUSE

Population increases in water-scant arid regions and inadequate supplies of water resources have led to the increased use of reclaimed (recycled) water for both potable and current nonpotable purposes. Recycled water is the liquid portion of municipal wastewater that has undergone a series of treatments usually involving a combination of physical, chemical and biological treatment technologies to remove suspended solids, dissolved solids, organic matter, nutrients, metals and pathogens (Jjemba et al., 2010). "Water reclamation" involves treatment of wastewater to make it reusable. "Water reuse" is the beneficial use of treated wastewater (Information Box 27.1). Recycled water is generally used for beneficial purposes including irrigation, industrial processes, toilet flushing or groundwater recharge of aquifers (WateReuse Association, 2011a). In addition, intentional reuse for the augmentation of potable supplies is also practiced.

The current extent of water reuse nationally in the U.S. is not precisely known, and in fact, a systematic analysis of the extent of effluent contributions to potable water supplies has not been made in the U.S. for over 30 years (NRC, 2012). However, in 2006 the U.S. EPA estimated that an average of at least 1.7 billion gallons of wastewater was reused daily (Brandhuber, 2006). Despite that, approximately 12 billion gallons of municipal wastewater effluent are still discharged each day into an ocean or estuary (NRC 2012). Currently, water recycling programs in Arizona, California, Florida and Texas account for about 90% of the water recycled in the U.S. However, new recycled water programs are emerging all over the country, including East Coast states such as Pennsylvania and Maryland (WateReuse Association, 2011b). As growth and population increase, sources of potable water become scarcer, and water recycling helps alleviate this need by replacing the use of potable water for applications that do not require water treated to such levels of quality. Thus, water recycling allows communities to become less reliant on ground and surface water sources.

Microorganisms play a critical role in the treatment of wastewater for reuse by transforming and removing

I.L. Pepper, C.P. Gerba, T.J. Gentry: Environmental Microbiology, Third edition. DOI: http://dx.doi.org/10.1016/B978-0-12-394626-3.00027-2
© 2015 Elsevier Inc. All rights reserved.

organic matter and chemical contaminants. In addition, most reuse applications require advanced treatment to further reduce exposure to waterborne pathogens. Finally, the growth and regrowth of water-based pathogens in the treated wastewater is also a potential concern.

27.2 TREATMENT TECHNOLOGIES TO PRODUCE RECYCLED WATER

Water utilities use a variety of well-tested and reliable treatment processes to treat effluent such that it can subsequently be reused. The four core stages of treatment are: preliminary treatment; primary treatment; secondary treatment; and tertiary or advanced treatment. The primary objective of conventional wastewater treatment is to reduce nutrient and contaminant loads to the environment, thereby maintaining the health of aquatic ecosystems. In recycled water applications, conventional wastewater treatment can be supplemented with additional processes to achieve a quality that is consistent with the intended use. At a minimum, recycled water will undergo some form of disinfection (WateReuse Association, 2011b). Examples of common treatment train processes are presented in Figure 27.1, while Table 27.1 outlines wastewater treatment processes. (See also Chapter 25).

> **Information Box 27.1   Terminology of Recycled Water**
>
> - Reclaimed water: Water that has been used more than one time before it is returned back into the natural water cycle.
> - Recycled water: Synonymous with reclaimed water.
> - Water reuse: The process of using water more than one time prior to environmental discharge. This term is frequently used with reference to potable uses.

27.3 RECYCLED WATER APPLICATION IN THE U.S.

Recycled water can be utilized as a resource in a variety of ways (Information Box 27.2), with the level of treatment being appropriate for specific end uses (see Figure 27.2).

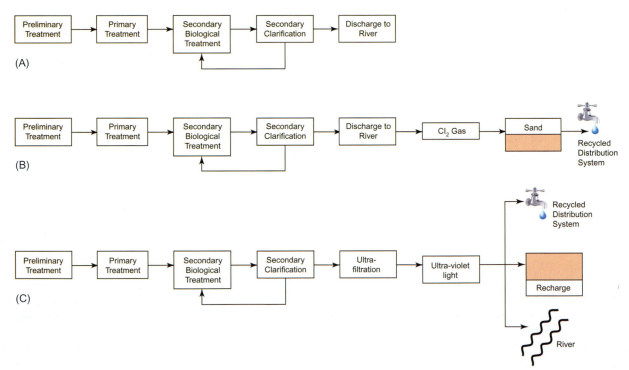

FIGURE 27.1 Wastewater treatment trains. (A) Preliminary treatment: screening of debris. Primary treatment: physical treatment in which solids settle out and are removed as primary sludge; fats, oils and greases (FDG) are skimmed off the surface. Secondary treatment: biological treatment via aerobic microbial degradation of organic matter which produces biological biomass on floc. Secondary clarification: biological floc settles out as secondary sludge. Discharge: secondary effluent is discharged to river. (B) Similar to treatment train (A) but with the addition of disinfection via chlorine gas, and sand filtration to remove particulates prior to distribution for irrigation. (C) Advanced treatment: utilizes ultrafiltration to remove particulates and ultraviolet light for water disinfection prior to discharge for irrigation, groundwater recharge or a surface water.

TABLE 27.1 Wastewater Treatment Processes: Purpose and Example Technologies

Treatment stage	Purpose	Technologies
Preliminary	Removal of large solids and grit particles	Screening; settling
Primary	Removal of suspended solids and some organic matter	Screening; sedimentation
Secondary	Biological treatment and removal of common biodegradable organic pollutants as well as inactivation of some microorganisms	Percolating or trickling filter; activated sludge; anaerobic treatment; waste stabilization ponds (oxidation ponds)
Tertiary (or advanced)	Removal of suspended particles and specific pollutants such as phosphorus, and removal/inactivation of bacteria, parasites and some viruses	Sand filtration; membrane bioreactor; microfiltration; ultrafiltration; reverse osmosis; chemical coagulation; UV; activated carbon

Modified from Wu et al. (2009).

Information Box 27.2   Recycled Water Applications in the U.S.

- Urban reuse
- Agricultural irrigation
- Industrial reuse
- Recreational reuse
- Groundwater recharge

FIGURE 27.2   The Santa Cruz River for much of the year has flow that is sustained by treated effluent discharge, south of Tucson, Arizona. Photo courtesy C. Rock.

27.3.1 Urban Reuse

Urban reuse is a term used to categorize a wide variety of different applications. Some of these applications have been utilized for many decades, especially golf course irrigation. The regulations and water quality standards required for any given application vary according to the potential for human contact with the recycled water. A large amount of recycled water is utilized for golf course and landscape irrigation. For example, in Arizona, the use of "secondary effluent," as it was then designated, was evaluated as a means for turf irrigation back in 1981 (Anderson *et al.*, 1981). Today, any new golf course brought into production in Arizona must use recycled water as a primary irrigation source. In addition, recycled water is now used extensively for irrigation of parks and playgrounds. A further benefit of such extensive reuse is that the practice is now widely accepted by the general public, and not subjected to the "yuck factor" (Information Box 27.3) associated with the "toilet-to-tap" perception. Other urban reuse applications include: fire protection; dust control; car washing; and toilet flushing. Generally, toilet flushing is restricted to commercial or industrial facilities, as in the case of Florida, where toilet flushing with recycled water is not permitted in residential homes.

27.3.2 Agricultural Irrigation

Agricultural irrigation is the oldest practice of recycled water use, and the largest end use by volume in the world. Depending on the type of crop, agricultural irrigation may be either unrestricted or restricted. Restricted irrigation applies to nonfood crops (e.g., cotton, switch grass) and requires less stringent treatment, whereas unrestricted irrigation for food crops (e.g., vegetables that are eaten raw) requires a very high level of treatment to produce appropriate water quality.

27.3.3 Industrial Reuse

Industrial reuse of water is increasing due to increased population growth and water conservation. For example,

---

Information Box 27.3 Recycled Water and the "Yuck Factor"

The concept of reusing water for potable purposes has for many people induced a fear and repugnance now colloquially known as the "yuck factor." The yuck factor was coined by University of Pennsylvania bioethicist Arthur Caplan to describe the instinctive adverse response to the concept of converting wastewater into drinking water (Schmidt, 2008). The yuck factor creates such strong feelings that it is difficult to overcome. In fact, the yuck factor creates feelings similar to the fear of eating genetically modified food crops, where opponents of such modified food exploited the gut reaction by calling it "Frankenfood" (Schmidt, 2008). Hence, even when presented with scientific facts that document the safety of recycled water, changing the opinions of the public is hard to do. Overall, many studies and projects have evaluated the safety of utilizing recycled water to augment potable sources, but there is no scientific documented adverse effect of such practices on human health.

Case Study 27.1 Tucson, Arizona, Recycled Water System

- Established in 1985
- In summer months daily deliveries of recycled water can exceed 30 mgd
- The recycled water distribution system consists of 160 miles of pipe
- Irrigation with recycled water serves 18 golf courses, 39 parks and 52 schools including the University of Arizona, with golf courses using about 60% of the recycled water annually
- Recycled water saves over a 6 billion gallons of drinking water annually, enough to serve 60,000 families for a year

---

the Curtis Stanton Energy Center in Orlando, Florida, uses 8 mgd (million gallons per day) of recycled water to cool the plant's boilers. Recycled water is also used as boiler feedwater to generate steam or hot water for thermal power stations. Such feedwater must be of very high quality to prevent boiler corrosion, scale and sediment deposits. Finally, recycled water is commonly used with flue-gas scrubbers found in waste incineration plants.

27.3.4 Environmental Enhancement

Recycled water can also be used for stream augmentation or the creation of artificial wetlands to serve as wildlife habitats and refuges. Use of recycled water for artificial wetlands is particularly attractive to municipalities since they serve several missions concurrently including: water treatment; creation of an urban wildlife habitat; and additional use as an outdoor classroom. At the Sweetwater Wetlands in Tucson, Arizona, secondarily treated effluent from a wastewater treatment plant enters recharge basins, filters through sediments beneath the basins and replenishes the local aquifer. This recycled wastewater is recovered by extraction wells during periods of high water demand and subsequently distributed and utilized for irrigation of golf courses, parks and other recreational areas (Case Study 27.1).

27.3.5 Recreational Reuse

An additional use of recycled water recreationally involves the creation of artificial lakes. These can be as simple as small water retention basins or ponds on golf courses; water-based recreational reservoirs with incidental human contact through fishing and boating; or full body contact involving swimming or wading. The recreational reuse can be restricted or unrestricted depending on the potential for public access, and the degree of body contact. Generally, higher degrees of treatment are utilized to provide better water quality when warranted. One of the primary issues with human-made lakes that utilize recycled water is the high inputs of nitrogen and phosphorus that can stimulate algal growth, turning the lake in question green. Controlling algal growth can involve application of chemicals such as copper sulfate, or aluminum phosphate, which sorb waterborne phosphate making it unavailable for algal growth. Advanced treatment in addition to aeration can also be used to reduce the amount of phosphate released from lake sediments.

27.3.6 Groundwater Recharge

In addition to utilizing recycled water for irrigation as a means to save potable water, recycled water can also be used to replenish aquifer water reserves. Multiple terms have been coined to describe groundwater replenishment with recycled water (Information Box 27.4). The process of using a shallow aquifer to treat recycled water is termed soil aquifer treatment (SAT) or aquifer recharge and recovery (ARR) (Missimer *et al.*, 2012). In contrast, aquifer storage and recovery (ASR) refers to the process of storing water during periods of abundance, and allowing for subsequent withdrawal and use in times of need. Groundwater recharge can also be used to prevent ground subsidence that can occur after large-scale groundwater withdrawals.

Regardless of the purpose, artificial groundwater recharge relies on the presence of gravels, sands and sites to remove contaminants. These include nutrients and trace

Information Box 27.4   Terminology Used to Describe Recycled Water Purification via Recharge Processes

- **Aquifer recharge**: The process of water movement from the land surface or unsaturated zone into the saturated zone or aquifer
- **Soil aquifer treatment (SAT)**: The process of using a shallow aquifer to treat recycled water
- **Aquifer recharge and recovery (ARR)**: Synonymous with SAT
- **Managed aquifer recharge (MAR)**: Synonymous with SAT
- **Aquifer storage and recovery (ASR)**: Injection of water into an aquifer for storage purpose prior to later withdrawal in times of need

organic chemicals, as well as pathogens, all of which can survive wastewater treatment (Missimer *et al.*, 2012). However, soils and vadose zone material can only adsorb finite amounts of phosphate, and due to the high phosphate content of most recycled waters, this increases the potential for phosphate contamination of groundwaters following long-term recycled water applications (20–25 years) (Moura *et al.*, 2011). Other compounds of concern that may be subject to leaching due to nonsorption include the pharmaceuticals diclofenac and ibuprofen, which also have relatively long half-lives of greater than 50 days due to low rates of microbial degradation (Lin and Gan, 2011). Other pharmaceuticals such as naproxen and trimethoprim demonstrate strong sorption to soil (Lin and Gan, 2011). If pharmaceuticals are sorbed, they generally do not degrade. Those that do not adsorb can be readily degraded by bacteria.

Microbial contamination of groundwater is also a concern since pathogenic organisms have been detected in groundwaters receiving wastewater (Levantesi *et al.*, 2010). This has posed the question of whether additional advanced tertiary treatment technology is needed prior to SAT. Despite these concerns, SAT, ARR and ASR have become more widely practiced in the twenty-first century in the U.S., Europe and Australia (Dillon *et al.*, 2008).

27.3.7 Potable Reuse

Potable reuse refers to the process of augmenting surface or groundwaters with recycled water to aid in water supply sustainability. This is practiced in many parts of the world including the U.S., Singapore, Australia, Saudi Arabia and the United Kingdom. Unplanned or incidental potable reuse occurs when wastewater is discharged from a wastewater treatment plant into a river, and is subsequently used as a drinking water source for a downstream

community. Such is the case for downstream communities on the Mississippi and Ohio rivers in the United States. In contrast to this, "planned" potable reuse can be direct or indirect.

27.3.8 Indirect Potable Reuse (IPR)

Planned indirect potable reuse (IPR) involves the intentional discharge of treated wastewater into bodies of water used as potable sources. Normally, this discharge occurs upstream of the drinking water treatment plant. Planned reuse indicates that there is an intention to reuse the water for potable use. The point of return could be into either a major water supply reservoir, a stream feeding a reservoir or a water supply aquifer (managed aquifer recharge or MAR). In the case of MAR, natural processes of filtration and dilution of the water with natural flows aim to reduce any real or perceived risks associated with eventual potable reuse. Locations of large planned indirect potable reuse facilities in the U.S. include San Diego, Tampa, Denver and the Orange County Water District. Abroad, examples include Australia, Singapore, Namibia and other water-scant areas. Many of these projects are large scale such as the Orange County Sanitation District IPR system known as the Groundwater Replenishment System (GWRS) (Case Study 27.2).

Following purification in the Orange County GWRS, approximately half of the high quality water (35 mgd) is pumped into injection wells to create a seawater intrusion barrier. The other half (35 mgd) is pumped into Orange County Water District's percolation basins in the city of Anaheim, where it flows through sand and gravel to deep underground aquifers, ultimately meeting the potable water needs of 600,000 residents. Upon withdrawal, the water is subjected to additional advanced treatment, then injected directly into the potable water distribution system without any recharge via SAT, or interaction with surface water.

27.3.9 Direct Potable Reuse

This is the so called toilet-to-tap perception, which has generally caused adverse public reaction known as the "yuck factor" (Information Box 27.3). A strict definition of direct potable reuse (DPR) is reclaimed water that is treated to potable water standards, and supplied pipe-to-pipe to consumers without an environmental buffer. In some countries, including Australia, the definition of DPR has been expanded to include: injection of recycled water directly into the potable water supply distribution system downstream of the water treatment plant, or into the raw water supply immediately upstream of the water treatment plant. Thus, injection could be either into a service

Case Study 27.2   The Orange County Groundwater Replenishment System (GWRS)

Wastewater is conventionally treated at the Orange County Sanitation District before it flows to the GWRS, where it undergoes advanced state-of-the-art purification process consisting of three technologies.

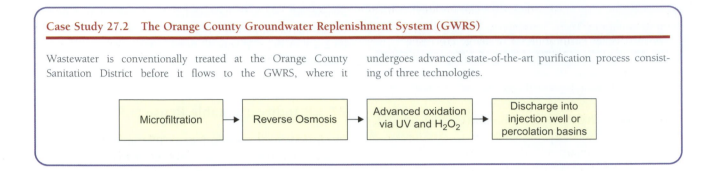

Case Study 27.3   NEWater in Singapore

Singapore is comprised of 63 small islands, and faced with water shortages. In 1974 the city-state began a water recycling program. By 1998 the Public Utilities Board (PUB) and the Ministry of the Environment and Water Resources in Singapore had instituted a water reclamation study which resulted in NEWater. NEWater is the result of advanced water treatment of municipal effluent using membrane filtration reverse osmosis and oxidative (UV) technologies. Today treated wastewater comprises 30% of Singapore's water supply. Much of the water is used for industrial purposes including wafer fabrication processes for the semiconductor industry and water for cooling towers, freeing up potable sources of water. However, some NEWater is added directly to potable water reservoirs. As a pioneer in indirect and direct potable reuse, the country is constantly evaluating new technologies to enhance the water quality of recycled water (Qin *et al.*, 2009).

reservoir or directly into a water pipeline. Therefore, the water used by consumers could be either undiluted or slightly diluted recycled water. In this definition, the key distinction with indirect potable reuse is that there is no temporal or spatial separation between the recycled water introduction and its distribution to consumers. Public perception of what extent of separation is required for reuse to become indirect may ultimately dictate the definition adopted. While most states in the U.S. consider DPR as a last resort scenario, there are a few examples where direct potable reuse has been practical. These locations include Windhoek, Namibia, in South Africa (duPisani, 2006), and the use of NEWater in Singapore (Case Study 27.3). Both of these locations were under such severe water stress that the communities had little choice but accept DPR as an alternative water supply. In these cases, adverse public perception issues were not readily apparent, illustrating that when potable water supplies are extremely scarce, the public is more accepting of toilet-to-tap.

27.4 RECYCLED WATER REGULATIONS

Of major concern in water reclamation is the removal of waterborne pathogens. Conventional sewage treatment (activated sludge) still contains significant concentrations of viruses and protozoan parasites. Treatment performance measures and fecal indicator organisms are both used to monitor the microbial quality of reclaimed waters. The greater the exposure to the public, the greater the needed treatment to reduce pathogens present in the wastewater. Usually, multiple barriers (e.g., disinfection and ultrafiltration) are employed as a redundancy plan to further reduce the risk, especially in indirect and direct potable reuse applications. Bacterial indicators are usually employed as treatment performance indicators, because of the lower cost and speed with which they can be detected compared to actual pathogen detection. Assessment of treatment processes for enteric viruses and protozoan parasites are sometimes required, depending upon local regulations.

Several states consider recycled water to be viable as a water source alternative, and have developed regulations with specific water quality requirements and/or treatment processes for a variety of reuse applications. In other states, water reuse regulations have been developed with the primary intent of providing a disposal alternative to surface water discharge. A few states have no specific regulations or guidelines on water reclamation and reuse, although programs may still be permitted with approval on a case-by-case basis.

27.4.1 Recycled Water Regulations and Water Quality Standards

To date, no federal regulations exist that govern water recycling in the U.S.; rather, such standards have been developed and implemented at the state government level. The lack of federal regulations and coordination between states has resulted in diverse standards for recycled water across the country. Despite the differing standards, the process of recycling water always involves a multi-barrier approach (i.e., physical, chemical and biological treatment processes). The U.S. EPA has developed *Guidelines for Water Reuse*, a comprehensive, technical document to encourage states to develop their own regulations (U.S. EPA, 2004).

The fundamental precondition for water recycling is that applications will not cause unacceptable public health risks (UNEP, 2004). Therefore, microbiological parameters have historically received the most attention in water reuse regulations and guidelines. Recently, the focus on microbiological parameters has shifted slightly to address contaminants of emerging concern, including pharmaceuticals and potential endocrine disrupting compounds (EDCs). However, microbial pathogens still pose the greater demonstrated risk in water recycling applications. Since monitoring for all pathogens is not practical, specific indicator organisms are monitored to minimize health risks (U.S. EPA, 2004). Indicator organisms (e.g., total coliforms, fecal coliforms) are generally not pathogenic, but their presence in a water sample signals the possible presence of fecal contamination, though recycled water guidelines may also mandate screening for disease-causing organisms (e.g., enteric viruses) (Table 27.2).

Updated in 2006, the World Health Organization (WHO) introduced guidance for the safe use of wastewater in 1971. The WHO guidelines are relatively less restrictive than water reuse regulations and guidelines adopted by the various states in the U.S. (Metcalf and Eddy 2007). The main intent of the WHO guidelines is to introduce some level of treatment of wastewater, and interrupt transmission of diseases, prior to food crop irrigation.

Since the U.S. EPA views it as a regional issue, regulations for water recycling have been developed and implemented at the state government level. According to U.S. EPA's 2004 *Guidelines for Water Reuse*, 26 states have adopted water recycling regulations, 15 states have guidelines or design standards and nine states have no regulations or guidelines.

The lack of federal regulations has resulted in varying regulations and guidelines. Among the states, Arizona, California, Colorado, Florida, Hawaii, Nevada, Oregon, Texas, Utah and Washington have developed regulations or guidelines specifying recycled water quality and treatment requirements. This allows for the full spectrum of reuse applications that allows for water reuse as a sustainable water conservation and management strategy.

27.5 MICROBIAL WATER QUALITY ASPECTS OF RECYCLED WATER

Recycled water is the end product of conventional wastewater treatment that receives additional advanced treatment prior to use. The level of treatment received varies depending on the targeted end use, and specific regulations that vary from state to state within the U.S. All classes of microorganisms can be found in wastewater including bacteria, virus, fungi, protozoa and helminths, but most studies have focused on bacteria. Here we describe relevant features of microbes found in recycled water. The types of organisms found in recycled water depend on the degree of treatment.

27.5.1 Bacteria

Recycled water normally contains large populations of bacteria that include heterotrophic plate count bacteria (HPC), but bacterial pathogens are generally reduced or eliminated. HPC concentrations within recycled water vary depending on the level of treatment but can be as high as 10^8 colony forming units (CFU) per 100 ml of water (Ajibode *et al.*, 2013). Standards require very low levels of indicator bacteria, which indicate the absence of waterborne bacterial pathogens in the treated water. However, regrowth of indicator bacteria may occur after treatment, and, in addition, *E. coli* has been detected in recycled water including O157:H7 (Jjemba *et al.*, 2009). Water-based pathogens including *Legionella* and *Mycobacterium* can also be found in recycled water

TABLE 27.2 Summary of Microbial Water Quality Parameters of Concern for Water Reuse

Parameter	Range in Secondary Effluents	Treatment Goal in Recycled Water	U.S. EPA Guideline
Total coliform	< 10 CFU/100 ml−10^7 CFU/100 ml	<1 CFU/100 ml−200 CFU/100 ml	−
Fecal coliform	< 1−10^6 CFU/100 ml	<1 CFU/100 ml−10^3 CFU/100 ml	14/100 ml for any one sample; 0/100 for 90% of samples
Helminth eggs	< 1/L−10/L	<0.1/L−5/L	−
Viruses	< 1/L−100/L	<1/50 L	−

Source: U.S. EPA (2004). Guidelines for Water Reuse.

Case Study 27.4 Influence of Residence Time in Water Distribution Systems on Recycled Water Quality

The distribution systems of two wastewater reclamation systems located in southwest Arizona were monitored over a 15-month period for microbial recycled water quality. The two plants' utilities produced recycled water of similar water quality except for nitrogen. Specifically, Utility A produced Class A+ water with a total nitrogen concentration of less than 10 mg/L, whereas Utility B produced Class A water with no nitrogen limitation. The two utilities also differed in their means of disinfection with Utility A utilizing chlorine and Utility B using ultraviolet light. Both distribution systems were monitored with increasing distance from the point of entry. Traditional waterborne indicators such as *E. coli* and *enterococci* were rarely detected and only at

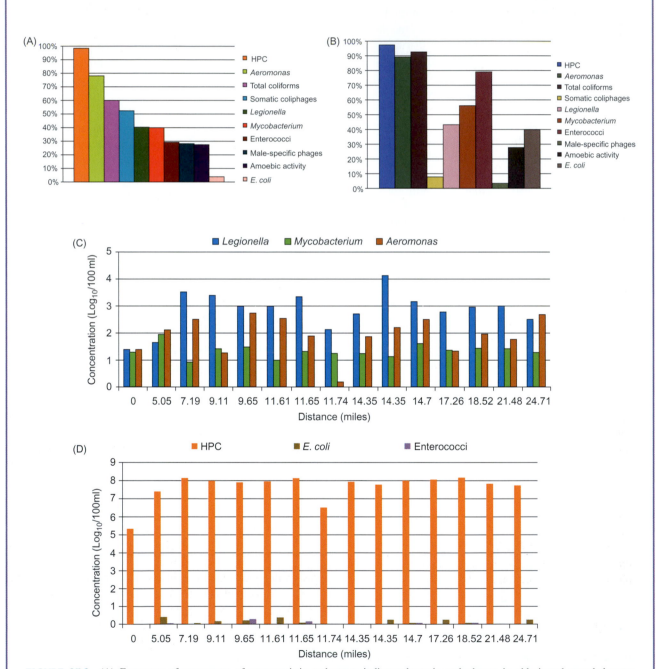

FIGURE 27.3 (A) Frequency of occurrence of opportunistic pathogens, indicator bacteria and phages in chlorinated recycled water at Utility A. (B) Frequency of occurrence of opportunistic pathogens and indicator bacteria and phages in UV disinfected recycled water at Utility B. (C) Influence of residence time on concentrations of water-based pathogens in recycled water distribution systems of Utility A. (D) Influence of residence time on microbial indicators in recycled water distribution systems of Utility A. *Source*: Ajibode *et al.* (2013).

low concentrations. In contrast, water-based pathogens such as *Legionella* or *Mycobacterium* were frequently detected (Figure 27.3A and B). Overall, waterborne indictor organisms such as *E. coli* were less prevalent in the chlorinated system (Utility A) than the UV system (Utility B). The impact of residence time on waterborne and water-based pathogens for plant A is shown in Figure 27.3C and D. Overall, there are numerous instances where significant concentrations of water-based pathogens were found, despite the absence of waterborne indicator organisms (*E. coli*). In fact, waterborne fecal indicators were not correlated with water-based pathogen incidence or concentration.

Regardless of the organism monitored, all opportunistic water-based pathogen concentrations increased fairly rapidly upon entry into the distribution system of either utility, perhaps due to rapid

dissipation of residual chlorine. This in turn was followed by reasonably constant concentrations of all organisms despite further increases in residence time. This is in contrast to the conventional view that microbial water quality decreases consistently as water age increases. The maintenance level of organisms may be due to assimilable organic carbon concentrations becoming constant as some organisms continued to grow, while others died and were lysed. Note that rechlorination of the distribution system at the 11.7 mile booster station in Utility A only reduced the concentrations of bacteria temporarily, resulting in the survival and regrowth of organisms including pathogens and indicators. Finally, amoebic activity was detected in approximately one-third of all samples from both systems, corroborating other studies with distribution systems cited by Thomas and Ashbolt (2011).

(Ajibode *et al.*, 2013). In contrast to bacteria, fungi have not been as well studied. However, *Aspergillus* spp. have been detected (Smit *et al.*, 2005).

27.5.2 Viruses

Viruses are more resistant to disinfection than bacteria, and are also more difficult to remove by filtration due to their small (nm) size. To reduce the level of viruses, extended chlorination or a combination of chlorine and UV light is routinely practiced. Soil aquifer treatment, ultra-filtration and membrane reverse osmosis also cause significant reductions in viruses.

27.5.3 Protozoa

Protozoa including *Cryptosporidium* and *Giardia* have been detected in recycled water despite prior filtration and disinfection (Jjemba *et al.*, 2009). Free-living amoebae can also be found in recycled water after treatment, particularly within biofilms found in distribution systems (Ajibode *et al.*, 2013). They are very resistant to chlorine in the biofilms, and live off of the bacteria in the biofilm. Of particular interest are *Acanthamoeba* spp., which act as a host for *Legionella*, and *Naegleria fowleri* and *Balamuthia mandrillaris*, which are known to cause brain encephalitis (see also Section 2.3.2.2).

27.5.4 Algae

Algae are photosynthetic microorganisms that can cause esthetic problems (taste and odor) and anoxic conditions in both open and closed reservoirs. However, they can also be found within recycled water distribution systems, where they can contribute to biodegradable carbon in the

distribution system, and create amoxic conditions that subsequently result in hydrogen sulfide production, creating foul odors (Jjemba *et al.*, 2010). Overall, blue–green algae or cyanobacteria are the most prevalent algae found in recycled water. Many of the cyanobacteria can produce anatoxins or endotoxins that cause flu-like symptoms following exposure via aerosols (Annadotter *et al.*, 2005).

27.6 INFLUENCE OF RESIDENCE TIME IN DISTRIBUTION SYSTEMS ON MICROBIAL WATER QUALITY

The chemical and microbial quality of recycled water is routinely monitored as it leaves the treatment plant, since this is frequently the point of compliance, where regulations require the water to be tested. However, due to the fact that the water must travel extensive distances through the distribution system to the point of use, there is potential for the quality of water to deteriorate within the distribution system, with residence time or water age being a key factor. Residence time will normally be a function of water demand and system operation and design. As residence time increases, both chemical and microbial water quality can be diminished (Weinrich *et al.*, 2010). The influence of residence time on microbial recycled water quality was recently evaluated in a study in Arizona. Specifically, two wastewater reclamation facilities were studied that utilized different treatment methods and different methods of disinfection. Recycled water from both facilities was monitored for an extended period of time and water quality was evaluated and compared (Case Study 27.4).

With increased residence time, concentrations of bacteria within distribution systems generally increased perhaps due mainly to two factors; the maintenance level of assimilable organic carbon; and the lack of residual chlorine in the distribution system. Since some of these bacteria are opportunistic pathogens they may be of concern and warrant additional evaluation.

QUESTIONS AND PROBLEMS

1. What is the largest use of recycled water?
2. What is the difference between indirect and direct potable reuse?
3. What is unplanned or incidental potable reuse?
4. What are the differences between water-based and waterborne pathogens?
5. Why do we use multiple barriers for the treatment of reclaimed waters for potable reuse? Look up the latest standards for water reuse in your state or country. Do they include bacterial standards? What are the microbial testing requirements? Is direct potable reuse allowed?
6. Why should water-based pathogens be of concern in recycled wastewater?

REFERENCES AND RECOMMENDED READING

Ajibode, O. M., Rock, C., Bright, K., McLain, J. E. T., Gerba, C. P., and Pepper, I. L. (2013) Influence of residence time of reclaimed water within distribution systems on reclaimed water quality. *J. Wat. Reuse & Desalination* **3**, 185–196.

Anderson, E. L., Pepper, I. L., and Kneebone, W. R. (1981) Reclamation of wastewater by means of a soil-turf filter. I. Nitrogen removal. *J. Wat. Pollut. Control Fed.* **53**, 1402–1407.

Annadotter, H., Cronberg, G., Nystrand, R., and Rylander, R. (2005) Endotoxins from cyanobacteria and Gram-negative bacteria as the cause of an acute influenza-like reaction after inhalation of aerosols. *Econ. Health* **2**, 209–221.

Brandhuber, P. (2006) EPA releases updated version of guidelines for water reuse. *Wat. Wastes Dig.* **46**, 1.

Dillon, P., Page, D., Vanderzalm, J., Pavelic, P., Toze, S., Bekele, E., *et al.* (2008) A critical evaluation of combined engineered and aquifer treatment systems in water recycling. *Wat. Sci. Technol.* **57**, 753–762.

duPisani, P. L. (2006) Direct reclamation of potable water at Windhoek's Goreangab reclamation plant. *Desalinization* **188**, 79–88.

Jjemba, P. K., Weinrich, L., Cheng, W., Givaldo, E., and LeChevallier, M. W. (2009) "Guidance Document on the Microbiological Quality and Biostability of Reclaimed Water Following Storage and Distribution," WateReuse Foundation Report WRF-05-002 WRF, Alexandria, VA.

Jjemba, P. K., Weinrich, L. A., Cheng, W., Giraldo, E., and LeChevallier, M. W. (2010) Regrowth of opportunistic pathogens and algae in reclaimed water distribution systems. *Appl. Environ. Microbiol.* **76**, 4169–4178.

Levantesi, C., Mantia, R. L., Masciopinto, C., Böckelmann, U., Ayuso-Gabella, M. N., Salgot, M., *et al.* (2010) Quantification of pathogenic microorganisms and microbial indicators in three wastewater reclamation and managed aquifer recharge facilities in Europe. *Sci. Total Environ.* **408**, 4923–4930.

Lin, K., and Gan, J. (2011) Sorption and degradation of wastewater-associated non-steroidal anti-inflammatory drugs and antibiotics in soils. *Chemosphere* **83**, 240–246.

Metcalf and Eddy (2007) "Water Reuse: Issues, Technologies, and Applications." McGraw-Hill, New York.

Missimer, T. M., Dreses, J. E., Amy, G., Maliva, R. G., and Keller, S. (2012) Restoration of Wadi aquifers by artificial recharge with treated wastewater. *Ground Water* **50**, 514–527.

Moura, D. R., Silveira, M. L., O'Connor, G. A., and Wise, W. R. (2011) Long-term reclaimed water application effects on phosphorus leaching potential in rapid infiltration basins. *J. Environ. Monit.* **13**, 2457–2462.

National Research Council (NRC) (2012) "WateReuse: Potential for Expanding the Nation's Water Supply through Reuse of Municipal Wastewater," National Academy Press, Washington DC.

Qin, J. J., Kekre, K. A., Oo, M. H., and Seah, H. (2009) Pilot study for reclamation of the secondary effluent at Changi Water Reclamation Plant. *Desalin. Wat. Treat.* **11**, 215–223.

Schmidt, C. W. (2008) The yuck factor when disgust meets discovery. *Environ. Health Perspect.* **116**, A524–A527.

Smit, L. A. M., Spaan, S., and Heederik, D. (2005) Endotoxin exposure and symptoms in wastewater treatment works. *Am. J. Indust. Med.* **48**, 30–39.

Thomas, J. M., and Ashbolt, N. J. (2011) Do free-living amoebae in treated drinking water systems present an emerging health risk? *Environ. Sci. Technol.* **45**, 860–869.

UNEP (2004) Water and wastewater reuse: an environmentally sound approach for sustainable urban water management. URL: <http://www.unep.or.jp/ietc/Publications/Water_Sanitation/wastewater_reuse/Booklet-Wastewater_Reuse.pdf>.

U.S. EPA (2004) *Guidelines for Water Reuse.* EPA/625/R-04/108. URL: <http://www.epa.gov/nrmrl/pubs/625r04108/625r04108.pdf>.

WateReuse Association (2011a) Sustainable solutions for a thirsty planet (Website). <http://watereuse.org/information-resources/about-water-reuse/faqs-o>; (accessed October 2011).

WateReuse Association (2011b) Sustainable solutions for a thirsty planet (Website). <http://athirstyplanet.com/be_informed/what_is_water_resuse/who-is-reusing>; (accessed August 2011).

Weinrich, L. A., Jjemba, P. K., Giraldo, E., and LeChevallier, M. S. (2010) Implications of organic carbon in the deterioration of water quality in reclaimed water distribution systems. *Wat. Res.* **18**, 5367–5375.

Wu, L., Weiping, C., French, C., and Chang, A. (2009) Safe Application of Reclaimed Water Reuse in the Southwestern United States. University of California Extension Publication #8357. URL: <http://anrcatalog.ucdavis.edu/>.

Chapter 28

Drinking Water Treatment and Distribution

Charles P. Gerba and Ian L. Pepper

Rivers, streams, lakes and underground aquifers are all potential sources of potable water. In the United States, all water obtained from surface sources must be filtered and disinfected to protect against the threat of microbial contaminants. Such treatment of surface waters also improves esthetic values such as taste, color and odors. In addition, groundwater under the direct influence of surface waters such as nearby rivers must be treated as if it were a surface supply. In many cases, however, groundwater needs either no treatment or only disinfection before use as drinking water. This is because soil itself has acted as a filter to remove pathogenic microorganisms, decreasing the chances of contamination of drinking water supplies.

At first, slow sand filtration was the only means employed for purifying public water supplies. Then, when Louis Pasteur and Robert Koch developed the germ theory of disease in the 1870s, things began to change. In 1881, Koch demonstrated in the laboratory that chlorine could kill bacteria. Following an outbreak of typhoid fever in London, continuous chlorination of a public water supply was used for the first time in 1905 (Montgomery, 1985). The regular use of disinfection in the United States began in Chicago in 1908. The application of modern water treatment processes had a major impact on water-transmitted diseases such as typhoid in the United States (Figure 28.1).

28.1 WATER TREATMENT PROCESSES

Modern water treatment processes provide barriers, or lines of defense, between the consumer and waterborne disease. These barriers, when implemented as a succession of treatment processes, are known collectively as a treatment process train (Figure 28.2). The simplest treatment process train, known as chlorination, consists of a single treatment process, disinfection by chlorination (Figure 28.2A). The treatment process train known as filtration entails chlorination followed by filtration through sand or coal, which removes particulate matter from the water and reduces turbidity (Figure 28.2B). At the next level of treatment, in-line filtration, a coagulant is added prior to filtration (Figure 28.2C). Coagulation alters the physical and chemical state of dissolved and suspended solids, and facilitates their removal by filtration. More conservative water treatment plants add a flocculation (stirring) step before filtration, which enhances the agglomeration of particles, and further improves the removal efficiency in a treatment process train called direct filtration (Figure 28.2D). In direct filtration, disinfection is enhanced by adding chlorine (or an alternative disinfectant, such as chlorine dioxide or ozone) at both the beginning and end of the process train. The most common treatment process train for surface water supplies, known as conventional treatment, consists

I.L. Pepper, C.P. Gerba, T.J. Gentry: Environmental Microbiology, Third edition. DOI: http://dx.doi.org/10.1016/B978-0-12-394626-3.00028-4
© 2015 Elsevier Inc. All rights reserved.

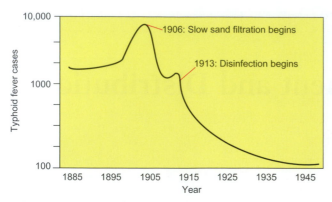

FIGURE 28.1 Impact of water filtration and chlorination on typhoid fever death rate in Albany, New York. From Logsdon and Lippy (1982).

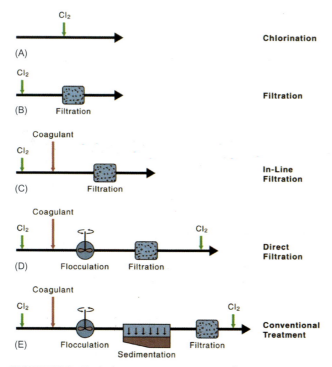

FIGURE 28.2 Typical water treatment process trains.

(usually polyacrylamides) are added after flocculation and sedimentation as an aid in the filtration step.

Coagulation can also remove dissolved organic and inorganic compounds. Hydrolyzing metal salts added to the water may react with the organic matter to form a precipitate, or they may form aluminum hydroxide or ferric hydroxide floc particles to which the organic molecules adsorb. The organic substances are then removed by sedimentation and filtration, or filtration alone if direct filtration or in-line filtration is used.

Flocculation is a purely physical process in which the treated water is gently stirred to increase interparticle collisions, thus promoting the formation of large particles. After adequate flocculation, most of the aggregates settle out during the 1 to 2 hours of sedimentation. Microorganisms are entrapped or adsorbed to the suspended particles and removed during sedimentation (Figure 28.3).

Sedimentation is another purely physical process, involving the gravitational settling of suspended particles that are denser than water. The resulting effluent is then subjected to rapid filtration to separate out solids that are still suspended in the water. Rapid filters typically consist of 50−75 cm of sand and/or anthracite, having a diameter between 0.5 and 1.0 mm (Figure 28.3). Particles are removed as water is filtered through the medium at rates of $4-24/min/10\ dm^2$. Filters need to be backwashed on a regular basis to remove the buildup of suspended matter. This backwash water may also contain significant concentrations of pathogens removed by the filtration process. Rapid filtration is commonly used in the United States. Another method, slow sand filtration, is also used. Employed primarily in the United Kingdom and Europe, this method operates at low filtration rates without the use of coagulation. Slow sand filters contain a layer of sand (60−120 cm deep) supported by a gravel layer (30−50 cm deep). The hydraulic loading rate is between 0.04 and 0.4 m/h. The buildup of a biologically active layer, called a *schmutzdecke*, occurs during the operation of a slow sand filter. This eventually leads to head loss across the filter, requiring removing or scraping the top layer of sand. Factors that influence pathogen removal by filtration are shown in Table 28.1.

Taken together, coagulation, flocculation, sedimentation and filtration effectively remove many contaminants as shown in Tables 28.2 and 28.3. Equally important, they reduce turbidity, yielding water of good clarity and hence enhanced disinfection efficiency. If not removed by such methods, particles may harbor microorganisms and make final disinfection more difficult. Filtration is an especially important barrier in the removal of the protozoan parasites *Giardia lamblia* and *Cryptosporidium*. The cysts and oocysts of these organisms are very resistant to inactivation by disinfectants, so disinfection alone cannot be relied on to prevent waterborne illness (see Case Study 22.1). However, because of their smaller size,

of disinfection, coagulation, flocculation, sedimentation, filtration and disinfection (Figure 28.2E)

Coagulation involves the addition of chemicals to facilitate the removal of dissolved and suspended solids by sedimentation and filtration. The most common primary coagulants are hydrolyzing metal salts, most notably alum $[Al_2(SO_4)_3 \cdot 14H_2O]$, ferric sulfate $[Fe_2(SO_4)_3]$ and ferric chloride $(FeCl_3)$. Additional chemicals that may be added to enhance coagulation are charged organic molecules called polyelectrolytes; these include high-molecular-weight polyacrylamides, dimethyldiallyl-ammonium chloride, polyamines and starch. These chemicals ensure the aggregation of the suspended solids during the next treatment step, flocculation. Sometimes polyelectrolytes

FIGURE 28.3 Drinking water treatment plant showing sand filter beds in the foreground and tanks containing alum flocculant in the background. Photo courtesy C.P. Gerba.

TABLE 28.1 Factors Effecting the Removal of Pathogens by Slow Sand Filters

Temperature
Sand grain size
Filter depth
Flow rate
Well-developed biofilm layer

viruses and bacteria can pass through the filtration process. Removal of viruses by filtration and coagulation depends on their attachment to particles (adsorption), which is dependent on the surface charge of the virus. This is related to the isoelectric point (the pH at which the virus has no charge) and is both strain and type dependent (see also Table 19.2). The variations in surface properties have been used to explain why different types of viruses are removed with different efficiencies by coagulation and filtration. Thus, disinfection remains the ultimate barrier to these microorganisms.

Generally, disinfection is accomplished through the addition of an oxidant. Chlorine is by far the most common disinfectant used to treat drinking water, but other oxidants, such as chloramines, chlorine dioxide and ozone, are also used (see Chapter 29). While ultraviolet can be used it does not leave a residual and usually a secondary disinfectant (i.e., chlorine) is added.

28.2 WATER TREATMENT REQUIREMENTS

Production of safe drinking water requires a holistic approach that considers the source of water, treatment processes and the distribution system. A multiple-barrier approach is used to ensure that if one barrier fails, the remaining barriers minimize pathogen presence in the water delivered to the consumer's tap. The essential barriers are:

- Source water protection
- Water plant processes
- Disinfection
- Distribution system residual disinfection
- Security

Source water protection means ensuring the highest water quality source possible before treatment by controlling use of the watershed, including minimizing sewage and domestic animal (e.g., cattle) contamination (Fox *et al.*, 2006). The treatment processes and disinfection must be adequate to ensure that the concentrations of pathogens are reduced to levels that minimize risk (see Chapter 26). This is dependent upon the concentrations expected in the water source. Thus, water sources which have significant sewage discharges or runoff from farm

TABLE 28.2 Coagulation, Sedimentation, Filtration: Typical Removal Efficiencies and Effluent Quality

Organisms	Coagulation and Sedimentation (% Removal)	Rapid Filtration (% Removal)	Slow Sand Filtration (% Removal)
Total coliforms	74—97	50—98	>99.999
Fecal coliforms	76—83	50—98	>99.999
Enteric viruses	88—95	10—99	>99.999
Giardia	58—99	97—99.9	>99
Cryptosporidium	90	99—99.9	99

From U.S. EPA (1988).

TABLE 28.3 Removal of Virus by Coagulation—Settling—Sand Filtration

Virus	Viral Assays, PFU (% Removal)		
	Input	Settled Water	Filtered Water
Poliovirus	5.2×10^7	1.0×10^6 (98)	8.7×10^4 (99.84)
Rotavirus	9.3×10^7	4.6×10^6 (95)	1.3×10^4 (99.987)
Hepatitis A virus	4.9×10^{10}	1.6×10^9 (97)	7.0×10^8 (98.6)

Adapted from Rao et al. (1988).

TABLE 28.4 Disinfection and Process Credits (Log Removal[a]) under the U.S. Environmental Protection Agency Surface Water Treatment Rule (U.S. EPA, 1991, 2003)

Process Credits	Viruses	Giardia	Cryptosporidium
Total log removal/ inactivation required	4.0	3.0	2.0 to 5.5[b]
Conventional treatment; sedimentation and filtration credit only	2.0	2.5	3.0
Disinfection required	2.0	0.5	0 to 2.5
Direct filtration credit	1.0	2.0	2.5
Disinfection required	3.0	1.0	0 to 3.5
No filtration	0	0	0
Disinfection required	4.0	3.0	2.0 to 5.5

[a]log 10 removal: each log is a 90% removal of the original concentration in the source water.
[b]Requirement depends on concentration of Cryptosporidium oocysts in source water.

areas with large numbers of cattle may need to include processes that can remove greater numbers of pathogens. In developing the Surface Treatment Rule, the United States Environmental Protection Agency recognized this issue, and now requires utilities to assess the concentration of pathogens in water sources (U.S. EPA, 2003). Minimum requirements for water treatment are based upon concentrations of pathogens in the raw water (Table 28.4). Under these rules, a system must remove at least two logs (99%) of Cryptosporidium oocysts. However, because of concerns that some water treatment plants may draw water from poor quality sources with elevated levels of Cryptosporidium, additional treatment is required if monitoring shows oocyst concentrations of 0.075/liter or greater. The actual treatment required depends upon how many oocysts are detected in the raw untreated water.

As the treated drinking water travels though the distribution pipe system to the consumer, the microbial quality slowly degrades. The degradation is caused by several factors including the loss of disinfectant residual, biofilm sloughing, stirred-up pipe sediments caused by rapid changes in flow, pipe breaks, intrusions of contaminants into the pipe network from pressure drops and cross connections. Regrowth of bacteria that survived the treatment processes, and growth of bacteria in biofilms on pipe walls and surfaces in storage tanks and reservoirs, can also occur (Fox et al., 2006). Heterotrophic bacterial growth or regrowth usually occurs when the free chlorine residual drops below 0.2 mg/L, the water temperature exceeds 10°C, and assimibile organic carbon (AOC) is greater than 50 µg/L. Because of this, chloramines are sometimes added to water to provide a residual disinfectant for water within the distribution system. In addition, new real-time monitoring systems for both biological and chemical contaminants are currently being developed not only to ensure water quality, but also to protect against water intrusion via terrorist activities.

28.3 WATER DISTRIBUTION SYSTEMS

28.3.1 Microbial Growth

Once drinking water is treated, it must often travel through many miles of pipe or be held in storage

TABLE 28.5   Problems Caused by Biofilms in Distribution Systems

Frictional resistance of fluids
Photoreduction of H_2S because of anaerobic conditions
Taste and odor problems
Colored water (red, black) from activity of iron- and manganese-oxidizing bacteria
Resistance to disinfection
Regrowth of coliform bacteria
Growth of pathogenic bacteria (i.e., *Legionella*)

reservoirs before it reaches the consumer. The presence of dissolved organic compounds in this water can cause problems, such as taste and odors, enhanced chlorine demand, and bacterial colonization of water distribution systems (Bitton, 2011). Bacterial concentrations in distribution system water vary from <1 colony forming units (CFU)/ml to as high as 10^5 to 10^6 CFU/ml in water from slow flow or stagnant areas in the distribution system. The density of bacteria in pipe biofilms, however, may be several orders of magnitude higher, 10^5 to 10^7 per sq cm (Berger *et al.*, 2012). Biofilms of microorganisms in the distribution are of concern because of the potential for protection of pathogens from the action of residual disinfectant in the water, and the regrowth of indicator bacteria such as coliforms (Table 28.5).

Biofilms may appear as a patchy mass in some pipe sections or as uniform layers (see Section 6.2.4). They may consist of a monolayer of cells in a microcolony or can be as thick as 10 to 40 mm, as in algal mats at the bottom of a reservoir (Geldreich, 1996). These biofilms often provide a variety of microenvironments for growth that include aerobic and anaerobic zones because of oxygen limitations within the biofilm. Growth of biofilms proceeds up to a critical thickness, at which nutrient diffusion across the biofilm becomes limiting. Biofilm microorganisms are held together by an extracellular polymeric matrix called a glycocalyx. The glycocalyx is composed of exopolysaccharides (EPS) including glucans, uronic acids, glycoproteins and mannans. The glycocalyx helps protect microorganisms from predation and adverse conditions (e.g., disinfectants) (see Section 2.2.4).

The occurrence of even low levels of organic matter in the distribution system allows the growth of biofilm microorganisms. Factors controlling the growth of these organisms are temperature, water hardness, pH, redox potential, dissolved carbon and residual disinfectant.

It has been demonstrated that biofilms can coexist with chlorine residuals in distribution systems (Geldreich, 1996). *Escherichia coli* is 2400 times more resistant to chlorine when attached to surfaces than as free cells in the water, leading to high survival rates within the distribution system (LeChevallier *et al.*, 1988a). The health significance of coliform growth in distribution systems is an important consideration for water utilities, because the presence of these bacteria may mask the presence of indicator bacteria in water supplies resulting from a breakdown in treatment barriers. Total numbers of heterotrophic bacteria growing in the water or biofilm may interfere with the detection of coliform bacteria. High heterotrophic plate counts (HPC) are also indicative of a deterioration of water quality in the distribution system. It has been recommended that HPC numbers not exceed 500 organisms per milliliter.

Biofilms in distribution systems are difficult to inactivate. Chlorine levels commonly used in water treatment are inadequate to control biofilms. Free chlorine levels as high as 4.3 mg/L have proved inefficient in eliminating coliform occurrence. It has been suggested that monochloramine may be more effective in controlling biofilms because of its ability to penetrate the biofilm (LeChevallier *et al.*, 1988b).

Water-based pathogens such as *Legionella* may also have the ability to grow in biofilms and the distribution system. *Legionella* is known to occur in distribution systems and colonize the plumbing, faucet fixtures and taps in homes (Colbourne *et al.*, 1988). *Legionella* are more resistant to chlorine than *E. coli*, and small numbers may survive in the distribution system. Hot-water tanks in homes and hospitals favor their growth. *Legionella* survives well at 50°C, and it is capable of growth at 42°C (Yee and Wadowsky, 1982). It has been found that sediments in water distribution systems and the natural microflora favor their survival. *Legionella* associated with hot-water systems in hospitals have caused numerous outbreaks of disease in immunocompromised patients. Sporadic cases in communities show a strong association with the colonization of household taps. Forty percent of sporadic cases of this illness may be related to household taps or showers.

28.3.2 Organic Carbon and Microbial Growth

Bacterial growth in distribution systems is influenced by the concentration of biodegradable organic matter, water temperatures, nature of the pipes, disinfectant residual concentration and detention time within the distribution system (Bitton, 2011). Bacteria such as *Pseudomonas aeruginosa* and *P. fluorescens* are able to grow in tap water at relatively low concentrations (μg/L) of low-molecular-weight organic substrates such as acetate, lactate, succinate and amino acids. The amount of biodegradable organic matter available to microorganisms is difficult to determine from data from dissolved organic carbon (DOC) or total organic carbon (TOC) measurements. These measurements capture only the bulk water portion of organic matter, of which only some is biodegradable (Figure 28.4).

Several bioassay tests have been proposed using either pure cultures of selected bacteria or mixed flora from the source water for assessment of the assimilable organic carbon (AOC) in water (Geldreich, 1996). Measurements

of bacterial action in the test sample over time are determined by plate counts, direct cell count, ATP, turbidity, etc. (see Example Calculation 28.1). It is estimated that the assimilable organic carbon in tap water is between 0.1 and 9% of the total organic carbon (van der Kooij et al., 2003), although this fraction may be higher if the treatment train involves ozonation, which breaks down complex organic molecules and makes them more available to microorganisms (Table 28.6 and Figure 28.5). In fact, rapid regrowth of heterotrophic plate counts usually occurs after ozonation of tapwater. Addition of a secondary disinfectant is required to control this regrowth.

Because determination of AOC levels based on *P. fluorescens* does not always appear to be a good indicator of the growth potential for coliforms, a coliform growth response (CGR) test has been developed (Geldreich, 1996) (Figure 28.5). This procedure uses *Enterobacter cloacae* as the bioassay organism. Changes in viable densities of this organism in the test over a 5-day period at 20°C are used to develop an index of nutrients available to support coliform biofilm growth. The CGR result is calculated by log transformation of the ratio of the colony density achieved at the end of the incubation period, to the initial cell concentration: Thus:

$$CGR = \log (N_5/N_0) \qquad \text{(Eq. 28.1)}$$

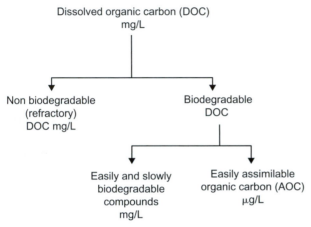

FIGURE 28.4   Fraction of organic matter in drinking water distribution systems. Based on Volk and LeChevallier (2000).

Example Calculation 28.1   Calculation of Assimilable Organic Carbon

Determination of assimilable organic carbon (AOC) involves a single bacterial species, usually either *Spirillum* NOX or *Pseudomonas fluorescens* P-17. The organism is inoculated into a sample of water which has been pasteurized by heat to kill the indigenous microflora and then the test bacterium in stationary phase is added. Growth is monitored (usually 7 to 9 days) until stationary phase is reached. The number of organisms at the stationary phase is assumed to be the maximum number of organisms that can be supported by the nutrients in the sample, and the yield (numbers of the test bacterium) on acetate carbon is assumed to equal the yield on naturally occurring AOC. When acetate is used as the carbon source in determination of yield, AOC concentrations may be reported as acetate-carbon equivalents. Reporting AOC as micrograms carbon per liter assumes that the yield (total number of bacteria after incubation for 7−9 days) on acetate is equal to the yield on naturally occurring AOC. In theory, the concentration of less than 1 μg carbon per liter can be detected. In practice, organic carbon contamination during glassware preparation and sample handling imposes a limit of detection of approximately 1−10 μg AOC/liter.

The AOC can be calculated as follows:

$$AOC\,(\mu g\ carbon/liter) = (N_{max} \times 1000)/Y$$

where

$$N_{max} = \text{maximum colony counts (CFU/ml) and}$$
$$Y = \text{yield coefficient(CFU/mg carbon).}$$

The AOC concentration is expressed as micrograms acetate-carbon equivalents per liter.

When using *P. fluorescens* strain P-17, $Y = 4.1 \times 10^6$ CFU/μg carbon. Thus, if the final yield of the test organism is 5×10^6/ml after 9 days of incubation,

$$AOC\ of\ sample = 1.22\ \mu g\ acetate\text{-}carbon\ equivalents/liter$$

where N_5 = number of CFU per milliliter at day 5 and N_0 = number of CFU per milliliter at day 0. Any sample that demonstrates a 1-log or greater increase is interpreted as supporting coliform growth. Calculated values between 0.51 and 0.99 are considered to be moderately growth supportive, and those less than 0.5 are regarded as not supportive of coliform growth.

It is important to note that the CGR test responds only to the concentrations of assimilable organic materials that support growth of coliforms characteristic of regrowth in biofilms. In fact, parallel assays comparing *E. coli* response with that of *Enterobacter cloacae* indicate a significant difference in growth response between these two coliforms. *Ent. cloacae* growth can occur in nutrient concentrations far below those required by *E. coli*.

Detecting any changes in the dissolved organic concentrations is another approach to obtaining a measure of assimilable organic carbon. Rather than using pure cultures, these procedures utilize the indigenous microflora of the raw surface-water source of the biomass washed from sand used in the sand filter during drinking water treatment (van der Kooij, 2003).

The biodegradable dissolved organic carbon (BDOC) is given by the following formula:

$$BDOC \ (mg/L) = initial \ DOC - final \ DOC \quad (Eq. \ 28.2)$$

The general approach is as follows. A water sample is sterilized by filtration through a 0.2-μm pore size filter, inoculated with indigenous microorganisms, and incubated in the dark at 20°C for 10−30 days until DOC reaches a constant level. BDOC is the difference between the initial and final DOC values (Servais *et al.*, 1987). The advantage of using the consortium of heterotrophic microorganisms that occurs in these aquatic habitats is in their acquired proficiency to degrade a diverse spectrum of dissolved organics that may be in test samples. Whether the inoculum is from the raw source water or the sand filter biofilm has not been found to be critical for optimal test performance.

TABLE 28.6 Concentrations of Assimilable Organic Carbon (AOC) in Various Water Samples

Source of Water	DOC (mg C/L)[a]	AOC (mg C/L)[a]
River Lek	6.8	0.062−0.085
River Meuse	4.7	0.118−0.128
Brabantse Diesbosch	4.0	0.08−0.103
Lake Yssel, after open storage	5.6	0.48−0.53
River Lek, after bank filtration	1.6	0.7−1.2
Aerobic groundwater	0.3	<0.15

Adapted from van der Kooij (2003).
[a]*DOC, dissolved organic carbon; AOC, assimilable organic carbon; mg C/L, milligrams carbon per liter.*

28.3.3 Microbial Community Structure

Genomic-based molecular methods have greatly increased our ability to understand microbial community dynamics in drinking water distribution systems. Recent studies indicate that these communities are complex and influenced by the source water (ground vs. surface), chemical properties of the water, treatment and type of disinfectant residual. Studies have shown Alphaproteobacteria, Betaproteobacteria or Gammaproteobacteria are in predominance (Hwang *et al.*, 2012). The abundance of different groups of bacteria has

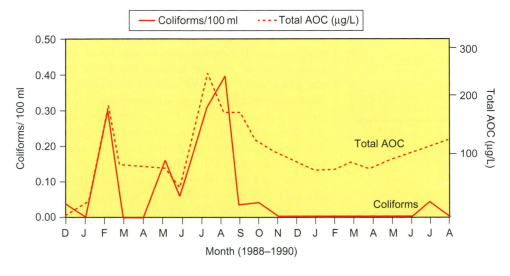

FIGURE 28.5 Relationship between mean coliform densities and total assimilable organic carbon (AOC) levels. From LeChevallier *et al.* (1992).

TABLE 28.7 Distribution of Opportunistic Pathogens in Distributions with Different Disinfection Residuals

Organism	Percent of Total in	
	Free Chlorine	Chloramine
Mycobacterium	1.29	19.65
Legionella	0.31	0.09
Amoeba	0.03	>0.0001

TABLE 28.8 Distribution of Members of Bacteria Domain Determined via Taxonomic Identifications of Annotated Proteins at the Class Level

Domain	Free Chlorine %[a]	Chloramine %
Actinobacteria	6.2	27.8
Cytophaga	0	2.3
Flavobacteria	0	2.3
Sphingobacteria	0	2.0
Chlamydiae	0.4	0.1
Chlorobia	1.4	0
Chloroflexi	1.3	0
Gloeobacteria	1.3	0
Cyanobacteria	9.0	0
Bacilli	4.1	0
Clostridia	5.6	0
Planctomycetacia	1.3	0
Alphaproteobacteria	35.1	22.5
Betaproteobacteria	6.2	24.1
Deltaproteobacteria	11.9	10.5
Gammaproteobacteria	0.5	0.1
Other classes representing <1%	15.4	8.4
Total	**100**	**100**

Source: Gomez-Alvarez et al. (2012).
[a]*Each number in brackets = % total sequences in each group.*

been found to vary between distribution systems that have a free chlorine residual and those that use chloramines (Gomez-Alvarez *et al.*, 2012). Such changes in community structure can be significant to protection of community

health as it was found that disinfection type can cause changes in abundance of opportunistic pathogens (Tables 28.7 and 28.8).

28.3.4 Intrusion Events

Intrusion of water into drinking water distribution systems can result in exposure to consumers after the water has been treated. Intrusion can occur from cross connection with water containing wastes, changes in water pressure or intentional addition from terrorist-motivated events. Transient negative pressure events (which create suction into the pipes) can occur in pipelines, and if leaks are present, provide a potential portal for water into the distribution systems. Karim *et al.* (2003) examined 66 soil and water samples immediately adjacent to drinking water pipes from eight utilities in six states of the United States. About 56% of the samples were found positive for human enteric viruses. In addition, total fecal coliform levels in some soil samples were greater than 1.6×10^4 CFU/100 g of soil, suggesting that the sampling locations were potentially under the influence of leaking sewage pipes. Several epidemiological studies have found that the greater the distance a person lives from the treatment plant, the greater the incidence of gastroenteritis. The use of point-of-use treatment devices at consumers' taps has also found a decrease in illness for gastroenteritis for the young and elderly. To better understand contamination in distribution systems, real-time monitoring systems are receiving increased attention.

28.4 REAL-TIME MONITORING OF MICROBIAL CONTAMINANTS IN WATER DISTRIBUTION SYSTEMS

Even in developed countries, microbial contamination of drinking water is a major issue, and over the past decade, the number of waterborne outbreaks associated with water distribution systems has increased. Contaminated drinking water in the distribution system can result from inadequate treatment, or from leaks or breaks in the distribution pipes, which result in accidental intrusion events that allow microbial or chemical contamiants to enter the potable water. In addition, EPA is concerned about deliberate intrusion events that could occur through bioterrorist activities. Traditionally, utilities have utilized indicator tests for fecal pollution to monitor for the potential presence of pathogens. However, such cultural assays can take up to 48 hours to complete, during which time contaminated water could be delivered to consumers.

Recently, utilities have been evaluating new online monitoring systems to augment traditional monitoring. Specifically, the goal has been to integrate software for

---

Case Study 28.1   Real-time Detection of *E. coli* and *Bacillus* Spores using Multi-angle Light Scattering

Multi-angle light scattering (MALS) is a technology that utilizes laser light scattering to detect intrusion of *E. coli* cells within a water distribution system (Miles *et al.*, 2011). Real-time detection was successful in tap water over a concentration range of $10^3 - 10^6$ cells per ml. Cell numbers as determined by MALS were similar to those obtained by conventional assays such as dilution and plating (see Chapter 10), and acridine orange direct counts (AODC) (see Chapter 9).

MALS was also utilized to detect *Bacillus thuringiensis* spores (a surrogate for *Bacillus anthracis* spores) introduced into a distribution system over a concentration range of $10^2 - 10^5$ spores per ml (Sherchan, 2013) (Figure 28.6).

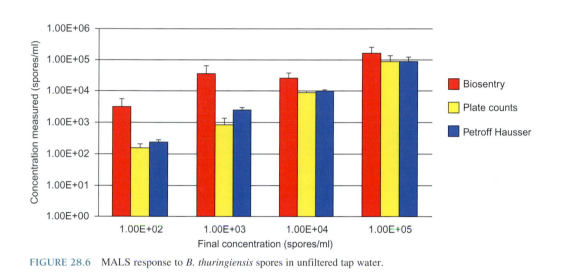

FIGURE 28.6   MALS response to *B. thuringiensis* spores in unfiltered tap water.

---

data management with new real-time sensor technologies to provide an early warning monitoring program via a supervisory control and data acquisition system (SCADA), installed at critical points within the distribution system. Use of such real-time technologies allows for a rapid response to contamination that safeguards the public from consuming contaminated water. Although utilities are concerned with both chemical and microbial contaminants, here we focus on real-time detection of microbial contaminants.

28.4.1 Real and Near Real-Time Technologies Form Monitoring Microbial Contaminants

Physical microbial characteristics can be used to detect microbial concentrations via several methodologies including vibrational spectroscopy and multi-angle light scattering technologies. Vibrational spectroscopy utilizes spectra that are emitted following excitation of molecules by laser light. These technologies include Raman spectroscopy and Fourier transformed infrared spectroscopy (Driskell *et al.*, 2005; Rule and Vikesland, 2009).

Multi-angle light scattering (MALS), which results from laser light that strikes particulates, can also be used to detect microorganisms. The specific light scattering that results allows for the differentiation of bacterial cells and spores in real time. Such technologies can also be used to detect increases in microbial cells in water, in essence functioning as a cell counter. To date, viruses cannot be detected in real time. In addition, the sensitivity of detection still needs to be improved to enable the technology to be used in water.

Near real-time assays can be defined as those assays that require approximately 2 hours to complete. These include quantitative PCR; ATP measurements via bioluminescence assays; antibody assays; flow cytometry; and matrix-assisted laser desorption/ionization mass spectrometry (MALDI-MS). Many of these assays are still in their infancy, and many still have issues related to sensitivity and specificity (Sherchan, 2013). Biosensors are also being developed that rely on recognition of specific biological targets including proteins or nucleic acid. Immunosensors rely on the interaction between antigens on target cells and antibodies (Shirale *et al.*, 2010). Viruses can be detected in near real time through the use of qPCR with a detection time of approximately 2 hours.

However, a major limitation of qPCR detection is that it does not discriminate between infectious and noninfectious virus. Very recently, aptamers have been utilized to detect microbes (Liu *et al.*, 2012). Aptamers are single-stranded nucleic acids (RNA or DNA) with defined tertiary structures for selective binding to target molecules. Aptamers can be readily produced by chemical synthesis and modified to result in aptamers that can bind to a broad range of biological targets with high affinity and specificity comparable to those of antibodies (Li *et al.*, 2012).

Overall, the field of real and near real-time detection of microbes is advancing rapidly, but many challenges remain, including the need for a real-time assay for viruses.

QUESTIONS AND PROBLEMS

1. Why is it important to reduce the amount of biodegradable organic matter and nutrients during water treatment?
2. Describe the major steps in the conventional treatment of drinking water.
3. What group of waterborne pathogens is most effectively removed by filtration? Why?
4. What methods can be used to assess the growth of bacteria in water?
5. Which pathogenic microorganisms are the most difficult to remove by conventional water treatment and why?
6. Why is coliform regrowth in distribution systems a problem?
7. Why does the HPC increase after ozonation of drinking water?
8. A water treatment plant is required to remove 3 $\log_{10}$ of *Cryptosporidium* oocysts from their water source. What treatment processes do you recommend? What if they have to remove 5 $\log_{10}$? See Chapter 25 for *Cryptosporidium* resistance to disinfectants.
9. To determine the AOC of drinking water, two 40 ml flasks are inoculated with 500 colony forming units (CFU) of *Pseudomonas fluorscens* strain P-17. After 9 days the concentration in one flask has reached 2.4×10^5 CFU/ml and 1.2×10^5 CFU/ml. Calculate the average AOC in this drinking water sample.
10. The type of residual disinfectant used in distribution systems affects the microbial community living in those systems. Why?

REFERENCES AND RECOMMENDED READING

Berger, P. S., Clark, R. M., Reasoner, D. J., Rice, E. W., and Santo Domingo, J. W. (2012) Drinking water. In "Topics in Ecological and Environmental Microbiology" (T. M. Schmidt, and M. Schaechter, eds.), Academic Press, San Diego.

Bitton, G. (2011) "Wastewater Microbiology," Wiley-Liss, New York.

Colbourne, J. S., Dennis, P. J., Trew, C., Berry, C., and Vesey, C. (1988) *Legionella* and public water supplies. *Wat. Sci. Technol.* **20**, 5–10.

Driskell, J. D., Kwarta, K. M., Lipert, R. J., Porter, M. D., Neill, J. D., and Ridpath, J. F. (2005) Low-level detection of virtual pathogens by a surface-enhanced Raman scattering based immunoassay. *Anal. Chem.* **77**, 6147–6154.

Fox, K. R., Reasoner, D. J., and Gertig, K. R. (2006) Water quality in source water, treatment, and distribution systems. In "Waterborne Pathogens," 2nd ed., American Water Works Association, Denver, CO, pp. 21–34.

Geldreich, E. E. (1996) "Microbial Quality of Water Supply in Distribution Systems," Lewis, Boca Raton, FL.

Gomez-Alvarez, V., Revetta, R. P., and Santo Domingo, J. W. (2012) Metagenomic analysis of drinking water receiving different disinfection treatments. *Appl. Environ. Microbiol.* **78**, 6095–6102.

Hwang, C., Ling, F., Andersen, G. L., LeChevallier, M. W., and Liu, W. T. (2012) Microbial community dynamics of an urban drinking water distribution system subjected to phases of chloramination and chlorination and chlorination treatments. *Appl. Environ. Microbiol.* **78**, 7856–7865.

Karim, M., Abbaszadegan, M., and LeChevallier, M. W. (2003) Potential for pathogen intrusion during pressure transients. *J. Am. Water Works Assoc.* **95**, 134–146.

LeChevallier, M. W., Becker, W. C., Schorr, P., and Lee, R. G. (1992) Evaluating the performance of biologically active rapid filters. *J. Am. Water Works Assoc.* **84**, 136–146.

LeChevallier, M. W., Cawthon, C. P., and Lee, R. G. (1988b) Inactivation of biofilm bacteria. *Appl. Environ. Microbiol.* **54**, 2492–2499.

LeChevallier, M. W., Cawthon, C. P., and Lee, R. G. (1988a) Factors promoting survival of bacteria in chlorinated water supplies. *Appl. Environ. Microbiol.* **54**, 649–654.

Li, L. L., Ge, P., Selvin, P. R., and Lu, Y. (2012) Direct detection of adenosine in undiluted serum using a luminescent aptamer sensor attached to a Terbium complex. *Anal. Chem.* **84**, 7852–7856.

Liu, G., Yu, X., Xue, F., Chen, W., Ye, Y., Yang, X., *et al.* (2012) Screening and preliminary application of a DNA aptamer for rapid detection of *Salmonella* 08. *Microchim. Acta* **178**, 237–244.

Logsdon, G. S., and Lippy, E. C. (1982) The role of filtration in preventing waterborne disease. *J. Am. Water Works Assoc.* **74**, 649–655.

Miles, S. L., Sinclair, R. G., Riley, M. R., and Pepper, I. L. (2011) Evaluation of select sensors for real-time monitoring of *Escherichia coli* in water distribution systems. *Appl. Environ. Microbiol.* **77**, 2813–2816.

Montgomery, J. M. (1985) "Water Treatment Principles and Design," John Wiley & Sons, New York.

Rao, V. C., Sumons, J. M., Wang, P., Metcalf, T. G., Hoff, J. C., and Melnick, J. L. (1988) Removal of hepatitis A virus and rotavirus by drinking water treatment. *J. Am. Water Works Assoc.* **80**, 59−67.

Rule, K. L., and Vikesland, P. J. (2009) Surface-enhanced resonance Raman spectroscopy for the rapid detection of *Cryptosporidium parvum* and *Giardia lamblia*. *Environ. Sci. Technol.* **43**, 1147−1152.

Servais, P., Billen, G., and Hascoet, M.-C. (1987) Determination of the biodegradable fraction of dissolved organic matter in waters. *Water Res.* **21**, 445−452.

Sherchan, S. P. (2013) Real-time monitoring of microorganisms in water. Ph.D. dissertation, The University of Arizona, Tucson, AZ.

Shirale, D. J., Bangar, M. S., Park, M., Yates, M. V., Chen, W., Myung, N. V., *et al.* (2010) Label-free chemiresistive immunosensors for viruses. *Environ. Sci. Technol.* **44**, 9030−9035.

U.S. EPA (1988) "Comparative Health Effects Assessment of Drinking Water Treatment Technologies," Office of Drinking Water, U.S. Environmental Protection Agency, Washington, DC.

U.S. EPA (2003) "National Primary Drinking Water Regulations: Long Term 2 Enhanced Surface Treatment Rule, Proposed Rule." Federal Register 40CFR Parts 141 and 142, 47639−47795.

U.S. EPA (1991) "Guidance Manual for Compliance with the Filtration and Disinfection Requirements for Public Water Systems using Surface Water Sources." March 1991 Edition, Office of Drinking Water, U.S. Environmental Protection Agency, Washington, DC.

van der Kooij, D. (2003) Managing regrowth in drinking water distribution systems. In "Heterotrophic Plate Counts and Drinking-Water Safety" (J. Bartram, M. Cotruvo, M. Extner, C. Fricker, and A. Glasmacher, eds.), IWA Publishing, London, pp. 199−232.

Volk, C., and LeChevallier, M. W. (2000) Assessing biodegradable organic matter. *J. Am. Water Works Assoc.* **92**, 64−67.

Yee, R. B., and Wadowsky, R. M. (1982) Multiplication of *Legionella pneumophila* in unsterilized tap water. *Appl. Environ. Microbiol.* **43**, 1130−1134.

Chapter 29

Disinfection

Charles P. Gerba

The destruction or prevention of growth of microorganisms is essential for the control of infectious disease transmission and preservation of foodstuffs and biodegradable materials. This is most commonly accomplished by heat, chemicals, filtration or radiation. Heat acts to kill or inactivate by denaturation of essential proteins (enzymes, viral capsids) and nucleic acids. Chemicals may act by many different means to kill organisms or prevent their growth, including destruction of membranes and cell walls, and interference with enzymic action and replication of nucleic acids (Table 29.1). Filtration is a process that acts to remove the organisms physically by size exclusion and does not result in destruction of the organism. Ultraviolet light and gamma radiation act directly on nucleic acids.

Sterilization is a process, physical or chemical, that destroys or eliminates all organisms. A sanitizer is an agent that reduces the number of bacterial contaminants to safe levels as judged by public health requirements. According to the official sanitizer test used in the United States, a sanitizer is a chemical that kills 99.9% of the specific test bacteria within 30 seconds under the conditions of the test (Block, 1991). A disinfectant is a physical or chemical agent that destroys disease-causing or other harmful microorganisms, but does not necessarily kill all microorganisms. Disinfectants are expected to kill more than 99.999% of the test organisms. Disinfectants are usually applied to water and inanimate objects (fomites) to control the spread of pathogenic microorganisms. They can also be used to treat foods and aerosols. A bacteriostat is usually a chemical agent that prevents the growth of bacteria but does not necessarily kill them. For example, silver is often added to activated carbon to prevent the growth of bacteria in home faucet-mounted water treatment devices.

29.1 THERMAL DESTRUCTION

The thermal destruction of microorganisms has been studied in great detail by the food industry because of the importance of this process in killing pathogenic bacteria and preventing foodborne spoilage. The thermal death of microorganisms is generally considered a first order relationship, i.e., linear with time. The time necessary to kill a given number of organisms at a specific temperature is called the thermal death time (TDT). The general procedure for determining TDT by these methods is to place a known number of organisms in a sufficient number of sealed containers to get the desired number of survivors for the test period. At the end of the heating period, the containers are quickly removed and cooled in cold water. Viability of the organism is assessed on standard culture media.

The TDTs of some foodborne and waterborne pathogens are shown in Table 29.2. The D value or decimal

I.L. Pepper, C.P. Gerba, T.J. Gentry: Environmental Microbiology, Third edition. DOI: http://dx.doi.org/10.1016/B978-0-12-394626-3.00029-6
© 2015 Elsevier Inc. All rights reserved.

TABLE 29.1 Mechanisms of Inactivation Used by Common Disinfectants

Target	Agent	Effect
Cell wall	Aldehydes Anionic surfactants	Interaction with $-NH_2$ groups Lysis
Cytoplasmic membrane	Quaternary ammonium compounds, biguanides, hexachlorophene	Leakage of low molecular weight material
Nucleic acids	Dyes, alkylating agents, ionizing and ultraviolet radiation	Breakage of bonds, cross-linking, binding of agents to nucleic acids
Enzymes or proteins	Metal ions (Ag, Cu) Alkylating agents Oxidizing agents (chlorine, hydrogen peroxide)	Bind to $-SH$ groups of enzymes Combine with DNA or RNA Damage of bacterial cell membranes; damage of proteins and nucleic acid

From Block (1991).

TABLE 29.2 Thermal Death Times of Water- and Food-borne Pathogenic Organisms

Organism	Temperature (°C)/time (min)	Reference
Campylobacter spp.	75/1	Bandres *et al.*, 1988
Escherichia coli	65/1	Bandres *et al.*, 1988
Legionella	66/0.45[a]	Sanden *et al.*, 1989
Mycobacterium spp.	70/2	Robbecke and Buchholtz, 1992
M. avium	70/2.3[a]	
Salmonella spp.	65/1	Bandres *et al.*, 1988
Shigella spp.	65/1	Bandres *et al.*, 1988
Vibrio cholerae	55/1[a]	Roberts and Gilbert, 1979
Cryptosporidium parvum	72.4/1	Fayer, 1994
Giardia lamblia	50/1[a]	Cerva, 1955
Hepatitis A virus	70/10	Siegl *et al.*, 1984
Rotavirus	50/30	Estes *et al.*, 1979

[a]*In buffered distilled water.*

reduction time is the time required to destroy 90% of the organisms. This value is numerically equal to the number of minutes required for the survivors as a function of time curve to traverse one log (Figure 29.1). This is equal to the reciprocal of the slope of the survivor curve, and is a measure of the death rate of an organism. The temperature at which the D value is determined is given as a subscript. For example, the D value for *Clostridium*

perfringens at 250°F is $D_{250} = 0.1$ to 0.2 (Jay, 1996). The z value refers to the degrees Fahrenheit required for the thermal destruction as a function of the temperature curve to traverse one log. This value is equal to the reciprocal of the slope of the TDT curve (Figure 29.2). Whereas D reflects the resistance of an organism to a specific temperature, z provides information on the relative resistance of an organism to different destructive temperatures; it allows the calculation of equivalent thermal processes at different temperatures. If, for example, 3.5 minutes at 140°F is considered to be an adequate process and $z = 8.0$, either 0.35 minutes at 148°F or 35 minutes at 132°F would be considered equivalent processes.

Spore-forming bacteria such as *Bacillus* and *Clostridium* are the most resistant to heat inactivation. Of the nonspore-forming waterborne and foodborne enteric pathogens, enteric viruses are the most heat resistant, followed by the bacteria and protozoa (Table 29.2). Parvoviruses are among the most heat-resistant viruses known (Eterpi *et al.*, 2009). In addition to the type of microorganism, factors that influence TDT in foods include water, fat, salts, sugars, pH and other substances. The heat resistance of microbial cells increases with decreasing humidity or moisture. Dried microbial cells are considerably more heat resistant than moist cells of the same type. Because protein denaturation occurs at a higher rate with heating in water than in air, it is likely that protein denaturation is closely associated with thermal death. The presence of fats and salts increases the heat resistance of some microorganisms. The effect of salt is variable and dependent on the kind of salt, concentration and cation. Cationic salts at molar concentrations greatly increase the thermal resistance of enteric viruses. For this reason, $MgCl_2$ is added to poliovirus vaccine to aid in extending its useful life. The presence of sugars causes an increase in the heat resistance of microorganisms, in part because of decreased water activity. Microorganisms are most

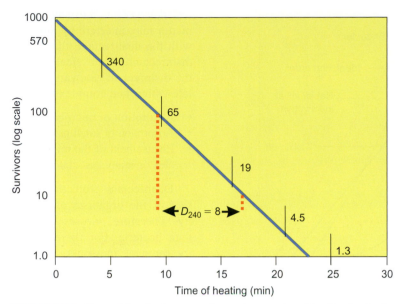

FIGURE 29.1 Thermal inactivation curve for a microorganism. The D value is the time required for inactivation of 90% of the organisms at a given temperature. In this case it required 8 minutes to kill 90% of the organisms at 240°F or $D_{240} = 8$ minutes.

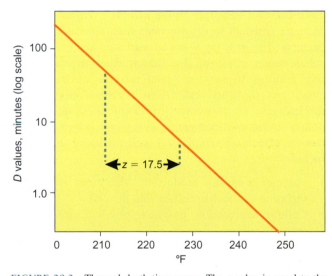

FIGURE 29.2 Thermal death time curve. The z value is equal to the degrees Fahrenheit required for the thermal destruction curve to traverse one log cycle.

resistant to heat at their optimal pH of growth, which is generally about 7.0. As the pH is lowered or raised from the optimal value, there is an increase in heat sensitivity. Thus, acid and alkaline foods require less heat processing than neutral foods. In water, suspended solids or organic matter increase heat resistance (Liew and Gerba, 1980).

29.2 KINETICS OF DISINFECTION

Inactivation of microorganisms is a gradual process that involves a series of physicochemical and biochemical steps. In an effort to predict the outcome of disinfection, various models have been developed on the basis of experimental data. The principal disinfection theory used today is still the Chick—Watson model, which expresses the rate of inactivation of microorganisms by a first-order chemical reaction.

$$N_t/N_0 = e^{-kt} \qquad \text{(Eq. 29.1)}$$

or

$$ln \ N_t/N_0 = -kt \qquad \text{(Eq. 29.2)}$$

where:

N_0 = number of microorganisms at time 0
N_t = number of microorganisms at time t
k = decay constant (1/time), and
t = time.

The logarithm of the survival rate (N_t/N_0) plots as a straight line versus time (Figure 29.3). Unfortunately, laboratory and field data often deviate from first-order kinetics. Shoulder curves may result from clumps of organisms or multiple hits of critical sites before inactivation. Curves of this type are common in disinfection of coliform bacteria by chloramines (Montgomery, 1988). The tailing-off curve, often seen with many disinfectants, may be explained by the survival of a resistant subpopulation as a result of protection by interfering substances (suspended matter in water), clumping or genetically conferred resistance.

In water applications, disinfectant effectiveness can be expressed as $C \cdot t$, where C is the disinfectant concentration and t the time required to inactivate a certain

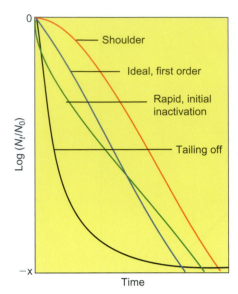

FIGURE 29.3 Types of inactivation curves observed for microorganisms.

percentage of the population under specific conditions (pH and temperature). Typically, a level of 99% inactivation is used when comparing $C \cdot t$ values. The lower the $C \cdot t$ value, the more effective the disinfectant. The $C \cdot t$ method allows a general comparison of the effectiveness of various disinfectants on different microbial agents (Tables 29.3 through 29.6). It is used by the drinking water industry to determine how much disinfectant must be applied during treatment to achieve a given reduction in pathogenic microorganisms. $C \cdot t$ values for chlorine for a variety of pathogenic microorganisms are shown in Table 29.3. The order of resistance to chlorine and most other disinfectants used to treat water is protozoan cysts > viruses > vegetative bacteria.

29.3 FACTORS AFFECTING DISINFECTANTS

Numerous factors determine the effectiveness and/or rate of kill of a given microorganism (Figure 29.4). Temperature has a major effect as it controls the rate of chemical reactions. Thus, as temperature increases, the rate of kill with a chemical disinfectant increases. The pH can affect the ionization of the disinfectant and the viability of the organism. Most waterborne organisms are adversely affected by pH levels below 3 and above 10. In the case of halogens such as chlorine, pH controls the amount of HOCl (hypochlorous acid) and $^-$OCl (hypochlorite) in solution (Figure 29.5). HOCl is more effective than $^-$OCl in the disinfection of microorganisms. With chlorine, the $C \cdot t$ increases with pH. Attachment of organisms to surfaces or particulate matter in water such as clays and organic detritus aids in the

resistance of microorganisms to disinfection. Particulate matter may interfere by either acting chemically to react with the disinfectant, thus neutralizing the action of the disinfectant, or physically shielding the organism from the disinfectant (Stewart and Olson, 1996). The particulate—microbial complex may be thought of as:

- Adsorption of microbes to larger particles
- Adsorption of small particles to the surface of the microbe
- Encasement of the microbe by one or more large particles or many associated small particles

Disinfectant protection is enhanced with decreasing size of the organism and increasing particle availability. Therefore, viruses are afforded greater protection than bacteria. For these reasons, particulate or turbidity removal in drinking water treatment is necessary to ensure the effectiveness of disinfection in the destruction of waterborne pathogens. Dissolved chemical substances that interfere with chemical disinfection include organic compounds, inorganic and organic nitrogenous compounds, iron, manganese and hydrogen sulfide.

Studies have demonstrated that pathogenic and indicator bacteria occurring in the natural environment may be more resistant to disinfectants than laboratory-grown bacteria. This resistance is cell mediated and physiological in nature, requiring that the organism develop adaptive features to survive under adverse environmental conditions (Stewart and Olson, 1996). Cell-mediated mechanisms of resistance to disinfectant agents are poorly understood compared with physicochemical protective effects. Examples of cell-mediated resistance include:

- Polymer or capsule production, which may act to limit diffusion of the disinfectant into the cell
- Cellular aggregation, providing physical protection to internal cells
- Cell wall and/or cell membrane alterations that result in reduced permeability to disinfectants
- Modification of sensitive sites, i.e., enzymes (Stewart and Olson, 1996)
- Efflux pumps to enhance removal of the substance from the bacteria (as in the case of metals and some antibiotics)
- Production of proteins to sequester metal ions

It has been speculated that many of these physiological events are a function of adaptation to low-nutrient conditions in the environment.

Repeated exposure of bacteria and viruses to strong oxidizing agents like chlorine may result in some selection for greater resistance (Bates *et al.*, 1977; Haas and Morrison, 1981). However, the enhanced resistance is not great enough to overcome concentrations of chlorine applied in practice.

TABLE 29.3 $C \cdot t$ Values for Chlorine Inactivation of Microorganisms in Water (99% Inactivation)[a]

Organism	°C	pH	$C \cdot t$
Bacteria			
Escherichia. coli	5	6.0	0.04
E. coli	23	10.0	0.6
Legionella. pneumophila	20	7.7	1.1
Mycobacterium avium (strain A5)	23	7.0	106
Mycobacterium avium (strain 1060)	23	7.0	204
Helicobacter pylori	5	6.0	0.12
Viruses			
Polio 1	5	6.0	1.7
Echo 1	5	6.0	0.24
Echo 1	5	7.8	0.56
Echo 1	5	10.0	47.0
Coxsackie B5	5	7.8	2.16
Coxsackie B5	5	10.0	33.0
Adenovirus 40	5	7.0	0.15
Protozoa			
Giardia lamblia cysts	5	6.0	54−87
Giardia lamblia cysts	5	7.0	83−133
Giardia lamblia cysts	5	8.0	119−192
Naegleria fowleri tropozoites	25	7.5	6
N. fowleri cysts	25	7.5	32.1
Encephalitozoon intestinalis spores	5	7.0	39
Cryptosporidium parvum oocysts	21−24	7.5−7.6	9740−11,300
Toxoplasma gondii	22	7.2	>133,920
Fungi			
Aspergillus fumigatus	25	7.0	946
A. terrus	25	7.0	1404
Penicillium citrirnum	25	7.0	959
Cladosporium tenuissimum	25	7.0	71

From Sobsey (1989); Rose et al. (1997); Gerba et al (2003); Wainwright et al. (2007). Shields et al. (2008) Sarkar and Gerba (2012); Pereira et al. (2013).
[a]*In buffered distilled water.*

29.4 HALOGENS

29.4.1 Chlorine

Chlorine and its compounds are the most commonly used disinfectants for treating drinking and wastewater. Chlorine is a strong oxidizing agent which, when added as a gas to water, forms a mixture of hypochlorous acid (HOCl) and hydrochloric acids:

$$Cl_2 + H_2O \equiv HOCl + HCl \qquad \text{(Eq. 29.3)}$$

In dilute solutions, little Cl_2 exists in solution. The disinfectant's action is associated with the HOCl formed. Hypochlorous acid dissociates as follows:

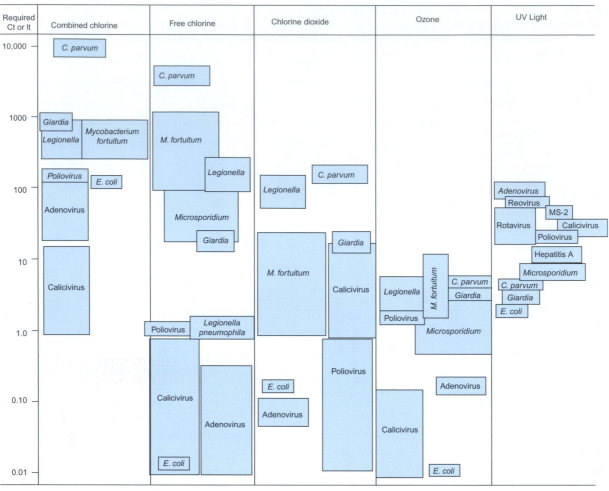

FIGURE 29.4 Overview of disinfection requirements for 99% inactivation of microorganisms. Ct = concentration of disinfectant × time. It = (μW s/cm^2) (time). Adapted from Jacangelo *et al.* (1997).

$$HOCl \equiv +H^+OCl^- \qquad \text{(Eq. 29.4)}$$

The ratio of hypochlorous acid and $^-$OCl (hypochorite ion) depends on the pH of the water (Figure 29.5). The amount of HOCl is greater at neutral and lower pH levels, resulting in greater disinfection ability of chlorine at these pH levels. Chlorine as HOCl or $^-$OCl is defined as free available chlorine. HOCl combines with ammonia and organic compounds to form what is referred to as combined chlorine. The reactions of chlorine with ammonia and nitrogen-containing organic substances are of great importance in water disinfection. These reactions result in the formation of monochloramine, dichloramine, trichloramine, etc.

$$NH_3 + HOCl \rightarrow \underset{\text{monochloramine}}{NH_2Cl} + H_2O \qquad \text{(Eq. 29.5)}$$

$$NH_2Cl + HOCl \rightarrow \underset{\text{dichloramine}}{NHCl_2} + H_2O \qquad \text{(Eq. 29.6)}$$

$$NHCl_2 + HOCl \rightarrow \underset{\text{trichloramine}}{NCl_3} + H_2O \qquad \text{(Eq. 29.7)}$$

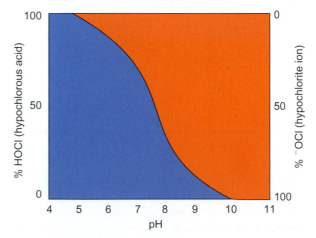

FIGURE 29.5 Distribution of HOCl and $^-$OCl in water as a function of pH. From Bitton (2011).

Such products retain some disinfecting power of hypochlorous acid, but are much less effective at a given concentration than chlorine.

Free chlorine is quite efficient in inactivating pathogenic microorganisms. In drinking water treatment, 1 mg/L or less for about 30 minutes is generally sufficient to reduce significantly bacterial numbers. The presence of interfering substances in wastewater reduces the disinfection efficacy of chlorine, and relatively high concentrations of chlorine (20–40 mg/L) are required (Bitton, 2011). Enteric viruses and protozoan parasites are more resistant to chlorine than bacteria (Table 29.3) and can be found in secondary wastewater effluents after normal disinfection practices. *Cryptosporidium* and *Toxoplasma* oocysts are extremely resistant to chlorine. A chlorine concentration of 100 mg/L is necessary to cause 99% inactivation of *Cryptosporidium* following a 100-minute contact time (Table 29.3). Chloramines are much less efficient than free chlorine (about 50 times less efficient) in inactivation of viruses.

Being a strong oxidizing agent, chlorine will react with any organic molecule including proteins, lipids, carbohydrates and nucleic acids to disrupt their structure. Bacterial inactivation by chlorine may result from (Stewart and Olson, 1996):

- Altered permeability of the outer cellular membrane, resulting in leakage of critical cell components
- Interference with cell-associated membrane functions (e.g., phosphorylation of high-energy compounds)
- Impairment of enzyme and protein function as a result of irreversible binding of the sulfyhydryl groups
- Nucleic acid denaturation

The actual mechanism of chlorine inactivation may involve a combination of these actions, or merely the effect of chlorine on a few critical sites. It appears that bacterial inactivation by chlorine is primarily caused by impairment of physiological functions associated with the bacterial cell membrane.

Chlorine may inactivate viruses by interaction with the viral capsid proteins or/and the nucleic acid (Figure 29.6). The site of action may also depend on the concentration of chlorine and the type of virus. It has been found that at free chlorine concentrations of less than 0.8 mg/L, inactivation of poliovirus RNA occurs without major structural changes, whereas chlorine concentrations in excess of 0.8 mg/L result in damage to the viral RNA and protein capsid (Alvarez and O'Brien, 1982) (Figure 29.11). The protein coat appears to be the target for the double-stranded RNA rotaviruses (Vaughn and Novotny, 1991). The protein involved in the binding to the host bacterium was found be involved with the loss of infectivity in MS 2 phage (Wigginton *et al.*, 2012) (Figure 29.7).

29.4.2 Chloramines

Inorganic chloramines are produced by combining chlorine and ammonia (NH_4) for drinking water

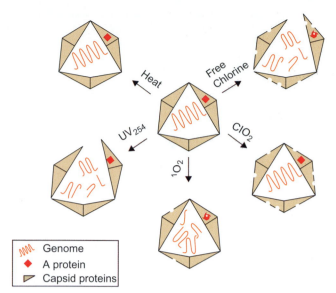

FIGURE 29.6   Mechanisms of MS 2 virus inactivation by disinfectants. Wigginton *et al.* (2012).

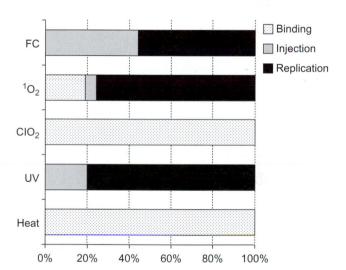

FIGURE 29.7   Relative contributions of MS 2 host binding protein, nucleic acid injection and replication to overall inactivation by different disinfectants. Wigginton *et al.* (2012).

disinfection. The species of chloramines formed (see Eqs. 29.5 through 29.7) depends on a number of factors, including the ratio of chlorine to ammonia-nitrogen, chlorine dose, temperature and pH. Up to a chlorine-to-ammonia mass ratio of 5, the predominant product formed is monochloramine, which demonstrates greater disinfection capability than other forms such as dichloramine and trichloramine. Chloramines are used to disinfect drinking water by some utilities in the United States, but because they are slow acting, they have mainly been used as secondary disinfectants when a residual in the distribution system is desired. For example, when ozone is used to treat drinking water,

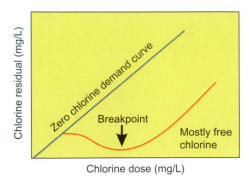

FIGURE 29.8 Dose—demand curve for chlorine.

no residual disinfectant remains. Because bacterial growth may occur after ozonation of tap water, chloramines are added to prevent regrowth in the distribution system. In addition, chloramines have been found to be more effective in controlling biofilm microorganisms on the surfaces of pipes in drinking water distribution systems because they interact poorly with capsular polysaccharides (LeChevallier *et al.*, 1990).

Because of the occurrence of ammonia in sewage effluents, most of the chlorine added is converted to chloramines. This demand on the chlorine must be met before free chorine is available for disinfection. As chlorine is added, the residual reaches a peak (formation of mostly monochloramine), and then decreases to a minimum called the breakpoint (Figure 29.8). At the breakpoint, the chloramine is oxidized to nitrogen gas in a complex series of reactions summarized in Eq. 29.8:

$$2NH_3 + 3HOCl \rightarrow N_2 + 3H_2O + 3HCl \qquad (Eq. 29.8)$$

Addition of chlorine beyond the breakpoint ensures the existence of free available chlorine residual.

Although numerous studies have been conducted to determine the mode of microbial inactivation by free chlorine, there have been fewer studies concerning chloramine inactivation mechanisms. It should be noted, however, that because of the poorly controlled experimental conditions employed by early investigators, many of the postulated chlorine inactivation mechanisms may have involved the action of chloramines rather than free chlorine. Research to date indicates that chloramines primarily inactivate microorganisms by irreversible denaturation of proteins (Stewart and Olson, 1996). Chloramine inactivation of bacteria is caused primarily by the oxidation of sulfyhdryl-containing enzymes, and to a lesser extent, a reaction with nucleic acid. In contrast to chlorine, there are no existing data to suggest that chloramines can modify the permeability state of the cell. Viral inactivation by chloramines is similar to the mechanism of inactivation by chlorine, in which primary targets consist of both capsid proteins and nucleic acid.

29.4.3 Chlorine Dioxide

Chlorine dioxide is an oxidizing agent that is extremely soluble in water (five times more than chlorine) and, unlike chlorine, does not react with ammonia or organic compounds to form trihalomethane, which is potentially carcinogenic. Therefore, it has received attention for use as a drinking water disinfectant. Chlorine dioxide must be generated on site because it cannot be stored. It is generated from the reaction of chlorine gas with sodium chlorite:

$$2NaClO_2 + Cl_2 \rightarrow 2ClO_2 + 2NaCl \qquad (Eq. 29.9)$$

Chlorine dioxide does not hydrolyze in water, but exists as a dissolved gas.

Studies have demonstrated that chlorine dioxide is as effective as, or more effective in inactivating bacteria and viruses in water than, chlorine (Table 29.4). As is the case with chlorine, chlorine dioxide inactivates microorganisms by: denaturation of the sulfyhdryl groups contained in proteins (Stewart and Olson, 1996); inhibition of protein synthesis (Bernarde *et al.*, 1967); denaturation of nucleic acid; and impairment of permeability control.

Studies with bacteriophage have suggested that the protein in the capsids is irreversibly damaged by chlorine dioxide (Figure 29.7). However, studies with poliovirus have suggested that the viral RNA is separated from the capsid during treatment (Vaughn and Novotny, 1991). The viricidal efficiency of chlorine dioxide increases as the pH is increased from 4.5 to 9.0 (Chen and Vaughn, 1990).

29.4.4 Bromine and Iodine

Bromine undergoes reactions in water similar to those of chlorine. However, its disinfecting capacity and mode of action differ from those of chlorine. The primary use of bromine is limited to hot tubs or spas and certain industrial applications (cooling towers). It is not as fast acting as chlorine, but is effective against bacteria (*Legionella*), viruses and protozoan parasites (*Entamoeba histolytica*). Bromine appears primarily to attach to the protein of viruses without causing structural damage (Keswick *et al.*, 1981). It does not appear to be able to penetrate the protein coat to inactivate the viral RNA.

Iodine has been used as a disinfectant primarily for small-scale water treatment needs such as those of campers, the space shuttle and small water treatment systems. On a comparative mg/L basis, more iodine than chlorine is required for a comparative bacterial kill. Iodine reacts in water as follows:

$$I + H_2O \approx \underset{\text{(iodine hydrolysis)}}{HOI} + H^+ + I^- \qquad (Eq. 29.10)$$

$$3I_2 + 3H_2O \approx \quad IO_3^- \quad + 5I^- + 6H^+ \quad \text{(Eq. 29.11)}$$
$$\text{(iodate formation)}$$

Iodide and iodate are primarily formed above pH 8.0, and have no viricidal action and little action against bacteria. Thus, at pH 9.0 HOI is the dominant form, whereas at pH 5.0, I_2 is dominant. At low pH iodine is more effective against some protozoan cysts because they are more sensitive to I_2 than to HOCl (Gottardi, 1991). This behavior is explained by the higher diffusibility of molecular iodine through the cell walls of cysts.

Iodine is not effective against all protozoa. For example, while *Giardia* cysts can be inactivated by iodine, *Cryptosporidium* oocysts are very resistant (Gerba *et al.*, 1997). In contrast to protozoa, viruses are more readily inactivated at pH levels above 7.0 because of the stronger oxidizing power of HOI. Iodine displays first-order inactivation kinetics, indicating single-site inactivation. Iodine oxidizes sulfhydryl groups and tryptophan and, perhaps more importantly, substitutes tyrosyl on histidyl moieties at neutral pH and room temperature. Structural changes in viral integrity have been noted by electron microscopy after treatment with iodine, and thus infectious RNA could be released into the environment.

29.5 OZONE

Ozone (O_3), a powerful oxidizing agent, can be produced by passing an electric discharge through a stream of air or oxygen. Ozone is more expensive than chlorination to apply to drinking water, but it has increased in popularity as a disinfectant because it does not produce trihalomethanes or other chlorinated byproducts, which are suspected carcinogens. However, aldehydes and bromates may be produced by ozonation, and may have adverse health effects. Because ozone does not leave any residual in water, ozone treatment is usually followed by chlorination or addition of chloramines. This is necessary to prevent regrowth of bacteria because ozone breaks down complex organic compounds present in water, into simpler ones that serve as substrates for growth in the water distribution system. The effectiveness of ozone as a disinfectant is not influenced by pH and ammonia.

Ozone is a much more powerful oxidant than chlorine (Tables 29.3 and 29.6). The $C \cdot t$ values for 99% inactivation are only $0.0011-0.2$ for enteric bacteria and $0.04-0.42$ for enteric viruses (Bitton, 2011). Ozone appears to inactivate bacteria by the same mechanisms as chlorine-based disinfection: by disruption of membrane

TABLE 29.4 $C \cdot t$ Values for Chlorine Dioxide Inactivation of Microorganisms in Water (99% Inactivation)

Microbe	ClO_2 Residual (mg/L)	Temperature (°C)	pH	% Reduction	$C \cdot t$
Bacteria					
Escherichia coli	0.3–0.8	5	7.0	99	0.48
B. subitilis spores		21	8.0	99	25
Viruses					
Polio 1	0.4–14.3	5	7.0	99	0.2–6.7
Rotavirus SA11					
Dispersed	0.5–1.0	5	6.0	99	0.2–0.3
Cell-associated	0.45–1.0	5	6.0	99	1.7
Hepatitis A virus	0.14–0.23	5	6.0	99	1.7
Adenovirus 40	0.1	5	7.0	99	0.28
Coliphage MS2	0.15	5	6.0	99	5.1
Protozoa					
Giardia muris	0.1–5.55	5	7.0	99	10.7
Giardia muris	0.26–1.2	25	5.0	99	5.8
Giardia muris	0.21–1.12	25	7.0	99	5.1
Giardia muris	0.15–0.81	25	9.0	99	2.7
Cryptosporidium parvum		21	8.0	99	1000

Adapted from Sobsey (1989); Rose et al. (1997); Charuet et al. (2001); Gerba et al. (2003).

TABLE 29.5 $°C \cdot t$ values for chloramine inactivation of microorganisms in water (99% inactivation)[a]

Microbe	°C	pH	$C \cdot t$
Bacteria			
Escherichia coli	5	9.0	113
Mycobacterium fortuitum	20	7.0	2667
Viruses			
Polio 1	5	9.0	1420
Echo 11	5	8.0	880
Hepatitis A	5	8.0	592
Adeno 2	5	8.0	990
Adeno 40	5	8.0	360
Coliphage MS2	5	8.0	2100
Rotavirus SA11			
Dispersed	5	8.0	4034
Cell-associated	5	8.0	6124
Protozoa			
Gardia muris	3	6.5–7.5	430–580
Gardia muris	5	7.0	1400
Cryptosporidium parvum	1	8.0	64,600
C. parvum	20	8.0	11,400

Adapted from Sobsey (1989); Rose et al. (1997); Driedger et al. (2001); Cromeans et al. (2010).
[a]*In buffered distilled water.*

TABLE 29.6 $C \cdot t$ values for Ozone Inactivation of Microorganisms in Water (99% Inactivation)

Organism	°C	pH	$C \cdot t$
Bacteria			
Escherichia coli	1	7.2	0.006–0.02
Viruses			
Polio 1	5	7.2	0.2
Polio 2	25	7.2	0.72
Rota SA11	4	6.0–8.0	0.019–0.064
Coxsackie B5	20	7.2	0.64–2.6
Adeno 40	5-7	7.0	0.02
Protozoa			
Giardia muris	5	7.0	1.94
Giardia lamblia	5	7.0	0.53
Encephalitozoon intestinalis	5	7.0	0.30.–0.04
Cryptosporidum parvum	1	—	40.0
C. parvum	7	—	7.0
C. parvum	22	—	3.5
Toxoplasma gondii	20	7.7–7.8	> 69

From Sobsey (1989); Rose et al. (1997) ; Gerba et al. (2003).

permeability (Stewart and Olson, 1996) (Figure 29.6); impairment of enzyme function and/or protein integrity by oxidation of sulfyhydryl groups; and nucleic acid denaturation. The effect of ozone on destruction of the cell wall of bacteria and protozoan oocysts is dramatic (Figures 29.9 and 29.10). *Cryptosporidium* oocysts can be inactivated by ozone, but a $C \cdot t$ of 1–3 is required. Viral inactivation may occur by breakup of the capsid proteins into subunits, resulting in release of the RNA, which may then be damaged (Figure 29.11).

29.6 METAL IONS

Heavy metals such as copper, silver, zinc, lead, cadmium, nickel and cobalt all exhibit antimicrobial activity; however, because of toxicity to animals, only copper and silver have seen widespread application. Copper and silver have seen use as swimming pool and hot tub disinfectants. Copper has been used to control the growth of *Legionella* in hospital distribution systems. Surfaces containing 65% or more copper have been approved as self-sanitizing surfaces (U.S. EPA, 2012). Silver has been used as a bacteriostat added to the activated carbon used in faucet-mounted water treatment devices for home use. Concentrations of copper used in water disinfection range from 200 to 400 µg/L. Silver exhibits greater antimicrobial action and concentrations of 40–90 µg/L give the same effectiveness. The effectiveness of metal ions is influenced by pH, presence of anions and soluble organic matter. Unlike halogens and other oxidizing disinfectants, metals remain active for long periods of time in water. The rate of inactivation is slow compared with oxidizing agents (Figure 29.12); however, their action is enhanced in the presence of low concentrations of oxidizing agents such as chloramines (Straub *et al.*, 1995). The enhanced rate of inactivation is due to a synergistic interaction of both disinfectants.

Metal ions may inactivate bacteria or viruses by reacting outside or inside the cell or virus either directly or indirectly. It has been suggested that the inactivating capacity of heavy-metal ions is due to their oxidation

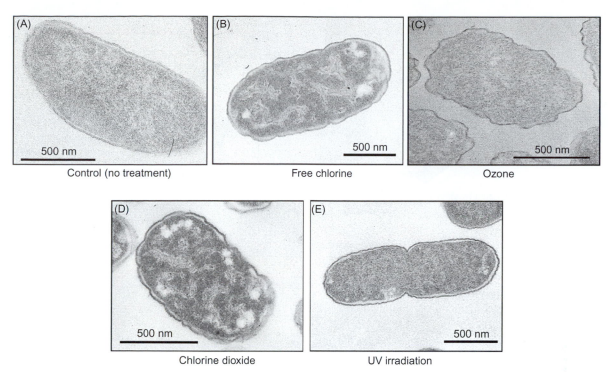

FIGURE 29.9   Transmission electron micrographs of *E. coli* before and after 90% inactivation by various disinfectants. Choi *et al.* (2010).

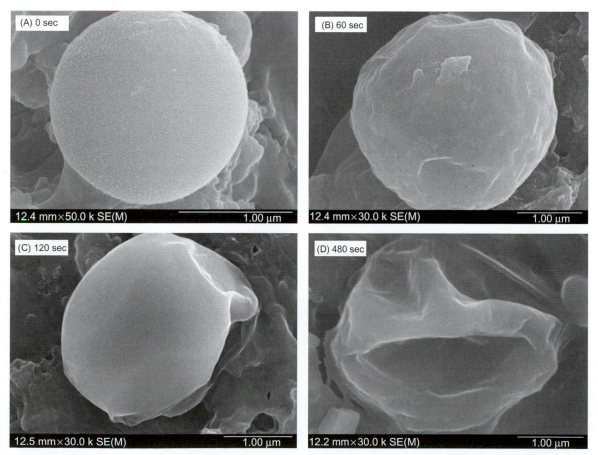

FIGURE 29.10   Scanning electron micrographs of *Cryptosporidium* oocysts after various time exposures to ozone. Ran *et al.* (2010).

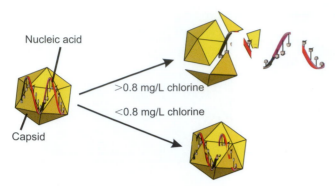

FIGURE 29.11   Virus inactivation by chlorine.

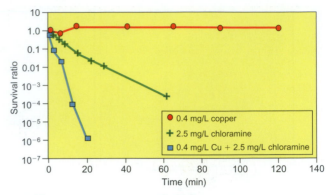

FIGURE 29.12   Synergistic inactivation of *Escherichia coli* by chloramines and copper. From Straub *et al.* (1995).

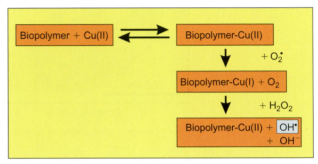

FIGURE 29.13   Modified site-specific Fenton mechanism. From Thurman and Gerba (1989).

power, and that a functional relationship exists between the inactivation rate and the oxidation potential of the ion (Thurman and Gerba, 1989). Inactivation of the macromolecules (proteins or nucleic acids) is thought to involve site-specific Fenton mechanisms. It is assumed that the metal ion binds to a biological target and is reduced by superoxide radicals or other reductants and subsequently reoxidized by H_2O_2, generating hydroxide radicals (Figure 29.13). Repeated cyclic redox reactions may result in multi-hit damage as radical formation occurs near the target site. Copper and silver may bind to proteins, interfering with the normal function of enzymes, resulting in cell death. Silver readily reacts with sulfyhydryl groups in proteins. Metals may also bind to the nucleic acids, forming complexes that interfere with replication.

The action of metals is slow and may be reversed by addition of chelating agents. For example, assay of samples containing silver for bacteria will give lower counts if the silver is not first neutralized by addition of sodium thiosulfate—sodium thioglycolate to inhibit the bacteriostatic effect of silver (Chambers *et al.*, 1962).

29.7 ULTRAVIOLET DISINFECTION

The use of ultraviolet disinfection of water and wastewater has seen increased popularity because it is not known to produce carcinogenic or toxic byproducts or taste and odor problems, and there is no need to handle or store toxic chemicals. Unfortunately, it has several disadvantages including higher costs than halogens, no disinfectant residual, difficulty in determining the UV dose, maintenance and cleaning of UV lamps, and potential photoreactivation of some enteric bacteria (Bitton, 2011). However, advances in UV technology are providing lower cost, more efficient lamps and more reliable equipment. These advances have aided in the commercial application of UV for water treatment in the pharmaceutical, cosmetic, beverage and electronics industries in addition to municipal water and wastewater application.

Microbial inactivation is proportional to the UV dose, which is expressed in microwatt-seconds per square centimeter (μW-s/cm^2) or:

$$\text{UV dose} = I \cdot t \qquad \text{(Eq. 29.12)}$$

where $I = \mu$W/cm^2 and t = exposure time.

In most disinfection studies, it has been observed that the logarithm of the surviving fraction of organisms is nearly linear when it is plotted against the dose, where dose is the product of concentration and time ($C \cdot t$) for chemical disinfectants or intensity and time ($I \cdot t$) for UV. A further observation is that constant dose yields constant inactivation. This is expressed mathematically as:

$$\log \frac{N_s}{N_i} = \text{function}(I_i t) \qquad \text{(Eq. 29.13)}$$

where N_s is the density of surviving organisms (number/cm^3) and N_i is the initial density of organisms before exposure (number/cm^3). Because of the logarithmic relationship of microbial inactivation versus UV dose, it is common to describe inactivation in terms of log survival, as expressed in Eq. 29.14. For example, if one organism in 1000 survived exposure to UV, the result would be a -3 log survival, or a 3 log reduction:

$$\log \text{survival} = \log \frac{N_s}{N_i} \qquad \text{(Eq. 29.14)}$$

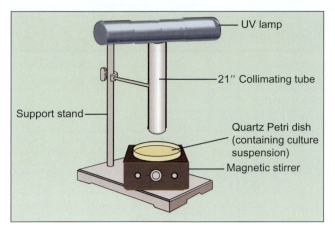

FIGURE 29.14   Collimating tube apparatus for UV dose application.

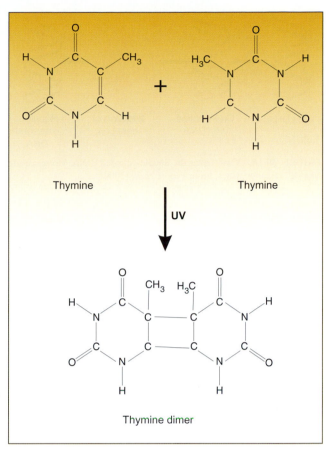

FIGURE 29.15   Formation of thymine dimers in the DNA of irradiated nonsporulating bacteria.

Determining the UV susceptibility of various indicator and pathogenic waterborne microorganisms is fundamental in quantifying the UV dose required for adequate water disinfection. Factors that may affect UV dose include cell clumping and shadowing, suspended solids, turbidity and UV absorption. UV susceptibility experiments described in the literature are often based on the exposure of microorganisms under conditions optimized for UV disinfection. Such conditions include filtration of the microorganisms to yield monodispersed, uniform cell suspensions, and the use of buffered water with low turbidity and high transmission at 254 nm. Thus, in reality, higher doses are required to achieve the same amount of microbial inactivation in full-scale flow through operating systems.

The effectiveness of UV light is decreased in wastewater effluents by substances that affect UV transmission in water. These include humic substances, phenolic compounds, lignin sulfonates and ferric iron. Suspended matter may protect microorganisms from the action of UV light; thus filtration of wastewater is usually necessary for effective UV light disinfection.

UV inactivation data are usually collected by placing a suspension of organisms in a stirred, flat, thin-layer dish in water with low UV light absorbance. In UV batch reactors, there are uniform UV intensities and contact time can be controlled. To deliver UV to these reactors, a collimating beam apparatus should be used (Figure 29.14). The light emitted at the end of the collimating beam is perpendicular to the batch reactor surface, thus creating a uniform, constant irradiation field that can be accurately quantified by means of a radiometer and photodetector calibrated for detecting 254-nm light. In general, the resistance of microorganisms to UV light follows the same pattern as the resistance to chemical disinfectants, i.e., double-stranded DNA viruses > MS 2 coliphage > bacterial spores > double-stranded RNA enteric viruses > single-stranded RNA enteric viruses > vegetative bacteria (Table 29.5).

Ultraviolet radiation damages microbial DNA or RNA at a wavelength of approximately 260 nm. It causes thymine dimerization (Figure 29.15), which blocks nucleic acid replication and effectively inactivates microorganisms. The initial site of UV damage in viruses is the genome, followed by structural damage to the virus protein coat. Viruses with high molecular weight, double-stranded DNA or RNA are easier to inactivate than those with low-molecular-weight, double-stranded genomes. Likewise, viruses with single-stranded nucleic acids of high molecular weight are easier to inactivate than those with single-stranded nucleic acids of low molecular weight. This is presumably because the target density is higher in larger genomes. However, viruses with double-stranded genomes are less susceptible than those with single-stranded genomes because of the ability of the naturally occurring enzymes within the host cell to repair damaged sections of the double-stranded genome, using the nondamaged strand as a template (Roessler and Severin, 1996) (Figure 29.16).

A phenomenon known as photoreactivation occurs in some UV light-damaged bacteria when exposed to visible wavelengths between 300 and 500 nm. The UV light

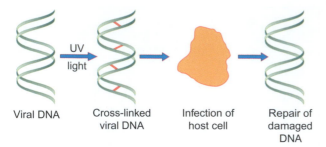

FIGURE 29.16 Viral repair in double-stranded DNA viruses using host cell repair enzymes. From Pepper *et al.* (2006).

TABLE 29.7 UV Dose to Kill Microorganisms

Organism	Ultraviolet Dose (μW-s/cm^2) Required for 90% Reduction
*Bacillus subtilis**	56,000
*Clostridium perfringens**	45,000
Campylobacter jejuni	1100
Escherichia coli	1300–3000
Klebsiella terrigena	3900
Legionella pneumophila	920–2500
Salmonella typhi	2100–2500
Shigella dysenteriae	890–2200
Vibrio cholerae	650–3400
Yersinia enterocolitica	1100
Adenovirus	23,600–56,000
Coxsackievirus	11,900–15,600
Echovirus	10,800–12,100
Poliovirus	5000–12,000
Hepatitis A	3700–7300
Rotavirus SA11	8000–9900
Coliphage MS-2	18,600
Cryptosporidium parvum	3000
Toxoplasma gondii	7000
Giardia	2000
Acanthamoeba	40,000
Naegleria fowleri (trohpozite)	6500
Naegleria fowleri (cyst)	31,500
Encephalitozoon intestinalis	2800

From Roessler and Severin (1996); John et al. 2003; Hijnen et al. (2006); Gerba et al. (2003); Sarkar and Gerba (2012).
**Environmental strains (spores)*

damage is repaired by activation of a photoreactivating enzyme, which binds and then splits the thymine dimers. DNA damage can also be repaired in the dark by a mechanism that excises dimerized pyrimidine base pairs, and allows the reinsertion of undimerized bases by other enzymes. The regenerative capacity of any organism is dependent on the type of organism. Total and fecal coliforms are capable of photoreactivation, but fecal streptococci are not. To prevent photoreactivation, sufficient doses must be applied or exposure to direct sunlight prevented.

A minimum dose of 16,000 μW s/cm^2 has been recommended for treating drinking water, as this results in a 99.9% reduction in coliforms. However, this level is not enough to inactivate enteric viruses and some protozoan cysts (Table 29.7) (Abbaszadegan *et al.*, 1997). *Cryptosporidium* oocysts and *Giardia* cysts are both very sensitive to UV light irradiation.

There are three types of UV light sources in use today. These include low pressure lamps, medium pressure lamps and pulsed UV light. Differences in the source lamp characteristics of these three types of UV result in different spectral outputs of UV light and photo densities that vary in their action on microorganisms. Low pressure UV lamps are the ones used most commonly for disinfection, and produce essentially monochromatic UV light at a wavelength of 253.7 nm. Medium pressure UV lamps emit polychromatic UV light ranging from 200 to 1400 nm with several peaks at 185 and 300 nm. Pulsed UV emits intense pulses of light in high photon densities, rather that the continuous, lower wavelength of low and medium pressure lamps. Since low pressure UV emits very near the 260 nm absorbance maximum for DNA, it inactivates microorganism largely by damaging their DNA/RNA. Medium and pulsed UV emit wavelengths which can damage other cellular components such as proteins, amino acids, lipids and small molecules such as carboxylic and ketone compounds (Eischeid *et al.*, 2011). An advantage to medium and pulsed UV is that it prevents photoreactivation of bacteria and adenoviruses, allowing the use of lower doses.

29.8 PHOTODYNAMIC INACTIVATION AND PHOTOCATAYLYSTS

The usefulness of photoreactive dyes for inactivating microorganisms and oxidizing toxic compounds and organic matter in wastewater has been demonstrated. Photodynamic action may be defined as the sensitization of microorganisms to inactivation by visible light through the action of certain dyes (e.g., methylene blue). The dye combines with the nucleic acid or another critical site, and the complex absorbs light energy and attains an excited energy state. The excited complex then combines with oxygen as the energy is released in a reaction that results in disruption of chemical bonds and loss of infectivity of the organism. Titanium dioxide is a photocatalyst that has a similar effect in the presence of UV light, causing strong oxidizing reactions at the surface of the metal oxide (Watts *et al.*, 1995). Different materials can be added to titanium dioxide to use visible light energy in the generation of free radicals.

29.9 OTHER CHEMICAL DISINFECTANTS

There are a few other chemical disinfectants that have seen widespread use primarily in consumer, institutional and industrial products. These include spray and wipe disinfectants available to the consumer and are widely used in the food industry. Quaternary ammonium compounds (Quats) are surfactants having both hydrophobic (water-repelling) and hydrophilic (water-attracting) properties. The basic structure of a quat is shown in Figure 29.17. The cation (positively charged) portion is a central nitrogen with four attached groups, which can contain a variety of structures, and is the functional part of the molecule. The anion (negatively charged) portion (X^-) is usually chlorine (Cl^-), and is linked to the nitrogen to form a quat salt. Benzalkonium chloride and cetylpyridinium chloride (Figure 29.18) are two of the most common basic quats structures in use. Benzalkonium chloride includes an aromatic ring, two methyl groups and a long chain ethyl ($CH_2^-CH_3$)/methyl chain, which can vary in length from C_{12} to C_{16}. Quats vary in their antimicrobial activity depending on the type and their formulations. They are effective against most common bacteria, but they are not sporicidal, although they may inhibit sporulation. They are effective against enveloped viruses (influenza), and specific formulations are effective against nonenveloped viruses (norovirus). Quats appear to act by adsorbing to and disrupting structure and function, eventually leading to leakage of cytoplasmic material (McDonnell, 2007). Direct interaction with viral and spore surface proteins may also cause prevention of growth, loss of function and disintegration. The presence of low-level residues after continued application may allow the selective development of bacterial strains with greater tolerance to quats over time (e.g., *Pseudomonas*). Tolerance is defined as the need for greater concentrations of an antimicrobial to kill the target organism, whereas resistance is defined as inability of the antimicrobial to kill the target organism. Thus, development of increased tolerance does not limit the practical application of a disinfectant, but may require increased concentrations to kill a target organism. It is difficult for microorganisms to develop resistance to disinfectants because they act nonspecifically on organic molecules, unlike

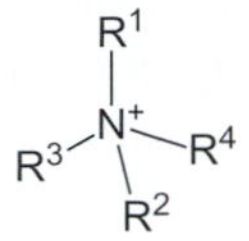

FIGURE 29.17 The basic structure of a quaternary ammonium compound.

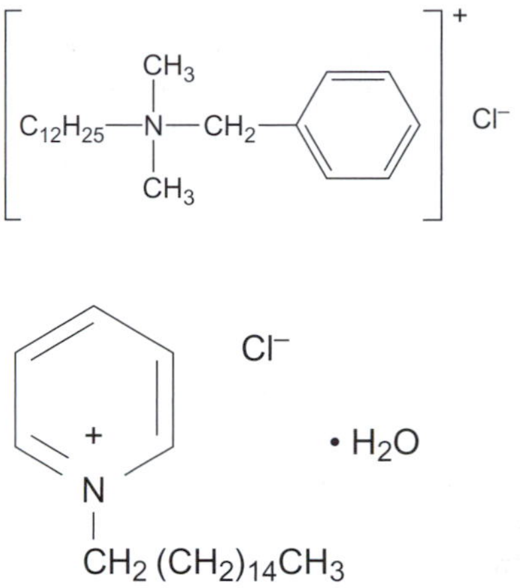

FIGURE 29.18 Structure of benzalkonium chloride (upper) and cetylpyridinium chloride (lower).

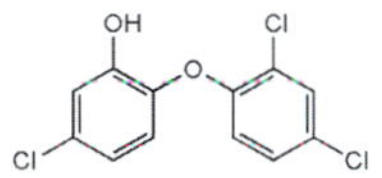

FIGURE 29.19 Triclosan.

antibiotics which act on specific sites in microorganisms. Thus, chlorine has been used for more than 100 years without microorganisms evolving a resistance to it (Rusin and Gerba, 2001).

Triclosan is also an antibacterial and fungal agent used in a wide variety of consumer products including hand soaps, mouth washes, shampoos and toothpastes, and via incorporation into materials (e.g., cutting boards). Triclosan is a bisphenol compound (Figure 29.19) and is known for its mildness to the skin. Mechanisms of triclosan action have received a great deal of study. Its action is much more specific than the other antimicrobials discussed in the chapter. Triclosan at concentrations used in products acts on multiple cytoplasmic and membrane targets (Russell, 2004). However, at lower concentrations, triclosan appears bacteriostatic, and is seen to target bacteria mainly by inhibiting fatty acid synthesis. Triclosan binds to bacterial enoyl–acyl carrier protein reductase enzyme (ENR). This binding increases the enzyme's affinity for nicotinamide adenine dinucleotide (NAD^+) resulting in the formation of a stable complex of ENR-NAD^+-triclosan, which is unable to participate in fatty acid synthesis. Fatty acids are necessary for reproducing and building cell membranes. Some bacterial species can develop low-level resistance to triclosan at its lower bacteriostatic concentrations, which results in a decrease of triclosan's effect on ENR-NAD$^+$ binding (Health *et al.*, 1999). Some bacteria have innate resistance to triclosan at low, bacteriostatic levels, such as *Pseudomonas aeruginosa*, which possesses multi-drug efflux pumps that "pump" triclosan out of the cell (Chuanchuen *et al.*, 2003). Other bacteria, such as some of the *Bacillus* genus, have alternative *FabI* genes (*FabK*) to which triclosan does not bind, and hence are less susceptible. Although increased tolerance to low levels of triclosan has been reported in numerous laboratory studies, this has not been enough to limit its use in practical applications.

29.10 GAMMA AND HIGH-ENERGY IRRADIATION

Ionizing radiation generated by radioactive materials such as cesium 127 or cobalt 60 and high-energy electron beams can inactivate microorganisms either directly or indirectly by production of free radicals. Nucleic acids are the main targets of ionizing radiation. Ionizing radiation has been studied in great detail for preservation of foods and for wastewater and sewage sludge treatment. Factors that influence the effectiveness of ionizing radiation include the type of organism (generally, the smaller the organism the more resistant); composition of the suspending medium (organic material offers protection); presence of oxygen (greater resistance in the absence of oxygen); and moisture (greater resistance of dried cells and radiolysis of water). The unit of dose is the rad, which is equivalent to the absorption of 100 ergs per gram of matter. A kilorad (krad) is equal to 1000 rads. Typical doses to produce a D value of 90% inactivation are shown in Table 29.8. Viruses are the most resistant to ionizing irradiation in water and sludge.

Sludge irradiators have been built in Europe and experimental electron beam irradiators in the United States. The electron beams are generated by a 750-kV electron accelerator. The unit treats a thin layer ($\approx$ 2 mm) of liquid sludge spread on a rotating drum. Such systems are costly for waste treatment and require thick concrete shielding.

QUESTIONS AND PROBLEMS

1. Of the non-spore-forming bacteria, which microbial group is the most resistant to thermal inactivation in water?

TABLE 29.8 Sludge Irradiation: D Values for Selected Pathogens and Parasites

Organism	D value (k rad)
Bacteria	
Escherichia coli	$< 22 - 36$
Klebsiella spp.	$36 - 92$
Enterobacter spp.	$34 - 62$
Salmonella typhimurium	$< 50 - 140$
Streptococcus faecalis	$110 - 250$
Viruses	
Poliovirus	350
Coxsackievirus	200
Echovirus	170
Reovirus	165
Adenovirus	150
Parasites	
Ascaris spp.	< 66

Modified from Ahlstrom and Lessel (1986).

2. What is thermal death time? D value?
3. Why are all microorganisms not inactivated according to first-order kinetics?
4. How long would you have to maintain a residual of 1.0 mg/L of free chlorine to obtain a $C \cdot t$ of 15? A $C \cdot t$ of 0.1?
5. Why is chlorine more effective against microorganisms at pH 5.0 than at pH 9.0?
6. Which chlorine compound is most effective against biofilms? Why?
7. What factors interfere with chlorine disinfection? Ultraviolet disinfection?
8. What is the main site of UV light inactivation in microorganisms? What group of microorganisms is the most resistant to UV light? Why?
9. At what pH is iodine most effective against protozoan parasites? Why?
10. What is photoreactivation? Are all microorganisms capable of photoreaction? If not, why?
11. What are two sources of ionizing radiation? How does ionizing radiation kill microorganisms?
12. Why does suspended matter interfere with the disinfection of microorganisms?
13. Chlorine has been in use for the disinfection of drinking and waste water for more than 100 years, yet no water- or foodborne bacteria or virus has developed resistance to chlorine. Why?

14. What dose of UV light would you need to kill 99.9% of the poliovirus in water?

15. What is a photocatalyst? How does it work?

REFERENCES AND RECOMMENDED READING

Abbaszadegan, M., Hasan, M. N., Gerba, C. P., Roessler, P. F., Wilson, B. R., Kuennen, R., *et al.* (1997) The disinfection efficacy of a point-of-use water treatment system against bacterial, viral and protozoan waterborne pathogens. *Water Res.* **31**, 574−582.

Ahlstrom, S. B., and Lessel, T. (1986) Irradiation of municipal shudge for pathogen control. In "Control of Sludge Pathogens" (C. A. Sorber, ed.), Water Pollution Contol Federation, Washington, DC.

Alvarez, M. E., and O'Brien, R. T. (1982) Effects of chlorine concentration on the structure of poliovirus. *Appl. Environ. Microbiol.* **43**, 237−239.

Bandres, J. C., Mawhewson, J. J., and Du Pont, H. L. (1988) Heat susceptibility of bacterial enteropathogens—implications for the prevention of travelers diarrhea. *Arch. Intern. Med.* **148**, 2261−2263.

Bates, R. C., Shaffer, P. T. B., and Sutherland, S. M. (1977) Development of poliovirus having increased resistance to chlorine inactivation. *Appl. Environ. Microbiol.* **33**, 849−853.

Bernarde, M. A., Snow, N. B., Olivieri, V. P., and Davidson, B. (1967) Kinetics and mechanism of bacterial disinfection by chloride dioxide. *Appl. Microbiol.* **15**, 257−265.

Bitton, G. (2011) "Wastewater Microbiology," 4th ed. Wiley-Liss, New York.

Block, S. S. (1991) "Disinfection, Sterilization, and Preservation," 4th ed. Lea & Febiger, Philadelphia, PA.

Cerva, L. (1955) The effect of disinfectants on the cysts of *Giardia intestinalis. Cesk. Parasitol.* **2**, 17−21.

Chambers, C. W., Procter, C. M., and Kabler, P. W. (1962) Bactericidal effect of low concentrations of silver. *J. Am. Water Works Assoc.* **54**, 208−216.

Chauret, C. P., Radziminski, C. S., Lepu, M., Creason, R., and Andrew, R. C. (2001) Chlorine dioxide inactivation of *Cryptosporidium parvum* oocysts and bacterial spore indicators. *Appl. Environ. Microbiol.* **67**, 2993−3001.

Chen, Y.-S., and Vaughn, J. (1990) Inactivation of human and simian rotaviruses by chlorine dioxide. *Appl. Environ. Microbiol.* **56**, 1363−1366.

Choi, M., Kim, J., Kim, J. Y., Yoon, J., and Kim, J. H. (2010) Mechanisms of *Escherichia coli* inactivation by several disinfectants. *Water Res.* **44**, 3410−3418.

Chuanchuen, R., Karkhoff-Schweizer, R. R., and Schweizer, H. P. (2003) High-level triclosan resistance in *Pseudomonas aerugginosa* is solely a result of efflux. *Am. J. Infect. Control.* **31**, 124−127.

Cromeans, T. L., Kahler, A. M., and Hill, V. R. (2010) Inactivation of adenoviruses, enteroviruses, and murine norovirus in water by free chlorine and monochloramine. *Appl. Environ. Microbiol.* **76**, 1028−1033.

Driedger, A. M., Rennecker, J. L., and Marinas, B. J. (2001) Inactivation of *Cryptosporidium* oocysts with ozone and monochloramine at low temperature. *Water Res.* **35**, 41−48.

Eischeid, A. C., Thurston, J. A., and Linden, K. A. (2011) UV disinfection of adenovirus: present state of research and future directions. *Crit. Rev. Environ. Sci. Technol.* **41**, 1375−1396.

Estes, M. K., Graham, D. Y., Smith, E. R., and Gerba, C. P. (1979) Rotavirus stability and inactivation. *J. Gen. Virol.* **43**, 403−409.

Eterpi, M., McDonnell, G., and Thomas, V. (2009) Disinfection against parvoviruses. *J. Infect. Hosp. Infect.* **73**, 64−70.

Fayer, R. (1994) Effect of high temperature on infectivity of *Cryptosporidium parvum* oocysts in water. *Appl. Environ. Microbiol.* **60**, 2732−2735.

Gerba, C. P., Johnson, D. C., and Hasan, M. N. (1997) Efficacy of iodine water purification tablets against *Cryptosporidium* oocysts and *Giardia* cysts. *Wilder. Environ. Med.* **8**, 96−100.

Gerba, C. P., Nwachuku, N., and Riley, K. R. (2003) Disinfection resistance of waterborne pathogens on the United States Environmental Protection Agency's Contaminant Candidate List (CCL). *J. Wat. Supply Res. Technol.* **52**, 81−94.

Gottardi, W. (1991) Iodine and iodine compounds. In "Disinfection, Sterilization, and Preservation" (S. S. Block, ed.), 4th ed., Lea and Febiger, Philadelphia, pp. 152−166.

Haas, C. N., and Morrison, E. C. (1981) Repeated exposure of *Escherichia coli* to free chlorine: production of strains possessing altered sensitivity. *Water Air Soil Pollut.* **16**, 233−242.

Health, R. J., Rubin, J. R., Holland, D. R., Zhang, E., Snow, M. E., and Rock, C. O. (1999) Mechanism of triclosan inhibition of bacterial fatty synthesis. *J. Biol. Chem.* **274**, 11110−11114.

Hijnen, W. A. M., Beerendonk, E. F., and Medada, G. J. (2006) Inactivation credit of UV radiation for viruses, bacteria and protozoan (oo)cysts in water: a review. *Water Res.* **40**, 3−22.

Jacangelo, J., Patania, N., Haas, C., Gerba, C., and Trussel, R. (1997) Inactivation of waterborne emerging pathogens by selected disinfectants. Report No. 442, American Water Works Research Foundation, Denver, Colorado.

Jay, J. M. (1996) "Modern Food Microbiology," 5th ed. Chapman & Hall, New York.

John, D. E., Nwachuku, N., Pepper, I. L., and Gerba, C. P. (2003) Development and optimization of a quantitative cell culture infectivity assay for *the microsporidium Encephalitozoon intestinalis* and application to ultraviolet light inactivation. *J. Microbiol. Methods* **52**, 183−196.

Keswick, B. H., Fujioka, R. S., and Loy, P. C. (1981) Mechanism of poliovirus inactivation by bromine chloride. *Appl. Environ. Microbiol.* **42**, 824−829.

LeChevallier, M. W., Lowry, C. H., and Lee, R. G. (1990) Disinfecting biofilm in a model distribution system. *J. Am. Water Works Assoc.* **82**, 85−99.

Liew, P., and Gerba, C. P. (1980) Thermo stabilization of enteroviruses on estuarine sediment. *Appl. Environ. Microbiol.* **40**, 305−308.

McDonnell, G. E. (2007) "Antisepsis, Disinfection, and Sterilization," ASM Press, Washington, DC.

Montgomery, J. M. (1988) "Water Treatment and Design," John Wiley & Sons, New York.

Pepper, I. L., Gerba, C. P., and Brusseau, M. L. (eds.) (2006) 2nd ed., Elsevier Science/Academic Press, San Diego, CA.

Pereira, V. J., Marques, R., Marques, M., Bernoliel, M. J., and Crespo, M. T. B. (2013) Free chlorine inactivation of fungi in drinking water sources. *Water Res.* **47**, 517−523.

Ran, Z., Li, S., Huang, J., *et al.* (2010) Inactivation of cryptosporidium by ozone and cell ultrastructures. *J. Environ. Sci. (China)* **22**, 1954−1959.

Robbecke, R. S., and Buchholtz, K. (1992) Heat susceptibility of aquatic Mycobacteria. *Appl. Environ. Microbiol.* **58**, 1869−1873.

Roberts, D., and Gilbert, R. J. (1979) Survival and growth of noncholera vibrios in various foods. *J. Hyg. Cambridge* **82**, 123–131.

Roessler, P. F., and Severin, B. F. (1996) Ultraviolet light disinfection of water and wastewater. In "Modeling Disease Transmission and Its Prevention by Disinfection" (C. J. Hurst, ed.), Cambridge University Press, Cambridge, UK, pp. 313–368.

Rose, J. B., Lisle, J. T., and LeChevallier, M. (1997) Waterborne Cryptosporidiosis: incidence, outbreaks, and treatment strategies. In "Cryptosporidium and Cryptosporidiosis" (R. Fayer, ed.), CRC Press, Boca Raton, FL, pp. 93–109.

Rusin, P., and Gerba, C. P. (2001) Association of chlorination and UV irradiation to increasing antibiotic resistance in bacteria. *Rev. Environ. Contamin. Toxicol.* **171**, 1–52.

Russell, A. D. (2004) Whither triclosan? *J. Antimicr. Chemother.* **53**, 693–695.

Sanden, G. N., Fields, B. S., Barbaree, J. M., and Feeley, J. C. (1989) Viability of *Legionella pneumophila* in chlorine-free water at elevated temperatures. *Curr. Microbiol.* **18**, 61–65.

Sarkar, P. S., and Gerba, C. P. (2012) Inactivation of *Naegleria fowleri* by chlorine and ultraviolet light. *J. Am. Water Works Assoc.* **104**, 95–96.

Shields, J. M., Hill, V. R., Arrowood, M. J., and Beach, M. J. (2008) Inactivation of *Cryptosporidium parvum* under chlorinated recreational water conditions. *J. Water Health* **6**, 513–520.

Siegl, G., Weitz, M., and Kronauer, G. (1984) Stability of hepatitis A virus. *Intervirology* **22**, 218–226.

Sobsey, M. D. (1989) Inactivation of health-related microorganisms in water by disinfection processes. *Water Sci. Technol.* **21**, 179–195.

Stewart, M. H., and Olson, B. H. (1996) Bacterial resistance to potable water disinfectants. In "Modeling Disease Transmission and Its Prevention by Disinfection," Cambridge University Press, Cambridge, UK, pp. 140–192.

Straub, T. M., Gerba, C. P., Zhou, X., Price, R., and Yahya, M. T. (1995) Synergistic inactivation of *Escherichia coli* and MS-2 coliphage by chloramine and cupric chloride. *Water Res.* **24**, 811–818.

Thurman, R. B., and Gerba, C. P. (1989) The molecular mechanisms of copper and silver ion disinfection of bacteria and viruses. *CRC Crit. Rev. Environ. Control* **18**, 295–315.

USEPA (2012) United States Environmental Protection Agency. Test Method for Efficacy of Copper Alloy Surfaces as a Sanitizer.

Vaughn, J. M., and Novotny, J. F. (1991) Virus inactivation by disinfectants. In "Modeling the Environmental Fate of Microorganisms" (C. J. Hurst, ed.), American Society for Microbiology, Washington, DC, pp. 217–241.

Wainwright, K. E., Miller, M. A., Barr, B. C., Gerdner, I. A., Meli, A. C., Essert, T., *et al.* (2007) Chemical inactivation of *Toxoplasma gondii* oocysts in water. *J. Parasitol.* **9**, 925–931.

Watts, R. J., Kong, S., Orr, M. P., Miller, G. C., and Henry, B. E. (1995) Photcatalytic inactivation of coliform bacteria and viruses in secondary wastewater effluent. *Water Res.* **29**, 95–100.

Wigginton, K. R., Pecson, B. M., Sigstam, T., Bosshard, F., and Kohn, T. (2012) Virus inactivation mechanisms: impact of disinfectants on virus function and structural integrity. *Environ. Sci. Technol.* **46**, 12069–12078.

Part VIII

Urban Microbiology

Chapter 30

Domestic and Indoor Microbiology

Charles P. Gerba and Ian L. Pepper

It has been suggested that perhaps 80% or more of all common infections including colds, flu, skin infections, gastroenteritis and diarrhea are acquired by exposure to our environment. Since humans in developed countries spend from 35 to 90% of their time indoors, sources of pathogens can be the air, food or water that enters a home. In addition, humans themselves that acquire an infection outside of the home can subsequently be a source of pathogens within the home. Once inside a home, pathogens can be transferred either via person–person contact, or via person–fomite–person. Other indoor environments including schools, workplaces and hospitals are also reservoirs of human pathogens. In this chapter we identify household and indoor sources of pathogens, and their fate and transport within the indoor environment.

30.1 HOUSEHOLD SOURCES OF PATHOGENS

30.1.1 Air

Microbial airborne pathogens occur as bioaerosols (see Chapter 5) and include bacteria, viruses, molds and spores. Molds are fungi that include species of *Cladosporium*, *Penicillium*, *Aspergillus* and *Alternaria*. Molds are highly prevalent in damp areas of homes. Since molds reproduce via spores that are easily wind-borne, they can easily enter households via open doors or windows. After landing, mold spores can colonize damp solid surfaces within 72 hours. Once they are established, it is difficult to remove molds, and frequently absorbent materials such as ceiling tiles or carpets need to be replaced. Molds cause a variety of health concerns including nasal stuffiness, eye irrigation, wheezing or skin irrigation. Some individuals have more serious allergies to molds, and high concentrations of molds can result in fever and/or shortness of breath. Threshold levels of molds are unknown and may vary with the type of mold. Health complaints, however, have been associated with concentrations of 2000 CFU m^{-3} of mixed mold populations in air samples (Reynolds, 2006).

House dust mites are microscopic organisms related to spiders and ticks (Figure 30.1). Allergens from dust mite feces contribute significantly to seasonal allergies and asthma. Approximately 20 million Americans are allergic to dust mites, which are ubiquitous and cause serious problems in approximately half of all U.S. homes (Reynolds, 2006). Dust mites feed on dead human skin, and thrive in household dust, bedding and carpeting, particularly in humid areas.

Another important household allergen is endotoxin (Figure 5.15). Endotoxin is derived from lipopolysaccharide contained within the outer cell wall of Gram-negative bacteria, and can cause a variety of health ailments. Most adverse health effects associated with endotoxin have been associated with occupational exposures such as grain houses, cotton dust or composting

I.L. Pepper, C.P. Gerba, T.J. Gentry: Environmental Microbiology, Third edition. DOI: http://dx.doi.org/10.1016/B978-0-12-394626-3.00030-2
© 2015 Elsevier Inc. All rights reserved.

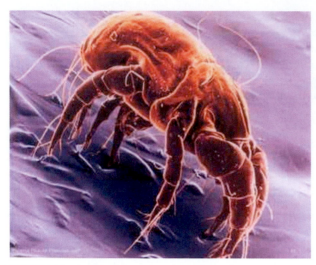

FIGURE 30.1   A dust mite.

FIGURE 30.2   Causes of illness in outbreaks from single food commodities in the United States, 1998—2010.

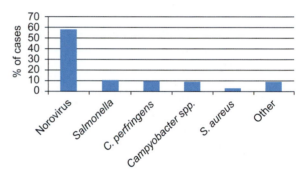

FIGURE 30.3   Top pathogens contributing to foodborne disease in the United States, 2011 (CDC, 2013). Created by C.P. Gerba.

plants. However, household dusts are also known to contain endotoxin, and the potential risks associated with routine, low-level exposures in common indoor environments are unknown (Reynolds, 2006).

Infectious pathogens, including viruses and bacteria, can also be found in household air. For example, influenza, colds and even chicken pox and tuberculosis are transmitted within households. In any given year up to 50 million Americans contract "the flu" and an average of 20,000 to 40,000 die each year. Overcrowded conditions and poor air circulation exacerbate the spread of infectious agents. Exposure to such pathogens can occur directly from humans through a sneeze or cough, or indirectly due to showering or toilet flushing. Home humidifiers are frequently a breeding ground for bacteria, protozoa or fungi, and have led to the use of the term "humidifier fever." Symptoms are similar to a short-term flu-like illness. Hot water systems and air conditioning units have been implicated in outbreaks of Legionnaires' disease and Pontiac fever caused by the bacterium *Legionella pneumophila*.

30.1.2 Food

It is estimated that there are about 47 million cases of foodborne illness every year in the United States (CDC, 2013). All types of foods can be associated with foodborne illness (Figure 30.2). Many common foods brought into homes routinely contain human pathogenic microbes including: *Salmonella*; *Campylobacter*; *Listeria monocytogenes*; *Staphylococcus aureus*; and *E. coli* O157:H7 (Chapter 22). *Salmonella* is often found in poultry and seafood, often in association with *Campylobacter*. Reng *et al.* (2007) detected *Salmonella*, *Campylobacter* or *Arcobacter* in 80% of 54 samples of duck. *Listeria* can be found in a variety of foods including raw milk, cheeses,

ice cream, raw vegetables, poultry and smoked fish. This organism causes a general group of disorders including meningitis and encephalitis. Of particular concern is its ability to grow at temperatures as low as 3°C, permitting multiplication even in refrigerated foods. *Staphylococcus aureus*, besides causing skin infections, can also cause staphylococcal food poisoning due to enterotoxins produced by some strains. *S. aureus* is more prevalent in foods that require excessive handling, and that are kept at slightly elevated temperatures after preparation. *E. coli* O157-H7, an enterohemorrhagic strain of *E. coli*, causes the acute disease hemorrhagic colitis, which can even be fatal. This strain of *E. coli* has been associated with undercooked hamburger. Outbreaks are also associated with produce such as lettuce and spinach which were contaminated before harvest of during handling (Delaquis *et al.*, 2007). Perfringens food poisoning is the term used to describe the common foodborne illness caused by *Clostridium perfringens*. Stored meat products are most commonly associated with *C. perfringens*. However, the major cause of most outbreaks are the human noroviruses, probably resulting from contamination of foods from handling by infected persons, from water or from other environmental sources of contamination (Figure 30.3).

All of these food products can result in human infections within households via two mechanisms: contamination of human hands and fomites during food preparation; and consumption of raw or undercooked foodstuffs. Contamination of fomites can subsequently result in further contamination of other fomite surfaces, particularly in the kitchen (Section 30.3).

30.1.3 Water

In developed countries, water is treated at public utilities and supplied to homes via utility distribution systems. Although water is disinfected prior to entering distribution systems, as the treated water travels through the distribution pipe system and into the home, the microbial water quality degrades and bacterial growth and regrowth can occur (see Chapter 28). Such microbial growth results in the development of biofilms, which further protects microbes (see Section 6.2.4.1). Thus, the number of heterotrophic bacteria (heterotrophic plate counts or HPC) in drinking water has been used as an estimate of the microbial water quality drinking water. Generally, a high HPC is equated with poor microbial water quality. The numbers of HPC would be expected to be elevated if surface waters are not adequately treated, or if a cross contamination event with a sewer line has occurred. However, it has been known for a long time that heterotrophic bacteria grow in distribution systems even in the presence of chlorine (Case Study 30.1).

Water distribution systems are clearly a source of microorganisms that humans are exposed to on a daily basis, but they appear to be no threat to normal healthy individuals. General groups of bacteria capable of specific biochemical transformations such as sulfate reduction and nitrification have been identified in tap water (Pepper et al., 2004). More recently, both cultural and molecular methods were used for microbial community analyses of drinking water from four United States cities (Case Study 30.2). Of organisms that grow in the distribution system Legionella pneumophila and Acanthamoeba spp. are the only ones that are commonly associated with illness. Legionella is associated with respiratory illnesses from exposure to warm water via showers, hot tubs, air humidifiers and water fountains.

Acanthamoeba infections are associated with persons who use tap water to wet their contact lenses. Both of these organisms grow in biofilms, and Legionella can actually grow inside Acanthamoeba amoeba, which protects it from the action of chlorine.

30.2 FOMITES: ROLE IN DISEASE SPREAD

In most discussions of disease spread, inanimate objects or fomites have been overlooked as agents of transmission. However, we continuously come into contact with a wide range of surfaces that may serve as vehicles or reservoirs of pathogenic microorganisms. Fomites can include doorknobs, sink taps, cutting boards, computer keyboards and of course the toilet seat. Because personal contact among nonrelated adults is limited in most cultures, fomites are believed to play a significant role in the transmission of some pathogens. What is often perceived to be person-to-person spread is actually person—fomite—person spread. For example, rhinovirus, the cause of the common cold, is readily transmitted by contact with virus-contaminated fingers brought to the nose or eyes (Figure 30.4) (Hendley et al., 1973). Fomites can also result in cross contamination from foods when raw meat contaminates a cutting board or a food handler, and then other foods such as spinach are prepared on the same board, and become contaminated and consumed raw. In addition, studies conducted in hospitals have demonstrated that fomites play a role in hospital or nosocomial infections.

Case Study 30.1   Heterotrophic Plate Count Bacteria in Source Waters and Household Taps

The concentrations of heterotrophic plate count (HPC) within water reaching consumer taps and from the water sources used by a major water utility were evaluated. The average HPC concentration in source waters ranged from 38 to 502 CFU/ml. The concentrations of HPC in a kitchen tap and other water containers are shown in Table 30.1. HPC in bathroom tap water are shown in Table 30.2. Clearly HPC in municipal waters were greater than the number of HPC in source waters, illustrating that bacterial growth had occurred. These data illustrate that water distribution systems contain living microbial communities that enter households.

TABLE 30.1   Heterotrophic Plate Counts from Seven Different Households[a]

	HPC (CFU/ml)			
	Kitchen Tap	Commercial Bottled Water	Sports Bottle	POU Device[b]
Range	$4-7 \times 10^7$	0–90,000	240–34,000	$4-1 \times 10^7$
Mean	399	1750	17,000	4000

Modified from Pepper et al. (2004).
[a]Assayed on trypticase soy broth at room temperature for 5 days.
[b]POU, point of use device mounted on the tap. These devices usually consist of activated charcoal to remove taste and odor.

TABLE 30.2 Heterotrophic Plate Count Bacteria in Bathroom Tap Water

	Overnight HPC (CFU/ml)			
	Before Flush		After Flush	
	House 1	House 2	House 1	House 2
Mean	2.4×10^3	2.4×10^3	1.5×10^2	1.4×10^2
S.D.	1.6×10^3	4.3×10^3	1.0×0^2	1.4×10^2

Modified from Pepper et al. (2004).

Case Study 30.2 Water Distribution Systems as Living Ecosystems: Impact on Taste and Odor

Six waters from different U.S. cities with known diverse taste and odor (TO) evaluations were selected for additional microbial characterization. All waters were subjected to microbial and cultural analyses, and four of the waters were further analyzed by cloning and sequencing of community 16S rRNA. The purpose of the study was to evaluate water distribution systems as living ecosystems, and the impact of these ecosystems on TO. All waters had total bacterial counts of at least 10^3 per ml. The water with lowest TO ranking had 10^6 total counts per ml. Community DNA sequence analysis identified diverse bacterial communities representing five different phyla and over 40 genera. Included in this diversity were heterotrophic and autotrophic species that were both aerobic and anaerobic. In addition, numerous opportunistic and nosocomial pathogens were identified (Table 30.3). Additionally, waters with the lowest TO evaluations contained significant sulfide concentrations, as well as bacteria associated with both the oxidation and reduction of inorganic sulfur compounds. Low redox conditions could have resulted in the reduced sulfur compounds and concomitant TO-related problems, and an increase in redox could help alleviate these problems. Overall, data show that water distribution systems contain living ecosystems that evolve based on specific environments within particular distribution systems that impact water TO.

From Scott and Pepper (2010).

Good hygiene practices have been shown to significantly reduce fomite transmission of pathogens. These practices include hand washing, use of hand sanitizers, cleaning and use of disinfectants. The importance of hand hygiene was demonstrated more than 100 years ago by Dr. Ignaz Semmelweis, an Austrian–Hungarian physician, who in 1847 discovered that the incidence of child bed fever (infections of the mother after delivering) could be drastically reduced by hand washing. Numerous studies have demonstrated that 30 to 50% reduction of illness can occur by providing adequate hand washing facilities

and by encouraging good hand washing practices (Van Curtis and Cairncross, 2003). Similar results have been obtained with alcohol gel sanitizers (White *et al.*, 2003; Vessey *et al.*, 2007).

Changes in lifestyles in the twenty-first century have increased our interactions with our indoor environments. This has increased the potential for fomite transmission of pathogens. For example, in the developed world most of us work in homes or offices, and spend most of our day indoors. We work in ever-larger buildings; vacation in larger hotels, resorts and cruise ships; visit stadiums of increasing size; visit health care centers rather than physicians' offices; fly in ever larger planes; and shop in large indoor shopping malls and super stores. All of these factors result in the increased sharing of fomites. Noroviruses provide an example of a virus easily spread by fomites that has resulted in the cancellation of vacation cruises and the closure of schools, hotels, gambling casinos, summer camps and hospital emergency rooms. This has created a need for a better understanding of pathogen spread by fomites and effective means for their control.

Enteric, respiratory and dermal pathogens have the greatest potential to be spread by fomites because they are released into the environment in large numbers via infected individuals. Also, enteric bacteria have the ability to grow in foods as well as some fomites such as sponges. The low number of viruses needed to cause infection makes fomite transmission more likely than for bacteria, which usually require contact with larger numbers of organisms to have a significant probability of infection (Chapter 22). Even blood-borne viruses can be spread by fomites. For example, an outbreak of hepatitis B virus was traced to computer cards, which infected small cuts when handled. Plantar warts, for which papovavirus is responsible, are generally contracted by walking barefoot in swimming areas, gyms, barracks or other public places. Some protozoa, such as *Giardia* and *Cryptosporidium*, may also be spread by this route, especially among young children.

30.2.1 Occurrence of Pathogens on Fomites

Fomites may become contaminated with pathogens by direct contact with bodily excretions/secretions (mucus, salvia, blood, feces). Alternatively, such fluids may be transferred from soiled hands to fomites, or airborne organisms may impinge or settle onto fomite surfaces. Fomites may also serve as a site for the replication of a pathogen, as in the case of enteric bacteria in household sponges or dishcloths. Until the development of molecular methods, such as PCR, data on the occurrence of pathogens on fomites were very limited because of the difficulty and cost associated with the isolation of pathogens

TABLE 30.3 Opportunistic and Nosocomial Pathogens Identified within Municipal Tap Water Collected from Four U.S. Cities

Genus/Species[a]	General Description
Microbacterium sp.	Aerobic heterotrophic bacterium found in soil and other environments. Some species are opportunistic pathogens.
Mycobacterium sp.	Aerobic heterotrophic bacterium associated with a variety of environments including household dust, drinking water and clinical specimens. Some species are pathogenic to humans.
Sphingobacterium multivorum	Aerobic heterotroph found in many environments including soil and clinical specimens. Opportunistic human pathogen.
Brevundimonas dimunitia	Aerobic, chemoorganotrophic to oligotrophic. Typically occupy aquatic habitats. Uncommon noscomial pathogen.
Methylobacterium sp.	Aerobic chemoheterotrophic and facultatively methylotrophic. Frequently airborne, found in dust and other environments such as freshwater. Some are opportunistic pathogens.
Acinetobacter johnsonii	Aerobic heterotroph found in a wide variety of environments. Opportunistic nosocomial pathogen.
Acinetobacter junii	Aerobic heterotroph. Opportunistic nosocomial pathogen.
Pseudomonas mendocina	Some strains reduce elemental sulfur. Very rarely found as opportunistic human pathogen.

Modified from Scott and Pepper (2010).
[a]*All clones had a sequence match identity of at least 98% as defined by Ribosomal Database Project 11 (Cole et al., 2007; Wang et al., 2007).*

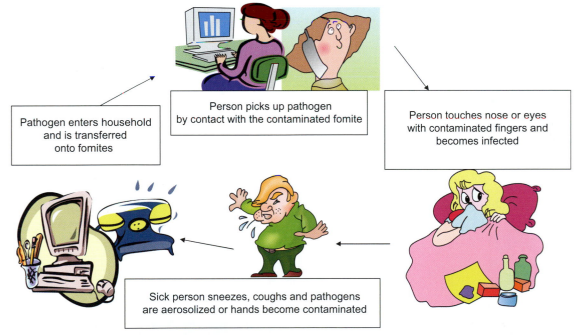

FIGURE 30.4 Role of fomites in respiratory disease transmission. Courtesy A. Moghe and C.P. Gerba.

from fomites. However, recent studies have shown that pathogens can rapidly contaminate the indoor environment. For example, it was found that in households with two children with influenza infections, influenza virus could be found on more than 50% of common fomites such as phones, TV remotes, faucets and doorknobs (Boone and Gerba, 2005). In a study of offices, parainfluenza could be isolated from one-third of all offices tested during the fall throughout the United States (Boone and Gerba, 2010). Rhinovirus, a major cause of the common cold, has been detected on 40 to 90% of the hands of adults with colds, and from 6 to 25% of selected fomites in rooms inhabited by persons with colds. During outbreaks, norovirus has been detected on toilet flush handles, gambling chips and doorknobs. Enteric bacteria such as *Salmonella* have been detected in likely places such as

---

Information Box 30.1   General Categories of Fomites

- **Reservoirs**—such as toilets, sinks, drains, clothing. These generally have high levels of contamination, including enteric bacteria, and have the potential environment for bacterial multiplication (Figure 30.5).
- **Reservoirs—disseminators**—cleaning tools such as sponges, dish cloths, vacuum cleaners and mops, on which there can be high levels of contamination, and the potential exists for bacterial multiplication. However, in addition, the potential exists for direct transfer of this contamination to surfaces whenever these items are used (Figure 30.6).
- **Hand and food contact surfaces**—such as kitchen counters, cutting boards, faucets, handles, laundry or fabrics on which there may be lower levels of contamination, but still the potential for the presence of pathogens, together with the constant potential for cross contamination to other crucial surfaces such as high-risk foods that are eaten raw or the hands (Scott, 1999).

Environmental Amplifiers

Food   Sink

Garbage Pail   Sponge

FIGURE 30.5 Locations in the household where bacteria can grow. Courtesy C.P. Gerba.

Disseminators

Hand   Washing Machine

Mop   Sponge

FIGURE 30.6 Objects involved in transfer of microorganisms in the household. Courtesy C.P. Gerba.

---

toilets, but also in household kitchen sinks, diaper hampers, vacuum cleaner dust, and cleaning tools, such as sponges, dishcloths and mops. Fomite contamination has been classified into three general categories of sites or surfaces on which the risk of contamination or cross contamination is greatest (Scott *et al.*, 1982) (Information Box 30.1).

Coliform bacteria in households are found in high concentration on kitchen sponges, and in sink areas relative to bathroom areas, which are more commonly associated with this group of bacteria (Figure 30.7). Enteric bacteria are brought into the home kitchen on raw meat and vegetables, where they can grow to large numbers in moist environments where food is available. This includes kitchen fomites and even cleaning tools—enteric bacteria can be spread around a home during normal cleaning of surfaces.

Public toilets have been shown epidemiologically to be responsible for outbreaks of *Shigella*, *Salmonella*, hepatitis A virus and norovirus. It has been demonstrated that viruses and bacteria are ejected to some degree when toilets are flushed, allowing for contamination of restroom areas adjacent to the toilet (Gerba *et al.*, 1975). The most common areas where fecal coliform bacteria are isolated in public restrooms include the floor, taps and sink drains (Figure 30.8), suggesting that these areas are more likely to be contaminated by pathogens originating from feces. Recent studies using pyrosequencing indicate that most bacteria in indoor environments originate from the human body (Figures 30.9 and 30.10).

30.2.2 Persistence of Pathogens on Fomites

The persistence of a pathogen on a fomite is dependent on a number of factors (Table 30.4). The rate of drying and temperature are the most important factors controlling survival. For most organisms, inactivation or death occurs most rapidly during the drying of the liquid in which it

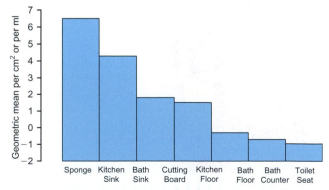

FIGURE 30.7 Concentration of coliform bacteria in the home. After Rusin *et al.* (1998).

was suspended. This is largely determined by the relative humidity of the air. The lower the relative humidity, the more rapid the drying takes place. Once drying is complete, the rate of organism die-off usually decreases (Figure 30.11). Some organisms survive better at lower relative humidity and others at high relative humidity (Table 30.5). In most indoor buildings with air handing systems, the relative humidity usually ranges from 40 to 60%. It is usually less in the winter because of indoor heating, which may favor the survival of some respiratory viruses such as influenza. Suspending media can also influence survival, for example rhinovirus in tryptose phosphate broth did not survive as well as in nasal

secretions (Sattar *et al.*, 2000). The nature of the virus and perhaps the route of transmission also play a role in the survival of viruses on fomites. While the survival of common respiratory viruses is usually a matter of hours to days, that of enteric viruses can be measured in days to weeks (Figures 30.12 and 30.13). The type of surface also may play a role, but a clear answer is not currently available because the efficient recovery of the organism from the different surfaces has not yet been determined.

30.3 TRANSFER OF PATHOGENS

Transfer of pathogens from an infected host to a fomite and pathogen transfer to a susceptible host are also important in understanding transmission. Clean hands can readily become contaminated when objects or surfaces are touched or handled. The reverse is also true. Individuals with rhinovirus colds were shown to deposit infectious rhinovirus particles on objects that they touch (Sattar and Springthorpe, 1996). The virus could also be recovered from the fingertips of volunteers who handled objects such as doorknobs previously touched by virus contaminated hands. Further studies using volunteers have demonstrated that placement of rotavirus-contaminated hands in the mouth or rhinovirus-contaminated hands in the nose also results in transmission of these viruses. Children frequently bring objects to their mouths. In fact, children less than 2 years of age have been observed to

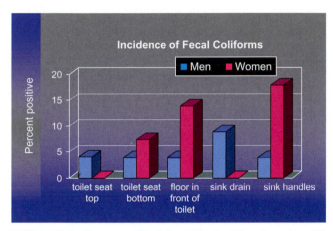

FIGURE 30.8   Frequency of isolation of fecal coliform bacteria at different locations in public restrooms. Courtesy C.P. Gerba.

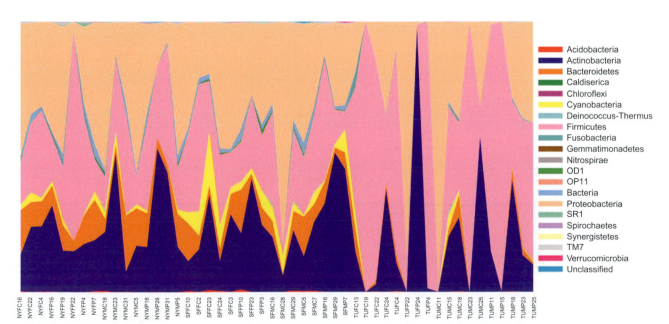

FIGURE 30.9   Relative abundance of bacterial divisions across samples. The abundances of various bacterial divisions (see color legend) in the 54 samples were based on multiplexed pyrosequencing of 16S rRNA gene sequences. The codes for each sample are presented along the X-axis and indicate the city (NY = New York, SF = San Francisco, TU = Tucson), gender of the office occupant (F = Female, M = Male) and site within the office from which the sample (C = Chair, P = Phone) was obtained, followed by sample number (Hewitt *et al.*, 2012).

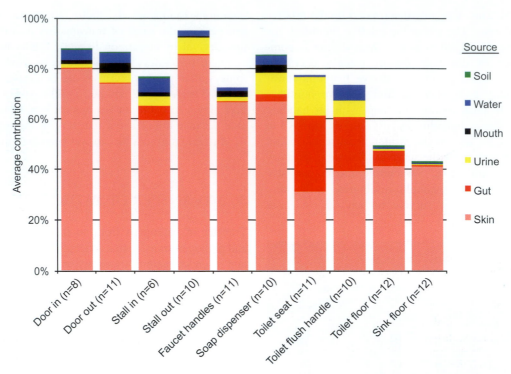

FIGURE 30.10 The average contributions of different sources to the surface-associated bacterial communities in 12 public restrooms. (The "unknown" source is not shown but would bring the total of each sample up to 100%.) (Flores *et al.*, 2011.)

TABLE 30.4 Environmental Factors Affecting the Survival of Microorganisms on Fomites

Factor	Primary Effect	Pathogen Example of Extreme Resistance
Temperature	Denaturization of proteins and nucleic acides; generally longer survival at lower temperatures	Bacterial spore formers Parvovirus
Solar irradiation	UV light causes cross-links along the nucleotides	Bacterial spore formers dsDNA viruses
Presence of organic matter	Can stabilize/destabilize proteins; protects against irradiation; neutralizes antagonistic substances; can serve as a nutrient source	
Interfaces	Greater stability at solid—water interfaces; less stability at air—water interfaces	Depends on nature of outer surface and resistance to denaturization
Dehydration	Loss of water causes denaturization of proteins	Spores
Relative humidity	Effects stability of proteins	Humidity range of least stability depends upon the organism

bring objects or their hands to their mouth an average of 81 times per hour (Tulv *et al.*, 2002). Studies have shown that the degree of virus transfer to the hand is related to:

- Age—an increase in the age of the individual reduces the relative amounts transferred probably because of less moisture in the skin
- The amount of pressure applied

- The application of friction which substantially increases the amount of virus transferred

The degree of transfer of any organism will depend on the nature and type of organism, nature of the surface and the amount of moisture (Figure 30.14). Higher bacterial transfer rates from fomite to the hand have been observed with hard nonporous surfaces (phone receiver, faucet) than with porous surfaces (clothing, sponges) (Rusin

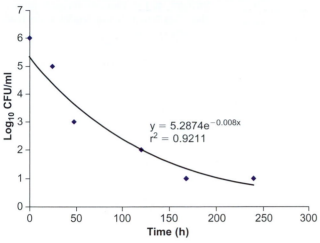

FIGURE 30.11 Survival of methicillin-resistant *Staphylococcus aureus* (MRSA) on stainless steel. Courtesy A. Moghe and C.P. Gerba.

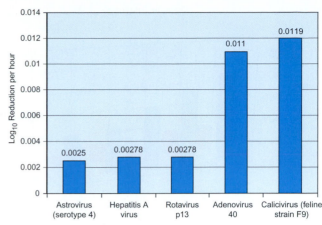

FIGURE 30.13 Inactivation rates of enteric viruses on fomites. From Boone and Gerba (2007). Reprinted with permission from the American Society for Microbiology.

TABLE 30.5 Examples of the Effects of Relative Humidity (RH) on Stability of Viruses

RH Range	Stability	Virus
>75	High	Vaccinia, reovirus
	Low	Adenovirus, poliovirus, foot and mouth disease, parainfluenza, measles
40–75%	High	Vaccinia, influenza
	Moderate	Poliovirus, MS 2 coliphage
	Low	Reovirus, T3 coliphage
<40%	High	Influenza, vaccinia
	Moderate	Reovirus, measles
	Low	Parainfluenza, T3 coliphage, poliovirus, adenovirus, foot and mouth disease

Selected from Spendlove and Fannin (1982).

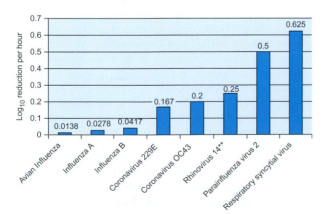

FIGURE 30.12 Inactivation rates of common respiratory viruses on fomites. From Boone and Gerba (2007). Reprinted with permission from the American Society for Microbiology.

et al., 2002). Although, greater numbers of bacteria were present in the wet sponge, the overall efficacy of transfer was less than from a stainless steel surface. When the same volunteers placed their fingers to their lips, 34 to 41% of the bacteria were transferred to their mouths depending on the type of bacteria. While such studies are useful for the demonstration of the potential role fomites play in disease transmission, they can also be used in risk assessment models to estimate the probability of disease transmission by fomites in a particular environment and the impact of interventions (Figure 30.15).

30.4 QUESTIONS AND PROBLEMS

1. Determine the time in hours for influenza A and hepatitis A virus to decrease in titer by 99% on a fomite.
2. Why would influenza virus be more likely to be transmitted in the winter by fomites than poliovirus?
3. What conditions would favor the growth of bacteria in/on fomites?
4. Determine the number of *Salmonella* a person will ingest if they touch a cutting board contaminated with 100,000 *Salmonella* per square centimeter. Assume that a fingertip has an area of one square centimeter and that only one finger touches the surface. List all your assumptions in a table. Using the risk model presented in Chapter 24 determine the probability of the individual becoming infected.
5. Look at the classroom you are sitting in and list three objects that would most likely become contaminated by a person infected with norovirus. Give your reasons why. Which object would have the greatest efficiency for transfer of a virus onto your hand?

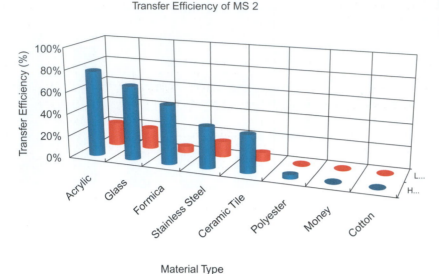

FIGURE 30.14 Transfer efficiency of coliphage MS 2 from various fomites to the hand. Courtesy G. Lopez.

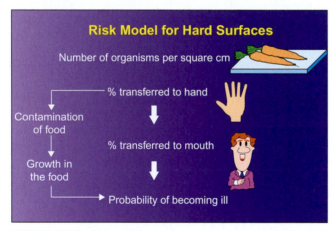

FIGURE 30.15 Risk model assessing pathogen transmission by hard surfaces. Courtesy C.P. Gerba.

REFERENCES AND RECOMMENDED READING

Boone, S. A., and Gerba, C. P. (2005) The occurrence of influenza A virus on household and day care center fomites. *J. Infect.* **51**, 103–109.

Boone, S. A., and Gerba, C. P. (2007) Significance of fomites in the spread of respiratory and enteric viral disease. *Appl. Environ. Microbiol.* **73**, 1687–1696.

Boone, S. A., and Gerba, C. P. (2010) The prevalence of human parainfluenza virus 1 on indoor office fomites. *Food Environ.Virol.* **2**, 41–46.

Centers for Disease Control (CDC, 2013) (2011) Estimates for foodborne illness in the United States. http://www.cdc.gov/foodborneburden/2011-foodborne-estimates.html

Cole, J. R., Chai, B., Farris, R. J., Wang, Q., Kulam-Shed-Mohideen, A. S., McGarrell, D. M., *et al.* (2007) The ribosomal database project (RDP-II): introducing *myRDP* space and quality controlled public data. *Nucl. Acids Res.* **35**, 169–172.

Delaquis, P., Bach, S., and Dinu, L. D. (2007) Behavior of *Escherichia coli* 0157-H7 in leafy vegetables. *J. Food Protect.* **70**, 1966–1974.

Flores, G. E., Bates, S. T., Knights, D., Lauber, C. L., Stombaugh, J., Knight, R., *et al.* (2011) Microbial biogeography of public restroom surfaces. *PLoS ONE* **6**, e28132. doi: 10.1371/journal.pone.0028132.

Gerba, C. P., Wallis, C., and Melnick, J. L. (1975) Microbiological hazards of household toilets-droplet production and fate of residual organisms. *Appl. Microbiol.* **30**, 229–237.

Hendley, J. O., Wenzel, R. P., and Gwaltney, J. M., Jr. (1973) Transmission of rhinovirus colds by self-inoculation. *New England J. Med.* **288**, 1361–1364.

Hewitt, K. M., Gerba, C. P., Maxwell, S. L., and Kelley, S. T. (2012) Office space bacterial abundance and diversity in three metropolitan areas. *PLoS ONE* **7**(5), e37849. doi: 10.1371/journal.pone.0037849.

Pepper, I. L., Rusin, P., Quintanar, D. R., Haney, C., Josephson, K. L., and Gerba, C. P. (2004) Tracking the concentration of heterotrophic plate count, bacteria from the source to the consumer's tap. *Int. J. Food Microbiol.* **92**(289–295).

Reng, V., Contzen, M., Drees, E., and Stegmanns, T. (2007) Study on the occurrence of *Salmonella* spp., *Campylobacter* spp. and *Arcobacter* spp. in duck breasts from retail trade prior to and after low-temperature cooking. *Arch. Lebensmittelhyg.* **58**, 170–174.

Reynolds, K. (2006) Indoor air quality. In "Environmental and Pollution Science" (I. L. Pepper, C. P. Gerba, and M. L. Brusseau, eds.), second ed., Elsevier, New York, NY.

Rusin, P., Maxwell, S., and Gerba, C. (2002) Comparative surface-to-hand and fingertip-to-mouth transfer efficiency of gram-positive bacteria, gram-negative bacteria, and phage. *J. Appl. Microbiol.* **93**, 585–592.

Rusin, P., Orosz-Coughlin, P., and Gerba, C. (1998) Reduction of faecal coliform, coliform and heterotrophic plate count bacteria in the household kitchen and bathroom by disinfection with hypochlorite cleaners. *J. Appl. Microbiol.* **85**, 819–828.

Sattar, S. A., Abebe, M., Bueti, A. J., Jampani, H., Newman, J., and Hua, S. (2000) Activity of an alcohol-based hand gel against adeno-, rhino-, and rotaviruses using the fingerpad method. *Infect. Contr. Hospital Epidemiol.* **21**, 516−519.

Sattar, S. A., and Springthorpe, V. S. (1996) Viral infections from animate and inanimate sources and infection control through chemical disinfection. In "Modeling Disease Transmission and its Prevention by Disinfection" (C. J. Hurst, ed.), University of Cambridge, UK, pp. 224−257.

Scott, B. A., and Pepper, I. L. (2010) Water distribution systems as living ecosystems: impact on taste and odor. *J. Environ. Sci. Health Part A* **45**, 890−900.

Scott, E. (1999) Hygiene issues in the home. *Am. J. Infect. Contr.* **27**, S22−S25.

Scott, E., Bloomfield, S. F., and Barlow, C. G. (1982) An investigation of microbial contamination in the home. *J. Hyg. (London)* **89**, 279−293.

Spendlove, J. C., and Fannin, K. F. (1982) Methods for the characterization of virus aerosols. In "Methods in Environmental Virology" (C. P. Gerba, and S. M. Goyal, eds.), Marcel Dekker, NY, pp. 261−329.

Tulv, N. S., Suggs, J. C., McCurdy, T., Hubal, E. A., and Moya, J. (2002) Frequency of mouthing behavior in young children. *J. Expo. Anal. Epidemiol.* **12**, 259−264.

Van Curtis, V., and Cairncross, S. (2003) Effect of washing hands with soap on diarrhea risk in the community: a systematic review. *Lancet Infect. Dis.* **3**, 275−281.

Vessey, J. A., Sherwood, J. J., Warner, D., and Clark, D. (2007) Comparing hand washing to hand sanitizers in reducing elementary school students' absenteeism. *Pediatr. Nurs.* **33**, 368−372.

Wang, Q., Garrity, G. M., Tiedje, J. M., and Cole, J. R. (2007) Naïve Bayesian classifier for rapid assignment of rRNA sequences into the new bacterial taxonomy. *Appl. Environ. Microbiol.* **73**, 5261−5267.

White, C., Kolble, R., Carlson, R., Lipson, N., Dolan, M., Ali, Y., *et al.* (2003) The effect of hand hygiene on illness rate among students in university residence halls. *Am. J.Infect. Contr.* **31**(36), 4−370.

Chapter 31

Global Emerging Microbial Issues in the Anthropocene Era

Ian L. Pepper, Charles P. Gerba and Terry J. Gentry

The anthropocene can be defined as the influence of human activities on the global ecosystem consisting of the lithosphere, the hydrosphere and the atmosphere, all of which potentially contain microorganisms. In this chapter we examine some of the major global microbial issues that have been impacted by human activities. Some of these issues are directly caused by adverse human activities that result in pollution, as in the case of the recent BP *Deepwater Horizon* spill, off the Gulf of Mexico. Other issues result indirectly, such as problems related to global warming, or antibiotic-resistant bacteria.

31.1 MICROBIAL CONTRIBUTIONS TO CLIMATE CHANGE

31.1.1 Nitrous Oxide Emissions from Soil

Microbial activities can influence the formation of greenhouse gases including N_2O, CH_4 and CO_2, and are a major cause of global warming (Table 31.1). The driving force for CO_2 emissions in the age of the anthropocene is thought to be fossil fuel use and land-use change (Inselsbacher *et al.*, 2011). In contrast, major N_2O and

CH_4 emissions may be due to agricultural practices (IPCC, 2007).

The agricultural practice of adding nitrogen, either as chemical or manure fertilizer, to soil is a major contributor to the gradual increase in nitrous oxide (N_2O) emissions to the atmosphere, although other sources of N_2O in the atmosphere include burning of biomass, combustion of fossil fuels and chemical manufacturing of nylon (Mosier *et al.*, 1996). The source of nitrogen in fertilizers is mainly ammonia or ammonium-producing compounds such as urea. However, normally only about half the total nitrogen applied to a field as fertilizer or manure is assimilated by the crop (Delgado and Mosier, 1996). The remaining nitrogen is lost through leaching, erosion, gaseous emissions and microbial activity. Due to microbial activity, about 75% of the total global anthropogenic emissions of N_2O are due to agricultural fertilizers (Jackson *et al.*, 2009). This activity includes two mechanisms, both of which have the potential to produce N_2O. The first is nitrification, a process in which ammonium is chemoautotrophically oxidized first to nitrite and then to nitrate (Chapter 2). It has been suggested that N_2O production is associated with low-oxygen conditions, in which nitrifying bacteria utilize nitrite as an electron

I.L. Pepper, C.P. Gerba, T.J. Gentry: Environmental Microbiology, Third edition. DOI: http://dx.doi.org/10.1016/B978-0-12-394626-3.00031-4
© 2015 Elsevier Inc. All rights reserved.

TABLE 31.1 Global Atmospheric Concentrations of Selected Greenhouse Gases[1]

	Microbially mediated/Anthropogenic (parts per million)			Anthropogenic only (parts per trillion)	
	CO_2	CH_4	N_2O	SF_6[a]	CFC[b]
Preindustrial	278	0.700	0.275	0	0
2004	377	1.789	0.319	5.22[c]	794
Atmospheric lifetime (years)	50–200	12	114	3200	45–100

Data used in this table are from the Carbon Dioxide Information Analysis Center, which is supported by the U.S. Department of Energy Climate Change Research Division, http://cdiac.ornl.gov/.
[a]SF_6 = sulfur hexafluoride.
[b]CFC = CFC-11 (trichlorofluoromethane) and CFC-12 (dichlorodifluoromethane).
[c]Value is from 2001.

acceptor in place of oxygen, as a result reducing NO_2^- to N_2O (Ambus *et al.*, 2006). The second step is denitrification, which is the utilization of nitrate (NO_3^-) as a terminal electron acceptor during anaerobic respiration of organic compounds (Chapter 15). N_2O is an intermediate in the reduction of NO_3^- to N_2, but can be the end point of nitrate reduction especially in environments where initial nitrate concentrations are low. Thus, N_2O can be produced as an intermediate both in denitrification and in nitrification. Repeated nitrogen fertilization of agricultural soils has been shown to increase N_2O emissions with a stimulation of both bacterial and archaeal ammonia oxidizers (Inselsbacher *et al.*, 2011). Interestingly, nitrogen fertilizers also lead to increased CO_2 emissions due to microbial respiration in the rhizosphere (Treseder, 2008).

Why is this a problem? Nitrous oxide gas that is released to the atmosphere from industrial and biogenic sources contributes to global warming as a greenhouse gas. Greenhouse gases are of concern because they absorb long-wave radiation from the sun after it hits the Earth and is reflected back into space. This effectively traps heat in the atmosphere. N_2O is of concern in several respects: it has a long residence time ($>$ 100 years) in the atmosphere; and it is highly efficient in absorbing long-wave radiation. One molecule of N_2O is equivalent in heat-trapping ability to about 200 molecules of CO_2. Therefore, small increases in atmospheric N_2O concentration can have a large impact on warming trends.

An additional concern with N_2O is that in the upper atmosphere, solar radiation can photolytically convert N_2O to nitric oxide (NO), which is a contributor to the depletion of the protective ozone layer (Figure 31.1). The ozone layer acts as a filter to remove biologically harmful ultraviolet (UV) light. Stratospheric ozone depletion occurs through a chemical interaction between sunlight, ozone and certain reactive chemical species, including nitrogen oxides and organohalogens such as chlorofluorocarbons (CFCs).

Depletion of the protective ozone layer can have serious ecological and human health consequences. Increased levels of UV radiation may be inhibitory to certain microorganisms, such as phytoplankton, and may also increase the incidence of skin cancer in humans. Ozone is depleted in the series of reactions shown in Figure 31.1, where light energy begins the reaction by splitting nitrous oxide into N_2 and singlet oxygen, in which one of the electrons is in a high-energy state. This singlet oxygen can react with nitrous oxide to form two molecules of nitric oxide. Nitric oxide in turn reacts with ozone (O_3) to produce nitrogen dioxide and oxygen. The nitrogen dioxide then reacts with singlet oxygen (O$^{\bullet}$) to produce oxygen and regenerate nitric oxide. The fact that nitric oxide is regenerated in this series of reactions means that for every nitrous oxide molecule released to the atmosphere, a large number of ozone molecules can be destroyed.

Strategies can be implemented to reduce biogenic N_2O emissions from agriculture. A primary factor in N_2O emissions is the low efficiency of utilization of nitrogen fertilizers, which leaves them subject to nitrification/denitrification processes. Several measures can be taken to increase fertilizer utilization efficiency. The most economical approach is simply to manage the amount and time of fertilizer application to a crop. Enough fertilizer must be added to meet crop needs, but overfertilization will result in increased nitrate formation and leaching. A second way to minimize nitrogen losses in irrigated croplands is through control of the timing and amount of irrigation. Remember, plants can take up nitrogen as either ammonia or nitrate, but nitrate is easily removed from the plant root area through leaching processes. Many states in the U.S. have studied and adopted best management practices (BMPs) for fertilizer application and irrigation. These BMPs are region-specific since climate and soil types change dramatically from region to region. Finally, two other approaches to minimizing

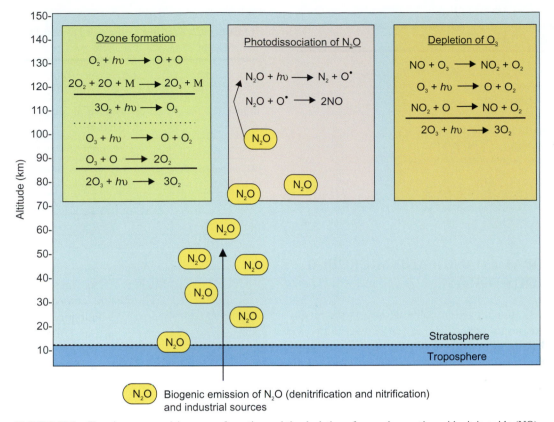

FIGURE 31.1 Equations summarizing ozone formation and the depletion of ozone by reaction with nitric oxide (NO). Solar UV radiation ($h\nu$) photodissociates molecular oxygen (O_2) into two oxygen atoms (O), which recombine with undissociated O_2 (in the presence of another chemical species, M) to form ozone (O_3). Ozone is then photodissociated back to molecular oxygen (O_2). The constant cycling between ozone and oxygen is important because it consumes harmful UV radiation in the stratosphere. Nitrous oxide (N_2O) emitted to the atmosphere is photodissociated to nitrogen and an oxygen atom in an electronically excited state, $O^\bullet$, which reacts with N_2O to produce nitric oxide (NO). Nitric oxide can react with ozone and, in a series of reactions, produce O_2 resulting in a net depletion of ozone.

nitrate formation and leaching are the use of slow-release fertilizers, which allow more controlled release of ammonia into the environment, and the application of nitrification inhibitors, e.g., N-serve (2-chloro-6-(trichloromethyl) pyridine), which suppress the formation of nitrate. However, application of slow-release fertilizers or nitrification inhibitors is costly, so neither approach is widely used.

31.1.2 Methane Emissions from Landfills

The formation of another global warming gas, methane (CH_4), is called methanogenesis and is predominantly a microbial process (see Chapter 16). The organisms responsible for this process are a group of obligately anaerobic Archaea called the methanogens. Many of these are commonly found in wetlands and marine sediments where anaerobic conditions prevail. However, methane emissions due to anthropogenic activities include emissions from landfills (Chiemchaisri *et al.*, 2012). Methane

is the third most important greenhouse gas after water vapor and CO_2, and has 25 times the global warming potential of an equal mass of CO_2. It has been estimated that one ton of deposited municipal solid waste results in $160-250 \, m^3$ of landfill gas, which consists of CH_4 (50–60% v/v) and CO_2 (40–45% v/v) (Scheutz *et al.*, 2004). Landfill gas has been shown to increase with increased moisture due to more optimal conditions for growth of anaerobic bacteria. In some modern landfills, anaerobic degradation and methanogenesis are actually encouraged in order to degrade organic materials thus freeing up valuable landfill space while generating CH_4, which can be captured to generate electricity. Examples of landfill methanogens identified through polymerase chain reaction (PCR) detection of the methyl coenzyme-M reductase gene (*mcrA*) are shown in Information Box 31.1. Overall, anthropogenic activities have significantly affected microbial activities, in many cases resulting in enhanced emissions of greenhouse gases and concomitant global warming.

Information Box 31.1   Species Identification of Methanogens from Two Landfills

- *Methanofollis liminatans, Methanospirillum hungatei*
- *Methanobacterium formicicum*
- *Methanocaldococcus jannaschi*
- *Methanobacterium formicicum mrtA*
- *Methanosarcina thermophile*
- *Methanocorpusculum bavaricum*
- *Methanocorpusculum aggregans, Methanocorpusculum parvum*

Source: Adapted from Luton *et al.* (2002).

31.2 GLOBAL CHANGE AND MICROBIAL INFECTIOUS DISEASE

In many cases, human activity directly affects the magnitude of exposure to pathogens that can be transmitted via the environment, as in the case of the voluntary risk associated with a decision to drink water that may or may not be fecally contaminated. However, in other situations, human activities are not directly involved such as when whole communities are exposed to pathogens due to accidental contamination of a drinking water supply. Urbanization and deforestation can also cause immense changes in environmental systems that impact exposure to pathogens. It is also increasingly clear that climate change will have an influence on our exposure to pathogens.

Here we discuss the influence of both indigenous environmental and animal-borne (including human) pathogens on human health and infectious disease, as well as the influence of anthropogenic activities on exposure to pathogens. Overall, microbial infectious diseases dwarf all other sources of human deaths annually (Table 31.2). But there is a disparity in where these deaths occur globally. In developing countries, microbial infectious diseases are clearly the leading agent of death (WHO, 2004) (Table 31.3). In developed countries, heart diseases and cancer are more important, although in the United States infectious diseases are still the third leading cause of death.

31.2.1 Global Warming and Microbial Infectious Disease

It is now well documented that global warming is occurring and that anthropogenic activities are at least partially to blame. Currently, human activities add about 7–9 gt (gigatons) of carbon to the atmosphere as carbon dioxide as well as other greenhouse gases that contribute to this process (see Chapter 16). Global warming is resulting in

TABLE 31.2 Leading Causes of Human Deaths Worldwide

Cause	Percent of deaths	Deaths per 100,000 per year		
		All	Male	Female
All causes	100.00	916.1	954.7	877.1
All microbial infectious disease	50.4	461.8	484.0	439.2
Cardiovascular diseases	29.34	268.8	259.3	278.4
Ischemic heart disease	12.64	115.8	121.4	110.1
Malignant neoplasms (cancers)	12.49	114.4	126.9	101.7
Cerebrovascular disease (stroke)	9.66	88.5	81.4	95.6
Respiratory diseases	6.49	59.5	61.1	57.9
Unintentional injuries	6.23	57.0	73.7	40.2
Chronic obstructive pulmonary disease	4.82	44.1	45.1	43.1
Perinatal conditions	4.32	39.6	43.7	35.4
Intentional injuries (suicide, murder, war, etc.)	2.84	26.0	37.0	14.9
Lung cancers	2.18	20.0	28.4	11.4
Road traffic accidents	2.09	19.1	27.8	10.4
Childhood diseases	1.97	18.1	18.0	18.2
Neuropsychiatric disorders	1.95	17.9	18.4	17.3
Diabetes mellitus	1.73	15.9	14.1	17.7
Hypertensive heart diseases	1.60	14.6	13.4	15.9
Suicide	1.53	14.0	17.4	10.6
Stomach cancer	1.49	13.6	14.1	13.1
Diseases of the genitourinary system	1.49	13.6	14.1	13.1
Cirrhosis of the liver	1.38	12.6	16.1	9.1
Nephritis/nephropathy	1.19	10.9	11.0	10.7
Colorectal cancer	1.09	10.0	10.3	9.7
Liver cancer	1.08	9.9	13.6	6.2

Adapted from WHO (2004).

changing weather patterns and the increased incidence of extreme weather events (Patz *et al.*, 2000). Severe weather events including prolonged droughts in arid areas, and excessive precipitation that results in flooding

TABLE 31.3 Leading Causes of Human Deaths: Developed vs. Developing Economics

Causes of death in developing countries	Number of deaths (annually)	Causes of death in developed countries	Number of deaths (annually)
HIV/AIDS	2,678,000	Ischemic heart disease	3,512,000
Lower respiratory infections	2,643,000	Stroke	3,346,000
Ischemic heart disease	2,484,000	Chronic obstructive pulmonary disease	1,829,000
Diarrhea	1,793,000	Lower respiratory infections	1,180,000
Cerebrovascular disease	1,381,000	Lung cancer	938,000
Childhood diseases	1,217,000	Car accident	669,000
Malaria	1,103,000	Stomach cancer	657,000
Tuberculosis	1,021,000	High blood pressure	635,000
Chronic obstructive pulmonary disease	748,000	Tuberculosis	571,000
Measles	674,000	Suicide	499,000

Adapted from WHO (2004).

Information Box 31.2   Influence of Global Warming on Microbial Infectious Disease

Mechanism	Impact	Example
Severe weather events (floods, droughts, heat waves)	Vector- and waterborne diseases increase	Most waterborne outbreaks occur after above normal rainfall events
Enhanced sea temperatures (El Niño)	Vector- and waterborne diseases increase	Enhanced growth of *Vibrio cholerae*, and increased malaria
Enhanced freshwater temperatures	Water-based pathogen numbers increase	Enhanced growth of the protozoan *Naegleria fowleri*
Decreased potable water sources	Waterborne disease increases	Use of source water of a lower quality

Information Box 31.3   Mosquito-borne Microbial Diseases

Disease	Type of Microbe	Genus
Malaria	Protozoan parasite	*Plasmodium*
Dengue fever	Virus	*Flavivirus*
Rift Valley fever	Virus	*Phlebovirus*
Encephalitis	Virus	West Nile virus

in other areas, can also affect the incidence of microbial infectious disease (Information Box 31.2).

Direct mechanisms for enhanced infectious disease due to global warming include enhanced incidence of vector-borne diseases, water-related pathogens and water-based pathogens. One example is mosquito-borne microbial diseases, which are particularly prevalent worldwide (Information Box 31.3). Malaria is an infectious disease carried by mosquitoes that occurs in tropical and subtropical regions including the Americas, Asia and Africa. The cause of the disease is a protozoan parasite in the genus *Plasmodium* (Snow *et al.*, 2005). Global warming is now believed to be increasing the

incidence of malaria in areas such as the East African highlands of Kenya, where previously high elevation and cooler temperatures limited the disease (Martens, 1999). Other vector-borne diseases including dengue fever and rift valley fever are also thought to be on the increase due to global warming and associated El Niño/Southern Oscillation (ENSO)-related climate anomalies. Rift valley fever is a viral zoonosis that primarily affects domestic livestock, but can be passed to humans resulting in a fever.

The El Niño/Southern Oscillation (ENSO) is a global-coupled ocean—atmospheric phenomenon. Specifically, ENSO is a prominent outcome of climate change that occurs due to enhanced temperature changes in surface waters of tropical areas of the Pacific Ocean. Currently, it is believed that global warming is enhancing El Niño effects, causing global climate anomalies and weather patterns including periods of excessive rainfall in some areas and enhanced drought in others. El Niño events are thought to be responsible for the enhanced incidence of several infectious diseases (Kovats *et al.*, 2003). El Niño rains resulted in enhanced cholera outbreaks in Uganda in

2002–2003 (Alajo *et al.*, 2006). Outbreaks of cholera in Peru were significantly correlated with elevated sea temperatures due to a 1997–1998 El Niño. Interestingly, the onset of cholera epidemics was linked not only to warmer sea temperatures, but also to plankton populations. When sea temperatures increase, phytoplankton populations first become abundant utilizing sunlight for energy. Subsequently, zooplankton populations increase with a concomitant increase in the cholera-causing bacterium *Vibrio cholerae* (Gil *et al.*, 2004). El Niño has also been linked to malarial epidemics in South America (Gagnon *et al.*, 2002). In the southwestern United States, the incidence of disease caused by hantavirus and West Nile virus is also thought to be related to El Niño events (Information Box 31.4).

In addition to clearly documented effects of El Niño on microbial infectious disease, other effects of global warming are now being noticed that have not yet been well studied. These include anecdotal reports of increased *Naegleria fowleri* infections in the United States, and increased Legionnaires' disease in the United Kingdom. In both instances, the water-based microorganisms involved in the infectious disease prefer warm, fresh water habitats for growth and reproduction. Whether or not these outbreaks are linked to global warming remains to be seen, but clearly from a broad perspective, global warming is now a major factor influencing the incidence of microbial infectious disease.

There are other ways in which humans impact the environment and influence infectious disease. For example, climate change-induced drought combined with increasing human populations and demand for potable water has strained the availability of potable water supplies. This has led to an increase in the consumption of fecally-contaminated water, particularly in developing countries. Floods can also enhance microbial infectious disease. In developed countries, excessive precipitation can result in storm water loads that overwhelm wastewater treatment and drinking water treatment plants, and wash sewage, animal manures and other

sources of pathogens into surface waters. Such was the case in the infamous Milwaukee, Wisconsin, U.S. *Cryptosporidium* outbreak in the early 1990s.

31.2.2 Urbanization and Deforestation

Urbanization can affect the incidence of microbial disease both positively and negatively. In developing countries, increased population growth within urban centers can lead to densely packed housing and inadequate water supplies. Such conditions can frequently lead to enhanced waterborne disease due to poor sanitation. Interestingly, the incidence of vector-borne disease has been shown to decrease during urbanization due to less available breeding sites for mosquitoes (Hay *et al.*, 2005). However, urbanization can also result in the creation of new pools of water such as dams, rice paddies or artificial wetlands, which serve as habitats that can enhance vector-borne disease including malaria.

Concomitant with urbanization are associated land cover changes that affect the incidence of infectious disease. Such anthropogenic land surface changes include: deforestation; road construction; agricultural encroachment; urban sprawl; and extractive industries such as mining, quarrying and oil drilling (NRC, 2007). Recently, deforestation, urban spawn and biodiversity loss were linked to an increase in the incidence of Lyme disease (Schmidt and Ostfeld, 2001). Lyme disease is now endemic in at least 19 states in the U.S. (Nadelman and Wormser, 2005). The disease is due to the spirochete bacterium *Borrelia burgdorferi*, which is transmitted to humans by ticks. Similarly, expansion and changes in agriculture were shown to be intimately associated with the emergence of Nipah virus in Malaysia (Lam and Chua, 2002). Thus, in many instances urbanization and deforestation can result in increased contact between humans and an infectious agent. Also of concern are the zoonotic pathogens of nonhuman origin with the potential for adaptation to cause human infections. It is thought that microbial agents causing infectious disease in humans often originate from disturbances of the natural environment. Emerging microbial diseases can be caused by infectious agents of wildlife that either recently adapted to infect humans, or were pre-adapted, and come into opportunistic contact with humans (Taylor *et al.*, 2001). Examples of such emerging pathogens include the HIV/AIDS virus, Ebola and West Nile virus.

In summary, anthropogenic activities that result in global climate change and changes in land use are influencing the incidence of known microbial infectious disease, and in some cases allowing for the discovery of new infections of microbial origin. As the human population continues to grow, these phenomena are unlikely to disappear.

Information Box 31.4   El Niño Enhanced Microbial Infectious Disease

Country/Region	Disease
Peru, Uganda	Cholera
Ecuador, Peru, Bangladesh, India	Malaria
Thailand, Brazil	Dengue fever
Southwestern United States	Hantavirus, West Nile virus

Adapted from Gagnon *et al.* (2002); Alajo *et al.* (2006).

31.3 MICROBIAL REMEDIATION OF MARINE OIL SPILLS

The effects of human activities on microorganisms in the age of the Anthropocene is explicitly illustrated by oil spill disasters such as the *Exxon Valdez* spill in 1989, and the BP *Deepwater Horizon* spill in 2010. Petroleum hydrocarbons occur naturally in all marine environments, with the result that numerous diverse microorganisms have evolved the capability to degrade similar hydrocarbons following an oil spill, as they utilize them as a source of carbon and energy for growth. Overall, there are hundreds of species of Bacteria, Archaea and Fungi that can degrade petroleum (Atlas and Hazen, 2011). Factors affecting the microbial degradation of petroleum hydrocarbons are shown in Information Box 31.5. Overall, oil-degrading indigenous microorganisms can play a significant role in the cleanup of marine oil spills. The *Exxon Valdez* oil spill in 1989 was notorious because of its impact on Alaskan wildlife and the pristine environments of Prince William Sound. The BP *Deepwater Horizon* spill became imprinted in the minds of the U.S. public due to the incessant nightly television images of oil gushing out of an oil wellhead far beneath the ocean surface for 87 days. The notoriety of both spills resulted in frequent comparisons, even though the two events were in fact quite different. However, in both cases biodegradation of oil by indigenous microorganisms significantly reduced the environmental impact of each disaster (Case Study 31.1).

Information Box 31.5   Factors Affecting Microbial Biodegradation Rates of Petroleum Hydrocarbon

Factor	Influence on Biodegradation
Type of oil	Lighter crude oils degrade more readily than heavier crudes or polycyclic aromatic hydrocarbons (PAHs)
Number and activity of oil-degrading microorganisms	Acclimation needed to enhance pre-spill degrader numbers
Essential nutrients	Lack of nitrates, phosphates and iron can limit rates of degradation
Environmental redox conditions	Aerobic conditions enhance degradation rates relative to anaerobic conditions
Oil surface area: volume ratio	Can limit degradation rates
Dispersants	Increase oil surface area and degradation rates

31.4 ANTIBIOTIC-RESISTANT BACTERIA

Bacteria are prokaryotic organisms with the ability to metabolize and replicate very quickly. They are also very adaptable genetically. When confronted with an antibiotic, there need only be one bacterial cell with a genetic or mutational change that confers resistance to that antibiotic to subsequently allow for the proliferation of antibiotic-resistant bacteria. Thus, the more antibiotics are used, the greater the likelihood of antibiotic-resistant strains developing. The greatest concern with antibiotic resistance is the potential for human pathogenic strains to become resistant to popular antibiotics, which subsequently cannot contain the infectious agent. Commonly used antibiotics and their mode of action are shown in Information Box 31.6. The widespread sometimes indiscriminant use of antibiotics has raised the question "Can antibiotic resistant genes be transferred from nonpathogenic bacteria to human pathogenic strains?" Here we address this issue by examining the incidence of antibiotic-resistant bacteria, and the potential for horizontal gene transfer between different groups of bacteria. In particular, we examine which anthropogenic practices are most likely to induce antibiotic-resistant bacteria.

31.4.1 Development of Bacterial Antibiotic Resistance

Antibiotic drug resistance is the acquired ability of an organism to resist the effects of a chemotherapeutic agent to which it is normally susceptible. In general, there are two ways in which a microbe can become resistant:

1. Mutational change (spontaneous): One way microbes can become more tolerant of an antibiotic is to alter the target of an agent within the cell. For example, spontaneous mutations in the genes encoding ribosomal RNAs can prevent antibiotics such as tetracycline from binding and blocking gene translation.
2. Introduction of foreign DNA: In this case, genes are obtained from other host microorganisms that offer protection against antibiotics. Well-studied examples here include efflux of antibiotics such as tetracycline, or inactivation of penicillins by β-lactamases. The genes encoding these resistances can be localized on a plasmid or on the chromosome. A gene responsible for tetracycline resistance encodes a membrane transporter responsible for pumping tetracycline out of the cell. The β-lactamases cleave the β-lactam ring of most penicillins thereby rendering them ineffective.

Because antibiotic-resistant genes can be transferred within and between bacteria, the issue of horizontal gene transfer becomes relevant to the discussion. Horizontal gene transfer (HGT) is the transfer of genetic material horizontally between bacteria of the same species, or other species or genera. The genetic material can consist of discrete pieces of DNA, whole genes or multiple genes contained within plasmids or transposons. The mechanisms

Case Study 31.1   The *Exxon Valdez* and BP *Deepwater Horizon* Oil Spills

In 1989 the *Exxon Valdez* oil tanker ran aground in Prince William Sound, Alaska, resulting in the second largest oil spill in U.S. history. Due to the nature of the heavy crude oil and concern about the use of dispersants, the decision was made that oil dispersion would not be attempted.

Tidal currents and winds then resulted in a significant portion of the oil floating ashore. Overall, approximately 1300 miles ($\cong$ 2100 km) of coastline in Prince William Sound and the Gulf of Alaska were contaminated with oil to some degree (Atlas and Hazen, 2011). Initially, physical washing and collection of the oil was the first cleanup strategy employed, but quickly it was realized that *in situ* bioremediation would likely be more effective. A graphic depiction of the *Exxon Valdez* spill and cleanup is shown in Figure 31.2.

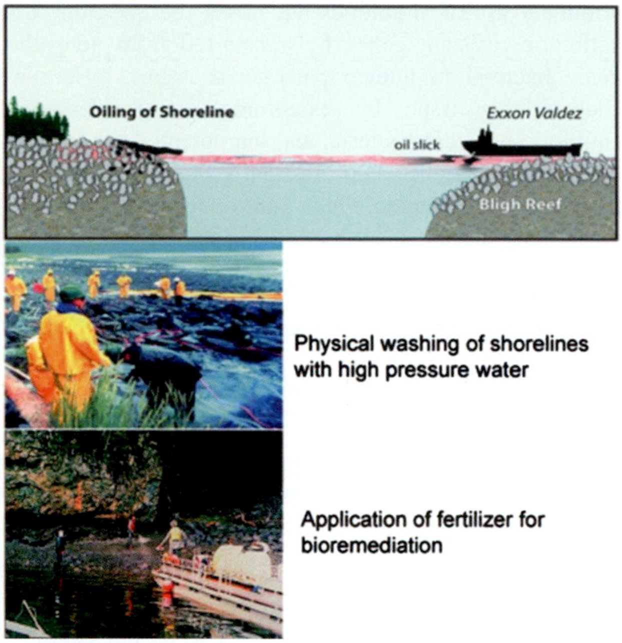

In 2010, the *Deepwater Horizon* oil rig exploded and sank, killing 11 people. Subsequent to this, a ruptured wellhead released oil into the Gulf of Mexico for almost 3 months 1500 m below the ocean surface. Unlike the *Exxon Valdez* spill, attempts were made to disperse the oil, since this was a light crude. The dispersant COREXIT 9500 was injected directly into the wellhead, resulting in a deepwater "cloud" of dispersed oil droplets, which remained beneath the ocean surface, and moved away from the wellhead. Overall, this resulted in fewer oil slicks forming at the surface of the ocean above the wellhead. The dispersion of the oil allowed for enhanced rates of oil biodegradation. A graphic depiction of the spill is shown in Figure 31.3. A comparison of the *Exxon Valdez* and BP *Deepwater Horizon* oil spills is shown in Table 31.4.

Bioremediation Aspects of the Two Spills

In the *Exxon Valdez* oil spill, enhanced *in situ* bioremediation was used to degrade oil on the contaminated shores. Rates of biodegradation were increased through the application of slow-release nitrogen fertilizers that also contained phosphate. Overall 107,000 lb (48,000 kg) of nitrogen were applied. Evidence of significant biodegradation of soil was shown by rapid increases in the numbers of naturally occurring oil-degrading bacteria from 103 CFU/ml of seawater ($\cong$ 1−10% of total heterotrophs) to 105 CFU/ml by late 1989 (up to 40% of total heterotrophs). The success of the bioremediation efforts was documented by National Oceanic and Atmospheric

Oiling of Shoreline

Physical washing of shorelines with high pressure water

Application of fertilizer for bioremediation

FIGURE 31.2 The *Exxon Valdez* oil spill. *Source:* Atlas and Hazen (2011).

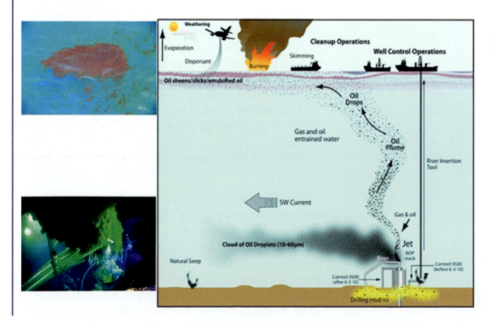

FIGURE 31.3 The BP *Deepwater Horizon* oil spill. *Source:* Atlas and Hazen (2011).

TABLE 31.4 Characteristics of the Two Major Oil Spills

Exxon Valdez	*BP Deepwater Horizon*
• 41.6 million liters (11 m gallons)	• 779 million liters estimated by the National Incident Command's Flow Rate
• North Slope Heavy Oil (API 29) tanker spill with known volume discharged	Technical Group (FRTG) ($\cong$ 200 m gallons)
	• Light Louisiana oil (API 35.2)
• Surface spill impacted cobble/rocky shorelines of nearby islands	• Large amounts of methane also released
	• Discharged at wellhead at a depth of 1500 m
• Bioremediation used extensively	• Spill occurred 77 km offshore
• Fate of oil remnants still studied more than 21 years after spill	• Impact by deep-sea cloud of fine droplets of low concentration oil, marshes affecting and sandy beach shorelines
• Scientific and operating experience gained is applicable to other spills	• Took 84 days to seal off leak
	• Aerial and subsurface dispersants used extensively
• No dispersants utilized	• Largest remediation and emergency response to an oil spill ever worldwide

Adapted from Atlas and Hazen (2011).

Administration (NOAA) surveys between 2001 and 2003, wherein 97.8% of all samples assayed had no oil residues (Short *et al.*, 2004). However, in localized areas some residual oil still persists, sequestered within fine-grained sediments, and even in 2013, the debate continues as to whether further bioremediation efforts should be conducted to remove all traces of the oil contaminants.

In the BP *Deepwater Horizon* oil spill, bioremediation efforts were also successful, although the situation was very different from the Prince William Sound spill. Rapid attenuation of oil has been observed in the aftermath of the BP spill, in part due to large populations of marine microorganisms capable of degrading oil. These populations are most likely due to adaptation following natural seeps of oil in the Gulf contributing more than 400,000 barrels of oil a year. Bioremediation was particularly significant in the "cloud" of dispersed oil, where bacterial density within the cloud (10^5 CFU/ml) was two orders of magnitude higher than outside of the cloud. Using a 16S rRNA microarray, it was shown that 16 classes of oil-degrading γ-*Proteobacteria* were significantly enriched within the cloud, with three families in the class *Oceanospirillales*

dominating (Hazen *et al.*, 2010). Subsequent studies utilizing metagenomic and metatranscriptomic sequencing revealed that genes for mobility, chemotaxis and aliphatic hydrocarbon degradation were significantly enriched and expressed in hydrocarbon plume water samples as compared to uncontaminated seawater. Sequencing of two *Oceanospirillales* single cells showed that both cells possessed genes coding for n-alkane and cycloalkane degradation (Mason *et al.*, 2012). Finally, note that significant populations of methanotrophs arose in contaminated seawater, presumably in response to the methane that was also released during the spill.

Multiple remedial actions were utilized following the BP spill including skimming, burning and *in situ* bioremediation. In addition, significant evaporation of alkanes with chain lengths up to C-20 occurred. Overall, it is estimated that 26% of the oil released still remains, much of it within shoreline and marsh sediments, where anaerobic conditions result in reduced rates of biodegradation (Atlas and Hazen, 2011). It is not known what the long-term environmental impacts of these oil remnants will be, and it will likely be decades before all effects are fully understood.

for gene transfer are shown in Information Box 31.7. Overall, acquisition of resistance via mutations has been shown to be less effective than complete gene acquisitions (Tomasz, 2006). Horizontal gene transfer in soils has been well documented. However, rates of gene transfer in soil are relatively low due to spatial separation of donors and recipients, and relatively low numbers of donors compared to nondonor organisms (Rensing *et al.*, 2002). Thus, rates of transfer of resistance genes to human pathogens in soil are likely to be low. However, public concern has still focused on the potential for increasing the number of antibiotic-resistant bacteria in the environment, which would increase the potential for transfer of resistance to human pathogenic bacteria. Despite this, a recent study has shown that antibiotic-resistant strains of bacteria are already indigenous and highly prevalent in soils (D'Costa

et al., 2006). In this study, *Streptomyces* isolates from diverse soil locations were tested for antibiotic resistance. All 480 isolates were resistant to at least six antibiotics, and some to as many as 20. The authors of this study therefore titled the article "Sampling the Antibiotic Resistance." Clearly, then, soils are a source not only of antibiotics but also of antibiotic resistance.

31.4.2 Anthropogenic Activities that Enhance the Incidence of Antibiotic-resistant Bacteria

Human activities such as land application of municipal biosolids or animal manures have been implicated as potentially increasing the concentrations of antibiotic-resistant bacteria within soil. However, in a

Information Box 31.6   Commonly Used Antibiotics and Their Mode of Action

Name	Spectrum	Mode of Action
Chloramphenicol	Broad spectrum	Inhibits protein synthesis by binding to 50S ribosomal subunit
Erythromycin	Mostly Gram-positive	Inhibits protein synthesis by binding to 50S ribosomal subunit
Tetracycline	Broad spectrum	Inhibits protein synthesis by binding to 30S ribosomal subunit
Streptomycin	Broad spectrum	Inhibits protein synthesis by binding to 30S ribosomal subunit
Polymyxin	Gram-negative bacteria, especially *Pseudomonas*	Disrupts cell membrane
Nalidixic acid	Gram-negative bacteria	Inhibits DNA synthesis
Novobiocin	Gram-negative bacteria	Inhibits DNA synthesis
Trimethoprim	Broad spectrum	Inhibits purine synthesis
Rifampicin	Gram-positive bacteria	Inhibits RNA synthesis
Penicillin	Mostly Gram-positive bacteria	Inhibits cell wall peptidoglycan synthesis

Information Box 31.7   Gene Transfer Mechanisms

- Conjugation: Transfer of genes from one prokaryotic organism to another by a mechanism involving cell-to-cell contact and a plasmid
- Transformation: Transfer of nucleic acid via uptake of free DNA. It does not require cell-to-cell contact
- Transduction: Transfer of nucleic acid mediated by a virus

recent study, concentrations of antibiotic-resistant bacteria did not increase even after 20 continuous years of land application of biosolids, when compared to neighboring soil that had not received biosolids (Brooks *et al.*, 2007). In contrast, soils receiving dairy manures have been shown to have enhanced concentrations (Esiobu *et al.*, 2002). Most concern over soil-borne antibiotic-resistant bacteria relates to the introduction of animal manures from swine, poultry or cattle fed antibiotics as part of their diet. For example, prophylactic dosing of antibiotics in confined animal feeding operations results in relatively high concentrations of antibiotic-resistant bacteria in animal manures compared to biosolids (Chee-Stanford *et al.*, 2009). Antibiotics are used as growth promoters to increase feed efficacy and subsequent daily growth of animals. Since many of these drugs are closely related to antibiotics used for treating human diseases, the transfer of these strains and the resistances they carry from animals to humans is of concern.

One area of concern is chickens which are fed Ciprofloxacin. Chickens are also frequently infected with *Salmonella* and/or *Campylobacter*. Therefore, there is the potential for resistance to Ciprofloxacin to increase, since microorganisms develop resistance to this antibiotic fairly easily. If a family consumes chicken that is not cooked properly, they may become infected with a Ciprofloxacin-resistant *Campylobacter*. Of additional interest is the fact that Ciprofloxacin is one of the primary drugs for the treatment of anthrax.

As another example, cattle are frequently fed a 90 to 100% grain diet prior to slaughter to enhance fat marbling in the meat. However, this diet is unnatural for cattle that are used to feeding on low-nutrition grasses, since they are ruminants. High grain diets can result in acid production, which causes ulcers in the cattle's stomachs. Bacteria can then migrate from the ulcers to the liver, where additional abscesses form. Because of this cattle are fed antibiotics to suppress the bacteria. Antibiotics also increase feed conversion efficiency.

However, new studies dispute the ease with which antibiotic resistance develops. A recent study examined 448 *Campylobacter* isolates from U.S. feedlot cattle for resistance to 12 antimicrobials (Englen *et al.*, 2005). These included: tetracycline; nalidixic acid; trimethoprim/sulfamethoxazone; and Ciprofloxacin. The study demonstrated only low levels of resistance to a broad range of commonly used antibiotics, relative to other recent studies. Therefore, the complete story on this emerging issue has yet to be written.

The major contributing factor for the selection and propagation of antibiotic-resistant strains is the use of antibiotics. In developed countries, many of the problems of antibiotic resistance are generated within hospitals because of the intensive use of antibiotics (Levy and Marshall, 2004). Methicillin-resistant *Staphylococcus aureus* (MRSA) is probably the most important antibiotic-resistant bacterium associated with hospital-acquired infections (HAI), since staphylococci are the most common pathogens causing bacteremia (blood infection). Bacteremia caused by MRSA are much more difficult to treat, and causes thousands of deaths every year in the United States.

Measures to manage and prevent the spread of drug resistance in hospitals include:

- isolation of infected patients that are potentially hazardous

- early identification and prompt implementation of control measures
- effective hand washing measures
- new therapeutic approaches

Overall, antibiotic-resistant bacteria appear to be widely dispersed throughout the environment, and yet it is only in specialized environments that their presence appears to adversely affect human health and welfare. In hospitals, antibiotic-resistant bacteria are selected due to high usage of antibiotics, and the fact that sterilized surfaces reduce biotic competition. Therefore, the potential for antibiotic-resistant pathogens is high as is the potential for nosocomial infections (hospital-acquired infections). The potential adverse effects from livestock feeding of antibiotics also warrants caution. In contrast, in environments such as soils or biosolids, the potential for gene transfer to human pathogenic bacteria would appear to be lower. Antimicrobial resistance has become one of the most important public health issues faced by the industrialized world. It is estimated that the annual cost of infections caused by antibiotic-resistant bacteria in the United States is U.S.\$4—5 billion.

QUESTIONS AND PROBLEMS

1. In your opinion, what is the greatest threat posed by global emerging microbial issues in the Anthropocene era?
2. Do soil microorganisms enhance or diminish global warming? Discuss.
3. Will we run out of new effective antibiotics? Discuss.

REFERENCES AND RECOMMENDED READING

Alajo, S.O., Nakavuma, J., and Erume, J. (2006). Cholera in endemic districts in Uganda during El Niño rains: 2002—2003. Central Public Health Laboratories, Ministry of Health, Wandegeya, Kampala, Uganda 6, 93—97.

Ambus, P., Zechmeister-Boltenstern, S., and Butterbach-Bahl, K. (2006) Sources of nitrous oxide emitted from European forest soils. *Biogeosciences* 3, 135—145.

Atlas, R. M., and Hazen, T. C. (2011) Oil biodegradation and bioremediation: a tale of the two worst spills in U.S. history. *Environ. Sci. Technol.* 45, 6709—6715.

Brooks, J. P., Maxwell, S. L., Rensing, C., Gerba, C. P., and Pepper, I. L. (2007) Occurrence of antibiotic-resistant bacteria and endotoxin associated with the land application of biosolids. *Can. J. Microbiol.* 53, 616—622.

Chee-Stanford, J. C., Koike, M. S., Krapac, I. G., Lin, Y., Yannarell, A. C., Maxwell, S., *et al.* (2009) Fate and transport of antibiotic residues and antibiotic resistance genes following land application of manure waste. *J. Environ. Qual.* 38, 1719—1727.

Chiemchaisri, C., Chiemchaisri, W., Kumar, S., and Wichramarachchi, P. N. (2012) Reduction of methane emission from landfill through microbial activities in cover soil: a brief review. *Crit. Rev. Sci. Technol.* 42, 412—434.

Delgado, J. A., and Mosier, A. R. (1996) Mitigation alternative to decrease nitrous oxide emissions and urea-nitrogen loss and their effect on methane flux. *J. Environ. Qual.* 25, 1105—1111.

D'Costa, V. M., McGrann, K. M., Hughes, D. W., and Wright, G. D. (2006) Sampling the antibiotic resistome. *Science* 311, 374—377.

Englen, M. D., Fedorka-Cray, P. J., Ladely, S. R., and Dargatz, D. A. (2005) Antimicrobial resistance patterns of Campylobacter from feedlot cattle. *J. Appl. Microbiol.* 99, 285—291.

Esiobu, N., Armenta, L., and Ike, J. (2002) Antibiotic resistance in soil and water environments. *Int. J. Environ. Hlth. Res.* 121, 133—144.

Gagnon, A. S., Smoyer-Tomic, K. E., and Bush, A. B. (2002) The El Niño southern oscillation and malaria epidemics in South America. *Int. J. Biometeorol.* 46, 81—89.

Gil, A. I., Louis, V. R., Rivera, I. N., Lipp, E., Hug, A., Lanata, C. F., *et al.* (2004) Occurrence and distribution of *Vibrio cholerae* in the coastal environment of Peru. *Environ. Microbiol.* 6, 699—706.

Hay, S. I., Guerra, C. A., Tatem, A. J., Atkinson, P. M., and Snow, R. W. (2005) Tropical infectious diseases: urbanization, malaria transmission and disease burden in Africa. *Nat. Rev. Microbiol.* 3, 81—90.

Hazen, T. C., Dubinsky, E. A., DeSantis, T. Z., Andersen, G. L., Piceno, Y. M., Singh, N., *et al.* (2010) Deep-sea oil plume enriches indigenous oil-degrading bacteria. *Science* 330, 204—208.

IPCC, (2007). Climate change 2007: synthesis report. *In* "Contribution of working groups I, II and III to the fourth assessment report of the intergovernmental panel on climate change" (Core Writing Team, R. K. Pachauri, and A. Reisinger, eds.), IPCC, Geneva.

Inselsbacher, E., Wanek, W., Ripka, K., Hackl, E., Sessitsch, A., Srauss, J., *et al.* (2011) Greenhouse gas fluxes respond to different N fertilizer types due to altered plant-soil-microbe interactions. *Plant Soil* 343, 17—35.

Jackson, J., Choudrie, S., Thistlewaite, G., Passant, N., Murrells, T., Watterson, J., *et al.* (2009) UK greenhouse gas inventory, 1990—2007 annual report for submission under the framework. In "Conventions on Climate Change," AEA Technology, p. 71.

Kovats, R. S., Bouma, M. J., Haiat, S., Worrall, E., and Haines, A. (2003) El Niño and health. *Lancet* 362, 1481—1489.

Lam, S. K., and Chua, K. B. (2002) Nipah virus encephalitis outbreak in Malaysia. *Clin. Infect. Dis.* 34, S48—S51.

Levy, S. B., and Marshall, B. (2004) Antibacterial resistance worldwide: causes, challenges and responses. *Nat. Med.* 10, S122—S129.

Luton, P. E., Wayne, J. M., Sharp, R. J., and Riley, P. W. (2002) The mcrA gene as an alternative to 16S rRNA in the phylogenetic analysis of methanogen populations in landfill. *Microbiology* 148, 3521—3530.

Martens, P. (1999) How will climate change affect human health. *Am. Sci.* 87, 534—541.

Mason, O. U., Hazen, T. C., Borglin, S., Chain, P. S. G., Dubinsky, E. A., Fortney, J. L., *et al.* (2012) Metagenome, metatranscriptome and single-cell sequencing reveal microbial response to Deepwater Horizon oil spill. *ISME J.* 6, 1715—1727.

Mosier, A. R., Duxbury, J. M., Freney, J. R., Heinemeyer, O., and Minami, K. (1996) Nitrous oxide emissions from agricultural fields: assessment, measurement and mitigation. *Plant Soil* 181, 95—108.

NRC (National Research Council) (2007) Earth Materials and Health. Washington, DC, National Academy of Sciences.

Nadelman, R. B., and Wormser, G. P. (2005) Poly-ticks: blue state versus red state for Lyme disease. *Lancet* **365**, 280.

Patz, J. A., Engelberg, D., and Last, J. (2000) The effects of changing weather on public health. *Am. Rev. Pub. Hlth.* **21**, 271–307.

Rensing, C., Newby, D. T., and Pepper, I. L. (2002) The role of selective pressure and selfish DNA in horizontal gene transfer and soil microbial community adaptations. *Soil Biol. Biochem.* **34**, 285–296.

Scheutz, C., Mosboek, H., and Kjeldsen, P. (2004) Attenuation of methane and volatile organic compounds in landfill soil covers. *J. Environ. Qual.* **33**, 61–70.

Schmidt, K. A., and Ostfeld, R. S. (2001) Biodiversity and the dilution effect in disease ecology. *Ecology* **82**, 609–619.

Short, J. W., Lindeberg, M. R., Harris, P. M., Maselko, J. M., Pellack, J. J., and Rice, S. D. (2004) Estimate of oil persisting on the beaches of Prince William Sound 12 years after the Exxon Valdez oil spill. *Environ. Sci. Technol.* **38**, 19–25.

Snow, R. W., Guerra, C. A., Noor, A. M., Myint, H. Y., and Hay, S. I. (2005) The global distribution of clinical episodes of *Plasmodium falciparum* malaria. *Nature* **434**, 214–217.

Taylor, L. H., Latham, S. M., and Woodhouse, M. E. (2001) Risk factors for human disease emergence. *Phil. Trans. R. Soc. Lond., Series B, Biol. Sci.* **356**, 983–989.

Tomasz, A. (2006) Weapons of microbial drug resistance abound in soil flora. *Science* **311**, 342–343.

Treseder, K. K. (2008) Nitrogen additions and microbial biomass: a meta-analysis of ecosystem studies. *Ecol. Lett.* **11**, 1111–1120.

WHO (World Health Organization) (2004) "The World Health Report".

Index

Note: Page numbers followed by "*b*", "*f*" and "*t*" refer to boxes, figures and tables, respectively.

D0225349